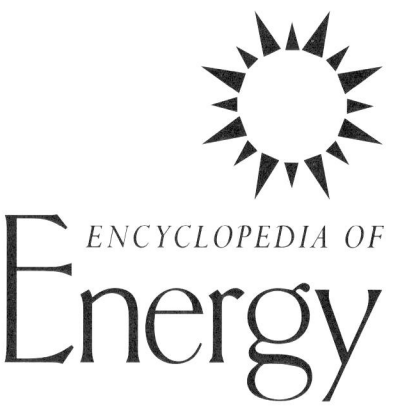

ENCYCLOPEDIA OF Energy

Gl – Ma
VOLUME 3

EDITORIAL BOARD

Editor-in-Chief

CUTLER J. CLEVELAND
Boston University, Boston, Massachusetts, United States

Associate Editors

ROBERT U. AYRES
Center for the Management of Environmental Resources, INSEAD, Fontainebleau, France, and Chalmers Institute of Technology, Gothenburg, Sweden

ROBERT COSTANZA
University of Vermont, Burlington, Vermont, United States

JOSÉ GOLDEMBERG
University of São Paulo, São Paulo, Brazil

MARIJA D. ILIC
Carnegie Mellon University, Pittsburgh, Pennsylvania, United States

EBERHARD JOCHEM
Swiss Federal Institute of Technology, Zurich, Switzerland, and Fraunhofer Institute for Systems and Innovation Research, Karlsruhe, Germany

ROBERT K. KAUFMANN
Boston University, Boston, Massachusetts, United States

AMORY B. LOVINS
Rocky Mountain Institute, Snowmass, Colorado, United States

MOHAN MUNASINGHE
Munasinghe Institute for Development, Colombo, Sri Lanka

R. K. PACHAURI
The Energy and Resources Institute (TERI), New Delhi, India

PER F. PETERSON
University of California, Berkeley, California, United States

LEE SCHIPPER
World Resources Institute, Washington, D.C., United States

CLAUDIA SHEINBAUM PARDO
Instituto de Ingeniería, Universidad Nacional Autónoma de México, Coyoacán, México

MARGARET SLADE
University of Warwick, Coventry, United Kingdom

VACLAV SMIL
University of Manitoba, Winnipeg, Manitoba, Canada

ERNST WORRELL
U.S. Department of Energy, Lawrence Berkeley National Laboratory, Berkeley, California, United States

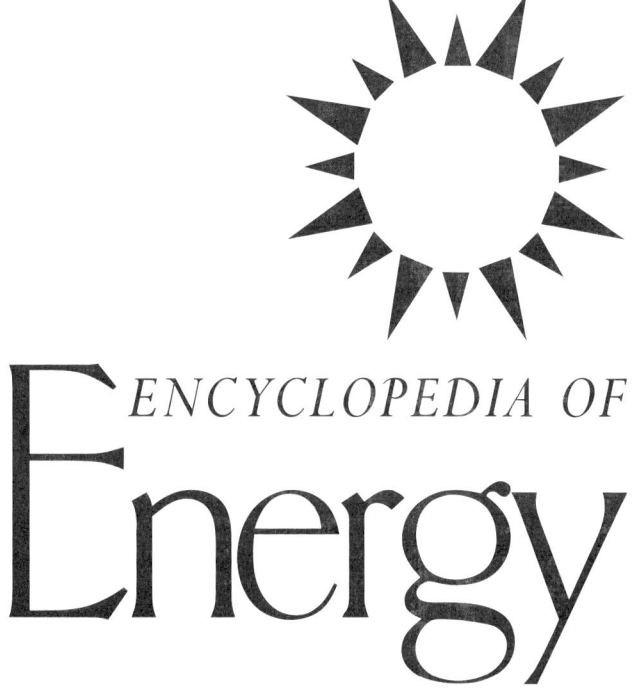

ENCYCLOPEDIA OF Energy

Editor-in-Chief

CUTLER J. CLEVELAND
Boston University, Boston, Massachusetts, United States

Gl – Ma
VOLUME 3

ELSEVIER
ACADEMIC
PRESS

Amsterdam   Boston   Heidelberg   London   New York
Oxford   Paris   San Diego   San Francisco   Singapore   Sydney   Tokyo

This book is printed on acid-free paper.

© 2004, Elsevier Inc.

All rights reserved.
No part of this publication may be reproduced or transmitted in any form or by any means, electronic or mechanical, including photocopy, recording, or any information storage and retrieval system, without permission in writing from the publisher.

Permissions may be sought directly from Elsevier's Science & Technology Rights Department in Oxford, UK: phone: (+44) 1865 843830, fax: (+44) 1865 853333, e-mail: permissions@elsevier.com. You may also complete your request on-line via the Elsevier homepage (http://www.elsevier.com), by selecting "Customer Support" and then "Obtaining Permissions."

Academic Press
An imprint of Elsevier Inc.
525 B Street, Suite 1900, San Diego, California 92101-4495, USA
http://www.academicpress.com

Academic Press
The Boulevard, Langford Lane, Kidlington, Oxford OX5 1GB, UK
http://www.academicpress.com

Library of Congress Catalog Card Number: 2003116610

International Standard Book Number: 0-12-176480-X (set)
International Standard Book Number: 0-12-176481-8 (Volume 1)
International Standard Book Number: 0-12-176482-6 (Volume 2)
International Standard Book Number: 0-12-176483-4 (Volume 3)
International Standard Book Number: 0-12-176484-2 (Volume 4)
International Standard Book Number: 0-12-176485-0 (Volume 5)
International Standard Book Number: 0-12-176486-9 (Volume 6)

Printed in the United States of America

04 05 06 07 08 09 MM 9 8 7 6 5 4 3 2 1

CONTENTS OF VOLUME 3

Contents of Volumes 1–6	ix
Contents by Subject Area	xvii
Foreword	xxvii
Preface	xxxi
Guide to the Encyclopedia	xxxiii

G

Glass and Energy Christopher W. Sinton	1
Global Energy Use: Status and Trends Linda E. Doman	11
Global Material Cycles and Energy Vaclav Smil	23
Goods and Services: Energy Costs Robert A. Herendeen	33
Green Accounting and Energy Peter Bartelmus	43
Greenhouse Gas Abatement: Controversies in Cost Assessment Mark Jaccard	57
Greenhouse Gas Emissions, Alternative Scenarios of Leo Schrattenholzer and Keywan Riahi	67
Greenhouse Gas Emissions from Energy Systems, Comparison and Overview Roberto Dones, Thomas Heck, and Stefan Hirschberg	77
Ground-Source Heat Pumps Arif Hepbasli	97

Gulf War, Environmental Impact of Farouk El-Baz	107

H

Hazardous Waste from Fossil Fuels Majid Ghassemi, Paul K. Andersen, Abbas Ghassemi, and Russell R. Chianelli	119
Heat Islands and Energy Haider Taha	133
Heat Transfer Fabio Gori	145
Heterotrophic Energy Flows Kenneth A. Nagy	159
Human Energetics William R. Leonard	173
Hunting and Gathering Societies, Energy Flows in Tim Bayliss-Smith	183
Hybrid Electric Vehicles John M. German	197
Hybrid Energy Systems J. F. Manwell	215
Hydrogen, End Uses and Economics David Hart	231
Hydrogen, History of Seth Dunn	241
Hydrogen Production Gene D. Berry	253

Contents of Volume 3

Hydrogen Storage and Transportation 267
Gene D. Berry, Joel Martinez-Frias,
Francisco Espinosa-Loza, and Salvador M. Aceves

Hydropower Economics 283
Brian K. Edwards

Hydropower, Environmental Impact of 291
Glenn Cada, Michael Sale, and Dennis Dauble

Hydropower, History and Technology of 301
John S. Gulliver and Roger E. A. Arndt

Hydropower Resettlement Projects,
Socioeconomic Impacts of 315
Adrian C. Sleigh and Sukhan Jackson

Hydropower Resources 325
Garold L. Sommers

Hydropower Technology 333
Peggy Brookshier

I

Indoor Air Quality in Developing Nations 343
Majid Ezzati

Indoor Air Quality in Industrial Nations 351
Adrian F. R. Watson

Industrial Agriculture, Energy Flows in 365
David Pimentel

Industrial Ecology 373
Amit Kapur and Thomas E. Graedel

Industrial Energy Efficiency 383
Wolfgang Eichhammer

Industrial Energy Use, Status and Trends 395
Ernst Worrell

Industrial Symbiosis 407
Marian R. Chertow

Inflation and Energy Prices 417
Paulo S. Esteves and Pedro D. Neves

Information Technology and Energy Use 425
Kurt W. Roth

Information Theory and Energy 439
Sven Erik Jørgensen

Innovation and Energy Prices 451
David Popp

Input–Output Analysis 459
Stephen D. Casler

Integration of Motor Vehicle and
Distributed Energy Systems 475
Timothy E. Lipman

Intelligent Transportation Systems 487
Susan A. Shaheen and Rachel Finson

Internal Combustion Engine Vehicles 497
K. G. Duleep

Internal Combustion (Gasoline and
Diesel) Engines 515
Robert N. Brady

International Comparisons of Energy End
Use: Benefits and Risks 529
Lee Schipper

International Energy Law and Policy 557
Thomas W. Wälde

Investment in Fossil Fuels Industries 583
Mark E. Fischer

L

Labels and Standards for Energy 599
Lloyd Harrington and Paul Waide

Land Requirements of Energy Systems 613
Vaclav Smil

Leisure, Energy Costs of 623
Susanne Becken

Life Cycle Analysis of Power
Generation Systems 635
Joule Bergerson and Lester Lave

Life Cycle Assessment and Energy Systems 647
Evert Nieuwlaar

Lifestyles and Energy 655
Kornelis Blok

Lithosphere, Energy Flows in 663
Seth Stein and Carol Stein

Livestock Production and Energy Use 671
David Pimentel

Lunar–Solar Power System 677
David R. Criswell

M

Magnetic Levitation 691
Donald M. Rote

Magnetohydrodynamics 705
Paul H. Roberts and Patrick H. Diamond

Manufactured Gas, History of 733
Joel A. Tarr

Marine Transportation and Energy Use 745
James J. Corbett

Market-Based Instruments, Overview 759
Stephen Farber

Market Failures in Energy Markets 769
Stephen J. DeCanio

Markets for Biofuels 781
Bengt Hillring

Markets for Coal 787
Richard L. Gordon

Markets for Natural Gas 799
Aad F. Correljé

Markets for Petroleum 809
M. A. Adelman and Michael C. Lynch

Marx, Energy, and Social Metabolism 825
Joan Martinez-Alier

Material Efficiency and Energy Use 835
Eberhard Jochem

Materials for Solar Energy 845
Claes G. Granqvist

Material Use in Automobiles 859
Sujit Das

CONTENTS OF VOLUMES 1–6

Volume 1

Acid Deposition and Energy Use

Aggregation of Energy

Aircraft and Energy Use

Air Pollution from Energy Production and Use

Air Pollution, Health Effects of

Alternative Transportation Fuels: Contemporary Case Studies

Aluminum Production and Energy

Aquaculture and Energy Use

Arid Environments, Impacts of Energy Development in

Batteries, Overview

Batteries, Transportation Applications

Bicycling

Biodiesel Fuels

Biomass, Chemicals from

Biomass Combustion

Biomass for Renewable Energy and Fuels

Biomass Gasification

Biomass: Impact on Carbon Cycle and Greenhouse Gas Emissions

Biomass Resource Assessment

Bottom-Up Energy Modeling

Business Cycles and Energy Prices

Carbon Capture and Storage from Fossil Fuel Use

Carbon Sequestration, Terrestrial

Carbon Taxes and Climate Change

Cement and Energy

City Planning and Energy Use

Clean Air Markets

Clean Coal Technology

Climate Change and Energy, Overview

Climate Change and Public Health: Emerging Infectious Diseases

Climate Change: Impact on the Demand for Energy

Climate Protection and Energy Policy

Coal, Chemical and Physical Properties

Coal Conversion

Coal, Fuel and Non-Fuel Uses

Coal Industry, Energy Policy in

Coal Industry, History of

Coal Mine Reclamation and Remediation

Coal Mining, Design and Methods of

Coal Mining in Appalachia, History of

Coal Preparation

Coal Resources, Formation of

Coal Storage and Transportation

Cogeneration

Combustion and Thermochemistry

Commercial Sector and Energy Use

Complex Systems and Energy

Computer Modeling of Renewable Power Systems

Conservation Measures for Energy, History of

Conservation of Energy Concept, History of

Conservation of Energy, Overview

Consumption, Energy, and the Environment

Conversion of Energy: People and Animals

Corporate Environmental Strategy

Cost–Benefit Analysis Applied to Energy

Crude Oil Releases to the Environment: Natural Fate and Remediation Options

Crude Oil Spills, Environmental Impact of

Cultural Evolution and Energy

Decomposition Analysis Applied to Energy

Depletion and Valuation of Energy Resources

Derivatives, Energy

Desalination and Energy Use

Development and Energy, Overview

Diet, Energy, and Greenhouse Gas Emissions

Discount Rates and Energy Efficiency Gap

Distributed Energy, Overview

District Heating and Cooling

Early Industrial World, Energy Flow in

Earth's Energy Balance

Easter Island: Resource Depletion and Collapse

Volume 2

Ecological Footprints and Energy

Ecological Risk Assessment Applied to Energy Development

Economic Geography of Energy

Economic Growth and Energy

Economic Growth, Liberalization, and the Environment

Economics of Energy Demand

Economics of Energy Efficiency

Economics of Energy Supply

Economic Systems and Energy, Conceptual Overview

Economic Thought, History of Energy in

Ecosystem Health: Energy Indicators

Ecosystems and Energy: History and Overview

Electrical Energy and Power

Electricity, Environmental Impacts of

Electricity Use, History of

Electric Motors

Electric Power: Critical Infrastructure Protection

Electric Power Generation: Fossil Fuel

Electric Power Generation: Valuation of Environmental Costs

Electric Power Measurements and Variables

Electric Power Reform: Social and Environmental Issues

Electric Power Systems Engineering

Electric Power: Traditional Monopoly Franchise Regulation and Rate Making

Electric Power: Transmission and Generation Reliability and Adequacy

Electromagnetic Fields, Health Impacts of

Electromagnetism

Emergy Analysis and Environmental Accounting

Energy Development on Public Land in the United States

Energy Efficiency and Climate Change

Energy Efficiency, Taxonomic Overview

Energy Futures and Options

Energy in the History and Philosophy of Science

Energy Ladder in Developing Nations

Energy Services Industry

Entrainment and Impingement of Organisms in Power Plant Cooling

Entropy

Entropy and the Economic Process

Environmental Change and Energy

Environmental Gradients and Energy

Environmental Injustices of Energy Facilities

Environmental Kuznets Curve

Equity and Distribution in Energy Policy

Ethanol Fuel

European Union Energy Policy

Evolutionary Economics and Energy

Exergoeconomics

Exergy

Exergy Analysis of Energy Systems

Exergy Analysis of Waste Emissions

Exergy: Reference States and Balance Conditions

Experience Curves for Energy Technologies

External Costs of Energy

Fire: A Socioecological and Historical Survey

Fisheries and Energy Use

Flywheels

Food Capture, Energy Costs of

Food System, Energy Use in

Forest Products and Energy

Forms and Measurement of Energy

Fuel Cells

Fuel Cell Vehicles

Fuel Cycle Analysis of Conventional and Alternative Fuel Vehicles

Fuel Economy Initiatives: International Comparisons

Fuzzy Logic Modeling of Energy Systems

Gas Hydrates

Gasoline Additives and Public Health

Geographic Thought, History of Energy in

Geopolitics of Energy

Geothermal Direct Use

Geothermal Power Generation

Volume 3

Glass and Energy

Global Energy Use: Status and Trends

Global Material Cycles and Energy

Goods and Services: Energy Costs

Green Accounting and Energy

Greenhouse Gas Abatement: Controversies in Cost Assessment

Greenhouse Gas Emissions, Alternative Scenarios of

Greenhouse Gas Emissions from Energy Systems, Comparison and Overview

Ground-Source Heat Pumps

Gulf War, Environmental Impact of

Hazardous Waste from Fossil Fuels

Heat Islands and Energy

Heat Transfer

Heterotrophic Energy Flows

Human Energetics

Hunting and Gathering Societies, Energy Flows in

Hybrid Electric Vehicles

Hybrid Energy Systems

Hydrogen, End Uses and Economics

Hydrogen, History of

Hydrogen Production

Hydrogen Storage and Transportation

Hydropower Economics

Hydropower, Environmental Impact of

Hydropower, History and Technology of

Hydropower Resettlement Projects, Socioeconomic Impacts of

Hydropower Resources

Hydropower Technology

Indoor Air Quality in Developing Nations

Indoor Air Quality in Industrial Nations

Industrial Agriculture, Energy Flows in

Industrial Ecology

Industrial Energy Efficiency

Industrial Energy Use, Status and Trends

Industrial Symbiosis

Inflation and Energy Prices

Information Technology and Energy Use

Information Theory and Energy

Innovation and Energy Prices

Input–Output Analysis

Integration of Motor Vehicle and Distributed Energy Systems

Intelligent Transportation Systems

Internal Combustion Engine Vehicles

Internal Combustion (Gasoline and Diesel) Engines

International Comparisons of Energy End Use: Benefits and Risks

International Energy Law and Policy

Investment in Fossil Fuels Industries

Labels and Standards for Energy

Land Requirements of Energy Systems

Leisure, Energy Costs of

Life Cycle Analysis of Power Generation Systems

Life Cycle Assessment and Energy Systems

Lifestyles and Energy

Lithosphere, Energy Flows in

Livestock Production and Energy Use

Lunar–Solar Power System

Magnetic Levitation

Magnetohydrodynamics

Manufactured Gas, History of

Marine Transportation and Energy Use

Market-Based Instruments, Overview

Market Failures in Energy Markets

Markets for Biofuels

Markets for Coal

Markets for Natural Gas

Markets for Petroleum

Marx, Energy, and Social Metabolism

Material Efficiency and Energy Use

Materials for Solar Energy

Material Use in Automobiles

Volume 4

Mechanical Energy

Media Portrayals of Energy

Microtechnology, Energy Applications of

Migration, Energy Costs of

Modeling Energy Markets and Climate Change Policy

Modeling Energy Supply and Demand: A Comparison of Approaches

Motor Vehicle Use, Social Costs of

Multicriteria Analysis of Energy

National Energy Modeling Systems

National Energy Policy: Brazil

National Energy Policy: China

National Energy Policy: India

National Energy Policy: Japan

National Energy Policy: United States

Nationalism and Oil

National Security and Energy

Natural Gas, History of

Natural Gas Industry, Energy Policy in

Natural Gas Processing and Products

Natural Gas Resources, Global Distribution of

Natural Gas Resources, Unconventional

Natural Gas Transportation and Storage

Net Energy Analysis: Concepts and Methods

Neural Network Modeling of Energy Systems

Nongovernmental Organizations (NGOs) and Energy

Nuclear Engineering

Nuclear Fission Reactors: Boiling Water and Pressurized Water Reactors

Nuclear Fuel: Design and Fabrication

Nuclear Fuel Reprocessing

Nuclear Fusion Reactors

Nuclear Power Economics

Nuclear Power, History of

Nuclear Power Plants, Decommissioning of

Nuclear Power: Risk Analysis

Nuclear Proliferation and Diversion

Nuclear Waste

Obstacles to Energy Efficiency

Occupational Health Risks in Crude Oil and Natural Gas Extraction

Occupational Health Risks in Nuclear Power

Ocean, Energy Flows in

Ocean Thermal Energy

Oil and Natural Gas Drilling

Oil and Natural Gas: Economics of Exploration

Oil and Natural Gas Exploration

Oil and Natural Gas Leasing

Oil and Natural Gas Liquids: Global Magnitude and Distribution

Oil and Natural Gas: Offshore Operations

Oil and Natural Gas Resource Assessment: Classifications and Terminology

Oil and Natural Gas Resource Assessment: Geological Methods

Oil and Natural Gas Resource Assessment: Production Growth Cycle Models

Oil Crises, Historical Perspective

Oil Industry, History of

Oil-Led Development: Social, Political, and Economic Consequences

Oil Pipelines

Oil Price Volatility

Oil Recovery

Oil Refining and Products

Oil Sands and Heavy Oil

Oil Shale

OPEC, History of

OPEC Market Behavior, 1973–2003

Origin of Life and Energy

Passenger Demand for Travel and Energy Use

Peat Resources

Petroleum Property Valuation

Petroleum System: Nature's Distribution System for Oil and Gas

Volume 5

Philanthropy and Energy

Photosynthesis and Autotrophic Energy Flows

Photosynthesis, Artificial

Photovoltaic Conversion: Space Applications

Photovoltaic Energy: Stand-Alone and Grid-Connected Systems

Photovoltaic Materials, Physics of

Photovoltaics, Environmental Impact of

Physics and Economics of Energy, Conceptual Overview

Plastics Production and Energy

Polar Regions, Impacts of Energy Development

Population Growth and Energy

Potential for Energy Efficiency: Developing Nations

Prices of Energy, History of

Public Reaction to Electricity Transmission Lines

Public Reaction to Energy, Overview

Public Reaction to Nuclear Power Siting and Disposal

Public Reaction to Offshore Oil

Public Reaction to Renewable Energy Sources and Systems

Radiation, Risks and Health Impacts of

Rebound Effect of Energy Conservation

Recycling of Metals

Recycling of Paper

Refrigeration and Air-Conditioning

Remote Sensing for Energy Resources

Renewable Energy and the City

Renewable Energy in Europe

Renewable Energy in Southern Africa

Renewable Energy in the United States

Renewable Energy Policies and Barriers

Renewable Energy, Taxonomic Overview

Renewable Portfolio Standard

Reproduction, Energy Costs of

Research and Development Trends for Energy

Resource Curse and Investment in Energy Industries

Reuse and Energy

Risk Analysis Applied to Energy Systems

Rocket Engines

Rural Energy in China

Rural Energy in India

Service and Commerce Sector, Energy Use in

Sociopolitical Collapse, Energy and

Solar Cells

Solar Cookers

Solar Cooling, Dehumidification, and Air-Conditioning

Solar Detoxification and Disinfection

Solar Distillation and Drying

Solar Energy, History of

Solar Fuels and Materials

Solar Heat Pumps

Solar Ponds

Solar Thermal Energy, Industrial Heat Applications

Solar Thermal Power Generation

Solar Water Desalination

Steel Production and Energy

Stock Markets and Energy Prices

Storage of Energy, Overview

Strategic Petroleum Reserves

Subsidies to Energy Industries

Suburbanization and Energy

Sun, Energy from

Sustainable Development: Basic Concepts and Application to Energy

System Dynamics and the Energy Industry

Volume 6

Tanker Transportation

Taxation of Energy

Technology Innovation and Energy

Temperature and Its Measurement

Thermal Comfort

Thermal Energy Storage

Thermal Pollution

Thermodynamics and Economics, Overview

Thermodynamic Sciences, History of

Thermodynamics, Laws of

Thermoregulation

Tidal Energy

Trade in Energy and Energy Services

Transitions in Energy Use

Transportation and Energy, Overview

Transportation and Energy Policy

Transportation Fuel Alternatives for Highway Vehicles

Turbines, Gas

Turbines, Steam

Ultralight Rail and Energy Use

United Nations Energy Agreements

Uranium and Thorium Resource Assessment

Uranium Mining: Environmental Impact

Uranium Mining, Processing, and Enrichment

Urbanization and Energy

Value Theory and Energy

Vehicles and Their Powerplants: Energy Use and Efficiency

War and Energy

Waste-to-Energy Technology

Wave and Tidal Energy Conversion

Wetlands: Impacts of Energy Development in the Mississippi Delta

Wind Energy Economics

Wind Energy, History of

Wind Energy Technology, Environmental Impacts of

Wind Farms

Wind Resource Base

Women and Energy: Issues in Developing Nations

Wood Energy, History of

Wood in Household Energy Use

Work, Power, and Energy

World Environment Summits: The Role of Energy

World History and Energy

CONTENTS BY SUBJECT AREA

Basics of Energy

Batteries, Overview
Cogeneration
Combustion and Thermochemistry
Conservation of Energy, Overview
Conversion of Energy: People and Animals
Electrical Energy and Power
Electric Motors
Electromagnetism
Entropy
Flywheels
Forms and Measurement of Energy
Fuel Cells
Heat Transfer
Internal Combustion (Gasoline and Diesel) Engines
Magnetohydrodynamics
Mechanical Energy
Refrigeration and Air-Conditioning
Rocket Engines
Storage of Energy, Overview
Sun, Energy from
Temperature and Its Measurement
Thermal Energy Storage
Thermodynamics, Laws of
Turbines, Gas
Turbines, Steam
Work, Power, and Energy

Coal

Coal Industry, Energy Policy in
Coal Industry, History of
Coal Mine Reclamation and Remediation
Coal Mining in Appalachia, History of
Clean Coal Technology
Coal, Chemical and Physical Properties
Coal Conversion
Coal, Fuel and Non-Fuel Uses
Coal Mining, Design and Methods of
Coal Preparation
Coal Resources, Formation of
Coal Storage and Transportation
Markets for Coal
Peat Resources

Conservation and End Use

Aircraft and Energy Use
Alternative Transportation Fuels: Contemporary Case Studies

xviii   Contents by Subject Area

Aquaculture and Energy Use
Batteries, Transportation Applications
Bicycling
Commercial Sector and Energy Use
Conservation Measures for Energy, History of
Conservation of Energy, Overview
Diet, Energy, and Greenhouse Gas Emissions
Discount Rates and Energy Efficiency Gap
Distributed Energy, Overview
District Heating and Cooling
Economics of Energy Efficiency
Energy Efficiency, Taxonomic Overview
Fisheries and Energy Use
Food System, Energy Use in
Fuel Cell Vehicles
Fuel Cycle Analysis of Conventional and Alternative Fuel Vehicles
Hybrid Electric Vehicles
Hydrogen, End Uses and Economics
Industrial Ecology
Industrial Energy Efficiency
Industrial Energy Use, Status and Trends
Information Technology and Energy Use
Integration of Motor Vehicle and Distributed Energy Systems
Intelligent Transportation Systems
Internal Combustion Engine Vehicles
International Comparisons of Energy End Use: Benefits and Risks
Lifestyles and Energy
Livestock Production and Energy Use
Lunar–Solar Power System
Magnetic Levitation
Marine Transportation and Energy Use
Obstacles to Energy Efficiency
Passenger Demand for Travel and Energy Use

Potential for Energy Efficiency: Developing Nations
Rebound Effect of Energy Conservation
Service and Commerce Sector, Energy Use in
Thermal Comfort
Transportation and Energy, Overview
Transportation Fuel Alternatives for Highway Vehicles
Ultralight Rail and Energy Use
Vehicles and Their Powerplants: Energy Use and Efficiency

Economics of Energy

Aggregation of Energy
Business Cycles and Energy Prices
Corporate Environmental Strategy
Derivatives, Energy
Economic Geography of Energy
Economic Growth and Energy
Economic Growth, Liberalization, and the Environment
Economics of Energy Demand
Economics of Energy Efficiency
Economics of Energy Supply
Economic Thought, History of Energy in
Energy Futures and Options
Energy Services Industry
Entropy and the Economic Process
Evolutionary Economics and Energy
Exergoeconomics
External Costs of Energy
Hydrogen, End Uses and Economics
Hydropower Economics
Inflation and Energy Prices
Innovation and Energy Prices

Investment in Fossil Fuels Industries

Market Failures in Energy Markets

Markets for Biofuels

Markets for Coal

Markets for Natural Gas

Markets for Petroleum

Marx, Energy, and Social Metabolism

Nuclear Power Economics

Oil and Natural Gas: Economics of Exploration

Oil Price Volatility

OPEC Market Behavior, 1973–2003

Petroleum Property Valuation

Physics and Economics of Energy, Conceptual Overview

Prices of Energy, History of

Rebound Effect of Energy Conservation

Resource Curse and Investment in Energy Industries

Stock Markets and Energy Prices

Subsidies to Energy Industries

Taxation of Energy

Thermodynamics and Economics, Overview

Trade in Energy and Energy Services

Wind Energy Economics

Electricity

Electrical Energy and Power

Electricity, Environmental Impacts of

Electricity Use, History of

Electric Motors

Electric Power: Critical Infrastructure Protection

Electric Power Generation: Fossil Fuel

Electric Power Generation: Valuation of Environmental Costs

Electric Power Measurements and Variables

Electric Power Reform: Social and Environmental Issues

Electric Power Systems Engineering

Electric Power: Traditional Monopoly Franchise Regulation and Rate Making

Electric Power: Transmission and Generation Reliability and Adequacy

Electromagnetic Fields, Health Impacts of

Electromagnetism

Hybrid Electric Vehicles

Public Reaction to Electricity Transmission Lines

Energy Flows

Conversion of Energy: People and Animals

Earth's Energy Balance

Ecosystem Health: Energy Indicators

Ecosystems and Energy: History and Overview

Environmental Gradients and Energy

Food Capture, Energy Costs of

Heat Transfer

Heterotrophic Energy Flows

Human Energetics

Industrial Agriculture, Energy Flows in

Industrial Symbiosis

Lithosphere, Energy Flows in

Migration, Energy Costs of

Ocean, Energy Flows in

Origin of Life and Energy

Photosynthesis and Autotrophic Energy Flows

Reproduction, Energy Costs of

Sun, Energy from

Thermoregulation

Environmental Issues

Acid Deposition and Energy Use

Air Pollution from Energy Production and Use

Air Pollution, Health Effects of

Aquaculture and Energy Use

Arid Environments, Impacts of Energy Development in

Biomass: Impact on Carbon Cycle and Greenhouse Gas Emissions

Carbon Capture and Storage from Fossil Fuel Use

Carbon Sequestration, Terrestrial

Clean Air Markets

Clean Coal Technology

Climate Change and Energy, Overview

Climate Change and Public Health: Emerging Infectious Diseases

Climate Change: Impact on the Demand for Energy

Climate Protection and Energy Policy

Coal Mine Reclamation and Remediation

Consumption, Energy, and the Environment

Crude Oil Releases to the Environment: Natural Fate and Remediation Options

Crude Oil Spills, Environmental Impact of

Desalination and Energy Use

Economic Growth, Liberalization, and the Environment

Ecosystem Health: Energy Indicators

Ecosystems and Energy: History and Overview

Electricity, Environmental Impacts of

Electric Power Generation: Valuation of Environmental Costs

Electric Power Reform: Social and Environmental Issues

Energy Efficiency and Climate Change

Entrainment and Impingement of Organisms in Power Plant Cooling

Environmental Change and Energy

Environmental Gradients and Energy

Environmental Injustices of Energy Facilities

Fisheries and Energy Use

Global Material Cycles and Energy

Greenhouse Gas Emissions, Alternative Scenarios of

Greenhouse Gas Emissions from Energy Systems, Comparison and Overview

Gulf War, Environmental Impact of

Hazardous Waste from Fossil Fuels

Heat Islands and Energy

Hydropower, Environmental Impact of

Indoor Air Quality in Developing Nations

Indoor Air Quality in Industrial Nations

Land Requirements of Energy Systems

Nuclear Power Plants, Decommissioning of

Nuclear Waste

Photovoltaics, Environmental Impact of

Polar Regions, Impacts of Energy Development

Thermal Pollution

Uranium Mining: Environmental Impact

Wetlands: Impacts of Energy Development in the Mississippi Delta

Wind Energy Technology, Environmental Impacts of

World Environment Summits: The Role of Energy

Global Issues

Climate Change and Energy, Overview

Cultural Evolution and Energy

Development and Energy, Overview

Economic Geography of Energy

Economic Growth and Energy

Geopolitics of Energy

Global Energy Use: Status and Trends

International Comparisons of Energy End Use: Benefits and Risks

International Energy Law and Policy

Nationalism and Oil

Nongovernmental Organizations (NGOs) and Energy

Nuclear Proliferation and Diversion

Population Growth and Energy

Technology Innovation and Energy

United Nations Energy Agreements

Women and Energy: Issues in Developing Nations

World Environment Summits: The Role of Energy

History and Energy

Coal Industry, History of

Coal Mining in Appalachia, History of

Conservation Measures for Energy, History of

Conservation of Energy Concept, History of

Early Industrial World, Energy Flow in

Economic Thought, History of Energy in

Ecosystems and Energy: History and Overview

Electricity Use, History of

Energy in the History and Philosophy of Science

Environmental Change and Energy

Fire: A Socioecological and Historical Survey

Geographic Thought, History of Energy in

Gulf War, Environmental Impact of

Hydrogen, History of

Hydropower, History and Technology of

Manufactured Gas, History of

Nationalism and Oil

Natural Gas, History of

Nuclear Power, History of

Oil Crises, Historical Perspective

Oil Industry, History of

OPEC, History of

OPEC Market Behavior, 1973–2003

Prices of Energy, History of

Sociopolitical Collapse, Energy and

Solar Energy, History of

Thermodynamic Sciences, History of

Transitions in Energy Use

War and Energy

Wind Energy, History of

Wood Energy, History of

World History and Energy

Material Use and Reuse

Aluminum Production and Energy

Cement and Energy

Forest Products and Energy

Glass and Energy

Global Material Cycles and Energy

Industrial Energy Efficiency

Material Efficiency and Energy Use

Materials for Solar Energy

Material Use in Automobiles

Plastics Production and Energy

Recycling of Metals

Recycling of Paper

Reuse and Energy

Steel Production and Energy

Uranium and Thorium Resource Assessment

Measurement and Models

Aggregation of Energy

Bottom-Up Energy Modeling

Computer Modeling of Renewable Power Systems

Cost–Benefit Analysis Applied to Energy

Decomposition Analysis Applied to Energy

Depletion and Valuation of Energy Resources

Ecological Risk Assessment Applied to Energy Development

Electric Power Generation: Valuation of Environmental Costs

Electric Power Measurements and Variables

Emergy Analysis and Environmental Accounting

Experience Curves for Energy Technologies

Forms and Measurement of Energy

Fuzzy Logic Modeling of Energy Systems

Green Accounting and Energy

Input–Output Analysis

Life Cycle Analysis of Power Generation Systems

Life Cycle Assessment and Energy Systems

Modeling Energy Markets and Climate Change Policy

Modeling Energy Supply and Demand: A Comparison of Approaches

Multicriteria Analysis of Energy

National Energy Modeling Systems

Net Energy Analysis: Concepts and Methods

Neural Network Modeling of Energy Systems

System Dynamics and the Energy Industry

Temperature and Its Measurement

Nuclear Power

Nuclear Engineering

Nuclear Fission Reactors: Boiling Water and Pressurized Water Reactors

Nuclear Fuel: Design and Fabrication

Nuclear Fuel Reprocessing

Nuclear Fusion Reactors

Nuclear Power Economics

Nuclear Power, History of

Nuclear Power Plants, Decommissioning of

Nuclear Power: Risk Analysis

Nuclear Proliferation and Diversion

Nuclear Waste

Occupational Health Risks in Nuclear Power

Public Reaction to Nuclear Power Siting and Disposal

Radiation, Risks and Health Impacts of

Uranium and Thorium Resource Assessment

Uranium Mining, Processing, and Enrichment

Oil and Natural Gas

Crude Oil Spills, Environmental Impact of

Gas Hydrates

Markets for Natural Gas

Markets for Petroleum

Natural Gas, History of

Natural Gas Processing and Products

Natural Gas Resources, Global Distribution of

Natural Gas Resources, Unconventional

Natural Gas Transportation and Storage

Occupational Health Risks in Crude Oil and Natural Gas Extraction

Oil and Natural Gas Drilling

Oil and Natural Gas: Economics of Exploration

Oil and Natural Gas Exploration

Oil and Natural Gas Leasing

Oil and Natural Gas Liquids: Global Magnitude and Distribution

Oil and Natural Gas: Offshore Operations

Oil and Natural Gas Resource Assessment: Classifications and Terminology

Oil and Natural Gas Resource Assessment: Geological Methods

Oil and Natural Gas Resource Assessment: Production Growth Cycle Models

Oil Crises, Historical Perspective

Oil Industry, History of

Oil-Led Development: Social, Political, and Economic Consequences

Oil Pipelines

Oil Price Volatility

Oil Recovery

Oil Refining and Products

Oil Sands and Heavy Oil

Oil Shale

OPEC, History of

Petroleum Property Valuation

Petroleum System: Nature's Distribution System for Oil and Gas

Public Reaction to Offshore Oil

Remote Sensing for Energy Resources

Strategic Petroleum Reserves

Tanker Transportation

Policy Issues

Carbon Taxes and Climate Change

City Planning and Energy Use

Clean Air Markets

Climate Protection and Energy Policy

Coal Industry, Energy Policy in

Corporate Environmental Strategy

Energy Development on Public Land in the United States

Equity and Distribution in Energy Policy

European Union Energy Policy

Fuel Economy Initiatives: International Comparisons

Geopolitics of Energy

Greenhouse Gas Abatement: Controversies in Cost Assessment

International Energy Law and Policy

Land Requirements of Energy Systems

Market-Based Instruments, Overview

National Energy Policy: Brazil

National Energy Policy: China

National Energy Policy: India

National Energy Policy: Japan

National Energy Policy: United States

National Security and Energy

Natural Gas Industry, Energy Policy in

Nuclear Proliferation and Diversion

Polar Regions, Impacts of Energy Development

Renewable Energy Policies and Barriers

Renewable Portfolio Standard

Research and Development Trends for Energy

Strategic Petroleum Reserves

Subsidies to Energy Industries

Taxation of Energy

Transportation and Energy Policy

Public Issues

City Planning and Energy Use

Climate Change and Public Health: Emerging Infectious Diseases

Consumption, Energy, and the Environment

Environmental Injustices of Energy Facilities

Hydropower Resettlement Projects, Socioeconomic Impacts of

Labels and Standards for Energy

Lifestyles and Energy

Media Portrayals of Energy

Motor Vehicle Use, Social Costs of

xxiv   Contents by Subject Area

Oil-Led Development: Social, Political, and Economic Consequences

Passenger Demand for Travel and Energy Use

Philanthropy and Energy

Population Growth and Energy

Public Reaction to Electricity Transmission Lines

Public Reaction to Energy, Overview

Public Reaction to Nuclear Power Siting and Disposal

Public Reaction to Offshore Oil

Public Reaction to Renewable Energy Sources and Systems

Suburbanization and Energy

United Nations Energy Agreements

Urbanization and Energy

Renewable and Alternative Sources

Alternative Transportation Fuels: Contemporary Case Studies

Biodiesel Fuels

Biomass, Chemicals from

Biomass Combustion

Biomass for Renewable Energy and Fuels

Biomass Gasification

Biomass Resource Assessment

Computer Modeling of Renewable Power Systems

Ethanol Fuel

Forest Products and Energy

Geothermal Direct Use

Geothermal Power Generation

Ground-Source Heat Pumps

Hybrid Energy Systems

Hydrogen, End Uses and Economics

Hydrogen Production

Hydrogen Storage and Transportation

Hydropower Economics

Hydropower Resources

Hydropower Technology

Lunar–Solar Power System

Materials for Solar Energy

Microtechnology, Energy Applications of

Ocean Thermal Energy

Photosynthesis, Artificial

Photovoltaic Conversion: Space Applications

Photovoltaic Energy: Stand-Alone and Grid-Connected Systems

Photovoltaic Materials, Physics of

Public Reaction to Renewable Energy Sources and Systems

Renewable Energy and the City

Renewable Energy in Europe

Renewable Energy Policies and Barriers

Renewable Energy in Southern Africa

Renewable Energy in the United States

Renewable Energy, Taxonomic Overview

Renewable Portfolio Standard

Solar Cells

Solar Cookers

Solar Cooling, Dehumidification, and Air-Conditioning

Solar Detoxification and Disinfection

Solar Distillation and Drying

Solar Energy, History of

Solar Fuels and Materials

Solar Heat Pumps

Solar Ponds

Solar Thermal Energy, Industrial Heat Applications

Solar Thermal Power Generation

Solar Water Desalination

Tidal Energy

Transportation Fuel Alternatives for Highway Vehicles

Waste-to-Energy Technology

Wave and Tidal Energy Conversion

Wind Energy Economics

Wind Energy, History of

Wind Farms

Wind Resource Base

Risks

Air Pollution from Energy Production and Use

Air Pollution, Health Effects of

Climate Change and Public Health: Emerging Infectious Diseases

Ecological Risk Assessment Applied to Energy Development

Electromagnetic Fields, Health Impacts of

Gasoline Additives and Public Health

Hazardous Waste from Fossil Fuels

Nuclear Power: Risk Analysis

Nuclear Proliferation and Diversion

Nuclear Waste

Occupational Health Risks in Crude Oil and Natural Gas Extraction

Occupational Health Risks in Nuclear Power

Radiation, Risks and Health Impacts of

Risk Analysis Applied to Energy Systems

Tanker Transportation

Society and Energy

Cultural Evolution and Energy

Early Industrial World, Energy Flow in

Easter Island: Resource Depletion and Collapse

Electric Power Reform: Social and Environmental Issues

Goods and Services: Energy Costs

Hunting and Gathering Societies, Energy Flows in

Hydropower Resettlement Projects, Socioeconomic Impacts of

Industrial Agriculture, Energy Flows in

Leisure, Energy Costs of

Lifestyles and Energy

Motor Vehicle Use, Social Costs of

Population Growth and Energy

Renewable Energy and the City

Sociopolitical Collapse, Energy and

Suburbanization and Energy

Urbanization and Energy

War and Energy

Sustainable Development

Development and Energy, Overview

Ecological Risk Assessment Applied to Energy Development

Economic Growth and Energy

Economic Growth, Liberalization, and the Environment

Energy Ladder in Developing Nations

Environmental Kuznets Curve

Indoor Air Quality in Developing Nations

Oil-Led Development: Social, Political, and Economic Consequences

Potential for Energy Efficiency: Developing Nations

Rural Energy in China

Rural Energy in India

Sustainable Development: Basic Concepts and Application to Energy

United Nations Energy Agreements

Women and Energy: Issues in Developing Nations

Wood in Household Energy Use

Systems of Energy

Aggregation of Energy

Complex Systems and Energy

Ecological Footprints and Energy

Economic Systems and Energy, Conceptual Overview

Emergy Analysis and Environmental Accounting

Entropy and the Economic Process

Exergoeconomics

Exergy

Exergy Analysis of Energy Systems

Exergy Analysis of Waste Emissions

Exergy: Reference States and Balance Conditions

Fuzzy Logic Modeling of Energy Systems

Information Theory and Energy

Life Cycle Assessment and Energy Systems

National Energy Modeling Systems

Neural Network Modeling of Energy Systems

Physics and Economics of Energy, Conceptual Overview

Risk Analysis Applied to Energy Systems

System Dynamics and the Energy Industry

Thermodynamics and Economics, Overview

FOREWORD

Energy generation and use are strongly linked to all elements of sustainable development: economic, social, and environmental. The history of human development rests on the availability and use of energy, the transformation from the early use of fire and animal power that improved lives, to the present world with use of electricity and clean fuels for a multitude of purposes. This progress built on basic scientific discoveries, such as electromagnetism and the inventions of technologies such as steam engines, light bulbs, and automobiles.

It is thus abundantly clear that access to affordable energy is fundamental to human activities, development, and economic growth. Without access to electricity and clean fuels, people's opportunities are significantly constrained. However, it is really energy services, not energy *per se* that matters. Yet, today some 2 billion people lack access to modern energy carriers.

In addition to the great benefits, the generation, transportation, and use of energy carriers unfortunately come with undesired effects. The environmental impacts are multifaceted and serious, although mostly less evident. Emissions of suspended fine particles and precursors of acid deposition contribute to local and regional air pollution and ecosystem degradation. Human health is threatened by high levels of air pollution resulting from particular types of energy use at the household, community, and regional levels.

Emissions of anthropogenic greenhouse gases (GHG), mostly from the production and use of energy, are altering the atmosphere in ways that are affecting the climate. There is new and stronger evidence that most of the global warming observed over the last 50 years is attributable to human activities. Stabilization of GHG in the atmosphere will require a major reduction in the projected carbon emissions to levels below the present.

Dependence on imported fuels leaves many countries vulnerable to disruption in supply, which might pose physical hardships and economic burdens; the weight of fossil fuel imports on the balance of payments is unbearable for many poorer countries. The present energy system of countries heavily dependent on fossil fuels geographically concentrated in a few regions of the world adds security of supply aspects.

From the issues indicated here it is clear that major changes are required in energy system development worldwide. At a first glance, there appears to be many conflicting objectives. For example, is it possible to sustain poverty alleviation and economic growth while reducing GHG emissions? Can urban areas and transport expand while improving air quality? What would be the preferable trade-offs? Finding ways to expand energy services while simultaneously addressing the environmental impacts associated with energy use represents a critical challenge to humanity.

What are the options? Looking at physical resources, one finds they are abundant. Fossil fuels will be able to provide the energy carriers that the world is used to for hundreds of years. Renewable energy flows on Earth are many thousands of times larger than flows through energy markets. Therefore, there are no apparent constraints from a resource point of view. However, the challenge is how to use these resources in an environmentally acceptable way. The broad categories of options for using energy in ways that support sustainable development are (1) more efficient use of energy in all sectors, especially at the point of end use, (2) increased use of renewable energy sources, and (3) accelerated development and deployment of new and advanced energy technologies, including next-generation fossil fuel technologies that produce near-zero harmful emissions. Technologies are available in these areas to meet the challenges of sustainable development. In addition, innovation provides increasing opportunities.

Analysis using energy scenarios indicates that it is indeed possible to simultaneously address the sustainable development objectives using the available natural resources and technical options presented. A prerequisite for achieving energy futures compatible with sustainable development objectives is finding ways to accelerate progress for new technologies along the energy innovation chain, including research and development, demonstration, deployment, and diffusion.

It is significant that there already exist combinations of technologies that meet all sustainable development challenges at the same time. This will make it easier to act locally to address pollution problems of a major city or country while at the same time mitigating climate change. Policies for energy for sustainable development can be largely motivated by national concerns and will not have to rely only on global pressures.

However, with present policies and conditions in the marketplaces that determine energy generation and use such desired energy futures will not happen. A prerequisite for sustainable development is change in policies affecting energy for sustainable development. This brings a need to focus on the policy situation and understand incentives and disincentives related to options for options for energy for sustainable development.

Policies and actions to promote energy for sustainable development would include the following:

- Developing capacity among all stakeholders in all countries, especially in the public sector, to address issues related to energy for sustainable development.
- Adopting policies and mechanisms to increase access to energy services through modern fuels and electricity for the 2 billion without.
- Advancing innovation, with balanced emphasis on all steps of the innovation chain: research and development, demonstrations, cost buy-down, and wide dissemination.
- Setting appropriate market framework conditions (including continued market reform, consistent regulatory measures, and targeted policies) to encourage competitiveness in energy markets, to reduce total cost of energy services to end-users, and to protect important public benefits, including the following:
  - Cost-based prices, including phasing out all forms of permanent subsidies for conventional energy (now on the order of $250 billion a year) and internalizing external environmental and health costs and benefits (now sometimes larger than the private costs).
  - Removing obstacles and providing incentives, as needed, to encourage greater energy efficiency and the development and/or diffusion of new technologies for energy for sustainable development to wider markets.
  - Recent power failures on the North American Eastern Seaboard, in California, London (United Kingdom), Sweden, and Italy illustrate the strong dependence on reliable power networks. Power sector reform that recognizes the unique character of electricity, and avoids power crises as seen in recent years, is needed.
- Reversing the trend of declining Official Development Assistance and Foreign Direct Investments, especially as related to energy for sustainable development.

This is a long list of opportunities and challenges. To move sufficiently in the direction of sustainability will require actions by the public and the private sector, as well as other stakeholders, at the national, regional, and global levels. The decisive issues are not technology or natural resource scarcity, but the institutions, rules, financing mechanisms, and regulations needed to make markets work in support of energy for sustainable development. A number of countries, including Spain, Germany, and Brazil, as well as some states in the United States have adopted successful laws and regulations designed to increase the use of renewable energy sources. Some regions, including Latin America and the European Union, have set targets for increased use of renewable energy. However, much remains to be done.

Energy was indeed one of the most intensely debated issues at the United Nations World Summit on Sustainable Development (WSSD), held in Johannesburg, South Africa, in August/September, 2002. In the end, agreement was reached on a text that significantly advances the attention given to energy in the context of sustainable development. This was in fact the first time agreements could be reached on energy at the world level! These developments followed years of efforts to focus on energy as an instrument for sustainable development that intensified after the United Nations Conference on Environment and Development in 1992.

The United Nations General Assembly adopted the Millennium Development Goals (MDG) in 2000. These goals are set in areas such as extreme poverty and hunger, universal primary education, gender equality and empowerment of women, child mortality, maternal health, HIV/AIDS, malaria and other diseases, and environmental sustainability. However, more than 2 billion people cannot access affordable energy services, based on the efficient use of gaseous and liquid fuels and electricity. This constrains their opportunities for economic development and improved living standards. Women and children suffer disproportionately because of their relative dependence on traditional fuels. Although no explicit goal on energy was adopted, access to energy services is a prerequisite to achieving all of the MDGs.

Some governments and corporations have already demonstrated that policies and measures to promote energy solutions conducive to sustainable development can work, and indeed work very well. The renewed focus and broad agreements on energy in the Johannesburg Plan of Implementation and at the 18th World Energy Congress are promising. The formation of many partnerships on energy between stakeholders at WSSD is another encouraging sign. A sustainable future in which energy plays a major positive role in supporting human well-being is possible!

Progress is being made on many fronts in bringing new technologies to the market, and to widening access to modern forms of energy. In relation to energy, a total of 39 partnerships were presented to the United Nations Secretariat for WSSD to promote programs on energy for sustainable development, 23 with energy as a central focus and 16 with a considerable impact on energy. These partnerships included most prominently the DESA-led Clean Fuels and Transport Initiative, the UNDP/World Bank-led Global Village Energy Partnership (GVEP), the Johannesburg Renewable Energy Coalition (JREC), the EU Partnership on Energy for Poverty Eradication and Sustainable Development, and the UNEP-led Global Network on Energy for Sustainable Development (GNESD).

With secure access to affordable and clean energy being so fundamental to sustainable development, the publication of the *Encyclopedia of Energy* is extremely timely and significant. Academics, professionals, scholars, politicians, students, and many more will benefit tremendously from the easy access to knowledge, experience, and insights that are provided here.

Thomas B. Johansson
Professor and Director
International Institute for
Industrial Environmental Economics
Lund University
Lund, Sweden

Former Director
Energy and Atmosphere Programme
United Nations Development Programme
New York, United States

PREFACE

The history of human culture can be viewed as the progressive development of new energy sources and their associated conversion technologies. Advances in our understanding of energy have produced unparalleled transformations of society, as exemplified by James Watt's steam engine and the discovery of oil. These transformations increased the ability of humans to exploit both additional energy and other resources, and hence to increase the comfort, longevity, and affluence of humans, as well as their numbers. Energy is related to human development in three important ways: as a motor of economic growth, as a principal source of environmental stress, and as a prerequisite for meeting basic human needs. Significant changes in each of these aspects of human existence are associated with changes in energy sources, beginning with the discovery of fire, the advent of agriculture and animal husbandry, and, ultimately, the development of hydrocarbon and nuclear fuels. The eventual economic depletion of fossil fuels will drive another major energy transition; geopolitical forces and environmental imperatives such as climate change may drive this transition faster than hydrocarbon depletion would have by itself. There is a diverse palette of alternatives to meet our energy needs, including a new generation of nuclear power, unconventional sources of hydrocarbons, myriad solar technologies, hydrogen, and more efficient energy end use. Each alternative has a different combination of economic, political, technological, social, and environmental attributes.

Energy is the common link between the living and non-living realms of the universe, and thus provides an organizing intellectual theme for diverse disciplines. Formalization of the concept of energy and identification of the laws governing its use by 19th century physical scientists such as Mayer and Carnot are cornerstones of modern science and engineering.

The study of energy has played a pivotal role in understanding the creation of the universe, the origin of life, the evolution of human civilization and culture, economic growth and the rise of living standards, war and geopolitics, and significant environmental change at local, regional, and global scales.

The unique importance of energy among natural resources makes information about all aspects of its attributes, formation, distribution, extraction, and use an extremely valuable commodity. The *Encyclopedia of Energy* is designed to deliver this information in a clear and comprehensive fashion. It uses an integrated approach that emphasizes not only the importance of the concept in individual disciplines such as physics and sociology, but also how energy is used to bridge seemingly disparate fields, such as ecology and economics. As such, this *Encyclopedia* provides the first comprehensive, organized body of knowledge for what is certain to continue as a major area of scientific study in the 21st century. It is designed to appeal to a wide audience including undergraduate and graduate students, teachers, academics, and research scientists who study energy, as well as business corporations, professional firms, government agencies, foundations, and other groups whose activities relate to energy.

Comprehensive and interdisciplinary are two words I use to describe the *Encyclopedia*. It has the comprehensive coverage one would expect: forms of energy, thermodynamics, electricity generation, climate change, energy storage, energy sources, the demand for energy, and so on. What makes this work unique, however, is its breadth of coverage, including insights from history, society, anthropology, public policy, international relations, human and ecosystem health, economics, technology, physics, geology, ecology, business management, environmental science, and engineering. The coverage and integration of the social sciences is a unique feature.

The interdisciplinary approach is employed in the treatment of important subjects. In the case of oil, as one example, there are entries on the history of oil, the history of OPEC, the history of oil prices, oil price volatility, the formation of oil and gas, the distribution of oil resources, oil exploration and drilling, offshore oil, occupational hazards in the oil industry, oil refining, energy policy in the oil industry, the geopolitics of oil, oil spills, oil transportation, public lands and oil development, social impacts of oil and gas development, gasoline additives and public health, and the environmental impact of the Persian Gulf War. Other subjects are treated in a similar way.

This has been a massive and extremely satisfying effort. As with any work of this scale, many people have contributed at every step of the process, including the staff of Academic Press/Elsevier. The project began through the encouragement of Frank Cynar and David Packer, with Frank helping to successfully launch the initiative. Henri van Dorssen skillfully guided the project through its completion. He was especially helpful with integrating the project formulation, production, and marketing aspects of the project. Chris Morris was with the project throughout, and displayed what I can only describe as an uncanny combination of vision, enthusiasm, and energy for the project. I owe Chris a great deal for his insight and professionalism. I spent countless hours on the phone with Robert Matsumura, who was the glue that held the project together. Chris and Robert were ably assisted by outstanding Academic Press/Elsevier staff, especially Nick Panissidi, Joanna Dinsmore, and Mike Early. Clare Marl and her team put together a highly effective and creative marketing plan.

At the next stage, the Editorial Board was invaluable in shaping the coverage and identifying authors. The Board is an outstanding collection of scholars from the natural, social, and engineering sciences who are recognized leaders in their fields of research. They helped assemble an equally impressive group of authors from every discipline and who represent universities, government agencies, national laboratories, consulting firms, think tanks, corporations, and nongovernmental organizations. I am especially proud of the international scope of the authors: more than 400 authors from 40 nations are represented from every continent and every stage of development. To all of these, I extend my thanks and congratulations.

Cutler Cleveland
Boston University
Boston, Massachusetts, United States

GUIDE TO THE ENCYCLOPEDIA

The *Encyclopedia of Energy* is a comprehensive and authoritative study of the subject of energy in all its various aspects, as well as the ways in which energy use involves or affects other areas such as environmental science, economics, public policy, international relations, and human development. The encyclopedia includes 380 different articles on various topics within the overall theme of energy, written by prominent experts from around the world who represent 40 different countries in all. The print version of this work consists of six separate volumes and about 5,400 pages.

Each entry in the encyclopedia provides a complete overview of the given topic, intended to inform a broad spectrum of readers, ranging from energy research professionals, energy policy makers, and scholars in energy-related fields to students and the interested general public. The entries are self-contained and can be read in isolation, but there is a general system linking related topics by means of cross referencing and by their placement in specific subject areas (see Organization, below).

In order that you, the reader, will derive the greatest possible benefit from your use of the *Encyclopedia of Energy*, we have provided this guide. It will explain how the encyclopedia has been formulated and how the information within it can be located.

ENTRY SELECTION

This encyclopedia was conceived with the goal of providing a complete description of all the issues contained within, or impacting upon, the field of energy. This approach defines energy not just in terms of its physical, chemical, and engineering aspects but with respect to all of its effects on society and the environment.

To that end, a thorough and systematic method of entry selection was devised for the work. To begin the selection process, the project's chief editor, Cutler Cleveland, prepared a thematic outline of the topic of energy. This thematic outline progressed through the entire scope of the subject from basic principles to peripheral issues, in the manner of a course curriculum or a textbook.

The reference staff of Academic Press/Elsevier then compared this original outline to a bibliography of leading source materials in the field, including books, journal articles, conference proceedings, Web sites, and so on. Professor Cleveland refined the original outline based on this research; at this point the number of possible entries was about 500, much larger than the eventual total would be for the published encyclopedia.

The outline was then cast in the form of a preliminary table of contents and was presented to members of the editorial board at a two-day conference in San Diego, California, United States. A number of proposed revisions for the contents list emerged from this forum, including suggestions for various topics worthy of being added and the identification of some existing topics that could be merged with others or dropped as nonessential.

Professor Cleveland then prepared another version of the thematic topic list in which he made certain revisions based on the editors' recommendations, and also other adjustments based on his own expert judgment. The Academic Press/Elsevier staff rearranged this thematic list into the alphabetical format that the actual encyclopedia would require and revised the wording of article titles as needed to accomplish this. The result was a working entry list of about 400 topics, which, after some attrition and the further combining of related topics, resulted in the final table of contents of 380 articles.

ORGANIZATION

For the purpose of this encyclopedia, the chief editor and the associate editors, in collaboration with the Academic Press/Elsevier staff, have defined the field of energy as consisting of 20 distinct subject areas, as follows:

> Basics of Energy
> Coal
> Conservation and End Use
> Economics of Energy
> Electricity
> Energy Flows
> Environmental Issues
> Global Issues
> History and Energy
> Material Use and Reuse
> Measurement and Models
> Nuclear Power
> Oil and Natural Gas
> Policy Issues
> Public Issues
> Renewable and Alternative Sources
> Risks
> Society and Energy
> Sustainable Development
> Systems of Energy

Every article in the encyclopedia is designated as part of one (or more) of these 20 subject areas. For example, various articles on energy pricing appear in the economics of energy subject area, and articles on climate change appear in the environmental issues section. (Please see p. xvii of this introductory section for a complete listing of the articles in the encyclopedia according to their subject area.) This table of contents by subject area is a good starting place for a reader who wants information on different facets of a particular issue, such as the relationship between energy use and sustainable development.

FORMAT

All the articles in the *Encyclopedia of Energy* are arranged in a single alphabetical sequence according to the wording of the article title. The placement of the articles by volume is as follows: articles whose titles begin with the letters A to Ea are in Volume 1, Ec to Ge in Volume 2, Gl to Ma in Volume 3, Me to Pe in Volume 4, Ph to S in Volume 5, and T to Z in Volume 6, along with the combined glossary, the appendix, and the subject index.

So that they can be easily located, article titles generally begin with the key word or phrase indicating the topic, with any generic terms following. Thus, for example, "Hydrogen, End Uses and Economics" is the article title rather than "End Uses and Economics of Hydrogen" and "Coal Mining, Design and Methods of" is the title rather than "Design and Methods of Coal Mining." This approach also allows related topics to be grouped together alphabetically, for example, a series of articles beginning with the word "Solar."

OUTLINE

Entries in the encyclopedia begin with a topical outline that indicates the general content of the article. This outline serves two functions. First, it provides a preview of the article, so that the reader can get a sense of what is contained there without having to leaf through the pages. Second, it serves to highlight important subtopics to be discussed within the article. For example, the article "Earth's Energy Balance" (by Kevin E. Trenberth) has this outline:

1. The Earth and Climate System
2. The Global Energy Balance
3. Regional Patterns
4. The Atmosphere
5. The Hydrological Cycle
6. The Oceans
7. The Land
8. Ice
9. The Role of Heat Storage
10. Atmosphere–Ocean Interaction: El Niño
11. Anthropogenic Climate Change
12. Observed and Projected Temperatures

The outline is intended as an overview and thus it lists only the major headings of the article. In addition, extensive second-level and third-level headings will be found within the article.

GLOSSARY

The glossary section appears before the beginning of the article text and is set off typographically from the narrative to follow. It contains terms that are important to an understanding of the article and that may be unfamiliar to the reader. Each term is defined within the context of the particular article in which it is used. Thus the same term may appear in more than one article with slightly varying definitions. The encyclopedia includes approximately 2,400 glossary terms. For example, the article "Oil and Natural Gas Liquids: Global Magnitude and Distribution" (by Thomas S. Ahlbrandt) includes the following glossary entries (among others):

continuous accumulations Petroleum that occurs in an extensive reservoir or reservoirs and is not necessarily related to conventional structural or stratigraphic traps. These accumulations of oil and/or gas lack well-defined down-dip petroleum/water contacts and thus are not localized by the buoyancy of oil or natural gas in water.

conventional accumulations Petroleum that occurs in structural or stratigraphic traps, commonly bounded by a down-dip water contact, and therefore affected by the buoyancy of petroleum in water.

crude oil A mixture of hydrocarbons that exists in a liquid phase in natural underground reservoirs and remains liquid at atmospheric pressure after passing through surface separation facilities. Crude oil may also contain some nonhydrocarbon components; referred to as oil in this article.

cumulative production Volumes of oil and natural gas liquids that have been produced.

endowment The sum of cumulative production, remaining reserves, mean undiscovered recoverable volumes, and mean additions to reserves by field growth.

field A contiguous area consisting of a single reservoir or multiple reservoirs of petroleum, all grouped on, or related to, a single geologic structural and/or stratigraphic feature.

future petroleum The sum of the remaining reserves, mean reserve growth, and the mean of the undiscovered volume. Cumulative production does not contribute to the future petroleum. The terms future oil, future liquid volume, or future endowment are sometimes used as variations of future petroleum to reflect those resources that are yet to be produced.

reserve The estimated quantities of petroleum expected to be commercially recovered from known accumulations relative to a specified date, under prevailing economic conditions, operating practices, and government regulations. Reserves are part of the identified (discovered) resources and include only recoverable materials.

reserve (field) growth The increases of estimated petroleum volume that commonly occur as oil and gas fields are developed and produced.

resource A concentration of naturally occurring solid, liquid, or gaseous hydrocarbons in or on the earth's crust, some of which is currently or potentially economically extractable.

DEFINING STATEMENT

The text of each article begins with a single introductory paragraph that precedes the body of the article. This introduction defines the topic under discussion and summarizes the content of the article. For example, the entry "Sociopolitical Collapse, Energy and" (by Joseph A. Tainter) begins with the following defining paragraph:

Collapse is the rapid simplification of a society. It is the sudden, pronounced loss of an established level of social, political, or economic complexity. Widely known examples include the collapses of Mesopotamia's Third Dynasty of Ur (ca. 2100–2000 BC), the Mycenaean society of Greece (ca. 1650–1050 BC), the Western Roman Empire (last emperor deposed 476 AD), and Maya civilization of the lowlands of Guatemala (ca. 250–800 AD). There are at least two dozen cases of collapse that are known from history, archaeology, or both. States and empires are not the only types of institutions that may rapidly simplify. The entire spectrum of societies, from simple foragers to extensive empires, yields examples of collapse. Since all but a few human societies existed before the development of writing, there may be dozens or even hundreds of cases that are not yet recognized archaeologically. Collapse is therefore a recurrent process, and perhaps no society is invulnerable to it.

CROSS-REFERENCES

All the articles in the encyclopedia have cross-references to other articles. These appear at the end of the article, following the end of the narrative text and preceding the further reading section. The encyclopedia contains about 3,300 cross-references in all. The cross-references indicate related articles that can be consulted for further information on the same topic, or for information on a related topic. For example, the article "European Union Energy Policy" (by Felix C. Matthes) has been provided with the following cross-references:

Equity and Distribution in Energy Policy • Fuel Economy Initiatives: International Comparisons • Geopolitics of Energy • National Energy Policy:

Brazil • National Energy Policy: China • National Energy Policy: India • National Energy Policy: Japan • National Energy Policy: United States • Research and Development Trends for Energy

FURTHER READING

The further reading section appears as the last element in an article. It consists of a selection of materials chosen by the author to provide readers with further information on the article topic. This section lists recent secondary sources to aid the reader in locating more detailed or more technical information. Review articles and research papers that are important to an understanding of the topic are also listed. For example, the article "OPEC, History of" (by Fadhil J. Chalabi) has the following suggested readings:

BP Amoco. (1970–2001). "Annual Statistical Reviews." BP Amoco, London.
Center for Global Energy Studies. (1990–2002). "Global Oil Reports." CGES, London.
Chalabi, F. J. (1980). "OPEC and the International Oil Industry: A Changing Structure." Oxford University Press, Oxford, UK.
Chalabi, F. J. (1989). "OPEC at the Crossroads." Pergamon, Oxford, UK.
Penrose, E. (1968). "The Large International Firm in Developing Countries: The International Petroleum Industry." Allen & Unwin, London.
Sampson, A. (1975). "The Seven Sisters." Hodder & Stroughton, London.

The further reading references are for the benefit of the reader; thus they consist of a limited number of entries. They do not represent a complete listing of all the sources consulted by the author in preparing the paper.

INDEX

A subject index is located at the end of Volume 6. This index is the most convenient way to locate a desired topic within the encyclopedia and thus it should be the first point of reference for any reader seeking to find a particular topic. The entries in the index are listed alphabetically and indicate the volume and page number where information on this topic can be found.

Glass and Energy

CHRISTOPHER W. SINTON
Alfred University
Alfred, New York, United States

1. Introduction
2. Types of Glass
3. Introduction to Commercial Glass Manufacturing
4. Energy

Glossary

container glass The largest glass sector; formed by a two-stage process of initial forming by pressing followed by blowing to obtain the finished hollow shape.
cullet Scrap or waste glass that is added to the batch materials.
float glass Flat glass that is made by floating molten glass on a pool of molten tin.
glass A material with a three-dimensional network of atoms that forms a solid lacking the long-range periodicity typical of crystalline materials.
optical fibers Very high-purity silica glass fibers used to transmit light for telecommunications.
oxy-fuel firing The replacement of the combustion air with oxygen in a glass furnace.
recuperator A continuous heat exchanger that preheats combustion air with the heat from the exhaust gases.
regenerator A stack of refractory material that absorbs the heat from exhaust gases. There are two regenerators in a regenerative furnace that are cycled over a specific time period.

Glass is a material that enjoys widespread use in many commercial applications, including buildings, beverage containers, automobiles, and telecommunications. However, the manufacture of glass is a very energy-intensive undertaking and the industry is constantly working to reduce its energy consumption. This article describes the use and flow of energy in commercial glass production. In order to understand the relationship between energy and glass manufacturing, this articles first defines what glass is and then describes how it is produced.

1. INTRODUCTION

The atomic or molecular structure of solid materials can generally be found in two states: ordered and disordered. If the atoms of a material are ordered in a repeated array throughout, the structure will be an ordered three-dimensional pattern. This is a crystalline material. Solid materials that lack long-range atomic order are noncrystalline or amorphous. When a material is cooled from a liquid state to a solid, its formation into a crystalline or amorphous solid depends on how easily the random atomic structures of the liquid can organize into an ordered state while cooling. Factors such as composition and cooling rate are important. Rapid cooling of a liquid enhances the formation of an amorphous structure since the material is given little time to reach an ordered state. This is a glass. Figure 1 shows a comparison of the ordered arrangement of crystalline silicon dioxide and the disordered arrangement of a silicon dioxide glass. A common definition of a glass was given by the physicist W. H. Zachariasen, who stated that glass is an extended, three-dimensional network of atoms that forms a solid lacking the long-range periodicity (or repeated, orderly arrangement) typical of crystalline materials.

Manufactured glass products are made by melting crystalline raw materials into a molten state. The melt is subsequently shaped into the desired product and cooled to make a rigid form. The composition of the glass depends on the desired properties of the glass, which in turn define which raw materials are needed. Conceptually, there are thousands of different combinations of raw materials that will produce a glass, but in practice there are a limited number of formulations that are used to make the common glass products that are used on a daily basis. These limitations are typically dictated by practical and economic considerations.

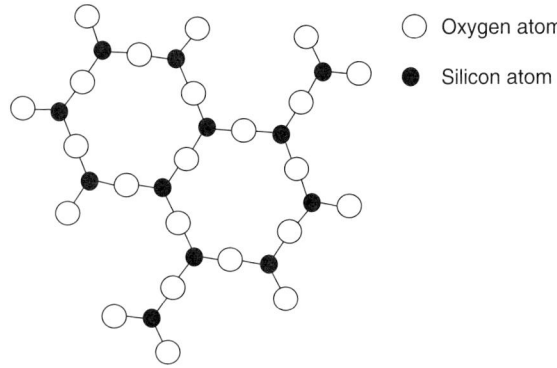

Crystalline silicon dioxide (SiO$_2$)

○ Oxygen atom
● Silicon atom

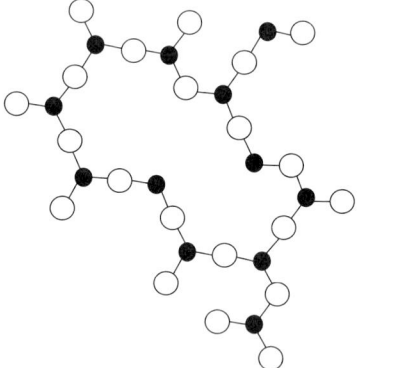

Noncrystalline silicon dioxide (SiO$_2$)

FIGURE 1 Schematic diagram showing the ordered arrangement of crystalline silicon dioxide and the disordered arrangement of a silicon dioxide glass.

2. TYPES OF GLASS

Glasses can be found in the natural environment, mostly in areas of volcanism where erupted magma is quickly cooled before crystallization can occur. Perhaps the best known example is obsidian, a black glass that consists mainly of oxides of silicon, aluminum, calcium, and sodium and/or potassium.

Commercially produced glasses differ widely in chemical composition and in physical and optical properties. Various materials can form a glassy or vitreous structure, such as the oxides of silicon and boron. Most commercial glasses are based on silica (SiO$_2$), which is the most abundant material on the earth's crust and is an excellent glass former. There are some nonsilicate glasses, but they make up only a very small fraction of commercially produced glass.

Pure vitreous silica glass can be made by melting quartz sand and cooling it quickly. Although silica makes an excellent glass for many applications, its high melting temperature ($>2000°C$) makes it too expensive for use in commodity glass products such as bottles and windows. The melting temperature can be reduced to a practical range by the addition of fluxes, the most common of which are alkali oxides (Li$_2$O, Na$_2$O, and K$_2$O). However, the addition of too much alkali can lead to the degradation of properties, such as the chemical durability of a glass (i.e., it will begin to dissolve when exposed to water). This can be compensated for by the addition of property modifiers, such as alkaline earth oxides (CaO and MgO) and aluminum oxide (Al$_2$O$_3$). The addition of these glass modifiers changes the glass structure, resulting in changes in the physical and chemical properties of the glass.

The composition used for a specific glass product depends on the desired final product performance, manufacturing characteristics (e.g., melting and forming), environmental considerations, and raw material/fuel costs. Although there are many different combinations of elements that will make a glass, there are some general compositional types that are widely used in commercial glass making. A glass made of silica, alkali, and alkaline earth elements is commonly called a soda-lime-silica glass and is the most common type. It has a relatively low melting temperature (begins melting at $\sim 800°C$ and is usually processed up to $1400°C$), has the proper viscosity characteristics for continuous production, and uses raw materials that are relatively inexpensive. It is the basis for making flat glass, containers, and consumer ware (e.g., drinking glasses).

Borosilicate glasses contain boron, which decreases thermal expansion, lowers viscosity, and allows faster fabrication speeds. Borosilicate glasses are used when resistance to thermal shock is desired, such as for cookware laboratory glass. Fiberglass products that are used for thermal insulation and structural reinforcement of composites contain boron.

Lead crystal and crystal glass are silicate glasses with added lead. The lead decreases viscosity and, hence, decreases the temperatures needed to work the glass and enhances the dissolution of refractory particles. Lead increases the refractive index of the glass, imparting a brilliance to the finished piece.

Specialty glasses include a variety of products, such as cathode ray tubes (CRTs), lighting glass (tubes and bulbs), laboratory and technical glassware, optical fibers, ceramic glasses (cookware), and glass for the electronics industry (e.g., LCD panels). Optical fibers are a specialty glass that is made of extremely high-purity silica. Optical fiber glasses cannot attain the needed purity levels produced by melting silica sand, so the glass is manufactured by precipitating chemically purified silica vapor.

3. INTRODUCTION TO COMMERCIAL GLASS MANUFACTURING

Glass manufacturing is essentially a commodity industry, with the majority of product sold to other industries, such as beverage, construction, and automobile manufacturers. In 1999, the U.S. glass manufacturing industry produced more than 18 million metric tons, worth approximately $17.6 billion. In the European Union, 1996 production was approximately 29 million metric tons (Table I).

3.1 Raw Materials

The overall goal of glass manufacturing is to convert crystalline raw materials into a homogeneous, flowing liquid that is free of visible defects that can be formed into a final product. This needs to be accomplished as quickly and as economically as possible and, at the same time, comply with all environmental regulations. Figure 2 shows a schematic diagram of the process of commercial glassmaking as described in this section. There are several methods of making and forming glass, but all begin with the mixing of raw materials or "batching." For most glasses, the most important glass former is silica. The raw material for this is quartz sand usually derived predominantly from sandstone deposits. Such deposits must contain few impurities, such as iron, which will affect the color of the glass. Glass property modifiers are introduced with a mix of limestone ($CaCO_3$) and dolomite ($CaMg[CO_3]_2$). The alkali flux is usually soda ash (Na_2CO_3), which comes from either a mined mineral ore called trona or the Solvay process of reacting sodium chloride with limestone. The soda and lime materials are carbonates so that upon heating the carbonate is driven off as volatile CO_2, leaving the metal oxide behind. Feldspar or nepheline syenite are aluminosilicate minerals that can be used in the batch to add alumina (Al_2O_3) and some alkali. Borosilicate glass batches can have 5–10% borate minerals, such a borax ($Na_2O \cdot 2B_2O_3 \cdot 5H_2O$) and colemanite ($Ca_2B_6O_{11} \cdot 5H_2O$) added to them. Mineral or rock wool is often made from slag that is a by-product from metal smelting.

Recycled glass, or cullet, is added to the batch to reuse waste glass from both within the plant and outside sources. Cullet also provides an early melting liquid phase that enhances reaction rates. The effect of cullet on energy use is discussed later. The main problem associated with cullet is the potential for contamination, particularly from pieces of metal, ceramic, or glass-ceramic. These contaminants can, at best, lead to rejected product and, at worst, damage the expensive refractory lining.

Minor additives include colorants, reducing agents, and fining agents. Colorants are added to control or add color and are usually oxides of transition metals or rare earth elements, such as cobalt, cerium, iron, and chrome. Fining agents are used to promote the removal of bubbles from the melt that, if not removed, will result in flaws in the final product. Common fining agents include sulfates,

TABLE I

Glass Sectors and Annual Production (Million Metric Tons)

Sector	European Union[a]	United States[b]
Flat glass	6.4	4.5
Container glass	17.4	8.7
Fiber	2.5	2.3
Speciality/Pressed/Blown	2.7	2.7
Total	29.0	18.3

[a] 1996 estimates.
[b] 1999 estimates.

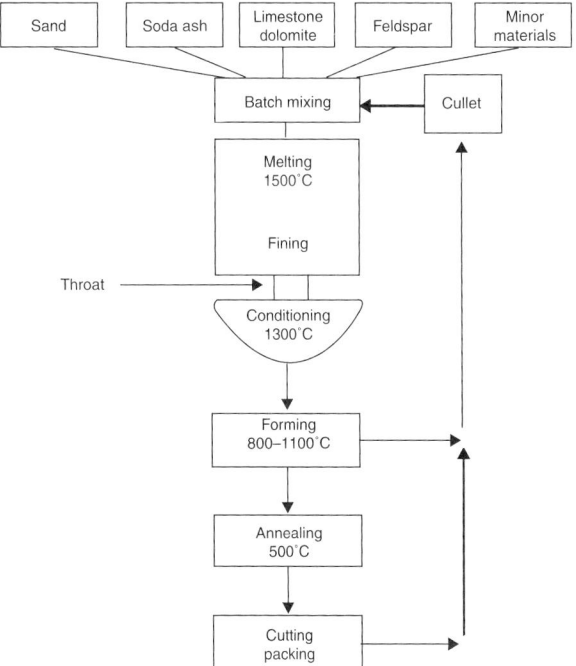

FIGURE 2 Schematic diagram of the process of commercial soda–lime–silica glassmaking showing raw materials, process sections, and approximate operating temperatures.

fluorides, nitrates, and oxides of arsenic or antimony. The most common combination for soda–lime–silica glasses is sodium sulfate and a reducing agent such as carbon.

The batching process consists of transferring raw materials directly from storage silos into a hopper and subsequently mixing the raw batch in a large-scale mixer. Cullet may be added after mixing. The mixed batch is placed in storage hoppers located directly adjacent to the "dog house," or the end of the glass furnace at which the batch is introduced and melting commences. The batch may be moistened with water or sometimes caustic soda to reduce particle segregation during transport to the furnace.

3.2 Glass Melting Furnaces

There are several types of melting furnaces used in the glass industry depending on the final product, raw materials, fuel choice, the size of operation, and economic factors. A melter can be either periodic or continuous. The latter is most widely used in large-scale operations. Continuous melters maintain a constant level by removing the glass melt as fast as raw material is added. Table II shows the three major types of furnaces and their advantages and disadvantages. The energy for melting comes from either the combustion of fossil fuels or electricity. The temperature necessary for melting ranges between 1300 and 1550°C. All furnaces are lined with high-temperature refractory materials to keep the corrosive melt from escaping and to insulate the melter.

Regenerative combustion furnaces recover heat from the exhaust stream to preheat the incoming combustion air by alternatively passing the exhaust and combustion air through large stacks of latticework refractory brick (regenerators or checkers). There are two sets of regenerators, so that as one is being preheated by the exhaust gases the other is transferring heat to the incoming combustion air (Fig. 3). The cycle is reversed approximately every

TABLE II
Advantages and Disadvantages of Different Furnace Types

Furnace type	Advantages	Disadvantages
Regenerative	Long furnace life	Higher NO$_x$ levels
	Long-term experience	Higher capital cost
Direct fired (mostly oxygen/fuel)	Reduction in fuel	Refractory corrosion
	NO$_x$ reduction	Short-term experience
	Lower capital cost	Cost of oxygen production
	Particulate reduction	
Electric	Efficient	High electricity cost
	No one-site pollution	Short furnace life

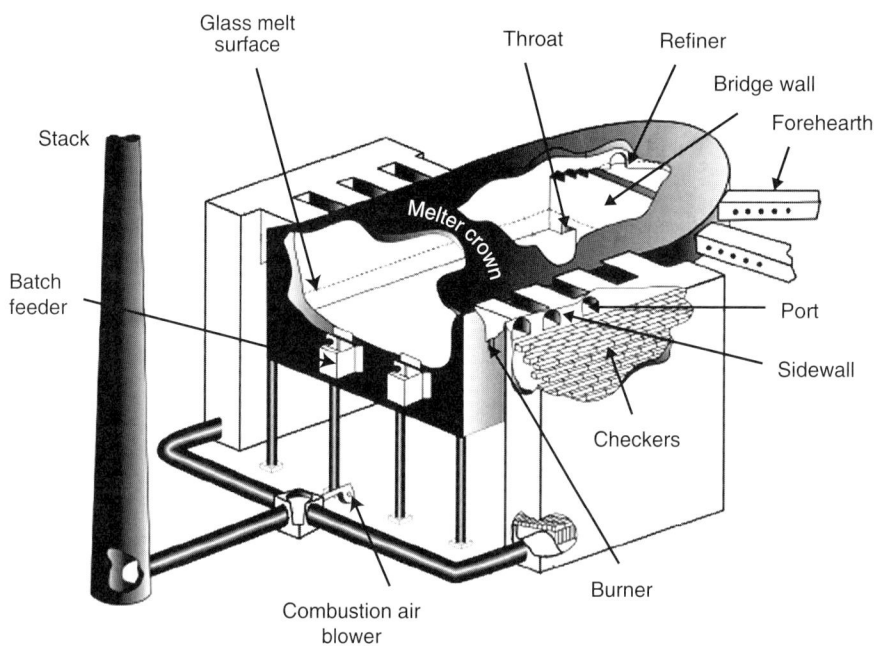

FIGURE 3 Diagram of a cross-fired regenerative furnace. This type is often used in large-volume production of container and flat glass. From U.S. Department of Energy (2002). "Energy and Environmental Profile of the U.S. Glass Industry."

20 min. Most glass-container plants have either end-fired (burners at each end) or cross-fired (burners on each side) regenerative furnaces, and all flat glass furnaces are cross-fired with five or six ports on each side with two burners for each port. Combustion air preheat temperatures of up to 1400 °C may be attained, leading to very high thermal efficiencies. A variant of the regenerative furnace is the recuperator, in which incoming combustion air is preheated continuously by the exhaust gas through a heat exchanger. Recuperative furnaces can achieve 800 °C preheated air temperatures. This system is more commonly used in smaller furnaces (25–100 tons per day). For large-capacity installations (>500 tons per day), cross-fired regenerative furnaces are almost always used. For medium-capacity installations (100–500 tons per day), regenerative end-port furnaces are most common.

A direct-fired furnace does not use any type of heat exchanger. Most direct-fired combustion furnaces use oxygen rather than air as the oxidizer. This is commonly called oxy-fuel melting. The main advantages of oxy-fuel melting are increased energy efficiency and reduced emission of nitrogen oxides (NO_x). By removing air, nitrogen is removed, which reduces the volume of the exhaust gases by approximately two-thirds and therefore reduces the energy needed to heat a gas not used in combustion. This also results in a dramatic decrease in the formation of thermal NO_x. However, furnaces designed for oxygen combustion cannot use heat-recovery systems to preheat the oxygen. Initially, furnaces that use 100% oxy-fuel were used primarily in smaller melters (<100 tons per day), but there is a movement toward using oxy-fuel in larger, float glass plants.

An electric furnace uses electrodes inserted into the furnace to melt the glass by resistive heating as the current passes through the molten glass. These furnaces are more efficient, are relatively easy to operate, have better on-site environmental performance, and have lower rebuild costs compared to the fossil-fueled furnaces. However, fossil fuels may be needed when the furnace is started up and are used to provide heat in the working end or forehearth. These furnaces are most common in smaller applications because at a certain size, the high cost of electricity negates the improved efficiency.

Some regenerative furnaces use oxygen enrichment or electric boost to optimize the melting process. Electric boosting adds extra heat to a glass furnace by using electrodes in the bottom of the tank. Traditionally, it is used to increase the throughput of a fossil fuel-fired furnace to meet periodic fluctuations in demand without incurring the costs of operating a larger furnace. It is also used to enhance the pull rate (the output at the forming end) of a furnace as it nears the end of its operating life.

3.3 The Melting Process

Regardless of the furnace type used, there are four basic steps to melting glass:

1. Conversion of the batch into a molten liquid free of undissolved crystalline material (called stones)
2. Removal of bubbles, or fining
3. Homogenizing the liquid so that it is chemically and thermally uniform
4. Conditioning the melt to bring it to a state in which it can be formed into the desired product

For a continuous furnace, these four steps occur simultaneously. The average residence time, from introduction of the batch to forming, is on the order of 24 h for container furnaces and up to 72 h for float glass furnaces. Glass melting tanks can contain 1500 tons or more of molten glass.

The melting of the batch consists of several steps from the moment it is introduced into the tank. Due to the low thermal conductivity of the batch materials, the initial melting process is quite slow. As the materials heat up, the moisture evaporates, some of the raw materials decompose, and the gases trapped in the raw materials escape. The first reaction is the decomposition of the carbonates at approximately 500 °C, which releases CO_2. The fluxes begin to melt at approximately 750 °C and begin to react with the silica. At 1200 °C, the silica sand and other refractory materials dissolve into the molten flux. As the remaining silica dissolves into the melt, its viscosity increases rapidly.

Bubbles are generated by the decomposition of raw materials, from reactions between the melt and the glass tank, and from air trapped between batch particles. These are removed as they rise buoyantly through the melt to the surface or as gases dissolve into the melt. Removal of the last few bubbles is the most difficult. Fining agents enlarge these bubbles to assist in their removal. Fining occurs as the melt moves forward through the throat of the melter to the conditioner, where melt homogenization and cooling take place.

Without homogenizing the melt, linear defects called cords (regions of chemical heterogeneity) will be seen in the final product. Furnaces are designed to promote recirculating convective currents within the melt to homogenize the glass. Convective mixing can be enhanced by physically stirring the melt with bubbles or mechanical stirring devices.

3.4 Forming Processes

After melting, fining, and homogenization, the melt passes through the throat of the furnace (Fig. 3) into a conditioning chamber, where it is cooled slowly to a working temperature between 900 and 1350°C. It is then ready to be delivered as a viscous mass through the forehearth to be formed into a specific product. The role of the forehearth is to remove temperature gradients and transfer the molten glass to forming operations. It consists of an insulated refractory channel equipped with gas-fired burners or electric heaters and an air cooling system.

Flat glass is most commonly formed by the float process, whereas patterned and wired glass are formed by rolling. Float glass and rolled glass are produced almost exclusively with cross-fired regenerative furnaces. The float process was developed by the Pilkington Brothers Glass Company of England in the 1950s, and it is the technique of pouring the melted glass onto a pool of molten tin. The glass floats on the tin and spreads out to form a uniform thickness. The glass is then mechanically stretched or constrained to attain the desired ribbon width and thickness. The tin bath is surrounded by a reducing atmosphere of nitrogen and hydrogen to keep the bath from oxidizing. The glass ribbon is pulled and modulated by water-cooled pulleys and exits the bath into the annealing lehr. Annealing is the process of heating and cooling the glass at a controlled rate to remove any residual stresses that could cause catastrophic failure while in use. After exiting the lehr, the glass is automatically cut into the desired sizes. The float method uses much less energy than preceding plate and sheet glass forming due to the use of more efficient furnaces and the elimination of the need for surface polishing and grinding. Rolled glass is formed by squeezing molten glass at approximately 1000°C between water-cooled steel rollers to produce a ribbon with a surface pattern or embedded wire.

Containers such as bottles and jars are made using a two-stage pressing and blowing process. This process is fully automated and begins with molten glass flowing from the furnace forehearth to a spout or set of spouts. The flow of glass is controlled by a mechanical plunger that produces a piece or gob of molten glass. The gob is then conveyed to the forming process, which produces a primary shape in a blank mold using compressed air or a metal plunger. The primary shape is placed into the finish mold and given its final shape by blowing compressed air into the container. During the forming process, the glass temperature is reduced by as much as 600°C to ensure that the containers are sufficiently rigid when placed on a conveyor. The cooling is achieved with high volumes of blown air against the molds. After forming, the containers are annealed by reheating to 550°C in a lehr and then cooled under controlled conditions. Larger operations can produce 500 bottles per minute.

Glass fibers can be found in nature where fountaining lavas are wind blown to produce long strands known as Pele's tears, after the Hawaiian goddess of fire. The manufacture of commercial glass fiber is similar, with some exceptions. Most fiber glass melts are produced with cross-fired recuperative furnaces, although there are several oxy-fuel fired furnaces in Europe. Short-strand glass wool is produced by spinning the melt in a crucible lined with holes. As the molten glass is ejected through the holes, high-velocity air blows the glass into fibers. For continuous glass fibers, the melt flows from the furnace through a series of refractory-lined channels to bushings. Bushings have several hundred calibrated holes or bushing tips that are electrically heated to precisely control the temperature and, thus, the viscosity and flow rate of the melt. The melt is drawn through the bushing tips by a high-speed drum to form continuous filaments. Once drawn from the bushing, the filaments are quickly cooled and then coated with a polymer binder.

4. ENERGY

Based on the previous description of glassmaking, it is evident that glass manufacturing is a very energy-intensive process. Energy accounts for 15–25% of the cost of glass products, depending on the sector and the country of manufacture. In 1999, the U.S. glass industry spent more than $1.3 billion on approximately 295 quadrillion joules (PJ) [280 trillion British thermal units (Btu)], including products from purchased glass. Taking into account electricity losses due to generation, transmission, and distribution, this figure increases to 417 PJ (395 trillion Btu) of total energy use.

By far the most energy-intensive components of glass manufacture are melting and refining, which together account for 70–80% of the total energy requirements. The remaining energy is used in the forehearths, tin bath (for float glass), forming/cutting processes, annealing, and other plant services. For example, in container manufacturing, the furnace accounts for 79% of the total energy used. The other areas of energy use are the forehearth (6%), compressed air (4%), mold cooling (2%), lehr (2%), and others (7%).

4.1 Sources of Energy

The sources of energy used to produce glass are natural gas, electricity, and, to a lesser extent, fuel oil. Fuel oil has been historically important, but today natural gas is the dominant fossil fuel used in glassmaking. Natural gas has higher purity, is relatively easy to control, and does not require on-site storage facilities. The higher purity of natural gas results in a reduction of sulfur dioxide emissions relative to fuel oil. Natural gas accounts for approximately 80% of energy use and more than 98% of purchased fuels in the U.S. glass industry. Although natural gas is mainly used to melt the raw materials, it is also used in emission control either to reduce NO_x in the exhaust stream or to incinerate toxic emissions from glass fiber production. Electricity is used in smaller resistance melters, fiberglass bushings, tin bath heating, annealing, and for powering motors to produce compressed air and operate conveyers and blowers.

4.2 Theoretical vs Actual Energy Use in Melting

The theoretical minimum energy required to transform the crystalline raw materials to a molten liquid is approximately 2.68 GJ/metric ton for a soda–lime–silica glass formulation. The calculation assumes all available heat is fully utilized and has three components:

- The heat of fusion to form the melt from the raw materials
- The heat, or enthalpy, required to raise the temperature of the glass from 20 to 1500°C
- The heat content of the gases released during melting.

The actual amount of energy required can be from 2 to 15 times the theoretical minimum, although for the majority of large-scale melters it is generally less than 8 GJ/ton. The amount of energy per unit of product produced can be referred to as the specific energy or energy intensity. As a comparison to other materials, the values for glass are similar to those of cement (5.4 GJ/ton) and lower than those of steel and iron (25.4 GJ/ton) and pulp and paper products (32.8 GJ/ton).

In addition to the specific product, the type of furnace and the desired quality of the glass affect the specific or per unit energy use. For instance, because flat glass needs to be completely free of imperfections, it generally requires longer residence times in the furnace than does container glass and therefore uses more energy per ton of glass produced. Table III shows the range of specific energy values for several different fossil fuel furnaces and glass products. Efficiencies in large regenerative furnaces can be as high as 55% and electric furnaces are 70–90% efficient (not considering the 25–30% efficiency of off-site electricity generation). Electric melters are more efficient because there are no heat losses in the stack or regenerator. Table IV shows the range of efficiencies for several types of furnaces.

The large differences between the theoretical and the actual energy used can be attributed to heat losses and other inefficiencies. The majority of heat loss is

TABLE III

Energy Consumption for Melting and Refining

Furnace type	U.S. average specific energy (GJ/metric ton)	EU average specific energy (GJ/metric ton)[a]
Flat glass		
Regenerative side-port	8.4	6.3
Electric boost side-port	6.2	—
Container glass		
Large regenerative	7.5	4.2
Electric boost	5.3	—
Oxy-fuel fired[b]	4.5	3.3
Small electric melter	2.7	—
Pressed and blown glass		
Regenerative	5.3	—
Oxy-fuel fired[b]	3.5	—
Electric melters	9.9	—
Recuperative	—	6.7
Regenerative TV tube	—	8.3
Insulation fiber glass		
Electric melters	7.2	—
Recuperative melters	6.7	4.3
Oxy-fuel fired[b]	5.4	—
Textile fiber		
Recuperative melters	10.1	—
Oxy-fuel fired[b]	5.4	—

[a] EU values assume 70% cullet for container glass, 40% cullet for television tube and tableware, and 20% cullet for flat.
[b] Does not include energy used in oxygen production.

TABLE IV
Comparison of Energy Efficiency Characteristics

Recuperative furnaces	20–40%
Regenerative furnaces	35–55%
Electric furnaces[a]	70–90%
Oxy-fuel[b] furnaces	40–55%
Regenerative w/batch/cullet preheater	50–65%
Oxy-fuel w/batch/cullet preheater[b]	50–65%

[a] Does not include 25–30% efficiency of off-site electricity generation.
[b] Does not include energy used in oxygen production.

from the furnace structure and through the removal of exhaust gases for fossil fuel melters. As described previously, modern melters that burn fuel employ a regenerator or recuperator to recover heat from the exhaust gases and preheat the incoming combustion air. This reduces the overall energy consumption by 50–70% compared to if it were direct-fired. The energy source, heating technique, and heat recovery method are central to the design of the furnace. The same choices also affect the environmental performance and energy efficiency of the melting operation.

4.3 Energy Reduction

The glass industry is constantly striving to close the gap between the theoretical and actual energy needed to melt raw materials into glass. Because melting and refining use the majority of energy in glass making, it is this part of the process that must be examined initially in any attempt to reduce per unit energy consumption. There are two general approaches toward reducing energy consumption: optimizing the existing system or completely replacing the current methods with and advanced melting system.

To reduce the amount of energy in conventional melting furnaces, the following may be addressed:

- Improvement of the combustion efficiency
- Reduction of the residence time in the tank
- Improvement of the insulation around the structure
- More effective use of the exhaust gas heat
- Oxy-fuel firing
- Increased cullet use

The melting technique chosen by a manufacturer is largely determined by economic considerations, and this can have a major effect on the energy efficiency. Although energy efficiency is an important aspect of the operating costs, these must be balanced by the other capital and operating costs, the desired production rate, and environmental performance (although energy efficiency and environmental performance are related).

Improvements in combustion efficiency have been driven by the need to reduce NO_x emissions. Reengineering of the burners and the control systems has also resulted in overall energy savings. Low NO_x burners reduce the volume of combustion air to close to stoichiometric levels, such that only the amount of air needed for combustion is introduced. Air that is introduced but not used in the combustion process would simply be heated and would carry that heat away in the exhaust.

The melt needs time (residence time) in the tank to produce a defect-free glass. However, the longer the melt remains in the tank, the more energy is expended in the process. Any reduction in the amount of time needed to melt, fine, or homogenize the melt will result in direct energy reductions. Residence times depend on the furnace temperature, composition of the batch, grain size of the batch ingredients, amount and grain size of cullet, and homogeneity of the batch. Improved batching techniques include ensuring optimal grain sizes and the addition of cullet. Because of the low viscosity of the initial melting fluxes, there may be a tendency for the batch to "unmix" and require excessive homogenizing times. Pelletizing or pre-reacting some of the batch components may reduce this effect. Furnace temperature can be increased to reduce melting and refining times, although this is done at the expense of the life of the refractory lining. Electric boost can be used in fossil fuel melters to increase the pull without changing the furnace size. In general, the electricity required to increase the pull rate by 1 ton is approximately 22–28 kW. The degree of homogeneity depends on the quality requirements, which vary considerably between and within glass sectors. Bubbling is a method of decreasing homogenization time in which air is injected through several nozzles installed at the bottom of the melting chamber, agitating the melt and promoting mixing.

As much as 30% of energy can be lost through the furnace structure. Structural heat losses are inversely proportional to the furnace size because of the change in melter surface area-to-volume ratio. Thus, larger furnaces have inherently lower heat losses. Regardless of the size of the melter, improvements in the insulation of the structure will increase energy efficiency. Most glass contact refractories that are used to line the furnace are dense, fusion cast materials that have high thermal conductivity and need to be insulated. The application of insulation to a furnace depends on the area of the furnace and the operating conditions, and not all parts can be insulated. Insulation increases the operating temperature of the refractory lining, which tends to shorten its life span. Overheated refractory material can also degrade and shed material into the glass, creating defects. Therefore, more efficient insulation needs to be balanced with the quality of the glass and lifetime of the refractories.

Advances in refractory material engineering have allowed furnaces to operate longer between rebuilds at higher levels of insulation. For example, super-duty, high-purity silica and alumina–zirconia–silica bricks have been developed for the crown to improve insulation. Sprayed fiber insulation can be applied to the regenerator structure, resulting in up to a 50% reduction in regenerator structural heat losses.

Up to 30% of the heat from melting is expelled through the exhaust gases. Furnaces equipped with regenerators or recuperators recover some of this heat, with end-fired regenerative furnaces generally more efficient than cross-fired regenerative furnaces. However, combustion control and furnace size are more limited for end-fired furnaces. The amount of energy recovered by regenerators may be increased by increasing the quantity of refractory bricks in the checkers using enlarged regenerator chambers or in separate but connected structures. However, as one approaches the physical maximum efficiency, there are limitations on the cost of the extra refractory bricks and the limitation of available space. Improved materials and designs such as corrugated cruciform bricks can improve heat transfer.

Gases exiting the regenerators or recuperators are 300–600°C, and there are other potential areas of use for the remaining waste heat, although too much heat loss in the exhaust stream will adversely affect the ability of the plant to vent the exhaust up the stack. One heat-recovery technique is to pass exhaust gases through a boiler to generate steam. The steam can be used for heating, on-site electricity generation, or to drive air compressors. Potential problems are fouling of the boiler tubes, high capital costs, and interconnection with the utilities (in the case of electricity generation).

Another potential use for the heat is preheating the batch or cullet, which is normally fed cold into the furnace. This can be accomplished by direct or indirect exposure of the batch to the exhaust gases. Direct preheating can be done by directing the hot exhaust gas opposite the batch as it falls from a feeder. This can raise the batch or cullet temperature up to 400°C. Indirect preheating systems use a plate heat exchanger and may consist of blocks of horizontal waste gas flow and vertical material funnels. This can lead to heating of the cullet or batch to 300°C. The exhaust gases exiting the preheaters will be cooled by approximately 270–300°C. Potential problems are increased particulate emissions and size sorting, although this is not as important in preheating cullet.

As mentioned previously, oxy-fuel firing can be more energy efficient than air-fuel furnaces. Most glass companies have incorporated some oxy-fuel firing into their operations, either as oxygen-enriched air or as 100% oxygen. A furnace that is equipped with oxygen-enriched air can produce the same amount of glass as with air combustion but at lower fuel input because of the reduced exhaust volume. However, when regenerative furnaces are converted to oxy-fuel firing, the heat-recovery checkers are eliminated. In addition, oxygen is expensive and to make oxy-fuel systems economical, a cost-effective and reliable source of oxygen is required. Several on-site systems can be used to fractionate oxygen from ambient air. Liquified oxygen can be used, but its use is limited by transportation costs. The energy required to use oxygen ranges from 250 to 800 kWh per ton of oxygen.

4.4 Cullet and Energy

Cullet has a lower melting energy requirement than the virgin raw materials because the heat of fusion of converting crystalline material to a liquid is not needed and its mass is 20% lower than that of the equivalent batch materials. General energy savings are 0.15–0.30% for each percent of cullet added to the batch. For example, a melter that uses 50% cullet will use 7.5–15% less energy than a batch with no cullet.

Most plants routinely recycle all internal cullet—that is, waste glass produced within the production process. For flat glass production, approximately 10–20% of the produced glass returns as internal cullet due to edge trimming, defects, product changes, and breakages. External sources of cullet (from consumer or external industrial sources) are also used, with the amount used depending on the product, specifications, availability, and price. The composition of external cullet can vary and the presence of impurities such as metal and ceramics may limit its use. Stringent final product quality requirements can restrict the amount of foreign cullet a manufacturer can use. Container glass manufacturing benefits from bottle recycling schemes and, depending on the country, this sector can consistently use 80% cullet in the batch. It is more difficult for flat glass manufacturers to find external cullet, although they are able to obtain scrap from the industries that purchase their product, such as architectural and automotive window manufacturers. All manufacturers keep a store of cullet on site in the event that the raw material batching system is offline and they need to melt at 100% cullet.

4.5 Advanced Melters

The next generation of glass melters needs to be more energy efficient than current melters, but they also must operate as rapidly and economically to produce the highest quality product. Currently, the most efficient glass melters are the large regenerative furnaces, but they are also the most capital intensive, have high rebuild costs, and are very inflexible in terms of varying or reducing output.

There are potential technologies that can make melting more efficient by making heat transfer to the batch more efficient. Submerged combustion melting uses natural gas and an oxidizer injected directly into the melt, which improves heat transfer and promotes rapid mixing. Plasma, which is a partially ionized gas that conducts electricity, can be very efficient at transferring heat directly to batch materials, but it is difficult to apply on a large scale. Microwaves can heat without direct contact, but different components of the batch have different susceptibilities to heating by microwaves.

One obvious approach to reducing the energy needs of the commercial glass industry is to reduce the residence time of the melt in the furnace. Rapid glass melting systems would allow the raw materials to be melted and homogenized without the need for the relatively slow, convective mixing that is currently practiced. The industry would like the current average 24- to 48-h residence time of a melt reduced to 1 h.

Combined melting and mixing can be achieved using a variety of mechanisms. Higher temperatures reduce viscosity and decrease mixing times, but the reduced time must be balanced by the increased energy input. Higher temperatures also adversely affect the refractory lining and can increase emissions of NO_x and volatile components of the batch. Improved mechanical mixing will decrease homogenization times. Decreased fining times can be achieved by controlling the atmosphere surrounding the melt or applying a vacuum.

There are several conceptual designs of fully reengineered methods of melting glass. Many of these segment the furnace into distinct sections for melting, fining, and homogenization. One segmented melting concept takes advantage of the different melting requirements of virgin batch materials and cullet. Virgin batch is charged into an all-electric premelting furnace that converts 75% of the raw material into glass. The premelted batch then moves into an enlarged doghouse where cullet is added (cullet comprises at least 60% of the raw material). The batch/cullet mix then enters the second melting chamber that uses oxy-fuel burners. The potential benefits are up to 25% improved energy efficiency and lower emissions. Another concept under development, referred to as the advanced glass melter, injects all batch materials into the flame of a natural gas-fired furnace, heating them rapidly and discharging them into the melt chamber. Because of lower flame temperatures, this system has potential for low NO_x emissions.

Acknowledgments

C.W.S. thanks Jim Shelby (Alfred University), Tony Longobardo (Guardian Industries), and Ernst Worrell (Lawrence Berkeley National Laboratory) for their assistance and reviews.

SEE ALSO THE FOLLOWING ARTICLES

Aluminum Production and Energy • *Cement and Energy* • *Industrial Energy Use, Status and Trends* • *Plastics Production and Energy* • *Steel Production and Energy*

Further Reading

Glass Manufacturing Industry Council (2001). "Glass Melting Technologies of the Future." Glass Manufacturing Industry Council, Westerville, OH.

Integrated Pollution Prevention Control (2000). "Reference Document on Best Available Practices in the Glass Manufacturing Industry." European Commission, Institute for Prospective Technological Studies, Seville.

Ruth, M., and Dell'Anno, P. (1997). An industrial ecology of the U.S. glass industry. *Resour. Policy* 23, 109–124.

Shelby, J. E. (1997). "Introduction to Glass Science and Technology." Royal Society of Chemistry, Cambridge, UK.

U.S. Department of Energy (2002). "Energy and Environmental Profile of the U.S. Glass Industry." U.S. Department of Energy, Washington, DC.

Varshneya, A. K. (1994). "Fundamentals of Inorganic Glasses." Academic Press, San Diego.

Woolley, F. E. (1991). Melting/fining. In "Engineered Materials Handbook: Ceramics and Glasses," Vol. 4, pp. 386–393. ASM International, Materials Park, OH.

Global Energy Use: Status and Trends

LINDA E. DOMAN
U.S. Department of Energy, Energy Information Administration
Washington, D.C., United States

1. Current Issues in Energy Demand
2. Today's Energy Use
3. Resources and Reserves
4. Drivers of Energy Demand
5. Regional Energy Use
6. Summary

Glossary

American Petroleum Institute (API) gravity A measure of specific gravity (in degrees) of crude oil or condensate. An arbitrary scale expressing the gravity or density of liquid petroleum products.
barrel A unit of volume equal to 42 U.S. gallons.
British thermal unit The quantity of heat required to raise the temperature of 1 lb of liquid water by 1°F at the temperature at which water has its greatest density (approximately 39°F).
fossil fuel An energy source formed in the earth's crust from decayed organic material. The common fossil fuels are petroleum, coal, and natural gas.
gas to liquids (GTL) A process that combines the carbon and hydrogen elements in natural gas molecules to make synthetic liquid petroleum products, such as diesel fuel.
generator capacity The maximum output, commonly expressed in megawatts, that generating equipment can supply to the system load, adjusted for ambient conditions.
liquefied natural gas (LNG) The liquefied product derived from natural gas (primarily methane) that has been reduced to a temperature of −260°F at atmospheric pressure.
megawatt One million watts of electricity.
oil resources The conventional oil resource base is defined by three categories: remaining reserves (oil that has been discovered but not produced); reserve growth (increases in reserves resulting mainly from technological factors that enhance a field's recovery rate); and undiscovered (oil that remains to be found through exploration).
pig iron The crude, high-carbon iron produced by reduction of iron ore in a blast furnace.
proved energy reserves The estimated quantities of energy sources that analysis of geologic and engineering data demonstrates with reasonable certainty are recoverable under existing economic and operating conditions. The location, quantity, and grade of the energy source are usually considered to be well established in such reserves.
tar sands The naturally occurring bitumen-impregnated sands that yield mixtures of liquid hydrocarbon and that require further processing other than mechanical blending before becoming finished petroleum products.
watt The unit of electrical power equal to 1 amp under a pressure of 1 V; 1 W is equal to 1/746 horsepower.
watt-hour The electrical energy unit of measure equal to 1 W of power supplied to, or taken from, an electric circuit steadily for 1 hr.

World energy use has increased steadily over the past several decades. Much of the growth in global energy consumption has been concentrated on the use of fossil fuels (oil, natural gas, and coal). This trend is expected to continue over the foreseeable future. Industrially mature nations will continue to rely on fossil fuels to meet their energy needs for all end uses, but the greatest rate of energy use is projected to occur in the emerging economies of the developing world. In these developing nations, fossil fuel use will be needed to feed growing industrial machinery, particularly the energy-intensive energy end uses that are needed to build industrial infrastructure, such as cement and steel-making, paper and pulp, and other energy-intensive processes. In this article, the trends of worldwide energy use by region are discussed. Also considered are the drivers of increased energy use, and the ramifications for future energy demand. These include the impact of increased energy demand on the environment and the role that advancing technologies may play on the world's future energy mix.

1. CURRENT ISSUES IN ENERGY DEMAND

Non-fossil-fuel energy sources (i.e., nuclear power and renewable energy sources) face difficulties in competing with fossil fuels. In the case of nuclear power generation, the industry has encountered political, economic, and social difficulties in many parts of the world, beginning with the nuclear accidents at Three Mile Island in the United States in 1979 and at Chernobyl in the Soviet Union in 1986. There is strong public sentiment against nuclear power in much of the industrialized world, based on concerns about plant safety, radioactive waste disposal, and the proliferation of nuclear weapons. Some growth in nuclear power may be expected among the developing countries of the world, particularly in developing Asia, where 17 of the 35 nuclear power reactors currently under construction worldwide may be found. However, the industrialized world is by and large expected to see a decline in nuclear power generation over the long term, as aging reactors are retired and are not replaced by new nuclear generators.

Renewable energy sources enjoy a great measure of popularity with the public—both because these sources of energy are cleaner burning compared to other energy sources and because they are not subject to resource depletion, perceived to be a problem for fossil fuels. However, renewable energy sources cannot yet compete economically with fossil fuel resources. With fossil fuel prices likely to remain moderate over the next quarter century, renewable energy sources will probably fail to gain market share relative to the fossil fuels in a business-as-usual scenario. In the absence of significant government policies aimed at reducing the impacts of carbon-emitting energy sources on the environment, it will be difficult to encourage a wide-scale expansion of renewable resources.

It should also be noted, however, that the renewable energy forecasts presented here deal only with commercial renewable energy sources, i.e., those projects connected to national electricity grids. The Paris-based International Energy Agency (IEA) has estimated that noncommercial, traditional biomass resources are the primary sources of heating and cooking for 2.4 billion people in the developing world. Unfortunately, sources of comprehensive data on the use of noncommercial fuels are not readily available on a consistent, worldwide basis, and so cannot be included in projections of global energy consumption. The IEA expects the developing populations of the world to begin to switch away from traditional fuels to commercial energy sources, but the reliance on the noncommercial fuels will clearly remain strong in the mid-term future given the high levels of current use.

Another trend that is expected to continue is a move toward improving energy intensity. As time goes by and energy technologies become more efficient and widespread, energy use becomes more and more productive. That is, by employing more efficient technologies, less energy is needed to produce the same amount of output. One part of the world where this is particularly obvious is among the nations of the former Soviet Union. Rich with fossil fuel resources and mired in dysfunctional political and economic systems, Russia and the Soviet satellite nations used energy very inefficiently for much of the past half century. With the fall of the Soviet Union in the early 1990s, these economies began to transition into a market-based economic system, and already the old, Soviet-era machinery and technologies are being replaced by more energy-efficient capital stock. This has meant an improvement in the region's energy intensity since the mid-1990s. As these trends continue, the transitional economies of Eastern Europe and the former Soviet Union will begin to see patterns of energy intensity that resemble those of the industrialized world.

The real growth in energy use is projected to occur in the developing economies of the world. Economic expansion is being accompanied by consumer demand for energy-consuming products. Growing demand for energy to fuel industrial end uses, as well as energy demand in the residential, commercial, and especially transportation and electric utility sectors, will have profound implications for the world's geopolitical relationships. There is tremendous opportunity for large increments of energy consumption in these economies, as nations strive to advance standards of living to match those of the industrialized world. In particular, many nations that are emerging economically will see consumer demand increasing for personal motor vehicles and a wide assortment of home appliances, including televisions and personal computers. National electrification has become an important goal of many of these nations, to improve both the productivity and the standards of living of their populations.

2. TODAY'S ENERGY USE

In the year 2001, 404 quadrillion British thermal units (Btu) of energy were consumed worldwide.

Petroleum products accounted for the largest share (39%) of the world's total energy consumption (Fig. 1). Oil products are particularly important for the transportation sector, and this will probably be true for the next several decades. The world's transportation infrastructure relies heavily on the combustion engine, which, in turn, largely relies on petroleum to operate. Currently, there are no truly commercially competitive alternatives to petroleum products in the transportation sector. Short of a substantial technological advance—such as hydrogen-fueled vehicles—that would offer competition to traditional petroleum-product-fueled automobiles, trucks, and airplanes, it will be difficult for any other fuel type to replace oil in this sector.

Although oil is predominant in the transportation sector, alternative fuels have displaced petroleum product use in other end-use sectors. In many parts of the world, for example, other fuels are increasingly used to fuel electricity generation and the importance of oil in this end-use sector has been decreasing over the past several decades. This trend is expected to continue into the future. Natural gas and nuclear power have been two energy sources that have taken market share away from oil in the electric utility sector, especially since the 1970s. Nuclear power generation advanced strongly in the 1970s. If it were not for the public concerns about the safety of nuclear power, because of accidents at Three Mile Island in 1979 (United States) and at the then-Soviet Chernobyl in 1986, nuclear generation may well have continued to experience significant growth and might have gained substantial market share among the electricity-generating energy sources.

Natural gas is now seen as the fuel of choice for new electricity generation worldwide. Combined cycle gas turbines have been demonstrated to be very efficient electricity generators. What is more, natural gas is widely regarded as a more cleanly burning fossil fuel compared to either coal or oil. In an age in which environmental concerns are increasingly important to the world's governments, natural gas is a desirable alternative to coal- or oil-fired power plants, and gas does not come with the negative aspects of nuclear power, such as concerns about safety and radioactive waste disposal. Where infrastructure and resources are available, natural gas has already gained significant share of the electricity sector. For instance, natural gas supplies more than half (54% in 2000) of the electricity generation in Russia, which, not coincidentally, possesses 31% of the world's total natural gas proved reserves.

Coal is the predominant source of electricity generation in many countries. In the United States, coal accounts for 52% of the country's total electricity-generating capacity. Although it is a relatively "dirty" form of energy (the combustion of coal emits more carbon dioxide than the same amount of natural gas and of most petroleum products), reserves of coal are also fairly widespread across the world and are plentiful. By some estimates, at current production levels, the world has enough coal to last for another 200 years or more. In two of the world's fastest growing developing economies, China and India, coal is an extremely important energy resource. These two countries alone are expected to account for more than 85% of the world increment in coal use over the next two and one-half decades (Fig. 2). Although incremental coal use over most of the world is expected to be for the generation of electricity, in China coal is also an important resource for

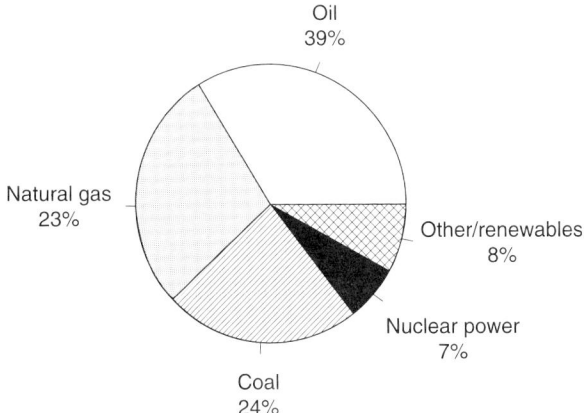

FIGURE 1 World energy use by fuel type, 2001. Data from Energy Information Administration (2003).

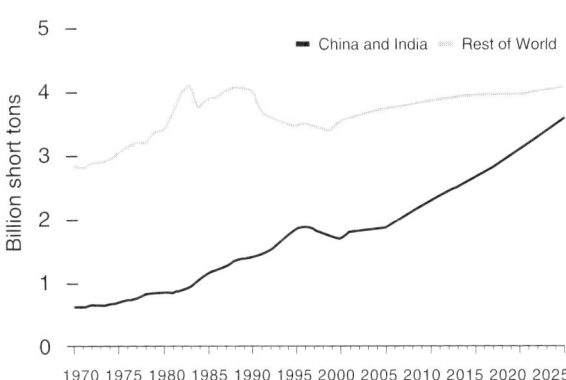

FIGURE 2 World coal use, 1970–2025. Data from Energy Information Administration (2003).

industrial purposes. China is the world's largest producer of both steel and pig iron. The country also uses coal for steam and direct heat for industrial applications, primarily in the chemical, cement, and paper and pulp industries. This heavy reliance on coal may lessen somewhat over the next two decades, but because coal is such a plentiful resource in China, it is difficult to envision a time when its importance will be eclipsed by any other currently available energy source.

As previously noted, nuclear generation showed great promise in the 1970s. Between 1970 and 1986, nuclear generation increased by an estimated 14% per year (Fig. 3). However, the nuclear accidents at Three Mile Island in 1979 and Chernobyl in 1986, in conjunction with the high capital costs associated with bringing nuclear generation into operation relative to coal- or natural-gas-fired generation, slowed the development of nuclear power. Despite problems associated with the disposal of highly radioactive waste, many analysts believed nuclear power would become an increasingly important source of the world's electric power mix. However, public wariness about the safety of nuclear power largely halted the growth of this sector in the 1990s, especially among the industrialized nations and the nations of Eastern Europe and the former Soviet Union. In fact, no orders for new nuclear generators have occurred in Austria, Hungary, Italy, Mexico, the Netherlands, Switzerland, or the United States since 1973. The only industrialized countries that still have plans to expand nuclear generation with new builds are Finland, France, and Japan.

Most of the growth in new nuclear generation is expected to occur in developing Asia (Fig. 3). Of the 35 reactors under construction in 2003, nearly half are located in the countries of developing Asia. This includes 8 reactors in India alone, 4 in China, 2 each in South Korea and Taiwan, and 1 in North Korea. These new reactors are expected to add some 12,000 MW of installed nuclear capacity to the existing 28,000 GW on-line in the region as of January 2003. Nuclear generation remains an attractive alternative to other forms of electricity in these countries, either because it will allow them to diversify their electricity-generating sources, such as in India and China, or because there are few indigenous resources that can be used to fuel electricity demand growth, such as in South Korea and Taiwan.

Many countries would like to expand the use of renewable energy resources. In the industrialized world, many governments view renewable energy resources (with the notable exception of large-scale hydroelectric power) as an attractive alternative to expanding electricity generation because of the lack of negative environmental impact, and renewable energy resources are often cited as a way for countries to reduce or limit emissions of carbon dioxide. In western Europe, many government incentive programs are in place to spur the growth of renewable resources, such as wind, solar, geothermal, and small hydroelectric generators, among others. Germany, Denmark, and Spain have made great strides in developing their wind generation capacity through government programs. That said, wind power and other nonhydroelectric forms of on-grid, commercial renewable energy resources remain fairly small portions of the electric generation mix of these nations.

As with any energy form, renewable energy does not come without its controversies. Sometimes populations faced with wind farms view them as eyesores, despoilers of the surrounding landscape, and there can be complaints about the noise that the whirring turbine blades can produce. Procuring site licenses has been particularly difficult in the United Kingdom. The Non Fossil Fuel Obligation—a program used in the United Kingdom to finance the development of renewable energy resources between 1989 and 2001, when it was replaced by the Renewable Obligation of the National Electricity Trading Arrangements program—approved power purchase contracts of wind capacity amounting to 2676 MW, but by the end of 2001 only 468 MW of wind capacity had been installed.

Hydroelectricity remains the largest and widest form of renewable energy worldwide. In the United States alone, hydroelectric power accounts for

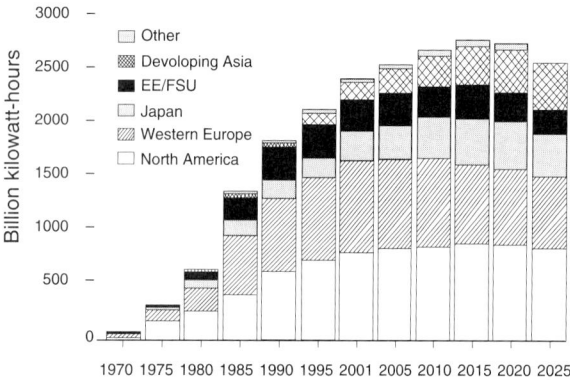

FIGURE 3 World nuclear power consumption by region, 1970–2025. EE/FSU, Eastern Europe/former Soviet Union. Data from Energy Information Administration (2003).

between 8 and 10% of the nation's total electricity generation, depending on the amount of rainfall in any given year. Many developing nations rely quite heavily on hydro power for their electricity supplies. In South America, Brazil, Peru, and Chile use hydroelectricity for 80% or more for their electricity needs. Large-scale hydroelectric facilities are often controversial because of their impact on the environment and because it is often necessary to displace surrounding populations in order to create the reservoirs needed to supply the hydro turbines with enough water to generate needed electricity.

Because most of the hydroelectric capacity in industrialized countries has either already been exploited or is located far from the population centers that could use the hydro-generated electricity, much of the growth in hydroelectricity is expected to be in the form of large-scale hydro projects in the developing countries of the world, especially those in developing Asia. Although the average size of a thermal electricity generator (i.e., oil, coal, or natural gas) is about 300 MW, China is currently constructing an 18,200-MW hydroelectric project on the Yangtze River, the Three Gorges Dam project. In Malaysia, the government is pressing forward to construct a 2400-MW Bakun Dam project. These projects and others in Vietnam, India, and Laos, among others, have met with staunch opposition from the international community because of their potentially detrimental impact on wildlife, plant life, and the indigenous populations that will be disrupted by the dams' construction. Still, governments persist with the developments of large-scale hydro projects because of the strong growth in demand for electricity and in an effort to diversify electricity-generating energy sources.

Noncommercial biomass is often not discussed as part of the world's energy use. Although the burning of wood and other plant life to cook and heat water is considered an important component of energy use in many rural parts of the developing world, it is also difficult to quantify. The International Energy Agency has been attempting to measure the amount of biomass used worldwide since 1994. Their estimates indicate that some 2.4 billion people in the developing world still rely on traditional, noncommercial biomass to meet their heating and cooking needs. Even when people in the developing world have access to electricity to meet their lighting needs, they often still rely on biomass for other residential uses. The heavy reliance on surrounding flora is accompanied by many problems. Because developing countries typically have high population growth rates relative to the industrialized world, there is a tendency to need more and more plant life to satisfy the populace's energy needs. The result is that a locale's biomass is often overharvested, resulting in the danger of deforestation, which can degrade the environment and make sustaining life difficult, if not impossible. The Tata Energy Research Institute of India noted that many women in rural South Africa carry 20 kg of fuel wood an average of 5 km each day. The effort required to carry out this task accounts for a large share of the gatherer's daily caloric intake, and as deforestation requires the person to go farther and farther to achieve the requisite fuel wood, the task will eventually not be sustainable.

The use of biomass, particularly fuel wood, dung, and charcoal, can also negatively impact the health of the people utilizing these products. People dependent on biomass resources are consistently exposed to indoor particulate and carbon monoxide levels that are far in excess of those the World Health Organization has deemed safe. Traditional stoves that burn dung or charcoal often emit large amounts of carbon monoxide and other noxious gases, resulting with acute respiratory illnesses and a shortened life span. Although the International Energy Agency has noted that the use of biomass has been declining steadily since biomass surveys were initiated, the agency also believes that it will be 30 years or more before the world will see a substantial reduction in the reliance that developing nations' populations have on this energy source.

3. RESOURCES AND RESERVES

In the 1970s, in an environment of oil shortages caused by the oil embargo staged by the Organization of Petroleum Exporting Countries (OPEC) cartel members, many analysts were concerned with estimating the point at which oil supplies and, indeed, all fossil fuels would be depleted. Gas rationing and high world oil prices associated with this time period further fueled this concern. In fact, work done by the U.S. geologist M. King Hubbert suggested that world oil production would ramp up and peak over only a few short decades and then just as rapidly decline.

Technology has helped to increase the volume of proved oil reserves over the past few decades. Three-dimensional drilling and seismic imaging have been enormously helpful in improving the ability of the oil industry to produce heretofore unrecoverable oil. In part as a result of these enhancements, most analysts today have a more measured view of the potential for world fossil fuel resources. In its most recent assessment of oil's long-term production potential, the U.S. Geological Survey identified at least 3 trillion barrels of ultimately recoverable conventional oil—that is, oil that can be recovered with current prices and technology—worldwide. Historically, it has been estimated that only about 25% of that oil that has been classified as "ultimately recoverable" has already been produced; rough calculations indicate that a likely peak in worldwide conventional oil production would occur at some point after 2020.

Analysts would agree that fossil fuels are a limited resource and, as such, are subject to depletion. Further, depletion would lead to scarcity and higher prices. There is, however, the matter of nonconventional resources. Nonconventional resources are defined as those resources that cannot be produced economically at current prices using current technology. This categorization shifts as the prices for these resources continue to rise, and the gap between conventional and nonconventional thus narrows. As a result, the combination of escalating prices and technological advancements will transform the nonconventional into the conventional. Much of the perceived pessimism regarding the finite quantities of resources has focused only on conventional resources as defined today. However, it was demonstrated in the 1990s that technological advances can help lower the costs of producing liquid fuels from several nonconventional sources, including heavy oils, tar sands, and natural gas.

Heavy oils generally have an American Petroleum Institute (API) gravity of less than 24° and will not flow on their own at standard reservoir temperatures. Tar sands are similar, but have more viscous oil that is located closer to the surface. More than 3.3 trillion barrels (oil in place) of heavy oil and tar sands are estimated to exist worldwide. The most significant deposits of these resources are found in Canada and Venezuela. There are two distinct ways to recover these resources. For deposits that are located close to the surface, mining can be used. For deposits located deeper under the surface, steam injection could be used to heat the oil and allow it to flow more easily, so that it behaves more like conventional resources. Although it is true that these heavy oils and tar sands will have to be cleaned and upgraded before they behave like conventional refinery feedstocks, higher prices for conventional sources would make the additional costs associated with the additional processes economical. A sustained world oil price of between $23 and $25 per barrel could make these nonconventional resources economical to develop and produce.

Significant portions of the world's natural gas resources are located in remote areas or in small deposits. Development of such sites is usually discouraged, because delivery by way of pipeline or in the form of liquefied natural gas (LNG) is not economical. However, natural gas molecules can be reconstituted as liquid synthetic petroleum products by employing an updated version of a technology that has been available since World War II, the so-called Fischer–Tropsch method. This gas-to-liquids (GTL) technology presents an attractive alternative for the small, remote deposits, because the infrastructure for petroleum products is already in place. GTL technology is also versatile enough to accommodate smaller gas deposits economically. By some estimates, GTL technology could be economically viable with sustained world oil prices of between $26 and $28 per barrel. It is possible to envision a scenario in which oil prices could climb even higher than those that would make nonconventional or GTL technologies economical. For instance, OPEC producers could adopt a conservative productive capacity strategy that might result in a doubling of prices, in real terms, over the next few decades. Sustained high world oil prices might further alter the supply side of the market even beyond those suggested by heavy oil, tar sands, and GTL. This might lead to the development of other nonconventional resources. Coal-to-liquids technologies and even shale oil could then economically be developed. Shale oil has enormous potential reserves that would dwarf those of conventional resources.

As with petroleum reserves, natural gas resources have in general increased in every year since the 1970s. As of January 1, 2003, proved natural gas reserves—that is, those quantities that can be recovered under present technologies and prices— were estimated by the *Oil & Gas Journal* at 5501 trillion cubic feet. Most of the increases in natural gas reserves in recent years have occurred in the developing world, and about 71% of the world's natural gas reserves are found in the Middle East and in the former Soviet Union (Fig. 4). The remaining reserves are spread fairly evenly among other regions of the world. Despite the high rates of increase in the use of natural gas worldwide, most regional reserves-to-production ratios have remained high. Worldwide, the reserves-to-production ratio is estimated at 62 years, but the former Soviet Union has a reserves-to-production ratio estimated at nearly 80 years; this

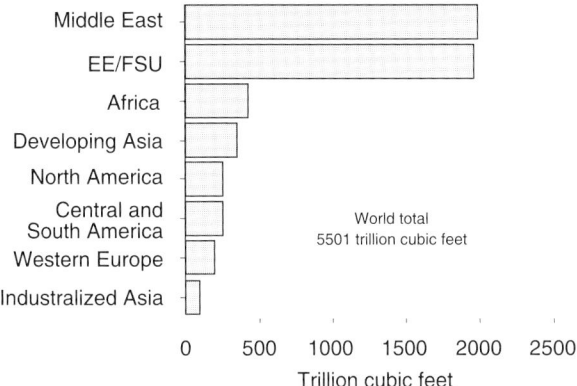

FIGURE 4 World natural gas reserves by region as of January 1, 2003. EE/FSU, Eastern Europe/former Soviet Union. Data from *Oil & Gas Journal*, Vol. 100, No. 52, December 23, 2002, pp. 114–115.

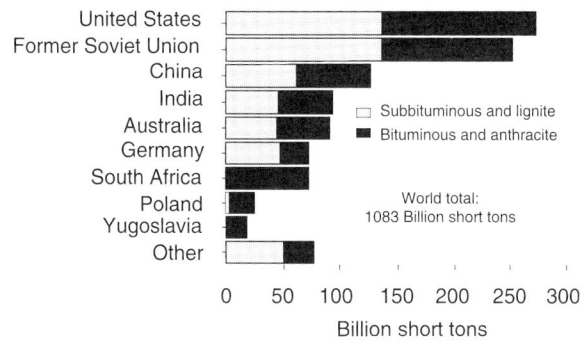

FIGURE 5 World recoverable coal reserves. Data from World Energy Council London, UK (2001) (for all countries except the United States; data for the United States provided by the Energy Information Administration); also based on unpublished data of the Coal Reserves Data Base (February 2002).

ratio in Africa and in the Middle East is about 90 years and over 100 years, respectively.

According to the U.S. Geological Survey's most recent assessment of world natural gas resources, there is a significant amount of natural gas that remains to be discovered. The mean estimate for worldwide undiscovered natural gas is 4839 trillion cubic feet. Of the natural gas resources that are expected to be added over the next 25 years, reserve growth accounts for 2347 trillion cubic feet. It is estimated that one-fourth of the undiscovered natural gas is located in undiscovered oil wells. Thus, it should come as no surprise that more than one-half of the mean undiscovered natural gas is expected to come from the Middle East, former Soviet Union, and North Africa. Although the United States has produced more than 40% of its total estimated natural gas endowment and carries less than 10% as remaining reserves, in the rest of the world reserves have been largely unexploited. Outside the United States, the world has so far produced less than 10% of its total estimated natural gas endowment and carries more than 30% as remaining reserves.

World coal reserves are also considered to be quite plentiful. Total recoverable reserves of coal are those quantities of coal that geological and engineering data indicate with reasonable certainty can be extracted in the future under existing economic and operating conditions; around the world, total recoverable reserves are estimated at 1083 billion short tons. At current consumption levels, there are still 210 years worth of sufficient coal reserves. Although coal deposits are widely distributed, 60% of the world's recoverable reserves is located in three regions; the United States holds an estimated 25% of world reserves, the former Soviet Union holds 23%, and China holds 12%. Another four countries, Australia, India, Germany, and South Africa, collectively account for another 29% (Fig. 5).

4. DRIVERS OF ENERGY DEMAND

There are several important drivers associated with the growth in energy demand. One driver, a nation's income, which can be measured by gross domestic product (GDP), can indicate how much energy a given nation will require to support the expanding economy. Potential growth in population will influence the demand for energy to support an expanding population. Another factor is the rate of efficiency gains a country makes in its energy consumption. Other factors, such as environmental policies, taxes, and subsidies, and other influences can also have an impact on the development of energy markets from country to country. An examination of these factors and how they are manifested in various parts of the world can help to explain the regional trends in energy consumption.

Although economic growth is certainly a key factor to determining the growth in energy demand, how energy is used relative to income can be equally important. Economic growth and energy demand are linked and the strength of that link varies among regions and their stages of economic development. In the industrialized world, history has shown the link to be a relatively weak one, with energy demand lagging behind economic growth. In developing countries, demand and economic growth have been closely correlated in the past, with demand growth tending to track the rate of economic expansion.

The historical behavior of energy intensity in the former Soviet Union (FSU) is problematic. Since World War II, the levels of energy intensity in the Soviet Republic economies have been higher than they have been in either the industrialized or the developing nations (Fig. 6). In the FSU, however, energy consumption grew more quickly than GDP until 1990, when the collapse of the Soviet Union created a situation in which both income and energy use were declining, but GDP fell more quickly; as a result, energy intensity increased. As the nations of the FSU continue to recover from the economic and social problems of the 1990s, energy intensity should begin to decline once again, as new, more efficient capital stock replaces inefficient Soviet-era equipment and GDP continues to expand.

The stage of economic development and the standard of living of individuals in a given region strongly influence the link between economic growth and energy demand. Advanced economies with high standards of living have relatively high per capita energy use, but they also tend to be economies in which per capita energy use is stable or changes very slowly, and increases in energy use tend to correlate with employment and population growth. In the industrialized world, there is a high penetration rate of modern appliances and motorized personal transportation equipment. To the extent that spending is directed to energy-consuming goods, it involves more often than not purchases of new equipment to replace old capital stock. The new stock tends to be more efficient than the equipment it replaces, resulting in a weaker link between income and energy demand. In the developing world, standards of living, while rising, tend to be low relative to those in more industrially advanced economies.

It is worth noting that, although economic growth and energy intensity impact the increases in energy demand, other factors may also influence the development of energy use. Individual countries may employ many devices to encourage consumer behavior. For instance, in the industrialized countries of western Europe, taxes are often levied in an effort to get consumers to use less energy. Motor gasoline and diesel fuel are heavily taxed in many western European countries. The government tries to use this disincentive to consume to lower congestion in urban areas and to lower the impact of motor vehicle emissions on the environment. In many industrialized countries, tax incentives, such as the wind production tax credit in the United States, are used to encourage consumers to use a more environmental, but not yet economically competitive, form of energy, in this case, wind. In the developing country, price subsidies are sometimes employed to make it possible for a population to increase the penetration of modern technologies. Many oil-producing countries, e.g., Indonesia and Nigeria, heavily subsidize their motor fuels to support fledgling motor vehicle ownership.

5. REGIONAL ENERGY USE

The current mix of the world's energy markets, the drivers associated with estimating increases in energy consumption, and the potential resources available to meet future demand have been considered; the regional trends in world energy use are now assessed.

At present, the industrialized world is responsible for the lion's share of the global energy use. The United States alone currently accounts for 25% of the total world energy use. In contrast, China, a country with a population of about four times that of the United States, presently consumes only 8% of the world's total energy production. Energy use among the industrialized countries is expected to grow, though more slowly in the future than it has over the past several decades. The industrialized countries will probably maintain the largest regional share of world energy consumption (Fig. 7). However, the growth in consumption is expected to be low relative to other parts of the world. Many industrialized nations—notably Japan and many of the countries of western Europe—are faced with slow population growth rates and thus energy markets will not need to accommodate a great deal of new populace. Further, most, if not all, of the energy end-use sectors

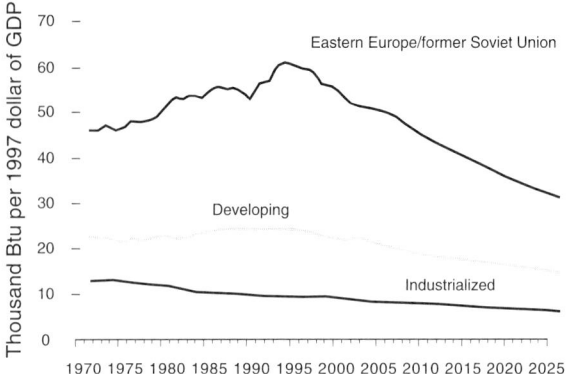

FIGURE 6 World energy intensity by region, 1970–2025. GDP, Gross domestic product. Data from Energy Information Administration (2003).

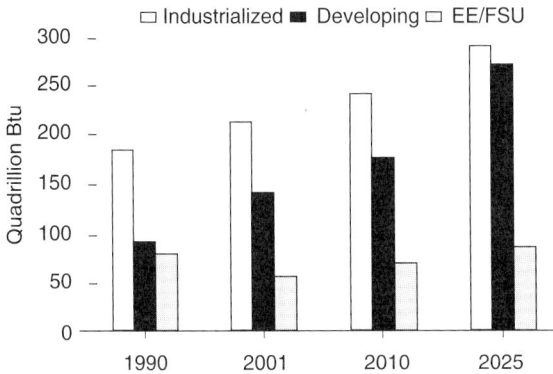

FIGURE 7 World energy consumption by region, 1990–2025. EE/FSU, Eastern Europe/former Soviet Union. Data from Energy Information Administration (2003).

TABLE I

Projected Growth in World Population by Region, 2001–2025[a]

Region	Average annual increase (%)
North America	0.9
Western Europe	−0.1
Industrialized Asia	0.2
Eastern Europe/former Soviet Union	−0.2
Developing Asia	1.1
Middle East	2.0
Africa	2.3
Central and South America	1.2
Total worldwide	1.1

[a]Data from United Nations (2001).

will not see large increments in energy demand. Residential and industrial consumers already possess well-established infrastructures, so most of the future energy demand will utilize more efficient equipment.

In comparison to the industrialized nations, there is enormous potential energy demand from burgeoning populations in the developing world. Africa, the Middle East, Central and South America, and developing Asia are all expected to experience average annual population increases of over 1.0% per year between 2001 and 2025 (Table I), with many individual countries among these regions seeing population growth in excess of 2% per year. A corresponding increment in energy consumption will be needed to ensure the prosperity of these economies. As these developing economies continue to emerge, there will be increasing consumer demand for electricity to run modern appliances such as refrigerators, stoves, ovens, washers, dryers, and even personal computers, to name a few, and also for personal motor vehicles. Motor vehicle ownership is considered to be among the first symbols of emerging prosperity in many developing countries, and as the industrial base expands, so too will the need to move goods and services to market.

The fastest growing energy markets will likely be centered in developing Asia. In the Energy Information Administration's *International Energy Outlook 2003*, economic growth in this region is projected to increase 5.1% per year between 2001 and 2025. This increment in the region's income will require a corresponding increase in energy use. The Energy Information Administration (EIA) expects developing Asian energy demand to grow by 3.0% per year over this time period. Many of the nations of this region have a relatively immature infrastructure. In China, for instance, there is great potential for the transportation sector to expand. In 1999, there were an estimated 12 motor vehicles per 1000 persons, compared to the 777 vehicles per 1000 persons recorded for the United States. Although a 2002 analysis by the EIA suggests that motorization levels in China might grow to as many as 52 vehicles per 1000 persons by 2020, China would still only have levels that are 6% that of the current U.S. motorization levels.

Growing transportation energy use could have a profound impact on the demand for world oil. With no economic alternatives to oil presently available, the pressures on world oil supplies will be great. In the case of China, should transportation energy use grow at a robust rate of, say, 5.0% per year between 2001 and 2025, an increment of 4.4 million barrels of oil per day would be required to fuel the Chinese transportation market. This is substantial, considering that total oil consumption in China is projected to grow by 5.9 million barrels per day over this time period. The geopolitical ramifications of such growth in demand could be significant. China became a net importer of oil in 1993, and in 2001 the country was importing 2.0 million barrels per day of oil, with 0.9 million barrels per day coming from the OPEC Persian Gulf. With the demand increases projected by the EIA, China's oil import dependence on the OPEC nations is apt to increase dramatically over the next two and one-half decades. In the EIA reference case, China will import 7.8 million barrels per day by 2025, most of which is expected to come from the Persian Gulf nations.

Energy demand could grow even more dramatically than analysts assume if developing markets manage to reach the consuming patterns of the industrialized countries more quickly than conventional wisdom suggests. This is entirely possible. Consider the case of South Korea, where energy consumption grew rapidly over the past three decades, by nearly 12% per year. South Korea embraced a market-based economy and has made tremendous strides in becoming an industrialized nation. Standards of living have grown substantially. In 1970, South Korea had a per capita GDP of about $2000 per person (in real 1997 U.S. dollars), but by 2001 this had climbed to $11,908 per year. Per capita energy use correspondingly increased, from 21 million Btu per person in 1970 to 172 million Btu in 2001. In contrast, China's per capita energy use was 31 million Btu in 2001. Should China's per person energy use achieve the current South Korea per capita energy use levels by 2025, another 163 quadrillion Btu of energy (or 80% more than projected for China in 2025 according to the EIA reference case forecast) would be needed to meet the additional demand. A similar paradigm could be made for many other developing countries.

The potential for new energy demand is also fairly high in the transitional economies of the former Soviet Union. Although energy demand plummeted in the early 1990s after the dissolution of the Soviet regime, by the late 1990s and into the 21st century, the economies of these nations began to recover. The growth in energy use is not expected to be as dramatic as that in the developing world, mostly because the rate of population growth is set to decline or will remain slow in most of these nations, and because energy, in the past, has been used very inefficiently. As these economies transition from planned economies to more market-based economies and as foreign investment in the private sectors begins to take hold, energy will be used more efficiently and with new, less energy-intensive equipment.

6. SUMMARY

World energy use is poised to increase substantially over the next few decades. How energy markets develop and what kinds of political, economic, or social changes might influence these changes are impossible to determine with any degree of certainty. Current understanding of fossil fuel resources suggests that there will be ample supplies of fossil fuels available to meet the growing demand for the foreseeable future, at relatively low consumer costs. It is also true that government intervention to encourage conservation or to support the consumption of nonfossil alternatives may decrease the need to exploit all of these resources. Further, there is an opportunity for technological innovation that may advance the use of other energy forms in the future, but this cannot yet be predicted. After all, the share of world energy use of nuclear power generation grew from less than 0.5% in 1970 to almost 7% in 1999. A similar advancement, say in automobiles that could economically run on hydrogen, could profoundly change the world reliance on petroleum products, albeit the penetration of such an advancement might require more than only a few decades without some incentive to make such a dramatic change to the infrastructure.

SEE ALSO THE FOLLOWING ARTICLES

Cultural Evolution and Energy • *Development and Energy, Overview* • *Earth's Energy Balance* • *Economic Geography of Energy* • *Economic Growth, Liberalization, and the Environment* • *Energy Ladder in Developing Nations* • *Geopolitics of Energy* • *Nationalism and Oil* • *Population Growth and Energy* • *Sustainable Development: Basic Concepts and Application to Energy* • *Technology Innovation and Energy* • *Transitions in Energy Use*

Further Reading

American Automobile Manufacturers Association. (1997). "World Motor Vehicle Data." AAMA, Detroit, Michigan.

Department of Trade and Industry. (2001). "New and Renewable Energy: Prospects for the 21st Century." Web site of the government of the United Kingdom of Great Britain and Northern Ireland, available at http://www.dti.gov/uk/renewable/consultations.htm.

Energy Information Administration. (1999). "International Energy Outlook 1999," DOE/EIA-0484(99). U.S. Department of Energy, Washington, D.C.

Energy Information Administration. (2002). "International Energy Outlook 2002," DOE/EIA-0484(2002). U.S. Department of Energy, Washington, D.C.

Energy Information Administration. (2003). "International Energy Outlook 2003," DOE/EIA-0484(2003). U.S. Department of Energy, Washington, D.C.

Energy Information Administration. (2003). "Annual Energy Outlook 2003," DOE/EIA-0383(2003). U.S. Department of Energy, Washington, D.C.

International Atomic Energy Agency. (2003). "Reference Data Series 2, Power Reactor Information System." Available on the Internet at http://www.iaea.org/.

International Energy Agency. (2002). "World Energy Outlook 2002." International Energy Agency, Paris, France.

International Energy Agency and PT Communications. (2002). "IEA Wind Energy Annual Report 2001." International Energy Agency and PT Communications, Boulder, Colorado.

National Energy Board. (2000). "Canada's Oil Sands: A Supply and Market Outlook to 2015." National Energy Board of Canada, Calgary, Alberta.

Porter, E. (1995). "Are We Running Out of Oil?" American Petroleum Institute Discussion Paper No. 081. American Petroleum Institute, Washington, D.C.

United Nations. (2001). "World Population Prospects: The 2000 Revision, Volume 1, Comprehensive Tables." United Nations, New York.

U.S. Geological Survey. (2000). "World Petroleum Assessment 2000." Web site of the U.S. Geological Survey. Available at http://greenwood.cr.usgs.gov/.

World Energy Council. (2001). "Survey of Energy Resources 2001." World Energy Council, London.

Worldwide look at reserves and production. (2002). *Oil Gas J.* **100**(52), 114–115.

Global Material Cycles and Energy

VACLAV SMIL
University of Manitoba
Winnipeg, Manitoba, Canada

1. Material Cycles
2. Geotectonic Cycle
3. Water Cycle
4. Carbon Cycle
5. Nitrogen Cycle
6. Sulfur Cycle

Glossary

carbon The sixth element of the periodical table and the key structural element of all living matter.
geotectonics The process of reforming the earth's surface that is driven primarily by creation and subduction of the ocean floor.
nitrogen The seventh element of the periodical table and an indispensable ingredient of proteins, the principal metabolic vehicles of life.
sulfur The 16th element of the periodical table whose presence in disulfide bonds makes the proteins three-dimensional.
water A triatomic molecule of hydrogen and oxygen with many unique physical and chemical properties that makes life possible.

The earth is a constant recipient of an enormous flux of solar radiation as well as considerable infall of the cosmic debris that also arrives periodically in huge, catastrophe-inducing encounters with other space bodies, particularly with massive comets and asteroids. Nothing else comes in, and the planet's relatively powerful gravity prevents anything substantial from leaving. Consequently, life on the earth is predicated on incessant cycling of water and materials that are needed to assemble living bodies as well as to provide suitable environmental conditions for their evolution. These cycles are energized by two distinct sources: Whereas the water cycle and the circulations of carbon, nitrogen, and sulfur, the three key life-building elements, are powered by solar radiation, the planet's grand tectonic cycle is driven by the earth's heat. For hundreds of thousands of years, our species had no discernible large-scale effect on these natural processes, but during the 20th century anthropogenic interferences in global biospheric cycles became a matter of scientific concern and an increasingly important, and controversial, topic of public policy debates.

1. MATERIAL CYCLES

Two massive material cycles would be taking place even on a lifeless Earth: (i) a slow formation and recycling of the planet's thin crust and (ii) evaporation, condensation, and runoff of water. As weathering breaks down exposed rocks, their constituent minerals are transported by wind in terrigenic dust and in flowing water as ionic solutions or suspended matter. Their journey eventually ends in the ocean, and the deposited materials are either drawn into the mantle to reemerge in new configurations 10^7–10^8 years later along the ocean ridges or in volcanic hot spots or they resurface as a result of tectonic movements as new mountain ranges. This grand sedimentary–tectonic sequence recycles not just all trace metals that are indispensable in small amounts for both auto- and heterotrophic metabolism but also phosphorus, one of the three essential plant macronutrients.

On human timescale (10^1 years), we can directly observe only a minuscule segment of the geotectonic weathering cycle and measure its rates mostly as one-way oceanward fluxes of water-borne compounds. Human actions are of no consequence as far as the geotectonic processes are concerned, but they have greatly accelerated the rate of weathering due to deforestation, overgrazing of pastures, and improper crop cultivation. Incredibly, the annual rate of this unintended earth movement, 50–80 billion metric tons (Gt), is now considerably higher than the aggregate rate of global weathering before the rise of agriculture (no more than 30 Gt/year).

In addition, we act as intentional geomorphic agents as we extract, displace, and process crustal minerals, mostly sand, stone, and fossil fuels. The aggregate mass moved annually by these activities is also larger than the natural preagricultural weathering rate. Our activities have also profoundly affected both the quantity and the quality of water cycled on local and regional scales, but they cannot directly change the global water flux driven predominantly by the evaporation from the ocean. However, a major consequence of a rapid anthropogenic global warming would be an eventual intensification of the water cycle.

In contrast to these two solely, or overwhelmingly, inanimate cycles, global circulations of carbon, nitrogen, and sulfur are to a large extent driven or mediated by living organisms, largely by bacteria, archaea, and plants. Another characteristic that these cycles have in common is their relatively modest claim on the planet's energy flows. Photosynthesis, carbon cycle's key terrestrial flux, requires only a tiny share of incident solar radiation, and the cycling of nitrogen and sulfur involves much smaller masses of materials and hence these two cycles need even less energy for their operation: Their indispensability for Earth's life is in renewing the availability of elements that impart specific qualities to living molecules rather than in moving large quantities of materials. Similarly, only a small fraction of the planet's geothermal heat is needed to power the processes of global geotectonics.

Although the basic energetics of all these cycles is fairly well-known, numerous particulars remain elusive, including not only specific details, whose elaboration would not change the fundamental understanding of how these cycles operate, but also major uncertainties regarding the extent and consequences of human interference in the flows of the three doubly mobile elements (carbon, nitrogen, and sulfur), whose compounds are transported while dissolved in water as well as relatively long-lived atmospheric gases. For millennia these interventions were discernible only on local and regional scales, but during the 20th century their impacts increased to such an extent that the effects of these interferences now present some historically unprecedented challenges even in global terms.

2. GEOTECTONIC CYCLE

The earth's surface is composed of rigid oceanic and continental plates. The former are relatively thin (mostly 5–7 km), short-lived [mostly less than 140 million years (Ma)], and highly mobile (up to 10–20 cm/year). The latter ones are thicker (only 35–40 km), long-lived (10^9 years), and some, most notably Asian and African plates, are virtually stationary. Plates ride on the nearly 3000-km-thick solid but flowing mantle, which is the source of hot magma whose outpourings create new seafloor along approximately 72,000 km of the earth-encircling ocean ridge system. Old seafloor is eventually recycled back into the mantle in subduction zones marked by deep ocean trenches (Fig. 1). There are substantial uncertainties concerning the actual movements of the subducted plates within the mantle: The principal debate is between the two-layer models and the whole-mantle flows.

Today's oceans and continents are thus transitory features produced by incessant geotectonic processes that are energized by three sources of the earth's heat: energy conducted through the lithosphere from the underlying hot mantle, radiogenic decay of heat-producing crustal elements, and convective transport by magmas and fluids during orogenic events. The relative importance of these heat sources remains in dispute, but the heat flow at the earth's surface can be measured with high accuracy. Its total is approximately 44 trillion watts (TW), prorating to approximately 85 mW/m² of the earth's surface. The average global heat flux of approximately 85 mW/m² is equal

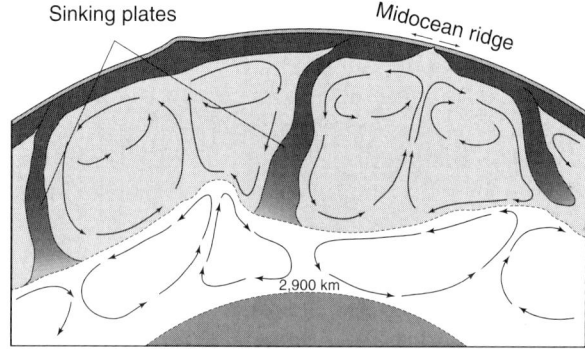

FIGURE 1 A cross section of the earth shows tectonic plates, an ocean ridge, a subduction trench, and, in this model, the circulation in mantle that is separated into two cells.

to a mere 0.05% of solar radiation absorbed by surfaces (168 W/m²), but acting over huge areas and across long timespans, this flow is responsible not only for creating new ocean floor, reshaping continents, and building enormous mountain chains but also for energizing earthquakes and volcanic eruptions.

Mean oceanic heat flow is approximately 100 mW/m², whereas the continental flow is only approximately half as high. The ocean floor transmits approximately 70% of the earth's total geothermal flux and nearly one-third of the oceanic heat loss occurs in the South Pacific, where the spreading rates, particularly along the Nazca Ridge, are faster than anywhere else on Earth. By far the highest heat flows are along midocean ridges, where approximately 3–3.5 km² of new ocean floor is created by hot magma rising from the mantle. The total heat flux at the ridges, composed of the latent heat of crystallization of newly formed ocean crust and of the heat of cooling from magmatic temperatures (approximately 1200°C) to hydrothermal temperatures (approximately 350°C), is between 2 and 4 TW.

Divergent spreading of oceanic plates eventually ends at subduction zones, where the crust and the uppermost part of the mantle are recycled deep into the mantle to reappear through igneous process along the ridges. Every new cycle begins with the breakup of a supercontinent, a process that typically takes approximately 200 Ma, and is followed first by slab avalanches and then by rising mantle plumes that produce juvenile crust. After 200–440 Ma, the new supercontinent is broken up by mantle upwelling beneath it.

During the past billion years (Ga), the earth has experienced the formation and breakup of three supercontinents: Rodinia (formed 1.32–1.0 Ga ago and broken up between 700 and 530 Ma ago), Gondwana (formed between 650 and 550 Ma ago), and Pangea (formed 450–250 Ma ago and began to break up approximately 160 Ma ago). Pangea spanned the planet latitudinally, from today's high Arctic to Antarctica; its eastern flank was notched by a V-shaped Tethys Sea centered approximately on the equator. Unmistakable signs of continental rifting can be seen today underneath the Red Sea and along the Great Rift Valley of East Africa: The process of continental breakup is still very much under way.

Inevitably, these grand geotectonic cycles had an enormous impact on the evolution of life. They made it possible to keep a significant share of the earth's surface above the sea during the past 3 Ga, allowing for the evolution of all complex terrestrial forms of life. Changing locations and distributions of continents and oceans create different patterns of global oceanic and atmospheric circulation, the two key determinants of climate. Emergence and diversification of land plants and terrestrial fauna during the past 500 Ma years have been much influenced by changing locations and sizes of the continents. The surface we inhabit today has been, to a large extent, fashioned by the still continuing breakup of Pangea. Two of the earth's massive land features that have determined the climate for nearly half of humanity—the Himalayas, the world's tallest mountain chain, and the high Tibetan Plateau—are its direct consequences. So is the northward flow of warm Atlantic waters carried by the Gulf Stream that helps to create the mild climate in Western Europe. Volcanic eruptions have been the most important natural source of CO_2 and, hence, a key variable in the long-term balance of the biospheric carbon. Intermittently, they are an enormous source of aerosols, whose high atmospheric concentrations have major global climatic impacts.

3. WATER CYCLE

Most of the living biomass is water, and this unique triatomic compound is also indispensable for all metabolic processes. The water molecule is too heavy to escape the earth's gravity, and juvenile water, originating in deeper layers of the crust, adds only a negligible amount to the compound's biospheric cycle. Additions due to human activities—mainly withdrawals from ancient aquifers, some chemical syntheses, and combustion of fossil fuels—are also negligible. Both the total volume of Earth's water and its division among the major reservoirs can thus be considered constant on a timescale of 10^3 years. On longer timescales, glacial–interglacial oscillations shift enormous volumes of water among oceans and glaciers and permanent snow.

The global water cycle is energized primarily by evaporation from the ocean (Fig. 2). The ocean covers 70% of the earth, it stores 96.5% of the earth's water, and it is the source of approximately 86% of all evaporation. Water has many extraordinary properties. A very high specific heat and heat capacity make it particularly suitable to be the medium of global temperature regulation. High heat of vaporization (approximately 2.45 kJ/g) makes it an ideal transporter of latent heat and helps to retain plant and soil moisture in hot environments. Vaporization of 1 mm/day requires 28 or 29 W/m², and with average daily evaporation of approximately

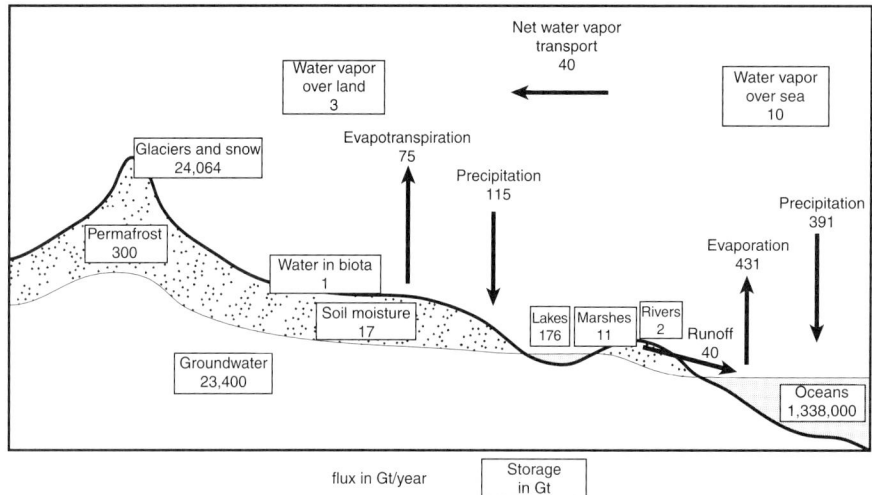

FIGURE 2 Annual flows of the earth's global water cycle. The totals are all in billion of metric tons.

3 mm, or an annual total of 1.1 m, the global latent heat flux averages approximately 90 W/m². This means that approximately 45 quadrillion watts (PW), or slightly more than one-third of the solar energy absorbed by the earth's surface and the planet's atmosphere (but 4000 times the amount of annual commercial energy uses), is needed to drive the global water cycle.

The global pattern of evaporation is determined not only by the intensity of insolation but also by the ocean's overturning circulation as cold, dense water sinks near the poles and is replaced by the flow from low latitudes. There are two nearly independent overturning cells—one connects the Atlantic Ocean to other basins through the Southern Ocean, and the other links the Indian and Pacific basins through the Indonesian archipelago. Measurements taken as a part of the World Ocean Circulation Experiment show that 1.3 PW of heat flows between the subtropical (latitude of southern Florida) and the North Atlantic, virtually the same flux as that observed between the Pacific and Indian Oceans through the Indonesian straits.

Poleward heat transfer of latent heat also includes some of the most violent low-pressure cells (cyclones), American hurricanes, and Asian typhoons. Oceanic evaporation (heating of equatorial waters) is also responsible for the Asian monsoon, the planet's most spectacular way of shifting the heat absorbed by warm tropical oceans to continents. The Asian monsoon (influencing marginally also parts of Africa) affects approximately half of the world's population, and it precipitates approximately 90% of water evaporated from the ocean back onto the sea. Evaporation exceeds precipitation in the Atlantic and Indian Oceans, the reverse is true in the Arctic Ocean, and the Pacific Ocean flows are nearly balanced.

Because of irregular patterns of oceanic evaporation (maxima up to 3 m/year) and precipitation (maxima up to 5 m/year), large compensating flows are needed to maintain sea level. The North Pacific is the largest surplus region (and hence its water is less salty), whereas evaporation dominates in the Atlantic. In the long term, oceanic evaporation is determined by the mean sea level. During the last glacial maximum 18,000 years ago, the level was 85–130 m lower than it is now, and satellite measurements indicate a recent global mean sea level rise of approximately 4 mm per year.

The difference between precipitation of evapotranspiration has a clear equatorial peak that is associated with convective cloud systems, whereas the secondary maxima (near 50°N and °S) are the result of extratropical cyclones and midlatitude convection. Regarding the rain on land, approximately one-third of it comes from ocean-derived water vapor, with the rest originating from evaporation from nonvegetated surfaces and from evapotranspiration (evaporation through stomata of leaves). Approximately 60% of all land precipitation is evaporated, 10% goes back to the ocean as surface runoff, and approximately 30% is carried by rivers (Fig. 2). Because the mean continental elevation is approximately 850 m, this riverborne flux implies an annual conversion of approximately 370×10^{18} J of potential energy to kinetic energy of flowing water, the principal agent of geomorphic denudation shaping the earth's surfaces. No more than approximately

15% of the flowing water's aggregate potential energy is convertible to hydroelectricity.

The closing arm of the global water cycle works rather rapidly because average residence times of freshwater range from just 2 weeks in river channels to weeks to months in soils. However, widespread construction of dams means that in some river basins water that reaches the sea is more than 1 year old. In contrast to the normally rapid surface runoff, water may spend many thousands of years in deep aquifers. Perhaps as much as two-thirds of all freshwater on Earth is contained in underground reservoirs, and they annually cycle the equivalent of approximately 30% of the total runoff to maintain stable river flows. Submarine groundwater discharge, the direct flow of water into the sea through porous rocks and sediments, appears to be a much larger flux of the global water cycle than previously estimated.

4. CARBON CYCLE

The key structural element of life keeps cycling on several spatial and temporal scales, ranging from virtually instantaneous returns to circulations that take 10^8 years to complete. At one extreme are very rapid assimilation–respiration flows between the plants and the atmosphere and less rapid but still fast assimilation–decomposition–assimilation circuits that move the element from the atmosphere to plant tissue of various ecosystems and return it to the atmosphere after carbon's temporary stores in dead organic matter (in surface litter and soils) are mineralized by decomposers (Fig. 3). The other extreme involves slow weathering of terrestrial carbonates, their dissolution, riverborne transport, and eventual sedimentation in the ocean, and the element's return to land by geotectonic processes. Human interest in the cycle is due not only to the fact that carbon's assimilation through photosynthesis is the foundation of all but a tiny fraction of the earth's life but also to the cycle's sensitivity to anthropogenic perturbations.

Photosynthesis, the dominant biospheric conversion of inorganic carbon from the atmospheric carbon dioxide (CO_2) to a huge array of complex organic molecules in plants, proceeds with very low efficiency. On average, only approximately 0.3% of the electromagnetic energy of solar radiation that reaches the earth's ice-free surfaces is converted to chemical energy of new phytomass. Short-term maxima in the most productive crop fields can be close to 5%, whereas even the tropical rain forests, the planet's most productive natural ecosystems, do not average more than 1%. Plant (autotrophic) respiration consumes approximately half of the gross photosynthesis, and it reduces the global net primary productivity [i.e., the new phytomass available every year for consumption by heterotrophs (from bacteria to humans)] to approximately 60 Gt of carbon on land and approximately 50 Gt of carbon in the ocean.

Virtually all this fixed carbon is eventually returned to the atmosphere through heterotrophic respiration, and only a tiny fraction that accumulates in terrestrial sediments forms the bridge between the element's rapid and slow cycles. After 10^6–10^8 years of exposure to high pressures and temperatures, these sediments are converted to fossil fuels. The most likely global total of biomass carbon that was transformed into the crustal resources of coals and hydrocarbons is in excess of 5 Tt of oil equivalent, or 2.1×10^{23} J. However, some hydrocarbons, particularly natural gas, may be of abiotic origin. Much more of the ancient organic carbon is sequestered in kerogens, which are transformed remains of buried biomass found mostly in calcareous and oil shales.

Intermediate and deep waters contain approximately 98% of the ocean's huge carbon stores, and they have a relatively large capacity to sequester more CO_2 from the atmosphere. The exchange of CO_2 between the atmosphere and the ocean's topmost (mixed) layer helps to keep these two carbon reservoirs stable. This equilibrium can be disrupted by massive CO_2-rich volcanic eruptions or abrupt climate changes. However, the rate of CO_2 exchange between the atmosphere and the ocean is limited because of the slow interchange between the relatively warm mixed layer ($>25°C$ in the tropics) and cold (2–4°C) deep waters. As is the case on land, only a small fraction of organic carbon is not respired and goes into ocean sediments (Fig. 3).

Analysis of air bubbles from polar ice cores shows that the atmospheric concentration of CO_2 has remained between 180 and 300 parts per million (ppm) during the past 420,000 years and 250–290 ppm since the rise of the first high civilizations during the 5000 years preceding the mid-19th century. Humans began changing this relatively stable state of affairs first by clearing natural vegetation for fields and settlements, but until the 19th century these releases of CO_2 did not make any marked difference. These conversions became much more extensive after 1850, coinciding with the emergence of fossil-fueled civilization.

Land use changes dominated the anthropogenic mobilization of carbon until the beginning of the 20th

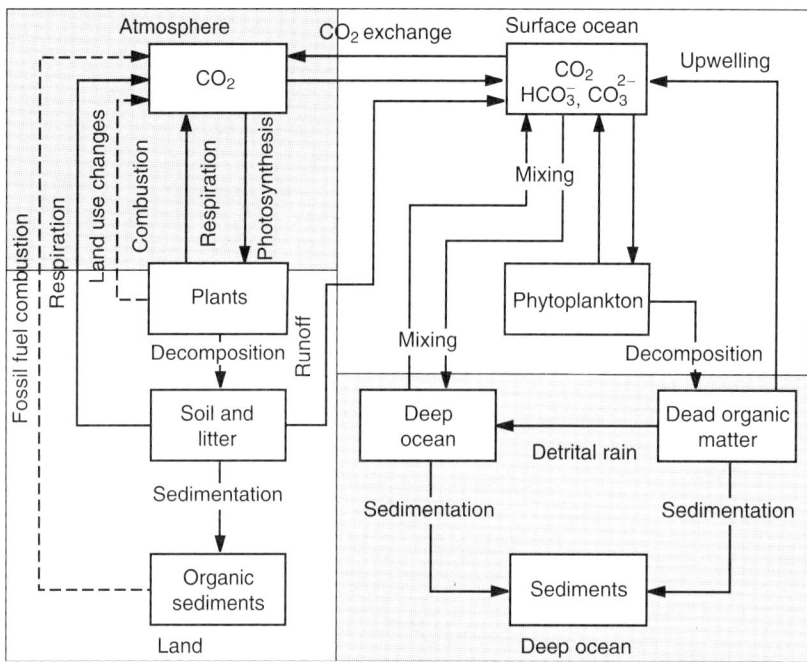

FIGURE 3 Principal reservoirs and flows of the biospheric carbon cycle.

century and they are now the source of 1.5–2 Gt of carbon per year to the atmosphere. Fossil fuel combustion had accelerated from an annual conversion of 22 EJ and emissions of less than 0.5 Gt of carbon in 1900 to more than 300 EJ and releases of more than 6 Gt of carbon by 2000. Every year, on average, approximately half of this anthropogenic flux of approximately 8 Gt of carbon stays in the atmosphere. As a result, atmospheric CO_2 levels that have been constantly monitored at Hawaii's Mauna Loa observatory since 1958 increased from 320 ppm during the first year of measurement to 370 ppm in 2000.

A nearly 40% increase in 150 years is of concern because CO_2 is a major greenhouse gas whose main absorption band of the outgoing radiation coincides with Earth's peak thermal emission. A greenhouse effect is necessary to maintain the average surface temperature approximately 33°C above the level that would result from an unimpeded reradiation of the absorbed insolation, and energy reradiated by the atmosphere is currently approximately 325 W/m^2. During the past 150 years, anthropogenic CO_2 has increased this flux by 1.5 W/m^2 and other greenhouse gases have added approximately 1 W/m^2.

The expected temperature increase caused by this forcing has been partially counteracted by the presence of anthropogenic sulfur compounds in the atmosphere, but another century of massive fossil fuel combustion could double the current CO_2 levels and result in rapid global warming. Its effects would be more pronounced in higher latitudes, and they would include intensification of the global water cycle that would be accompanied by unevenly distributed changes in precipitation patterns and by thermal expansion of seawater leading to a gradual rise in sea level.

5. NITROGEN CYCLE

Although carbon is the most important constituent of all living matter (making up, on average, approximately 45% of dry mass), cellulose and lignin, the two dominant molecules of terrestrial life, contain no nitrogen. Only plant seeds and lean animal tissues have high protein content (mostly between 10 and 25%; soybeans up to 40%), and although nitrogen is an indispensable ingredient of all enzymes, its total reservoir in living matter is less than 2% of carbon's huge stores. Unlike the cycles of water or carbon that include large flows entirely or largely unconnected to biota, every major link in nitrogen's biospheric cycle is mediated by bacteria.

Nitrogen fixation—a severing of the element's highly stable and chemically inert molecule (N_2) that makes up 78% of the atmosphere and its incorporation into reactive ammonia—launches this great cascade of biogenic transformations. Only a very limited number of organisms that possess nitrogenase, the enzyme able to cleave N_2 at ambient temperature and at normal pressure, can perform this conversion. Several genera of rhizobial bacteria live symbiotically with leguminous plants and are by far the most important fixers of nitrogen on land, whereas cyanobacteria dominate the process in freshwaters as well as in the ocean.

There are also endophytic symbionts (bacteria living inside stems and leaves) and free-living bacterial fixers in soils. The current annual rate of biofixation in terrestrial ecosystems is no less than 160 million metric tons (Mt) of nitrogen, and at least 30 Mt of nitrogen is fixed in crop fields. Biofixation is a relatively energy-intensive process, with 6–12 g of carbon (delivered by the plant to bacteria as sugar) needed for every gram of fixed nitrogen. This means that the terrestrial biofixation of approximately 200 Mt of nitrogen requires at least 1.2 Gt of carbon per year, but this is still no more than 2% of the net terrestrial primary productivity.

Subsequently, there are four key links in the nitrogen cycle (Fig. 4). After nitrogen fixation is nitrification, the conversion of ammonia to more water-soluble nitrates, which is performed by a small number of bacterial species living in both soils and water. Assimilation is the incorporation of reactive nitrogen, be it as ammonia (NH_3) or as nitrate (NO^{-3}), by autotrophic organisms into amino acids that are then used to synthesize proteins. Ammonification is enzymatic decomposition of organic matter performed by many prokaryotes that produces ammonia for nitrification. Finally, denitrification is the closing arm of the cycle in which bacteria convert nitrates into nitrites and then into N_2.

Human interference in the nitrogen cycle is largely due to the fact that heterotrophs, including people, cannot synthesize their own amino acids and hence must digest preformed proteins. Because nitrogen is the most common yield-limiting nutrient in crop cultivation, all traditional agricultures had to resort to widespread planting of leguminous crops and to diligent recycling of nitrogen-rich organic wastes. However, the extent of these practices is clearly limited by the need to grow higher yielding cereals and by the availability of crop residues and human and animal wastes. Only the invention of ammonia synthesis from its elements by Fritz Haber in 1909 and an unusually rapid commercialization of this process by the BASF company under the leadership of Carl Bosch (by 1913) removed the nitrogen limit on crop production and allowed for the expansion of the human population supplied by adequate nutrition (both chemists were later awarded Nobel Prizes for their work). Currently, malnutrition is a matter of distribution and access to food, not of its shortage.

These advances have a considerable energy cost. Despite impressive efficiency gains, particularly since the early 1960s, NH_3 synthesis requires, on average, more than 40 GJ/t, and the conversion of the compound to urea, the dominant nitrogen fertilizer, raises the energy cost to approximately 60 GJ/t of nitrogen. In 2000, Haber-Bosch synthesis produced approximately 130 Mt of NH_3 (or 110 Mt of nitrogen). With production and transportation losses of approximately 10%, and with 15% destined for industrial uses, approximately 82 Mt of nitrogen was used as fertilizer at an aggregate embodied cost of approximately 5 EJ, or roughly 1.5% of the world's commercial supply of primary energy.

The third most important human interference in the global nitrogen cycle (besides NH_3 synthesis and planting of legumes) is the high-temperature combustion of fossil fuels, which released more than 30 Mt of nitrogen, mostly as nitrogen oxides (NO_x), in 2000. Conversion of these oxides to nitrates, leaching of nitrates, and volatilization of ammonia from fertilized fields (only approximately half of the nutrient is actually assimilated by crops) and from animal wastes are the principal sources of nitrogen leakage into ecosystems. Excess nitrogen

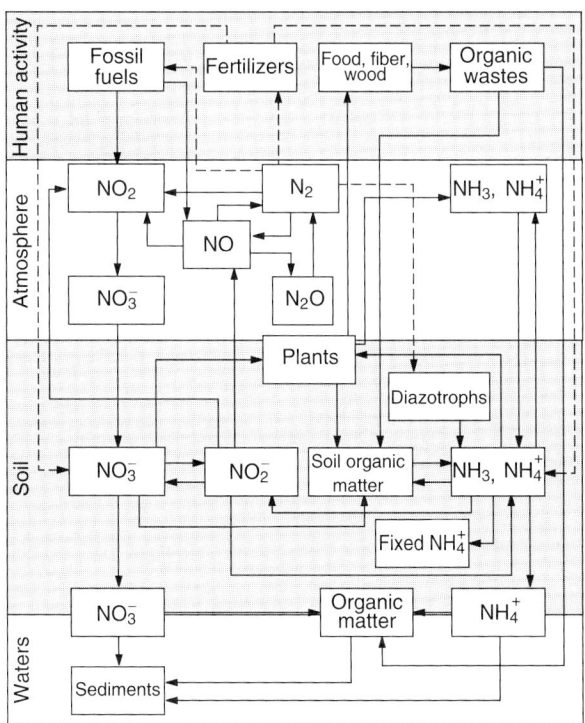

FIGURE 4 Biogeochemical nitrogen cycle centered on plants.

causes eutrophication of fresh and coastal waters, and it raises concerns about the long-term effects on biodiversity and productivity of grasslands and forests. NO_x are key precursors of photochemical smog, and rain-deposited nitrates contribute to acidification of waters and soils. Nitrous oxide, a potent greenhouse gas, released during imperfect denitrification is another long-term environmental concern.

6. SULFUR CYCLE

Sulfur is an even rarer component of living molecules than is nitrogen. Only 2 of the 20 amino acids that make up proteins (methionine and cysteine) have the element as part of their molecules, but every protein needs disulfide bridges to make long three-dimensional polypeptide chains whose complex folds allow proteins to be engaged in countless biochemical reactions. The largest natural flux of the element is also the most ephemeral: Sea spray annually carries 140–180 Mt of sulfur into the atmosphere, but 90% of this mass is promptly redeposited in the ocean. Some volcanic eruptions are very large sources of SO_2, whereas others have sulfur-poor veils; a long-term average works out to approximately 20 Mt of sulfur per year, and dust, mainly desert gypsum, may contribute the same amount (Fig. 5).

Biogenic sulfur flows may be as low as 15 Mt and as high as 40 Mt of sulfur per year, and they are produced on land by both sulfur-oxidizing and sulfate-reducing bacteria present in water, mud, and hot springs. Emitted gases are dominated by hydrogen sulfide, dimethyl sulfide (DMS), propyl sulfide, and methyl mercaptan. DMS is also generated in relatively large amounts by the decomposition of algal methionine in the ocean, and these emissions have been suggested to act as homeostatic controllers of the earth's climate. Higher concentrations of DMS would increase albedo by providing more condensation nuclei for increased cloudiness, the resulting reduced insulation would lower planktonic photosynthesis, and diminished DMS emissions would provide fewer condensation nuclei and let in more solar radiation. Later studies showed that not just the magnitude but also the very direction of this feedback are questionable.

Human intensification of the sulfur cycle is due largely to the combustion of coals (containing typically approximately 2% sulfur) and crude oils (sweet crudes have less than 0.5% and sour ones more than 2% sulfur). Some natural gases are high in sulfur, containing it mainly as hydrogen sulfide (H_2S), but it is easily removed before they enter pipelines. Fossil fuel combustion accounts for more than 90% of all anthropogenic sulfur emissions, whose global total increased from 5 Mt in 1900 to approximately 80 Mt in 2000. The remainder is emitted largely during the smelting of color metals (mainly Cu, Zn, and Pb), and a small share originates in chemical syntheses.

Our interest in the sulfur cycle is mainly due to the fact that sulfur dioxide (SO_2) emitted from fossil fuel combustion and nonferrous metallurgy is rapidly oxidized to sulfates, whose deposition is the leading source of acid deposition, both in dry and in wet forms. Before their deposition, sulfur compounds remain up to 3 or 4 days in the lowermost troposphere (average, <40 h), which means that they can travel several hundred to more than 1000 km and bring acid deposition to distant ecosystems. Indeed, the phenomenon was first discovered when SO_2 emissions from central Europe began to affect lake biota in southern Scandinavia. Because of the limits on long-distance transport (atmospheric residence times of DMS and H_2S are also short-lived), sulfur does not have a true global cycle as does carbon (via CO_2) or nitrogen (via denitrification), and the effects of significant acid deposition are limited to large regions or to semicontinental areas.

Deposited sulfates (and nitrates) have the greatest impact on aquatic ecosystems, especially on lakes with low or no buffering capacity, leading to the decline and then the demise of sensitive fish, amphibians, and invertebrates and eventually to profound changes in a lake's biodiversity. Leaching of alkaline elements and mobilization of toxic aluminum from forest soils is another widespread problem in acidified areas, but acid deposition is not the sole, or the principal, cause of reduced forest productivity. Since the early 1980s, concerns about the potential long-term effects of acidification have led to highly successful efforts to reduce anthropogenic SO_2 emissions throughout Europe and in eastern North America.

Reductions of sulfur emissions have been achieved by switching to low-sulfur fuels and by desulfurizing fuels and flue gases. Desulfurization of oils, although routine, is costly, and removing organic sulfur from coal is very difficult (pyritic sulfur is much easier to remove). Consequently, the best control strategy for coal sulfur is flue gas desulfurization (FGD) carried out by reacting SO_2 with wet or dry basic compounds (CaO and $CaCO_3$). FGD is effective,

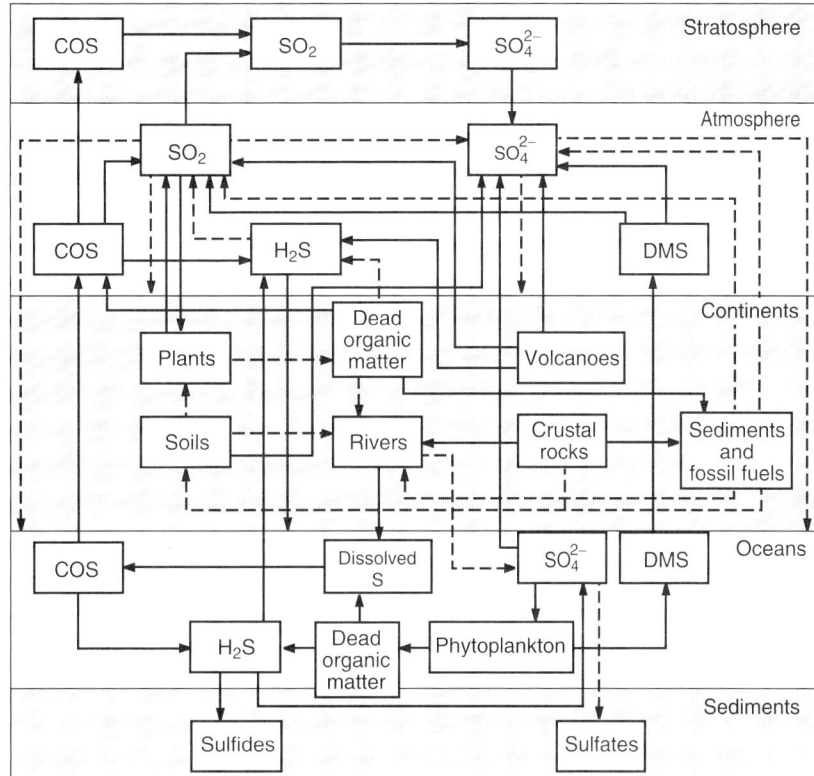

FIGURE 5 Sulfur's biogeochemical cycle extends from the stratosphere to marine sediments.

removing 70–90% of all sulfur, but the average U.S. cost of $125/kW adds at least 10–15% to the original capital expense, and operating the units increases the cost of electricity production by a similar amount. Moreover, FGD generates large volumes of wastes for disposal (mostly $CaSO_4$).

Airborne sulfates also play a relatively major role in the radiation balance of the earth because they partially counteract the effect of greenhouse gases by cooling the troposphere: The global average of the aggregate forcing by all sulfur emissions is approximately $-0.6 W/m^2$. The effect is unevenly distributed, with pronounced peaks in eastern North America, Europe, and East Asia, the three regions with the highest sulfate levels.

A 16-fold increase in the world's commercial energy consumption that took place during the 20th century resulted in the unprecedented level of human interference in global biogeochemical cycles. Thus, we must seriously consider the novel and a very sobering possibility that the future limits on human acquisition and conversion of energy may arise from the necessity to keep these cycles compatible with the long-term habitability of the biosphere rather than from any shortages of available energy resources.

SEE ALSO THE FOLLOWING ARTICLES

Biomass: Impact on Carbon Cycle and Greenhouse Gas Emissions • Conversion of Energy: People and Animals • Diet, Energy, and Greenhouse Gas Emissions • Earth's Energy Balance • Economic Geography of Energy • Ecosystem Health: Energy Indicators • Environmental Gradients and Energy • Land Requirements of Energy Systems • Sun, Energy from • War and Energy • World History and Energy

Further Reading

Browning, K. A., and Gurney, R. J. (eds.). (1999). "Global Energy and Water Cycles." Cambridge Univ. Press, New York.
Condie, K. C. (1997). "Plate Tectonics and Crustal Evolution." Oxford Univ. Press, Oxford, UK.
Davies, G. F. (1999). "Dynamic Earth: Plates, Plumes and Mantle Convection." Cambridge Univ. Press, Cambridge, UK.

Galloway, J. N., and Cowling, E. B. (2002). Reactive nitrogen and the world: 200 years of change. *Ambio* **31,** 64–71.

Godbold, D. L., and Hutterman, A. (1994). "Effects of Acid Precipitation on Forest Processes." Wiley–Liss, New York.

Houghton, J. T., *et al.* (eds.). (2001). "Climate Change 2001. The Scientific Basis." Cambridge Univ. Press, New York.

Irving, P. M. (ed.). (1991). "Acidic Deposition: State of Science and Technology." U.S. National Acid Precipitation Assessment Program, Washington, DC.

Smil, V. (1985). "Carbon Nitrogen Sulfur." Plenum, New York.

Smil, V. (1991). "General Energetics: Energy in the Biosphere and Civilization." Wiley, New York.

Smil, V. (2000). "Cycles of Life." Scientific American Library, New York.

Smil, V. (2001). "Enriching the Earth: Fritz Haber, Carl Bosch, and the Transformation of World Food Production." MIT Press, Cambridge, MA.

Smil, V. (2002). "The Earth's Biosphere: Evolution, Dynamics, and Change." MIT Press, Cambridge, MA.

Goods and Services: Energy Costs

ROBERT A. HERENDEEN
Illinois Natural History Survey
Champaign, Illinois, United States

1. Introduction
2. Methods to Determine Energy Cost
3. Illustrations/Applications of Energy Cost
4. Conclusions

Glossary

embodied energy The energy consumed "upstream" to facilitate a flow of goods or services (units = energy).
energy cost of living The energy required to facilitate a household's consumption of goods and services (units = energy/year).
energy intensity (ε) The energy required to facilitate a flow of one unit of a specified good or service (units = energy/gloof).
gloof Generic term to cover the range of goods and services; for example, a pound of fertilizer, the dollar value of an airline ticket, or a liter of water.
indirect effect Necessary costs not considered "direct." For example, auto fuel is usually considered direct energy, whereas energy to build roads is considered indirect. It depends on the definition and system boundary.
input–output (I–O) analysis A subdiscipline of economics that explicitly determines indirect effects.
net energy analysis A comparison of the energy costs and the energy produced by an energy technology such as a coal mine or photovoltaic panel.
power Energy per unit time (units = energy/time).
system boundary The limit up to which costs, benefits, impacts, and consequences are considered; can refer to spatial, temporal, or conceptual issues.
trophic position (TP) An indicator of dependence of an organism or group of organisms on solar-driven photosynthesis in an ecosystem. If the system has a linear food chain, it is called trophic level (dimensionless).
vertical analysis (also known as process analysis) A method to determine energy cost by tracing back through the production process.

Energy analysis determines the total energy required to produce a good or service, including the indirect effects through the chain of production and delivery. This energy is commonly called energy cost, although it does not mean the monetary cost of energy. Once the energy cost is known, the energy requirements of different patterns of production/consumption of goods and services can be analyzed. In this article, the imprecise term energy is used, although the preferred, thermodynamically correct term is free energy.

1. INTRODUCTION

Most of us know intellectually that energy—solar and fossil—supports all of our life-support systems, from agriculture to transportation, commerce, and medicine. Some of us know this more viscerally as well; ecologists and farmers see every day that "all flesh is grass." Readers may remember the U.S. east coast blackout of 1965, the oil crisis of 1973 (long waiting lines at gas stations), or even the California electricity shortages during the summer of 2001. I first "got it" while sitting in traffic in 1971 in Oak Ridge, Tennessee. Ahead of me was a car towing a trailer on which there was an aluminum pleasure boat. Because there was increasing controversy in the Tennessee Valley regarding electricity powered by strip-mined coal and nuclear reactors and also from dams, energy was on my mind. I knew that aluminum is an energy-intensive metal, and I realized that I was looking at "embodied" energy, and hence embodied environmental damage and controversy.

On that day, I started 30 years of work in energy analysis. When I see bread, autos, baby carriages, lawn sprinklers, hospitals, playgrounds, orchestral concerts, haircuts, blue jeans, orthodontic braces, street artists, aircraft carriers, or banjoes, I see energy. The question was and is the following: How much embodied energy? In this article, I present methods used to quantify energy cost and illustrate how it is used in several applications.

Energy cost is an example of an indirect effect. Once we realize that a good or service "embodies" an input that is not evident, we are free to calculate the energy cost of an orange, the orange cost of energy, the water cost of hamburger, or how much pollution is released when a Sierra Club book (or this encyclopedia) is published. However, then we are vexed with how far to go in the calculation. For example, how much energy is required to allow a car to travel 1 mile? Here are possible answers to the question:

1. The fuel burned
2. Plus the energy to extract, refine, and transport the fuel
3. Plus the energy to manufacture the car (prorated to 1 mile's use)
4. Plus the energy to produce tires, replacement parts, etc.
5. Plus the energy to build and maintain roads.
6. Plus the energy to maintain auto repair shops, government regulation and registration services, etc.
7. Plus the energy to produce and maintain that portion of the health system used to care for accident victims and sufferers of auto-related health problems.
8. Plus …

Where to stop depends on how we bound the issue, and in the end this decision should be made by the users of the result.

Also, there are additional steps. A similar expanding wave of concern and impact ripples out from, for example, the energy to make the car, which could include

1. The energy consumed at the assembly plant
2. Plus the energy used to make the steel, glass, rubber, etc.
3. Plus the energy used at the iron mine, sand pit, and sulfur mine
4. Plus …

even including the cars used by the iron mine so that the process runs in circles, although successive steps become smaller.

2. METHODS TO DETERMINE ENERGY COST

2.1 Vertical Analysis

The previous process is called vertical analysis or process analysis; I call it following your nose. Fig. 1 indicates the stepwise process of producing a product

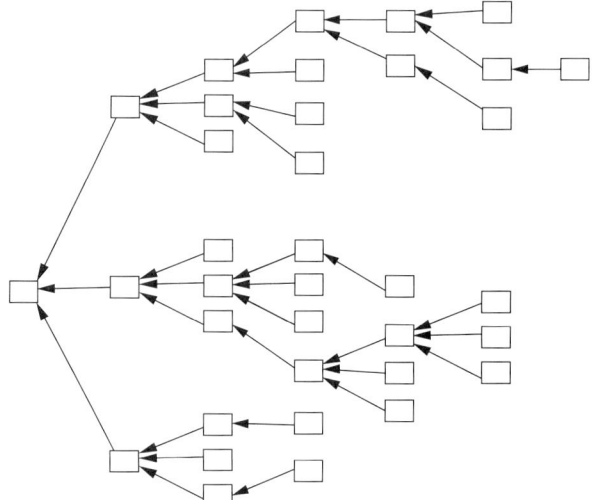

FIGURE 1 Schematic representation of vertical analysis (also called process analysis) to determine energy cost.

such as a car. The auto factory is the box on the far left. To produce an auto, the factory receives inputs from three other sectors—for example, steel, glass, and rubber. Each of these receives inputs from other sectors, and so on. Vertical analysis traces this web of inputs backward. At each node, it evaluates the amount of the desired input (e.g., energy) at that point and prorates it on the basis of that sector's output. Given the complexity of a modern economy, vertical analysis is labor-intensive and data-demanding.

2.2 Input–Output Analysis

Economic input–output (I–O) analysis is a well-established method to trace indirect effects in monetary terms. The U.S. Department of Commerce prepares detailed information on the dollar transactions among 350–500 sectors covering the U.S. economy. Many other nations have similar programs. Using several assumptions, one can use the analytical machinery of I–O to combine these data with supplementary information on energy use to determine energy cost.

We start with an economy with N sectors, each with an output flow measured in units of gloof/time. Gloof can be almost anything: tons of steel, pounds of butter, hours in a dentist's chair, dollars worth of day care, etc. It can even vary from compartment to compartment (as in the following example). The assumptions are as follows:

1. Every flow in the system that we wish to count (units = gloof/time) has an associated energy

intensity according to its source compartment (units = energy/gloof).

2. Embodied energy flows everywhere there are flows that we consider important, and the magnitude of the embodied energy flow (units = energy/time) is the product of the energy intensity (units = energy/gloof) times the original flow (units = gloof/time). This assumes that the flow variable is a good surrogate for embodied energy, which is a judgment call by the user.

3. Embodied energy is conserved in each compartment—that is, the amount "in" equals the amount "out."

Assumption 3 is stated by Eq. (1) and illustrated in Fig. 2.

$$\sum_{i=1}^{N} \varepsilon_i X_{ij} + E_j = \varepsilon_j X_j, \qquad (1)$$

where X_{ij} is the actual gloof flow from compartment i to compartment j (units = gloof$_i$/time), X_j is the sum of all output flows from compartment j (units = gloof$_j$/time), E_j is the actual energy input flow to compartment j from outside the system (units = energy/time), and ε_j is the energy intensity of output of compartment j (units = energy/gloof$_j$). Equation (1) is assumed to hold for each of the N compartments, yielding N equations in N unknowns. We know the X_{ij}'s and E_j's, so we can solve for the ε_j's. Equation (1) formalizes the allocation of indirect effects in a self-consistent manner and can be used to allocate indirect anything, not just energy.

2.3 An Example of the Calculation of Energy Cost

Figure 3A shows a two-sector economy burning oil to produce steel and cars. We use Eq. (1) to determine the energy intensities:

Steel sector: $10.8 + 1\varepsilon_{car} = 12\varepsilon_{steel}$

Car sector: $10\varepsilon_{steel} + 2 = 10\varepsilon_{car}$

Solving these two equations gives $\varepsilon_{steel} = 1$ barrel oil/ton steel and $\varepsilon_{car} = 1.2$ barrel oil/car. The embodied energy flows are obtained by multiplying the flows in Fig. 3A by the intensities, giving the result shown in Fig. 3B. Each sector is in embodied energy balance, as we assumed in Eq. (1). In addition, the whole system is in balance: 12.8 barrels of oil enters and is burned, and 12.8 barrels of embodied oil is contained in the shipments of steel and cars to final consumption. Furthermore, only 2/12 of the energy

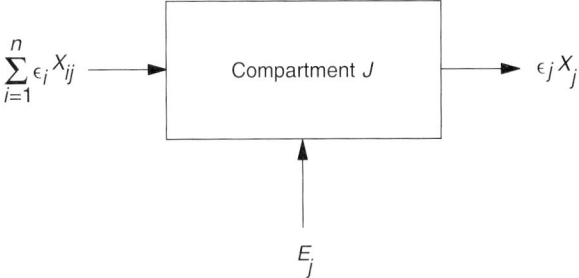

FIGURE 2 Input–output approach to determine energy cost. It is assumed that the embodied energy in = embodied energy out. E_j is the energy flow into sector j (units = energy/time), X_{ij} is the flow of product i to sector j (units = gloof/time), X_j is the total output of sector j (units = gloof/time), and ε_j is the energy intensity of product j (units = energy/gloof).

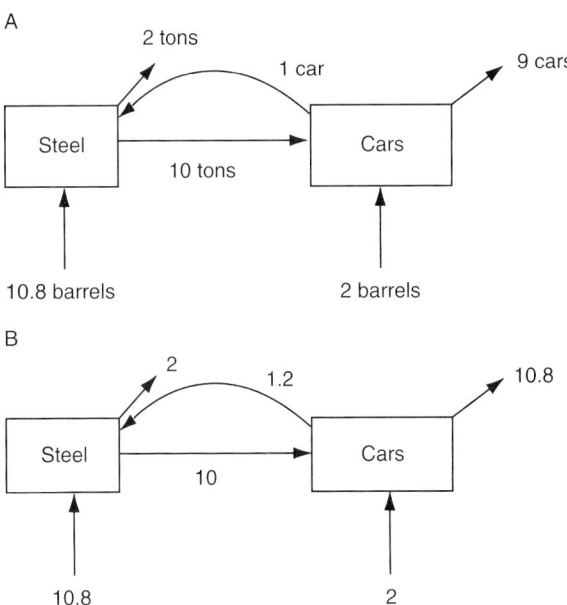

FIGURE 3 A two-sector economy illustrating the calculation of energy cost. (A) Original flows per unit time. Diagonal arrows represent shipments to final consumption. Energy inputs are in barrels of oil. (B) Embodied energy flows (units = barrels/unit time). All flows are assumed steady over time.

to produce a car is used directly (i.e., burned in the auto factory).

This is the I–O approach for two sectors. Calculation for systems with more sectors is operationally the same but tediously complicated—a job easily handled by computers. Typical results are shown in Table I. In addition, one may show that a vertical analysis of Fig. 3A yields the same results as the I–O approach. Because of the feedback of cars to steel, the analysis has an infinite number

TABLE I

Energy Intensities for Personal Consumption Categories, 1972–1973[a]

Consumption category	Energy intensity (10³ Btu/$1972)	Consumption category	Energy intensity (10³ Btu/$1972)
Food at home	54	Auto insurance	19
Food away from home	44	Auto registration and fees	0[a]
Alcoholic beverage	51	Local bus and train, commuting	56
Tobacco	30	Local bus and train, school	56
Rented dwelling	14	Local bus and train, other	56
Owned dwelling	12	Transportation to school away from home	132
Other shelter, lodgings	21	Plane, trip outside commuting area	167
Fuel oil and kerosene	890	Train and bus, trip outside commuting area	68
Coal and wood	795	Ship, trip outside commuting area	129
Other fuels	795	Limousine and taxi, trip outside commuting area	56
Gas main	915		
Gas in bottles and tanks	999	Car rental	38
Electricity	616	Health insurance	19
Water, sewage, trash	26	Health care	47
Telephone, telegraph, cable	24	Personal care	48
Domestic service	0[a]	Owned vacation home	12
Household textiles	59	Other transportation costs on vacation	91
Furniture	45	All-expense tours, summer camps	31
Floor coverings	71	Other vacation expenses	33
Major appliances	66	Boats, aircraft, motorcycles	65
Small appliances	65	Television	48
Housewares	57	Other recreation	41
Miscellaneous house furnishings	48	Reading	43
Dry cleaning, laundry	44	Private education	35
Clothing	44	Public education	37
Vehicle purchase	68	Miscellaneous consumption expenditure	41
Vehicle purchase finance charges	27	Personal insurance and pensions	20
Gasoline and oil	443	Gifts and contributions	41
Tires and lubrication	69	Increase in savings	47
Batteries	59	Housing purchase and improvement	40
Auto repair and service	37	Increase in investment	47

[a] Labor and government services are assumed to have zero energy intensity relative to the consumer. For labor, this is done to avoid double counting. For government, the energy requirement is not lost but rather charged against government expenditures.

of steps, but a computational trick makes this manageable.

For typical energy studies, the following assumptions have usually been made: Energy inputs are coal, petroleum, natural gas, and a "fossil equivalent" of hydro, wind, photovoltaic, and nuclear electricity. Solar energy input to silviculture and agriculture is usually considered a free input and not counted for economic systems. However, solar energy is usually counted in energy analysis of ecosystems.

3. ILLUSTRATIONS/APPLICATIONS OF ENERGY COST

3.1 Auto Manufacture and Use

Energy analysis showed that approximately 12% of the energy to produce and market a U.S. automobile was consumed in the automobile sector vs. 37% in the primary metals sectors. In addition, the fuel in the tank was only approximately 60% of the energy to provide auto transport; the other 40% was car

TABLE II

Energy Cost of Corn Production in Illinois, 1996

Type of agiculture:	Conventional
Crop:	Yellow corn
Field size:	76 acres
Yield:	168 bushels/acre
Calorific energy of crop	5.13×10^9 Btu (402×10^3 Btu/bushel)
Crop energy/input energy:	8.4

	$1996 Cost or Btu	Energy intensity (10^3 Btu/$1996 or Btu/Btu)	Energy (10^6 Btu)	% of total
Seed	$2284	26.2	59.8	9.8
Fertilizer	$4725	769.7	363.7	59.6
Herbicide	$2179	26.2	57.0	9.3
Pesticide	0	26.2	0	0.0
Machines	$2017	10.7	43.0	7.0
Farm fuel	52×10^6 Btu	1.20[a]	62.3	10.2
Custom application	$2129	11.6	24.7	4.0
Total			610.4	100.0

[a] Twenty percent extra energy is consumed in extracting, refining, and transporting fuel to point of use. This is an experimental plot; in commercial practice, typically some pesticides would be applied.

manufacture (8%), energy cost of energy (9%), tires, maintenance, road construction, parking, insurance, etc. This is an old result from the 1970s, but it is still compelling.

3.2 Grain Production

The ecologist Howard T. Odum said we eat "potatoes made of oil," a reflection of the energy demands of modern mechanized agriculture. Table II shows typical energy requirements for a corn crop. In the rich soil and favorable climate of Illinois, the fossil energy inputs sum to only one-eighth of the crop's calorific value (the sun is not counted). Note that fertilizer and herbicides account for 2/3 of total energy input, while farm fuel is only 10%. It is debatable whether fossil fuel inputs should be compared with the calorific content of food. More compelling are studies that show that the fossil energy input per unit of food output has increased during the past several centuries. Likewise, if we move to the right on any of these spectra— good→marginal soil and climate, grain→flesh, low→high processing and packaging, near→distant—the energy inputs typically increase. For the average food as delivered to American mouths, the fossil energy input is 5–10 times the calories delivered.

3.3 Household Energy (Energy Cost of Living)

Once obtained, energy intensities can be used to link everyday personal consumption patterns with their impact in resource use, as indicated in Eq. (2):

$$\text{Energy} = \sum_{i=1}^{N} (\text{energy intensity of consumer good } i) \times (\text{consumption of consumer good } i). \quad (2)$$

As with much energy analysis, effort peaked in the 1970s and 1980s, but here I present an update.

3.3.1 Energy Intensities

Very few vertical analyses have been performed on specific products, such as automobiles and agricultural chemicals. The remaining energy intensities have been calculated for approximately 350 sectors covering the full range of economic expenditures in the U.S. I–O economic tables. One assumes that dollars are a satisfactory numeraire for allocating embodied energy. Energy data compatible with I–O data are provided by the U.S. Department of Energy, Energy Information Administration. These are used with a version of Eq. 1.

One immediate question is whether the energy intensities do vary among sectors. If not, the energy

TABLE III

Energy Intensities of Consumer Expenditures, Updated (by Approximations) to 1999

Expenditure category	Energy intensity (Btu/$1999)
Food	9095
Alcohol and tobacco	5457
Housing	5457
Residential energy	105000
Gasoline and motor oil	115000
Auto purchase, maintenance	10914
Public transportation	20000
Apparel	8025
Health, personal care	8560
Entertainment and communication	5457
Education and reading	7276
Miscellaneous	7490
Contributions	7490
Insurance, pension	3638
Asset change	8560

consequences of spending a dollar for any good or service would be the same, and the details of expenditures would be irrelevant. We already know that the primary metals industry is energy intensive compared with a service industry, but we are interested in consumption options open to individuals and households, which generally covers finished products. Even so, there is a significant variation, as shown in Table I. For example, the energy intensity of gasoline was 443,000 Btu/$ 1972, whereas that of private education was 35,000 Btu/$ 1972, a variation of a factor of 13. Among nonenergy, nontransportation categories, there was a variation of a factor of 5, from rental dwellings (14,000 Btu/$ 1972) to floor coverings (71,000 Btu/$ 1972). In Table III, the consumption categories have been aggregated to 15 and updated to 1999.

3.3.2 Household Consumption Data

The U.S. Bureau of Labor Statistics (BLS) periodically performs detailed surveys of consumption "market baskets" of U.S. households. A total of 19,000+ households were surveyed in 1972 and 1973, with expenditures broken down into 61 categories as given in Table I. In the updated analysis here, I use results of a smaller BLS survey for 1999. The system boundary thus is defined by what the household purchases.

Combining expenditures and intensities using Eq. (2) yields Fig. 4, which shows the energy cost taken back to the mine mouth or wellhead, of the market basket of average households in lowest and highest expenditure quintiles and for the overall average household. Of the 15 consumption categories, 2 are direct energy expenditures (residential fuel and electricity and auto fuel). For the average households, these amount to 5% of monetary expenditures (Fig. 4A) but 44% of the household's total energy cost (Fig. 4B). Figure 4B also shows that for the lowest quintile, the direct energy requirement is approximately 57% of the total, and for the top quintile it is approximately 33%. There seems to be an upper limit on how much auto fuel a household can use but not on football tickets, clothing, furniture, plane tickets, second homes, etc. For the average expenditure level and above, one would miss more than half of the energy cost of living by examining only direct energy.

Under certain assumptions, the energy cost of living can be used for two more speculative purposes: to predict the economic effects of energy price increases on different households (this has been done), and in the analysis of sprawl issues (because more auto use and larger residences are associated with living further from urban centers).

3.4 Appliances: Energy to Operate vs Energy to Produce/Maintain/Dispose

What fraction of the energy use for a refrigerator or a toaster is used to manufacture, market, maintain, and mash it, and what fraction is used to operate it? The answer is useful in evaluating different strategies to reduce energy use, such as increasing operational efficiency, decreasing energy use in manufacture, or increasing the device's lifetime. This is an example of life cycle cost in energy terms. Consider all the energy associated with an appliance:

$$E_{\text{life}} = \text{manufacturing energy} + \text{maintenance energy} + \text{disposal energy} + \text{operational energy}.$$

The average total power is the lifetime energy divided by the lifetime, T:

$$p_{\text{tot}} = \frac{E_{\text{life}}}{T}$$
$$= \left(\frac{\text{manufacturing} + \text{maintenance} + \text{disposal}}{E_{\text{life}}} + 1\right) \times p_{\text{operation}}. \quad (3)$$

For electric appliances, energy is measured in kilowatt-hours, and power is measured in kilowatts.

Table IV summarizes energy use for selected household appliances. Most appliances tend to have

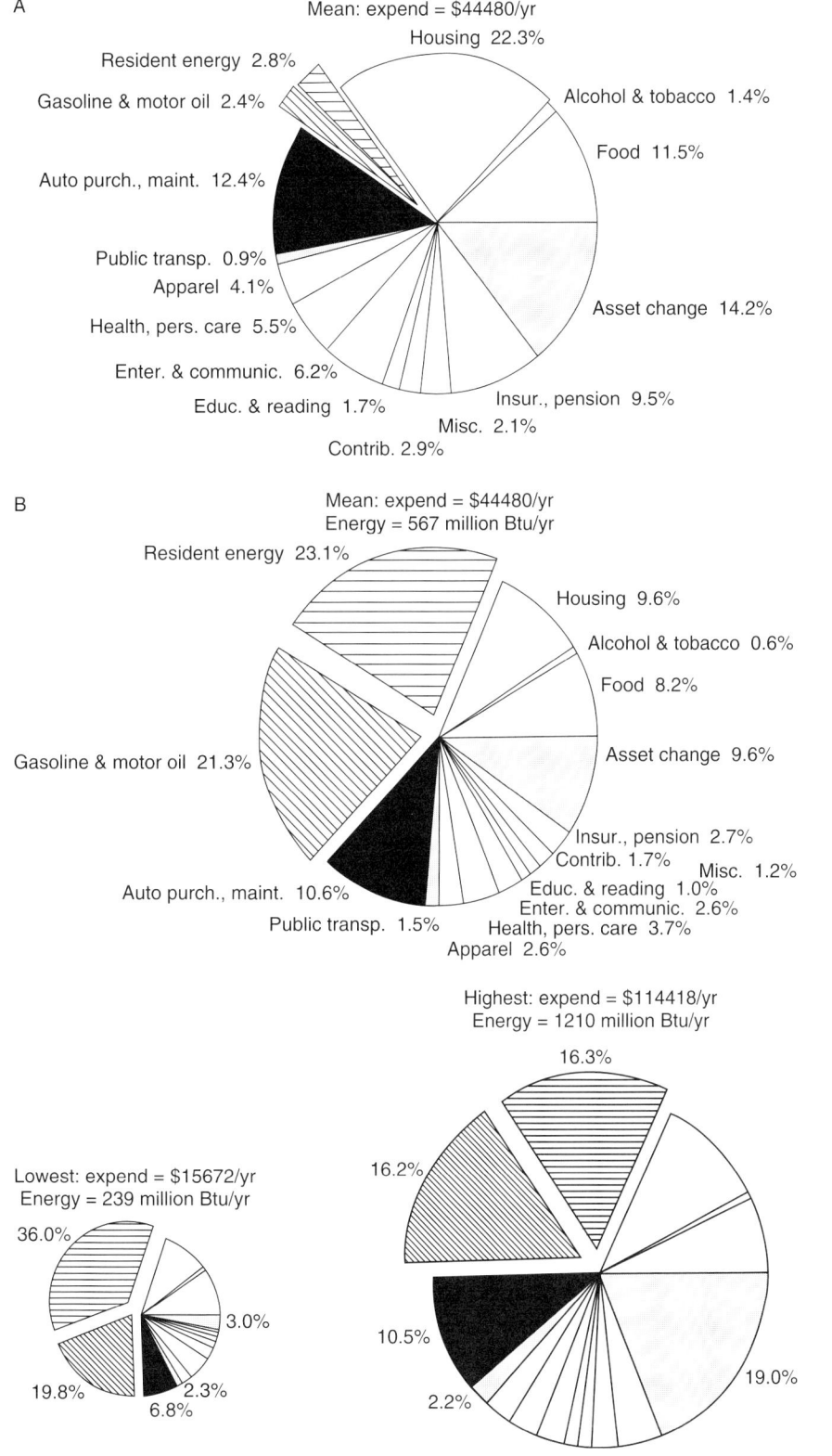

FIGURE 4 Energy cost of living, 1999. (A) Annual expenditures for average U.S. household. (B) Energy costs of expenditures in Fig. 4A and for lowest and highest expenditure quintiles as well. Figure 4B is obtained from Fig. 4A by multiplying energy intensities by expenditures (Eq. 2). In Fig. 4B, the area of the circle is proportional to the total household energy impact.

TABLE IV

Comparison of Operating and (Manufacuring + Maintenance + Disposal) Energy for Selected Appliances

Electric appliance	$p_{\text{operation}}$ (kWh/year)	T(year)	Indirect fraction (manufacturing – maintenance + disposal) as fraction of life energy	Classification ($p_{\text{operation}}$/ indirect fraction)[b]
Blender	17	14	0.34	Low/high
Mixer	14	14	0.35	Low/high
Refrigerator	750	14	0.13	High/low
Water heater	4700	8	0.01	High/low
75-W incandescent bulb[b]	657	0.1	0.02	High/low
17-W compact fluorescent bulb[b]	146	1.14	0.08	High/low
Coffee maker	83	8	0.07	Low/low
Toaster	43	8	0.19	Low/low
Personal computer (at work)[c]	160	7	0.5	High/high

[a] Classification: $p_{\text{operation}}$: <100 kWh/year = low; ≥ 100 kWh/year = high. Indirect fraction: <0.25 = low; ≥ 0.25 = high.
[b] Assumes bulbs are on continuously. These two bulbs provide approximately equal light levels.
[c] Assumes the computer is used 8 h/day, 200 days/year.

high operation power and a low indirect energy fraction [i.e., (manufacturing + maintenance + disposal)/$E_{\text{life}} < 1$] or vice versa. For the appliances usually considered large energy users, such as refrigerators, freezers, water heaters, and light bulbs, the fraction is ≤ 0.25. For smaller appliances, it is ≥ 0.25. This implies, but does not prove, that to save energy one should concentrate on operational energy efficiency for the first group, whereas for the second there is likely potential in both operational and manufacturing efficiency.

The personal computer, if used 8 h a day, 5 days a week, has a high operation power (160 kWh/year) and a high indirect fraction (≈ 0.5).

3.5 Energy Cost in Ecosystems

There are several indicator quantities that summarize indirect energy effects in ecosystems. Two that are especially appropriate for quantifying the solar dependence of all life (with the negligible exceptions of deep-sea vent communities, which are based on geochemical energy) are energy intensity and trophic position. Trophic structure refers to the number of energy "transactions" separating the sun (which is assigned trophic position 0) from the compartment in question. Originally, energy flow in ecosystems was visualized in terms of straight food chains, for which trophic positions (called trophic levels in this case) are successive integers (Fig. 5A). However, most ecosystems have web-like energy flows (Fig. 5B). Equation (1) applies to webs, so calculating energy intensities requires no new technique. On the other hand, trophic position needs to be defined. The trophic position of a compartment, TP, is the energy-weighted sum of TPs of inputs + 1. Using the language we used for calculating energy intensities, for each compartment,

$$\text{TP}_j = \sum_{i=1}^{N} \left(\frac{X_{ij}}{\sum_{i=1}^{N} X_{ij}} \right) \text{TP}_i + 1. \quad (4)$$

Equation (4) is similar to Eq. (1), but it has two important differences. First, the flows X_{ij} must be in terms of biomass energy because trophic ecology is by definition concerned with energy flow. Second, the factors in the bracket sum to 1 because they are normalized with respect to the input flows, whereas in Eq. (1) normalization is with respect to total output and the factors X_{ij}/X_j need not sum to 1. Equation (4) represents N equations in N unknowns.

For the (idealized) food chain in Fig. 5A, the weighting factors are all 1, and Eq. (4) gives the trophic positions 1–4. For the (real) food web in Fig. 5B, TPs are 1, 3.43, 3.90, and 4.90. Because plants receive input only from the sun, $\text{TP}_{\text{plants}} = 1$. Because decomposers gets all their input from detritus, $\text{TP}_{\text{decomposers}} = \text{TP}_{\text{detritus}} + 1$. Animals and detritus have more than one input, and their TPs are mixtures. In this accounting, decomposers (e.g., bacteria) are on top of the energy pyramid and the food web.

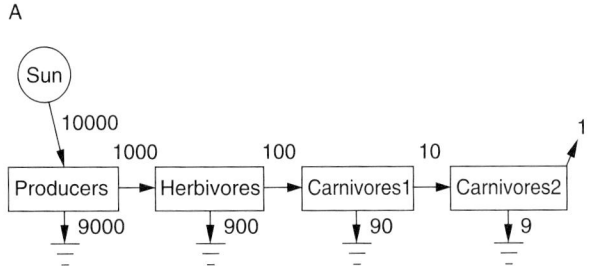
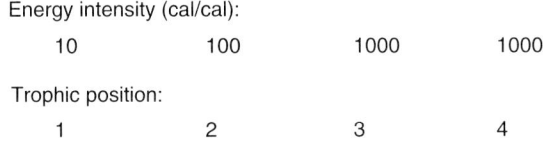
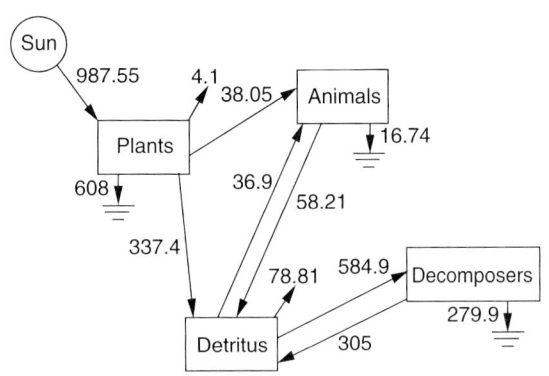

Compartment	Energy intensity (cal/cal)	Trophic position
Plants	2.60	1.00
Animals	9.56	3.43
Detritus	12.40	3.90
Decomposers	23.80	4.90

FIGURE 5 Energy intensities and trophic positions for food chain and web. (A) Idealized food chain. (B) Boreal bog ecosystem food web. Units for both = g fixed carbon m^{-2} year^{-1}. Flows to ground symbols are metabolic losses. Detritus is dead material the origin of which is not visually identifiable. Decomposers include bacteria.

4. CONCLUSIONS

Energy is important, and the idea of energy cost makes intuitive sense. The details of obtaining and using energy cost complicate its usefulness, but when done transparently, it can provide insight and help in decision making. To do it transparently, we require at the least a procedure, a system boundary, and many assumptions. For brevity, I have not discussed most of the assumptions, and I have included only a few applications of energy cost. Others include comparing energy embodied in imported and exported goods and services (energy balance of trade), comparing embodied energy and labor costs to determine the employment effects of energy policy, determining equity effects of energy taxes, and comparing the energy cost with the energy output of an energy source technology such as a power dam. The latter, called net energy analysis, is presented in a separate article in this encyclopedia.

SEE ALSO THE FOLLOWING ARTICLES

Cost–Benefit Analysis Applied to Energy • *Energy Services Industry* • *External Costs of Energy* • *Global Energy Use: Status and Trends* • *Industrial Ecology* • *Industrial Energy Efficiency* • *Industrial Energy Use, Status and Trends* • *Input–Output Analysis* • *Leisure, Energy Costs of*

Further Reading

Bullard, C., and Herendeen, R. (1975). The energy costs of goods and services. *Energy Policy* **3**, 268–278.

Bullard, C., Penner, P., and Pilati, D. (1978). Net energy analysis: Handbook for combining process and input–output analysis. *Resour. Energy* **1**, 267–313.

Finn, J. T. (1976). Measures of ecosystem structure and function derived from analysis of flows. *J. Theor. Biol.* **56**, 115–124.

Hall, C., Cleveland, C., and Kaufmann, R. (1986). "Energy and Resource Quality: The Ecology of the Economic Process." Wiley, New York.

Herendeen, R. (1988). Net energy considerations. *In* "Economic Analysis of Solar Thermal Energy Systems" (R. West and F. Kreith, Eds.), pp. 255–273. MIT Press, Cambridge, MA.

Herendeen, R. (1998). "Ecological Numeracy: Quantitative Analysis of Environmental Issues." Wiley, New York. [See Chap. 8].

Herendeen, R., and Fazel, F. (1984). Distributional aspects of an energy conserving tax and rebate. *Resour. Energy* **6**, 277–304.

Herendeen, R., Ford, C., and Hannon, B. (1981). Energy cost of living, 1972–1973. *Energy* **6**, 1433–1450.

Odum, H. T. (1996). "Environmental Accounting." Wiley, New York.

Schulze, P. (ed.). (1999). Measures of Environmental Performance and Ecosystem Condition. National Academy of Engineering, Washington, D.C.

Slesser, M. (ed.). (1974). Energy analysis workshop on methodology and conventions, August 25–30, Guldsmedhyttan, Sweden. International Federation of Institutes of Advanced Study, Stockholm, Sweden. [Reprinted in Energy accounting as a policy analysis tool, Committee print for the 94th Congress, 2nd session, Serial CC, 68–391. U.S. Government Printing Office, Washington, DC, 1976]

Green Accounting and Energy

PETER BARTELMUS
Wuppertal Institute for Climate, Environment and Energy
Bergische Universität Wuppertal, Germany

1. Accounting for Sustainability
2. The Concepts and Methods of Integrated Environmental and Economic Accounting
3. Accounting for Energy and Related Impacts
4. Outlook: Policy Use of Accounting Indicators
   Appendix: Monetary Valuation in the SEEA

Glossary

capital maintenance Keeping produced and natural capital intact as a principle of sustainability of economic activity, applied in environmental accounting.

dematerialization Reduction of material inputs into, and residual outflows from, the production and consumption of goods and services as a means of diminishing environmental pressure from economic activity.

energy accounts Accounts and balances, which measure energy inputs, transformations and uses; sometimes extended to include material flows in embodied energy values.

energy theory of value Rationale for the quantitative evaluation of natural resources and goods and services in terms of energy expended in their production.

environmental accounting (1) National accounting: short for integrated environmental and economic accounting, as advanced by the System of Environmental and Economic Accounts (SEEA). (2) Corporate accounting: usually refers to environmental auditing but also includes environmental costing and life-cycle (cradle-to-grave) assessments of products.

environmental valuation Different techniques of estimating a monetary value for natural assets or asset changes in environmental accounting, including market valuation of natural resources, maintenance costing of environmental impacts, and damage valuation of environmental effects.

environmentally adjusted (net) capital formation (ECF) Environmental accounting indicator that deducts the cost of natural capital consumption from net capital formation.

environmentally adjusted (net) domestic product (EDP) Environmental accounting aggregate, obtained by deducting the cost of natural capital consumption from net domestic product.

green accounting Popular term for environmental accounting, including material flow accounts.

material flow accounts Physical accounts and balances, which present inputs of materials into the economy, their accumulation, and their outflow to the natural environment and other economies; usually measured in mass (weight) units.

natural capital Natural assets in their role of natural resource supply and provision of environmental services (notably of waste/residual disposal) for economic production and consumption.

System of Environmental and Economic Accounts (SEEA) Extended national accounts system, advanced by the United Nations and other international organizations; incorporates natural assets and asset uses to measure the interactions between environment and economy in physical and monetary units.

total material requirement (TMR) Key indicator of material flow accounts that measures the total material base of the economy; includes besides direct material inputs, "hidden" movements of materials associated with imports and extraction of natural resources.

Green accounting is a popular term for environmental accounting or, more precisely, integrated environmental and economic accounting at national and corporate levels. At both levels, the purpose is to capture the elusive notion of long-term sustainability of economic performance, hampered by environmental impacts of production and consumption. Long lists of environmental and sustainable development indicators failed to assess sustainable economic growth and development in a comprehensive and transparent manner. The more systematic greening of the national accounts succeeds in defining and measuring the environmental sustainability of economic activity. This article describes the concepts and methods of environmental accounting at the national level and addresses specifically the role of energy in accounting and sustainability analysis.

1. ACCOUNTING FOR SUSTAINABILITY

1.1 Concepts and Indicators of Sustainable Development

Conspicuous pollution incidents in the 1960s and the energy crises of 1973 and 1974 brought about environmental gloom. Well-known titles like Carson's "silent spring" or Loraine's "death of tomorrow" alerted to risks for human survival. The Club of Rome presented in 1972 a computerized and hence seemingly objective world model, which predicted the collapse of the industrial system and Malthusian population decline within this century. Nonrenewable resource depletion was the main reason for this business-as-usual scenario.

William Kapp can be credited with alerting mainstream economists to the social costs of environmental concerns. Neoclassical economics had largely ignored environmental phenomena as external to economic planning and budgeting. In Kapp's view the staggering social costs of environmental damage had reached a magnitude beyond the capacities of welfare analysis, national income accounting, and compartmentalized social and natural sciences. Based on a similar assessment of compartmentalized institutions, the World Commission on Environment and Development (WCED) advanced in 1987 the notion of sustainable development. The commission's report on "Our Common Future" described the new development paradigm as an integrative approach to tackling the interlocking environmental and developmental crises. Two Earth Summits, convened by the United Nations in 1992 in Rio de Janeiro and 2002 in Johannesburg, confirmed sustainable development as a global paradigm of equal importance for industrialized and developing countries.

The WCED also offered the popular definition of sustainable development as "development that meets the needs of the present without compromising the ability of future generations to meet their own needs." The definition is vague: it does not specify the categories of human needs, does not give a clear time frame for analysis and does not indicate particular roles for the environment or social concerns in long-term development. The paradigm needs to be operationalized, and indeed numerous indicators for and of sustainable development have since been advanced.

Indicators *for* sustainable development typically include long lists of measures of environmental, social, economic, and institutional aspects of development. Relatively loose frameworks generate some useful categorization but do not in general link or aggregate these indicators. The main drawback of indicator lists is the difficulty of comparing and evaluating data expressed in different units of measurement. Such indicators may support the management of particular areas of environmental, social, or economic concern but are less useful for the overall planning and implementation of sustainable development.

Proposals of indicators, or indices rather, *of* sustainable development has been the response to the comparability and aggregation problem. The approach is to average or otherwise combine selected indicators, or to correct conventional economic aggregates of wealth, income, and consumption. Compound measures of sustainable or human development or social progress deserve our attention since they all try to take the environment "into account." Well-known examples are the following:

- Human Development Index (HDI), developed by the United Nations Development Programme. The index is an average of per capita income, life expectancy, and education, intended to provide a more complete (than GDP) assessment of human achievements.
- Environmental Sustainability Index (ESI), initiated by the World Economic Forum. An average high performance of 20 indicators, made up of 68 variables, defines environmental sustainability.
- Genuine Progress Indicator (GPI) for the United States, later applied and modified in other countries. The GPI subtracts defensive expenditures (of environmental protection, defense, and other welfare maintaining—rather than increasing—expenses) and adds or subtracts positive and negative environmental and other social costs and benefits.
- World Bank's National Wealth aggregate of produced assets, natural capital, and human resources.

All these indices suffer from a more or less arbitrary selection of partially correlated indicators, a questionable mix of weighting and valuation techniques, or inconsistency with standard accounting concepts and conventions. The idea seems to be to demonstrate the nonsustainability of economic growth, rather than to support policy making. For instance, the authors of the GPI present their indicator explicitly as a means to discredit GDP as a measure of social progress.

1.2 Green Accounting: Assessing the Environmental Sustainability of Economic Growth

The elusive notion of sustainability needs to be operationalized in a more transparent and systematic manner. To this end, natural science offers basic principles for the nature-economy interface, and economics provides the tool of accounting for this interface. Thermodynamic laws of matter and energy conservation and dissipation govern the use of natural resources. Formal double-entry accounting can then be applied to assess the use (input) and dispersion (output) of these resources from/to the natural environment.

Figure 1 illustrates the "throughput" of materials (including energy carriers) through the economy. The figure also shows the two basic functions of the environment: (1) resource supply of raw materials, space, and energy to the economy (source function) and (2) waste assimilation (sink function). The latter represent an important ecological service. Other ecosystem services such as habitat for species, soil protection, or local climate control are sometimes categorized as indirect benefits for humans; they are probably better assessed by regionalized environmental and ecological statistics.

The fortunate coincidence of physical laws that ensure the equality of inputs and outputs of energy and materials, and input-output based economic accounting points to an obvious approach of applying the latter to the former. The result is green accounting—that is, the extension of conventional economic accounts into the physical world of the use and abuse of nature by economic activity.

The application of accounting tools requires aggregation of physical environmental data by means of a common measuring rod. Environmentalists and economists—to make a crude distinction between ecological and economic views of the human environment—disagree on whether to use physical measures, or prices and costs, for weighting the importance of environmental impacts. Environmental economists seek to "internalize" the costs of environmental impacts (environmental "externalities") into the decisions of households and enterprises. Environmentalists, on the other hand, refute such commodification of an indivisible (public) good on whose value markets should not have a say.

Two operational sustainability concepts can thus be distinguished according to the economic and ecological outlook:

- Economic sustainability makes use of the established requisite for economic growth, capital maintenance, and extends the (produced) capital concept to include nonproduced natural capital.
- Ecological sustainability considers material throughput through the economy as a pressure on the carrying capacities of natural systems, and aims to reduce this pressure to tolerable levels by dematerializing the economy.

Two accounting systems attempt respectively to capture these notions of sustainability. Material flow accounts (MFA) and related physical input-output tables (PIOT) measure material flows in physical—weight—units. The United Nations System for integrated Environmental and Economic Accounting (SEEA) embraces both physical and monetary accounts, seeking compatibility with the worldwide adopted System of National Accounts (the SNA). Considering the integrative power of the SEEA, the

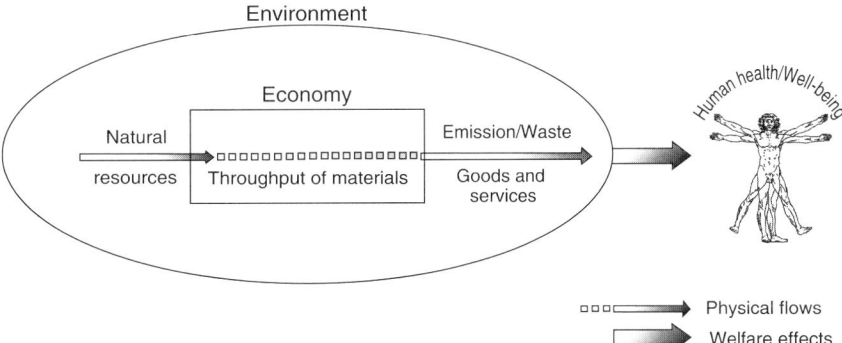

FIGURE 1 Environment-economy interaction and effects. Flows of material throughput from the environment into the economy and out of the economy into the environment. Outputs from the economy and the environment affect human welfare. From Bartelmus (2001) Fig. 1, p.1723, with permission from EOLSS publishers.

1992 Earth Summit called for its application by all member States of the United Nations in its Agenda 21. An international group of experts and accountants has now revised the SEEA, which the United Nations and other international organizations will issue as "Integrated Environmental and Economic Accounting 2003."

The narrow focus on environmental (at the expense of social and other concerns) sustainability of economic activity by both types of accounts does not do justice to the much broader paradigm of sustainable development. Development itself encompasses, besides economic objectives, social, ecological, cultural, and political ones. A corresponding broad sustainability notion would therefore have to incorporate all these objectives. Measurability is the main problem in accounting for these concerns, even when more narrowly focusing on the costs of human, social, and institutional capital formation and use in economic growth.

1.3 Energy: The Ultimate Limit?

Several energy analysts and ecological economists have deplored the neglect of energy in mainstream economic analysis. In their view, energy can be seen as the ultimate source of life and at the same time as its ultimate limit. The reasons for this neglect include a relatively low share of energy costs in production due to incomplete energy accounting and the underpricing of energy resources. Prices that ignore the social costs of energy depletion and pollution from energy production and use create a delusion of abundance and might undermine the sustainability of economic activity. Subsidies, notably of labor-intensive fossil fuel production, may increase employment but aggravate underpricing and environmental impacts. Addressing the pressures on and scarcities of natural resources, the WCED seems to concur that the ultimate limits to global development might indeed be determined by the impacts of energy use.

The significance of energy production and consumption for the overall sustainability of economic development can, however, only be assessed in a broader context of accounting, comprehensively and comparably, for economic benefits, natural resource use, and waste generation by different economic sectors. Section 2 describes the general concepts and methods of such accounting.

Section 3 then indicates how environmental accounting applies to energy production and consumption and related environmental impacts. It does not elaborate on the different techniques of narrowly defined energy balances and analyses. Rather, the section shows how energy fits into the environmental accounting frame and how such accounting may contribute to assessing the sustainability of economic performance and growth.

2. THE CONCEPTS AND METHODS OF INTEGRATED ENVIRONMENTAL AND ECONOMIC ACCOUNTING

2.1 Material Flow Accounts: Assessing the Physical Metabolism of the Economy

Figure 2 can be viewed as the accounting interpretation of the generic environment-economy interface of Fig. 1. It represents the material flow accounts (MFA), which Robert Ayres and others have characterized as picturing the physical metabolism of society. The MFA show material inputs from abroad and the natural environment, their accumulation as stock in infrastructures and other durable goods, and their output of wastes and residuals into the environment and (as exports) to other countries. The MFA were developed in Austria and Germany and mostly applied in industrialized countries. The statistical office of the European Union, Eurostat, prepared guidelines on economy-wide MFA for improving the international comparability of the accounts.

Adding up the material flows (in tons) obtains various indicators at different stages of throughput as shown in acronyms in Fig. 2. Table I elaborates on the definition and contents of these indicators. Perhaps best known is the total material requirement (TMR). The indicator is defined as the sum of primary material inputs, including their ecological rucksacks. These rucksacks are hidden flows, which do not become a physical part of any product but are associated with its production, use, and disposal (e.g., soil erosion from agriculture or forestry, or gangue from mining).

Scholars of the German Wuppertal Institute, who focused their work on compiling and analyzing material flow accounts, have argued that their indicators measure actual and potential, and possibly still unknown, environmental impacts. A downward trend of the TMR, together with the usual upward trend of economic output (GDP), would thus indicate a decoupling of environmental pressure from economic growth. This can be seen as a process of dematerialization of the economy, and hence of ecological sustainability, as defined previously.

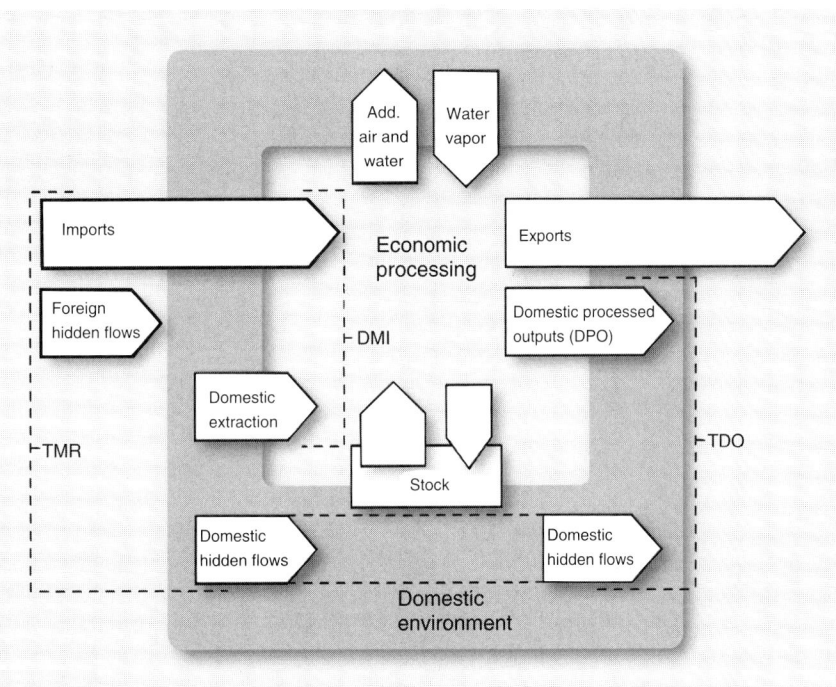

FIGURE 2 Material flow accounts and indicators. Adding up material flows at various stages of throughput through the economy obtains different indicators of material input, accumulation, and output. From Matthews *et al.* (2000), "The Weight of Nations: Material Outflows from Industrial Economies," Fig. 1, p. 5, World Resources Institute, Washington, DC, with permission.

The question is, how much dematerialization makes economic growth sustainable? Since physical material flows are not expressed in monetary units and are therefore not directly comparable to economic output or income, we can answer this question only by setting standards or targets for desirable rates of economic growth and necessary throughput reduction. The former president of the Wuppertal Institute, Ernst von Weizsäcker advanced the "Factor 4" standard of halving material input into the economy while doubling wealth over a period of 3 to 5 decades. Factor 4 and other standards of even a higher order can be interpreted as dematerialization targets for global sustainable development.

Overall indicators like the TMR are relatively crude measures of environmental impact since they use the weight of material inputs for weighting the significance of potential environmental effects. Table I shows therefore also various output indicators of wastes and emissions. They do weigh impacts by weight but are closer to specific environmental pressures and impacts. Their decline is indicative of the detoxification of the economy as they address the sink/pollution concern of sustainability.

2.2 Linking Physical and Monetary Accounts

The MFA were not developed in full consistency with economic concepts and measures. Their purpose was, at least initially, to provide an alternative to misleading—with regard to the human quality of life—accounting aggregates such as GDP. As a result, the MFA differ from standard and extended (see Section 2.3) economic concepts and indicators, notably with regard to the following:

- The narrow focus on material flows, excluding stocks and stock changes of produced and natural assets
- A corresponding inability to assess permanent losses (depletion and degradation beyond regeneration) of natural assets by flow indicators such as the TMR—an essential concern of sustainability
- A focus on different categories of materials by the MFA, rather than on the economic sectors responsible for the consumption of natural resources and the emission of residuals
- The assumption that material inputs reflect actual and potential environmental impacts, as opposed

TABLE I
Material Flow Balance and Derived Indicators[a]

Inputs (origin)	Outputs (destination)
Domestic extraction	Emissions and wastes
• Fossil fuels (coal, oil, etc.)	• Emissions to air
• Minerals (ores, gravel, etc.)	• Waste landfilled
• Biomass (timber, cereals, etc.)	• Emissions to water
	Dissipative use of products
Imports	• (Fertilizer, manure, compost, seeds, etc.)
Direct material inputs (DMI)	Domestic processed output (DPO) to nature
Unused domestic extraction	Disposal of unused domestic extraction
• From mining/quarrying	• From mining/quarrying
• From biomass harvest	• From biomass harvest
• Soil excavation	• Soil excavation
Total material input (TMI)	Total domestic output (TDO) to nature
	Exports
	Total material output (TMO)
Upstream flows associated with imports	
Total material requirement (TMR)	Net additions to stock (NAS)
	• Infrastructures and buildings
	• Other (machinery, durable goods, etc.)
	Upstream flows associated with exports

Source. Bringezu (2002). Material flow analysis—unveiling the physical basis of economies. In "Unveiling Wealth—On Money, Quality of Life and Sustainability." (P. Bartelmus, Ed.), Table IV. 2, p.119. Kluwer Academic, Dordrecht, Boston, and London.

[a] Excludes water and air flows (unless contained in other materials).

to the (cost value of) actual impacts recorded for a specific accounting period
- The use of physical (mass) units of measurement rather than the monetary (price and cost) values of economic accounts.

Facing mounting criticism of the conventional accounts and pressure to modify (to "green") these accounts for environmental impacts, national accountants took first steps toward incorporating readily available environmental statistics into their accounting frameworks. To this end, they made use of—physical, Leontief-type—input-output tables whose monetary counterparts are already a part of the national accounts system. The result is a relatively weak integration of environmental data either into a purely physical input-output table (PIOT) or as an add-on of physical indicators into the conventional accounts. The latter generates a hybrid physical-monetary National Accounting Matrix including Environmental Accounts (NAMEA).

German statisticians compiled a comprehensive PIOT for 1990 and updated the tabulations for 1995 (but did not publish them so far). The German PIOT is basically a sectoralization of the compound MFA: it specifies the material and substance flows as inputs and outputs in a detailed sectoral breakdown. Stock/asset accounts of produced capital goods and natural resources are omitted, as are difficult-to-assess-and-allocate ecological rucksacks. Supplementary sub-tables present energy balances in calorific values (terajoules) and in terms of inputs, outputs and transformations of energy carriers.

The Dutch NAMEA goes further by linking physical with monetary data. It stops short of any monetary valuation of environmental impacts by juxtaposing environmental statistics and indicators next to the economic activities responsible for environmental impacts. Like the PIOT, the NAMEA incorporates impacts on the source and sink functions of the environment in its supply (output) and use (input) accounts, as shown in Table II. In addition, the NAMEA presents some further weighting (beyond the use of tons) and aggregation for national environmental themes. The themes include energy (in calorific values), acidification, eutrophication, greenhouse effect, ozone layer depletion, and loss of natural resources (of gas and oil).

Table II is a simplified structure of a hybrid accounting system, based on the NAMEA approach. It presents the conventional economic accounts of production, consumption, capital accumulation, income generation, and international trade. Below and to the right of these accounts (separated by dashed lines), material and substance flow accounts are added to the economic transaction rows and columns of the economic accounts. The NAMEA can thus be seen as a PIOT linked to a monetary input-output table with some further supplementary indicators for selected environmental policy themes.

2.3 SEEA: Integrating Physical and Monetary Accounts

Physical accounts do not have the integrative power of the monetary national accounts. Weighting

TABLE II
Simplified Structure of a NAMEA

Supply \ Use	Monetary accounts				Physical accounts	
	Industries 1, 2, ...	Households (public and private)	Capital	Rest of the world (ROW)	Material flows (residuals, natural resources)	Policy themes, loss of natural resources
Products	Intermediate consumption	Household consumption	Gross capital formation	Exports	Emissions from industries	
Consumption of households					Emissions from households	
Capital	Capital consumption			Capital transfer from ROW	Other volume changes (discoveries etc.)	
Income generation	Value added, NDP					
Rest of the world (ROW)	Imports		Capital transfer to ROW	Balances	Transboundary pollutants to ROW	
Material flows (residuals, natural resources)	Natural resource inputs		Contributions to policy themes, loss of natural resources	Transboundary pollutants to ROW		Contributions to policy themes, loss of natural resources

Source. Adapted from Keuning and de Haan (1998), Table 1, pp. 146–147.

environmental impacts by weight of substance and resource flows may alert us to trends of overall environmental pressures and serve the management of particular natural resources; physical accounts cannot, however, compare the significance of environmental problems with the environmental and economic costs and benefits, as perceived by economic agents. Calculating ratios of resource productivity (GDP/TMR) or ecoefficiency (environmental impacts per GDP) is the most one can do to relate physical impacts to economic output. Monetary valuation appears to be the only possibility of fully integrating environmental concerns into the economic accounting system while ensuring consistency and comparability of "greened" with conventional economic indicators.

Pricing "priceless" environmental source and sink functions that are not traded in markets has its own problems. National accountants, especially from industrialized countries, have therefore been quite recalcitrant to implementing environmental accounts in monetary terms. They seem to believe that they might lose some of their longstanding goodwill if they let in new controversial concepts and valuations even through supplementary, so-called satellite accounts. While some accountants now favor the incorporation of the cost of depletion of economic natural resources (like timber, oil, or fish; see the appendix to this article) into the conventional accounts, many still consider the costing of environmental externalities a matter of modeling.

Nonetheless, the relatively new system for integrated environmental and economic accounting (SEEA), advanced by the author and his colleagues at the United Nations Statistics Division, did tackle the valuation problem. The objective was to incorporate environmental concerns into a comprehensive, integrated national accounts system. The basic methods of market valuation, maintenance costing, and contingent (damage) valuations are briefly described in the appendix. Their objective is to overcome major flaws of the conventional (monetary) accounts, in particular the following:

- Neglect of new scarcities of natural resources, which are a threat to sustained productivity of the economy
- Exclusion of environmental degradation as an externality of economic activity
- Accounting for environmental protection expenditures as national income and product when such outlays could be considered as a maintenance cost of society.

Figure 3 shows how environmental functions can be consistently integrated in the national accounts by doing the following:

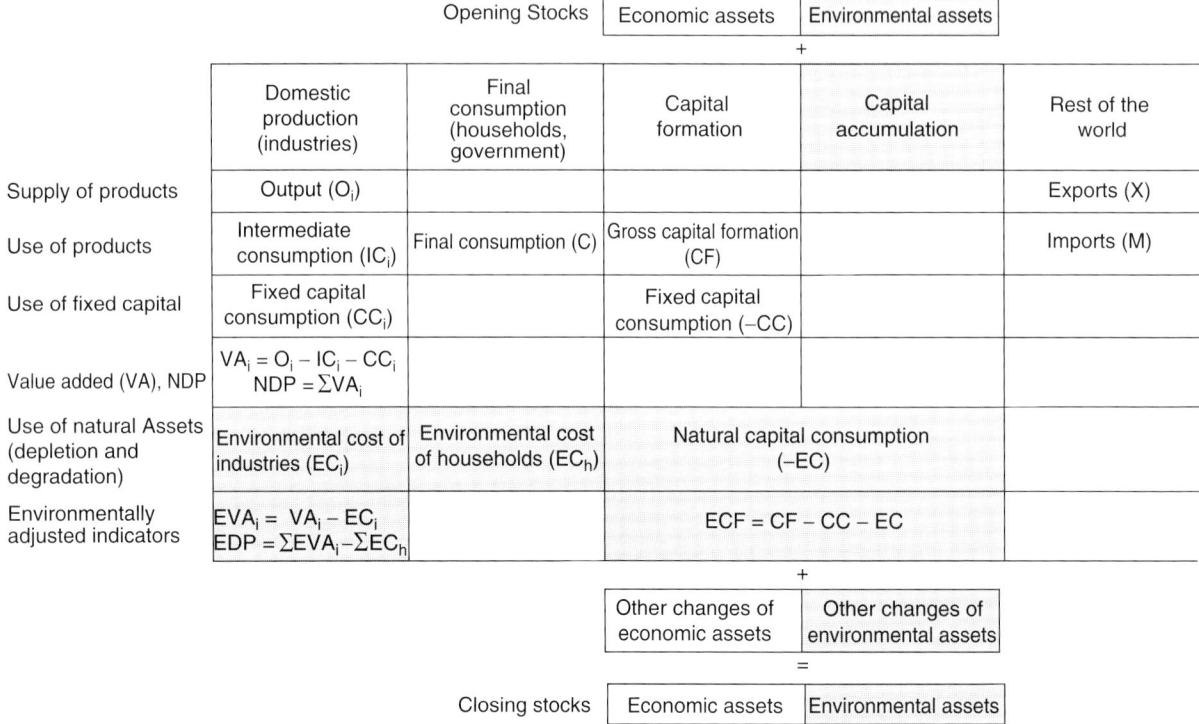

FIGURE 3 SEEA: structure and main indicators. The integrated accounting system incorporates environmental and natural – economic assets into the conventional flow and stock accounts and defines new environmentally adjusted ("green") economic indicators. From Bartelmus (1999), "Greening the National Accounts, Approach and Policy Use," DESA Discussion Paper No. 3, Fig. 3, p.7. United Nations, New York.

- Incorporating environmental assets as natural capital—in addition to produced (fixed) economic capital—into the asset accounts of the SNA
- Costing natural capital consumption (EC) in addition to and in consistency with fixed capital consumption (CC) in both the asset, and supply and use accounts
- Maintaining the accounting identities for the definition and calculation of the key economic (value added, VA, net domestic product, NDP, capital formation, CF), and environmentally adjusted (EVA, EDP, and ECF) indicators.

Environmental protection expenditures could be shown explicitly as "thereof" items of the supply and use of products in Fig. 3. The SEEA identifies these expenditures in appropriate classifications but does not suggest their deduction from GDP as a defense against welfare losses.

Accounting for natural capital consumption and maintenance (through reinvestment) extends the sustainability notion, as already built into the conventional net indicators of value added and capital formation, into the field of environmental services for the economy. In analogy to the wear and tear (i.e., the ultimate destruction) of capital goods in the production process, natural capital loss is defined as the permanent loss of natural resource stocks and waste absorption capacities. This definition clarifies the content of physical depletion and degradation—beyond natural regeneration and replenishment: the regeneration of nature can be seen as a cost-free natural repair process rather than as a capital loss.

Moreover, the costing of capital maintenance avoids controversial damage (to health and ecosystems) valuation (see the appendix). It can be argued, from an economic, utility maximizing optimality point of view, that such welfare effects are the correct values for environmental cost internalization into the budgets of households and enterprises. However, the purpose of national accounting is not welfare analysis or modeling of optimal behavior but the ex-post assessment of economic performance.

The revised SEEA takes a different, somewhat ambiguous view on welfare and other valuations. On the one hand, the monetization of the use of environmental assets is deemed to be a case of modeling a so-called green GDP; on the other hand, damage valuation, as applied in cost-benefit analyses, is extensively discussed. As a consequence, the revised SEEA trades much of its original systemic character for flexibility in a modular, issue-oriented approach.

3. ACCOUNTING FOR ENERGY AND RELATED IMPACTS

Energy resource scarcities and environmental impacts from energy production and use are sometimes presented as the principal environmental problem; however, they do not tell the whole sustainability story. The previously described, accounting systems tell a more comprehensive story. This section illustrates, therefore, how the part of energy can be assessed within the full network of economic-environmental interactions. A detailed technical description of energy statistics and balances is thus beyond this article and would also be more relevant for the management of selected energy categories than for overall environment and sustainability policies.

3.1 Physical Energy Accounts

Physical energy balances are an established tool of energy statistics. They present the origin and destination of energy flows in calorific values or other energy equivalents. In accounting terms, such balances are "supply and use accounts," which can be made compatible with the national accounts. Typically they ignore, however, a number of environmental concerns, including the following:

- Important new and renewable energy sources, as well as nutritional energy flows, needed for getting the full picture of what Helmut Haberl calls "society's energetic metabolism"
- Environmental and social impacts of energy production and use such as resource depletion, emissions, land use changes, resettlement for hydropower development, and accidents.

To some extent, the previously described physical accounts incorporate the depletion of energy sources and the environmental impacts of energy production and use by appropriate coverage and classifications of energy carriers and substance flows. Perhaps best suited for detailed energy accounting in this regard are the physical input-output tables (PIOT). Table III illustrates the approach taken in supplementary energy tables of the PIOT. The table can also be interpreted as an elaboration of the physical part of the hybrid NAMEA (Table II) with regard to energy inputs and outputs, measured in calorific values. However, the PIOT do not relate emissions to energy supply and use but list them separately for the standard economic sectors. As in the NAMEA, the flows of residual substances could of course be added as outputs of energy production and consumption and as inputs into nature and the rest of the world.

The flow accounts of the PIOT do not show stocks of produced or natural capital, but stock changes only, as accumulation. As a consequence, periodic inputs and outputs balance only in the nonaccumulative flows for industries and households. The last row of Table III indicates an imbalance for produced and nonproduced natural assets and foreign trade (i.e., a net accumulation in the inventories of energy products and cultivated natural assets) and for nonproduced energy raw materials. The trade balance is negative, reflecting a net import of goods and services.

Physical flow accounts do not measure the ultimate objective of energy use, namely the production of energy services like heating, transportation or work. As a proxy, the energy PIOT includes various categories of final, or useful, energy. Most energy accounts exclude, however, the nutritional energy supply to humans and animals, needed for work and drive power, as well as hidden energy flows generated outside the boundary of the particular energy system. Inclusion of these flows would permit a comprehensive assessment of the energetic metabolism of societies as a complement to the industrial metabolism measured by material flow accounts.

3.2 Integrated Physical and Monetary Energy Accounts

Section 2.3 and Fig. 3 described the integration of physical and monetary accounts in the SEEA. The revised SEEA focuses largely on the physical side, at least as far as (noneconomic) environmental assets (see the appendix) are concerned. Its flow accounts thus cover energy supply and use in a similar fashion as PIOT and NAMEA. Both systems play in fact a dominant part in the revised SEEA. Considering, however, that the main reasons for greening the (monetary) national accounts have been environment-related deficiencies of the established accounts, the following brief discussion of energy stocks and flows is based on the more integrative original SEEA and its operational handbook.

TABLE III

PIOT for Energy Balances[a]

Supply	Industries 1, 2 etc.	Accumulation		Household consumption	Rest of the world (ROW)
		Produced assets	Nonproduced natural assets		
Inputs					
Raw materials[b]	x[c]				
Products[d]	x	x[e]		x	x
Waste for incineration	x				
Useful energy to nature[f]			x		
Energy losses to nature			x		
Outputs					
Raw materials[b]			x		
Products[d]	x	x[g]			x
Waste for incineration					
Useful energy[f]	x			x	
Energy losses	x			x	
Inputs minus outputs		x(+)	x(+)		x(−)

Source. Based on a case study for Germany (Stahmer, Kuhn, and Braun, 1998, Tables 4 and 5).

[a] Unit of measurement: calorific value, for example, joule or ton of oil equivalent (toe).
[b] Energy carriers and hydropower.
[c] Energy production (electricity, mining).
[d] Fuelwood, electricity, gas, coal, oil, natural gas, chemical and petroleum products, etc.
[e] Change in stocks.
[f] Light, heat.
[g] Cultivated natural assets (forestry, fishery).

The main advantages of such accounting over physical energy balances are the following:

- Coverage, definition, and valuation of energy stocks (oil deposits, standing fuel wood, etc.)
- Clear definition and costing of changes in stocks, notably the depletion of renewable (biotic and aquatic) and nonrenewable (fossil) energy resources as capital consumption
- Comparability of energy stocks and flows (stock changes)
- Comparability of conventional economic indicators of the energy sector (capital formation, value added, cost, income, etc.) with environmentally (cost-)adjusted indicators.

Assessing the current and future availability of energy resources, and hence of energy stocks and consumption, is a key concern of any sustainability analysis from the "capital maintenance" point of view (see Section 1.2). A number of categories and definitions used in the SEEA cater to this concern and are particularly relevant to energy carriers. For mineral and energy (subsoil) resources, the so-called McKelvey box categorizes physical stocks according to the probability of their existence and the economic feasibility of their extraction. The national—economic—accounts include proven reserves only (as opposed to probable, possible, or speculative resources), which are deemed profitable to exploit. The SEEA adopts the same definition of proven reserves to ensure consistency in (depletion) cost accounting and corresponding net indicators. Separate tables could of course present asset stocks weighted by probability (of existence) factors.

Subsoil energy carriers are nonrenewable resources. Any extraction reduces by definition the total availability of the resource, regardless of any potential discoveries of new deposits, and hence depletes the resource. The U.S. Bureau of Economic Analysis (in a one-time case study) disagreed, however, and treated discoveries as production (development) of a capital good. This approach offsets largely the extraction of the nonrenewable resource, turning it statistically into a renewable one. As a result, the total value of stocks of mineral reserves has not changed much in the United States since 1958 (in constant prices). Further noteworthy issues, related to the accounting for nonrenewables and dealt with in the forthcoming revised SEEA, are the treatment of decommissioning costs (e.g., of nuclear power plants) as anticipated ("leveled-off") capital consumption, and of subsidized assets (with negative rents) as noneconomic, zero-value stocks.

The assessment of the depletion of renewable (energy) resources is more complex, since processes of natural regrowth of biotic (fuel wood) and replenishment of cyclical (water) resources, as well as capital formation (afforestation) need to be accounted for. The idea is to obtain the net amount of the permanent loss of the resource—beyond its sustainable yield—which should be costed as nonsustainable capital consumption. The determination of sustainable yield requires complex modeling for specific yield uses such as fuel wood from timber production, or water use for hydropower or as a coolant. In practice, the overall change in the physical amount and value of a resource between opening and closing stocks is frequently taken as a proxy for its depletion in physical and monetary terms. For example, the sustainable cut of forests is usually measured as the net increase in forest biomass, accounting for felling, natural losses from disease or fire, natural regrowth, and reforestation. This simplification neglects, however, changes in the composition and age structure of the forest that may affect future yields and economic returns.

Similarly, changes in the assimilative capacity of environmental media, owing to natural and human-made restoration, should be taken into account when estimating the net emissions—from energy production and use—that cannot be safely absorbed by natural systems. Only these emissions overload nature's existing assimilative capacities and should be costed as (nonsustainable) environmental externalities (see the appendix).

As discussed earlier, the physical and monetary environmental accounts consider only the permanent loss of natural capital as capital consumption. Most new and renewable energy sources like wind, sun, or ocean, have not experienced such loss, and their *(in situ)* economic value can be assumed to be zero. Of course, this does not mean that there are no investment and operating cost for harnessing these energy sources. However, these are conventional expenses (capital formation, capital consumption, and operating cost) that are covered by the established national accounts. The SEEA, and notably its revised version, identifies these environmental activities as natural resource management and exploitation by means of additional regrouping and classification. These environmental expenditures are important for the management of the different resources but do not assess the core environmental concerns of depletion and degradation.

The purpose of monetary estimates of economic (private) and environmental (social) costs is to assess the environmental sustainability of economic activity. However, some energy analysts and ecological economists appear to share Malcolm Slesser's view that monetary valuation is "after all nothing more than a highly sophisticated value judgement." Instead, they suggest the use of energy values or potentials for measuring the natural or absolute value of goods and services—independent of volatile market preferences. Different valuation principles for material and energy flow accounting include the following:

- Embodied energy, the direct and indirect energy required to produce goods and services
- Exergy, the potential work content of materials, energy carriers, and residuals
- E*m*ergy, the energy memory of the total amount of energy of one kind, required directly and indirectly to make a product or service (advanced in the writings of the late Howard Odum)

The main critique of these energy valuations has been, in turn, on the neglect of individual value judgements or preferences of economic agents. Additional, more practical problems are the data and knowledge requirements for a myriad of different production and transformation processes that use energy and generate emissions and discharges. Energy value accounting does therefore not seem to be capable of capturing the overall interaction of environment and economy. Such accounting could provide useful supplementary information for energy policies and management; it is less likely, though, to become a tool of overall assessments of economic or ecological sustainability.

4. OUTLOOK: POLICY USE OF ACCOUNTING INDICATORS

Rapther than reviewing the wide range of uses of a multipurpose accounting system, this section leads us back to the initial main concern of green accounting—the assessment of sustainability.

Table IV provides a framework for sustainability accounting and analysis with regard to the two key operational, economic and ecological, sustainability concepts. The table sets out from the rationale and strategies of reducing material throughput through dematerialization and of sustaining economic growth through capital maintenance. The corresponding physical and monetary accounting tools and indicators measure relatively strong and weak sustainability, respectively. Compliance with minimum standards for the preservation of critical (irreplaceable) capital is to attain strong sustainability. On the other hand, monetary accounts and their adjusted economic indicators imply the substitutability of different capital categories and production factors. They aim to assess the—weak—sustainability concept of overall capital maintenance.

Reducing material flows into the economy, as expressed in particular by their sum total, TMR, aims at decoupling economic growth from the generation of environmental impacts. A collaborative study, published by the World Resources Institute, shows that TMR per capita has been leveling off for selected industrialized countries (Germany, Netherlands, the United States) at about 80 tons per annum, except for Japan at 45 tons (because of its low per capita energy use). Given that GDP per capita has been rising in the high-TMR countries, there is some—relative—delinkage from growth, albeit far from the sustainability targets of Factor 4 (see Section 2.1). At least in these countries, ecological sustainability has not been in sight during the accounting periods (1975–1993).

Unfortunately, there does not seem to be an automatic solution to delinking environmental impacts or pressures from economic growth as suggested by the Environmental Kuznets Curve hypothesis. The corresponding inverted U-curve correlates low environmental impacts with low levels of economic development (per capita income), increasing impacts with industrialization and decreasing impacts with affluence. The possible transition to a dematerialized service economy might explain the automaticity, but empirical confirmation (except for a few pollutants) is elusive. Environmental regulation, possibly supported by fiscal incentives (subsidies for resource-saving technologies and disincentives such as material consumption taxes) will therefore need to be applied to ensure the preservation of critical capital.

Costing produced and natural capital consumption is a means to reserve funds for reinvestment and hence capital maintenance. At the macroeconomic level, the deduction of these costs from GDP obtains EDP, a net indicator whose decrease would indicate nonsustainable economic growth. Case studies of the SEEA, as presented, for instance, by Kimio Uno and Peter Bartelmus, did not show a reversal of (upward) GDP trends by EDP. However, this does not necessarily confirm sustainability because of the relatively short time series available.

A more pertinent way of looking into the sustainability of economic performance would be

TABLE IV

Concepts, Indicators, and Policies of Environmental Sustainability

	Ecological sustainability: Dematerialization/detoxification	Economic sustainability: Produced and natural capital maintenance
Rationale	Reducing throughput below carrying capacities	Sustaining economic growth
Strategy	Decoupling economic growth from environmental pressure	Maximizing economic growth while keeping produced and natural capital intact
Accounting tools	Material flow accounts (MFA), physical input-output table (PIOT), national accounting matrix including environmental accounts (NAMEA)	System for integrated environmental and economic accounting (SEEA)
Key indicators	TMR, TDO, DMI, TMC, resource productivity: GDP/TMR	"Green" economic aggregates: EDP, EVA, ECF, total wealth
Strength of sustainability	Strong: reduction of material and substance flows to meet sustainability standards (Factor X targets), maintenance of critical capital	Weak: overall capital maintenance, allowing substitution between produced and natural capital and other production factors
Policy instruments	Regulation and standard setting (Factor X, emission standards), fiscal (dis)incentives for material-saving innovation	Use of green indicators and variables in economic analysis and policy; market instruments of environmental cost internalization

Source. Adapted from Bartelmus, P. (2003), abridged version of Table 1, p.68, with permission.

to measure a nation's ability to generate new capital after taking produced and natural capital consumption into account. The Environmentally adjusted Net Capital Formation (ECF) indicator showed a significant difference between some developing and industrialized countries. For example, Indonesia, Ghana, and Mexico at times exhibited nonsustainability by using up more capital than generating new one (i.e., they showed an overall disinvestment in produced and natural assets). Industrialized countries, on the other hand, did increase their capital build-up and met thus one necessary condition for sustaining their economies.

Past overall capital maintenance or increase hide, however, the fact that in the long run complementarities of natural capital might make it impossible to maintain production and consumption patterns and growth rates. The empirical testing of this assumption should be an important field of sustainability research.

At the microeconomic level, one important use of environmental (social) cost accounting is to determine the level at which market instruments of cost internalization such as eco-taxes should be initially set for different economic agents according to polluter/user-pays-principles. The ultimate effects of environmental cost internalization on production and consumption patterns and overall economic growth and employment need, of course, be modeled with the usual restrictions and assumptions. The results of such modeling might point to the need of resetting the original levels of market instruments according to their predicted efficacy.

Greened national accounts and PIOTs are a particularly useful tool for integrative environmental-economic modeling since they generate a high degree of transparency with regard to the underlying data and indicators. This applies also to the manifold energy scenarios (and related climate change models), which have predicted or denied reaching the limits of energy supply and, possibly, economic growth. They have also generated a wide range of policy options according to differing coverage, modeling techniques, and policy targets. The International Institute for Applied Systems Analysis (IIASA) and the World Energy Council have presented concise overviews of energy scenarios on their Web sites.

As already mentioned, greened national accounts can only capture the immediate interaction between environment and economy. A broader assessment of sustainability would have to take account of social, institutional, and political aspects of sustainable development. In the field of energy, the disparities in the distribution of and access to scarce nonrenewable resources have triggered aggressive policies of securing affordable energy supply. Assessing natural resource availability, ownership, and trends of consumption in different regions and nations could alert us to looming resource scarcities and to possible international conflicts in resource use. It might also be the first step toward more rational policies of ensuring energy security, and perhaps security, for all.

APPENDIX: MONETARY VALUATION IN THE SEEA

Market valuation uses prices for natural assets, which are observed in the market. It is usually applied to economic assets of natural resources. The SNA defines economic assets as "entities (a) over which ownership rights are enforced ... and (b) from which economic benefits may be derived by their owners." Figure 3 displays therefore part of natural capital consumption under the column of economic assets.

Where market prices for natural resource stocks, such as oil deposits or fish in the ocean are not available, the economic value of these assets can be derived from estimates of the—discounted—sum of life-long net returns, obtained from their potential use in production. It is at this value that a natural asset would be traded if a market existed for the asset. Market valuation techniques are also applied to changes in asset values, caused in particular by depletion (i.e., their nonsustainable use). These value changes represent losses in the income-spinning capacity of an economic asset. Depletion cost allowances reflect thus a weak sustainability concept, calling for the reinvestment of environmental cost in any income-generating activity.

Maintenance valuation permits the costing of the degradation of noneconomic—environmental—assets, in particular from emissions. Typically these externalities are not traded in markets. The SEEA defines maintenance cost as those that "would have been incurred if the environment had been used in such a way as not to have affected its future use." Maintenance costs refer to best-available technologies or production processes with which to avoid, mitigate, or reduce environmental impacts. They are used to weight actual environmental impacts, generated during the accounting period by different economic agents. As with depreciation allowances for the wear and tear of produced capital, such costing can be seen as the funds required for reinvesting in capital maintenance.

The SEEA also discusses contingent and related damage valuations. These valuations are applied in cost-benefit analyses of particular projects and programs but are hardly applicable at the national level. They refer to the ultimate welfare effects (damages) of environmental impacts, which are incompatible with the market price and cost conventions of the national accounts. The original SEEA considered damage valuation in environmental accounting as experimental, and its operational handbook discourages its application.

SEE ALSO THE FOLLOWING ARTICLES

Cost–Benefit Analysis Applied to Energy • *Development and Energy, Overview* • *Economic Growth and Energy* • *Environmental Change and Energy* • *Environmental Kuznets Curve* • *Goods and Services: Energy Costs* • *Sustainable Development: Basic Concepts and Application to Energy* • *World Environmental Summits: The Role of Energy*

Further Reading

Ayres, R. U., and Ayres, L. W. (Eds.). (2002). "A Handbook of Industrial Ecology." Edward Elgar, Cheltenham, United Kingdom, and Northampton, MA.

Bartelmus, P. (2001). Accounting for sustainability: Greening the national accounts. *In* "Our Fragile World, Challenges and Opportunities for Sustainable Development" (M. K. Tolba, Ed.), Forerunner to the Encyclopedia of Life Support Systems, Vol. II. EOLSS, Oxford.

Bartelmus, P. (2003). Dematerialization and capital maintenance: Two sides of the sustainability coin. *Ecol. Econ.* 46, 61–81.

Eurostat (2001). "Economy-wide Material Flow Accounts and Derived Indicators, a Methodological Guide." European Communities, Luxembourg.

Haberl, H. (2001). The energetic metabolism of societies, part I: Accounting concepts. *J. Industrial Ecology* 5(1), 11–33.

Kapp, K. W. (1971, and previous editions: 1950 and 1963). "The Social Costs of Private Enterprise." Schocken Books, New York.

Keuning, S. J., and deHaan, M. (1998). Netherlands: What's in a NAMEA? Recent results. *In* "Environmental Accounting in Theory and Practice" (K. Uno and P. Bartelmus, Eds.). Kluwer Academic, Dordrecht, Boston, and London.

Odum, H. T. (1996). "Environmental Accounting, Energy and Environmental Decision Making." Wiley, New York.

Slesser, M. (1975). Accounting for Energy. *Nature* 254, 170–172.

Stahmer, C., Kuhn, M., and Braun, N. (1998). "Physical Input-output Tables for Germany 1990." Eurostat Working Paper No. 2/1998/B/1. Eurostat, Luxembourg.

United Nations (1993). "Integrated Environmental and Economic Accounting." United Nations, New York.

United Nations (2000). "Integrated Environmental and Economic Accounting—An Operational Manual." United Nations, New York.

United Nations, *et al.* (1993). "System of National Accounts 1993." United Nations and others, New York and other cities.

United Nations, *et al.* (in prep.). "Integrated Environmental and Economic Accounting 2003." United Nations and others, New York and other cities.

Uno, K., and Bartelmus, P. (Eds.). (1998). "Environmental Accounting in Theory and Practice." Kluwer Academic, Dordrecht, Boston, and London.

World Resources Institute (WRI), *et al.* (1997). "Resource Flows: The Material Basis of Industrial Economies." World Resources Institute, Washington, DC.

Greenhouse Gas Abatement: Controversies in Cost Assessment

MARK JACCARD
Simon Fraser University
Vancouver, British Columbia, Canada

1. Introduction
2. Alternative Definitions of Cost
3. Uncertainties and Assumptions about Technology and Preference Change
4. Assessment

Glossary

autonomous energy efficiency index (AEEI) The rate at which the economy becomes more or less energy intensive, with the price of energy remaining constant relative to the prices of other inputs. Indicates productivity improvements in energy use that are not caused by increases in energy prices.

bottom-up analysis (model) An analysis (or model) that focuses on the financial costs of technologies that, if widely adopted to meet the energy service needs of firms and households, would lead to reductions in greenhouse gas emissions. Tends to estimate low costs of greenhouse gas abatement.

computable general equilibrium (CGE) model A top-down model of the economy that includes all of its major components (consumers, producers, government, external agents) and markets (capital, labor, materials, energy, interindustry trade, final goods and services, government revenues and expenditures, international trade) and the relationships between them. Such a model can compute the new general equilibrium that results after a perturbation to one component or market, as would be caused by a greenhouse gas tax that affected the price of energy and perhaps other goods and services.

consumers' surplus A premium that consumers are willing to pay for a good or service above what they actually pay for it (its market price).

economies of learning The reductions in financial costs (capital and operating) of technologies (especially newer technologies) as manufacturers and operators gain experience with them with greater production. The rate of learning, and thus cost decline, is sometimes referred to as an "experience curve."

economies of scale The reductions in financial costs (capital and operating) of technologies as larger units are built or a greater number of units are manufactured on a regular basis. Can occur with old and new technologies.

elasticities of substitution (ESUB) Changes in the relative use of inputs (capital, labor, energy, materials) in response to changes in their relative costs. A high elasticity implies an easy ability to switch away from an input, such as energy, as its relative price increases, and a low elasticity implies the opposite.

financial costs The monetary costs that are used by engineers and financial planners to estimate the initial and ongoing capital outlay required by an investment. Unless preceded by a term such as "expected," financial cost estimates exclude consideration of financial risks, or intangible costs and benefits (e.g., consumers' surplus).

hybrid analysis (model) An analysis (or model) that combines the technological explicitness of the bottom-up approach with estimation of consumer and firm behavior of the top-down approach.

option value The increase in expected net benefits resulting from the decision to delay an irreversible investment under uncertainty. While this definition is predominant in energy efficiency literature, there are alternative definitions for this term.

top-down analysis (model) An analysis (or model) that focuses on aggregate relationships between inputs and outputs of the economy, and the interplay of these through price feedback mechanisms. The relative costs and uses of inputs (capital, labor, energy, and materials) are portrayed in production functions for firms and in consumption or utility functions for households. A "computable general equilibrium model" (CGE) is a type of top-down model. Top-down models tend to estimate high costs of greenhouse gas abatement, although this depends on the values of their elasticities of substitution and autonomous energy efficiency index parameters.

In deciding how and by how much to reduce greenhouse gas emissions, policy makers need to know the potential cost. But there are alternative definitions of cost, the evolution of technologies and preferences is a complex and highly uncertain process that can profoundly influence estimates of long-run cost. For these and other reasons, independent researchers and interest groups have produced widely divergent estimates of costs, and policy makers do not know whom to believe. The different analytical and modeling approaches are discussed here in order to explain their relative contribution to the differences in greenhouse gas reduction cost estimates. Each contrasting cost definition and contrasting assumption of technical and preference evolution informs the estimates. The focus is on energy-related technologies, because these produce the greatest anthropogenic contribution to rising atmospheric concentrations of greenhouse gases, but the issues are equally relevant to all technologies and management practices that reduce greenhouse gas emissions, whether involving forestry, agriculture, treatment of municipal solid waste, or capture and storage of CO_2.

1. INTRODUCTION

In 1991, M. Grubb referred to "the missing models" in arguing for the development of a new generation of models that would better inform policy makers of the critical issues in estimating greenhouse gas (GHG) reduction costs. To be useful to policy makers, models for costing GHG reduction need to explain the key factors behind divergent cost estimates. But over a decade later, cost estimates still diverge widely, in part because of different assumptions about technology and preference evolution (which is perhaps inevitable) but also because of how costs are defined and depicted in GHG costing models. "Bottom-up" models tend to ignore or underrepresent important constraints to the adoption of new equipment, whereas "top-down" models incorporate these constraints but inadequately represent the potential for technological innovation and preference changes that can reduce costs in the long run. Newer "hybrid" models attempt to incorporate key strengths of both top-down and bottom-up approaches in order to be more useful to policy makers who face critical decisions about the magnitude and timing of GHG reduction targets.

2. ALTERNATIVE DEFINITIONS OF COST

An action to reduce GHG emissions is defined as a change in equipment choice, equipment use rate, lifestyle, or resource management practice that changes GHG emissions from what they otherwise would be. Examples are: choosing more efficient light bulbs, turning off unused lights, telecommuting some days of the week, and practicing forestry or agriculture differently. The cost of an action is the difference in costs between the business-as-usual scenario and a world in which the action is undertaken. Unfortunately, analysts apply alternative definitions of these costs.

2.1 Bottom-Up Cost Definition and Critique

Bottom-up analysis, applied frequently by engineers, physicists and environmental advocates, estimates how changes in energy efficiency; fuel type; emission controls on equipment, buildings, and infrastructure; and even land-use practices might lead to different levels of GHG emissions. Technologies (furnaces, light bulbs, electric motors, vehicles) that provide the same energy service (space heating, lighting, industrial motive force, personal mobility) are generally assumed to be perfect substitutes except for differences in their financial costs and their emissions of GHGs and other pollutants. When their financial costs (capital and operating) in different time periods are converted into present value using a social discount rate, many current and emerging technologies available for reducing GHG emissions appear to be profitable or just slightly more expensive relative to existing equipment. Bottom-up analyses often show, therefore, that substantial GHG emission reduction can be profitable or low cost if the low-emission technologies were to increase from their small market share to achieve market dominance.

Although they are all characterized by a focus on technologies and by cost definitions that are restricted to financial costs, bottom-up models can vary in terms of how they depict the interconnections between different decisions in the economy. A simple bottom-up analysis would assess each energy service individually (home heating, commercial lighting) and conduct a financial analysis to determine the least-cost technology for each. A fully integrated, optimizing bottom-up model would determine the least-cost technology for all services simultaneously. This integrated approach may indicate a very different outcome if the model shows, for example, that dramatic reductions in GHG emissions from the electricity sector are inexpensive. If these are pursued, then a drop in the GHG content of electricity will reduce the GHG cost-effectiveness of electricity conservation actions in the end-use sectors. A well-known, integrated bottom-up linear programming model used in North America, Europe, and increasingly elsewhere is the MARKAL (market allocation) optimization model.

An indication of the bottom-up approach is provided by researchers assessing the costs to the United States of achieving the target it negotiated (but has not ratified) under the Kyoto Protocol of 1997. The U.S. government agreed to reduce its GHG emissions to 7% below their 1990 level by 2010. According to forecasts at the time, this represented a reduction of almost 30% from expected emission levels for 2010. Brown and coauthors at five major energy research laboratories funded by the U.S. government found that the United States could reduce to its 1990 level of emissions with no net cost to the U.S. economy, and thus virtually no impact on gross domestic product (GDP). This could be achieved with a host of policies, including a GHG tax of no more than $25/ton of carbon ($25/t C). All of the reductions would be attained domestically, even though the United States has the capability under the Kyoto agreement to purchase reduction credits from other countries.

Many economists, however, criticize the bottom-up approach in general, and this study in particular, for the assumption that a single, *ex ante* (anticipated) estimate of financial cost differences indicates the full social cost of switching technologies. Economists such as A. Jaffe and R. Stavins argue that technologies may differ to consumers and businesses in ways that are not represented by this single financial value. The mundane example of purchasing a light bulb illustrates these differences.

- New technologies usually have a higher chance of premature failure than do conventional technologies and therefore pose greater financial risk. There is some probability that *ex poste* (realized) financial costs will exceed *ex ante* financial costs for new technologies. New compact fluorescent light bulbs have exhibited higher rates of premature failure compared to conventional incandescent light bulbs, requiring a higher than expected financial outlay because of early replacement in some cases.

- Technologies with longer payback periods (relatively high capital costs) are riskier if the cumulative probability of failure or accident, or undesired economic conditions, increases over time. Because of the higher purchase cost of compact fluorescent light bulbs, the chance of accidental breakage prior to paying back the initial investment is higher than for an incandescent light bulb.

- Two technologies may appear to provide the same service to an engineer but not to the consumer. Many people find compact fluorescent light bulbs to be less than perfect substitutes for incandescent light bulbs in terms of attractiveness of the bulb, compatibility with fixtures, quality of light, and timing to reach full intensity, and would pay more to maintain high levels of these nonfinancial attributes.

- Not all firms and households face identical financial costs: acquisition, installation, and operating costs can vary by location and type of facility. Compact fluorescents may be more expensive to acquire in nonmetropolitan areas where the market is smaller. This heterogeneity means that a comparison of single-point estimates of financial costs may exaggerate the benefits of market domination by a low-GHG technology.

The first two differences relate to "option value," which is the expected gain from delaying or avoiding an irreversible investment while waiting for new information that might lead to a better decision. The third difference refers to what economists call "consumers' surplus," or the extra value that consumers realize above the financial cost of a particular technology. The fourth difference simply acknowledges "market heterogeneity," i.e., that a technology may be cheaper than its competitor in one circumstance but more expensive in another.

When consumers and businesses are asked, induced, or forced to switch away from a technology they would otherwise have chosen, economists say that the social cost of this switch is the difference in financial costs plus or minus any intangible costs related to option value, consumers' surplus, and market heterogeneity. By ignoring these values, bottom-up analysts may overestimate the willingness of firms and households to take actions that reduce GHG emissions, and thus underestimate the social cost. Politicians, in contrast, seem to be instinctively aware of these other values, and tend to question claims that GHG emissions can be reduced at little or no cost. In providing only part of the cost picture, the bottom-up approach is less helpful to policy makers than it could be.

Another challenge with the bottom-up approach is that its technology-specific focus hinders the ability to assess broader macroeconomic effects, notably the trade and structural repercussions resulting from changes in energy prices and costs in the economy. In this sense, bottom-up models usually provide only a "partial equilibrium analysis" of the response to GHG reduction policies. If, however, the cost of GHG reduction is as low as this approach often concludes, or if governments are unwilling to pursue rapid and costly decreases in GHG emissions, then macroeconomic feedbacks may be small enough that this partial equilibrium assessment is sufficient. But even if energy-focused GHG policies are too modest to elicit a significant macroeconomic change, they may cause unintended responses in energy markets that are outside the purview of a conventional bottom-up model. For example, subsidized or voluntary decisions to acquire more energy-efficient equipment (cars, lights, appliances, industrial machinery) reduce energy service operating costs, and may thus encourage greater consumption of energy services. The magnitude of this "rebound effect" is disputed, but certainly all sides agree that at some point lower operating costs foster higher use. Returning to the lightbulb example, the installation of outdoor decorative and security lighting, often using high-efficiency lightbulbs, is increasing in North America. Most bottom-up models do not, however, link the demand for energy services to the cost of the service, and thus would not simulate a rebound effect.

2.2 Top-Down Cost Definition and Critique

The alternative, top-down analysis of economists usually relies on historical and cross-sectional market data to estimate aggregate relationships between the relative costs and the relative market shares of energy and other inputs to the economy, and links these to sectoral and total economic output in a broader, equilibrium framework. Elasticities of substitution (ESUB) indicate the substitutability between any two pairs of aggregate inputs (capital, labor, energy, materials) and between energy forms (coal, oil, gas, renewables) within the energy aggregate. Another key parameter in such models, the autonomous energy efficiency index (AEEI), indicates the rate at which price-independent technological evolution improves energy productivity. Relatively high parameter values for AEEI and for energy-related ESUB (a large degree of substitutability between energy and capital, and between GHG-intensive and non-GHG-intensive forms of energy) equate to a relatively low cost of GHG emission reduction. Because these parameters are estimated from real market behavior, as energy prices and energy consumption have changed historically, they are said to reveal the actual preferences of consumers and businesses—and therefore implicitly incorporate losses or gains in option value and consumers' surplus, as well as reflect the market heterogeneity of real-world financial cost conditions.

With these key response parameters estimated, economists simulate the economy's reaction to an economic signal (a GHG tax, tradable GHG permits, technology subsidies, or some combination of these) that increases the relative cost of GHG-intensive technologies and behavior. The magnitude of the economic signal necessary to achieve a given GHG reduction target provides the implicit marginal cost of that target. Because top-down models usually integrate energy supply and demand within a total economy model, this marginal cost includes the feedback between energy supply and demand, on the one hand, and between these components and the economy as a whole, on the other. Top-down approaches are, therefore, more often associated with a "general equilibrium analysis" of policy effects; models designed expressly to close all key economic feedback loops are referred to as computable general equilibrium models (CGE models). Because movement away from current technologies includes all intangible value losses, top-down analysis usually generates substantially higher estimates of GHG reduction costs.

The top-down parameters, ESUB and AEEI, are usually estimated from historical data. Historical data generate relatively low values for the economy's willingness to substitute (ESUB) away from GHG-intensive technologies and activities, in part because full long-run price responses are difficult to detect in an ever-changing world. Experts can, however, judgmentally change the AEEI or ESUB parameter values to generate low-cost estimates, simply by portraying consumers and businesses as more prone to switch technologies in response to changes in financial costs—in effect, reducing the estimated option value losses, consumers' surplus losses, and market heterogeneity that otherwise explain technological inertia. But setting these parameters judgmentally undermines to some extent the very argument in favor of the top-down approach, by forgoing its real-world empirical basis. The future value of AEEI, being highly uncertain, is hotly disputed.

Top-down models are relatively easy to construct, and this may explain why there are so many variants in contrast to the dominance that MARKAL has attained among integrated bottom-up modelers. Some of the top-down models in the United States are G-Cubed, MERGE, SGM, and MIT-EPPA; all of these are computable general equilibrium models. Top-down models have also been applied to estimating the costs to the United States of achieving its Kyoto target, and these provide a contrast with the bottom-up estimate. Weyant and Hill summarized the results of a multimodel evaluation of the Kyoto protocol by several top-down models, especially the CGE type. Of the 11 participating models, 8 models found that the United States would require a carbon tax of at least $150/t C, and of these, 4 models required a tax higher than $250/t C (one higher than $400/t C). GDP impacts ranged from modest levels to the loss of over 3% of economic growth.

Like the bottom-up approach, however, the top-down approach is also vulnerable to the criticism of being unhelpful to policy makers. Substantial reduction of GHG emissions, in order eventually to stabilize atmospheric concentrations, is a long-run objective that requires dramatic technological change over a lengthy period in which consumer preferences are also likely to evolve. Policy makers need to know if and how their policies can influence long-run technological change, and thus the *ex ante* and *ex poste* financial costs of technologies, as well as the long-run evolution of consumer preferences, in ways that reduce these intangible losses from switching to less GHG-intensive technologies.

Even if the confidence intervals of the estimated ESUB and AEEI parameters are narrow, meaning that they are relatively robust at explaining past changes in energy demand, there is no guarantee that parameter values derived from past experience will remain valid into the future. Financial costs and consumer preferences change over time, and the magnitude and direction of this change may be influenced by policy. For example, until recently, there was no incentive to design and commercialize low-GHG technologies. Now, such technologies are under development worldwide. Azar and Dowlatabadi note that as research, development, and production expand, aided by government policy, "economies of scale" and "economies of learning" reduce financial costs and can reduce option values, perhaps dramatically. The market penetration of these new technologies and their falling costs imply higher future values for ESUB and AEEI, reflecting the economy's increasing ability to reduce GHG emissions at lower costs.

A similar logic applies to historical estimates of consumers' surplus losses. A typical pattern in new technology commercialization is an initial phase of slow market penetration, limited to acquisition by early adopters, followed by a market transformation phase in which expanding consumer awareness and acceptance precipitate a feedback loop of product improvement, intensified marketing effort, and accelerating dissemination. This shift in consumer preferences means lower consumers' surplus losses from technology switching, implying that the long-run social costs of GHG reduction may be partially influenced by policy. Again, this would cause higher long-run values for the all-encompassing ESUB and AEEI parameters of top-down models, and thus lower costs of GHG reduction. Returning to the light bulb example, the growing cumulative production of compact fluorescent light bulbs has reduced their production cost and market price over the past decade, which in turn has shortened investment paybacks and decreased the financial risks from premature breakage. Manufacturers have also improved the reliability and attractiveness of compact fluorescents (appearance, size, hue) so that they are now closer substitutes for incandescent light bulbs. Does this mean that the household market for light bulbs is on the verge of a market transformation in which compact fluorescent light bulbs overtake and marginalize incandescent light bulbs? It might. But an important caution is that low-GHG technologies are not guaranteed lower financial costs and growing consumer acceptance. Experienced marketers of new products are well aware of the difficulty of predicting financial cost reductions and future consumer preferences; the experience of every new technology is different. If the general trend of technological evolution, however, is for new technologies to increase the ability and willingness of consumers and businesses to opt for low-GHG choices, as is likely, then AEEI and ESUB values will increase and the long-run costs of GHG emission reduction will fall.

The importance of technological change for estimating future costs raises another challenge for the top-down approach in terms of its usefulness to policy makers. Because conventional top-down analysis conceives of technological change as an abstract, aggregate phenomenon, characterized by ESUB and AEEI parameter values, it only helps policy makers assess top-level policy instruments, such as taxes and tradable permits. But policy makers often prefer technology- and building-specific policies in the form of tax credits, subsidies, and regulations because these policies avoid the dramatic (and unpopular) increases in energy prices that would otherwise be required to cause the degree of technological transformation for substantial GHG emission reductions. A model is thus more useful to policy makers if it can assess the combined effect of these focused policies with perhaps relatively modest economy-wide applications of taxes and tradable permits, but conventional, aggregated top-down models are unable to do this.

2.3 Compromise Hybrid Costing Approaches

In summary, the top-down and bottom-up approaches offer competing cost definitions, both of which have their failings in terms of usefulness to policy makers. The bottom-up approach ignores option values and consumer preferences. The top-down approach incorporates these cost factors, at an aggregate level, but ignores the potential for long-run changes to them as technology-specific and economy-wide policies stimulate technical change and perhaps shifting consumer preferences.

Researchers have tried to address these different strengths and weaknesses of top-down and bottom-up analysis by designing policy costing models that are both technologically explicit and behaviorally realistic—simulating the economy's long-run building and equipment stock turnover as a function of policies that target technological innovation and business and consumer preferences, while incorporating feedbacks between energy supply and demand, and between the energy system and the structure and output of the economy. Bohringer and others refer to this as a "hybrid modeling" approach, because it attempts to bridge the top-down/bottom-up schism. An example of a so-called hybrid model is the National Energy Modeling System (NEMS) of the U.S. government. Like a bottom-up model, NEMS is highly disaggregated, having substantial technological detail in its representation of energy supply and of the individual energy demand sectors (transportation, residential, commercial/institutional, and industry). But, unlike a typical bottom-up model, technology choices within these sectors are simulated based on behavioral parameters that are mostly estimated from real-world empirical data. Alternative vehicle choice in the NEMS transportation module, for example, is based on parameters estimated from surveys that asked consumers about vehicle options. Also, the model is close to general equilibrium in that a policy simulation equilibrates energy supply and demand, on the one hand, and this supply–demand composite with the macroeconomy as a whole, on the other hand.

With the hybrid modeling approach, the policy analyst can help the policy maker sort out the importance of differing definitions of cost for the estimates of GHG reduction costs. A hybrid model can first be applied in bottom-up mode, focusing just on financial costs and treating these in a deterministic manner, ignoring risk factors and intangible costs such as consumers' surplus losses. Then, its parameters can be adjusted to incorporate the estimated effects of these other factors. The policy maker has the opportunity to see the importance of these different cost definitions for the different policy cost estimates, and to test the cost assumptions that seem most plausible. For example, the U.S. Energy Information Administration applied NEMS in 1998 to estimate the costs of U.S. compliance with the Kyoto Protocol. It estimated that a carbon tax of $294/t C was required to reach the target with domestic actions alone and that this would cause a 3% loss in GDP growth. These results are within the range of top-down models in the multimodel evaluation, but distant from the bottom-up study of the U.S. national research laboratories. This is not surprising given that the 12-year period between 1998 and the Kyoto deadline year of 2010 provides insufficient time for policies to influence the evolution of new technology costs or consumer preferences. Because hybrid models and top-down models have the same definition of cost, they are likely to show similar costs of GHG reduction over a short-period simulation. Over a 20- or 30-year time frame, however, the cost estimates from a hybrid model could diverge significantly from those of a typical top-down model, if the hybrid included dynamic influences on technology costs and consumer preferences, as is explained in the following section.

3. UNCERTAINTIES AND ASSUMPTIONS ABOUT TECHNOLOGY AND PREFERENCE CHANGE

Even if the definition of costs were not in dispute, the challenge of GHG emission reduction presents enormous uncertainties that can result in widely divergent cost estimates. For a given country, three broad categories of uncertainty stand out.

- Uncertainty about external factors creates profound uncertainty for the GHG reduction costs of a given country. Will other countries make and keep emission reduction targets, and if so at what cost and with what implication for competitive trade conditions? To what extent will a given country be allowed to utilize international permits and international actions to meet its commitment? What international emission permit trading mechanism will exist and what will be the price of permits?

- Uncertainty about technological innovation and its implications for future financial costs increases as the time horizon is extended. Over a 20- to 30-year time frame, dramatic technological change may occur with equally dramatic implications for the financial cost of GHG reduction policies. For example, innovations that reduce the cost of converting fossil fuels into hydrogen, while sequestering safely underground the CO_2 by-product, raise the prospect of accelerated commercialization of GHG-free, hydrogen-based technologies such as fuel cells. The financial costs of such technologies are not independent of public policy, although the relationship is highly uncertain.

- Uncertainty about consumer and business preferences increases as the time horizon is extended. How might these preferences change as environmental awareness, income, available technologies, and policy alter over that same 20- to 30-year time frame? Thus, policies that mandate the development and commercialization of hydrogen-based fuel cell vehicles are likely to influence the marketing efforts of vehicle manufacturers, with a possible effect on the preferences of consumers. Again, the magnitude of this effect is highly uncertain.

The importance of external uncertainty is usually assessed by constructing and simulating scenarios that show contrasted but plausible future values for uncertain factors, such as the price of internationally traded GHG emission permits or crude oil. If, however, a country has a significant influence on the market of internationally traded energy commodities or GHG permits, as is the case for the United States, it should also model the relationship between its domestic policies and these international prices. The NEMS model of the U.S. government has modules that link developments in the U.S. market to international energy prices and could link to a module of internationally traded GHG permits. Top-down models, being more aggregated, have dominated the exploration of this uncertainty so far, although there have been significant advances in recent years in linking together the MARKAL bottom-up models from several countries for this type of analysis.

Given the long-run nature of an environmental challenge such as climate change risk, and the profound technological change that is required to achieve substantial GHG emission reductions, the uncertainties associated with technological innovation and consumer and business preferences are of great significance for the estimation of costs. For addressing these uncertainties, bottom-up, top-down, and hybrid approaches offer different capabilities.

3.1 Top-Down Approaches to Long-Run Uncertainty, and Critique

Although top-down models have not traditionally been applied to extremely long-run periods in which the technological context is likely to change, some economists have used them for 50- and even 100-year simulations of the policy response to climate change risk. But as the time frame extends further into the future, the character of technical change plays an increasingly important role in determining the costs of GHG reduction, yet technical change in top-down models is simply portrayed via the ESUB and AEEI parameters. These parameters are typically estimated from past technological and preference changes associated with price changes for energy and other inputs.

How confident can the modeler, and the policy maker, be in the stability of these parameter values in future time periods? For example, how does the top-down modeler translate new information about the cost of renewable electricity generation or high-efficiency light bulbs into their parameter estimates for AEEI and ESUB? How does the modeler show the policy maker the relationship between technology-focused research and development policies and these same parameters? To be useful to policy makers, models need to indicate the direction, probability, and impact of technological changes that influence both the financial costs and consumers' surplus costs of GHG emission reduction—and the extent to which these changes might be influenced by policy. Parameters estimated from historical conditions may have little ability to anticipate the future technology and preference dynamics in response to policies seeking to induce long-run, profound technological change.

3.2 Bottom-Up Approaches to Long-Run Uncertainty, and Critique

In recognition of this challenge, bottom-up modelers have focused on applying lessons from past technological innovations to new technologies, linking cumulative manufacture and expanded application of a technology to reductions in its financial costs—the "economies of learning" or "experience curves" discussed earlier. For example, as global production of windmills has grown over the past 20 years, the cost of windmill-generated electricity fell quickly at first and today continues gradually down toward an asymptote, possibly in the range of (U.S.) $0.04/kWh or even lower. Researchers suggest that zero-emission, hydrogen-fueled vehicles could exhibit a similar declining cost pattern over the next 20–30 years.

In this bottom-up framework, the analyst can make projections about the declining financial costs of emerging technologies as they penetrate the market, assume a rate of diffusion, and then calculate the resulting GHG reductions and financial costs. Bottom-up modelers can also link the future financial costs of these new technologies to policy initiatives, such as research and development subsidies, in what Grubler, Nakicenovic, and Victor, among others, refer to as "endogenous modeling of induced technical change." But the very criticisms of bottom-up modeling that motivated the development of hybrid modeling approaches apply equally to the exploration of long-run technical change. Uncertainty about emerging technologies concerns not just the diversity of views on their future *ex ante* financial costs (the slope of their experience curves), but also uncertainty about the willingness of consumers and businesses to embrace any particular technology as its financial costs fall and awareness of it increases.

An example is the choice of technology for providing personal mobility in the urban areas of industrialized countries. Emerging substitutes for the conventional vehicle driven by a gasoline-fueled internal combustion engine include replacing gasoline with ethanol, adopting battery–gasoline vehicles, adopting hydrogen-fuel-cell vehicles, and switching to public transit. Bottom-up modeling of induced technical change could show a substantial decrease in the *ex ante* financial cost of public transit relative to the alternatives, but this would mislead policy makers if it had dramatically underestimated the reluctance of urbanites in wealthy countries to switch away from personal vehicles. A similar concern applies to many of the new technologies that are now being applied to estimate the long-run costs of GHG reduction.

3.3 Hybrid Efforts to Address Long-Run Uncertainty

Uncertainty internal to the system being modeled, therefore, is related both to the long-run dynamic between technology commercialization and financial costs, on the one hand, and between technology commercialization and preferences, on the other. To be useful to policy makers, simulations of policies to induce long-run technical change should provide an indication of how reductions in the *ex ante* financial costs of technology options might translate into rates of market penetration, the critical issue overlooked by most bottom-up modeling.

To address this uncertainty, researchers of technological change policy, such as Train, increasingly apply "discrete-choice models" to relate past preferences to certain technology attributes in what is called "revealed preference research." Estimated confidence intervals provide an indication of uncertainty. But for new technologies there is no past market behavior, so discrete-choice researchers must instead ask consumers and businesses to make choices between new or even hypothetical technologies with different bundles of attributes, what is called "stated preference research." An example is to ask consumers to make a choice from among a set of vehicles that vary in terms of acceleration, refueling time, fuel type, emissions, and range. Confidence intervals from this research indicate uncertainty, but may be biased or imprecise given that the choices are hypothetical.

Hybrid models are well suited for incorporating the estimates from discrete-choice modeling into their behavioral parameters. First, discrete-choice research is focused at the technology level, the same level of resolution at which bottom-up and hybrid models operate. Second, hybrid models have behavioral parameters at the level of detail required for simulating technology choices—decision-specific discount rates, and technology-specific costs that can be adjusted to reflect consumers' surplus differences. Research in this direction is still at an early stage, but recent advances are showing a good deal of promise. The transportation module of the NEMS hybrid model, for example, already contains in its behavioral parameters the results of revealed and stated preference research on alternative vehicle choices. Other approaches to addressing technical uncertainty involve subjective adjustment of top-down model parameters (ESUB and AEEI) to incorporate lessons from technology-focused research, and the introduction of behavioral elements into the technology-choice algorithms of bottom-up models.

4. ASSESSMENT

Bottom-up models can be useful to policy makers by showing the technological possibilities for advancing toward an environmental target, such as reducing GHG emissions. However, these models can also be misleading in that their definition of cost is restricted to a single *ex ante* estimate of financial costs that ignores possible losses of option value and consumers' surplus by businesses and consumers. Most bottom-up models are also not able to estimate the overall effect of energy-impacting policies on the macroeconomy. Top-down models, in contrast, can be useful to policy makers because they incorporate the preferences of businesses and consumers—in that their parameters are usually estimated from real-world behavior—and connect the energy system to the macroeconomy. Bottom-up models tend to estimate low costs of GHG reduction relative to top-down models. Different types of comparative analysis, some of it using hybrid models that combine technological explicitness with behavioral realism, can help policy makers understand the role of these different definitions of cost on the divergence in cost estimates between the two approaches.

A second key challenge for GHG reduction cost estimation is due to uncertainties, and hence different assumptions, about technology and preference changes over very long periods. Recent research with bottom-up models attempts to explore how financial costs may decrease as new technologies experience the benefits of learning and perhaps mass production. However, these models lack the behavioral component that is critical in assessing the extent to which the decreasing cost of a new technology will drive its adoption by consumers or firms. Models with hybrid characteristics of technological explicitness and behavioral realism offer some hope to improve the usefulness of models for decision makers and thus address Grubb's challenge of more than a decade ago to develop "the missing models."

SEE ALSO THE FOLLOWING ARTICLES

Biomass: Impact on Carbon Cycle and Greenhouse Gas Emissions • Bottom-Up Energy Modeling • Diet, Energy, and Greenhouse Gas Emissions • External Costs of Energy • Goods and Services: Energy Costs • Greenhouse Gas Emissions, Alternative Scenarios of • Greenhouse Gas Emissions from Energy Systems, Comparision and Overview • Modeling Energy Markets and Climate Change Policy • Multicriteria Analysis of Energy

Further Reading

Azar, C., and Dowlatabadi, H. (1999). A review of technical change in assessment of climate policy. *Annu. Rev. Energy Environ.* **24**, 513–544.

Bohringer, C. (1998). The synthesis of bottom-up and top-down in energy policy modeling. *Energy Econ.* **20**, 233–248.

Brown, M., Levine, M., Romm, J., Rosenfeld, A., and Koomey, J. (1998). Engineering-economic studies of energy technologies to reduce greenhouse gas emissions: Opportunities and challenges. *Annu. Rev. Energy Environ.* **23**, 287–385.

Grubb, M. (1991). Policy modeling for climate change: the missing models. *Energy Policy* **21**, 203–208.

Grubler, A., Nakicenovic, N., and Victor, D. (1999). Modeling technological change: Implications for the global environment. *Annu. Rev. Energy Environ.* **24**, 545–569.

Jaccard, M., Nyboer, J., Bataille, C., and Sadownik, B. (2003). Modeling the cost of climate policy: Distinguishing between alternative cost definitions and long-run cost dynamics. *Energy J.* **24**, 49–73.

Jaffe, A., and Stavins, R. (1994). Energy-efficiency investments and public policy. *Energy J.* **15**, 43–65.

Train, K., and Atherton, T. (1995). Rebates, loans, and customers' choice of appliance efficiency level: Combining Stated- and revealed-preference data. *Energy J.* **16**, 55–69.

U.S. Energy Information Administration. (1998). "Impacts of the Kyoto Protocol on U.S. Energy Markets and Economic Activity." U.S. Department of Energy, Washington, D.C.

Weyant, J., and Hill, J. (1999). Introduction and overview (special issue on the costs of the Kyoto Protocol: A multi-model evaluation). *Energy J.* vii–xliv.

Greenhouse Gas Emissions, Alternative Scenarios of

LEO SCHRATTENHOLZER and KEYWAN RIAHI
International Institute for Applied Systems Analysis
Laxenburg, Austria

1. Greenhouse Gas Emissions Scenarios
2. The Difference between Scenarios, Projections, and Forecasts
3. Scenarios, Uncertainty, and Probability
4. Energy–Economy–Environment Scenarios
5. Why Build Scenarios of Greenhouse Gas Emissions?
6. How are Scenarios of Greenhouse Gas Emissions Built?
7. Driving Variables of Scenarios
8. Scenarios and E3 Models
9. Model Inputs and Outputs
10. Published Scenarios
11. Overall Ranges of Greenhouse Gas Emissions and Their Driving Forces
12. Three Groups of Scenarios
13. Integrated Assessment of Climate Change
14. Scenarios and Policymaking

Glossary

E3 models Energy–economy–environment models.
final energy The form of energy that is bought and sold in the market to consumers.
greenhouse gas (GHG) A gas that can trap heat in the atmosphere by absorbing the longwave radiation emanating from Earth's surface.
Kyoto Protocol An agreement between industrialized countries (the so-called Annex B countries), obliging them, individually or jointly, to ensure that their average annual carbon dioxide equivalent emissions of six greenhouse gases (CO_2, CH_4, N_2O, SF_6, CF_4, and halocarbons) will be 5% below 1990 levels in the commitment period 2008 to 2012.
primary energy Naturally occurring form of energy.
reference energy system (RES) A schematic and aggregated representation of all energy conversion technologies and the flows between them. A RES is not uniquely defined, but rather depends on the level of aggregation considered in a particular analysis.
scenario A qualitative and/or quantitative description of (the development) of a system, based on a set of internally consistent assumptions concerning the main driving forces.
stochastic models Representations that include uncertainty by replacing certainty (of parameters and functional relations) with probability distributions.
storyline In connection with a scenario, a storyline is a qualitative, textual description of how the future might evolve.

This article reports on global carbon emissions projected with long-term energy–economy–environment (E3) scenarios. Many of the prominent emission scenarios have been published under the auspices of the Intergovernmental Panel on Climate Change (IPCC). After briefly discussing methodological aspects of scenario building—in particular its driving forces and other inputs—the article proceeds to systematically discuss scenario results for three groups of scenarios, that is, high-impact, CO_2 mitigation, and sustainable development scenarios. For these groups, ranges are given for the most important inputs and outputs. For mitigation scenarios, typical mitigation costs are reported.

1. GREENHOUSE GAS EMISSIONS SCENARIOS

In the context of this article, the term "scenario" means the qualitative and/or quantitative descriptions of (the development of) a system, based on a set of internally consistent assumptions concerning the main driving forces. A scenario of greenhouse gas (GHG) emissions then depicts the development of GHG emissions over time for a given geographical region. As the greenhouse gas problem is a global problem, the focus here is describing global scenarios.

In addition to including a geographical specification, the system used to describe GHG emissions usually also includes so-called driving forces. These driving forces comprise socioeconomic, demographic, and technology variables that determine the evolution of the energy, agricultural, and other systems, the major sources of anthropogenic GHG emissions. GHG emission scenarios include one or more greenhouse gases, but in most cases, one of the greenhouse gases included is carbon dioxide (CO_2). For most GHG emission scenarios, the six gases included in the Kyoto Protocol are the maximum number of gases considered. Analogous to the glass roofs of actual greenhouses, greenhouse gases are restricted in the narrow sense to mean gases that trap the heat that emanates from Earth's surface, but other gases and aerosols (such as SO_x, water vapor, black carbon, and others) influence the heat balance of Earth (for instance, but not always, by shading, which has a cooling effect). The so-called precursor gases are yet another important factor for the dynamic heat balance of the atmosphere. They are not greenhouse gases, but they are involved in atmospheric chemical reactions that produce or absorb greenhouse gases.

The most important use of greenhouse gas emission scenarios is to serve as inputs for models that calculate the climate impact of given global emission pathways. Conceptually, the climate impact of GHG emissions is described by the chain: emissions, atmospheric concentrations, radiative forcing (a measure of the amount of heat trapped per unit area, e.g., per square meter), temperature change, and specific impacts (e.g. land losses due to sea level rise). Consecutive links of this chain are connected through a complex system that is not yet completely understood. Any evaluation of the causal chain between GHG emissions and impact is therefore surrounded by considerable uncertainty. To give an example, according to recent estimates from the Intergovernmental Panel on Climate Change (IPCC), the temperature increase that is expected to result from a doubling of atmospheric GHG concentrations (relative to preindustrial levels) may range between 1.7 and 4.2 K.

2. THE DIFFERENCE BETWEEN SCENARIOS, PROJECTIONS, AND FORECASTS

It is important to distinguish a scenario from forecasts and, even more so, from predictions, the most important difference being that forecasts and (to an even higher degree) predictions are meant to portray particularly likely future developments. In contrast, scenarios may very well include elements that may not be considered the most likely development. For example, the task of perhaps the most prominent efforts to generate GHG emission scenarios, the IPCC Special Report on Emission Scenarios (SRES), edited by Nakićenović and Swart in 2000, was to portray the entire range of plausible greenhouse gas emissions during the 21st century. Another example of scenarios that are not constructed for the purpose of portraying particularly likely developments are those scenarios that describe a consistent set of circumstances that lead to a desired outcome. If models are used to generate such a scenario, the model is often run "backward." Scenarios designed to generate a particular outcome are called "normative." In contrast, scenarios that describe the consequence of assuming alternative future states of the world, which usually are meant to be particularly likely, are called "descriptive." This distinction is conceptual, and borderline cases exist that could be classified either way.

3. SCENARIOS, UNCERTAINTY, AND PROBABILITY

A common way of treating uncertainty is to define probability distributions, according to which an uncertain event is distributed. Many probability distributions are well understood, and the interpretation of medians, percentiles, variances, and other indicators is a common practice in research. Perhaps less common is the insight that probabilities such as those describing uncertainties about the evolvement of the climate system are subjective and that they therefore cannot be uniquely determined. As a practical definition of probability (in contrast to a strictly formal mathematical definition), de Finetti (1974) proposed "degree of belief." Of course, belief is always in the eyes of the beholder, and it is therefore consistent with this definition to postulate that all probabilities are subjective and that probability theory is the science of analyzing and processing given (subjective) probability distributions.

This way of looking at probabilities is far from commonplace, and the perception and interpretation of probabilistic statements by the interested public are therefore unpredictable. Moreover, it can easily happen that two scientists attach different degrees of belief to one and the same scenario, even when they have produced that scenario jointly. Publishing different probabilities for the same scenario would then almost certainly cause confusion. This was the reason that Nakićenović and Swart, and the other authors of the IPCC SRES report, refrained from estimating probabilities or the relative likelihood of each of the 40 GHG emission scenarios that they generated and published. Of course, there are scientific disciplines in which subjective probabilities of most scientists match or in which it is acceptable to work with probability distributions such as Gaussian (normal) distributions, the parameters of which have been estimated from empirical data. In such cases, the use of probability distributions and, possibly, of stochastic models is well established, and the interpretation of the results is straightforward, at least conceptually. This is not the case for GHG emission scenarios, for which subjective probabilities of different scientists and policymakers alike may vary quite substantially. Instead of hiding these differences and then wondering how it is possible that different conclusions are drawn from identical scenarios, it would appear worth trying to be more specific about differences of subjective probabilities, particularly of the catastrophic consequences of climate change. Doing so could be a way of transferring disagreement on the mitigation measures to disagreement on probabilities, which could lead to agreement on how to manage uncertainty surrounding many aspects of climate change.

4. ENERGY–ECONOMY–ENVIRONMENT SCENARIOS

Many scenarios of greenhouse gas emissions belong to the family of energy–economy–environment (E3) scenarios. There is a close relationship between the three systems. E3 scenarios usually focus on the energy system, including the economy as a driving force and the environment to map the environmental impact of energy production, conversion, and end use. Although global E3 scenarios now usually include the climate issue as one of the most important global environmental problems, the global climate need not be the main theme of such scenarios. More and more scenario builders address the issue of sustainable development, which is a more general concept that includes the issue of climate change.

5. WHY BUILD SCENARIOS OF GREENHOUSE GAS EMISSIONS?

Generally speaking, GHG emission scenarios are built to analyze the consequences of policy action (or non-action) on the development of greenhouse gas emissions, the costs of their abatement, and the impact on the economy. National scenarios can serve the purpose of designing national policies to fulfill a GHG emission reduction commitment such as the targets set by the Kyoto Protocol. Global scenarios can serve as inputs to climate models. In this way, many GHG emission scenarios can collectively describe the range of plausible emissions, with climate models describing the range of possible consequences on climate. It is still too early for climate models to calculate numerical estimates of climate impacts within reasonably narrow bounds. Should this be possible at some future date, GHG emission scenarios can be the first step of an analysis that compares abatement costs with avoided damage. Until that date, damage scenarios will have to serve as placeholders for accurate quantitative estimates.

6. HOW ARE SCENARIOS OF GREENHOUSE GAS EMISSIONS BUILT?

The process of scenario building depends on the subject of the scenarios and the issues to be analyzed with them. An illustration of a typical and widespread method to build long-term GHG emission scenarios is the method that was used to build the scenarios described in the IPCC Special Report on Emission Scenarios (SRES).

As already mentioned, the purpose of the SRES was to gauge the range of plausible global GHG emissions in the 21st century. This required thinking of limiting cases on the high and low ends of the plausible range. The first step in defining the limiting cases—and, in fact, defining any case—was to think of general circumstances that would be consistent with and lead to the targeted outcome in terms of

GHG emissions. High levels of emissions, for example, could be the consequence of high rates of population growth and economic growth, high carbon-intensity of gross domestic product (GDP), poor technological progress, and other factors. Such qualitative assumptions are then formulated in the "storyline." A storyline is a qualitative, textual description of how the future might evolve. It gives the rationale and narrative behind a specific combination of driving forces aiming at directing the development path of aggregate variables of a scenario (for instance, relative to their past trends). For example, "high economic growth rate" means that assumptions of economic growth in a scenario should be on the high side of past trends. Such qualitative characterizations included in storylines are later quantified and transformed into model input numbers.

7. DRIVING VARIABLES OF SCENARIOS

The most important of the "driving variables" of energy demand, which in turn is a major determinant of carbon emissions, are GDP and energy intensity of GDP. Disaggregating one step further, GDP can be thought of as the composite of population and GDP per capita. Energy intensity, together with the carbon intensity of energy supply, can be aggregated to the carbon intensity of GDP. Looking only at the logic of the variables and the problem, the carbon intensity of GDP might seem to be the most obvious driving variable to determine carbon emissions of an economy. However, the point of disaggregating, creating even more variables, is eventually to arrive at those variables that can be most plausibly projected into the future. This criterion depends on the expertise of the scenario builders, and different groups therefore use different driving variables for the projection of carbon emissions.

8. SCENARIOS AND E3 MODELS

The route from storylines, via driving forces, to scenarios is through formal (mathematical) models. Although this discussion is not on modeling, it appears appropriate to give a description of some relevant modeling terms here.

There are several criteria to classify models. One way is to classify them by the overall purpose of model usage, into normative and descriptive. This distinction is particularly important for models that project carbon emissions. For instance, scenarios projecting low-level carbon emissions are often built for the purpose of specifying a set of conditions under which certain environmental goals can be reached. In contrast, descriptive models typically describe the development of a system as the consequence of few assumptions, which are often meant to be particularly likely. An important class of descriptive scenarios is the group of "business-as-usual" scenarios, which aim at projecting past trends in a straightforward way into the future. Sometimes, such scenarios are also referred to as "dynamics-as-usual" scenarios if the point is to be made that systematic changes of trends have been incorporated in them and that incorporating them leads to distinctly different results, compared to mere trend extrapolations.

The core of energy models aiming at projections of greenhouse gases usually incorporates an abstract image of energy flows from primary-energy to final consumption. This image is usually called a reference energy system (RES). Conceptually, the RES has primary energy carriers on its input end and energy demands (and by-products such as pollutants) on the output end. The two ends are connected by a set of technologies that extract, convert, transport, and distribute various forms of energy. One important characterization of an energy model is the degree of detail with which its RES is described in the model. Another way of characterizing models is by its so-called system boundaries. For energy models, one important issue with respect to the system boundary of the model is whether energy demand is substitutable by other factors of production, such as capital and labor, and whether it is price responsive. This issue concerns the degree of detail in which the overall economy is described in the energy model.

9. MODEL INPUTS AND OUTPUTS

As previously stated, the driving forces determine model output such as greenhouse gas emissions. A most important qualifier to this general statement is the amendment "within the given constraints." Using the concept of the reference energy system, such constraints concern primary-energy recoverability, the availability and performance of technologies, and energy demand. Of these groups, energy demand is the one that is shaped the most by the driving forces discussed thus far, because it reflects GDP, population, GDP per capita, and energy intensity. Energy intensity is also related to the conversion efficiencies of the technologies included in the reference energy system. So is the carbon intensity, which, in addition to being related to the performance of non-carbon-emitting technologies, is also related to the model assumptions on the recoverability of hydrocarbon energy resources.

Generally speaking, technological progress can therefore be considered as another driving force of greenhouse gas emissions. It overlaps with other driving forces, but in many studies, the role of technological development is central. In comparison to the other driving forces, this may well be regarded by policymakers as one of the driving forces that can be steered particularly effectively (for instance, by supporting research and development of energy technologies). As it will be substantiated later, during the discussion of the policy relevance of long-term GHG emission scenarios, it can be argued that such a policy would be aiming at influencing the driving forces in a much more sustainable way, compared to energy taxes or "command-and-control" measures. This, of course, is not meant to say anything against the latter, just to emphasize the importance of technology policy as an important additional way of driving greenhouse gas emissions in the right direction. In fact, the mitigation of greenhouse gases to volumes that would lead to a "stabilization of greenhouse gas concentrations in the atmosphere at a level that would prevent dangerous anthropogenic interference with the climate system," as postulated by the United Nations Framework Convention on Climate Change, would suggest combining all available policies and measures for the given purpose.

10. PUBLISHED SCENARIOS

One recent comprehensive effort to explore alternative GHG emission paths projected for the 21st century led to the publication of the IPCC Special Report on Emission Scenarios (SRES). The SRES scenarios were developed to represent the range of driving forces and emissions in the scenario literature so as to reflect current understanding and knowledge about underlying uncertainties. Hence, plausibility, size, and location of the emissions range were defined by emission scenarios that had been published by the time the SRES work began. Note that the SRES scenarios do not include additional climate initiatives, which means that no scenarios are included that explicitly assume the implementation of climate policies, such as the United Nations Framework for Climate Change (UNFCC) or the emissions targets of the Kyoto Protocol.

The first step of SRES work thus consisted of compiling what has become also known as the SRES database, produced by Morita and Lee in 1998. This database includes the results of some 400 E3 scenarios, which are described in terms of the most important variables characterizing the long-term development of the E3 system either globally or for major world regions. The variables include population, economic growth, energy demand, carbon emissions, and others. Although not all scenarios in the database report on all variables, the scenarios included can be regarded as representative of the range of possibilities still regarded as plausible by the global modeling community 5 years after they were collected in the SRES database.

One way of interpreting the ranges of the projected carbon emissions and their driving forces would be to see them as quantifications of the uncertainty surrounding the evolvement of the variables during the course of the present century. To what extent the ranges and frequency distributions quantify the uncertainty in any reliable way (i.e., to what extent they can be interpreted as probabilities) is open for speculation and individual judgment. In the overview of scenario results given in the following section, such an interpretation of frequencies as probabilities is withheld, and the focus is instead on the description of statistical indicators, such as medians, variances, the minimum, and the maximum.

11. OVERALL RANGES OF GREENHOUSE GAS EMISSIONS AND THEIR DRIVING FORCES

For the summary presentation of ranges of greenhouse gas emissions and their driving forces, a common way of graphically presenting values and ranges of a number of variables is to use a regular polygon and (zero-based) axes between the center and each vertex. A heptagon representation with seven variables is chosen to summarize the global scenarios of the SRES database; five variables describe the values of scenario variables in the year 2100, one describes a cumulative figure up to that year, and one describes a growth rate. The five variables with values for 2100 are (1) CO_2 emissions, expressed in billions of tons (Gt, gigatons; 10^9 tons)

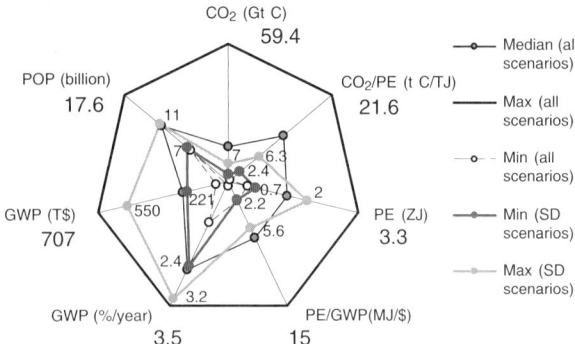

FIGURE 1 Global carbon emissions and their main driving forces. The minimum and maximum of the values for the Special Report on Emissions Scenarios database, as well as for the sustainable-development (SD) scenarios only, are shown on seven axes of the heptagon, and they are connected. The seven axes show ranges for indicators across the scenarios in 2100 (see text for discussion). PE, Primary energy; GWP, gross world product; T$ trillions of dollars; POP, population. IIASA, International Institute for Applied Systems Analysis.

of carbon, (2) specific carbon emissions per unit of primary energy (PE), expressed in tons of carbon per terajoule, (3) total primary energy demand up to the year 2100, expressed in zetajoules (10^{21} joules), (4) specific primary energy consumption of economic output, expressed as gross world product (GWP) in megajoules per U.S. dollar (1990 purchasing power), and (5) gross world product, expressed in trillions (10^{12}) of U.S. dollars (1990 purchasing power); (6) population, expressed in billions (10^9) of people; and (7) growth of gross world product, expressed as the average annual growth rate of the GWP between 1990 and 2100. The resulting heptagon is presented in Fig. 1.

The outer, symmetric heptagon in Fig. 1 represents the maximum values for the group of all SRES database scenarios for each of the seven variables. The smallest irregular heptagon inside this envelope represents the respective minimum values for all SRES database scenarios. Another irregular heptagon shows the respective seven median values for all SRES database scenarios. Preempting the later discussion of sustainable development (SD) scenarios, the figure also shows, for comparison, two heptagons that connect the seven minimum and maximum values for the group of SD scenarios for each of the seven variables.

In interpreting the heptagons, it should be remembered that any one of them is unlikely to connect points that belong to the same scenario. For instance, it would not appear logical if either a minimum or a maximum heptagon connected values for the same scenario. Take economic growth and primary-energy intensity, for example. One common assumption in E3 scenarios is that with higher economic growth, technological progress can occur faster. As a consequence, the maximum primary energy intensity should not coincide with maximum economic growth. Figure 1 also illustrates that CO_2 emissions in the year 2100 cover a range between zero, i.e., no net carbon emissions, to almost 60 billion tons (gigatons) of carbon (Gt C), with the median at approximately 18 Gt C, significantly below halfway. In contrast, world GDP growth and population both have a "bias" toward the maximum. For economic output, this means that the SRES database does not include scenarios with zero or low economic growth during the 21st century.

12. THREE GROUPS OF SCENARIOS

The overall ranges of global GHG emissions and their driving forces provide a first overview of possibilities that have been considered by modelers. To better understand the sometimes considerable differences between GHG emissions (and between other variables) projected in different scenarios, three separate groups of scenarios are now described: high-impact, CO_2 mitigation, and sustainable-development scenarios. The description of these three groups also includes scenarios that have contributed to neither SRES scenarios nor the SRES database. The reason is that the mandate for the authors of the SRES was to survey carbon emissions in scenarios that did not include explicit climate policies. Although scenarios that included more general environmentally benign strategies, such as sustainable development, were included, explicit climate mitigation scenarios were not included in the SRES process. Rather, they were included in another IPCC publication, the Third Assessment Report, compiled by Metz et al. in 2001.

Another preface is required before we can meaningfully describe SD scenarios. We follow a definition of SD E3 scenarios as proposed by the Environmentally Compatible Energy Strategies (ECS) Project at IIASA. According to this definition, all scenarios that satisfy the following four criteria will be referred to as sustainable-development scenarios:

1. Economic growth (in terms of GDP per capita) is sustained throughout the whole time horizon of the scenario.

2. Socioeconomic inequity among regions, expressed as the world-regional differences of gross domestic product per capita, is reduced significantly

over the 21st century, in the sense that by 2100, the per capita income ratios between all world regions are reduced to ratios close to those prevailing between Organization for Economic Cooperation and Development (OECD) countries today.

3. Long-term environmental stress is mitigated significantly. In particular, carbon emissions at the end of the century are approximately at or below today's emissions. Other GHG emissions may increase, but total radiative forcing, which determines global warming, is on a path to long-term stabilization. Other long-term environmental stress to be mitigated includes impacts to land use (e.g., desertification). Short- to medium-term environmental stress (e.g., acidification) may not exceed critical loads that threaten long-term habitat well being.

4. The reserves-to-production (R/P) ratios of exhaustible primary-energy carriers do not decrease substantially from today's levels.

The proponents of this definition argue that it has been inspired by and is consistent with the "Brundtland spirit," which is quantitative enough to serve the purpose of classifying long-term energy–economic scenarios, and that it comes close to the definition of weak economic sustainability, complemented with some environmental and social constraints.

The quantitative characterizations of the three groups of E3 scenarios in this section are based on 34 scenarios that were developed and published by the IIASA's ECS Project since 1998. These include scenarios developed with the World Energy Council (WEC) in 1998, scenarios developed for the IPCC Special Report on Emissions Scenarios, as well as scenarios developed for the impact assessment for the Working Group III of the IPCC Third Assessment Report. As to the ranges of carbon emissions and driving forces, the 34 IIASA ECS scenarios are representative of the scenarios in the SRES database.

12.1 High-Impact Scenarios

Most scenarios of the SRES database include a high-level environmental impact in at least one respect. For the quantitative summary in this subsection, the eight IIASA ECS scenarios that are representative of these high-impact scenarios have been selected. One class of scenarios in this group consists of the business-as-usual scenarios. In particular, the SRES-B2 scenario, which was designed in such a way that its driving variables stay in close proximity to the medians of the SRES database, is classified as high impact. Another important function of high-impact scenarios is to serve as a reference, relative to which mitigation and other policies can be quantified. One example is the SRES-A1C scenario ("C" stands for coal), in which atmospheric CO_2 concentrations in the year 2100 reach 950 parts per million by volume (ppmv), more than three times the preindustrial level. SRES-A1C is also characterized by a high rate of economic growth. The projected global GDP in 2100 is 550 trillion (10^{12}) U.S. dollars (in 1990 prices). Relative to global GDP in 1990, 20.9 trillion, this corresponds to an average annual growth rate of just below 3%. The lowest economic growth rate in this group of eight scenarios is projected in the IIASA WEC B scenario, one of the business-as-usual scenarios. There the projected average annual growth rate is 2%.

Global population does not show characteristic features in the high-impact scenarios. Projections range from 7.1 billion (after an earlier peak) to 15.1 billion in the year 2100. Global primary-energy consumption in the high-impact scenarios ranges from 1500 to 2700 exajoules (EJ) in 2100. Relative to the global consumption in 1990, 350 EJ, this range implies annual growth rates between 1.3 and 1.9%.

12.2 CO_2 Mitigation Scenarios

Mitigation scenarios explore cases in which the global atmospheric CO_2 concentration is stabilized at different levels. The following description is based on 19 IIASA ECS scenarios. These scenarios were constrained to stabilize atmospheric CO_2 concentrations at levels of 450, 550, 650, and 750 ppmv in 2100, and they used high-impact scenarios as references. The criterion that classifies scenarios in this group raises the question as to whether constraining atmospheric CO_2 concentrations in this way leads to sustainable-development scenarios. The answer is "not necessarily." Of the 19 scenarios included in this group, only 8 would also qualify as SD scenarios. Although this observation is based on a sample that might be considered small, it suggests that sustainable development is a more general strategic goal compared to climate mitigation.

Making a reference-mitigation pair of scenarios comparable suggests leaving crucial assumptions on general technological development and resource availabilities the same in both scenarios. At the same time, it could be argued that it would appear implausible to assume exactly the same kind of technological progress as in the reference and stringent emission mitigation, and at the same time. There is no clear-cut solution to this issue, but to keep rules simple, the same set of technology data and the same set of resource availabilities as for the corresponding baseline scenarios were assumed for the mitigation cases. To respond to the "technological response" argument, sustainable-development scenarios were built (these are described in the following subsection).

The main mechanisms to reduce emissions relative to the reference cases can therefore be summarized as supply shifts toward primary-energy carriers, with less carbon content and price responsiveness of final energy demand. As a general characterization of the carbon emission trajectories in the mitigation cases, it can be said that they are similar to those derived by other modelers for the same concentration limits, in particular the popular "WRE" trajectories described by Wigley, Richels, and Edmonds in 1996. Very often, the first question about mitigation concerns costs. Choosing typical examples of mitigation aiming at atmospheric CO_2 concentrations of 550 ppmv (corresponding to roughly twice the preindustrial level), and quantifying the difference in global GDP in 2100 between the mitigation scenarios and the corresponding reference cases, obtains reductions of 7 trillion U.S. dollars for the SRES-A2 scenario (236 instead of 243) and 8 trillion for the SRES-A1C scenario (542 instead of 550). These correspond to relative reductions by 3 and 1.5%, respectively. At the average annual growth rates of economic output assumed in the two scenarios, this would mean delaying the achievement of the given output by 1.5 and 0.5 years for the A2 and A1C scenarios, respectively. In comparison with the reductions of global economic output in the mitigation cases, the difference in primary-energy consumption is relatively higher. In A2-550, primary-energy consumption in 2100 is 18% lower than that of the reference A2 scenario. In A1C, this difference is lower (6%).

The comparison of these two pairs of scenarios clearly shows that the relative economic impact of mitigation is higher where lower growth had been projected in the corresponding reference case. Nonetheless, reaching the mitigation goal of 550 ppmv in the higher growth scenario A1C reduces atmospheric CO_2 concentrations in the year 2100 by a larger amount (from 950 ppmv), compared to the lower growth scenario A2 (from 780 ppmv).

12.3 Sustainable-Development Scenarios

By the definition used here and in comparison to most other scenarios, sustainable-development scenarios project low levels of GHG emissions. The following description is based on seven IIASA ECS scenarios. Three distinct general qualitative strategic ideas are reflected in this group of scenarios. One idea can be dubbed "small is beautiful," and it leads to the scenarios of the SRES-B1 family. These scenarios are characterized by low energy intensity of GDP, which is dominated by the service sector. Energy supply is highly decentralized and relies to a high degree on renewable energy. In contrast, the SRES-A1T scenario describes a high-technology interconnected world in which technological progress (in particular, in the field of carbon-free sources, which include nuclear energy) is key to creating an environmentally compatible global energy system.

Built some years earlier than the SRES scenarios, the two sustainable-development scenarios of the IIASA WEC set (IIASA-WEC C1 and IIASA-WEC C2) describe a "global family" in which the industrialized countries transfer substantial amounts of resources to the developing countries and regions, to aid development in a way that preserves the global environment. Even without involving the assumption of massive international transfers of wealth, all scenarios in this group assume a fully cooperative world with efficient markets and without constraints such as trade barriers, monopolies, and the like. This assumption can be regarded mostly normative in the sense of the distinction between normative and descriptive, and substantial amounts of goodwill would be required to implement any of the sustainable-development scenarios.

Comparing projected GDP of sustainable-development scenarios with the projections of all scenarios in the SRES database (Fig. 2) shows that projected GDP in SD scenarios tends to cover the higher part of all GDP projections. The crosshatched area in Fig. 2 includes average annual GDP growth rates between 2 and 3%. This suggests that sustainable development tends to require wealth rather than to prevent it. According to the definition of sustainable-development scenarios, projected global carbon emissions in this group in 2100 are below today's emissions. Accordingly, atmospheric CO_2 concentrations in the year 2100 are between 445 and 560 ppmv. Figure 3 illustrates the carbon emission reductions projected in sustainable-development scenarios, comparing the carbon intensity of primary energy in SD scenarios with that of all scenarios in the SRES database. Figure 3 shows that the highest projected average decarbonization rates in the database are near 3.3% per year between 1990 and 2100.

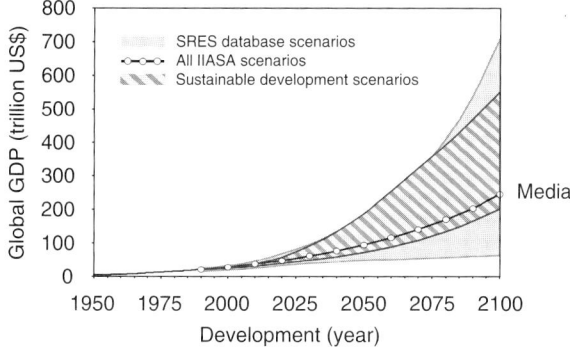

FIGURE 2 Global economic product (GDP) in trillions of 1990 U.S. dollars; development from 1950 to 1990 and in the scenarios [193 of which are from the Special Report on Emission Scenarios (SRES) database] from 1990 to 2100. Historical data are from two United Nation studies published in 1993; the database is from Morita and Lee (1998). IIASA, International Institute for Applied Systems Analysis.

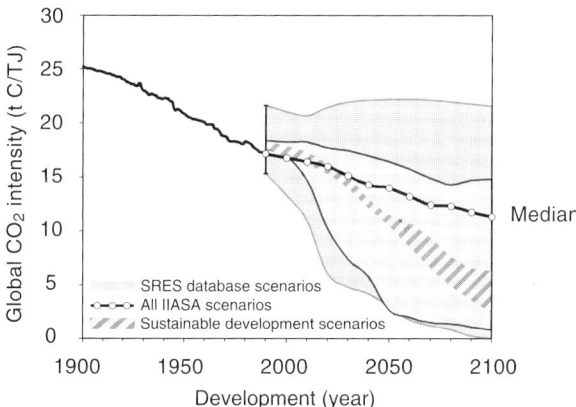

FIGURE 3 Global carbon intensity of primary energy; actual development from 1900 to 1990 and in the scenarios [140 of which are from the Special Report on Emission Scenarios (SRES) database] from 1990 to 2100. Historical data are from Nakićenović (1996); the database is from Morita and Lee (1998). Note that net carbon emissions are used to calculate the carbon intensities (i.e., bioenergy is excluded). The differences in the carbon intensities for the base year are due to different accounting conventions and data problems in general. IIASA, International Institute for Applied Systems Analysis.

This means a reduction of energy-related carbon emissions per unit of primary energy by a factor of 40 during this time horizon, which, in turn, leads to an energy system with almost no carbon emissions. Decarbonization in the sustainable-development scenarios leads to projected carbon intensities for the year 2100 between 2.4 and 6.3 C/TJ, both down from 18 C/TJ in 1990.

13. INTEGRATED ASSESSMENT OF CLIMATE CHANGE

Analyses of emission reduction, explicitly or implicitly, mainly address the cost component of climate abatement. For a more complete view of the problem, however, it is necessary also to get an estimate of the benefits of emission mitigation in terms of avoided damage. This is the task addressed by models for the integrated assessment of climate change. In view of the substantial uncertainties surrounding the impacts of climate change, estimates of avoided damage vary within wide ranges, even for given emissions. Still, a range of estimates can be better than no estimate, and any damage estimate is good enough to analyze the cost-effectiveness of policies and measures (that is, the cost of a unit of emission reduction). This aspect of cost-effectiveness was one motivation for the ongoing efforts to study more greenhouse gases than CO_2 only, although CO_2 is the most important anthropogenic greenhouse gas and therefore has been studied most intensively. The most obvious non-CO_2 greenhouse gases to analyze in addition to CO_2 are those included in the Kyoto Protocol of the UNFCC (CH_4, N_2O, SF_6, CF_4, and halocarbons). The six "Kyoto gases" and their abatement are the subject of the Stanford-based Energy Modeling Forum study (EMF-21), which involves modelers from North America, Europe, and Asia. The study is still ongoing in 2003, but the first preliminary results show that considering GHGs other than CO_2 leads to significantly reduced mitigation costs.

14. SCENARIOS AND POLICYMAKING

Mitigation scenarios incur mitigation costs in terms of GDP lost, but such costs for sustainable-development scenarios have not been reported here. The reason is that the mitigation scenarios discussed herein were defined relative to reference cases, whereas SD scenarios were defined in absolute terms. This may seem just like a curious coincidence, but it is not, and it appears worth emphasizing the policy implications of this difference, which is not just a methodological peculiarity, but by deliberate design. This difference means to express the idea that a strategy aiming directly at sustainable development could be more effective than conceptually constraining the evolvement of a system that is implicitly assumed to be preferable, if not for the risk of catastrophic consequences of climate change. This has been recognized in the past for strategies to mitigate local pollution in cases in which the issue is to limit pollution either by so-called end-of-pipe technologies or by avoiding it from the beginning. It has been argued, for instance, that given enough notice (time), progress in automobile technology could have achieved the same emission reduction as catalytic converters, by improving the internal combustion engine instead of treating the exhaust gases. It is the idea of sufficient notice that may provide for an analogy to GHG emissions. If it holds, this analogy suggests that aiming for sustainable development might be more promising than imposing carbon constraints, given that sufficient time is available for action. In turn, this means that action should be initiated sooner rather than later.

The projections of GHG emissions and their driving forces reported here vary over wide ranges. Carrying the analysis further, to atmospheric GHG concentrations, radiative forcing, temperature change, and impacts, in a sense increases the size of these ranges, as a consequence of the uncertainties surrounding the evolvement of the global climate, which is not understood well enough to predict the consequences of given GHG emissions within narrow limits. Whatever the overall strategy to avoid catastrophic consequences of climate change, this uncertainty will play a major role in policymaking. What will be called for is uncertainty management at unprecedented scales, and flexibility of strategies will be an obvious plus.

SEE ALSO THE FOLLOWING ARTICLES

Biomass: Impact on Carbon Cycle and Greenhouse Gas Emissions • *Carbon Capture and Storage from Fossil Fuel Use* • *Carbon Taxes and Climate Change* • *Climate Change and Energy, Overview* • *Climate Protection and Energy Policy* • *Diet, Energy, and Greenhouse Gas Emissions* • *Greenhouse Gas Abatement: Controversies in Cost Assessment* • *Greenhouse Gas Emissions from Energy Systems, Comparison and Overview*

Further Reading

De Finetti, B. (1974). "Theory of Probability," Vol. 1. John Wiley & Sons, London and New York.

Houghton, J. T., Ding, Y., Griggs, D. J., Noguer, M., van der Linden, P. J., and Xiaosu, D. (eds.). (2001). "Climate Change 2001: The Scientific Basis." Contribution of Working Group I to the Third Assessment Report of the Intergovernmental Panel on Climate Change (IPCC). Cambridge Univ. Press, Cambridge, UK.

Klaassen, G., Miketa, A., Riahi, K., and Schrattenholzer, L. (2002). Targeting technological progress towards sustainable development. *Energy Environ.* **13**(4/5), 553–578.

Metz, B., Davidson, O., Swart, R., and Pan, J. (eds.). (2001). "Climate Change 2001: Mitigation." Contribution of Working Group III to the Third Assessment Report of the Intergovernmental Panel on Climate Change. Cambridge Univ. Press, Cambridge UK.

Morita, T., and Lee, H.-C. (1998). "IPCC SRES Database, Version 0.1." Emission Scenario Database, prepared for IPCC Special Report on Emissions Scenarios. Available on the Internet at http://www.cger.nies.go.jp.

Nakićenović, N. (1996). Freeing energy from carbon. *Daedalus* **125**(3), 95–112.

Nakićenović, N., Grübler, A., and McDonald, A. (eds.). (1998). *Global Energy Perspectives.* Cambridge University Press, Cambridge UK, pp. 299, ISBN 0521645697.

Nakićenović, N., and Swart, R. (eds.). (2000). "Emissions Scenarios." Special Report of the Intergovernmental Panel on Climate Change (IPCC). Cambridge Univ. Press, Cambridge UK.

Rao, S., and Riahi, K. (2003). "Long-term Multigas Mitigation Strategies Using MESSAGE." Presentation at the International Energy Workshop, 25 June. International Institute for Applied Systems Analysis (IIASA), Laxenburg, Austria.

Riahi, K., and Roehrl, R. A. (2000). Greenhouse gas emissions in a dynamics-as-usual scenario of economic and energy development. *Technol. Forecast. Social Change* **63**(3), 195–205.

Riahi, K., and Roehrl, R. A. (2000). Energy technology strategies for carbon dioxide mitigation and sustainable development. *Environ. Econ. Policy Stud.* **3**(2), 89–123.

United Nations (UN). (1993). "Macroeconomic Data Systems, MSPA Data Bank of World Development Statistics, MSPA Handbook of World Development Statistics." MEDS/DTA/1 and 2 June. UN, New York.

United Nations (UN). (1993). "UN MEDS Macroeconomic Data Systems, MSPA Data Bank of World Development Statistics." MEDS/DTA/1 MSPA-BK.93, Long-Term Socioeconomic Perspectives Branch, Department of Economic and Social Information and Policy Analysis. UN, New York.

Wigley, T. M. L., Richels, R., and Edmonds, J. A. (1996). Economic and environmental choices in the stabilization of atmospheric CO_2 concentrations. *Nature* **379**, 240–243.

Greenhouse Gas Emissions from Energy Systems, Comparison and Overview

ROBERTO DONES, THOMAS HECK, and
STEFAN HIRSCHBERG
Paul Scherrer Institut
Villigen, Switzerland

1. Introduction
2. Fuel-Specific Greenhouse Gas Emission Factors
3. Energy Chain-Specific Greenhouse Gas Emissions
4. Comparison of Greenhouse Gas Emissions

Glossary

Central European Electricity Interconnection (CENTREL) Regional group of four transmission system operator companies of the Czech Republic, Hungary, Poland, and Slovak Republic.

energy chain All stages associated with energy supply by means of specific energy carriers including mining, processing, and conversion of an energy carrier, as well as the treatment and disposal of the produced waste.

mixed oxides fuel (MOX) Oxides of plutonium and uranium.

Organization for the Nordic Electric Cooperation (NORDEL) Includes Denmark, Finland, Norway, Sweden, and Island.

Union for the Coordination of Transmission of Electricity (UCTE) Members in year 2002 are Austria, Belgium, Bosnia-Herzegovina, Czech Republic, Croatia, Denmark (associated member), France, Germany, Greece, Hungary, Italy, Luxembourg, Macedonia, the Netherlands, Poland, Portugal, Slovak Republic, Slovenia, Spain, Switzerland, and Serbia and Montenegro. The CENTREL countries Czech Republic, Hungary, Poland, and Slovak Republic have officially joined UCTE in 2001.

According to the Intergovernmental Panel on Climate Change, the earth's climate system has changed, globally and regionally, with some of these changes being attributable to human activities resulting in emissions of greenhouse gases. Energy supply and the fossil systems in particular are the dominant contributors to the emissions of these gases. This article provides an overview and comparison of greenhouse gas emissions associated with fossil, nuclear, and renewable energy systems. In this context, both the direct technology-specific emissions as well as the contributions from full energy chains within the life cycle assessment framework are considered. This discussion also provides examples illustrating the differences between industrialized and developing countries and explores the impact of technological advancements.

1. INTRODUCTION

1.1 Global Greenhouse Gas Emissions from Energy Systems and Other Sources

According to the Intergovernmental Panel on Climate Change (IPCC), the anthropogenic greenhouse gas (GHG) emissions contributing most clearly to global warming in terms of relative radiative forcing are CO_2, CH_4, halocarbons, and N_2O. Radiative forcing is the change in the net vertical irradiance (expressed in watts per square meter: Wm^{-2}) at the boundary between the troposphere and the stratosphere. Compared to the preindustrial era 250 years ago, additional radiative forcing due to increases of the well-mixed greenhouse gases is estimated to be $2.43\,Wm^{-2}$. CO_2 contributes most ($1.46\,Wm^{-2}$), followed by methane ($0.48\,Wm^{-2}$), halocarbons ($0.34\,Wm^{-2}$), and N_2O ($0.15\,Wm^{-2}$).

Other possible factors influencing the global climate are less well understood and thus quantitatively more uncertain. Among them are stratospheric ozone (cooling), tropospheric ozone (warming), sulphate (cooling), black carbon and organic carbon (warming or cooling), biomass burning (cooling), mineral dust (warming or cooling), aerosol indirect effects (cooling), land use (change of albedo, i.e., share of reflected sun light), and solar variation.

Table I shows the global emissions of the major greenhouse gases and the contribution of anthropogenic sources for the late 1990s. CO_2 emissions originate mainly from combustion of fossil fuels (i.e., from energy systems). These CO_2 emissions are quite well known. For CH_4 and N_2O, the total emission rates are much more uncertain and it is difficult to assess global emissions from single sources. Halocarbons are molecules containing carbon and either chlorine, fluorine, bromine, or iodine. Among the halocarbons are several refrigerants that are used as working fluids for cooling or heating (i.e., for transport of thermal energy). Many refrigerants interesting for energy uses (in refrigerators, heat pumps, air conditioners, etc.) have very high global warming potentials (GWP) on a per kg emitted basis. Energy scenarios for reduction of CO_2 emissions often include increased use of heat pumps substituting for fossil fuel heating systems. It has to be kept in mind that refrigerant emissions due to leakages counteract to a certain extent in the total GHG balance. Halocarbon emissions are almost completely made by humans. The table does not include refrigerants like CFC-11 or CFC-12, which have been banned because of their high ozone-depleting potential; they show low emission rates but are still abundant in the atmosphere due to emissions in the past.

TABLE I

Annual Emissions of Important Greenhouse Gases in the Late 1990s

	Annual emissions (Mt/year)	Lifetime (years)	GWP 100-year	CO_2-equiv. 100-year (Mt/year)
II. CO_2	29,000		1	29,000
Fossil fuels	22,400			22,400
Cement production	700			700
Land use, etc.	6000(3000–9000)			6000
III. CH_4	600	8.4–12	23	13,800
Energy	100(89–110)			2300
Biomass burning	40(23–55)			900
Other anthropogenic sources	230			5300
Natural sources	230			5300
IV. N_2O	26	120	296	7700
Automobiles	0.3(0.2–0.4)			90
Industry, including energy	2.0(1.1–2.8)			600
Other anthropogenic sources	8.5(7.6–9.6)			2500
Natural sources	15			4500
V. HFC refrigerants				
HFC-23	0.007	260	12,000	84
HFC-134a	0.025	13.8	1300	33
HFC-152a	0.004	1.4	120	0.5
VI. Other halocarbons				
Perfluoromethane (CF_4)	0.015	>50,000	5700	86
Perfluoroethane (C_2F_6)	0.002	10,000	11,900	24
Sulphur hexafluoride (SF_6)	0.006	3200	22,200	133

Emissions are given in real mass per year and in CO_2-equiv. mass per year. GWP, global warming potential; HFC, hydrofluorocarbon. *Source.* IEA (1999); rounded values.

1.2 Methodological Basis and Scope of Comparisons

The most straight-forward accounting of GHG emissions can be based on emission factors associated with combustion of the various fossil fuels. This approach can be used for the establishment of the overall national emission inventories but is not practical when the goal is to account fully for emissions associated with the use of specific technologies. While uses of nuclear and renewable energy sources exhibit practically negligible emission levels for GHGs at the stage of power generation, the same is not necessarily true for other stages of the corresponding energy chains. In addition, emissions may arise when manufacturing the components for the plants, transporting fuels and other materials, or in the decommissioning stage.

Life cycle assessment (LCA), an approach utilizing process chain analysis specific to the types of fuels used in each process, allows for the full accounting of all such emissions, also when they take place outside of the national boundaries. Thus, LCA considers not only emissions from power plant construction, operation, and decommissioning but also the environmental burdens associated with the entire lifetime of all relevant upstream and downstream processes within the energy chain. This includes exploration, extraction, processing, and transport of the energy carrier, as well as waste treatment and disposal. The direct emissions include releases from the operation of power plants, mines, processing factories, and transport systems. In addition, indirect emissions originating from manufacturing and transport of materials, from energy inputs to all steps of the chain, and from infrastructure are covered.

An alternative, not process-oriented approach, is the input/output (I/O) method; it divides an entire economy into distinct sectors and based on the inputs and outputs between the sectors generates the energy flows and the associated emissions. A hybrid approach is also frequently employed, combining LCA and I/O methods; the I/O method is then used exclusively for assessing processes of secondary importance.

This article provides comparisons of GHG emissions, based on several approaches mentioned earlier. The largest part of the data cited here has been derived using a full scope LCA implementation. Due to consistency and other practical reasons it is not feasible to account for results from all major published studies addressing the GHG emissions.

As the main reference work, the most recent results of the comprehensive Swiss study addressing LCA-based environmental inventories in Europe has been used. This is supplemented by some results from studies carried out by researchers in and for other countries. More details on the methodological aspects of LCA as applied in the Swiss study will be provided in Section 3.

2. FUEL-SPECIFIC GREENHOUSE GAS EMISSION FACTORS

For national accounting of direct CO_2 emissions from fuel combustion, the International Energy Agency (IEA) within OECD uses the IPCC methodology. Neither GHGs other than CO_2 nor non-energy emission sources are included in the IEA statistics. Table II summarizes the fuel-specific emission factors.

The direct emissions based on the factors provided in Table II can be combined with appropriate contributions from the production chain of fuels. Table III shows the emission factors per unit of mass from the upstream chain of selected fuels in the Swiss LCA study. The large variations for coal products depend on the differences between the producing regions (mainly the energy needed for mining and the coalbed methane emissions) and between supply mixes (transport energy). The variations for coke and coke by-products depend on the lower average efficiency assumed for the world average production versus coke making in Central Europe.

A Japanese LCA study addressed burdens from fossil fuel extraction, processing, and transportation, associated with uses of the fuels in Japan. The supplying countries were addressed separately. In Table IV, only the average emission factors are shown. In the Japanese study, CO_2 and CH_4 emissions are given separately; here they are summed using for methane the GWP according to IPCC. The average values for coal are in the lower end of those estimated by the Swiss study, partly because of the different distribution of the origin of the coal and the transport distances, partly due to the different assumptions on the energy uses at mine and methane emissions. The Japanese study did not address the refinery step, which explains the differences for oil. The results for CO_2 emissions for a single supplying country may change by a factor of four for coke, three for steam coal, two for crude oil and LPG, but only by about 20% for LNG.

TABLE II

Carbon Conversion Factors for Fuels[a]

	Fuel	Carbon emission factor (tC/TJ)	Carbon emission factor (tCO$_2$/TJ)
Liquid fossil	Primary fuels		
	Crude oil	20	73
	Orimulsion	22	81
	Natural gas liquids	17.2	63
	Secondary fuels/products		
	Gasoline	18.9	69
	Jet kerosene	19.5	72
	Other kerosene	19.6	72
	Shale oil	20	73
	Gas/diesel oil	20.2	74
	Residual fuel oil	21.1	77
	LPG (liquid petroleum gas)	17.2	63
	Ethane	16.8	62
	Naphta	20	73
	Bitumen	22	81
	Lubricants	20	73
	Petroleum coke	27.5	101
	Refinery feedstocks	20	73
	Refinery gas	18.2	67
	Other oil	20	73
Solid fossil	Primary fuels		
	Anthracite	26.8	98
	Coking coal	25.8	95
	Other bituminous coal	25.8	95
	Subbituminous coal	26.2	96
	Lignite	27.6	101
	Oil shale	29.1	107
	Peat	28.9	106
	Secondary fuels/products		
	Patent fuel and brown coal/peat briquettes (BKB)	25.8	95
	Coke oven/gas coke	29.5	108
	Coke oven gas	13	48
	Blast furnace gas	66	242
Gaseous fossil	Natural gas (dry)	15.3	56
Biomass	Solid biomass	29.9	110
	Liquid biomass	20	73
	Gas biomass	30.6	112

[a] The factors are given in terms of tonne C and tonne CO$_2$ to facilitate application. Values are rounded.
Source. Revised 1996 IPCC Guidelines for National Greenhouse Gas Inventories (1997), IPCC/OECD/IEA, Paris.

TABLE III
Greenhouse Gas Emissions from the Upstream Chains of Fossil Fuels Used in Europe Calculated in the Swiss LCA Study

		kg CO_2-equiv./kg fuel	
		Minimum	Maximum
Hard coal	At producing region[a]	0.040	0.343
	Country-specific supply mix[b]	0.188	0.322
	Supply mix to UCTE	0.270	
	Coke[c]	0.492	0.655
	Coke oven gas[c]	0.612	0.814
	Benzene[c]	0.685	0.911
	Tar[c]	0.637	0.847
	Briquettes	0.336	
Lignite	At mine	0.017	
	Briquettes	2.223	
	Pulverized	0.812	
Oil[d]	Heavy fuel oil	0.423	
	Light fuel oil	0.480	
	Bitumen	0.402	
	Diesel	0.482	
	Diesel low sulphur	0.495	
	Kerosene	0.477	
	Naphta	0.417	
	Petroleum coke	0.488	
	Petrol low sulphur	0.769	
	Propane/butane	0.573	
	Refinery gas	0.558	
Natural gas	Western European high-pressure grid	0.491	

[a] Minimum: South America; maximum: Western Europe.
[b] Minimum: Poland; maximum: Germany.
[c] Coke oven gas, benzene, and tar are by-products of coke making. Minimum: Central Europe; maximum: estimation for world average.
[d] Values for Western Europe. The analysis performed for the Swiss-related chain resulted in greater values by a factor 1.2 to 1.4.
Source. Dones *et al.* (2003).

3. ENERGY CHAIN-SPECIFIC GREENHOUSE GAS EMISSIONS

This section provides chain-specific results obtained in the Swiss LCA study for Switzerland and Europe, supplemented by representative result examples from a number of published analyses by researchers in other countries. Some additional basic features of the LCA methodology as applied in the Swiss

TABLE IV
LCA Greenhouse Gas Emissions from the Upstream Chain of Fossil Fuel Mixes Used in Japan

Fuel	kg CO_2-equiv./kg fuel
Steam coal	0.168
Coke	0.118
LNG	0.640
Crude oil (for refinery)	0.122
Crude oil (for power plant)	0.153
LPG total[a]	0.319
NGL total[b]	0.258

[a] Weighted over the two processes producing liquid petroleum gas (LPG) from liquefied natural gas (LNG) (0.560 kg CO_2-equiv./kg) and producing LPG at oil fields (0.258 kg CO_2-equiv./kg).
[b] Weighted over the three processes producing natural gas liquids (NGLs) during LNG liquefaction (0.214 kg CO_2-equiv./kg to 0.481 kg CO_2-equiv./kg) and directly at the oil field (0.254 kg CO_2-equiv./kg).
Source. Kato *et al.* (2000).

applications are summarized here; most of these principles also apply to other state-of-the-art studies, which, however, may differ in terms of scope, level of detail, specific assumptions, and methodology applied (some results are based on hybrid approaches). The most important features are as follows:

- Energy systems, transport systems, material manufacturing, production of chemicals, waste treatment and disposal, as well as agricultural products, have been assessed using detailed process analysis developed under consistently defined common rules.
- Electricity inputs were modeled using production technology or supply mix as close as feasible to the actual situation. In case of lack of specification, the UCTE mix was used as an approximation.
- Allocation criteria were developed for multipurpose processes.

The results provided here focus on electricity supply, but selected results are also given for heat generation and for cogeneration systems.

3.1 Fossil Energy Chains

3.1.1 Coal

3.1.1.1 Hard Coal In the Swiss study European country-specific average power plants operating around year 2000 were analyzed. For the estimation of the infrastructure of the plants, two power levels of 100 and 500 MW were considered; a mix with a share of 10 and 90%, respectively, was defined for the reference plant. The reference unit was assumed to be used for middle load with 4000 hours of operation per year at full capacity, and operate 150,000 hours during its lifetime. Emissions of CO_2 were estimated on the basis of the coal characteristics (e.g., lower heating values in the range of 22.1 to 26.6 MJ/kg), and the average net efficiencies of single units operating in Europe (29 to 40%), while emissions of CH_4, N_2O, and CO are averages taken from the literature.

Average coal from eight supplying regions was considered: West and East Europe, North and South America, Australia, Russia, South Africa, and Far East Asia. Specific average data were used for coal characteristics (e.g., upper heating values in the range of 20 MJ/kg to 28 MJ/kg), share of open pit and underground mines, methane emissions, land use, and energy uses for each of these regions. Import mixes in year 2000 for all European countries with coal plants have been defined.

The average methane emissions from coal mining in the eight modelled regions range between 0.16 g CH_4/kg (typical emission from open pit coal mines in the United States) and 13.6 g CH_4/kg product coal (West Europe).

The results for GHG GWP 100a from the chains associated with the average hard coal in European countries is between 949 g CO_2-equiv./kWh for the NORDEL countries and 1280 g CO_2-equiv./kWh for the Czech Republic (including several cogenerating plants, where the emission is entirely allocated to the electricity generation). The average for UCTE countries (excluding CENTREL) in year 2000 is 1070 g CO_2-equiv./kWh. Methane contributes nearly 7%, and N_2O 0.8% to the total GHG emissions for the UCTE average hard coal chain; CO_2 emissions practically cover the rest.

The upstream chain contributes between 8% (Portugal) to 12.5% (Germany) to total GHG. The total GHG associated with production regions varies between 0.04 kg CO_2-equiv./kg coal at production in South America up to 0.288 kg CO_2-equiv./kg coal at production in Russia.

Figure 1 provides a comparison between the range of average estimates of normalized GHG emissions for UCTE-countries, the average for Japan and United States, the range obtained for the coal chain in the Province Shandong in China, and the range according to a worldwide survey carried out in 1997. The Japanese and U.S. results are on the lower side of the ranges for UCTE countries and in the 1997 survey; the same applies to the lower range of the estimates from the study for China. The higher range from the China study reflects the low efficiencies characteristic in particular for small units in China as well as large contributions from mining; the potentially substantial but difficult to estimate additional GHG emissions from uncontrolled coal fires have not been included.

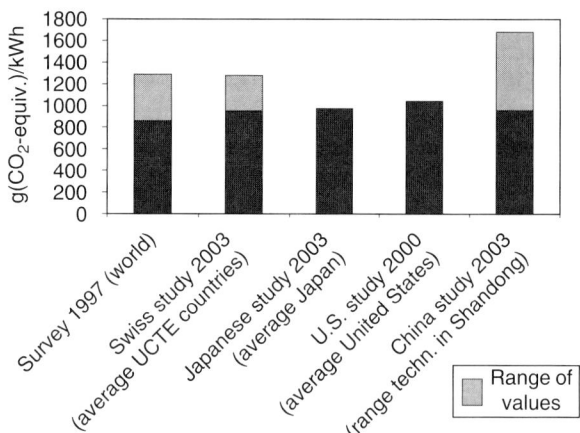

FIGURE 1 Greenhouse gas emissions from coal power plants. The figure shows ranges of greenhouse gas emissions in CO_2-equivalents per kWh electricity from coal power plants and associated fuel cycles according to different studies. From Survey (1997), Van de Vate, J. F., *Energy Policy* 25, pp. 1–6; Swiss study (2003), Dones *et al.*; Japanese study (2003), Hondo (2003); U.S. study (1999), Spath *et al.* (1999); China study (2003), Dones *et al.* (2003).

3.1.1.2 Lignite The reference plant used for lignite in the Swiss LCA study has similar characteristics as for hard coal but a larger share of plants with low rated power was used. The reference unit is assumed to be used for base load with 6000 hours of operation per year at full capacity, and work 200,000 hours during its lifetime. Emissions of CO_2 are estimated on the basis of the characteristics of the average lignite (e.g., lower heating values in the range of 5.2 to 16.6 MJ/kg) and the average efficiencies of single units operating in Europe (range for country-average is between 23 and 40%), while emissions of CH_4, N_2O, and CO are UCTE averages taken from the literature.

Considering that lignite plants are mine-mouth, only an average European open-pit mine has been modeled on the basis of information limited to only few mines. Only 0.23 g CH_4/kg lignite are assumed to be emitted during mining operations.

The results for GHG GWP 100a from the chains associated with the average lignite power plants in

European countries is between 1060 g CO_2-equiv./kWh for Austria and 1690 g CO_2-equiv./kWh for the Slovak Republic. The average for UCTE countries (excluding CENTREL) in year 2000 is calculated as 1230 g CO_2-equiv./kWh. Methane contributes about 0.6%, and N_2O 0.5% to total GHG emission for the UCTE average lignite chain; CO_2 emissions practically cover the rest. Mining contributes marginally between 0.9% (France) to 2.6% (Greece) to total GHG.

3.1.2 Oil

Since the role of oil in electricity generation is decreasing, only few key facts are provided from the Swiss study. The European country averages for GHG emissions of oil chains range from 519 to 1200 g CO_2-equiv./kWh 100-year. The UCTE average around the year 2000 was about 880 g CO_2-equiv./kWh 100-year, of which about 88% or 775 g/kWh are the direct emissions during power plant operation.

For the fuel oil supply, emissions occur during crude oil exploration, long-distance transport (e.g., in transoceanic tankers), oil processing in refineries, and local distribution. For an average oil power plant in Europe, the highest contributions to the upstream GHG emissions occur in the oil exploration phase and in the heavy oil processing in refineries.

3.1.3 Gas

3.1.3.1 Natural Gas Chain For natural gas, like for the other fossil fuel fired electricity or heating systems, the dominant contributor to GHG emissions are the CO_2 emissions from the power plant or boiler. Natural gas is transported in pipelines over long distances. Since natural gas consists mainly of methane (i.e., natural gas itself is a greenhouse gas), leakages in the pipelines can contribute significantly to the total GHG emissions. For European countries, the methane emissions can make up to about 10% of the total GHG emissions in the full chain depending on the location of the natural gas power plant or the boiler. Together with CO_2 and other GHG emissions in the upstream chain, the emissions other than directly from the power plant can constitute more than 10% of the total GHG emissions for European natural gas power plants (about 17% for the UCTE average in year 2000).

The country averages of the full chain GHG emissions of natural gas power plants in Europe range from 485 to 991 g CO_2-equiv./kWh 100-year. The UCTE average around the year 2000 was about 640 g CO_2-equiv./kWh 100-year. Therein are about 530 g direct CO_2-equiv./kWh emissions during operation of the power plants.

For the modeling of the best technology combined-cycle gas power plant, data from the new 400 MW power plant Mainz-Wiesbaden (Germany) were used. According to the operators, this is the natural gas power plant with the highest net electric efficiency (58.4%) worldwide (year 2001). Because the efficiency depends also on the local conditions (the Mainz-Wiesbaden plant is located directly at the river Rhine, which yields good cooling conditions), it was assumed that a comparable plant at an average location in Europe would have a net efficiency of about 57.5%. The full chain GHG emissions of the best combined-cycle power plant (about 420 g/kWh) are much lower than those of an average gas power plant.

Figure 2 compares the range of average estimates of normalized GHG emissions from the gas chain for UCTE-countries, with the average for LNG in Japan, results for combined cycle plants in Europe, Japan and the United States, and the range according to the worldwide survey carried out in 1997. The upper range in the survey probably represents a plant using a gas mix rather than pure natural gas.

3.1.3.2 Industrial Gas Industrial gas covers blast furnace gas from steel production and coke oven gas.

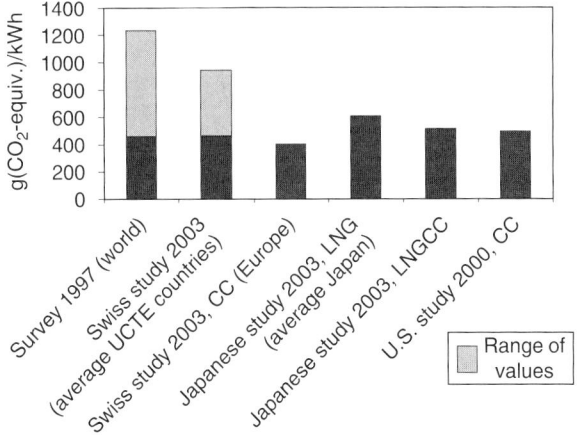

FIGURE 2 Greenhouse gas emissions from gas power plants. The figure shows ranges of greenhouse gas emissions in CO_2 equivalents per kWh electricity from gas power plants according to different studies. From Survey (1997), Van de Vate, J. F., *Energy Policy* 25, pp. 1–6; Swiss study (2003) Dones *et al.*; Japanese study (2003), Hondo (2003); U.S. study (2000), Spath, P. L., and Mann., M. K., "Life Cycle Assessment of a Natural Gas Combined-Cycle Power Generation System," National Renewable Energy Laboratory, Golden, CO.

The results cited here are based on the mix of blast furnace gas and coke oven gas used for electricity production in UCTE around year 2000. Due to its high CO_2 content, blast furnace gas has high CO_2 emission factors of 90 to 260 g/MJ burned gas. Emission factors of coke oven gas range between 40 and 90 g/MJ gas burned. This leads to exceptionally high total GHG emissions, exceeding 1700 g CO_2-equiv./kWh for the European average industrial gas mix. The methane contribution comes mainly from the coke oven gas production chain.

3.1.4 Heating and Cogeneration

3.1.4.1 Heating Two hard coal heating systems were modeled in the Swiss LCA study: an industrial furnace with capacity in the range 1 to 10 MW, and a stove of about 5 to 15 kW. The thermal efficiency of the furnace is 80%, the one of the stove 70%. The industrial furnace is assumed to be fuelled with the average Western European hard coal supply mix, the stove either with coke or with briquettes.

Assuming that all CO is oxidized into CO_2, CO contributes about 10% to the total GHG emissions associated with the stove. Direct CH_4 emissions from burning briquettes are 20 times higher than from burning coke in the stove. In case of the briquette, direct methane emissions from the stove are about 50% of the total methane emissions calculated for the chain. Due to a lower carbon content per unit of energy, burning briquettes has lower direct CO_2 emissions than burning coke.

Condensing gas and oil boilers use the heat from combustion and the heat from condensation of the water in the flue gas. Modern condensing natural gas boilers can reach annual net efficiencies of about 102%, modern oil boilers about 98%. (The ratio refers to the lower heating value of the fuel therefore efficiencies of more than 100% are possible for condensing boilers.) High efficiency reduces the CO_2 emissions. Direct CO_2 emissions of a modern condensing natural gas boiler <100 kW are about 56 g/MJ, the GHG emissions of the full chain for a condensing natural gas boiler in Central Europe in year 2000 add up to about 71 g/MJ CO_2-equiv. 100-year. For a 10 kW condensing nonmodulating oil boiler with direct CO_2 emissions of about 74 g/MJ, the full chain GHG emissions for location Central Europe are about 89 g/MJ CO_2-equiv. 100-year.

Figure 3 shows a comparison of GHG emissions for the energy chains based on various heating technologies of different capacity, utilizing hard coal, natural gas, and oil. The lowest emissions are for the natural gas systems, followed by oil systems.

3.1.4.2 Cogeneration Figure 4 shows a comparison of CO_2 emissions per kWh_e from modern small cogeneration plants of different capacity and technology, located in Switzerland. Allocation of emissions to the products is in this case based on exergy. The higher the capacity, the higher the electric efficiency and the lower the CO_2 emissions for electricity. The total efficiency is approximately constant for the different plants shown. The CO_2 emissions per MJ fuel burned are the same for all natural gas plants but higher for the diesel plant because of the higher emission factor of diesel oil.

3.2 Nuclear Energy Chain

The amount of GHG emissions from the nuclear chain associated with light water reactors (LWRs) is controlled by few parameters: the nuclear cycle considered, the average enrichment and burnup at discharge of the fuel, the lifetime of the plants, especially of the power plant, and, most important, the enrichment process used and the electricity supply to the enrichment diffusion plant (if its services are used).

The Swiss LCA study on energy systems addresses the Swiss, French, German, and average UCTE nuclear chains, separately modeling boiling water reactor (BWR) and pressurized water reactor (PWR) power plant technologies with reference to the two 1000 MW class units installed in Switzerland.

The chain was decomposed into several steps: uranium mining (open pit and underground), milling, conversion, enrichment (diffusion and centrifuge), fuel fabrication, power plant (light water reactors), reprocessing, conditioning (encapsulating spent fuel), interim storage of radioactive waste, and final repositories (geological for high and intermediate level waste).

The study assumes partial or total reprocessing of the fuel, according to the conditions in West European countries. In particular, 40% of the total spent fuel produced at Swiss and German plants during their lifetime was assumed to be reprocessed, whereas for France this is most likely the fate of 100% of the fuel; for UCTE a weighted average of 80% of total spent nuclear fuel was assumed to undergo reprocessing. Mixed oxides fuel (MOX) was modeled in the study as MOX is

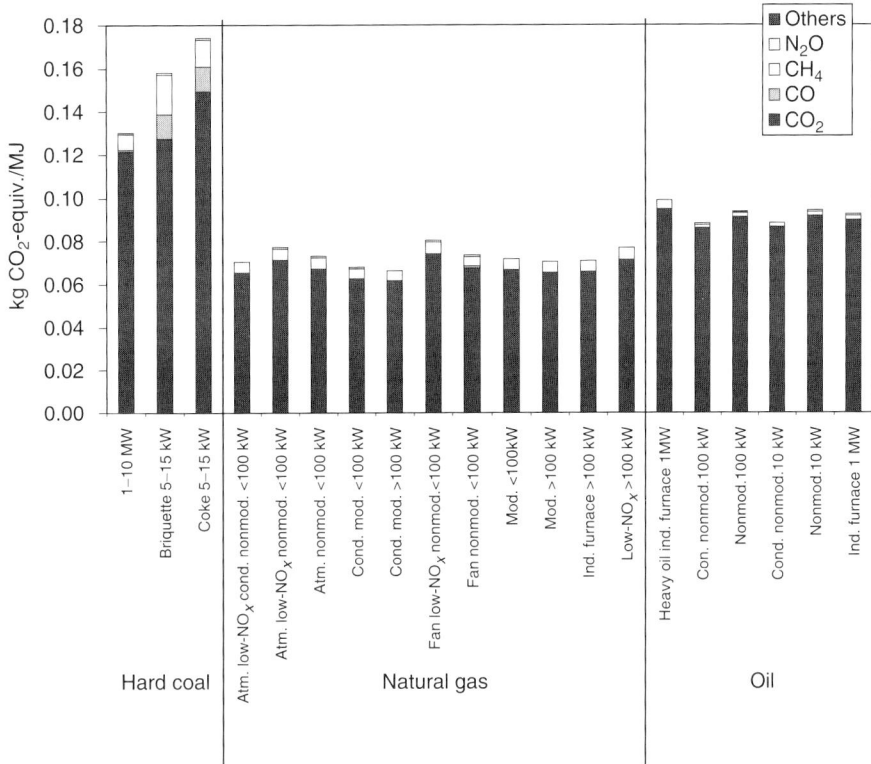

FIGURE 3 Greenhouse gas emissions from fossil fuel heating systems. CO_2-equivalent emissions per MJ heat are shown for different fuels and different technologies. Atm., atmospheric boiler; Fan, fan burner; Mod., modulating boiler; Cond., condensing boiler; Ind. furnace, industrial furnace. Low-NO_x refers to boilers built in early 1990s. All emission factors estimated for boilers operating around year 2000 in Central Europe; modulating boilers refer to the most modern technology on the market. From Dones *et al.* (2003).

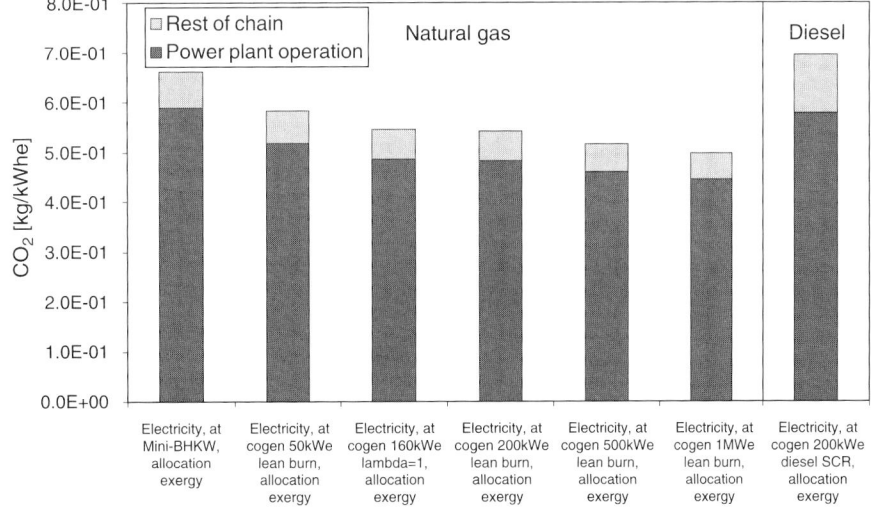

FIGURE 4 CO_2 emissions per kWhe from modern small cogeneration plants of different capacity and technology. Lambda = 1: cogeneration unit with three-way catalyst. Mini-BHKW is a small device comparable to a boiler for a small house with about 2 kWe. SCR, selective catalytic reduction. The allocation to electricity is based on the exergy of electricity and heat assuming an upper temperature of about 80°C for heat. From Dones *et al.* (2003).

widely used in Western Europe, for which recycled plutonium is mixed with depleted uranium from enrichment.

Lifetime of 40 years was assumed for the power plants. Average U-235 enrichment ranges between 3.8% for French PWR and Swiss BWR and 4.2% for Swiss PWR. The corresponding average burnup of spent fuel ranges between nearly 43 MW$_{th}$ day/kgU for the French plants and 52 MW$_{th}$ day/kgU for the Swiss PWR. The study includes a description of enrichment supplies and relevant electricity input, based on a literature search and assumptions on the worldwide enrichment market and country-specific enrichment services.

The total emissions of GHG for the modeled nuclear chains are calculated between approximately 6 and 11 g CO_2-equiv./kWh, where the minimum is estimated for the French nuclear chain, assuming 10% MOX, and 100% use of enrichment at the diffusion Eurodif plant in Tricastin (supplied by nuclear plants and using no CFCs as refrigerants but water). The maximum was calculated for Germany, under the assumption of about 13% MOX and a mix of enrichment services including 10% of the USEC diffusion plant which is assumed to be supplied by coal power plants; nearly 70% of the enrichment services are assumed to be supplied by URENCO facilities (one is based in the north of Germany), based on centrifugal technology, which is about 65 times less energy intensive than the USEC plant.

Calculated total GHG is prevalently from CO_2 emissions, between 90 and 93%. Methane contributes between 3 and 6%, whereas N_2O is between approximately 1 and 3%. The hydrofluoro- and hydrochlorocarbon emissions from the enrichment stage are below 5%.

The GHG associated with the power plant ranges between approximately 1 and 1.3 g CO_2-equiv./kWh, and the waste management (back-end or downstream) between approximately 0.6 and 1 g CO_2-equiv./kWh. The upstream chain makes the rest, which may substantially change according to the main assumptions for the cycle.

For comparison, the total GHG emissions for the Chinese reference chain were estimated at 9 g CO_2-equiv./kWh using centrifuge technology for all enrichment services. Conversely, taking the extreme assumption of only diffusion enrichment powered by coal plants, the highest GHG emission was calculated at nearly 80 g CO_2-equiv./kWh. When electricity mixes with mixed fuels (also including gas and nuclear) are used, this amount could be halved to about 45 g CO_2-equiv./kWh.

3.3 Renewable Energy Chains

3.3.1 Biomass

For biomass burning boilers and cogeneration systems modeled in the Swiss LCA study, only untreated wood was considered (i.e., no waste wood combustion was accounted for). The emission of direct, biogenic CO_2 from combustion was calculated with a carbon content in dry wood of 0.494% for all types of wood fuels. All carbon absorbed by the trees and contained in the wood that is eventually burned was assumed to be emitted during the combustion, either as CO_2 or as CO. Furthermore, all CO is assumed to be fully oxidized to CO_2 in the atmosphere and as such accounted in the total GHG.

Infrastructure and emission data are corresponding to average operation of modern wood boilers, available on the central European market around year 2000. The wood fuel supply chain represents average central European conditions.

Several classes of heating technologies were modeled, namely 6, 30, and 100 kW for wood log furnace, and 50, 300, and 1000 kW for wood chips furnace, using wood directly from forest or residual wood from industrial processes. Boiler operation was modeled for hardwood (mainly beech), softwood (mainly spruce), and the Swiss commercial residual wood mix (72% softwood, 28% hardwood). Wood pellet furnaces of 15 and 50 kW were also modeled.

For the industrial wood chips, practically all burdens of the processes have been allocated to commercial wood products on the basis of an economic evaluation, rather than to the residues eventually burned in furnaces. This zero-allocation assumption might not be applicable if wood chips would become more important for heat production.

Figure 5 shows the comparison of the results of the Swiss study with an Austrian LCA study, the latter analyzing wood heating systems in Central Europe. Only the systems with high efficiency modeled in the Austrian study are used here, as these have comparable efficiencies as assumed in the Swiss study.

The capacities of the logs and chips boilers used for this comparison are between 6 and 1000 kW for the Swiss study and 10 to 50 kW for the Austrian study. The pellet boilers have similar capacities in the Swiss and Austrian studies. The electric power rates of the CHP plants are between 335 and 400 kWe for the Swiss study and 210 kWe to 36 MWe for the Austrian one.

The total efficiencies of log/chips and pellet boilers are between 68 and 85%. The electric efficiencies of

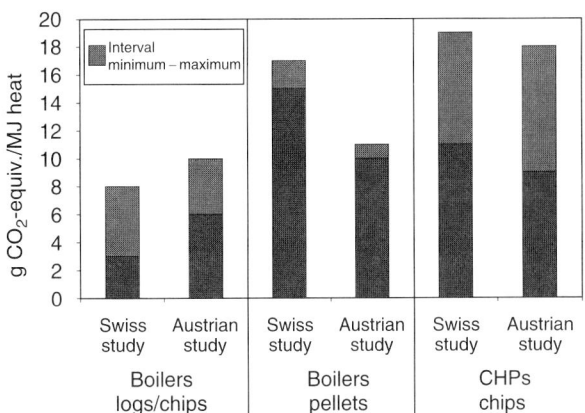

FIGURE 5 Greenhouse gas emissions for wood heating systems. The figure shows the comparison of results of two studies, Swiss and Austrian, analyzing wood heating systems in Central Europe, using LCA. From Swiss study, Dones et al. (2003); Austrian study, Jungmeier et al. (1999), Treibhausgasbilanz der Bioenergie. Vergleich der Treibhausgasemissionen aus Bioenergie-Systemen und fossilen Energiesystemen. Joanneum Research, Institut für Energieforschung, Graz. Bericht Nr.: IEF-B-06/99.

the cogeneration (CHP) plants in the Swiss study are between 0.9 and 2.3%, thermal efficiencies are between 76 and 78%. The Austrian study reports electric efficiencies of the CHP plants between 8 and 39%, and thermal efficiencies between 51 and 80%.

All values for net GHG emissions of the compared systems are within a relatively small range between 3 g CO_2-equiv./MJ and 19 g CO_2-equiv./MJ. There is no obvious correlation between system capacity and net GHG emissions. Direct CH_4 and N_2O emissions are usually lower at a higher capacity, but the indirect GHG emissions may be higher.

In general, the net GHG emissions from wood chips and wood logs heating systems are not significantly different. Due to incomplete combustion, burning wood logs in automatically controlled furnaces produces more CO and CH_4 than burning chips. Compared to all other heating systems, the (modern) 6 kW fireplaces are emitting relatively high amounts of CO and CH_4.

Compared to other wood boilers, pellet furnaces have higher net GHG emissions, between 15 and 17 g CO_2-equiv./MJ. More than 90% are CO_2 emissions of fossil origin. About 65% of these are due to material and energy uses in the fuel supply chain, and 22% are due to material and energy uses in fuel transport. The main reason is that energy requirements for the pellet production are higher than for chips and logs. Additionally, the transport distances of the pellets to the boilers are usually higher than for chips and logs.

Two wood cogeneration systems of 6400 and 1400 kW$_{th}$ installed in Switzerland and burning industrial chips have been assessed in the Swiss LCA study, each of them with two alternative modes, with multicyclone filter and with enhanced emission control (electric filter + selective noncatalytic reduction (SNCR), assumed to operate with urea). Both systems are mainly designed for heat production (thermal efficiency of nearly 77%); the electricity (efficiency between 1 and 3%) is mostly used for self-consumption.

Allocating all emissions to heat production, 11 g CO_2-equiv./MJ are calculated for the technology with multicyclone-filter and 19 g CO_2-equiv./MJ for the technology with emission control. The reason for the higher GHG emissions of the second technology is that the urea in the SNCR is partially transformed to N_2O. However, the emission factor of this N_2O emission is quite uncertain, and only little information is available in the literature. The interval of N_2O emissions is between 4 and 41 mg/MJ wood input.

3.3.2 Hydro

3.3.2.1 Swiss and European Hydro Power Plants

To reflect the situation in Switzerland and Europe, two main types of hydropower plants (storage and run-of-river) and additionally pumped storage plants were analyzed in the Swiss LCA study. A representative sample of four run-of-river power plants in Switzerland and one in Austria was used. Lifetime was assumed to be 80 years for the fixed structures and 40 years for other parts. Net average efficiency is 82% (best efficiency can be 88%).

A representative sample of 52 Swiss concrete dams with a height of more than 30 m was considered for calculating the average. Lifetime was assumed to be 150 years for the dam, 40 years for turbines and pipes, and 80 years for the rest of materials. Net average efficiency, including pipe losses, is 78% (best efficiency can be 84%). The data refer to plant construction of a mix of dam types built between 1945 and 1970; therefore they might not be fully representative for more modern construction, or for a specific dam type, or for an individual unit. The data have been extrapolated to preliminary describe dam mixes in other alpine countries (Austria, Italy, and France) and nonalpine European countries.

No studies seem to exist for CH_4 and N_2O emissions from hydroelectric storage plants in the Alpine region. Hence, the methane emission has been roughly extrapolated from data for natural Swiss lakes, which has been estimated as $4 \cdot 10^{-3}$ kg m^{-2} a^{-1}. Applying this value as first guess to all Swiss reservoirs, the emission factor would be calculated at 0.014 g CH_4/kWh or 0.32 g CO_2-equiv./kWh. Measurements for N_2O releases from a natural lake in Switzerland have been used for a rough estimation of this emissions from reservoirs, given at $7.7 \cdot 10^{-5}$ g N_2O/kWh or 0.023 g CO_2-equiv./kWh. CO_2 emissions were not accounted for, due to the lack of information on the net balance (see below) for alpine regions.

The Swiss LCA study used available figures for direct GHG emissions from Norwegian and Swedish reservoirs, averaging around 6 g CO_2-equiv./kWh, also for all non-Alpine regions except Finish reservoirs, for which 30 g CO_2-equiv./kWh were assumed.

The results for GHG GWP 100a show that for alpine regions the construction of the dams contributes the most to the total emission of 4 to 5 g CO_2-equiv./kWh. The material (mostly concrete) contributes about 70% to total. With the above assumptions on direct CH_4 emissions, other European reservoirs may exhibit average emissions of 10 g CO_2-equiv./kWh and Finland around 34 g CO_2-equiv./kWh. By increasing GHG fluxes, the emissions associated with the construction of the dam decrease in importance. In case of run-of-river, results are of the order of 3 g CO_2-equiv./kWh.

A survey of several LCA studies performed in 1997 concluded that the indirect contributions approximately range 1 to 10 g CO_2-equiv./kWh for reservoir as well as for run-of-river plants, the difference depending on the plant and site characteristics (type of dam, height/width of dam, capacity of reservoir, location, installed electric capacity, load factor, dedicated transmission lines). However, small hydro may have somewhat higher GHG emission factors.

Pumped storage environmental burdens are a function of the input electricity source and the total efficiency, assumed to be 70%. Hence, results for GHG are strongly depending on the assumption for the input electricity mix. Results of the Swiss LCA study span from the lowest value for Norway (0.027 kg CO_2-equiv./kWh—mix based prevalently on hydro) to 1.62 kg CO_2-equiv./kWh for Poland (mix based mostly on coal). However, these results should be compared only with other systems providing the same service as pumped storage (i.e., peak load).

3.3.2.2 Issues Related to Direct GHG Emissions from Hydroelectric Reservoirs Direct GHG emitted during the operation of a reservoir due to the decomposition of organic matter should be counted for additionally to indirect emissions related to the construction, maintenance, and decommissioning of the plant, in order to compare hydropower with other electricity systems.

Since the late 1980s, research has been invested in the assessment of direct GHG emissions from the air-water interface of reservoirs. Key parameters mentioned for the GHG emission factors are climate (in tropical reservoirs the biodegradation is faster); amount of flooded biomass; nature of the flooded soils; depth of reservoir, which controls methane oxidization; and the ratio energy production to surface.

The state-of-the-art in 2003 for the estimation of GHG emissions from reservoirs comprises the consideration of the emissions from flooded land area only, with contributions from CO_2 diffusive emissions for the initial 10-year period after first filling, CH_4 bubbling and diffusive emissions, N_2O diffusive emissions, and degassing emissions for all previously mentioned species. Ongoing research is aiming at modeling the net emissions, including the entire catchment area and the various phases in the life of the reservoirs in the boreal, temperate, and tropical regions, including pre-impoundment natural GHG emissions, which may be substantial.

Diffusive flow is generally leading to greater GHG emissions than bubbling flow. It is reported that about 99% of CO_2 is released through diffusion, whereas for CH_4 this mechanism releases from 14 to 90% of the total flow. Canadian researchers report that diffusive CO_2 accounts for 80% of total emissions in boreal and temperate sites, whereas methane bubbling makes 60% of total emissions in shallow tropical sites but less than 10% in deep reservoir. In general, N_2O contributes only a few percentage points.

Canadian researchers estimated total GHG emissions (using IPCC 1996 GWP) of 265 g CO_2-equiv. $m^{-2} a^{-1}$ (± 150) from boreal and temperate reservoirs on the basis of measurements of diffusive emission fluxes at the water-air interface of 15 different hydroelectric developments. Data for six tropical reservoirs were summarized giving 4300 g CO_2-equiv. $m^{-2} a^{-1}$ (± 2300) for reservoirs with large shallow water areas (average depth below 10 m) and 1700 g CO_2-equiv. $m^{-2} a^{-1}$ (± 400) for deeper reservoirs (average depth >25 m). The Canadian researchers concluded that reservoirs in tropical regions emit between 5 and 20 times more GHG than in boreal and temperate regions.

Brazilian researchers reported the results of two surveys of measurements in several Brazilian reservoirs located in different parts of the country. By extrapolating these two surveys to 1-year emissions, they obtained for total CO_2 plus CH_4 the approximate range 1050 to 3930 g CO_2-equiv. $m^{-2} a^{-1}$ (using IPCC 1996 GWP). Methane contributed 7 to 86% of the total GHG emissions. Using the average capacity factor for Brazilian hydroelectric plants of approximately 50%, an interval of 12 to 2077 g CO_2 equiv./kWh from direct emissions in the year of the measurements can be derived for different reservoirs. The average GHG emission factor for direct CO_2 and CH_4 emissions from seven plants, weighted by the energy generation, would be around 340 g CO_2 equiv./kWh. However, it would be inappropriate to use such estimation for the lifetime of the plants, because variations of the fluxes may occur.

On the basis of measurements by several teams of direct emissions in boreal and tropical reservoirs of direct emissions, and of the expected electricity production assuming a load factor of 0.6 and a lifetime of 100 years, Canadian researchers have estimated ranges of GHG emission factors. Boreal and temperate reservoirs would have average GHG emission factors between 0.01 and 0.06 kg CO_2 equiv./kWh, whereas emissions from tropical reservoirs would range between 0.2 and 3 kg CO_2 equiv./kWh. The highest specific emissions were found in shallow (average depth between 4 and 10 m) tropical hydroelectric developments where the ratio of energy production to flooded area is low. A 1997 survey reported for the La Grande Complex in Canada a direct GHG emission of 34 g CO_2-equiv./kWh, based on measurements; this would reduce to 15 g CO_2-equiv./kWh including all sources, assuming gradual return to zero net emissions after 50 years of operation.

Besides decomposition of organic matter, there is evidence of the strong influence of the flooded peat soil for methane emissions in Finn reservoirs, especially if they are shallow. Finnish sources based on direct measurements reported an average range of 65 to 72 g CO_2-equiv./kWh for such reservoirs.

The issue of the net GHG emissions deserves further research in various regions of the world considering the potential developments of hydropower.

3.3.3 Wind

The production of electricity by wind turbines generates no direct GHG emissions. Indirect emissions are following the construction and assembling of the wind power plant, the production and assembling of its materials, transports of materials, and waste disposal processes.

The Swiss LCA study analyzed several wind turbines for Swiss-specific and European average conditions. Plants with 30 kW, 150 kW, 600 kW, and 800 kW were modeled for onshore, and a unit of 2 MW for offshore (based on data from the wind park Middelgrunden, near to Copenhagen). Actual capacity factors in Switzerland range from 8 to 14%. The average for Europe was assumed 20% for onshore and 30% for offshore. The wind mix in Switzerland is mostly composed of 600 and 800 kW turbine, making 57 and 40% of wind electricity in 2002, respectively.

The GHG emissions are decreasing with increasing installed capacity of the turbine. For Swiss conditions, one 800 kW unit emits 20 g CO_2-equiv./kWh, while for average European conditions it decreases to 14 g CO_2-equiv./kWh, for both onshore and offshore. Other studies estimated a typical release of GHG associated to wind turbines in areas with optimal wind speed on the order of 10 g CO_2-equiv./kWh.

For the 800 kW plant, about 72% of GHG emissions stems from the production of materials; about 15% are caused by material processing, 8% by waste disposal, and the rest by transports, final assembling, and installation. The fixed parts (foundation and tower) cause 23% of the CO_2 emissions, the moving parts the rest. For the 2 MW offshore plant, about 81% of the emissions originate from the production of materials. The fixed parts make 44% of the total CO_2 emissions.

3.3.4 Solar

Mono- (mc-Si) and polycrystalline (pc-Si) silicon cell technologies utilized in photovoltaic (PV) panels were analyzed in the Swiss LCA study for Swiss conditions (yield of 885 kWh $kW_{peak}^{-1} a^{-1}$). The plants considered were grid-connected 3 kW_{peak} units installed on buildings in two configurations (i.e., integrated and nonintegrated slanted-roof); also façade and flat-roof were considered. The production chain has been decomposed into detailed steps. The efficiency of the PV cells was assumed to be 16.5% for mc-Si and 14.8% for pc-Si.

Use of electronic-grade silicon from the European electronic industry was assumed as input to the PV industry. The production of metallurgical grade silicon is assumed to occur in Norway (hydro-based electricity mix) and the purification in Germany, with use of a mix of hydropower and a highly efficient combined heat and power plant. To reflect market conditions, besides electronic grade also a 50% share of off-grade silicon has been considered for the production of the wafers. Energy uses for purification were allocated on the basis of economic value of these two coproducts plus silicon tetrachloride. Changing these assumptions may lead to substantial change of the calculated emissions. In the future, solar grade silicon will probably be used, with associated reduced energy inputs for purification. This production was also modeled in the Swiss study, assuming an increase of cell efficiencies to 17.5% for the mc-Si and 15.7% for the pc-Si systems. Thirty years lifetime was considered for all conditions.

These assumptions lead to a decrease of total GHG emissions compared to previous analyses. Nonintegrated slanted-roof mc-Si panels exhibit about 73 g CO_2-equiv./kWh, whereas for pc-Si technology this value is 59 g CO_2-equiv./kWh. Integrated types have slightly lower LCA emissions. For future conditions, the emissions may decrease to 46 g CO_2-equiv./kWh and 39 g CO_2-equiv./kWh for the mc-Si and pc-Si technologies, respectively.

3.3.5 Heat Pumps

An attractive alternative to fossil fuel fired boilers is the use of heat pumps. Heat pumps can gain heat either from the ambient air, from the ground, or from groundwater. The seasonal performance factor (SPF) of an electric heat pump is the ratio between the annual heat energy output and the annual electric energy input. The SPF depends on the efficiency of the heat pump (coefficient of performance, COP), but also on the local climatic conditions and on the integration of the heat pump into the building (e.g., low- or high-temperature hydronic heat distribution system). Field measurements showed that the average SPF in Switzerland in the year 1998 was about 3.2. Air/water heat pumps using the ambient air had an average SPF of 2.8, brine/water heat pumps using ground heat had an average SPF of 3.9.

The total GHG balance of an electric heat pump depends strongly on the SPF of the heat pump and on the electricity mix used for operation. A heat pump is almost CO_2-free if an almost CO_2-free source of electricity (renewables or nuclear) can be used. But heat pumps can also improve the GHG balance of heating when electricity from fossil fuels is used. A combination of a heat pump with a seasonal performance factor of 3.5 and a cogeneration plant with an electric efficiency of 35%, a total efficiency of about 90%, and electric transmission losses of 2.5% yields about 174% heat relative to the fossil fuel input. The most modern natural gas combined-cycle power plants in Europe can reach a net electric efficiency of 57 to 58%. A combination of a 57.5% CC power plant with an SPF 3.5 heat pump yields a heat efficiency of about 200%.

On the negative side, there are the emissions of refrigerants from heat pumps (besides the indirect emissions from material production and maintenance valid also for other systems). The HFC refrigerant emissions from heat pumps over the lifetime are roughly 2 to 5 mg/MJ. For a heat pump with refrigerant R134a (HFC-134a), which has a 100-years global warming potential of 1300, this corresponds to GHG emissions of about 3–6 g/MJ CO_2-equiv. 100-year. Compared to the CO_2 emissions of a natural gas boiler in the order of 60 g/MJ, heat pumps still can provide a GHG reduction potential if the electricity supply is favorable from the GHG point of view.

Assuming a market penetration of 30% in the building sector using presently available technology, the IEA Heat Pump Centre estimated the CO_2 reduction potential of heat pumps being about 6% of the total global CO_2 emissions from fossil fuels in 1997. This does not include a full LCA analysis and insofar has to be viewed as a rough estimate.

4. COMPARISON OF GREENHOUSE GAS EMISSIONS

The energy sector plays a dominant role with regard to worldwide emissions of GHGs, in particular CO_2. About 77% of yearly CO_2 emissions worldwide originate from fossil fuels combustion. The contribution of the energy sector to the emission of other most important GHGs (i.e., CH_4 and N_2O) is much smaller in relative terms but still important. Based on the total GWP of all GHGs emitted yearly worldwide, the overall share of the energy sector in emissions relevant for global warming is of the order of 50%.

4.1 Current Technologies

The results presented in Section 3 for the individual electricity systems are summarized here. This focuses on the results of the Swiss LCA study for Switzerland and Europe, which cover a broad spectrum of options in a consistent manner. Some conclusions are provided. For a number of detailed comparisons of results for specific chains with other studies and for special issues, refer to Section 3.

Figure 6 shows the overall results with contribution of the various GHGs. In Figure 7 the same overall results are provided but with differentiation between contributions from power plant and rest of the chain. Figures 6 and 7 are complemented by the ranges of GHG emissions from European country-specific energy systems provided in Table V.

1. *Electricity systems ranking based on GHG emissions.* GHG emissions per unit of generated electricity are typically highest for industrial gas, followed by lignite, hard coal, oil, and natural gas. Hydro exhibits very low GHG emissions in boreal and temperate regions, in most cases two orders of magnitude lower than coal. However, hydroelectric developments in tropical regions may emit during

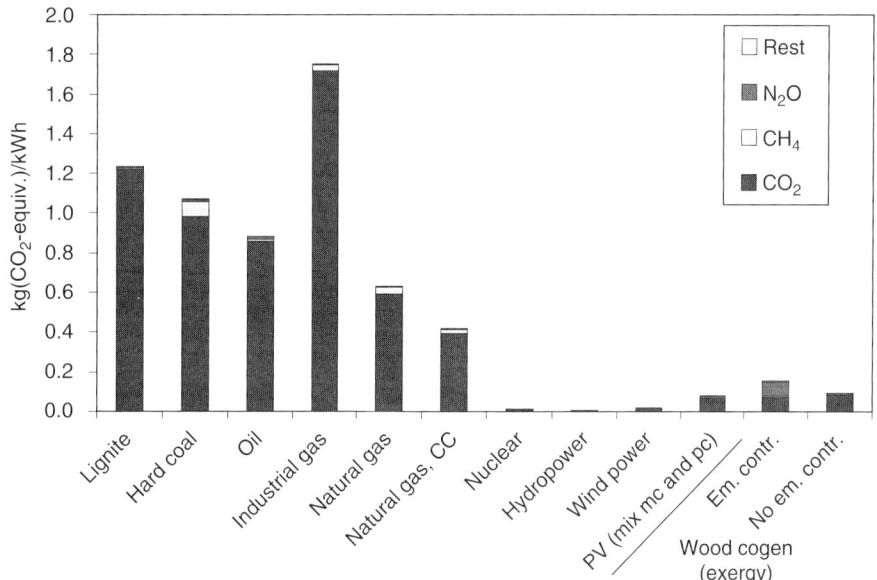

FIGURE 6 Overview of full chain greenhouse gas emissions for different electricity generation systems. The figure shows the contributions of the different greenhouse gases in kg CO_2-equivalents (100 years) per kWh_e. Lignite, hard coal, oil, industrial gas, natural gas, and nuclear refer to energy systems associated with average UCTE power plants around year 2000. Natural gas CC: Combined-cycle power plant of currently best technology, gas supplied by the European high-pressure grid. The renewables (hydropower, wind power, photovoltaic, wood) refer to location Switzerland. From Dones *et al.* (2003).

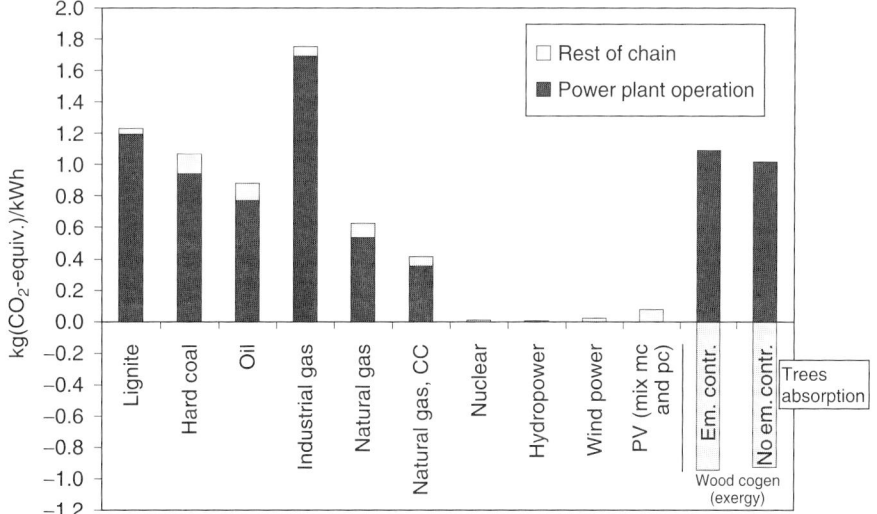

FIGURE 7 Greenhouse gas emissions from power plant operation and from the rest of chain for different electricity generation systems. For explanations see Fig. 6 and text.

TABLE V

Ranges of GHG Emissions per kWh Electricity from European Country-Specific Energy Sources, using GWP 100 years

	Minimum (kg CO_2-equiv./kWh)	Maximum (kg CO_2-equiv./kWh)
Lignite	1.060	1.690
Hard coal	0.949	1.280
Oil	0.519	1.190
Industrial gas	0.865	2.410
Natural gas	0.485	0.991
Nuclear power	0.008	0.011
Hydropower[a]	0.003	0.027
Wind power[b]	0.014	0.021
PV (mix mc and pc)[c]	0.079	—
Wood cogeneration[d]	0.092	0.156

[a] Mix of reservoirs and run-of-river plants.
[b] Calculated for average conditions of Switzerland (maximum) and Western Europe (minimum).
[c] Calculated for average Swiss conditions.
[d] 6400 kW_{th}/400 kW_e plant; allocation exergy; no emission control (minimum); emission control SNRC (maximum).
Source. Dones et al. (2003).

operation between 5 and 20 times more GHGs, which at the higher range is comparable to emissions from fossil sources. Taking into account full energy chain contributions including also the grey ones, GHG emissions from nuclear and wind energy (under favorable wind conditions) are in the same low range as typical for hydro. The corresponding net emissions for biomass are in the middle range (i.e., one order of magnitude lower than coal and one order of magnitude higher than nuclear and wind). The same applies to older solar photovoltaic systems in countries with climate conditions such as in central Europe, while current photovoltaic technologies exhibit even smaller differences compared to nuclear and wind; still better performance can be achieved in southern countries.

2. *Differences between industrialized and developing countries.* Based on quite limited comparative material for fossil systems, a pattern can be distinguished, namely the GHG emissions are significantly higher in the developing countries. This is due to use of technologies characterized by lower efficiency as well as distinct features of fuel cycles (relatively large methane emissions from mines and from gas pipelines; coalfield fires in China).

3. *Dominant GHGs.* CO_2 dominates GHG emissions from energy systems. In the case of hard coal and natural gas, the methane contribution in terms of CO_2-equivalents may reach the level of 10% due to high leakage rates from coal mines and gas pipelines. N_2O may be of high relative importance for wood cogeneration, if pollutant emission controls employing catalysts generating N_2O are applied.

4. *Role of GHG emissions other than from power plant.* For fossil systems power plant GHG emissions are clearly dominant. It can be noted that in the case of major pollutants, the situation may be quite different—that is, emissions from power plants are higher than the other contributions, but the latter may be very substantial. For European conditions the upstream part of the chain typically contributes about 10% of the total GHG emissions for the hard coal chain, slightly more than 10% for oil chain, and can approach 20% for conventional natural gas systems. For nuclear, wind and solar photovoltaic, the GHG emissions other than from power plant are totally dominant in relative terms. The same applies to typical hydro plants, with the exception of dams operating under special conditions resulting in very large direct emissions from reservoirs.

For comparisons of heating and cogeneration systems, refer to Figs. 3, 4, and 5. Gas boilers are as expected better performers than oil boilers when modern systems are compared. Biomass exhibits best GHG performance along with heat pumps, provided that electricity input to the latter comes from zero or low carbon sources. Cogeneration based on biomass is clearly favorable in terms of GHG emissions compared to the corresponding gas and oil systems. Evaluation of cogeneration is subject to variability, depending on which allocation scheme is employed.

4.2 Electricity Mixes

Calculating and presenting GHG emission data for single technologies is important for comparisons and decision making. However, it is also essential to generate the total burdens associated with the electricity mixes. These can then be used in LCA studies of products.

Two approaches are used to generate the relevant mixes (i.e., marginal and average mix approach). The marginal approach suggests to use the last available technology or mix of technologies (e.g., the cheapest available for base load, or the technology[ies], which will most probably cover an increase in demand, or the specific energy technology which will be chosen for a process/industry), whereas with the average approach only average mixes supplying the grid are considered.

The Swiss LCA study addressed national production mixes for all European countries and the three major associations of electric transmission operators UCTE, CENTREL, and NORDEL in year 2000. GHG results for these electricity mixes are summarized in Table VI.

Depending on the shares of technologies, the total GHG emissions can vary by several orders of magnitude from the hydro-based Norwegian mix to the coal-based Polish mix. Additionally, the study addressed the issue of the supply mix. This was modeled by merging all imports to the domestic production mix and thus recalculating the share of technologies to be used both for the domestic supply mix as well as for the export mix. This model may somewhat overestimate the effective import shares and may not describe seasonal exchanges properly. The results for Switzerland, a country with intense exchanges of electricity with neighboring countries due to its central position in Europe, are illustrated in Table VII.

The GHG-free production mix based on hydro and nuclear (58 and 38%, respectively, in year 2000) changes to a supply mix with quite significant fossil components due to imports from Germany. Another issue concerns the influence of the grids (losses and material use) to the GHG associated to the electricity services. The analysis performed for the Swiss conditions leads to increases of GHG by 2.5, 5, and 18% with respect to the supply mix at the busbar of plants in the case of high, medium, and low voltage levels, respectively. With equal network losses, the increases may be lower in percent in case of country supply mixes with higher fossil share.

A Japanese study covered the electricity production mixes for 10 Japanese electric companies. Besides direct emissions from fossil plants, the authors included the emissions from the upstream chains, but excluded construction and decommissioning of the various facilities and transport systems. Only CO_2 and CH_4 were considered. Using GWP from IPCC 2001, the results are in the range of 296 to 795 g CO_2-equiv./kWh for nine companies, whose average is calculated at 468 and 1079 g CO_2-equiv./kWh for Okinawa. Remarkably, the UCTE (2000) and the Japanese production mixes have practically the same GHG intensity.

4.3 Advanced Technologies and Sequestration

4.3.1 Advanced Technologies

The LCA methodology has been originally developed and primarily applied for operating systems. Consequently, the input is normally based on the actual experience. Furthermore, the standard approach is static and its application to future systems requires extensions, extrapolations, and a number of additional assumptions on technological advancements and structural changes.

Few studies have implemented extensions of LCA applications to future systems, based on literature as well as direct information from the industry and on expert judgment. The result-driving parameters in the LCA study are emissions, efficiencies, material intensities (for construction and operation), and transportation requirements. The relative importance of these parameters varies significantly between energy chains.

The most important expected changes towards improvements of ecological performance identified in

TABLE VI

LCA-Based GHG Emissions for European Electric Associations and Countries

Electricity mix	kg CO_2-equiv./kWh
Production mix (at busbar)	
UCTE	0.470
CENTREL	0.932
NORDEL	0.132
Country-specific production mix[a]	0.007–1.13
Country-specific supply mix[a,b]	0.010–1.11

[a] Minimum: Norway; maximum: Poland.
[b] The used model adds all electricity imports to the domestic production, recalculates the shares of technologies on this basis, and applies this portfolio to domestic supply and exports.
Source. Dones et al. (2003).

TABLE VII

LCA-Based GHG Emissions for Swiss Electricity Mixes

Electricity mix	kg CO_2-equiv./kWh
Production mix (at busbar)	0.018
Supply mix (with import/export)[a]	0.120
High voltage	0.123
Medium voltage	0.126
Low voltage	0.142

[a] See note b in Table VI.
Source. Dones et al. (2003).

the 1996 Swiss study on evolutionary future electricity generating technologies were as follows:

- *Gas systems.* Reduction of gas leakage, improvements of power plant burner performance characteristics and of overall power plant efficiency.
- *Hard coal systems.* Increased methane recovery in underground mining, improvements in power plant abatement technology, and of overall power plant efficiency.
- *Nuclear systems.* Reductions of electricity consumption in enrichment by replacement of diffusion by centrifuges or laser technologies, power plant improvements (particularly extended life time and increased burn-up).
- *Hydro systems.* Overall power plant efficiency improvements (turbine).
- *PV systems.* Improvements in the manufacturing of mono-crystalline-silicon (m-Si) and amorphous-silicon (a-Si) solar cells (yield, electricity consumption), and in cell efficiencies.

Some of the improvements mentioned have already been implemented in the best technologies available around year 2000, which are not yet widely disseminated.

For electricity inputs needed for the LCA modules external to Switzerland, a European mix for year 2010 was defined, based on a forecast by the International Energy Agency. The mix reflects the expansion of gas, reduction of oil shares, and a relatively small but significantly increased contribution of renewables other than hydro. Coal, hydro, and nuclear remain at about the same level percent. Since the new systems generally show better performance and lower emissions than the old ones, assumptions needed to be made with respect to the market penetration of the new ones. This was done individually for each energy source, taking into account their specificity and the expected developments reflected by the assumed overall contribution of each energy carrier to the mix.

Future electricity generation systems based on fossil chains have the potential to reduce GHG emissions by 20 to 40% in comparison with the corresponding present average European fossil systems. However, they remain the largest GHG producers: two orders of magnitude higher than hydro in Alpine regions, wind and nuclear, and one to two orders higher than photovoltaic.

4.3.2 Sequestration

Since CO_2 emissions into the atmosphere should be reduced, storage of CO_2 in geological reservoirs or in the ocean is a possible option under discussion. The sequestration procedure includes capture of CO_2 from the fuel combustion device, dehydration, compression, transport, and injection into the final reservoir.

The fate of CO_2 in ocean is not yet fully understood. Storage in geological reservoirs seems to be a more promising method, but also in this case there are open issues to be investigated like the permanency of the storage. Possible environmental risks of storage in saline aquifers include leakages, catastrophic releases (although considered unlikely), changes in geochemistry, water quality, and ecosystem health caused by pH shifts due to carbonic acid, or contamination of potable water due to displacement of brine.

Table VIII shows estimates of the storage capacities of possible geological reservoirs. Depleted gas and oil fields could store only about 40% of the CO_2 emissions expected between year 2000 and year 2050 (though this does not include fields that are not yet operating). The actual storage capacity of saline aquifers under reasonable economic conditions is highly uncertain.

The main techniques for CO_2 capture are postcombustion capture, precombustion capture, and oxy-fuel combustion. Postcombustion capture of CO_2 from the flue gas usually involves chemical solvents like monothanolamine (MEA). In precombustion capture, the fuel is converted into CO_2 and hydrogen; the hydrogen is then combusted. Oxy-fuel combustion means an increase of CO_2 concentration in the flue gas by using concentrated oxygen (which is, e.g., produced by cryogenic air separation) for combustion.

CO_2 capture reduces the net efficiency of a power plant and thus increases the fuel consumption per kWh delivered to grid. Using current technology, the

TABLE VIII
Global CO_2 Storage Capacities of Geological Reservoirs

CO_2 emissions and storage capacities	Gt CO_2
Emissions 2000–2050 (IPCC business as usual scenario)	2000
Theoretical storage capacities of geological reservoirs	
Depleted gas fields	690
Depleted oil fields	120
Deep saline aquifers	400–10,000
Unminable coal seams	40

Source. Gale (2002). (Storage capacities based on injection costs of up to 20 US $/ton CO_2; gas and oil fields exclude fields that are not yet producing.)

changes would be significant. For a natural gas combined cycle power plant, the fuel consumption is expected to increase by 16 to 28% due to CO_2 capture. The respective figures for coal power plants have been estimated to be 22 to 38% for pulverized coal and 16 to 21% for integrated gasification combined cycle (IGCC). Capital costs would increase by 300 to 800 US$/kW, or about 30 to 50% for IGCC, about 70 to 80% for pulverized coal, and about 80 to 100% for natural gas CC.

SEE ALSO THE FOLLOWING ARTICLES

Acid Deposition and Energy Use • *Biomass: Impact on Carbon Cycle and Greenhouse Gas Emissions* • *Carbon Capture and Storage from Fossil Fuel Use* • *Carbon Sequestration, Terrestrial* • *Climate Change and Energy, Overview* • *Diet, Energy, and Greenhouse Gas Emissions* • *Ecosystem Health: Energy Indicators* • *Greenhouse Gas Abatement: Controversies in Cost Assessment* • *Greenhouse Gas Emissions, Alternative Scenarios of*

Further Reading

Bartlett, K. B., *et al.* (1990). Methane flux from the Amazon River Floodplain. *J. Geophys. Res.* Sept. 20, 16773–16788.

Dones, R., Zhou, X., and Tian, C. (2003). Life cycle assessment. *In* "Integrated Assessment of Sustainable Energy Systems in China—The China Energy Technology Program" (Eliasson, B., and Lee, Y. Y., Eds.), pp. 320–344. Kluwer.

Dones, R., Bauer, C., Bolliger, R., Burger, B., Faist Emmenegger, M., Frischknecht, R., Heck, T., Jungbluth, N., and Röder A. (2003). Life cycle inventories of energy systems: Results for current systems in Switzerland and other UCTE countries. Final report ecoinvent 2000 No. 5. Paul Scherrer Institut, Villigen, and Swiss Centre for Life Cycle Inventories, Duebendorf, Switzerland. Online version under www.ecoinvent.ch.

Dones, R., Gantner, U., Hirschberg, S., Doka, G., and Knoepfel, I. (1996). Environmental inventories for future electricity supply systems for Switzerland. PSI Report No. 96-07, Villigen, Switzerland.

Duchemin, E., Lucotte, M., St-Louis, V., and Canuel, R. (2002). Hydroelectric reservoirs as an anthropogenic source of greenhouse gases. *World Resource Rev.* 14(3), 334–353.

Fearnside, P.M. (2002). Greenhouse gas emissions from a hydroelectric reservoir (Brazil's Tucuruí dam) and the energy policy implications. *Water, Air, and Soil Pollution* 133, 69–96.

Gale, J. (2002). Proceedings of "Workshop on carbon dioxide capture and storage," Nov. 18–21. IPCC. Regina, Canada.

Hondo, H. (2003). Life cycle GHG emission analysis of power generation systems: Japanese case. Submitted to *Energy—The International Journal*.

International Energy Agency (IEA) (1999). CO_2 emissions from fuel combustion 1971–1997. OECD/IEA, Paris.

Intergovernmental Panel on Climate Change (IPCC) (2001). "Climate Change 2001: The Scientific Basis." Contribution of Working Group I to the Third Assessment Report of the Intergovernmental Panel on Climate Change (Houghton, J. T., *et al.*, Eds.). Cambridge Univ. Press, Cambridge, UK.

Jungmeier *et al.* (1999). Jungmeier, G., Canella, L., Spitzer, J., and Stiglbrunner, R. Treibhausgasbilanz der Bioenergie. Vergleich der Treibhausgasemissionen aus Bioenergie-Systemen und fossilen Energiesystemen. Joanneum Research, Institut für Energieforschung, Graz. Bericht Nr.: IEF-B-06/99.

Kato, Y., Inaba, A., and Mayasuki, S. (2000). The CO_2 emissions for fossil fuels from producing countries to Japan. Proceedings of APEC/AIST/NEDO symposium "LCA for APEC Member economies," H12.11.1-2, pp. 87–92. Tsukuba, Japan.

Matsuno, Y., and Betz, M. (2000). Development of life cycle inventories for electricity grid mixes in Japan. *Int. J. LCA* 5(5), 295–305.

Pinguelli Rosa, L., dos Santos, A., Matvienko, B., and Sikar, E. (2002). Carbon dioxide and methane emissions from Brazilian hydroelectric reservoirs. First Brazilian Inventory of Anthropogenic Greenhouse Gas Emissions—Background Reports, Alberto Luiz Coimbra Institute for Graduate Studies and Research in Engineering COPPE, Brazilian Ministry of Science and Technology, Brasilia.

Spath, P., Mann, M., and Kerr, D. (1999). Life-cycle assessment of coal-fired power production, National Renewable Energy Laboratory, NREL/TP-570-25119, Golden, CO.

Van de Vate, J. F., and Gagnon, L. (1997). Greenhouse gas emissions from hydropower—The state of research in 1996. *Energy Policy* 25(1), 7–13.

Van de Vate, J. F. (1997). Comparison of energy sources in terms of their full energy chain emission factors of greenhouse gases. *Energy Policy* 25(1), 1–6.

Ground-Source Heat Pumps

ARIF HEPBASLI
Ege University
Izmir, Turkey

1. Ground-Source Heat Pumps
2. Worldwide Ground-Source Heat Pump Installations
3. Main Components of a Ground-Source Heat Pump System
4. Design Types of Ground Heat Exchangers
5. Soil Characteristics
6. Groundwater Heat Pumps
7. Surface Water Heat Pumps
8. Hybrid Ground-Source Heat Pump Systems
9. Performance Evaluation of Ground-Source Heat Pumps

Glossary

Air Conditioning and Refrigeration Institute (ARI) The nonprofit organization that sets the standards for testing and comparisons for the heating, air-conditioning, and refrigeration industries.

aquifer A water-bearing geological formation; a large body of underground water.

coefficient of performance (COP) The ratio of heat energy delivered or extracted to work to operate the equipment.

compressor The device that, through compression, increases the pressure and temperature of entering refrigerant, and discharges a hot dense gas.

condenser A heat-exchange coil within a mechanical refrigeration system; used to reject heat from the system.

energy efficiency ratio (EER) The ratio of the rated cooling capacity (in British thermal units per hour) divided by the amount of electrical power used (in watts) for any given set of conditions.

entering water temperature (EWT) The temperature of the heat transfer fluid leaving the ground heat exchanger.

evaporator A heat-exchange coil within a mechanical refrigeration system; used to absorb heat into the system.

ground heat exchanger (GHE) A device that exchanges ground heat; thermal energy is extracted from (or rejected into, depending on mode of operation) the ground.

heat exchanger A device used to transfer heat from a fluid (liquid or gas) to another fluid; the two fluids are physically separated (usually by metal tubing).

heating seasonal performance factor (HSPF) The ratio of the total amount of heat energy supplied to the space for the heating season with respect to the total electrical energy required by the heating system for the same season.

heat pump A device that uses a refrigerant cycle to take low-quality heat, concentrate it to a higher quality heat, and move that heat to another location in the cycle.

heat sink The area or media where heat is deposited (inside a building, etc.).

heat source The area or media from which heat is removed (water, air, etc.).

heat transfer fluid A fluid (water, or a mixture of water and antifreeze) that is circulated through the ground heat exchanger via a circulating pump.

refrigerant A fluid (liquid or gas) that picks up heat by evaporating at a low temperature and pressure.

reversing valve A valve that changes the direction of refrigerant flow in a heat pump.

seasonal energy efficiency ratio (SEER) The average energy efficiency ratio over the entire cooling season.

Ground-source heat pumps (GSHPs), also known as geothermal heat pumps (GHPs), use electricity to move heat, not to generate it, and are recognized to be outstanding heating, cooling, and water-heating systems. The primary benefit of GSHPs compared to other systems is the increase in operating efficiency, which translates to reduced heating and cooling costs, but there are additional advantages. One notable benefit is that GSHPs are classified as renewable energy technology, although they are electrically driven. According to the U.S. Environmental Protection Agency (EPA), well-designed and properly installed high-efficiency GSHP systems produce less environmental harm as compared to any other alternative space-conditioning technology currently available. On a full fuel cycle basis, these systems are the most efficient technology available, with the lowest CO_2 emissions for minimum greenhouse warming impact.

1. GROUND-SOURCE HEAT PUMPS

Ground-source heat pumps are not a new idea. Patents on the technology date back to 1912, in Switzerland. One of the oldest GSHP systems in the United States, in the United Illuminating headquarters building in New Haven, Connecticut, has been operating since the 1930s.

There are two main types of GSHP systems: ground-coupled (vertical or horizontal) closed loop and water-source (groundwater) open loop, as shown in Figs. 1 and 2, respectively. Ground-coupled heat pumps (GCHPs) are known by a variety of names. These include ground-source heat pumps, earth-coupled heat pumps (ECHPs), earth energy heat pumping systems, earth energy systems, ground-source systems, geothermal heat pumps, closed-loop heat pumps, solar energy heat pumps, geoexchange systems, geosource heat pumps, and a few other variations. In marketing ground-source heat pumps, to cope with the diverse terminology, sales personnel may wish to connect GCHPs to renewable energy sources (solar, geothermal), to connect them to environmental awareness (earth energy), and to dissociate them from air heat pumps (ground-source systems). However, two terms are commonly used to describe the technology in general: geothermal heat pumps and ground-source heat pumps. The former term is typically used by individuals in marketing and government, and the latter is used by engineering and technical types. In addition, ground-coupled, groundwater, and surface water heat pumps are referred to as subsets of ground-source heat pumps. The systems will be referred to as ground-source heat pumps throughout this text.

The closed-loop systems use a buried earth coil as the ground heat exchanger (GHE) through which a heat transfer fluid, typically an antifreeze solution, is circulated. The GHE, installed either vertically in borings or horizontally in trenches, exchanges heat with the ground (Fig. 1). In open-loop systems, the groundwater is pumped into the heat pump unit, where heat is extracted from (or rejected into, depending on mode of operation) the water; the water is then disposed of in an appropriate manner. If possible, releasing the water into a stream, river, lake, pond, ditch, or drainage tile, known as open discharge, is the easiest and least expensive disposal method. A second means of water discharge is using a return well that returns the water to the ground aquifer (Fig. 2).

2. WORLDWIDE GROUND-SOURCE HEAT PUMP INSTALLATIONS

Among the worldwide applications of geothermal energy for direct utilization, GSHPs have had the largest rate of growth since 1995, almost 59%, representing 9.7% annually. Most of this growth has occurred in the United States and Europe, though interest is developing in other countries, such as Japan and Turkey. The installed capacity is 6875 thermal-derived megawatts (MW_t), and the annual energy use is 23,287 terajoules (TJ)/year, or 6453 gigawatt-hours (GWh)/year, at the beginning of 2000 in 27 countries, as shown in Table I. It is estimated that the actual number of installed units is around 500,000, and the equivalent number of 12-kW units installed is slightly over 570,000. It is estimated that 450,000 units are presently installed in the United States; thus, this rate of installation would add an additional 1.1 million units for a total of about 1.5 million units by 2010.

The world's largest GSHP installation is the Galt House East Hotel and Waterfront Office Buildings in Louisville, Kentucky. A 4700-ton GSHP system is used to meet the heating and cooling needs of the complex. The 70,000 m² (750,000 ft²) Galt House East Hotel was completed in 1984 and uses a

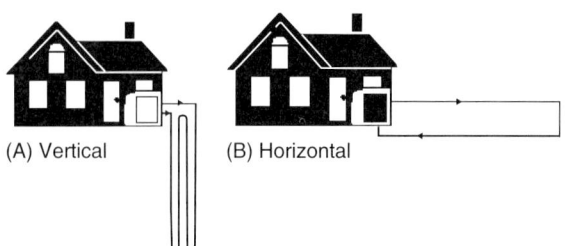

FIGURE 1 Ground-coupled (closed-loop) systems. Reproduced by permission from the Geo-Heat Center, Oregon Institute of Technology.

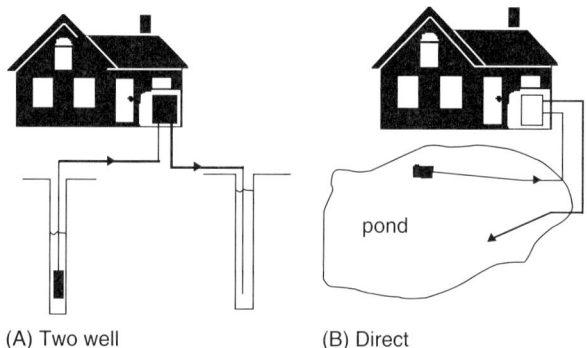

FIGURE 2 Groundwater (open-loop) systems. Reproduced by permission from the Geo-Heat Center, Oregon Institute of Technology.

TABLE I

Worldwide Geothermal Heat Pump Installations in 2000

Country	MW_t[a]	TJ/year	GWh/year	Actual no.	Equiv. no. (12 kW)
Australia	24	57.6	16.0	2000	2000
Austria	228	1094	303.9	19,000	19,000
Bulgaria	13.3	162	45.0	16	1108
Canada	360	891	247.5	30,000	30,000
Czech Republic	8.0	38.2	10.6	390	663
Denmark	3	20.8	5.8	250	250
Finland	80.5	484	134.5	10,000	6708
France	48	255	70.8	120	4000
Germany	344	1149	319.2	18,000	28,667
Greece	0.4	3.1	0.9	3	333
Hungary	3.8	20.2	5.6	317	317
Iceland	4	20	5.6	33	33
Italy	1.2	6.4	1.8	100	100
Japan	3.9	64	17.8	323	323
Lithuania	21	598.8	166.3	13	1750
Netherlands	10.8	57.4	15.9	900	900
Norway	6	31.9	8.9	500	500
Russia	1.2	11.5	3.2	100	100
Poland	26.2	108.3	30.1	4000	2183
Serbia	6	40	11.1	500	500
Slovak Republic	1.4	12.1	3.4	8	117
Slovenia	2.6	46.8	13.0	63	217
Sweden	377	4128	1146.8	55,000	31,417
Switzerland	500	1980	550.0	21,000	41,667
Turkey	0.5	4.0	1.1	23	43
UK	0.6	2.7	0.8	49	53
USA	4800	12,000	3333.6	350,000	400,000
Total	6875.4	23,286.9	6453.1	512,678	572,949

Source. Data from Lund (2000).

[a] Thermally derived megawatts.

1700-ton GSHP system. The installation cost was $1500 per ton. In comparison, a conventional system would have cost between $2000 and $3000 per ton. The hotel complex energy use is approximately 53% of a similar non-GSHP system in an adjust unit, and the system saves about $25,000 monthly in reduced energy costs. The Waterfront Office Buildings, built in 1994, add about 89,000 m² (960,000 ft²) of office space and almost 3000 tons of GSHP capacity to the project, making this the world's largest commercial GSHP project. It is reported that Galt House East has been operating for 15 years with no system problems and that the GSHP system has performed even better than expected.

One of the largest commercial GSHP systems running today is at Stockton College in Pomona, New Jersey, where 63 GSHPs with a total capacity of 1655 tons (5825 kW) are connected to a GHE, which consists of 400 wells, each 129 m (425 ft) deep.

3. MAIN COMPONENTS OF A GROUND-SOURCE HEAT PUMP SYSTEM

GSHPs use the relatively constant temperatures, typically 7° to 21°C (45° to 70°F), of the soil and water beneath the frost line to provide efficient heating and cooling all year long. The efficiencies achieved by these systems are impressive and allow commercial users to save up to 50% over conventional heating and cooling systems, plus they reduce maintenance costs. GSHPs use the stable temperature of the ground as a heat source to warm buildings in winter and as a heat sink to cool them in summer.

A GSHP system for space heating and cooling of houses consists of three distinct subsystems (see Fig. 3):

1. Earth connection or ground heat exchanger. Many types of GHE designs are possible, but for GSHP systems, only two basic design concepts have been implemented so far. They are direct-expansion GHEs, which consist of a buried copper piping network through which refrigerant is circulated, and secondary fluid circulating GHEs, through which a water or a water–antifreeze solution is circulated. The GHE is buried in the ground near the building to be conditioned. Heat transfer fluid is circulated through the GHE via a circulating pump (stage 1, in Fig. 3).

2. Heat pump unit. This device is composed of five primary components (Fig. 3): a compressor (stage 2), a refrigerant–air (or water) heat exchanger (stage 3), an expansion valve (stage 4), and a heat exchanger between the refrigerant and GHE fluid (stage 5); there is also a reversing valve (stage 6). In the heating mode, heat is extracted from the ground by the GHE. This heat is transferred to the refrigerant in the refrigerant–fluid heat exchanger, is upgraded in the heat pump cycle, and is supplied to the house by the refrigerant–air heat exchanger, which acts as the condenser. In the cooling mode, the process is reversed to extract heat from the house and reject it to the ground. The most widely used type of unit is a water-to-air heat pump, which circulates a water or water–antifreeze solution through a liquid-to-refrigerant coil and a GHE. Air to be heated or cooled is circulated through a conventional finned-tube air-to-refrigerant coil and duct system. Water-to-water heat pumps are also used by replacing the forced air system with a hydronic loop. Systems using water-to-air and water-to-water heat pumps are often referred to as GCHPs with secondary solution loops, to distinguish them from direct-expansion GCHPs.

3. Heat distribution. There are two basic ways to distribute the energy extracted from the earth and water, namely, forced-air systems and hydronic systems. The simplest way to cool a building is to circulate air by using fans to blow the air through coils. This kind of system is not suitable for heating, however, unless the earth or water source is usually warm. In addition to space conditioning, GSHPs can be used to provide domestic hot water when the system is operating.

4. DESIGN TYPES OF GROUND HEAT EXCHANGERS

The type of GHE used will affect heat pump system performance (therefore, the heat pump energy consumption), auxiliary pumping energy requirements, and installation costs. Choice of the most appropriate type of GHE for a site is usually a function of specific geography, available land area, and life cycle cost economics. Primarily, only two types of GHE designs are used, vertical and horizontal.

4.1 Vertical Ground Heat Exchangers

The vertical ground heat exchanger (VGHE), illustrated in Fig. 1A, is generally considered when land surface is limited. The VGHE consists mainly of two small-diameter high-density polyethylene (PE) tubes that have been placed in a vertical borehole that is subsequently filled with a solid medium. The tubes are thermally fused at the bottom of the bore to a close-return U-bend. Vertical tubes range from 0.75 to 1.5 inches in nominal diameter, and bore depths vary from 15.2 to 183 m (50 to 600 ft), depending on local drilling conditions and available equipment. Multiple wells are typically required, with well spacing not less than 4.6 m (15 ft) in northern climates and not less than 6.1 m (20 ft) in southern climates to achieve the total heat transfer requirements. A minimum base separation distance of 6.1 m

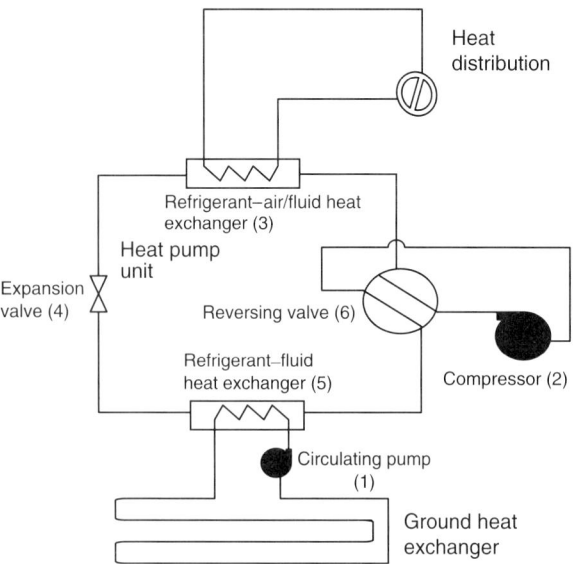

FIGURE 3 A schematic of a ground-source heat pump system, showing stages 1 through 6 (see text for discussion).

(20 ft) is recommended when loops are placed in a grid pattern. This distance may be reduced when the bores are placed in a single row, when the annual heating load is much greater than the annual cooling load, or when vertical water movement mitigates the effect of heat buildup in the loop field.

4.2 Horizontal Ground Heat Exchangers

Horizontal ground heat exchangers, illustrated in Fig. 1B, are most often used for small installations where space is not a premium. Pipes are buried in trenches, typically at a depth of 1.2–3.0 m (4–10 ft). Depending on the specific design, as many as six pipes may be installed in each trench, with about a foot separating the pipes, and trenches typically being 3.7–4.6 m (12–15 ft) apart. The "slinky," a spiral loop comprising overlapping coils of pipe, can be used to maximize the number of feet of pipe buried per foot of trench, but slinkys require more feet of pipe per ton of thermal energy needed, compared to conventional horizontal systems.

Another variation of the spiral-loop system involves placing the loops upright in narrow vertical trenches. The spiral-loop configuration generally requires more piping, typically 43.3–86.6 m/kW per system cooling ton (500–1000 ft), but less total trenching as compared to the multiple horizontal-loop systems just described. For the horizontal spiral-loop layout, trenches are generally 0.9–1.8 m (3–6 ft) wide; multiple trenches are typically spaced about 3.7 m (12 ft) apart. For the vertical spiral-loop layout, trenches are generally 15.2 cm (6 inches) wide; the pipe loops stand vertically in the narrow trenches. In cases in which trenching is a large component of the overall installation costs, spiral-loop systems are a means of reducing the installation cost. As noted with horizontal systems, slinky systems are also generally associated with lower tonnage systems when land area requirements are not a limiting factor.

The key issue that makes GSHPs unique is the design of the GHE system. Most operational problems with GSHPs stem from the performance of the GHE system. There are several ways of sizing a GHE:

1. Using rules-of-thumb. In the residential sector, many systems have been designed using rules-of-thumb and local experience, but for commercial-scale systems, such practices are ill advised. For this purpose, heat extraction/rejection rates obtained from experience are used. For all but the most northern climates, commercial-scale buildings will have significantly more heat rejection than extraction. This imbalance in heat rejection/extraction can cause heat buildup in the ground to the point where heat pump performance is adversely affected and hence system efficiency and possibly occupant comfort suffer.

2. Short method 1. This method uses the building-block load, estimates of average equivalent full-load operating hours, average values of the equipment energy efficiency ratio (EER) and coefficient of performance (COP), and some values to estimate the temperature penalty due to long-term heat exchange in the GHE. Besides this, this computation also requires obtaining ground thermal properties and thermal resistances.

3. Short method 2. Besides the energy parameters for the heat pump (heating and cooling loads, and run time), the highest supply water temperature acceptable for cooling mode and the lowest temperature acceptable for heating mode must be known. These values, along with local ground temperature, establish the design temperatures for sizing a GHE. Using three simple tables and heat transfer equations, the required calculation may be easily done.

4. Using design software. The abbreviated computation described in short methods 1 and 2 is still very time consuming. More importantly, it is difficult to evaluate the impact of design alternatives, because most of the calculations have to be repeated. The procedure lends itself well to computer-aided design.

Proper design for commercial-scale systems almost always benefits from the use of design software. Today, software tools are available to support the design of any GHE system that meets the needs of designers and installers. Software for commercial-scale GSHP system design should consider the interaction of adjacent loops and predict the potential for long-term heat buildup in the soil. These tools are available from several sources, including the International Ground-Source Heat Pump Association (IGSHPA). In addition, several manufacturers have designed their own proprietary tools, and these are more closely tuned to particular system requirements.

5. SOIL CHARACTERISTICS

Soil characteristics are an important concern for any GHE design. The operation of the heat exchanger induces a simultaneous heat and moisture flow in the surrounding soil. The transfer of heat between the GHE and adjoining soil is primarily by heat conduction and to a certain degree by moisture migration. Therefore, transfer depends strongly on the soil type, temperature, and moisture gradients. With horizontal GHEs, the soil type can be more easily determined because the excavated soil can be inspected and tested.

The soil thermal diffusivity (α_s) is a defined property and is the ratio of the thermal conductivity (k_s) and the heat capacity ($\rho_s C_s$). Therefore, the three soil properties, k_s, ρ_s, and C_s, must be known or estimated to predict the thermal behavior of GHEs. Obtaining accurate values for thermal properties of the soil requires a detailed site survey. In order to estimate the thermal properties of granular soils (sand, clay, silt) it is necessary to determine the sand and clay content, dry density, and moisture content. The following analytical equations are used to calculate the thermal properties of the soil based on soil dry density and moisture content.

The thermal conductivity of the soil, k, in watts/meter degree Kelvin, may be calculated from Eqs. (1) and (2):

$$k_s = 0.14423(0.9 \log \omega - 0.2)10^{0.000624\rho_{sd}} \quad (1)$$

for silt and clay soils

$$k_s = 0.14423(0.7 \log \omega + 0.4)10^{0.000624\rho_{sd}} \quad (2)$$

for sand soils where ω is moisture content in percentage by weight and ρ_{sd} is dry density in kilograms/cubic meter.

Thermal diffusivity is sensitive to moisture content because the specific heat of water (4.1868 kJ/kg K) is far greater compared to the typical values for soils and rock (0.84 to 1.05 kJ/kg K). Both the dry specific heat and the dry density of soil, C_{psd} and ρ_{sd}, must be corrected in order to determine α_s as follows:

$$\alpha_s = \frac{k_s}{C_{psc} \rho_{sc}} \quad (3)$$

with

$$C_{psc} = [\omega C_{pw} + (100 - \omega)C_{psd}]/100 \quad (4)$$

and

$$\rho_{sc} = [\omega \rho_{pw} + (100 - \omega)\rho_{sd}]/100 \quad (5)$$

where C_{psc} is the corrected specific heat of soil, ρ_{sc} is the corrected density of soil, C_{pw} is the specific heat of water (4.1868 kJ/kg K), and ρ_{pw} is the density of water (1000 kg/m^3).

6. GROUNDWATER HEAT PUMPS

Groundwater heat pump (open-loop) systems are the oldest and most well established of the GSHP systems, with the first examples of large commercial applications installed in the late 1940s. Common design variations include direct (groundwater used directly in the heat pump units), indirect (building loop isolated with a plate heat exchanger), and standing column (water produced and returned to the same well), as shown in Fig. 2A. Direct systems are typically limited to the smallest applications. Standing-column systems are employed in hard-rock geology sites where it is not possible to produce sufficient water for a conventional system.

Although seemingly simple in nature, these systems require careful consideration of well design, groundwater flow, heat exchanger selection, and disposal, even for very small commercial applications. The direct groundwater system is very susceptible to problems induced by poor water quality, the most common of which is scaling of the refrigerant-to-water heat exchangers. This design is recommended in only the smallest applications when practicality or economics precludes the use of an isolation heat exchanger and/or when groundwater quality is excellent (the determination of which requires extensive testing). The standing-column system has been installed in many locations in the northeast portion of the United States. Like the direct groundwater system, it, too, is subject to problems induced by poor water quality. In general, water quality in the area where most of the systems have been installed is extremely good, with low pH and low hardness (little scaling potential). Standing-column systems are used in locations underlain by hard-rock geology, where wells do not produce sufficient water for conventional open-loop systems and where water quality is excellent. Well depths are often in the range of 304.8 to 457.2 m (1000 to 1500 ft) and the systems operate at temperatures between those of open- and closed-loop systems.

In most commercial applications, the optimum design dictates a flow of 0.045–0.054 liters/sec per kW, or 2.5–3.0 gallons per minute (gpm)/ton, on the building-loop side of the exchanger and 0.018–0.045 liters/sec per kW (1–2.5 gpm/ton) on the groundwater side. As a result of this, the approach, or minimum, temperature difference between the two flows occurs at the building-loop return (heat pump "leaving" water) and groundwater leaving end of the exchanger. Groundwater below 60°F may be circulated directly through hydronic coils in series or in parallel with heat pumps.

There are two basic options for water disposal from an open-loop system: surface and injection. Both options are subject to regulatory oversight and permitting. Surface disposal, the most common method used in the past, is less expensive, but requires that the receiving body be capable of accepting the water over a long period. Injection is more complex and costly but offers the certainty that the groundwater aquifer will not be adversely affected (aquifer decline) by the operation of the system over the long term, because the water is "recycled." For surface disposal, it may be advisable to place a pressure-sustaining valve on the end of the system to maintain full piping when the pump is not operating. Some designers prefer simply to place a motorized valve at this point in the system and interlock it with the pump (through an end switch).

The optimum design of the groundwater heat pump (GWHP) is to realize optimal cooperation for all components, and to consider the running costs and first investment simultaneously under present technical and economic conditions.

7. SURFACE WATER HEAT PUMPS

Surface water heat pump (SWHP) systems can be either closed-loop systems similar to GCHPs or open-loop systems similar to GWHPs (Fig. 2B). However, the thermal characteristics of surface water bodies may be quite different from those of the ground or groundwater. Some unique applications are possible, but special precautions may be warranted.

Closed loops of pipe can be positioned at the bottom of a body of water at sites where the water is deep enough and has significant flow. Such sites utilize the circulating fluids to effect excellent heat exchange with the surrounding water, and limit the need for excavation. SWHP systems are a viable and relatively low-cost GSHP option. Lakes, streams, and bays can be very good heat sources and sinks for these systems, if properly utilized. Many successful systems are currently in operation.

Open systems are restricted for use in warmer climates or for buildings in colder climates with cooling-only applications (or very small heating loads). In colder climates, lake temperatures may range from 38° to 40°F. Pumping at a rate of 0.054 liters/sec per nominal kilowatt (3 gpm/nominal ton) will result in an outlet temperature 3.3°C (6°F) below the heat pump inlet. Even in warmer climates, caution is necessary to ensure that pumping rates are adequate and that the reservoir is large and deep enough to prevent the water temperature from falling below 5.6°C (42°F).

8. HYBRID GROUND-SOURCE HEAT PUMP SYSTEMS

Ground-source heat pump systems offer an attractive alternative for both residential and commercial heating and cooling applications due to their higher energy efficiency compared to conventional systems, but their higher first cost has been a significant drawback to wider acceptance of the technology. This is especially true in commercial and institutional applications, where the vertical GHE is commonly preferred. These types of buildings are generally cooling dominated and therefore reject more heat to the ground than they extract on an annual basis (Fig. 4). As a result, the required GHE length is significantly greater than the required length if the annual loads were balanced. One option to reduce the size of the GHE, and therefore the first cost of the system, is to balance the ground thermal loads by incorporating a supplemental heat rejecter into the system. GSHP systems that incorporate a supplemental heat rejecter have been referred to as hybrid GSHP systems.

Supplemental heat rejection can be accomplished with a cooling tower, fluid cooler, cooling pond, or pavement heating system. Currently suggested design methods for hybrid GSHP systems attempt to size the GHE based on the annual heating load and then size

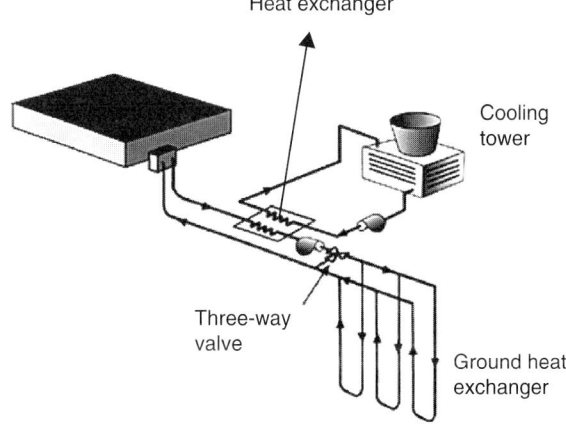

FIGURE 4 Cooling-tower supplemental for cooling-dominated loads.

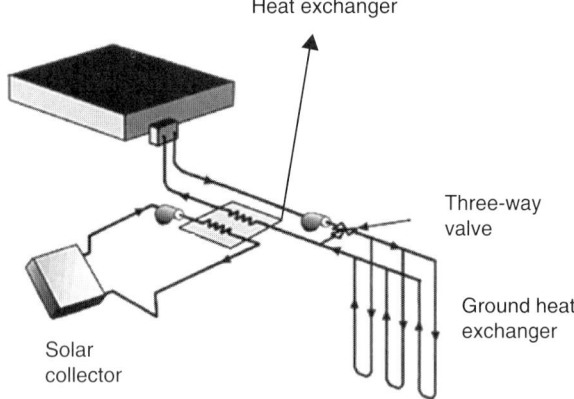

FIGURE 5 Solar-assisted system for heating-dominated loads.

the supplemental heat rejecter to balance the annual ground loads. However, the design of the system components also depends on the strategy used to control the supplemental heat rejecter. A smaller supplemental heat rejecter operated for more time may reject the same amount of heat as a larger supplemental heat rejecter operated for less time. Hence, a balance between the size of GHE, the size of the supplemental heat rejecter, and the control strategy is required to achieve the best economic alternative.

In northern climates, where the heating load is the driving design factor, supplementing the system with solar heat can reduce the required size of a closed-loop ground-coupling system. Solar panels, designed to heat water, can be installed into the ground-coupled loop (by means of a heat exchanger or directly), as illustrated in Fig. 5. The panels provide additional heat to the heat transfer fluid. This type of variation can reduce the required size of the ground-coupled system and increase heat pump efficiency by providing a higher temperature heat transfer fluid.

9. PERFORMANCE EVALUATION OF GROUND-SOURCE HEAT PUMPS

The energy performance of a GSHP system is influenced by three primary factors: the heat pump unit, the circulating pump or well pumps, and the GHE. Designers often devote more energy and resources to the GHE because it is a novel component to most heating, ventilating, and air-conditioning (HVAC) engineers. When the GHEs are correctly designed and installed and high-efficiency extended range heat pump units are specified, GSHPs are extremely efficient. Thus, the high demand and energy use of pumps that may result from unnecessary oversizing, restrictive piping, and poor control are especially noticeable.

9.1 Heat Pump Unit

The heat pump is the largest single energy consumer in the system. Its performance is a function of the efficiency of the machine and the water temperature produced by the GHE or well (either in the heating or cooling mode). The most important strategy is to start with an efficient heat pump. It is difficult and expensive to enlarge a GHE to improve the performance of an inefficient heat pump.

The purpose of rating the efficiency of a heat pump unit is to indicate the relative amount of energy needed to provide a specific heating/cooling output. The more efficient the equipment, the less energy will be used to do the same job. Many of the terms used in the heating, air-conditioning, and refrigeration industry convey performance and efficiency. Among these, four terms, the coefficient of performance, the energy efficiency ratio, the seasonal energy efficiency ratio (SEER), and the heating seasonal performance factor (HSPF), are widely used for these purposes in the United States and are currently rated by the American Refrigeration Institute (ARI). Although codes and standards identify minimum efficiencies, they do not fully communicate the energy efficiency that is achieved by today's heat pumps. A review of manufacturers' literature on commercially available equipment indicates that cooling efficiencies (EERs) of 12.0 to 16.8 Btu/Wh and heating efficiencies (COPs) of 3.0 to 4.3 are readily available (to convert the EER rating to the COP, divide the EER rating by 3.413). In the United States, heat pumps are rated on tonnage (i.e., 1 ton of cooling power produced by 1 ton of ice is equal to 12,000 Btu/hour, or about 3.51 kW).

When comparing equipment efficiencies, it is important to make an appropriate comparison. The efficiency of any heating and cooling equipment varies with application, load, and related heat-source and heat-sink temperatures. Furthermore, standard ratings are for specific conditions and operating conditions. It is highly recommended that the ratings systems be used only to compare one type of equipment to another. The performance of equipment rated in one standard should not be compared with the performance of units rated in another standard. The ratings of water-to-air heat pumps should not be compared with equipment rated by the air-source heat pump standard, which is much more complex and optimistic.

The actual performance of the equipment is a function of the water temperature produced by the GHE, which is a function of the ground temperature, the pumping energy, and the design of the GHE. In other words, the entering water temperature (EWT) of the heat pump unit is perhaps the single most representative parameter of the GSH effectiveness and heat pump loading. For example, in a region where the local ground temperature is 16°C and the GHE is designed for the customary 11–14°C above ground temperature, a heat pump rated at an EER of 16.8 would actually operate at an EER of 14.2 under peak load conditions. A poorly designed GHE that forces the unit to operate at 17°C above ground temperature would reduce the value to less than 13.0. These are for the cooling operation, which is the dominant load in commercial applications. However, the same relationship holds for heating operations. Besides this, systems with inadequate GHE capacity will have excessively high EWTs in summer and excessively low EWTs in winter.

9.2 Circulating Pump

The energy performance of the system is also influenced by the pumping energy required to circulate the fluid through the heat pump and the GHP. In the design of GSHP systems, the benchmarks given in Table II are proposed for pumping power for commercial GCHP systems. These benchmarks are for judging the effectiveness of a pumping and piping system design for a minimum of $0.162 \, m^3/kWh$ of cooling, with optimum pumping

TABLE II
Benchmarks for GSHP System Pumping Efficiency[a]

Watts (input)		Performance	
Per ton	Per kilowatt	Efficiency	Grade
≤50	≤14	Efficient systems	A: Excellent
50–75	14–21	Acceptable systems	B: Good
75–100	21–28	Acceptable systems	C: Mediate
100–150	28–42	Inefficient systems	D: Poor
>150	>42	Inefficient systems	E: Bad

[a] Required pump power to achieve cooling capacity.
Source. Copyright 2003, American Society of Heating, Refrigerating and Air-Conditioning Engineers, Inc. Reprinted by permission from Kavanaugh *et al.* (1997).

flow rates ranging from 0.162 to $0.192 \, m^3/kWh$ of cooling. Higher rates require excessive pump power, and lower rates limit the heat pump. In addition, pumping energy should range between 6 and 7% of the total system energy used.

To put the values given in Table II into perspective, consider an office building with a 175-kW cooling load and heat pump units selected to operate at an EER of 14 under peak conditions. With an efficient circulating pump design (10 W/kW), the energy demand of the circulating pump would amount to $175 \, kW \times 10 \, W/kW = 1750 \, W$. Combining the heat pump demand with the heat pump unit demand results in a system EER of 13.5. The same building and equipment coupled to a poorly designed pumping system consuming 34 W/kW would yield a system EER of only 12.2, thus compromising the premium paid for the higher efficiency equipment.

9.3 Ground Heat Exchanger

The design of GHEs is complicated by the variety of geological formations and properties that affect thermal performance. Proper identification of materials, moisture content, and water movement is important. Careful consideration of the interaction between these three components and their impact on system performance is necessary in order to minimize operating costs for the building owner.

The key parameter for GHE layout is the specific performance, i.e., the heat extraction rate during the heating season or the heat rejection rate during the cooling season in watts per meter of borehole (or meter of borehole per kilowatt) for vertical GHEs (or length of buried pipe loop for horizontal GHEs). The heat rejection rate ranges from 40 to 100 W/m, with a typical average of 55–70 W/m in mid-European countries. Based on a study that covered the 22 GSHP systems in the United States, for closed-loop systems, the GHE pipe length ranged from 20.5 to $52.0 \, m/kW_t$, with an average of $39.3 \, m/kW_t$. Of those with vertical bores, the range is 14.4–17.7 m of bore per kW_t.

The impact soil type and moisture content on the performance of GSHPs has recently been reported by some investigators. The length of a GHE can be reduced by using sand or thermally improved bentonite grout instead of standard bentonite grout. It can be concluded that using thermally enhanced materials instead of standard ones decreases the required GHE length by 15–20%. For applications in combination with other heat sources, the specific performance may differ considerably. Therefore, it is essential to obtain a second parameter, i.e., the specific heat extraction per meter of borehole length for a year. This value ranges between 50 and 200 kWh/m per year for mid-European facilities.

GHEs and associated components have significantly different characteristics compared to other types of HVAC heat exchangers; these differences impact the selection of flow rates, allowable head losses, and piping materials. GSHP piping design may be affected by the following special factors:

1. The ground provides by far the greatest resistance to heat flow; thus, highly conductive pipe materials, extended surfaces, and high internal fluid velocities are of little value.

2. Because the fluid must travel through both the building and GHE, rust inhibitor solutions must be acceptable for in-ground use. Thus, a material that does not require toxic inhibitors is necessary.

3. When design guides are followed, high fluid velocities are unnecessary for good heat transfer in the GHE and head losses through the ground coils can be relatively small [<6 ft (2 m) of water].

4. Because high fluid velocities are unnecessary for good heat transfer in the GHE, moderate flow imbalances will have little impact on overall heat exchange.

5. High-efficiency, extended-range water-to-air heat pumps do not require precise water flow control (or high head loss) to operate near maximum efficiency. A $\pm 33\%$ change in water flow rate results in less than a $\pm 2\%$ change in cooling capacity.

6. Because the water coils of extended-range, high-efficiency heat pumps are much larger than water-loop heat pump coils, head loss is relatively small [<12 ft (4 m) of water].

7. Because the combined head losses in the ground coil and heat pump are small [<18 ft (5.5 m) of water], the lengths of the headers, control valve losses, and fitting restrictions have a significant impact on overall system head loss.

8. Because the cost of the recommended piping material (thermally fused high-density polyethylene) is low relative to labor costs, low friction losses can be economically justified as a method of minimizing pump head.

9. Corrosion- and leak-resistant piping material with a minimal number of in-ground joints is critical to high system reliability and life.

SEE ALSO THE FOLLOWING ARTICLES

Geothermal Direct Use • *Geothermal Power Generation* • *Ocean Thermal Energy* • *Thermal Energy Storage*

Further Reading

Bose, J. E., Parker, J. D., and McQuiston, F. C. (1985). "Design/Data Manual for Closed-Loop Ground-Coupled Heat Pump Systems." American Society of Heating, Refrigerating, and Air-Conditioning Engineers, Inc. Atlanta.

Federal Technology Alerts (FTA). (2001). "Ground-Source Heat Pumps Applied to Federal Facilities," 2nd Ed. Available on the Internet at http://www.pnl.gov.

Healy, P. F., and Ugursal, I. (1997). Performance and economic feasibility of ground source heat pumps in cold climate. *Int. J. Energy Res.* **21**, 857–870.

Hepbasli, A. (2002). Performance evaluation of a vertical ground-source heat pump system. *Int. J. Energy Res.* **26**, 1121–1139.

Huttrer, G. W. (1997). Geothermal heat pumps: An increasingly successful technology. *Renew. Energy* **10**(2/3), 481–488.

Kavanaugh, S. (1991). "Ground and Water Source Heat Pumps: A Manual for the Design and Installation of Ground Coupled, Ground Water and Lake Water Heating and Cooling Systems in Southern Climates." Energy Information Services, Tuscaloosa, Alabama.

Kavanaugh, S. P., and Rafferty, K. (1997). "Ground-Source Heat Pumps: Design of Geothermal Systems for Commercial and Institutional Buildings." American Society of Heating, Refrigerating and Air-Conditioning Engineers, Inc., Atlanta.

Lund, J. W. (2000). Ground-source (geothermal) heat pumps. *In* "Course on Heating with Geothermal Energy: Conventional and New Schemes Course textbook of World Geothermal Congress 2000" (P. J. Lienau, Ed.), pp. 209–236. Kazuno, Tohoku District, Japan.

Lund, J. W. (ed.). (2001). Geothermal heat pumps—An overview. *Q. Bull.* **22**(1), 1–2.

Lund, J. W. (2003). Direct-use of geothermal energy in the USA. *Appl. Energy* **74**, 33–42.

Rafferty, K. (1999). Design issues in the commercial application of GSHP systems in the U.S. *In* "Proceedings of International Summer School" (K. Popovski, J. Lund, D. Gibson, and T. Boyd, Eds.), pp. 153–157. Geo-Heat Center, Oregon Institute of Technology, Klamath Falls, Oregon.

Gulf War, Environmental Impact of

FAROUK EL-BAZ
Boston University
Boston, Massachusetts, United States

1. Definition of War Impacts
2. General Setting
3. Preinvasion Conditions
4. Postwar Effects
5. Remediation Approaches
6. Future Monitoring

Glossary

desert pavement Natural residual concentration of wind-segregated and closely packed pebbles and other rock fragments mantling a desert surface.

Gulf War The international military campaign of "coalition forces" led by the United States to liberate Kuwait, during January–November 1991, from the occupation by Iraq's forces that began in August 1990. "Gulf" refers to the Arabian Gulf, which is also known as the Persian Gulf.

Landsat A series of U.S. satellites designed to obtain and transmit images of the surface of the earth up to 15 m in resolution (detail).

sabkha Flat, low area covered by clay, silt, and sand and often encrusted with salt; it forms in restricted coastal areas of arid and semiarid environments.

sand dune An accumulation of loose sand heaped up by the wind, where there is an abundant source of sand and constant wind.

sand sheet A large plain of wind-blown sand lacking the discernible sides or faces that distinguish sand dunes.

Systeme pour L'Observation de la Terre (SPOT) A French satellite that obtains and transmits images of the earth up to 5 m in resolution (detail).

tarcrete A hardened surface layer composed of surface sand, gravel, and other rock fragments cemented together by petroleum droplets and soot from the plumes of oil well fires (in Kuwait).

Thematic Mapper An advanced imaging system of Landsat 4, 5, and 7 that collects surface radiation in multispectral bands up to 30 m in resolution (detail) and panchromatic (black-and-white) bands of 15-m resolution.

The Gulf War of 1990–1991 resulted in severe environmental effects, which were analyzed using satellite images followed by field observations. Major disturbances were caused by the disruption of the "desert pavement," a layer of pebbles and gravel that protects the fine-grained soil beneath. The disruption was caused by the activities of military personnel including the digging of trenches and pits, building of berms and sand walls, and the traffic of heavy military vehicles. The exposure of fine-grained soil caused the increase of dust storms and the formation of numerous sand dunes that encroached on roads, installations, and farms. In addition, the explosion of over 600 oil wells caused other major environmental hazards. These included the formation of 240 oil lakes in topographically low areas. Also, plumes from the well fires deposited oil droplets and soot, which hardened on the desert surface as a layer of "tarcrete," affecting desert animals and natural vegetation. Nearly one-third of the land surface of Kuwait was impacted by the war. Although some remediation was performed, long-term monitoring is required to fully understand and ameliorate the impacts on the desert environment.

1. DEFINITION OF WAR IMPACTS

Iraq invaded Kuwait on August 2, 1990. On January 17, 1991, international coalition forces led by the United States began a military campaign for the liberation of Kuwait that became known as the Gulf War. These events resulted in numerous environmental impacts on the desert surface of Kuwait. The impacts were identified based on the study and analysis of satellite images. The most significant of these impacts were the results of exploding the oil wells, most of which caught fire. Plumes of smoke and oil droplets were carried by the southward wind and deposited on the desert surface. The resulting layer of hardened tarcrete disrupted natural vegetation. Some of the exploded oil wells did not catch fire and the oil seeped to form vast lakes. The oil-soaked lake beds continue to be an environmental hazard more than a decade later.

In addition to oilfield-related environmental impacts, the desert surface was disrupted by military troop actions. Disruption of the surface layer occurred due to the passage of heavy military vehicles, digging of trenches, building of berms, etc. All these activities exposed fine-grained material from below the protective layer of pebbles known as desert pavement. Exposure of the fine-grained soil to wind increased dust storms as well as the number and size of sand dunes.

An understanding of the nature of the desert surface in Kuwait is necessary to appreciate these environmental impacts. For this reason, a description of the general setting of Kuwait and the characteristics of its surface is given, followed by an assessment of the prewar conditions based on comparisons of satellite images obtained before and after the Gulf War. The major impacts are described with this background in mind. Recommendations are given for monitoring the changes and remediation of the environmental problems.

2. GENERAL SETTING

Kuwait is an arid country that lies along the northwestern corner of the Arabian (Persian) Gulf, bordering Saudi Arabia to the south, Iraq to the north and west, and the Gulf to the east (Fig. 1). The surface topography of its 17,818 km is marked by flat to undulating gravel-covered plains that slope gently eastward from a maximum elevation of 284 m (above mean sea level) in the west.

During the past three decades, rainfall ranged from less than 25 mm in 1964 to 375 mm in 1972, with an average of 115 mm per year. The precipitation infiltrates rapidly, leaving no permanent surface water, although some wadis fill temporarily with water from winter rains. The largest of these is Wadi Al-Batin, which is 5 km in width at the western border

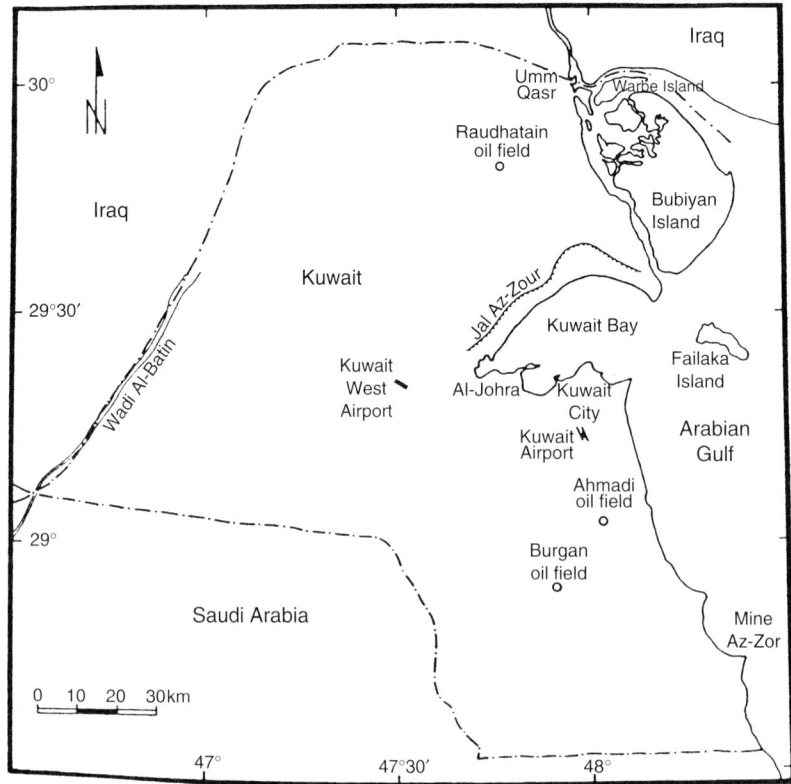

FIGURE 1 Sketch map of Kuwait showing its major features.

of the country. This wadi is part of a partially buried, water-dug channel whose tributaries drain part of the Hijaz Mountains 850 km to the southwest in Saudi Arabia. It was proposed to name the complete channel the Arabia River, whose delta deposits form much of the land surface of Kuwait (Fig. 2).

As mapped by the U.S. Geological Survey, Kuwait consists of flat-lying Tertiary rocks overlying gently folded Cretaceous and Jurassic formations. Exposed rock types include the Eocene Dammam formation, a white, fine-grained cherty limestone that shows some karst development at its contact with the overlying Miocene and Pliocene age Fars and Ghar formations, which are primarily calcareous sandstones, sandy limestones, clay, and sand. Completing the section above the Fars and Ghar formations is the Dibdiba formation, which is composed of sandstone. Quaternary deposits include lag gravels that blanket much of the interior of Kuwait and coastal deposits including unconsolidated marine sands, mudflats, and supratidal sabkha surfaces.

Structurally, Kuwait lies on the Arabian Shield, an area noted for its stability since Cambrian times. The Shield tilts slightly to the northeast, giving rise to sedimentation of the Arabian Shelf consisting of a sequence of laterally extensive but thin limestones, marls, shales, sandstones, and evaporites. North–south trending anticlinal structures, developed with gentle warping and uplift, serve as traps for Kuwait's oil reserves. In addition to this warping, there is evidence of salt diapirs with magnetic low signatures on the anticlinal structures.

Topographically, Kuwait is a flat sandy desert that may be divided into two parts. The north is a vegetated stone desert with shallow depressions and low hills running northeast to southwest and ending near an area of underground water storage. The principal hills in the north are Jal Az-Zor (145 m) and the Liyah Ridge. The southern region is a bare plain covered by sand; the Ahmadi Hill (125 m) is the sole exception to the flat terrain. In addition to Wadi Al-Batin, the only other valley of note is Ash-Shaqq, a portion of which lies within the southern part of the country and the rest in Saudi Arabia.

Morphologically, east of Wadi Al-Batin (along the borders between Kuwait and Saudi Arabia; Fig. 1) the desert surface is composed of a calcretic plain. This unit is composed of rock debris surrounding hills and gravel ridges. Often, parts of the unit are covered with a crust of saline or alkaline soils. The largest morphological unit of Kuwait lies east of the calcrete unit and northeast of Wadi Al-Batin. In this unit, the

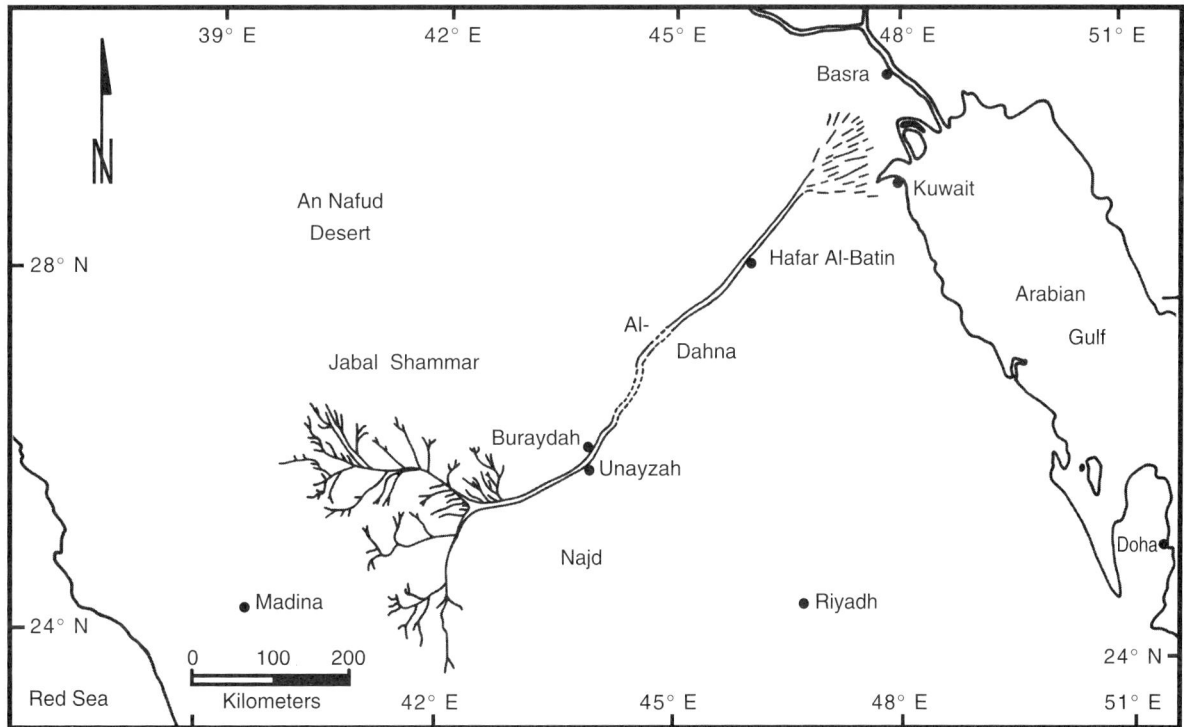

FIGURE 2 Schematic illustration of the tributaries and partly sand-buried channel of the "Arabian River." Note that western Kuwait is part of the ancient delta of the river.

terrain is rugged and composed mainly of a vegetated sand sheet, locally referred to as the Al-Huwaimliyah plain. The ruggedness is caused by the accumulation of sand in the lee of the wind behind desert shrubs. Spaces between the sand piles are composed of a thick bed of wind-blown sand. Therefore, destruction of the vegetation by extended droughts, overgrazing, or vehicular overuse results in the mobilization of sand in the otherwise stable sand sheet.

Northern Kuwait is covered by a vast plain with particles ranging in size from 4 to 10 mm in diameter. This unit is considered a desert pavement of lag gravels, which are fragments of rock too large to be moved by the wind and therefore "lag" behind the finer particles that move with the wind. Their concentration varies from scattered pebbles to densely packed accumulations of pebbles and granules. The predominance of the unit in northern Kuwait is largely attributed, in the literature, to the occurrence of the gravelly Dibdiba formation, which outcrops in northern Kuwait and southern Iraq. However, I propose that the gravel originated in western Saudi Arabia and was transported through the channel of the Arabia River through Wadi Al-Batin (Fig. 2). This is largely based on an apparent provenance of the gravel; rocks from which the gravel may have originated abound in the Higaz Mountains of western Arabia (Fig. 2).

From the characteristics of the desert pavement of Kuwait, it is clear that gravels were first moved by water currents and deposited in a sheet together with finer fractions, which were later blown away by deflation. As a result of the continuous deflation, pebbles and granules were concentrated on the surface until a deflation armor or desert pavement was acquired. The gravel ridges were formed by reversed topography as a result of the preferential deflation of flood plains dissected by fluvial channels in the fan-shaped delta of Wadi Al-Batin.

The coastal zone of Kuwait is composed mainly of sabkha deposits. These are flat areas of clay, silt, and sand that are often encrusted with salt. Along the coasts of Warba and Bubiyan islands, the supratidal flats are covered by thin laminae of sandy mud encrusted by a thin lithified crust. Large areas of sabkha are composed of extensive sheets of mud with almost no relief and flourishing mangroves.

3. PREINVASION CONDITIONS

During the two decades before 1990, developing land for agriculture had increased dramatically in Kuwait. This increase in agricultural activity resulted in the disruption of the desert pavement in developed areas and the exposure of soil to wind action. This, in turn, increased the amount of dust in the air and the movement of sand and its accumulation in mobile dunes. In addition, increases in dust and sand storms in the northern Gulf region were noticed in the 1970s due to major agricultural expansion in the Western Desert of Iraq. Satellite images clearly show that dust storms in southwestern Iraq spread over the eastern part of the Arabian Peninsula.

Similarly, growth of urban communities results in the disturbance of the stable desert surface. Most of the environmental effects of urbanization occur during digging, when soils are exposed to eolian erosion. Such efforts were particularly obvious during the expansion of Al-Jahrah, a community west of Kuwait City (Fig. 1).

Farther inland, disruption of the desert surface from the oil industry resulted from exploration and development activities, including geophysical surveys, drilling operations, building of roads, and laying of pipelines. These activities usually require a large number of vehicles, which criss-cross the desert surface affecting both the desert pavement and the scant desert vegetation.

For example, the aforementioned morphological unit of Al-Huwaimliyah plain, the vegetated sand sheet, extended into the Burgan oil field in south central Kuwait (Fig. 1). This area was badly degraded due to excessive vehicular activities and the desert surface was disturbed. Because disturbances of this type mobilize the sand, vehicular traffic in the open desert in Burgan field had been prohibited during the 6 years prior to the invasion of 1990. Due to this prohibition, vegetation began to flourish and the sand sheet was once again stabilized.

Military installations in the open desert also had some impact on the environment of Kuwait's desert prior to the Gulf War. Military bases have greatly suffered from the effects of destabilizing the desert surface in terms of having to constantly alleviate the effects of sand accumulations around buildings and their entrances. The study of sand deposits within Ali Al-Salem and Ahmad Al-Jabir bases provided insights into accumulations on roads. Three cases of pavement profiles on a slightly raised bed were considered: a flat pavement, one in which the underlying bed was slightly elevated on the upwind side of the road, and another with the elevation on the downwind side of the road (Fig. 3).

In all three cases, sand accumulated on either side of the raised roadbed; regardless of its slight elevation, a roadbed represents a topographic

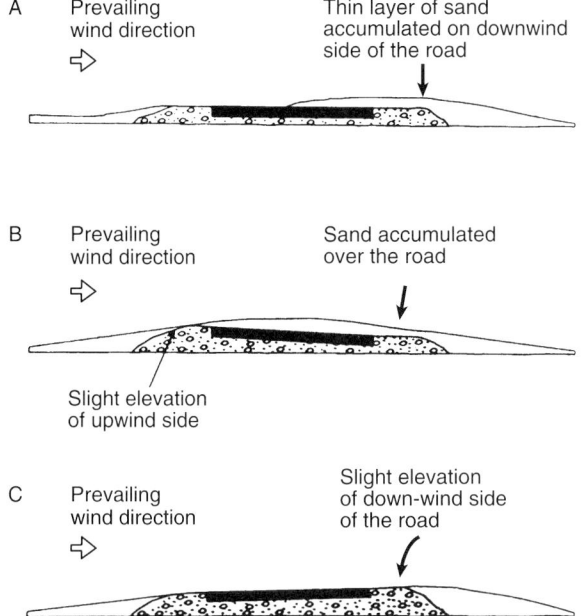

FIGURE 3 Deposition of sand along a road with a horizontal profile (A), an upwind elevation (B), and a downwind elevation (C). Modified from Fikry Khalaf.

impediment to the wind. In the case of the flat roadbed and pavement, sand accumulated on the downwind side of the road. The road with a slight elevation upwind was completely covered by sand because vorticity of the wind as it was confronted by the raised roadbed caused deposition of its sand load. In the case of the slight elevation of the roadbed and pavement on the downwind side, the road remained free of sand. In this case, the wind energy was increased as it moved upslope and carried the sand with it across the road to be deposited beyond the downwind elevation of the roadbed.

In the future, similar precautions and protective measures will have to be taken to safeguard installations in the open desert, such as roads, airport runways, helicopter pads, oil wells and storage tanks, checkpoints, and residential sites in the oil fields of Kuwait. Each site will have to be considered separately because remedies may differ due to local topographic characteristics, the nature of the desert surface, and the potential of particle transport by the wind.

4. POSTWAR EFFECTS

The physical effects of the Gulf War on the desert surface of Kuwait are numerous. Satellite images were used to identify these effects and to quantify their extent. NASA's Landsat Thematic Mapper images with a 30-m resolution were used in conjunction with Systeme pour L'Observation de la Terre (SPOT) images at 10-m resolution.

The variety and extent of environmental damage due to the Gulf War were characterized mainly through "change detection." In this process, an image of a specific area is compared to another image of the same area taken at a different time; the change from one image to another is computed. In this case, four Landsat images of Kuwait obtained in 1989 (prior to the invasion) were compared to four taken in 1992, upon cessation of the last oil well fire.

Most people are familiar with the horrific images of burning oil wells, which resulted from the wanton explosion of Kuwait's 700 oil wells by the Iraqi troops prior to their withdrawal. Naturally, this tragedy resulted in severe environmental effects, particularly spewing smoke and oil particles in the atmosphere for nearly 1 year. There were numerous other environmental effects, which are discussed in the following sections.

4.1 Vehicle Tracks

The movement of a great number of tanks and mechanized vehicles on the natural surface of Kuwait adversely impacted the desert pavement, particularly in the western part of the country. During the invasion, columns of Iraqi military equipment moved along parallel lines both inside and just west of the border of Kuwait. Such heavy equipment forces the surface cover of pebbles and cobbles deep into the sand and excavates large amounts of soil on either side of the tracks.

The net result of vehicle movement on a naturally packed surface is the decrease in the closeness of packing of sand grains. This phenomenon increases the potential of grain movement by the wind, resulting in the mobilization of previously protected fine particles. The spread of newly exposed soil on either side of the tracks allows for their detection on satellite images, even if the resolution of the images is less than the width of the tracks (Fig. 4).

During occupation, most of the troop movements were along paved roads. However, some areas were trampled by military vehicles, causing damage to the desert pavement. This is particularly true in the southwestern and northwestern parts of the country as Iraqi soldiers raised defensive fortifications. In areas covered by a sand sheet, such as in southeastern Kuwait, the tracks of military vehicles have already been covered by drifting sand. On the other hand, areas covered by a desert pavement remained deeply

FIGURE 4 Vehicle tracks on the desert surface as depicted on three successive Landsat images of Kuwait's southwestern corner, just east of Wadi Al-Batin. Courtesy of Earthsat Corporation.

FIGURE 5 A trench dug in the desert surface north of Jal Az-Zor that was partly filled by sand less than 1 year from the end of the Gulf War.

scarred, such as north of the Jal Az-Zor escarpment (Fig. 1).

4.2 Trenches and Pits

Iraq's armored forces spent much of their time during the fall of 1990 digging trenches in the desert of Kuwait (Fig. 5). This is particularly true along Kuwait's coastline. Beyond the trenches, pits were dug as hiding places for military equipment and storage areas of ammunition.

After the passage of a windy (shamal) season, most pits and trenches had been filled by drifting sand. The ammunition pits that were covered by sand represent a potential danger in the future. These may be identified by the use of ground-penetrating radar so they can be excavated and destroyed. Such procedures would again mobilize the sand for downwind movement if the dug-up areas are not stabilized, for example, by spreading pebbles on top of the exposed soil.

4.3 Berms and Sand Walls

Berms were built in Kuwait's desert to slow the potential advance of liberation forces. The major berms were constructed in the south along the Saudi border, in the central part (south of the Burgam oil field), and in the northern part of Kuwait (north and west of the Jal Az-Zor escarpment). The berm-building activity continued throughout the occupation phase. Iraqi military strategists knew that the sand walls would be eroded by the consistent wind. Therefore, berms were sprayed with crude oil by pipelines that were laid specifically for this purpose from Kuwait's nearby oil fields. These berms represent a source of shifting sand in the long term. Currently, the soil in these berms is stabilized by the hardened oil crust (Fig. 6). However, this crust will gradually be dismantled by wind erosion and the fine-grained sand will be released regularly.

The effects of such berms on the spread of sand on the desert surface are exemplified by the case of the sand wall in Saudi Arabia just south of its border with Kuwait, which was constructed to stop potential smugglers, mostly of liquor. The berm is less than 3 m high and a few meters wide at the base. However, its sand was mobilized and spread over such a large area (up to 1 km) so as to make it clearly visible on satellite images of 30-m resolution (Fig. 7).

In addition to the berms built by Iraqi military personnel, an extensive berm was constructed during the summer of 1992 by Kuwait's security forces in the northern part of the country. This berm was intended to stop a wave of smugglers and political agitators from Iraq following the cessations of hostilities. Since wind erodes the fine-grained material of the berm, it is constantly maintained by bulldozers. This berm is a source of drifting sand and must be monitored.

4.4 Mines and Ordnance Detonation

The Iraqi military forces planted between 500,000 and 2.5 million land mines in the desert and coastal areas of Kuwait. The mining was done during the occupation phase to hamper the expected liberation of Kuwait by military action. Numerous defense lines were established particularly along the border of Saudi Arabia and the coastline of the Arabian Gulf. In addition, mines were laid in areas of high troop concentrations, particularly in the triborder area near the western corner of Kuwait.

The laying of land mines is a process that results in the disturbance of stabilized soil and exposure of fine-grained material to the action of wind. This is particularly obvious in a minefield along the northern coast of Kuwait Bay (Fig. 1). In this case, the mines

FIGURE 6 An oil-covered berm in northern Kuwait built by Iraqi forces. Crude oil was used to stabilize the sand.

FIGURE 7 Effects of wind erosion on a berm in Saudi Arabia just south of its border with Kuwait. The berm is on the sharp northern edge of the white zone; the wind blowing to the southeast caused the jagged southern boundary due to sand transport downwind. This is part of a SPOT image acquired on May 25, 1992.

were laid in a soil that was stabilized by salt crust. The process of digging the land to plant the mines destroyed the drier crust, resulting in the deflation of the soil layer and exposure of the mines (Fig. 8).

During liberation, several passages were made through the minefield along the southern border of Kuwait. Breakthroughs were made by exploding the mines in groups. This process also disturbed additional stretches of desert pavement. In addition, bombing of Iraqi defensive positions by coalition forces resulted in cratering of the desert surface where bombs missed their targets. Ejecta from the craters exposed much fine-grained soil to the wind. However, because of the accurate targeting these craters are few and far between.

Much of the disruption of the desert surface in the minefields occurred during the postliberation phase as a result of mine detection and detonation. Crews of mine detectors were followed by mine sweepers and/or detonation equipment. The raking of the land and the explosion of mines and undetonated cluster bombs and other ordnance further disturbed the desert surface by exposing fine-grained soil to the wind.

4.5 Oil Crust

As shown in Fig. 9, deposits of oil and soot particles from the plumes of oil well fires were clearly visible in satellite images. When observed in the field, this deposit formed a hardened crust on top of the desert surface. It varied in thickness, with an average of 6 cm; the closer to the center of the deposit, the thicker the crust. Minor variations in the exact orientation of shamal winds caused the spread of the oil rain downwind of the well fires. Naturally, the thicker the deposit, the darker the surface. Minor variations in the spectral reflectance of such deposits were easily depicted on Landsat Thematic Mapper images, and it was possible to correlate the reflectance values to the thickness of the oil crust. Because it represented a new rock formation, this deposit was named tarcrete and defined as a conglomerate consisting of surface sand and gravel cemented together into a hard mass by petroleum droplets and soot.

The total surface area of Kuwait covered by tarcrete was approximately $943 \, km^2$. Particulates in this area had been basically stabilized. Petroleum sprays had been used in the past to stabilize dune sands. Thus, from the standpoint of sand mobility, the tarcrete had a positive impact.

The effects of the tarcrete on plant life must be considered separately. Some plants died from the oil spray. Other species flourished but were not palatable to grazing animals. Furthermore, the salinity of the soil in areas covered by the tarcrete may change with time as sulfur compounds react with moisture to increase soil acidity. For this reason, monitoring of the pH values and other soil characteristics is recommended.

FIGURE 8 Mines exposed by wind erosion of sand along the coast of Kuwait. The minefield was supposed to prevent a marine landing of coalition forces (photograph by Jassim Al-Hassan).

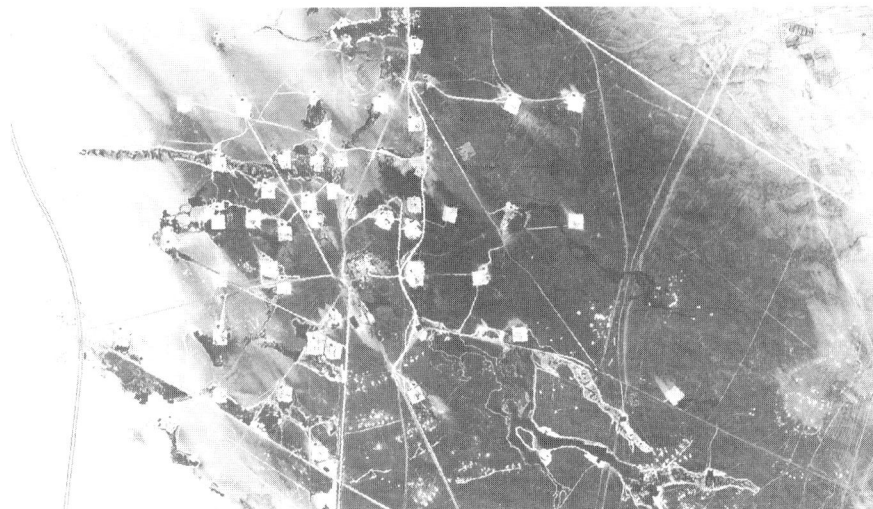

FIGURE 9 Part of a SPOT image acquired on March 25, 1992, of the Raudhatain oil field showing the new dirt roads (white lines) and sand pits around well heads (white squares) that were created during the firefighting operations. Note the oil lakes, which are oriented roughly east–west, within and along the western edge of the darkened surface.

4.6 Oil Lakes

Some of the exploded oil wells did not catch fire, and the seeped oil accumulated in low areas forming lakes (Fig. 9). Smaller oil lakes were also created during the firefighting phase. These lakes were identified using satellite images. The production of accurate maps of oil lakes necessitated the study of Kuwait by both Landsat Thematic Mapper and SPOT as well as maps that were based on field observations. Change detection techniques were used to study the characteristics of oil lakes in Thematic Mapper images. Also, geographic information system correlations of all data sets allowed their accuracy to be checked and the mapped lake boundaries to be confirmed. It was concluded that after a full understanding of the spectral characteristics of the oil lakes in Thematic Mapper images, their accurate mapping could be improved by using the higher resolution SPOT data.

Disregarding patches smaller than $1\,km^2$, the surface area covered by oil lakes in the desert of Kuwait was $35.4\,km^2$ in the fall of 1991, as mapped by the Kuwait Institute for Scientific Research. The total area of oil lakes was $27.5\,km^2$ as mapped from SPOT images obtained on March 25, 1992. The difference between the two measurements is easily ascribed to the shrinkage of the lakes due to the evaporation of aromatic compounds and the seepage of some of the oil into the underlying soil during the intervening months. The larger figure must be used in the estimation of remediation needs because all oil-saturated soil must be treated or removed, replaced, and stabilized.

5. REMEDIATION APPROACHES

It is clear from the previous discussion that the Gulf War has had severe short- and long-term effects on the desert surface of Kuwait and neighboring areas. Short term effects included the following:

1. Evaporation of aromatic compounds such as benzine from the surfaces of the oil lakes created a hazard to humans, animals, and plants.

2. Increased dust storm activity due to the release of fine silt and clay in areas previously protected by course desert pavement increased hazards to aviation in the whole Gulf region. In addition, mineral aerosols in the air constituted a hazard to humans if lodged in the throat, lungs, and eyes.

3. Movement of sand and its accumulation over minefields and ordnance pits hampered the detection of mines and unexploded bombs. In early 1992, several explosions caused severe injury to people and death to grazing animals.

4. The movements of heavy military vehicles as well as the oil and soot from the well fireplumes severely reduced plant cover in the open desert.

Long-term effects include the following:

1. An increase in sand storm activity due to the mobilization of sand-sized particles. Such storms

interfere with all modes of transportation and cause traffic accidents.

2. The formation and movement of sand dunes, which encroach on roads, farms, oil installations, and settlements in the desert. Once a dune is formed, it continues to move as long as the wind blows. For this reason, sand dune encroachment represents a longest term impact on the desert surface of Kuwait.

3. An increase in soil acidity due to the interaction of moisture with remnants of the oil on the desert surface. This increase hampers the rejuvenation of some natural plants.

4. Potential pollution of the groundwater resources from seepage, through fractures, of oil from the stagnating pools of crude petroleum. This represents a long-term impact particularly since agriculture in Kuwait depends solely on groundwater reservoirs.

For these reasons, mitigation of the effects of the Gulf War on the desert surface of Kuwait will require decades for remediation and monitoring of its effects. Remediation approaches include the following:

1. Filling of empty trenches using recently mobilized sand to limit the exposure of sand-sized particles to the action of wind.

2. Leveling of berms and other sand barriers and walls because these structures represent a source of sand for future activation and mobilization by wind.

3. Stabilizing the flattened surface either by increasing desert plant cover or, more efficiently, by spreading gravel on top of the soil to mimic the natural desert pavement.

4. Removing the oil-saturated soil from areas covered by oil lakes to limit potential long-term effects on soil acidity or groundwater pollution.

5. Replacing the removed soil with clean sand to return the land to the original flat contours.

6. Stabilizing the new surfaces by spreading gravel to form the armor against the erosive action of wind.

7. Seeding of areas of the desert that lost vegetation cover using, as much as possible, seeds of natural plants.

6. FUTURE MONITORING

The sources of environmental impacts on the desert surface due to the Gulf War are related to the invasion of Kuwait by Iraqi forces, the activities that persisted during the occupation phase, the initiation of liberating Kuwait by coalition forces and the events that accompanied the retreat of Iraqi troops, and the environmental impact of fighting the oil well fires and the spread of the oil lakes on the desert surface of Kuwait.

Some remediation of the environmental effects of the Gulf War was accomplished during the past decade. This is particularly true in the case of oil lakes. Most of the oil was pumped and bioremediation of the soil was accomplished. However, the more subtle impacts, such as those of sand movement and plant growth, have not been given equal priority. Therefore, the need exists for a long-term effort to monitor these factors and study the best ways to limit their impact on the natural environment.

These impacts can be monitored by use of satellite images, including National Oceanic and Atmospheric Administration satellites to monitor the initiation of dust storms, Landsat Thematic Mapper and SPOT images to monitor sand accumulation and movement, and higher resolution (1 m) satellite images to monitor changes to the topsoil. Interpretations of satellite image data need to be checked and confirmed in the field at specific field stations. At these stations, instruments should be installed to measure the activities discussed in the following sections.

6.1 Dust Storms

The exposure of vast amounts of soil to the action of wind causes an increase in the frequency and ferocity of dust storms. These hamper visibility on roads and at airports and cause respiratory ailments. The need exists to monitor the initiation of dust storms upwind for advance warning of the impending danger.

6.2 Sand Accumulation

The wind carries enormous amounts of sand in the flat terrain of Kuwait. The sand begins to accumulate where the wind regime is affected by topographic impediments. Such accumulations hamper transport and operations in the oil fields and other installations in the desert. Thus, a monitoring system should be established to alleviate the impact of encroaching sand dunes by stabilizing the moving sand with desert vegetation.

6.3 Soil Rejuvenation

The oil rain has caused much damage to the natural desert soils in Kuwait. It is feared that additional harm to the soil might be caused by an increase in acidity. A program needs to be instituted to monitor soil conditions and characteristics so that a remedy to the potential problem can be carried out in a sustained manner. This can be accomplished by monitoring soil acidity and its treatment wherever pH levels require it.

6.4 Plant Growth

Degradation of the soil has caused much damage to natural desert plants in Kuwait. Some plants have died in areas covered by the oil soot and droplets, and other species have appeared that are not palatable to grazing animals. The growth of plants should be monitored in view of the drastic changes in the desert environment.

SEE ALSO THE FOLLOWING ARTICLES

Arid Environments, Impacts of Energy Development in • Crude Oil Releases to the Environment: Natural Fate and Remediation Options • Crude Oil Spills, Environmental Impact of • Hazardous Waste from Fossil Fuels • Nationalism and Oil • National Security and Energy • Oil Crises, Historical Perspective • War and Energy

Further Reading

Al Doasari, A. (2000). Analysis of the changes in the tarcrete layer on the desert surface of Kuwait using satellite imagery and cell-based modeling. Ph.D. dissertation, Boston University, Boston, MA.

Al-Hassan, J. M. (1992). "The Iraqi Invasion of Kuwait, an Environmental Catastrophe." Fahad & Al Marzouk, Kuwait.

El-Baz, F., and Al-Sarawi, M. (1996). Kuwait as an alluvial fan of a paleo-river. *Z. Geomorphol. N. F. Suppl.-Bd.* **103**, 49–59.

El-Baz, F., and Al Sarawi, M. (2000). "Atlas of the State of Kuwait from Satellite Images." Kuwait Foundation for the Advancement of Sciences, Kuwait.

El-Baz, F., and Makharita, R. M. (2000). "The Gulf War and the Environment." Gordon & Breach, Lausanne, Switzerland.

Holden, C. (1991). Kuwait's unjust desserts: Damage to its desert. *Science* **251**, 1175.

Khalaf, F. I., Gharib, I. M., and Al-Kadi, A. S. (1982). Sources and genesis of the Pleistocene gravelly deposits in northern Kuwait. *Sedimentary Geol.* **31**, 101–117.

Koch, M., and El-Baz, F. (1998). Identifying the effects of the Gulf War on the geomorphic features of Kuwait by remote sensing and GIS. *Photogrammetic Engineers Remote Sensing* **64**(7), 739–747.

United Nations (1991). Kuwait. Report to the secretary-general on the scope and nature of damage inflicted on the Kuwait infrastructure during the Iraqi occupation, Security Council Document No. S/22535. United Nations, New York.

U.S. Geological Survey (1963). Geologic map of the Arabian Peninsula. U.S. Geological Survey (Washington, D.C.) and Arabian-American Oil Company (Aramco, Dhahran, Saudi Arabia). [U.S. Geological Survey Misc. Geol. Inv. Map I-270; scale 1:2,000,000]

Hazardous Waste from Fossil Fuels

MAJID GHASSEMI
New Mexico Institute of Mining and Technology
Socorro, New Mexico, United States

PAUL K. ANDERSEN and ABBAS GHASSEMI
New Mexico State University
Las Cruces, New Mexico, United States

RUSSELL R. CHIANELLI
University of Texas at El Paso
El Paso, Texas, United States

1. Overview of the Fossil Fuel Cycle
2. Wastes Associated with Petroleum and Its Products
3. Waste Associated with Coal and Its Production
4. Summary

Glossary

asphaltenes The part precipitated by addition of a low-boiling paraffin solvent such as normal pentane and benzene-soluble fraction, derived from carbonaceous sources such as petroleum, coal, or oil shale.

boiler slag A black granular material that is coarser than conventional fly ash.

BTEX The BTEX chemicals (benzene, toluene, ethylbenzene, and xylenes) are volatile monoaromatic hydrocarbons that are commonly used in crude petroleum and petroleum products.

characteristic hazardous waste Waste solids, liquids, or containerized gases that exhibit ignitability, corrosivity, reactivity, or toxicity characteristics.

coal bottom ash The coarse, granular, incombustible by-product that is collected from the bottom of furnaces that burn coal.

coal combustion by-products (CCBs) Fly ash, bottom ash, boiler slag, and flue gas emission that are produced when coal is burned.

drilling mud Waste stream associated with drilling operations, including oil-based mud (OBM), synthetic-based mud (SBM), and water-based mud (WBM).

drilling wastes Wastes that are associated with oil exploration and production (E&P).

exploration wastes Wastes that are primarily related to drilling and well completion.

fly ash The fine powder formed from the mineral matter in coal, consisting of the noncombustible matter in coal plus a small amount of carbon that remains from incomplete combustion.

flue gas desulfurization (FGD) gypsum Also known as scrubber gypsum; it is the by-product of an air pollution control system that removes sulfur from the flue gas in calcium-based scrubbing systems.

fossil fuels The biomass that produces coal, gas, oil, and tar sands.

heavy crude Crude oil that contains a relatively high portion of residuum.

listed hazardous wastes Wastes from nonspecific sources and wastes from specific sources and discarded commercial chemicals.

methyl tertiary-butyl ether (MTBE) MTBE is a gasoline additive and is highly water soluble.

naphtha The petroleum fraction ranging from low-boiling C_4 hydrocarbons to those boiling as high as approximately 220°C.

polycyclic aromatic hydrocarbons (PAHs) Often referred to as polynuclear aromatics (PNA); they are a class of very stable organic molecules composed of only carbon and hydrogen.

produced water Water produced in association with crude oil.

proppants/frac sand wastes Semisolid sludge consisting of aluminum silicate beads and formation sand.

residuum The petroleum fraction boiling above 343°C.

Superfund A commonly used name for the federal Comprehensive Environmental Response, Compensation and Liability Act (CERCLA).

tank bottom wastes Sediment that accumulates in the bottom of oil field vessels and pipelines when fluid turbulence is low.

waste oil Oil arising as a waste product of the use of oils in a wide range of industrial and commercial activities.

workover and completion wastes Wastes from operations in which an oil well's head is partially open to the atmosphere and is filled with a water-base fluid that maintains pressure on the formation to prevent blowout.

The world uses two basic fossil fuels for everyday life. Coal production is used mainly for the production of electricity and petroleum products are mainly used for transportation. The population of the United States uses an amount of fossil fuel in one year that took nature approximately one million years to produce! There are major environmental problems involved with the production, transportation, and consumption of these fossil fuels. Petroleum production has environmental problems at every stage of production, transportation, and consumption. During production, the waste products include drilling mud, drill cuttings, chemical additives, and produced water. In fact, there are 7.5 barrels of produced water for one barrel of oil. The main environmental problem of petroleum refining and storage include oil process wastes and leakage of the storage tanks. Pipeline leaks are also among major environmental issues for transportation of petroleum products. The production of coal requires mining operations. The initial mining virtually ravages the land. The processing of the coal leaves large amounts of tailings and other by-products. Additionally, the use of coal causes major problems with air quality. The fact that the United States uses one billion tons of coal annually for electricity alone is an extremely large environmental problem.

1. OVERVIEW OF THE FOSSIL FUEL CYCLE

Energy fosters human activities and is a necessity for modern life. Energy heats and cools buildings, powers industries, fuels transportation system, etc. All energy on Earth comes from the sun, which contributes to the formation of biomass. The biomass produces decay, which becomes buried in sediments, thereby produces coal, gas, oil, and tar sands. These are called fossil fuels. Figure 1 illustrates a detailed fossil fuel cycle. Fossil fuels are nonrenewable. In 1 year, mankind expends an amount of fossil fuel that took nature approximately 1 million years to produce.

Energy production from fossil fuels can create major environmental problems at every stage of production, transportation, and consumption. Fossil fuels pose risks to the environment. In this article, we discuss the wastes that are associated with fossil fuel in its production, transportation, and consumption.

2. WASTES ASSOCIATED WITH PETROLEUM AND ITS PRODUCTS

Petroleum is derived from Latin *petra*, meaning rock or stone, and Latin *oleum*, meaning oil. Crude oil is a natural occurring material extracted from formations beneath the earth via established drilling production processes. Crude oil refining operations involve extracting useful petroleum products that include naphtha, middle distillate, and residuum. The petroleum fraction ranging from low-boiling C_4 hydrocarbons to those boiling as high as approximately 220°C is called naphtha, which is used primarily in gasoline production. The petroleum fraction boiling from approximately 150 to 343°C is called middle distillate and is used for jet fuel, diesel fuel, and home and commercial heating oils. The petroleum fraction boiling above 343°C is called residuum. The portion that boils from approximately 343 to approximately 566°C is called vacuum gas oil and has no natural market. Material boiling above 566°C is called vacuum residuum, which is a heavy and tarry substance containing a heavier asphaltene fraction.

If crude oil contains a relatively high portion of residuum, it is called heavy crude. Residuum is burned as an industrial fuel in power plants and in ship boilers or is asphalted to make road material. Since gasoline is in great demand, petroleum processes convert lighter fractions (reforming processes) and heavier fractions (cracking processes) to gasoline-range products.

The petroleum industry produces a large percentage of the United State's hazardous waste sites. Various wastes are generated during petroleum exploration, production, and refining. During exploration and production, large quantities of low-toxicity solid waste and wastewater are produced.

In the United States, petroleum production and transportation result in the leakage of approximately 280 million barrels of petroleum (crude oil) each, creating serious environmental problems. Petroleum refinery wastes result from processes designed to

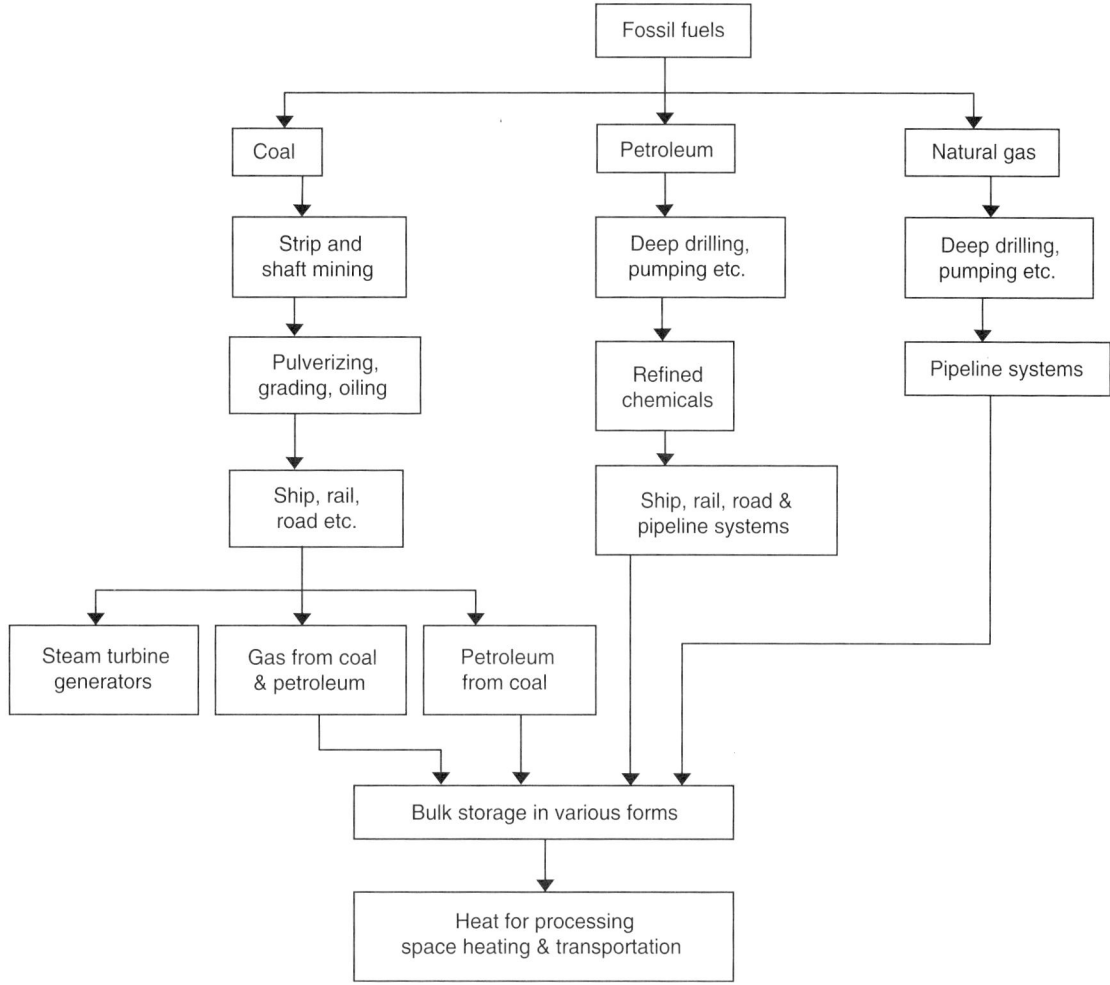

FIGURE 1 Detail of the fossil fuel cycle.

remove naturally occurring contaminants in the crude oil, including water, sulfur, nitrogen, and heavy metals. The waste streams from refineries are smaller but are considerably more toxic, containing listed hazardous waste and characteristic hazardous waste. In addition, poor housekeeping practices during exploration, production, and refining can produce excessive waste.

2.1 Waste from Exploration, Production, and Refining

Oil exploration and production (E&P) operations generate large volumes of drilling waste every year.

2.1.1 Waste from Petroleum Exploration

Exploration wastes are primarily related to drilling and well completion. Table I lists oil production, drilling, and exploration waste streams. Between 1986 and 1995, an average of 313 wells were drilled each year. Drilling operations generate more than 1000 tons of drilling waste per well. This means that more than 300,000 tons of drill cuttings may be produced each year. In addition, the American Petroleum Institute (API) survey of onshore and coastal operators in the United States estimates that approximately 149 million barrels (bbl) of drilling waste (17.9 million bbl of produced water and 20.6 million bbl of other associated wastes) were generated from E&P operations in 1995. The API has also estimated that the average volume of produced water was approximately 7.5 barrels of water per barrel of oil in 1995.

Drilling wastes include drilling mud, drill cuttings, wash water, and chemical additives (Table I). Drilling mud comprises the largest waste stream associated with drilling operation. There are three types of

TABLE I
Oil Production, Drilling, and Exploration Waste Streams[a]

	Produced water
Benzene	Hydrocarbons
Toluene	Radioactive pollutants Ra^{226} and Ra^{228}
	Miscellaneous surface water
Snowmelt and precipitation	Engine coolant
Runoff from cleaning vehicles and shops	Unused cement slurry
Rigwash	
	Drilling mud
Heavy metals	Chlorides (including potassium chloride, sodium chloride, calcium chloride, and magnesium chloride)
Polynuclear aromatic hydrocarbon components (OBM only)	
Barium, gypsum, barites	Clays (bentonite, kaolinities)
	Additives (biocides, scale inhibitors, corrosion inhibitors)
	Drill cuttings
Pieces of rocks and sand	Soil and wet sludge containing oil and water
	Workover and completion waste
Well completion, treatment and stimulation fluids	Fracturing media
Surfactants	Used filters
Weighting agents/viscosifiers	Biocides
Thinners	Packing fluids
Muds (WBM or OBM associated with workovers)	Detergents
Produced water (associated with workovers)	Defoamers
Crude oil	Paraffin
Acidizing agents	Sludges
Corrosion inhibitors	Pieces of downhole equipments
Gels	Intermaterial from downhole mechanical repair (produced sands, formation or pipe scale, cement cutting, and slurries)
Solvents	
Paraffin solvents and dispersants	
	Proppants/frac sand
Semisolid sludge containing the aluminum silicate beads along with formation sand	
	Bottom waste
Solids	Accumulated heavy hydrocarbons (asphaltic or paraffin)
Sands	Drilling mud
Emulsions	Semisolid waste
Pigging wastes	
Produced water	
	Oily debris, spent filters, and spent filter media
Rags	Sand
Sorbent materials	Diatomaceous earth
Water with surfactant (backwash)	Particulates
Oil	Socks
Coal	Cartridges
Gravel	Canisters
	Hydrocarbon wastes
Dirty diesel	Organic halides
Hydrocarbon-bearing soil	Hydrocarbon solids
	Camp wastes
Surplus paints	Wood cribbing supporting temporary pipelines
Batteries	Thread protectors from pipe ends
Light bulbs	Drums for chemical storage

continues

Table I continued

Radioactive smoke detector elements	Oil and air filters on vehicles
Wood pallets	Vehicle lube oil
Fabric pit liners	
	Ozone-depleting chlorofluorocarbon
Halons	Freon
	Naturally occuring radioactive materials
Spent filters	Produced water
Contaminated soil	Completion waste
Camp waste	Hydrocarbon waste

[a] Abbreviations used: OBM, oil-based mud; WBM, water-based mud.

drilling mud: oil-based mud (OBM), synthetic-based mud (SBM), and water-based mud (WBM). Most onshore wells are drilled with WBM or OBM. The key constituents of OBM, SBM, and WBM contaminated cuttings are summarized in Table I. The leachability, runoff, and seepage of heavy metals, chlorides, and the total polyaromatic hydrocarbons degrade the soil and contaminate groundwater, fish, and shellfish. Through the consumption of contaminated fish and shellfish, they can also endanger human health.

2.1.2 Waste from Petroleum Production

Production operations generate massive quantities of produced water. Produced water—water produced in association with crude oil—is by far the largest waste stream in most oil fields. The volume of produced water far exceeds the volume of drilling wastes (approximately 60 times more). The produced water is discharged into surface waters, injected underground, or transported to a commercial oil field waste disposal facility. The majority of produced water is disposed of underground through injection wells, which is allowed by the U.S. Environmental Protection Agency (EPA).

Produced water and drilling fluids could damage agricultural land, crops, ephemeral streams, and livestock and threaten endangered species, fish, and other aquatic and bird life. Metals and polycyclic aromatic hydrocarbon (PAH) discharge cause contamination of aquatic and bird life. Also, improper functioning of injection wells can degrade groundwater. In addition, oil field waste disposal facilities pose significant risk to migratory birds and other wildlife.

Workover and completion wastes result from operations in which an oil well's head is partially open to the atmosphere and is filled with a water-based fluid that maintains pressure on the formation to prevent blowout. Excluding the fluids that were disposed via on-site underground injection, the API calculates that U.S operators generated approximately 5.6 million bbl of completion and workover fluids in 1985. More current data are not available.

The environmental impact related to injection of workover and completion fluids is the potential migration of wastes to underground drinking water sources. Also, seepage of native brines from improperly plugged, unplugged, and abandoned wells could damage groundwater, agricultural land, and domestic and irrigation water.

Tank bottom wastes are a type of sediment that accumulates in the bottom of oil field vessels and pipelines when fluid turbulence is low. Tank bottom wastes are generated continuously at nearly all oil production operations. Oily debris saturated with crude oil comes from oil spill cleanups (minor and major) and spent filter and filter media waste comes from filtering crude oil. According to the API, U.S operators generated a total of 1.5 million bbl of bottom tank wastes and oily debris in 1985. From this total, 1,232,000 bbl (42 U.S. gallons per barrel) were bottom tank wastes and 1,261,000 bbl oily debris.

Potential environmental impact of tank bottom wastes and oily debris may come from landspreading, roadspreading, on-site pits, and on-site burial. The potential impact of landspreading and roadspreading may be a risk of off-site migration of constituents of concern due to surface water run-on and runoff. In addition, off-site transportation could result in high oil and grease concentration wastes to soil and/or vegetation. Also, on-site pits and on-site burial could potentially impact groundwater due to downward migration of constituents of concern and also surface water and soils.

Proppants/frac sand wastes are related to the aluminum silicate beads that are injected into wells to hold formation fractures open, along with formation sand, that together form a semisolid sludge. Hydrocarbon wastes such as dirty diesel fuel are contaminated from pressure testing pipelines. Camp wastes are equivalent in quantity to those of small municipalities.

Ozone-depleting chlorofluorocarbons (CFCs) are used in oil and gas production as a fire and explosion suppressant. Naturally occurring radioactive material is produced in solution with oil field brines and deposited as scale on the inside of oil field vessels and piping. Its radioactivity (rated as "low specific activity" by the Nuclear Regulatory Commission) poses a hazard if it is inhaled as a dry dust during vessel and pipe cleaning or repair.

2.1.3 Waste from Petroleum Refining

Table II shows the waste generation data for petroleum refineries as given by the API survey of U.S. refineries. The major sources of wastewater in petroleum refinery are cooling tower blowdown, boiler blowdown, oily wastes, and other process wastes. Solid wastes from refineries are also a significant problem.

In 1976, Congress enacted the Resource Conservation and Recovery Act (RCRA) for reduction of hazardous waste. This early hazardous waste legislation imposed "cradle to grave" liability on waste generators for the waste that they produce and dispose of. In 1984, Congress passed the Hazardous and Solid Waste Amendments (HSWA), which amended RCRA in an attempt to reduce groundwater contamination by banning or regulating landfill disposal of specific listed hazardous wastes. The refining sector of the petroleum industry generates both HSWA listed and other hazardous wastes. Five refinery solid waste streams are listed hazardous wastes by the EPA and subject to HSWA regulations: dissolved-air-flotation floating solids (K048), slop-oil-emulsion solids (K049), heat-exchanger-bundle cleaning solids (K050), API separator sludge (K051), and leaded-gasoline-tank bottoms (K052).

Several refinery, petrochemical, waste treatment, recycling, and disposal sites have been listed as Superfund sites. An example of a refinery and petrochemical Superfund site is the Sikes Disposal Pits Superfund site. According to the EPA, numerous refineries and petrochemical companies have disposed of chemical and oil-based wastes at the Sikes site. Other examples are the Baldwin Waste Oil site and the Falcon Refinery site, also known as the National Oil Recovery Corporation. The R&H Oil Company site, also known as the Eldorado Refining and Marketing, Inc. site and Tropicana Energy site, and the Dubose Oil Products Company Superfund site, which is a waste treatment, recycling, and disposal facility, are other examples of Superfund sites.

A major portion of refining sector hazardous wastes are not HSWA listed but are characterized as hazardous waste because testing has shown that they have one or more characteristics of hazardous waste (i.e., toxicity, ignitability, reactivity, and corrosivity). These wastes include primary treatment plant sludge, secondary treatment plant sludge, biological treatment sludge, cooling tower sludge, ion exchange regenerant, fluidized catalytic cracking catalysts, and hydrocracking catalysts.

Direct releases from a refinery, especially air emission, can pose significant risk to public, employee, and environmental health. These releases may involve process, storage, transport, fugitive, and secondary emissions.

2.2 Waste from Petroleum Products and Petroleum Storage

Oils originating from petroleum crude can be processed, blended, and formulated to produce numerous products for many applications. Users of

TABLE II
Refining Residual Stream

Oily sludge and other organic residuals	
American Petroleum Institute separator sludge	Primary sludge (F037)
Dissolved-air-flotation unit float	Primary sludge (F038)
Slop oil emulsion solids	Nonleaded tank bottoms
Leaded tank bottom	Waste oils/spent solvents
Pond sediments	Other oily sludge/organic residuals
Contaminated soils and solids	
Heat exchanger bundle cleaning sludge	Residual/waste sulfur
Contaminated soils/solids	Other contaminated solids
Residual coke/carbon/charcoal	
Spent catalysts	
Fluid cracking catalyst or equivalent	Other spent catalyst
Hydroprocessing catalyst	
Aqueous residuals	
Biomass	Spent stretford solution
Oil-contaminated water (not wastewater)	Treatment, storage and disposal leachate (F039)
High-Ph/low-Ph waters	Other aqueous residuals
Spent sulfide solution	
Chemicals and inorganic residuals	
Spent caustics	Residual amines
Spend acids	Other inorganic residuals
Other residuals	

oils and oil products include not only industrial and commercial interests but also the individual citizen through the use of gasoline, lubricants, and fuel oils. These users of petroleum products produce large quantities of waste. Waste oils, due to the nature of their use, contain various contaminants, including heavy metals, combustion by-products, and substances arising from the original use (e.g., PCBs).

The principal uses of oils are as fuels, lubricants, hydraulic and transmission fluids, heat transfer fluids, and insulants. As fuels, oils cover a wider spectrum from light, highly mobile liquids to heavy fuel oils of such low viscosity at ambient temperatures that they must be heated to enable their use. The transportation industry is a major user of oil as a fuel for aircraft, shipping, and many land-based transport systems.

Petroleum products (gasoline, diesel fuel, crankcase oil, and waste oil) are a complex mixture of hundreds of chemicals with individual toxic effects. Those that are most often measured are referred to as BTEX (benzene, toluene, ethylbenzene, and xylenes). The BTEX chemicals are volatile monoaromatic hydrocarbons that are commonly found together in crude petroleum and petroleum products. They are considered one of the major causes of environmental pollution because of widespread occurrences of leakage from underground petroleum storage tanks and spills at petroleum production wells, refineries, pipelines, and distribution terminals. It has been estimated that 35% of the 1.4 million gasoline storage tanks in the United States are leaking. Contamination of groundwater with the BTEX compounds is difficult to remedy because these compounds are relatively soluble in water and can diffuse rapidly once introduced into an aquifer.

Some of the petroleum waste products are gasoline, oil, lubricants, machining or cutting oils, leakage from storage tanks, aqueous waste, sulfur dioxide, and PAHs.

2.2.1 Gasoline

When gasoline or another petroleum product is released into the environment, primary concerns generally include the possibility of contaminating soil and ground- and surface water. Gasoline spill and leaking underground storage tank contaminants are known to be a source of groundwater contamination. The United States uses approximately 16 million bbl of oil and oil by-products, such as gasoline, each day (245 billion gallons per year). Much of this material is stored in tanks at some point during transport from source to point of use. There are approximately 6 million tanks in the United States for storing petroleum, petroleum by-products, and petroleum-related chemicals. Many of the storage tanks are underground. According to the EPA, approximately 1 out of every 4 underground storage tanks may be leaking. Even a small gasoline leak of one drop per second can result in the release of approximately 400 gallons of gasoline into the groundwater in 1 year. A few quarts of gasoline in the groundwater may be enough to severely pollute a farmstead's drinking water.

As gasoline seeps through the soil and encounters groundwater, some components will dissolve and travel with the groundwater. After seeping underground, the petroleum products remain near the top of the water table because they are lighter than water. When wells draw water from the upper portions of the water table, they collect the contaminants during periods of low water availability and high consumption.

A plume of gasoline contaminants has been found in groundwater due to spills and leaking underground storage tanks. Petroleum fuels contain a number of potentially toxic compounds, including common solvents, such as benzene, toluene, and xylene, and additives, such as methyl tertiary-butyl ether (MTBE), ethylene dibromide (EDB), and organic lead compounds. The gasoline additive MTBE has contaminated many thousands of drinking water supplies and wells. MTBE causes widespread groundwater contamination due to its high water solubility and other unique properties. The strong, terpene-like odor of MTBE makes its presence easy to detect, even in minute quantities.

These contaminants (petroleum fuel toxic compounds) are all confirmed or potential carcinogens and could pose a significantly increased cancer risk. MTBE may cause cancer and myriad other health problems, such as asthma and respiratory illnesses. EDB and benzene are also considered carcinogens (cancer-causing).

2.2.2 Crude Oil (Petroleum)

Oils are not especially toxic, but contaminants such as additives, breakdown products, and other substances that may have become mixed with oils during their legitimate use may be much more so. Oils, however, do have considerable potential to cause environmental damage by virtue of their persistence and their ability to spread over large areas of land or water. Films or coverings of oil substances may reduce or prevent air from reaching life-forms of all types within an area of land or sea and can rapidly result in significant degradation of environmental quality in those media. Waste acid sludge arising from the acid/clay process, a common method of recycling waste oils, represents a serious environmental threat.

2.2.3 Lubricants

Lubricants embrace an even wider spectrum of products with many different viscosities and viscosity ranges and the extensive use of performance-enhancing additives. A major source of waste oils is from lubrication. Lubricating oils need replacing at intervals, and although improved designs of engines and better oil grades may reduce the frequency of change, there is little prospect of the need for oil changes being avoided altogether. Major contributors to this market are cars, commercial vehicles, aircraft, railway locomotives, and other large pieces of machinery. Oils from these sources usually contain not only contamination from their use but also a variety of additives intended to improve performance. For example, lubricants are used for propulsion engines. Motor vehicles of modern design have several different non-fuel oil applications, such as engine crankshaft lubrication, gearbox lubrication, and hydraulic fluids for brake systems. Motor vehicles and their use impinge on most aspects of life.

2.2.4 Machining or Cutting Oils

Machining or cutting oils are used extensively for metal cutting, machining of components on lathes, and general engineering activities. In many instances, depending on the metal being machined, oils for this purpose are used as an emulsion in water with detergents, biocides, chlorinated paraffins, halogenated and nonhalogenated additives, and additives included to improve performance. The waste, which can be substantial in volume, comprises an emulsion of typically 2–5% oil in water. The waste may also contain metal dusts and similar contaminates.

Unintentional releases of process fluid from equipment are another source of waste. Any equipment that can potentially leak (e.g., pump, valve, and flanges) is a source of pollution. An oil refinery may have 250,000 such pieces of equipment.

2.2.5 Storage Tanks

Another source of pollution is leakage from storage tanks. At refineries and other oil/product handling facilities, there are many aboveground crude oil and refined product storage tanks in service. API Standard 653 requires tanks to be taken out of service periodically and to be visually inspected for leaks. However, this standard is often ignored since inspection can be very costly. Buried pipelines within refinery boundaries represent another source of leaks, and according to the API they have caused more groundwater contamination than the more visible storage tanks.

Tank washings and sludge from storage tanks and refineries, including those from the cleaning of oil storage tanks, can be a source of contamination since they contain breakdown products. In addition, tank washings can generate substantial volumes of waste, usually with medium to low oil contents. Typical sources are oily bilge water, oily ballast water, water from the jetting and cleaning of tanks both onboard ships and at land-based installations, and true tank-cleaning residues. The range of oils encountered in the generation of these wastes can be broad and can include accumulated oil sludge and sediment, tank scale, and even grit from grit blasting and related cleaning work.

2.2.6 Aqueous Waste

Many oil wastes are highly aqueous in nature and must first be subject to treatment to separate the waste into an oil-rich fraction that may be usable in some way and an aqueous waste stream that may require further treatment prior to final disposal. Processes of this type usually yield sludges that, along with other oil waste-based sludges, have no beneficial use and will require final disposal. Aqueous wastes are unlikely to be entirely free from oil and may need further treatment. Refining processes such as coking sulfur recovery, steam cracking, hydro-cracking, and crude desalting create large volumes of oily wastewater. Wastewater facilities in particular are considered to be a source of significant atmospheric hydrocarbon emissions. Wastewaters constitute more than 80% of total hazardous waste mass at refineries. Also, aqueous wastes may contain other contaminants, depending on the original process, including surfactants, solvents, and toxic metals. PAHs are of particular concern because of their carcinogenic properties.

2.2.7 Sulfur Dioxide

All fossil fuels, including crude oil, contain organic sulfur compounds. When the fuels are burned, sulfur oxides are emitted. Sulfur oxides can cause respiratory problems, but emissions of these compounds are a concern primarily because they react in the atmosphere, forming acids. Over time, deposition of these acids through precipitation and diffusion can increase the acidity of natural bodies of water and change local ecosystems. Crude oil contains materials such as soils, metals, and sulfur- and nitrogen-containing compounds. Removing these materials during refining results in large volumes of waste.

2.2.8 Polycyclic Aromatic Hydrocarbons

Polycyclic aromatic hydrocarbons, often referred to as polynuclear aromatics, are a class of very stable organic molecules composed of only carbon and hydrogen. These molecules are flat, with each carbon having three neighboring atoms, much like graphite. PAHs are released into the atmosphere as gases or particles during the incomplete combustion of organic material. PAHs have a number of sources, including cars, trucks, ships, and aircraft; industrial-power generation; oil refining and waste incineration; and primarily domestic combustion for heating and cooking, especially solid fuel heaters using wood and coal.

Of the 1410 hazardous waste sites listed on the National Priorities List (NPL) (Superfund or NPL sites), 352 have been contaminated by PAHs found in soil, groundwater, and sediment. Exposure to PAHs usually occurs by breathing air contaminated by wildfires or coal tar or by eating foods that have been grilled. The Department of Health and Human Services has determined that some PAHs may reasonably be expected to be carcinogens.

The smallest member of the PAH group is naphthalene, a two-ring compound that is a gas at normal temperatures. Three- to five-ring PAHs occur as either gases or particles in air. PAHs with five or more rings tend to be solids that attach to the surface of other particulate matter in the atmosphere.

2.3 Pipeline Leaks

Pipelines have historically been a primary source of oil spills both offshore and onshore. Pipeline leaks are not a rare occurrence. When they do occur somewhere within the 160,000 miles of liquid fuel-carrying pipelines that criss-cross the nation, they can have disastrous consequences. In 2000, gasoline leaking from an Olympic Pipe Line Company line in Bellingham, Washington, caused an explosion that killed three people. During the 1990s, major reportable spills, those involving at least 2100 gallons, occurred approximately four times a week and totaled approximately 6 million gallons per year. In 1996, a rupture in a 36-in. Colonial pipeline released more than 1 million gallons of diesel fuel into the Reedy River near Fork Shoals, South Carolina, creating a slick that spread for more than three-fourths of a mile down the waterway. A 1993 leak

TABLE III
Sources of Oil Spills from Inland Transportation and Distribution System, 1987–1995

Inland source	No. of spills	Volume spilled (millions of gallons)	% of total volume spilled
Fixed facility	43,188	79.5	62
Pipeline	13,771	33.7	26
Highway transport	13,280	6.2	5
Rail transport	2483	3.7	3
Above ground storage tank	921	4.4	3
Underground storage tank	795	0.5	<1
Total	74,438	128	100

TABLE IV
Causes of Oil Spills from Inland Transportation and Distribution System, 1987–1995

Cause	No. of spills	Volume spilled (millions of gallons)	% of total volume spilled
Equipment failure	11,476	22.8	50
Operator error	4235	4.3	9
Transportation accident	3187	3.0	7
Natural causes	583	1.7	4
Dumping	2250	1.0	2
Other	2584	2.3	5
Unknown	934	10.4	23
Total	25,249	45.5	100

from a Colonial pipeline in Virginia dumped more than 300,000 gallons of oil into a tributary of the Potomac, forcing Fairfax to temporarily switch to an alternate water supply. A 1991 leak from a Colonial pipeline near Spartanburg, South Carolina, spewed an estimated 562,000 gallons of oil into a tributary of the Enoree River, forcing two towns to close their water-treatment plants. they Tables III and IV summarize the sources and causes of oil spills from the inland transportation and distribution system.

2.4 Noise Contamination

Noise in refineries originates from rotating machinery, such as cooler fans, turbines, compressors, engines, and motors. High-velocity flow of fluids through valves, nozzles, and piping also contributes to the general noise level.

3. WASTE ASSOCIATED WITH COAL AND ITS PRODUCTION

Coal waste products are similar to those produced during petroleum exploration, production, refining, and use. Coal is a solid with a complex hydrocarbon structure and a higher carbon-to-hydrogen ratio than petroleum. Coal is the world's most plentiful fossil fuel energy resource, and its use is likely to quadruple by 2020. Coal has a wide range of moisture content (2–40%), sulfur content (0.2–8%), and ash content (5–40%), causing environmental problems from its use. Burning of coal creates approximately 60% of soot-creating sulfur dioxide emission in the United State. It is also a major source of smog-forming nitrogen oxide pollution, mercury contamination, and carbon dioxide, releasing more than 51 tons of mercury and 2 billion tons of carbon dioxide into the air every year. Sulfur dioxide and nitrogen oxide pollution also creates fine particles, which can cause significant increases in acute and chronic respiratory diseases. At every stage of production, transportation, and consumption, coal poses risks to the environment.

Coal is dirty fossil fuel: Mining it can ravage the land, and burning it can generate a large amount of carbon dioxide and other pollutants. The mining/cleaning of coal creates a significant amount of waste. Wastes that are associated with mine extraction (i.e., surface mining, open-pit mining, and underground mining) and processes of coal include liquid effluents (mine tailings containing fine materials that are discharged as slurry to impoundment areas), solid wastes (coarse materials typically >0.5 mm), and coal dust. According to a survey conducted by the U.S. Department of Energy, approximately 2.5–3 billion tons of fine coal has been deposited in tailing impoundments that typically have a life of up to 20 years.

Mining of coal has left large volumes of mining overburden throughout the United States. Mine overburden and waste soils (commonly referred to as mine tailings) are the waste products generated during mining and milling operations. Mine overburden is any material, consolidated or unconsolidated, excluding topsoil, that lies above a natural deposit of coal. It is also the material after removal from its natural state in the process of surface coal mining.

A complex set of chemical and biological reactions and conditions occur over time in mine overburden. Surface deposition of overburden waste products results in atmospheric exposure and subsequent weathering of sulfide minerals present in the rock, precipitation, and ground- or surface water. Oxidation of sulfide minerals by chemical and microbial processes produces sulfuric acid and heavy metal ions that find their way into surface and groundwaters. The release of these components into natural waters is a widespread and persistent problem in areas of the country that have a long history of coal or precious metal mining. Many of these tailings are causing problems with water quality and revegetation.

Mine overburden piles have acid-producing potential as well as acid-neutralization potential. Generally, the rate of acid production exceeds the rate of acid neutralization. Degradation of water quality of surface and groundwaters in contact with acid-producing mine tailings piles is the major environmental problem associated with acid mine drainage. The pH of the water is lowered due to the introduction of sulfuric acid as a result of sulfide mineral weathering. Acidic mine drainage laden with heavy metals and the precipitation of "yellow boy" has a devastating effect on the ecology of freshwater streams.

Heavy metal ions are released into the water as a result of mineral weathering. Depending on the composition of the mine overburden, the pollutants released into surface and groundwaters typically include any of the following heavy metal ions: iron, zinc, aluminum, manganese, cobalt, nickel, copper, arsenic, selenium, cadmium, and lead.

Mine overburden piles also contain *Thiobacillus ferrooxidans*. *Thiobacillus ferrooxidans* is responsible for the oxidation of iron and inorganic sulfur compounds in mine overburden areas. *Thiobacillus ferrooxidans* is acidophilic (acid loving, capable of surviving at low pH) and has a physiology that is well suited for growth in an inorganic mining environment. It obtains carbon by fixing atmospheric carbon dioxide and is strictly autotrophic. *Thiobacillus ferrooxidans* catalyzes the acid and heavy metal and produces reactions that occur in mine tailings resulting in the degradation of water quality downstream of mining operations.

Approximately 1 billion tons of coal is burned annually to generate electricity in the United States; as a result, more than 97 million tons of coal combustion by-products (CCBs) are produced. CCBs are considered a product by the American Coal Ash Association and are referred to as coal combustion products (CCPs). In 1998, CCPs ranked as the third largest nonfuel mineral commodity produced in the United States. Only one-fourth of these are utilized, whereas the rest (80% CCBs) are disposed of mainly in landfills and surface impoundments (Table V).

With shrinking landfill space and increasing amounts of ash generation, there is an urgent need to utilize these materials in an environmentally safe manner. One potential use of these materials is waste stabilization. Intrinsic properties of these products allow immobilization of hazardous nuclear, organic, and toxic metal wastes for safe and effective environmental disposal. Additionally, coal ash contains plant nutrients and can be used to amend highly acidic soils.

CCBs produced during coal-fired electric power production are fly ash, bottom ash, boiler slag, and flue gas desulfurization (FGD) material. The physical characteristics and trace element composition of fly ash, bottom ash, boiler ash, and FGD are given in the Tables VI and VII, respectively.

3.1 Fly Ash

Fly ash is the finest of coal ash particles. It is called fly ash because it is transported from the combustion chamber by exhaust gases. Fly ash is the fine powder formed from the mineral matter in coal, consisting of the noncombustible matter in coal and a small amount of carbon that remains from incomplete combustion. Fly ash is generally light tan in color and consists mostly of silt-sized and clay-sized glassy spheres. Properties of fly ash vary significantly with coal composition and plant operating conditions. In the United States, approximately 50 million tons of fly ash is reused annually.

Fly ash is referred to as either cementations or pozzolanic. A cementations material is one that hardens when mixed with water. A pozzolanic material will also harden with water but only after activation with an alkaline substance such as lime. These cementations and pozzolanic properties make some fly ashes useful for cement replacement in concrete and many other building applications. Fly ash is used in concrete and cement products, road base, oil stabilizer, clean fill, filler in asphalt, metal recovery, and mineral filler.

3.2 Coal Bottom Ash

Coal bottom ash and fly ash are quite different physically, mineralogically, and chemically. Bottom ash is a coarse, granular, incombustible by-product collected from the bottom of furnaces that burn coal for the generation of steam, the production of electric power, or both. Bottom ash is coarser than fly ash, with grain sizes ranging from that of fine sand to that of fine gravel. The type of by-product produced depends on the type of furnace used to burn the coal. Bottom ash is used as construction and railroad fill material, abrasive blasting grit, granules on asphalt roofing shingles, aggregate for concrete and masonry blocks, substitute for sand for traction on icy roadways, and soil amendment to increase permeability.

TABLE V
Amount of Coal Combustion By-products (CCBs) Produced in the United States

Material	Metric tons $\times 10^6$
Crushed stone	1500
Sand and gravel	1020
CCBs	97.7
Cement	85.5
Iron ore	62

TABLE VI
Summary of Physical Characteristics and Engineering Properties of Coal Combustion By-products

Physical characteristic	Fly ash	Bottom ash/boiler ash	Flue gas desulfurization material	
			Wet	Dry
Particle size (mm)	0.001–0.1	0.1–10.0	0.001–0.05	0.002–0.074
Compressibility (%)	1.8	1.4		
Dry density (lb/ft^3)	40–90	40–100	56–106	64–87
Permeability (cm/s)	10^6–10^4	10^{-3}–10^{-1}	10^{-6}–10^{-4}	10^{-7}–10^{-6}
Shear strength	0–170	0		
Cohesion (psi)	24–45	24–45		
Angle of internal friction (°)				
Unconfined compressive strength (psi)			0–1600	41–2250

TABLE VII
Trace Element Composition of Coal Combustion By-products

| | Fly ash | | | | | | | |
| | Mechanical | | Esp/baghouse | | Bottom ash/boiler slag | | Dry flue gas desulfurization material | |
Element (mg/kg)	Range	Median	Range	Median	Range	Median	Range	Median
Arsenic	3.3–160	25.2	2.3–279	56.7	0.50–168	4.45	44.1–186	86.5
Boron	205–714	258	10–1300	371	41.9–513	161	145–418	318
Barium	52–1152	872	110–5400	991	300–5789	1600	100–300	235
Cadmium	0.40–14.3	4.27	0.10–18.0	1.60	0.1–4.7	0.86	1.7–4.9	2.9
Cobalt	6.22–76.9	48.3	4.90–79.0	35.9	7.1–60.4	24	8.9–45.6	26.7
Chromium	83.3–305	172	3.6–437	136	3.4–350	120	16.9–76.6	43.2
Copper	42.0–326	130	33.0–349	116	3.7–250	68.1	30.8–251	80.8
Fluorine	2.50–83.3	41.8	0.4–320	29.0	2.5–104	50.0	—	—
Mercury	0.008–3.0	0.073	0.005–2.5	0.10	0.005–4.2	0.023	—	—
Manganese	123–430	191	24.5–750	250	56.7–769	297	127–207	167
Lead	5.2–101	13.0	3.10–252	66.5	0.4–90.6	7.1	11.3–59.2	36.9
Selenium	0.13–11.8	5.52	0.6–19.0	9.97	0.08–14	0.610	3.6–15.2	10.0
Silver	0.08–4.0	0.70	0.04–8.0	0.501	0.1–0.51	0.20	—	—
Strontium	396–2430	931	30–3855	775	170–1800	800	308–565	432
Vanadium	100–377	251	11.9–570	248	12.0–377	141	—	—
Zinc	56.7–215	155	14–2300	210	4.0–798	99.6	108–208	141

3.3 Boiler Slag

Boiler slag is coarser than conventional fly ash and is formed in cyclone boilers, which produce a molten ash that is cooled with water. Boiler slag is generally a black granular material with numerous engineering uses.

3.4 FGD Gypsum

FGD gypsum is also known as scrubber gypsum. FGD gypsum is the by-product of an air pollution control system that removes sulfur from the flue gas in calcium-based scrubbing systems. It is produced by employing forced oxidation in the scrubber and is composed mostly of calcium sulfate. FGD gypsum is most commonly used for agricultural purposes and for wallboard production.

4. SUMMARY

Worldwide energy demand increases even as knowledge of how fossil fuel use threatens the global environment grows. Environmental impacts arise at all stages in the life cycle of fossil fuel sources. These impacts are as follows:

- Fossil fuel extraction: coal mining and oil extraction
- Fossil fuel distribution: transport of fuels by road, rail, water, and pipelines; transmission of electricity
- Fossil fuel processing: petroleum refining and coal preparation
- Fossil fuel combustion in the domestic, commercial, industrial, and transport sectors
- Waste and residue management: coal ash and refinery residues

The environmental burdens associated with these activities are many and varied, including:

- Emissions of pollutants to air, mainly from fossil fuel combustion but also from methane leakage from coal and oil extraction and distribution
- Discharges of pollutants to rivers, estuaries, and seas from coal mining and oil extraction
- Groundwater pollution from coal mining
- Visual intrusion, noise, and loss of wildlife habitat

SEE ALSO THE FOLLOWING ARTICLES

Acid Deposition and Energy Use • Air Pollution from Energy Production and Use • Air Pollution, Health Effects of • Clean Air Markets • Clean Coal Technology • Crude Oil Spills, Environmental Impact of • Ecosystem Health: Energy Indicators • Ecosystems and Energy: History and Overview • Electricity, Environmental Impacts of • Gasoline Additives and Public Health • Gulf War, Environmental Impact of

Further Reading

Aldridge, C. L., and Bearden, R. (1985). "Studies on Heavy Hydrocarbon Conversion." Exxon Research and Development Laboratories, Baton Rouge, Louisianna.

Allen, D. T., and Rosselot, K. S. (1997). "Pollution Prevention for Chemical Processes." Wiley, New York.

Fries, M. R., Zhou, J., Chee-Sanford, J., and Tiedje, J. M. (1994). Isolation, characterization, and distribution of denitrifying toluene degraders from a variety of habitats. *Appl. Environ. Microbiol.* **60**(8), 2802–2810.

Gary, J. H., and Handwerk, G. E. (1994). "Petroleum Refining Technology and Economics." Dekker, New York.

Ghassemi, A. (2002). "Handbook of Pollution Control and Waste Minimization: Minimization and Use of Coal Combustion By-products (CCBs): Concepts and Applications." Dekker, New York.

Gibbons, J. H., Blair, P. D., and Gwin, H. L. (1990). "Managing Planet Earth; Strategies for Energy Use." Freeman, New York.

Harwood, C. S., and Gibson, J. (1997). Shedding light on anaerobic benzene ring degradation: A Process Unique to Prokaryotes. *J. Bacteriol.* **179**(2), 301–309.

Higgins, T. E. (1995). "Pollution Prevention Handbook." Lewis, New York.

Pacific Northwest Pollution Prevention Research Center (1993). "Pollution Prevention Opportunities in Oil and Gas Production, Drilling, and Exploration." Pacific Northwest Pollution Prevention Research Center. Seattle, Washington.

Salisbury Post (2001, June). Better monitoring needed—Pipeline spill no rare event.

Smith, M., Manning, A., and Lang, M. (1999, November). Research on the re-use of drill cuttings onshore, Report No. Cordah/COR012/1999 for Talisman Energy (UK) Limited.

U.S. Environmental Protection Agency (1995). "Data obtained from EPA Emergency Responsive Notification System," Inland spill data. U.S. Environmental Protection Agency, Washington, DC.

U.S. Environmental Protection Agency (2000, January). "Special Wastes: Crude Oil and Natural Gas: Associated Waste Report; Executive Summary." U.S. Environmental Protection Agency, Washington, DC.

World Bank Group (1998). "Pollution Prevention and Abatement Handbook." World Bank Group, Geneva.

Heat Islands and Energy

HAIDER TAHA
Altostratus, Inc.
Martinez, California, United States

1. Introduction
2. Definition of Heat Islands
3. Causes and Effects of Urban Heat Islands
4. Typical Magnitudes of UHIs
5. Impacts of UHIs on Energy
6. Mitigation of UHI Effects to Save Energy

Glossary

albedo (a) For a flat surface, albedo is generally defined as hemisphere- and wavelength-integrated reflectivity. Thus, whereas reflectivity is the ratio of the amount of reflected to total incident radiation at a particular wavelength, albedo can, by definition, encompass a range of wavelengths of interest. In this article, the interest is in solar albedo with an integral over the portion of the solar spectrum between approximately 0.28 and 2.8 μm. Effective albedo is the albedo resulting from geometrical effects (e.g., multiple reflections) in addition to the effects of the surface albedo. Of course, the definition of albedo is not restricted to surfaces.

boundary layer (BL) (e.g., planetary BL and canopy BL) A layer in the fluid (e.g., air) whose momentum, mass, or heat characteristics are affected or dictated by the fluxes of respective quantities at the underlying surface. Thus, generically, a BL is a layer in which the effects of the surface are "felt."

Bowen ratio (β) The ratio of sensible heat flux (H) to latent heat flux (λe) from a surface or plane, land cover, or region of interest: $\beta = H/\lambda e$. The definition can be applied to any scale.

cost of conserved energy (CCE) In this article, CCE is defined as the compounded additional cost per saved kWh that must be invested in order to implement an energy-saving strategy or feature. The CCE of interest is that of increased roof or urban albedo. CCE = $[\Delta I/\Delta E] \times [i/\{1-(1+i)^{-n}\}]$, where ΔI is additional investment (\$), ΔE is the annual energy savings (kWh/year), i is interest or compounding rate (e.g., 6%), and n is the number of years.

degree-days (DD) A measure of the magnitude and duration of temperature deviation from a certain given threshold over a certain period of time (e.g., annual, seasonal, and monthly). Thus, for example, DD = $\Sigma(T_{avg}-T_t)$, where T_t is any threshold (reference) temperature [e.g., 18.3°C (65°F)], and T_{avg} is the 24-h average temperature. The summation can be taken over any period of interest (e.g., 1 year).

efficiency Coefficient of performance (COP), energy efficiency ratio (EER), and seasonal energy efficiency ratio (SEER) are measures of the ratio of work provided by a system (e.g., cooling or heating) to the power input to that system. COP is dimensionless, but EER and SEER typically have units of Btu/h W^{-1}.

evaporative cooling The process of cooling the air by evaporation of water, in which case the heat of vaporization is taken from the air. Thus, evaporative cooling is a constant enthalpy process.

view factor (e.g., sky view factor) The ratio of the amount of radiation (e.g., long-wave flux) leaving a surface (e.g., street level) through the sky aperture "seen" by the surface to the radiation leaving the surface in all directions (e.g., to the hemisphere). The view factor is thus a purely geometrical definition and can be qualitatively defined as the ratio of sky area or aperture (solid angle) seen from a certain surface (or point) to the area (solid angle) of the entire hemisphere (or sphere) seen by that same surface (or point).

Weather conditions and local meteorology, e.g., urban climates, have significant impacts on regional and local energy use intensities and patterns. An important phenomenon of urban climates, which has direct bearings on energy use, is the urban heat island (UHI). The UHI itself is partly a result of energy use in urban areas. When a UHI is present, air temperature can be several degrees higher than in the surrounds and in summer the urban-scale peak electric demand for cooling can increase by up to 10%.

1. INTRODUCTION

Urban heat islands (UHIs) and energy are interrelated in two complementary ways. First, energy conversion and use can contribute to creating UHIs via increased anthropogenic heating of ambient air by machinery, equipment, air conditioners, cooling towers, motor vehicles, and other sources of heat (e.g., combustion). Although this is only one of many UHI causative factors, it can be significant in urban areas, where such sources are highly concentrated. Energy use and conversion can also result in increased emissions of air pollutants, CO_2, nitrous oxide (N_2O), and water vapor (motor exhausts, refineries, chimneys, electricity generators, etc.), which can enhance local (e.g., urban) greenhouse effects. From a relatively more extreme point of view, it can also be argued that passive and active energy conversion systems (e.g., solar and, to a much lesser extent, wind systems) can also add to the UHI. This would occur, for example, if a vast array of photovoltaic cells were installed in an urban area such that the effective urban albedo is decreased and heat transfer from the cells to the air becomes larger. However, this effect is small since photovoltaics and other solar equipment (e.g., solar water heaters) are sparse and cover a small surface area. Thus, although the albedo of a photovoltaic cell may be approximately 0.05, the impact on overall urban albedo is very small. Also, most high-concentration solar energy conversion facilities (e.g., solar power plants and wind farms) are usually located in desert and rural areas, not in urban regions. Thus, they do not contribute to UHIs directly.

Second, UHIs directly affect energy use in buildings; heating, ventilating, and air-conditioning (HVAC) systems and equipment; motor vehicles; industrial processes; and so on. In buildings, more cooling and less heating are generally needed when UHIs are present. In addition, equipment efficiencies (coefficient of performance, energy efficiency ratio, and seasonal energy efficiency ratio) of HVAC, industrial, and other equipment are to a certain extent functions of ambient temperature and thus are affected by UHIs, even if only by small amounts.

2. DEFINITION OF HEAT ISLANDS

A heat island is simply an area, volume, or region in which the temperature is higher than that of the surroundings. This very general definition can obviously apply to any scale and environment, from micrometers to thousands of kilometers. That is, a heat island can exist on a scale of a few millimeters, as a bubble of hot air surrounded by colder air, or on a scale of thousands of kilometers, such as a high-pressure weather system moving across a relatively cooler air mass. It can also be thought of as a more global or universal phenomenon. Obviously, we would like to put some bounds on this definition (for the purpose of this article) and speak only of the UHI. Thus, a UHI is the difference between the temperature in a certain urban location and that at a given reference point in a nonurban location (e.g., an upwind rural location). Obviously, the choice of urban and reference points or a collection of points can affect the magnitude and the characteristics of a reported UHI.

Keeping the urban scale in mind, there can be a surface temperature heat island and an air temperature heat island (Fig. 1). The latter can be further broken down using various scales and criteria, but some common classifications include canopy-layer heat islands and the more generic boundary-layer heat islands. The former generally occur below effective roof level (quite qualitatively) and the latter extend up to the boundaries of the urban heat plume.

Figure 1 is a conceptual diagram of a 3.5°C UHI. In reality, UHI attributes are very site specific and meteorology dependent. Figure 1 shows a hypothetical temperature-difference profile (difference from a rural reference temperature), indicated with a thick dashed line. Of course, the intensity of the UHI (0–3.5°C) is purely for illustration purposes. Note the downwind displacement (advection) of the location of peak air temperature difference (i.e., the location of maximum UHI, denoted by "max").

3. CAUSES AND EFFECTS OF URBAN HEAT ISLANDS

In the simplest situations, air temperature is dictated by energy balance at the underlying surface. The following is the basic form of a surface energy balance equation:

$$\alpha_s Q + \alpha_s q + \alpha_l L + Q_f = \varepsilon \sigma T_o^4 + h_c(T_o - T_a) + k dT_o/dz|_{z=0} + \lambda e,$$

where α_s and α_l are the absorptivities for short- and long-wave radiation, respectively; Q and q are the direct and diffuse short-wave radiative fluxes, respectively; L is incoming long-wave flux; Q_f is anthropogenic heat flux; ε is emissivity; σ is the Stefan–Boltzmann constant; T_o and T_a are surface and air temperatures, respectively; h_c is the convective heat transfer coefficient; k is thermal conductivity; λ is latent heat of vaporization; and e is

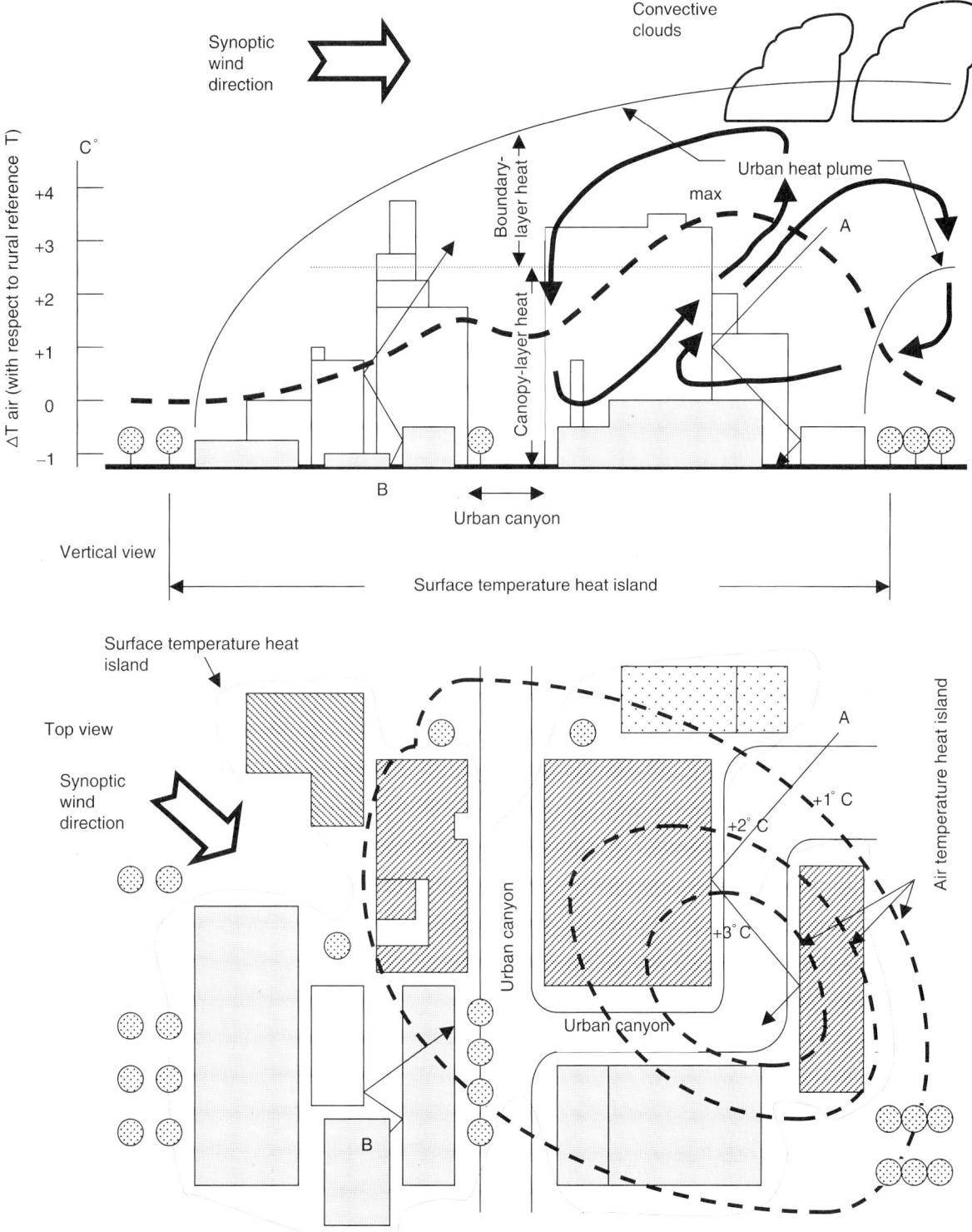

FIGURE 1 Schematic of an urban heat island: (top) vertical view (cross section); (bottom) corresponding projection top view. The thick broken lines show the air temperature UHI intensity read off the scale (°C) at left in the vertical view. The thick arrows show a UHI convective cell advected downwind of the urban center by the synoptic wind. The boundaries of an urban heat plume (also advected downwind) are shown. Other aspects shown are the boundaries of the surface temperature UHI, canopy and boundary layer UHI, convective clouds, incoming and outgoing radiative beams (A and B), and urban canyon.

evaporation rate (mass flux). Thus, the first and second terms on the left side represent the total absorbed short-wave (e.g., solar) radiation flux at the surface (both direct and diffuse), and the third term is the absorbed long-wave (e.g., near infrared) flux. The fourth term on the left is the anthropogenic heat flux (e.g., from motor vehicles, building equipment, and other heat sources) into the surface or near-surface atmospheric layer. On the right side of the equation, the first term represents long-wave radiative flux from the surface, the second term is the sensible heat flux, the third term is ground heat flux (heat conduction through the surface), and the last term is latent heat flux. Some of the terms in this equation are discussed directly or indirectly in the following sections.

3.1 Causes of UHIs

When the energy balance of an urban area is altered so that there is slower cooling relative to that in rural areas (e.g., after sunset), a classic nighttime UHI results. However, when there is significant heat storage in the urban canopy layer, or any additional sources of heat (e.g., anthropogenic), a UHI can arise at any time. Heat islands can appear as a result of a single causative factor or, more likely, a combination of several factors. The following are the main causative factors of the UHI:

Thermal capacity and urban geometry: Urban areas in general are abundant in materials that have high thermal capacity (e.g., concrete, brick, stone, pavements, and asphalt). In addition, urban regions have much more total surface area (exposed to the sun) per horizontal area than do rural regions [surface-to-area ratio (SAR)]. Rural areas tend to have a SAR close to 1. In residential neighborhoods, the SAR is on the order of 2 or 3, whereas in urban cores it is approximately 5 or higher. In cities such as Manhattan, the SAR can easily reach 10 or higher. As a result of the combination of these two factors (increased thermal capacity and increased SAR), solar radiation is captured and stored more efficiently in urban areas. The release of this extra heat, stored in all types of urban structures and materials, contributes directly to warming the air and the creation of a UHI. For example, see beams A and B in Fig. 1.

Sky view factor (SVF): The smaller SVF in urban areas, relative to that in open surrounds, helps create a UHI, especially the nighttime type. This is because radiative cooling of the urban surfaces (and canopy-layer air) to the cooler sky is hindered by the obstructions in urban areas (walls, structures, etc.). In Fig. 1, the beam at point B undergoes multiple reflections that decrease its intensity and warm the surfaces and air in contact with them. The number of reflections may increase when the urban "canyon" is surrounded by taller structures. As a result, a UHI is created due to the differential in cooling rates between the urban area and that of the rural surrounds.

Albedo and effective albedo: The albedo of building materials, pavements, and other urban structures may be lower than that of vegetation or barren land (which is abundant in rural areas). However, in some cases, the reverse is true. For example, low-altitude aircraft measurements have shown that urban albedo can reach 0.20 when the rural vegetated surroundings have an albedo of 0.15. In general, a UHI can arise when urban albedo is lower than that of the rural surrounding areas. In addition to the fact that many urban surfaces (roofs, walls, pavements, streets, structures, etc.) have low albedos, urban areas also create a lower effective albedo compared to that of rural or surrounding areas. The effect of geometry (i.e., urban canyons) is to increase multiple reflections of incident solar radiation and therefore increase the probability that a photon will be absorbed by the canyon surfaces (instead of escaping back), thus resulting in lower effective albedo. This can be seen by the simplified path of beam A in Fig. 1. After several reflections, the beam is less intense and some photons never make it to the canyon floor.

Bowen ratio: As a result of a relatively less dense vegetation cover in urban areas, the partitioning of incoming solar radiation into sensible and latent heat fluxes (among others) is altered. Thus, in urban areas, the Bowen ratio (β) is larger and this results in higher air temperatures. In rural areas or suburbs in which there is typically more vegetation, evaporation, and evaporative cooling, β is smaller, which can help keep air temperatures relatively lower. Typical values for β are approximately 4 or 5 in urban areas and approximately 0.8–1.5 in vegetative canopies. By comparison, β is approximately 0.1 for oceans and in the tropical forests it is approximately 0.2.

Anthropogenic heating: Urban areas alter the energy balance not only "passively," such as through effects of urban geometry, thermal capacity, decreased effective albedo, and lack of vegetation, but also "actively" via injection of heat directly into the air and, to a smaller extent, into the surface. This is typically referred to as furnace heat, anthropogenic heat, or man-made heat. Typical sources of anthropogenic heat include motor vehicles, stacks and chimneys, HVAC in motor vehicles and buildings, industry, refineries and processing plants, and power plants. Because it is difficult to directly measure anthropogenic heat flux, it is usually indirectly estimated or derived from energy use profiles at sources or in areas of interest. Typical values of anthropogenic heat flux density in residential areas are on the order of 10–20 W m^{-2} and in dense urban cores in the range of 50–100 W m^{-2}. Flux densities larger than 100 W m^{-2} are thought to occur at certain times in extremely dense urban cores, such as Lower Manhattan (approximately 120–150 W m^{-2}). In Japan, flux densities of up to 500 W m^{-2} have been reported, although this is highly uncommon. For comparison, the maximum incoming solar radiation flux density at street level on a cloudless summer day can range between 800 and 1000 W m^{-2} around solar noon. Outside of the atmosphere, the incoming solar radiation flux density (solar constant) is approximately 1350 W m^{-2}.

Of course, a UHI can be caused by any one of the previously discussed causative factors or, most commonly, by a varying degree of combinations of such. As a result, the characters of UHIs and their extents can be very diverse and different from one location or time to another.

3.2 Effects of UHIs

UHIs can alter the meteorology, air pollutant emission rates, pollutant transport, and photochemistry in the area in which they exist. Of course, they also affect energy use. Briefly, the nonenergy effects of UHIs are

Creation of a wind pattern or circulation, which is typically a convective cell located in the UHI vicinity: Depending on the strength of the background winds (e.g., synoptic) relative to that of the UHI winds, the convective cell can be asymmetrical and be advected downwind (see, for example, the thick arrows in Fig. 1). In general, the larger and denser a city (larger population), the stronger its UHI circulation.

Increased cloudiness and rainfall as a result of enhanced convective activity within the heat plume associated with UHIs (see hypothetical clouds in Fig. 1, downwind of the UHI maximum and near the location of divergence of the UHI circulation): Of course, UHIs produce increased convective clouds or precipitation only when the right conditions are met (e.g., temperature, winds, moisture, and condensation nuclei). Long-term observations suggest that UHIs can cause up to an 8% increase in cloudiness, 14% more rainfall, and 15% more thunderstorms near urban areas compared to their surroundings.

Increased ambient temperatures, the most obvious effect of UHIs: On average, typical UHIs are on the order of 2°C, but extremes as high as 8°C have been recorded. Figure 1 shows a hypothetical UHI of 3.5°C downwind of the highest urban concentration. It is obvious that the choice of urban and rural reference points can dictate the "observed" UHI intensity.

Increased mixing of pollutants in the boundary layer (BL): Depending on stability, depth of the BL, and convective activity and strength, pollutants emitted into the urban atmosphere can be well mixed and dispersed within the BL, resulting in lower apparent concentrations compared to a situation in which the BL is shallower and pollutant mixing is inhibited.

Increased temperature-dependent emissions of air pollutants: For example, the emission of biogenic hydrocarbons (e.g., isoprene from vegetation) is highly temperature dependent. It can double or triple for every 6–10°C increase in air temperature in the range of 25–37°C. In addition, most fugitive emissions of hydrocarbons (e.g., from fuel storage tanks, on-road motor vehicles, motor engines hot-soak, and evaporative losses) are temperature dependent. Direct emissions of particulate matter (e.g., of interest PM_{10} and $PM_{2.5}$) are also temperature dependent to a certain extent. Increased precursor emissions generally lead to an increased rate of smog formation, decreased visibility, and poorer air quality.

Photochemical production of smog generally positively correlated with temperature: This is true at least for some of the most influential photochemical reactions. Thus, a UHI can act as a catalyst for accelerating the formation of tropospheric/urban ozone, which poses a significant health problem in major urban areas. Because higher temperatures can exacerbate ozone formation while at the same time increase the mixing of pollutants in the BL, there may appear to be a competing effect. However, most data and observations suggest that the net effect of a UHI in terms of ozone concentrations is a negative impact on air quality.

4. TYPICAL MAGNITUDES OF UHIs

Because UHIs form as a result of several factors that differ in space, time, location, geography, and with combinations of such, it is obvious that their intensities differ not only in absolute terms but also in location, time, duration, and an array of spatio-temporal characteristics. For example, some UHIs peak at night (most common), others during the day, and others have multiple diurnal peaks. In infrequent cases, a UHI can also be roughly constant during a 24-h period. In addition to hourly and subhourly fluctuation in intensity, UHIs also have diurnal and seasonal fluctuation components.

Thus, it is difficult to generalize and characterize UHI intensities with one or two indicators. However, there may be some observed upper bounds and common (frequently observed) intensities. The upper bounds seem to be related to city size (population) in some ways. Urban heat islands of 1 or 2°C are typical of the North American continent (the United States and Canada), where they can peak up to 4°C or more. Observations and numerical simulations of a number of U.S. metropolitan areas suggest that UHIs on the order of 2°C are typical in Los Angeles, Chicago, Atlanta, Washington, DC, Philadelphia, and New York. UHIs of 1–1.5°C are found in Houston, Dallas, Phoenix, and Miami. In Japan and China, UHIs on the order of 3°C have been frequently reported. Larger UHIs have been reported in Brazil (as high as 8°C in Sao Paolo) and Britain (as high as 6 or 7°C in London). In Mexico, similar UHIs have been observed. As discussed later, a UHI of 1 or 2°C is a significant factor in energy use.

5. IMPACTS OF UHIs ON ENERGY

The impacts of UHI on energy can be seen on regional, urban, and building scales. In terms of energy supply, conversion, and use, these scales roughly correspond to utility service territory and power plants, city-scale grids, and building scales. Of course, urban and building energy use are not the only contributors to utility-scale loads, but most often they are a major component. As previously mentioned, a UHI can cause several meteorological perturbations, but perhaps the most relevant from an energy perspective is that of increased ambient air temperature. Temperature can also be thought of as a surrogate (implicit indicator) for other local weather parameters, such as available sunshine, cloudiness, precipitation, and winds.

The typical energy impact of UHIs is to increase cooling loads in summer and decrease heating loads in winter. The net effect, of course, depends on the characteristics of a region's climate (summer dominated, winter dominated, both, or none) and general meteorological features (e.g., wind patterns, available sunshine, precipitation, and topographically induced flows and phenomena). It also depends on local energy-specific aspects, such as population density, dominant sectors of energy use, types of buildings, age, distribution, general building envelope characteristics, saturation of HVAC equipment, and local cost of fuel and electricity rates. However, in general, a UHI has relatively larger impacts on (increasing) cooling loads than on (decreasing) heating loads. Also, the former is an unwanted effect, whereas the latter is generally a welcomed one. Thus, the remainder of this section focuses more on the cooling energy impacts of UHIs than on their heating energy implications.

A first estimate of the effects of UHIs on energy (e.g., heating and cooling loads) would involve quantifying their impacts on heating and cooling degree-days (HDD and CDD, respectively). If a certain reference temperature is selected [e.g., 18.3°C (65°F)], we find that most urban areas in the United States see a typical increase in CDD of 15–35% as a result of UHIs, with some extremes as high as 90%. A typical decrease in HDD is approximately 5–14%, with extremes as high as approximately 30%. Table I shows the impact of UHIs on HDD and CDD in selected regions.

However, one CDD is not equivalent to one HDD when converted to energy costs, for example, if cooling is done with electricity and heating mainly with natural gas. Thus, in addition to the fact that electricity and gas costs are different, the HDD/CDD weight also depends on how cooling and heating are achieved locally (i.e., HVAC systems, primary and secondary systems, power plants, and overall system

TABLE I

Impacts of UHI on Annual HDD and CDD in Selected Cities (Base 18.3°C)[a]

City	Impact of UHI on HDD (%)	Impact of UHI on CDD (%)
Los Angeles	−30	+90
Washington, DC	−5	+20
St. Louis	−5	+10
New York	−5	+25
Baltimore	−15	+35
Seattle	−15	+55
Detroit	−5	+15
Chicago	−5	+25
Denver	−10	+20

[a] Rounded to the nearest 5%.

efficiencies). Thus, the conversion from HDD/CDD to energy use and costs depends on local factors and energy rates. In the United States, the average electricity cost is approximately $0.08/kWh and for gas approximately $0.65/therm (1 therm = 10^5 Btu), but deviations from these averages (by geographical location and time/season) can be quite significant. Thus, the HDD/CDD data given in Table I are for qualitative comparison purposes.

At the building scale, the energy impacts of UHIs are rather straightforward to understand, even though building characteristics, thermal integrity, occupancy schedules, and thermal comfort preferences may be complicating factors. In general, residential or small nonresidential buildings are significantly affected by the UHI. In larger buildings, in which internal loads (heat gain from occupants, lighting, machinery, and equipment) are larger than the heat gain or loss through the envelope, the sensitivity to air temperature (UHI) is relatively smaller. Table II gives some orders of magnitudes for expected peak cooling electric demand change with temperature in North America. These ranges are for illustration purposes and not exact numbers.

Thus, for example, the effect of a 3°C UHI would be to increase peak demand for cooling on a summer afternoon by approximately 5% of the daily average demand in large office buildings (relative to whole-building electricity consumption). For small offices, the increase would correspond to approximately 10% and for residential buildings to approximately 15%.

The impacts of UHIs on building energy use have also been studied and analyzed with numerical simulations in an attempt to isolate the effect of UHIs from other factors. Computer modeling is needed in such sensitivity studies since it is nearly impossible in the real world to find exactly identical buildings some within and others outside of a UHI with everything else similar so that the impacts of UHIs can be isolated. Advanced computer simulations show that in terms of system cooling peak, UHIs can cause increases in the range of 5–40% depending on weather, location, and building type. In terms of heating peaks, the UHIs can cause a decrease on the order of 4 or 5%. The simulations also show that in some regions and climates, the effect of UHIs is small or negligible. In addition to HVAC-related energy use, there are other aspects in buildings that are indirectly related to UHIs, such as indoor thermal comfort, indoor environmental quality, natural ventilation, and the performance of passive energy systems.

At the urban scale, data suggest a sensitivity in the peak electric demand of 2–5%/°C in the United States. Values of 2.5–3%/°C are common in cities such as Los Angeles, Chicago, Washington, DC, and New York. Values of approximately 5%/°C are found in cities with relatively hotter climates, such as Atlanta, Houston, Dallas, Phoenix, and Miami. The increase in electric demand occurs when the daily maximum temperature is above a certain threshold (e.g., 20°C in summer). In terms of absolute values, these percentages translate into a range of effects. In areas such as the Los Angeles basin, this corresponds to approximately 130 MW/°C, whereas in hotter regions, such as California's Central Valley or the state of Texas, this is equivalent to approximately 300 MW/°C. Thus, depending on the UHI and the regional climate, the actual increase in demand can be significant. On the other hand, in some cold-weather regions, the sensitivity of the electric demand may be negative, indicating savings (rather than penalties) with increased air temperatures as a result of decreased electricity use for heating. This occurs in the northeastern United States and along the Canadian border.

For the purpose of this discussion, it is possible to identify four climate groups in the United States that correspond to certain ranges in the sensitivity of the peak electric demand to temperature. Such climate types are cold, moderate, warm, and hot, as shown in Table III.

According to various numerical and observational studies, UHIs are typically in the range of 1–3°C in North America. Thus, in Los Angeles, Chicago, Washington, DC, New York, and Phoenix, approximately 5% of the peak electric demand is a result of UHIs. In Houston and Dallas, this is approximately

TABLE II

Magnitudes for Typical Temperature Sensitivity of Peak Electric Demand for Cooling at the Building Scale

	Total floor area (× 1000 m^2)	kW °C^{-1}	Total electricity usage (kWh m^{-2} year^{-1})
Large office buildings	50–100	50–110	100–130
Small commercial buildings	1–1.8	1.2–3	70–120
Residential buildings	0.1–0.4	0.2–0.35	50–100

TABLE III

Sensitivities of Peak Electric Demand by Climate Types in the United States

Climate	Sensitivity of peak (%/°C)	States
Cold	−1	Maine, Vermont, Montana
Moderate	1	Massachusetts, Virginia, the Dakotas, Michigan
Warm	2–3	Alabama, Tennessee, California
Hot	3–5	Florida, Louisiana, Texas, Arizona

8%, in Atlanta it is approximately 11%, whereas in Philadelphia it is approximately 3%. In general, it is estimated that 5–10% of the urban peak electric demand in summer-dominated large U.S. urban areas is a result of the UHI effect. Of course, the higher temperatures associated with a UHI not only increase the demand for cooling but also decrease the efficiency of the air-conditioning/cooling systems in meeting the loads.

Finally, UHIs also indirectly affect other aspects of energy use. Warmer conditions, for example, induce an increase in watering needs (e.g., for urban vegetation, nearby crops, and parks) and thus an increased need for pumping water and distributing it. Higher air temperatures also cause an increased need for air-conditioning use in motor vehicles. The additional air-conditioner load requires more energy (increased fuel consumption). For motor vehicle air-conditioner compressors, the "on time" fraction increases by approximately 5%/°C (i.e., the compressor is on 5% longer for every 1°C increase in ambient air temperature above a certain threshold). Another aspect of UHI is health and thermal comfort, which entail indirect use of energy as well. Heat waves, for example, can be exacerbated with UHIs and thus heat-related hospital admissions increase and health care becomes a serious concern during the summertime. Although difficult to quantify, these aspects involve increased energy usage in health facilities.

6. MITIGATION OF UHI EFFECTS TO SAVE ENERGY

Just as the UHI imparts effects on energy at various scales (e.g., building and urban scales), its mitigation (reversing or compensating for its effects) can also be addressed at such scales.

Thus, at the building scale, the goal is to cool the building envelope or cool the ambient air in the immediate vicinity or both. More accurately, the goal is to prevent the envelope and ambient air from heating up as much as they would without UHI control. The current thinking is that cooling the building envelope should be done using high-albedo (reflective) materials so that the amount of solar radiative flux absorbed, and hence the heat conducted through the structure, is reduced. Typical building materials (e.g., roofing materials) have albedos in the range of 0.08–0.25. For example, the majority of asphalt shingles typically have an albedo of 0.10–0.15. Roofing membranes, such as black single-ply roofing, have a typical albedo of 0.06 and gravel roofs between 0.12 and 0.34, depending on the color of the gravel, but most tend to be approximately 0.15.

A hypothetical example of conventional roofing materials with an albedo of 0.25 is shown by the lower (thick) line in Fig. 2. Roof albedos can be increased rather easily with high-albedo (reflective) materials or coatings so that the new albedos attain the range of 0.4–0.6 after accounting for effects of weathering, aging, and soiling. Some reflective materials have an initial albedo of up to 0.85 or higher but require maintenance to keep it at that high value. A hypothetical example of a high-albedo, light-colored material (with an albedo of 0.70) is shown by the upper (thin) line in Fig. 2. However, high-albedo materials need not be lightly colored or white. For example, a hypothetical material shown by the middle line (thick dashed) in Fig. 2 has an albedo of 0.5, but it will appear dark to the human eye because of its low reflectivity in the visible range.

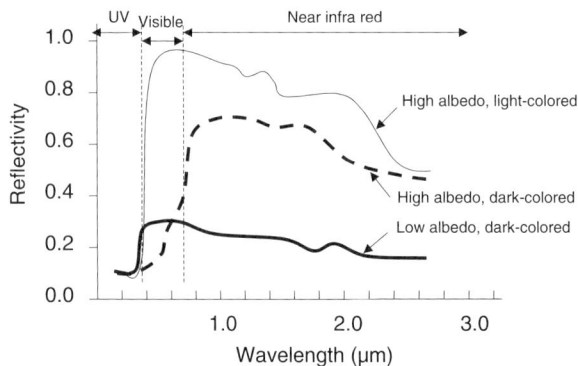

FIGURE 2 Hypothetical spectral reflectivities (albedo) of various materials. These hypothetical materials have albedos of 0.25, 0.50, and 0.70, respectively, from bottom to top.

Materials for increasing roof albedo exist in a variety of colors and textures. In the United States, their incremental cost over traditional roofing materials is $0–2/m^2. Incremental cost is defined here as the extra costs incurred as a result of selecting a reflective roof instead of a traditional one. Of course, if the selection is done at the time of maintenance reroofing or during the initial construction phases, there may be no incremental costs. The implementation of high-albedo roofs can offset or more than offset the local effects of UHIs on energy use in buildings. Roof surfaces with higher albedo are typically 5–15°C (up to 30°C) cooler than their low-albedo counterparts. Reflective roofs warm up at approximately one-third the rate of more conventional roofs. Thus, whereas conventional roofs can be as much as 40–45°C warmer than the overlying air, high-albedo roofs are typically 10–15°C warmer than the air, assuming similar underlying construction and materials. During certain times, the surface temperature of reflective roofs can be very close to that of ambient air.

To cool the ambient air around a building, the current thinking is to increase vegetative cover so as to promote evaporative cooling. The tree canopy also has the beneficial effect of shading the building structure, thus significantly decreasing solar heat gain by the envelope. Another energy benefit of increased urban vegetation, particularly of the evergreen type, is wind shielding. Although this is only indirectly related to heat islands, it has a significant energy effect. In winter, for example, trees shield buildings from the cold winds, reducing the infiltration of colder air to the indoor spaces and thus helping to reduce heating energy needs, especially at higher latitudes. Of course, this effect is important only for one or two-storied buildings, for which trees can shield the wind from the structures. Upper floors in taller buildings do not benefit from ground-level tree canopies. Urban vegetation has other benefits, such as improving air quality by trapping and depositing pollutants from the air. It also has aesthetics values and qualities.

At the urban scale, the main concept is to implement high-albedo materials not only on buildings (e.g., roofs) but also on pavements, roadways, parking lots, sidewalks, and other built-up surfaces. The strategy would also include a reforestation program in which a large number of trees are planted throughout urban areas. The effects of such large-scale implementation of urban albedo and reforestation programs would be to reduce the urban heat island, completely offset it, or sometimes more than offset it. Simulations show that regional air temperature can be decreased by 4°C or more at certain locations and times as a result of increased urban albedo. These modeling studies also show that increasing vegetative canopy cover by 30% can potentially decrease air temperature by 3°C or more. However, a reduction in air temperature in urban areas of 1 or 2°C is more typical according to mesoscale meteorological modeling work. Thus, in most cities discussed previously, the UHI can be offset by large-scale implementation of high-albedo materials and urban forest. In terms of utility-scale impacts of such strategies, a decrease of 2°C in the UHI would be equivalent to savings of 4–10% from the afternoon peak electric demand in summer.

Extrapolation of numerical simulation results to regional and national scales suggests that the savings from implementing an aggressive reflective roof and pavement program can be significant, even after accounting for the wintertime heating energy penalties that can inadvertently result from increased reflectivity to incoming solar radiation. Typically, however, the heating penalties are smaller than the cooling benefits because in wintertime, solar radiation flux density is smaller (lower sun altitude angle), cloud cover is larger, overcast events are more frequent, and snow and rain can cover building and built-up surfaces (e.g., pavements and parking lots). Thus, the impacts of increased albedo is minimal under such conditions.

Table IV provides some simulation results obtained at the Lawrence Berkeley National Laboratory. The first row gives estimated net energy savings (in millions of dollars per year) from high-albedo roofs for each city. These results were based on the assumptions that residential reflective roofs have an albedo of 0.5 and that nonresidential reflective roofs have an albedo of 0.7 (both having an initial albedo of 0.2).

Table IV also gives the CDD and HDD (base 18.3°C) to show the climate traits for each region. Obviously, the regionwide savings also relate to city size and urban population density, not just to climate type. In addition, savings in cooling electricity and increases in natural gas needs for heating are given per 100 m^2 per year for these selected cities. The savings and penalties are given for both residential and nonresidential buildings and the net effect, integrated over the region, is shown. Extrapolating these results to nationwide scale, the net savings from an aggressive high-albedo roofs program in the United States is estimated to be on the order of $700 million per year.

TABLE IV
Energy Savings from Complete Implementation of High-Albedo Roofs

	City							
	Los Angeles	Phoenix	Dallas	New Orleans	Atlanta	Miami	WDC/Baltimore	New York
$M year^{-1} (net savings)	35	37	20	9	9	20	8	16
CDD, base 18.3°C (65°F)	940	3815	2415	2620	4920	4130	1045	1000
HDD, base 18.3°C (65°F)	1300	1155	2305	1680	3020	140	5235	5090
kWh year^{-1} 100 m^{-2}, residential	−182	−314	−166	−199	−153	−259	−137	−104
Therms year^{-1} 100 m^{-2}, residential	+4	+1	+3	+3	+4	0	+6	+10
kWh year^{-1} 100 m^{-2}, nonresidential	−350	−409	−224	−287	−239	−340	−221	−211
Therms year^{-1} 100 m^{-2}, nonresidential	+3	+2	+4	+2	+6	0	+9	+9

The cost of conserved energy (CCE) has been calculated for a range of building types, roof types, insulation levels, roof life spans, various levels of increased albedo, and a range of weather conditions. The target, of course, is to keep the CCE under $0.08/kWh, which is the U.S. national average cost for electricity. To achieve this target, the incremental cost for reflective roofs should be kept under $0.6 m^{-2} if it is assumed that the life spans of reflective and nonreflective roofs are equal. This is not true, however, since the life span of reflective roofs can be double that of nonreflective roofs, and if the cost of replacement for nonreflective roofs is factored in, then the cutoff incremental cost for reflective roofs can be much larger (to keep CCE under $0.08/kWh). This is quite significant since other benefits of reflective roofs are not factored in (e.g., air quality and health benefits).

Increased vegetative cover cools the ambient air within and under a moist, evaporating canopy by up to 3°C. For example, parks are typically 0.5–1.5°C cooler than surrounding areas in higher latitudes. Observations in Canada show that urban parks can be 2.5°C cooler than their immediate built-up surroundings. In relatively lower latitudes (e.g., the southwestern United States), they have been observed to be 1–3°C cooler. Measurements in California's Central Valley show that orchards can be up to 6°C cooler than surrounding nonvegetated land but are typically only 2°C cooler. Studies in Japan show urban parks to be 1.5°C cooler than their surroundings in the summertime. However, it has also been shown that trees can sometimes increase air temperature within or above the canopy, especially if they are drier. The increase can be on the order of 1°C. Thus, the net effects of urban reforestation can be more complex to quantify than the effects of increased urban albedo, which are relatively more straightforward to understand.

For vegetation, the costs vary widely depending on tree species, age, size, planting location, etc. In the United States, trees can cost less than $5 when small and young. However, some 5-m trees cost $500 or more if purchased individually. Labor (planting) and maintenance (watering, pruning, and trimming) costs should also be factored in for an accurate analysis.

SEE ALSO THE FOLLOWING ARTICLES

City Planning and Energy Use • Ecological Footprints and Energy • Economic Growth and Energy • Land Requirements of Energy Systems • Population Growth and Energy • Suburbanization and Energy • Urbanization and Energy

Further Reading

American Society of Heating, Refrigerating and Air-Conditioning Engineers (1998). Energy savings of reflective roofs. *ASHRAE Tech. Data Bull.* **14**(2).

Arya, S. P. (1988). "Introduction to Micrometeorology." Academic Press, New York.

Department of Energy (1990). Energy and climate change, Report of the DOE Multi-Laboratory Climate Change Committee. Lewis, Chelsea, MI.

Landsberg, H. (1981). "The Urban Climate." Academic Press, New York.

Linder, K. P., and Inglis, M. R. (1989). The potential effects of climate change on regional and national demands for electricity. *In* "The Potential Effects of Global Climate Change on the United States," Report No. EPA-230-05-89-058, Appendix H, pp. 1–38. Environmental Protection Agency, Washington, DC.

Oke, T. R. (1987). "Boundary-Layer Climates." Methuen, London.

Pomerantz, M., Akbari, H., Berdahl, P., Konopakci, S., and Taha, H. (1999). Reflective surfaces for cooler buildings and cities. *Philos. Magazine B* **79**(9), 1457–1476.

Taha, H. (ed.). (1997a). *Energy Buildings Special Issue Urban Heat Islands* **25**(2).

Taha, H. (1997b). Modeling the impacts of large-scale albedo changes on ozone air quality in the south coast air basin. *Atmos. Environ.* **31**(11), 667–1676.

Taha, H., Konopacki, S., and Gabersek, S. (1999). Impacts of large-scale surface modifications on meteorological conditions and energy use: A 10-region modeling study. *Theor. Appl. Climatol.* **62**, 175–185.

Heat Transfer

FABIO GORI
University of Rome "Tor Vergata"
Rome, Italy

1. Historical Survey
2. Modality of Heat Transfer
3. Heat Conduction
4. Thermal Radiation
5. Heat Transfer by Convection
6. Dimensional Analysis

Glossary

blackbody A body absorbing all the thermal radiation impinging on it.
boundary The geometric contour dividing the thermodynamic system from the exterior.
conservation equations Equations expressing the balance of a prescribed quantity (e.g., mass, momentum, and energy).
convection Heat transfer at the boundary between a solid wall and a moving fluid.
exterior The part of space external to the thermodynamic system.
first law of thermodynamics The physical principle relating heat transfer to mechanical work.
Fourier equation Basic constitutive equation of heat conduction relating heat transfer to temperature gradient.
heat conduction Heat transfer in a substance, solid, liquid, or gas without macroscopic movements of the matter (not involving electromagnetic waves).
heat transfer Energy exchange across the boundary of a thermodynamic system due to a temperature difference.
Newton equation Basic constitutive equation of a fluid relating shear stress to velocity gradient.
non-Fourier equation Constitutive equation of heat conduction for special phenomena (e.g., low temperatures).
non-Newtonian equation Constitutive equation for shear stress of special fluids.
scale of temperature In the International System, the unit of measurement of the temperature is the degree Kelvin (K), which is 1/273.16 of the temperature of the triple point of water, and the dimension is $[\theta]$.

second law of thermodynamics The physical principle establishing a limit in the conversion of heat into work.
temperature The fundamental quantity measured by a thermometer.
thermal radiation Heat transfer connected to electromagnetic waves.
thermodynamic system Part of space under investigation.

Heat transfer is the energy interaction between a thermodynamic system and the exterior, due to a temperature difference, across the boundary. Heat transfer is measured, according to the first law of thermodynamics, as the work exchanged by a closed system in a cyclic process. The unit of heat transfer is Joule (J) (i.e., equal to the work unit), and the dimensions are $[ML^2T^{-2}]$. No instrument can directly measure heat transfer, and only the thermometer can measure temperature.

1. HISTORICAL SURVEY

Experiments on heat transfer were done in ancient times and several kinds of heat machines have been invented. The first proposal to use a thermoscope to measure temperature is attributed to Galileo Galilei. The Galileo scientific heritage was continued by the Cimento Academy, founded in Florence in 1657 by the Granduke Ferdinand II and Prince Leopold of Medici. The Academy promoted experiments that allowed modern science to make the jump from the heat transfer empiricism of the past to the modern heat transfer scientific design. The most significant scientific contributions of the Cimento Academy, in its 10 years of existence, were the proposal to use several instruments to measure temperature and experiments in barometry and vacuum, which allowed other physical quantities to be defined, such as pressure, which plays a significant role in heat transfer.

2. MODALITY OF HEAT TRANSFER

There are three modalities of heat transfer: heat conduction, thermal radiation, and convection. Heat conduction is the kind of heat transfer, associated with the internal energy of matter, that occurs within a substance (gas, liquid, or solid) without macroscopic movement of its parts. No heat conduction is present in vacuum. Thermal radiation is the kind of heat transfer that is present also in vacuum. Convection is a combined type of heat transfer, occurring in a fluid (liquid or gas), which includes heat conduction and fluid motion.

2.1 Heat Conduction

This kind of heat transfer is present within a substance and is associated with the energy of the molecules, atoms, and their components. Heat conduction is related to the microscopic behavior of the matter. On the other hand, the basic, or constitutive, equation of heat conduction is related to the temperature gradient. The fundamental law of heat conduction must be consistent with the second law of thermodynamics—that is, heat must be exchanged between parts of a body or bodies at different temperatures. In contrast, no heat transfer is present between parts at the same temperature. The first law of thermodynamics, or the energy conservation equation, states that energy is conserved in the absence of heat sources and sinks. In general, the temperature at a certain point on the matter is a function of the space coordinate and time. An isothermal surface is a geometric surface on which temperature is the same throughout. If the temperature along this surface is constant in time, the isotherm can be said to be in steady state. Assume two isothermal surfaces with a temperature difference equal to dT distant in space dn, where dn is measured perpendicularly to the isothermal surface. The heat conduction per unit time and area, Q/A, has been proposed by J. B. Fourier to be proportional to a macroscopic physical property of the substance k according to the equation

$$q = Q/A = -k\, dT/dn = -k\nabla T. \qquad (1)$$

The property k, called thermal conductivity of the substance, is defined by Eq. (1) and has been found to be a physical property (i.e., it is a function of temperature and pressure). The units of k are W/m K and the dimensions are $[M\, L\, T^{-2}\, \theta^{-1}]$. The thermal conductivity can be related to the microscopic behavior of the substance. The coupling of the Fourier equation, also called the constitutive equation for heat conduction, with the first law of thermodynamics, or the energy conservation equation, allows one to write a differential equation, in a three-dimensional rectangular domain, in unsteady state as

$$\rho c\, \partial T/\partial t = \partial/\partial x(k\, \partial T/\partial x) + \partial/\partial y\, (k\, \partial T/\partial y) \\ + \partial/\partial z\, (k\, \partial T/\partial z) + q_g, \qquad (2)$$

where q_g is the heat generated per unit time and volume, $k = f(x, y, z, t)$, $\rho = f(x, y, z, t)$, and $c = f(x, y, z, t)$. Equation (2) is valid for isotropic, heterogeneous media. The assumption of constant thermal conductivity allows Eq. (2) to be written as

$$\partial T/\partial t \\ = k(\rho c)\, (\partial^2 T/\partial x^2 + \partial^2 T/\partial y^2 + \partial^2 T/\partial z^2) + q_g/(\rho c) \\ = \alpha \nabla^2 T + q_g/(\rho c). \qquad (3)$$

The new physical property of the substance, $\alpha = k/\rho c$, called thermal diffusivity, has been introduced because of the unsteady state; it has the units (m^2/s) and the dimensions $[L^2 T^{-1}]$. Equation (3) has a parabolic form, which implies an infinite velocity of propagation of the effect or, in other words, that the effects of a thermal disturbance propagate instantaneously to an infinite distance from the source. Equation (3) has been verified for phenomena at ambient and high temperatures, and experimental results have been found to be in agreement with the theory. For very low-temperature phenomena, Eq. (1) has been modified and the following is one of the proposals:

$$\alpha/C^2\, \partial q/\partial t + q = -k\nabla T, \qquad (4)$$

and the corresponding conservation equation becomes

$$1/C^2 \partial^2 T/\partial^2 t + 1/\alpha\, \partial T/\partial t = \nabla^2 T, \qquad (5)$$

where C is the velocity of propagation of heat transfer, given by $C^2 = \alpha/t_o$, and t_o is a relaxation time. The heat transfer with finite propagation velocity has been confirmed experimentally at liquid helium II temperatures. A similar effect of the liquid helium II has been found in materials of great purity and at low temperatures. Equation (5) is a hyperbolic equation.

2.2 Thermal Radiation

Thermal radiation is energy transfer in the form of electromagnetic waves. The microscopic mechanism can be related to the energy transport by photons released from molecules and atoms. The physical parameters that describe thermal radiation are the photon or wave velocity, c, the wavelength, λ, and the frequency, ν. The photon energy is given by the relation

$$E = h\nu, \qquad (6)$$

where h is the Planck constant, $h = 6.6256 \times 10^{-34}$ Js. Frequency and wavelength are related by

$$\lambda \nu = c. \qquad (7)$$

The thermal radiation velocity in vacuum is equal to $c_0 = 2.997925 \times 10^8$ m/s. The wave velocity c in a medium is connected to c_0 by the relation

$$n_0 = c_0/c, \qquad (8)$$

where n_0 is the refraction index of the medium traveled by the electromagnetic waves. Electromagnetic waves are classified by the wavelength λ, which is usually measured on the scale of 10^{-6} m = 1 μm. Visible light is in the range 0.4–0.7 μm. The wavelengths generated by heated bodies are in the range 0.3–10 μm and are the thermal radiations of common interest. Radiation with a wavelength larger than the visible one is called infrared, whereas that with a wavelength smaller than the visible is called ultraviolet. In all energy applications except cryogenics, the characteristic dimensions of the system are large compared to the wavelengths of the thermal radiation. In cryogenics, because of the low temperatures, the radiation wavelengths are large. In the following, it is assumed that the energy transfer by radiation occurs along straight lines, excluding scattering or refraction. The radiant energy flux per unit time, $d\Phi$, can be evaluated for an area element dA along a direction with an angle β toward the surface normal. The radiant energy flux $d\Phi$ contained in a solid angle $d\omega$ within the frequency range $d\nu$ is given by

$$d\Phi = K_\nu \cos\beta \, dA \, d\omega \, d\nu = K_\lambda \cos\beta \, dA \, d\omega \, d\lambda, \qquad (9)$$

where K_ν and K_λ are the monochromatic intensities of the radiant flux. In a nonemitting and nonabsorbing medium the intensity K_ν is constant along a ray, whereas it varies if it emits or absorbs radiation, and the radiant energy flux increase $d^2\Phi_e$, which is an order of magnitude smaller than $d\Phi$, is

$$d^2\Phi_e = j_\nu \, \rho dV \, d\omega \, d\nu, \qquad (10)$$

where j_ν is the energy emitted per unit time and unit mass into a unit solid angle and within a unit frequency range. The energy flux absorbed along a path length ds is

$$d^2\Phi_a = \chi_\nu \, \rho ds \, d\Phi, \qquad (11)$$

where χ_ν is the coefficient of absorption. The decrease in the radiant energy flux due to scattering is

$$d^2\Phi_s = -\sigma_\nu \, \rho ds \, d\Phi, \qquad (12)$$

where σ_ν is the coefficient of scattering. Thermal radiation is present when there is local thermodynamic equilibrium in a medium. Application of the laws of thermodynamics to media in thermodynamic equilibrium allows the following conclusions:

The monochromatic intensity emitted, $K_{\nu e}$, is given by $K_{\nu e} = j_\nu/\chi_\nu$.

A radiation beam, traveling toward the interface between two media, with an angle β toward the surface normal is partly reflected with the same angle, with a ratio given by the reflectance or reflectivity, ρ_ν, and it partly penetrates, with an angle β' according to the law $\sin\beta'/\sin\beta = c'/c$.

The assumption of constant velocities, c' and c, allows one to find the final relation: $j_\nu c^2/\chi_\nu = j'_\nu c'^2/\chi'_\nu$ = constant, or a universal function, independent of the medium.

The absorbance of a medium, α_ν, is the ratio between the radiant energy flux absorbed and that approaching the interface.

The intensity of the radiation emitted by a medium to another medium, $i_{\nu e}$, is found in relation to the previous parameters as $i_{\nu e} = \alpha_\nu K_{\nu e}$.

A medium with $\alpha_\nu = 1$ is a blackbody, and no radiation impinging on that body is reflected or transmitted but is only absorbed. For a blackbody, the intensity $i_{\nu b}$ becomes $i_{\nu b} = K_{\nu e}$.

One of the conclusions of the electromagnetic theory of Maxwell was that the radiation reflected from an interface between two media exerts a pressure on that surface. The radiation pressure, which the solar radiation exerts on the surface of Earth, is very small (4×10^{-6} N/m^2). An important application of radiation pressure is associated with space flight, for which it has been proposed as a means of propelling space vehicles. The monochromatic intensity has been derived by Planck from the quantum theory as

$$i_{\nu b} = 2h\nu^3/(c_0^2(\exp(h\nu/kT) - 1)), \qquad (13)$$

where $k = 1.38054 \times 10^{-23}$ J K, the Boltzmann constant is universal. From Eq. (7), written in vacuum, it is found that

$$i_{\nu b} = 2C_1/[\lambda^5(\exp(C_2/kT) - 1)], \qquad (14)$$

where $C_1 = 0.59548 \times 10^{-16}$ W m^2 and $C_2 = 1.43879$ cm K. The maximum of $i_{\nu b}$ occurs according to the relation

$$\lambda_{\max} T = C_3, \qquad (15)$$

where $C_3 = 0.28978$ cm K. The total heat radiated by a blackbody, into a unit solid angle, in the wavelength range from 0 to ∞ is

$$i_b = \int_0^\infty i_{\lambda b} d\lambda = \sigma T^4/\pi. \qquad (16)$$

The emissive power e_b is found by integration over the whole hemispherical space (i.e., the solid angle ω),

$$e_b = i_b \int \cos \beta d\omega = \pi i_b = \sigma T^4, \qquad (17)$$

where $\sigma = 5.6697 \times 10^{-8}$ W/m^2 K^4. The previous equation is a good approximation of the behavior of the radiation emitted from a blackbody into a gas.

2.3 Heat Transfer by Convection

Convection occurs in a fluid during its flow at the boundary with a solid wall. The temperature of the fluid distant from the wall can be assumed to be constant and the gradient of temperature is mainly concentrated in a relatively narrow layer, δ, of fluid, close to the wall. A thin film of fluid adheres to the wall and temperature variation takes place across the layer δ. The temperature is constant outside the layer δ due to mixing of the fluid. In the layer δ of fluid, the heat is transferred by pure conduction as in a solid body. In the oversimplified case of a linear temperature variation across the layer, it is possible to write

$$Q = k_f A(T_f - T_w)/\delta, \qquad (18)$$

where k_f is the thermal conductivity of the fluid, A is the surface involved in the heat transfer, T_f is the fluid temperature, and T_w is the wall temperature. The layer δ depends on the velocity of the fluid, the shape and the structure of the surface, and other factors. In the practical application, the ratio k_f/δ is substituted by a new coefficient h, the convective heat transfer coefficient, such that Eq. (18) becomes

$$Q = hA(T_f - T_w). \qquad (19)$$

The coefficient h is not a property of the fluid, as is k in the constitutive equation of heat conduction, but it depends on the fluid dynamics conditions established in the fluid flow. It is important to use this coefficient in heat conduction problems as thermal boundary conditions. Later, the link between the heat transfer coefficient and the fluid flow is discussed. In convective studies, the concept of boundary layer, shear layer, and turbulence must be introduced. Several forces are involved in the movement of a fluid: body forces (e.g., gravity), electric or magnetic forces, centrifugal or Coriolis force, surface forces due to the influence of the surrounding fluid, and tensile and shear (or viscous) forces. The surface forces per unit area are the stresses. In a fluid, the tensile and shear stresses are related to the time rate of the deformation, and at rest only the tensile stress (i.e., pressure) is present. The shear stress must be related to the rate of deformation. For a simple flow situation in which the velocities, variation is only in the direction y, normal to the velocity, Newton proposed the proportional relation between the shear stress τ, in the flow direction, to the velocity gradient

$$\tau = \mu du/dy, \qquad (20)$$

where the property μ is the dynamic viscosity, with units in kg/m s and dimensions $[M\ L^{-1}\ T^{-1}]$. The kinematic viscosity, ν, also called momentum diffusivity, is defined as

$$\nu = \mu/\rho, \qquad (21)$$

with units m^2/s and the dimensions $[L^2\ T^{-1}]$. If the fluid obeys Eq. (20), it is Newtonian; otherwise, it is non-Newtonian. Several equations have been proposed for non-Newtonian fluids, but only some of them are shown here:

$$\text{Ostwald–de Waele fluid}: \tau = m(du/dy)^n, \qquad (22)$$

where m is the consistency of the fluid and n is the index.

$$\text{Eyring fluid}: \tau = A \sin h^{-1}(du/dy\ 1/B), \qquad (23)$$

where A and B are constants.

$$\text{Bingham plastic}: \tau = \mu du/dy + \tau_o \text{ if } \tau > \tau_o \text{ and}$$
$$du/dy = 0 \text{ if } \tau < \tau_o. \qquad (24)$$

Fluid particles in contact with a surface at rest have zero velocity, but velocity increases rapidly in the thin layer of thickness δ, called the boundary layer. The Reynolds experiment demonstrated the existence of two forms of flow—laminar and turbulent. In laminar flow, the individual streamlines run orderly side by side, whereas in turbulent flow they are interwoven with each other in an irregular manner. The motion of fluid particles in turbulent flow fluctuates around some mean flow path, and vortices of different size and orientation occur. The measurement of the velocity component, u, in turbulent flow at a location versus time results in the sum of a time-averaged value, $\bar{u}$, and an instantaneous value, u', of the timewise fluctuation. Assuming that the mean velocity is only in the x direction of u and that $\bar{v}$ and $\bar{w}$ are zero, the turbulent intensity J is defined as

$$J = \sqrt{1/3(\bar{u}'^2 + \bar{v}'^2 + \bar{w}'^2)}/\bar{u}, \quad (25)$$

where each of the three terms is the time mean of the square of the velocity fluctuations. Heat transfer is influenced by the nature of the turbulence, as described by the length scale of the turbulent fluctuations and the vortices. A measure of the length scale of the turbulent fluctuations and vortices is given by the length L,

$$L = \int_0^\infty R(y) dy, \quad (26)$$

where the parameter R, given by

$$R = (\bar{u}'_1 \bar{u}'_2) / \sqrt{\overline{u_1'^2}} \sqrt{\overline{u_2'^2}} \quad (27)$$

is obtained by hot-wire measurements, which determine the instantaneous velocity fluctuations simultaneously but at a varying distance y. Heat transfer by convection is enhanced by turbulent flow compared to the laminar one.

2.4 Heat Transfer Applications

There are many applications of heat transfer. The main applications are discussed here. Heat transfer is important in many practical applications relating to different branches of engineering but with opposite goals. Indeed, for mechanical and civil engineering, when energy must be conserved, and heat transfer to the exterior must be reduced, e.g., heat transfer losses in buildings during the winter. Energy savings from a building's heating and cooling systems can be obtained with the use of heating networks and cogeneration plants. One very important application is for very low-temperature situations such as cryogenics, which uses superconducting systems, in which heat transfer control with the exterior is crucial. On the other hand, in electrical and electronic engineering, because of the Joule effect, heat transfer to the exterior must be increased to maintain a constant temperature or to decrease the temperature of the system.

The problem of dissipating heat is an important issue in many electrical and electronic applications, such as in electrical wires and cables. Increased heat dissipation can be obtained with finned surfaces, which are of great importance in many energy systems, such as economizers in steam power plants, convectors for steam and hot water heating systems, electric transformers, and aircraft engines. In electrical machines, dissipation of electric energy requires an appropriate design, which is also required for chemical and nuclear fields in which heat sources are present. Heat transfer with change of phase is important in melting and solidification processes, combustion, frost penetration into the ground, and in aerodynamic ablation phenomena such as the protective devices of atmospheric reentry vehicles. The presence of a moving heat source or sinks in heat conduction is important in arc welding, surface hardening, continuous casting, or quenching. Radiation heat transfer phenomena are important in conventional, nuclear, and fusion power plants; gas turbines; and propulsion devices for aircraft, missiles, satellites, and space vehicles. Radiation pressure, in connection with space flight, has been proposed as a means of propelling space vehicles. Energy exchange by radiation is important in the calculation of heat transfer between surfaces in enclosures, as applied to satellites or space probes, moving in vacuum. The correct evaluation of the shape factor is important in several engineering applications, such as the radiation cooling surfaces in the furnace of a steam boiler or the tungsten spiral in an electric light bulb. Gas radiation is important in the heat transfer evaluation of furnaces and steam boilers. It is also important in plasma technology and the reentry phase of satellites or space vehicles through the atmosphere, when the gas is at very high temperatures. Flame radiation is important in the evaluation of heat transfer within tube bundles of water-tube boilers. Heat transfer analysis is useful for evaluating the temperature of a fluid flowing in a tube if the fluid temperature differs greatly from the ambient temperature. An additional systematic error in the temperature measurement is due to the heat exchange by radiation with the surrounding solid surfaces. Solar radiation evaluation is another important application of heat transfer because it can be employed for house heating and cooling.

3. HEAT CONDUCTION

3.1 Steady State

3.1.1 Composite Plane Wall

Consider a composite wall with three layers of different materials with thermal conductivities k_1–k_3, thickness x_1–x_3, and temperatures at the two extremes equal to T_{w1} and T_{w2}. The solution to Eq. (2), using the one-dimensional assumption and steady state, gives the heat flux through the surface A:

$$Q/A = (T_{w1} - T_{w2})/(x_1/k_1 + x_2/k_2 + x_3/k_3). \quad (28)$$

The temperature variation within the layers is linear if the thermal conductivity is constant versus the temperature. If the boundary conditions at the two extreme walls are not given with assumed values of temperature but are defined according to Eq. (19) (i.e., the heat flux is specified in the form of the two convective heat transfer coefficients, h_i and h_e), and the two temperatures of the fluids are known as T_e and T_i, the heat flux is given by

$$\frac{Q}{A} = \frac{(T_i - T_e)}{(1/h_i + x_1/k_1 + x_2/k_2 + x_3/k_3 + 1/h_e)}. \quad (29)$$

3.1.2 Composite Cylindrical Wall

Consider a metallic tube with internal radius R_1, external radius R_m, and thermal conductivity k_m covered by an insulating material with external radius R_i and thermal conductivity k_i. Assume the internal wall of the metallic tube is T_{w1} and the external wall of the insulating material is T_{w2}. The solution to the conservation equation, obtained using the one-dimensional cylindrical assumption and the steady state, gives the heat transfer per unit length L of the cylinder:

$$\frac{Q}{L} = \frac{(T_{w1} - T_{w2})}{[\ln(R_m/R_1)/(2\pi k_m) + \ln(R_i/R_m)/2\pi k_i)]}. \quad (30)$$

If the boundary conditions are of specified heat fluxes, with the convective heat transfer coefficients h_i and h_e and the two fluid temperatures (the internal T_i and the external T_e), then the heat transfer is

$$Q/L = (T_i - T_e)/(1/(2\pi R_1 h_i) + \ln(R_m/R_1)/(2\pi k_m) \\ + \ln(R_i/R_m)/(2\pi k_i) + 1/(2\pi R_i h_e)). \quad (31)$$

For a large radius of the insulation, R_i, the previous equation shows a decrease in heat transfer with the increase in the radius R_i. In the case of small values of R_i the previous equation presents a maximum depending on R_i. The radius of the maximum heat transfer is equal to

$$R_{crit} = k_i/h_e. \quad (32)$$

The practical application is for the insulation of electrical wires, where the objective is to provide an electric insulation and the maximum heat transfer for wire cooling.

3.1.3 Extended Surfaces

Assume that a solid body needs to dissipate more heat and the temperature of the cooling fluid cannot be modified. The only possibility is to increase the surface of heat transfer between the solid body and fluid. The simplest case is that of a thin rod with constant cross section A, perimeter C, and length L that is exchanging heat with a fluid of temperature T_f. With the assumptions that the temperature of the rod is only dependent on the distance from the body, is equal to T_0 at the base of the rod near the wall, is constant across the section A, and is in steady state, Eq. (3) can be solved to give the following temperature variation along the rod:

$$\frac{[T(x) - T_f]}{[T_0 - T_f]} \\ = \frac{[\cosh(m(L-x)) + (h_2/(mk))\sinh(m(L-x))]}{[\cosh(mL) + (h_2/(mk))\sinh(mL)]}, \quad (33)$$

and the heat flux at the boundary between rod and body is

$$\frac{Q}{A} = \frac{mk(T_0 - T_f)[h_2/(mk) + \tanh(mL)]}{[1 + (h_2/(mk))\tanh(mL)]}, \quad (34)$$

where $m^2 = hC/(kA)$, h is the convective heat transfer along the perimeter and h_2 at the final section of the rod, and k is the thermal conductivity of the metal rod. If the temperature of a fluid flowing in a tube is measured by a thermometer located in a well welded into the tube wall, and the tube wall has a lower temperature than the fluid, heat will be exchanged by conduction from the well to the tube wall and the indicated temperature will not be the true fluid temperature. The error can be calculated using the previous equations. Finned surfaces are of great importance in many energy systems, such as economizers in steam power plants, convectors for steam and hot water heating systems, electric transformers, and in aircraft engines. Practical finned surfaces differ from the thin rod, and the problem becomes more complicated from a mathematical standpoint. The finned surfaces of practical interest are rectangular fins, straight fins of triangular profile, and circumferential fins, for which the problems are the minimum weight, the fin effectiveness, and the fin arrangement.

3.1.4 Walls with Heat Sources

When heat sources are present in heat conducting bodies, it is necessary to calculate the temperature distribution within the bodies. In electrical machines and apparatus, the dissipation of electric energy requires an appropriate design. In the chemical and nuclear fields, similar heat sources are present. In a plane wall $2L$ wide, with heat sources surrounded by a fluid at temperature T_e, Eq. (2) becomes, in steady state

$$d^2(T - T_e)/dx^2 + q_g/k = 0, \quad (35)$$

and the solution is

$$(T - T_e) = -q_g x^2/(2k) + C_1 x + C_2, \quad (36)$$

where the two constants C_1 and C_2 can be calculated by the boundary conditions imposed on the surfaces. With the convective heat transfer boundary condition at the two walls, and the convective heat transfer coefficient given by h, the final solution is

$$(T - T_e) = q_g(L^2 - x^2)/(2k) + q_g L/h. \quad (37)$$

The temperature distribution along the plane wall is parabolic with the maximum on the middle of the plane.

3.2 Unsteady State

The temperature of a body exchanging heat with a fluid of constant temperature is variable with time in unsteady state. Assume that the body has a large thermal conductivity compared to that of the fluid and the temperature difference is low in the body compared to that of the fluid. In this case, the first law of thermodynamics allows one to write the equivalence between the body internal energy variation (per unit time) and the heat transfer by convection with the fluid:

$$\rho c V dT/dt = hA(T_f - T), \quad (38)$$

where V is the volume and A is the surface of the body. With the initial condition of the body temperature, T_0, the following solution is found:

$$(T - T_f)/(T_0 - T_f) = \exp - (hA/\rho c V t)$$
$$= \exp - [(hL/k)(\alpha t/L^2)], \quad (39)$$

where $L = V/A$ is a characteristic dimension that depends on the form of the body. The group of variables hL/k is called the Biot number and is represented by B_i. The group of variables $\alpha t/L^2$ is called the Fourier number and is represented by F_o.

3.3 Moving Boundaries

In many engineering heat transfer problems, there is a change of phase in the conducting medium. Examples include melting and solidification processes, combustion, frost penetration into the ground, and aerodynamic ablation applications such as in the protective devices of atmospheric reentry vehicles. The interface between the two regions of different physical properties moves as a function of time and latent heat is rejected or absorbed at its boundary. The formation of ice in water at $0\,°C$ is a problem that allows an exact solution, neglecting the heat capacity in the ice. The ice layer at time t is $x(t)$, the temperature in the liquid phase is equal to the phase change temperature T_{cp} and is constant throughout the liquid layer, and the solid phase temperature changes from the phase change value T_{cp} to that of the external surface T_s. The latent heat of fusion, Q_L, released by the change of phase of a layer dx has to be removed by conduction through the solid layer according to the following equation:

$$\rho Q_L dx/dt = k(T_{cp} - T_s)/x, \quad (40)$$

where ρ and k are relative to the solid. Imposing the boundary condition that the ice layer is zero at the beginning time, the analytical solution is

$$x = \sqrt{2k(T_{cp} - T_s)t/(\rho Q_L)}. \quad (41)$$

Consider the solidification of a liquid that has a temperature T_2 at a long distance from the surface. Heat is removed from the surface, which is maintained at the constant temperature T_1. In addition to the latent heat of fusion, solidification can take into account that the thermal conductivities, specific heats, and densities are different for the two phases. With the method developed by Neumann, it is found that the two temperatures, T_S and T_L, respectively in the solid (ice) and liquid (water) phases, are given by

$$\frac{(T_{cp} - T_L)}{(T_{cp} - T_1)} = 1 - \frac{\operatorname{erf}(x/2\sqrt{\alpha_1 t})}{\operatorname{erf}(K\beta/2)\sqrt{\alpha_1}} \quad (42)$$

$$\frac{(T_{cp} - T_s)}{(T_{cp} - T_2)} = 1 - \frac{\operatorname{erfc}(x/2\sqrt{\alpha_2 t})}{\operatorname{erfc}(K/2\sqrt{\alpha_2})}. \quad (43)$$

The interfaces move according to the laws

$$X_1 = K\beta\sqrt{t} \quad (44)$$

$$X_2 = K\sqrt{t}, \quad (45)$$

where $\beta = \rho_2/\rho_1$ takes into account the density decrease of ice and K is a constant to be determined numerically. In the water–ice system, an adequate approximation of K is

$$K_0 = \sqrt{2(T_{cp} - T_1)k_1/(Q_L \rho_1)} \quad \text{for } \beta = 1 \quad (46)$$

$$K_\beta = \sqrt{2(T_{cp} - T_1)k_1/(Q_L \rho_1 \beta^2)} \quad \text{for } \beta > 1. \quad (47)$$

In the process of ablation, the melting fluid can be assumed to be completely removed by aerodynamic forces, the surface recedes with time, the surface temperature remains constant at the phase change temperature, and a temperature distribution exists only in the remaining solid.

Assuming the semiinfinite solid is heated by a constant heat flux, q_0, at the surface $x = 0$, the melted material is removed, the boundary of the solid and melt line is located at $X(t)$, the temperature distribution in the solid penetrates to a depth $\delta(t)$, and the temperature for $x > \delta(t)$ is constant and equal to T_S. The variation of the melt line X with time t is represented parametrically by

$$\Omega = -1/3[\zeta - 2 + 2(1+v)\ln\{(2(1+v) - \zeta)/(2v)\}], \quad (48)$$

where $\Omega = q_0^2 t/(\rho Q_L k \theta_P)$, $v = Q_L/(c\theta_P)$, $\zeta = q_0(\delta - X)/(\theta_P k)$, and $\theta_P = T_{cp} - T_S$, and also by

$$\lambda = -1/3[\zeta - 2 + 2v \ln\{(2(1+v) - \zeta)/(2v)\}], \quad (49)$$

where $\lambda = q_0 X/(K\theta_P)$.

The presence of a moving heat source or sinks in heat conduction is important in arc welding, surface hardening, continuous casting, or quenching. If a point source moves at velocity U in x direction in an infinite medium, with temperature T_f and a heat generated by the point source equal to q_g, the temperature distribution is given by

$$T - T_f = q_g \exp(-(U/(2\alpha))(r + \xi))/(4\pi k r), \quad (50)$$

where $r = \sqrt{(\xi^2 + y^2 + z^2)}$ and $\xi = x - Ut$.

4. THERMAL RADIATION

Several processes occur at high temperatures in engineering, and the proper knowledge of the radiation heat transfer phenomena is important for the correct design of equipment. Applications are relate to conventional, nuclear, and fusion power plants; gas turbines; and propulsion devices for aircraft, missiles, satellites, and space vehicles.

4.1 Radiation of Strongly Absorbing Media

For strongly absorbing media, the parameters can be related to the surface conditions neglecting the differences in the thermodynamic state within the thin layer, which contributes to absorption or emission. A new term, monochromatic emittance ε_λ, is introduced by the equation

$$\varepsilon_\lambda = i_\lambda/i_{\lambda b} = \alpha_\lambda, \quad (51)$$

which shows that the monochromatic absorptance is equal to the monochromatic emittance. A radiant beam, incident on a medium, can be absorbed in the medium, transmitted through it, or reflected. Reflectance, ρ_λ, is the ratio of the reflected to the incident energy; absorbance, α_λ, is the ratio of the absorbed to the incident energy, and transmittance, τ_λ, is the ratio of the transmitted to the incident energy. According to energy conservation, the following relation can be written:

$$\rho_\lambda + \alpha_\lambda + \tau_\lambda = 1, \quad (52)$$

which for a strongly absorbing medium, in which the transmitted energy is small, becomes

$$\rho_\lambda + \alpha_\lambda = 1. \quad (53)$$

Most solids absorb radiation very strongly and all the incident radiation is absorbed in a very thin layer of the surface. Exceptions are a few solid substances like glass, quartz, rock salt, and most liquids in the visible and near-infrared range. However, all the mentioned substances and liquids absorb very strongly in the bulk of the infrared.

4.1.1 Enclosures with a Nonparticipating Medium

Energy exchange calculations between surfaces in an enclosure with a nonparticipating medium can be applied to satellites or space probes moving in vacuum. The surfaces of the enclosure are assumed to be strongly absorbing media. The surfaces emit with an intensity independent of direction and with perfectly diffuse or specular reflection. The shape factor determines the influence of the local arrangement of the surfaces involved. Considering two area elements, dA_1 and dA_2, with dimensions small compared to the distance, s, between them, the shape factor of dA_1 toward dA_2, $dF_{dA_1-dA_2}$, is given by the ratio between the heat flux incident on dA_2 to the heat flux leaving dA_1 as

$$dF_{dA_1-dA_2} = \cos\beta_1 \cos\beta_2 \, dA_2/(\pi s^2). \quad (54)$$

The shape factor depends only on the angular relationship describing the mutual position of the two area elements. It can be shown that

$$dA_1 dF_{dA_1-dA_2} = dA_2 dF_{dA_2-dA_1}, \quad (55)$$

which is called the reciprocity relation. The shape factor for a finite area A_2 is finally given by

$$F_{dA_1-A_2} = \int_{A_2} (\cos\beta_1 \cos\beta_2 \, dA_2/(\pi s^2)) \quad (56)$$

and the reciprocity relation for finite areas is

$$A_1 F_{A_1-A_2} = A_2 F_{A_2-A_1}. \quad (57)$$

The shape factor is 1 for two surfaces infinite in extent and for a surface inside an enclosure. For other configurations, the integral must be solved. For a surface A_i, completely surrounded by n surfaces, A_k, the following relation holds:

$$\sum_k F_{A_i-A_k} = 1. \quad (58)$$

The correct evaluation of the shape factor is important in several engineering applications, such as the radiation cooling surfaces in the furnace of a steam boiler or the tungsten spiral in an electric light bulb.

4.1.2 Enclosures with Black Surfaces

The radiation energy flux leaving the surface i, with A_i, is $A_i e_{bi}$, where $e_{bi} = \sigma T_i^4$. The radiation energy flux leaving a surface k in the direction toward surface i is $A_k F_{k-i} e_{bk}$; it arrives at surface i and is absorbed by it. The total energy flux absorbed by the surface i is $\sum_k A_k F_{k-i} e_{bk}$, and the net heat flux, exchanged per unit time, is

$$Q_i = A_i e_{bi} - \sum_k A_k F_{k-i} e_{bk} \quad (59)$$

Using the reciprocity law and Eq. (54), we obtain

$$Q_i = A_i \sum_k F_{i-k}(e_{bi} - e_{bk}), \quad (60)$$

and this equation can be written for each surface. The system of equations can be solved once the boundary conditions are prescribed, usually knowing the temperature, the emissive power, and the shape factor. Rapid solution of these equations can be carried out with analogy to the electric networks.

4.1.3 Enclosures with Gray Surfaces

Surfaces that emit radiation with an intensity independent of direction and that are perfect diffuse reflectors allow the introduction of radiosity, B, as the total heat flux leaving the surface per unit area and time. Radiosity is then independent of whether radiation is generated by emission or reflection. Irradiance, H, is the heat flux arriving at a surface per unit area and time. The surfaces are also assumed as gray and non-transparent, with $\alpha_i = \varepsilon_i$ and $\rho_i + \alpha_i = 1$. The heat loss of the surface by radiation is

$$Q_i = (B_i - H_i) A_i. \quad (61)$$

The heat flux H_i arriving from the surrounding surfaces is

$$H_i = \sum_k F_{i-k} B_k. \quad (62)$$

The radiosity B_i, due to emission and reflection, is given by

$$B_i = \varepsilon_i e_{bi} + \rho_i H_i. \quad (63)$$

The term H_i can be eliminated from the previous equations, with the assumption $\rho_i = 1 - \varepsilon_i$, giving

$$Q_i = \sum_k A_i F_{i-k}(B_i - B_k) \quad (64)$$

$$Q_i = A_i(e_{bi} - B_i)\varepsilon_i/(1 - \varepsilon_i). \quad (65)$$

Similar equations can be solved for each surface with the same boundary conditions as those for the enclosure with black surfaces, and the similar electrical analogy can be carried on. An additional resistance has been introduced into each connection of a node without influence on the potential of this node. In other words, the temperature of an adiabatic wall is independent of the absorption of the wall. The application of the procedure to two parallel gray surfaces of very large extent gives the heat transfer

$$Q_1/A_1 = (e_{b1} - e_{b2})/(1/\varepsilon_1 + 1/\varepsilon_2 - 1), \quad (66)$$

whereas for a surface A_1, enclosed in A_2, it gives

$$Q_1/A_1 = (e_{b1} - e_{b2})/(1/\varepsilon_1 + A_1/A_2(1/\varepsilon_2 - 1)). \quad (67)$$

If the surface 2 is much larger than 1, Eq. (67) becomes

$$Q_1/A_1 = (e_{b1} - e_{b2})\varepsilon_1. \quad (68)$$

4.2 Radiation of Weakly Absorbing Media (Gases)

Scattering is neglected in gases. Knowledge of the coefficient of absorption χ_λ is sufficient for the analysis because of the Kirchhoff equation,

$$j_\lambda = \chi_\lambda K_{\lambda e} = \chi_\lambda i_{\lambda b}. \quad (69)$$

The coefficient of absorption χ_λ can be measured from emission and absorption measurements in a medium of dimension S using the simplified equation of the monochromatic emissivity, ε_λ:

$$\varepsilon_\lambda = 1 - \exp(-\rho \chi_\lambda S) = \alpha_\lambda = 1 - \tau_\lambda. \quad (70)$$

Heat transfer by radiation can be calculated in an enclosure filled with an absorbing and emitting gas, with the assumptions that the intensity of the walls of the enclosure are independent of direction and that they are diffuse reflectors and gray surfaces. The gas filling the enclosure is gray. The radiation heat flux leaving the surface k to the surface i is $A_i F_{i-k} B_k$, and the part arriving on surface i is $A_i F_{i-k}(1-\alpha_{i-k,g}) B_k$, where $\alpha_{i-k,g}$ is the average absorbance between surfaces i and k. The radiation heat flux arriving at surface i is $A_i \varepsilon_{i,g} e_{bg}$, where $\varepsilon_{i,g}$ is the emissivity of the gas. The radiation heat flux leaving i is $A_i B_i$, and the net radiation heat loss of the surface i is

$$Q_i = A_i[B_i - \sum_k F_{i-k}(1-\alpha_{i-k,g})B_k - \varepsilon_{i,g}e_{bg}]$$
$$= A_i\varepsilon_{i,g}(B_i - e_{bg})$$
$$+ \sum_k A_i F_{i-k}(1-\alpha_{i-k,g})(B_i - B_k), \quad (71)$$

and the net heat loss of the gas due to the radiation heat is

$$Q_g = \sum_i A_i \varepsilon_{i,g}(e_{bg} - B_i). \quad (72)$$

The last three equations are sufficient to calculate the heat loss for surfaces with known temperature, the temperature for adiabatic surfaces, and the heat loss of the gas in the enclosure. Gas radiation is important in the heat transfer evaluation of furnaces and steam boilers. It is also important in plasma technology and in the reentry phase of satellites or space vehicles through the atmosphere when gases are at very high temperatures. In this condition, radiation has its largest contribution from visible or near-ultraviolet wavelength range. Flame radiation is important in the evaluation of heat transfer within the convective surfaces of tube bundles in water-tube boilers.

5. HEAT TRANSFER BY CONVECTION

5.1 Conservation Equations

The conservation equations are written for a volume of fluid, in a system of Cartesian coordinates, of dimensions dx, dy, dz. The mass conservation for a fluid system without chemical reactions is

$$\partial \rho/\partial t + \partial(\rho u)/\partial x + \partial(\rho v)/\partial y + \partial(\rho w)/\partial z = 0, \quad (73)$$

and introducing the notation

$$D/dt = \partial/\partial t + u\partial/\partial x + v\partial/\partial y + w\partial/\partial z \quad (74)$$

it can be written as

$$D\rho/dt + \rho(\partial u/\partial x + \partial v/\partial y + \partial w/\partial z) = 0. \quad (75)$$

The conservation of momentum for the control volume states that the sum of all the forces acting on the mass must equal the time derivative of the momentum of the fluid mass. The fluid momentum changes due to the momentum flux, both out of and in the control volume, or due to the time change of the momentum of mass of the control volume. The surface forces per unit area, or the stresses, represent the action of the surrounding fluid on the control volume fluid. The stresses have two indices, as, for example, p_{xy} and p_{xz}. The first index indicates the direction normal to the surface and the second the direction of the stress component. The shear stresses are, for example, p_{xy} and p_{xz}. The normal stress $(p_{xx}-p)$ is composed of a viscous stress p_{xx}, which is a tensile stress that vanishes when the velocity becomes zero, and the fluid pressure p. The minus sign for the pressure takes into account the fact that it is assumed positive when it is compression stress (i.e., it acts from outside). The momentum conservation in the x direction is

$$\partial(\rho u^2)/\partial x + \partial(\rho vu)/\partial y + \partial(\rho wu)/\partial z + \partial(\rho u)/\partial t$$
$$= \partial(p_{xx}-p)/\partial x + \partial p_{yx}/\partial y + \partial p_{zx}/\partial z + \rho g_x. \quad (76)$$

Using the mass conservation equation, it becomes

$$\rho u\partial u/\partial x + \rho v\partial u/\partial y + \rho w\partial u/\partial z + \rho\partial u/\partial t$$
$$= -\partial p/\partial x + \partial p_{xx}/\partial x + \partial p_{yx}/\partial y + \partial p_{zx}/\partial z + \rho g_x, \quad (77)$$

and with notation similar to that of the mass conservation equation, that is,

$$Du/dt = \partial u/\partial t + u\partial u/\partial x + v\partial u/\partial y + w\partial u/\partial z, \quad (78)$$

it becomes

$$\rho Du/dt = -\partial p/\partial x + \partial p_{xx}/\partial x$$
$$+ \partial p_{yx}/\partial y + \partial p_{zx}/\partial z + \rho g_x. \quad (79)$$

It is similarly obtained in the other two directions:

$$\rho Dv/dt = -\partial p/\partial y + \partial p_{xy}/\partial x$$
$$+ \partial p_{yy}/\partial y + \partial p_{zy}/\partial z + \rho g_y \quad (80)$$

$$\rho Dw/dt = -\partial p/\partial z + \partial p_{xz}/\partial x + \partial p_{yz}/\partial y$$
$$+ \partial p_{zz}/\partial z + \rho g_z. \quad (81)$$

At this point, the empirical relations connecting the stresses with the deformation of the fluid volume, also called constitutive equations, must be introduced, but only for Newtonian fluids. The shear stresses for a Newtonian fluid have been derived by Stokes and are the following

$$p_{xx} = 2\mu \partial u/\partial x - \mu'(\partial u/\partial x + \partial v/\partial y + \partial w/\partial z) \quad (82)$$

$$p_{yx} = p_{xy} = \mu(\partial v/\partial x + \partial u/\partial y) \quad (83)$$

for the x direction and similarly for the other directions. The viscosity μ' has the value of $2\mu/3$ for monoatomic molecules and is the only one considered here. Introducing the previously mentioned constitutive equations, the three momentum conservation equations become

$$\begin{aligned}\rho Du/dt =& -\partial p/\partial x + 2\partial(\mu \partial u/\partial x)/\partial x \\ &+ \partial[\mu(\partial v/\partial x + \partial u/\partial y)]/\partial y \\ &+ \partial[\mu(\partial u/\partial z + \partial w/\partial x)]/\partial z \\ &- \partial[\mu'(\partial u/\partial x + \partial v/\partial y + \partial w/\partial z)]/\partial x \\ &+ \rho g_x \end{aligned} \quad (84)$$

$$\begin{aligned}\rho Dv/dt =& -\partial p/\partial y + 2\partial(\mu \partial v/\partial y)/\partial y \\ &+ \partial[\mu(\partial v/\partial z + \partial w/\partial y)]/\partial z \\ &+ \partial[\mu(\partial v/\partial x + \partial u/\partial y)]/\partial x \\ &- \partial[\mu'(\partial u/\partial x + \partial v/\partial y + \partial w/\partial z)]/\partial y \\ &+ \rho g_y \end{aligned} \quad (85)$$

$$\begin{aligned}\rho Dw/dt =& -\partial p/\partial z + 2\partial(\mu \partial w/\partial z)/\partial z \\ &+ \partial[\mu(\partial w/\partial x + \partial u/\partial z)]/\partial x \\ &+ \partial[\mu(\partial w/\partial y) + \partial v/\partial z)]/\partial y \\ &- \partial[\mu'(\partial u/\partial x + \partial v/\partial y + \partial w/\partial z)]/\partial z \\ &+ \rho g_z. \end{aligned} \quad (86)$$

These equations are referred to as Navier–Stokes equations. The energy conservation equation is derived on the basis of the first law of thermodynamics applied to the fluid contained in the control volume. The time variation of the total energy contained in the control volume, $e_t = u + V^2/2$, which is the sum of the internal and kinetic energy, $dE_t/dt = d(e_t M)/dt$, is equal to the sum of the work transfer per unit time, dW/dt, of the energy exchange by convection, dE_c/dt, of the energy transfer by molecular movements, dE_m/dt (i.e., heat conduction) but also by diffusion:

$$dE_t/dt = d(e_t M)/dt = dW/dt + dE_c/dt + dE_m/dt. \quad (87)$$

For the control volume of dimensions dx, dy, dz, the total energy becomes

$$d(e_t M)/dt = \partial[(\rho e_t)dxdydz]\partial t. \quad (88)$$

The work exchanged in the x direction is positive when the force is acting in the flow direction and negative in the opposite direction and is

$$\begin{aligned}dW/dt =& [\partial(up)/\partial x - \partial(up_{xx})/\partial x - \partial(up_{yx})/\partial y \\ &- \partial(up_{zx})/\partial z]dxdydz + \rho u g_x dxdydz. \end{aligned} \quad (89)$$

The energy flux by convection in the x direction is

$$dE_c/dt = -\partial[\rho u e_c dxdydz]/\partial x. \quad (90)$$

The energy transfer by molecular movements in the x direction is

$$dE_m/dt = -\partial[e_{mx}dxdydz]/\partial x. \quad (91)$$

Using similar expressions for the other two directions, with the help of the mass conservation equation and the substantial derivative of the total energy, the total energy conservation equation is obtained as follows:

$$\begin{aligned}&\rho De_t/dt + \partial(up)/\partial x + \partial(vp)/\partial y + \partial(wp)/\partial z \\ &+ \partial(e_{mx})/\partial x + \partial(e_{my})/\partial y + \partial(e_{mz})/\partial z \\ &= \partial(up_{xx})/\partial x + \partial(up_{yx})/\partial y + \partial(up_{zx})/\partial z \\ &+ \partial(vp_{xy})/\partial x + \partial(vp_{yy})/dy + \partial(vp_{zy})/\partial z \\ &+ \partial(wp_{xz})/\partial x + \partial(wp_{yz})/\partial y + \partial(wp_{zz})/\partial z \\ &+ \rho u g_x + \rho v g_y + \rho w g_z. \end{aligned} \quad (92)$$

For a one-component fluid, the energy flux e_m is due to the Fourier equation.

5.2 Forced Convection

5.2.1 Laminar Flow

In a two-dimensional steady-state flow, the previous equations, restricted to the boundary layer, reduce to

$$\partial u/\partial x + \partial v/\partial y = 0 \quad (93)$$

$$\rho(u\partial u/\partial x + v\partial u/\partial y) = -\partial p/\partial y + \mu \partial^2 u/\partial y^2 \quad (94)$$

$$\partial p/\partial y = 0 \quad (95)$$

$$\rho c_p(u\partial T/\partial x + v\partial T/\partial y) = k\partial^2 T/\partial y^2, \quad (96)$$

considering negligible the influence of the gravitational body forces and assuming a fluid with constant properties.

5.2.2 Turbulent Flow

Considering a constant-property fluid in a two-dimensional flow and dividing the turbulent flow in a time-averaged value, $\bar{v}$ and a fluctuating value, v', for each variable, $u = \bar{u} + u'$, $v = \bar{v} + v'$, $p = \bar{p} + p'$, $T = \bar{T} + T'$, using the time average rules it is possible to obtain the following equations for the conservation of mass and momentum:

$$\partial \bar{u}/\partial x + \partial \bar{v}/\partial y = 0 \quad (97)$$

$$\rho(\partial\bar{u}/\partial t+ \bar{u}\partial\bar{u}/\partial x + \bar{v}\partial\bar{u}/\partial y + \partial\bar{u}'^2/\partial x + \partial\,\bar{u}'\bar{v}'/\partial y$$
$$= -\partial\bar{p}/\partial x + \mu\partial^2\bar{u}/\partial y^2 \quad (98)$$

$$\rho\partial\,\bar{v}'^2/\partial y = -\partial\,\bar{p}/\partial y. \quad (99)$$

Integration of Eq. (99) gives $p_s = \bar{p} + \rho\bar{v}'^2$, which simplifies Eq. (98) giving the following equations, including the energy conservation equation:

$$\partial\bar{u}/\partial x + \partial\bar{v}/\partial y = 0 \quad (100)$$

$$\rho(\partial\bar{u}/\partial t + \bar{u}\partial\bar{u}/\partial x + \bar{v}\partial\bar{u}/\partial y)$$
$$= -\partial p_s/\partial x + \mu\partial^2\bar{u}/\partial y^2 - \rho\partial\bar{u}'\bar{v}'/\partial y \quad (101)$$

$$\rho c_p(\partial\bar{T}/\partial t + \bar{u}\partial\bar{T}/\partial x + \bar{v}\partial\bar{T}/\partial y)$$
$$= k\partial^2\bar{T}/\partial y^2 - \rho c_p\partial\bar{T}'v'/\partial y. \quad (102)$$

The momentum equation differs from the laminar counterpart for the term containing the fluctuating quantities, which can be interpreted as resulting from turbulent stresses. The turbulent–boundary-layer equations contain too many unknowns and it is necessary to relate the turbulent stresses to the mean time values. Boussinesq first proposed the following:

$$\tau_t = -\rho\,\bar{u}'\,\bar{v}' = \rho\varepsilon_m\partial\,\bar{u}/\partial y, \quad (103)$$

where ε_m is the turbulent diffusivity of momentum. Similarly, the corresponding term of the energy equation has been proposed as given by

$$-\rho c_p\,\bar{T}'\,\bar{v}' = \rho c_p\varepsilon_h\partial\,\bar{T}/\partial y, \quad (104)$$

which introduces ε_h, the turbulent diffusivity of heat. The ratio of the two turbulent diffusivities has been defined as the turbulent Prandtl number:

$$Pr_t = \varepsilon_m/\varepsilon_h. \quad (105)$$

5.3 Natural Convection

The laminar boundary-layer equations under body force due to gravitational acceleration in low-velocity conditions, when the effect of viscous dissipation can be neglected, can be written for plane flow as well as for steady rotationally symmetric flow, as

$$\partial(\rho u r_0^n)/\partial s + \partial(\rho v r_0^n)/\partial y = 0 \quad (106)$$

$$\rho(u\partial u/\partial s + v\partial u/\partial y) = -dp/ds$$
$$+ \partial(\mu\partial u/\partial y)/\partial y + \rho g_s \quad (107)$$

$$\rho c_p(u\partial T/\partial s + v\partial T/\partial y) = \partial(k\partial T/\partial y)/\partial y, \quad (108)$$

where s is the coordinate along the curved surface and y is normal to it. The exponent n is zero for two-dimensional flow and 1 for rotationally symmetric flow. Outside the boundary layer, the viscous effects are negligible and the momentum equation reduces to

$$dp/ds = \rho_e g_s - \rho_e u_e du_e/ds, \quad (109)$$

where e refers to the conditions outside the boundary layer. The momentum equation becomes

$$\rho(u\partial u/\partial s + v\partial u/\partial y) = g_s(\rho - \rho_e) + \rho_e u_e du_e/ds$$
$$+ \partial(\mu\partial u/\partial y)/\partial y, \quad (110)$$

and introducing the thermal expansion coefficient

$$\beta = 1/v(\partial v/\partial t)_p = -1/\rho(\partial\rho/\partial t)_p, \quad (111)$$

the following equation is obtained:

$$\rho(u\partial u/\partial s + v\partial u/\partial y) = -\beta\rho g_s(T - T_e) + \rho u_e du_e/ds$$
$$+ \partial(\mu\partial u/\partial y)/\partial y. \quad (112)$$

6. DIMENSIONAL ANALYSIS

For some heat transfer phenomena, the basic understanding of the physical processes is not enough to permit the writing of the fundamental equations. The alternative is to perform experiments and to generalize the results by dimensional analysis. For other heat transfer applications, an approximate knowledge of the equations is possible, and this allows one to start with these equations and make them dimensionless. One example of this approach is the case of flow separation on cylinders. The writing of the basic Navier–Stokes equations allows the conclusion that the velocity fields and pressure fields around cylinders are physically similar, provided the Reynolds number has the same value for all the considered cases. If the cylinder is kept at a temperature T_w and the fluid is at T_0, the Navier–Stokes equations can be written dimensionless and the dimensionless temperature can be found as

$$\theta^* = (T - T_0)/(T_w - T_0) = f(x/d,\ Re,\ Pr,\ Ec), \quad (113)$$

where x is the Cartesian coordinate and d is the cylinder diameter. The Reynolds number, Re, is defined as

$$Re = \rho u_0 d/\mu, \quad (114)$$

where u_0 is the fluid velocity outside the boundary layer. The Prandtl number, Pr, is defined as

$$Pr = \mu c_p/k. \quad (115)$$

The Eckert number, Ec, is defined as

$$Ec = u_0^2/c_p\theta_w. \quad (116)$$

The convective heat transfer coefficient at the wall between fluid and cylinder, h, can be put in dimensionless form by introducing the Nusselt number, Nu, which is defined as

$$Nu = hd/k = (\partial\theta^*/\partial(x/d)). \quad (117)$$

Considering also a fluid with variable properties, T_w/T_0, the final relation of the average Nusselt number on the total surface is

$$Nu_{average} = f(Re, Pr, Ma, T_w/T_0, \alpha, \gamma), \quad (118)$$

where the Mach number, Ma, is defined as

$$Ma = u_0/a_0, \quad (119)$$

where u_0 is the velocity of the fluid outside the boundary layer and a_0 is the velocity of sound for a gas with c_p, $\gamma = c_p/c_v$, $k = C_k T^\alpha$, and $\mu = C_\mu T^\alpha$:

$$a_0 = \sqrt{(\gamma-1)c_p T_0} = \sqrt{\gamma R\,T}. \quad (120)$$

The following relation holds between Mach and Eckert numbers:

$$Ec = (\gamma - 1)Ma^2/(T_w/T_0 - 1). \quad (121)$$

In the practical applications, Eq. (118) is usually stated as the sum of exponential terms of the form

$$X = C_1 + C_2 Y^a Z^b W^c, \quad (122)$$

where the values of C_1, C_2, a, b, and c have to be found by specific experiments. The effect due to the rarefaction of gas is taken into account by the dimensionless Knudsen number, Kn, which is defined as

$$Kn = \lambda/L = Ma/Re\,\sqrt{\gamma\pi/2}, \quad (123)$$

where λ is the mean free molecular path and L is a body dimension characteristic in the flow field. In natural or free convection flow (i.e., when the flow and heat transfer are present because of a constant body force due to the gravitational effect, g) two additional dimensionless numbers are defined: the Grashof number, Gr, and the Rayleigh number, Ra, defined as

$$Gr = \beta g(T_w - T_0)L^3/\nu^2 \quad (124)$$

$$Ra = GrPr. \quad (125)$$

Another dimensionless number that is often defined in heat transfer problems is the Stanton number, St, defined as

$$St = Nu/(RePr). \quad (126)$$

SEE ALSO THE FOLLOWING ARTICLES

Combustion and Thermochemistry • Conservation of Energy Concept, History of • Energy in the History and Philosophy of Science • Forms and Measurement of Energy • Mechanical Energy • Temperature and Its Measurement • Thermal Energy Storage • Thermodynamic Sciences, History of • Work, Power, and Energy

Further Reading

Carslaw, H. S., and Jaeger, J. C. (1970). "Conduction of Heat in Solids." Oxford Univ. Press, New York.

Eckert, E. R. G., and Drake, R. M., Jr. (1972). "Analysis of Heat and Mass Transfer." McGraw-Hill, New York.

Manuscripts (1666–1667). "Essays of Natural Experiences Done in the Cimento Academy." National Library of Florence, Florence, Italy.

Heterotrophic Energy Flows

KENNETH A. NAGY
University of California, Los Angeles
Los Angeles, California, United States

1. Importance of Heterotrophic Energy Flow
2. Energy Budget of an Animal
3. Energy Intake
4. Energy Use
5. Efficiency
6. Regulation of Energy Metabolism
7. Summary

Glossary

anaerobic metabolism Cellular production of energy-containing adenosine triphosphate in the absence of oxygen.

basal metabolic rate The lowest rate of heat production by an adult mammal or bird (endothermic or "warm-blooded" animal), measured when body temperature is normal and costs for activity, temperature regulation, digestion, and other expenses are low or zero.

chemical potential energy Energy that is contained in the foodstuffs and that is released as heat or as other forms of chemical energy (e.g., adenosine triphosphate) during cellular oxidative or anaerobic metabolism.

efficiency The portion of energy used or converted out of the total energy input, usually expressed as a ratio (out/in) or as a percentage (100 * out/in).

feeding rate The amount of food, usually in grams of dry matter of the diet, that an animal eats each day.

field metabolic rate The rate of heat production by an animal living undisturbed in its natural habitat.

homeostasis Regarding energy flow, the maintenance over time of energy balance, where in = out and energy storage is zero (none accumulated or depleted over time).

oxidative metabolism Cellular production of energy-containing adenosine triphosphate from foodstuffs via an oxygen-using process.

production The synthesis of new biochemicals that contain chemical potential energy and their accumulation as stored fatty tissue, new body tissue (growth), or reproductive tissues (e.g., embryos, eggs, milk).

standard metabolic rate The lowest rate of heat production by a reptile, amphibian, fish, or invertebrate (ectothermic or "cold-blooded" animal) measured at a specified body temperature and when costs for activity, temperature regulation, digestion, and other expenses are low or zero.

Autotrophic organisms, including most plants and some micro-organisms, "eat" energy in the form of photons (radiant energy as sunlight) or as energy-containing bonds in inorganic chemicals. Heterotrophs eat autotrophs, thereby obtaining their energy in the form of organic chemicals (carbohydrates, fats, and proteins). Nearly all animals are strict heterotrophs, dining either on plants, other animals, or the remains of these organisms. Thus, life on earth depends on the flow of energy, primarily from the sun, through plants, through herbivores, through carnivores, and finally through detritivores.

1. IMPORTANCE OF HETEROTROPHIC ENERGY FLOW

1.1 Money Analogy

Energy is intimately involved in living processes at every level of biological organization, from the organelles in cells to ecosystems. The central importance of energy as a common currency for individual animals and within ecosystems is somewhat analogous to the importance of money in human societies. The main goal of working each day is to obtain this resource (food or money) in sufficient quantities to meet the needs of living that day. Daily needs are not fixed but rather vary with lifestyle and especially with behavior. If income exceeds daily needs, the excess can be stored or used for growth or reproduction. Too little of this resource can be fatal.

1.2 Competition for Food, Selection, and Adaptation

Thomas Malthus observed that populations reproduce faster than their food supply, and Charles

Darwin realized that this should cause intense competition for food within a species. The critical component of the food is presumed to be its energy. Hence, energy availability should limit population density. It follows that there should be intense selection favoring individuals that manage their energy budgets more effectively. Effective management may involve (1) getting more energy from the environment each day, (2) digesting food more thoroughly, (3) reducing daily metabolic expenditures, and/or (4) increasing the "efficiency" of reproduction by producing more successful offspring from the same energy intake. If energy is the principal limiting resource for wild animals, selection pressure on biological properties that are directly concerned with energy metabolism should be more intense than selection pressure on other properties of animals. Thus, energy should be evolutionarily paramount, and most of the important features of living animals should be understandable from the perspective of energetics.

Unfortunately, studies indicate that this paradigm may be too simplistic an explanation of the biology of wild animals given that only rarely can food supply be shown to be an important limiting resource. When food energy is in excess, there is no pressure to conserve energy. If food is abundant, animals can simply eat more and overcome any shortcomings of energy management or inefficiencies within their bodies. Food shortages probably do limit populations at times, but such bottlenecks may be decades or centuries apart, making them difficult to study. During intervals between energy bottlenecks, other causes of mortality, such as predation, may predominate. Nevertheless, the energy bottlenecks are probably quite severe when they occur and should have a large selective effect on the characteristics of subsequent generations. Thus, animals may display phenotypes that reflect adaptations to energy limitations even though energy might not be limiting at the time. If so, knowledge of the energetics of a species should yield a better understanding of its biology, ecology, and evolution in general.

1.3 Energy Costs of Living

Living animal cells are chemically different from their environment, which is the extracellular fluid within the animal. These differences across the cell membrane tend to disappear due to passive diffusion down concentration gradients, and maintenance of constant intracellular conditions requires a continuous expenditure of energy by the cells. Additional energy is used by groups of differentiated cells (organs and organ systems) in carrying out the higher level functions they perform within animals. At the level of the whole animal, more energy is used for integrated actions such as social interchanges, predator avoidance, and food acquisition. Population energy flow is the sum of energy use by individuals, and community energy flow incorporates the effects of interactions among species of producers, consumers, and detritivores. Whole animals also must spend extra energy, beyond that spent by their cells on intracellular maintenance, to maintain internal chemical homeostasis in their extracellular fluids. Additional energy is needed for animals to move, obtain and digest their food, interact socially, grow, and reproduce. A complete energy budget for an animal would include all of the avenues of heat gain and loss and chemical potential energy gain and loss. Such a budget could be given in units of daily energy intake, loss, and storage in kilojoules per day (kJ/day). These units are rates reflecting the dynamic nature of animal energetics. This article focuses on chemical potential energy intake as food and follows the fate of that food energy.

2. ENERGY BUDGET OF AN ANIMAL

Animals fall into two main metabolic categories: endotherms (primarily mammals and birds that are sometimes termed "warm-blooded") and ectotherms (fish, amphibians, reptiles, and invertebrates). Endotherms use their high metabolic rates to produce the internal heat that allows them to maintain relatively high and constant body temperatures compared with their surroundings. Ectotherms have quite low rates of metabolic heat production, too low to warm their bodies much at all. However, some species of ectotherms, especially lizards, can achieve and maintain high body temperatures for hours by using external heat sources such as sunlight and sun-warmed rocks. Endotherms have much more metabolic machinery (enzymes) in their cells, and this makes their cellular operating costs much higher than those of ectotherms. The ability of endotherms to keep themselves warm allows them to occupy colder habitats such as those at high latitudes and altitudes. Ectotherms, which have good fasting endurance due to their low energy needs, can persist in habitats such as deserts and the ocean's depths, where food may be scarce periodically or continuously.

The only source of chemical potential energy for most animals is their food. The food can be taken in by mouth, or nutrients can be absorbed across body walls as in some parasites and marine organisms. Some of the food energy is digested and absorbed (assimilated energy), and the indigestible food residues are voided by the animals as feces. Feces actually contain additional substances such as bacteria from the hindgut that grow and feed on the undigested dietary residues as well as sloughed off gut cells and proteins (e.g., mucus) that contain energy that was assimilated earlier but escaped redigestion and reassimilation. Thus, the term "apparent assimilation efficiency" (AE) is used to describe the following ratio: (food energy in) − (fecal energy out)/(food energy in). Some assimilated chemical potential energy leaves animals as urine, and this energy is also not available for metabolism. Most of the urinary energy is in nitrogen waste products (e.g., urea, uric acid), but some is in mucus. The difference between energy ingested as food and that excreted as urine and feces is called metabolizable energy, and "apparent metabolizable energy efficiency" (MEE) is used to describe the following ratio: (food energy in) − (feces and urinary energy out)/(food energy in) (Fig. 1).

Metabolizable energy is used primarily for oxidative metabolism in most animals, and all of this energy ultimately leaves the animal as heat (Fig. 1). If any metabolizable energy is left over after paying the energetic costs of the essential physiological processes, it can be used for production (growth, storage, or reproduction). The essential energetic costs of living include the minimum (idling) cost, called basal metabolic rate (BMR) in endothermic animals and standard metabolic rate (SMR) in ectotherms, along with the additional costs of obtaining, ingesting, digesting, absorbing, and biochemically processing food as well as the costs of being alert, interacting socially, maintaining body temperature, being active (e.g., locomoting), and responding to diseases or parasites. Energy savings resulting from reduced oxidative metabolism may occur in endotherms that can lower their body temperatures for hours or days (torpor) or weeks (hibernation). Total metabolic cost also may be lower than the sum of its (separately determined) components if an animal can substitute heat produced from one process (e.g., activity cost, digestion cost) for simultaneous requirements for heat needed for maintenance of body temperature (thermoregulation cost).

When an animal eats just enough food to provide the metabolizable energy it burns in oxidative

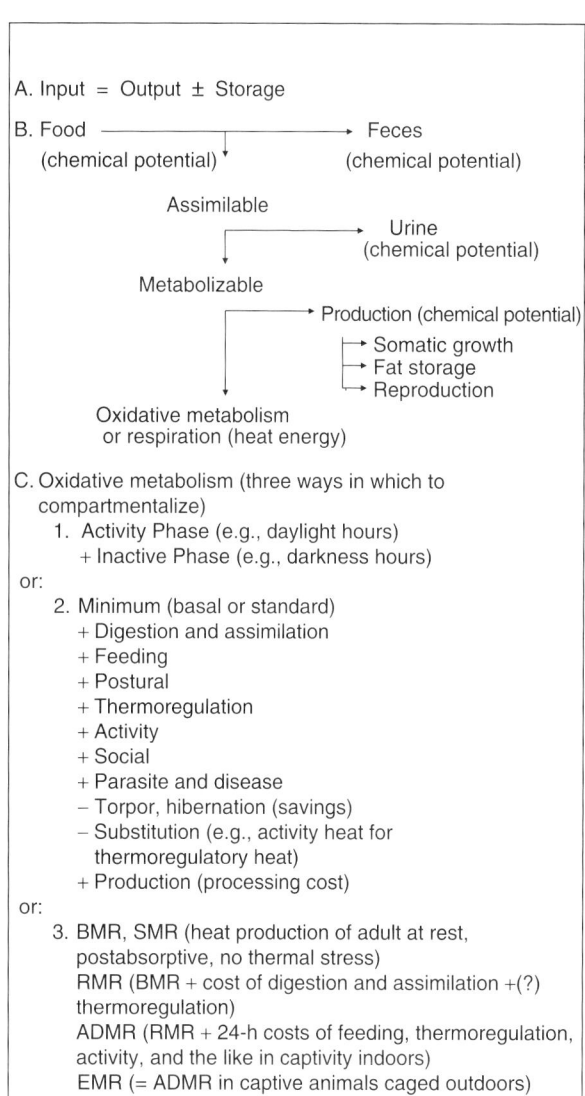

FIGURE 1 Three versions of a daily energy budget. Version A is in the form of a simple equation, and Version B accounts for the substances (food, feces, and urine) that move in and out of an animal. Version C shows three ways of subdividing the daily metabolic rate (daily heat production) portion of the energy budget of an animal, reflecting the different ways of measuring component parts. BMR, basal metabolic rate; SMR, standard metabolic rate; RMR, resting metabolic rate; ADMR, average daily metabolic rate; EMR, existence metabolic rate; FMR, field metabolic rate; DEE, daily energy expenditure.

metabolism each day, it is maintaining energy balance. The chemical potential energy content of its body will not be changing from day to day, although during a given day it will have a positive energy budget when absorbing its food and a negative energy budget when the assimilation rate declines below its metabolic rate. When an animal consumes more or less metabolizable energy than it burns in a day, its body will store the excess or contribute the deficit from body stores of energy. Most animals must have balanced energy budgets over relatively short periods of time (days, weeks, or even just hours for small endotherms such as some shrews and hummingbirds) to maintain life.

There are several ways in which to subdivide, or "itemize," an energy budget (Fig. 1). In theory, the various costs of living can be identified and listed as separate entries in the budget. Such a theoretical budget has intellectual value and appeal, but the practical difficulties of measuring each component of such a budget usually lead to its modification. For example, a theoretical budget might include the energetic expenses of hunting for and obtaining food (foraging cost), looking for and courting a mate (reproductive cost), and watching for and avoiding predators (antipredation cost). Using a rodent such as a desert kangaroo rat as an example, let us assume that it is in its burrow for 22 h and out of the burrow and active for 2 h each day. Let us also assume that we were able to measure the rat's daily metabolic expenditure and its metabolism only while it was in its burrow and that the difference between these expenditures (45 kJ/24 h–31 kJ/22 h) is 14 kJ/2 h active. The first problem in partitioning these 14 kJ into their theoretical budget categories is deciding whether basal metabolism, thermoregulatory and digestion costs, and the like, which the rat is also paying while it is outside its burrow doing other things, should be subtracted off to determine the pure cost of foraging, antipredation, and the like. This correction requires assumptions about energy allocation in active animals that have not yet been verified. The second problem is deciding, say from detailed observations of the animal's behavior while it was above ground, what portion of those 2 h the rat spent foraging, looking for mates, and avoiding predators and then subdividing the 14 kJ appropriately. This requires reading the mind of the kangaroo rat to determine whether it was looking for food, predators, or mates while it was sitting still in an alert, eyes-open posture during much of its time above ground. In fact, active animals probably do all three things simultaneously. The conservative choice is to label the 14 kJ simply as "metabolic cost while above ground." Creative experimentation will be required to increase the resolution of energy metabolism measurements of wild animals to evaluate the components of theoretical energy budgets. For now, researchers are settling for budget categories that reflect measurable parameters such as active and inactive periods of the day or basal (BMR), resting (RMR), average daily (ADMR), existence (EMR), and field (FMR) metabolic rates (Fig. 1).

Metabolizable energy that is not burned is available for production (synthesis of new biomass). Excess energy can simply be converted to fat and stored somewhere in the body, where it can be used for somatic growth (mostly as the protein and lipids needed for cell division and development) or for reproduction (growth of eggs or embryos as well as production of milk by mammals). Production involves formation of new biomass from foodstuffs. There are metabolic costs involved in these biochemical processes that add to the total rate of oxidative metabolism (heat production) of an animal producing new biomass. Animals are using energy continuously, and it is important to view energetics as an assemblage of processes, each having its own rate that varies through time. A useful analogy is that of a leaky bucket under an intermittently opened faucet. Various holes in the bucket can represent the components of oxidative metabolism that drain water (energy) out, lowering the water level. The intermittent faucet represents feeding events that add water (energy) to the bucket periodically. If more water is added than drains out each day, the volume of water in the bucket will increase (grow) over time. However, unlike the holes in the bucket, animals can influence the rates at which they use energy for various purposes, and they also influence their rate of energy intake (feeding rate).

We are accustomed to thinking about energy as packets, such as food items, that are measured in units of calories or joules. These are static units that do not represent rates of energy flow. The ratio of energy units to time units (joules/day or watts) does represent flow, and it is used herein. It is also important to pay attention to the time unit in discussions of animal energetics. Animals can spend energy very quickly over short time intervals (seconds or minutes, as during sprinting), and their maximum rates of energy expenditure can be far higher than their maximum rates of energy intake. However, over the course of a day, they may achieve energy balance by interspersing periods of activity with periods of resting, so that total energy expenditures do not exceed income. Likewise, some seasons may be difficult for animals and other seasons may be benevolent to animals, but over a whole year, the animals may achieve energy balance or even a positive energy budget that allows for growth or reproduction.

3. ENERGY INTAKE

The rate of metabolizable ("usable") energy intake that an animal achieves depends on many factors, including food availability in the habitat, feeding rate, diet composition and digestibility, capacities of the digestive tract, and appetite.

3.1 Food Availability

The availability of food in a habitat is different from food present in a habitat. Some of the potential food items are not available for a variety of reasons. For example, chemical potential energy in plant tissues might not be available to herbivorous animals because it is unreachable (underground or high on a shrub or tree), inedible (woody trunk of a tree or spine-covered stem or leaf), distasteful (containing antifeedant chemicals), toxic (containing poisons), or nutritionally inadequate (poor balance of nutrients or full of protein digestion inhibitors such as tannins). Some plant parts may be nutritionally poor or even toxic by themselves but can be edible and nutritious when eaten along with other plant parts that counteract toxins or make up for nutritional deficiencies. Similarly, plant items that are nutritious for an animal species having certain detoxification or digestive capabilities may be detrimental to another animal species that does not have those capabilities. Moreover, food availability also includes effects of patchy food distribution in space and time, and the ability of various animal species to exploit a given patch depends on the characteristics of the animals (e.g., small vs large, ability to catch insects). Thus, "food availability" must be defined from the perspective of the animal of concern and not simply as a fixed property of a habitat.

3.2 Food Composition and Digestibility

The chemical composition of diets that animals choose varies greatly. The plant matter in the diets of herbivores can consist largely of cell wall components that include cellulose, hemicellulose, and lignins. The chemical potential energy in these substances (collectively called "fiber") is generally unavailable to many herbivorous animals because they do not have the enzymes needed to digest fiber. Thus, less than half of the potential energy in the food may be available to these herbivores. On the other hand, animals that eat other animals are ingesting a food that has a chemical composition very similar to that of their own bodies, and most of those chemicals are digestible.

Diet digestibility, or digestive efficiency (DE), refers to the percentage of the dry matter (not energy, which is called assimilation efficiency or AE) of food that is digested and absorbed across the gut: DE = 100 * [(dry matter in) − (dry matter out as feces)/(dry matter in)]. DE varies greatly among different animals and diets. DE can be as low as 0.5% in earthworms (detritivores) eating soil with little organic matter in it (Table I) or as high as essentially 100% in hummingbirds or butterflies eating nectar, which is pure sugar dissolved in water. Herbivorous animals fall into two main categories: fermenters and nonfermenters. The most specialized fermenters are the ruminant mammals (e.g., cattle, antelope, sheep), which have a four-chambered stomach that contains an ecosystem of micro-organisms capable of digesting plant cell walls. The fiber is broken down by cellulases produced by microbes and is fermented by those microbes into volatile fatty acids that can be used by the animals for energy. The microbes use fiber fermentation and most of the soluble chemicals in the food (e.g., sugars, starch, proteins, lipids from the plant cells) for their own metabolism

TABLE I

Average Digestibilities and Energy Yields for Some Representative Diet/Consumer Combinations

Dietary category	Example species	Apparent dry matter digestibility (percentage)	Energy yield (metabolizable kJ/g dry matter)
Herbivore, monogastric	Gopher, jackrabbit	30–75	7–13
Herbivore, fermenter	Sheep, antelope	60–75	10–13
Carnivore	Insectivorous lizard	70–85	16–19
Granivore	Seed-eating rodent	90	16.9
Nectarivore	Hummingbird, butterfly	100	18
Soil eater	Earthworm	0.5–5.0	0.7

and production. Then, the microbes are digested and absorbed later in the digestive tract (beginning in the fourth stomach chamber, which is acidic). Thus, ruminants "cultivate" and "eat" microbes (mainly bacteria) and also eat plants. Nonruminant vertebrate herbivores ("monogastrics") have the standard acidic stomach and are able to digest only the soluble components in plant matter in their stomachs and small intestine. Thus, they obtain much less of the total energy in the diet than do fermenters (Table I), although in the field monogastrics often compensate by selecting young vegetation that has more soluble and less fiber components. However, many species of monogastric herbivores have some fermentation occurring somewhere in their gut, either in a somewhat modified portion of the stomach (e.g., macropod kangaroos, some primates) or in the large intestine (e.g., horses, many primates), and these species gain some of the energy locked up in plant fiber. Rabbits and some herbivorous rodents ingest their own feces (coprophagy) and thereby derive nutrition from the fermentative microbes bodies, as do ruminants. Carnivores digest most of the dry matter in their food (Table I). Granivores and nectarivores have the highest gain of energy from their food. They eat products of plants with which they have coevolved, wherein the plant benefits from pollination or seed dispersal and the animals benefit by obtaining food of high energetic quality. It is not surprising to find large benefits to consumers in circumstances where those producers that favor consumption have a selective advantage.

3.3 Stomach Capacity

Daily food consumption may be influenced by maximum stomach capacity. If predation or competitive pressures are high, animals that can obtain more food during a single foraging trip should have greater fitness. However, the information needed to test this hypothesis is largely unavailable. Early attempts to measure stomach volume by removing stomachs during autopsy, filling them with water, and weighing them are of questionable relevance to the situation in a live animal. In general, the capacities (volumes) of organs increase as a constant percentage of body mass as body size increases, but there is much room for interspecific differences. This is a topic for future research to address.

3.4 Appetite

Appetite is a major factor influencing energy intake rates, but little is known about appetite in wild animals. There is much literature on laboratory rats and mice, as well as on humans in relation to obesity and control of body mass, in both the physiological and psychological arenas. The scanty information on wild animals comes from measurements of feeding rate under various conditions rather than from measurements of hunger or satiation, which are the components of appetite, probably because of the difficulties in evaluating hunger and satiation directly in wild animals. Nutrient deprivation and hormone manipulation experiments indicate that food consumption rates of wild animals are determined primarily by energy needs rather than by needs for specific nutrients such as certain amino acids, fatty acids, vitamins, minerals, and/or carbohydrates or by needs for protein, carbohydrates, lipids, and/or vitamins in general. Animals eating diets deficient in specific vitamins, minerals, or amino acids generally do not increase food consumption beyond that needed to satisfy their energy requirements, but new foods become more attractive to them. However, animals deprived of water do develop an appetite for water (thirst), and sodium deprivation leads to "salt hunger" in some species. In general, appetite is influenced primarily by the energy requirements of the animal. This is also an area where research is needed.

3.5 Rates of Feeding

Feeding rates that are actually achieved in the field are influenced not only by intrinsic factors, such as endothermy or ectothermy, activity level, reproductive status, and digestive physiology, but also by extrinsic factors, such as season, food availability, competitive and social interactions with other individuals of the same or other species, and reduction of feeding opportunities due to the presence of predators or inclement environmental conditions (e.g., excessive midday heat, darkness, rainstorms). Over a period of several days or weeks, animals are usually able to compensate for short-term difficulties in getting food and are able to obtain approximately enough food to maintain energy balance. Body size is the most important determinant of feeding rate; bigger animals eat more food than do smaller animals. However, this is not a one-to-one relationship. An animal weighing 10 times more than another animal does not eat 10 times more food each day; rather, it eats only approximately 5 to 6 times more, in accordance with the scaling of metabolic rate. When differences in body mass are accounted for, feeding rates still vary by more than 25 times among species during their activity seasons. For example, a representative insectivorous lizard that weighs 100 g consumes about 0.7 g of food (dry matter intake [DMI]) per day, whereas a typical 100-g bird living in a marine habitat consumes about 18 g of dry matter per day (Table II). Both may be living in the same seashore habitat, eating a similar diet (arthropods), and maintaining the same body temperature during the day, but the lizard's metabolic energy expenditures over a 24-h period are only about 4% those of the bird, so the lizard's food needs are proportionately lower as well. Within various groups of endothermic vertebrates, there is a 230% variation in feeding rates among same-sized animals. Desert-dwelling mammals and birds have relatively low feeding rates, and marine birds have relatively high feeding rates. Desert mammals and birds are known to have lower metabolic rates (basal and field) than do related species living in other habitats, so their food requirements should be correspondingly lower. The high feeding rates of herbivorous mammals, relative to other mammals, most likely result in part from their low MEE (Table I), meaning that relatively more plant food must be eaten to obtain a given daily rate of metabolizable energy intake.

4. ENERGY USE

4.1 Metabolic Rate

Oxidative metabolism is the avenue of energy use that powers the daily costs of living (analogous to the overhead of human financial budgets such as housing, food, and transportation). This portion of the energy budget, also known as respiratory metabolism or metabolic rate, ultimately ends up as heat, which is produced primarily inside the animal during biochemical reactions or by friction of moving parts but also outside the animal, mainly by friction with, and work done on, the environment. Although most animals are capable of fueling these energy expenditures by anaerobic metabolism for short periods, either because of a lack of oxygen during breath holding (e.g., forced diving) or because the muscles are called on to spend adenosine triphosphate (ATP) energy faster than it can be produced by oxidative processes, the "oxygen debt" that is built up is paid off ultimately by aerobic processes. However, a few animals live most or all of their lives in anaerobic conditions (e.g., worms in anoxic mud and unstirred water in swampy areas, some intestinal parasites), and anaerobic metabolism is their main source of ATP energy. Nevertheless, most animals have access to oxygen and produce ATP primarily by oxidative metabolism.

Energy needs for oxidative metabolism apparently get the highest priority for metabolizable energy. When an animal is eating extra food, it will either store fat, grow in mass, or reproduce with the extra metabolizable energy (Fig. 2). However, as the feeding rate is diminished, these productive processes decline to zero while the metabolic rate remains nearly constant. At feeding levels below those needed for energy balance (where the rate of metabolizable energy intake = the rate of oxidative metabolic expenditure), production is "negative" in that the animal draws on chemical potential energy stored in its body to make up the deficit between intake from food and metabolic expenses (Fig. 2) and body mass declines. An animal's rate of oxidative metabolism does not remain exactly constant over a range of feeding rates. Metabolic rate increases as food intake increases above the weight maintenance level of intake, apparently due to greater costs of digestion and assimilation and to added

TABLE II

Feeding Rates of Wild Vertebrates, Summarized in Allometric Equations Derived from Field Measurements of Energy Metabolism, for Various Groups of Mammals, Birds, and Reptiles

Animal group	a	b	Expected feeding rate (g dry matter/day) for a 100-g animal
Eutherian mammals (58 species)	0.299	0.767	10.2
Marsupial mammals (20 species)	0.483	0.666	10.4
Herbivorous mammals (26 species)	0.859	0.628	15.5
Desert mammals (25 species)	0.192	0.806	7.9
Birds (95 species)	0.638	0.685	15.0
Passerine birds (39 species)	0.630	0.683	14.6
Desert birds (15 species)	0.407	0.681	9.4
Marine birds (36 species)	0.880	0.658	18.2
Reptiles (55 species)	0.011	0.920	0.77
Herbivorous reptiles (9 species)	0.033	0.717	0.91
Insectivorous lizards (27 species)	0.011	0.914	0.73

Note. The equations have the following form: $y = ax^b$, where y = feeding rate (in grams dry matter consumed per day), x = body mass (in grams), a is the intercept at body mass = 1 g, and b is the allometric slope.

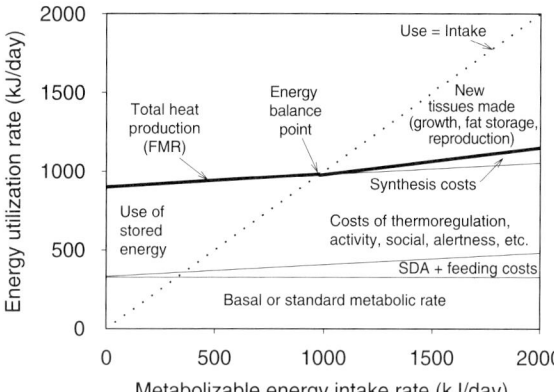

FIGURE 2 Fate of metabolizable energy input at various feeding rates in a representative 1-kg mammal. As daily feeding rate increases, metabolizable energy intake (dotted line) increases to a point where intake rate = utilization rate (energy balance point), and stored energy (e.g., fat) in the body is no longer used to make up the deficit (deficit indicated by the triangular space between dotted line [intake rate] and heavy solid line [metabolic rate] to the left of the energy balance point). At higher intake rates, extra energy intake (indicated by the triangular space above the heavy solid line to the right of the energy balance point) ends up in new biomass (production) either as fat, new somatic tissues, or reproductive tissues. Specific dynamic action (SDA, the cost of digesting food) and feeding costs increase as more food is consumed in a day, and costs of synthesizing new tissues increase as food consumption exceeds the energy balance point.

biosynthesis costs for production. The decline in metabolic rate as the food intake rate drops farther below the maintenance level is due to reduced digestion costs and probably to changes in behavior (reduced activity) and physiology (e.g., lower peripheral and even core body temperatures) that often accompany various degrees of starvation.

Rates of energy metabolism in free-living wild vertebrates are known as field metabolic rates (FMRs) or as daily energy expenditures (DEEs) and are expressed in units of kilojoules per day. There is much variation in FMRs of different animals, with measured values ranging from 0.2 kJ/day ($\sim$2.5 mW given that 1 W = 86.4 kJ/day) in a small African gecko to 67,700 kJ/d (784 W) in a male elephant seal weighing half a metric ton. This difference of nearly six orders of magnitude (10^6) in FMR is due largely to differences in body mass; bigger animals need more energy each day to pay their daily costs of living. FMRs increase exponentially (rather than linearly) with body mass, as do many other physiological and morphological variables. The slopes of the log–log (allometric) relationships for FMRs in various taxonomic, dietary, and habitat groupings of vertebrates range from 0.5 to near 1.0 and average approximately 0.7 (Table III). A slope of 0.7 indicates that an animal that is 10 times larger than another animal will have an FMR approximately five times greater.

By solving the allometric equations in Table III for a 100-g animal, we can adjust FMRs for body size effects and then compare the FMRs of various taxa. It is obvious that birds and mammals are "high-energy" machines in comparison with reptiles. The endotherms need 15 to 25 times more energy per day to pay their costs of living (overhead) as do lizards of the same body mass. This enormous difference in the daily cost of living is due in part to the much higher concentrations and activities of metabolic enzymes, higher mitochondrial densities and cellular membrane surface areas, and higher thyroid activities in endotherms. It is also due to metabolic responses to the daily body temperature regimes of the two groups. Lizards become cool at night and their metabolism decreases, but endotherms stay warm and their metabolism may increase rather than

TABLE III

Field Metabolic Rates of Wild Vertebrates, Summarized in Allometric Equations Derived from Field Measurements of Energy Metabolism, for Various Groups of Mammals, Birds, and Reptiles

Animal group	a	b	Expected FMR (kJ/day) for a 100-g animal
Eutherian mammals (58 species)	4.21	0.772	147
Marsupial mammals (20 species)	10.1	0.590	153
Rodents (30 species)	5.48	0.712	145
Desert rodents (15 species)	9.68	0.487	91
Herbivorous mammals (26 species)	7.94	0.646	156
Birds (95 species)	10.5	0.681	242
Passerine birds (40 species)	10.4	0.680	238
Marine birds (36 species)	14.25	0.659	296
Desert birds (15 species)	6.35	0.671	140
Reptiles (55 species)	0.196	0.889	11.8
Desert lizards (16 species)	0.177	0.935	13.1
Herbivorous reptiles (8 species)	0.232	0.813	9.8

Note. The equations have the following form: $y = ax^b$, where y = feeding rate (in grams dry matter consumed per day), x = body mass (in grams), a is the intercept at body mass = 1 g, and b is the allometric slope.

decrease at night to provide the needed heat. Among the endotherms, daily costs are similar for eutherian mammals and marsupial mammals but are higher for birds. One group of birds, those that live in marine habitats, has a very high cost of living, being about twice that of similar-sized mammals. As a group, desert-dwelling mammals and birds have relatively low FMRs, and this may benefit them in their sparsely provisioned habitat. FMRs also change with seasons, as indicated by the doubling in FMRs of desert kangaroo rats from summer to winter (Fig. 3A). This is probably related to the higher thermoregulatory costs during the cooler months (Fig. 3B). To the contrary, FMRs of desert tortoises decline during winter because the cold weather inhibits activity in these ectothermic animals and they hibernate during winter (Fig. 3).

What determines FMRs? Itemizing daily field energy expenses can reveal the mechanisms underlying daily costs of living and can provide explanations for differences in FMRs.

4.2 Production

Metabolizable energy consumed in excess of that required by oxidative metabolism can be used to create new biomass in the form of somatic growth, storage, or reproduction—or even two or all three simultaneously (Fig. 2). There are two issues here. The first concerns the conditions under which an animal consumes more energy than it needs to maintain energy balance, and the second concerns how extra ingested energy will be allocated. Both issues depend on species of the animal, its age, its sex, its endocrinological status, and the food supply in its habitat, among other things.

Many animal species (e.g., most reptiles, many invertebrates) seem to eat whenever food is available to them even though they may have already eaten enough to meet their immediate metabolic needs. These species have indeterminate growth and continue to increase in size throughout life via somatic growth. Although the extra energy is allocated mainly to growth, fat may be accumulated during times of great food abundance. For example, some snakes have greater fat contents just before winter begins as well as during occasional rodent population explosions, and filter-feeding bivalve mollusks have high fat contents during autumn as well as during algal blooms that may occur in some years and not others. These species do accumulate fat at certain times of the year, usually before seasons of low food availability, but the regulation of mode of production is poorly understood in many species. Some indeterminate growers, such as marine crustaceans, bivalves, and coelenterates, are able to reverse the growth process and become smaller while oxidizing their own somatic tissues during periods of food shortage.

Determinate growers, such as birds and many insects and mammals, have a fixed adult size range, and individuals do not grow much after reaching adult size. In fact, most birds reach adult size before they fledge and leave the nest. For adults of these species, appetite usually limits food intake to just that needed for oxidative metabolism even though food may be abundant in the habitat. In general, wild

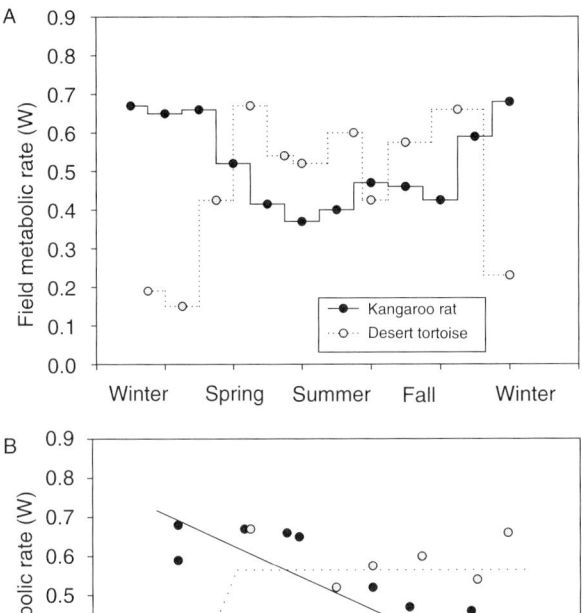

FIGURE 3 Field metabolic rates throughout a year in two desert animals: the kangaroo rat (*Dipodomys merriami*) and the desert tortoise (*Gopherus agassizii*). (A) The endotherm (rat) had higher metabolic rates during the cold seasons than during the warm seasons, but the reverse was true for the ectotherm (tortoise), which had higher FMRs during the warm seasons. (B) The ambient temperature effects are more easily seen when the FMR data are replotted versus average ambient air temperature. Interestingly, the annual energy expenditures of the two species are similar because the tortoise, a "low-energy" ectotherm, is about 30 times larger than the kangaroo rat, a "high-energy" endotherm. Thus, one individual of either species will have a similar impact on the food resources in its habitat.

vertebrates do not become obese, in contrast to some species of domestic animals and humans. Adults of species having determinate growth tend to have similar body fat contents in the field during a given season, and very fat individuals are rare. Many of these species are capable of increasing appetite and accumulating much adipose tissue, but they do not get fat until just before the period of need (e.g., winter hibernation, migration). One could speculate that obese animals would be quite attractive to predators due to their high energetic reward as well as to their lowered agility and consequent increased vulnerability. If so, selection would act to restrict periods of obesity to times of need only.

During the breeding season, both indeterminate and determinate growers allocate energy-containing chemicals to reproductive tissues. In many species, the males' energy investment in reproductive tissues (testes and accessory gland enlargement and sperm production) is minor compared with that of females. (An exception is found in male crickets and other insect species where males pass a large spermatheca to females, which use some of the chemical energy in the spermatheca to produce eggs.) Depending on species and body size, females may synthesize reproductive tissues (e.g., eggs, embryos, placentas, milk) at very high rates, and their feeding rates may double or even triple (in some species of small rodents) during some phases of the breeding period. For example, a female deer mouse may produce a litter that, at weaning time, may weigh several times her own body mass—all accomplished in a few weeks.

The chemicals needed for synthesizing the new biomass apparently are taken from the pool of substrates circulating in the blood. Hormones (i.e., growth hormone and thyroid hormones in the case of somatic growth, reproductive hormones in the case of reproductive tissues) trigger anabolic metabolism in certain tissues, and this apparently depletes blood substrates. This is presumably detected by the liver and other organs, which then add substrates to the blood to restore homeostasis. On the other hand, fat deposition appears to be a consequence of higher than normal blood levels of nutrients, favoring production and storage of lipids in adipose tissue. This process is influenced by the pancreatic hormones insulin and glucagon. It is not clear whether increased appetite or hormonal stimulation of fat deposition is the primary stimulus for the rapid deposition of fat in some small mammals just before they hibernate and in some birds just before they migrate.

It is possible for both "negative" and "positive" production to occur simultaneously in an animal. For example, lactating female mammals may draw on their own tissues to produce milk for their babies when food intake is inadequate. Similarly, female desert tortoises use protein and fats from their own soma, along with newly consumed foodstuffs, to synthesize eggs during spring.

It is important to recognize that neither food energy supply nor energy intake is the primary determinant of production in many animals. Food supply may be high, but that does not necessarily mean that an animal will consume more and then grow or reproduce. Although fat deposition may occur as a result of a high food intake in response to high food availability, somatic growth (in determinate growers) and reproduction are usually triggered by factors other than energetics, and it is these factors that increase appetite. Appetite seems to be determined largely by hormonal conditions in the animal that are responding to factors such as season, photoperiod, and age and sex of the animal as well as by energetic parameters in the animal and its environment.

In the field, an animal that is producing new biomass is paying not only additional energy for biosynthesis costs but also the extra costs associated with getting the additional food needed for growth. These costs include the extra activity involved in foraging longer, avoiding predators longer, thermoregulating longer, and interacting socially longer while getting the extra food as well as the higher specific dynamic action required to process the extra food. These added costs should also be considered as costs of production.

5. EFFICIENCY

Not all of the chemical potential energy in the food can be used by animals because some is indigestible and is lost as feces, and some that is digested and absorbed leaves the body as urinary wastes (which in some animals is mixed inseparably with feces when voided). The fraction of the food's energy that is absorbed across the gut and not lost in feces is called assimilated energy, and the percentage of dietary energy that this amounts to is the apparent AE. Apparent MEE, or the percentage of dietary energy that is usable for metabolism or production, is assimilated energy corrected for urinary energy losses. Production efficiency can be expressed in a variety of ways. Here, it is given as net production efficiency (NPE), which is the percentage of metabolizable energy that appears as chemical potential energy in new biomass (e.g., fat, growth, eggs, embryos, milk).

TABLE IV

Typical Energetic Efficiencies (Percentages) in Animals Having Various Diet

Diet type	Assimilation efficiency	Metabolizable energy efficiency	Net production efficiency
Herbivore, monogastric	30–77	22–73	20–55
Herbivore, ruminant	60–75	50–70	23
Herbivore, hindgut fermenter	60–72	55–68	23
Granivore	77–90	74–81	22–24
Carnivore	70–88	62–81	30–58
Omnivore	80	74	23–58
Lactivore	95	87	45
Detritivore	12–44	12–42	53
Parasites	77–100	71–95	50

Note. AE (assimilation efficiency) = [(I−L) / I]100; MEE (metabolizable energy efficiency) = [(I−L−U)/I]100; NPE (net production efficiency) = [P/(I−L−U)]100; I, food energy intake; L, feces energy loss; U, urinary energy loss, P, energy allocated to production of biomass.

AE and MEE vary greatly among diets (Table IV). This variation is due mainly to the relatively indigestible organic components of the various diets: cellulose in plants, keratin in vertebrates, and chitin in arthropods. Within the herbivores, the type of digestive system an animal has also influences AE. For a given diet, herbivores that ferment foods, either at the beginning of the gut in specialized stomachs (e.g., ruminants) or at the end of the gut (cecum or colon fermenters), can digest and assimilate more of that food's energy. These fermenters obtain some of the energy in cellulose, which is broken down by cellulases produced by anaerobic microbes in their fore- or hindgut. This allows ruminants to live on lower quality plant foods, such as dry grasses (straw), and reduces competition for food among herbivore types. Monogastric herbivores often choose high-quality foods (e.g., young leaves, flowers, buds) when available, and they may attain high AE and MEE (up to 73–77% [Table IV]) by diet selection, thereby avoiding the low AE (22–30%) they would have if they ate dry grasses.

Young suckling mammals (lactivores), parasites, and animals that eat seeds (granivores), nectar (nectarivores), or other animals (carnivores) all have relatively high AE and MEE. Coevolution of plants along with the animals that pollinate their flowers and disperse their seeds probably has favored their cooperation and the ready satisfaction of those animals' energy needs from those plants' products. On the other hand, herbivores are essentially predators on plants, and evolution has favored plants that developed defenses against herbivory such as spines, woody tissues, poisons, crystalline inclusions (silica spines), and antifeedant chemicals (secondary plant compounds). Moreover, plant foods often are unbalanced in regard to the nutrient proportions (e.g., energy, protein, minerals, vitamins) required by the consumer. Thus, some herbivores have extraordinary modifications of their feeding and digestive apparati that improve their ability to survive on vegetation (e.g., plant toxin avoidance or resistance by some caterpillars and beetles, coprophagy by rabbits, rumination by sheep and antelope). Animals that eat other animals (carnivores and parasites) do not face the problems of food quality that the herbivores face because the composition of these animals' food is essentially the same as their own. However, carnivores often are challenged by problems of food quantity in that their prey may be hard to obtain.

Production efficiency can be measured only when an animal is producing new biomass, and this may be a fleeting event in the lives of some wild animals, especially birds and many mammals and insects. The values for NPE in Table IV are mainly from domestic, farm, or captive wild animals, and they represent near maximum efficiencies that occur during peak production periods. Efficiencies will be lower in animals that forage for their own food because extra energy is required to obtain the extra food needed to support growth. Nevertheless, some generalizations seem warranted from studies on domesticated animals. Ectotherms generally have higher peak NPE than do endotherms, apparently due to the higher metabolic expenditures for oxidative metabolism paid by endotherms. Where a range of values for NPE is given in Table IV, the highest efficiencies are for ectotherms and the lowest are for endotherms. Diet type has a large effect on NPE, with high-protein foods having lower efficiencies than high-fat and high-starch diets. Finally, the type of production occurring influences NPE, with milk production and somatic growth having efficiencies substantially higher than those for fattening. There is still controversy about whether NPE declines at higher rates of food intake, so the "law of diminishing returns," derived from experiments on adding increasing amounts of fertilizer to crops, might not be applicable to animal energetics.

Studies on five species of farm animals, selected for good production capabilities, show a modest and surprisingly uniform maximum rate of food energy consumption (maximum ME intake = 4.4–5.7 BMR). A subsequent evaluation of maximum observed feeding rates in domestic and wild mammals (8 species) and birds (11 species) indicates that maximum ME intake ranges from three to six times BMR. Thus, animals may limit their food intake to rates that their guts can deal with effectively. The role of appetite is clearly critically important but still poorly understood for wild animals.

Energetic efficiency values are useful for evaluating the quality of various foods and for identifying feeding and digestive adaptations of animals eating a variety of food types. It is tempting to hypothesize that animals should be selected for having higher efficiencies, but this is unlikely for three reasons. First, efficiency is a ratio of two rate processes, and there is no evidence that natural selection responds to such an abstract derived variable as a ratio. Second, some important efficiencies are already near the maximum of 100%, leaving little variation on which selection might operate. Third, an animal can increase its daily rate of metabolizable energy intake much more easily and dramatically by eating more food and keeping AE constant than by keeping feeding rate constant and improving its digestive physiology enough to increase MEE substantially.

6. REGULATION OF ENERGY METABOLISM

Most physiological processes and systems are closely regulated, so it is reasonable to ask whether and how oxidative metabolism of animals is regulated. Metabolic rates of marine and aquatic invertebrates and water-breathing vertebrates may be regulated by oxygen availability in their environments. Metabolism of terrestrial and aquatic invertebrates and ectothermic vertebrates that live in cold climates may be reduced dramatically when these animals freeze during winter. For air-breathing vertebrates (birds, mammals, reptiles, and most amphibians), current knowledge indicates that aspects of energy budgets (e.g., thermoregulation, reproductive output) are regulated but that overall daily energy expenditure (FMR) itself is probably not directly controlled. Instead, the FMR appears to be a consequence of its component parts. For example, an ectothermic vertebrate has a total daily expense for its resting metabolic costs that will be determined by its daily body temperature excursion, which is most likely determined by the animal's minute-by-minute decisions to hide, forage, bask, display, or escape predators and not by some metabolic control center in its central nervous system. This animal may spend additional energy for activity associated with basking, feeding, interacting, and escaping. These costs are also consequences of other phenomena, as are the added costs of digesting and processing food, producing new biomass, and remaining alert and responsive while above ground. Thus, the total daily expense (FMR) seems to be the sum of costs incurred for tasks not directly related to energy metabolism. The situation for an endotherm is similar except that its minimum metabolic rate (BMR) is fixed and its thermoregulatory costs are determined by ambient temperatures.

Animals do some things that may reflect a crude ability to regulate (mainly reduce) FMRs. Ectotherms may choose to remain inactive in cool retreats for days, especially during periods of environmental stress (e.g., drought, low food availability), and this behavior reduces the daily use of energy and other nutrients. Similarly, some small endotherms may use torpor or hibernate, and their low body temperatures reduce FMRs greatly, but this also renders them relatively unresponsive and vulnerable to predation. However, for normally active animals, there is no known mechanism that might sense the rate of energy expenditure or the accumulated daily total energy expenditure and then limit the amount allocated to thermoregulation, activity, digestion, and the like during the rest of the day (regulate output). Instead, it seems that animals spend whatever energy is called for to get through the day and then regulate intake, by means of altered appetite, to achieve energy balance. Similarly, energy allocation to growth and reproduction appears to be a consequence of hormonal changes associated with regulation of those processes, with appetite being the "cart" following the "horses" of growth and reproduction.

Even though the rate of oxidative metabolism might not be under direct physiological control, there are upper limits that constrain energy expenditures.

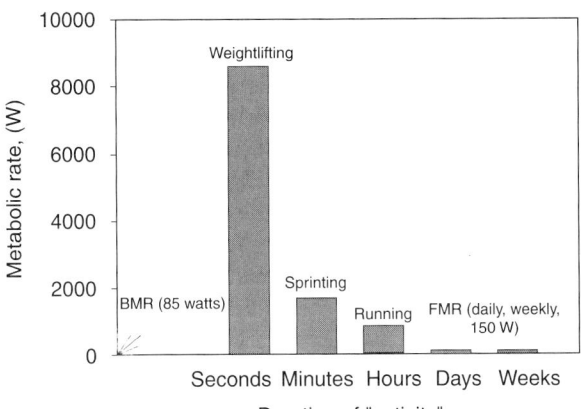

FIGURE 4 Rates of energy expenditure achieved by human athletes during bouts of activity of various kinds. Weightlifters can use ATP very quickly during a lift (100 times their BMR), but this cannot be sustained for very long due to buildup of oxygen debt. Sprinters also use anaerobic sources of ATP and cannot sustain sprinting for longer than a few minutes. Runners (at 20 times BMR) sustain their activity aerobically but are not maintaining energy balance while running because their intake is zero. Maximum sustainable metabolic rates, where humans are in energy balance, are only 5 times BMR. Typical FMRs of nonathlete humans are 1.5 to 2.5 times BMR.

The level of these limits depends on the amount of time involved in the determination of the limit. For example, a champion weightlifter can put out power during a lift that is more than 100 times his BMR, but this effort draws on and uses up energy-containing chemicals and oxygen stored in muscles, so it cannot be sustained for more than a few seconds (Fig. 4). Similarly, sprinters use aerobic as well as anaerobic metabolism to power their running, so sprinting cannot be sustained for more than a few minutes, but sprinters can run faster (use energy faster) than can long-distance runners, who are entirely aerobic. However, continuous running is not a sustainable activity either due to the needs to eat and sleep periodically. The highest sustainable FMR measured in humans is approximately 480 W (in Tour de France cyclists) or approximately 5.6 times BMR. Maximums measured in wild animals are approximately 6.9 times BMR in a small marsupial mammal, 5.0 times BMR in a rodent (eutherian mammal), 6.7 times BMR in a breeding seabird, and 4.6 times estimated field SMR in a small reptile.

7. SUMMARY

Chemical potential energy, in the form of food, is a centrally important resource for animals (heterotrophs) and has been termed the "primary currency of life." Knowledge of the ways in which animals gain, use, and store energy, and of their rates or "flows," may reveal much about the biology of various species, and such knowledge is essential in understanding energy flows in populations and through communities. The input side of a chemical potential energy budget is simple, being made up only of food intake. Energy outputs, corrected for loss of chemical energy in excreta, include heat resulting from oxidative metabolism and biomass (production of new tissues in the form of adipose tissue [fat storage], somatic growth, and reproductive tissues [sperm, embryos, eggs, placentas, and milk]). Different kinds of foods yield widely different amounts of energy; soil and detritus contain the least, plant tissues are intermediate, animal matter is higher, and seeds and nectar are highest. However, the availability of these food types in the field is not uniform in space or time. Endothermic animals require much more (15–25 times more) energy each day than do same-sized ectothermic animals. Large animals need more food energy than do small animals, but the increased food requirement is only about half that expected from the difference in body masses (a 10-fold mass increase requires only a 5- to 6-fold food intake increase). Most or all of the available food energy (metabolizable energy) is used for oxidative metabolism (metabolic rate), which has highest priority. Excess metabolizable energy can be used for production, but many animals do not consume excess food, despite its availability, except during certain seasons (e.g., breeding, premigratory, prehibernation). Daily energy needs appear to be influenced mainly by energy use for resting costs, temperature regulation, and activity. Although there appear to be upper and lower limits to rates of energy expenditure and food intake, actual daily energy needs seem to be determined more by lifestyle, season, and local circumstances than by precise regulation of energetic parameters themselves. Digestibility, AE, and MEE are determined mainly by diet type, although there is much variation among digestive capabilities of different herbivores. Net production efficiency is usually higher in ectotherms than in endotherms. Although daily energy expenditures by animals have operational upper and lower limits, actual expenditures do not seem to be regulated directly; rather, they are subservient to other aspects of the animals' biology such as thermoregulation needs, foraging expenses, reproductive and migration costs, and costs of interactions with other organisms (e.g., conspecifics, predators) in their surroundings. Appetite, a poorly understood factor in wild animals, apparently matches income with expenditures, so that most animals are approximately in energy balance much of the time. The evolution and success of birds and mammals, which have extraordinarily high costs of living compared with those of ectotherms, still needs a compelling explanation.

SEE ALSO THE FOLLOWING ARTICLES

Conversion of Energy: People and Animals • Earth's Energy Balance • Ecosystems and Energy: History and Overview • Food Capture, Energy Costs of • Human Energetics • Migration, Energy Costs of • Photosynthesis and Autotrophic Energy Flows • Reproduction, Energy Costs of • Thermoregulation

Further Reading

Blaxter, K. (1989). "Energy Metabolism in Animals and Man." Cambridge University Press, Cambridge, UK.
Kleiber, M. (1975). "The Fire of Life." R. E. Krieger, Huntington, NY.
Louw, G. N. (1993). "Physiological Animal Ecology." Longman Scientific and Technical, Essex, UK.
McNab, B. K. (2002). "The Physiological Ecology of Vertebrates: A View from Energetics." Cornell University Press, Ithaca, NY.
Nagy, K. A. (2001). Food requirements of wild animals: Predictive equations for free-living mammals, reptiles, and birds. *Nutr. Abstr. Rev.* **71**, R21–R32.
Nagy, K. A., Girard, I. A., and Brown, T. K. (1999). Energetics of free-ranging mammals, reptiles, and birds. *Annu. Rev. Nutr.* **19**, 247–277.
Robbins, C. T. (1993). "Wildlife Feeding and Nutrition." 2nd ed. Academic Press, San Diego.

Human Energetics

WILLIAM R. LEONARD
Northwestern University
Evanston, Illinois, United States

1. Introduction
2. Calorimetry
3. Sources of Food Energy
4. Determinants of Daily Energy Needs
5. Human Variation in Energy Dynamics

Glossary

aerobic respiration (metabolism) Energy producing reactions that require oxygen (O_2).
basal metabolic rate (BMR) The minimum level of energy required to sustain basic biological function of the body.
body mass index (BMI) A weight-for-height index used to assess physical nutritional health, calculated as weight(kg)/height(meters)^2.
calorimetry The measurement of heat transfer.
kilocalorie (kcal) The amount of heat required to raise the temperature of 1 kilogram (or 1 liter) of water by 1°C. This is the unit most commonly used to measure the energy content of food and human energy requirements.
macronutrient Chemical components of food that are used as sources of energy. These include carbohydrates, fats, and proteins.
physical activity level (PAL) The ratio of total daily energy expenditure to basal metabolic rate (TDEE:BMR).
total daily energy expenditure (TDEE) The total energy requirements of a person over the course of an entire day.

The dynamic relationship between energy intake and expenditure has important consequences for human health, survival, and reproduction. Human energy requirements are determined by a variety of factors, including age, sex, body size, and activity levels. Energy availability, on the other hand, is strongly shaped by one's social, economic, and physical environments. Worldwide variation in nutritional health directly reflects differences in long-term energy balance. Among rural populations of the developing world, conditions of negative energy balance and undernutrition are common due to high levels of daily energy expenditure and marginal food availability. Conversely, in the industrialized world rates of obesity and chronic metabolic diseases are increasing at an alarming rate due to the abundance of dietary energy and increasingly sedentary lifestyles.

1. INTRODUCTION

Sufficient dietary energy is essential to the survival and health of all animals. From the perspective of human biology and health, energy is particularly important for a number of reasons. First, food and energy represent critical points of interaction between humans and their environment. The environments in which we live determine the range of food resources that are available and how much energy and effort are necessary to procure those resources. Indeed, the dynamic between energy intake and energy expenditure is quite different for a subsistence farmer of Latin America than it is for an urban executive living in the United States. Beyond differences in the physical environment, social, cultural, and economic variations also shape aspects of energy balance. Social and cultural norms are important for shaping food preferences, whereas differences in subsistence behavior and socioeconomic status strongly influence food availability and the effort required to obtain food.

Additionally, the balance between energy expenditure and energy acquired has important adaptive consequences for both survival and reproduction. Obtaining sufficient food energy has been an important stressor throughout human evolutionary history, and it continues to strongly shape the biology of traditional human populations.

This article examines aspects of energy expenditure and energy intake in humans. First it explores the principles of calorimetry, the methods used to measure energy expenditure in humans and other animals. The discussion then examines sources of energy in our diets, the macronutrients—carbohydrates, fats, and proteins. Next, the of the physiological basis of variation in human energy requirements is evaluated, specifically the different factors that determine how much energy a person must consume to sustain himself of herself. Finally, the discussion considers patterns of variation in energy expenditure and intake among modern human populations, highlighting the different strategies that humans use to fulfill their dietary energy needs and the health consequences associated with the increasingly sedentary lifeways of modern society.

2. CALORIMETRY

The study of energy relies on the principle of calorimetry, the measurement of heat transfer. In food and nutrition, energy is most often measured in kilocalories (kcal). One kilocalorie is the amount of heat required to raise the temperature of 1 kilogram (or 1 liter) of water, 1°C. Thus, a food item containing 150 kcal (two pieces of bread, for example) contains enough stored chemical energy to increase the temperature of 150 liters of water by 1°C. Another common unit for measuring energy is the joule or the kilojoule [1 kilojoule (kJ) = 1000 joules]. The conversion between calories and joules is 1 kcal is equal to 4.184 kilojoules.

Techniques for measuring energy expenditure involve either measuring heat loss directly (direct calorimetry) or measuring a proxy of heat loss such as oxygen consumption (O_2) or carbon dioxide (CO_2) production (indirect calorimetry). Direct calorimetry is done under controlled laboratory conditions in insulated chambers that measure changes in air temperature associated with the heat being released by a subject. Although quite accurate, direct calorimetry is not widely used because of its expense and technical difficulty.

Thus, methods of indirect calorimetry are more commonly used to quantify human energy expenditure. The most widely used of these techniques involve measuring oxygen consumption. Because the body's energy production is dependent on oxygen (aerobic respiration), O_2 consumption provides an accurate indirect way of measuring a person's energy expenditure. Every liter of O_2 consumed by the body is equivalent to energy cost of approximately 5 kcal. Consequently, by measuring O_2 use while a person is performing a particular task (for example, standing, walking, running on a treadmill), the energy cost of the task can be determined.

Figures 1 and 2 show alternative methods for measuring of oxygen consumption. Figure 1 shows the classic Douglas bag method for measuring O_2 uptake. With this technique, subjects breathe through a valve that allows them to inhale room air and exhale into a large collection bag. The volume and the O_2 and CO_2 contents of the collected air sample are then measured to determine the total amount of oxygen consumed by the subject. Recent advances in computer technology allow for the

FIGURE 1 Measurement of resting energy expenditure using the Douglas Bag method of indirect calorimetry. The subject breathes into the collection bag for a specified amount of time. The volume of the collected air sample and its oxygen and carbon dioxide content will then be measured to determine the amount of energy the subject has used.

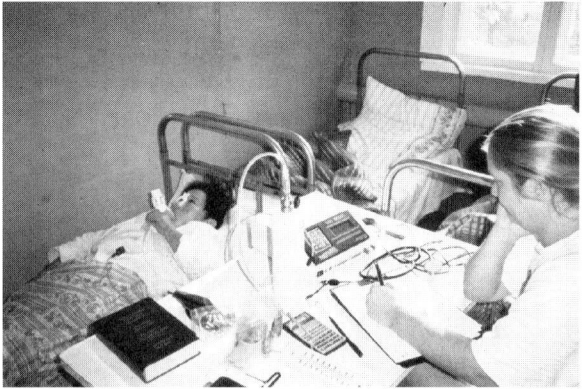

FIGURE 2 Measurement of energy expenditure using a computerized metabolic system (Aerosport TEEM 100). As with the Douglas bag method, this system measures the volume of expired air along with its oxygen and carbon dioxide contents to determine rates of energy expenditure.

determination of O_2 consumption more quickly without directly collecting the expired air samples. Figure 2 shows an example of a computerized system for measuring oxygen consumption. As with the Douglas bag method, this system determines energy costs by measuring the volume and the O_2 and CO_2 concentrations of expired air samples.

The most recently developed method of indirect calorimetry is the doubly labeled water (DLW) technique that involves the use of stable isotopes. With the DLW method, subjects drink a small amount of water that has been enriched with isotopes of both hydrogen (^2H, deuterium) and oxygen (^{18}O). After the labeled water has been consumed, subjects then provide urine samples for the subsequent 10 to 14 days. Over this period, the concentrations of both the ^2H and ^{18}O isotopes decrease, because they are being lost in the products of aerobic respiration, H_2O and CO_2. The concentration of ^{18}O in the body decreases more rapidly than that of ^2H, because the oxygen is contained in both metabolic waste products (H_2O and CO_2), whereas hydrogen is contained in only one (H_2O). Figure 3 shows a graph of the rates of decline (washout rates) of the two isotopes over a 14-day period. The differences in the washout rates provide an accurate measurement of CO_2 production and O_2 consumption over the period of study. These estimates of CO_2 production and O_2 consumption are then converted to kilocalories to provide a measure of total energy expenditure of the subject over the entire study period.

Validation studies of the DLW method against direct calorimetry have shown that the technique is quite accurate (± 2 to 3% error for individuals). Hence, since the late 1980s, DLW has become the most accurate technique for measuring human energy expenditure in real-world (i.e., nonlaboratory) situations. The main limitations of the technique remain its expense, about $800 to $1000 per subject.

3. SOURCES OF FOOD ENERGY

The main chemical sources of energy in our foods are carbohydrates, protein, and fats. Collectively these three energy sources are known as macronutrients. Vitamins and minerals (micronutrients) are required in much smaller amounts and are important for regulating many aspects of biological function.

Carbohydrates and proteins have similar energy contents; each provides ~4 kilocalories of metabolic energy per gram. In contrast, fat is more calorically dense; each gram provides about 9 to 10 kilocalories. Alcohol, although not a required nutrient, also can be used as an energy source, contributing ~7 kcal/g. Regardless of the source, excess dietary energy can be stored by the body as glycogen (a carbohydrate) or as fat. Humans have relatively limited glycogen stores (about 375 to 475 grams), in the liver and muscles. Fat, however, represents a much large source of stored energy, accounting for approximately 13 to 20% of body weight in men and 25 to 28% in women.

3.1 Carbohydrates

The largest source of dietary energy for most humans is carbohydrates (~45 to 50% of calories in the typical American diet). The three types of carbohydrates are monosaccharides, disaccharides, and polysaccharides. Monosaccharides, or simple sugars, include glucose, the body's primary metabolic fuel; fructose (fruit sugar); and galactose. Disaccharides, as the name implies, are sugars formed by a combination of two monosaccharides. Sucrose (glucose and fructose), the most common disaccharide, is found in sugar, honey, and maple syrup. Lactose, the sugar found in milk, is composed of glucose and galactose. Maltose (glucose and glucose), the least common of the disaccharides, is found in malt products and germinating cereals. Polysaccharides,

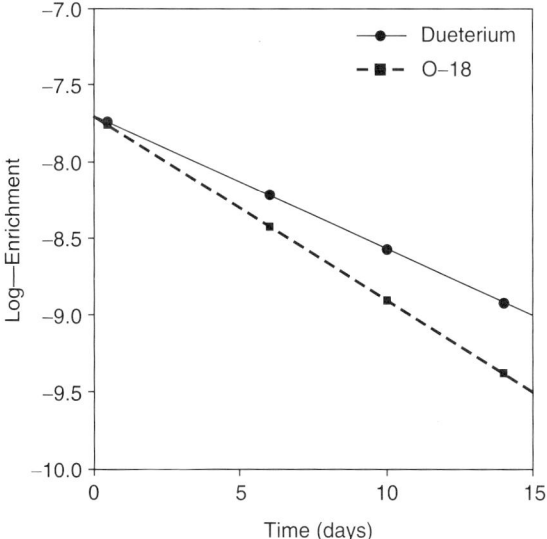

FIGURE 3 Graph showing the rates of decline (washout rates) in the two key isotopes (^{18}O and ^2H [deuterium]) used to measure daily energy expenditure in the "doubly labeled water" method. The difference in the washout rates of the two isotopes provides a measure of carbon dioxide (CO_2) production, which allows for the determination of energy expenditure.

or complex carbohydrates, are composed of three or more simple sugar molecules. Glycogen is the polysaccharide used for storing carbohydrates in animal tissues. In plants, the two most common polysaccharides are starch and cellulose. Starch is found in a wide variety of foods, such as grains, cereals, and breads, and provides an important source of dietary energy. In contrast, cellulose—the fibrous, structural parts of plant material—is not digestible to humans and passes through the gastrointestinal tract as fiber.

3.2 Fats

Fats provide the largest store of potential energy for biological work in the body. They are divided into three main groups: simple, compound, and derived. The simple, or neutral fats, consist primarily of triglycerides. A triglyceride consists of two component molecules: glycerol and fatty acid. Fatty acid molecules, in turn, are divided into two broad groups: saturated and unsaturated. These categories reflect the chemical bonding pattern between the carbon atoms of the fatty acid molecule. Saturated fatty acids have no double bonds between carbons, thus allowing for the maximum number of hydrogen atoms to be bound to the carbon (that is, the carbons are saturated with hydrogen atoms). In contrast, unsaturated fatty acids have one (monounsaturated) or more (polyunsaturated) double bonds. Saturated fats are abundant in animal products, whereas unsaturated fats predominate in vegetable oils.

Compound fats consist of a neutral fat in combination with some other chemical substance (for example, a sugar or a protein). Examples of compound fats include phospholipids and lipoproteins. Phospholipids are important in blood clotting and insulating nerve fibers, whereas lipoproteins are the main form of transport for fat in the bloodstream.

Derived fats are substances synthesized from simple and compound fats. The best known derived fat is cholesterol. Cholesterol is present in all human cells and may be derived from foods (exogenous) or synthesized by the body (endogenous). Cholesterol is necessary for normal development and function because it is critical for the synthesis of such hormones as estradiol, progesterone, and testosterone.

3.3 Proteins

Proteins, in addition to providing an energy source, are also critical for the growth and replacement of living tissues. They are composed of nitrogen-containing compounds known as amino acids. Of the 20 different amino acids required by the body, 9 (leucine, isoleucine, valine, lysine, threonine, methionine, phenylalanine, tryptophan, and histidine) are known as essential because they cannot be synthesized by the body and thus must be derived from food. Two others, cystine and tyrosine, are synthesized in the body from methionine and phenylalanine, respectively. The remaining amino acids are called nonessential because they can be produced by the body and need not be derived from the diet.

4. DETERMINANTS OF DAILY ENERGY NEEDS

A person's daily energy requirements are determined by several different factors. The major components of an individual's energy budget are associated with resting or basal metabolism, activity, growth, and reproduction. Basal metabolic rate (BMR) represents the minimum amount of energy necessary to keep a person alive. Basal metabolism is measured under controlled conditions while a subject is lying in a relaxed and fasted state.

In addition to basal requirements, energy is expended to perform various types of work, such as daily activities and exercise, digestion and transport of food, and regulating body temperature. The energy costs associated with food handling (that is, the thermic effect of food) make up a relatively small proportion of daily energy expenditure and are influenced by amount consumed and the composition of the diet (for example, high-protein meals elevate dietary thermogenesis). In addition, at extreme temperatures, energy must be spent to heat or cool the body. Humans (unclothed) have a thermoneutral range of 25 to 27°C (77 to 81°F). Within this temperature range, the minimum amount of metabolic energy is spent to maintain body temperature. Finally, during one's lifetime, additional energy is required for physical growth and for reproduction (for example, pregnancy, lactation).

In 1985, the World Health Organization (WHO) presented recommendations for assessing human energy requirements. The procedure used for determining energy needs involves first estimating BMR from body weight on the basis of predictive equations developed by the WHO. These equations are presented in Table I. After estimating BMR, the total daily energy expenditure (TDEE) for adults (18

TABLE I

Equations for Predicting Basal Metabolic Rate (BMR) Based on Body Weight (Wt, in Kilograms)

	BMR (kcal/day)	
Age (years)	Males	Females
0–2.9	60.9 (Wt)−54	61.0 (Wt)−51
3.0–9.9	27.7 (Wt)+495	22.5 (Wt)+499
10.0–17.9	17.5 (Wt)+651	12.2 (Wt)+746
18.0–29.9	15.3 (Wt)+679	14.7 (Wt)+496
30.0–59.9	11.6 (Wt)+879	8.7 (Wt)+829
60+	13.5 (Wt)+487	10.5 (Wt)+596

Source. FAO/WHO/UNU (1985).

TABLE II

Physical Activity Levels (PALs) Associated with Different Types of Occupational Work Among Adults (18 Years and Older)

	PAL			
Sex	Minimal	Light	Moderate	Heavy
Male	1.40	1.55	1.78	2.10
Female	1.40	1.56	1.64	1.82

Source. FAO/WHO/UNU (1985).

TABLE III

Energy Constants and PALs Recommended for Estimating Daily Energy Requirements for Individuals Under the Age of 18 Years Old

Age (years)	Males	Females
	Energy constant (kcal/kg body weight)	
<1.0	103	103
1.0–1.9	104	108
2.0–2.9	104	102
3.0–3.9	99	95
4.0–4.9	95	92
5.0–5.9	92	88
6.0–6.9	88	83
7.0–7.9	83	76
8.0–8.9	77	69
9.0–9.9	72	62
	PAL	
10.0–10.9	1.76	1.65
11.0–11.9	1.72	1.62
12.0–12.9	1.69	1.60
13.0–13.9	1.67	1.58
14.0–14.9	1.65	1.57
15.0–15.9	1.62	1.54
16.0–16.9	1.60	1.52
17.0–17.9	1.60	1.52

Source. FAO/WHO/UNU (1985); James and Schofield (1990).

years old and older) is determined as a multiple of BMR, based on the individual's activity level. This multiplier, known as the physical activity level (PAL) index, reflects the proportion of energy above basal requirements that an individual spends over the course of a normal day. The PALs associated with different occupational work levels for adult men and women are presented in Table II. The WHO recommends that minimal daily activities such as dressing, washing, and eating are commensurate with a PAL of 1.4 for both men and women. Sedentary lifestyles (for example, office work) require PALs of 1.55 for men and 1.56 for women. At higher work levels, however, the sex differences are greater. Moderate work is associated with a PAL of 1.78 in men and 1.64 in women, whereas heavy work levels (for example, manual labor, traditional agriculture) necessitate PALs of 2.10 and 1.82 for men and women, respectively.

In addition to the costs of daily activity and work, energy costs for reproduction also must be considered. The WHO recommends an additional 285 kcal/day for women who are pregnant and an additional 500 kcal/day for those who are lactating.

Energy requirements for children and adolescents are estimated differently because extra energy is necessary for growth and because relatively less is known about variation in their activity patterns. For children and adolescents between 10 and 18 years old, the WHO recommends the use of age- and sex-specific PALs. In contrast, energy requirements for children under 10 years old are determined by multiplying the child's weight by an age- and sex-specific constant. The reference values for boys and girls under 18 years old are presented in Table III.

5. HUMAN VARIATION IN ENERGY DYNAMICS

5.1 Energy Consumption

Compared to most other mammals, humans are able to survive and flourish eating a remarkably wide range of foods. Human diets range from completely vegetarian (as observed in many populations of South Asia) to ones based almost entirely on meat and animal foods (for example, traditional Eskimo/Inuit populations of the Arctic). Thus, over the course of our evolutionary history, humans have developed a high degree of dietary plasticity. This ability to utilize a diverse array of plant and animal resources for food is one of the features that allowed humans to spread and colonize ecosystems all over the world.

Table IV presents information on the percentage of energy derived from plant and animal foods for subsistence-level (that is, food producing) and industrial human societies. The relative contribution of animal foods varies considerably, ranging from <10% of dietary energy in traditional farming communities of tropical South America, to more than 95% among traditionally living Inuit hunters of the Canadian Arctic.

Subsistence-level agricultural populations, as a group, have the lowest consumption of animal foods. Among hunting and gathering populations, the contribution of animal foods to the diet is variable, partly reflecting the environments in which these populations reside. For example, the !Kung San, who live in arid desert environments of southern Africa, have among the lowest levels of animal food consumption among hunter-gatherers. In contrast, hunters of the arctic rely almost entirely on animal foods for their daily energy. Foragers living in forest and grassland regions of the tropics (for example, the Ache and the Hiwi) have intermediate levels of animal consumption.

The amount of energy that humans consume also varies widely. Table V shows per capita energy consumption (kcal/person/day) for selected countries around the world. Daily energy intakes in the United States (3699 kcal) and France (3518 kcal) are more than double those found in Afghanistan (1745 kcal). Overall, individuals from industrialized countries consume about 55% more energy per day than their counterparts from the developing world (3311 versus 2140 kcal/day).

These broad worldwide differences in daily energy consumption are mirrored in differences in under- and over-nutrition. As discussed Section 5.3, undernutrition remains a persistent problem among populations in the developing world, despite the fact that worldwide food production is sufficient to meet the energy needs of all people. In contrast, populations of the industrialized world, most

TABLE IV

Percentage of Dietary Energy Derived from Animal and Plant Foods in Selected Human Populations

Population	Percentage energy from animal foods (%)	Percentage energy from plant foods (%)
Hunter-gatherers		
!Kung (Botswana)	33	67
Ache (Paraguay)	56	44
Hiwi (Venezuela)	68	32
Inuit (Canada)	96	4
Pastoralists		
Turkana (Kenya)	80	20
Evenki (Russia)	31	69
Agriculturalists		
Quechua (highland Peru)	5	95
Coastal Ecuador	7	93
Yapú (lowland Colombia)	11	89
Industrial societies		
France	38	62
Japan	20	80
United Kingdom	31	69
United States	27	73

TABLE V

Per Capita of Dietary Energy Supply (kcal/Person/Day) for Selected Countries

Country	Per capita energy supply (kcal/day)
Developing countries	
Afghanistan	1745
Bolivia	2174
Botswana	2183
Columbia	2597
Ethiopia	1858
Haiti	1869
Pakistan	2476
Phillipines	2366
Tanzania	1995
Average	**2140**
Industrialized countries	
France	3518
Canada	3119
Israel	3278
Japan	2932
Norway	3357
Spain	3310
United Kingdom	3276
United States	3699
Average	**3311**

Source. Millstone and Lang (2003).

notably the United States, are now heavier and fatter than ever before. Just as there is a dramatic economic divide between rich and poor populations around the world, there is also a stark nutritional divide.

5.2 Energy Expenditure

Humans also show considerable variation in levels of energy expenditure. Recent comparative analyses indicate that daily energy expenditure in human groups typically ranges from 1.2 to 5.0 x BMR (that is, PAL = 1.2–5.0). The lowest levels of physical activity, PALs of 1.20 to 1.25, are observed among hospitalized, and nonambulatory populations. In contrast, the highest levels of physical activity (PALs of 2.5 to 5.0) have been observed among elite athletes and soldiers in combat training. Within this group, Tour de France cyclists have the highest recorded daily energy demands of 8050 kcal/day (a PAL = 4.68)!

Table VI presents data on body weight, total daily energy expenditure, and PALs of adult men and women from selected humans groups. Men of the subsistence-level populations (that is, foragers, herders, and agriculturalists) are, on average, 20 kilograms (45 pounds) lighter than their counterparts from the industrialized world and yet have similar levels of daily energy expenditure (2897 versus 2859 kcal/day). The same pattern is true for

TABLE VI

Weight (kg), Total Daily Energy Expenditure (TDEE; kcal/Day), Basal Metabolic Rate (BMR; kcal/Day), and Physical Activity Level (PAL) of Selected Human Groups

Group	Sex	Weight (kg)	TDEE (kcal/day)	BMR (kcal/day)	PAL (TDEE/BMR)
Industrialized populations					
18–29 years	M	75.6	3298	1793	1.84
	F	69.2	2486	1480	1.68
30–39 years	M	86.1	3418	1960	1.74
	F	67.9	2390	1434	1.67
40–64 years	M	77.0	2749	1673	1.64
	F	70.0	2342	1386	1.69
65–74 years	M	76.4	2629	1650	1.59
	F	60.2	2055	1267	1.62
75 and older	M	72.6	2199	1434	1.53
	F	48.3	1458	980	1.48
Average	M	77.5	2859	1702	1.67
	F	63.1	2146	1309	1.63
Subsistence populations					
!Kung foragers	M	46.0	2319	1383	1.68
	F	41.0	1712	1099	1.56
Ache foragers	M	59.6	3327	1531	2.17
	F	51.8	2626	1394	1.88
Inuit hunters	M	65.0	3010	1673	1.80
	F	55.0	2350	1305	1.80
Evenki pastoralists	M	58.4	2681	1558	1.75
	F	52.7	2067	1288	1.63
Aymara agriculturalists	M	54.6	2713	1355	2.00
	F	50.5	2376	1166	2.03
Highland Ecuador, agriculturalists	M	61.3	3810	1601	2.38
	F	55.7	2460	1252	1.96
Costal Ecuador, agriculturalists	M	55.6	2416	1529	1.58
	F	47.8	1993	1226	1.63
Average	M	57.2	2897	1519	1.90
	F	50.6	2227	1247	1.78

women; those from subsistence-level populations are 12.5 kilograms (~28 pounds) lighter than women of industrialized societies but have higher levels of daily energy expenditure (2227 versus 2146 kcal/day).

Thus, when we express daily energy needs relative to BMR, we find that adults living a modern lifestyle in the industrialized world have significantly lower physical activity levels than those living more traditional lives. Among men, PALs in the industrialized societies average 1.67 (range = 1.53 to 1.84), as compared to 1.90 (range = 1.58 to 2.38) among the subsistence-level groups. Physical activity levels among women average 1.63 in the industrialized world (range = 1.48 to 1.69) and 1.78 (range = 1.56 to 2.03) among the subsistence-level societies.

The differences in daily energy demands between subsistence-level and industrialized populations are further highlighted in Fig. 4, which shows daily energy expenditure (kilocalories/day) plotted relative to body weight (in kilograms). The two lines denote the best-fit regressions for both groups. These regressions show us that at the same body weight, adults of the industrialized world have daily energy needs that are 600 to 1000 kilocalories lower than those of people living in subsistence-level societies.

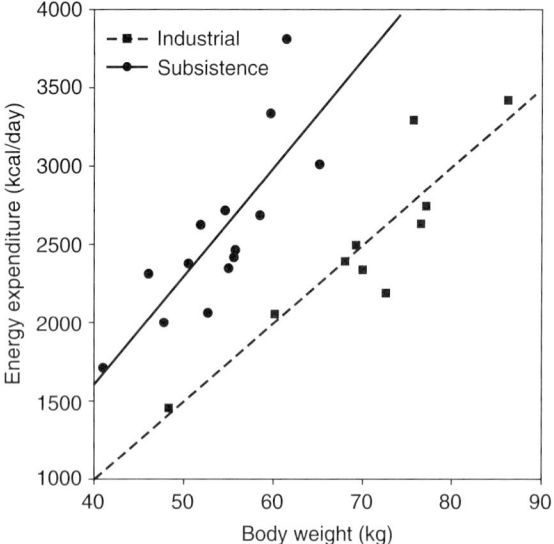

FIGURE 4 Daily energy expenditure (kcal/day) versus body weight (kg) for adult men and women of industrial and subsistence-level populations (from data presented in Table V). At a given body weight, individuals from subsistence-level groups have levels of energy expenditure that are 600 to 1000 kcal greater than those of the industrialized populations.

5.3 Energy Balance and Its Health Consequences

Ultimately, energy balance, the dynamic relationship between energy consumed and expended, is critical for determining nutritional health and well-being. In the industrialized world, low levels of physical activity coupled with relative abundance of food create a net surplus of energy, or a positive energy balance. Conversely, among rural food-producing societies of the developing world, limited energy availability and heavy workloads produce widespread under-nutrition (negative energy balance).

To assess long-term energy balance (physical nutritional status), nutritionists typically measure body weight and composition (i.e., fat and muscle). One of the most widely used measures of physical nutritional status is the body mass index (BMI). The BMI is a weight-for-height measure that is calculated as weight (in kilograms) divided by height (in meters) squared. Because the BMI is strongly correlated with levels of body fatness, it provides good long-term measure of energy balance.

Table VII shows the BMI cutoffs for different levels of nutritional health for adults (individuals 18 years and older). BMIs between 18.5 and 24.9 kg/m^2 are considered healthy because this range is associated with the lowest mortality risks. Individuals with BMIs under 18.5 kg/m^2 are considered to be underweight. Mild and moderate under-nutrition is associated with BMIs of 17.0 to 18.4 and 16.0 to 17.0 kg/m^2, respectively. BMIs of less than 16 kg/m^2 are indicative of starvation conditions (i.e., severe and acute under-nutrition). BMIs of 25 and above are indicative of over-nutrition. Individuals with BMIs between 25.0 and 29.9 kg/m^2 are considered overweight, and those with BMIs of 30.0 kg/m^2 or more are classified as obese.

In the United States, rates of obesity have increased dramatically (see Table VIII). As of 2001, 21% of the

TABLE VII

BMI Categories for Assessing Physical Nutritional Status in Adults

BMI Range	Classification
<16.0	Severe under-nutrition
16.0–16.9	Moderate under-nutrition
17.0–18.4	Mild under-nutrition
18.5–24.9	"Healthy"
25.0–29.9	Overweight
≥ 30	Obese

TABLE VIII

Percentage of U.S. Adults Who Are Obese (BMI ≥ 30 kg/m^2), 1991–2001

Sample	Percent obese (BMI ≥ 30)					
	1991	1995	1998	1999	2000	2001
Men	11.7	15.6	17.7	19.1	20.2	21.0
Women	12.2	15.0	18.1	18.6	19.4	20.8
Race, ethnicity						
White, non-Hispanic	11.3	14.5	16.6	17.7	18.5	19.6
Black, non-Hispanic	19.3	22.6	26.9	27.3	29.3	31.1
Hispanic	11.6	16.8	20.8	21.5	23.4	23.7
Other	7.3	9.6	11.9	12.4	12.0	15.7
Educational level						
Less than High school	16.5	20.1	24.1	25.3	26.1	27.4
High school degree	13.3	16.7	19.4	20.6	21.7	23.2
Some college	10.7	15.1	17.8	18.1	19.5	21.0
College or above	8.0	11.0	13.1	14.3	15.2	15.7

Source. National Center for Chronic Disease Prevention and Health Promotion, Behavioral Risk Factor Surveillance System (1991–2001); www.cdc.gov/nccdphp/dnpa/obesity/trend/prev_char.htm.

adult American population was considered to be obese, up from ~12% in 1991. Obesity rates are particularly high in minority groups—31% in African Americans and 24% among Hispanics—linked to conditions of poverty, class, and level of education.

These rising rates of obesity in the United States are also associated with increased rates of other cardiovascular problems, including high cholesterol, high blood pressure, and diabetes. This cluster of chronic health problems is referred to as the metabolic syndrome, because all of the individual features reflect a surplus of energy.

In the developing world, under-nutrition has declined; however, it remains a persistent problem. As shown in Table IX, rates of under-nutrition (BMIs < 18.5 kg/m^2) are relatively higher in Asia and Africa than in Latin America. In China and India, the levels are 12% and 49%, respectively whereas those in the Sahelian region of Africa fall between 15% and 20%. The prevalence of under-nutrition in Latin American nations is typically less than 10%. These national-level statistics can be misleading in that they mask large urban-rural differences in nutritional health. Sadly, it has generally been the impoverished rural populations of the developing world that have shown the smallest improvements in health and nutritional status.

While under-nutrition remains a severe problem in rural regions of the Third World, increasing rates of obesity are being found among urban populations of the developing world. This trend has been documen-

TABLE IX

Percentage Under-Nutrition (BMI < 18.5 kg/m^2) among Adults in Selected Countries

Region/Country	Percentage under-nutrition
Africa	
Ghana	20
Mali	16
Asia	
China	12
India	49
Latin America	
Brazil	6
Cuba	7
Peru	3

Source. Shetty and James (1994).

ted in North Africa (e.g., Morocco and Tunisia) and South America (e.g., Brazil and Chile) and is likely attributable to the Westernization of dietary habits and more sedentary lifestyles. As in the United States, diabetes and other cardiovascular diseases have increased as a consequence of the rise in obesity. It thus appears that much of developing world is now undergoing a rapid nutritional transition such that obesity and under-nutrition exist as parallel problems.

Addressing the emerging obesity epidemic in the United States and throughout the world will require changes in consumption patterns as well as broader lifestyle modifications. The wider availability of inexpensive market foods that are high in fats and simple carbohydrates is certainly a primary cause of the obesity problem. However, the other half of energy balance equation—reduced energy expenditure associated with increasingly sedentary lifestyles—is also a critical and yet often overlooked part of the problem. Consequently, it is not surprising the most effective nutritional interventions are those that promote increased exercise and activity levels as well dietary modifications.

SEE ALSO THE FOLLOWING ARTICLES

Complex Systems and Energy • *Conversion of Energy: People and Animals* • *Earth's Energy Balance* • *Easter Island: Resource Depletion and Collapse* • *Ecosystems and Energy: History and Overview* • *Emergy Analysis and Environmental Accounting* • *Food Capture, Energy Costs of* • *Food System, Energy Use in* • *Goods and Services: Energy Costs* • *Heterotrophic Energy Flows* • *Hunting and Gathering Societies, Energy Flows in* • *Thermoregulation*

Further Reading

Black, A. E., Coward, W. A., Cole, T. J., and Prentice, A. M. (1996). Human energy expenditure in affluent societies: An analysis of 574 double-labelled water measurements. *European J. Clinical Nutrition* **50**, 72–92.

Durnin, J. V. G. A., and Passmore, R. (1967). "Energy, Work and Leisure." London: Heineman.

Food and Agriculture Organization, World Health Organization, and United Nations University (FAO/WHO/UNU) (1985). Energy and protein requirements. Report of Joint FAO/WHO/UNU Expert Consultation. WHO Technical Report Series No. 724. World Health Organization, Geneva.

Gibson, R. S. (1990). "Principles of Nutritional Assessment." Oxford University Press, Oxford.

James, W. P. T., and Schofield, E. C. (1990). "Human Energy Requirements: A Manual for Planners and Nutritionists." Oxford University Press, Oxford.

Kleiber, M. (1975). "The Fire of Life: An Introduction to Animal Energetics." 2nd Ed. Krieger. Huntington, NY.

Leonard, W. R. (2000). Human Nutritional Evolution. *In* "Human Biology: A Biocultural and Evolutionary Approach" (S. Stinson, B. Bogin, R. Huss-Ashmore, and D. O'Rourke, eds.), pp. 295–343. Wiley-Liss, New York.

Leonard, W. R. (2002). Food for thought: Dietary change was a driving force in human evolution. *Scientific American* **287**(6), 106–115.

McArdle, W. D., Katch, F. I., and Katch, V. L. (2001). "Exercise Physiology: Energy, Nutrition and Human Performance." 5th Ed. Lippincott Williams & Wilkins, Philadelphia.

McLean, J. A., and Tobin, G. (1987). "Animal and Human Calorimetry." Cambridge University Press, Cambridge, UK.

Millstone, E., and Lang, T. (2003). "The Penguin Atlas of Food." Penguin Books, New York.

Montoye, H. J., Kemper, H. C. G., Saris, W. M. H., and Washburn, R. A. (1996). "Measuring Physical Activity and Energy Expenditure." Human Kinetics, Champaign, IL.

Prentice, A. (1990). "The Doubly-Labelled Water Method for Measuring Energy Expenditure: A Consensus Report by the IDECG Working Group." International Atomic Energy Agency, Vienna.

Shetty, P. S., and James, W. P. T. (1994). Body mass index: A measure of chronic energy deficiency. FAO Food and Nutrition Paper, no. 50. FAO, Rome.

Ulijaszek, S. J. (1995). "Human Energetics in Biological Anthropology." Cambridge University Press, Cambridge, UK.

Hunting and Gathering Societies, Energy Flows in

TIM BAYLISS-SMITH
University of Cambridge
Cambridge, United Kingdom

1. Introduction
2. Problems in Defining Hunting and Gathering
3. Problems in Generalizing from Modern HG Societies
4. Energy Flow in Contrasting HG Environments
5. Energy Flow and Optimal Foraging Theory
6. Conclusion

Glossary

collecting A hunter–gatherer strategy in which task groups move out from a residential base to harvest, process, and transport seasonal concentrations of resources for future use.

foraging A hunter–gatherer strategy for exploitation of resources in the vicinity of the residential base, to harvest and consume resources that are continuously available in low concentrations.

Holocene The period of Earth history that began 10,000 years ago, at the end of the last glacial period.

hominid A primate of the order Hominidae, which includes modern humans (*Homo sapiens sapiens*), earlier human subspecies, and their direct ancestors.

logistic mobility The pattern of spatial movements associated with collecting resources from a residential base.

megafauna Mammals, reptiles, and birds that belong to the largest size categories.

optimal foraging theory (OFT) Models of hunter–gatherer behavior that predict the selection of resources, foraging behavior within patches of resources, or the mobility of the residential base. OFT uses microeconomic theory to assess what outcome will be the most efficient for the individual or for society—for example, by maximizing energy gain.

Pleistocene A period of Earth history that began 2.5 million years ago, and characterized by repeated glacial/interglacial fluctuations in climate.

residential mobility The pattern of frequent moves of the residential base made by hunter–gatherer groups that exploit resources by foraging.

sedentism A sedentary settlement pattern, through which hunter–gatherers settle down to live in fixed locations and in more permanent dwellings.

Hunter–gatherer (HG) societies have existed in all continents, and today survive mainly in tropical forests, semidesert areas, boreal forests, the tundra, and the Arctic, all environments where agricultural colonization and other forms of displacement have not yet occurred. These societies are defined principally according to their dependence on nondomesticated resources, accessed through hunting, trapping, fishing, and gathering. In many societies, cultivation or herding coexists with the HG mode, and exchange often provides people with some items not available from local sources. Typically, these societies are small scale, egalitarian, and live at low population densities, although not in all cases. Their patterns of energy flow, including diet, labor use, mobility, and optimizing behavior, have been used as models for understanding past HG societies in areas that are now agricultural, and also to illuminate the behavior of other hominids, because HG subsistence has been dominant for about 90% of the history of *Homo sapiens sapiens* and about 99% of the time since hominids first evolved.

1. INTRODUCTION

Extrapolating data back in time from a few modern HG societies must be done cautiously. The geographical distribution of modern hunter–gatherers is highly skewed; modern groups live in environments marginal for farming or pastoralism and in places remote from colonialism, modernity, and the nation state. Because of the disappearance of megafauna, they exploit a range of resources that are much depleted compared to the Pleistocene norm. Hunter–gatherers encompass such a range of different adaptations, particularly in relation to variations in the seasonality in energy supply and variations in the spatial concentration of food energy, that generalizing from particular case studies is very hazardous. Problems in defining HG societies need first to be addressed (Section 2), before confronting this ecological diversity and achieving a meaningful classification of HG societies (Section 3). Only then is it possible to make valid generalizations about the patterns of energy flow in HG societies (Sections 4 and 5), to generate findings that might be applicable to other times and places.

2. PROBLEMS IN DEFINING HUNTING AND GATHERING

2.1 Origins of the Hunter–Gatherer Concept

The concept of "hunter–gatherer" became necessary when, in the 19th century, Western scholars began to classify the different modes of subsistence that existed in the world. Sedentary agriculturalists were distinguished from pastoralists, people primarily dependent on their herds of domesticated animals. But what of those societies whose food resources were not domesticated, who lived instead by hunting wild animals, fishing, and gathering wild plant foods? Such peoples lived in an immense range of different environments and in rather strongly contrasted societies, in places as remote from each other as African deserts, South American jungles, and the Arctic tundra. Even so, for convenience, all were put in the one category, hunter–gatherer, for reasons that perhaps were partly political. By seeing HG societies as a homogeneous category with a preagricultural and therefore "primitive" technology, it became easier to justify their displacement following the spread of European settlement and colonial expansion.

Despite its unsatisfactory origins, the HG concept became even more firmly established when the long prehistory of the genus *Homo* was demonstrated by comparison with the rather short history of plant and animal domestication. Whereas agriculture was recognized as extending no further back than the Neolithic, perhaps 6000 years ago, the origins of *Homo sapiens* lay a long way back in the Pleistocene. In 1966, a symposium was held in Chicago on the theme "Man the Hunter," at which Richard Lee and Irven DeVore claimed that "cultural man has been on earth for some 2,000,000 years, and for more than 90% of this time he has lived as a hunter–gatherer."

Although modern archaeology would not extend the origins of *H. sapiens sapiens* (anatomically modern humans) beyond about 130,000 years ago, research has also confirmed that no clear evidence exists for full-scale agriculture anywhere until well after the Holocene began 10,000 years ago. Nor is pastoralism any older; indeed, most pastoral societies postdate the formation of settled farming communities, with which they typically exchange products (e.g., cattle for grain). From various perspectives, therefore, it became important for scholars to understand more about the economies, societies, and cosmologies of these preagricultural and prepastoral HG societies, because they were now seen as representing such a large part of the human heritage. Yet we inherit many prejudices and assumptions about such societies. Prejudice in Western thought can be traced back at least to the time of Thomas Hobbes (1651), who typecast hunter–gatherers as not merely the primeval state of humanity, but people who lived lives that he famously described as "solitary, poor, nasty, brutish and short." The influence of this stereotype undoubtedly lingered until the mid-20th century, but it did so alongside another more romantic vision of hunter–gatherers. This alternative view can be traced back to John Dryden, who in 1670 depicted hunter–gatherers as living in a state of grace from which the rest of humanity had fallen, coining the equally famous phrase "the noble savage" to describe them.

The stereotypes of both Hobbes and Dryden influenced Western attitudes and policies towards HG societies during the colonial period, and were not finally dispelled until the 1960s when the results of scientific research became more widely known, including studies of the energy efficiency of hunting and gathering. Ironically, by then most HG societies were disappearing, displaced by colonial genocide and modernity or absorbed by the expansion of neighboring groups and by the nation state.

2.2 Sources of HG Food Energy

Hunter–gatherers are usually identified today by means of a "package" of traits, including economic, ecological, sociocultural, and ideological features, but of these the HG economy is the most important characteristic. The *Cambridge Encyclopaedia of Hunters and Gatherers* offers the following definition of the HG mode: "Subsistence based on hunting of wild animals, gathering of wild plant foods, and fishing, with no domestication of plants and no domesticated animals except the dog." As well as their food energy, the fuels, fibers, and tools of hunter–gatherers also come from nondomesticated, noncultivated sources, although iron tools, pottery, and a few other items may be acquired by exchange.

In most HG societies, there are five categories of food energy procurement: (1) hunting, involving the active pursuit of mammals, reptiles, or birds; (2) trapping, which is the passive capture of prey (mammals, birds, or fish) by means of prior investments in pitfalls, traps, nets, weirs, and other technical means; (3) fishing, using a wide range of active techniques; and (4) gathering, which involves the acquisition of passive animal prey/products such as shellfish, snails, insect larvae, birds' eggs, and honey, and also plant foods of all kinds. The final category, (5) exchange with neighboring groups, should be added, in almost all modern cases, although the extent to which food energy is traded, rather than other goods, varies a lot.

The relative importance of the various sources of subsistence is highly variable between different HG societies, dependent on environmental opportunities and constraints. However, to prioritize hunting in the definition of such societies is misleading. Only in a minority of cases (for example, in the Arctic, in boreal forests, and in places where fish and sea mammals are especially abundant) do hunting, trapping, and fishing contribute more to the diet than does the gathering of plant foods. Meat is generally the preferred food in all societies, but gathered plant foods are more abundant in a majority of cases and, crucially, are more dependable. "Gatherer–hunters" might, therefore, be a more accurate term.

Other definitions of HG subsistence could be cited, but what they all have in common is the absence of direct human control over the sources of food energy on which people depend, including the reproduction, the distribution, and the abundance of the various plant and animal species. Lack of direct control in HG societies is contrasted to what agriculturalists achieve, including (1) a transformation of ecosystems using energy-intensive means and (2) a partial or total domestication of plants and animals, with either minor or major genetic changes to the various species involved.

2.3 Energy Foraging versus Energy Collecting

For understanding spatial behavior in HG societies, the archaeologist Lewis Binford made an important distinction between foraging and collecting food energy. These represent alternative strategies for solving problems of resource distribution. Foraging takes place where resources are widespread and continuously available, and where frequent residential mobility enables hunter–gatherers to move their camps to the resource they wish to use. In this way, they can forage for what they need by means of immediate return technologies and social practices. The storage of food energy is minimal in such societies, which are typical of HG societies in the semideserts and forests of tropical and midlatitudes (Fig. 1).

In contrast, logistical mobility takes place where collecting of resources is needed for future consumption. From the residential base, task groups are sent out to hunt or gather particular foods, undertake the initial processing, and return to base with a surplus that then provides a future energy store. This alternative has been called by James Woodburn a strategy of delayed return, and it is essential where periods of resource shortage can be anticipated. The northwest coast of North America, for example, was

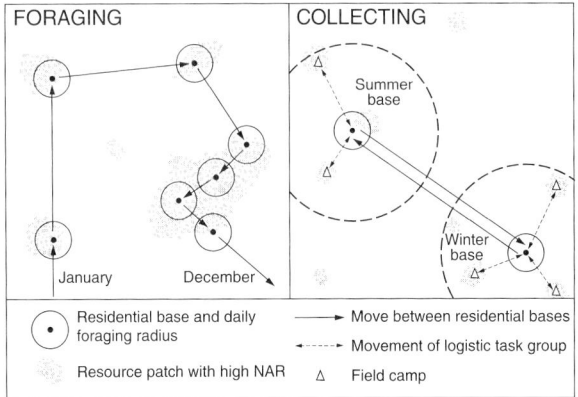

FIGURE 1 Schematic maps contrasting the foraging and the collecting strategies of hunter–gatherers. In both cases, residential bases are located in places with good access to resource patches with high net acquisition rates. A net acquisition rate (NAR) is defined as food energy output divided by work energy input, per hour of foraging/collecting activity. The residential bases of the foragers are each occupied for much shorter periods, compared to those of the collectors. In both cases, virtually no stored food energy is in the possession of the society at the end of the more difficult season of the year, but foragers consume energy on an immediate-return basis in the intervening period, whereas collectors have a delayed-return consumption pattern made possible by their logistic strategy.

an environment where salmon and waterfowl were superabundant in season, and where a single sturgeon, sea lion, or whale could provide enough food for many days or weeks of consumption. However, according to anthropologist Wayne Suttles, "abundance consisted only of certain things at certain times and always with some possibility of failure." To cope with abundance, in a situation where for some foods (e.g., salmon) more effort was needed to store the food than to collect it, complex technologies and social institutions had be developed that could guarantee delayed energy return, and so prevent the society from being destroyed by famine at the end of winter.

Most HG societies need to have a mix of foraging and collecting strategies, being residentially mobile in some seasons (e.g., summer or dry season) but in other seasons (e.g., winter or wet season) operating from fixed base camps by means of logistical mobility. A primary emphasis on delayed-return energy use seems to be associated with the emergence of more stratified ("complex") societies, such as the ranked societies of chiefs, freemen, and slaves that existed on the northwest coast of North America. Ian Hodder has argued that the ideologies associated with delayed-return HG subsistence paved the way for agriculture, but other archaeologists disagree.

Using these concepts, HG societies can be classified according to their adaptations to two critical aspects of their energy resources: (1) seasonal and year-to-year fluctuations, which control the scale of energy storage required, and (2) variations in the spatial concentration of critical resources (food, water, fuel), which permit different degrees of sedentism. The four main categories of HG society are shown in Table I, together with examples. In practice, it is rare today to find societies that depend solely on hunting, fishing, and gathering. For their subsistence, most documented communities engage in a limited amount of cultivation, herding, and/or trading, in addition to their main HG focus.

2.4 Use of Energy Subsidies

Unlike farmers, hunter–gatherers do not depend on energy subsidies (e.g., from fossil fuels), and their impact on the landscape is small-scale apart from localized fuelwood consumption. However, HG societies do exert indirect control over energy flows in the ecosystem, particularly by using fire to assist with hunting and to improve the grazing. In what is today the United States, driving game with fire was practiced almost everywhere in pre-Columbian times. Open areas of grassland found today in the apparent Yosemite wilderness are in fact the legacy of former HG hunting practices. In Australia, the use of fire by HG societies transformed vegetation over a much longer period (perhaps 50,000 years), reducing

TABLE I

A Classification of Hunter–Gatherer Societies[a]

Spatial concentration of main sources of food energy[b]	Different mobility patterns	Example	
		Seasonal and year-to-year fluctuations in resource abundance[c]	
		Low	High
Low	Residential mobility	Inland Australian Aborigines Kalahari San (e.g., !Kung)	No examples known
Intermediate	Seasonal base camps with logistic mobility	Coastal Australian Aborigines (e.g., Gidjingali)	Sub-Arctic maritime societies (e.g., Coastal Sámi, Varanger Fjord, Norway)
		Arctic tundra (e.g., Tuluaqmiut Inuit, north Alaska)	Boreal forest hunters (e.g., Cree, northern Ontario)
		Sub-Arctic inland (e.g., Mountain Sámi, northern Sweden)	Torres Strait islanders, Australia
High	Sedentism	No examples known	Northwest Coast Indians (e.g., Kwakiutl, Salish, Chinook)

[a] Based on the extent of seasonal and year-to-year fluctuations in available food energy and the extent of spatial concentration in food energy, with ethnographic examples.
[b] Correlates with different mobility patterns.
[c] Correlates with (1) extent of capital investments in storage to safeguard future energy supplies and (2) with institutions for ownership and defense of territories.

evergreen rainforest to small patches and greatly enlarging the open woodlands dominated by *Eucalyptus*. Far from being "natural," *Eucalyptus* woodland is now interpreted as an anthropogenic fire-climax vegetation.

Some anthropologists studying hunting practices in Aboriginal Australia and among North American Indians are so impressed by the importance of fire as an indirect tool for ecosystem management that they use the phrase "firestick farming." In the savannas of northern Australia, for example, in the early part of the dry season when cool nights dampen the ground with dew, the Aborigines use small and controlled bush fires that clear away undergrowth but leave the trees undamaged. The growth of new grass in such areas attracts game and makes hunting much easier. Because the modern state has imposed a ban on firestick farming, infrequent but catastrophic and far more destructive forest fires take place.

2.5 Social Differences in Energy Use

A consequence of the low intensity of management of food energy sources by HG societies is that they live at very low population densities, often in small bands of 25–50 persons, with temporary or seasonal settlements and frequent long-distance mobility. In most respects, these societies are strongly egalitarian, with food sharing and an absence of social stratification and only weak institutions for political control. However, there is a strong gender division of labor, with males doing most of the hunting and females gathering the plant foods. Fishing and the collecting of shellfish and small game may be an activity of both sexes, and children participate in the food quest from an early age.

Hierarchy is most strongly developed in semisedentary HG societies where seasonally abundant resources make possible large-scale food energy storage, as among Northwest Coast Indians such as the Kwakiutl in British Columbia or the Chinook in Washington state. Lee and DeVore, in the classic text *Man The Hunter*, wrote "we make two basic assumptions about hunters and gatherers, (1) they live in small groups, and (2) they move around a lot." Perhaps these "simple," residentially mobile HG societies have always coexisted with more sedentary, more complex, and more hierarchical ones.

As regards their attitudes to energy consumption, these are not societies in which maximizing energy flow finds ideological support. Far more important is maximizing the security of energy supplies, simultaneously minimizing the energy expended in work so that more time is spent enjoying leisure and strengthening social bonds with others—itself, arguably, a useful risk-minimizing strategy.

3. PROBLEMS IN GENERALIZING FROM MODERN HG SOCIETIES

3.1 Can We Extrapolate from HG Societies to Hominids?

There has been much interest in the possibility that studies of energy flow in HG societies can be used to achieve an understanding of our hominid ancestors. As Robert Foley put it, "until very recently being a hominid, being a human and being a hunter–gatherer were very nearly the same thing." Do studies of the energetic efficiency of food-gaining among the !Kung San of the Kalahari, or the energy storage strategies of sub-Arctic maritime hunting/fishing communities, have general implications for the behavior of humans or hominids who are remote from these societies in time and space? Perhaps it is the intentionality of the human subject that distinguishes his or her behavior from the superficially similar food quest of the nonhuman HG groups, as well as the human capacity for a symbolic expression of hunting experiences, as shown in the cave art of the Upper Palaeolithic and the shamanism of modern HG groups.

Remember also that modern HG groups exploit a depleted set of resources compared to their Pleistocene ancestors, because of megafaunal extinctions. The capacity clearly existed in the past for rapid population growth and a relationship with nature that is far removed from the stable and rather harmonious one that HG societies have exhibited in modern times. Moreover, these societies are not a random sample of those that existed on Earth before the spread of agriculture. Not only are they restricted to environments less suited ecologically to agriculture or pastoralism, but also they are no longer "pure" in their HG lifestyle, interacting with farmers or herdsmen when opportunity arises.

3.2 The Skewed Distribution of HG Societies Today

Using ethnographic sources, we can map the distribution of 115 HG societies that existed in the 20th century (Fig. 2). Based on the environments they occupy, five groups can be distinguished:

FIGURE 2 Map of 115 hunter–gatherer societies that were ethnographically documented in the 19th and 20th centuries.

- Hot deserts and tropical savannas (19 societies).
- Tropical rain forests and coral reefs (30 societies).
- Boreal forest and tundra zone (38 societies).
- Cool temperate maritime regions (13 societies).
- Other environments (15 societies).

It is apparent that in modern times HG societies have been absent from almost all environments that are prime farming land. Therefore, those that still survive are a very biased sample of the HG mode of the past. The temperate deciduous forests, for example, have not been the domain of hunter–gatherers in Europe since the Neolithic 5000–6000 years ago. In eastern North America and Chile, such forests were used for a mixed HG/agricultural strategy before being taken over by European colonists some 400 years ago; the same process removed Aboriginal HG societies from the temperate forests of Tasmania in the 19th century. While being careful not to label the environments of modern HG societies as necessarily "marginal," which they are not except from the point of view of agriculture, we should also bear in mind the predominance of hot deserts, jungles, and seasonally frozen environments in the world list.

4. ENERGY FLOW IN CONTRASTING HG ENVIRONMENTS

4.1 Managing the Daily Risks of Energy Shortage

4.1.1 General Principles
In some societies, it might make sense to understand behavior in terms of profit or surplus (i.e., maximizing energy output), but most anthropologists now believe that for HG societies maximizing security (i.e., minimizing risk) makes more sense. Security is increased when the risks of food shortage are successfully managed. From her studies of the San in southern Africa, Polly Wiessner identified four general ways in which risk is reduced: (1) prevention of energy deficit, (2) transfer of deficit, (3) pooling of resources, and (4) storage strategies. To prevent deficits on a daily basis, hunter–gatherers need appropriate tools for killing prey or gathering plants; expert knowledge of the probabilities of success and how it can be achieved; and intragroup cooperation in tracking, processing, and transporting foods. The transfer of deficit is achieved by sharing within the

group and by exchange relationships, whereas the pooling of resources is another social strategy that provides for an individual some insurance against failure in food energy acquisition.

Storage strategies for food and water are also potentially useful, but except in the short term these are difficult to achieve in hot and humid climates. Fuel storage is essential only for HG societies coping with extreme winter conditions. Unless food storage can be achieved effectively, for the hunter or gatherer to maximize short-term gain by producing as much energy as possible is not a sensible strategy. Where energy storage is problematic, the social strategies of sharing, cooperation, and exchange are more effective, but all require some form of investment. Investing in friendships and exchange, activities that require leisure time, might prove to be a better strategy for risk reduction, compared to killing more and more animals.

4.1.2 Example: Anbarra Gidjingali Aborigines

The ways in which risk and productivity can interact are well demonstrated by the example of the Anbarra, a subgroup of about 40 people that are part of the Gidjingali (population 350) of Arnhem Land, northern Australia. The economy of the Anbarra was studied by Rhys Jones and Betty Meehan at a time (early 1970s) when a large part of the Anbarra diet still derived from HG resources. Anbarra territory is at the mouth of the Blyth river, in the subhumid tropical savanna zone. There is seasonal access to the rich resources of open woodlands, grassy plains, freshwater swamps and lagoons, sand dunes, and mangroves, as well as the open sea, using a semisedentary pattern of settlement. In the wet season, people live by the coast and depend largely on fish and shellfish. Later in the year, they range inland to take advantage of cycad fruits, wild yams, and small game, and they move further inland across the grassy plains for wallaby hunting in the latter part of the dry season. The seasonal availability of foods is one obvious constraint over their pattern of energy consumption.

Another important factor, however, is the probability of success in the food-gaining quest. There is an inverse relationship between the productivity of labor and the probability of a successful trip. Activities such as wallaby hunting provide on average a high yield of food energy per unit of time invested, but are also the most risky activities, with a 75% chance of complete failure of a hunting trip. At the other end of the spectrum are plant-gathering activities, for which success is virtually guaranteed but the returns to labor are considerably less attractive. The Anbarra's overall strategy is to combine high-yield, high-risk activities (e.g., wallaby hunting or spear fishing) with low-yield, low-risk tasks (e.g., collecting shellfish and gathering of plant foods such as yams and cycads). Table II shows data for risk and yield for four of the major Anbarra food resources.

A viable long-term strategy for energy acquisition therefore requires a blend of different activities. Hunting is men's work, and to ensure against their failure the women must engage in less rewarding but more secure tasks such as gathering cycads and other plants and collecting shellfish. In this way, Anbarra

TABLE II

The Trade-off between Productivity and Risk in Northern Australia[a]

Hunter–gatherer activity	Sex of persons involved[b]	Gross acquisition rate (MJ)[c]	Risk[d]	Importance[e]
Buffalo hunting	M	8400	0.001	8
Wallaby hunting	M	210	0.25	7
Fishing with spear	M	146	0.4	3
Trapping fish	M	146	0.55	5
Fishing with hook and line	M and F	25	0.65	1
Cycad gathering and processing	F	5.4	1.0	6
Yam gathering	F	5.0	0.9	4
Shellfish gathering	F	4.2	0.9	2

[a] The gross acquisition rate for various Anbarra activities during successful trips, compared to the probability of success per attempted trip. Calculations based on 1972–1973 data from Jones (1981).
[b] M, Male; F, female.
[c] Megajoules of food energy per person-hour on a successful trip.
[d] Probability of success per attempted trip.
[e] Ranked according to percentage of days in the year when the activity was attempted.

society maximizes utility in terms of risk avoidance, secures a balanced diet, and maintains social solidarity and mutual aid between the sexes.

4.2 Managing the Energy Cost of Mobility

4.2.1 General Principles

Rather than managing plants and animals to achieve a spatially focused concentration of food energy, which is the strategy of agriculturalists and pastoralists, hunter–gatherers must depend on the natural distribution patterns in time and space of the resources on which they depend. In circumstances in which there is a natural concentration of resources in time and space, sedentism is encouraged, so that people and their possessions do not have to keep shifting from place to place. Even if temporary, sedentism is likely to be attractive to HG societies for many reasons: (1) it confers many social benefits; (2) it allows long-term storage of food energy to be an option, thereby increasing future security; (3) it reduces the direct energy expenditures involved in walking, carrying infants, and transporting possessions from one residential base to another; and (4) it reduces the indirect costs of high mobility, such as higher infant mortality, less equipment, and smaller energy stores, if all food must be carried. However, unless food energy sources are spatially concentrated, sedentism also reduces the accessible territory to a local area, which is soon likely to become overexploited. In arid areas, water may also become scarce. The HG lifestyle can be seen as the outcome of a perpetual compromise between the various costs of mobility (many of them measurable as energy costs) and the costs of sedentism. The disadvantages of sedentism will become energy costs once carrying capacity is exceeded, because when hunting yields decline and other resources are depleted, labor productivity will fall. As a result, people must work for longer periods, and/or journey times from home base to food source must increase, so as to compensate.

4.2.2 Example: The Dobe!Kung

The Dobe subgroup of the !Kung San were studied by Richard Lee in the 1960s, and provide an example of these interacting constraints. For the Dobe, water is the overriding factor that dictates the location of camps. The camp serves as the home base for 30–40 people, who move out each morning in groups of 2 or 3 to collect plant foods (especially mongongo nuts) or hunt game. All groups must return before evening if they wish to share the common meal. After moving to a new camp, the food resources of the local area (defined by the Dobe as the area within a 2-hour hike) are exploited first. By fanning out in all directions and moving at about 5 km/hour with pauses for rest and work, an area within a 10-km radius can be worked in the space of a 6-hour day.

Once the low-productivity/low-risk resources within this local area have been depleted (i.e., the plant foods and easily collected small game), it is more efficient for the group to move to a new camp than it is for people to spend more and more time walking. A division of labor becomes evident, with younger, more active people making longer trips to mongongo woodlands and the older people and children staying close to the camp and searching for less desirable foods. Any who are not prepared to walk further for the staple mongongo must eat foods such as bitter melons, roots, acacia gum, and the heart of the ivory palm. Richard Lee showed that for mongongo nut foraging, the round-trip distance was the best measurement of the time and energy "cost" of obtaining this desirable food. Camps are moved quite frequently (five or six times in the year), but never very far. A rainy season camp in the mongongo woodland is seldom more than 16–20 km from the previous one, but this is sufficient to shift the spatial focus of exploitation to a new and almost unexploited area.

A 10-km radius is a relatively small territory for hunting the larger game animals, which can move over long distances. Potentially, Dobe hunters must be more mobile than the gatherers and small-game hunters, and the Dobe hunters do sometimes range up to 30 km from a home base using overnight stops, in pursuit of high-yield but high-risk game. However, to do so involves the carrying of much equipment, including water, and the return journey with an animal carcass can be equally arduous. Even for these hunters it is generally more energy efficient to focus on those game available within a 10-km radius.

4.2.3 The Original Affluent Society?

For HG groups, any risk of subsistence crisis will be reduced where (1) food energy derives from a diverse set plant and animals and (2) aggregate food resources are continuously available during the year, even if certain foods are more abundant in certain seasons. If a preferred food such as meat is temporarily scarce, then, through more intensive efforts and high spatial mobility, other less desirable foods (e.g., edible plants) can usually be accessed elsewhere. As a result of these strategies, basic supplies of food energy can be secured throughout the year, and what is produced is almost immediately shared and consumed.

Both the Anbarra Gidjingali (Section 4.1.2) and the Dobe!Kung San (Section 4.2.2) are examples of HG societies that can exploit such circumstances, but with somewhat different mobility patterns. Dobe campsites must be moved frequently, whereas the Anbarra can be semisedentary, with long-term seasonal residence at the boundary between land and estuary. But even in their arid "desert wasteland" (judged from an agropastoralist standpoint), the Dobe!Kung strategy can be judged as highly efficient in maximizing leisure time. According to Lee's data, they devoted only 12–19 hours per week to food-gaining "work" to achieve a state of subsistence energy balance (nobody hungry, but no storage). The surprisingly low level of energy input resulted in the San (and, by extension, all HG societies) being characterized by Marshall Sahlins as "the original affluent society." Agriculturalists, by contrast, usually have to work harder, enjoy less adequate diets, and sometime suffer harvest failure.

In this situation, there is no urgent reason for an HG society to produce a food surplus even if circumstances permit, and there is also little incentive to invest in techniques for food storage. Such "hand-to-mouth" strategies characterize the relatively nonseasonal environments of tropical forests, midlatitude deserts, and productive high-latitude coasts such as the Aleutian islands, Tierra del Fuego, and Tasmania. HG societies that occupy such environments share a number of features in common. Robin Torrence has demonstrated that the technologies employed by such societies involve relatively few types of tools. Hunting focuses on the most rewarding methods of securing prey, and gathering is equally focused on plant foods with the highest returns per unit of gathering effort. New tools and new techniques are added to the HG repertoire in order to reduce the probability of failure, but this only tends to happen in more uncertain environments where the costs of concentrating efforts on only the "best" (i.e., most rewarding) food sources can be very high.

4.3 Coping with Extreme Seasonality in Energy Resources

4.3.1 General Principles
Higher levels of risk are typical of more seasonally fluctuating environments, especially where winter conditions of darkness and extreme cold make hunting difficult and the option of constant mobility becomes extremely arduous. Midwinter sedentism is virtually obligatory in sub-Arctic and Arctic environments, where food storage is not so technically difficult (stored food will not spoil in the freezing conditions), but where substantial quantities of food energy must be amassed if the group's winter survival is to be safeguarded. HG societies in such environments typically invest much energy in strategies that enable them to cope with extreme seasonality—in particular, mass harvesting and mass storage.

4.3.2 Mass Harvesting
To exploit seasonal gluts of animal food requires first an efficient technology and social organization, so that by a concentration of effort a huge surplus of food energy can be amassed. Hunting wild reindeer/caribou, for example, requires leadership, communal effort, and the creation of decoys, barriers, and pitfall traps covering extensive areas, so that many animals can be killed during the window of opportunity that opens up during the brief period of migration in the fall. At Rondane in central Norway, wild reindeer were hunted on a large scale during the period 1000–1200 AD. On the high plateaus, large trapping systems were designed using 3-km lines of wooden stakes to guide the animals into a steep gully, where they could be slaughtered in a funnel-shaped corral. Ørnulf Vorren has found and mapped the remains of 1700 stakes, each supported on a cairn of stones and requiring timber to be transported from pine forests located at least 10 km away. It was obviously hard work to set up and to maintain this system, and the hunt needed the coordinated efforts of many people, but the reward was a large stockpile of meat and furs. When the snows came, this stockpile could be transported to winter base camps on sledges, using tame reindeer as draft animals.

4.3.3 Mass Storage
The limits to energy storage capacity have often been set less by people's capacity to collect food than by their capacity to store it. For HG groups to take advantage of seasonal runs of fish, for example, they require large investments in weirs, nets, and traps, and these must be in place before harvesting begins. Equally important, however, are the investments in processing, transport, and secure storage of such foods. For groups such as the Chinook, Salish, or Kwakiutl, which occupy the big rivers of the northwest coast of North America, a heavy fall run of

salmon represented a crisis of abundance. How much food could be stored depended mainly on the number of hands available for the work of cutting and skewering the fish, how many drying racks had been prepared in houses, the availability of fuel for smoke, and the number of containers available. Preserving food was largely women's work, and the key role of women was reflected in contested rights over them, high bride prices, and the institution of slavery.

Not all fish keep equally well, and many salmon species cannot be preserved for the winter simply by drying. On the Columbia River, for example, the Chinook dried the filleted salmon and then pulverized it with berries. Tons of this preserved food was then stored in wooden boxes sealed with oil rendered from sea mammal fat. On the Great Plains, buffalo meat was prepared in a similar way with berries and grease to make pemmican, which, if made properly, could be stored for years, although most was consumed within a single year.

Techniques to preserve food energy, such as salting, drying, and smoking, all represent major investments. Like the technology of mass harvesting, these storage technologies represent forms both of stored energy and of social capital to provide the reserves without which winter survival would be impossible. A consequence of the investments in traps and storage is that a form of property is created by the group or individual responsible, and this property and the territories that serve as catchments (pastures, bays, rivers) may also be defended by groups against outsiders. In such circumstances, too, a more stratified and hierarchical HG society tends to emerge.

4.3.4 Example of Mass Harvesting/Storage: The Varanger Sámi

The Coastal Sámi of Varanger fjord in north Norway are today largely assimilated into the modern society and economy of Norway. They are farmers, commercial fishermen, and reindeer herders rather than hunters and gatherers. In the past, however, this large, treeless Arctic fjord, with its hinterland of stunted birch forests and mountain tundra, was the site of one of Europe's last hunter–gatherer societies. Their seasonal pattern of resource use has been reconstructed by the archaeologist Knut Odner (Fig. 3).

Until the 1700s, the main basis for both subsistence and lifestyle was fishing (cod, coalfish, and haddock), hunting wild reindeer, and hunting whales and seals. All of these foods could be plentiful in season, and a range of minor foods such as seabirds and their eggs, grouse, hare, and freshwater salmon

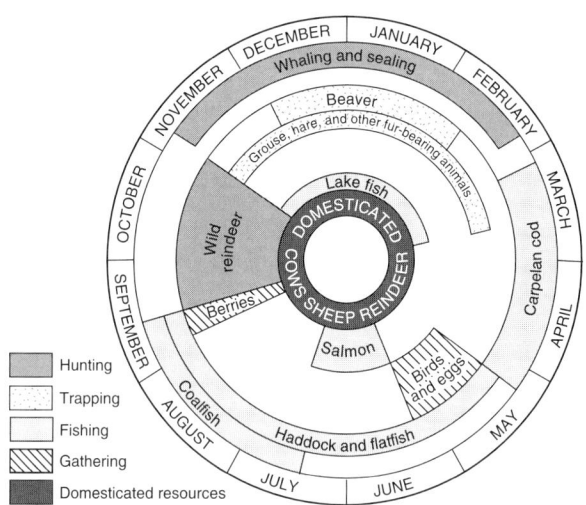

FIGURE 3 Seasonality of the food energy sources available the Varanger Sámi before circa 1700, as reconstructed by Knut Odner.

was also available for brief periods. The problem was to devise effective delayed-return strategies to enable the society to cope with fluctuations, particularly the very sparse resources available in the late-winter season, after the end of whaling and sealing in mid-February and before the arrival of shoals of cod at the end of March.

Of all the various resources, sea fishing for resident species was the most dependable year-round source of energy (low productivity but low risk). Migratory shoals also made possible a storable surplus, particularly the spring cod. On average, a fisherman could expect in one day to land about 100 kg of cod, and he could maintain this level of productivity for about 10 weeks. Each family might therefore expect to be able to dry about 7000 kg of fish in the season. Drying racks, women's labor, and access to boats, fish hooks, and lines were all essential for success. In an environment where fuel energy was scarce, air drying was the main preserving technique in spring, until warmer summer weather caused the fish to become fly-blown.

Until the wild reindeer became extinct, the Varanger Sámi employed various techniques for mass harvesting during their fall migration. The most effective were systems of fences and corrals using stones or poles stuck in the ground, and also lines of pitfalls. The remains of about 5000 hunting pits can be seen around Varanger fjord today. Probably all families contributed hunters to the task group, and the carcasses would have been equally divided. The frozen meat and the skins provided a source of food and clothing for the community in the winter months.

By 1700, growing trade and reindeer domestication made possible an increased sedentism, the animals playing a crucial role in transporting meat and fuel energy from the hinterland. A more sedentary lifestyle in turn made it possible for cattle and sheep to be kept on a small scale, despite the high energy costs that foddering and housing them entailed. The animals and their wool, milk, and cheese made a small but critical contribution to the seasonal energy gap that existed in the late winter, especially after the disappearance of wild reindeer as a significant resource. In these ways, the dependence of the Varanger Sámi on HG energy sources diminished, as they diversified to become semi-agropastoralists.

5. ENERGY FLOW AND OPTIMAL FORAGING THEORY

5.1 Resource Selection

Biologists would argue that we owe our current capacity to consume and enjoy a great variety of foodstuffs to the wide-ranging omnivory of our primate ancestors. That omnivory implies that hominid foragers consistently faced multiple choices in the subsistence quest. Almost all environments contain more species that are edible by humans than can be effectively be harvested by them. Resource selection models are designed to analyze this situation, by asking the question "which of the available resources will an efficient forager seek to harvest?" By comparing the predicted pattern of resource selection, which represents maximum efficiency, with the actual behavior and diet of HG groups, we can perhaps obtain some measure of how far the groups conform to efficiency predictions, rather than being led by other preferences.

The situation that is modeled is where a searching forager randomly encounters edible resources and is then faced with a decision. Should he/she take time to harvest (hunt, gather) this item, or, alternatively, is it better to continue searching in the hope of finding something more rewarding to harvest? Anthropologists have devised a measure of profitability termed the net acquisition rate (NAR), which is a measure of how much net energy is acquired (food energy output divided by work energy input) per hour of time devoted to the harvest. Some examples of the data necessary for these calculations are shown in Table II for the Anbarra. In the Anbarra case, the gross yield per person-hour for shellfish collecting averaged a modest 900 kcal, whereas spearfishing generated 14,000 kcal/hour and wallaby hunting much more. Knowing that fish suitable for spearing might be encountered later in the day, or a wallaby might be found, does the forager pause at the shellfish beds and harvest them, or does he or she move on in the hope of better things? The dilemma is that stopping to harvest what is encountered precludes harvesting something else—in other words, there are opportunity costs involved in making any choice. (In practice, mindful of the risks of failure in seeking high-risk resources with a high NAR, probably the woman forager will concentrate on harvesting the reliable but low-NAR shellfish and plant foods, whereas the male hunter will gamble on finding game.)

Bruce Winterhalder has used this approach to show how an optimal foraging strategy can be modeled. The model predicts the total set of resources that will be exploited if foragers have perfect knowledge (or gradually acquire it) regarding the benefits (the perceived NAR) of making certain choices. These modeling exercises suggest that resources with very low profitability, however abundant, will always be avoided if higher ranked species are available, because the opportunity costs of exploiting low-profit resources are too high. As Winterhalder says, "pennies in a field of sufficient dollars will always be ignored whatever their numbers." For Cree Indian foragers living in the boreal forests of northern Ontario, Canada, the model predicted that the introduction of steel traps, snares, and rifles should have led to an increase in profitability (energy gain per hour). As a result, this new technology should have expanded the range of species that the forager decided was worth harvesting. By contrast, the more recent introduction of the snowmobile so greatly increased the rate of encountering items that the efficient forager would now focus on a narrower set of resources. These predicted shifts were verified from the historical record, suggesting that this approach might have a wide potential for understanding the diversity of HG foraging behavior.

5.2 The Use of Patches

It is also possible to model foraging behavior within patches of resources. Again, the assumption (derived from microeconomics) is that foragers make decisions based on their knowledge of alternatives. A decision to stop consumption of one resource, or to cease one harvesting activity in favor of another one, is made by comparing the value of the last unit gained to the anticipated value of the alternative.

Thus the marginal value of harvesting one resource is compared to the average value of moving on.

Patches are discrete, localized concentration of resources, located such that a mobile forager might encounter several in the course of a day. Is it better to strip a patch of all the items that are encountered, or should the optimum forager move on the next patch quickly, having only taken the most accessible items? Maximum efficiency requires the forager to abandon a patch when its declining marginal rate of return equals the average NAR of many visits to many patches. This application of optimum foraging theory predicts that foragers will nearly always depart from a patch before it has been depleted of all its resources, which matches anecdotal evidence from HG practice. However, as with all these models, testing the predictions against real-world data is far from easy.

5.3 Central Place Foraging

Ethnographic studies show that few HG societies are wholly sedentary, and that residential relocation occurs with a frequency of up to 60 moves per year in extreme cases. The distance of relocation is normally in the 5- to 10-km range, but relocation is occasionally up to 70 km in one move. If we consider these base camps as central places, then can "central place theory" of human geography help to explain the diversity of HG residential mobility patterns? The theory predicts that to economize on the energy costs of resource acquisition, central places will be chosen either (1) adjacent to a critical resource that is rare and bulky, such as water, or more generally (2) at a spot chosen because it is at the center of gravity of the key resources that need to be transported there, such as water, fuel, and food. From this location, adjacent resources are exploited first. The net acquisition rate declines as the forager moves outward, reflecting the energy costs of walking to the resource and then transporting the items that are harvested back to base. Later, as the inner zone becomes depleted, the group experiences lower and lower NAR levels, until a point is reached when the marginal return from continuing to live in a central place is lower than the energy costs of moving elsewhere.

Winterhalder has shown that manipulation of the model predicts that camps will relocate more often (1) if relocation costs are low (for example, where the next suitable central place is not too far away); (2) if depletion of zones close to the central place is rapid, which will happen if patch density is low; and (3) if alternative sites offer a high initial rate of return. The very high mobility of foraging bands in the initial phase of colonization of North America at the end of the Ice Age is consistent with the predictions of the model, in a situation whereby the big game herds were abundant but easily depleted.

6. CONCLUSION

6.1 Energy and HG Theory

Energy provides a convenient way in which costs and benefits can be quantified, at a range of spatial and temporal scales. It is clear that the enormous diversity of HG societies makes it impossible to construct simple generalizations about energy flow, but on the other hand we can use concepts of efficiency to make some progress in understanding the various adaptations that can be observed. So far, theoretical work has focused on optimal foraging behavior, wherein "optimum" is defined in terms of maximizing energy yield or minimizing the direct energy expenditures involved in human labor.

Meanwhile, ethnographic studies of HG communities emphasize risk avoidance as well as energy efficiency as the dominant motives underlying economic behavior. The avoidance of energy deficit during critical seasons when resources are unavailable, and insurance against subsistence failure, are seen by anthropologists as key factors. These concepts can help to explain much of the variation that exists between different HG societies in their residential mobility, their focus on energy storage, and their defense of territories, and the social institutions that underpin these adaptations. Concepts of energy flow may help to provide a unifying theory of HG subsistence behavior.

6.2 The Future of HG Research

Ironically, as HG societies disappear or are transformed by modernity and the market economy, the opportunities for new research will become increasingly limited to the most marginal, and therefore the most atypical, HG groups. As our interest in mankind's own past grows, the desire to know more about energy flows in preagricultural societies is not likely to diminish. The surge of new information from palaeontology and human genetics about the remote origins of *Homo sapiens* is likely to arouse even more curiosity about this vanishing mode of food energy acquisition, given the dependence of all hominids on HG subsistence techniques. Only by placing modern HG societies firmly into both their ecological and historical contexts can we use 21st century knowledge to illuminate patterns of ecosystem management in the remote Pleistocene era.

SEE ALSO THE FOLLOWING ARTICLES

Cultural Evolution and Energy • Early Industrial World, Energy Flow in • Easter Island: Resource Depletion and Collapse • Fire: A Socioecological and Historical Survey • Food Capture, Energy Costs of • Human Energetics • Industrial Agriculture, Energy Flows in • Transitions in Energy Use • Wood Energy, History of • World History and Energy

Further Reading

Harris, D. R. (1996). Introduction: themes and concepts in the study of early agriculture. *In* "The Origins and Spread of Agriculture and Pastoralism in Eurasia" (D. R. Harris, Ed.), pp. 1–9. UCL Press, London.

Ingold, T., Riches, D., and Woodburn, J. (eds.). (1987–1988). "Hunters and Gatherers. Vol. 1, History, Evolution and Social Change"; "Hunters and Gatherers. Vol. 2. Property, Power and Ideology." Berg, Oxford, and St Martin's Press, New York.

Jones, R. (1981). Hunters in the Australian coastal savanna. *In* "Human Ecology in Savanna Environments" (D. R. Harris, Ed.), pp. 107–146. Academic Press, London.

Kelly, R. L. (1995). "The Foraging Spectrum: diversity in Hunter–Gatherer Lifeways." Smithsonian Institution Press, Washington, D.C.

Lee, R. B., and DeVore, I. (eds.). (1968). "Man The Hunter." Aldine, New York.

Lee, R. B., and Daly, R. (eds.). (1999). "The Cambridge Encyclopaedia of Hunters and Gatherers." Cambridge Univ. Press, Cambridge.

Odner, K. (1992). "The Varanger Saami, Habitation and Economy AD 1200–1900." Scandinavian Univ. Press, Oslo.

Panter-Brick, C., Layton, R. H., and Rowley-Conwy, P. (eds.). (2001). "Hunter–Gatherers: An Interdisciplinary Perspective." Cambridge Univ. Press, Cambridge.

Winterhalder, B., and Smith, E. A. (eds.). (1981). "Hunter–Gatherer Foraging Strategies: Ethnographic and Archaeological Analyses." Univ. Chicago Press, Chicago.

Hybrid Electric Vehicles

JOHN M. GERMAN
American Honda Motor Company, Inc.
Ann Arbor, Michigan, United States

1. Introduction: Effects of Hybrid Electric Vehicles
2. History
3. Hybrid Electric System Design
4. Real-World Hybrid Examples
5. Hybrid Advantages and Disadvantages
6. Future Technology Development
7. Future Hybrid Markets

Glossary

energy density Energy storage per unit weight. Energy is the power rate multiplied by the length of time the power is maintained; measured as the amount of energy (such as kilowatt-hour) per unit of weight (such as kilogram or pound).

hybrid electric A system that combines a combustion engine (e.g., gasoline, diesel, or turbine) and an electric motor and energy storage (e.g., a battery or ultracapacitor).

hybrid propulsion A combination of two different power sources for vehicle propulsion. One source derives its power from fuel, such as an internal combustion engine or a fuel cell. The second source is a device that stores and reuses energy. This can be electrical energy or electromechanical energy, such as a hydraulic accumulator or a flywheel.

load Forces that must be overcome to accelerate a vehicle and maintain cruising speed. These include air drag, tire rolling resistance, mechanical resistances in the drivetrain, vehicle mass resistance to acceleration, and accessory loads from pumps, air conditioning, and power steering, for example.

power density The power delivered per unit weight; measured as the amount of power, such as kilowatts or horsepower, per unit of weight, such as kilograms or pounds.

regenerative braking The capture of energy from a vehicle's inertia when the vehicle slows down. A large amount of the energy that is used to accelerate a vehicle to operating speeds is normally lost when the vehicle slows down or brakes. Regenerative braking captures part of this energy and uses it to recharge an energy storage device.

ultracapacitors Higher specific energy and power versions of electrolytic capacitors; devices that store energy as an electrostatic charge.

The internal combustion engine has dominated the car and light-truck market for over 100 years. Although remarkable improvements have been made over the past 30 years to reduce air pollutants to nearly zero and to almost double vehicle efficiency, increasing concerns about global warming and energy security are pushing vehicles toward even greater efficiency. Adding a high-power electric motor and electric storage capacity to an internal combustion engine offers significant fuel savings. The engine can be shut off at idle to avoid wasting fuel, the motor can be driven in reverse when braking to capture energy that would otherwise be lost, and the boost from the electric motor allows use of a smaller, more efficient, engine. In the long term, the electrical power from the motor can be used to replace existing mechanical accessory drives with more efficient electrical devices, as well as to provide power for additional features desired by customers. Against these benefits must be weighed the additional complexity and costs of the hybrid electric power system.

1. INTRODUCTION: EFFECTS OF HYBRID ELECTRIC VEHICLES

In the early days of the automobile, around the turn of the 20th century, there was spirited competition between vehicles powered by electricity and by internal combustion engines. The internal combustion engine won primarily because of the high amount of energy in liquid fuels, which allows vehicles to travel farther without refueling than is possible on an electric charge. Ten gallons of gasoline weighs only about

28 kg (62 lb), but contains about 330 kWh (kilowatt-hours) of energy (1.1 million Btu). By comparison, a modern lead-acid battery weighing the same 28 kg provides only about 1.1 kWh. This overwhelming energy advantage of liquid fuel has ensured the dominance of the internal combustion engine for the past 100 years, despite its relatively low efficiency.

Problems with the use of internal combustion engines and fossil fuels are well documented, including low efficiency, air pollution, fossil fuel use, energy security, dependence on foreign oil suppliers, lead poisoning, and leaking storage tanks. To combat these problems, many alternatives to the gasoline internal combustion engine have been proposed over the years, such as steam power, turbine engines, electric vehicles, and the use of alternative fuels such as methanol, ethanol, compressed natural gas, and propane. The latest contender is the fuel cell, powered by hydrogen. However, so far, the internal combustion engine has won over all proposed alternatives. It has accomplished this because the incremental advantages of switching to another propulsion system have been less than the cost of switching to an entirely new infrastructure. In addition, every time the internal combustion engine has been challenged, it has responded with enough improvement to keep the alternative off the market. For example, with the development of modern computer controls and catalysts, air pollutants have been reduced to levels undreamed of even 10 years ago. Fuel efficiency has also increased greatly due to advances in technology. The average car today achieves almost twice the fuel economy of the average car 30 years ago, in addition to accelerating much faster. However, even with the efficiency increases, the average efficiency of a gasoline internal combustion engine in typical in-use operation is still only about 15%. The other 85% is lost to engine heat, heated exhaust gases, aerodynamic drag, tire rolling resistance, driveline losses, and braking.

Adding an electric motor and energy storage to the internal combustion engine can significantly improve efficiency in a variety of ways, depending on how the system is designed:

- Engines are least efficient when operating at low loads. The electric motor can be used to supply part or all of the propulsion energy at low speeds and loads, minimizing engine use under inefficient conditions.
- The electric motor can assist the engine during acceleration. This allows use of a smaller engine without any loss in overall performance. For a given load, smaller engines have better efficiency, due to lower frictional and heat losses.
- The high-power electric motor allows rapid engine restarts. This allows the engine to be shut off at idle and avoids fuel consumption while the vehicle is stopped.
- The electric motor can be used to capture regenerative braking energy. This is done by using the vehicle's inertia to drive the electric motor in reverse while the vehicle is slowing down, thus creating free electric energy.
- The additional electric power from the motor/generator can be used to replace mechanical and hydraulic devices and pumps with more efficient electric versions.

The advantages of hybrid systems have long been recognized. The first hybrid vehicle was built in 1898 and several manufacturers sold hybrid vehicles in the early 1900s. However, hybrid vehicles also have significant problems. They require two propulsion systems, which take up room, add weight, and greatly increase the cost. Another problem is that careful coordination of the operation of the engine and the motor is necessary to achieve much of the efficiency benefits and to avoid driveability problems. This was not possible with mechanical controls, and the production of hybrid vehicles did not survive continued development of the internal combustion engine in the early 1900s.

Renewed interest in hybrid vehicles has coincided with development of computer controls and improved batteries, combined with increasing concerns about the contribution of carbon dioxide emissions to global warming. Sophisticated computer controls allow maximum efficiency benefits while providing smooth, seamless coordination of the two propulsion systems. Advanced batteries, such as those made with nickel–metal hydride (NiMH), provide higher energy density and much longer cycle life, which is important to customer acceptance of hybrid technology. Hybrid vehicles offer a way to reduce fuel use significantly, with corresponding reductions in global warming gases and fuel cost to consumers, plus modest reductions in criteria air pollutants. The primary concern with hybrid systems remains the cost of the additional components.

2. HISTORY

2.1 Early Hybrid Electric Vehicles

From about 1890 through 1905, electric vehicles, internal combustion-powered vehicles, and steam cars were all competitively marketed and sold in the

United States. Electric vehicles had an early development lead in the United States due to the work of electricity pioneers, such as Edison, Tesla, and Westinghouse. Also, the limited driving range of electric vehicles was sufficient for the small U.S. cities in the late 19th and early 20th centuries, and the great distances between cities had roads that were largely inadequate for vehicle travel. However, it was obvious from the beginning that batteries severely limited the range and utility of electric vehicles. Although most developers went straight to internal combustion-powered vehicles, some did try to marry the advantages of the electric vehicle and the internal combustion engine.

Justus Entz probably built the first hybrid vehicle in 1898 for the Pope Manufacturing Co. in Connecticut. The vehicle caught fire during the initial test drive and Pope Manufacturing abandoned the concept. The first production hybrid vehicles were made in Europe in the early 1900s. Camile Jenatzy demonstrated a parallel hybrid system at the 1901 Paris Auto Show, with a 6-hp engine and a 14-hp motor/generator. In 1903, the Lohner–Porshe group brought out a series hybrid system, using a 20-hp engine to drive a 21-kW direct current (DC) generator and an electric motor in each of the two front wheels. A few years later, the Mercedes Company teamed with Mixte to introduce the Mercedes-Mixte. To improve the limited range of electric vehicles, two well-known U.S. electric car manufacturers, Baker and Woods, independently developed hybrid vehicles in 1917. However, after some initial development and vehicle operation, both concluded that the hybrid approach added too much complexity, weight, and cost.

2.2 Modern Development of Hybrid Vehicles

Air pollution concerns in the late 1960s and early 1970s spurred renewed interest in electric vehicles and hybrid electric vehicles. At that time, many people believed the internal combustion engine was inherently highly polluting and that it would need to be replaced in order to clean up the air. Battery–electric vehicles were a logical candidate because they have no emissions while in operation, although there is some offset from increased emissions from electric power plants. However, the short range and high cost of batteries continued to be insurmountable problems. To extend the range of primarily electric vehicles, a substantial number of concepts were introduced at the time, with small gasoline engines that could be turned on when needed. However, the large battery packs required for this type of vehicle were too expensive and degraded performance too much to be commercially viable.

The oil crises of 1973 and 1979 turned the auto industry on its head. Gasoline shortages, fuel price spikes, and predictions of continued oil shortages and increasing fuel prices caused the public to suddenly demand higher efficiency vehicles. Congress responded in 1975 with the Energy Policy and Conservation Act (EPCA). The cornerstone of EPCA was Corporate Average Fuel Economy (CAFE) standards, which mandated that cars double their average fuel economy by 1985 and that the Department of Transportation set cost-effective standards for light trucks.

As fuel prices went down in the early 1980s and stayed down the entire decade, the buying public gradually forgot about fuel economy. By 1990, vehicle purchasers were back to demanding other attributes they valued more highly than fuel savings, such as luxury, performance, and utility. The average fuel economy of new cars and light trucks, combined, peaked in 1987 and has been slowly declining ever since. Against the declining public concern with fuel economy was increasing concern about a new problem, global warming. Every gallon of gasoline burned produces about 20 lb of carbon dioxide, the primary greenhouse gas accused of increasing global temperatures. In addition to global warming concerns, the Gulf War of 1991 provoked concerns about national energy security.

Faced with such concerns, several attempts were made in Congress to raise CAFE standards beyond the levels set during the early energy crisis years. However, in Congress, there was also significant opposition to raising the CAFE standards; resistance came from a coalition of members concerned about the health of the domestic auto industry and general antiregulatory interests, and from those who did not believe that global warming is a significant concern.

2.3 Supercar

Rather than try to push a mandatory program through Congress, President Clinton and Vice-President Gore struck a deal with General Motors (GM), Ford, and Chrysler. If the domestic automakers would promise to try to build ultraefficient cars, the Clinton administration would back off from its campaign promise to raise CAFE standards. In the fall of 1993, the President announced that the U.S. government and the American auto industry would work together to build a family car that got 80 mpg, or about three times the mileage of an average car. Officially named the Partnership for a New Generation of Vehicles (PNGV), most people soon dubbed the program "Supercar." The partnership spent 9 years and about $3 billion, roughly half each from the government and industry, on developing a super-efficient hybrid vehicle. General Motors, Ford, and Chrysler each developed a working hybrid demonstration vehicle, as illustrated by GM's 80-mpg Precept concept vehicle (Fig. 1). Although the program spurred advances in materials and hybrid design, it did not produce a vehicle that the automakers felt could be mass-produced at a price consumers would be willing to pay.

The primary problem with the Supercar program was the arbitrary goal of 80 mpg. This goal forced the automakers to choose options that were not cost-effective or were not marketable, such as diesel engines, lightweight materials, and extreme aerodynamic improvements. However, the program did spur significant advances in hybrid design that are being incorporated into future hybrid vehicles.

The announcement of the PNGV program also helped spur Toyota into designing its own hybrid vehicle, the Prius, in secret. Toyota's goal was a more realistic 55 mpg, or twice the mileage of an average car, using a highly efficient gasoline–electric hybrid design. The initial design was tailored to congested urban driving conditions in Japan and performance was not adequate for the U.S. market. However, it was sold in Japan while Supercar was still establishing concepts. Toyota redesigned the hybrid system on the Prius 2 years later to increase performance to levels acceptable to U.S. customers. In the meantime, Honda developed its own hybrid vehicle, the Insight, and beat Toyota to the U.S. market in late 1999.

FIGURE 1 The GM Precept concept hybrid vehicle.

3. HYBRID ELECTRIC SYSTEM DESIGN

The amount of power needed to propel a vehicle varies greatly, depending on the driving requirements. The vehicle must be designed to meet the most demanding requirements, such as hard accelerations under fully loaded conditions. However, this amount of power is needed infrequently. The large majority of driving requires only a small fraction of the available power from the engine. With a hybrid electric system, a smaller, more efficient engine can be used, with an electric motor added to restore performance.

Other types of hybrid systems have been considered. For example, Chrysler built a prototype vehicle in the early 1990s that combined a combustion engine with a flywheel that stores mechanical energy and provides power to the wheels. More recently, Ford and the U.S. Environmental Protection Agency (EPA) cooperated on a hybrid that combines an internal combustion engine with a hydraulic/nitrogen gas energy storage system. This system can provide considerable power from a stop and may find an application for helping to launch heavy-duty trucks.

For the large majority of applications, only the hybrid electric vehicle offers improved fuel economy and adequate performance at a cost customers might be willing to pay. In addition, the high amount of electrical power available on hybrid electric vehicles can be used to meet the increasing electrical power demands associated with other vehicle features, potentially adding value beyond just higher powertrain efficiency.

3.1 Types of Hybrid Electric Systems

Although the ways in which the electric motor and the engine can be designed, and how they interact, are virtually limitless, the two basic configurations are series and parallel hybrids.

3.1.1 Series Hybrid
In the series hybrid, the engine is connected to a generator, which in turn is used to supply the motor and/or charge the battery (Fig. 2). Only the electric motor is directly used to drive the wheels. The battery pack can be charged by the engine and generator set and by using the generator to capture energy normally lost to braking. The principal advantage of this configuration is that the engine can be run in more efficient operating zones and shut off when it is not needed. The system also eliminates the need for

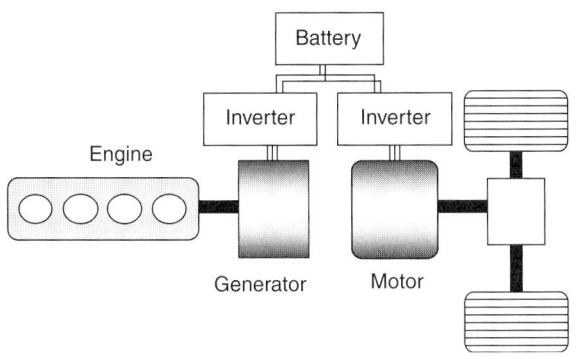

FIGURE 2 Series hybrid system diagram. Courtesy of Toyota Motor Company.

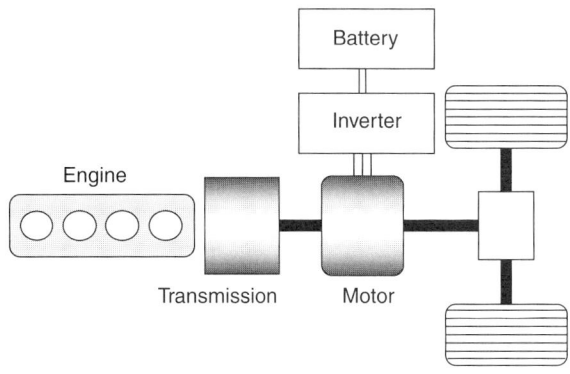

FIGURE 3 Parallel hybrid system diagram. Courtesy of Toyota Motor Company.

clutches and conventional transmissions. The need to convert the output of the engine to electricity before driving the motor creates significant inefficiencies if the engine is used as the primary power source, so this configuration is usually set up with a small engine that provides a range booster to a large battery pack. This configuration has very low emissions and has excellent efficiency during slow stop-and-go driving. However, it requires a large, heavy battery pack, which significantly increases cost and reduces vehicle performance. It is also much less efficient during high-speed driving, due to losses in converting the mechanical power from the engine to electricity and in charging and discharging of the battery. In addition, the internal combustion motor must still be large enough to meet minimum acceleration needs should the battery pack become depleted.

3.1.2 Parallel Hybrid

The parallel hybrid uses an engine and a motor, which are both connected directly to the drivetrain (Fig. 3). The highest efficiency range of each is selected and used, depending on operating condition. A parallel system is more complicated than a series system and there are additional challenges in integration of the two power sources. The major advantage is that large efficiency gains can be realized even with a relatively small, light battery pack. This minimizes the additional cost of the battery pack and the performance penalty from the weight of the batteries.

3.1.3 Variations on the Basic Design

The simplest way to arrange a parallel hybrid is to mount the electric motor in line with the engine. Another feasible arrangement is to power one set of wheels (either front or rear) with the conventional engine and the other set of wheels with an electric motor or motors. This system is referred to as a "through-the-road" parallel hybrid, because the two power sources are run in parallel by independently driving different wheels. The primary advantage of this system is that it adds part-time all-wheel-drive (AWD) capability. It also allows more flexibility in the mounting location of the electric motor. However, the system adds considerable complexity and cost. In addition, the AWD capability is limited to the energy in the battery pack and cannot be maintained indefinitely.

Another option is to add "plug-in" capacity. The basic hybrid system is self-contained, with all of the electricity generated by the on-board motor/generator while the vehicle is in operation. A vehicle with a battery pack that can be optionally recharged while the vehicle is parked, by plugging it into an electrical output, is commonly referred to as a grid-connected hybrid. Grid-connected hybrids can be used with either parallel or series design. The advantage of a grid-connected system is that it allows significant driving range in all-electric mode, without turning on the engine. Grid-connected hybrids have been promoted as a way to extend the range of a true electric vehicle, while still relying on the all-electric operation for urban operation with no combustion emissions at the vehicle. However, the value of electric-only operation has greatly diminished as ultraclean gasoline engines have become available. More importantly, the electric-only vehicle also suffers from many of the problems of a pure electric vehicle, such as the need for a large, heavy, and expensive battery pack and long recharge times.

3.2 Hybrid Components

3.2.1 Electric motor

Because the electric motor is used frequently, both for acceleration and for regenerative braking, efficiency and high torque are important attributes. If the motor is placed between the engine and the transmission, then a compact design and high heat resistance are also important. There are currently two basic types of electric motors. One is an induction alternating current (AC) motor. This design uses part of the electric current to generate the magnetic field, within which the motor windings turn. The second design is a permanent magnet direct current motor. This type of motor generally uses three-phase windings and switching modules to control the flow of electricity sequentially to each phase. The AC design is simpler and cheaper. However, brushless DC (BLDC) permanent magnet motors are more efficient, more compact, and lower weight. The most efficient BLDC motors use magnets sintered with rare earth metals, such as neodymium. It is the cost of the magnet that makes DC motors more expensive.

Another important factor is the design voltage of the hybrid system. Higher motor power capacities are best met at higher voltages, because this reduces current and allows smaller wires and more efficient motors and inverters. The operating voltage can affect the choice of battery and the type of power electronics, which has cost implications. If the electric motor is located between the flywheel and the transmission clutch, it becomes very important to make the motor as thin as possible to minimize increases in powertrain length. This is especially important for transverse-mounted front-wheel-drive applications, in which the distance between the wheels is limited. Fig. 4. illustrates the placement of a compact motor design on the Civic Hybrid. The BLDC electric motor uses concentrated windings to conserve space (Fig. 5).

3.2.2 Energy Storage

Battery packs are the primary type of energy storage for hybrid electric vehicles. The only reasonable alternative is the ultracapacitor. Unless significant all-electric range is desired, energy requirements for hybrid electric vehicles are fairly modest. Energy storage is primarily used for acceleration, regenerative braking, and low-speed, low-power operation. At higher loads, it is more cost-efficient to use the engine to power the vehicle than it is to increase the size of the battery pack. Thus, hybrid vehicle electric power requirements are either brief or low load. Although some energy reserve is needed for extended hill climbs, energy storage demands are relatively small.

On the other hand, power requirements are substantial. If significant engine downsizing is to be

FIGURE 4 Electric motor placement on the Honda Civic Hybrid. Courtesy of Honda Motor Company.

FIGURE 5 Thin electric motor designed to fit between the engine and transmission on the Honda Civic Hybrid. Courtesy of Honda Motor Company.

achieved, the electric motor must provide at least a 50% torque boost at low engine speeds. Although the electric motor can be biased to provide its maximum power at relatively low engine speeds, this still requires an electric motor that has at least 15% of the power output of the engine. Even for a small car, the electric motor needs to be at least 10 kW (about 13 hp). The electric motor power requirements scale up with vehicle size and with the use of electric-only operation. Thus, batteries for hybrid vehicles must be able to provide or absorb high power rates to support acceleration and to recapture high amounts of energy during brief braking events.

Batteries developed for electric vehicles in the 1990s were optimized for energy storage, because they needed to maximize all-electric vehicle range. These batteries are not suitable for hybrid vehicles. Batteries were redesigned for hybrid vehicles to provide much higher power density, with the trade-off of lower energy density. Even with the redesign, most battery packs still must be sized to deliver maximum power without overheating. This results in more energy storage than is needed for the vast majority of driving conditions, which increases the cost and weight of the battery pack.

Ultracapacitors have very high power density. They excel at rapid transfer of power and offer the potential for higher durability and lower cooling requirements. They are ideal for assisting acceleration and at recapturing regenerative braking energy. Unfortunately, they have very low energy density and cannot store enough energy to cover longer load demands, such as hill climbs. The energy density is so low that no manufacturers are considering their use for hybrid vehicles, although some manufacturers are using them to provide momentary assist on fuel cell vehicles.

3.2.2.1 Lead-Acid Battery The primary advantage of the lead-acid battery is low cost. Being developed for hybrid vehicles are advanced lead-acid batteries with higher power, such as the valve-regulated lead-acid (VRLA) battery. However, even VRLA batteries have relatively low energy density and short cycle life. The short cycle life is a particular concern, because periodic replacement raises the effective cost of the system. It is uncertain how customers will react to the cost and hassle of periodically replacing the hybrid battery pack. Despite the concerns, lead-acid batteries will be used in some relatively low-power hybrid vehicle applications due to their low cost.

3.2.2.2 Nickel–Metal Hydride Battery NiMH batteries have higher power and energy density and a much longer life cycle compared to lead-acid batteries. They are also completely safe and their power output is not affected by the battery state of charge. The main concern with nickel–metal hydride batteries is that they are very expensive. Other concerns are high self-discharge rates, low-temperature operation, and higher cooling requirements. NiMH batteries are used in higher power hybrid designs, such as the Toyota Prius, the Honda Insight and Civic Hybrid, and the upcoming small sport utility hybrid vehicles from Ford and General Motors.

3.2.2.3 Lithium-Ion Battery Lithium-ion batteries have higher energy and power density compared to NiMH batteries. They also have better low-temperature performance and a low self-discharge rate. However, they need improvements in durability and cost before they will be commercially viable for hybrid electric vehicles.

3.2.3 Power Electronics

Direct current motors generally use three-phase motor windings. Three switching modules are needed to control current flow to the proper windings as the motor turns. These switches must be able to turn large amounts of current on and off rapidly and accurately, as well as to control the direction and phase of the current. This operation produces a substantial amount of heat and the power control unit must be cooled to keep the heat from affecting the electronics. The size and cost of the power electronics are substantial, due to the high power demands of the system and the need for cooling.

The power control unit includes a DC–DC converter to supply 12 V for the conventional electrical system. Improvements to the switching method are being incorporated to reduce switching losses and increase the efficiency over the output range of the converter.

3.2.4 Cooling System

Both the battery pack and the power electronics must be cooled. Charging and discharging a battery is a chemical reaction, which generates heat as a by-product. High battery temperatures degrade performance and shorten battery life. Power electronics generate a lot of heat and do not function properly if they get too hot. Efficient cooling systems are needed for both the battery pack and the power electronics. Due to the lower heat generated by newer power electronics, it is now possible to cool the power electronics with air discharged from the battery. Combining the battery pack and power electronics into a single unit with an integrated cooling system offers substantial reductions in overall size and weight.

3.3 Fuel Efficieny Benefits

3.3.1 Engine Downsizing

The greatest demands on horsepower and torque occur while accelerating and climbing grades. Minimal power is needed to maintain a vehicle's speed while cruising on a level road. By using an electric motor to provide a power boost to the engine when appropriate, a smaller engine can be used. Other things being equal, a smaller engine is more efficient for a given load, because it has lower frictional losses, less heat loss to the engine block and cylinder head, and larger throttle openings, which reduce the energy lost in forcing intake air past the throttle (commonly referred to as pumping losses).

3.3.2 Regenerative Braking

A large amount of energy is lost in conventional vehicles when the vehicle slows down or brakes. This energy is primarily consumed by braking and engine friction, which includes mechanical friction and pumping losses. The electric motor can be used as a generator to capture this energy. When the vehicle is slowing down or braking, the electric motor captures—or regenerates—electrical energy from the vehicle's forward momentum that would otherwise be wasted. This energy is stored in the battery pack until it is reused during acceleration, when power from the gasoline engine is boosted by power generated by the electric motor. This process is referred to as "regenerative braking." If the electric motor is bolted to the engine, then some of the potential regenerative braking energy is lost to engine friction. Installation of a clutch between the engine and the electric motor eliminates engine friction during braking and increases the amount of energy available for recharging the battery. However, the clutch adds cost and increases packaging problems, especially for front-wheel-drive vehicles.

3.3.3 Idle Off

Hybrid systems allow the engine to be shut off at idle, eliminating idle fuel consumption and emissions. The powerful electric motor can restart the engine far more quickly than can a conventional starter motor. In most cases, the motor is powerful enough to spin the engine up to normal idle speed in only about 0.2 sec. This allows smooth, rapid restarts that are acceptable to the driver. In addition, fuel injection usually does not start until after the engine is already up to normal idle speed, which virtually eliminates unburned fuel and its attendant hydrocarbon emissions. The average vehicle sits and idles about 20% of the time, thus turning off the engine at idle can reduce fuel consumption by about 5–10%.

3.3.4 Engine Efficiency

Proper integration of the electric motor and the internal combustion engine can have a major impact on the efficiency of the engine during different driving conditions. Optimization of the overall system allows the engine to operate in more efficient modes. For example, the high torque of the electric motor allows the engine to turn more slowly, and hence more efficiently, during highway driving while still maintaining adequate acceleration. Another example is careful integration with transmission operation, to keep the engine running at higher efficiency speed and load points.

The boost from the electric motor can also facilitate use of innovative engine designs, such as the Atkinson cycle gasoline engine. The Atkinson cycle uses a longer expansion stroke to extract more energy from the combustion process and boost efficiency. The downside of the Atkinson cycle is that it generates much lower peak torque and horsepower. This is more acceptable on a hybrid vehicle because the electric motor can provide power boost during acceleration.

3.3.5 Electrical Accessories

The additional electric power can be used to improve the efficiency of engine accessories, such as the air conditioner, power steering pump, and water pump. Currently, most accessories are driven by mechanically connecting them to the engine with belts, gears, or chains. The mechanical efficiencies are often very low. Another problem is that the accessory speed goes up and down with the engine speed. This is very wasteful, because the accessory must be sized to provide adequate operation at low engine speed. Using electric motors to power accessories allows much more efficient designs, because they can be operated independently of the engine and only as needed.

On conventional vehicles, the use of electric accessories is limited by alternator load, which is already reaching the limits of a 12-V alternator. The vehicle electric demand has increased from a few hundred watts 10 years ago to over a kilowatt today. Inducing high currents in a 12-V system results in high efficiency losses and requires large, heavy, and expensive electrical cables. The high power and higher voltage available from the hybrid electric motor can enable use of electric air-conditioning compressors and electric pumps for power steering and circulation of fluids.

3.4 Design Considerations

Design of hybrid systems is an exercise in trade-offs. Larger electric motors and battery packs allow more acceleration assist and smaller engines, recapture more regenerative braking energy, and enable some all-electric operation to replace the most inefficient engine operation modes. Very large battery packs and motors allow off-board recharging from the electric grid. However, battery packs are very expensive and their additional weight affects performance. Thus, it is highly desirable to limit the size of the battery pack. The type of battery also has a profound impact on durability, size, and cost. Finally, the design and sophistication of the hybrid system, including individual components and how they are integrated, can have a greater impact on fuel efficiency than the simple inclusion of individual hybrid features. Juggling the relative benefits, costs, and packaging of all the different options is difficult. Each manufacturer has a different idea of how to balance the factors. Ultimately, the market will determine what works and what does not.

Hybrid vehicles can incorporate different fuel-efficient techniques, such as idle off, regenerative braking, engine downsizing, motor-only operation, and off-board recharging. Except for idle-off, each of these features can be utilized to different levels.

3.4.1 Integrated Starter Generator

The most basic hybrid type is the integrated starter generator. These are conventional vehicles that have a higher power electric motor that can turn the engine off when the vehicle is stopped and restart it as needed. The starter generator motor can be mounted on the driveshaft or it can be connected with a belt. Most systems under development use a 42-V lead-acid battery pack, although 12-V systems are also being evaluated. These systems are relatively inexpensive and can provide some additional power for accessories. However, the fuel economy improvement is generally only 5–10%. An integrated starter generator system generally is not classified as a hybrid vehicle unless its electrical motor also provides propulsion assistance.

3.4.2 Full-Function Hybrid

The next step is to use a larger, full-function electric motor and a high-voltage battery pack to provide assist on all accelerations. This design significantly increases the cost of the system, but it greatly expands the amount of regenerative braking, enables engine downsizing, and allows integration of the engine and motor functions. Depending on the sophistication of the system and the amount of engine downsizing, fuel economy improvements can range from about 20 to 50%.

Trade-offs between fuel economy and performance are also possible. One interesting design concept is to add an electric motor without downsizing the engine. This can dramatically increase vehicle performance, especially at low speeds, when the electric motor has very high torque output, while still providing significant fuel savings from idle-off and regenerative braking. The result can be thought of as an environmental supercharger. Because performance is valued more highly than fuel economy by most U.S. new-vehicle purchasers, this may be a way to help public acceptance of hybrid vehicles.

3.4.3 Motor-Only Operation

The next step in design sophistication is to add motor-only operation. These systems use the electric motor to propel the vehicle at low speeds and loads, eliminating engine operation in its most inefficient modes. Unfortunately, the incremental efficiency gains from motor-only operation are modest. There is little efficiency benefit from motor-only operation above about 20 mph, because the engine can operate reasonably efficiently at higher speeds. Even below 20 mph the engine must be used for harder accelerations, so the primary efficiency benefits are in congested, low-speed urban driving, such as those encountered in Japan. In addition, large amounts of motor-only operation require larger battery packs, driving up cost and weight. Limited use of motor-only operation at low speeds can achieve some efficiency improvement at reasonable cost, but the overall impact under typical U.S. driving conditions is relatively small.

3.4.4 Grid-Connected Hybrids

The final step in hybridization extends the motor-only range by allowing the battery to be recharged from the electric grid. However, the value of electric-only operation has greatly diminished as ultraclean gasoline engines have become available. In any case, the very high cost and weight of the large battery packs needed for these systems generally preclude their use in production.

4. REAL-WORLD HYBRID EXAMPLES

In the real world, engineers can get creative about applying hybrid concepts to actual vehicles. Hybrid vehicles in production or being readied for production in the United States illustrate a number of innovative hybrid techniques.

4.1 Toyota Prius

The Prius was the first modern hybrid to be placed in production, starting sales in Japan in late 1997. A version for the United States was introduced in 2000 (Fig. 6). Rather than choose between series and parallel systems, Toyota developed a new type of hybrid system that combines some of the advantages of both (Fig. 7). The Prius powertrain is based on the parallel type. However, it uses a planetary gear system and a separate generator to optimize the engine's operation point and to allow series-like operation. Toyota's hybrid system requires a power split device and an extra generator and inverter. However, it offers great flexibility in engine operation and calibration. The system also acts as a continuously variable transmission (CVT), eliminating the need for a conventional transmission.

The Prius used a relatively small NiMH battery pack, rated at 288 V and about 2 kWh of storage capacity. The electric motor is rated at 44 hp, allowing limited acceleration and cruise at light loads and low speeds on just the electric motor. The Prius also uses an Atkinson cycle engine for maximum efficiency. The overall design excels at low-speed fuel efficiency and the city rating of 52 mpg is much higher than the highway rating of 45 mpg. These values represent a fuel economy improvement of about 80% on the U.S. EPA city driving cycle, although less than 20% on the U.S. EPA highway cycle. On the Japanese driving cycle, involving relatively low speeds and many starts and stops, the Prius offers about a 100% improvement over a car of comparable size and performance.

4.2 Honda Insight and Civic Hybrid

Honda was the first company to market a modern hybrid vehicle in the United States, introducing the Honda Insight in late 1999 (Fig. 8). Honda developed an electric motor assist system for the Insight (Fig. 9), which integrates a high-torque, high-efficiency brushless DC motor between the engine and the transmission. This 10-kW (13-hp) motor is only 60 mm (2.4 in) thick and is connected directly to the engine's crankshaft. This is a straightforward method for packaging a parallel hybrid system and minimizes the weight increase.

The Insight's 144-V NiMH battery pack is rated at about 1 kWh of storage and weighs only about 22 kg (48 pounds). This helps to maintain in-use performance and efficiency while maintaining most of the hybrid system benefits. The Insight also incorporates a wide range of other efficiency technologies, such as an all-aluminum body, aerodynamic improvements, and a 1.0-liter three-cylinder engine with variable valve timing, reduced friction, and lean burn. With a manual transmission, the fuel economy ratings are 61 mpg in the city and 68 mph on the highway, the

FIGURE 6 The Toyota Prius. Courtesy of Toyota Motor Company.

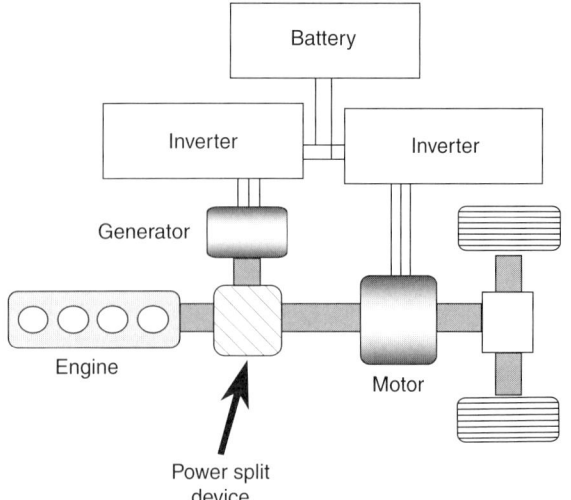

FIGURE 7 Hybrid system for the Toyota Prius, illustrating the planetary gear system and separate motor and generator. Courtesy of Toyota Motor Company.

FIGURE 8 The Honda Insight. Courtesy of Honda Motor Company.

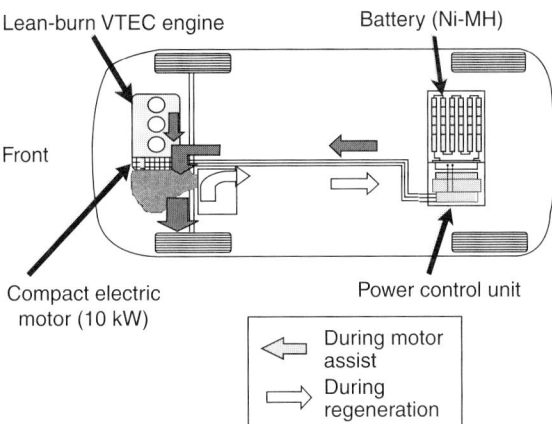

FIGURE 9 Hybrid system layout for the Honda Insight, with a variable valve-timing and lift electronic control (VTEC) engine and a nickel–metal hydride (NiMH) battery. Courtesy of Honda Motor Company.

best fuel economy ratings ever recorded by the EPA for a production vehicle.

In 2003, Honda expanded the integrated motor-assist (IMA) system to the Civic. This was the first time a hybrid system was offered as a powertrain option on a conventional vehicle. A number of improvements were made to increase motor torque, integrate many of the functions of the battery pack and power electronics, and improve battery pack efficiency. The Civic Hybrid also has a cylinder-idling system that deactivates valves on three of the four cylinders during deceleration and braking, cutting engine friction by 50% and allowing more power to be recaptured by the electric motor. The Civic hybrid with a manual transmission is rated at 46 mpg city and 51 mpg highway. With the automatic continuously variable transmission (CVT) option, fuel economy is 48 mpg city and 47 mpg highway. These values represent fuel economy improvements of about 55% (average of manual and automatic) on the U.S. EPA city driving cycle and about 30% on the U.S. EPA highway cycle.

4.3 Toyota Estima Minivan

Introduced in 2002 and sold only in Japan, the Toyota Estima small minivan is the first through-the-road parallel hybrid vehicle. The front powertrain combines a gasoline engine with a 216-V, 13-kW, belt-driven integrated starter generator. A second 216-V electric motor, rated at 18-kW, provides on-demand power to the rear wheels for temporary four-wheel-drive operation. The computer can choose any combination of the three power sources and either of the motor/generators can be used to recharge the NiMH battery pack.

4.4 Future Compact Sport Utility Vehicle Hybrids

Ford has announced plans for a hybrid versions of their small sport utility vehicle (SUV), the Ford Escape, in 2005. The hybrid system includes a 300-V NiMH battery pack, an efficient electric air-conditioning compressor that operates even when the engine is turned off at idle, and relatively large electric motors that provide motor-only operation at low speeds and loads. The drive motor is rated at 65 kW (87 hp), plus it has a 28-kW motor/generator. The Escape hybrid combines the electric motor, separate generator, power electronics, and planetary gear into a single unit that takes the place of a conventional transmission. This arrangement provides flexible power combinations and is similar in concept to the planetary gear system used in the Toyota Prius.

Toyota plans to produce the first through-the-road parallel hybrid system for the U.S. market based on its Lexus RX330 compact SUV in 2005. A V6 gasoline engine powers the front wheels while two electric motors, one in the front and one in the back, provide the vehicle with all-wheel-drive capability. This vehicle also illustrates a hybrid electric design optimized for performance, providing the vehicle with very high acceleration ability along with a significant boost in fuel economy. DaimlerChrysler also plans to offer a through-the-road hybrid system on future models, such as the Jeep Liberty and Dodge Durango.

4.5 Hybrids as Electric Generators

Both General Motors and DaimlerChrysler plan to offer similar 42-V hybrid systems on full-size pickup trucks in 2004. An AC induction motor/generator is integrated into the drivetrain and is used for idle-stop and limited acceleration assist. Both use 42-V lead-acid batteries, which are recharged with regenerative braking. The fuel economy improvement is relatively modest, about 10–15%, because the engines are not downsized, in order to preserve cargo-hauling and trailer-towing abilities. However, the fuel economy improvement is secondary to the on-board electrical-generating capability provided from the on-board engine and generator on these vehicles. The GM pickups provide continuous 20-amp power through each of two 110-V outlets. The Dodge Ram can provide up to 20 kW of either 110- or 220-V power. These systems are primarily targeted at construction site work, but they may also appeal to campers and outdoor enthusiasts and to farmers and homeowners for backup power in case of a power failure.

4.6 Military Applications

Military actions have some unique characteristics that make hybrid vehicles very desirable. For example, fuel may need to be transported hundreds, or even thousands, of miles in overseas military action. Fuel is the single largest supply component in a military action and the price of delivering a gallon of fuel to a battlefield can easily exceed $30 per gallon. Thus, even a 10–15% improvement in fuel efficiency is extremely desirable to the military, as long as it does not impact performance and utility.

The military has to furnish its own electricity in a hostile country and the use of electronic controls and components for weapons, intelligence, and communication is accelerating. Currently, the only way to meet these rising electrical needs is to haul around generators, which could be eliminated by generating electricity directly from hybrid vehicles. Importantly, the military is also very interested in the stealth capability of hybrid powertrains. The battery pack can supply electricity to power on-board electronics for substantial periods, eliminating the noise and heat signature from a running engine. Military vehicles would also have the ability to travel limited distances on the electric motor alone, making it easier to move equipment without detection.

4.7 Urban Buses and Delivery Vehicles

Urban buses and delivery vehicles are ideal applications for hybrid vehicles. The electric motor is very good at launching a vehicle and accelerating at slow speeds, plus the frequent stops in urban operation offer much larger benefits from regenerative braking. In addition to larger fuel savings, brake maintenance is also reduced because the regenerative braking system helps to slow the vehicle down. Because brake repairs are the single largest maintenance item for urban buses, this is a significant cost saving.

As of early 2003, roughly 200 urban buses with hybrid electric powertrains had been placed in demonstration fleets. Results from the demonstration fleets have been positive, and Orion Bus Industries has a contract to deliver 325 production buses with a series hybrid system, diesel engine, and lead-acid battery pack to New York City Transit.

The first urban delivery hybrid application is FedEx's recent purchase from Eaton Corporation of 20 delivery trucks with parallel hybrid systems. The trucks use a smaller, four-cylinder diesel and fuel economy is expected to increase by 50%. Eaton is using lithium-ion batteries, which will be the first market experience with the durability of lithium-ion batteries in a hybrid vehicle. If these trucks meet project goals, the program has the potential to replace FedEx's 30,000 medium-duty trucks with hybrid vehicles over the next 10 years.

5. HYBRID ADVANTAGES AND DISADVANTAGES

5.1 Advantages

Hybrids have a number of positive features that are desired by customers, in addition to the obvious fuel savings:

- Hybrids use widely available fuel, thus there are no concerns about creating a new infrastructure to support fueling or recharging.
- The customer benefits from extended range and fewer trips to the gas station.
- The vehicle is quieter inside; both at idle when the engine is turned off and on the highway, where electric motor assist allows the engine to run at slower speeds.
- Hybrid systems have good synergy with other fuel-economy technologies. For example, not only could the hybrid system supply electrical power for fully adjustable electromechanical valves, but the motor and the valve timing could also be coordinated to optimize the benefits from the adjustable valves.
- Emissions can be reduced in a number of ways. Upstream emissions from fuel production and distribution are reduced in direct proportion to the reductions in fuel consumption. Increased driving range reduces evaporative emissions from refueling. Hybrid systems also reduce engine-out hydrocarbon and NO_x emissions. A conventional starter spins the engine up to only about 200–250 rpm and fuel must be combusted to propel the engine up to normal idle speeds. Combustion is very unstable at such low engine speeds and much of the fuel during the starting process goes out the tailpipe unburned, as hydrocarbon emissions. The greater power available from the hybrid electric motor can

quickly spin up the engine and fuel is not injected until after the engine has reached normal idle speed. This allows instantaneous combustion and reduced tailpipe hydrocarbon emissions. Hybrids also offer reductions in engine-out NO_x emissions. NO_x is formed when nitrogen and oxygen in the air are compressed at high temperatures. NO_x formation correlates closely with engine load, such that far more NO_x is generated during acceleration than it is during cruises. The hybrid system reduces the load on the engine during acceleration, with a corresponding reduction in NO_x formation. These reductions in engine-out emissions make it easier to achieve the ultralow levels that will be required starting in 2004.

- There is little impact on how the vehicle operates. The vehicle drives and operates similar to a conventional vehicle.
- Brake life is extended, because the electric motor captures some of the energy that is usually dissipated in the brakes.
- Part-time all-wheel drive can be added, if desired, by using a through-the-road hybrid system.
- If desired, the motor can be used to boost performance instead of reducing engine displacement.
- The motor can supply electricity for features desired by customers, such as off-board electricity generation, heated seats and windows, brake- and steer-by-wire systems, multimedia components, dynamic ride control, and radar and camera safety/warning systems. This feature is greatly underappreciated because electrical technology and accessories do not currently exist to take advantage of the high power from hybrid vehicles. A growth in customer demand for high-power features could be the "ace in the hole" for hybrid system popularity.

5.2 Potential Barriers

Traditionally, customer distrust of new technology has slowed market penetration. However, public reaction to introduction of hybrids in the United States has been surprisingly positive. Both the Insight and the Prius exceeded their original sales targets and public surveys indicate widespread knowledge and acceptance of hybrid technology. For example, a survey conduced by J. D. Power and Associates in 2002 found that 60% of new vehicle buyers would strongly consider a hybrid electric vehicle.

There are claims that development resources would be better spent on fuel cells. Certainly fuel cells are a promising long-term option. Hydrogen fuel cells have virtually no emissions and are extremely efficient. Large-scale production of hydrogen would use resources other than oil. However, fuel cells still have a lot of issues to resolve. Cost and size must be dramatically reduced, plus major breakthroughs are needed to enable practical on-board hydrogen storage. Durability must also be proved. Even after all these problems are solved, the problems and cost of creating an entirely new infrastructure must be addressed. Thus, fuel cell vehicles, at best, will need a long time for development and may yet turn out to be a dead end. Hybrids can help fill the gap until fuel cells and their related infrastructure are ready. In fact, hybrids might become the preferred technology until nonfossil sources of hydrogen are ready, especially because widespread adoption of hybrid vehicles will reduce the benefits of switching to fuel cell vehicles. Hybrids are also a useful hedge in case fuel cell development hits an insurmountable barrier. Finally, hybrid vehicles are a good path toward fuel cell vehicles. Many of the hybrid technologies can likely be incorporated into fuel cell systems.

Room must found for the electric motor, battery pack, and power electronics. This is not a problem on larger vehicles, but can pose challenges for smaller cars. A small reduction in interior space can be expected on smaller vehicles. The additional hybrid components also add some additional weight to the vehicle, which reduces performance. For example, the hybrid system adds about 100 lb to the Civic. However, this effect is relatively minor compared to the performance and efficiency boosts from the electric motor.

Many of the benefits of reducing fossil fuel use are shared by all of society, but the new vehicle purchaser must bear the burden of high first cost. Experience with vehicle emission standards has shown that, despite widespread support for pollution controls that raise the cost for everyone, few customers are willing to pay more individually for vehicles with lower emissions. New vehicle purchasers do respond rationally to the cost of putting fuel in their vehicles. This means that they severely discount expected lifetime fuel savings, both due to the time value of money and due to expected sale of the vehicle in a few years. It also means that most new vehicle purchasers are relatively indifferent to technologies that enhance fuel economy. The cost of driving a mile today is only about half what it was in the early 1970s before the first oil crisis, because fuel prices, adjusted for inflation, are the lowest in history and new vehicles are far more efficient than they were 30 years ago. Given the multitude of other factors affecting purchase decisions, it is no surprise that efficiency ranks very low in the car-buying process.

Increased costs due to long-term maintenance and repair of hybrid components are another potential barrier to acceptance of hybrid electric vehicles. Although the electric motor and power electronics are very reliable and durable, there is a potential concern with battery packs. Batteries in hybrid vehicles should be far more durable compared to the batteries in electric vehicles. This is because most battery deterioration occurs at the lowest 20% and highest 20% of the battery state-of-charge. These conditions are unavoidable with electric vehicles, which already suffer from limited range. However, the batteries on hybrid vehicles are carefully controlled to always stay in the middle 60% range, allowing much longer battery life. Still, hybrid batteries are not expected to last the life of the vehicle and are expensive to replace. NiMH batteries last a lot longer than do lead-acid batteries, but cost a lot more when they do finally need to be replaced.

The bottom line is that there is really only one factor holding back hybrid systems—cost, including battery replacement costs. Unfortunately, hybrid systems are not cheap. Manufacturers are understandably reluctant to discuss the cost of their hybrid systems, so it is difficult to determine a realistic cost. The Electric Power Research Institute (EPRI), in a study in 2002 on hybrid costs, estimated that the incremental lifetime cost for a high-volume, full-function hybrid system, including in-use battery replacement, would be about $2500 to $3600 for a compact car, $4000 to $5500 for a midsize SUV, and $4500 to $6300 for a large SUV. To put this estimated hybrid cost into context requires looking at what customers might be willing to pay in exchange for the fuel savings, both in the United States and worldwide. To do this necessitates making a few assumptions. The most critical is customer discounting of fuel savings.

In the March 2002 issue of *The Power Report*, J. D. Power reported some findings from their research on consumer interest in hybrid vehicles. One was that about one-third of potential hybrid buyers would still buy a hybrid even if the savings from reduced fuel costs during their ownership period would be less than the extra cost of purchasing the hybrid option. This is encouraging, in the sense that there are some customers who are willing to pay extra for hybrids. However, if this statement is turned around, it implies that two-thirds would not buy a hybrid vehicle if the extra cost of purchasing it were more than the fuel savings just during their ownership period. This is roughly equivalent to assuming that most customers value the fuel savings only for about the first 50,000 miles of driving.

The following formula, using miles per gallon (mpg), is used to calculate the fuel savings from hybrid vehicles:

$$\left[\frac{50,000 \text{ miles}}{\text{baseline mpg}}\right] - \left[\frac{50,000 \text{ miles}}{\text{baseline mpg} \times (1 + \% \text{ mpg increase})}\right] \times \text{fuel price}.$$

Estimates using this formula were made for three different vehicle sizes: small cars, midsize cars, and large trucks. Estimates were also made for two different fuel prices: $1.50 per gallon to represent the United States and $4.00 to represent Japan and many countries in Europe. A reasonable factor for the fuel economy improvement from just the hybrid system and corresponding engine size reduction is probably about 30–40% over EPA's combined cycles. Sensitivity cases of 20% (for very mild hybrids) and 80% (for advanced hybrid systems) are also shown in Table I. The results in Table I are supported by the J. D. Power survey on hybrids, which also found that consumers would be willing to pay about $1000 to $1500 more for a hybrid powertrain.

From a societal view, the fuel savings over the full life of the vehicle (which are about three times the values in Table I) are likely to justify the additional cost of hybrid systems. However, the typical new vehicle customer does not value the fuel savings enough to pay for the incremental cost. To address the cost differential, hybrid system costs need to come down or the price of gasoline must increase significantly.

Some customers who drive a lot or value the benefits and features more highly are willing to pay

TABLE I

Customer Value of Hybrid Fuel Savings[a]

Hybrid FE increase[b]	Fuel cost	Small car (34 mpg baseline)	Midsize car (23 mpg baseline)	Large truck (14 mpg baseline)
+20%	$1.50/gal	$368	$543	$893
	$4.00/gal	$980	$1449	$2381
+40%	$1.50/gal	$630	$932	$1531
	$4.00/gal	$1681	$2484	$4082
+80%	$1.50/gal	$980	$1449	$2381
	$4.00/gal	$2614	$3865	$6349

[a] Savings for the first 50,000 miles.
[b] FE, fuel economy.

the premium for a hybrid vehicle. Hybrid sales will increase as the costs come down. However, hybrids will probably not break into the mainstream market in the United States until the cost of hybrid systems comes down to less than $1500 and/or gasoline prices increase to over $3.00 per gallon. In the long term, increased demand for high-power features could increase the price customers are willing to pay for hybrid vehicles and make it easier to achieve the necessary cost reductions.

5.3 Practical Hybrid Designs

Because battery costs dominate customer acceptance of hybrid vehicles, it is not practical to design a hybrid system with a large battery pack for cars and light trucks. All light-duty hybrid vehicles will continue to be powered primarily by the internal combustion engine for the foreseeable future. Although manufacturers are introducing a wide variety of hybrid systems, they all share the basic design of a parallel system, with electric motors used primarily to assist the internal combustion engine and recapture regenerative braking energy. Some manufacturers choose to add motor-only operation at low-load, low-speed conditions and series hybrid operation under some conditions, but the efficiency gains are limited by the need to minimize the size and cost of the battery pack.

6. FUTURE TECHNOLOGY DEVELOPMENT

6.1 Electric Motor

The permanent magnet is one of the primary costs of a brushless DC electric motor. Because magnet production is a long-established and optimized technology, only modest cost reductions can be expected on magnets. Indeed, the electric motor in general is a mature technology. However, integration into a hybrid electric vehicle poses some unique packaging challenges, so the first hybrid electric vehicles are paying some additional premiums to cover development and tooling costs. These can be expected to come down in the future as production volumes increase. In the long run, switched reluctance motors may offer advantages over the brushless DC motor, because they provide good efficiency and reliability with reasonable cost and weight. They may offer a modest decrease in cost in the long term.

6.2 Energy Storage

Energy storage costs are the single largest long-term obstacle to widespread market acceptance of hybrid electric vehicles. Although lead-acid batteries are likely to be used initially in many hybrid vehicles due to their relatively low cost, their limited power density and short cycle life will force manufacturers to turn to advanced batteries in the future.

Nickel–metal hydride batteries are the near-term choice, due to their higher power density, longer cycle life, and better response to high-power pulses. Unfortunately, nickel–metal hydride batteries are expensive. Although some cost reduction is expected in the future, costs are likely to remain at high levels because of inherently high material costs. Nickel–metal hydride batteries also have high cooling requirements, due to high heat generation.

Lithium-ion batteries need improvements in cycle life and cost before they will be commercially viable for hybrid electric vehicles. However, lithium-ion battery use is rapidly growing in consumer markets, such as in laptops, and their cycle life and cost are expected to improve. Eventually, lithium-ion batteries should provide a lighter and slightly lower cost alternative to nickel–metal hydride, as well as better performance at low temperatures. Although development of lithium-ion batteries will improve the situation, their costs will likely still be a barrier to widespread acceptance of hybrid systems. A breakthrough in energy storage is needed if hybrid electric vehicles are to become universally accepted.

One possible long-term solution might be to combine a small battery pack with an ultracapacitor. Because of their relatively low power density, battery packs are sized to handle peak power demands during acceleration and regenerative braking. This leads to more energy storage than is necessary to support efficient hybrid operation. Ultracapacitors have very high power density, but can store only a small amount of energy. In theory, combining a battery pack with an ultracapacitor would allow the ultracapacitor to provide peak power for short acceleration and regenerative braking events, while the much smaller battery pack could recharge the ultracapacitor after acceleration and adsorb excess energy after regenerative braking. In essence, the ultracapacitor would act as a buffer for the battery pack, reducing the battery size requirements and energy storage system costs. This system is still in early stages of development and the savings in battery cost may prove to be less than the additional complexity and cost of the ultracapacitor.

Another possible technology is the solid-state battery, such as lithium-polymer batteries. Because they do not rely on liquids as the storage medium, solid-state batteries have the potential to be much smaller than conventional batteries. However, these batteries are still in their infancy and currently have lower energy density, shorter cycle life, and higher manufacturing cost compared to lithium-ion batteries.

6.3 Power Electronics

As with all electronics, advances in power electronics should continue steadily into the future, continuously improving performance and reducing cost. Higher temperature components are being developed that require less cooling, further improving performance and reducing cooling costs and packaging problems. The different components and modules are also being integrated to reduce size and complexity. Hybrid systems can also enable secondary improvements, by providing the electrical power needed to replace mechanical devices and pumps with superior electronic equivalents. This offers significant opportunities to reduce costs and provide features desired by customers.

7. FUTURE HYBRID MARKETS

Hybrid vehicles will continue to expand in niche applications. They offer a number of desirable features, in addition to fuel savings, and will appeal to a limited number of car and light-truck purchasers who value these features or the fuel savings enough to pay the increased price. Another niche market is through-the-road all-wheel-drive hybrid systems, which will appeal to some drivers in snowy climates. On-board electric generating will appeal to a small segment of contractors, farmers, and recreational users. Performance hybrids, in which the electric motor is used to boost acceleration instead of downsizing the engine, have not been tested in the market yet, but may appeal to another niche market of performance enthusiasts. A hybrid designed with both high-performance and part-time AWD might be especially appealing to SUV and luxury vehicle customers, because such a design could allow customers to have all the features they want, while still improving fuel economy by about 20%. Finally, although they do not represent huge markets, military, urban bus, and urban delivery applications are very promising and hybrids may come to dominate these applications in about 10 years.

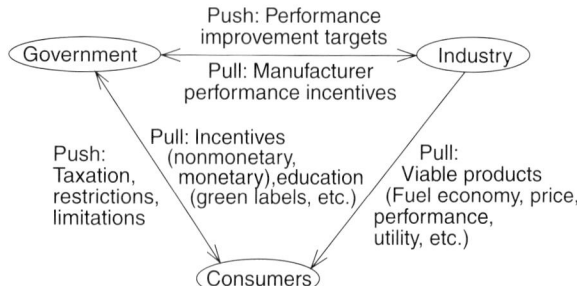

FIGURE 10 Schematic emphasizing the importance of interaction between consumers, government, and industry.

Hybrid system costs will come down as the market expands. The lower costs will, in turn, draw in more customers. J. D. Power predicts that hybrid vehicles will account for about 1%t of the market by 2005, 3% by 2009, and nearly 5% by 2013. However, in the long term, it is not clear if hybrid vehicles will stagnate at 5–10% of the market or if costs will eventually drop enough for hybrid vehicles to break into the mass market. An unknown factor is how much customer demand will increase for high-power features in the long term. Strong demand for high-power features would greatly help market acceptance of hybrid vehicles.

The industry can provide a market "pull" by providing products desired by the consumer. But customers cannot be "pushed" into buying vehicles they do not want. Government programs to stimulate demand, provide incentives, and educate the customer could dramatically affect acceptance of hybrids and market penetration (Fig. 10).

SEE ALSO THE FOLLOWING ARTICLES

Alternative Transportation Fuels: Contemporary Case Studies • *Batteries, Transportation Applications* • *Biodiesel Fuels* • *Biomass for Renewable Energy and Fuels* • *Energy Efficiency, Taxonomic Overview* • *Ethanol Fuel* • *Fuel Cell Vehicles* • *Fuel Cycle Analysis of Conventional and Alternative Fuel Vehicles* • *Fuel Economy Initiatives: International Comparisons* • *Internal Combustion (Gasoline and Diesel) Engines* • *Transportation Fuel Alternatives for Highway Vehicles*

Further Reading

An, F., Stodolsky, F., and Santini, D. (1999). "Hybrid Options for Light-Duty Vehicles, SAE Paper No 1999-01-2929." Society of Automotive Engineers, Warrendale, Pennsylvania.

Aoki, K., Kenji, N., Kajiwara, S., Sato, H., Yamamoto, Y. (2000). "Development of Integrated Motor Assist Hybrid System," SAE Paper 2000-01-2059. Society of Automotive Engineers, Warrendale, Pennsylvania.

DeCicco, J. M. (2000). Hybrid Vehicles in Perspective: Opportunities, Obstacles, and Outlook. American Council for an Energy Efficient Economy, Washington, D.C.

Electric Power Research Institute. (2002). "Comparing the Benefits and Impacts of Hybrid Electric Vehicle Options for Compact Sedan and Sport Utility Vehicles," 1006892. EPRI, Palo Alto, California.

German, J. M. (2001). Hybrid vehicles go to market. *TR News* **213**(March–April), 15–23.

Itazaki, H. (1999). "The Prius that Shook the World." Nikkan Kogyo Shimbun, Ltd., Tokyo, Japan.

Wakefield, E. H. (1998). "History of the Electric Automobile Hybrid Electric Vehicles." Society of Automotive Engineers, Warrendale, Pennsylvania.

Hybrid Energy Systems

J. F. MANWELL
University of Massachusetts
Amherst, Massachusetts, United States

1. Introduction to Hybrid Energy Systems
2. Characteristics of Hybrid Energy Systems
3. Technology Used in Hybrid Energy Systems
4. Energy Loads
5. Renewable Energy Resource Characteristics
6. Design Considerations
7. Economics
8. Trends in Hybrid Energy Systems

Glossary

biomass Any organic material that is used as a fuel.
deferrable load An electrical load that must be supplied a certain amount of energy over a specified time period in an isolated electrical network, but that has some flexibility as to exactly when.
diesel generator A device for producing electricity through the use of a diesel engine.
dump load A device used to dissipate power in order to maintain stability in an isolated hybrid energy system.
fuel cell A device for continuously producing electricity through a reaction between hydrogen and oxygen.
hybrid energy system A combination of two or more energy conversion devices (e.g., electricity generators or storage devices), or two or more fuels for the same device, that, when integrated, overcome limitations that may be inherent in either.
load (1) An energy consuming device; (2) the cumulative power requirement due to a number of devices connected to an electrical network.
maximum power point tracker A DC-DC converter whose function is to maximize the power produced by a variable voltage DC generator, such as PV panels or a wind turbine, connected to a fixed voltage bus.
optional load An electrical load which can be supplied by excess power in an isolated electrical network but need not be supplied if no excess power is available.
penetration The instantaneous power from the renewable generator divided by the total electrical load being served in an isolated electrical network.
photovoltaic panel (PV) A semiconductor-based device that converts sunlight to electricity.
renewable energy source An energy source that is derived ultimately from the sun, moon, or earth's internal heat.
supervisory control A central control system whose function is to ensure proper operation of all the devices within a hybrid energy system.
weak grid An electrical network whose operational characteristics can be affected by the presence of a generator or load connected to it.
wind turbine A device that converts the power in the wind to electrical or mechanical power.

The term hybrid energy system refers to those applications in which multiple energy conversion devices are used together to supply an energy requirement. These systems are often used in isolated applications and normally include at least one renewable energy source in the configuration. Hybrid energy systems are used an alternative to more conventional systems, which typically are based on a single fossil fuel source. Hybrid energy systems may also be used as part of distributed generation application in conventional electricity grid. The most general definition is the following: "Hybrid energy systems are combinations of two or more energy conversion devices (e.g., electricity generators or storage devices), or two or more fuels for the same device, that when integrated, overcome limitations that may be inherent in either." This definition is useful because it includes a wide range of possibilities and the essential feature of the multiplicity of energy conversion. This article focuses on stationary power systems, where at least one of the energy conversion devices is powered by a renewable energy source (which, in the context of this article, is one based ultimately on the sun).

1. INTRODUCTION TO HYBRID ENERGY SYSTEMS

A considerable interest has emerged in combined or 'hybrid' energy systems. In the context used here, that refers to an application in which multiple energy conversion devices are used together to supply an energy requirement. These systems are often used in isolated applications and normally include at least one renewable energy source in the configuration. Hybrid systems are used as an alternative to more conventional systems, which typically are based on a single fossil fuel source.

1.1 Definitions

Hybrid energy systems have been defined in a number of ways. The most general, and probably most useful, is the following:

"Hybrid energy systems are combinations of two or more energy conversion devices (e.g., electricity generators or storage devices), or two or more fuels for the same device, that when integrated, overcome limitations that may be inherent in either."

This definition is useful because it includes a wide range of possibilities and the essential feature of the multiplicity of energy conversion. Note that this broad definition does not necessarily include a renewable energy based device and allows for transportation energy systems. In the present discussion, however, we focus on stationary power systems, where at least one of the energy conversion devices is powered by a renewable energy source (which is one based ultimately on the sun, moon, or earth's internal heat).

For the purpose of comparison, it is useful to consider briefly the nature of conventional energy systems that are normally used where hybrid system might be used instead. There are basically three types of conventional systems of interest: (1) large utility networks, (2) isolated networks, and (3) small electrical load with dedicated generator.

Large utility networks consist of power plants, transmission lines, distribution lines and electrical consumers (loads). These networks are based on alternating current (AC) with constant frequency. Such networks are frequently assumed to have an infinite bus. This means that the voltage and frequency are unaffected by the presence of additional generators or loads.

Isolated electrical networks are found on many islands or other remote locations. They are similar in many ways to large networks, but they are normally supplied by one or more diesel generators. Generally, they do not have a transmission system distinct from the distribution system. Isolated networks do not behave as an infinite bus and may be affected by additional generators or loads.

For many small applications, it is common to supply an electrical load with a dedicated generator. This is the case, for example, at construction sites, highway signs, and vacation cabins. These systems are also normally AC but have no distribution system.

1.2 Applications for Hybrid Energy Systems

There are numerous possible applications for hybrid power systems. The most common examples are (1) remote AC network, (2) distributed generation applications in a conventional utility network, and (3) isolated or special purpose electrical loads.

The classic example of the hybrid energy system is the remote, diesel-powered AC network. The basic goal is to decrease the amount of fuel consumed by diesel generators and to decrease the number of hours that they operate. The first addition to "hybridize" the system is to add another type of generator, normally using a renewable source. These renewable generators are most commonly wind turbines or photovoltaic panels. Experience has shown, however, that simply adding another generator is not sufficient to produce the desired results. Accordingly, most hybrid systems also include one or more of the following: supervisory control system, short-term energy storage, and load management. Each of these will be described in more detail. An example of a typical hybrid energy system, in this case a wind/diesel system, is illustrated in Fig. 1.

During the 1990s, management of many large electrical networks changed so that it is now possible

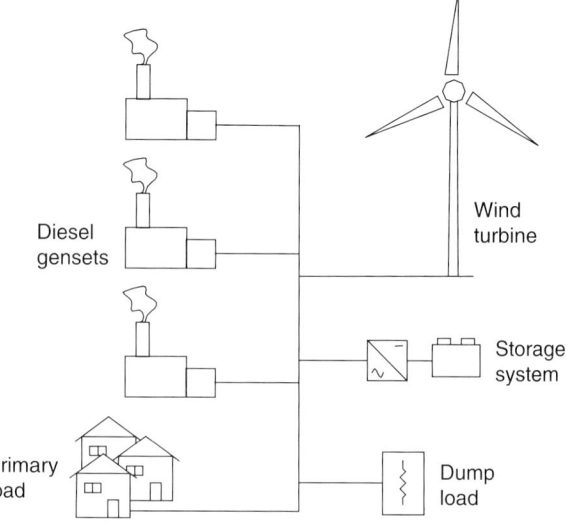

FIGURE 1 Schematic of wind diesel system with storage.

for individuals or businesses to add generation in the distribution system of the utility. This is known as distributed generation (DG). Distributed generation can have a variety of purposes, but in any case, it is sometimes desirable to combine a number of energy conversion devices together with the distributed generator. This results in a hybrid energy system in the distributed generation application.

Hybrid systems can be of particular value in conjunction with an isolated or special-purpose application. One example is that of using photovoltaic panels, together with some battery storage and power electronic converters to supply a small amount of energy to a load in a remote location. Other examples include water pumping or water desalination. In these applications, electrical loads may take a range of forms. They may be conventional AC, DC, or even variable voltage and variable frequency.

1.3 Impetus for Hybrid Energy Systems

There are a variety of reasons why a hybrid system might be used. An important reason is the reduction of fossil fuel use. Fossil fuel can be costly, especially in remote locations where the cost of transporting the fuel to the site must be considered. In many remote locations fuel oil must be stored, often for an entire winter. Reducing the amount of fuel use can reduce storage costs.

An isolated hybrid power system can be an alternative to power line construction or power line upgrade. Hybrid power systems, because they include so many components, tend to be relatively expensive. When they can be used as an alternative to power line construction, the savings can help compensate for the cost of the hybrid system.

Under some conditions a power line already exists, but is not capable of carrying the desired current. This could occur, for example, when a new load is added. In this case, the hybrid system would be a type of distributed generation. The hybrid system would only need to provide the difference between the total load and what the existing lines could carry. The presence of the existing line could simplify the design of the hybrid system, in that it need not necessarily be required to set the power line frequency and might not have either storage or dump loads (q.v.).

Distributed generation has advantages in a number of situations, and, in some cases, the overall benefit is enhanced by using a hybrid system. One common application for distributed generation is combined heat and power (CHP). In this case, waste heat associated with fossil fuel combustion is used for space or process heating, thereby increasing the overall energetic efficiency of the fuel use.

Use of hybrid systems can result in local environmental benefits. In particular, diesel generators (in isolated applications) should run less often and for shorter periods. This will reduce locally produced air pollution and noise, as well as reducing risks associated with fuel transport.

2. CHARACTERISTICS OF HYBRID ENERGY SYSTEMS

The characteristics and components of a hybrid system depend greatly on the application. The most important consideration is whether the system is isolated or connected to a central utility grid.

2.1 Central Grid Connected Hybrid Systems

If the hybrid system is connected to a central utility grid, as in a DG application, then the design is simplified to a certain degree and the number of components may be reduced. This is because the voltage and frequency are set by the utility system and need not be controlled by the hybrid system. In addition, the grid normally provides the reactive power. When more energy is required than supplied by the hybrid system the deficit can be in general be provided by the utility. Similarly, any excess produced by the hybrid system can be absorbed by the utility. In some cases, the grid does not act as an infinite bus, however. It is then said to be "weak." Additional components and control may need to be added. The grid connected hybrid system will then come to more closely resemble an isolated one.

2.2 Isolated Grid Hybrid Systems

Isolated grid hybrid systems differ in many ways from most of those connected to a central grid. First, they must be able to provide for all the energy that is required at any time on the grid or find a graceful way to shed load when they cannot. They must be able to set the grid frequency and control the voltage. The latter requirement implies that they must be able to provide reactive power as needed. Under certain conditions, renewable generators may produce energy in excess of what is needed. This energy must be dissipated in some way so as not to introduce instabilities into the system.

There are basically two types of isolated grid hybrid systems which include a renewable energy generator among their components. These are known as low penetration or high penetration. In this context, "penetration" is defined as the instantaneous power from the renewable generator divided by the total electrical load being served. Low penetration, which is on the order of 20% or less, signifies that the impact of the renewable generator on the grid is minor, and little or no special equipment or control is required. High penetration, which is typically over 50% and may exceed 100%, signifies that the impact of the renewable generator on the grid is significant and special equipment or control is almost certainly required. High-penetration systems may incorporate supervisory control, so-called dump loads, short-term storage, and load management systems.

Two important considerations in an isolated system are whether the system can at times run totally on the renewable source (without any diesel generator on) and whether the renewable source can run in parallel with (i.e., at the same time as) the diesel generator. It is most common for one or the other to be possible (and normal). It is less common that both modes of operation are possible. This latter system offers the greatest fuel savings but is more complicated.

2.3 Isolated or Special Purpose Hybrid Systems

Some hybrid systems are used for a dedicated purpose, without use of real distribution network. These special purposes could include water pumping, aerating, heating, desalination, or running grinders or other machinery. Design of these systems is usually such that system frequency and voltage control are not major issues, nor is excess power production. In those cases where energy may be required even when renewable source be temporarily unavailable, a more conventional generator may be provided. Renewable generators in small isolated systems typically do not run in parallel with a fossil fuel generator.

3. TECHNOLOGY USED IN HYBRID ENERGY SYSTEMS

A wide range of technology may be used in a hybrid energy system. This section describes some of them in more detail than was done in previous sections. Devices to be discussed include energy consuming devices (loads), rotating electrical machinery, renewable energy converters, fossil fuel generators (often, but not always, diesels), energy storage devices, power converters, control systems, and load management devices. Some of the various possible devices and arrangements that may be found in hybrid energy system are illustrated in Fig. 2. The primary focus here is on isolated network hybrid systems, but much of the technology applies to other types of hybrid systems as well.

3.1 Energy Consuming Devices

Hybrid energy systems typically use the same types of energy consuming devices that are found in conventional systems. These include lights, heaters, motors, and electronic devices. The combined energy requirement of all the devices is know as the total load, or just load. The load will typically vary significantly over the day and over the year. An example of a load varying over a year on an isolated island is shown in Fig. 3.

3.2 Rotating Electrical Machinery

Rotating electrical machinery is found in many places in a hybrid energy system. Most such machines can function as either motors or generators, depending on the application. This section focuses on the generating function.

3.2.1 Induction Generators
The induction generator has been the most common type of generator used in wind turbines. They are also occasionally used with other prime movers, such as hydro turbines or landfill gas fueled internal combustion engines. More information on induction machines is provided elsewhere in this encyclopedia, so they will not be discussed in detail here. As far as hybrid systems, however, there are two important considerations. First of all, induction machines require a significant amount of reactive power. This is not a fatal problem, but it does affect the design of the system in a number of ways.

The second consideration is starting. When an induction machine is brought on line from stand still, it requires much higher current than when operating normally. Provision must be made to ensure that the hybrid system has the capability of starting any induction motors or generators that it connected to it.

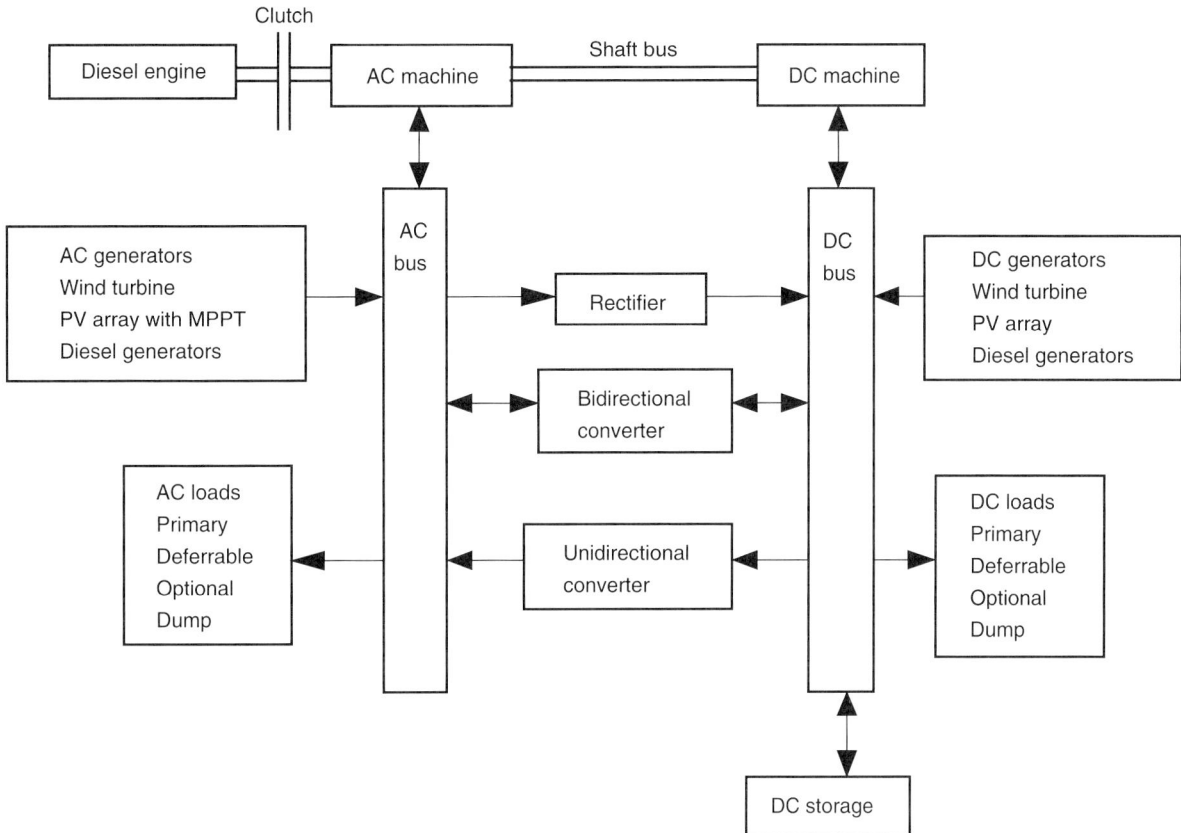

FIGURE 2 Devices and arrangements in hybrid energy systems.

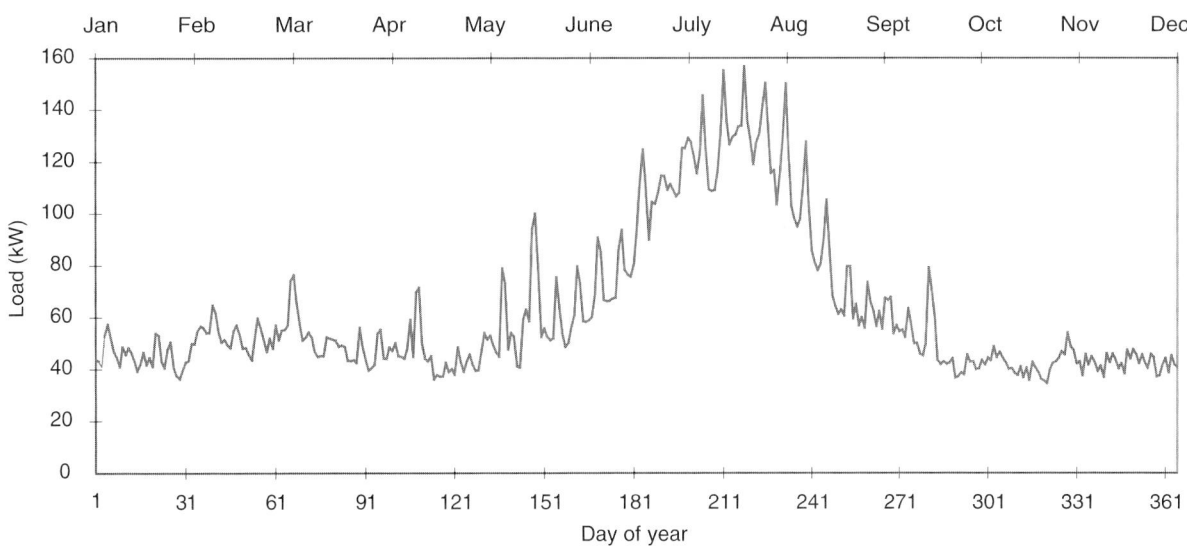

FIGURE 3 Daily electrical loads on Cuttyhunk Island, Massachusetts.

3.2.2 Synchronous Generators

Synchronous generators (SG) may also be used with a variety of prime movers. These generators are discussed in detail elsewhere in this encyclopedia. In hybrid system applications SGs are of one of two types: those with electromagnetic fields and those using permanent magnets. The first type, operating with a voltage regulator, can maintain the voltage on the electrical network and provide reactive power required by other devices in the system. Permanent magnet SGs are often used in small wind turbines. Such generators cannot maintain voltage and so are normally used in conjunction with power electronic converters.

A synchronous machine can be brought on line so that it will run in parallel with other generators. When this is done, particular attention must be taken so that the various generators are in phase with each other.

3.3 Renewable Energy Generators

Renewable energy generators are devices that convert energy from its original form in the renewable energy source into electricity. Renewable energy generators that are most likely to be found in hybrid energy systems include wind turbines and photovoltaic panels. Some hybrid energy systems use hydroelectric generators, biomass fueled generators, or fuel cells. It should be noted that many renewable energy generators include rotating electrical machines acting in the generating mode, which is also called a generator. It should be clear from the context what is meant.

3.3.1 Wind Turbines

Wind turbines are devices that convert the energy in wind into electricity. A typical wind turbine is shown in Fig. 4. The main parts of a wind turbine are the rotor, the drive train (including the generator), main frame, tower, foundation, and control system. The rotor consists of the blades and a hub. The blades serve to convert the force of the wind to a torque that ultimately drives the generator.

The two most important features of a wind turbine as far as a hybrid energy system is concerned are the type of generator and the nature of the rotor control. Most wind turbines use induction generators, although some use synchronous generators. In either case the generator may be connected directly to the electrical network or it may be connected indirectly through a power electronic converter.

There are two main forms of rotor control on wind turbines: stall control and pitch control. The most important function of rotor control is to protect the wind turbine from high winds. In hybrid systems, rotor control is also important because it affects on how energy flows are regulated within the hybrid system. Under some conditions (with pitch control), the rotor can be controlled to facilitate start up or to reduce production when full output is not required.

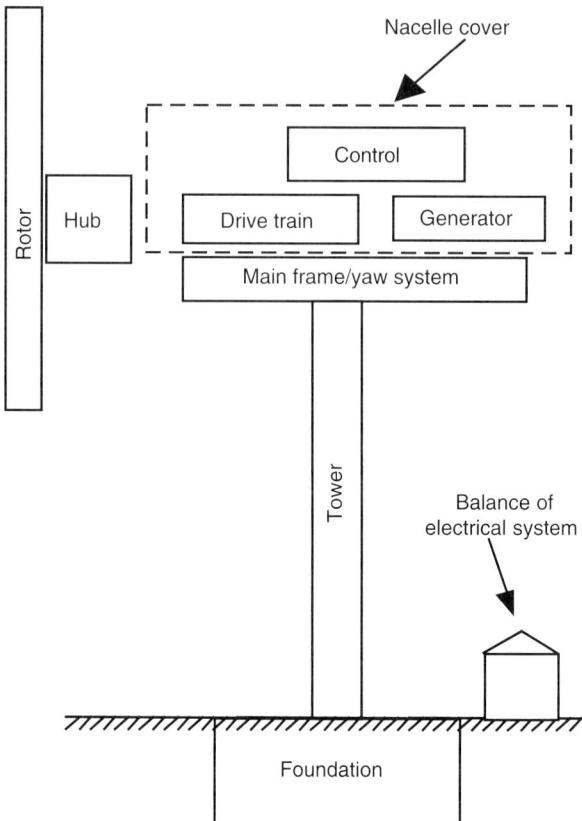

FIGURE 4 Components of typical wind turbine.

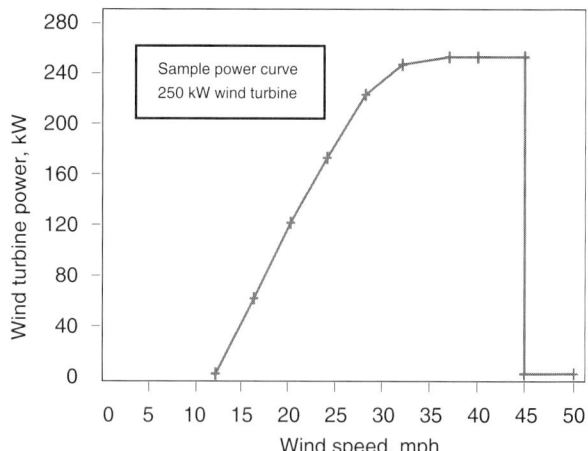

FIGURE 5 Typical wind turbine power curve.

Details on rotor control may be found elsewhere in this encyclopedia or in Manwell *et al*.

The power generated by a wind turbine varies in relation to the wind speed. That relation is summarized in a power curve, a typical example of which is illustrated in Fig. 5.

3.3.2 Photovoltaic Panels

Photovoltaic (PV) panels are used to produce electricity directly from sunlight. PV panels consist of a number of individual cells connected together to produce electricity of a desired voltage. Photovoltaic panels are inherently DC devices. To produce AC, they must be used together with an inverter.

Most PV cells are made from crystalline silicon. PV cells produce current in proportion to the solar radiation level (up to a certain voltage). The current/voltage relation of a typical silicon cell at a fixed level of solar radiation is shown in Fig. 6. Since power is proportional to the product of current and voltage, the power from a PV cell will continue to increase until the current begins to drop.

Because the maximum voltage from individual cells is less than 1 V, multiple cells are connected together in series on a PV panel.

The actual radiation level at any given time at a particular spot on the earth's surface will vary significantly over the year and over the day. See Section 5 for more details. More information on PV panels themselves is provided elsewhere in this encyclopedia.

3.3.3 Hydro Turbines

Hydropower is one of the oldest forms of electricity production and is used in large electricity networks as well as small ones. Small hydroelectric plants are used in isolated systems in many parts of the world. Hybrid systems, which include hydroelectric plants with other forms of generation, are relatively rare, but they do occur. The most common of these are in locations where the resource varies significantly over the year, and there is not enough water in certain seasons. A diesel generator is then used instead of the hydroelectric generator during those seasons. The primary requirements of a hydroelectric plant are a continuous source of water and a change in elevation. More information on hydroelectric generators is provided elsewhere in this encyclopedia.

3.3.4 Biomass Fueled Generators

Biomass fueled generators are occasionally used in hybrid energy, though most often when they are employed in isolated networks they are the only power plants used. Biomass fueled generators are quite similar to conventional coal or oil fired generating plants, except for the combustors themselves and the fuel handling equipment.

Biomass is any organic material which is used as a fuel. For purposes of power generation, the most common sources of biomass include wastes from wood products or the sugar cane industry. Other agricultural wastes are sometimes used as well.

Biomass generators are sometimes fueled with landfill gas. Such generators most commonly employ an internal combustion engine as a prime mover. Occasionally fuel cells have been used.

3.3.5 Fuel Cells

Fuel cells can be used in hybrid power systems, and are expected to become more common as their costs drop. Fuel cells ultimately run on hydrogen, but in many cases the input fuel is some other gas, such as natural gas or landfill gas.

Fuel cells can be thought of as a continuous battery. The fundamental reaction is between hydrogen and oxygen, the product of which is water and electric current. The reaction takes place in the vicinity of a membrane, which serves to separate the various components of the electrolyte and the electrodes. The electrodes are the terminals of the fuel cell and carry the current into an external circuit. Like batteries, fuel cells are inherently DC devices. Fuel cells can be used in an AC network if their output is converted to AC via an inverter.

When natural gas or landfill gas is used as the fuel, it must first be used to produce hydrogen. This is done in a reformer.

There are a number of different designs for fuel cells. They include proton exchange membrane (PEM), solid oxide, and molten carbonate. More information on fuel cells is presented elsewhere in this encyclopedia.

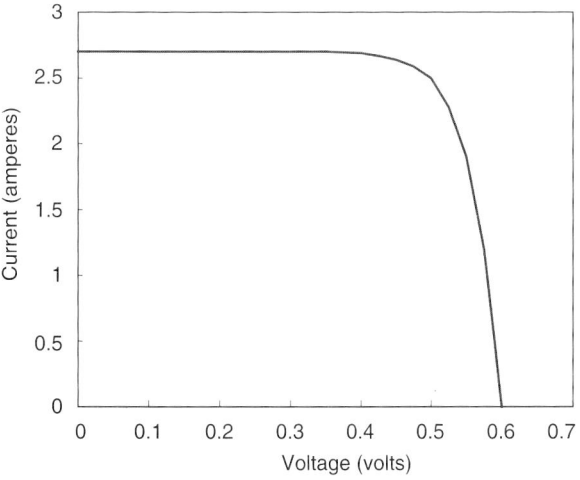

FIGURE 6 Photovoltaic cell current versus voltage.

3.4 Fossil Fuel Generators

Fossil fuel generators are commonly used in hybrid energy systems. In fact, most isolated power systems at the present time are based on fossil fuel, using internal combustion engines as prime movers. Most medium-sized and larger isolated systems use diesel engine/generators. The smallest systems sometimes use gasoline. Some very large isolated power systems sometimes use conventional oil-fired steam power plants. They will not be discussed here, however.

3.4.1 Diesel Engine/Generators

Diesel engine generators are discussed elsewhere in this encyclopedia, so they will only be summarized here. Emphasis is given to those aspects of relevance to hybrid energy systems.

Diesel generators typically consist of three main functional units: a diesel engine, a synchronous generator with voltage regulator, and a governor.

The fuel injection system is an important part of the diesel engine. Its function is to inject the proper amount of fuel into the proper cylinder at the appropriate time in the cycle. The timing of the injection is determined by the design of the engine itself; the amount of fuel is determined by the governor.

The diesel engine is normally connected directly to a synchronous generator. A voltage regulator ensures the proper voltage is produced. The frequency of the AC power is directly proportional to the engine speed, which in turn is controlled by the governor.

Historically diesel generators have employed "droop" type governors. A set of flyballs is driven by the engine, so the speed varies in proportion to the engine speed. Through a set of linkages, the amount of fuel that can be injected at any time is varied according to how far the operating speed differs from (or "droops") from nominal. As the electrical load on the generator increases, the droop increases and more fuel is injected. Diesel generators are now more likely to have electronic governors without droop, but the function is the same.

An important consideration regarding diesel generators is their fuel consumption, both at full load and part load. Diesel fuel consumption is frequently described in terms of electricity produced per gallon of fuel consumed. Full load values ranging from a low of 8 kWh/gallon to 14 kWh/gallon have been reported.

Diesel engine generators are often called on to follow the load. That means that their output must be equal to the system load (or to the system load less

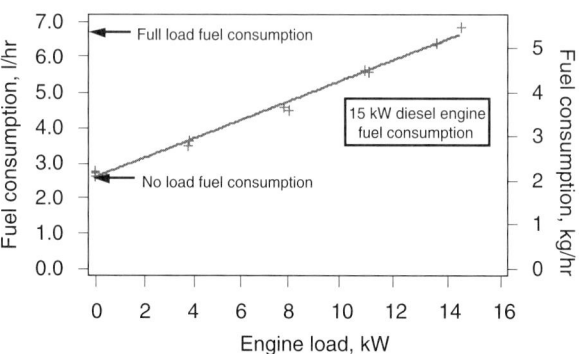

FIGURE 7 Typical diesel engine generator fuel curve.

the production of any other generators that might be on; this is called the net load). As the load may go up and down, so must the electricity generated. This is known as part load operation. Generally, the conversion efficiency is less at part load than at full load. Fuel consumption over the full range of operation is summarized in fuel curves. In these curves, fuel consumption is graphed against power. Figure 7 shows a typical example.

Regardless of efficiency considerations, manufacturers normally recommend that diesel generators not be run below some specified minimum power level, known as the minimum load. Typically, the minimum recommended load is between 25% and 50% of rated. Engines run for long periods at levels below the minimum recommended can experience a number of problems.

3.4.2 Gasoline Generators

Gasoline generators are sometimes used for very small hybrid energy systems applications. These generators have the advantage that they are readily available throughout the world and are relatively expensive. They are similar in many ways to diesel generators, except that use spark ignition, and have lower compression ratios. They also typically use carburetors rather than fuel injection. The main disadvantage of gasoline generators is that they are less efficient than diesel.

3.5 Energy Storage

Energy storage is often useful in hybrid energy systems. Energy storage can have two main functions. First of all, it can be used to adapt to a mismatch between the electrical load and the renewable energy resource. Second, it can be used to facilitate the control and operation of the overall system. There are basically two types of energy storage, convertible and end use. Convertible storage is that which can readily be converted back to electricity. End-use storage can be applied to a particular end-use requirement but may not readily be converted back to electricity.

3.5.1 Convertible Storage

There are a number of convertible storage media, although only a few of them have been used frequently in hybrid energy systems. The most commonly used form of convertible storage is the battery. Less commonly used, but frequently discussed, forms include pumped hydroelectric, flywheels, compressed air, and hydrogen.

3.5.1.1 Batteries Batteries are the most commonly used form of convertible storage for hybrid energy systems. They have been used both for short term (less than 1 hour) and long term (more than 1 day) storage. A number of types of batteries have been developed. The most common type of storage battery for hybrid applications is the lead acid battery. Nickel cadmium has also been used occasionally. Batteries are discussed elsewhere in this encyclopedia, so only those aspects most relevant to hybrid energy systems are summarized here.

As far as hybrid energy systems are concerned, there are five important performance characteristics of batteries: (1) voltage, (2) energy storage capacity, (3) charge/discharge rates, (4) efficiency, and (5) battery lifetime.

Batteries by their nature are DC. Individual batteries are made up by a number of cells in series, with each cell nominally two volts. Complete batteries are typically 2, 6, 12, or 24 Volts. The actual terminal voltage will depend on three factors: (1) state of charge, (2) whether the battery is being charged or discharged, and (3) the rate of charge or discharge.

The energy storage capacity is primarily a function of battery voltage and the amount of charge it can hold and then return. Charge is measured in units of current times time (Ampere-hours). The amount of charge that is stored in a battery at any particular time is often described with reference to its full state by the term "state of charge" (SOC). Discharging and charging back to a given level (normally fully charged) is referred to a cycle. The total amount of charge that a battery can hold is primarily a function of the amount of material used in the construction.

Battery capacity is normally specified with reference to a specific discharge rate. This is because the apparent capacity of batteries actually differs with charge and discharge rate. Higher rates result in smaller apparent capacities.

As energy storage media, batteries are not 100% efficient. That is, more energy is expended in charging than can be recovered. Overall efficiencies are typically in range of 50 to 80%.

An important characteristic of batteries is their useful lifetime. Experience has shown that the process of using batteries actually decreases their storage capacity until eventually the battery is no longer useful. The important factors in battery life are the number of cycles and the depth of discharge in the cycles. Depending on the type of battery, the number of deep cycles to which a battery can be subjected ranges from a few thousand down to hundreds or even tens of cycles. The cycle life of a typical battery is illustrated in Fig. 8.

3.5.1.2 Pumped Hydro One form of convertible storage that has been applied in some hybrid energy systems is pumped storage. In this case, water is pumped from one reservoir at a low elevation up to one at a higher elevation. The amount of energy that can be stored is a function of the size of the reservoir and the difference in elevation. The overall efficiency of the storage is a function of the efficiency of the pumps and turbines (which may be the same devices) and the hydraulic losses in the pipes connecting the two reservoirs. The use of pumped storage is limited by the lack of sites where such facilities can be installed at a reasonable price.

3.5.1.3 Flywheels Flywheels can be used to store energy in a hybrid system. A flywheel energy storage system consists of the following components: (1) the flywheel itself, (2) an enclosure, usually evacuated to minimize frictional losses, (3) a variable speed motor/generate to accelerate and decelerate the

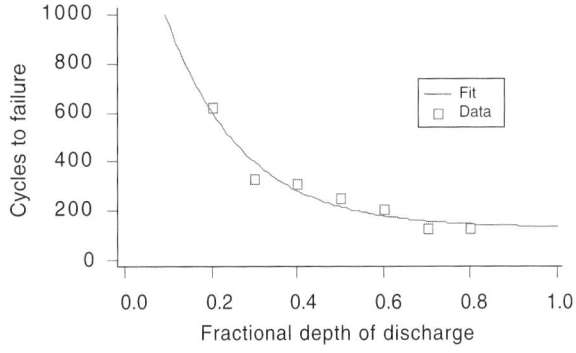

FIGURE 8 Battery cycle life.

wheel, and (4) a power electronic converter. Energy is absorbed when the wheel is accelerated up to its maximum speed and is released when the wheel is decelerated to some lower speed. A power electronic converter accompanies the flywheel and motor/generator because the input and output power to and from the motor/generator is typically variable voltage and variable frequency AC. Flywheels typically store relatively small amounts of energy, but they can absorb or release the energy at high rates. Thus in hybrid power systems they can be used to smooth short term fluctuations in power (on the order of seconds or minutes) and they can facilitate system control.

3.5.1.4 Compressed Air Compressed air can be used for storage in hybrid systems, and some experimental prototypes have been built. Efficiency is relatively low, however, and this method of storage has not been widely used.

3.5.1.5 Hydrogen Hydrogen can be produced by the electrolysis of water, using a renewable energy source for the electricity. The hydrogen can be stored indefinitely and then used in an internal combustion engine/generator or a fuel cell to generate electricity again. This method of storage appears to have a lot of potential, but it is still quite expensive.

3.5.2 End-Use Storage
End-use storage, as opposed to convertible storage, refers to the situation in which some product is created through the use of electricity when it is available. The product is then stored and made available when it is needed. Load management schemes often incorporate some end-use storage. Three examples are given next.

3.5.2.1 Thermal Energy A common form of end-use storage is thermal energy, most often in the form of hot water. Hot water can be used for space heating applications, domestic hot water, swimming pools, and so on. Hot water can be stored relatively inexpensively in insulated water tanks. Depending on the location of the water tanks relative to the end use, the efficiency of the storage can be quite high.

3.5.2.2 Pumped Water Pumped water is another form of end-use storage. This form of storage has a long history of use in conjunction with windmills. In this application, water for any plausible purpose is pumped into a reservoir or storage tank, usually elevated, from which it can be released as needed. When the storage is elevated, then the water can flow by gravity to the point of use, so no further input of externally produced energy is required.

3.5.2.3 Pure Water Another, less common form of end-use storage is the production and then storage of pure water from salty or brackish water. Depending on the quality of water at the input of the process, ultrafiltration, reverse osmosis, or vapor compression evaporation may be used to produce the pure water. All of these are energy intensive, so the implicit energy density of the storage is high.

3.6 Power Converters

For any hybrid energy system to function properly, it is common that one or more power converter be incorporated into the system. These are either electromechanical or electronic devices.

3.6.1 Electromechanical Power Converters
There are at least two types of electromechanical power converters that have been used in hybrid energy systems. They include the rotary converter and the synchronous condenser.

3.6.1.1 Rotary Converter A rotary converter is an electromechanical device that converts AC to DC or vice versa. When it is converting AC to DC it is a rectifier. When operating the other way it is an inverter. The rotary converter consists of two rotating electrical machines that are directly connected together. One of them is a DC machine; the other is an AC machine. Either of them can run as a motor or generator depending on the intended direction of power flow. The AC machine can be either an induction machine or a synchronous machine. Which is used will depend on the requirements of the system.

Rotary converters have the advantage that they are a rugged and well-understood technology that has been around for many years. Their disadvantage is that their costs are high and their efficiencies are lower than are electronic devices that can serve the same purpose.

3.6.1.2 Synchronous Condenser Synchronous condenser is the name given to a synchronous machine that is connected into an electrical network to help in maintaining the system voltage. The synchronous machine in this case is essentially a motor to which no load is connected. A voltage regulator is connected to the synchronous machine and functions in the same way as described previously. Synchronous condensers are used in hybrid energy systems when there are, for at least some of the time, no other synchronous machines connected. This is often the case in those systems that include diesel generators but that are intended to allow all the diesels generators to be turned off under some circumstances. For example, a synchronous condenser could be used in a wind/diesel system in which the wind turbines used induction generators and could at times supply all of the electrical load. The diesel could then be turned off, but the synchronous machine would be needed to supply the required reactive power to the induction generators and maintain system voltage in the process.

3.6.2 Electronic Power Converters

A wide range of electronic devices has been developed and adapted for use in hybrid energy systems. Many of them serve similar functions to the electromechanical devices described previously, but have a number of advantages, such as lower cost, higher efficiencies, and greater controllability. The devices of greatest interest include rectifiers, inverters, dump loads, and maximum power point trackers.

3.6.2.1 Rectifiers A rectifier is a device that converts AC to DC. The primary functional elements are diodes, which only let current pass one way. By suitable layout of the diodes AC in a single or three-phase circuit is converted to a rippling, but single direction, current. Capacitors may be used to smooth the resulting current.

3.6.2.2 Inverter An inverter is a device that converts DC to AC. The primary switching elements are either silicon controlled rectifiers (SCRs) or power transistors (IGBTs). They are arranged in a bridge circuit and switched on (and off, in the case of transistors) in such a way that an oscillating waveform results. Some inverters operate in conjunction with other devices that set the system frequency. These are referred to as line commutated. Other inverters have the capability to set frequency themselves. These are called self-commutated.

3.6.3 Maximum Power Point Trackers

Another electronic device that may be used in hybrid energy systems, particularly ones with photovoltaic panels is the maximum power tracker. This device is DC-DC converter than can be partially thought of as a DC transformer. Its function to provide a particular desired output voltage to the rest of the system, while adjusting the voltage at the input to allow the maximum power production by the generator that is connected to it. In the case of photovoltaic panels, the voltage at the input will be the maximum power point voltage corresponding to the incident solar radiation level. Maximum power point trackers have also been used with small wind turbines.

3.7 Dump Loads

A dump load is device that is used to dissipate power in order to maintain stability in an isolated hybrid energy system. Dump loads are used primarily to maintain power balances. They may also be used to control frequency. The most common use of a dump load is in isolated wind/diesel systems where the penetration of the wind energy generation is such that instantaneous power levels sometimes exceed the system load less than minimum allowed diesel power level. The dump load control can sense the excess power and dissipate whatever is required to ensure that the total generated power is exactly equal to the actual system load plus that which is dissipated. The dump load is virtually a requirement in high penetration wind/diesel systems using stall controlled wind turbines, because adjusting the power is impossible. When pitch controlled turbines are used it may be possible to dispense with the dump load, since power can be limited by pitching the blades.

3.8 Supervisory Controller

Many hybrid systems, especially the more complex ones, have a supervisory controller to ensure proper operation of all the devices within the system. The possible functions of a supervisory controller are illustrated in Fig. 9. The controller itself consists of three main functional units: (1) sensors, (2) logical unit, and (3) control commands.

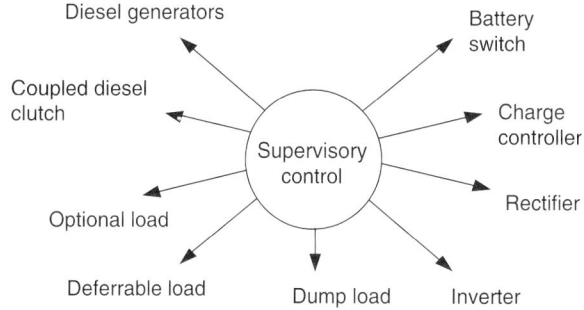

FIGURE 9 Supervisory controller functions.

Sensors are distributed throughout the system. They provide information on power levels, operating conditions, and so on. The information gathered from the sensors is directed to the logical unit. The logical unit is based on a computer or microprocessor. It will make decisions based on an internal algorithm and the data from the sensors. Algorithms can be fairly simple or quite complex.

The decisions made by the logical unit are referred to as dispatch decisions, since their function relates to dispatching of the various devices in the system. Dispatch in this sense refers to turning a device on or off, or in some cases to setting its power level. The dispatch instructions from the supervisory controller are sent to controllers of the various devices in the system.

4. ENERGY LOADS

The energy supplied by a hybrid system can be categorized in a variety of ways. The first has to do whether the energy supplied is electricity or heat. Within the electrical category, electricity loads are often divided into primary and secondary loads. Primary loads are those that must be served immediately. Secondary loads are associated with load management, and they may be further divided in what are known as deferrable and optional loads. Deferrable loads have some flexibility in when they are served, while optional loads are those which are only served if there happens to be sufficient excess energy available to do so. Regardless of type, it may be noted that loads frequently vary significantly from one season to the next as well as over the week and over the day, as was discussed earlier.

5. RENEWABLE ENERGY RESOURCE CHARACTERISTICS

It is worth considering briefly the time varying nature of the various renewable resources that might be used in a hybrid system, because that nature will affect the design and operation of the system.

5.1 Wind

The wind resource is ultimately generated by the sun, but it tends to be very dependent on location. Over most of the earth, the average wind speed varies from one season to another. It is also likely to be affected by general weather patterns and the time of day. It is not uncommon for a site to experience a number of days of relatively high winds and for those days to be followed by others of lower winds. The daily and monthly average wind speed for an island off the coast of Massachusetts, illustrating these variations, is illustrated in Fig. 10. The wind also exhibits short-term variations in speed and direction. This is known as turbulence. Turbulent fluctuations take place over time periods of seconds to minutes.

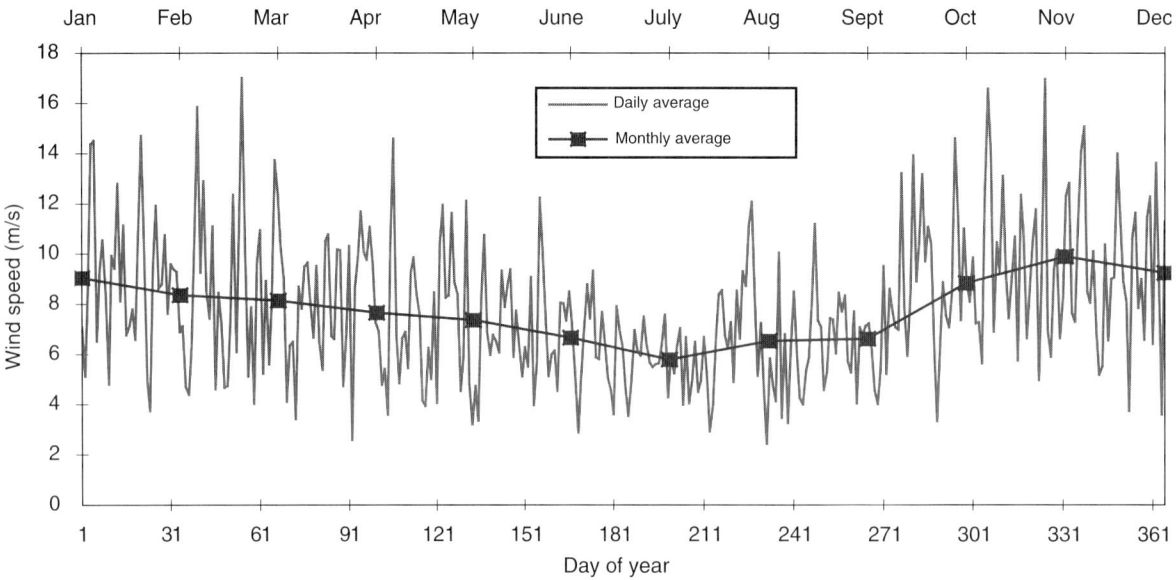

FIGURE 10 Daily and monthly wind speeds on Cuttyhunk Island, Massachusetts.

5.2 Solar Radiation

The solar radiation resource is fundamentally determined by the location on the earth's surface, the date, and the time of day. Those factors will determine the maximum level of radiation. Other factors, such as height above sea level, water vapor or pollutants in the atmosphere, and cloud cover, decrease the radiation level below the maximum possible. Solar radiation does not experience the same type of turbulence that wind does, but there can be variations over the short term. Most often, these are related to the passage of clouds. Figure 11 illustrates the solar radiation over a 5-day period in December in Boston, Massachusetts.

5.3 Hydropower

The hydropower resource at a site is a function of the amount of flowing water available (discharge) in a river or stream and the change in elevation (head). The head is usually relative constant (affected only by high water during storms), but the amount of water available can vary significantly over time. The average discharge is determined by rainfall and the drainage area upstream of the site on which the rain falls. Discharge will increase after storms and decrease during droughts. Soil conditions and nature of the terrain can also affect the discharge. Short-term fluctuations are normally insignificant.

5.4 Biomass

Sources of biomass are forest or agricultural products. The resource is ultimately a function of such factors as solar radiation, rainfall, soil conditions, temperatures, and the plant species that can be grown.

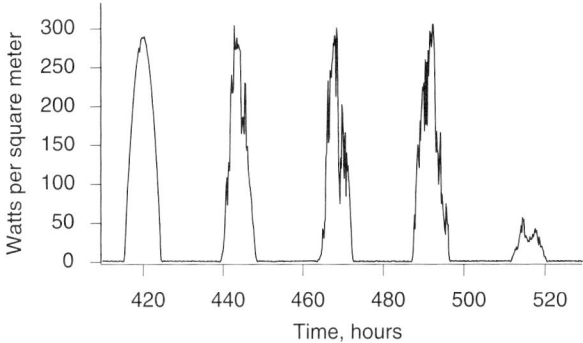

FIGURE 11 Hourly solar for 5 days radiation in Boston, Massachusetts.

6. DESIGN CONSIDERATIONS

The design of a hybrid energy system will depend on the type of application and the nature of the resources available. The primary consideration is whether the system will be isolated or grid connected. Other important considerations include (1) how the various generator types will be intended to operate with each other, (2) excess power dissipation, (3) use of storage, (4) frequency control, (5) voltage control, (6) possible dynamic interactions between components. Many of the possible technologies were discussed previously and will not be repeated here.

6.1 Load Matching

In the design of an isolated hybrid system with renewable energy source generators one matter of particular concern is known as load-matching. This refers to coincidence in time (or lack thereof) of the renewable resource and the load. The match between the two will affect the details of the design and will also determine how much of the energy that can be generated from the renewable energy source can actually be used. When too much is available, some of it may have to be dissipated. When too little is available, either storage or a conventional fueled generator will be needed.

Under ideal circumstances, the energy requirement will match the available resource. Often, however, that is not the case. In many islands of the world, for example, the largest loads are during the summer tourist season, but the wind resource is greatest in the winter. Figures 3 and 10 above illustrated a typical example of this for the load and wind, respectively.

When there is a mismatch between the resource and the load, the system must be designed to function properly under all conditions. For example, a wind/diesel system designed for the situation shown in Figs. 3 and 10 would most likely be designed so that wind energy could supply most of the load in the winter, keeping most or all of the diesel generators off for long periods, but in the summer, the diesels would continue to run and the wind turbines would serve as fuel savers. Figure 12 illustrates the possible energy flows for a 50-second period for a hypothetical wind/diesel system on an isolated island. The system illustrated has a primary load, an optional heating load, a dump load, a wind turbine, and multiple diesel engine/generators, at least one of which must always remain on.

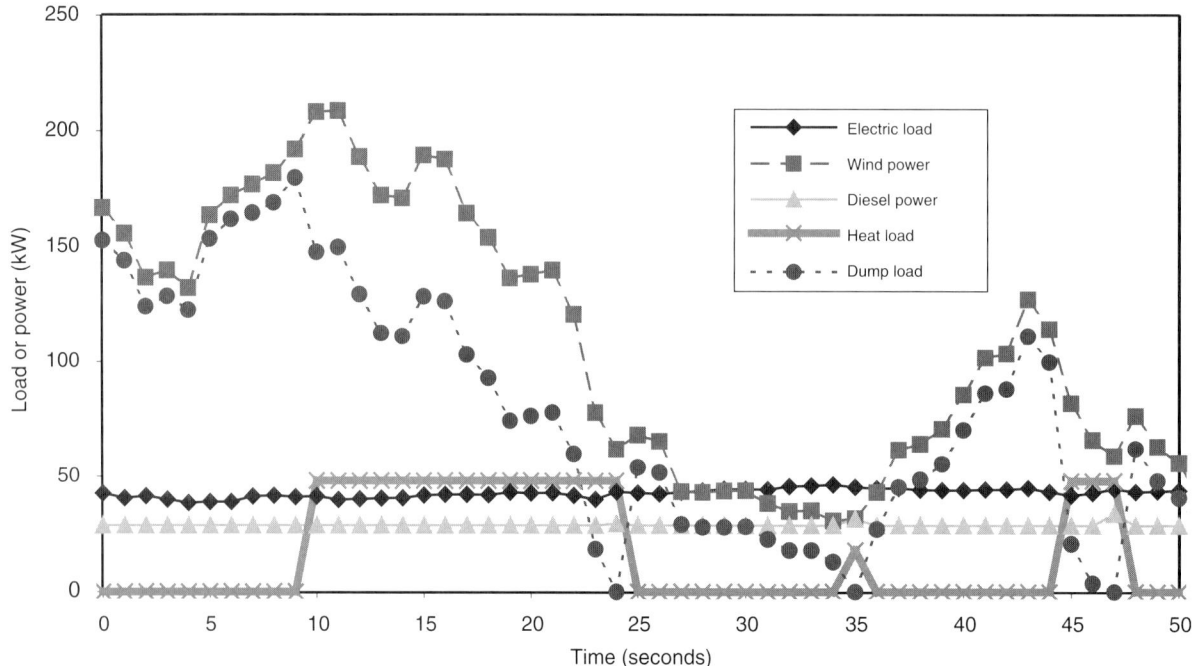

FIGURE 12 Power flows in hypothetical wind/diesel system.

7. ECONOMICS

Whether or not a hybrid energy system is actually used, and what it looks like, is strongly related to the economics of the system. In particular, will the cost of energy from the hybrid system be lower than that of a more conventional alternative? The cost of energy of a hybrid system is determined primarily by two factors: the cost of the system and the amount of useful energy that is produced. Other factors that are also important include the value of the energy, the cost of conventional energy, lifetime of the system, maintenance costs, and financial costs.

The cost of a hybrid energy system is first of all affected by the cost of the individual components that make up the system. Installation of the components and integrating them into a functioning unit will also contribute to the total cost. Generally, the renewable energy generators themselves are the most expensive items. For example, at the present time, wind turbines cost on the order of $1000 to $2000/kW, depending on the size. Photovoltaic panels cost on the order of $6000 to $8000/kW. Diesel generators are also costly, but considerably less so than wind turbines. Costs of $250 to $500/kW are typical. Lead acid batteries cost $100 to $200/kWh.

The energy that can be produced by a renewable energy generator will depend on the type of generator, its productivity at different levels of the resource, and the distribution of the occurrences of those levels. For example, a wind turbine will produce different amounts of power depending on the wind speed. If the average wind speed each hour over a year is known, this information can be combined with the data in the turbine power curve to predict the total energy that the turbine could produce. Similar curves describe the output of photovoltaic panels. These curves, combined with hourly data on the solar resource, can be used to predict the total annual energy from the panels.

When a renewable energy generator is operating in a hybrid energy system, predicting the useful energy is not as simple as it would appear, based on the previous discussion. First of all, because of a likely mismatch between the available resource and the load, it is may be that not all of the energy can be used when it is available, except in very low penetration systems. When not all of the load is supplied by the renewable generator, the rest is made up a conventional generator. Because of the difficulty in determining the amount of useful energy produced by the renewable energy generators in a hybrid system and predicting the actual fuel savings, detailed computer models are frequently used to facilitate the process.

The value of energy in a hybrid energy system is related to the nature of the energy produced, the cost of the alternatives, and the vantage point of the operator of the system. For example, the operator of an island diesel power system, considering installing a wind turbine, would want to know that the cost of the turbine would be offset by the decrease in diesel fuel purchased.

Economics of hybrid energy systems are typically evaluated by a technique known as life cycle costing. This method takes account of the fact that hybrid energy systems have relatively high initial costs but long lifetimes and low operating costs. These factors, together with various financial parameters, are used to predict present value costs, which can then be compared with costs from the conventional alternatives.

8. TRENDS IN HYBRID ENERGY SYSTEMS

Hybrid energy systems are still an emerging technology. It is expected that technology will continue to evolve in the future, so that it will have wider applicability and lower costs. There will be more standardized designs, and it will be easier to select a system suited to particular applications. There will be increased communication between components. This will facilitate control, monitoring, and diagnosis. Finally, there will be increased use of power electronic converters. Power electronic devices are already used in many hybrid systems, and as costs go down and reliability improves, they are expected to be used more and more.

SEE ALSO THE FOLLOWING ARTICLES

Biomass for Renewable Energy and Fuels • *Cogeneration* • *Economics of Energy Supply* • *Flywheels* • *Fuel Cells* • *Geothermal Power Generation* • *Hydrogen Production* • *Hydropower Technology* • *Photovoltaic Energy: Stand-Alone and Grid-Connected Systems* • *Solar Thermal Power Generation* • *Waste-to-Energy Technology* • *Wind Farms*

Further Reading

Hunter, R., and Elliot, G. (1994). "Wind-Diesel Systems." Cambridge University Press, Cambridge, United Kingdom.

Lundsager, P., Binder, H., Clausen, H-E., Frandsen, S. Hansen, L., and Hansen, J. (2001). "Isolated Systems with Wind Power." Riso National Laboratory Report, Riso-R-1256 (EN), Roskilde, Denmark, www.risoe.dk/rispubl/VEA/veapdf/ris-r-1256.pdf.

Manwell, J. F., Rogers, A. L., and McGowan, J. G. (2002). "Wind Energy Explained: Theory, Design and Application." John Wiley & Sons, Chichester, United Kingdom.

Hydrogen, End Uses and Economics

DAVID HART
Imperial College London
London, United Kingdom

1. Introduction to Hydrogen Energy End Uses
2. Transport End Uses
3. Stationary Heat and Power End Uses
4. Other Hydrogen End Use Technologies
5. Economics of Hydrogen Systems
6. Conclusions

Glossary

external costs of energy Costs, such as health damage from pollution, that can be directly attributed to energy use but are not included in its price.

fuel cell Device in which hydrogen can be electrochemically combusted to produce electricity and heat.

hydrogen embrittlement Reduction in material strength due to atomic hydrogen interaction with material.

hydrogen energy Hydrogen used for production of power, heat, or another energy service.

liquid hydrogen Hydrogen cooled to −253°C, where it changes from a gaseous state to a liquid state.

Hydrogen is an appropriate fuel for many energetic end uses, from the production of heat in buildings to the generation of power for transport. It can be burned in place of conventional fuels, but it can also be electrochemically oxidized rather easily in comparison with most hydrocarbons. This allows it to be used efficiently in fuel cells and broadens the available scope of end use applications. In addition, hydrogen provides a potentially valuable form of energy storage for intermittent renewable energy systems such as wind and solar power. Currently, hydrogen energy is used to any great extent only within the space industry, where its combustion with pure oxygen provides propulsion, typically for secondary rocket stages. However, chemical hydrogen is in widespread use in the oil industry, where it is used to clean, refine, and upgrade petroleum products for energy uses. Many small hydrogen energy systems are in development, including hydrogen vehicles using both internal combustion engines and fuel cells; stationary power and heat production plants, and small-scale electrical and electronic systems. None is yet commercially available outside very specific cases because the use of hydrogen is typically more costly than the use of conventional commercial fuels. The economics of hydrogen energy will improve as production, storage, and end use technology become more widespread and cheaper. If environmental benefits such as low emissions are valued, hydrogen should become even more competitive with conventional energy.

1. INTRODUCTION TO HYDROGEN ENERGY END USES

Hydrogen has been used as an industrial chemical since its discovery and naming during the late 18th century, but rarely for its energetic properties. The use of hydrogen as an energy carrier was proposed early on by Jules Verne, among others, but hydrogen is found only in compound form on earth, and so readily available conventional fuels, such as coal and oil, were preferred. Interest in hydrogen energy increased sharply toward the end of the 20th century. This increased interest was prompted by growing global concerns about (1) emissions of carbon dioxide (CO_2) and local pollutants such as NO_x and (2) potential reductions in the availability of sources of primary fossil energy. Hydrogen energy offers the possibility of solving these problems. CO_2 emissions can be reduced because no carbon atoms are present in the hydrogen, local pollutants can be reduced because hydrogen can be burned cleanly, and the wide range of possible hydrogen sources allows for flexibility in production. The extent to which these solutions are possible depends on all the elements of the hydrogen fuel chain: production, storage and transportation, and end use.

As with all fuels, the end use of hydrogen energy technologies is that of providing a service—illumination, warmth, and mobility. Various technologies may be used to achieve this, and various categorizations may be applied to the end uses.

Hydrogen end uses are frequently categorized in one of two ways: by type of energy output (e.g., heat, light, electricity) or by end use type (e.g., transport, stationary power generation). In many cases, these categorizations become blurred, for example, as the coproduction of heat and power becomes more widespread. The economics of each end use depend strongly on a wide variety of factors, including the source of the hydrogen but also the cost, lifetime, and efficiency of the end use technology. Hydrogen end use technologies are generally conventional heat and power production technologies with modifications to accommodate the various characteristics of the hydrogen fuel.

Importantly, hydrogen energy technologies are in their infancy, and little is published about the costs of many of them. The costs can be expected to change rapidly and probably over a large range.

Table I shows technologies that can be used with hydrogen in a variety of generic applications. The potential for providing transport, heat, stationary power, or power for small-scale electronics is indicated. A range of hydrogen end use technologies is shown, although other conversion devices, such as Stirling engines, may also be used.

1.1 Hydrogen End Use Technologies

As indicated in Table I, many hydrogen technologies can be applied to different end uses, either unchanged or with minor modifications. This is also common to other fuels.

The hydrogen used in transport can be used in various prime movers for most applications. For example, automotive companies are developing hydrogen vehicles based on both internal combustion engine and fuel cell technology. Aircraft will be powered primarily by gas turbines, although fuel cell aircraft are under consideration. Ships may use modified versions of diesel engines. The various prime movers offer different characteristics, as do those used with conventional fuels.

Similar technologies will be deployed in stationary power. Fuel cells are viewed as a promising technology with development work still to be done, and conventional machinery, such as turbines and gas engines, may be used effectively in stationary applications. Fuel cells are also the only hydrogen technology likely to be deployed in commercial electronics applications.

Hydrogen, especially when used in a very pure state, can embrittle certain materials and cause them to become weak. Its high flame speed may also cause problems. Therefore, in some instances, especially in the development of hydrogen turbines, new or advanced material developments have been required. Hydrogen in small-scale electronics would almost certainly be stored as a compound, probably in a metal hydride.

1.1.1 Spark Ignition, Diesel, and Wankel Engines

Hydrogen can be burned in a standard spark ignition engine with marginal modification, but only with a performance penalty. This comes about due to the large differences between the flame speed and flammability range of hydrogen and those of hydrocarbon fuels. The fuel flow and ratio must be optimized for hydrogen by using a pressure regulator (used for all gaseous fuels) rather than a liquid pump and by modifying the timing of the spark. Larger fuel-mixing equipment and fuel injectors are also necessary because hydrogen is less energy dense than a liquid fuel and a greater volume of it must be introduced for combustion. Then, there is less room for air in the cylinder, the expansion is reduced, and consequently so is the output power—by as much as 30% relative to the original gasoline engine.

If a dedicated hydrogen engine is used, its performance may be better than a comparable gasoline engine. The compression ratio of the cylinder can be increased, as is done with natural gas engines, by forcing more of the fuel mixture into the available space before combustion occurs or by using a turbocharger or supercharger to force more air into the cylinder. If the mixture is cooled, it is also possible to increase the amount introduced. Adopting one of these measures can allow the hydrogen engine to operate at a power up to 25% greater than the equivalent petroleum engine.

The wide flammability range of hydrogen (mixtures of between 4 and 75% in air) means that it is possible to operate the engine at a very lean mixture, increasing fuel efficiency and ensuring complete fuel combustion. Fuel efficiency is also improved because the mixing of gaseous hydrogen with air is better than that of a liquid fuel. The thermal efficiency of a hydrogen engine can be on the order of 25% better than a gasoline equivalent, although the benefit in miles per gallon is not as simple to calculate because it also depends on the driving cycle being undertaken. The combustion temperature is lower with a lean mixture than with a rich one, reducing NO_x formation.

Hydrogen can also be used in modified diesel engines, given the introduction of a suitable ignition process, because hydrogen is not a compression ignition fuel. The Wankel rotary engine, operating on a slightly different principle from conventional piston engines, is particularly suitable for hydrogen fueling. The engine itself relies on combustion of the fuel/air mixture within an unusually shaped chamber known as an epitrochoid, which drives the rotary equivalent of a piston. The Wankel rotary engine has a separate intake and combustion chamber, allowing the hydrogen mixture to be taken in without backfire problems and enabling smoother running.

Backfiring has been an issue associated with the use of pure hydrogen in internal combustion engines in addition to knocking. The use of water injection reduces these tendencies, as does the injection of the hydrogen in liquid form into the chamber. Preignition of the hydrogen may occur if a mixture of hydrogen and air comes into contact with a hot surface. This can be avoided by mixing the fuel and air only within the cylinder and ensuring that this is done only just before the spark plug fires so as to leave no opportunity for preignition.

Use of hydrogen in internal combustion engines of any type in stationary applications is likely to be simpler than that in vehicles. Hydrogen storage is unlikely to be a critical parameter, and the system is more likely to run under steady-state conditions. Costs for engines are expected to be similar to those for conventional machines, but a premium of 10 to 20% may be attached to early examples.

1.1.2 Turbines

Hydrogen-burning turbines for vehicles are being investigated primarily for aircraft. Again, emissions from conventional kerosene fuels are a driver. Major development work is being undertaken to examine the potential for a passenger or cargo airliner, but work on hydrogen aircraft dates back to the 1950s when military projects were conducted.

The need to maintain range means that aircraft are nearly certain to use liquid hydrogen, and turbine designs are based around this expectation. In an aircraft engine, the liquid hydrogen would be passed through a heat exchanger to be vaporized before driving a turbine and a fan to compress incoming air to be driven through the engine. The hydrogen would then be burned in the afterburner to provide thrust.

Stationary hydrogen gas turbines are also under development, notably in Japan as part of a major government research project. Hydrogen's combustion properties offer advantages but have disadvantages as well. The fact that hydrogen burns cleanly is advantageous because no ash particles and residues that can produce sediment or corrosion on the turbine blades arise, unlike the burning of fossil fuels. In addition, the inlet temperature of the turbine can be raised much higher than would be possible with fossil fuels, producing higher thermal efficiencies under the constraints of the Carnot limit. However, very high temperatures and operation under pure hydrogen/oxygen conditions cause great stresses in materials, and so early turbine tests were conducted with nitrogen dilution. While reducing materials problems, this solution may result in production of high levels of NO_x, and that may be unacceptable.

Limited research is being conducted on hydrogen-burning turbines, although a 500-MW combined cycle turbine system burning liquid hydrogen and oxygen is under investigation in Japan. A thermal efficiency of 60% is predicted.

1.1.2.1 Economics of Hydrogen Engines The costs of dedicated and converted hydrogen combustion engines are not public because is available commercially. However, the minimal modification required, and the comparatively standard materials used, means that costs are expected to vary little from standard engine costs of between U.S. $30/kW for mass-produced vehicle engines and $500/kW for more specialized stationary machinery. Operating costs will depend not only on efficiency but also on the cost of fuel. Hydrogen supplied in large quantities can be competitive with gasoline, but until it becomes widespread, higher costs (perhaps double or triple) are likely.

Hydrogen turbines are even further from commercial costing, although capital costs of between U.S. $1000 and $2000/kW have been suggested.

1.1.3 Fuel Cells

The fuel cell works on the principle of electrochemical combustion and offers the opportunity for highly efficient conversion of hydrogen and an oxidant (usually air) to electrical energy. Polymer electrolyte membrane fuel cells (PEMFCs) are undergoing major development and deployment programs for automotive use, and the first demonstration fuel cell buses using hydrogen were put into service during the

1990s. Cars using fuel cell technology and hydrogen fuel were introduced in very limited numbers late in 2002, although fully commercial vehicles are not expected until around 2010.

A fuel cell-powered vehicle is an electric vehicle. The fuel cell produces electricity from the hydrogen reaction, and this is then channeled through a power conditioning system into either one electric motor that takes the place of the internal combustion engine or smaller motors mounted independently on each wheel. No transmission linkage or gearbox is required.

Fuel cell vehicle development has been driven particularly by emissions regulations such as the California zero emissions vehicle (ZEV) mandate, which has forced automotive manufacturers to develop vehicles that produce no regulated pollutants. Only battery and hydrogen fuel cell electric vehicles currently satisfy the criteria of the mandate, and it appears that fuel cell development may outpace that of batteries, meaning that fuel cell vehicles are more likely to come into widespread use in the future. However, suitable performance and cost reduction must be demonstrated for this to happen. In addition, a refueling infrastructure for hydrogen must be developed if that is the fuel chosen.

Fuel cells are equally applicable in stationary end uses, where they are likely to be used not only for electricity generation but also for heat. A range of fuel cells in addition to the PEMFC can also be used for stationary end uses, with each offering specific advantages and drawbacks. Low-temperature fuel cells are theoretically more efficient at producing electricity from hydrogen than are high-temperature fuel cells. However, they do not produce useful heat except perhaps for space heating applications. Many high-temperature fuel cells will operate directly using a hydrocarbon fuel source such as natural gas rather than running on hydrogen. Although they convert the natural gas to hydrogen on the anode before oxidation, they cannot strictly be termed "hydrogen technologies" given that only a proportion of fuel cells for stationary power and heat production are expected to operate purely on hydrogen.

1.1.3.1 Economics of Fuel Cells The economics of a fuel cell system are highly unclear. Capital costs are high, as the technology is still generally in laboratory-scale development, and are quoted in the wide range U.S. $500 to $10,000/kW over the range of technologies. Operating and maintenance costs, lifetime, and durability still remain to be proven.

1.1.4 Catalytic Burners
A catalytic burner uses a coating of a catalyst, such as platinum, to electrochemically combust a fuel over a wide range of fuel/air ratios. A pure catalytic burner uses only the electrochemical reaction between fuel and oxidant to generate heat, whereas a hybrid system uses a catalyst to enhance a conventional combustion reaction. Hydrogen is fed to the catalytic surface, where it reacts with oxygen in the air to produce heat and water vapor. A catalytic burner is self-starting; the reaction begins when hydrogen is fed to the catalyst, and it ceases again as the flow stops. The temperature can be controlled over a wide range to enable applications as diverse as catalytic cooking stoves and low-intensity heating panels that could be hung, uncovered, on the wall of a room. Emissions from catalytic combustion systems will typically be zero.

1.1.5 Gas Boilers
Hydrogen can be burned in a boiler in the same way as can any fuel gas. Hydrogen fuel and air are fed to the combustion zone, where they are ignited to produce heat.

Using hydrogen in a natural gas boiler is much the same as using natural gas. A number of small modifications to the boiler are required, the most important of which is a reduction of the orifice size to control the flow of the gas. The flow rate of the input gas must also be adjusted to take care of the fact that hydrogen is much less dense than natural gas but has a higher calorific value.

Emissions of NO_x may be slightly increased initially, but optimizing the flow rate and the introduction of boilers dedicated to hydrogen use should solve this problem, whereas carbon monoxide (CO) and hydrocarbon emissions both will be reduced.

1.1.6 Liquid and Gaseous Hydrogen and Hydrogen Mixtures
Apart from the technologies themselves, hydrogen will be used in various applications in different forms.

Aircraft apart, hydrogen can be stored on-board most vehicles in liquid or gaseous form or in metallic or other compounds. Vehicles using both gaseous and liquid systems have been demonstrated successfully and continue to be developed. Liquid hydrogen systems typically offer the advantage of greater energy density than does gaseous hydrogen, allowing greater range for less space occupied within the vehicle. However, very high-pressure storage systems of 10,000 psi/700 bar offer similar energy densities to liquid hydrogen. But both combustion engines and fuel cells will use the hydrogen in a gaseous state.

In addition, hydrogen may be mixed with other fuels to improve combustion characteristics. This has been the subject of experiment since German engineer Rudolf Erren conducted early conversions of internal combustion engines during the 1930s. However, it is difficult to mix hydrogen with a liquid fuel effectively, and methane (or natural gas) has often been chosen. Mixtures of hydrogen in methane varying from 15 to 50% by volume have been examined. The effect of this mixing is to generally improve the combustion characteristics of the natural gas, which has a low flame speed. Increasing this speed assists in reducing hydrocarbon formation at the cylinder walls, thereby lowering production of CO, but also enables the engine to be run on a leaner mixture, cutting down NO_x emissions. CO_2 emissions from the combustion process will also be reduced because less carbon is contained in the fuel. Nevertheless, it appears likely that pure hydrogen will be preferred in automotive applications.

In stationary power systems, pure hydrogen will be used in remote areas where an integrated renewable power system can be put into place. The hydrogen will be used as an energy storage buffer to compensate for the intermittency of the system, but excess hydrogen could also be used for other purposes (e.g., as a transport fuel). Pure hydrogen may also be used in many other applications, but hydrogen/methane mixtures are being considered as a way of reducing CO_2 emissions from natural gas networks. Up to 20% hydrogen by volume can be introduced into the natural gas transmission system in many countries, without requiring end use technologies such as gas ovens to be adjusted and while remaining within gas standard specifications. CO_2 emissions from end use are reduced by the same amount as the hydrogen introduced into the system.

2. TRANSPORT END USES

Hydrogen is being explored for use in transport in a wide variety of applications. Motivated primarily by regulations to reduce emissions of CO_2 and regulated pollutants, all major automotive manufacturers have hydrogen research and development (R&D) under way for cars and some for buses. Smaller vehicles (e.g., scooters) and larger vehicles (e.g., ships, airplanes) are also being developed. Hydrogen's low density creates particular problems for transport applications because the range of vehicles using hydrogen fuel may be short, but it may be mixed with other fuels to give some of the advantages of each.

2.1 Hydrogen Vehicle Types

Hydrogen could be used to power vehicles in all sectors, although it is expected to be used first in smaller scale applications such as buses and cars and only later in replacing the fuel for very large engines such as marine diesels.

Geographical locations that are sensitive to emissions may be the first to adopt hydrogen vehicles, and these may be specific to the needs of the location. For example, many cities in Southeast Asia suffer from high levels of pollution generated by two-wheeled vehicles with two-stroke engines, and so hydrogen scooters, both internal combustion engine and fuel cell, are under development. Conversely, hydrogen buses are seen as a good early technology for European cities.

2.1.1 Niche Vehicles: Scooters, Forklifts, Battery Replacements

Scooters and small motorcycles can be run on either fuel cells or internal combustion engines, and developments of each are under way. Hydrogen for these applications is typically stored either in compressed gas form or in a metal hydride, providing a range on the order of 100 km and very low emissions.

Forklift trucks conventionally operate on batteries in warehouse and food industry environments where emissions must be zero to meet health and safety regulations. Hydrogen used in fuel cells can also meet these regulations without the long recharge periods required by many batteries.

Many other specialist vehicles, including aircraft tugs at airports, operate using batteries. Hydrogen used in fuel cells is under consideration as a replacement fuel for these vehicles.

2.1.2 Common Vehicles: Cars, Buses, Vans, Trucks

The largest potential for hydrogen vehicles may come in the main automotive markets, although the cost and technology targets for hydrogen cars are very challenging. However, the use of hydrogen in cars is viewed as one leading option to enable continued mobility while reducing some of its environmental impacts. Analysis carried out in Germany has suggested that hydrogen offers the best long-term combination of low emissions, renewable feedstock potential and vehicle performance, and should be seen as the main long-term strategic fuel for the country.

Significant debate is ongoing as to the best way in which to introduce hydrogen into the car market. Both buses and cars using liquid hydrogen burned in internal combustion engines are being demonstrated in North America, Japan, and Europe. The advantage over fuel cell vehicles is that the engine technology is close to maturity in terms of both cost and technical development. However, liquid hydrogen is bulky to carry and is subject to losses through "boil-off," where a small rise in temperature causes evaporation and a subsequent pressure buildup within the storage vessel and gas is then vented. In addition, the efficiency of the internal combustion engine is lower than that of a fuel cell, so both vehicle range and fuel cost can be worse than with a fuel cell vehicle.

Fuel cell vehicles have also been demonstrated using hydrogen stored as a liquid, as compressed gas, or in metal hydrides. The vast majority of these have used PEMFCs, although some early bus systems used phosphoric acid fuel cells (PAFCs). As of early 2003, some tens of buses and cars had been tested or were within demonstration projects. Fuel cell vehicles are entering into limited precommercial demonstration, but the technology must still be proven to be reliable under the same conditions as conventional engines, costs must be reduced, and fueling infrastructure must be provided. Fuel cell vehicles are not expected to be available commercially until around 2010.

Fuel cell and hydrogen internal combustion engine light duty vans are also undergoing tests, but limited work is ongoing in developing hydrogen fueling for heavy-duty vehicles such as long-distance trucks. The characteristics of these vehicles are (1) high efficiency brought about by the use of heavy diesel engines and numerous gears to allow the engine to operate at close to optimal efficiency at all times and (2) very long engine life (typically hundreds of thousands of hours). Fuel cell technology is not yet at a stage where it can compete with these characteristics, and hydrogen storage is not yet advanced enough to allow sufficient range for the use of hydrogen in internal combustion engines in trucks.

2.1.3 Large/Specialist Vehicles: Locomotives, Boats, Ships, Planes, Submarines, Aerospace

Large-scale vehicles using hydrogen have also been considered, and some have been put into practice. In submarines and spacecraft, the use of hydrogen is comparatively common. In the former, the use of hydrogen in fuel cell systems means that emissions-free propulsion can be provided with only a very low heat and noise signature. These characteristics are invaluable for a submarine operating in "stealth" mode, where it aims to remain undetected. Typically, the hydrogen is stored in metal hydrides, which are both chemically stable and heavy, providing ballast for the submarine. Oxygen is also stored on-board, and the exhaust of the fuel cell—pure water—can be vented directly into the sea water, where it produces no telltale bubbles. The fuel cell is used not as the primary drive but rather as a supplementary power source.

Spacecraft use liquid hydrogen and liquid oxygen as fuel for rocket boosters. The mix contains approximately 40% more power per unit mass than do other rocket fuels, enabling greater payloads to be transported. A typical specific impulse for a rocket engine using liquid hydrogen/liquid oxygen is approximately 420 s, as compared with a liquid oxygen/kerosene rating of perhaps 250 s. Hydrogen has also been used in fuel cells on-board spacecraft, from the Gemini mission to the Space Shuttle, to produce both electrical power for the equipment and drinking water for the astronauts.

Aircraft have been designed to use hydrogen, and several small-scale tests have been conducted, including a Russian experiment where one of the turbines of a Tupolev aircraft was converted and operated successfully on liquid hydrogen during a test flight, but no complete aircraft yet exists. Liquid hydrogen contains less energy per unit volume than does kerosene, meaning that larger fuel tanks will be required in hydrogen aircraft to give them the same range as in conventional aircraft. These tanks cannot be situated in the wings, unlike in conventional airplanes, because the shape of the tanks is not suitable for maintaining liquid hydrogen at very low temperatures and the fluid will not act as ballast in the same way as will kerosene. Instead, the use of large tanks in the top of the main body of the aircraft has been suggested. The volume of hydrogen required will be on the order of four times that of kerosene for equivalent energy content. However, hydrogen is very light, and this will enable hydrogen aircraft either to be lighter at takeoff than their conventional counterparts or to carry greater payloads.

The use of locomotives using hydrogen has been suggested, but limited investigation has been carried out. Erren conducted early work during the 1930s, operating a few locomotives on hydrogen mixtures. Similarly, little analysis has been conducted on ships, particularly on a large scale. Several small boats operating on rivers have been converted to hydrogen and fuel cell use, and the potential exists for further exploitation of this area.

3. STATIONARY HEAT AND POWER END USES

In addition to transport, hydrogen can be used effectively in producing heat, power, or a combination of the two. Hydrogen can be used in place of other fuels in a wide range of different technologies, as suggested earlier.

3.1 Hydrogen for Heat Production

Hydrogen's generally high cost means that it is unlikely to be used as a fuel simply for heating because other fuels are effective for this purpose. However, if hydrogen is available cheaply, boiler, catalytic heater, or internal combustion engine technologies may be used to produce heat. Hydrogen used in fuel cells will also produce heat, and even low-temperature fuel cell technologies such as PEMFCs can prove to be sufficient for space heating and hot water requirements.

3.2 Hydrogen for Power Production

Hydrogen is typically seen as a form of energy storage and clean fuel, and it is most frequently considered in the context of power production.

The most efficient and cleanest technology for producing power from hydrogen is the fuel cell, but high cost and limited availability mean that few have been introduced outside of demonstration markets. The first small backup PEMFC generator (1.2 kW) was made available late in 2002 at a price of U.S. $6000, including a hydrogen supply stored in a metal hydride within the package. Other fuel cell-based hydrogen power systems have typically been on the order of several or tens of kilowatts, although a pure hydrogen-using PAFC operates in the city of Hamburg, Germany, and produces 200 kW of power along with a similar amount of heat for a local district heating system. The largest fuel cell system run to date was an 11-MW PAFC system in Japan that operated on hydrogen produced from steam-reformed natural gas to produce electrical power and heat.

3.2.1 Remote Integrated Hydrogen Power Systems

Many islands and other geographically remote areas suffer both from high imported fuel costs and from pollution caused when that fuel is used in providing heat and power. A typical example might be the use of diesel oil imported to many small islands and used in old generator sets for power production. However, these islands frequently have high levels of indigenous renewable energy resources such as solar, wind, and wave power. Using renewable resources to produce electricity as required, with any excess electricity subsequently used in electrolysis to produce hydrogen, can provide an integrated power system with no polluting emissions.

A basic system would include renewable electricity production, electrolysis and hydrogen storage (typically as a compressed gas), and an end use technology such as a fuel cell or combustion engine. Sizing the system is a complex process that depends on the fluctuation between renewable power availability and local demand, both diurnally and annually, and the volume of hydrogen storage required may be large to meet all fluctuations between supply and demand. The system may also be used in conjunction with a conventional power supply to meet either baseload or peaking power production. The economics of the system will also depend significantly on local factors, and electricity produced in this way may be expensive, possibly on the order of U.S. $1.00/kWh. However, this is frequently competitive with imported diesel while avoiding the pollution associated with the use of hydrocarbon fuels.

Hydrogen systems may be equally integrated with biomass production of hydrogen, making them both flexible and appropriate for use in a wide range of countries and circumstances. Rural electrification in many areas, including developing countries, may be enhanced through the use of these technologies.

3.2.2 Large-Scale Hydrogen Power Production

As hydrogen energy becomes more widespread, it may also be used for large-scale power production, with end use technologies of tens of megawatts or even gigawatts. Large-scale hydrogen combustion turbines could potentially be sited closer to urban areas than can fossil fuel-powered turbines because they would produce no pollution and, hence, would require no smokestacks. A 500-MW liquid hydrogen-powered turbine is under development in Japan, with the expectation that it will be used in the future as part of that country's commitment to hydrogen energy.

Hydrogen could also be used in fuel cells for this purpose, with several smaller fuel cells (perhaps on the order of 1 MW) connected to provide a modular power production opportunity of greater size. However, it might be that the modular nature of the fuel cells, and the potentially distributed nature of hydrogen production resources, offers better opportunities for decentralized hydrogen power systems.

4. OTHER HYDROGEN END USE TECHNOLOGIES

A small number of technologies not directly associated with heat, power, or transport production may also operate on hydrogen. In general, these are expected to be technologies in which hydrogen plus a fuel cell replaces a battery. Modeling suggests that under optimal conditions, a hydrogen and fuel cell combination could provide both more power and more energy from the same weight and volume than do many advanced batteries. However, such small hydrogen fuel cell systems have not yet been optimized, and so batteries continue to offer the most convenient solution.

Possible end uses include military power packs for communications, laptop computer power, and battery replacement for mobile phones. However, in the smallest applications, such as mobile phones, methanol or another liquid fuel may be more suitable than hydrogen due to its higher volumetric energy density.

5. ECONOMICS OF HYDROGEN SYSTEMS

The fundamental economics of hydrogen energy systems are the same as those of conventional energy systems. However, hydrogen energy systems suffer from two peculiarities in comparison with the majority of conventional technologies. The first is that hydrogen is typically one of the more expensive fuels to transport, so that the economics become significantly worse with long-distance fuel movement. In conventional fossil fuel systems, fuel transport costs are generally a marginal factor in the overall economics. In addition, hydrogen technologies are still largely under development, and so costs may vary widely even in similar applications.

5.1 An Introduction to the Costs of Hydrogen Systems

The basic economics of hydrogen systems are dependent, as is any energy system, on three primary factors: (1) fuel (hydrogen) cost, (2) equipment capital cost, and (3) equipment performance (e.g., lifetime, maintenance, efficiency).

As has been discussed, the basic cost of hydrogen is higher than that of the raw materials from which it is produced. Losses are inherent in any conversion process, and the conversion equipment also has a cost associated with it. However, the cost of the end use service is what should be considered as far as possible—the cost per unit of luminosity, of mobility, or of warmth. Because of this, the efficiency and other characteristics of the end use technology have a critical role.

High fuel cost can be compensated, to a greater or lesser extent, by high end use efficiency. As a simplistic example, if fuel costs are twice as high for hydrogen but the efficiency of end use is doubled in comparison with a conventional fuel, the cost of the end use due to fuel will remain unchanged. Other costs, such as those for maintenance and capital, may still alter the final delivered service cost. For this reason, although fuel cells suffer very high capital costs in the near term due to their early development status, they can offer some expected economic benefits related to efficiency. In addition, maintenance costs are expected to be low in the long term, although this has yet to be demonstrated.

5.2 Environmental Costs

An increasingly important element to be considered in the economics of any energy system is that of environmental costs, both internal and external. A primary criticism of fossil energy systems is that they have severe environmental impacts associated with most stages of the fuel chain, from discovery and extraction of raw materials to emissions from processing and final conversion. A significant advantage of many hydrogen systems is the absence or reduction of these environmental impacts.

Emissions from end use are one of the effects most commonly considered. In conventional systems, these tend to have the greatest health impact, although emissions from refining and processing must also be taken into account. Ideally, to ensure valid comparisons, emissions from entire fuel chains should be evaluated.

These emissions can also be considered as part of the economic equation for evaluation of an energy system. To do so requires their representation as costs using principles of externality costing that are not trivial. However, this enables expensive fuels that produce little or no pollution to be compared, on a somewhat more equal basis, with cheap fuels that have a high health impact cost not included in the price of the fuel. Hydrogen can be considered one of the more expensive but less polluting fuels.

6. CONCLUSIONS

Hydrogen energy end use technologies are not yet mature, unlike conventional technologies. However, they offer potentially significant advantages in terms of low or zero emissions and flexibility in fuel sources. As immature technologies, costs are high and reliability and durability are not yet proven. Nevertheless, early demonstrations of hydrogen vehicles using internal combustion engines and fuel cells are under way and expected to become more widespread.

Similarly, hydrogen energy systems for power and heat production are operating under demonstration conditions in many areas. The possible integration of hydrogen energy storage with renewable energy sources offers the prospect of economically efficient remote power systems and reductions in the external costs of energy associated with many fossil fuels.

Over the long term, hydrogen energy is likely to be used in many everyday situations, and hydrogen energy technologies will replace many conventional technologies. The economics of such hydrogen technologies will improve as they enter widespread use, and the added value of low pollution should further enhance the value of such systems.

SEE ALSO THE FOLLOWING ARTICLES

Alternative Transportation Fuels: Contemporary Case Studies • *Arid Environments, Impacts of Energy Development in* • *Fuel Cells* • *Fuel Cell Vehicles* • *Hydrogen, History of* • *Hydrogen Production* • *Hydrogen Storage and Transportation* • *Internal Combustion (Gasoline and Diesel) Engines* • *Transportation Fuel Alternatives for Highway Vehicles.*

Further Reading

Hoffmann, P. (2001). "Tomorrow's Energy: Hydrogen, Fuel Cells, and the Prospects for a Cleaner Planet." MIT Press, Cambridge, MA.

Ogden, J. M. (1999). Prospects for building a hydrogen energy infrastructure. *Annu. Rev. Energy Environ.* 24, 227–279.

Padró, C. E. G., Putsche, V. (1999). "Survey of the Economics of Hydrogen Technologies," NREL/TP-57027079. National Renewable Energy Laboratory, Golden, CO.

Hydrogen, History of

SETH DUNN

Worldwatch Institute
New Haven, Connecticut, United States

1. Discovery of Hydrogen
2. Origins of Interest in Hydrogen Energy
3. Roots of the Hydrogen Economy Concept
4. Globalization of the Hydrogen Movement
5. Future of Hydrogen History

Glossary

electrolysis The process of splitting water into its components, hydrogen and oxygen, by means of an electrical current.
energy carrier A form of matter that carries energy from one point to another, as distinct from an energy source, for which hydrogen is often mistaken.
fuel cell An electrochemical device that combines hydrogen and oxygen, producing an electrical current and water as a by-product.
hydrogen economy A term first introduced in the 1970s to describe the concept of an energy system based primarily on the use of hydrogen as an energy carrier.
proton-exchange membrane One type of fuel cell, featuring a membrane through which protons but not electrons can pass, forcing the latter to move along an electrode and generate a current.

The history of hydrogen as an energy carrier begins with its discovery and the comprehension that it is one of the basic building blocks of the universe. When the early Greek philosophers named water, air, fire, and earth to their list of elements, they did not realize that water in fact consisted of two elements. One of these elements is oxygen, the respiratory prerequisite for human life. The other is hydrogen, which modern science has revealed as the lightest and most abundant element in the universe.

1. DISCOVERY OF HYDROGEN

By the most recent estimates, hydrogen accounts for approximately 75% of the mass of the entire universe and accounts for more than 90% of all molecules. According to the Harvard astrophysicist Steven Weinberg, between 70 and 80% of the observable universe is composed of hydrogen, with the rest attributable to helium, the universe's second lightest element. This is why hydrogen is called *Wasserstoff*—the stuff of water—in German.

The first recorded production of hydrogen took place in the 15th century, when the Middle Age physician Theophrastus Paracelsus dissolved iron in the acid vitriol, producing hydrogen gas. Paracelsus did not note that hydrogen was flammable; this was left to Turquet de Mayeme and Nicolas Lemery, French scientists who, in the 17th century, mixed sulfuric acid with iron, yet did not think that the resulting gas could be an element unto itself, only a burnable form of sulfur.

The actual classification and formal description of hydrogen would have to wait until the 18th century. As the early Greeks would have no doubt appreciated, the modern discovery of hydrogen was tied to advances in identifying its fellow component in water, oxygen. Such advances would, of course, require first that scientists reconsider the conventional notion of air as a basic element. The first to question this longstanding characterization was Herman Boerhaave, a Dutch doctor and naturalist who believed that air contained an ingredient that made breathing and combustion possible. Wrote Boerhaave in 1732, "The chemist will find out what it actually is, how it functions, and what it does; it is still in the dark. Happy he who will discover it." Boerhaave had a modicum of intellectual support in the 17th-century writings of British scientists Robert Boyle and John Mayo, who suspected that some substances in the air were responsible for the process of combustion. But this ran counter to the phlogiston theory, another 17th-century idea. The phlogiston theory, one of the first attempts to explain the process of combustion, was first published in 1697. Its origins lay with a German scientist named Georg Ernst Stahl, who contended that a substance known as phlogiston gave all matter the ability to burn. Because phlogiston disappeared from material during the combustion process, and because it was considered impossible to reduce the substance to a pure state, this was not a theory easily disproved. Modern chemistry has since shown that Stahl had it precisely backwards: the process of burning arises from adding a substance—oxygen. The idea that combustion consisted of the release of phlogiston served as an impediment to the recognition of oxygen and hydrogen as gases.

Fortunately for scientific progress (but rather unfortunately for Stahl), the late 18th century would witness a scientific race of sorts to successfully isolate oxygen. Among those who discovered, but did not name, the element were Joseph Priestley, a British minister; Carl Wilhelm Scheele, a doctor of Swedish–German descent; and Antoine Laurent Lavoisier, by then France's leading chemist. Although Scheele was the first to produce pure oxygen, between 1771 and 1772, the other two beat him to the press, publishing their findings in 1774. Because the scientists studying air were also investigating water, the isolation and identification of hydrogen proceeded in parallel fashion. But most chemists, such as Boyle, still saw the hydrogen they produced as another type of air. That would change with the English nobleman Henry Cavendish, the first to discover and describe many of the important characteristics of hydrogen. Ironically, Cavendish did not name hydrogen because he was not fully free from the phlogiston theory—he thought he had discovered the substance in its pure state. But his key breakthrough was to identify two kinds of "factitious air": "fixed air," or carbon dioxide, and "inflammable air," or hydrogen. In a 1766 paper to the Royal Society of London, Cavendish provided exact measurements of the weights and densities of both gases, revealing the inherent lightness of hydrogen. (Recognition of this led the French physicist Jacques Alexandre Cesar Charles to fly a hydrogen-filled balloon to a height of close to 2 miles in 1783.) In his *Experiments on Air* treatise of the 1780s, Cavendish showed it was possible to mix hydrogen with air, ignite the mixture with a spark, and produce water.

Cavendish's findings attracted Lavoisier, who tried his experiments in reverse. Splitting water molecules into hydrogen and oxygen in an experiment considered to be a breakthrough, Lavoisier combined hydrogen and oxygen to produce 45 g of water (which are still preserved in the French Academy of Science). In February 1785, Lavoisier conducted, before a large group of scientists, definitive experiments proving that hydrogen and oxygen are the fundamental constituents of water. His greatest work, *The Method of Chemical Nomenclature*, labels the "life-sustaining air" oxygen and the "inflammable air" hydrogen.

The first large-scale production of hydrogen was helped along by France's historical circumstances: the storming of the Bastille in 1789 and the subsequent warfare of the French Revolution. Guyton de Norveau, chemist and member of the Comité de Salut Public (Committee for Public Salvation), proposed that the army use hydrogen-filled captive balloons as observation platforms, a proposal that was approved after he and Lavoisier repeated the latter's 1783 experiment. The scientist Jean Pierre Coutelle built the first hydrogen generator; this was a furnace with a cast iron tube, containing iron fillings, with steam piped in at one end and hydrogen emerging at the other. A standard generator was later developed and included temperature-control systems, precursors to the coal gas generators that would later be used for lighting and heating in the early 19th century; coal gas would often be misnamed "hydrogen gas."

2. ORIGINS OF INTEREST IN HYDROGEN ENERGY

In 1800, six years after Lavoisier went to the guillotine in the melee of the French Revolution, another important scientific discovery was made. This was electrolysis, the splitting of water into hydrogen and oxygen by the passage of an electric current through it. William Nicholson and Sir Anthony Carlisle, British scientists, had discovered electrolysis, only a few weeks after Alessandro Volta built the first electric cell. As the 19th century progressed, other proposals to put hydrogen to practical use emerged, and from intriguing sources—including the clergy. In 1820, Reverend W. Cecil, a Fellow of Magdalen College and of the Cambridge Philosophical Society, read "On the Application of Hydrogen Gas to Produce Moving Power in Machinery." His treatise described the space and time limitations of water-driven and steam engines, and the advantages of a hydrogen engine in providing motive force in any location and with little preparation needed. The transactions of the society provide considerable detail of the engine, but no evidence as to whether it was ever built by Cecil.

Hydrogen, and its unusual properties, became a topic of growing interest among not only scientists but also science fiction writers. One of the most well-known fictional discussions of hydrogen is found in Jules Verne's *The Mysterious Island*, one of the writer's final works, published in 1874, almost precisely one century before modern hydrogen research took off. Set in the American Civil War, the book at one point depicts five characters speculating on the future of the Union, and the impact on commerce and industry were coal supplies to dwindle. Whereas four of the characters express fear about the loss of coal-powered machinery, railways, ships, manufacturing, and other forms of modern civilization, the other character, the engineer Cyrus Harding, asserts that they will burn "water decomposed into its primitive elements":

> Yes, my friends, I believe that water will one day be employed as fuel, that hydrogen and oxygen which constitute it, either singly or together, will furnish an inexhaustible source of heat and light, of an intensity of which coal is not capable. Some day the coalrooms of steamers and the tenders of locomotives will, instead of coal, be stored with these two condensed gases, which will burn in the furnaces with enormous calorific power. There is, therefore, nothing to fear. As long as the earth is inhabited it will supply the wants of its inhabitants, and there will be no want of either light or heat as long as the productions of the vegetable, mineral, or animal kingdoms do not fail us. I believe, then, that when the deposits of coal are exhausted we shall heat and warm ourselves with water. Water will be the coal of the future.

Verne may have skipped two energy transitions along the path from coal to hydrogen—oil and natural gas—and therefore he missed what energy expert Robert Hefner would in 2000 depict as the "age of energy gases" (Fig. 1). Nor did Verne mention what the primary source of the hydrogen might be. But it is a prescient passage, given the state of late-19th-century science.

Another adventure story, *The Iron Pirate*, authored by Max Pemberton and published in England in 1900, featured speedy battleships using hydrogen engines. The book influenced the British scientist W. Hastings Campbell, as he related to colleagues at Britain's Institute of Fuel in 1933. His reference was noted in the *Journal of the Institute of Fuel* as "another instance of the very annoying persistence with which art always seemed to anticipate discoveries."

Interest in hydrogen as a fuel had begun to blossom in the wake of the First World War, prompted in part by concern about access to energy supplies. At first, interest was strong in Germany and England, as well as in Canada. The early 20th century had seen the emergence of several new companies with a stake in hydrogen production, such as the Canadian Electrolyser Corporation Limited (now Stuart Energy Systems, one of the world's leading electrolytic hydrogen plant manufacturers). The firm arose from the interest of its founder, Alexander Stuart, in utilizing the nation's considerable excess hydroelectric capacity. The first commercial uses of its electrolyzers were not, however, to make hydrogen fuel, but rather to make hydrogen and oxygen for steel-cutting procedures. Other uses included fertilizer production, and in 1934 Ontario Hydro constructed a 400-kW electrolysis plant. But World War II and growing use of natural gas in Canada halted the project.

Back in Britain, a young Scottish scientist, J. B. S. Haldane, prophesied, in a lecture at Cambridge University in 1923, hydrogen as the fuel of the

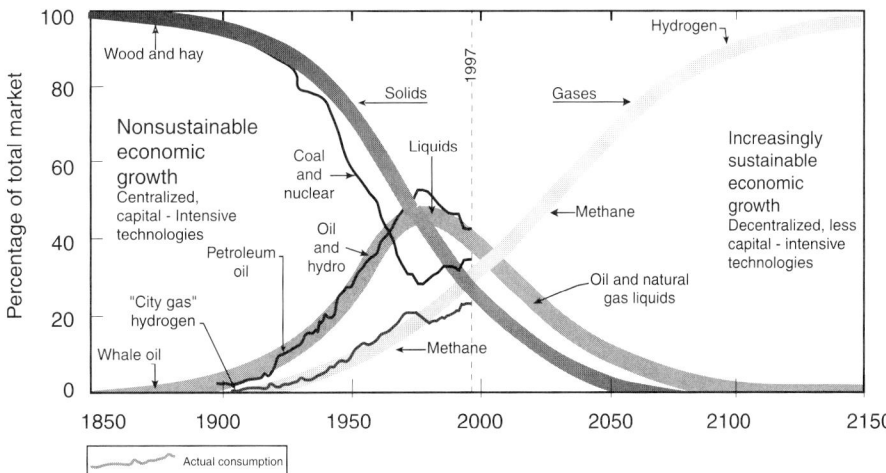

FIGURE 1 In the "age of energy gases," the transition in global energy systems. Reprinted from Dunn (2001), with permission.

future—derived from wind power, liquefied, and stored. Future British energy needs would be met by metallic windmills that decomposed water and stored the gases in vast underground reservoirs for later use. Britain's other early researchers and proponents of hydrogen as a fuel included Harry Ricardo, one of the developers of the internal combustion engine, and R. O. King, a member of the British Air Ministry Laboratory.

Germany counted numerous hydrogen supporters among its engineers and scientists, some of them influenced by Verne. Franz Lawaczeck, a turbine designer, sketched concepts for car, trains, and engines and is considered the first to propose the transport of hydrogen energy by pipeline. More whimsical was Hermann Honnef, who envisioned huge steel towers with windmills producing hydrogen. Use of liquid hydrogen in aircraft attracted attention in Italy, the United States, and Germany in the late 1930s. Lecturing before the American Institution of Electrical Engineers, Igor Sikorski cited the fuel's potential for improving the speed and performance of long-distance aircraft. But the most intriguing application was for the German zeppelins, in wide use in the 1920s and 1930s. As zeppelins lost weight through fuel consumption, they typically blew off hydrogen; engineers proposed burning the hydrogen as extra fuel, increasing output and saving energy. Although tests in the Mediterranean were promising, there is no evidence that the principles were applied to typical flights.

A prominent figure in early hydrogen advocacy was the German Rudolf Erren. Erren spent much of the 1930s in Britain, where he developed advanced combustion processes that would allow the use of hydrogen as a fuel or "booster" additive. In the mid-1930s, he and the engineer Kurt Weil proposed to the Nazi government that it convert from the internal combustion engines to these multifuel systems. At the time, Germany was interested in economic self-sufficiency and reducing dependence on imported liquid fuel. Eventually, between 1000 and 4000 cars and trucks were converted to the new system. Erren's efforts extended elsewhere. In Dresden, a hydrogen rail car ran for several years. In England, delivery vans and buses were tested; a prototype plane was readied and hydrogen-propelled submarines and torpedoes drew public interest in the early 1940s. But Erren's possessions were confiscated during World War II, after which he was repatriated to Germany. None of his engines survived the war years.

The war spurred direct interest in hydrogen in areas of the world where fuel supplies were in danger of being cut off. As Australia's wartime fuel needs grew and supplies from Borneo were lost to Japan, the government of the state of Queensland authorized construction of a hydrogen plant in Brisbane, using off-peak electricity. But the Allied victory in 1945 and the reversion to cheap oil and gasoline halted progress—as may be said about postwar hydrogen progress more generally.

3. ROOTS OF THE HYDROGEN ECONOMY CONCEPT

The early 1950s saw a resumption of interest in hydrogen. The British scientist Francis Bacon developed the first practical fuel cell. Though invented in 1836, the fuel cell had yet to see significant applications outside the laboratory, a situation that would soon change with the U.S. space program.

In the 1960s, the idea of hydrogen as a medium for energy storage spread. German physicist Eduard Justi proposed in 1962 the recombination of hydrogen and oxygen in fuel cells, and in 1965 the production of solar hydrogen along the Mediterranean, for later piping to Germany and other nations. John Bockris, an American electrochemist from Australia, proposed in 1962 the energizing of U.S. cities with solar energy from hydrogen, an idea he expanded on in his 1975 book, *Energy: The Solar-Hydrogen Alternative*. Bockris traces the coining of the term hydrogen economy to a 1970 discussion at the General Motors (GM) Technical Center in Warren, Michigan. As a consultant to GM, Bockris was discussing possible alternatives to gasoline in the wake of growing environmental awareness, and the group agreed that "hydrogen would be the fuel for all types of transports." GM began experimental work on hydrogen, but would later be eclipsed by overseas carmakers.

Various visions of the hydrogen economy began to emerge. The Italian scientist Cesare Marchetti, an influential advocate in Europe, promoted the large-scale production of hydrogen from the water and heat of nuclear reactors. Derek Gregory and Henry Linden of the Institute of Gas Technology, leaders in early U.S. hydrogen research and development, were motivated by the prospect of hydrogen as a substitute for natural gas. Space scientists and engineers continued to pursue liquid hydrogen potential. Despite the limited efforts of GM, Ford, and Chrysler, interest among automotive engineers grew. The Army, Navy, and Air Force pursued liquid

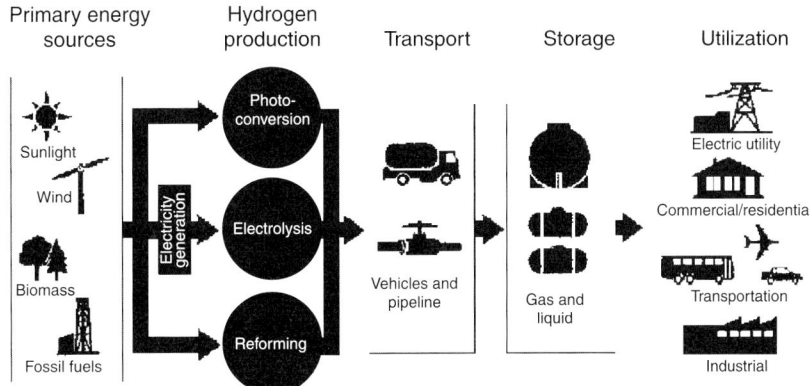

FIGURE 2 The components of a hydrogen energy system. Reprinted from Dunn (2001), with permission.

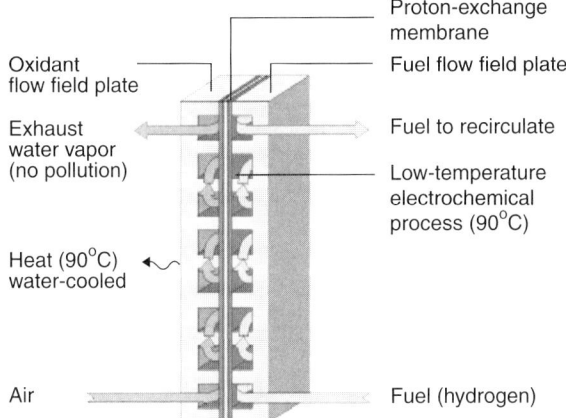

FIGURE 3 A proton-exchange membrane fuel cell. Reprinted from Dunn (2001), with permission.

hydrogen and oxygen applications, their experiments spilling over into the space program. During the 1970s, aided by the two oil shocks, both environmental concerns and energy security accelerated interest in hydrogen. Popular journals such as *Business Week*, *Fortune*, and *Time* ran stories on hydrogen development. Researchers formed groups such as the International Association for Hydrogen Energy (which currently publishes the *International Journal of Hydrogen Energy*).

In the 1980s and 1990s, governments and international organizations began to pay additional attention to the idea of a hydrogen system and its components of production, delivery, storage, and use (Fig. 2). The United States, West Germany, European Community, and Japan began to lay out funding for hydrogen research and development in the mid-1970s, but would cut back as the effects of the oil shock waned. Nevertheless, the institutionalization of hydrogen into energy policy and strategy was underway. The 1980s were a paradoxical period for the hydrogen prospect. Official government support by and large waned, particularly within the U.S. Administration, which significantly cut funding for alternative fuel budgets to make room for its expanding Cold War weapons buildup. One unintended consequence of this benign neglect was the decision of a creative but frustrated engineer named Geoffrey Ballard, then under the employ of the Department of Energy, to leave Washington, D.C. for the west coast of Canada. In Vancouver, British Columbia, Ballard and several like-minded scientists decided to focus on developing electric batteries—a focus that soon switched to the proton-exchange membrane (PEM) fuel cell (Fig. 3).

Although the fuel cell had, by the 1980s, enjoyed several decades of use in the U.S. space program, efforts to bring the technology down to earth had been hampered by issues of cost and efficiency. Most fuel cells relied on significant amounts of the expensive metal platinum and were large and bulky, minor considerations for space programs but major problems for terrestrial commercial applications. Through a combination of ingenuity, government seed money (from the Canadian defense agency), venture capital, and persistence, Ballard's company achieved a 20-fold reduction in platinum requirements for its PEM fuel cell, with commensurate cost reductions. In 1993, Ballard rolled out the first PEM fuel cell bus at its headquarters near Vancouver. That same year, it entered into a cooperative agreement with Daimler Benz (later to become DaimlerChrysler) to jointly develop fuel cells for cars and buses, a partnership that was joined by Ford in 1997, with an overall estimated investment of $300 million. Daimler announced a $1 billion, 10-year commitment to fuel cell development; Daimler's founders were the first to bring the internal combustion engine car to market, and the company appears intent on doing the same for the fuel cell vehicle.

The late 1990s were characterized by an auto race of sorts, as companies paraded demonstration fuel cell vehicles at auto shows and the media buzzed over whether and when carmakers would unveil their first commercial vehicles. Executives also seemed to be staging another competition based on who could sound most bullish, even a bit visionary, about the hydrogen prospect. Echoing the GM engineers of three decades ago, Executive Director Robert Purcell announced in 2000, at the annual meeting of the National Petrochemical and Refiners Association, that "our long-term vision is of a hydrogen economy." As with the introduction of fuel-efficient vehicles in the 1980s, and hybrid electric vehicles in 2000–2002, the Japanese automakers quietly poured money into efforts that threatened to put them ahead of their U.S. and European competitors. In late 2002, Honda and Toyota announced they would be the first to introduce fuel cell cars to the U.S. market.

Energy suppliers were also giving more attention to hydrogen as the century came to a close. In 1999, Royal Dutch/Shell established a core business, Shell Hydrogen, to pursue business opportunities related to the hydrogen economy—an announcement that caused British Petroleum, ChevronTexaco, and ExxonMobil to follow with similar but smaller investments—and advocacy. Testifying before the U.S. Congress, Frank Ingriselli of Texaco (soon to become ChevronTexaco) asserted that "greenery, innovation, and market forces are propelling us inexorably towards hydrogen. Those who don't pursue it, will rue it." Shell Hydrogen would become involved in one of the most interesting hydrogen developments of the late 20th century: the announcement by Iceland that it planned to become the world's first hydrogen economy. The idea of using hydrogen in Iceland was first proposed in the 1970s by a chemistry graduate student named Bragi Arnason as a way to utilize the island's abundant geothermal and hydro resources to reduce oil import reliance. As oil price volatility resurfaced in the 1990s, it coincided with the nation's struggle to address greenhouse gas emissions and with growing global interest in fuel cells and hydrogen. Acting on the recommendations of a panel chaired by Arnason (who had by then earned the nickname "Professor Hydrogen"), the Icelandic government formed in 1999 a consortium, Icelandic New Energy, with Shell, DaimlerChrysler, and the Norwegian power company Norsk Hydro, to investigate the potential for transforming the nation's energy economy to hydrogen between 2030 and 2040. The consortium planned to convert gradually, from petroleum to hydrogen, the buses of the capital, Reykjavik, followed by the island's entire fleets of buses, cars, and fishing boats.

Other islands, likewise vulnerable to oil price volatility, soon followed Iceland's lead. Hawaii approved in early 2001 a jump-start grant to support a public/private partnership in hydrogen research and development, intended eventually to make Hawaii a major player in hydrogen exports. The South Pacific island of Vanuatu launched a feasibility study for transitioning to hydrogen, and plans to be completely renewable-energy-based by 2020. Cuba is also contemplating a move to hydrogen.

These developments shed useful light on the converging forces that are renewing interest in hydrogen. Technological advances and the advent of greater competition in the energy industry are part of the equation. But equally important motivations for exploring hydrogen are the energy-related problems of energy insecurity, air pollution, and climate change—problems that are collectively calling into question the fundamental sustainability of the current energy system. In the 21st century, it is likely that there will be five drivers for hydrogen—technology, competition, energy security, air pollution, and climate change. They are far more powerful drivers than those that existed during the 20th century, and they help to explain why geographically remote islands, stationed on the front lines of vulnerability to high oil prices and climate change, are in the vanguard of the hydrogen transition.

4. GLOBALIZATION OF THE HYDROGEN MOVEMENT

Iceland and other nations represent just the tip of the iceberg in terms of the changes that lie ahead in the energy world. The commercial implications of a transition to hydrogen as the world's major energy currency will be staggering, putting a $2 trillion energy industry through its greatest tumult since the early days of Edison, Ford, and Rockefeller. During 2002, over 100 companies were aiming to commercialize fuel cells for a broad range of applications, from cell phones, laptop computers, and soda machines to homes, offices, and factories, to vehicles of all kinds. Hydrogen was also being researched for direct use in cars and planes. Particularly active were fuel and auto companies, who were spending between $500 million and $1 billion annually on hydrogen. Leading energy suppliers had created hydrogen divisions, and major carmakers continued to pour billions of dollars into a race to put the first fuel cell vehicles on the market between 2003 and 2005. Buses were being test-driven in North America, Europe, Asia, and Australia. In California, 23 auto, fuel, and fuel cell companies and seven government agencies had partnered to fuel and test drive 70 cars and buses. Hydrogen and fuel cell companies had captured the attention of venture capital firms and investment banks anxious to get into the hot new space known as or energy technology (ET).

The geopolitical implications of hydrogen are enormous as well. Coal fueled the 18th- and 19th-century rise of Great Britain and modern Germany; in the 20th century, oil laid the foundation for the unprecedented economic and military power of the United States, of which the fin-de-siècle superpower status, in turn, may be eventually eclipsed by countries that harness hydrogen as aggressively as the United States tapped oil a century ago. Countries that focus their efforts on producing oil until the resource is gone will be left behind in the rush for tomorrow's prize. As Don Huberts, Chief Executive Officer of Shell Hydrogen, has noted, "The Stone Age did not end because we ran out of stones, and the oil age will not end because we run out of oil." Access to geographically concentrated petroleum has also shaped world wars, the 1991 Gulf War, and relations between and among Western economies, the Middle East, and the developing world. Shifting to the plentiful, more dispersed hydrogen could alter the power balances among energy-producing and energy-consuming nations, possibly turning today's importers into tomorrow's exporters.

The most important consequence of a hydrogen economy may be the replacement of 20th-century "hydrocarbon society" with something far better. The 20th century saw a 16-fold increase in energy use—10 times more than the energy used by humans in the 1000 years preceding 1900. This increase was enabled primarily by fossil fuels, which account for 90% of energy worldwide. Global energy consumption is projected to rise by close to 60% over the next 20 years. Use of coal and oil is projected to increase by approximately 30 and 40%, respectively.

Most of the future growth in energy is expected to take place in transportation, where motorization continues to increase and where petroleum is the dominant fuel, accounting for 95% of the total. Failure to develop alternatives to oil would heighten growing reliance on oil imports, raising the risk of political and military conflict and economic disruption. In industrial nations, the share of imports in overall oil demand will rise from roughly 56% today to 72% by 2010. Coal, meanwhile, is projected to maintain its grip on more than half the world's power supply. Continued increases in coal and oil use will exacerbate urban air problems in megacities such as Delhi, Beijing, and Mexico City, which experience thousands of pollution-related deaths each year. And prolonging petroleum and coal reliance in transportation and electricity would increase global carbon emissions from 6.1 to 9.8 billion tons of carbon by 2020, accelerating climate change and the associated impacts of sea level rise, coastal flooding, and loss of small islands; extreme weather events; reduced agricultural productivity and water availability; and the loss of biodiversity.

Hydrogen cannot, on its own, entirely solve each of these complex problems, which are affected not only by fuel supply but also by factors such as population, over- and underconsumption, sprawl, congestion, and vehicle dependence. But hydrogen could make a major dent in addressing these issues. By enabling the spread of appliances, more decentralized "micropower" plants, and vehicles based on efficient fuel cells (of which the only by-product is water), hydrogen use would dramatically cut emissions of particulates, carbon monoxide, sulfur and nitrogen oxides, and other local air pollutants. By providing a secure and abundant domestic supply of fuel, hydrogen would significantly reduce oil import requirements, providing the independence and energy security that many nations crave.

Hydrogen has also been increasingly recognized by the scientific community as facilitating the transition from limited nonrenewable stocks of fossil fuels to unlimited flows of renewable sources, playing an essential role in the "decarbonization" of the global energy system needed to avoid the most severe effects of climate change (Fig. 4). The United Nations 2001 *World Energy Assessment* emphasized "the strategic importance of hydrogen as an energy carrier"; the accelerated replacement of oil and other fossil fuels with hydrogen could help achieve "deep reductions" in carbon emissions and avoid a doubling of preindustrial CO_2 concentrations in the atmosphere, a level at which scientists expect major, and potentially irreversible, ecological and economic disruptions. Hydrogen fuel cells could also help address global energy inequities, providing fuel and

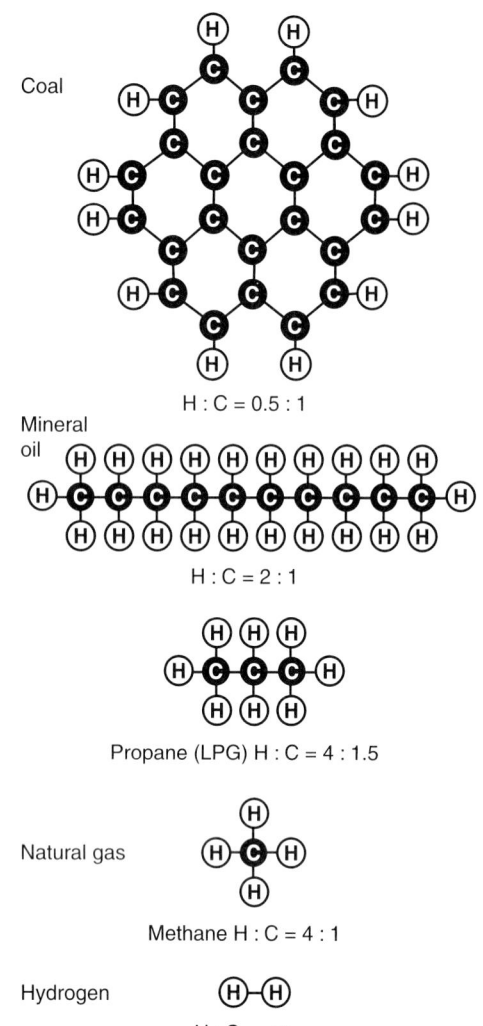

FIGURE 4 The atomic hydrogen:carbon ratios of selected energy sources. LPG, Liquid propane gas. Reprinted from Dunn (2001), with permission.

power and spurring employment and exports in the rural regions of the developing world, where nearly 2 billion people lack access to modern energy services.

Despite these potential benefits, and despite early movements toward a hydrogen economy, its full realization has faced an array of technical and economic obstacles. The feasibility of production, delivery, and use all awaits further improvement. Hydrogen is only beginning to be piped into the mainstream of the energy policies and strategies of governments and businesses, which still tend to aim at preserving the hydrocarbon-based status quo and expanding fossil fuel production. Market structures have been typically tilted toward expanding fossil fuel production. Subsidies to these energy sources, in the form of direct supports and the "external" costs of pollution, were estimated at $300 billion annually in 2001. The perverse signals in the energy market that lead to artificially low fossil fuel prices and encourage their production and use make it difficult for hydrogen and fuel cells (for which the production, delivery, and storage costs are improving but look high under such circumstances) to compete with the entrenched gasoline-run internal combustion engines and coal-fired power plants. This could push the broad availability of fuel cell vehicles and power plants a decade or more into the future. Unless the antiquated rules of the energy economy, aimed at keeping hydrocarbon production cheap by shifting the cost to consumers and the environment, are reformed, hydrogen will be slow to make major inroads.

One of the most significant obstacles to realizing the full promise of hydrogen is the prevailing perception that a full-fledged hydrogen infrastructure—the system for producing, storing, and delivering the gas—would immediately cost hundreds of billions of dollars to build, far more than a system based on liquids such as gasoline or methanol. This is tied to the "chicken-and-egg" dilemma that confronts any new infrastructure. Automakers have been hesitant to mass-produce fuel cell vehicles without assurance of a sufficient fueling network. Energy suppliers, meanwhile, have been resistant to ramping up hydrogen production without assurance of a sufficient number of vehicles to use the fuel. Because of this dilemma, auto and energy companies have been investing millions of dollars into the development of reformer and vehicle technologies that would derive and use hydrogen from gasoline and methanol, keeping the current petroleum-based infrastructure intact. To some analysts, this incremental path—continuing to rely on the dirtiest, least secure fossil fuels as a bridge to the new energy system—represents a costly wrong turn, both financially and environmentally. If manufacturers "lock in" to mass-producing inferior fuel cell vehicles just as a hydrogen infrastructure approaches viability, trillions of dollars of assets could be wasted. Furthermore, by perpetuating petroleum consumption and import dependence and the excess emission of air pollutants and greenhouse gases, this route would deprive society of numerous benefits. By the late 1990s, fossil fuels were the source of some 95% of the hydrogen being produced worldwide—approximately 400 billion cubic meters in 1999, primarily not for energy but for the refining of petroleum and the manufacture of resins, plastics, solvents, and other industrial commodities. Over the long run, critics argued, this proportion would need to be shifted toward renewable sources, not maintained, for hydrogen production to be sustainable.

The "fuel choice" question heated up in the 1990s, complicated by competing life-cycle and "well-to-wheels" studies by government, industry, and nongovernmental organizations. A growing number of scientists in government and industry openly challenged the conventional wisdom of the incremental path. Their research suggested that the direct use of hydrogen would be the quickest and least costly route—to the consumer and the environment—toward a hydrogen infrastructure. Their studies pointed to an alternative pathway that would initially use the existing infrastructure for natural gas (the cleanest fossil fuel and the fastest growing in terms of use) and employ fuel cells in niche applications to buy down their costs to competitive levels, spurring added hydrogen infrastructure investment. As the costs of producing hydrogen from renewable energy fell, meanwhile, hydrogen would evolve into the major source of storage for the limitless but intermittent flows of the sun, wind, tides, and Earth's heat. The end result would be a clean, natural hydrogen cycle, with renewable energy used to split water into hydrogen, which would be used in fuel cells to produce electricity and water—which then would be available to repeat the process (Fig. 5). Some of the experts argued that there were no major technical obstacles to the alternative path to hydrogen. As one researcher put it, "If we really wanted to, we could have a hydrogen economy by 2010." But the political and institutional barriers appeared formidable. Both government and industry had devoted more resources toward the gasoline- and methanol-based route than to the direct hydrogen path, though they were open to the latter: remarked one industry executive, "everyone is placing bets on several horses."

Meanwhile, hydrogen still received a tiny fraction of the research funding that was allocated to coal, oil, nuclear, and other mature, commercial energy sources. Within energy companies, the hydrocarbon side of the business continued to argue that oil will be dominant for decades to come, even as other divisions prepare for its successor. Hydrogen also posed a management challenge, that of looking beyond continuous, incremental improvement of existing products and processes, and toward more radical, "disruptive" technological change. And little had been done to educate people about the properties and safety of hydrogen, even though public acceptance, or lack thereof, would in the end make or break the hydrogen future. The societal and environ-

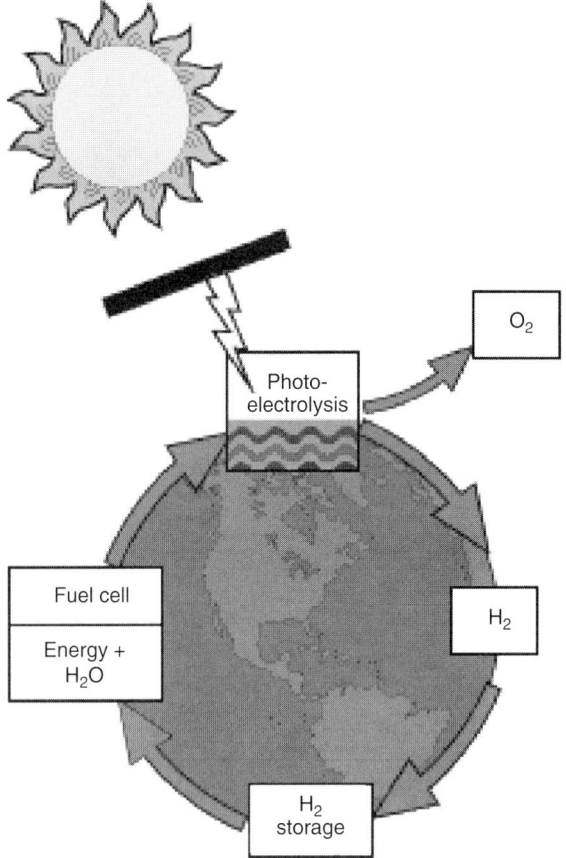

FIGURE 5 A renewable-hydrogen cycle. Reprinted from Dunn (2001), with permission.

mental advantages of the cleaner, more secure path to hydrogen illustrated the essential—and underrecognized—role for government in the hydrogen transition. Indeed, without aggressive energy and environmental policies, the hydrogen economy is likely to emerge along the more incremental path, and at a pace that is inadequate for dealing with the range of challenges posed by the incumbent energy system. Neither market forces nor government fiat could, in isolation, move society down the more direct, more difficult route. The challenge appears to be for government to guide the transition, setting the rules of the game and working with industry and society toward the preferable hydrogen future.

This catalytic leadership role, some advocates have argued, would be analogous to that played by government in launching another infrastructure in the early years of the Cold War. Recognizing the strategic importance of having its networks of information more decentralized and less vulnerable to attack, the U.S. government undertook critical research, incentives, and public/private collaboration into the development of what we now call the Internet. A similar case could be made for strategically laying the groundwork for a hydrogen energy infrastructure that best limits vulnerability to air pollution, energy insecurity, and climate change. Investments made at present would heavily influence how, and how fast, the hydrogen economy emerges in coming decades. As with the Internet, putting a man on the moon, and other great human endeavors, the dawning of the hydrogen age was unlikely to result from business as usual.

5. FUTURE OF HYDROGEN HISTORY

While its effects continue to ripple outward, the events of September 11, 2001 will become an important reference point in the history—written and unwritten—of hydrogen's development as an energy source. Interest in hydrogen surged in the aftermath of the 2001 terrorist attacks on the United States. Subsequent discussions of the appropriate military response revealed the extent to which dependence on Middle East oil had complicated U.S. foreign policy choices in that region. By chance, Shell had planned to unveil its latest long-term energy scenarios in New York City the following month. Their exposition of uncertainties, discontinuities, the "unthinkable," and the "unknowable" struck a strong chord at a moment when an unthinkable event had just happened—one that, to more business and political decision makers, made the vision of a hydrogen economy even more desirable than before.

Presenting the scenarios, Shell chairman Philip Watts presented two scenarios, "Dynamics as Usual" and "Spirit of the Coming Age," and three factors that would drive changes in the energy system: resource scarcity, new technology, and shifting social and personal priorities. Under the first scenario, there is an evolutionary progression or "carbon shift" from coal to gas to renewable energy. The second scenario involves "something more revolutionary: the potential for a truly hydrogen economy, growing out of new and exciting developments in fuel cells, advanced hydrocarbon technologies and carbon dioxide sequestration." The prominence of hydrogen in this scenario—not a prediction, but a tool for helping managers plan for different possible futures and make better strategic decisions—is one indication of the growing attention being given to hydrogen by the world's leading multinationals.

Hinting at hydrogen's ascendancy as well is *Tomorrow's Energy* by Peter Hoffmann, a journalist who had followed hydrogen and fuel cell developments since the 1970s and who edited the well-respected industry periodical *Hydrogen and Fuel Cell Letter*. His work originally began as an update of his 1981 book *The Forever Fuel*. But it soon became a new book, due to the progress in hydrogen developments that had since been made. Among the late-20th-century advances reported by Hoffmann was the latest discovery related to one of hydrogen's biggest public relations problems: the Hindenburg. The prospect of hydrogen as a fuel often raises safety questions, in part because of its association with the German airship that exploded in 1937, taking 36 lives. For years, it was widely believed that the cause of the explosion was the ignition of the hydrogen gas used for lifting the ship. Bain Addison, a retired National Aeronautics and Space Administration (NASA) scientist who doubted this explanation, carefully studied film footage and documents relating to the incident. In 1997, he publicized his surprising finding that the zeppelin's cover had been painted with iron oxides and aluminum, compounds used in rocket fuel. Addison also uncovered evidence that the Nazi government was aware of the design flaw but suppressed this information.

Beyond the Hindenburg, the safety of hydrogen can be compared with that of other fuels in use today. Hydrogen does have a wide range of limits for flammability and detonability, which means that a broad range of mixtures of hydrogen in air can lead to a flame or explosion. However, the lower limits are most relevant to transportation uses, and in this regard, hydrogen is comparable to, or better than, gasoline and natural gas. At this lower flammability limit, the ignition energy, i.e., the energy in a spark to ignite a fuel mixed in air, is about the same for hydrogen and methane. Hydrogen is also nontoxic, unlike methanol or gasoline in higher concentrations. And hydrogen is very buoyant, escaping quickly from leaks, whereas gasoline puddles, causing its fumes to build up. The prevention, detection, and management of hydrogen leaks comprise an important safety issue, requiring that areas where hydrogen is stored and dispensed be well ventilated.

Although the chemical industry routinely handles large quantities of hydrogen safely, the question is whether this safety record will be transferred to hydrogen vehicle and refueling systems. Several studies in the 1990s explored this question. A 1994 report by researchers at Sandia National Laboratories stated that "hydrogen can be handled safely, if its unique properties—sometimes better, sometimes worse, and sometimes just different from other fuels—are respected." A 1997 study by Ford Motor Company concluded that, with proper engineering, the safety of a hydrogen fuel cell vehicle would potentially be better than that of a gasoline or propane vehicle. To ensure safe and standardized practices for using hydrogen, several national and international organizations have begun to develop codes and standards for hydrogen and fuel cells.

Another indicator of the increased attention given to hydrogen is the publication of *The Hydrogen Economy* by social critic Jeremy Rifkin. Rifkin, who has written extensively on the social and economic implications of scientific and technological trends, writes that hydrogen and the communications revolution will create a relationship that "could fundamentally reconfigure human relationships in the 21st and 22nd centuries. Since hydrogen is everywhere and is inexhaustible if properly harnessed, every human being on Earth could be 'empowered,' making hydrogen energy the first truly democratic energy regime in history." The resulting hydrogen energy web "will be the next great technological, commercial, and social revolution in history." Rifkin's utopian vision has engendered some justifiable skepticism from the corporate world. A review in *Business Week* opined that "the hydrogen hype...may be too much, too soon. While hydrogen's abundance and cleanliness make it the favorite as the fuel of the future, fossil fuels still have a cost advantage. And it's not obvious that even when hydrogen does come into widespread use, democracy will follow." The article argues that sharp cost declines in the cost of renewable energy, not a shortage of oil, will be "the factor that forces the transition to a hydrogen economy....The hydrogen economy will dawn when systems based on renewables and hydrogen beat hydrocarbons in straight competition for consumers' dollars."

Indeed, hydrogen still faces considerable cost hurdles at all stages of the energy system, from production to transportation to storage to use. In addition to major advances in renewable energy, improvements in storage technologies such as carbon nanotubes and metal hydrides, development of hydrogen-compatible pipelines, and further strides with fuel cells are among the prerequisites to making hydrogen competitive with hydrocarbons. The most economic method of producing hydrogen at present, via the reformation of natural gas, has been mostly limited to feedstock production; extracting hydrogen from electrolysis from solar or wind power costs three to five times more.

At the same time, and as recognized to varying degrees by business and government, renewables and hydrogen do not face a level playing field today, making concerted public action essential to the hydrogen transition. Japan had set the pace in the global race to hydrogen with its 1993 announcement of a 10-year, $2 billion commitment to promote hydrogen internationally through its World Energy Network (WE-NET) Project. But the stakes had been raised by 2002. The European Commission (EC) unveiled a $2 billion, 4-year investment in hydrogen technology research. In announcing the initiative, EC President Romano Prodi argued that the program would be as important to Europe as the space program was for the United States in the 1960s. "It's like going to the moon in a series of steps....We expect an [even] better technological fallout."

The U.S. government, often criticized for supporting hydrogen less aggressively compared to its European and Japanese counterparts, also moved forward in 2002, announcing a "Freedom Car" partnership with the "big three" automakers to develop fuel cell vehicles. The National Academy of Sciences commissioned a panel study to examine the technical potential of transitioning to a hydrogen economy. Soon after, the Department of Energy unveiled an ambitious hydrogen road map, a blueprint crafted by public and private experts to coordinate the long-term development of a hydrogen economy. In an introduction to the report, Secretary Spencer Abraham sounded almost reminiscent of Jules Verne: "To talk about the 'hydrogen economy' is to talk about a world that is fundamentally different from the one we now know. A hydrogen economy will mean a world where our pollution problems are solved and where our need for abundant and affordable energy is secure...and where concerns about dwindling resources are a thing of the past."

It has taken more than a century, but the idea of hydrogen as an energy source has moved off the pages of science fiction and into the speeches of business and political leaders. In other words, hydrogen energy has evolved from a matter of scientific and technological curiosity to a major strategic issue for political and corporate decision makers. Whether it continues to sustain such interest will determine whether a hydrogen economy truly emerges or whether hydrogen remains, as enthusiasts have called it for several decades, "tomorrow's energy" or "the fuel of the future."

The history of hydrogen as an energy source is in many respects only beginning, with its future path and direction full of many possibilities. However it unfolds, the story of hydrogen in the 21st century promises to be as exciting as *The Prize*, Daniel Yergin's epic about oil in the 20th century. Whether, how, and how fast the hydrogen economy is realized hold enormous consequences for humanity's social, economic, political, and environmental future. That, in turn, will hinge on whether hydrogen becomes not only an interesting possibility, but also a compelling imperative. The hydrogen endeavor is a truly an historic one, bringing to mind President John F. Kennedy's impassioned appeal for the Apollo space program: "There are risks and costs to a program of action, but they are far less than the long-range risks and costs of comfortable inaction."

SEE ALSO THE FOLLOWING ARTICLES

Coal Industry, History of • *Combustion and Thermochemistry* • *Electricity Use, History of* • *Hydrogen, End Uses and Economics* • *Hydrogen Production* • *Hydrogen Storage and Transportation* • *Manufactured Gas, History of* • *Natural Gas, History of* • *Nuclear Power, History of* • *Oil Industry, History of*

Further Reading

Dunn, S. (2000). The hydrogen experiment. *World Watch* **13**(6), 14–25.

Dunn, S. (2001). "Hydrogen Futures: Toward a Sustainable Energy System." Worldwatch Pap. 157. Worldwatch Institute, Washington, D.C.

Hoffmann, P. (2001). "Tomorrow's Energy: Hydrogen, Fuel Cells, and the Prospects for a Cleaner Planet." MIT Press, Cambridge, Massachusetts.

Jensen, M. W., and Ross, M. (1999). The ultimate challenge: Building an infrastructure for fuel cell vehicles. *Environment* **42**(7), 10–22.

Koppel, T. (1999). "Powering the Future: The Ballard Fuel Cell and the Race to Change the World." John Wiley & Sons, New York.

McNeill, J. R. (2000). "Something New Under the Sun: An Environmental History of the Twentieth-Century World." W. W. Norton & Co., New York.

Ogden, J. M. (1999). Prospects for building a hydrogen infrastructure. *Annu. Rev. Energy Environ.* **24**, 227–279.

Rifkin, J. (2002). "The Hydrogen Economy." Penguin Putnam, New York.

Royal Dutch/Shell Group. (2001). "Energy Needs, Choices and Possibilities: Scenarios to 2050." Royal Dutch/Shell, Amsterdam.

Veziroglu, T. N. (2000). Quarter century of hydrogen movement, 1974–2000. *Int. J. Hydrogen Energy* **25**, 1143–1150.

U.N. Development Programme, U.N. Department of Economic and Social Affairs, and World Energy Council. (2000). "World Energy Assessment Report." United Nations, New York.

U.S. Department of Energy. (2002). "National Hydrogen Energy Roadmap." Department of Energy, Washington, D.C.

Yergin, D. (1991). "The Prize: The Epic Quest for Oil, Money, and Power." Simon & Schuster, New York.

Hydrogen Production

GENE D. BERRY
Lawrence Livermore National Laboratory
Livermore, California, United States

1. Introduction
2. Hydrogen Production Using Fossil Energy
3. Thermodynamics of Water Splitting
4. Multiple-Step Thermochemical Cycles for H₂O Decomposition
5. Electrolysis of Water and Steam
6. Overall Prospects for Hydrogen Production

Glossary

Carnot efficiency The maximum efficiency with which thermodynamic work can be produced from thermal energy flowing across a temperature difference; the Carnot efficiency is given by $\eta_{\text{Carnot}} = 1 - \frac{T_{\text{low}}}{T_{\text{high}}}$.

Gibbs free energy The minimum thermodynamic work (at constant pressure) to drive a chemical reaction (or, if negative, the maximum work that can be done by the reaction); the standard Gibbs free energy necessary to decompose H_2O into H_2 and O_2 is 237 kJ/mol. The Gibbs free energy decreases with temperature, but increases with the pressure of hydrogen (and oxygen) produced.

lower heating value (LHV) Thermal energy released when a fuel is burned with air (oxygen) at ambient temperature and pressure, assuming that water in the exhaust remains as vapor, where 1 kg of hydrogen has an LHV of 120 MJ, nearly the same as a gallon (3.785 L) of gasoline; the higher heating value (HHV) of hydrogen is 142 MJ/kg, the thermal energy released when hydrogen reacts with air (oxygen) to produce liquid water exhaust.

Nernst equation The thermodynamic minimum voltage (i.e., in equilibrium) of an electrochemical reaction; it is given by the Gibbs free energy per mole of electrons involved in the reaction divided by Faraday's constant (F), defined as 96,500 Coulombs of charge per mole of electrons.

overvoltage The additional voltage, in excess of the Nernst voltage, required to overcome electrochemical reaction kinetics and resistance to the flow of ions and electrons necessary to drive electrochemical reactions at finite reaction rates; water electrolysis has a Nernst voltage of 1.23 V under ambient conditions.

steam reforming The dominant method of commercial hydrogen production; it is based on reacting methane with water to produce hydrogen and carbon dioxide according to the (endothermic) reaction: $CH_4 + 2H_2O \rightarrow 4H_2 + CO_2$.

thermochemical (hydrogen) cycle A multiple-step chemical reaction cycle that sums to the overall production of hydrogen (and oxygen) by water decomposition; the Carnot efficiency of the highest temperature step places a theoretical limit on the overall hydrogen production efficiency of such a cycle.

Composed of just one proton and an accompanying electron, and accounting for 90% of all matter in the universe, hydrogen (H_2) is by far the simplest, most abundant, and lightest element. H_2 offers distinct advantages as an alternative fuel. Its extreme buoyancy allows H_2 to disperse rapidly when accidentally released or spilled, ultimately escaping the earth's atmosphere or returning as water (H_2O) when combined with atmospheric oxygen (O_2). H_2 can be produced from H_2O using virtually any energy source(s), even electricity, allowing strategic economic and environmental flexibility, especially if generated by electrolysis. H_2 is ecologically ideal because it is carbon free, forming H_2O when burned (or otherwise reacted with O_2) to release chemical energy. Among alternative fuels, only H_2 is likely to be light enough for aircraft and, therefore, the only alternative transportation fuel capable of being universal.

1. INTRODUCTION

The widespread use of hydrogen (H_2) faces three fundamental obstacles. First, as the lightest known molecule, H_2 possesses extreme physical properties that make compressing, liquefying, and storing H_2 costly and energy intensive. Second, high-pressure gaseous H_2, or even liquid hydrogen (LH_2), has a very low energy density relative to polyatomic hydrocarbon fuels (e.g., gasoline [$\sim C_8H_{18}$], natural gas [CH_4], ethanol [CH_3CH_2OH], methanol [CH_3OH]), whose molecules contain more hydrogen atoms than does H_2. Finally, like electricity, H_2 is an energy carrier that must be generated from high-value nuclear, solar, or fossil energy (almost certainly by decomposing water [H_2O]). H_2 consequently entails an energetic and economic premium proportional to the energy consumed in H_2 production and storage. This premium will force H_2 to be more costly than fossil fuels (per unit energy) until nonfossil energy sources cost significantly less than fossil fuels.

Higher fuel economy can compensate for the high cost of H_2 in applications where heavier fossil fuels are at a disadvantage (e.g., spacecraft, aircraft). H_2 can also be used with the greatest efficiency of any fuel (1) in conventional combustion engines, (2) by flameless oxidation on catalytic surfaces, or (3) by electrochemical oxidation to directly generate electricity in fuel cells. Future high efficiency (~ 80 mpg, 34 km/L equivalent) H_2-fueled hybrid electric or fuel cell automobiles will travel 400 miles (640 km) on less than 5 kg of H_2 fuel (energy equivalent to 5 gallons of gasoline).

A full-fledged H_2 economy will likely require not just high fuel efficiency but also full realization of (and capitalization on) the unique benefits of H_2 as an alternative fuel. These include enduring ecological advantages when made from carbonless energy; a broad spectrum of H_2 production energy sources, methods, scales and related storage options and distribution infrastructures; and universal utility throughout transportation (land, sea, and air), energy storage, and industrial sectors.

H_2 fuel can be generated by decomposing H_2O using thermal, electrolytic, or chemical methods, perhaps in combination (Fig. 1). The various H_2 production approaches are characterized by extreme contrasts in technological maturity, energetics, ecological potential, and their implications for feasibility on a global scale.

It is the author's belief that the worldwide replacement of fossil fuels by H_2 will be driven chiefly by the potential scope, efficiency, and energetic costs of H_2 production technologies as they approach the thermodynamic limits of H_2O decomposition. Consequently, this article does not consider H_2 production methods that are ultimately inefficient pathways for harnessing solar energy or are likely to have limited applicability (e.g.,

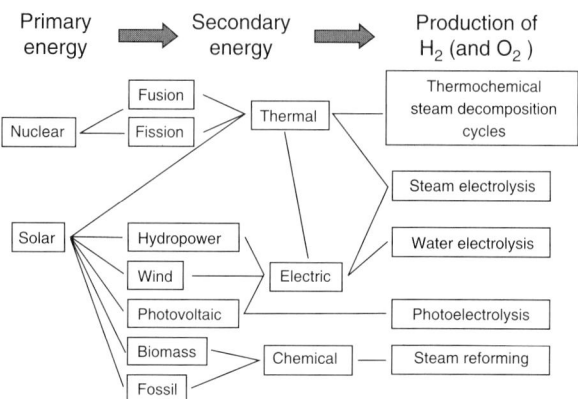

FIGURE 1 Flowchart of hydrogen production pathways and methods. Hydrogen is not an energy source but rather a carrier of energy from nuclear or solar primary energy sources (or indirectly through fossil, biomass, wind, and/or hydroelectric pathways). Energy can be stored in hydrogen by decomposing water using chemical, thermal, and/or electrical energy by steam reforming, electrolysis, photoelectrolysis, or thermochemical cycles.

biomass gasification; biological, photobiological, and photochemical production). Instead, this article surveys historical and current H_2 production using fossil energy, the thermodynamics of producing H_2 exclusively from H_2O, and the two chief technological options for carbonless H_2 production on a global scale: multiple-step thermochemical cycles and electrolysis.

2. HYDROGEN PRODUCTION USING FOSSIL ENERGY

Following Lavoisier's identification of hydrogen as an element of water in 1783, large amounts of H_2 gas were first produced in 1794 by reacting H_2O with iron (Fe) for suggested use in military observation balloons in France. H_2 was produced as a heating fuel during the late 19th century by reacting coal (carbon) with H_2O to generate mixtures of H_2 and carbon monoxide (CO)—known as town gas, blue gas, or blue water gas. Pipelined town gas is still used in parts of Europe, but these gases were ultimately displaced by natural gas (chiefly CH_4) in the United States following World War II, although the oil shocks of the 1970s revived interest in coal gasification to generate H_2 as a precursor to synthetic natural gas (SNG).

Natural gas prices have since fallen and economics now favor the reverse, that is, producing H_2 from natural gas (CH_4) by the process known as steam-methane reforming, essentially harvesting H_2 from H_2O using natural gas by the following overall (endothermic) reaction:

$$2H_2O(l) + CH_4(g) + 245 kJ \rightarrow CO_2(g) + 4H_2(g). \quad (1)$$

Steam methane reforming can be 70% efficient based on the lower heating value (LHV) of hydrogen (120 MJ or 33.3 kWh/kg), with substantially lower capital investment than other H_2 production methods. Purge gases, diverted feed methane, and even nonfossil sources (e.g., solar, nuclear) can provide the additional thermal energy to drive steam reforming at peak temperatures of approximately 1100 K. The product H_2 gas is typically separated by pressure swing adsorption (PSA) relying on the weak adsorption properties of H_2 relative to other larger molecules. CO is the principal contaminant in H_2 from this process. The vast majority of commercial H_2 is currently produced by steam reforming of natural gas at or near the point of use. This H_2 is used in refinery operations (upgrading crude oil) or chemical manufacture (e.g., NH_3, CH_3OH), with only approximately 3% transported as compressed gas or cryogenic liquid (LH_2) in the "merchant" H_2 market.

Unfortunately, steam reforming, or other fossil-fueled H_2 production methods, generate carbon dioxide (CO_2) and are generally not environmentally advantageous. The (high) energy intensity of producing and storing H_2 can actually increase net energy use, attendant pollution, and greenhouse gas emissions associated with "fossil" H_2 relative to direct use of CH_4 or other fossil fuels.

In general, the (high) energy intensity of H_2 production and storage can be justified in applications where H_2 can be used with far greater fuel economy than can hydrocarbon fuels and electricity is not a viable substitute (e.g., supersonic intercontinental aircraft). More recently, the potential ease of CO_2 sequestration has been advanced as a justification for fossil-fueled H_2 production because CO_2 emissions would generally be available to sequester without diluent nitrogen (N_2) from the atmosphere. CO_2 sequestration will be limited to large centralized H_2 production facilities, implying a substantial H_2 pipeline infrastructure and/or energy-intensive cryogenic LH_2 distribution. Sequestration is also likely to be relevant only for H_2 production using coal given that conventional natural gas and petroleum are projected to be minority contributors to global CO_2 emissions over the 21st century.

3. THERMODYNAMICS OF WATER SPLITTING

The desire to use nonfossil energy to efficiently produce H_2 underlies interest in decomposing H_2O into its elements: H_2 and O_2. So-called water-splitting approaches are broadly divisible into thermochemical and electrochemical methods, which can differ dramatically in terms of primary energy sources, scale, intermittent operability, and associated infrastructure implications. However, both approaches fundamentally share the same thermodynamic problem, that is, decomposing H_2O.

H_2 is an energetic and attractive fuel due to an extreme chemical affinity for oxygen (O_2). The earth's atmosphere is so O_2 rich that in thermodynamic equilibrium, even a single H_2 molecule would not remain free within it. The hydrogen–oxygen bond in H_2O is so strong that 1000 cubic meters (m^3) of pure steam (1 atm) in thermodynamic equilibrium will contain just 10 dissociated molecules on average.

Producing H_2 fuel from H_2O requires substantial energy to oppose these thermodynamic forces. The theoretical minimum energy input (work) is given by

$$\text{Minimum Work} = \Delta G_{\text{std}}(T) + RT \ln \left(\frac{[H_2][O_2]}{[H_2O]} \right)^{1/2} \quad (2)$$

where ΔG_{std} is the standard Gibbs free energy of water decomposition, R is the universal gas constant (8.314 J/mol Kelvin), T is absolute temperature, and $[H_2]$, $[O_2]$, and $[H_2O]$ are the partial pressures (in atmospheres) of hydrogen, oxygen, and water, respectively.

The minimum work to decompose saturated water vapor in equilibrium with liquid water and atmospheric air (0.2 bar O_2, 300 K) is 235 kJ/mol H_2O. Thermodynamic tables presume a higher O_2 concentration (1 bar, 300 K), raising the standard minimum work of H_2O decomposition to 237 kJ/mol, equivalent to 1.229 V, the well-known theoretical minimum voltage for water electrolysis.

Furthermore, H_2O decomposition generates entropy (i.e., positive ΔS) by creating three molecules (two H_2 and one O_2) from every two H_2O molecules and, therefore, is endothermic (i.e., absorbs heat), making H_2O decomposition thermodynamically favored by higher temperatures. This is reflected in the reduced work (ΔG) requirement but larger thermal energy ($T\Delta S$) demand at higher temperatures (Fig. 2). Overall, the thermodynamic requirements

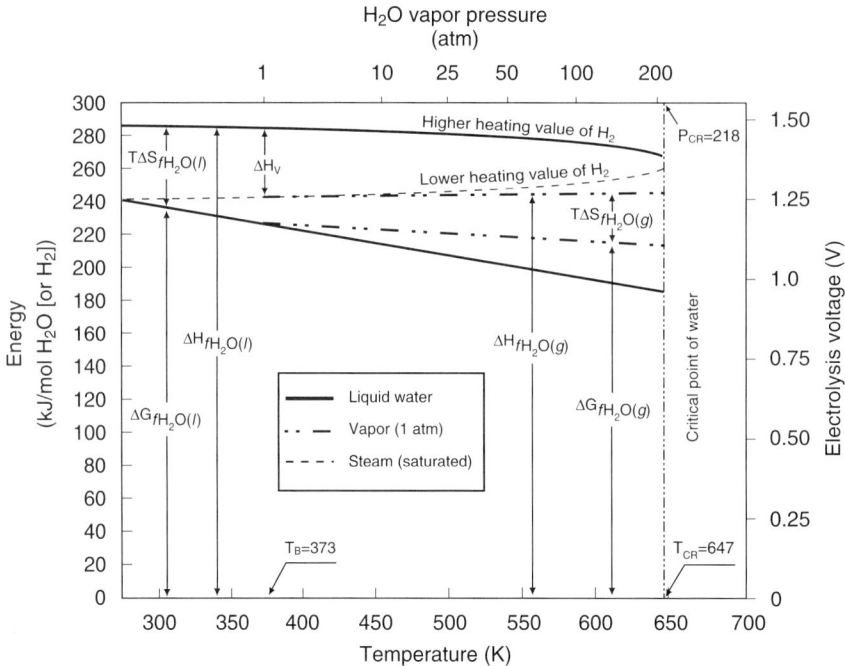

FIGURE 2 Theoretical energetics of water decomposition to produce hydrogen. Generating 1 atm of hydrogen (and oxygen) from liquid water (thick lines) requires a total energy input corresponding to the enthalpy of formation of water $(\Delta H_{fH_2O(l)})$ equivalent to the higher heating value of hydrogen (286 kJ/mol). This energy can be supplied as work or as a combination of work $(\Delta G_{fH_2O(l)})$ and thermal energy $(T\Delta S_{fH_2O(l)})$. The total energy to electrolyze steam at 1 atm (dash double dot) $(\Delta H_{fH_2O(g)})$ is considerably lower due to the thermal energy absorbed (ΔH_v) by the vaporization of water, representing the difference between the higher and lower heating values of hydrogen. This difference approaches zero as the thermodynamic properties of saturated steam (dashed line) and liquid water converge at the critical point (647 K, 218 atm). The work required to produce hydrogen from 1 atm steam $(\Delta G_{fH_2O(g)})$ decreases with temperature due to a corresponding increase in absorbed thermal energy $(T\Delta S_{fH_2O(g)})$. The work required to produce hydrogen from high-pressure (top-axis) liquid water $(\Delta G_{fH_2O(l)})$ decreases more sharply with temperature, but the thermal energy requirement $(T\Delta S_{fH_2O(l)})$ is higher as well. The right-hand axis indicates the energy equivalent voltage to generate hydrogen by electrolysis of water or steam.

for H_2 production from liquid H_2O are minimums of heat and work that sum to the decomposition enthalpy (ΔH) of 286 kJ/mol H_2O, equivalent to the thermoneutral electrolysis voltage of 1.48 V and the corresponding energy content of H_2 fuel as defined by its higher heating value (HHV).

4. MULTIPLE-STEP THERMOCHEMICAL CYCLES FOR H_2O DECOMPOSITION

At sufficiently high temperatures (>3000 K) the thermodynamic work of steam decomposition (ΔG) will shrink to zero and H_2O will decompose spontaneously, theoretically permitting H_2 production using only thermal energy. The conceptual simplicity of thermal decomposition is potentially attractive because it circumvents the intermediate step of electricity generation prerequisite for H_2 production by water electrolysis. However, steam decomposes only slightly at temperatures up to 1200 K, forming approximately 5 ppm H_2 (and 2.5 ppm O_2). Steam decomposition increases exponentially at higher temperatures, reaching 4% H_2 at 2500 K (1 atm), but at these temperatures H_2 and O_2 molecules themselves begin to decompose (Fig. 3), forming hydroxyl (OH) molecules and even individual atoms (O and H), by reactions that generate even greater entropy than does steam decomposition.

This presents fundamental difficulties for direct thermal production of H_2 at very high temperatures (T > 2500 K). Not only are H_2 and O_2 molecules unstable at these temperatures, but most other materials are as well. The related practical issues of long-term material stability, bulk diffusion, reaction with very hot gases or monatomic species, and separation of product gases are generally thought to preclude H_2 production by single-step steam decomposition.

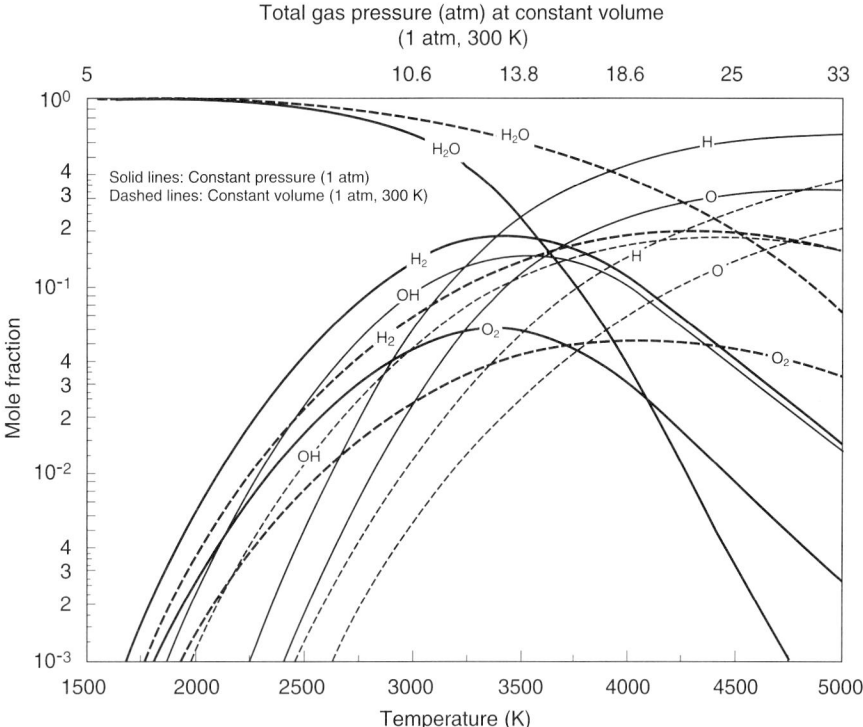

FIGURE 3 Equilibrium thermodynamics of water vapor decomposition. Water vapor decomposes when heated under conditions of constant pressure (solid lines) or volume (dashed lines). Hydrogen and oxygen are the major decomposition products at low temperatures, where less than 1% of water vapor decomposes. At higher temperatures, a greater fraction of water decomposes, but the thermodynamic equilibrium becomes complex and hydroxyl (OH) molecules are also generated in significant amounts. Above the very high temperatures (~ 2500 K) necessary for significant H_2O decomposition, H_2 and O_2 molecules themselves begin to decompose into individual atoms, complicating direct water decomposition schemes. The upper axis indicates the total gas pressure for the constant volume case. At constant volume, pressure increases as the temperature and number of molecules increase. The initial condition is 1 atm, 300 K.

First explored by Funk in 1966, interest in producing H_2 by multiple-step thermochemical H_2O decomposition cycles skyrocketed during the oil shocks of the 1970s. In theory, reduced reaction temperatures (T < 1200 K) compatible with advanced nuclear heat sources, or perhaps concentrated solar energy, can be achieved by dividing H_2O decomposition into multiple steps chosen for optimal properties and operated under varied conditions but that, overall, still sum to H_2O decomposition.

The simplest illustrative example is a notional two-step thermochemical decomposition cycle in which a conceptual oxide material AO absorbs heat, decomposing at a high temperature (e.g., 1173 K), releasing O_2, and later reacts with H_2O to liberate H_2, releasing heat near room temperature (373 K):

$$\text{Heat} + 2AO \rightarrow 2A + O_2 \ (1173\,K) \quad (3)$$

$$2H_2O + 2A \rightarrow 2AO + 2H_2 + \text{Heat} \ (373\,K). \quad (4)$$

Because both reactions sum to steam decomposition, the thermodynamic properties (i.e., entropy and enthalpy) of the individual reactions also sum to the properties of steam decomposition:

$$\Delta S_1 + \Delta S_2 = \Delta S_{H_2O} \quad (5)$$

$$\Delta H_1 + \Delta H_2 = \Delta H_{H_2O}. \quad (6)$$

These relationships, in combination with the temperature difference ($T_{high} - T_{low}$) between reaction steps, enable calculation of the theoretical thermodynamic requirements (i.e., $\Delta G = 0$ for each reaction) sufficient for both to be driven by heat alone:

$$\Delta G_1(T_{high}) = 0 = \Delta G_2(T_{low}) \quad (7)$$

$$\Delta H_{high} = T_{high} \Delta S_{high} \quad (8)$$

$$\Delta S_{high} = \frac{[\Delta G_{H_2O}(T_{low})]}{[T_{high} - T_{low}]}. \quad (9)$$

Assuming that the second reaction step is run at 373 K, the theoretical minimum reaction entropy needed for a high temperature step at 1373 K is 225 J/mol $H_2 \cdot$ K, a value that is five times as high as steam decomposition itself (45 J/mol $H_2 \cdot$ K). This must be compensated by a low-temperature step with a correspondingly large negative reaction entropy (-180 J/mol $H_2 \cdot$ K). Unfortunately, it is doubtful whether chemical reactions based on practical (i.e., simple) substances that generate such large changes in entropy can be found. Candidate reactions for a theoretical two-step thermochemical cycle with a peak temperature below 1000 K have not been proposed to date. Instead, cycles with additional reaction steps have been examined.

Dividing steam decomposition into more than two steps relaxes the thermodynamic requirements on each reaction step, permitting lower peak reaction temperatures. Temperature limitations on nuclear heat sources stimulated theoretical consideration of more than 100 thermochemical cycles using three, four, and five reaction steps during the 1970s, but interest waned during the 1980s, with work carried out only in Japan. Multiple-step reaction cycles complicate thermochemical water splitting substantially, reducing practical efficiencies and increasing equipment cost. Recent work has narrowed to practical consideration of two quite different thermochemical reaction cycles envisioned for use with high-temperature ($\sim$1200 K) nuclear reactors, which can be considered prototypical: the sulfur–iodine (SI) and UT-3 cycles.

The SI cycle is a three-step thermochemical water decomposition cycle conducted entirely during the fluid phase (Fig. 4). The SI cycle relies principally on the high-temperature (and high-entropy) decomposition of the complex molecule hydrogen sulfate (H_2SO_4) into many simpler ones (H_2O, SO_2, O_2), from which O_2 is removed. A second lower temperature reaction step regenerates H_2SO_4 by adding H_2O and captures the liberated hydrogen as hydrogen iodide (HI), which is then decomposed in a third step, producing H_2 at pressures of up to 20 atm. The iodine (I_2) is recycled for use in the second reaction, closing the thermochemical cycle conceptually expressed by the following reaction sequence:

$$H_2SO_4 \rightarrow SO_2 + H_2O + 1/2\ O_2\ (1100\ K) \quad (10)$$

$$SO_2 + 2H_2O + I_2 \rightarrow 2HI + H_2SO_4\ (400\ K) \quad (11)$$

$$2HI \rightarrow I_2 + H_2\ (625\ K). \quad (12)$$

Development of the SI cycle was pursued at General Atomics in San Diego during the 1970s, with bench scale experiments predicting H_2 production efficiencies as high as 44% (LHV). Potential improvement of the second reaction step appears

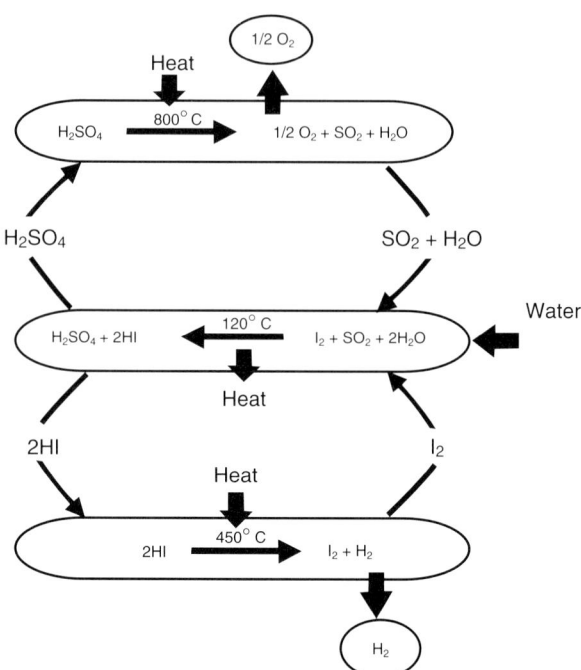

FIGURE 4 The sulfur–iodine cycle. The vertical schematic shows the leading thermochemical hydrogen production cycle using sulfuric acid and iodine. Sulfuric acid is decomposed in a high-temperature step, rejecting liberated oxygen. A second step in the cycle regenerates this sulfuric acid through the addition of water. The hydrogen from this additional water is removed as hydrogen iodide, which is decomposed in a third step liberating iodine for reuse and producing hydrogen for removal from the cycle. The thermal efficiency of this cycle is thought to be capable of approaching 50% on a higher heating value of hydrogen (142 MJ/kg) basis.

possible by avoiding separation of HI from excess I_2. Interest in the cycle has recently been renewed at General Atomics.

The high-temperature, gas-cooled nuclear reactors (HTGR) envisioned to power the SI (or other) thermochemical cycle(s) have been commercially demonstrated, producing electricity up to scales of 330 MWe. These reactors are projected to be capable of 48% efficiency, with capital costs of $1100/kWe and operating costs of $0.015/kWh for fuel, maintenance, and the like, resulting in busbar electricity costs of approximately $0.03/kWh.

Reactors designed to generate H_2 would rely on helium (He) coolant, transporting heat to the thermochemical cycle at approximately 1150 K, requiring additional capital investment for the thermochemical hydrogen plant (even with a credit for elimination of the electrical generation equipment). Assuming that 75 kWh of nuclear heat could produce 1 kg of H_2 (44% LHV efficiency) and that annual operations and plant maintenance costs average 7% of capital investment ($315/kW$_{thermal}$), HTGR SI plants are projected to be capable of producing H_2 for approximately $1.50/kg, equivalent to $0.045/kWh$_{thermal}$ or $12.50/GJ H_2 (LHV). These costs would be lower than those of other nonfossil H_2 production approaches and would be competitive with costs of steam reforming if natural gas prices rise in the future.

An alternative H_2O decomposition cycle, known as the UT-3 cycle, has been under development for decades at the University of Tokyo. This cycle differs substantially from the General Atomics SI cycle in that it relies entirely on gas–solid reactions, with batched forward and reverse reactions occurring in stationary beds instead of recycling fluid reactants and products. The UT-3 cycle can be described thermodynamically by the following sequence of four reactions:

$$CaBr_2(s) + H_2O(g) \rightarrow CaO(s) + 2HBr(g) \quad (1030\,K) \quad (13)$$

$$3FeBr_2(s) + 4H_2O(g) \rightarrow Fe_3O_4(s) + 6HBr(g) + H_2(g) \quad (835\,K) \quad (14)$$

$$Fe_3O_4(s) + 8HBr(g) \rightarrow Br_2(g) + 3FeBr_2(s) + 4H_2O(g) \quad (490\,K) \quad (15)$$

$$Br_2(g) + CaO(s) \rightarrow CaBr_2(s) + 1/2\,O_2(g) \quad (945\,K). \quad (16)$$

The chemistry of this cycle has been tested on the pilot plant scale. Operation consists of two sets of identical stationary beds. Two beds are calcium (Ca), cycled between the oxide (CaO) and bromide ($CaBr_2$). The other two beds cycle iron (Fe) between the oxide (Fe_3O_4) and bromide ($FeBr_2$). Hydrogen bromide (HBr) and H_2O are circulated throughout the beds, and subatmospheric oxygen and hydrogen are extracted from this vapor stream by membranes.

The first bed decomposes input H_2O (steam) over $CaBr_2$ at the peak reaction temperature of approximately 1030 K, yielding CaO and HBr. Similarly, the $FeBr_2$ bed is oxidized by H_2O, forming Fe_3O_4 and $H_2/H_2O/HBr$ mixture, from which H_2 is extracted. The remaining H_2O and HBr regenerate both the $FeBr_2$ from Fe_3O_4 and the $CaBr_2$ from CaO, ultimately producing an O_2/H_2O mixture from which O_2 is extracted and additional H_2O is added to repeat the cycle.

The UT-3 cycle has a projected H_2 production efficiency of 34% (LHV), producing H_2 at subatmospheric pressures. The non-steady-state nature of the cycle could potentially reduce efficiency in long-term operation due to attrition of solids in the bed.

H_2 production by thermochemical cycles faces fundamental challenges beyond the details of any given cycle. To achieve commercial success, every step of the cycle must be accomplished with virtually perfect material recycling to avoid pollution and/or raw material cost issues given the enormous quantities of H_2 (billions of tonnes [metric tons] annually) needed in a mature hydrogen economy. The development of much simpler two-step cycles would facilitate material recycling greatly but necessarily require higher temperatures (T > 1000 K) and the new development of high-temperature heat sources.

Higher temperatures would also raise the theoretical limit on thermochemical H_2 production given by the Carnot efficiency:

$$\eta_{thermal} = \left(1 - \frac{T_{low}}{T_{high}}\right) \cdot \frac{\Delta H_f}{\Delta G_f}, \quad (17)$$

where ΔH_f and ΔG_f are the enthalpy and Gibbs free energy of formation for liquid water under standard conditions, a ratio of 1.20.

5. ELECTROLYSIS OF WATER AND STEAM

Electrolysis is a conceptually simple and elegant method of decomposing water (and other exceptionally stable chemical compounds). Electrolysis was discovered by Nicholson and Carlisle shortly after the invention of the voltaic cell by Alessandro Volta in 1800. Early 19th-century scientists studied electrolysis for decades in advance of knowledge of the chemical elements, knowledge of thermodynamics, or the advent of the internal combustion engine. The ecological and global implications of producing H_2 by industrial scale electrolysis of water were first noted by Jules Verne in his 1874 novel *The Mysterious Island*. Following scientific groundwork by Faraday, Arrhenius, Gibbs, and Nernst, the first commercial electrolyzer was built in 1902.

The fundamental unit of an electrolyzer is an electrochemical cell consisting of three essential elements: an anode, a cathode, and an intervening electrolyte. Electrons flow from the anode (where oxidation occurs) to the cathode (where reduction occurs). The intervening electrolyte conducts ions between the anode and cathode (electrodes), maintaining overall charge neutrality of an electrochemical reaction. Water electrolysis technology is easily classified into three approaches, with each approach conducting one of three ionic species across the electrolyte: liquid alkaline electrolytes, which transport hydroxide (OH^-) ions; proton exchange membranes, typically polymers that transport protons (H^+); and ceramic solid oxide membranes, which conduct oxygen (O^{2-}) ions (Fig. 5).

In electrolysis, the current is directly proportional to the H_2 production rate, which is accelerated by the voltage applied in excess of the thermodynamic minimum. Well-designed electrolyzers avoid parasitic side reactions and achieve Faradaic (current) efficiencies of virtually 100%. Consequently, the electricity consumption (and efficiency) of an electrolysis cell is a function only of the applied voltage necessary to achieve a satisfactory H_2 production rate (e.g., an electrolysis current density of 0.1–1.0 amp/cm^2).

Decomposing water requires a minimum voltage determined by the chemical potential of gases (or liquids) at each electrode, and these variables, in turn, are functions of temperature and relative gas concentrations. These variables influence the overall Gibbs free energy difference (ΔG) between products and reactants. When this is divided by the moles of electrons (n) transferred per mole of H_2 generated and Faraday's constant (F), it gives the electrolysis voltage (E) using the Nernst equation:

$$E = \frac{\Delta G}{nF}. \qquad (18)$$

Under 1 atm each of H_2 and O_2, the minimum work (voltage) necessary to decompose water at 300 K is 237 kJ/mol H_2O or 1.23 V, corresponding to 32.6 kWh/kg of H_2 produced. If liquid, H_2O absorbs substantial heat ($\sim$7 kWh/kg H_2) from the surroundings when electrolyzed into H_2 and O_2.

Thermodynamic irreversibilities (resistance to ion flow through the electrolyte, electron flow through electrodes and wires, and additional overvoltage necessary to accelerate electrode reactions) currently generate waste heat greater than 7 kWh/kg H_2, making water electrolysis exothermic in today's commercial electrolyzers. The transition from endothermic to exothermic occurs (Fig. 6) when the electrolysis voltage exceeds 1.48 V, the thermoneutral voltage corresponding to the HHV of hydrogen (142 MJ/kg H_2) or, equivalently, the enthalpy of water decomposition (286 kJ/mol H_2O). Water electrolysis efficiencies have been customarily quoted based on this thermoneutral (enthalpic) voltage (1.48 V) rather than on the lower Gibbs free energy

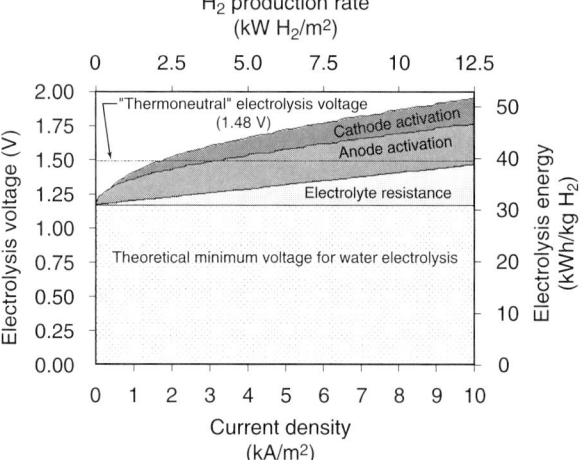

FIGURE 6 Electrolysis voltage and overvoltage mechanisms versus current density. The chart shows operating voltage versus current density for an advanced alkaline electrolyzer producing 1 atm of hydrogen and oxygen at 363 K (90°C), using aggressive technical assumptions. The thermoneutral voltage for water electrolysis is 1.48 V. If excess heat is available, the theoretical minimum (i.e., Nernst) voltage for water electrolysis can be much lower (1.18 V), corresponding to electricity energy consumption of only 31.3 kWh/kg H_2. Typically, excess heat is generated by overvoltage losses due to resistance to ion flow in the electrolyte, activation voltage to accelerate O_2 production at the anode (+), and activation voltage to accelerate H_2 evolution at the cathode (–). Electrolyte resistance is linear with electrolysis rate (current), whereas activation overvoltages rise more slowly and are shown as logarithmic. The chart indicates that the sum of the Nernst voltage and overvoltages (at 10,000 A/m^2) can rise to 1.9 V, corresponding to an electrolysis energy consumption of approximately 51 kWh/kg H_2 or 65% electrolysis efficiency (lower heating value). For reference, 1 gallon of gasoline contains approximately 33 kWh energy equivalent to 1 kg of hydrogen fuel.

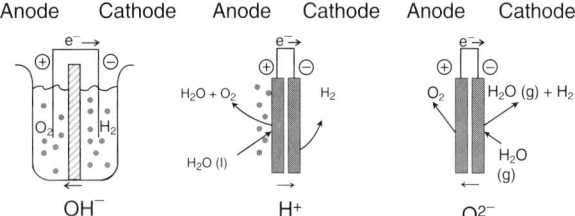

FIGURE 5 The three classes of electrolysis technology. The three classes of electrolysis technology are most easily differentiated by electrolyte. Electrons in a conventional alkaline electrolysis cell (left) generate hydrogen and hydroxide ions that diffuse through an aqueous potassium hydroxide (KOH) electrolyte, where they are converted to oxygen and water at the anode, giving up electrons to the external circuit. The more recently developed proton exchange membrane (PEM) (middle) replaces the corrosive KOH electrolyte with a polymeric membrane that, when hydrated, conducts protons (H^+) from the anode to the cathode, where they combine with electrons to generate hydrogen gas. Steam electrolysis (right) relies on solid oxide ceramics (usually heavily doped ZrO_2) to conduct oxygen ions extracted from steam at the cathode and release them as oxygen gas at the anode, returning electrons to the external circuit.

voltage (1.23). However, this measure is obviously flawed given that, by the measure, the maximum "efficiency" of a perfect electrolyzer could be 120% (1.48/1.23). The correct definition of electrochemical efficiency uses the minimum reversible thermodynamic electrolysis voltage under standard conditions (1.23 V) divided by the actual voltage applied during electrolysis.

The thermodynamics of water decomposition discussed earlier (Fig. 2) apply equally well to thermochemical or electrolytic methods. H_2O decomposition has a positive reaction entropy, reducing the thermodynamic minimum electrolysis voltage at warmer temperatures and potentially allowing for greater use of the excess heat evolved in practical electrolyzers. Warmer temperatures sharply reduce the minimum electrolysis voltage for liquid water, declining from 1.23 V at 300 K to 1.17 V at 373 K. The effect is less dramatic for steam electrolysis (1 atm) above the boiling point of water (from 1.17 V at 373 K to 1.11 V at 573 K) because water vapor has a higher entropy than does liquid water.

Elevated temperatures also increase the effect of steam, hydrogen, and oxygen pressure(s) on minimum electrolysis voltages. High steam pressures reduce (and high hydrogen and oxygen pressures increase) electrolysis voltage logarithmically according to the following (approximate) equation:

$$E = \frac{\Delta G_{std}(T) + RT \ln \frac{[pH_2][pO_2]^{1/2}}{[pH_2O]}}{nF}. \quad (19)$$

A rigorous thermodynamic treatment recognizes the minimum electrolysis voltage as the difference in chemical potential of oxygen (or hydrogen) between the equilibrium gas concentrations at the anode and cathode of any electrolytic cell. This results in the minimum work of electrolysis given by

$$\text{work}_{minimum}(\text{mole of } H_2O) \\ = RT \ln(pO_2^{0.5}) + \Delta H_{std} - T\Delta S_{std} \\ + RT\ln\left(\frac{pH_2}{pH_2O}\right) \quad (20)$$

when oxygen ions are transported across the electrolyte.

Alternatively, for a proton-conducting electrolyte, the minimum work is given by

$$\text{work}_{minimum}(\text{mole of } H_2O) = RT \ln(pH_2) \\ + \Delta H_{std} - T\Delta S_{std} + RT\ln\left(\frac{pO_2^{0.5}}{pH_2O}\right). \quad (21)$$

The major distinction to be drawn from these two equations is that although both equations yield identical voltages under the same overall H_2O, H_2, and O_2 pressures, the ideal voltage (work) requirement of oxygen ion-based electrolysis depends on the absolute pressure (concentration) of O_2 at the cathode but only on the relative H_2/H_2O ratio at the anode. Conversely, ideal electrolysis voltages across proton-conducting membranes depend instead on the absolute H_2 anode pressure (concentration) and the relative O_2/H_2O ratio at the cathode.

At finite reaction rates (i.e., current density), two kinetic mechanisms (ionic transport across the electrolyte and activation of the rate limiting step at each electrode) raise electrolysis voltages above the ideal thermodynamic minimums, reducing efficiency (Fig. 6). First, ions flowing between electrodes and through the electrolyte encounter ohmic resistance that can be minimized only by using thinner electrolytes, more conductive electrolytes, or larger electrolyte cross-sectional areas (reducing current density). Second, electrons traveling from the external circuit to the cathode, and from the anode to the external circuit, also encounter ohmic resistance, which is typically lower than electrolyte resistance and is, to some extent, a design issue rather than fundamental. Overall, the resistance(s) to ion and electron transport during electrolysis is determined essentially by cell geometry and interconnections, varying with temperature as electrode and electrolyte resistivities change.

The ultimate limits to electrolysis efficiency are controlled by a more fundamental factor: the activation overpotential(s) necessary to accelerate the electrochemical reactions at the anode and cathode. Fortunately, electrolysis reactions accelerate exponentially with activation overpotential (η) according to the Butler–Volmer equation:

$$I = I_0(T) \cdot \left(\exp\left(\frac{(1-\beta)F\eta}{RT}\right) - \exp\left(\frac{-\beta F\eta}{RT}\right)\right). \quad (22)$$

The factors determining the activation overpotential (η in volts) necessary to achieve a given current density at an electrode are physical constants (ideal gas constant R and Faraday's constant F), environment (absolute temperature [T]), and electrochemical surface chemistry fundamentals (exchange current density I_o [T] and a reaction symmetry factor β that is usually ~ 0.5). The most important factor is the exchange current density I_o, which describes the speed with which adsorbed molecule, ion, and electron populations transform into one another on the surface of each electrode in dynamic equilibrium (i.e., when no net electrolysis current is flowing).

The electrode kinetics that determine exchange current density (I_o) are also influenced by temperature, partial and absolute activities (pressures) of reactants and products, and the density, reactivity, and location of reaction sites. It is thought that electrochemical reactions on electrodes take place almost exclusively near the triple phase boundary, where electrons from the electrode, ions from the electrolyte, and surface-adsorbed gas molecules all exist in close proximity and can readily participate in reactions. If so, structure and geometry issues at the electrode/electrolyte interface can limit electrolysis efficiency. However, beyond issues of cell and electrode geometry, fundamental material, electronic, ionic, and defect chemistry properties, particularly at the surface, drive electrode kinetics and define the ultimate limits to electrolysis efficiency.

The three major electrolysis techniques are based on three classes of electrolytes (aqueous alkaline solution, proton exchange membrane, and solid oxide) at very different stages of commercialization and development. Because the electrolyte determines the operating temperature needed for sufficiently high ionic conductivity, choice of electrolyte is fundamental to an electrolyzer.

Aqueous alkaline electrolysis is the most mature electrolysis technology developed since the 19th century. Alkaline electrolyzers use an aqueous potassium hydroxide (KOH) electrolyte with metallic electrodes in solution. Proton exchange membrane (PEM) electrolyzers, which entered space and defense applications beginning in the 1970s and are now in commercial use as laboratory hydrogen generators, rely instead on a polymer electrolyte membrane with high protonic conductivity. Steam electrolysis at very high temperatures (>1000 K) was researched alongside solid oxide fuel cells (SOFCs) beginning in the 1960s, was developed in Germany during the 1980s, and is still under development today. Virtually all steam electrolysis concepts rely on solid oxide ceramic materials that conduct oxygen ions at very high temperatures (>1000 K).

Alkaline water electrolysis has developed slowly since commercialization during the early 20th century. A few very large plants (~100 MW H_2 output power or 3000 kg H_2/h) were built through the mid-20th century near large hydroelectric sources to produce H_2 for NH_3 synthesis. Developments have included electrolysis at elevated pressures to minimize electrolyte resistance from gas bubbles and a "zero-gap" design in which the electrolyte is essentially absorbed in an asbestos diaphragm sandwiched between the electrodes. Alternative diaphragm materials more resistant to corrosion than asbestos have been researched, but none has yet appeared to simultaneously satisfy the economic and technical criteria as well as asbestos. Nearly all alkaline electrolyzer manufacturers use the so-called filter-press electrolyzer design linking electrolysis cells in series to construct compact, higher pressure electrolyzers. Lurgi Corporation manufactures electrolysis units capable of electrolyzing water at pressures of 3 MPa (~500 psi). Only Electrolyzer Corporation in Canada uses a unipolar electrolyzer design linking electrolysis cells in parallel, reducing the consequences of any individual cell failure in exchange for higher cell resistance and lower H_2 production pressures.

The fundamental reactions at the electrodes of an alkaline electrolyzer are as follows:

$$\text{Cathode}: \quad 4H_2O + 4e^- = 4H_2 + 4OH^- \quad (23)$$

$$\text{Anode}: \quad 4OH^- = O_2 + 2H_2O + 4e^-. \quad (24)$$

The high concentration of hydroxyl (OH^-) ions in an alkaline electrolyte such as potassium hydroxide (KOH) promotes rapid reaction kinetics and, therefore, smaller activation overvoltages, especially for oxygen evolution at the anode. Higher temperatures would improve electrode kinetics and reduce electrolyte resistance. But corrosion issues restrict alkaline electrolysis to moderate temperatures (<90°C). All commercial alkaline electrolyzers use aqueous KOH electrolyte to minimize corrosion.

PEM electrolyzers eliminate the corrosion issues of alkaline electrolysis by using very pure water and replacing the KOH electrolyte with a proton-conducting polymer membrane. Electrochemical devices using PEMs began during the 1950s. DuPont's introduction of Nafion, a perfluorinated ionomer, improved the durability of electrolyte membranes, enabling long-life applications. The basic process of PEM electrolysis is described by the two electrode half-reactions:

$$\text{Cathode}: \quad (4H^+ + 4e^- = 2H_2) \quad (25)$$

$$\text{Anode}: \quad (2H_2O = 4H^+ + 4e^- + O_2). \quad (26)$$

PEM electrolysis differs fundamentally from alkaline electrolysis in that it is acidic rather than basic. Hydrated protons, rather than hydroxide ions, are conducted across the polymer electrolyte. Proton mobility is very high in perfluorosulfonic acid polymers because protons jump between sulfonate groups attached in high concentrations to a backbone polymer chain. Proton conductivity is controlled by the concentrations of sulfonic groups but increases relatively slowly with temperature, allowing operation near room temperature. PEM electrolysis requires very active noble metal catalysts to achieve high exchange current densities at room temperature and, therefore, to achieve rapid reaction rates.

The development of commercial PEM electrolysis has its roots in military and aerospace applications, principally for O_2 generation rather than H_2 generation. During the 1980s, PEM membranes also entered industrial use for the production of chlorine by electrolysis of brine. The principal advantages of PEM electrolysis are the use of pure water and high-pressure operation, potentially higher than 200 atm ($\sim$3000 psi) without mechanical compression. PEM electrolyzers for H_2 generation have begun to be commercialized only recently, in the wake of much larger development efforts devoted to closely related stationary and automotive PEM fuel cell technology.

Similarly, steam electrolysis development using doped ZrO_2 electrolytes is an offshoot of larger efforts in ceramic SOFC development. The basic principle of steam electrolysis using solid oxide electrolytes is given by the following reactions:

$$\text{Cathode: } H_2O(g) + 2e^- = H_2(g) + O^{2-} \qquad (27)$$

$$\text{Anode: } O^{2-} = 1/2\, O_2(g) + 2e^-. \qquad (28)$$

These reactions occur at very elevated temperatures (1000–1200 K), necessary for adequate conduction of oxygen ions (O^{2-}) through thin layers of ZrO_2 doped with Y_2O_3 to produce large concentrations of oxygen ion vacancies. High temperatures narrow the range of materials that can be used in steam electrolyzer construction, complicated by gas-tight sealing of brittle ceramic materials. But these very high temperatures lower theoretical electrolysis voltages dramatically, accelerate electrode reactions, and raise ionic mobility in the electrolyte, resulting in substantially lower electricity use for steam electrolysis than for any other electrolysis technology.

Steam electrolysis differs fundamentally from other approaches in that the electrode reactions occur between gas phase reactants and crystalline ionic solids rather than aqueous solutions or saturated polymers. A key issue is maximizing the electrode/electrolyte interfacial area in contact with reactant and/or product gases because it is thought that virtually all electrolysis occurs in this region. Contact resistance to ion flow across this interface is also thought to limit the ultimate high current density performance of high-temperature steam electrolysis.

Steam electrolysis was pursued in Germany throughout the 1980s, resulting in pilot plant designs based on ceramic tubes of steam electrolysis cells connected in series. This led to predicted electrolysis requirements of approximately 35 kWh/kg H_2, far lower than the alkaline or polymer water electrolysis requirements of 50 to 60 kWh/kg H_2. However, high-temperature steam electrolysis requires substantial additional thermal energy to generate steam for operation at hot temperatures. Recent solid oxide electrolysis research in the United States has grown out of SOFC development. Technology Management in Cleveland, Ohio, is pursuing steam electrolysis using excess thermal energy from an SOFC.

A long-term possibility for steam electrolysis is the replacement of oxygen ion-conducting ceramic electrolyte materials with proton-conducting ceramics. In principle, this would have the advantage of permitting the production of pure H_2 rather than an H_2/H_2O mixture on the anode side of a steam electrolyzer. The electrode reaction kinetics may differ as well, providing potential opportunities.

In general, high-temperature steam electrolysis appears to offer significant potential to improve electrolysis efficiency but is currently in embryonic development relative to alkaline and PEM electrolysis technologies. Both of the "advanced" electrolysis technologies (PEM and steam electrolysis) would likely develop far more rapidly if PEM fuel cells and/or SOFCs achieve mass commercialization, implying superb underlying electrochemical performance for either or both of these technologies.

6. OVERALL PROSPECTS FOR HYDROGEN PRODUCTION

Currently, the most economic method of H_2 production ($1.00/kg H_2) is steam reforming of natural gas, but future cost reductions are unlikely. In addition, the economic and energetic effectiveness of producing H_2 fuel by consuming natural gas must be compared with direct use of natural gas instead. Because the energy efficiency of direct use may be greater, using natural gas directly may both cost less and produce lower overall greenhouse gas emissions than does using H_2 produced by steam reforming.

Coal gasification to produce H_2 may become competitive if natural gas prices rise, but coal gasification has a higher capital, energy, and carbon intensity than does steam reforming of natural gas. If

CO_2 proves to be sequesterable on a global scale at low cost, both coal and natural gas could power H_2 production well into the 21st century, but at a higher cost than today's approximately $1.00/kg H_2.

It is likely that thermochemical cycles powered by future fission or (eventually) fusion heat sources can produce low-cost H_2 (e.g., $1.50/kg) at moderate pressures, continuously, and on large scales (>200 MW). Very large scales (>2000 MW) may be required in the case of fusion-driven thermochemical H_2 production. The characteristics of thermochemical cycles could have advantages for H_2 production integrated with NH_3 synthesis or LH_2 production for aircraft.

Beyond these two applications, the economic and energetic effectiveness of dedicating dispatchable (i.e., nonintermittent) nonfossil energy sources (e.g., nuclear) to H_2 generation is unclear. Generating electricity instead of H_2 from these sources is generally easier, more efficient, and likely to be more profitable. Therefore, thermochemical H_2 production schemes based on nuclear energy face a strategic question: are high-temperature nonfossil heat sources better applied, in economic and/or environmental terms, to dedicated thermochemical H_2 production or electricity generation, from which any surplus (i.e., low-value) electricity could generate H_2 by electrolysis? Answers to this question depend in part on electric generation mixes and pricing structures. More fundamentally, they depend on the efficiency of electrolysis technology.

Production of H_2 by electrolysis eliminates the need to dedicate primary energy sources to generating only hydrogen or electricity. For moderate capital costs (~$500/kW LHV H_2), electrolysis permits flexible use of primary energy sources to coproduce both electricity and H_2, potentially in response to real-time supply/demand fluctuations and corresponding price signals. Electrolytic H_2 production (either for storage or for transportation fuel) is likely to be the highest value use of surplus electricity from intermittent renewables (e.g., solar, wind), and widespread electrolysis is consequently necessary for these nonfossil energy sources to become dominant.

Electrolysis would also permit a broad range of H_2 refueling infrastructures and onboard storage technologies. H_2 vehicles equipped with metal hydride onboard storage could be refueled overnight in a garage using low-pressure electrolysis. High-pressure electrolysis is the most efficient method of H_2 compression and would allow refueling of vehicles at H_2 pressures of at least 200 atm (~3000 psi). Refueling pressures could be boosted, if necessary, by mechanical compression to more than 350 atm (~5000 psi).

However, the strategic and infrastructure advantages of electrolysis are counterbalanced by higher H_2 production cost due nearly entirely to the price of electricity. If future nonfossil electricity prices drop as low as $0.05/kWh, conventional electrolytic H_2 will cost $2.50 to $3.00/kg (energy equivalent to $2.50–$3.00/gallon of gasoline) for the electricity alone.

Although these prices can be easily tolerated by 80-mpg equivalent fuel cell automobiles, which need only approximately 150 kg H_2 annually ($450/year), they are not economically competitive for H_2-fueled freight trucks and aircraft, which will consume an increasing share of transportation energy in the future. The economic viability of H_2 for these applications will require lower electricity prices and/or lower electrolysis voltages. High-temperature steam electrolysis, whether based on oxygen ion- or proton-conducting ceramic electrolytes, has the greatest prospects for achieving very low electrolysis voltages (~1.0 V) but might not easily withstand intermittent operation, requiring either a dispatchable high-temperature heat source (e.g., nuclear) or thermal storage. Steam electrolysis may be best suited for cost-sensitive continuous operations such as production of NH_3 or LH_2 fuel for aircraft. If the electricity to produce H_2 is not continuously available, such as when wind and solar power are widespread or when electricity transmission and distribution are strained during peak hours, the flexibility of alkaline or PEM technology may offset the necessarily higher electrolysis voltages. Assuming that surplus electricity is available in the future for $0.05/kWh, improving electrolysis voltages from the current 1.9 to 2.0 V to near the thermoneutral voltage (~1.5 V) could reduce the electricity cost of H_2 to approximately $2.00/kg.

SEE ALSO THE FOLLOWING ARTICLES

Alternative Transportation Fuels: Contemporary Case Studies • *Hydrogen, End Uses and Economics* • *Hydrogen, History of* • *Hydrogen Storage and Transportation* • *Natural Gas Processing and Products* • *Oil Refining and Products*

Further Reading

Berry, G., and Lamont, A. (2002). Carbonless transportation and energy storage in future energy systems. *In* "Innovative Energy Strategies for CO_2 Stabilization" (R. Watts, Ed.), pp. 181–210. Cambridge University Press, Cambridge, UK.

Bockris, J. O. (1976). "Energy: The Solar–Hydrogen Alternative." Halsted Press, New York.

Casper, M. S. (1978). "Hydrogen Manufacture by Electrolysis, Thermal Decomposition, and Unusual Techniques." Noyes Data Corporation, Park Ridge, NJ.

Cox, K., and Williamson, K. (1977). "Hydrogen: Its Technology and Implications." vols. 1–5. CRC Press, Cleveland, OH.

Hoffmann, P. (2001). "Tomorrow's Energy." MIT Press, Cambridge, MA.

Ogden, J., and Williams, R. (1989). "Solar Hydrogen: Moving beyond Fossil Fuels." World Resources Institute, Washington, DC.

Winter, C. J., and Nitsch, J. (1988). "Hydrogen as an Energy Carrier." Springer-Verlag, New York.

Hydrogen Storage and Transportation

GENE D. BERRY, JOEL MARTINEZ-FRIAS,
FRANCISCO ESPINOSA-LOZA, and SALVADOR
M. ACEVES
Lawrence Livermore National Laboratory
Livermore, California, United States

1. Introduction
2. Hydrogen in the Context of Fuels and Energy Storage Modes
3. Historical Approaches to Hydrogen Storage and Transportation
4. Liquid Hydrogen
5. Compressed Hydrogen
6. Compressed Cryogenic Hydrogen
7. Hydrogen Storage by Chemical Absorption in Metal Hydrides
8. Hydrogen Storage by Physical Adsorption on Carbon
9. Hydrogen Storage in Chemical Carriers
10. Future Prospects for Hydrogen in Transportation

Glossary

adiabatic A process (e.g., expansion of a gas) in which no heat is transferred to (or from) the external environment; in a reversible adiabatic expansion, as a gas cools, its internal energy is reduced by the amount of work done by the gas on the environment.

critical point The highest temperature and pressure at which the liquid and gas phases of a substance coexist in equilibrium; the critical point of hydrogen is 33 K and 13 atm.

dormancy The duration for which a cryogenic (liquid) hydrogen vessel can absorb heat from the surroundings without venting hydrogen vapor, thereby preventing the vessel internal pressure from exceeding the maximum allowable working pressure.

lower heating value (LHV) Thermal energy released when a fuel is burned with air at ambient temperature and pressure, assuming that the water exhaust remains as vapor, where 1 kg of hydrogen has an LHV of 120 MJ, nearly equivalent to the LHV of 1 gallon (3.785 L) of gasoline.

multiple-layer vacuum insulation Thermal insulation consisting of multiple layers of highly reflective metallized plastic located between the walls of an evacuated vessel.

steam reforming The dominant method of commercial production of hydrogen from methane and water according to the following reaction: $CH_4(g) + 2H_2O(l) + 245 kJ \rightarrow 4H_2(g) + CO_2(g)$.

Composed of a single proton and an accompanying electron, hydrogen is the simplest and lightest element. As a diatomic molecule, hydrogen gas (H_2) is the lightest substance known, nearly 15 times lighter than air and able to escape the earth's atmosphere. Terrestrially, hydrogen is found chemically bound only within heavier substances. In addition to its abundance in the world's oceans and biological matter, millions of years of photosynthetic decomposition of water (H_2O) and atmospheric carbon dioxide (CO_2) is generally believed to be responsible for the creation of the world's fossil fuels, sequestering sizable reserves of hydrogen in hydrocarbon form. This "fossil hydrogen" accounts for more than 30% of the fuel energy in oil and more than 50% of that in natural gas (chiefly CH_4). Today, H_2 is used principally in fertilizer manufacture and as a rocket fuel and chemical. Nearly all commercial H_2 is essentially harvested from CH_4 by reaction with H_2O in a process known as steam methane reforming, which produces H_2 at low cost (if low-cost natural gas is available). Unfortunately, it also generates the by-product CO_2, a greenhouse gas.

1. INTRODUCTION

Recently both industry interest and public awareness of H_2 as a future automobile fuel has increased dramatically. Hydrogen (H_2) differs from other fuels in that it can be produced without carbon dioxide (CO_2) by simple decomposition of water (H_2O) using electricity and/or heat from solar, wind, fission, or fusion power sources. Therefore, H_2 is a versatile and universal carbonless energy carrier, a necessary element for future energy systems aimed at being free of air pollution, CO_2, and other greenhouse gases. If generated from renewable energy, H_2 becomes the crucial link in an inexhaustible global fuel cycle based on the cleanest, most abundant, natural, and elementary substances: H_2, oxygen (O_2), and H_2O.

The physical and chemical properties of hydrogen make its use superior to fossil fuels. H_2 is a simple non-toxic molecule that generates power cleanly and efficiently, even silently and without combustion if desired. However, widespread use of H_2 has been challenging due to its low energy density relative to conventional (hydrocarbon) fuels. Energy density fundamentally drives the feasibility of H_2 fuel by determining the capital, materials, volume, and energy needed for onboard storage.

The economic and physical consequences of energy density, in turn, become strategic, placing limits on the scale, efficiency, cost, range, and shape of H_2 vehicles and refueling infrastructure(s) linking the production of H_2 with its use onboard vehicles. Consequently, improvements in the technology, economics, and energetics of onboard hydrogen storage will likely do the most to speed (or slow) and shape the transition to H_2 as a universal transportation fuel.

2. HYDROGEN IN THE CONTEXT OF FUELS AND ENERGY STORAGE MODES

Hydrogen's position in the energy density hierarchy of fuels and energy storage modes is dictated by physical and chemical fundamentals. Gravitational energy storage is relatively weak and at the base of the energy density hierarchy (Fig. 1). The pneumatic energy of air compressed to 700 atm ($\sim$10,000 psi) is roughly 100 times greater than the gravitational energy of an equal volume of water elevated 370 m. Energy density 10 times greater still is present in chemical bonds. Mechanical energy storage (e.g., kinetic energy of a flywheel) can be on the order of chemical bond energies but can also be constrained by geometry. Batteries generate electricity directly (if slowly) from the energy in chemical bonds but are relatively heavy because they must retain their reaction products to be recharged.

Fuels quickly (e.g., combustion) release the energy in chemical bonds but sacrifice efficiency. Fuels also have very high energy densities when they can rely on the atmosphere to provide reactants (O_2) and absorb exhaust products (H_2O and/or CO_2). On a lower heating value (LHV) basis, 3.75 gallons (1 kg) of liquid hydrogen (LH_2) is energy equivalent (33.3 kWh, 120 MJ) to 1 gallon of gasoline (2.6 kg), making H_2 by far the lightest, but also the least compact, fuel due chiefly to the extreme simplicity of the H_2 molecule.

Unlike polyatomic hydrocarbon fuels such as gasoline (C_8H_{18}), H_2 consists of just two monovalent atoms. This unique structure gives H_2 extreme combustion characteristics (e.g., low ignition energy, low emissivity, wide flammability limits, high flame speed) and physical properties (e.g., high diffusivity, thermal conductivity, buoyancy, incompressibility). An extremely low boiling point (20.3 K), second only to helium, gives LH_2 a very large coefficient of thermal expansion, expanding nearly 40% between 20 and 30 K.

Particularly striking is the very low atomic density of H_2 both as a compressed gas and as a cryogenic liquid. There are fewer hydrogen atoms by volume in LH_2 than in metal hydrides, most fuels, and common molecular liquids such as octane (C_8H_{18}), ethanol (CH_3CH_2OH), methanol (CH_3OH), cryogenic liquid methane (LCH_4), ammonia (NH_3), hydrazine (N_2H_4), hydrogen peroxide (H_2O_2), and even H_2O (Fig. 2).

The low intrinsic energy density of H_2 can be counterbalanced by its efficient utilization. Although automotive scale fossil-fueled heat engines are limited by variance in output and the balance between thermal and frictional losses to efficiencies of approximately 40%, hydrogen-powered fuel cells can circumvent the Carnot cycle and directly generate electricity at approximately 60% efficiency (or higher at partial load). This improves the comparison between H_2 and other fuels on a deliverable energy basis (Fig. 1). The elemental simplicity of H_2 also allows the inverse process (i.e., H_2 generation by electrolysis) to be elegant and efficient, even on a small scale. Thermochemical production of H_2 is simpler, and therefore more efficient, than production of synthetic hydrocarbon fuels (e.g., CH_3OH), which typically involve H_2 as an intermediate.

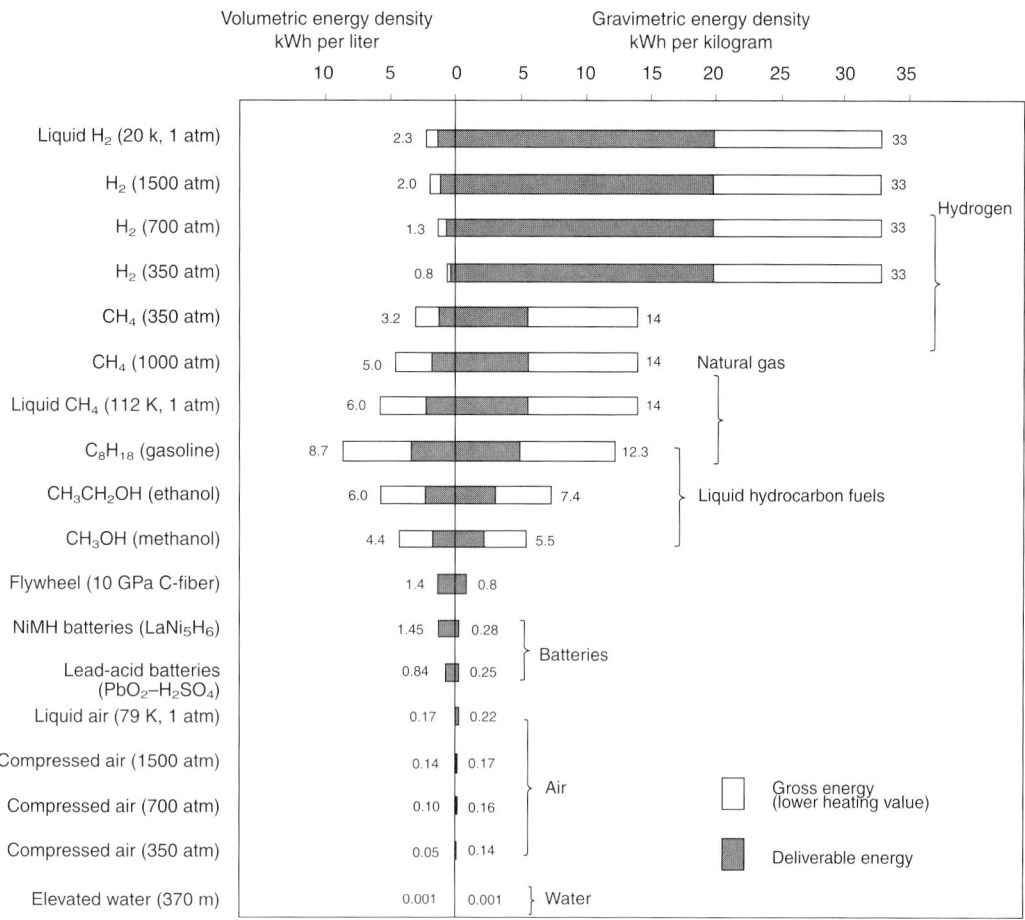

FIGURE 1 Theoretical intrinsic gravimetric and volumetric energy densities of fuels (on an LHV basis) and energy storage technologies. Compressed air and gaseous fuels are shown at 300 K and various pressures, and liquids are shown at 300 K or (if cryogenic) their boiling point. The mass and volume of containment vessels are excluded. Energy densities for flywheels and batteries include only active mass. The values given for compressed air represent maximum possible system work. Numerical values indicate gross energy densities. Maximum deliverable energy densities are indicated graphically for assumed efficiencies of 100% for all energy storage systems, 60% for hydrogen-powered fuel cells, and 40% for hydrocarbon fuels used in automotive scale combustion engines.

3. HISTORICAL APPROACHES TO HYDROGEN STORAGE AND TRANSPORTATION

During the first half of the 20th century, hydrogen was stored infrequently due in part to its low density. At the time, the most important use of H_2 was NH_3 synthesis. Convenience and economics favored producing H_2 by electrolysis, where inexpensive hydroelectric power was available, or else by fossil-fueled thermochemical methods.

H_2 was first used as a fuel in combination with carbon monoxide (CO), distributed by pipeline, and sometimes stored in underground caverns. These H_2/CO mixtures, known as "town gas," were produced by reaction of H_2O with coal and are still distributed by pipeline in Europe today. Pure H_2 pipelines have been operated for decades in the chemical industry, and it should be possible to distribute H_2 through existing natural gas (CH_4) pipeline systems, albeit with reduced energy flow rates and higher compression energies due to the low energy density of H_2. Consideration must also be given to H_2-induced embrittlement of carbon steel pipeline walls. Pipelines and underground storage will be the most economical method of distribution and storage of gaseous H_2 if and when the demand for H_2 becomes large and stable.

When H_2 demand is small, variable, or localized, on-site production by electrolysis or truck transport of compressed hydrogen is viable. Because the high capital and transportation costs of heavy metallic cylinders are magnified by the low energy density of

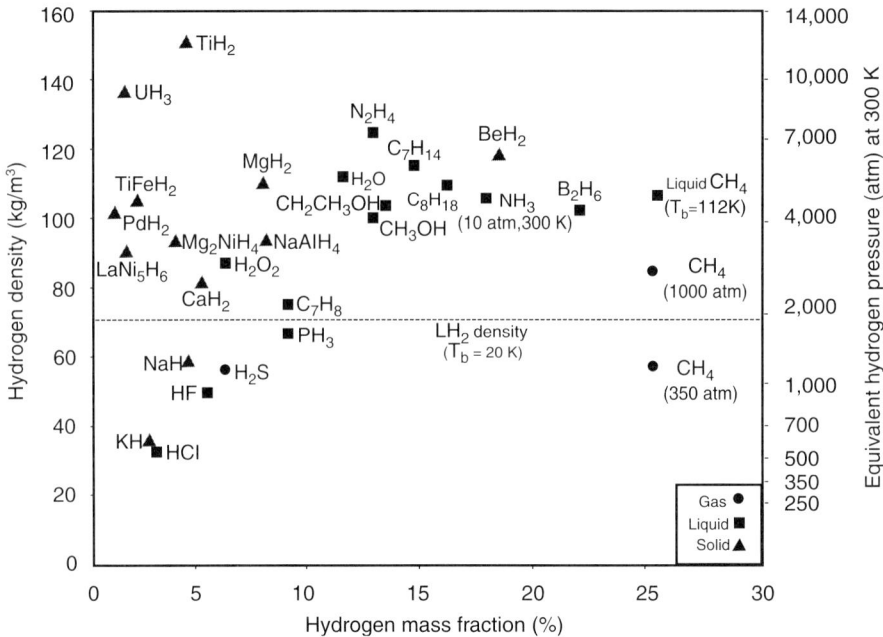

FIGURE 2 Density and mass fraction of hydrogen contained in substances spanning a spectrum of ionic, covalent, and metallic (both high- and low-temperature) hydrides. Hydrocarbon fuels are also shown, including compressed methane (CH_4) at 300 K. The right-hand scale (nonlinear) shows the pressure needed to achieve an equivalent density of compressed hydrogen gas at 300 K. The dashed line indicates that the density (70 kg/m^3) of liquid hydrogen (LH$_2$) at its boiling point (20 K) is substantially lower than the atomic hydrogen density of many compounds.

gaseous H_2, even moderately sized gaseous H_2 end uses (e.g., fuel stations) are more economically met by lightweight LH$_2$ transport by truck, on-site steam reforming, or electrolysis.

H_2 was first liquefied by James Dewar in 1898, but liquefaction did not begin on an industrial scale until the 1950s. LH$_2$ capacity grew during the 1960s to supply the U.S. space program, resulting in large commercial LH$_2$ plants (60,000 kg/day). Smaller LH$_2$ plants in Europe (5,000–20,000 kg/day) are less energy efficient. U.S. LH$_2$ capacity (~250,000 kg/day) is substantially greater than in other parts of the world, and LH$_2$ is distributed exclusively by tanker trucks with capacities of up to 4000 kg.

4. LIQUID HYDROGEN

LH$_2$ use has become routine for launching spacecraft, but liquefying H_2 remains very energy intensive despite decades of industrial experience. Hydrogen liquefaction is complex and energy intensive relative to the other bulk gases (CH_4, N_2, and O_2). Even the noble gases (except helium) are more easily liquefied than is H_2. One complication is that unlike nearly all gases, H_2 has a negative Joule–Thomson coefficient at room temperature, warming when expanded by throttling. H_2 must instead be cooled through work on the environment done by expansion of the gas (e.g., through a turbine). Theoretically, H_2 could be liquefied in a single step if expanded adiabatically from approximately 20,000 atm (to 1 atm). In practice, H_2 is precooled by a closed-loop LN$_2$ refrigerator, cyclically compressed to moderate pressures (<100 atm), and expanded before liquefying by throttling through a Joule–Thomson expansion valve.

An additional complexity is that H_2 molecules, whether liquid or gaseous, exist with two different nuclear spin arrangements. The higher energy form (spins aligned) is orthohydrogen. The lower energy form (spins opposite) is parahydrogen. The relative amounts of each form are in dynamic equilibrium, determined solely by temperature. The equilibrium fraction of parahydrogen is only 25% at 300 K but grows to 99% at 20 K. Spontaneous conversion of the higher energy orthohydrogen to the more stable parahydrogen form at cryogenic temperatures releases enough thermal energy (~500 kJ/kg H$_2$) to vaporize 10% of a given quantity of LH$_2$ within 10 h and 50% within 1 week.

The conversion of orthohydrogen to parahydrogen is intentionally accelerated in LH$_2$ plants, generating additional heat that must be removed.

This increases the theoretical work of H_2 liquefaction from 3.25 to 3.92 kWh/kg LH_2. In industrial practice, producing 1 kg of pure para-LH_2 requires 10 to 14 kWh, equal to 33 to 42% of the fuel energy (LHV) in LH_2. Magnetocaloric refrigeration, employing cyclic magnetization of tailored materials in very high magnetic fields, may ultimately be capable of reducing liquefaction energy requirements to perhaps 7 kWh/kg LH_2.

Despite the energy intensity of cryogenics, LH_2 is the favored method of large-scale hydrogen storage (e.g., ~270,000 kg LH_2 at Cape Kennedy, energy equivalent to 9 million kWh). LH_2 is particularly suited for aircraft, where the higher gravimetric energy density of hydrogen (120 MJ/kg for LH_2 vs 45 MJ/kg for jet fuel) counterbalances its roughly four times greater volume. The first successful test of LH_2 was in one engine of a modified B-57 aircraft during the late 1950s. In 1988, the Soviet Union reported successful testing of hydrogen in one engine of a modified civilian airliner.

Several design studies by Lockheed examined LH_2 airliners during the 1970s, evaluating lightweight fuel tanks that could store two to three times their own weight in fuel (25,000 kg LH_2). These studies indicate that the most favorable locations for LH_2 tanks are at both ends of the fuselage rather than within or supported from the wings. LH_2 aircraft would likely be 10% longer than fossil-fueled aircraft but would compare favorably in terms of fuel economy, wingspan, engine size, and takeoff noise, weight, and distance.

An LH_2 airliner designed to fly 400 passengers 10,000 km would weigh 75% as much as a conventional aircraft at takeoff and use 90% of the fuel energy. If designed for an 18,000-km range, an LH_2 aircraft could weigh only 50% as much as a fossil-fueled airliner at takeoff and use 65% of the fuel energy. It is generally agreed that LH_2 aircraft will compare even more favorably to conventional aircraft for supersonic air travel. LH_2 aircraft may also be able to use the cooling capacity of LH_2 to chill the wings and other surfaces so as to maintain laminar airflow and reduce aerodynamic drag.

The rationale for LH_2 storage onboard automobiles is similar to the case for aircraft but is not as clear-cut. The advantages of LH_2 originate from the fact that it is relatively compact (70 kg H_2/m^3), minimizing the capital investment necessary to centrally produce, liquefy, distribute, and store LH_2 throughout the refueling infrastructure and onboard automobiles. Automotive LH_2 issues have been researched for more than two decades, principally by BMW. Through five generations of prototype vehicles, BMW has developed engines and onboard LH_2 storage sufficient to drive a 2100-kg sedan 580 km (360 miles) using 13.5 kg (50 gallons) of LH_2 (a gasoline equivalent fuel economy of 28 mpg or 11.3 km/L). BMW has built and tested a fleet of 15 dual-fuel LH_2 sedans with reduced fuel economy and H_2 range (320 km or 200 miles). Manual and automated LH_2 stations can now refuel vehicles in less than 3 min. The world's first public LH_2 filling station opened in Munich, Germany, in 1998, but automotive LH_2 storage issues remain.

Automotive LH_2 tanks, storing only 5 to 15 kg LH_2, have surface/volume ratios 200 to 400 times larger than those for aircraft (25,000 kg LH_2), sharply curtailing capital cost and volumetric efficiency advantages. Typical automotive LH_2 tanks occupy a volume 1.3 to 1.5 times greater, and weigh 5 to 10 times more, than the LH_2 they contain. One fundamental difficulty is the extreme thermal expansion coefficient of LH_2 at temperatures and pressures below the critical point (33 K, 13 atm) where LH_2 and H_2 vapor coexist. In some cases, as LH_2 warms inside a tank, H_2 vapor actually condenses to accommodate LH_2 thermal expansion. Therefore, LH_2 tanks are fueled only 85 to 90% full to prevent LH_2 spills. This intrinsic volume penalty makes heat transfer reduction without increasing vessel volume critical. LH_2 vessels are typically double-walled aluminum or steel, sandwiching approximately 2 cm (70 layers) of multiple-layer insulation *in vacuo*.

Multiple-layer insulation can reduce heat flow into automotive LH_2 vessels to less than 1 W, a significant fraction of which is conducted through plumbing and supports suspending the inner vessel to avoid crushing the insulation. LH_2 tanks are designed to vent H_2 when warmed to 27 to 28 K (where vapor pressure is 6–8 atm). Most materials have very small heat capacities at such low temperatures. The heat capacity of aluminum is 50 times lower at 30 K than at 300 K. Below 30 K, the thermal inertia of an LH_2 tank is due principally to the LH_2 itself. Warming 140 L (5 kg) of LH_2 from 21 to 31 K requires nearly the same thermal energy as does melting 2 L of ice. A continuous 1-W heat leak will cause a 5-kg LH_2 vessel to "boil off" H_2 vapor at a rate equivalent to approximately 300 W of lost fuel energy after only 5 days. This is probably the greatest challenge facing onboard LH_2 storage for automobiles.

Dormancy times and boil-off rates of LH_2 tanks can significantly degrade effective fuel economy unless vehicles are driven far enough, on average, to consume H_2 faster than it evaporates. To avoid venting H$_2$ vapor, an 80-mpg (34-km/L) equivalent LH$_2$ vehicle fueled with 5 kg LH$_2$ must be driven an average of approximately 15 miles (25 km) daily to offset 1 W of heat transfer (Fig. 3). If parked, the vehicle could remain dormant (i.e., no venting of H$_2$ vapor) for only 5 days. If future LH$_2$ tanks absorbed heat at only approximately 0.25 W (representing "perfect" insulation, magnetic vessel suspension, or perhaps active refrigeration), 80-mpg equivalent vehicles driven an average of only 4 miles (6.5 km) daily would never vent H$_2$, and 5-kg LH$_2$ tanks could remain dormant for 20 days.

However, without such advances, high(er) fuel economy vehicles will likely lose some fuel to evaporation unless they are routinely driven more than approximately 15 miles daily, reducing effective fuel economy. This could create an incentive for drivers to refuel often so as to keep their fuel tanks cold, or it could encourage excess driving. Vehicles venting H$_2$ can raise safety issues in unventilated areas (e.g., garages), and LH$_2$ vehicles parked for long periods of time (e.g., at airports) might not retain enough fuel to reach an LH$_2$ station.

In the final analysis, LH$_2$ storage is the most energy-intensive method of onboard H$_2$ storage, but it is also the lowest capital cost, and likely the most compact and perhaps safest, method. LH$_2$ storage may be most appropriate for vehicles storing more than 5 kg of H$_2$ fuel. The low-capital but energy-intensive cost structure of LH$_2$ storage is best suited for vehicles that are refueled infrequently but are occasionally driven long distances. For vehicles driven and refueled frequently (e.g., 12,000 miles/year for a 60-mpg equivalent H$_2$ vehicle with 5 kg LH$_2$ onboard), the capital cost advantage (perhaps ~$1000 saved for each vehicle) of LH$_2$ storage may be offset by the higher (energy) cost (~$0.50/kg) of LH$_2$ versus compressed H$_2$.

Therefore, to be economically compelling, it appears that LH$_2$ tank technology will need to be capable of about 2 weeks of unrefueled moderate (20 miles/day on average) driving in high-fuel economy H$_2$ vehicles. Improved LH$_2$ tanks with heat transfer rates of 0.5 W or lower would be energetically efficient (venting no H$_2$ vapor) at this level of vehicle usage and would provide dormancy of 10 days for 5-kg LH$_2$ tanks.

5. COMPRESSED HYDROGEN

Currently, the principal alternative to LH$_2$ is to compress gaseous hydrogen (GH$_2$). Use of gaseous fuels onboard vehicles is fairly well established. There are approximately 250,000 compressed natural gas (CNG) vehicle cylinders currently used in the United States. There is also relatively broad experience with GH$_2$ in many demonstration vehicles, especially buses, which can accommodate large storage vessels. Under ambient conditions (1 atm, 300 K), H$_2$ has a density of only 0.0818 kg/m^3. Compression of H$_2$ up to pressures of 250 atm, a standard pressure for CNG vehicles, is thought to be insufficient for GH$_2$ vehicles. GH$_2$ at 250 atm is four times bulkier than LH$_2$ and 15 times bulkier than the energy equivalent volume of gasoline.

Because the theoretical (isothermal) work of compressing hydrogen rises only logarithmically with pressure, ambient temperature compression is the most energy-efficient method for densifying H$_2$ in both theoretical (Fig. 4) and practical terms. Compressing GH$_2$ to 250–1000 atm requires 1.5–2.0 kWh/kg theoretically and 2.5–4.0 kWh/kg in practice, substantially less energy than is required for conventional H$_2$ liquefaction (10–14 kWh/kg). Electrolysis at high pressure can compress GH$_2$ at near theoretical efficiency.

Compressed H$_2$ storage has historically been limited by the strength, and especially the weight, of pressure vessel materials. Current high-strength (1.5–3.0 GPa ultimate strength) composites permit lightweight vessels to approach the fundamental limits of GH$_2$ storage (Fig. 5) imposed by the

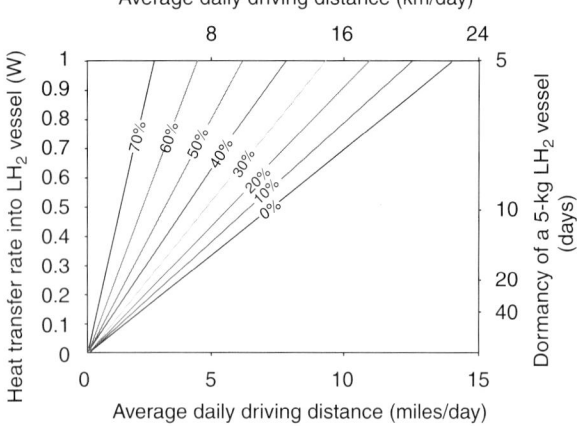

FIGURE 3 Cumulative LH$_2$ fuel lost to venting (boil off) of H$_2$ vapor due to continuous heat transfer (left axis) into the LH$_2$ tank of an 80-mpg equivalent H$_2$ vehicle as a function of its average daily driving distance (horizontal axes). The vehicle is assumed to refuel 84% full with 21 K LH$_2$ and vent hydrogen as necessary to keep the internal H$_2$ vapor pressure at or below 6 atm. The right axis indicates the dormancy (days before venting any H$_2$ vapor) of a 5-kg LH$_2$ tank if the vehicle is parked immediately on refueling.

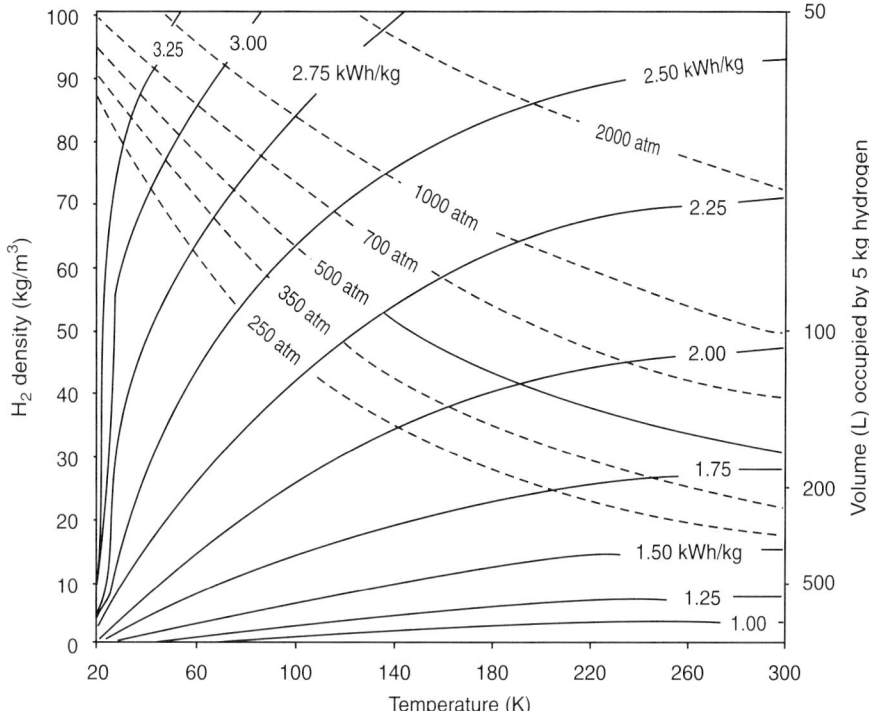

FIGURE 4 Theoretical minimum work to densify hydrogen from the reference gas density (0.0818 kg/m^3) at 300 K and 1 atm pressure to the densities (left axis) and temperatures (bottom axis) shown. The corresponding volume occupied by 5 kg of hydrogen gas, energy equivalent to 5 gallons (19 L) of gasoline, is shown on the right-hand axis. Dashed lines indicate the pressure needed to achieve a given hydrogen density as a function of temperature. The density that can be achieved for a given amount of work (solid lines) declines sharply at cryogenic temperatures. The work necessary to convert orthohydrogen to parahydrogen is not included but can be important for hydrogen stored more than a few days at temperatures below 60 K. The ideal work of converting normal (75% ortho) LH$_2$ to pure para-LH$_2$ is 0.7 kWh/kg at the boiling point (20 K), where LH$_2$ has a density of 70 kgH$_2$/m^3.

nonideal gas behavior of H$_2$. Even at 300 K, H$_2$ is not an ideal gas. It is less compressible and becomes more so at higher pressures, reaching only approximately 80% of ideal gas density at 350 atm (~5000 psi) and 60% at 1000 atm. Even for idealized infinitely thin pressure vessels, this incompressibility limits the hydrogen storage density achievable by high-pressure GH$_2$ storage.

For a given storage pressure, the strength of practical wall materials places a lower limit on vessel wall thickness. Thicker vessel walls occupy volume otherwise available for H$_2$ gas storage. Theoretical minimum wall thicknesses are more than doubled in practice to provide a burst pressure safety factor of 2.25 times the actual H$_2$ storage pressure. Future pressure vessel designs, using onboard sensors, may be allowed to use burst pressure safety factors of 1.8.

The net impact of H$_2$ gas behavior, burst pressure safety factor, and vessel wall strength is best shown by considering the amount of H$_2$ gas that can be stored in a vessel of fixed external volume (Fig. 6).

The volume between the rear seat and the trunk of an automobile is approximately 250 L (66 gallons). This volume could accommodate H$_2$ storage in a 54-cm diameter cylindrical pressure vessel 120 cm in length with ellipsoidal end caps. Although actual vessel configurations will vary, H$_2$ storage density depends little on vessel aspect ratio or size. However, the density of H$_2$ gas itself depends nonlinearly on pressure. A 250-L volume (300 K) contains 5.9 kg H$_2$ at 350 atm or 10 kg H$_2$ at 700 atm but only 12.3 kg H$_2$ at 1000 atm, representing the ideal maximum H$_2$ storage capacities of (infinitely thin) pressure vessels with 100% volumetric efficiency.

Ambient temperature pressure vessels made of high-strength materials (1.5–3.0 GPa or 220,000–440,000 psi ultimate strength) achieve 90 to 95% volumetric efficiency (hydrogen volume/vessel volume) at pressures up to 350 atm. Above 350 atm, as thicker walls become necessary, wall strength and safety factor are significant influences on volumetric efficiency and, therefore, on storage capacity. For

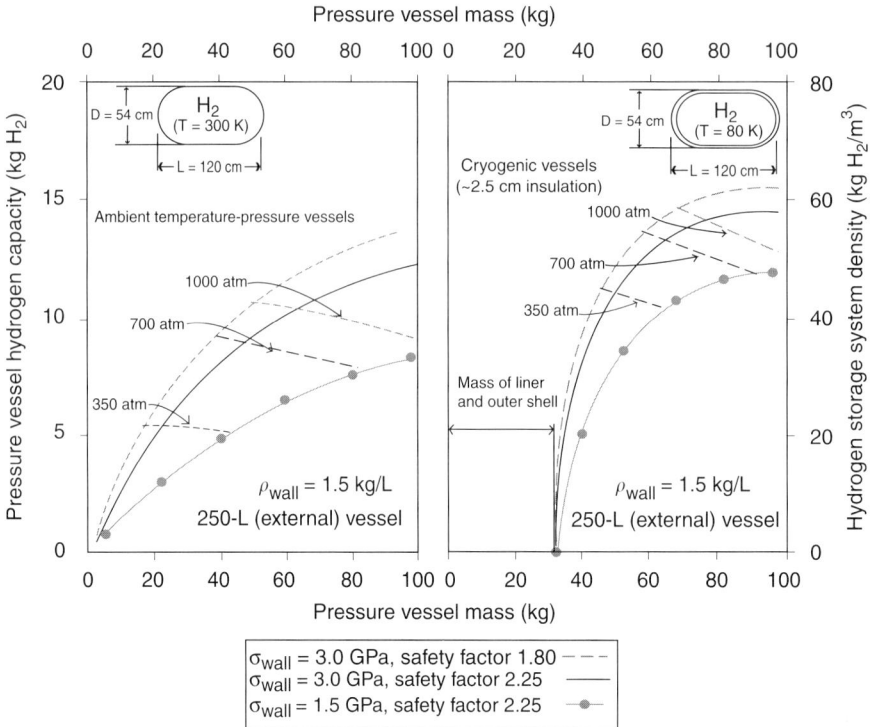

FIGURE 5 Mass of three idealized carbon composite pressure vessels as a function of H2 storage capacity at 300 K (left) and 80 K (right). All vessels shown are cylindrical with ellipsoidal end caps and identical external volumes (250 L), and dimensions (120 cm long, 54 cm diameter). Vessel wall thickness calculations include thick-wall effects. The heaviest vessels are designed with a burst pressure safety factor of 2.25 and wall stress of 1.5 GPa. The intermediate-weight vessels in both panels have twice the wall stress (3.0 GPa). The lightest vessels have identical wall stress (3.0 GPa) but are designed to a (lower) burst pressure safety factor of 1.8. The cryogenic (80 K) pressure vessels include an additional mass of approximately 30 kg for 3 mm aluminum liner, 2 mm outer shell, and 2.5 cm multiple-layer cryogenic insulation. Overall hydrogen storage system density is shown on the right-hand axis.

example, a 700-atm (∼10,000-psi), 250-L (external volume) pressure vessel made of 1.5-GPa material with a safety factor of 2.25 can store only 7.6 kg H_2 instead of the theoretical 10 kg H_2. The same 250-L vessel made of materials with twice the strength (3.0 GPa) would have thinner walls and could store 8.6 kg H_2 at 700 atm.

The disadvantage of low-strength wall materials is compounded by the nonlinear relationship between the density of H_2 gas and pressure. Reducing the wall strength of a 250-L vessel storing this same 8.6 kg H_2 from 3.0 to 1.5 GPa increases wall thickness, displacing available H_2 storage volume requiring a 40% higher pressure (1000 atm) to overcome. At higher pressures (>1000 atm), an ultimate H_2 storage density is reached as hydrogen gas becomes more incompressible and vessel walls must be ever thicker. These two factors ultimately limit maximum H_2 storage density to pressures of approximately 10% of vessel wall material strength and 60% volumetric efficiency, whereas maximum economic GH_2 pressures are likely to occur at approximately 5% of vessel wall material strength.

The attractiveness of automotive compressed H_2 storage will grow with the strength of the materials available for pressure vessel construction. Pressure vessels for H_2 vehicles have been certified up to 10,000 psi (∼700 atm), enabling a 250-L pressure vessel to store 8 kg H_2, adequate for a 650-mile (1100-km) range in an 80-mpg equivalent H_2 vehicle. Vessels made of very strong materials (3.0 GPa) could store more than 10 kg H_2 using higher pressures (>1000 atm). Future vessel materials stronger than 6.0 GPa (900,000 psi) and vessel pressures exceeding 1500 atm would be needed for GH_2 storage densities (at 300 K) comparable to LH_2 tanks (13 kg H_2 in 250 L).

One complication of all gaseous fuel storage, even at moderate pressures, is refueling overpressure. When a pressure vessel is quickly filled with gas, the temperature rises due to the compression of gas within the vessel by subsequent gas molecules.

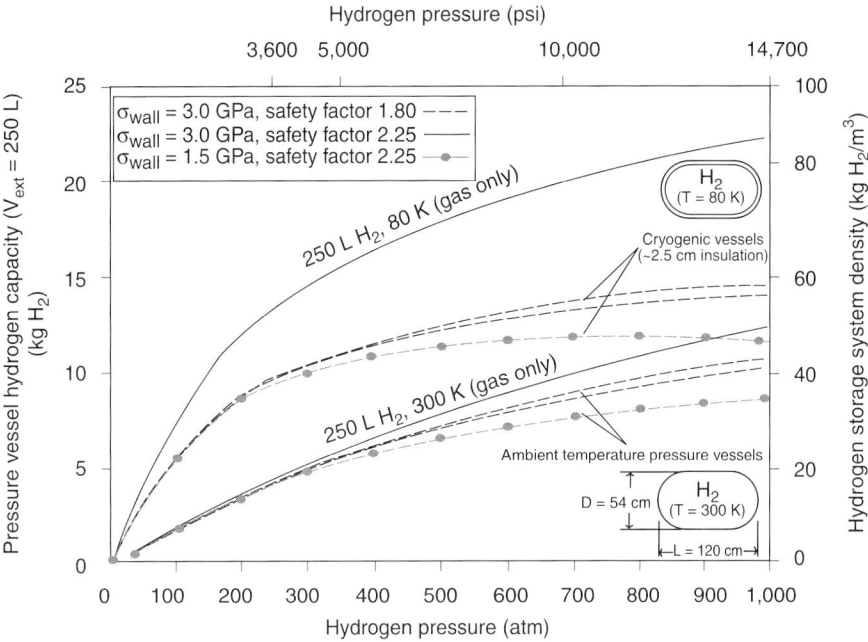

FIGURE 6 Hydrogen storage capacity versus pressure for idealized pressure vessels of varying wall stress and burst pressure safety factor at ambient (300 K) and cryogenic (80 K) temperatures. All vessels shown are cylindrical with ellipsoidal end caps and identical external volumes (250 L) and dimensions (120 cm long, 54 cm diameter). Vessel wall thickness calculations include thick-wall effects. Three vessels are shown for each temperature case: 3.0 GPa wall stress with a safety factor of 1.80, 3.0 GPa wall stress with burst pressure safety factor of 2.25, and 1.5 GPa wall stress with burst pressure safety factor of 2.25. Ambient vessels assume negligible (i.e., metallized polymer) liner thickness. Cryogenic vessels have 2.5 cm multiple-layer insulation, 3 mm internal metal liner, and 2 mm outer shell. Corresponding hydrogen storage system densities are shown on the right-hand axis. The quantities of H2 gas contained in a 250-L volume as a function of pressure at 80 and 300 K are also shown for reference.

Quickly filling an empty vessel to 350 atm with H_2 will theoretically warm it from 300 to 440 K, reducing storage capacity by 30% and requiring a 50% overpressure (525 atm) to overcome. Fortunately, in practice, pressure vessels absorb some of this thermal energy, and so warming to only approximately 350 K is observed. This reduces storage capacity by only 12%, and the storage capacity can be restored with 15% overpressure (400 atm). Overpressure effects vary little with filling pressure between 350 and 1000 atm and are lessened because H_2 is easier to compress as it warms. CH_4 exhibits opposite physical properties, becoming less compressible as it warms. The storage capacity reduction from warming CH_4 by 25 K is equivalent to H_2 warmed by 50 K. Therefore, although hydrogen vehicles will have warmer fast fill temperatures, fast fill overpressures may be comparable for H_2 and CNG vehicles.

Safety, both real and perceived, is an often raised criticism of compressed H_2 storage. However, the safety risks of storing compressed gases are not a simple function of pressure. The overall safety of pressure vessels can be counterintuitive, for although vessel wall strength and impact resistance increase directly with storage pressure, the maximum mechanical energy released by sudden expansion (e.g., in a vessel rupture) of the stored gas (hydrogen) does not. That is, 1 kg H_2 stored at 70 atm (300 K) will release a theoretical maximum mechanical energy of 0.55 kWh if suddenly expanded to atmospheric pressure (cooling substantially in the process). This maximum energy release increases only slightly if H_2 is stored at much higher pressures (Fig. 7). Raising pressure from 70 to 1000 atm (a 1400% increase) increases the maximum (theoretical) mechanical energy release by only 10% while shrinking vessel volume 83% and strengthening the vessel wall many times over. Over the likely range of onboard GH_2 storage pressures (350–1000 atm), the maximum mechanical energy release is nearly constant at 0.6 kWh/kg GH_2. Therefore, compact, higher pressure vessels with thick walls are likely safer than larger, lower pressure vessels with thinner walls.

In addition to pressure, the crash-worthiness of compressed H_2 storage depends on vessel shape,

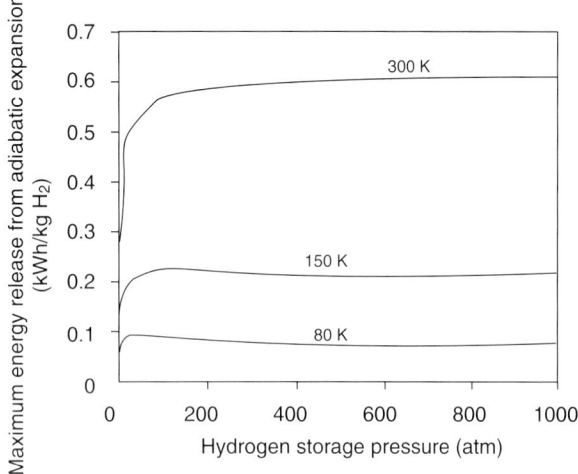

FIGURE 7 Maximum mechanical energy (per kilogram of hydrogen) released on instantaneous expansion of H2 gas (e.g., from a pressure vessel) as a function of initial storage pressure at 80, 150, and 300 K. This mechanical energy is the theoretical maximum available work based on reversible adiabatic expansion from the pressure shown to 1 atm, calculated from internal energy differences of H2 gas before and after isentropic expansion. For comparison, the chemical energy of hydrogen fuel is 33.3 kWh/kg.

orientation, volume, location, and likely accident scenarios. Vehicles with GH_2 stored onboard have been dropped from cranes, simulating a 30-mph rear-end collision, without releasing any hydrogen gas. It is generally thought that the best location to store H_2 is probably underneath/behind the rear seat of a vehicle. This sets an upper bound on vessel volume, aspect ratio, and thereby automotive onboard H_2 storage capacity given the limits of pressure and material properties (Fig. 6). Current carbon composite pressure vessels are designed to withstand pressures 2.25 times higher than their operating pressure and to leak before bursting. They are also subjected to a variety of safety tests, most notably bonfire, gunfire, and drop tests.

6. COMPRESSED CRYOGENIC HYDROGEN

Compressed H_2 is much denser at cryogenic temperatures (Fig. 4), potentially reducing storage volume dramatically, although cryogenic vessels incur volume inefficiencies similar to LH_2 tanks due to an inner liner, multiple-layer cryogenic insulation *in vacuo*, and a metallic outer shell. Built to withstand higher pressures than do LH_2 tanks, cryogenic pressure vessels could reduce H_2 vapor boil-off rates and extend dormancy when filled with LH_2. Alternatively, cryogenic pressure vessels can lower energy intensity if fueled with cryogenic GH_2 rather than LH_2. Cooling GH_2 to 80 K requires less energy than does liquefying H_2 at 20 K (Fig. 4). Additional energy could be saved by avoiding conversion of orthohydrogen to parahydrogen given that the equilibrium fraction of parahydrogen (25% at 300 K) is only 50% at 80 K. Flexibly fueled cryogenic vessels could reduce energy intensity and fuel cost further if ambient (300 K) GH_2 were used for urban driving and cryogenic (80 K) GH_2 or LH_2 were used only for occasional long trips.

Flexibly refueled cryogenic vessels will need to store H_2 at 200 to 400 atm to compare favorably with conventional ambient temperature pressure vessels. An insulated pressure vessel of moderate strength (1.5 GPa) material with an external volume of 250 L could contain 2.5 kg H_2 at 200 atm and 300 K, sufficient for a 200-mile (320-km) range in an 80-mpg equivalent H_2 vehicle. If filled with 80 K GH_2, the same 250-L, 200-atm vessel would hold 8.5 kg H_2 (680-mile or 1100-km range). Pressures of 1000 atm would be needed to store the same 8.5 kg H_2 at 300 K in a conventional pressure vessel of the same strength wall material (1.5 GPa) (Fig. 6). This dramatic relative advantage of flexibly fueled cryogenic vessels declines sharply in excess of 200 atm. Doubling H_2 pressure (to 400 atm) nearly doubles ambient H_2 capacity of a cryogenic vessel (to 5 kg GH_2 sufficient for a 400-mile or 660-km driving range). However, cryogenic (80 K) H_2 density over the same pressure range improves only 30% (Fig. 4). A conventional 250-L GH_2 vessel could store 5.5 kg H_2 at 400 atm (Fig. 6), sufficient for a 440-mile (725-km) range without the cost and complexity of cryogenics.

Cryogenic pressure vessels could also be designed to operate exclusively on LH_2, accepting the full energy intensity of liquefying H_2 but improving thermal endurance over low-pressure LH_2 tanks, enabling very compact storage for high-fuel economy vehicles requiring small amounts of fuel (kg LH_2). Pressure vessels can be filled more completely than can LH_2 tanks because the former can withstand the high pressures generated by LH_2 thermal expansion. Whereas conventional LH_2 tanks fueled 85% full must vent H_2 when warmed to 27 to 28 K, a 100-atm, LH_2-fueled pressure vessel could remain dormant until warmed to 33 K (45 K if 85% full). When fueled with para-LH_2, cryogenic vessels could be cooled substantially if catalysis of the endothermic conversion of cold (40–80 K) parahydrogen vapor to orthohydrogen were feasible.

Carbon composite pressure vessels lined with approximately 3 mm of aluminum have passed high-pressure ambient and cryogenic cycling tests representative of a 240,000-km vehicle life without any degradation in burst pressure. They have also passed a number of certification tests similar to those for ambient pressure vessels (e.g., gunfire, bonfire).

A potentially significant safety distinction between cryogenic and ambient tanks is that cryogenic gases release much lower amounts of mechanical energy when suddenly expanded from high pressure. The theoretical maximum mechanical energy that can be released by sudden (adiabatic) expansion of hydrogen from 350 atm is nearly eight times smaller at 80 K than at 300 K (Fig. 7).

Although cryogenic pressure vessels may ultimately offer safety and other technical advantages over ambient pressure vessels, and they may offer energy intensity or dormancy advantages over LH_2 tanks, cryogenic vessels are a far less mature technology. To date, preliminary evaluations of cryogenic pressure vessels versus ambient high-temperature tanks indicate comparable or lower safety risks in the former, but insulation performance, the use of cryogenic pressure vessels on vehicles, and flexible refueling have yet to be demonstrated.

7. HYDROGEN STORAGE BY CHEMICAL ABSORPTION IN METAL HYDRIDES

Although H_2 vehicles can feasibly use LH_2, GH_2, or cryogenic GH_2 storage, H_2 has higher storage costs (both capital and energy) and occupies greater volume in each of these forms than do competing alternative fuels, especially CH_4 (natural gas), even when H_2 is stored at higher pressures and/or lower temperatures. H_2 can also be stored by unique methods, circumventing the need for high pressures and/or cryogenics.

The most mature of these approaches is the reversible chemical reaction of metals with H_2 to form what are known as metal hydrides. The roots of this approach date back to Graham's 1866 discovery of H_2 absorption in palladium. Many metals react with and absorb hydrogen, releasing heat. Later, this hydrogen can be recovered on moderate heating. Metal hydrides have the advantage of storing H_2 at high density but are typically very heavy, storing only 1 to 2 wt% H_2. Most metal hydrides and other compounds have atomic hydrogen densities of 90 to 110 kg H_2/m^3 (Fig. 2)—30 to 60% denser than LH_2. However, hydride storage systems usually achieve only approximately 50% volumetric efficiency because the hydrides are in powder form and expand when chemically absorbing H_2.

Chemisorption of H_2 by virtually every element has been studied without finding an ideal candidate for H_2 fuel storage. Most are too heavy, bond too strongly to hydrogen, or are insufficiently abundant. Magnesium is sufficiently light to store 7.6 wt% H_2, forming MgH_2, the elemental hydride closest to automotive viability. But the strong magnesium–hydrogen bond requires significant energy (75 kJ/mol H_2 at temperatures near 550 K) for H_2 release, making MgH_2 (and related alloys) unattractive for current fuel cell vehicles with exhaust temperatures of only approximately 380 K.

The field of candidate H_2 storage materials was expanded from elements to compounds with the discovery of the reversible hydride $ZrNiH_3$ by Libowitz in 1958. Entire classes of intermetallic hydrides were subsequently developed. These are known as the A_2B, AB, AB_2, and AB_5 hydrides, in which "A" denotes elements that strongly absorbs hydrogen (e.g., Mg, Ti, La) and "B" denotes elements that do not, usually lighter transition metals in the first row of the periodic table (e.g., V, Cr, Fe, Co, Ni, Cu). During the late 1960s, two prototype practical hydrides, $LaNi_5H_6$ and $TiFeH_2$, were developed to adsorb H_2 interstitially, with low enough bond energies ($\sim$30 kJ/mol H_2) to release H_2 between 273 and 373 K and to adsorb H_2 quickly at low pressures (1–10 atm). Many variations on these materials have been studied, with some finding application in nickel metal hydride batteries, but all known low-temperature hydrides store at most 2 wt% H_2. Practical automotive hydride systems will need additional mass, reducing effective storage densities to 1.1 to 1.5 wt% H_2, corresponding to a 330- to 450-kg system mass for 5 kg of onboard H_2 storage capacity (energy equivalent to 5 gallons of gasoline). Daimler–Benz built and tested a fleet of 10 hydride vans and station wagons during the 1980s. They were successfully operated over 240,000 km in West Berlin over 4 years, proving automotive hydride storage to be feasible, if perhaps too heavy. The most significant issue for hydride storage is probably refueling speed. H_2 absorption during refueling generates heat that, if not removed quickly, warms the hydride, requiring higher pressures to drive further hydrogen absorption, undercutting an advantage (low pressure) of metal hydrides.

More recent hydride research has ventured beyond the classical crystalline intermetallic hydride compounds, focusing on solid solution alloys as well as disordered, amorphous, and nanocrystalline materials. However, such hydrides may be metastable or susceptible to degradation over many absorption/desorption and thermal cycles.

The greatest research progress has arguably been made with hydride complexes, in which hydrogen and a metal atom form a complex ion, stabilized by another element. Magnesium nickel hydride (Mg_2NiH_4) is a prototypical example, with Mg stabilizing an $(NiH_4)^{-2}$ ion. Complex hydrides can be formed (e.g., Mg_2FeH_6) from metals that do not themselves form intermetallic compounds, further expanding the range of potential hydride storage materials. Complex hydrides made of light metals (e.g., Li, Be, B, Na, Al) contain 5 to 14 wt% H_2 but typically do not absorb H_2 directly. The hydride with the highest known ratio of hydrogen to metal atoms, storing 2.7 wt% H_2 at twice the density of LH_2, is the complex hydride $BaReH_9$, but it is irreversible and is synthesized by aqueous techniques.

A breakthrough in hydride research occurred during the mid-1990s when Bogdanovic reported that sodium alanate ($NaAlH_4$), a lightweight complex hydride of low-cost materials, could reversibly store and release H_2 at temperatures of 400 K if catalyzed by titanium and/or zirconium. $NaAlH_4$ is more complex than the intermetallic hydrides. Hydrogen atoms reside on lattice sites in $NaAlH_4$ and are released through multiple decomposition steps involving metal atom diffusion and, therefore, warmer temperatures.

When $NaAlH_4$ is warmed, it initially releases half of its contained H_2, decomposing into aluminum (Al) and an aluminum-poor complex, Na_3AlH_6, which contains strongly bound hydrogen. On further warming, Na_3AlH_6 releases additional H_2 and decomposes into Al and sodium hydride (NaH), retaining a quarter of its original hydrogen atoms.

Based on these reactions $NaAlH_4$ contains 7.46 wt% H_2, of which 5.60 wt% is theoretically recoverable (3.73 wt% H_2 if only the low-temperature decomposition is used). The decomposition of Na_3AlH_6 is endothermic (47 kJ/mol H_2), releasing 1 atm H_2 at 380 K. When the cycle is reversed for refueling, H_2 is absorbed by Al and NaH, generating temperatures in excess of 455 K (the melting point of $NaAlH_4$). These high temperatures can drive refueling pressures as high as 100 atm but are needed to accelerate the sluggish refueling kinetics of $NaAlH_4$.

The tightly bound hydrogen in NaH reduces the effective maximum H_2 storage density of $NaAlH_4$ to approximately 70 kg H_2/m^3 (comparable to the density of LH_2). This may be offset by potential hydride bed engineering advantages. Volumetric efficiency of 75% would be needed for $NaAlH_4$ systems to achieve H_2 storage density comparable to that of other hydride systems. Future automotive (5 kg H_2) $NaAlH_4$ systems could perhaps occupy 100 L and weigh 170 kg.

8. HYDROGEN STORAGE BY PHYSICAL ADSORPTION ON CARBON

A second approach is to physically adsorb and store H_2 molecules on the surface of an adsorbent (e.g., carbon) rather than to dissociate H_2 molecules on a metal surface and absorb hydrogen atoms throughout the body of a metal, forming a metal hydride. Physical adsorption of H_2 molecules onto high-surface area adsorbents has been studied for decades. Initial research began during the 1960s and continued through the early 1990s, examining cryogenic H_2 adsorption onto high-surface area carbon materials. However, the advantage of cryogenic adsorbents was found to decline with pressure because the volume of the adsorbent itself displaces ever larger quantities of H_2. At pressures exceeding approximately 200 atm, removing the adsorbent usually increases cryogenic H_2 storage density.

Recent research has focused more directly on storage, pursuing H_2 adsorption at low pressures and ambient temperatures, where the relative advantage of adsorbents is greater. Ambient temperature H_2 adsorption has required engineered carbon materials, with atomic scale porosity and high surface areas. Graphite nanofibers are a class of engineered carbon materials that received significant attention, with experimental claims of extraordinary gravimetric H_2 storage densities (>50 wt% H_2). Subsequent work has been unable to duplicate these claims but also has produced widely varying results.

Carbon nanotubes are a subsequent class of engineered adsorbents. They are atomic scale tubes with graphite structure walls and internal diameters of 20 to 100 Å. Intercalating lithium or potassium in multiple-wall carbon nanotubes has been reported to enable dry adsorption of approximately 2 wt% w H_2.

Single-wall carbon nanotubes are a further refinement, capable of higher purity ($\sim 50\%$) fabrication, with diameters of approximately 10 Å, nearly the 2.9-Å kinetic diameter of H_2. Single-wall nanotubes form compact hexagonal bundles. Early experiments on single-wall nanotubes measured an adsorption enthalpy of approximately 20 kJ/mol H_2, substantially larger than the 4 kJ/mol H_2 seen with graphite adsorbents. Experiments on higher purity nanotubes measured desorption of H_2 from two classes of adsorption sites: one active near 300 K and another active only at higher temperatures. Current evidence indicates that 4 to 8 wt% H_2 adsorption is possible at 300 K, which for a bulk nanotube density of approximately 1 g/cm^3 would equate to intrinsic maximum storage densities of 40 to 80 kg H_2/m^3.

Current understanding of H_2 adsorption within carbon nanotubes (or other engineered adsorbents) is embryonic. Theoretical calculations and simulation results vary depending on methodology and assumptions. The reality of H_2 adsorption within a nanotube array may be very complex, depending on tube diameter distributions, interstitial sites, nanotube electronic and structural properties, defects, and changes caused by adsorbed H_2 molecules. The ideal adsorbent would cause H_2 molecules to adsorb, order, and condense at room temperature, akin to a phase change, representing the ultimate H_2 storage technology. Whether this ideal is possible within the theoretical limits of H_2 adsorption or the capabilities of economically fabricated carbon nanotube adsorbents needs to be known before their ultimate H_2 storage potential can be evaluated.

9. HYDROGEN STORAGE IN CHEMICAL CARRIERS

In addition to hydride and adsorbent storage, H_2 can be stored by chemical absorption in liquids. Aromatic molecules with carbon–carbon double bonds are the leading candidates for liquid chemical hydrogen storage. For example, methylcyclohexane (C_7H_{14}) releases three H_2 molecules when heated and becomes toluene (C_7H_8). Using this reaction, a theoretical 6.16 wt% H_2 can be stored at a density of 49 kg H_2/m^3. However, significant thermal energy (a theoretical minimum of ~ 70 kJ/mol H_2) at temperatures of 650 K or higher is needed to release this H_2. Therefore, H_2 storage in aromatic liquids is probably best suited to compete with LH$_2$ for stationary long-term storage of H_2 in bulk, where the energy and capital costs of H_2 release are offset by the very low capital costs of storing liquid chemicals.

Methanol (CH_3OH) is a liquid chemical H_2 carrier containing 12.5 wt% H_2 at high density (100 kg H_2/m^3). H_2 accounts for 75% of the fuel value (19.92 MJ/kg) of CH_3OH, which is produced on a large scale with efficiencies as high as 75% from H_2/CO/CO_2 mixtures derived from natural gas. CH_3OH can be decomposed to H_2 and either CO or CO_2 (with H_2O). Both reaction pathways are very endothermic (65 and 40 kJ/mol H_2, respectively), making H_2 release from CH_3OH energy intensive unless waste heat is available to vaporize the methanol and/or steam. The CO_2 reaction product, toxicity, and high solubility in (ground) water cloud the ecological attractiveness of CH_3OH as an H_2 carrier. Methanol synthesis could be CO_2 neutral if atmospheric CO_2 or perhaps surplus biomass were used as the necessary carbonaceous feedstock.

NH_3 is the most attractive carbonless liquid chemical H_2 carrier. NH_3 boils at 240 K but can be condensed at 300 K under moderate pressure (10 atm) containing 17.6 wt% H_2 with an atomic hydrogen density of 105 kg H_2/m^3—150% denser than LH$_2$. Decomposing NH_3 into H_2 (and nitrogen [N_2]) is thermodynamically easier (30 kJ/mol H_2) than other liquid chemical H_2 carriers but is kinetically complicated by the N_2 triple bond. Rapid NH_3 synthesis is currently practical only at elevated temperatures (and pressures) in large plants (1000 tonnes [metric tons] NH_3/day) achieving efficiencies of 75 to 80% from CH_4.

Higher efficiencies are possible using atmospheric N_2 and nonfossil H_2 sources because NH_3 formation is exothermic. NH_3 has been proposed as an alternative fuel in its own right but has two major drawbacks: chemical reactivity with certain metals and toxicity. Although NH_3 can be lethal in concentrations of 1000 to 5000 ppm, it is easily detectable by smell at less than 20 ppm and is lighter than air. NH_3-based H_2 storage onboard vehicles would likely require avoiding any NH_3 emissions through nearly perfect decomposition to N_2 and H_2. NH_3 may hold greater immediate promise for H_2 distribution and transportation to refueling stations, avoiding the lower energy density, higher capital cost, and higher energy intensity of LH$_2$ transport, distribution, and storage.

10. FUTURE PROSPECTS FOR HYDROGEN IN TRANSPORTATION

Unlike other fuels, hydrogen possesses a broad array of possible storage approaches and attendant vehicle and refueling infrastructure implications. Each differs dramatically in capital and energy intensity, maturity, and perceived or potential business and safety risks. For some applications, a given H_2 storage approach is clearly best. For example, LH$_2$ storage is clearly the only option for H_2 aircraft, especially those designed for intercontinental and perhaps future hypersonic air travel. Economies of scale and infrequent refueling argue for the use of LH_2 in ships and trains as well.

Since H_2 fuel began receiving serious interest in the 1970s, technological advances have weakened the case for LH_2 in light-duty vehicles and perhaps even in large trucks. Hybrid electric vehicles have been commercially introduced, and at the time of this writing, it is quite possible that fuel cell vehicles will be introduced before 2005. Either technology allows the production of H_2 automobiles that achieve fuel economies equivalent to 60 to 100 mpg (25–42 km/L). This leap in fuel economy permits H_2 vehicles using as little as 5 kg H_2 onboard storage to achieve a 300- to 500-mile driving range. However, these fuel economy improvements would be insufficient to weaken the case for LH_2 without the tripling in strength of lightweight composite materials since the 1970s. If future higher strength materials ($\sim$ 7 GPa or 1 million psi) become routinely available, compressed H_2 may rival the density of LH_2 storage.

Flexibly refueled cryogenic pressure vessels would permit light-duty vehicles to compromise between LH_2 and GH_2 and be compatible with a multiple-mode refueling infrastructure: centralized LH_2 production and delivery to stations alongside a spectrum of on-site GH_2 production technologies at stations and perhaps garage electrolysis.

Decentralized H_2 production, and particularly slow (e.g., overnight) refueling, would be ideal for metal hydride storage, easing refueling times and reducing electrolysis pressures. The lighter higher temperature hydrides (e.g., $NaAlH_4$) are not yet sufficiently developed for vehicles, but the heavier ambient temperature hydrides (e.g., $TiFeH_2$) will be increasingly feasible as H_2 vehicle fuel economy increases. In addition, electrolytic garage refueling of onboard hydride storage, or perhaps moderate-pressure GH_2 storage, would conceivably permit early H_2 light-duty vehicles to circumvent the greatest strategic barrier facing any alternative fuel vehicle: the initial lack of refueling infrastructure.

The ultimate shape of the refueling infrastructure and hydrogen energy economy may depend on whether LH_2 or GH_2 is the dominant fuel for future large commercial trucks. If gaseous hydrogen and/or hydride storage turns out to be sufficient for light-duty vehicles, the choice of hydrogen storage for trucks could determine whether LH_2 will be an everyday road vehicle fuel or principally used only in aircraft. The use of LH_2 would favor a centralized H_2 infrastructure for efficient liquefaction, whereas a GH_2 infrastructure would very likely have lower ultimate costs due to reduced energy intensity. A GH_2 infrastructure could also be more easily decentralized, relying on electrolysis at filling stations, in garages, or ultimately onboard passenger vehicles.

The factors determining the onboard H_2 storage technology best suited for commercial heavy trucks are very different from those for light-duty vehicles. A future H_2-fueled light-duty vehicle will refuel with 4 to 5 kg H_2 approximately 40 times annually. A liquefaction energy cost premium of $0.50/kg for LH_2 would amount to only $80 to $100/year for such vehicles.

H_2-powered commercial heavy trucks will consume 100 to 200 times as much fuel as will light-duty vehicles, with the former refueling with 75 to 100 kg of H_2 daily and being much more sensitive to fuel costs. Potential GH_2 fuel cost savings of as little as $0.25/kg relative to LH_2 would probably justify large investments ($30,000–$40,000) in onboard GH_2 storage for trucks. However, large amounts (100 kg) of high-pressure GH_2 storage on heavy trucks would benefit little from economies of scale and could raise overall safety and volume concerns. Composite vessels pressurized with 700 to 1000 atm GH_2 would be 6.0 to 7.5 times larger than equivalent energy 100-gallon (380-L) gasoline tanks.

For heavy trucks, cryogenic pressure vessels may best balance hydrogen storage volume, capital costs of onboard storage, and energy intensity-related fuel costs. Safety issues of cryogenic vessels could be eased by the substantially reduced maximum mechanical energy release at cryogenic temperatures. Dormancy, boil-off, and insulation requirements for cryogenic vessels would be minimal for heavy trucks. Cryogenic vessels store H_2 at approximately twice the density of ambient vessels at the same pressure, or they can achieve comparable H_2 storage density with far lower pressures, vessel wall strength, and likely capital costs. Finally, although the energy intensity of an 80 K H_2 infrastructure has yet to be demonstrated, its theoretical energy intensity is significantly lower than that of LH_2 (Fig. 4) and likely easier to approach than is the case with current H_2 liquefaction technology (10–14 kWh/kg LH_2).

If developed successfully, $NaAlH_4$ would allow hydride storage for heavy trucks with moderate desorption temperatures. However, a 100-kg H_2 storage system for a heavy truck would still weigh approximately 3000 kg (six times heavier than two full 80-gallon diesel tanks currently used on heavy trucks).

Whichever blend of hydrogen storage technologies ultimately develops for both light-duty and heavier vehicles, it is likely that the most critical characteristic over the long term will be the flexibility, versatility, and efficiency of the hydrogen production and utilization infrastructure each storage technology permits. Hydrogen storage technologies with greater flexibility (cryogenic pressure vessels, hydrides, and perhaps engineered adsorbents) offer benefits that may be significant over the long run by permitting an H_2 production, storage, and utilization infrastructure that easily adapts.

A versatile hydrogen infrastructure that can benefit from or even encourage technological change (e.g., scale efficiencies, vehicle fuel economies), dynamically balance energetic and capital costs, and offer intangible benefits (e.g., energy infrastructure security, vehicle safety, sustainability) may ultimately be of greater value than the sum of its individual H_2 production and storage components.

SEE ALSO THE FOLLOWING ARTICLES

Coal Storage and Transportation • Hydrogen, End Uses and Economics • Hydrogen, History of • Hydrogen Production • Natural Gas Transportation and Storage • Storage of Energy, Overview.

Further Reading

Berry, G., and Lamont, A. (2002). Carbonless transportation and energy storage in future energy systems. *In* "Innovative Energy Strategies for CO_2 Stabilization" (R. Watts, Ed.), pp. 181–210. Cambridge University Press, Cambridge UK.

Cox, K., and Williamson, K. (1977). "Hydrogen: Its Technology and Implications Vols. 1–5." CRC Press, Cleveland, OH.

Hoffmann, P. (2001). "Tomorrow's Energy." MIT Press, Cambridge, MA.

Ogden, J., and Williams, R. (1989). "Solar Hydrogen: Moving Beyond Fossil Fuels." World Resources Institute, Washington, DC.

Peschka, W. (1992). "Liquid Hydrogen." Springer-Verlag, New York.

Sandrock, G., Suda, S., and Schlapbach, L. (1992). Applications. *In* "Hydrogen in Intermetallic Compounds II" (L. Schlapbach, Ed.), pp. 197–246. Springer-Verlag, New York.

Winter, C. J., and Nitsch, J. (1988). "Hydrogen as an Energy Carrier." Springer-Verlag, New York.

Hydropower Economics

BRIAN K. EDWARDS
Brian K. Edwards Associates
Downers Grove, Illinois, United States

1. Introduction
2. Development and Regulation of Hydroelectric Power in the United States
3. Policy Issues Relating to Dams and Hydroelectric Power
4. Generating Hydroelectric Power
5. Optimal Provision of Hydroelectric Power
6. Summary and Conclusions

Glossary

dam head The distance between the elevation of a reservoir and the top of the penstock.
elevation The number of feet the surface of a reservoir is above sea level.
forebay Where water is stored behind a dam prior to release for generation.
hydroelectric power The electric power generated by the flow of water.
hydro-shifting The practice of storing water behind a dam during off-peak hours for release during on-peak hours.
penstock A tube through which water flows, from the afterbay of a dam to the turbines, for generation.

With few exceptions, the academic research relating to hydroelectric power has come from the power systems (engineering) literature. Some published works have addressed hydroelectric power from an economics perspective, but many of these articles have addressed environmental issues associated with water use; these have included valuing damages to ecosystems and valuing the recreational benefits of water resources, and some have estimated the benefits of changes in downstream river flows sometimes attributable to changes in how dams are operated. In short, issues relating to dam operations have largely been relegated to the periphery of discussions of other, though certainly not unimportant, topics. Thus, the focus of this article will be on the use of water to generate hydroelectric power, by presenting a simple model of how a dam operator decides to use the water resource to generate that power.

1. INTRODUCTION

One of the difficulties associated with any economic analysis of hydroelectric power is how widely the functions of dams vary, not only within individual countries, but between countries. Studies by the International Commission on Large Dams and the World Commission on Dams reveal a great diversity in dam use. For example, one-third of all countries depend on hydropower for over one-half of their electricity. One-third of European dams are used to generate hydroelectric power and another 20% are used for irrigation. Nearly two-thirds of Asian dams are used for irrigation, but only 7% are used to generate hydroelectric power. In North and Central America, dam use is almost equally divided between irrigation, hydroelectric power, water supply, flood control, and recreation functions. In the United States, hydroelectric generation accounts for approximately 10% of total electric generation. In South America, nearly one-fourth of the dams are used for hydroelectric generation. Nearly three-fourths of African dams are used for irrigation and water supply. In Australasia, nearly 50% of the dams are used for water supply and another 20% are used to generate hydroelectric power.

These statistics suggest that the author of any article on dams and hydroelectric power must take care to avoid making too general an analysis of dams, or at least must provide adequate warning to readers expecting a complete and self-contained presentation on dams and hydroelectric power, covering every possible area of interest. Thus, the primary focus here is on the use of water to generate hydroelectric power; a simple model of a dam operator is presented in that regard, and other issues are ignored, i.e., those relating to operational restrictions that limit the rate at which water can be released, hydrothermal generation (i.e., the joint provision of hydroelectric power with other kinds of electrical power), hydraulically coupled dams (i.e., dams connected by a common river), and the recreational impacts of dam operations. However, some of the analysis presented here will relate to these other uses of water, but only marginally. For example, the operating restrictions often imposed on dam operators are motivated by environmental, recreational, and water use considerations. However, there is no attempt to integrate these other uses of water explicitly into this analysis. Nevertheless, the model presented here could be extended to consider alternative water uses explicitly, and these extensions would indeed represent fruitful extensions of the work presented herein. For example, the model could be extended to include explicit agricultural or recreational uses of the water that supplement the treatment of the water resource as an input in electricity generation. There could also be an attempt to integrate the benefits and costs of hydroelectric power generation into a broader analysis that also considers the recreational benefits of alternative flow regimes explicitly. In this case, trade-offs between using water to generate hydroelectric power and the benefits of imposing restrictions on changes in water release rates to improve the value of downstream recreational activities could be examined.

A brief legal history of hydroelectric power with particular emphasis on the United States is presented in Section 2. An outline of some of the energy, environmental, and economic development issues surrounding dams and hydroelectric power is presented in Section 3. In Section 4, a simple economic model of hydroelectric power generation is presented. In this model, a decision maker will decide on the hourly pattern of water releases that maximizes net receipts subject to an equation of motion that describes the dynamics of water supply.

2. DEVELOPMENT AND REGULATION OF HYDROELECTRIC POWER IN THE UNITED STATES

Although the development and regulation of hydroelectric power has a long history of federal oversight, the demand for and supply of such power have largely been regulated at the state and local levels. Unless a dam has been constructed specifically for federal purposes such as flood control or navigation, hydroelectric power generation has to comply with state proprietary water laws. Under the Federal Pollution Control Act, as amended by the Clean Water Act Amendments of 1977 (Clean Water Act), unless state and local environmental laws conflict with federal law, hydroelectric projects must also comply with state and local environmental laws.

In the beginning of the 20th century, most hydroelectric power development was private, developed primarily through holding companies, so that by 1916, most of the 87 hydroelectric projects in the United States were in private hands. During this period, the federal government increased its role in hydroelectric power development primarily through federal reclamation projects by the United States Bureau of Reclamation (BOR) and water resource projects by the United States Army Corps of Engineers (Corps). Over the years, the Corps has been the largest single producer of hydroelectric power in the United States, and currently operates 75 dams with an installed capacity of over 21,000 MW.

The BOR was created by the Reclamation Act of 1902. This act authorized the Secretary of the Interior to develop irrigation and hydropower projects in 17 western states. The increased involvement by the BOR was a natural consequence of the need to manage scarce water supplies in the western United States. Despite much of the emphasis on hydroelectric power generation, the development of dams resulted from the need to manage water supplies effectively in a predominantly arid region. Moreover, hydroelectric power was often used during the construction phases of these projects to provide electrical power for the processing of materials and the running of sawmills, concrete plants, and construction equipment, and allowed for operation of construction activities during the night. After construction was finished, hydroelectricity was used to power drainage pumps that moved water to higher elevations than was possible with gravity-flow canals. Many of these hydroelectric power facilities found that they had surplus power, thus it was necessary to establish the terms under which such surplus power could be sold. To that end, the Towns Sites and Development Act (1906) gave the Secretary of the Interior authority to lease surplus power or power privileges. As a result, surplus power was sold to existing power distribution systems, benefiting local industries, towns, and farming communities.

Indeed, the Bureau of Reclamation's first hydroelectric facility was built to support construction of the Theodore Roosevelt Dam, located on the Salt River northwest of Phoenix, Arizona. Although the primary purpose of the hydroelectric power was to power construction equipment, the surplus power generated by this plant was sold to the local community. Public support for additional power led to an expansion of the capacity of this dam, so that in 1909 five generators were placed in operation, providing power for irrigated agriculture and the Phoenix area. This expansion of hydroelectric power encouraged the economic growth of the Phoenix area, including expanding electrically irrigated agriculture to cover more than 10,000 acres, and providing all of the residential and commercial power requirements of the area.

In 1920, the Federal Power Act established the Federal Power Commission (FPC), giving them authority to issue licenses for developing hydroelectric projects in public lands and on navigable waters. In 1928, the Boulder Canyon Project Act authorized construction of the Hoover Dam. Even though initiated before the New Deal, the construction of the Hoover dam has come to symbolize many of the large-scale public works projects that came to characterize the New Deal. Until the 1930s, the FPC took a fairly *laissez-faire* approach to requests for licenses for power projects, granting most of these requests with little federal review or oversight. It was not until the administration of President Franklin Roosevelt that the role that the federal government currently plays would be established. The Public Utility Act of 1935 expanded the FPC's jurisdiction, which hitherto had been limited to licensing hydroelectric generation, to include setting wholesale electricity rates. Under the Public Utility Act, however, regulation of consumer rates remained with state public utility commissions. The administration of Franklin Roosevelt also ushered in the "big dam" period, which, in addition to the Hoover Dam, saw the construction of many other large multipurpose hydroelectric projects, including the Grand Coulee Dam on the Columbia River in Washington state (authorized by the Rivers and Harbors Act of 1935) and the Central Valley Project in California. The Tennessee Valley Authority (TVA) was created in 1933 and the Bonneville Project Act of 1937 created the Bonneville Power Administration (BPA). The construction of the dams had much to do with the migration to the west that occurred during the 1930s. During World War II, cheap hydroelectric power encouraged firms involved in the defense effort to locate in the west.

From the legal developments, two federal policies emerged: (1) primary federal regulation of the construction and use of private hydroelectric projects and (2) federal construction and control of the largest such projects. With the exception of increased use of federal subsidies and tax incentives designed to encourage small hydroelectric power development at both existing and new facilities, these two policies remain in effect.

Even in the 1930s, the environmental impacts of hydroelectric power generation, including the lowering of water tables brought on by the rapid growth of irrigated agriculture, began to be recognized and would eventually lead to legislation that addressed these environmental impacts. Most of the early legislation laid the groundwork for the construction and operation of hydroelectricity projects, but it would not be until 1968, with the passage of the Wild and Scenic Rivers Act, that these environmental concerns would begin to be addressed. That Act protected rivers in their natural state by excluding them from consideration as hydroelectric power generation sites. One year later, the National Environmental Policy Act (NEPA) required federal agencies to take into account environmental considerations in the construction and operation of hydroelectricity generating facilities. In 1973, the Endangered Species Act listed endangered species and their critical habitats, many of which are directly affected by the construction and operation of dams. The Fish and Wildlife Coordination Act was first amended in 1946, to ensure that fish and wildlife would receive equal consideration for protection by federal agencies. Sections 401 and 404 of the Clean Water Act ensured that federal and nonfederal entities would comply with state water quality standards by requiring state certification as a condition to Federal Energy Regulatory Commission (FERC) license approval.

Other regulations addressed the financing and administration of dams. The Federal Water Power Act of 1920 regulated hydroelectric development of navigable waterways. The Reclamation Project Act of 1939 extended the contract term to 40 years for sale of power or lease of power privileges, giving preference to public entities. The Flood Control Act of 1944 gave the Secretary of the Interior authority to market power from Army Corps of Engineers projects and authorized the Pick–Sloan Missouri Basin Program. The Federal Columbia River Transmission Act of 1974 authorized the Bonneville Power Administration to issue revenue bonds.

In 1977, the United States Department of Energy was created by the Department of Energy Organization Act. This Act transferred the existing Power Marketing Administrations to the newly created

Department of Energy and created the Western Area Power Administration. The Public Utility Regulatory Policies Act (PURPA) of 1978 encouraged small-scale power production facilities; exempted certain hydroelectric projects from federal licensing requirements, and required utilities to purchase—at "avoided cost" rates—power from small production facilities that use renewable resources.

3. POLICY ISSUES RELATING TO DAMS AND HYDROELECTRIC POWER

The past few years have seen many changes in the energy industries in the United States and other countries. The British energy industry has already undergone deregulation. Deregulation in the United States, at least at the retail level, has largely proceeded on a state-by-state basis, and wholesale bulk power markets are in an advanced stage of deregulation. The hopes of deregulation are to increase competition and eventually lower energy prices to consumers, but also to allow pricing signals to direct energy resources to their most efficient use and provide incentives to users to conserve energy.

However, there is much more at stake than economic efficiency. Because energy and air pollution are often joint products, socially unacceptable levels of air pollution can result regardless of whether markets are used to ration electricity. Proponents of hydroelectric power point out that hydro generation produces no emissions of acid rain or ozone precursors, and dams have also not created the sort of not-in-my-backyard (NIMBY) concerns that have stigmatized nuclear power facilities and proposed nuclear waste storage facilities.

Despite the potential for displacing emissions, however, hydroelectric facility construction is virtually irreversible. Although some smaller dams have been torn down, it is difficult to fathom how the Hoover Dam could be decommissioned and taken apart without causing energy and economic disruptions in the American West. Moreover, dam construction alone can change the environment dramatically, even before any hydroelectric power is generated, the Xiaolangdi and Three Gorges dams in China being two recent cases in point.

Once built, dams can adjust their rate of generation quickly, but this very flexibility causes rapid changes in the downstream flow of rivers, which can erode downstream sandbanks, change the deposition of downstream river sediment, thereby altering the ecological makeup of the downstream environment, and degrade the quality of downstream recreational activities such as fishing. Unfortunately, reducing these fluctuations can cause other environmental problems by mitigating the positive effects of annual spring floods. These floods can replenish downstream sandbanks eroded by the river flows over the course of the year. The United States Bureau of Reclamation has been experimenting with periodic sustained high releases from Glen Canyon Dam to mimic the spring flooding that used to occur there naturally. The argument that hydroelectric power displaces air pollution precursors has been challenged. Storage dams can become silted with vegetation that can rot, emitting carbon dioxide and methane, two greenhouse gases. These increased greenhouse gas emissions could, in some cases, offset any fossil fuel emission reductions from thermal generation.

The salmon populations that Lewis and Clark observed in the Columbia River in abundance during the early years of the 19th century have fallen dramatically since then, perhaps a result of the current system of dams on that river. Hydroelectric power threatens fish species throughout the world by disrupting the movement of many fish species to their spawning sites. In 2000, more than 1500 fishermen and farmers occupied the Pak Mun hydroelectric dam in the rural northeast of Thailand, demanding that the floodgates of these dams be opened and the dams be demolished. Stocks of sturgeon, striped bass, Atlantic salmon, and alewives have increased, water quality has improved, and more animals, plants, and birds now occupy the banks of the Kennebec River in Maine since the bulldozing of the Edwards Dam in 1999. In the United States, over 40 dams have either been removed, or slated for removal, and dam-removal advocates have recently pointed their attention at the Snake River, a main tributary of the Columbia River, where four dams have virtually stopped the flow of salmon.

Dams can also have unforeseen impacts even when they accomplish their intended purposes. Flood control dams might encourage people to live closer to rivers, believing that the risks of flooding are mitigated by the dam. This can be catastrophic, as demonstrated in 1993, when floods in Mississippi overwhelmed the dam system, causing billions of dollars in damages to communities along that river. Presumably, these damages would have been less severe had people chosen to locate further from the river. To the extent that irrigation dams are successful, they can also encourage the wasteful use of water, as evidenced by farmers switching from drip and other less water-using irrigation methods to more wasteful spray methods of irrigation. To the extent that hydroelectric power is sold at below-market rates, as is done in many parts of the United States, electrical irrigation methods are encouraged, possibly inducing farmers to specialize in more water-intensive crops, and possibly leading to more rapid depletion of aquifers. What else can account for the high share of American cotton grown in Arizona, or, for that matter, the preponderance of golf courses in the Phoenix area?

Hydroelectric power is often touted as a reliable source of electricity, but its availability often depends on the whims of nature. Years of low rainfall in Brazil, a country that depends on hydroelectric power for more than 90% of its electricity needs, have lead Brazil to approve the licensing of more than 20 new gas-fired generating plants and to consider the construction of additional nuclear capacity. In the meantime, Brazil faces the possibility of rolling blackouts to bring electricity consumption in line with their generating capacity. Shortages of water in the western United States have revealed weaknesses in how the state of California deregulated its power industry. Sufficient water might have allowed enough power to be generated to prevent, or at least forestall, the energy problems the state has faced.

4. GENERATING HYDROELECTRIC POWER

Water is stored behind a dam in what is called a forebay. The quantity of water in the forebay can be expressed in terms of acre-feet of volume, but can also be measured in terms of elevation, which refers to the number of feet the surface of the reservoir is above sea level. The elevation of the reservoir will determine an important attribute of a dam that has a direct bearing on the amount of hydroelectric power than can be generated, namely, the head of the dam, which refers to the distance between the elevation of the reservoir and the top of the penstock, which is a tube through which water flows to the generator. The higher the head, the faster the water will fall through the penstocks, and hence the greater the amount of kinetic energy that the generator will eventually convert into electrical energy. The technical relationship between water release, dam head, and generation can be expressed as $q = (\text{CFS} \times \text{head})/\alpha$, where q is hydroelectric output, CFS is the water release rate (measured in cubic feet or cubic meters per second), and head is the head of the dam (usually measured in feet or meters). The coefficient α is a factor that converts the product of water releases and dam head to hydroelectric generation. For example, if we use a figure of $\alpha = 8.8$ and release water at a rate of 5000 CFS from a dam with a 400-foot head, we generate 22,773 hp of electricity. Dividing this figure by 746 and converting (from watts to megawatts) yields 170 MW of hydroelectric generation.

To facilitate integrating the technical relationship between these factors into hydroelectric generation for use in the models that follow, we combine the relationship between elevation, content, and head into the following simpler equation that we will use as our production function for hydroelectric generation:

$$q_t^h = q_t^h(r_t, W_t). \tag{1}$$

In this equation, hydroelectric power, q_t^h, is generated in period t by releasing water at a rate given by r_t. The argument for reservoir content in the production function, W_t, will capture the combined effects that changes in content have on elevation, which in turn influences dam head, which finally influences the productivity of water releases. If we were to take a more formal approach, we would define the hydroelectric generation production function as $q_t^h = \tilde{q}_t^h\{r_t, h_t[e_t(W_t)]\}$. In this equation, we define head as a function of elevation, $h_t = h_t(e_t)$, and elevation as a function of content, $e_t = e_t(W_t)$. We derive Eq. (1) by successively substituting the head and elevation equations. Nevertheless, we assume that Eq. (1) is continuously differentiable in both arguments, that both first derivatives are positive, and that both second derivatives are negative. Accordingly, hydroelectric generation will be higher, for a given dam content, the higher the rate of water release. We also assume that hydroelectric generation will be higher, the higher dam content, for a given rate of water release.

5. OPTIMAL PROVISION OF HYDROELECTRIC POWER

Our simple model of hydroelectric generation assumes that the hydroelectric power dispatcher generates power and sells it in a large market, with the hydroelectric power sold in a market at a price given by the inverse demand function $p_t(Q_t)$, where Q_t represents the quantity of electricity sold during period t. We express industry output as $Q_t = \tilde{Q}_t + q_t^h(r_t, W_t)$, where, $\tilde{Q}_t$ represents power sold by the other generators or energy marketers serving this particular market. Water release costs are given by

$c_t^b(r_t)$ which we assume to be continuously differentiable with positive first and second derivatives. Accordingly, profits in each period are given by the following equation:

$$\pi_t = p_t(Q_t)q_t^b(r_t, W_t) - c_t^b(r_t). \quad (2)$$

The only constraints that the hydroelectric power operator faces in our simple model is an equation of motion, $W_t = W_{t-1} - r_{t-1} + f_{t-1}$, which has reservoir content determined by water releases and inflows into the reservoir (f_{t-1}), and a requirement that the dispatcher operate within the operating characteristics of the generator. For now, we will assume that the dispatcher satisfies this constraint implicitly.

The dispatcher will choose rates of water release to maximize the discounted present value of profits subject to the equation of motion over t periods ($t = 1,\ldots T$). We use a finite-time horizon because dam operations are often subject to operating plans promulgated by regulatory agencies that specify water release and other requirements over annual, monthly, and other finite time horizons. Nevertheless, the resulting discrete-time Hamiltonian function is given by

$$\begin{aligned}H = \sum_{t=1}^{T-1} &[p_t(Q_t)q_t^b(r_t, W_t) - c_t^b(r_t) \\ &+ \gamma_{t+1}(W_{t+1} - W_t + r_t - f_t)] \\ &+ P_T(Q_T)q_T^b(r_T, W_T) - c_T^b(r_T) \\ &+ \gamma_T(W_T + r_T - f_T). \quad (3)\end{aligned}$$

In Eq. (3), we define γ_{t+1} as the discrete-time costate associated with the equation of motion and define γ_T as the final period costate on the equation of motion.

The first-order conditions on the control variable are given by

$$\frac{\partial H}{\partial r_t} = p_t \frac{\partial q_t^b}{\partial r_t} + \frac{\partial p_t}{\partial q_t^b} \frac{\partial q_t^b}{\partial r_t} q_t^b - \frac{\partial c_t^b}{\partial r_t} - \gamma_{t+1} \leq 0 \quad (4)$$

$$r_t \frac{\partial H}{\partial r_t} = 0. \quad (5)$$

Equations (4) and (5) indicate the conditions that govern hourly water release rates. Equation (4) can be rearranged to form the following equation:

$$\frac{\partial H}{\partial r_t} = p_t \frac{\partial q_t^b}{\partial r_t} + \frac{\partial p_t}{\partial q_t^b} \frac{\partial q_t^b}{\partial r_t} q_t^b \leq \frac{\partial c_t^b}{\partial r_t} + \gamma_{t+1}. \quad (6)$$

The first part of the right-hand side of Eq. (6) is the marginal benefit of releasing water in period t ($p_t \partial q_t^b/\partial r_t + (\partial p_t/\partial q_t^b)(\partial q_t^b/\partial r_t)q_t^b$). The second part of the right-hand side of Eq. (6) is the marginal cost of releasing water in period t ($\partial c_t^b/\partial r_t + \gamma_{t+1}$) and includes two components. The first part ($\partial c_t^b/\partial r_t$) is interpreted as the marginal cost of releasing water. The second part (γ_{t+1}) is the shadow price of water and captures the impact that releasing water now has on future net benefits. We include it in the marginal cost component of Eqs. (4)–(6) because it reflects the opportunity cost of using the water resource to generate hydroelectric power in the current period.

Nevertheless, Eqs. (4) and (5) are the Kuhn–Tucker necessary conditions for an optimum and can be interpreted as follows. If Eq. (4) is strictly negative, and the marginal benefits of releasing water in period t are less than the marginal costs, Eq. (5) will then hold with equality and water releases in that period will be zero. Otherwise, water is released at a rate up to the point where Eq. (4) holds with equality, and will still satisfy Eq. (5). We can take the first part of Eq. (4) and rearrange it into terms more familiar in microeconomics:

$$P_t \frac{\partial q_t^b}{\partial r_t} + \frac{\partial p_t}{\partial q_t^b} \frac{\partial q_t^b}{\partial r_t} q_t^b = p_t \frac{\partial q_t^b}{\partial r_t}\left(1 + \frac{1}{\eta_t}\right), \quad (7)$$

where η_t is the elasticity of demand for electricity. This allows us to express the right-hand side of Eq. (7) as $MR_t(1 + 1/\eta_t)$. On making this substitution, and assuming an interior solution, our first-order condition becomes

$$p_t \frac{\partial q_t^b}{\partial r_t}\left(1 + \frac{1}{\eta_t}\right) = \frac{\partial c_t^b}{\partial r_t} + \gamma_{t+1}. \quad (8)$$

According to this equation, the operator releases water at that rate which equates the marginal benefit of releasing water to the marginal cost of releasing water.

In this model, water has two values. The first, which we call the flow value, captures the role that plays a direct input in hydroelectric power generation. The second, which we call the stock value, reflects the effect that stored water has on the generating efficiency of the dam in future periods through effects that storing additional water have on the head of the dam. *Ceteris paribus*, higher release rates in the current period increase hydroelectric generation in the current period, but reduce the potential future reservoir elevation, thereby reducing the generating efficiency of future releases.

For a given reservoir size, releasing water in the current period increases revenue by $p_t\{\partial q_t^b(r_t, W_t)/\partial r_t\}$ and costs the hydroelectric power dispatcher $\partial c_t^b/\partial r_t$. The net impact of releasing water in the current period is $p_t\{\partial q_t^b(r_t, W_t)/\partial r_t\} - \partial c_t^b/\partial r_t$.

However, direct water release costs are invariant with respect to reservoir elevations, so additional water in the reservoir provides additional generation without incurring higher operating expenses. If we differentiate Eq. (3) with respect to the reservoir content, we obtain an equation that describes the dynamic conditions governing the change in the marginal cost of water.

$$\frac{\partial H}{\partial W_t} = p_t \frac{\partial q_t^b}{\partial W_t} + \frac{\partial p_t}{\partial q_t^b} \frac{\partial q_t^b}{\partial W_t} q_t^b + \gamma_{t+1} - \gamma_t = 0. \quad (9)$$

We can use the previous result involving the elasticity of demand to restate Eq. (9) as

$$p_t \frac{\partial q_t^b}{\partial W_t} \left(1 + \frac{1}{\eta}\right) = \gamma_{t+1} - \gamma_t = 0. \quad (10)$$

In Eq. (10), the $p_t(\partial q_t^b / \partial W_t)(1 + 1/\eta)$ term represents the marginal benefit of additional stored water measured in terms of the value of additional hydroelectric power generation for a given rate of water release. The term defining the change in the shadow price of the water, $\gamma_{t+1} - \gamma_t$, reflects the net contribution of the water in future periods. Rearranging Eq. (10) as a difference equation (for $\Delta \gamma = \gamma_{t+1} - \gamma_t$) and solving it yields Eq. (11), which expresses the current-period shadow value of water in terms of the value that the stored water has on remaining future period revenues, less the cost of the additional hydroelectric generation in these periods.

$$\tilde{y}_t = \sum_{t=1}^{T-1} p_t \frac{\partial q_t^b}{\partial W_t}\left(1 + \frac{1}{\eta_t}\right) + p_T \frac{\partial q_T^b}{\partial W_T}\left(1 + \frac{1}{\eta_T}\right). \quad (11)$$

According to this equation, the benefits of storing water will decrease over time. In the last period, water has value only for its generating value in that period. We also see that γ_t is the sum of these benefits for the remaining periods, which decline as we approach the end of the time horizon.

We now present a simulation example of a simple version of the above model. In order to implement the simulation, we will make a number of simplifying assumptions. First, we will run the simulation for 24 periods. We will also assume that spot electricity prices are determined exogenously and vary over the course of the simulation. We can easily think of the 24 periods as hours, and so we can easily imagine that our "day" is characterized by periods of lower demand (and hence lower spot prices) and periods of higher demand (and hence higher spot prices). The prices will be expressed in terms of dollars per megawatt and will follow the time path given in Fig. 1. According to Fig. 1, prices are low during the early and late hours of the day. We can easily think of these as off-peak hours. As the hours progress, however, prices rise until they reach their highest point sometime during the middle of the day. We easily think of these as on-peak hours. We will also assume that generating hydroelectric power costs a constant $15/megawatt to generate.

We assume that the reservoir begins with 120,000 acre-feet of content and that hourly inflows fluctuate randomly at a rate between 200 and 800 CFS. Generation will occur according to $q_t^b = 0.0005 r_t h_t$ (where h_t is the head of the dam in period t). We also assume that dam head depends on content, according to $h_t = 0.002 W_t$. We assume finally that the operator must release a total of 2000 acre-feet of water over the 24 periods. The operator will choose rates of water release in each of the 24 periods to maximize the sum of net receipts subject to the equation of motion. For our simulation, we will use a discount rate of 5% (0.05). Finally, we will assume that generator capacity limits hourly release rates to 1500 CFS (a small dam!).

In Fig. 2, we graph hydroelectric generation (which is measured along the left-vertical axis) in each of the 24 periods. Along the right-vertical axis, we measure the rate of water release in each of the 24

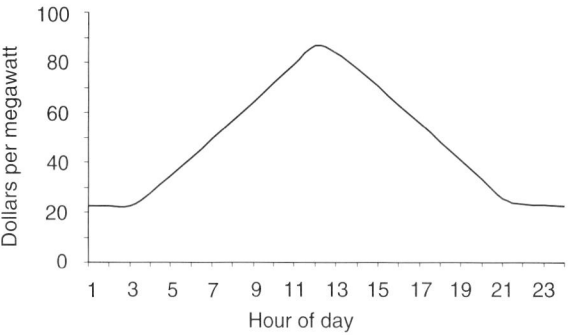

FIGURE 1 Hypothetical spot electricity prices for simulation.

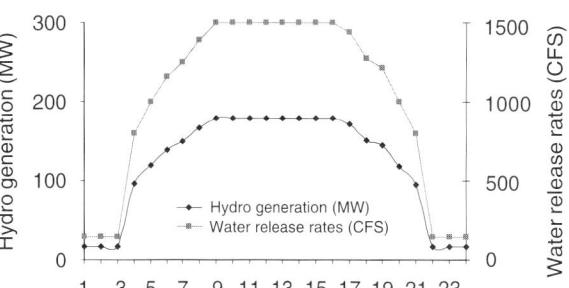

FIGURE 2 Water releases and hydroelectric generation. CFS, cubic feet/second.

periods. Because generation and transmission costs are constant throughout the day, but spot prices rise during the on-peak hours, it should come as no surprise that release rates (and generation) follow the pattern displayed by prices. From the perspective of power generation, this behavior typifies load following, whereby generation follows demand (as reflected in the higher on-peak period prices). In terms of hourly water release patterns, this illustrates what we term hydro-shifting, which refers to storing water behind the dam during off-peak hours and releasing the water for generation during on-peak hours.

6. SUMMARY AND CONCLUSIONS

The preceding discussion has presented a brief introduction to the economics of dams and hydroelectric power. Although the models and their results are simple, such a simplified approach represents a foundation for subsequent research and analysis of more complex issues involving dams and hydroelectric power. These models also couch the entire discussion of hydroelectric power firmly in the natural resource economics, and more specifically, in the water management domain, one that reflects the view of the author that operating dams and generating hydroelectric power require decisions involving how best to manage water resources.

SEE ALSO THE FOLLOWING ARTICLES

Electricity, Environmental Impacts of • Electric Power Generation: Fossil Fuel • Electric Power: Transmission and Generation Reliability and Adequacy • Hydropower, Environmental Impact of • Hydropower, History and Technology of • Hydropower Resettlement Projects, Socioeconomic Impacts of • Hydropower Resources • Hydropower Technology

Further Reading

Disheroon, F. R. (1993). Hydroelectric power. *In* "Environmental Law: From Resources to Recovery" (C. B. Campbell-Mohn and J. W. Futrell, Eds.), pp. 527–557. West Publishing Co., St. Paul, Minnesota.

Edwards, B. K. (2003). "The Economics of Hydroelectric Power." Edward Elgar, Cheltenham, UK.

Edwards, B. K., Flaim, S. J., and Ancrile, J. D. (1992). Maximizing the value of thermally integrated hydroelectric generating facilities. *In* "Modeling and Simulation, Proceedings of the Twenty-Third Annual Pittsburgh Conference on Modeling and Simulation (1992)" (W. G. Vogt and M. H. Mickle, Eds.), pp. 463–470.

Edwards, B. K., Howitt, R. E., and Flaim, S. J. (1996). Fuel, crop, and water substitution in irrigated agriculture. *Resource Energy Econ.* **18**(3), 311–331.

Edwards, B. K., Flaim, S. J., and Howitt, R. E. (1999). Optimal provision of hydroelectric power under environmental constraints. *Land Econ.* **75**(2), 267–283.

Ferguson, C. E. (1972). "Microeconomic Theory." Richard D. Irwin, Inc., Homewood, Illinois.

Francfort, J. (1997). "Hydropower's Contribution to Carbon Dioxide Emission Reduction," Idaho National Engineering and Environmental Laboratory (November). U.S. Department of Energy, Idaho Falls, Idaho.

International Commission on Large Dams (ICOLD). (1998). "World Register of Dams, 1998." ICOLD, Paris, France.

United States Department of the Interior, Bureau of Reclamation. (2001). "The History of Hydropower Development in the United States." Available on the Internet at http://www.usbr.gov.

Varian, H. R. (1992). "Microeconomic Analysis." W. W. Norton, New York.

World Commission on Dams. (2000). "Dams and Development: A New Framework for Decision-Making." Earthscan Publ. Ltd., London and Sterling, Virginia.

Hydropower, Environmental Impact of

GLENN CADA and MICHAEL SALE
Oak Ridge National Laboratory
Oak Ridge, Tennessee, United States

DENNIS DAUBLE
Pacific Northwest National Laboratory
Richland, Washington, United States

1. Introduction
2. Water Quality Effects
3. Changes in Sediment and Nutrient Balance
4. Alteration of Terrestrial Habitats
5. Alteration of Aquatic Habitats
6. Interference with Movements of Aquatic Organisms
7. Greenhouse Gas Emissions

Glossary

anadromous fish Fish that spend most of their lives in the sea and then migrate to fresh water to breed (e.g., salmon).
catadromous fish Fish that spend most of their lives in fresh water and then migrate to the sea to breed (e.g., eels).
entrainment The process by which an aquatic organism is drawn into the intake of a power plant along with the inflowing water.
epilimnion The upper, wind-mixed layer of a thermally stratified lake or reservoir; because of its exposure and turbulent mixing, dissolved gases are freely exchanged with the atmosphere.
eutrophication The process by which a lake or reservoir receives a large (often excessive) supply of nutrients, resulting in a high production of organic matter by plants and animals.
fishway A constructed device that allows fish to migrate upstream past dams, waterfalls, and rapids.
flushing rate The average length of time that water resides in a lake or reservoir, ranging from several days in small impoundments to many years in large lakes. The flushing rate is calculated by dividing the volume of water passing through the lake per unit time by the lake volume.
gas bubble trauma A condition in which fish tissues become supersaturated with dissolved gases that subsequently come out of solution and form bubbles in the gills, blood vessels, and other tissues.
hypolimnion The bottom-most layer of a thermally stratified lake or reservoir; it is isolated from wind mixing, often too deep for sunlight penetration and plant photosynthesis, and typically cooler and less oxygenated than the epilimnion.
metalimnion The middle or transitional layer of water in a reservoir, between the epilimnion and hypolimnion, in which the temperature exhibits the greatest difference in a vertical direction.
plankton Floating or weakly swimming organisms in the open water of a reservoir or river.
supersaturation of gases A state in which gases, such as nitrogen and oxygen, are held in solution in water in quantities exceeding the saturation level of the water (i.e., greater than 100% saturation) for a given temperature and pressure.
thermal stratification The temporary separation of a lake or reservoir into distinct vertical layers due to temperature differences.
trap efficiency The ability of a reservoir to trap and retain sediments and nutrients; it is usually expressed as a percentage of the inflowing sediments or nutrients that are retained in the reservoir.

Although free of many of the harmful emissions associated with other energy sources, hydroelectric power plants can affect the environment by impounding water, flooding terrestrial habitats, and creating a barrier to the movements of sediments, nutrients, and aquatic organisms. The ability of hydropower projects to alter the flow of water can affect both aquatic and terrestrial habitats downstream from dams.

1. INTRODUCTION

Hydroelectric power production is largely free of several major classes of environmental impacts associated with nonrenewable energy sources (Table I). Unlike fossil-fueled power plants, hydroelectric generation does not emit toxic contaminants (e.g., mercury) or sulfur and nitrogen oxides that can cause acidic precipitation. Although construction of the hydropower project may result in temporary emissions from dust and equipment, these impacts are at least two orders of magnitude less than the continuing emissions from a coal-fired power plant. Inundation and decomposition of vegetation by the reservoir can cause the release of greenhouse gases (e.g., methane, carbon dioxide), but the significance of this source depends greatly on the age, size, and geographic location of the reservoir and on the amount of vegetation and soil carbon flooded. Hydroelectric power provides an opportunity to reduce the production of air pollutants and greenhouse gases when compared with power produced by the combustion of fossil fuels or biofuels.

Unlike most other forms of energy production, hydroelectric power plants generate few solid wastes. Land may be required for the disposal of material dredged from reservoirs or waterborne debris, but the amounts of land needed are very small compared with those needed for the continuing disposal of coal ash and slag. Similarly, hydropower production does not create hazardous or radioactive wastes that require safe long-term storage facilities. Many other environmental impacts associated with the overall fuel cycles of other energy sources are minor or nonexistent for hydroelectric power. These externalities include impacts associated with resource extraction (e.g., coal mining, oil drilling), fuel preparation (e.g., refining), and transportation (e.g., oil spills, other accidents).

Although hydroelectric power plants have many advantages over other energy sources, they also have potential environmental impacts. Most of the adverse impacts of dams are caused by habitat alterations. Reservoirs associated with large dams can inundate large areas of terrestrial and river habitats and displace people from their homes. Diverting water from the stream channel or curtailing reservoir releases to store water for future electrical generation can dry out streamside (riparian) vegetation. Insufficient water releases degrade habitat for fish and other aquatic organisms in the river below the dam. Water in a reservoir is stagnant compared with that in a free-flowing river. Consequently, waterborne sediments and nutrients can be trapped, resulting in the undesirable proliferation of algae and aquatic weeds (eutrophication). In some cases, water spilled from high dams may become supersaturated with nitrogen gas, resulting in gas bubble disease in aquatic organisms inhabiting the tailwaters.

Hydropower projects can also affect aquatic organisms directly. The dam can block upstream movements of fish, and this can have severe consequences for anadromous fish (e.g., salmon,

TABLE I

Potential Environmental Benefits and Adverse Impacts of Hydroelectric Power Production

Benefits	Adverse impacts
No emission of sulfur and nitrogen oxides	Inundation of wetlands and terrestrial vegetation
Few solid wastes	Emissions of greenhouse gases from flooded vegetation at some sites
Minimal impacts from resource extraction, preparation, and transportation	Conversion of a free-flowing river to a reservoir
Flood control	Replacement of riverine aquatic communities with reservoir communities
Water supply for drinking, irrigation, and industry	Displacement of people and terrestrial wildlife
Reservoir-based recreation	Alteration of river flow patterns below the dam
Reservoir-based fisheries	Loss of river-based recreation and fisheries
Enhanced tailwater fisheries	Desiccation of streamside vegetation below the dam
Improved navigation on inland waterways below the dam	Retention of sediments and nutrients in the reservoir
	Development of aquatic weeds and eutrophication
	Alteration of water quality and temperature
	Interference with upstream and downstream passage of aquatic organisms

steelhead, American shad), catadromous fish (e.g., American eels), and riverine fish that make seasonal migrations to spawn (e.g., sturgeon, paddlefish). Fish moving downstream may be entrained (i.e., drawn into the power plant intake flow). Entrained fish are exposed to physical stresses (e.g., pressure changes, shear, turbulence, strike) as they pass through the turbine, and these may cause disorientation, physiological stress, injury, and/or mortality.

This article describes the potential environmental impacts associated with hydropower production as well as measures that are employed to mitigate or eliminate these impacts.

2. WATER QUALITY EFFECTS

Construction of hydropower facilities (dam, powerhouse, roads, and transmission lines) will result in the disturbance and movement of soil. Water quality would be degraded if soil erosion increases the amounts of suspended material in the water (turbidity) or blankets the bottom of the river (sedimentation). Oils, greases, and chemical wastes may accidentally be spilled into the water during project construction. Well-known engineering practices, including soil stabilization techniques and storm-water retention dikes, can reduce impacts to water quality during construction. In most cases, the continuing effects of operation of the hydropower project are of greater concern. The remainder of this article considers the impacts of hydropower operation.

2.1 Temperature

The most common water quality issues associated with hydroelectric power production are changes in temperature and dissolved oxygen concentrations in the water released from the dam. Reservoirs in temperate regions often become thermally stratified, especially if they are deep or sheltered from the wind. When reservoir water begins warming during the spring, a warm surface layer floats over the colder bottom waters. In time, three distinct layers of water form: a warm upper layer (epilimnion), a narrow zone of rapidly decreasing water temperature (metalimnion), and a cool bottom layer (hypolimnion) (Fig. 1). This stratified condition continues throughout the summer until the surface waters cool during the autumn and top-to-bottom mixing occurs. Although thermal stratification is common in temperate regions, not all natural lakes and artificial reservoirs stratify. In general, reservoirs that are

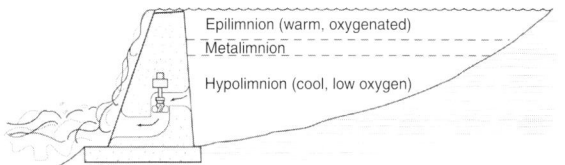

FIGURE 1 Thermal stratification of a hydropower reservoir.

most likely to stratify are deep, have a low surface area/volume ratio and flushing rate, and are protected from mixing by their orientation to wind and the surrounding topography.

Many hydroelectric turbine intakes withdraw water from the depths of the reservoir, and summer water temperatures in the discharged water are lower than would be the case in the free-flowing river. The opposite situation may occur during the winter. Because the large volume of water in the reservoir cools more slowly than does an unimpounded river, winter discharges from a hydropower plant may be warmer than natural. Altered water temperatures in the tailwaters have profound effects on aquatic organisms. Cold-water discharges during the summer can slow the growth rates and reduce productivity of fish and aquatic invertebrates, but they can also allow the establishment of cold-water fisheries in areas where the natural rivers are too warm. Conversely, warm-water discharges during the winter can speed the metabolic rate of aquatic insects and fish eggs, so that they develop and emerge before the appropriate season.

Mitigation of temperature effects is not difficult but can be expensive. Devices such as propellers have been used to break up stratification in small reservoirs. For large reservoirs, multiple-level intakes allow water to be withdrawn and mixed from different depths, so that water of the appropriate temperature can be discharged into the tailwaters.

2.2 Dissolved Oxygen

Thermal stratification can also decrease the concentrations of dissolved oxygen in the water discharged from the reservoir. The relatively stagnant hypolimnion is isolated from the processes of atmospheric diffusion and wind mixing that replenish dissolved gases in the epilimnion. Plant and animal respiration, bacterial decomposition of organic matter, and chemical oxidation all can act to progressively remove dissolved oxygen from bottom waters. This situation may be exacerbated by the input of high levels of oxygen-consuming organic materials that enter the reservoir from the watershed upstream. In temperate regions, the decline in dissolved oxygen concentrations begins at the onset of stratification during the spring and continues until either all of the oxygen is used or reoxygenation occurs during the autumn turnover of the water body. Turbines that withdraw from the hypolimnion may discharge water that is low in dissolved oxygen, affecting the aesthetic qualities (taste, odor, and appearance), communities of aquatic organisms, and waste assimilation capacity of the tailwaters.

Low dissolved oxygen concentrations create chemically reducing conditions, which in turn may affect other water quality parameters. For example, decomposition of sulfur and nitrogen compounds in the absence of dissolved oxygen may result in the buildup of toxic hydrogen sulfide and ammonia in the hypolimnion. Chemically reducing conditions also increase the solubility of iron, manganese, and some heavy metals. These elements may be mobilized from the sediments and suspended particulate material and may be discharged into the tailwaters. Hypolimnetic discharges from hydropower facilities that have low dissolved oxygen concentrations and elevated levels of iron and other metals, manganese, ammonia, and sulfides may have adverse effects on downstream aquatic organisms and other water users. Adequate levels of dissolved oxygen (≥ 5 mg/L in most waters) are necessary not only for aesthetic qualities (taste and odor) but also to support a balanced community of aquatic organisms. Hydrogen sulfide creates odor problems and may cause downstream fish kills. High levels of iron and manganese increase the cost of domestic water treatment because they impart objectionable tastes to drinking water and can stain laundry and plumbing fixtures.

There are a number of structural measures that can be employed to increase dissolved oxygen concentrations in the discharge water (e.g., aerating turbines). Dissolved oxygen levels can also be increased through modifications in dam operations, including fluctuating the timing and duration of flow releases, spilling surface water from the top of the dam, and flow mixing with multiple-level intakes.

2.3 Gas Supersaturation

Dissolved gas concentrations may be raised as well as lowered by the presence of a hydroelectric dam. Spilling water over the crest of the dam (Fig. 2) often increases the concentration of total dissolved gases because entrained air bubbles dissolve under pressure when the water plunges into deep tailwater pools. As this water surfaces in the shallow river channel

FIGURE 2 Spilling water from the Bonneville Dam on the Columbia River in the United States.

downstream, it becomes supersaturated ($>100\%$ saturation) with gases relative to the lower water pressures at the surface. If the subsequent rate of equilibration is slow, gas supersaturation may persist in flowing waters for long time periods and distances from the dam.

Dissolved gas saturations of 110% or greater are considered potentially damaging to aquatic organisms. Water that has become supersaturated with dissolved nitrogen can cause gas bubble trauma in fish, a potentially lethal condition similar to the "bends" experienced by deep divers who return to the surface too quickly. Gas supersaturation leading to gas bubble trauma has been most commonly reported at large dams in the Columbia River basin in the United States. Historically, gas supersaturations in the range of 115 to 143% have occurred in the Columbia and Snake rivers during periods of high spilling, but the problem has declined during recent years as a result of the implementation of mitigation measures.

The necessary conditions for nitrogen gas supersaturation at hydropower projects include a deep plunge pool (so that air bubbles will dissolve under elevated pressures) and a high dam (so that spilled water has sufficient velocity to reach the bottom of the pool). One successful method for preventing nitrogen gas supersaturation is to install flip lips, that is, structures at the base of the spillway that direct the spilled water horizontally downstream instead of deep into a plunge pool. By keeping spilled water (with entrained air bubbles) near the surface, there is less chance for excess nitrogen gas to become dissolved.

2.4 Eutrophication

When a free-flowing river enters a reservoir, the water slows down, spreads out, and warms up, often creating excellent conditions for the growth of plants. Submerged rooted vegetation and emergent vegetation (e.g., cattails) proliferate in shallow water areas. The relatively low-flow conditions also favor the growth throughout the reservoir of microscopic planktonic plants (algae). In tropical areas, they also favor the growth of floating plants such as water hyacinth and water lettuce.

Depending on the circumstances, the proliferation of aquatic vegetation in the reservoir can be either beneficial or damaging. Rooted vegetation provides a valuable habitat for aquatic invertebrates and fish, and algae provide an essential food source for small planktonic animals and, in turn, plankton-eating fish. Often, the production of these organisms in a reservoir may far exceed that in a free-flowing river. On the other hand, the excessive undesirable growth of plants, known as eutrophication, can be a serious problem. Although many species of planktonic algae are food sources, some species that thrive in warm nutrient-rich waters release chemicals that are toxic to animals. Large plants can interfere with the operation of a hydropower project by clogging the intake structure, increasing the evapotranspiration loss of water from the reservoir, reducing the effective storage volume of the reservoir, and trapping more sediments and nutrients (and thereby decreasing the useful life of the reservoir). The decomposition of excessive growths of algae and large aquatic plants can degrade water quality in the reservoir and in the tailwaters downstream, impairing alternative water uses (e.g., fishing, swimming, drinking).

Surface water released from an eutrophic reservoir often has high concentrations of algae and low concentrations of nutrients compared with the inflowing river. This encourages the production of plankton-feeding organisms in the tailwaters. On the other hand, bottom water may have relatively high levels of nutrients due to the decomposition of organic material in the reservoir sediments. Discharges of nutrient-rich bottom water may stimulate excessive production of attached plants in the river channel below the dam, and this often has adverse impacts on habitats of fish and other aquatic organisms.

Mitigation of alterations in the nutrient balance of the river and reservoir is possible, but it is often costly and complicated. Excess growth of large aquatic plants can be controlled by mechanical harvesting or the introduction of herbivorous fish. Microscopic planktonic algae are more difficult to control. It is often easier to reduce the input of nutrients from the watershed or flush nutrients from the reservoir to limit algal production.

3. CHANGES IN SEDIMENT AND NUTRIENT BALANCE

Reservoirs may act as traps for sediments carried by the inflowing river. A stream entering a reservoir spreads out, reducing the water velocities and sediment-carrying capacity. A delta forms at the entrance to the reservoir, with the size and location of the delta depending primarily on the size of the sediment particles in the river and the velocity of flow through the reservoir. Coarse-grained heavy sediments carried by the stream are deposited immediately. Fine sediments remain in suspension and can be deposited farther into the reservoir or carried out of the reservoir in the outflow.

The ability of a reservoir to trap and retain sediments is known as the trap efficiency, expressed as the percentage of the total inflowing sediment that is deposited. Sediment trapping can be beneficial in cases where the reservoir retains soil that eroded from the watershed due to improper land use practices. However, reservoirs with high trap efficiencies will also tend to have the shortest useful life spans. The deposited sediment fills up the reservoir and, unless removed by dredging, reduces the storage capacity needed for hydropower production. Sediment trapping can also have biological effects. Compared with a turbid river, sedimentation will increase the clarity of water, benefiting aquatic plants and sight-feeding animals. Gravel deposits at the entrance to a reservoir will often provide an excellent fish-spawning habitat. On the other hand, fine sediments deposited elsewhere in the reservoir may blanket habitat and smother clams, fish eggs, and other nonmobile organisms.

Reservoirs also trap inflowing nutrients by two processes: (1) sedimentation of nutrient-bearing particulate matter and (2) transformation of dissolved nutrients into particulate material, which is then subject to sedimentation. Dissolved nutrients may be converted to particulate forms by physicochemical processes (adsorption and precipitation) or through uptake by plants and bacteria. The impacts of nutrient trapping may be beneficial or harmful, depending on the site. Nutrients trapped in the reservoir can support increased production of aquatic plants and fish. However, eutrophication caused by increased nutrient availability often diminishes water quality.

Removal of sediments and nutrients by the reservoir generally results in the tailwaters being "starved" of these materials, relative to the stream above the hydropower project. Clear water discharged from the reservoir has erosive power to pick up available sediment particles from the riverbed and stream banks below the dam. The river channel below the dam may be eroded, and scour may damage riverside structures (e.g., bridges, culverts, road embankments). The eroded sediments will be deposited farther downstream. Similarly, nutrients may be trapped in the reservoir sediments, leaving the discharged water nutrient poor compared with the water flowing into the reservoir. This can be beneficial in rivers where excessive nutrients are a form of water pollution. However, particularly in large rivers, reductions in nutrient flow may drastically lower the productivity of important fisheries.

The simplest way in which to mitigate for adverse sediment and nutrient trapping is to dredge the reservoir as needed. Numerous mechanical and hydraulic dredging techniques are available. Sediments in some reservoirs can be flushed out through pipes or notches in the dam. Large reservoirs impound enough water, so that sediments can be flushed out at any time, whereas in smaller reservoirs sediments might be able to be flushed only during floods and other high stream flow events.

4. ALTERATION OF TERRESTRIAL HABITATS

Creation of a reservoir behind a dam will flood terrestrial vegetation and displace associated wildlife and human populations. The significance of this impact depends on the size and location of the reservoir. A small reservoir will not cause major losses of terrestrial habitats and, in fact, may actually have benefits. For example, a hydropower reservoir typically creates a larger water surface than the original river channel that it flooded. Consequently, the reservoir can provide more habitat for waterfowl and, in arid regions, a more permanent source of drinking water for wildlife. Human populations may benefit from additional uses for hydropower reservoirs: flood control, recreation, fisheries, and a reliable source of water for drinking, industry, and agriculture. On the other hand, a large hydropower reservoir may flood villages and towns, displace large numbers of people, displace or drown terrestrial wildlife, and eliminate important floodplain vegetation. The decomposition of flooded vegetation may degrade water quality and increase the emissions of greenhouse gases.

Road construction and site clearance for the hydropower dam, powerhouse, and associated facilities may eliminate wildlife habitat, and construction noise and activities could interfere with the nesting of birds or migratory movements of terrestrial mammals.

Downstream from the reservoir, stream flow changes during construction and operation of the hydropower project can affect streamside (riparian) vegetation and associated animals. As with the reservoir, these effects may be positive or negative, depending on the site, design of the project, and mode of operation. For example, relatively constant releases from the reservoir associated with hydropower production could provide reliable flows in rivers that naturally experience wide variations in stream flow. This could encourage the establishment of riparian vegetation. On the other hand, some riparian communities require the seasonal flooding that is associated with natural unimpounded rivers, and they may decline under more uniform controlled flow releases. Some hydropower projects are operated in a "pulsing" mode, with water released from powerhouses only during portions of the day when electricity is most needed or valuable and with discharges from powerhouses being halted for the remainder of the day. By alternating flooding and drying within a single day, pulsing projects create an unnatural flow regime that can have serious impacts on both aquatic organisms and riparian communities.

5. ALTERATION OF AQUATIC HABITATS

Many, but not all, hydropower projects are associated with dams and reservoirs that are constructed either to store water or to increase the height, or hydraulic head, of water above the power plants. The water stored behind hydropower dams is used to shift river flows from times of surplus water to other times when water use is more valuable, for example, from the spring season when river flows are high to the summer season when they are low.

When dams and reservoirs are constructed on rivers that were previously free flowing, the aquatic habitats can be changed substantially, both in the reservoirs and downstream from the dams. The degree of habitat change depends on the reservoir operation (how and when the active storage is used to detain flows) and the physical dimensions of the reservoir. Some hydropower projects divert water out of the natural river channel but do not have large impoundments. Such a diversion project usually has a bypassed reach of the river that has a substantially reduced water flow and modified aquatic habitat as well as a pipeline or canal that routes the diverted water to a powerhouse downstream.

5.1 Habitat Change in the Reservoir

In addition to inundation of terrestrial habitat, riverine habitats are altered by the creation of reservoirs. Natural riverine habitats are generally characterized by relatively shallow water with fast currents and a substrate of larger sand, gravel, boulders, and/or exposed bedrock. Reservoirs impounded behind dams are usually deeper and slower moving than free-flowing rivers. Because water velocities in these impoundments are greatly reduced, finer sediments that would have been transported downstream tend to accumulate. Silt covers the gravel and boulders that formed the bottom substrates in the former stream channel. Silt and sand habitats often support less varied invertebrate and fish communities. Conversely, the quiet waters and muddy bottoms of reservoirs may allow the development of beds of aquatic plants that are very productive habitats for some aquatic animals and waterfowl. Because of these habitat changes, animals and plants that are characteristic of river ecosystems are replaced by animals and plants that are adapted to lake-like conditions.

If the reservoir is deep and stagnant enough and the climatic conditions are conducive, the deep colder water may become isolated from the warmer surface water, a condition known as thermal stratification. The deep-water zone (hypolimnion) provides a very different habitat for aquatic organisms than does the unimpounded river. Depending on the transparency of the water and the depth of the reservoir, light might no longer reach the bottom, preventing aquatic plants from growing there. Storage and release of water for power production alternately floods and dries out the shoreline areas, further limiting the growth of aquatic plants along the margins of the reservoir.

5.2 Altered Flows below the Reservoir

A hydropower storage reservoir can retain enough water that the river below the dam dries up. Similarly, a diversion project withdraws water from the river, passes it through a canal or pipeline, and returns it to the river downstream from the powerhouse. In both types of hydroelectric projects, stream flows and aquatic habitats in a portion of the river are greatly or entirely reduced. This may eliminate fish and invertebrates that were residing in the affected reach and may constrain the movements of migratory animals.

Releasing a predetermined amount of water down the river channel is often required to sustain the other instream uses of water, including maintenance of fish and wildlife communities, streamside vegetation, recreation, aesthetics, water quality, and navigation. Providing flows downstream from a storage reservoir or hydroelectric diversion is simple; water can be spilled from the dam instead of diverted to a pipeline or stored in a reservoir. Because releasing water to support instream uses below the dam frequently makes that water unavailable for generation of electricity, hydropower operators are interested in providing sufficient, but not excessive, releases. Methods have been developed to ascertain the instream flow requirements of many of these uses, but the needs of biological resources are often difficult to assess. A variety of instream flow assessment methodologies are available to help determine how much water needs to be released to maintain aquatic and riparian communities.

6. INTERFERENCE WITH MOVEMENTS OF AQUATIC ORGANISMS

Hydropower projects can affect aquatic organisms directly by impeding their migration, blocking their movement upstream and/or downstream, and indirectly by altering the flow patterns and water quality conditions to which they respond. Fish are more likely to be affected than are other aquatic organisms because fish have more complex behaviors. Probably the most important change to physical habitats that influences fish movements is reduced water velocity upstream of dams. Flowing streams are changed to reservoirs, and the average depth and cross section are increased markedly. Higher water temperatures may also result from the increased surface area and reduced water velocities in reservoirs. The rate of downstream transport of aquatic invertebrates and other organisms can be slowed when water velocities upstream of hydroelectric dams are reduced substantially or when a large proportion of the stream is diverted for hydroelectric generation.

As discussed earlier, project operations such as spilling surface water can result in increased concentrations of dissolved gas, whereas storage practices may cause lower concentrations of dissolved oxygen to be present than is the case under riverine conditions. Water temperature may also be altered due to storage practices. For example, in rivers with extensive hydroelectric development, there may be a phase shift where the timing of seasonal maximum or minimum water temperature is altered. Because aquatic organisms require a certain range of conditions to survive, grow, and reproduce, any major changes in water quality can affect their relative abundance and distribution. The following discussion focuses on the effects of hydropower projects on the upstream and downstream movements of migratory fish.

6.1 Upstream Passage

Most hydropower dams pose a physical barrier to upstream-migrating fish. Fish migrate upstream for a variety of reasons, including to return to natal areas to spawn (e.g., adult salmon, steelhead, American shad), to complete their freshwater rearing periods (e.g., juvenile American eel), and to feed (e.g., sturgeon, sucker). Many freshwater fish move seasonally between large rivers and their tributary streams. These migration routes may be blocked by dams, eliminating large areas from production of fish.

Many hydroelectric projects have ways in which to assist the upstream movement of fish, including fish ladders, trap-and-haul operations, and fish elevators. In all cases, fish are slowed while they search for and decide whether to use the upstream passage facilities. Delay of adult salmon and steelhead at Columbia River dams is one factor contributing to their mortality during migration from the Pacific Ocean to spawning areas. However, adult salmon generally migrate faster through reservoirs than they do through nonimpounded or natural sections of rivers that have higher water velocities against which the fish must swim.

Critical features of fishway design that affect passage success of fish include water velocity, depth, slope, and availability of resting areas. The entrance design is also an important consideration. Some fish species have a difficult time in locating passageway openings or auxiliary flows provided to attract those seeking upstream passage.

The majority of fishways in North America have been designed for efficient upstream passage of salmonids (Fig. 3). However, other species of fish might not respond similarly to these fishways due to their unique behaviors. For example, adult white sturgeons are often too large to navigate certain designs. The upstream passage of adult lampreys is also poor relative to the number of fish that approach hydropower projects.

6.2 Downstream Passage

Fish that migrate downstream past a hydropower project have three primary routes of passage. They may (1) be drawn into the power plant intake flow (entrainment) and pass through the turbine, (2) be diverted via bypass screens into a gatewell and to a collection facility or the tailrace, or (3) pass over the

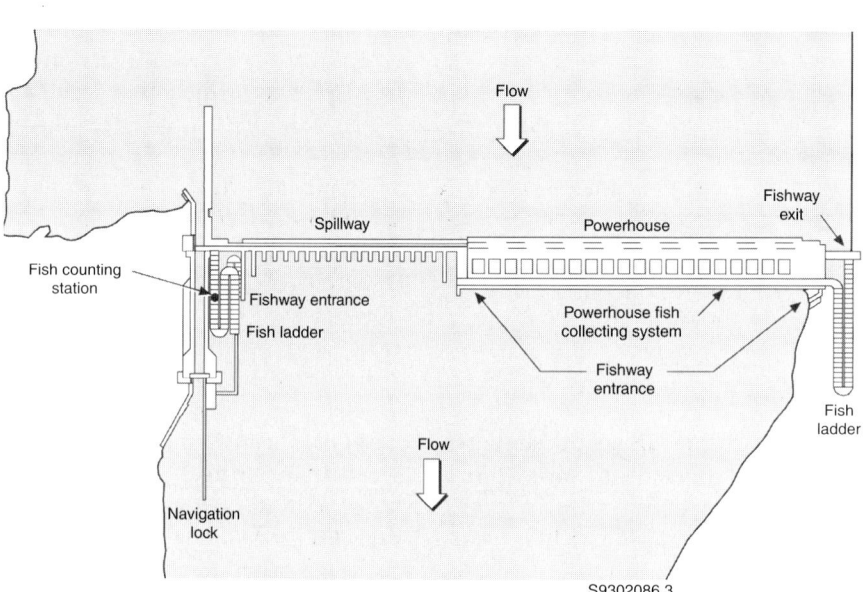

FIGURE 3 Generic design of a hydroelectric facility in the Columbia River system showing the location of fishways.

dam in spilled water. Recent modifications made to dams to decrease numbers of turbine-passed fish include guiding migrating fish toward spillbays and using surface bypass systems (Fig. 4) and behavioral guidance walls. Ice and trash sluiceways have also been modified to provide a surface passage route for migrating fish.

Entrained fish are exposed to physical stresses (e.g., rapid change in pressure, shear, turbulence, blade strike) that may be injurious. In the best existing turbines, up to 5% of turbine-passed fish may be injured or killed, and mortalities in some turbines may be 30% or more. Several new design concepts are under development with a goal of reducing mortality of turbine-passed fish to 2% or less.

Nonturbine passage routes pose some risk to fish as well. Weak swimmers, such as juvenile lampreys and small resident fish (e.g., minnows, sunfish), may be impinged or injured on contact with bypass screens. Design features that influence the ability of a diversion screen to guide fish safely include screen type (e.g., rotating, fixed), screen material, size of openings, length relative to intake opening size, angle, and approach velocity. The design and location of outfalls from fish bypasses are also critical to minimizing exposures of bypassed fish to predatory fish in the dam tailrace. Fish that pass via the spill of a high-head dam are subjected to extremely high and variable water velocities, may be abraded by contact with the dam face, and may collide with submersed structures below the dam, including those designed to dissipate the high energy of spilled water. However, project spill may reduce the residence time of migrating fish immediately above the dam during which they are vulnerable to predators in the reservoir. Thus, design of the entire project, from forebay entrance to tailrace exit, affects the safe and efficient passage of fish.

6.3 Effects of Water Quality Alterations on Movements

Adverse water quality conditions resulting from operation of hydroelectric projects may influence the movement of aquatic organisms. For instance, dissolved gases in excess of 100% saturation can occur under conditions of high project spill, which in turn can lead to gas bubble trauma. Some storage reservoirs may experience low dissolved oxygen concentrations and/or thermal stratification. Fish could avoid or be attracted to these conditions, depending on their life history requirements, relative sensitivity of their sensory systems, and their physiological sensitivity. The movement and distribution of aquatic organisms may be influenced by water temperatures and dissolved gas changes caused by hydropower development. Hydropower operators can modify project structures or operations to mitigate for high total dissolved gas levels and improve passage conditions for fish.

7. GREENHOUSE GAS EMISSIONS

Concern about the climatic effects of increased emissions of greenhouse gases, primarily carbon dioxide and methane, is a relatively recent issue for electric power production, and the contribution of hydroelectric plants is still being clarified. It had been assumed that hydropower projects emit few, if any, greenhouse gases compared with those emitted by fossil-fueled power plants. However, decomposition of inundated vegetation and other organic matter within the reservoir can result in greenhouse gas emissions that may continue for decades after the initial flooding. The amounts of greenhouse gases released from a hydropower reservoir vary widely, depending on geography, altitude, latitude, water temperature, reservoir size and depth, depth of turbine intakes, hydropower operation, carbon input from the river basin, and reservoir construction (e.g., whether vegetation was cleared from the reservoir before the dam was closed). Recent studies suggest that greenhouse gas emissions from reservoirs in higher latitudes (e.g., Canada, Scandinavia) are low but that they may be a greater concern in tropical areas. For example, studies of 30 reservoirs in Brazil

FIGURE 4 Surface view (looking downstream toward the project) of the prototype surface collector at the Bonneville I powerhouse on the Columbia River. This structure was designed to provide a surface passage route for juvenile salmon and steelhead that would otherwise pass through turbines.

found a 500-fold difference in greenhouse gas emissions among dams. The lowest emissions from Brazilian reservoirs were similar to those from Canadian lakes and reservoirs, whereas the highest emissions were within the range of gross emissions from fossil-fueled power plants. Thus, hydropower plants cannot be automatically assumed to release less greenhouse gases than do fossil-fueled sources of electricity. Emissions must be determined on a case-by-case basis.

SEE ALSO THE FOLLOWING ARTICLES

Electricity, Environmental Impacts of • *Entrainment and Impingement of Organisms in Power Plant Cooling* • *Hydropower Economics* • *Hydropower, History and Technology of* • *Hydropower Resettlement Projects, Socioeconomic Impacts of* • *Hydropower Resources* • *Hydropower Technology* • *Wind Energy Technology, Environmental Impacts of*

Further Reading

Baxter, R. M. (1977). Environmental effects of dams and impoundments. *Annu. Rev. Ecol. Systematics* **8**, 255–283.

Mattice, J. S. (1991). Ecological effects of hydropower facilities. *In* "Hydropower Engineering Handbook" (J. S. Gulliver and R. E. A. Arndt, Eds.), pp. 8.1–8.57. McGraw–Hill, New York.

Petts, G. E. (1984). "Impounded Rivers: Perspectives for Ecological Management." John Wiley, New York.

World Commission on Dams. (2000). "Dams and Development: A New Framework for Decision-Making," report of the World Commission on Dams. Earthscan Publications, London. www.dams.org.

Hydropower, History and Technology of

JOHN S. GULLIVER and ROGER E. A. ARNDT
University of Minnesota
Minneapolis, Minnesota, United States

1. Introduction
2. Hydropower Potential
3. What Nature Gives Us
4. How Hydropower is Captured
5. Hydroturbines
6. Ecological Considerations
7. Advantages and Disadvantages of Hydropower Development

Glossary

axial-flow turbine A collective term for turbines with axial flow through the runner blades axially to the turbine shaft; both propeller turbines and Kaplan turbines are axial-flow turbines.

base load Typically, the minimum load over a given period of time.

capacity The greatest load that a piece of equipment can safely serve.

dam A massive wall or structure built across a valley or river for storing water.

design head The head at which the turbine is designed to operate at maximum efficiency.

draft tube The diffuser that regains the residual velocity energy of the water leaving the turbine runner.

energy The power of doing work for a given period, usually measured in kilowatt-hours.

forebay The upstream part of the bay-like extension of the river for the location of a powerhouse.

Francis turbine A radial-inflow reaction turbine where the flow through the runner is radial to the shaft.

generator A machine powered by a turbine that produces electric current.

gross head The difference between headwater level and tailwater level at the powerhouse.

headrace The portion of the power canal that extends from the intake works to the powerhouse.

headwater The water upstream from the powerhouse, or generally the water upstream from any hydraulic structure creating a head.

headwater elevation (or headwater level) The height of the headwater in the reservoir.

hydraulic efficiency An efficiency component of the turbine, expressing exclusively the power decrement due to hydraulic losses (e.g., friction, separation, impact), including the losses in the scroll case and the draft tube.

hydroelectric power The electric current produced from water power.

hydroelectric power plant A building in which turbines are operated, to drive generators, by the energy of natural or artificial waterfalls.

hydropower plant (or hydropower development) The comprehensive term for all structures (one powerhouse and pertaining installations) necessary for using a selected power site.

hydropower station A term equivalent to the powerhouse and sometimes including the structures situated nearby.

hydropower system Two or more power plants (and therefore two or more powerhouses) that are cooperating electrically through a common network.

intake (or intake works or headworks) A hydraulic structure built at the upstream end of the diversion canal (or tunnel) for controlling the discharge and preventing silt, debris, and ice from entering the diversion.

Kaplan turbine An axial-flow reaction turbine with adjustable runner blades and adjustable guide vanes.

kinetic energy Energy that a moving body has because of its motion, dependent on its mass and the rate at which it is moving.

load The amount of electric energy delivered at a given point.

load demand A sudden electrical load on the generating units, inducing the rapid opening of the turbines.

load factor The ratio of the annually produced kilowatt-hours and of the energy theoretically producible at installed capacity during the whole year.

load rejection A sudden cessation of electrical load on the generating units, inducing the rapid closure of the turbines.

main shaft The rotating element that transmits torque developed by the turbine runner to the generator rotor or that transmits torque developed by the motor to the pump impeller.

mechanical efficiency An efficiency component of the turbine, expressing exclusively the power losses of the revolving parts due to mechanical friction.

needle valve A streamlined regulating body moving like a piston in the enlarged housing of the valve.

net head The part of the gross head that is directly available for the turbines.

nozzle (or jet nozzle) A curved steel pipe supplied with a discharge-regulating device to direct the jet onto the buckets in impulse runners.

peak load The greatest amount of power given out or taken in by a machine or power distribution system during a given time period.

Pelton turbine The main type of turbine used under high heads.

penstock (or pressure pipe) A pressurized pipeline conveying the water in high-head developments from the headpond or the surge tank to the powerhouse.

plant discharge (or plant discharge capacity) The maximum discharge that can be used by all turbines of the power plant with full gateage (i.e., the entire discharging capacity of the turbines).

pondage That rate of storage in run-of-river developments that can cover daily peaks only.

power The rate at which work is done by an electric current or mechanical force, generally measured in watts or horsepower.

powerhouse The main structure of a water power plant, housing the generating units and the pertaining installations.

propeller-type turbine The collective term for axial-flow reaction turbines; in this terminology, it denotes two types: fixed-blade propeller turbines and adjustable-blade propeller turbines (i.e., Kaplan turbines).

pumped-storage development A combined pumping and generating plant; it is not a primary producer of electrical power, but by means of a dual conversion, it stores the superfluous power of the network and returns it during peak load periods, as would a battery.

reaction turbine A collective term for turbines in which the water jet enters the runner under a pressure exceeding the atmospheric value. The water flowing to the runner still has potential energy, in the form of pressure, that is converted into mechanical power along the runner blades.

reservoir An artificial lake into which water flows and is stored for future use.

run-of-river plant A development with little or no pondage regulation such that the power output varies with the fluctuations in the stream flow.

runner The rotating element of the turbine that converts hydraulic energy into mechanical energy; for reversible pump-turbines, the element is called an impeller and converts mechanical energy into hydraulic energy for the pump mode.

scroll case (or spiral case) A spiral-shaped steel intake guiding the flow into the wicket gates of the reaction turbine.

semi-scroll case (or spiral case) A concrete intake directing flow to the upstream portion of the turbine with a spiral case surrounding the downstream portion of the turbine to provide uniform water distribution.

setting The vertical distance between the tailwater level and the center of a turbine runner.

specific speed A universal number that indicates the machine design: impulse, Francis, or axial.

tailrace The portion of the power canal that extends from the powerhouse to the recipient watercourse.

tailwater The water downstream from the powerhouse, generally the water downstream from any hydraulic structure creating a head.

tidal power plant (or tidal power station) A power station that uses the potential hydraulic power originating from the tidal cycles of the sea.

Turbine A device that produces power by diverting water through blades of a rotating wheel that turns a shaft to drive generators.

turbine discharge capacity The maximum flow that can be discharged by a single turbine at full gateage.

turbine efficiency The entire efficiency of the turbine (i.e., the product of hydraulic mechanical and volumetric efficiencies).

water power A general term used for characterizing both power (kilowatts) and energy (kilowatt-hours) of watercourses, lakes, reservoirs, and seas.

Hydropower has a long and rich history, being important to the development of modern industrial society and essential to the electrification of the world. This article reviews some of this history and the hydropower potential that remains in the world today. Next, the water cycle and how hydropower is extracted from this cycle will be discussed. Finally, some of the ecological considerations and the advantages and disadvantages of hydropower development will be addressed.

1. INTRODUCTION

Falling or flowing water has been used to perform work for thousands of years, with the particular uses varying with the social and political conditions of the times. Although the Greeks and Romans knew of waterwheels since the 3rd century BC, these labor-saving devices were not used extensively until the 14th century. Early tasks included grinding grain, sawing wood, powering textile mills, and (later) operating manufacturing plants. Mills or factories were located at the hydropower sites to directly use the available energy. By the end of the 18th century, there were approximately 10,000 waterwheels in New England alone. The power output of these early plants, usually limited to about 100 kW, is compared with other power sources in Fig. 1.

During the 19th century, hydropower became a source of electrical energy, although some form of hydroelectric turbine development can be traced back as far as 1750. Benoit Fourneyron is credited with developing the first modern hydroelectric turbine in 1833. The first hydroelectric plant is documented as coming online September 30, 1882, in Appleton, Wisconsin, and is still functioning. However, there is some dispute over this; Merritt cited the Minneapolis Brush Electric Company as beginning operation of a hydroelectric plant some 25 days earlier. The generation of electricity from falling water expanded the need for larger hydroelectric plants because the energy did not need to be used on-site. The transmission of power over long distances became economical in 1901 when George Westinghouse installed alternating current equipment at Niagara Falls in New York, further expanding the potential uses of hydropower.

As Fig. 1 indicates, the power capabilities of water turbines became larger as the need grew. During the 1930s, large dams and ever-increasing turbine capacities became the norm. The power capacity of steam turbines was also increasing rapidly, and the relative cost of electricity continued to fall. Finally, during the period between 1940 and 1970, the cost of operating and maintaining older, smaller hydroelectric plants became greater than the income they could produce, and many were retired. This is seen in Fig. 2, where small hydropower capacity decreased as overall hydropower capacity climbed rapidly in the United States. A similar trend occurred in European countries. Hydropower development in other parts of the world was insignificant before 1930, as indicated by world hydropower production in Fig. 3.

The largest hydropower facility in the world is currently the Itaipu Dam on the Parana River, located between Brazil and Paraguay. This hydroplant's 12,600-MW capacity is the equivalent of 12 large thermal powerplants. It is upstream from the incomparable Iguasu Falls. China is building an even larger project at Three Gorges. The world's largest capacity hydroelectric plants in 2000 are listed in Table I.

There are two other basic sources of hydropower besides that which is extracted from the world's rivers: tidal power and wave power. Small tidal mills to provide mechanical power existed hundreds of years ago. However, the first hydroelectric tidal plant

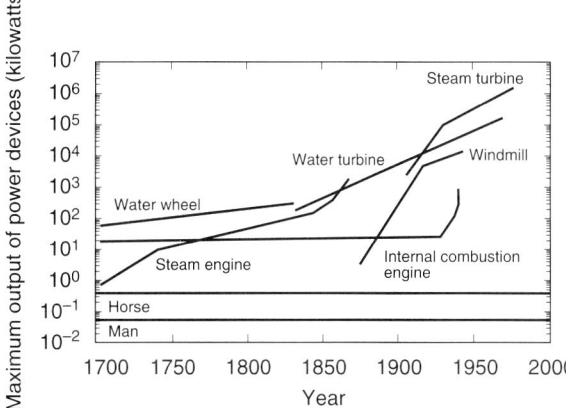

FIGURE 1 The maximum power output of selected power devices over the period from 1700 to 1970. From Chappel, 1984.

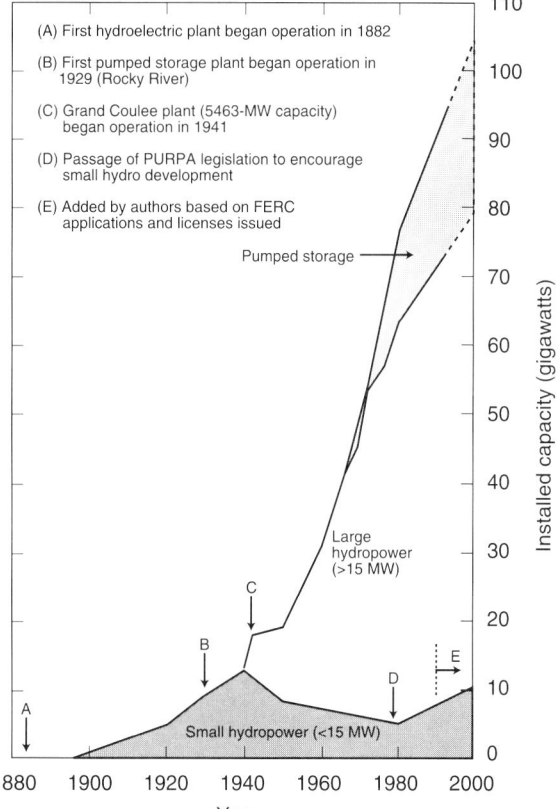

FIGURE 2 Installed hydroelectric capacity in the United States for the period from 1882 to 2000. FERC, Federal Energy Regulatory Commission. Assembled from Federal Energy Regulatory Commission, 1992, and International Waterpower and Dam Construction annual yearbooks.

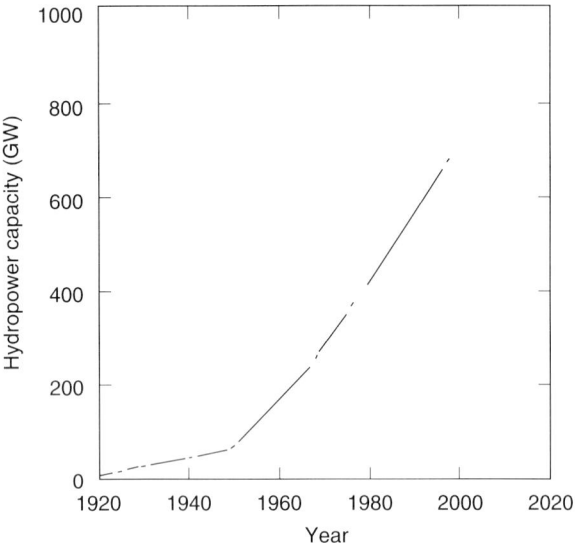

FIGURE 3 World hydropower production. Assembled from International Water Power and Dam Construction annual yearbooks.

was developed during the 1960s on the La Rance estuary in northern France. Producing up to 240 MW of power, this tidal plant uses a dam across a cove mouth to form a pond. Sluice gates open to let water flow in during the rising tide and then close with the returning tide as water is directed through a standard hydroturbine.

The feasibility of tidal power depends on the range of tide experience and on finding a location where an inordinately long dam does not need to be built. Thus, only a handful of tidal plants have been developed since the La Rance plant: a 10-MW plant in 1986 and a number of smaller plants in China; an 18-MW plant in 1984 at an existing flood structure at Annapolis Royal in Nova Scotia, Canada; and a 400-kW plant in the Soviet Union. The turbines at Annapolis Royal are of interest because they can turn in both directions, capturing energy from both the incoming and outgoing tides.

The United Kingdom has studied the feasibility of building a dam across the Severn estuary, generating up to 7000 MW of power with 192 hydroturbines. Canada has conducted a similar study on the Bay of Fundy.

All tidal power projects are low head, up to 36 feet (11 m). Gulliver and Dotan found that the cost per kilowatt produced is approximately proportional to $H^{-0.53}$, where H is the hydraulic head across the dam. Obviously, any low-head hydropower project must be planned carefully.

2. HYDROPOWER POTENTIAL

On a worldwide basis, hydropower represents roughly one-quarter of the total electrical energy generated. There is approximately 3800 GW of technically feasible hydropower potential existing in the world that could produce 14,370,000 GWh per year. Of this amount, 2100 GW (8,082,000 GWh/year) is feasible under current economic conditions and 700 GW (2,645,000 GWh/year) was installed in 2001. The technically feasible potential

TABLE I

World's Largest Capacity Hydroelectric Plants in 2000

		Rated capacity (MW)		
Name of dam	Location	Present	Ultimate	Year of initial operation
Itaipu	Brazil/Paraguay	12,600	14,000	1983
Guri	Venezuela	10,000	10,000	1986
Grand Coulee	Washington State (United States)	6494	6494	1942
Sayano-Shushensk	Russia	6400	6400	1989
Krasnoyarsk	Russia	6000	6000	1968
Churchill Falls	Canada	5428	6528	1971
La Grande 2	Canada	5328	5328	1979
Bratsk	Russia	4500	4500	1961
Moxoto	Brazil	4328	4328	n.a.
Ust-Ilim	Russia	4320	4320	1977
Tucurui	Brazil	4245	8370	1984

Source: Mermel, T. W. (2000). The world's major dams and hydro plants. "International Water Power and Dam Construction," 2000 Handbook. Wilmington, Kent, UK.

is that considered developable based on physical considerations without considering current economics or other environmental issues. It represents 39% of the total energy in the world's rivers. The available and developed power in 2001 is given by continent in Fig. 4. The tremendous potential in Asia, Africa, and South America is apparent, amounting to more than half of the world's total hydropower potential. On the other hand, hydropower developed or under construction in Europe and North American is at 73 and 70% of the economically feasible potential, respectively. The future hydropower development in these two continents is likely to be in the form of upgrades in power and pumped storage.

3. WHAT NATURE GIVES US

Hydroelectric power, similar to many of humanity's activities, is produced by insertion into the hydrologic cycle, where water evaporates from the oceans, seas, lakes, and the like into the atmosphere, forms clouds of condensed water drops, falls as precipitation (e.g., rain, sleet, snow), and flows into rivers and eventually into the ocean. This cycle has significant variability given that floods and droughts are common. A dam placed in a river forms a reservoir with a given hydraulic head or a difference in elevation between the water behind the reservoir and the water below it. This hydraulic head is used to create power according to the relation

$$P = \eta \gamma Q H, \qquad (1)$$

where P is the power produced (in watts), H is in meters, Q is the discharge (flow) routed through the hydroelectric turbine (in cubic meters/second), γ is the specific weight of water (in Newtons/cubic meter), and η is the overall efficiency of the hydroelectric facility (fraction between 0 and 1).

The parameter that is highly variable in Eq. (1) is the discharge, Q. At a typical dam and reservoir on a river, the discharge over the dam during floods can be 1000 times greater than that during droughts. Predicting this discharge for a typical year represents much of the risk in developing a hydroelectric facility because it typically requires 30 years of records to reduce the uncertainty in mean discharge to $\pm 15\%$.

A watershed or drainage basin is the region contributing flow to a given location such as a prospective hydropower site on a stream. Any watershed is composed of a wide variety of foliage, soils, geological formations, streams, and the like and, therefore, is unique. Because no two watersheds are the same, comparisons of runoff from two separate watersheds and generalized relationships for watershed runoff are not accurate. The data obtained from stream gauges in the watersheds of interest should be used whenever possible.

The ratio of stream runoff to precipitation within a given watershed depends primarily on seven factors that are illustrated in Fig. 5 and described as follows:

- *Interception.* Interception is the precipitation stored on vegetative cover and then evaporated. Interception accounts for between 10 and 20% of annual precipitation in a well-developed forest. The interception of crops varies greatly. However, some approximate values for a 25-mm storm are as follows: cotton, 33%; small grains, 16%; tobacco, 7%; corn, 3%; alfalfa, 3%; meadow grass, 3%.

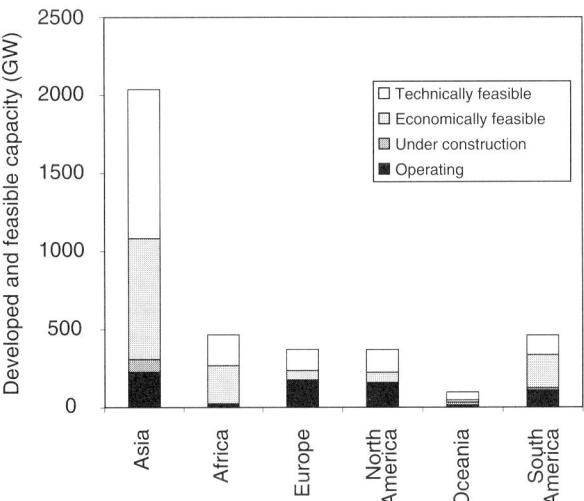

FIGURE 4 World hydropower resources by continent. Assembled from "International Hydropower and Dams," 2001 Handbook.

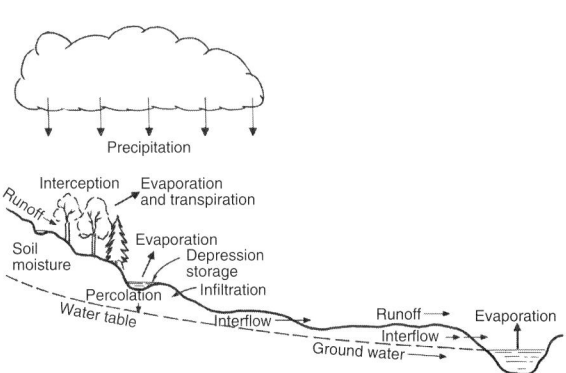

FIGURE 5 Processes affecting the relationship between precipitation and runoff.

- *Depression storage.* Depression storage is the rainwater retained in puddles, ditches, and other depressions in the surface. It occurs when the rainfall intensity exceeds infiltration capacity. At the end of the storm, water held in depression storage is either evaporated or infiltrated into the soil. The depression storage capacity of most drainage basins is between 13 and 50 mm of precipitation. A specific type of depression storage that is handled separately is a blind or self-enclosed drainage basin, that is, a portion of the drainage basin that does not drain into the stream network but rather is self-enclosed, usually with a lake, marsh, or bog at the center. Blind drainage basins are normally excluded from the hydrologic analysis of a hydropower site.

- *Surface runoff.* Surface runoff is the precipitation that moves downslope along the soil surface until it reaches a stream or lake. It is primarily associated with flood events, although in larger watershed basins the effects of surface runoff can be felt for up to a month. A small portion of any precipitation event will fall directly into the channel or stream. This immediately becomes runoff. It is grouped together with runoff from impervious areas, such as parking lots, streets, and buildings, because this runoff occurs rather quickly after precipitation.

- *Infiltration.* Infiltration is the passage of water through the soil surface. It usually implies percolation, which is the movement of water through unsaturated soil. Infiltration capacity depends on soil porosity, moisture content, and vegetative cover. Sandy or highly organic soils will have a greater infiltration due largely to increased porosity. A wet soil will have a lower infiltration capacity. Vegetation cover increases infiltration by retarding surface flow, increasing soil porosity with the root system, and reducing rain packing of the soil surface.

- *Soil moisture.* The precipitation that infiltrates into the soil will first be used to replenish soil moisture. Over time, the soil moisture is taken by plant root systems and eventually is transpired from plant foliage as part of the photosynthetic process. Water used to replenish soil moisture will not appear as stream flow. Soils with a high percentage of decayed plant material have a large capacity to retain moisture.

- *Interflow.* Interflow is water that infiltrates through the soil and moves laterally in the upper soil layers until it reemerges as surface runoff. A thin soil surface covering rock, hardpan, or plow bed will usually have large quantities of interflow. Interflow often emerges as a spring in riverbanks. It will not usually affect flood peaks but will increase stream flow at a steady rate for some time after the peak.

- *Groundwater flow.* If the infiltrated water percolates downward until it reaches the water table (i.e., zone saturated with water), it will eventually reach stream as groundwater flow, which is the primary source of base flow for streams. Groundwater flow influences stream flow on a seasonal, rather than a weekly, time scale.

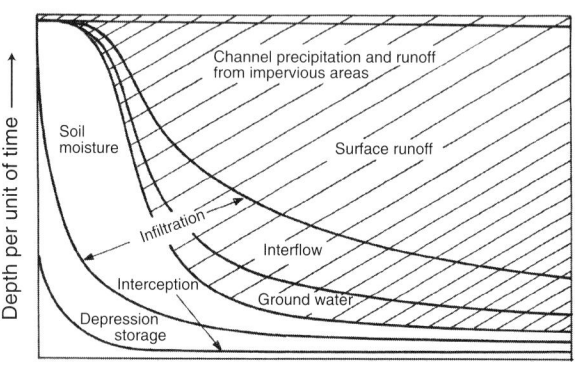

FIGURE 6 Schematic diagram of the segmentation of rainfall for an extensive storm in a relatively dry basin. The shaded area indicates precipitation that eventually results in a runoff.

A schematic diagram of the segmentation of rainfall for an extensive storm in a relatively dry basin is given in Fig. 6. The shaded area indicates the quantity of rainfall that will eventually become stream flow. The general order in which the various types of flow reach the stream is as follows: channel precipitation, surface runoff, interflow, ground water.

The seven parameters that influence the runoff process, which depends on foliage, soil type, geological formation, and watershed geomorphology, can vary greatly within a given watershed and will certainly vary between two distinct watersheds. Therefore, it should be no surprise to discover that two adjacent watersheds of similar drainage areas have entirely different runoff characteristics.

4. HOW HYDROPOWER IS CAPTURED

The broad variety of natural conditions at hydropower sites has resulted in development of many different types of hydropower schemes. This is a disadvantage so far as engineering costs are concerned, but it provides a great deal of stimulus to those involved in hydropower development. A longitudinal section of a typical scheme is provided in Fig. 7.

In the run-of-river development, a short penstock or dam directs the water through the turbines. The powerhouse is often an integral part of the dam, as shown in Fig. 8. The natural flow of the river remains relatively unaltered. A more complex development occurs at diversion or canal projects, where the water is diverted from the natural channel into a canal or long penstock, as shown in Fig. 9. This results in a significant change in the flow of water in a given reach of the river, sometimes for a considerable distance.

Storage regulation developments are defined as those in which an extensive impoundment at the power plant, or at the reservoir upstream of the power plant, allows for regulation of the flow downstream through storage. Water is stored during high-flow periods and is used to augment the flow during low-flow periods. This allows for a relatively constant supply of energy over the course of the year.

Pump storage facilities are normally large developments in which water is pumped from a lower reservoir during off-peak hours when the cost of energy is low. The pumps are run in reverse as turbines during the hours of peak demand to augment the power supplied from other sources. Thus, the pumped storage facility serves the same function as a large battery. It is used to allow large base load facilities to operate at continuous power output in cases where it is uneconomical to allow the power output of a large plant to fluctuate.

Tidal power plants have also been developed or considered. These are located in areas where there

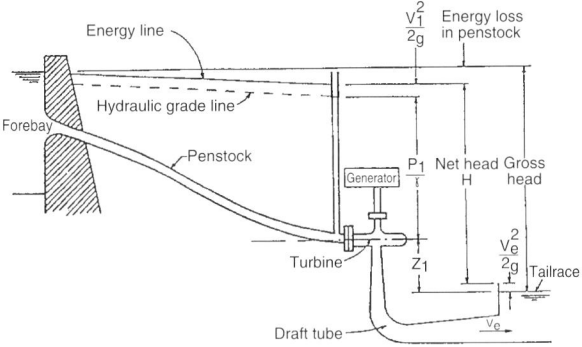

FIGURE 7 Schematic of a hydropower installation. $V_1^2/2g$ is head lost before the turbine, P_1/γ is the positive head extracted by the turbine, Z_1 is the suction head on the turbine, and $V_e^2/2g$ is the velocity head remaining at the draft tube outlet.

FIGURE 8 The 8.4-MW St. Cloud hydropower facility is a run-of-river development with the powerhouse composing a portion of the dam. Net head is 5.2 m. (A) Upstream view of the spillway and intakes into gates and powerhouse. (B) Downstream view of the powerhouse and spillway. Courtesy of M. A. Mortenson Company.

FIGURE 9 The Mayfield project on the Cowlitz River in Washington. It is composed of an arch dam and spillway (background), a power tunnel leading to surge tanks (left), and penstocks leading to a 122-MW powerhouse (center front). Notice how the layout maximizes efficient use of the terrain. Courtesy of Harza Engineering.

are large tidal fluctuations. Low-head turbines are used to harness the energy in the tidal cycle, and an entire bay or estuary is enclosed by a low dam.

Hydropower schemes can either be single purpose, having as their only purpose the production of electricity, or multipurpose, having hydropower production as just one aspect of the total use of the facility. Multipurpose facilities include those in which hydropower is developed in conjunction with irrigation, flood control, navigation, and water supply. Hydropower plants are also categorized by the type of use. For example, a base load plant is one in which the power is used to meet all or part of a sustained and constant portion of the electrical load. Energy from these plants that is available at all times is referred to as "firm power." The need for power varies during the course of the day, and power requirements over and above the base load requirement are met by peak load facilities. These are plants in which the electrical production capacity is relatively high and the volume of water discharged through the units can be changed readily to meet peak demands. Storage or pondage of the water supply is necessary for these load demands because such plants can be started and stopped more rapidly and economically than can fossil fuel and nuclear power plants.

A typical hydropower facility consists of the following components:

- The dam or diversion structure with the associated reservoir
- The powerhouse structure and its foundation
- Hydraulic conveyance facilities, including the head race, headworks, penstock, gates and valves, and tailrace
- The turbine-generator unit, including the guide vanes or wicket gates, turbine, draft tube, speed increaser, generator, and speed-regulating governor
- Station electrical equipment, including the transformer, switch gear, automatic controls, conduit, and grounding and lightning systems
- Ventilation, fire protection, communication, and bearing cooling water equipment
- Transmission lines.

The selection of the proper turbine for a given site depends on the head and flow available as well as the use that is to be made of the facility. Because there is such a broad variation in available conditions in the field, many different turbine configurations have been developed and are available. It is essential that a hydropower engineer understand in detail the rationale for selection of hydroturbines. The other items listed, although not unique to hydropower production, are also crucial to a properly operating facility.

5. HYDROTURBINES

The water turbine has a rich and varied history. It has been developed as a result of a natural evolutionary process from the waterwheel. Originally used for direct drive of machinery, the use of the water turbine for the generation of electricity is a comparatively recent activity. Much of its development occurred in France, which did not have the cheap and plentiful sources of coal that England had and that sparked the industrial revolution during the 18th century. During the 19th century, France found itself with its most abundant energy resource being water. To this day, *Houille Blanche* (literally "white coal") is the French term for water power.

In 1926, the Société d'Encouragement pour l'Industrie Nationale offered a prize of 6000 francs to anyone who would "succeed in applying at large scale, in a satisfactory manner, in mills and factories, the hydraulic turbines or wheels with curved blades of Bélidor." Bélidor was an 18th-century hydraulic and military engineer who, between 1737 and 1753, authored a monumental four-volume work, *Architecture Hydraulique*, a descriptive compilation of hydraulic engineering information of every sort. The waterwheels described by Bélidor departed from convention by having a vertical axis of rotation and being enclosed in a long cylindrical chamber approximately 1 m in diameter. Large quantities of water were supplied from a tapered sluice at a tangent to the chamber. The water entered with considerable rotational velocity. This preswirl, combined with the weight of water above the wheel, was the driving force. The original tub wheel had an efficiency of only 15 to 20%.

Water turbine development proceeded on several fronts from 1750 to 1850. The classical horizontal axis waterwheel was improved by engineers such as John Smeaton (1724–1792) of England and J. V. Poncelet of France (1788–1867). This resulted in waterwheels having efficiencies in the range of 60 to 70%. At the same time, reaction turbines (somewhat akin to modern lawn sprinklers) were being considered by several workers. The great Swiss mathematician Leohard Euler (1707–1783) investigated the theory of operation of these devices. A practical application of the concept was introduced in France in 1807 by Monnoury de Ectot (1777–1822). His machines were, in effect, radially outward-flow machines. The theoretical analyses of Burdin (1790–1893), a French professor of mining engineering who introduced the word "turbine" into engineering terminology, contributed much to our understanding of the principles of turbine operation and underscored the principal requirements of shock-free entry and exit with minimum velocity as the basic requirements for high efficiency. A student of Burdin, Benoit Fourneyron (1802–1867), was responsible for putting his teacher's theory to practical use. Fourneyron's work led to the development of high-speed outward-flow turbines with efficiencies on the order of 80%. The early work of Fourneyron resulted in several practical applications and the winning of the coveted 6000-franc prize in 1833. After nearly a century of development, Bélidor's tub wheel had been officially improved. Fourneyron spent the remaining years of his life developing some 100 turbines in France and Europe. Some turbines even found their way to the United States, the first around 1843.

As successful as the Fourneyron turbines were, they lacked flexibility and were efficient over only a narrow range of operating conditions. This problem was addressed by S. Howd and U. A. Boyden (1804–1879). Their work evolved into the concept of an inward-flow motor as a result of the work of James B. Francis (1815–1892). The modern Francis turbine, shown in Fig. 10, is the result of this line of development. At the same time, European engineers addressed the idea of axial-flow machines, which today are represented by "propeller" turbines of both fixed-pitch and Kaplan types, as shown in Fig. 11.

Just as the vertical-axis tub wheels of Bélidor evolved into modern reaction turbines of the Francis and Kaplan types, development of the classical horizontal-axis waterwheel reached its peak with the introduction of the impulse turbine. The seeds of development were sown by Poncelet in 1826 with his description of the criteria for an efficient waterwheel. These ideas were cultivated by a group of California engineers during the late 19th century, one of whom was Lester A. Pelton (1829–1908), whose name is given to the Pelton wheel (shown in Fig. 12), which consists of a jet or jets of water impinging on an array of specially shaped buckets closely around the periphery of a wheel.

Modern impulse units are generally of the Pelton type and are restricted to relatively high-head applications (Fig. 3). One or more jets of water impinge on a wheel containing many curved buckets. The jet stream is directed inward, sideways, and outward, thereby producing a force on the bucket, which in turn results in a torque on the shaft. All kinetic energy leaving the runner is "lost." A draft tube is generally not used because the runner operates under approximately atmospheric pressure and the head represented by the elevation of the unit above tailwater cannot be used. Because this is a high-head device, the loss in available head is relatively unimportant. The Pelton wheel is a low-specific speed device. Specific speed can be increased by the addition of extra nozzles, with the specific speed increasing by

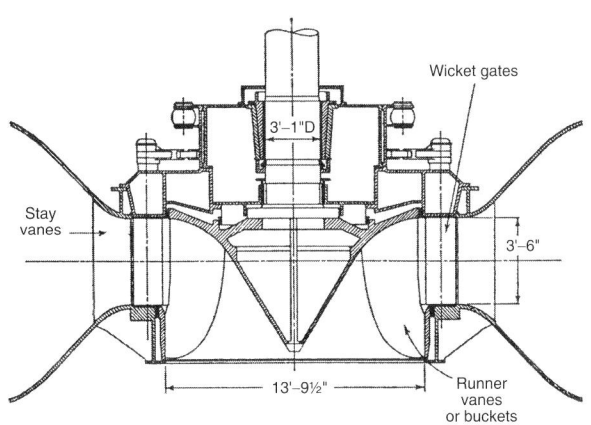

FIGURE 10 Francis turbine. From Daily (1950).

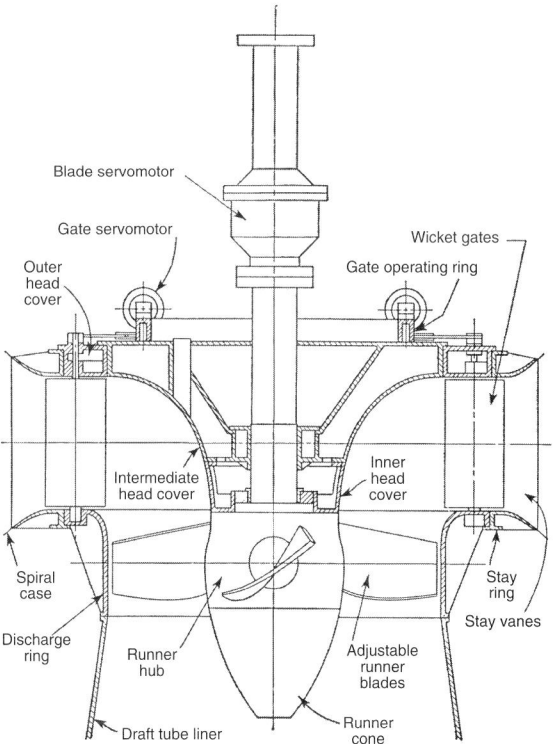

FIGURE 11 Smith–Kaplan axial-flow turbine with adjustable-pitch runner blades. From Daily (1950).

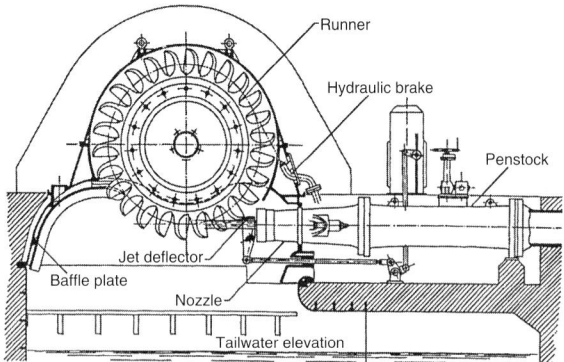

FIGURE 12 Cross section of a single-wheel, single-jet Pelton turbine. This is the third highest head Pelton turbine in the world ($H = 1447$ m, $n = 500$ rpm, $P = 35.2$ MW, $N_s \sim 0.038$). Courtesy of Vevey Charmilles Engineering Works. Adapted from Raabe, J. (1985), "Hydro Power: The Design, Use, and Function of Hydromechanical, Hydraulic, and Electrical Equipment." VDI-Verlag, Dusseldorf, Germany.

the square root of the number of nozzles. Specific speed can also be increased by a change in the manner of inflow and outflow. Special designs such as the turbo and crossflow turbines are examples of relatively high-specific speed impulse units.

Over more than 250 years of development, many ideas were tried. Some were rejected, whereas others were retained and incorporated into the design of the hydraulic turbine as we know it today. This development has resulted in highly efficient devices, with efficiencies as high as 96% in the larger sizes. In terms of design concept, these fall into roughly three categories: reaction turbines of the Francis design, reaction turbines of the propeller design, and impulse wheels of the Pelton type.

Most Pelton wheels are mounted on a horizontal axis, although newer vertical-axis units have been developed. Because of physical constraints on orderly outflow from the unit, the maximum number of nozzles is generally limited to six or less. Although the power of a reaction turbine is controlled by the wicket gates, the power of the Pelton wheel is controlled by varying the nozzle discharge by means of an automatically adjusted needle, as illustrated in Fig. 12. Jet deflectors, or auxiliary nozzles, are provided for emergency unloading of the wheel. Additional power can be obtained by connecting two wheels to a single generator or by using multiple nozzles. Because the needle valve can throttle the flow while maintaining essentially constant jet velocity, the relative velocities at entrance and exit remain unchanged, producing nearly constant efficiency over a wide range of power output.

5.1 Performance Comparison

The physical characteristics of various runner configurations are summarized in Fig. 13. It is obvious that the configurations change with speed and head. Impulse turbines are efficient over a relatively narrow range of specific speed, whereas Francis and propeller turbines have a wider useful range. An important consideration is whether or not a turbine is required to operate over a wide range of load. Pelton wheels tend to operate efficiently over a wide range of power loading due to their nozzle design. In the case of reaction machines that have fixed geometry, such as Francis and propeller turbines, efficiency can vary widely with load. However, Kaplan turbines can maintain high efficiency over a wide range of operation conditions. The decision of whether to select a simple configuration with a relatively "peaky" efficiency curve or take on the added expense of installing a more complex machine with a broad efficiency curve will depend on the expected operation of the plant and other economic factors.

Note in Fig. 13 that there is an overlap in the range of application of various types of equipment. This means that either type of unit can be designed for good efficiency in this range but that other factors, such as generator speed cavitation and cost, may dictate the final selection.

5.2 Speed Regulation

The speed regulation of a turbine is an important and complicated problem. The magnitude of the problem varies with size, type of machine and installation, type of electrical load, and whether or not the plant is tied into an electrical grid. It should also be kept in mind that runaway or "no load" speed can be higher than the design speed by a factor as high as 2.6. This is an important design consideration for all rotating parts, including the generator.

The speed of a turbine has to be controlled to a value that matches the generator characteristics and the grid frequency,

$$n = \frac{120f}{N_p}, \qquad (2)$$

where n is the turbine speed in revolutions per minute (rpm), f is the required grid frequency in hertz, and N_p is the number of poles in the generator. Typically, N_p is in multiples of 4. There is a tendency to select higher speed generators to minimize weight and cost. However, consideration has to be given to speed regulation.

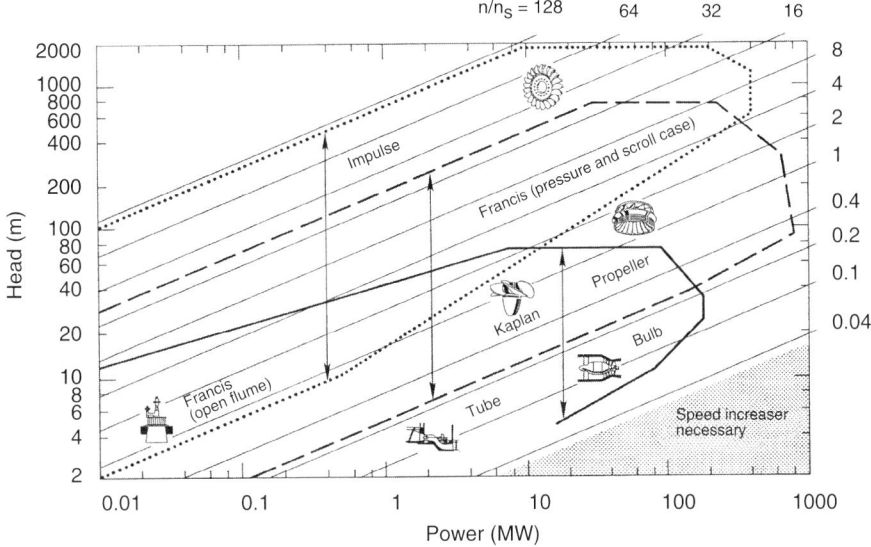

FIGURE 13 Application chart for various turbine types. n/n_s ratio of turbine speed in revolutions per minute n to specific speed defined in the system; $n_s = nP^{1/2}/H^{3/4}$ (with P in kilowatts). From Arndt, R. E. A. [1991]. Hydraulic turbines. *In* "Hydropower Engineering Handbook" [J. S. Gulliver and R. E. A. Arndt, Eds.]. McGraw–Hill, New York.

It is beyond the scope of this subsection to discuss the question of speed regulation in detail. Regulation of speed is normally accomplished through flow control. Adequate control requires sufficient rotational inertia of the rotating parts. When load is rejected, power is absorbed, and this accelerates the flywheel; when load is applied, some additional power is available from deceleration of the flywheel. Response time of the governor must be carefully selected because rapid closing time can lead to excessive pressures in the penstock.

A Francis turbine is controlled by the opening and closing of the wicket gates, which vary the flow of water according to the load. The actuator components of a governor are required to overcome the hydraulic and frictional forces and to maintain the wicket gates in a fixed position under steady load. For this reason, most governors have hydraulic actuators. On the other hand, impulse turbines are more easily controlled. This is due to the fact that the jet can be deflected or an auxiliary jet can bypass flow from the power-producing jet without changing the flow rate in the penstock. This permits long delay times for adjusting the flow rate to the new power conditions. The spear or needle valve controlling the flow rate can close quite slowly, say in 30 to 60 s, thereby minimizing any pressure rise in the penstock.

Several types of governors are available that vary with the work capacity desired and/or the degree of sophistication of control. These vary from pure mechanical, to mechanical–hydraulic, to electrohydraulic. Electrohydraulic units are sophisticated pieces of equipment and would not be suitable for remote regions. The precision of governing required will depend on whether the electrical generator is synchronous or asynchronous (i.e., induction type). There are advantages to the induction type of generator. It is less complex and, therefore, less expensive; however, it typically has slightly lower efficiency. Its frequency is controlled by the frequency of the grid it is feeding into, thereby eliminating the need for an expensive conventional governor. It cannot operate independently and can only feed into a network, and it does so with a lagging power factor that may or may not be a disadvantage, depending on the nature of the load. For example, long transmission lines have a high capacitance, so in this case the lagging power factor may be an advantage.

Speed regulation is a function of the flywheel effect of the rotating components and the inertia of the water column of the system. The start-up of the rotating system, t_s, is given by

$$t_s = \frac{I\omega^2}{P} \qquad (3)$$

where I is the moment of inertia of the generator and turbine (in kilogram-meters), ω is the rotational speed of the unit (in radians/second), and P is the power generated (in watts).

The starting-up time of the water column, t_p, is given by

$$t_p = \frac{\Sigma L V}{gH} \qquad (4)$$

where L is the length of water column segment, V is the velocity in each segment of the water column, and g is the acceleration of gravity.

For good speed regulation, it is desirable to keep $t_s/t_p > 4$. Lower values can also be used, although special precautions are necessary in the control equipment. It can readily be seen that higher ratios of t_s/t_p can be obtained by increasing I or decreasing t_p. Increasing I implies a larger generator, which also results in high costs. The start-up time of the water column can be reduced by reducing the length of the flow system, by using lower velocities, or by adding surge tanks, which essentially reduce the effective length of the conduit. A detailed analysis should be made for each installation because, for a given length, head, and discharge, the flow area must be increased to reduce t_p, leading to associated higher construction costs.

5.3 Cavitation and Turbine Setting

Another factor that must be considered prior to equipment selection is the evaluation of the turbine with respect to tailwater elevations. Hydraulic turbines are subject to pitting due to cavitation. For a given head, a smaller, lower cost, high-speed turbine runner must be set lower (i.e., closer to tailwater or even below tailwater) than does a larger, higher cost, low-speed turbine runner. Also, atmospheric pressure or plant elevation above sea level is a factor, as are tailwater elevation variations and operating requirements. This is a complex subject that can be accurately resolved only through model tests. Every runner design will have different cavitation characteristics. Therefore, the anticipated turbine location or setting with respect to tailwater elevations is an important consideration in turbine selection.

Cavitation is not normally a problem with impulse wheels. However, by the very nature of their operation, cavitation is an important factor in reaction turbine installations. The susceptibility to cavitation is a function of the installation and the turbine design. This can be expressed conveniently in terms of Thoma's sigma, defined as

$$\sigma_T = \frac{H_a - H_v - z}{H}, \qquad (5)$$

where H_a is the atmospheric pressure head, H_v is the vapor pressure head (generally negligible), and z is the elevation of a turbine reference plane above the tailwater (Fig. 7). Draft tube losses and the exit velocity head have been neglected.

The parameter σ_T must be above a certain value to avoid cavitation problems. The critical value of σ_T is a function of specific speed. The Bureau of Reclamation suggested that cavitation problems can be avoided when

$$\sigma_T > 0.26 N_S^{1.64} \qquad (6)$$

where N_S is the specific speed of the turbine or $N_S = n(\text{rpm}) \, P(\text{kW})^{1/2}/H^{5/4}$ (feet). Equation (6) does not guarantee total elimination of cavitation; it guarantees only that cavitation is within acceptable limits. Cavitation can be totally avoided only if the value σ_T at an installation is much greater than the limiting value given in Eq. (6). The value of σ_T for a given installation is known as the plant sigma σ_P. Equation (6) should only be considered as a guide in selecting σ_P, which is normally determined by a model test in the manufacturer's laboratory. For a turbine operating under a given head, the only variable controlling σ_P is the turbine setting z. The required value of σ_P then controls the allowable setting above tailwater:

$$z_{allow} = H_a - H_v - \sigma_p H \qquad (7)$$

It must be borne in mind that H_a varies with elevation. As a rule of thumb, from the sea level value of 10.3 m, H_a decreases by 1.1 m for every 1000 m above sea level.

6. ECOLOGICAL CONSIDERATIONS

The ecological impacts of hydroelectric power production that are perceived depend on the scale at which attention is focused. At the global environmental scale, hydropower is still seen as a clean form of energy. It is a replacement of fossil fuel power production, which is an inherently dirty source of energy. Terrestrial ecology is disturbed by mining activities, transportation of the fuel, and plants that burn the fuel. Water from rivers and lakes that is used to cool power generation systems returns hotter than normal, altering the ecology of these water bodies. Finally, atmospheric impacts of fossil fuel power production are substantial, including the production of nitrous oxides, sulfur compounds, heavy metals, and carbon dioxide. The future state of the earth and its atmosphere is affected by our power production through burning fossil fuels.

Although hydroelectric power production on the global scale is a relatively clean alternative to power produced through fossil fuels, on the local scale (i.e., the river system) the ecological impacts of hydropower can be significant. The dam and reservoir, which most hydroelectric power facilities require, significantly alter the aquatic habitat and the species present. A dam will typically hinder migrating fish species in their spawning activities and cause water quality problems such as dissolved gas bubble disease. The reservoir used for storage can alter flows, such that floods and droughts are less frequent, and will catch most of the sediment. This will alter the river morphology in the downstream reach because the flood flows are important to forming river morphology. In tropical climates, parasite transmission can increase below a dam (e.g., below the Aswan High Dam). With extensive development of dams, a river will become a series of reservoirs and the river habitat will have disappeared entirely, such that many clams or mussel species will no longer exist and the dominant fish species will have been altered.

There are only selected locations on a river that are appropriate for dams and reservoirs. These locations are also the best for recreational activities such as canoeing, rafting, and fishing. Therefore, there is significant public interest in any proposed dam/hydropower facility that is proposed. As the prospective locations for dams and hydropower facilities are developed, there are fewer remaining locations for the recreational activities that need to compete for these sites. From Fig. 4, it can be seen that between 70 and 73% of the economic hydropower potential of Europe and North American has been developed. The recreational and environmental interests of the two continents need to be considered, and it is unlikely that much of the remaining hydropower potential of these two continents will be developed. However, in Asia, Africa, and South America, there is substantial hydropower remaining, and the power needs of the 21st century will be significant. The building of new dams and hydropower facilities in these regions is more likely.

7. ADVANTAGES AND DISADVANTAGES OF HYDROPOWER DEVELOPMENT

It is obvious that hydropower development is a renewable energy source, and this is a significant advantage. When hydropower development occurs at existing dam sites, the environmental impact is often minimal. However, significant changes can occur when an existing dam site is developed for power generation. For example, water that would normally flow over a spillway, where a large amount of aeration would take place, is now funneled through turbines, where little or no aeration occurs. The resulting substantial difference in the rate of aeration at a given point in the river can have a notable influence on the dissolved oxygen content of a considerable reach of river. In certain areas, especially those near large municipalities, the environmental impact can be significant.

Although the fuel costs of hydropower plants are negligible, their construction and capital equipment costs are usually substantially greater per unit of installed capacity than those of thermal power plants. This can be seen as either a benefit or a detriment. The economic feasibility of hydropower development is very sensitive to the difference between a discount rate, used to bring future income and costs to present value, and an assumed escalation rate, used to predict the future cost of electricity. The discount rate is taken as the load interest rate, which is usually fixed. The escalation rate is an estimate, which can vary greatly. For example, during a period of escalating inflation, a hydroelectric project that is already in place looks like a very good investment.

When considering hydropower development within a given region, it is also important to look at the overall economics of that region. For example, it should be noted that hydropower development means that a substantially larger percentage of the investment capital can stay within a given region because much of the developmental work can be done by local engineers and contractors. The more sophisticated coal-fired and nuclear power plants are designed and built by specialized contractors, and this often means that large amounts of capital leave the local economy. In many instances, the same is true for the amount of capital necessary for fuel for thermal power plants. This substantial drain on the economy can be very significant. In addition, hydropower facilities require minimal maintenance and do not have the same requirements for skilled personnel as do the more sophisticated thermal power plants.

There are other advantages of hydropower, and especially of small hydropower suitable as appropriate technology in less developed economic regions. Many future possibilities in small hydropower development will depend on the economic climate. There is a definite market for small turbine technology. If this market is developed, one could expect that significant improvements, both in operational characteristics of turbines and in reduced cost, can lead to small hydropower facilities being more cost-effective. There are many other aspects of small hydropower that have not yet been explored. In many cases in developing nations, small hydropower stations becomes the catalysts for the development of small manufacturing facilities. One case in point is electrolytic manufacturing of fertilizer. It should also be noted that the current miniprocessor technology has developed to the point where it is feasible to operate a system of small hydropower plants on a completely automated basis, with only a traveling crew of workers needed for maintenance. In some cases, the number of small sites developed could supply the same amount of power as could one large nuclear power plant, without the safety or security hazards normally involved with a large-scale development. If the total amount of power is distributed over several small plants, overall reliability can increase because it is very unlikely that all plants would suffer outages at the same time.

However, there are also several disadvantages to small hydropower development. The most obvious is the fact that economy of scale does not prevail. This results in high initial cost for a relatively low installed capacity. In many cases, these plants are run-of-river; that is, their capability for generating power fluctuates wildly with the seasons, and this prevents a system of small power plants from acting as an equivalent base load plant. In many areas of the world, peak power is available during the late spring, whereas peak demand occurs during the midsummer or midwinter. This mismatch of power need and availability can be quite serious.

In addition to the lost economy of scale, there are other disadvantages that relate to the head available. For example, low-head facilities are those in which the available head is less than approximately 20 m. Because available power is proportional to the product of flow and head, larger amounts of flow must be handled to generate a given power level at lower head. Thus, the size of the machine increases, producing a disproportionate increase in cost for the amount of power developed.

SEE ALSO THE FOLLOWING ARTICLES

Coal Industry, History of • *Electricity Use, History of* • *Hydropower Economics* • *Hydropower, Environmental Impact of* • *Hydropower Resettlement Projects, Socioeconomic Impacts of* • *Hydropower Resources* • *Manufactured Gas, History of* • *Natural Gas, History of* • *Nuclear Power, History of* • *Oil Industry, History of* • *Wind Energy, History of* • *Wood Energy, History of*

Further Reading

Armstrong, E. L. (1985). The global outlook for additional hydropower use. In "Waterpower '85." American Society of Civil Engineers, New York.

Arndt, R. E. A. (1981). Cavitation in fluid machinery and hydraulic structures. *Annu. Rev. Fluid Mech.* **13**, 273–328.

Chapman, D. W. (1986). Salmon and steelhead abundance in the Columbia River in the nineteenth century. *Trans. Am. Fish. Soc.* **155**, 662–670.

Chappel, J. R. (1984). "The Future of Hydropower." EG&G Idaho, Idaho Falls, ID.

Daily, J. W. (1950). Hydraulic machinery. *In* "Engineering Hydraulics" (H. Rouse, Ed.). John Wiley, New York.

Deudney, D. (1981). "Rivers of Energy: The Hydropower Potential, Worldwatch Paper 44." Worldwatch Institute, Washington, DC.

Federal Energy Regulatory Commission. (1992). "Hydroelectric Power Resources of the United States, Developed and Undeveloped," FERC-0070. FERC, Washington, DC.

Gulliver, J. S., and Arndt, R. E. A. (1991). "Hydropower Engineering Handbook." McGraw-Hill, New York.

Gulliver, J. S., and Dotan, A. (1984). Cost estimates for hydropower at existing dams *J. Energy Eng.* **110**, 204–214.

International Electrotechnical Commission. (1963). "International Code for the Field Acceptance Tests of Hydraulic Turbines," Publication 41.

Isom, B. G. (1969). The mussel resource of the Tennessee River. *Malacologia* **7**, 397–425.

Merritt, R. H. (1979). "Creativity, Conflict, and Controversy: A history of the St. Paul District," U.S. Army Corps of Engineers, 008-022-00138-7. U.S. Government Printing Office, Washington, DC.

Rouse, H., and Ince, S. (1963). "History of Hydraulics." Dover, New York.

Shea, C. P. (1988). "Renewable Energy: Today's Contribution, Tomorrow's Promise," Worldwatch Paper 81. Worldwatch Institute, Washington, DC.

Smith, N. (1980 January). The origins of the water turbine. *Sci. Amer.* **242**(1).

Van der Schalie, H. (1972). World Health Organization Project Egypt 10: A case history of a schistosomiasis control project. *In* "The Careless Technology: Ecology and International Development" (J. T. Farver and J. P. Milton, Eds.). National History Press, Garden City, NY.

Wirth, B. D. (1997). Reviewing the success of international flooding of the Grand Canyon. *Hydro Rev.* **16**(2), 10–16.

Hydropower Resettlement Projects, Socioeconomic Impacts of

ADRIAN C. SLEIGH.
National Centre for Epidemiology and Population Health, Australian National University
Canberra, Australian Capital Territory, Australia

SUKHAN JACKSON
School of Economics, University of Queensland
Brisbane, Queensland, Australia

1. Introduction
2. Hydropower for Energy: Past and Present
3. Purposes of Hydrodams
4. The Height and Might of Dams
5. Opposition to Dams
6. Proliferation of Large Dams around the World
7. Winners and Losers in Hydropower Projects
8. Displaced, Host, Upstream and Downstream Communities
9. Construction Workers
10. Indigenous People and Gender Effects
11. Planning Blights and Intergenerational Differences
12. Health Impacts
13. Health Impact Assessment
14. Conclusion

Glossary

anadromous Fish born in fresh water that descend to live in the ocean before ascending the river again to spawn and reproduce.
eutrophic Nutrient-enriched (water); for reservoirs this is usually due to the large biomass of vegetation trapped within the reservoir when it filled.
head For a hydrodam, the vertical distance between the water surface in the reservoir and the exit of water from the power-generating turbines.
oustees A word used by social scientists to describe people forced to leave their habitats and livelihoods by the reservoirs and related works of dams or other large projects.
reparation Compensation for a loss. For involuntary resettlers, this should cover in money or kind (or both) what they have lost and what they need to start again.
salinization Accumulation of salt in soil or water.
silt Sediment deposited by rivers and composed of fine soil particles that are usually rich in nutrients needed for agriculture.
waterlogging Saturation of soil with water.

In November 2000 the World Commission on Dams (WCD) released a landmark report, the first rigorous and inclusive global review of large dams. Worldwide, up to a third of these dams are for hydropower and another third are multipurpose. This article focuses on the social impact of such dams, especially the associated involuntary resettlement that has affected millions of people and will affect millions more.

1. INTRODUCTION

The World Commission on Dams (WCD) was formed in 1998 by a joint initiative of the World Conservation Union (IUCN) and the World Bank (WB) after a historic meeting of leading dam proponents and opponents in Switzerland, with proceedings reported by Dorcey and others. The WCD was asked to discover the truth about the cost, effect, and benefit of large dams and it functioned independently for 2 years at a cost of $10 million. The money was well spent, and at the end of that period the WCD produced a comprehensive report and numerous support documents, and then dissolved itself. The report and support documents remain on the Internet available to all, and they have changed forever the debate about dams. Unfounded assertions have been replaced with independent observations that constitute the largest multifaceted knowledge base on dams. The balanced analysis, development-oriented core values, and suggested guidelines are already influencing decisions about dams. Social impacts are now prominent elements in rights-and-risks assessments suggested for decision making about all future dams. Some governments have rejected this approach, but the overall views about dams and development have surely changed.

At the end of this article is a list of books, articles, reports, and other documents that support and expand on all the issues raised. The most useful general source of additional information is the "Dams and Development" report of the World Commission on Dams and many other documents available on its Web site. Both McCully and Smith give extensive accounts of the history of dams. Several references listed address aspects of India's Sardar Sarovar and other Narmada river dams, as well as China's Three Gorges dam; these are especially informative case studies for the political economy and social impacts of very large dams in the countries most committed to dam-based development. The works of Cernea, Scudder, McCully, Brody, Dorcey, Goodland, and Jackson have broader application for social impacts of large dams. Health impacts are covered by works cited for Hunter, Jobin, McCully, Sleigh, and the World Health Organization.

2. HYDROPOWER FOR ENERGY: PAST AND PRESENT

Water manipulation dates from the dawn of civilization. Mesopotamians built dams and irrigation canals at least 8000 years ago and town water supply systems operated from 3000 BC in Jordan. Ancient water systems and dams have been found in Asia and South America. Humans began to capture energy from flowing water using wheels with buckets to lift water in ancient Egypt and Sumeria. The Romans used watermills to grind corn 2100 years ago. A thousand years ago, England already had thousands of watermills. Nineteenth-century Britain led the charge to develop large dams and built nearly 200 above 15 m, usually to store water for the growing industrial cities. In 1832, Benoit Fourneyron had developed a water turbine to capture the potential energy of falling water, far more useful than flow-dependent waterwheels. Engineers later used turbines to generate electricity, and the first hydro plant began producing in Wisconsin in 1882. Hydropower integrates four technologies—dam building, water turbines, electric generators, and electricity transmission to other locations—enabling human society to produce energy from water at one location and use it at another. Such a useful invention soon spread and the size of dams and power stations increased from 30-m heads (in 1900) to 200-m heads (1930s). Modern dams began to spread to the developing world. In 1902, the British built the Low Aswan Dam to regulate the Nile and irrigate cotton fields and eventually India and China began to build many large dams. After 1950 these two countries invested huge sums in dams, foregoing other potential developments. From 1950 to 1980, the Chinese built about 600 such dams per year, but quality was poor. Many dams burst, some catastrophically. For example, in 1975 more than 250,000 people were killed in Henan when dams burst after an extraordinary rainfall. The early dam failure rate in China was much higher than elsewhere.

The World Bank, multilateral development banks, the Food and Agriculture Organization (FAO) and UN Development Program (UNDP) played a big role in promoting large dams and irrigation schemes in developing countries. Bilateral funds through the U.S. Agency for International Development (USAID) and the British Overseas Development Administration (ODA) also helped to plan and fund many dams. Since the 1970s, aid has supported construction companies in rich countries once the work at home had declined. Every year about $20 billion is spent on large dams, and 100 to 200 are commissioned.

3. PURPOSES OF HYDRODAMS

The water levels of most of the world's rivers fluctuate with seasonal rains, melting snows, or droughts, and in the arid zones there are flash floods. It makes sense to capture and store the abundant waters and later release it during the low flow periods or use it for power. So dams have been built for the purposes of irrigation, municipal water supply, and hydroelectricity generation. Flood control and transportation benefits are also important.

Cities are growing all over the developing world, and they require water supplies and reservoirs. Since the beginning of the 20th century, hydropower has steadily increased and it now supplies 19% of worldwide electricity.

Many large dams are multipurpose. For example, the Hoover Dam in the United States provides water for irrigation in California's Imperial Valley, piped water for the residents of Los Angeles and central Arizona, hydroelectricity to consumers in the region, and it controls flood in the lower Colorado River Valley.

4. THE HEIGHT AND MIGHT OF DAMS

Dams are described as major, large, or small. The size is usually based on height, the reservoir area, the generation capacity, or the area under irrigation. The International Commission on Large Dams (ICOLD) defines a large dam as one that is more than 15 m from foundation to crest (taller than a four-story building). But tallness is not always an indication of the dam's impact on ecology and on the number of displaced people. A dam is defined as major (or a super dam) on the basis of either (1) a height of at least 150 m, volume at least 15 million cubic meters, and reservoir storage at least 25 cubic kilometers or (2) a generation capacity of at least 1000 megawatts. It is difficult to calculate the number of large and major dams throughout the world partly because of unreliable information from China and the former USSR. Pearce estimated in 1992 that there were then more than 100 super dams with a height of more than 150 m, of which three quarters were built in the previous 35 years. Since then, others have been built and many more are planned or under way.

5. OPPOSITION TO DAMS

Since the 1970s, natural scientists, hydrologists, economists, and even engineers have become ever more concerned about the negative impacts of large and major dams. The Hoover Dam and the Grand Coulee Dam may have been regarded as successes, but thousands of other large dams performed poorly and were abject failures as instruments of development. First the economists began to criticize the cost overruns and schedule delays of many dams (1960s). Then environmentalists began to report on adverse effects on biodiversity and riparian ecology (1970s).

Ecologists now point to the natural advantages of silt-depositing annual floods that recharge soils, groundwater, wells, and wetlands. The reservoirs created by large dams extinguish wildlife, fish, herbs, and fruits, natural bounties that sustain the rural poor and indigenous populations. The environmental focus is now on sustainability of power generation systems, including hydropower.

By the 1980s social scientists began to publicize the negative impacts on oustees, communities hosting those resettled, construction workers attracted to the project, and downstream fishers and farmers. Michael Cernea was especially important for bringing these social concerns into the World Bank, and he has written widely on the topic. For several decades, health researchers have been reporting a great variety of adverse health impacts, especially infectious diseases associated with tropical dams. These reports are well summarized by the works of Hunter and others, Jobin, and the World Health Organization. Finally the project-affected people have globally united their dissenting voices (1990s) culminating in their Declaration of Curitiba. They demand that dam builders put an end to violent expulsions, pay reparations for livelihood losses, and grant negotiating rights to all those affected.

Impact assessments are now obligatory for virtually all dam planners but are often subject to manipulation, censorship or delay until finance, design, or construction decisions become irreversible. Some well-documented examples include the grievously flawed impact assessments for India's Narmada dams and China's Three Gorges Dam. But many politicians and dam builders remain so confident of their product that few can even imagine that dams could do more harm than good. Yet for many large dams, the harm inflicted on people and nature is irreversible.

6. PROLIFERATION OF LARGE DAMS AROUND THE WORLD

After World War II, many rich nations began to export their 20th century model of economic development to the poor nations of the world, including the emphasis on building large dams. Now at least 45,000 such dams exist, nearly half constructed in China since 1949. Nearly 40,000 of all large dams were commissioned after 1950. The WCD estimated that the total cost to governments exceeded $2 trillion, and most dams had substantial cost overruns and have performed below functional expectations. Despite such enormous investments, systematic reviews of large dam performance were never undertaken and even crude economic outcomes were rarely evaluated. The failure to report negative outcomes, and the apparent (but usually unmeasured) benefits of some large hydropower, flood control, or multipurpose projects in North America and Europe, contributed to the post–World War II determination of many Asian, South American, and African leaders to build large dams for national development, with activity peaking in the decades after 1970. The 20th century leaders of dam-based development, in order of number of dams, were the United States, Spain, Canada, the United Kingdom, Italy, Norway, and Germany. The followers included China (now with about half the world total), India, Japan, South Korea, Turkey, Brazil, South Africa, and Mexico.

The WCD reported that expected benefits were invariably exaggerated by dam proponents, and this had helped create an unstoppable development momentum. About 90% of the global inventory of large dams is located in the 15 countries listed, but 125 other nations have at least one such dam and many more are planned.

Brody, in a theme paper prepared for the WCD, notes that pro-dam sentiment in the United States in the 1930s, 1940s, and 1950s influenced world opinion profoundly. This enthusiasm is captured perfectly by the American folk musician Woody Guthrie, who wrote "The Grand Coulee Dam" (for the Bonneville Power Authority in the Pacific Northwest) and set it to the tune of a famous lament (Lead Belly's immortal song "Irene Goodnight").

> *Uncle Sam took the challenge in the year of 'thirty-three,*
> *For the farmer and the factory and for all of you and me,*
> *He said 'Roll along, Columbia, you can ramble to the sea,*
> *But river, while you're rambling, you can do some work for me.'*
>
> *Now in Washington and Oregon you can hear the factories hum,*
> *Making chrome and making manganese and light aluminum,*
> *And there roars the flying fortress now to fight for Uncle Sam,*
> *Spawned upon the King Columbia by the big Grand Coulee Dam.* ©

Now that we know the adverse impact of large dams on many ecosystems, and millions of poor people worldwide, Guthrie's words seem rather naïve. But they must be read in the context of his bitter experience of the great 1930s drought, the resulting dustbowl in rural Oklahoma and neighboring states, and the forced exodus of millions of impoverished farmers after the bank foreclosures that resulted. The dams he supported on the Columbia river system were sources of hope to many Americans, and they did stimulate the north-west economy.

By the 1970s, the main locus of large dam building had moved to the developing world. Since then, the World Bank has provided the core of the massive finance needed and has led the development of guidelines to assess, monitor, and mitigate environmental and social impacts. Over that period, the social impacts became more noteworthy because dam building affected the poorer and more densely populated parts of the world at a time of increasing interest in human rights and environmental protection. Global reporting and access to information also improved. The adverse social and environmental impacts of large dams were widely (but haphazardly) reported and such information fed an ever-growing, and increasingly vocal, global opposition. Now most such dams are being built in developing countries. Meanwhile, in the United States, decommissioning dams has become more frequent than building them, a trend expected to spread to most of the developed world in the early 21st century. So the social impact of large dams is now largely an issue for the developing world, and the pro-dam lobbies in China, India, Iran, Turkey, Brazil, and many other countries remain very powerful. Anti-dam dissent is poorly tolerated in most of those countries, and intimidation and violence are frequent.

Now the World Bank is far less enthusiastic than before. Since the 1980s, it was stung by strident criticism of its support for dams, which included reference to its failure to follow its own guidelines for mitigating social impacts and even an accusation of a coverup of a state-supported massacre of nearly 400 indigenous people who refused to move for a WB-supported dam in Guatemala. McCully gives a full account of these criticisms in his book *Silenced Rivers*, published in 2001. In the 1990s, the World Bank began to withdraw from or avoid the most troublesome projects, such as India's Sardar Sarovar dam or China's Three Gorges dam. These dams are proceeding without World Bank finance. Even private and government international dam-finance agencies are now subject to increased scrutiny, and some have withdrawn from controversial projects.

By the late 1990s, the dam debate had reached an impasse, and only a substantial and independent global inquiry into large dams could bridge the gulf between those who opposed such dams and those who financed and built them. The resulting World

Commission on Dams included engineers, environmentalists, anti-dam activists, civil servants responsible for water resources, lawyers, scientists, and others drawn from all parts of the world. Most observers expect that fewer dams will be built in the future, planning and cost-benefit estimations will be more honest and comprehensive, and the development emphasis will focus on the rights, risks, and livelihoods for affected people. As well, technological advances in power generation, especially gas turbines (at first), then wind and solar power (next 10 years), and finally fuel cell (hydrogen-based and renewable) should make future hydropower projects less frequent. Hydroelectric power and central electrification will become progressively more expensive relative to other options, despite the fact that hydrodams performed closer to their estimated economic and production targets than did most irrigation dams in the WCD assessment.

Asia has so far exploited only about 11% of its hydropower capacity and Africa about 3%, so the hydropower supply momentum in China and India, and perhaps Africa too, will persist for some time. China, India, Turkey, and Russia expressed grave doubts about the negative WCD findings and pro-dam sentiment remains powerful in other countries as well. The need to improve domestic and industrial water supplies, to electrify cities and rural areas, and to irrigate land to produce food for the growing world population are expected to be the key demand drivers of most future large dams.

7. WINNERS AND LOSERS IN HYDROPOWER PROJECTS

The winners in hydropower and other large dam projects are those who gain electricity, domestic water supplies, sanitation, drought relief, sustainable irrigation, flood control, improved water transport, industrial and employment opportunities, food security, better roads, telecommunications, and other benefits. As well, other economic multipliers may arise and whole regions may flourish. This is an impressive list and is the very basis of modernization and liberation from the drudgery of traditional low-technology rural life. Electricity, radio, and domestic water supplies completely transform the lives of those who gain these goods, and better health and education are the most obvious outcomes.

But the WCD (and many other reports before and after) found that the benefits of large dams were rarely so comprehensive, and the common river-basin disbenefits included waterlogging and salination of irrigated land, unseasonal reversal of downstream flows, dangerous river flows at unpredictable times, loss of vital silt deposits, encroachment of seawater in river deltas, and obstruction of migrating anadromous fish. Subtle changes included the switch to cash cropping, frequently with negative consequences for the nutrition of vulnerable women, children, and the elderly, and concentration of land ownership.

More direct and even more negative effects occur for those forced to leave their homes and land, for communities that are broken up, and for other communities that host the oustees. Migrants arrive to construct the dam or to take advantage of ecological changes they are better equipped to exploit, such as lake fishermen after reservoirs form. Funds distributed in compensation for losses are often embezzled or subject to nepotism. All these changes produce social friction and fights are common. Many dam-affected groups become sick due to the water impoundment or its ecological or socioeconomic effects. In many well documented cases (and others not documented), the evictions of oustees have been violent and sometimes lethal. Details of such ill-treatment are given in the publications of the WCD and McCully and in many other sources. All these groups described previously are the losers. They are not beneficiaries of development dams. Sometimes winners and losers coexist in a single community or even within one family.

In his paper on social impacts of large dams, prepared for WCD, Brody listed several important principles to consider, including the following:

1. Dams take one set of resources, a river and the lands along its banks, and transform them into another set of resources, a reservoir, hydropower, and irrigation. The means of production and livelihood for one group of people are changed into benefits for people living elsewhere.
2. Powerful economic and political interests attach to such undertakings. The larger the dam, the more costly the project, the greater its construction momentum.
3. The larger the project and the greater the momentum, the greater the risks to the social well-being and human rights of those directly affected.
4. For consultation and participation to be authentic and effective, they must take place in a way and at a time when they can influence

decisions about the project and mitigation of its impacts.

8. DISPLACED, HOST, UPSTREAM AND DOWNSTREAM COMMUNITIES

The adverse human outcomes of blocking the world's rivers have been even more dramatic than the ecological ones. But social impacts vary enormously within and among these groups. Upstream communities are either displaced directly, often with inadequate or no reparation, or become hosts for some of the oustees. The latter situation causes frequent conflict, especially (as is often the case) if land is scarce and resettlers arrive with more available capital than the hosts. The same tensions often arise for downstream communities that are forced to host oustees.

Upstream communities displaced are often the worst affected. In most dams studied by the WCD, and according to numerous other reports, compensation for oustees is almost always inadequate and sometimes is not offered at all. Dams have flooded 40 million to 80 million people off their land, with 16 million to 38 million displaced in India (differentially indigenous and always poor) and more than 10 million peasants displaced in China (just for the Yangtze basin). It seems probable (but usually unstudied) that many people never recovered from the ordeal, either economically or psychologically, and this trauma has been passed to the next generation. Displaced communities were usually fragmented and destroyed as cultural groups. Many more people lost their fisheries, the irrigation of annual floods, and the usufruct of flooded lands. Lost archeological and cultural sites could not be replaced.

Dam resettlement authorities often state they aim to offer land-for-land to people displaced. In fact, this is often not possible, but even if it is done the problems for those resettled are not over. The replacement land is likely to be less valuable, less fertile, less watered, and quite different to the land they occupied before. Essential infrastructure is often lagging or not provided at all. As the new land may be higher up, and sloping, water will be the first problem and may be an insurmountable obstacle to successful agricultural activity. The next obstacle will be to choose suitable crops, and these are likely to be unfamiliar to the farmer. Without scientific agricultural support, results will be poor. Even with that support, the first few crops may be disastrous. So resettled farmers need to be supported and monitored, and this must continue for several years. It rarely does. But unless resettlers livelihoods are established successfully and return to the level they had before or better, they will be long-term losers and may never recover.

Upstream communities not displaced must learn to live with a changed environment. Travel across the landscape is obstructed by the reservoir and new and unfamiliar industries and occupations arise. Inflation may affect the area due to the influx of money to build the dam. For very large reservoirs, the climate may change with relative humidity and temperature increasing and fog may become more frequent. There are other positive impacts as well. Transport may improve and journeys that once took days (by road or foot paths) may now take only 1 to 2 hours across the water. If the dam authority develops a boat service or creates a new township at the top of the reservoir to provide municipal services, people are better off.

Downstream communities are affected by large dams for hundreds of kilometers. The riparian impacts are greatest but the effects of trade, transport, and food supplies may be felt far away from the river basin. The benefits of electricity and domestic water supplies are often downstream. Some downstream communities are winners and others are losers. Farmers and fishers are at high risk of adverse impacts due to retention of silt by the dam, unseasonal water flows, loss of water to distant irrigators, and destruction of fisheries. Irrigation can be a mixed blessing; if not done well, it leads to salination and water logging, noted by the WCD to affect 20% of such areas.

9. CONSTRUCTION WORKERS

Large dams attract construction workers in large numbers. A temporary workforce of 10,000 to 30,000 is not uncommon and the peak of work may last 4 to 8 years. Sometimes whole families are employed, but often marginalized men, or women, are the workers recruited. Conflict is common and injuries and accidental death are frequent. Illnesses due to introduced or endemic infectious diseases, including sexual infections and AIDS, are expected. Health services are often well developed for this important group whose welfare is crucial to completion of the dam. Infrequently these services are made available to the local population or become the basis for boosting regional health services after the dam is completed. If the latter occurs, the whole community benefits.

10. INDIGENOUS PEOPLE AND GENDER EFFECTS

The World Bank has expressed great concern for indigenous people affected by dams. These groups are so dependent on their local environments they cannot recover once they are removed. In India and South America, indigenous people have borne disproportionate shares of the displacement burdens. Many instances of violence and social catastrophes have been recorded. It is now thought that if indigenous persons are going to be affected, dams should not proceed unless equivalent lands can be found. Such substitute lands are quite improbable.

The gender effects of dam displacement have rarely been studied and were a special focus of the WCD report. The commission noted that in any situation where females were already disadvantaged dams often make things worse. Compensation, if paid, is usually given to husbands or sons, even if the group is matrilineal. This reflects the male gender bias of the state. Common lands lost are not compensated even if they were the location of food and herb gathering, small game hunting, and other social activities conducted exclusively by women. Women also suffer when violence erupts, which is more likely among the displaced. Once a family is impoverished, unequal females may become even worse off and may turn to prostitution and crime to survive. Gender-sensitive planning has been very rare for large dams and other infrastructure projects.

11. PLANNING BLIGHTS AND INTERGENERATIONAL DIFFERENCES

For many years and sometimes many decades before a large dam is built, the project is widely discussed. During this long period, all investment in the area is substantially curtailed by both the state and the local residents. Consequently, the people are poorer than they would have been if no dam was contemplated, and public services and infrastructures are run down or not provided at all. This planning blight should be considered when calculating compensation for oustees, but almost invariably it is ignored.

The most vulnerable groups for social impacts of large dams are the aged, children, and the infirm. Within families, a young couple might see the cash compensation and resettlement in a new house elsewhere, or even a chance to move to a city, as a golden opportunity to escape a dreary rural life. But their parents could be devastated by the same choices, become disoriented and depressed, and lose all hope. They would sink into poverty while their children rose to become relatively wealthy; in such circumstances, family harmony breaks down or parents are even abandoned. Other groups may be selectively disadvantaged, such as those who have no official residence in the place they live and must leave. This arises for up to ten per cent of displaced populations in China, and they receive nothing at all even if they lose a long-established business such as a restaurant or noodle bar. Others may be forced into the cash economy and quickly stripped of their life savings by the predatory behavior of swindlers they encounter in their new town. For many displaced persons—uncounted millions since the 1950s—the final result of their large dam has been loss of community, culture, burial sites, identity, habitat, and livelihoods. This leads to poverty and urban or rural squatting without adequate housing, water, electricity, food, or employment. They have no voice and are forgotten.

12. HEALTH IMPACTS

Dams built to stimulate development should not produce disease. But they do. Most information on the health impacts of large dams comes from studies conducted in Africa and the Caribbean. Vector-borne diseases have been the most noticeable disease effect, especially epidemics of parasitic and viral infections around large (or small) dams and irrigation systems. These begin soon after reservoirs form once the dams are closed. First among these is bilharzia, also known as snail fever, water belly, or schistosomiasis. The disease is caused by one or more pairs of blood flukes (schistosomes) whose eggs pass out in human urine or faeces, infect snails, multiply, and produce free swimming larvae that reinfect human hosts. Several species of schistosome infect humans, including one in China that also infects many domestic mammals. The disease is present in 73 countries including most of Africa, parts of the Middle East, some Caribbean islands, Brazil and several other South American countries, China, Laos, Cambodia, the Philippines, and Indonesia. There are numerous examples of entire communities becoming infected due to large dams; average life spans are shortened and many people are debilitated and impoverished as a result. Attempts to prevent snail breeding in reservoirs and canals by engineering modifications or chemicals have met with limited success and are described in detail in Jobin's book *Dams and Disease*.

The next most reported dam-induced infection is malaria and this is spread even more widely than schistosomiasis. Other mosquito-borne infections include the nematode worm that causes elephantiasis, also known as filariasis, and several serious viral diseases, including Rift Valley fever, Japanese encephalitis, and dengue. Black fly populations can breed on dam spillways in parts of Africa and Latin America, and transmit the nematode worm that causes river blindness, also called onchocerciasis.

Much less reported, but perhaps of even greater importance, are the diseases of poverty and poor sanitation because they are made worse by inadequate resettlement. These problems include maternal anaemia, death in childbirth, childhood malnutrition, and infections such as diarrhoea, pneumonia, and tuberculosis. Because they were present before the dam was built and are relatively common, they are not often noted as problems exacerbated by large dams.

Even less well recognized and studied are the mental health problems produced by involuntary loss of habitat and livelihood. There is no doubt that they exist, but little is known about their consequences. Some indigenous people are so deeply affected that they die soon after displacement. Other oustees become depressed or suicidal. Violence and social breakdown are also serious problems and are inevitable if communities are displaced and not reformed with adequate support.

Poisoning and injury are well documented consequences of large dams. Methyl-mercury is formed in the reservoir and ascends the food chain, reaching humans via the fish. Other heavy metals may be collected in the runoff after flooding certain industrial plants. Organic solvents and other poisons can pollute reservoirs also. Biological toxins pollute the water if algae bloom or cyanobacteria proliferate, which is very likely in eutrophic waters typical of tropical reservoirs soon after they fill. Injury is a constant risk for construction workers and many are injured or killed by falling or after being struck by flying objects at the dam site or in the powerhouse or other related works. Drowning is also a risk after the reservoir forms, especially in tropical areas where people may start recreational swimming with no safety measures in place and little experience of deep water. Alcohol makes this risk higher. Inebriated revelers may swim out into reservoirs and be struck by boats around busy jetties or further out, or they may simply drown.

Finally there is a risk of a catastrophe inherent for all large dams. They can collapse and when overtopped or breached due to a geological fault, flood, or a landslide-induced giant wave, the results have sometimes been disastrous. Landslides may be made more likely by the earthquakes induced by the weight of water stored behind high dams. Accounts of such catastrophes, and seismic activity induced by impounded water, are given in the WCD report and in McCully's book *Silenced Rivers*. The most spectacular landslide example occurred in the Italian Alps in 1963. It caused a huge wave that rose 110 m above the 261-m wall of Vaiont Dam, overtopping the dam and killing 2600 people in villages located downstream. The most lethal example is the multiple dam collapse in Henan, China, in 1975, which killed at least 250,000 people downstream.

13. HEALTH IMPACT ASSESSMENT

Health impact assessment is not often requested by dam planners and builders. It involves a baseline (pre-dam) study of existing risks and an estimation of new risks that will or could arise. The at-risk community should then be followed and steps taken to reduce or prevent the identified health risks. In some instances, the risks are too high and cannot be controlled and the dam should not proceed. This almost never happens because such risks are not identified in time. Usually, at best, a health assessment representative is placed on an environmental or social panel to monitor the outcomes and is not resourced or supported with the needed array of public health expertise (epidemiology, nutrition, infectious diseases, entomology, malacology, toxicology). It is not possible to have much effect in these circumstances. In some cases the dam builder has investigated the risks well but not integrated the process with other preventive activity. This was done for China's Three Gorges Dam, but the geographers led by Chen and others made assessments without being connected to the health system, and the health personnel in the area were not involved. No follow-up activity resulted, and no mitigation of risks occurred. More than 1 million people are being displaced and many millions more affected along the 600-km reservoir or in the densely populated middle reaches of the river stretching hundreds of kilometers below the dam.

14. CONCLUSION

Many social benefits flow from well-planned, economically justified, and equitably executed large dams. But very few dams have been completed that way. In future, the rights-and-risks approach recommended by the World Commission on Dams, with full participation of project-affected people, is likely to be a requirement before attracting international finance for a new dam. Until then we have to address the enormous adverse social and health impacts created by many of the existing 45,000 large dams, a legacy of 20th-century enthusiasm that was often misplaced. Some of these dams have harmed a great many people, and some were successful as agents of development. It would be unfortunate if future dams did not take all this experience into account, and most people would agree that large dams should not be built if they make people sick.

Acknowledgments

Lyrics to "The Grand Coulee Dam" (words and music by Woody Guthrie) are used by permission. TRO—© Copyright 1958 (renewed), 1963 (renewed), 1976 Ludlow Music, Inc., New York, NY.

SEE ALSO THE FOLLOWING ARTICLES

Electricity, Environmental Impacts of • Hydropower Economics • Hydropower, Environmental Impact of • Hydropower, History and Technology of • Hydropower Resources • Hydropower Technology • Oil-Led Development: Social, Political, and Economic Consequences

Further Reading

Alvares, C., and Billorey, R. (1988). "Damming the Narmada." Third World Network/Appen, Penang, Malaysia.

Barber, M., and Ryder, G. (eds.). (1993). "Damming the Three Gorges. What Dam Builders Don't Want You to Know." Earthscan, London.

Brody, H. (2000). Social impacts of large dams equity and distributional issues (*Thematic Rev.*, World Commission on Dams), www. dams. org/docs/kbase/contrib/soc192.pdf (accessed September 24, 2003).

Cernea, M. M. (1988). Involuntary resettlement in development projects: Policy guidelines in World Bank-financed projects. *World Bank Technical Paper*, 80. World Bank, Washington, DC.

Cernea, M. M. (1999). "The Economics of Involuntary Resettlement: Questions and Challenges." World Bank, Washington, DC.

Cernea, M. M., and McDowell, C. (eds.). (2000). "Risks and Reconstruction: Experiences of Resettlers and Refugees." World Bank, Washington, DC.

Chen, Y., Shi, M., Zhao, M., Chen, M., Huang, X., Luo, X., Lie, B., Wu, Y., Yang, L., Yan, J., Wang, J., Zhong, D., Yao, S., Fan, F., Wang, D., and Han, A. (eds.). (1990). "Atlas of the Ecology and Environment in the Three Gorges Area of the Changjiang River." Science Press, Beijing.

Declaration of Curitiba (1997). "Affirming the Right to Life and Livelihood of People Affected by Dams." Approved at the First International Meeting of People Affected by Dams. Curitiba, Brazil, March 14, 1997, www. irn. org/programs/curitiba. html (accessed September 24, 2003).

Dorcey, T., Steiner, A., Acreman, M., and Orlando, B. (eds.). (1997). "Large Dams. Learning from the Past, Looking at the Future." Gland, Switzerland: World Conservation Union (IUCN) and World Bank Workshop Proceedings, April 11–12.

Fearnside, P. M. (1994). The Canadian feasibility study of the Three Gorges Dam proposed for China's Yangtzi River: A grave embarrassment to the impact assessment profession. *Impact Assessment* **12**, 21–53.

Goodland, R. (1997). Environmental sustainability in the hydro industry. *In* "Large Dams. Learning from the Past, Looking at the Future." (T. Dorcey, A. Steiner, M. Acreman, and B. Orlando, Eds.), pp. 69–102. Gland, Switzerland: World Conservation Union (IUCN) and World Bank Workshop Proceedings, April 11–12.

Hunter, J. M., Rey, L., Chu, K. Y., Adekolu-John, E. O., and Mott, K. E. (1993). "Parasitic Diseases and Water Resource Development." World Health Organization, Geneva.

Jackson, S., and Sleigh, A. (2000). Resettlement for China's Three Gorges Dam: Socio-economic impact and institutional tensions. *Communist and Post-Communist Studies* **33**, 223–241.

Jackson, S., and Sleigh, A. C. (2001). Political economy and socio-economic impact of China's Three Gorges Dam. *Asian Studies Review* **25**, 57–72.

Jobin, W. (1999). "Dams and Disease: Ecological Design and Health Impacts of Large Dams, Canals and Irrigation Systems." E & FN Spoon, New York.

McCully, P. (2001). "Silenced Rivers. The Ecology and Politics of Large Dams." Zed Books, London.

Morse, B., and Berger, T. (1992). "Sardar Sarovar: Report of the Independent Review." Resource Futures International, Ottawa.

Pearce, F. (1992). "The Dammed: Rivers, Dams and the Coming World Water Crisis." The Bodley Head, London.

Scudder, T. (1997). Social impacts of large dam projects. *In* "Large Dams. Learning from the Past, Looking at the Future" (T. Dorcey, A. Steiner, M. Acreman, and B. Orlando, eds.), 41–68. Gland, Switzerland: World Conservation Union (IUCN) and World Bank Workshop Proceedings, April 11–12.

Sleigh, A., and Jackson, S. (1998). Public health and public choice: Dammed off at China's Three Gorges? *The Lancet* **351**, 1449–1450.

Sleigh, A. C., and Jackson, S. (2001). Dams, development and health: A missed opportunity. *Lancet* **357**, 570–571.

Smith, N. A. F. (1971). "A History of Dams." Peter Davis, London.

World Commission on Dams (2000). "Dams and Development. A New Framework for Decision Making." Earthscan, London, www.dams.org (accessed September 24, 2003).

World Health Organization (1999). "Human Health and Dams." World Health Organization, Geneva (submission to the World Commission on Dams), www.dams.org/docs/kbase/working/health.pdf (accessed September 24, 2003).

Hydropower Resources

GAROLD L. SOMMERS
Idaho National Engineering and Environmental Laboratory
Idaho Falls, Idaho, United States

1. The Resource
2. How Hydropower Resources are Assessed
3. Global Distribution
4. Current Resources
5. Potential of Hydropower Resources

Glossary

benefit/cost ratio (B/C) Ratio of the present value of the economical benefit to the present value of the project cost, computed for comparable price level assumptions.

capability The maximum load that a generating unit, generating station, or other electrical apparatus can carry under specified conditions for a given period of time without exceeding approved limits of temperature and stress.

capacity The maximum power output or load for which a turbine generator, station, or system is rated.

dam A wall or structure built across a valley or river for storing water.

dependable capacity The load-carrying ability of a hydropower plant under adverse hydrologic conditions for a specified time interval and period.

distribution system The portion of an electric system dedicated to delivering electric energy to an end user; the distribution system "steps down" power from high-voltage transmission lines to a level that can be used in homes and businesses.

diversion dam A dam used in conjunction with long tunnels, canals, or pipelines to divert water to a powerhouse located a distance from the dam.

drawdown The distance that the water surface of a reservoir drops from a given elevation as the result of withdrawal of water.

energy The capacity for performing work; the electrical energy term generally used is kilowatt-hours and represents power (kilowatts) operating for some time period (hours).

forebay The water intake area for a canal, penstock, or turbine designed to reduce water velocity and turbulence so as to settle suspended material and keep it from entering the system.

generator A machine that converts mechanical energy into electric energy.

gigawatt (GW) Unit of electric power equal to 1 million kW.

head The difference in elevation between the headwater surface above and the tailwater surface below a hydroelectric power plant under specified conditions.

horsepower A unit of the rate of doing work equal to 33,000 foot pounds per minute or 745.8 W (Britain), 746 W (United States), or 736 W (Europe).

hydroelectric plant or hydropower plant An electric power plant in which the turbine generators are driven by falling water.

hydroelectric power Electric current produced from waterpower.

hydrology The scientific study of the properties, distribution, and effects of water on the earth's surface, in the soil and underlying rocks, and in the atmosphere.

hydrologic cycle Water constantly moving through a vast global cycle in which it evaporates from lakes and oceans, forms clouds, precipitates as rain or snow, and then flows back to the ocean; the energy of this water cycle, which is driven by the sun, is tapped most efficiently with hydropower.

kilowatt (kW) Unit of electric power equal to 1000 W or approximately 1.34 horsepower; for example, it is the amount of electric energy required to light 10 100-W light bulbs.

kilowatt-hour (kWh) The unit of electrical energy commonly used in marketing electric power; it is the energy produced by 1 kW acting for 1 h.

kinetic energy Energy that a moving body has due to its motion, dependent on its mass and the rate at which it is moving.

megawatt (MW) A unit of power equal to 1 million W; for example, it is the amount of electric energy required to light 10,000 100-W bulbs.

megawatt-hour (MWh) Unit of electric power equal to 1000 kWh.

microhydropower A hydroelectric plant with a rated capacity of 100 kW or less.

peaking capacity That part of a system's capacity that is operated during the hours of highest power demand.

penstock A closed conduit or pipe for conducting water to a powerhouse.

plant factor The ratio of the average output to the installed capacity of the plant, expressed as an annual percentage.

power (electric) The rate of generation or use of electric energy, usually measured in kilowatts.

pumped-storage hydroelectric plant A plant that usually generates electric energy during peak-load periods by using water previously pumped into an elevated storage reservoir during off-peak periods when excess generating capacity is available to do so; when additional generating capacity is needed, the water can be released from the reservoir through a conduit to turbine generators located in a power plant at a lower level.

reservoir An artificial lake into which water flows and is stored for future use.

turbine A machine for generating rotary mechanical power from the energy of a stream of fluid (e.g., water, steam, hot gas); turbines convert the kinetic energy of fluids to mechanical energy through the principles of impulse and reaction or a mixture of the two.

watt (W) The unit used to measure production/usage rate of all types of energy; the unit for power.

watt-hour (Wh) The unit of energy equal to the work done by 1 W in 1 h.

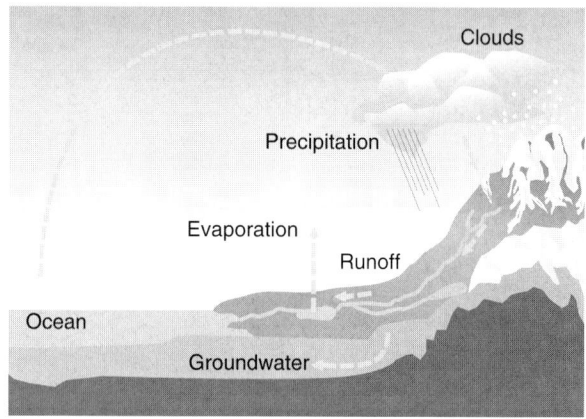

FIGURE 1 The hydrologic cycle.

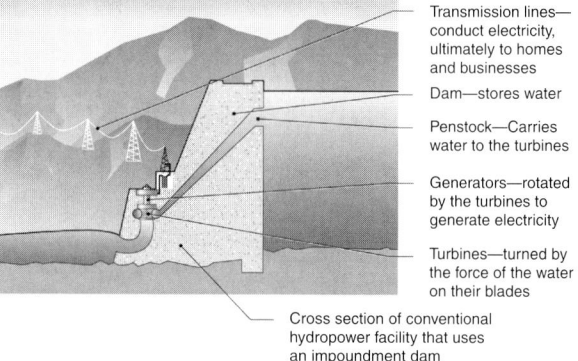

FIGURE 2 Impoundment hydropower project.

This article provides a brief overview of the hydropower resource—how this resource is assessed, current resources worldwide, and future potential development.

1. THE RESOURCE

Hydroelectric power comes from water at work. To generate electricity, water must be in motion. This kinetic energy turns the blades of a water turbine, which changes the kinetic energy to mechanical (machine) energy. The turbine shaft turns a generator, which then converts this mechanical energy into electricity. Because water is the initial source of electrical energy, this technology is called hydroelectric power or "hydropower" for short.

Water constantly moves through the hydrologic cycle (Fig. 1), which is driven by solar energy. In the hydrologic cycle, atmospheric water reaches the earth's surface as precipitation. Some of this water evaporates, but much of it either percolates into the soil or becomes surface runoff. Water from rain and melting snow eventually reaches ponds, lakes, reservoirs, or oceans where evaporation is constantly occurring. The hydrologic cycle ensures that water is a renewable resource.

Hydropower plants can be located on rivers, streams, and canals, but dams are needed for a reliable water supply. Dams store water for later release for purposes such as irrigation, domestic and industrial use, and power generation. The reservoir acts much like a battery, storing water to be released as needed to generate power. The dam also creates a head, that is, a height from which water flows. A pipe (penstock) carries the water from the reservoir to the turbine.

Typical conventional hydropower projects are of three types: impoundment, diversion, and run-of-river. Impoundment projects (Fig. 2) use a dam to store water. Water is released to meet water use demands, including generation of electricity. Diversion projects (Fig. 3) channel a portion of the river through a canal or penstock to produce electrical energy. A small diversion dam may be required to channel a portion of the water from the river. Run-of-river projects (Fig. 4) use the flow of water within the natural range of the river. They may require a small impoundment to develop head. Run-of-river projects generally blend in with the natural river system.

FIGURE 3 Diversion project.

FIGURE 4 Run-of-river project.

Unconventional hydropower projects include pumped storage. These projects pump water from a lower reservoir to an upper reservoir during times when demand for electricity is low. During periods of high electrical demand, the water is released back to the lower reservoir to generate electricity.

2. HOW HYDROPOWER RESOURCES ARE ASSESSED

The first step in assessing a potential water resource for hydropower is to calculate how much power can be produced. The estimated power is determined by the volume of water available (flow) and the vertical distance the water falls (head). A given amount of water falling a given distance will produce a certain amount of energy. In the English system of measurement, flow is commonly measured in cubic feet per second (cfs). Head is measured in feet.

The capacity of hydropower plants is measured in kilowatts or megawatts. Potential capacity can be calculated as follows:

$$kW = Q \times H/11.8,$$

where

- kW = electrical power;
- Q = flow rate in cubic feet per second;
- H = head in feet; and
- 11.8 = a constant (conversion of foot pounds to horsepower and then horsepower to kilowatts).

Note: the equation does not include efficiency losses in the turbine, generator, penstock, and so on.

Power potential can be converted to potential annual energy production, which is expressed in kilowatt-hours as follows:

$$kWh = kW \times 8760 \times PF,$$

where

- kWh = annual energy;
- kW = project capacity;
- 8760 = 24 h per day × 365 days per year; and
- PF = plant factor.

Annual energy production is based on installed capacity and available water for power production. In most cases, water flow varies throughout the year. During low-flow periods, energy production is reduced. Therefore, when calculating annual energy production, a plant factor must be applied. If the actual plant factor cannot be determined based on flow conditions, an average can be used to estimate energy production. The average plant factor for hydropower projects in the United States is approximately 50% (0.5).

The potential installed capacity for a specific project is used to evaluate development potential. In general, a river with a large flow (Q) or potential drop in elevation (H) will be desirable for producing hydropower. However, specific project evaluations include issues such as project economics and accommodating the necessary legal, institutional, and environmental constraints. The sum of the installed capacity of all the projects in a region or country represents that region's or country's potential resource.

Specific issues can be applied to each project, further reducing its potential resource. Some projects cannot be developed because they are located in excluded areas. Excluded areas include wild and scenic rivers, national wilderness areas, and national parks, monuments, preserves, refuges, and historic areas. Other projects may not be developed, or developed to full potential, due to environmental, legal, and/or institutional issues. Environmental concerns may include releasing water for nonpower production to improve water quality or fish passage. Other issues could be cultural, historical, geological, recreational, scenic, loss of productive land, and displacement of people.

Undeveloped hydropower resources are generally reported by potential capacity. This capacity can be evaluated based on the various conditions and constraints that require specific site information and characteristics.

3. GLOBAL DISTRIBUTION

In 1998, hydroelectricity provided approximately 21.6% of the worldwide electricity capacity and 18.8% of the worldwide generation of electricity. Figure 5 illustrates the top hydroelectric generating countries in the world. Table I shows the installed hydropower capacity and energy production by major geographical regions worldwide.

In 1999, the developed hydropower capacity in the United States was approximately 79,700 MW. In 2000, the electrical energy production was approximately 269,000 GWh/year. This is approximately 7.1% of the total electricity in the United States.

4. CURRENT RESOURCES

Water is one of the most valuable resources, and hydropower makes further use of this renewable treasure. The United States has invested more than $150 billion (in 1993 dollars) in hydropower facilities. This investment does not include the cost to build most of the large federal dams in the western part of the country, but it does include the hydropower portion of these projects. The U.S. hydropower capacity is more than 94,000 MW (includes pumped storage), enough to supply the electrical power needs of approximately 28 million households.

Hydroelectric facilities have many characteristics that favor their use and encourage upgrading existing plants and developing new projects. Hydroelectric facilities offer several benefits, including the following:

- use a renewable resource to generate power;
- are highly reliable and have low operating costs; and
- can start up quickly and have the capability to adjust their output (load following) capability and peak capacity.

As an added benefit, reservoirs have scenic and recreational value for campers, fishermen, and water sports enthusiasts. Water is the home for fish and wildlife as well. Dams add to domestic water supplies, control water quality, provide irrigation for agriculture, and avert flooding. Dams can actually improve the quality of downstream conditions by allowing mud and other debris to settle out.

Just as there is a wide variety of natural settings for hydropower sites, there is a variety of developmental schemes that have various benefits and impacts. The most common distinction is between large and small dams. Dams are often referred to as high-head and low-head dams, reflecting the

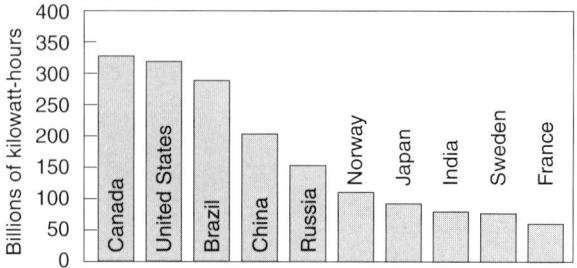

FIGURE 5 Top hydroelectric-generating countries in 1998. From the Energy Information Administration. (2000). "Annual Energy Review 1999," Table 11.15. EIA, Washington, DC.

TABLE I

Worldwide Hydropower Capacity and Energy Production

Geographic region	Installed capacity (~MW)	Energy production (~GWh/year)
Africa	20,651	80,575
Asia (including Russia and Turkey)	241,624	793,045
Australia/Oceania	13,271	42,000
Europe	175,625	593,377
North and Central America	158,000	700,000
South America	111,459	531,168
World totals	720,630	2,740,165

Source. Aqua~Media International Westmead House. (2002). "The International Journal on Hydropower and Dams: World Atlas and Industry Guide" (pp. 13–15). Aqua~Media International Westmead House, Sutton, UK.

concept that hydraulic head (the height from which water drops before reaching the turbine) is a key aspect of electrical generation. As a general rule of thumb, the electrical generating capacity of large hydropower projects ranges from approximately 25 to more than 10,000 MW. Small hydropower projects range from 1 to 25 MW. Smaller yet are the minihydropower projects (less than 1 MW) and microhydropower projects (less than 100 kW). Mini- and microhydropower projects are often located on very small streams to provide decentralized electrical power in remote locations. Many of the most controversial aspects of hydroelectric development, such as the social impacts of human resettlement or inundation of terrestrial habitat, arise from the creation of large reservoirs associated with large dams. However, large dams are far less numerous than small dams. For example, the International Commission on Large Dams estimates that there are approximately 40,000 large dams (dams with a height of 15 m or more or shorter dams with high discharges) and 800,000 small dams worldwide.

Existing power plants can be upgraded, or new power plants can be added at current dam sites, without significant additional effect on the environment. Only approximately 2400 of the nation's 80,000 dams are currently used to generate power (Fig. 6). In addition, new facilities can be constructed with consideration of the environment. For instance, dams can be built at remote locations; power plants can be placed underground; and selective withdrawal systems can be used to control the water temperature released from dams. Facilities can incorporate features that aid fish and wildlife such as salmon runs and resting places for migratory birds. For a history of hydroelectricity's progression in the United States, see Table II.

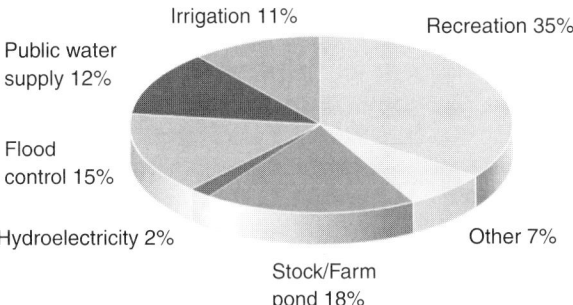

FIGURE 6 Primary benefits of U.S. dams. From the U.S. Army Corps of Engineers, National Inventory of Dams.

5. POTENTIAL OF HYDROPOWER RESOURCES

The potential of hydropower resources is generally evaluated in two steps. The first looks at the resources that are technically feasible. The site physical characteristics are evaluated and generally include site access, water flow data, potential development schemes, potential head, potential installed capacity and energy production, and special site issues. Technically feasible projects require a water resource that can be developed using known engineering and construction techniques. The second step looks at other conditions that apply to developing the resource, including economic factors, environmental issues, legal issues, and institutional concerns. In most cases, economic screening is the second step in the evaluation. The economic factors include investment costs (total cost of the project), interest rate (cost of borrowing money), debt service life (duration to

TABLE II
History of Hydroelectricity's Progression in the United States

- July 1880: Michigan's Grand Rapids Electric Light and Power Company generates electricity by a dynamo belted to a water turbine at the Wolverine Chair Factory; it lights 16 brush-arc lamps.
- 1881: City of Niagara Falls street lamps powered by hydropower.
- 1886: Approximately 45 water-powered electric plants exist in the United States and Canada.
- 1887: San Bernardino, California, is first hydroelectric plant in the western states.
- 1889: Approximately 200 electric plants in the United States use waterpower for some or all generation of electricity.
- 1901: First federal Water Power Act is passed.
- 1907: Approximately 15% of electric generating capacity in the United States is provided by hydropower.
- 1920: Approximately 25% of U.S. electrical generation is hydropower.
- 1920: Federal Power Act establishes Federal Power Commission authority to issue licenses for hydro development on public lands.
- 1935: Federal Power Commission's authority is extended to all hydroelectric projects built by utilities engaged in interstate commerce.
- 1938: Bonneville Dam is first federal dam on the Columbia River.
- 1940: Approximately 40% of electrical generation is hydropower.
- Conventional capacity in the United States triples between 1921 and 1940 and nearly triples again between 1940 and 1980.
- Currently, approximately 7% of U.S. electricity comes from hydropower; approximately 80,000 MW is conventional capacity and 14,000 MW is pumped storage.

TABLE III

Worldwide Hydropower Potential

Geographical region	Technically feasible potential (∼GWh/year)	Economically feasible potential (∼GWh/year)
Africa	1,750,000	1,100,000
Countries with major resources:		
Angola	90,000	65,000
Cameroon	115,000	103,000
Congo, Democratic Republic of	774,000	419,000
Ethiopia		260,000
Madagascar	180,000	49,000
Asia (including Russia and Turkey)	6,800,000	3,600,000
Countries with major resources:		
China, People's Republic of	2,200,000	1,270,000
India	660,000	
Japan	135,000	114,000
Russian Federation	1,670,000	852,000
Turkey	215,000	123,000
Vietnam	100,000	80,000
Australia/Oceania	270,000	107,000
Countries with major resources:		
Australia		30,000
New Zealand	77,000	40,000
Europe	1,035,000	791,000
Countries with major resources:		
Austria	54,000	50,000
France	72,000	70,000
Iceland	64,000	44,000
Italy	69,000	54,000
Norway	200,000	187,000
Spain	70,000	41,000
Sweden	200,000	130,000
North and Central America	1,663,000	1,000,000
Countries with major resources:		
Canada	981,000	536,000
Mexico	49,000	32,000
United States	528,000	376,000
South America	2,700,000	1,600,000
Countries with major resources:		
Argentina	130,000	
Bolivia	126,000	50,000
Brazil	1,300,000	764,000
Chile	162,000	
Colombia	200,000	140,000
Ecuador	133,000	106,000
Venezuela	261,000	100,000
World Totals	14,218,000	8,198,000

Source. Aqua∼Media International Westmead House. (2002). "The International Journal on Hydropower and Dams: World Atlas and Industry Guide" (pp. 13–15). Aqua∼Media International Westemead House, Sutton, UK.

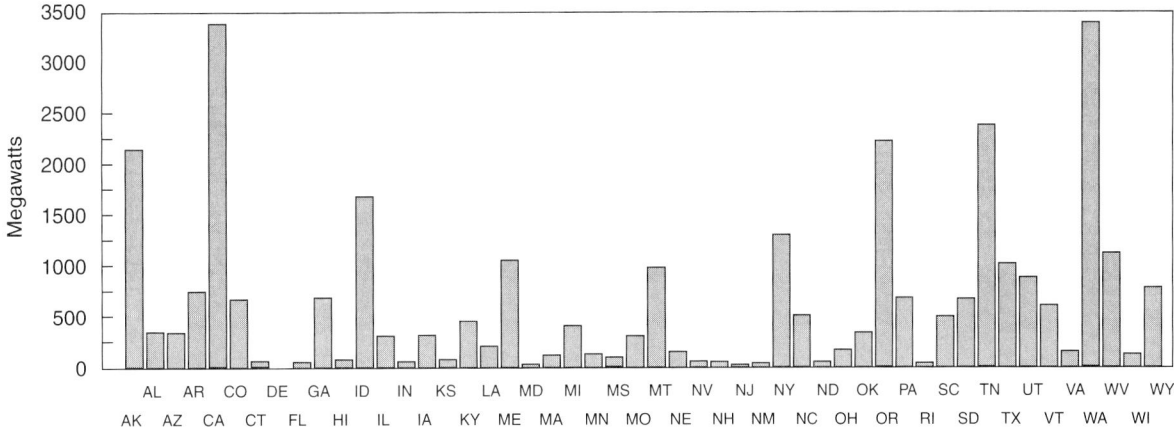

FIGURE 7 Undeveloped hydropower potential by state. From the Hydroelectric Power Resources Assessment database (FERC) and Hydropower Evaluation Software (INEEL). DOE has modeled the undeveloped conventional hydropower potential in the United States. This does not include developed capacity. Various state agencies have reviewed the modeled results and provided input. The 50-state undeveloped conventional hydropower potential is approximately 30,000 MW. The model includes environmental, legal, and institutional constraints to development.

repay the investment with interest), energy value (value or income from the energy produced), operations and maintenance cost (annual cost to operate and maintain the plant), and escalation rates (future value of energy and future operation/maintenance costs). These engineering economic factors can be used to determine the cash flow and benefit/cost ratio. If the resource appears feasible after this step, additional screening for other conditions can be applied to further evaluate the resource. Conditions are unique to each project, and evaluation requires site-specific information. The additional conditions may also affect economics, so a new economic evaluation might be appropriate after the other conditions have been determined.

In some cases, the other screening issues can be applied after the technically feasible resources have been identified. For example, if the environmental issues prevent a resource from being developed, no further evaluation is needed.

Table III shows the additional hydropower potential worldwide by major geographical area, based on technically and economically feasible resources.

The Federal Energy Regulatory Commission (FERC) has provided estimates of the undeveloped hydropower resources in the United States in its Hydropower Resource Assessment (HPRA) database. Other agencies, such as the Army Corps of Engineers, the Bureau of Reclamation, and the Power Marketing Administration (Alaska Power Administration, Bonneville Power Administration, West Area Power Administration, Southwestern Power Administration, and Southeastern Power Administration), have conducted regional HPRA studies.

In June 1989, the U.S. Department of Energy (DOE) initiated development of a National Energy Strategy to identify the energy resources available to support the expanding demand for energy in the United States. Public hearings conducted as part of the strategy development process indicate that the undeveloped hydropower resources were not well defined. One of the issues was that some assessments did not estimate the undeveloped hydropower capacity based on site characteristics, stream flow data, and available hydraulic heads.

Therefore, the DOE developed uniform criteria to perform HPRA studies. They used the FERC HPRA database as the basis for reevaluations. Information from other databases was factored into the assessments. The information was also reviewed by various state agencies. After the review process, the specific site information was computer modeled to further screen development potential base on environmental, legal, and institutional constraints.

The DOE HPRA studies involved 5677 sites and initially determined a capacity of approximately 70,000 MW. This computer modeling of these undeveloped hydropower resources based on environmental, legal, and institutional constraints resulted in an estimated total undeveloped capacity of approximately 30,000 MW. Figure 7 illustrates the undeveloped hydropower potential by state.

The DOE continues to conduct assessments of hydropower resources in the United States.

SEE ALSO THE FOLLOWING ARTICLES

Biomass Resource Assessment • Hydropower Economics • Hydropower, Environmental Impact of • Hydropower, History and Technology of • Hydropower Resettlement Projects, Socioeconomic Impacts of • Hydropower Technology • Oil and Natural Gas Liquids: Global Magnitude and Distribution • Wind Resource Base

Further Reading

Aqua~Media International Westmead House. (2002). "The International Journal on Hydropower and Dams: 2002 World Atlas and Industry Guide." Aqua~Media International Westmead House, Sutton, UK.

Conner, A.M., et al., (1998). "U.S. Hydropower Resource Assessment Final Report." (DOE/ID–10430.2). U.S. Department of Energy, Idaho Operations Office, Idaho Falls, ID.

Gulliver, J. S., and Arndt, R. E. A. (eds.). (1991). "Hydropower Engineering Handbook." McGraw–Hill, New York.

Hall, D. G., et al. (2002). "Low Head/Low Power Hydropower Resource Assessment of the Arkansas White Red Hydrologic Region." U.S. Department of Energy, Idaho Operations Office, Idaho Falls, ID.

Hall, D. G., et al. (2002). "Low Head/Low Power Hydropower Resource Assessment of the Pacific Northwest Hydrologic Region." U.S. Department of Energy, Idaho Operations Office, Idaho Falls, ID.

Hydrologic Engineering Center and Institute for Water Resources. (1979). "Feasibility Studies for Small Scale Hydropower Additions: A Guide Manual." U.S. Army Corps of Engineers, Hydrologic Engineering Center, Davis, CA.

U.S. Bureau of Reclamation, Power Resources Office. (2001). Hydroelectric power. www.usbr.gov/power/index.html.

U.S. Department of Energy. (1997). "Hydropower: America's Leading Renewable Energy Resource" [brochure]. U.S. Department of Energy, Idaho Operations Office, Idaho Falls, ID.

U.S. Department of Energy. (2001). "Hydropower: Partnership with the Environment" [brochure]. U.S. Department of Energy, Idaho Operations Office, Idaho Falls, ID.

Hydropower Technology

PEGGY BROOKSHIER
U.S. Department of Energy
Idaho Falls, Idaho, United States

1. Brief History
2. Hydrologic Cycle
3. Hydropower Plant Components and Functions
4. General Description of the Types of Hydraulic Turbines
5. Past Research and Development Efforts
6. Current R&D
7. Future of R&D

Glossary

alternating current (AC) Electric current that reverses direction many times per second.
ancillary services Operations provided by hydroelectric plants that ensure stable electricity delivery and optimize transmission system efficiency.
cavitation Noise or vibration causing damage to the turbine blades as a result of bubbles that form in the water as it goes through the turbine, causing a capacity loss, head loss, and efficiency loss; the bubbles collapse when they pass into higher regions of pressure.
direct current (DC) Electric current that flows in one direction.
draft tube A water conduit that can be straight or curved, depending on the turbine installation, and that maintains a column of water from the turbine outlet and the downstream water level.
efficiency A percentage obtained by dividing the actual power or energy by the theoretical power or energy; it represents how well the hydropower plant converts the energy of the water into electrical energy.
flow Volume of water, expressed as cubic feet or cubic meters per second, passing a point in a given amount of time.
head Vertical change in elevation, expressed in either feet or meters, between the headwater level and the tailwater level.
headwater The water level above the powerhouse.
low head Head of 66 feet or less (as defined by the U.S. Department of Energy).
penstock A closed conduit or pipe for conducting water to the powerhouse.
runner The rotating part of the turbine that converts the energy of falling water into mechanical energy.
scroll case A spiral-shaped steel intake guiding the flow into the wicket gates located just prior to the turbine.
small hydro Projects that produce 30 MW or less (as defined by the U.S. Department of Energy).
tailrace The channel that carries water away from a dam.
tailwater The water downstream of the powerhouse.
ultra low head Head of 10 feet or less (as defined by the U.S. Department of Energy).
wicket gates Adjustable elements that control the flow of water to the turbine passage.

Hydropower, also known as hydroelectric power, is the use of water to produce power. Harnessing water to perform work has been going on for thousands of years. The Greeks used waterwheels for grinding wheat into flour more than 2000 years ago. Besides grinding flour, the power of the water was used to saw wood and to power textile mills and manufacturing plants. This article looks briefly at how hydropower began, why it is considered renewable, the parts of a hydropower plant, types of turbines and when they are used, and what research and development is occurring.

1. BRIEF HISTORY

The evolution of the modern turbine began during the mid-1700s when a French hydraulic and military engineer, Bernard Forest de Bélidor, wrote *Architecture Hydraulique*. In this four-volume work, he described using a vertical axis versus a horizontal axis machine. During the 1700s and 1800s, water turbine development continued. In 1880, a brush arc light dynamo driven by a water turbine was used to provide theater and storefront lighting in Grand Rapids, Michigan, and in 1881, a brush dynamo connected to a turbine in a flour mill provided street lighting in Niagara Falls, New York. These two used direct current (DC) technology. Alternating current (AC) is used today. That breakthrough came when the electric generator was coupled to the turbine, resulting in the first hydroelectric plant in the United States—and in the world—located in Appleton, Wisconsin.

2. HYDROLOGIC CYCLE

The hydrologic cycle, shown in Fig. 1, depicts the continuous water cycle from evaporation, to precipitation, and back to the water source. This is why hydropower is considered a renewable energy.

A rough rule of thumb for determining potential power is as follows:

$$P(kW) = \underbrace{[eH(ft)Q(ft^3/s)]/11.82}_{\text{English}}$$
$$= \underbrace{[eH(m)Q(m^3/s)] \times 9.81}_{\text{Metric}},$$

where P = generator output (power) in kilowatts (kW), e = overall plant efficiency, H = head (foot [ft] or meter [m] depending on which equation is used), and Q = flow (cubic feet or cubic meters per second depending on which equation is used. This equation assumes a constant water density of $62.4 \, lb/ft^3$ or $9806 \, N/m^3$.

3. HYDROPOWER PLANT COMPONENTS AND FUNCTIONS

There are many reasons why dams are built. Figure 2 shows the primary purpose of dams in the United States. Many dams were built for other purposes, and hydropower was added later.

Worldwide, there are approximately 840,000 dams. However, only a small percentage of them produce power. For example, the United States has approximately 80,000 dams, but only 2400 produce power. The rest are for recreation, stock/farm ponds, flood control, water supply, and/or irrigation.

In general, the purpose of dams is to control water by impounding it, raising its level, or diverting it. Some dams are built to contain sediments that are carried in the water (e.g., mining wastes).

There are three major categories of dams:

- *Embankment:* constructed from the natural materials soil and rock (two types of embankment dams: earthfill and rockfill)
- *Concrete:* constructed from concrete (four types of concrete dams: arch, gravity, combination arch dam with gravity tangents, and buttress)
- *Composite:* generally consisting of concrete gravity or buttress sections in combination with earthfill or rockfill sections

Determining which category and type of dam to build is dependent on many factors, including topography, geological conditions, availability of materials, spillway size and location, and environmental considerations.

There are three types of hydropower facilities:

- *Impoundment.* This type requires a dam to store water. Water may be released either to meet changing electricity needs or to maintain a constant reservoir level.
- *Diversion.* This type channels a portion of the river through a canal or penstock, which may or may not require a dam.
- *Pumped storage.* This type pumps water, when demand for electricity is low, from a lower reservoir to an upper reservoir. During periods of high demand, the water is released back to the lower reservoir to generate electricity.

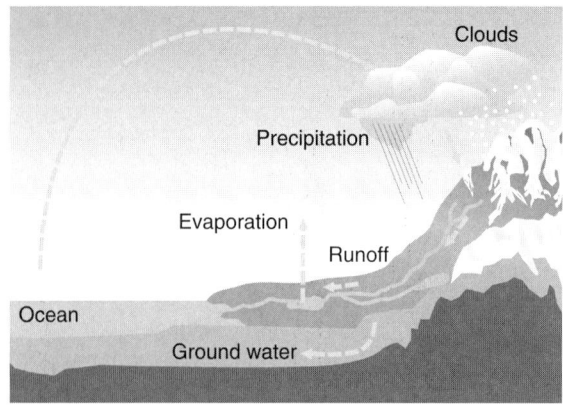

FIGURE 1 Hydrologic cycle.

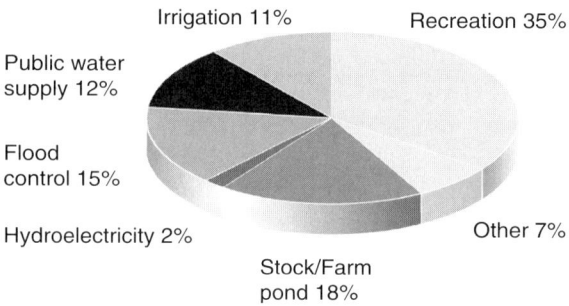

FIGURE 2 Primary purposes or benefits of U.S. dams. From the U.S. Army Corps of Engineers, National Inventory of Dams.

Figure 3 illustrates the components of a typical hydropower facility that requires a dam. Figure 4 illustrates a facility that does not use a dam. Figure 5 depicts a pump storage facility. Figure 6 depicts a typical reversible pump.

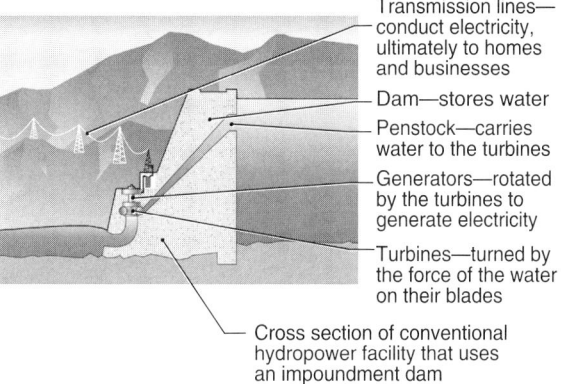

FIGURE 3 Components of a typical hydropower facility.

FIGURE 4 Tazimina Hydropower Project—diversion.

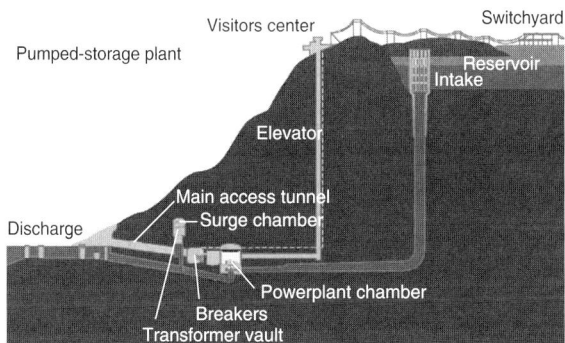

FIGURE 5 Cutaway diagram of a typical pump storage facility. From the Tennessee Valley Authority.

4. GENERAL DESCRIPTION OF THE TYPES OF HYDRAULIC TURBINES

Hydraulic turbines have two main classifications: impulse and reaction. The impulse turbine generally uses the velocity of the water to move the runner and discharges to atmospheric pressure. The water stream hits each bucket on the runner. There is no suction on the down side of the turbine, and the water flows out the bottom of the turbine housing after hitting the runner. An impulse turbine is generally suitable for high-head, low-flow applications.

A reaction turbine develops power from the combined action of pressure and moving water. The runner is placed directly in the water stream flowing over the blades rather than striking each one individually. Reaction turbines are generally used for lower head, higher flow applications in comparison with impulse turbines. However, as shown in Fig. 7, there is a lot of overlap.

4.1 Impulse Turbine

4.1.1 Pelton
A pelton wheel has one or more free jets discharging water into an aerated space and impinging on the buckets of a runner. Draft tubes are not required for an impulse turbine because the runner must be located above the maximum tailwater to permit operation at atmospheric pressure. Figure 8 shows a cutaway of a Pelton.

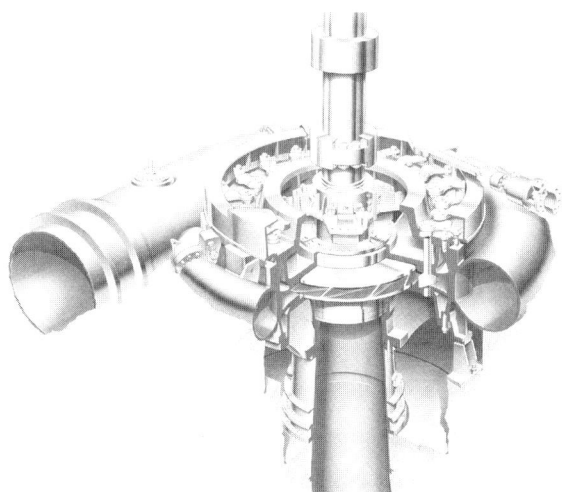

FIGURE 6 Cutaway diagram of a reversible pump used at pump storage facilities. It pumps to an upper reservoir and acts as a turbine when water flows to the lower reservoir. From GE Hydro.

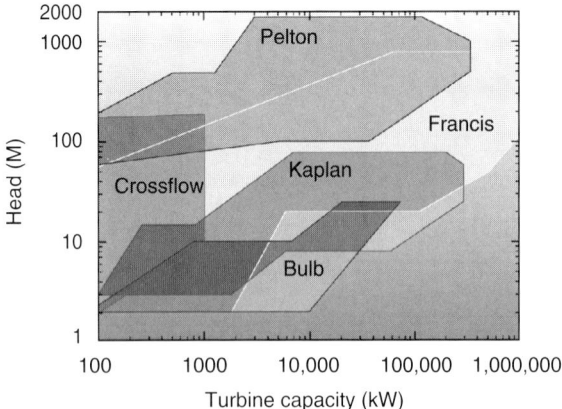

FIGURE 7 Head and flow ranges for various types of turbines.

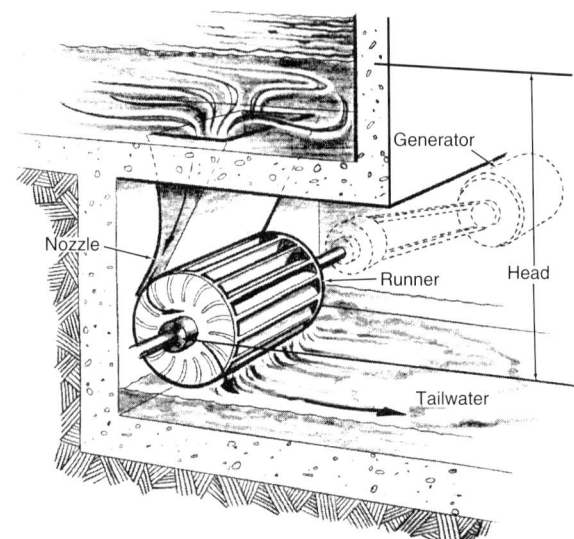

FIGURE 9 Cutaway of a typical cross-flow turbine.

to accommodate larger water flows and lower heads than is the case with the Pelton. Figure 9 shows a cutaway of a cross-flow turbine.

4.2 Reaction Turbine

4.2.1 Propeller

A propeller turbine generally has a runner with three to six blades in which the water contacts all of the blades constantly. One may picture a boat propeller running in a pipe. Through the pipe, the pressure is constant; if it were not constant, the runner would be out of balance. The pitch of the blades may be fixed or adjustable. The major components, besides the runner, are a scroll case, wicket gates, and a draft tube. There are several different types of propeller turbines:

- *Bulb turbine.* The turbine and generator are a sealed unit placed directly in the water stream. Figure 10 shows a cutaway of a bulb turbine.
- *Straflo.* The generator is attached directly to the perimeter of the turbine.
- *Tube turbine.* The penstock bends just before or after the runner, allowing a straight line connection to the generator.
- *Kaplan.* Both the blades and the wicket gates are adjustable, allowing for a wider range of operation. Figure 11 shows a cutaway of a Kaplan turbine.

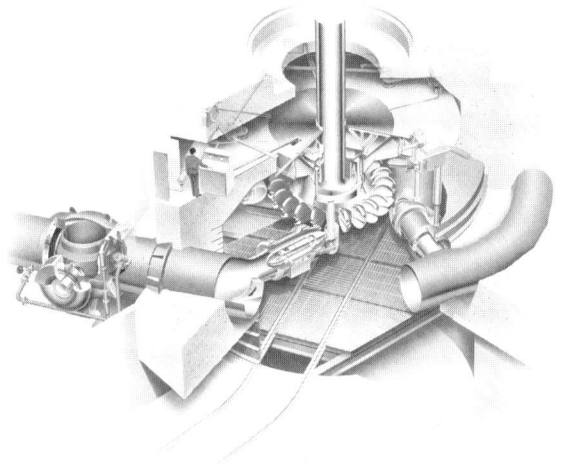

FIGURE 8 Cutaway of a typical Pelton turbine. From GE Hydro.

A turgo wheel is a variation on the Pelton and is made exclusively by Gilkes in England. The Turgo runner is a cast wheel whose shape generally resembles a fan blade that is closed on the outer edges. The water stream is applied on one side, goes across the blades, and exits on the other side.

4.1.2 Cross-Flow

A cross-flow turbine is drum shaped and uses an elongated, rectangular section nozzle directed against curved vanes on a cylindrically shaped runner. It resembles a "squirrel cage" blower. The cross-flow turbine allows the water to flow through the blades twice. The first pass is when the water flows from the outside of the blades to the inside, and the second pass is from the inside back out. A guide vane at the entrance to the turbine directs the flow to a limited portion of the runner. The cross-flow was developed

4.2.2 Francis

A Francis turbine has a runner with a number of fixed buckets (vanes)—usually nine or more. Water is introduced just above the runner and all around it

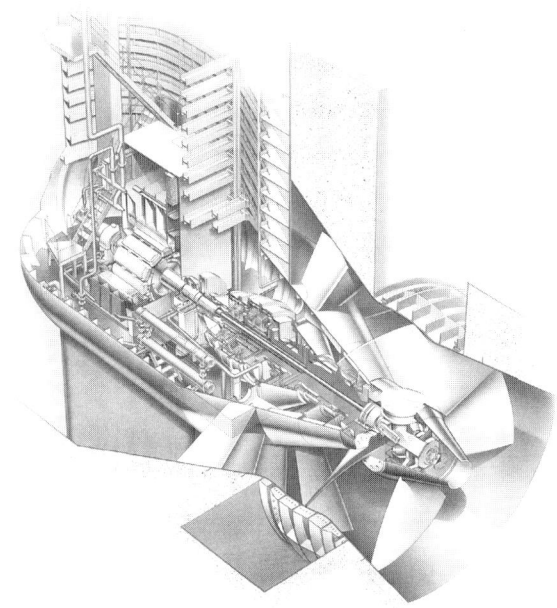

FIGURE 10 Cutaway of a typical bulb turbine. From GE Hydro.

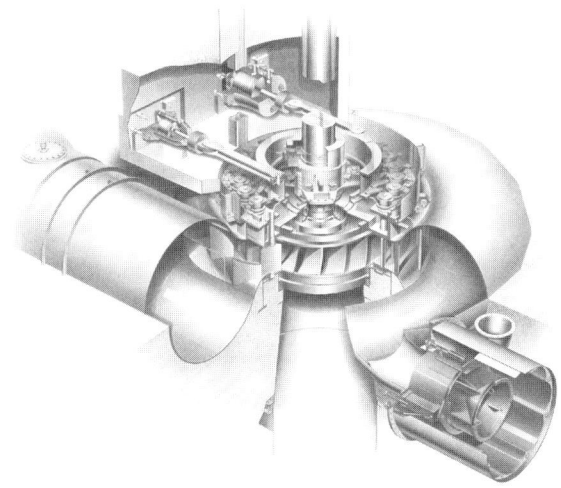

FIGURE 12 Cutaway of a typical Francis turbine. From GE Hydro.

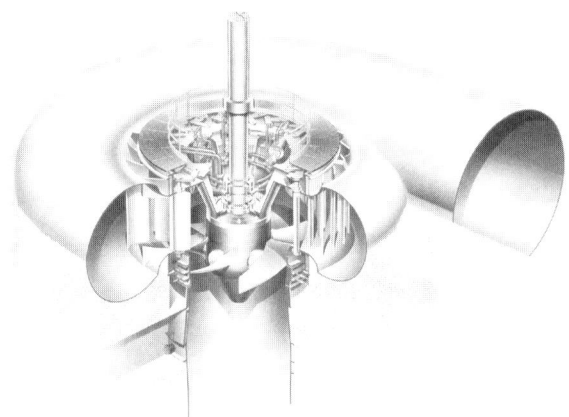

FIGURE 11 Cutaway of a typical Kaplan turbine. From GE Hydro.

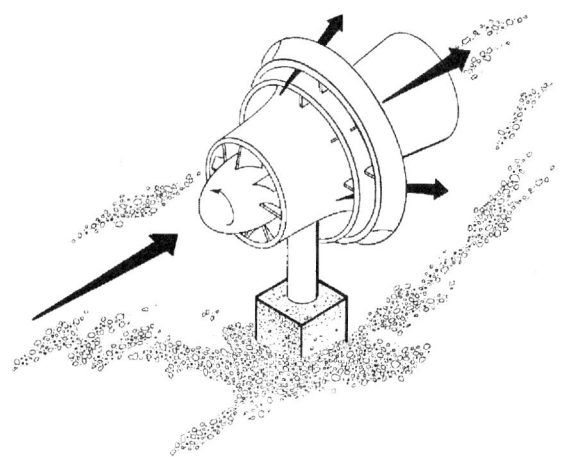

FIGURE 13 Ducted free-flow turbine.

and then falls through, causing it to spin. Besides the runner, the other major components are the scroll case, wicket gates, and draft tube. Figure 12 shows a cutaway of a Francis turbine.

4.2.3 Kinetic

Kinetic energy turbines, also called free-flow turbines, generate electricity from the kinetic energy present in flowing water rather than from the potential energy from the head. The systems may operate in rivers, manmade channels, tidal waters, or ocean currents. Kinetic systems use the water stream's natural pathway. They do not require the diversion of water through manmade channels, riverbeds, or pipes, although they might have applications in such conduits. Kinetic systems do not require large civil works; however, they can use existing structures such as bridges, tailraces, and channels. Figure 13 shows a ducted free-flow turbine.

Turbine selection is first based on head and flow. As can be seen in Fig. 7, many different types of turbines overlap. Other deciding factors include how deep the turbine must be set, efficiency, and cost.

5. PAST RESEARCH AND DEVELOPMENT EFFORTS

Up to around 1980, research and development (R&D) efforts focused mainly on improving turbine efficiency, reducing cavitations, and increasing generation. Whereas older units had efficiency ratings as low as 60%, the new units have efficiency ratings of approximately 90%. The U.S. Department of Energy (DOE) also looked at new technologies for developing low-head, ultra low-head, and small hydro projects. These technologies included the following:

- *Pumps used as turbines.* This technology uses either actual off-the-shelf pumps or pumps slightly modified by the pump manufacturer as a turbine. Water is now flowing down through the pump rather than up. In some cases, the efficiency is as high as 85%.
- *Float-in powerhouses.* This concept is where the powerhouse is constructed in modules at a shipyard and then floated, acting as a barge, to an existing dam, where it would be sunk into place and the voids would be filled with concrete. This is very similar in concept to modular homes.
- *Marine thruster.* This technology uses thrusters, normally used to maneuver ships while docking, to generate power. Initial tests showed an efficiency of approximately 60%.
- *Free-flow turbines.* This technology uses the flow of the water only to generate power. Therefore, no dam or reservoir is needed. This type of turbine can be anchored to the bottom of a river or hung from bridges. This concept has now been applied to ocean currents.
- *Hydraulic air compressor.* This technology uses entrainment of air bubbles in a penstock full of water, which is allowed to flow from the headwater to a deep chamber, where the then compressed air is collected for use in a gas turbine. The method of using the compressed air determines the overall plant efficiency, which is approximately 85%. The extraction of the work from the compressed air by direct expansion through a gas turbine will produce shaft power that may be used to generate electricity. In addition, the exhaust can be very cold and may be used for cooling.
- *Inexpensive cross-flow turbines.* This concept looked at designing and building an inexpensive cross-flow turbine by using cheaper materials and building in a conventional machine shop.
- *Siphon penstock.* This technology is a penstock that goes over a dam rather than through the dam. Thus, the integrity of the dam is not jeopardized, and the cost of installation is less than what it would be going through the dam. This would be useful at an existing dam with no hydropower. However, it has a lift limitation of 32 feet at sea level.

6. CURRENT R&D

Although hydropower turbine manufacturers have incrementally improved turbine technology to improve efficiencies, the basic design concepts have not changed for decades. These late 19th- and early 20th-century designs did not consider environmental effects because little was known about environmental effects of hydropower at the time.

During the 1980s, the environmental concerns in the United States became more important in hydropower projects, both existing and planned. This trend has been slowly spreading across the globe. In 1996, DOE looked at the hydropower R&D being conducted by federal agencies as well as the private sector. Approximately $63.5 million was spent on hydro R&D, with approximately 84% of that going toward fish passage, behavior, and response-type R&D.

Hydropower has many competitors for its source of power, that is, water. Water is used for drinking, irrigation, recreation, fish and wildlife, navigation, power, cooling, and the like. As shown earlier in Fig. 2, dams also have multiple purposes. With so many competing uses for both the water and the dams, it is understandable why an existing hydropower project, let alone a planned project, would face many barriers and issues.

The hydropower industry recognizes that hydropower plants have an effect on the environment. The industry also recognizes that there is a great need to bring turbine designs into the 21st century. The industry visualizes innovative turbines designed from a new perspective. This perspective would look at the "turbine system" (which could include everything except the dam and powerhouse) intended to balance environmental, technical, and economic considerations.

6.1 Environmental Challenges

Although hydroelectric power plants have many advantages over other energy sources, the potential environmental impacts are also well known. Most of the adverse impacts of dams are caused by habitat alterations. Reservoirs associated with large dams can inundate large amounts of terrestrial and river habitats and displace human populations. Diverting water out of the stream channel (or curtailing reservoir releases so as to store water for future electrical generation) can desiccate riparian (streamside) vegetation. Insufficient stream flow releases degrade habitat for fish and other aquatic organisms in the affected river reach below the dam. Water in the reservoir is stagnant compared with that in a free-flowing river; consequently, waterborne sediments and nutrients can be trapped, resulting in the undesirable proliferation of algae and aquatic weeds. In some cases, water spilled from high dams may become supersaturated with nitrogen gas and cause gas bubble disease in aquatic organisms inhabiting the tailwaters.

A hydropower project can also affect aquatic organisms directly. The dam can block upstream movements of migratory fish (e.g., salmon, steelhead, American shad, sturgeon, paddlefish, eels). Downstream-moving fish may be drawn into the power plant intake flow and pass through the turbine. These fish are exposed to physical stresses (e.g., pressure changes, shear, turbulence, strike) that may cause disorientation, physiological stress, injury, and/or mortality.

R&D is currently under way to help fishery biologists and turbine designers better understand what is happening in the turbine passage. Biological tests are being conducted that will quantify the physical stresses that cause injury or mortality to fish. In addition to these tests, tools are being developed to help both engineers and biologists. These tools include developing a "sensor fish," which is similar to a crash dummy fish. It will be able to measure the physical stresses in a turbine passage and be used instead of live fish to gather information. Another tool is the development of a computational fluid dynamics program that models potential fish behavior in the turbine passage. The test results and tools will help turbine manufacturers to design a more environmentally friendly turbine that will reduce the physical stresses to which fish are exposed. New products such as greaseless bearings eliminate the possibility of petroleum products being released into the water.

6.2 Turbine

During the mid-1990s, DOE began the Advanced Hydropower Turbine System (AHTS) program as part of the Hydropower Program. The goal of the AHTS program is to develop environmentally friendly turbines. DOE funded the conceptual designs of four turbine types: three by Voith–Siemens and one by the team of Alden Research Laboratory (ARL) and Northern Research and Engineering Corporation (NREC). Voith–Siemens looked at redesigning a Kaplan and Francis turbine and developing a dissolved oxygen-enhancing turbine. ARL redesigned a pump impeller, used in the food processing industry to pump tomatoes and fish, to be used as a turbine. Figure 14 shows what the ARL–NREC design would look like.

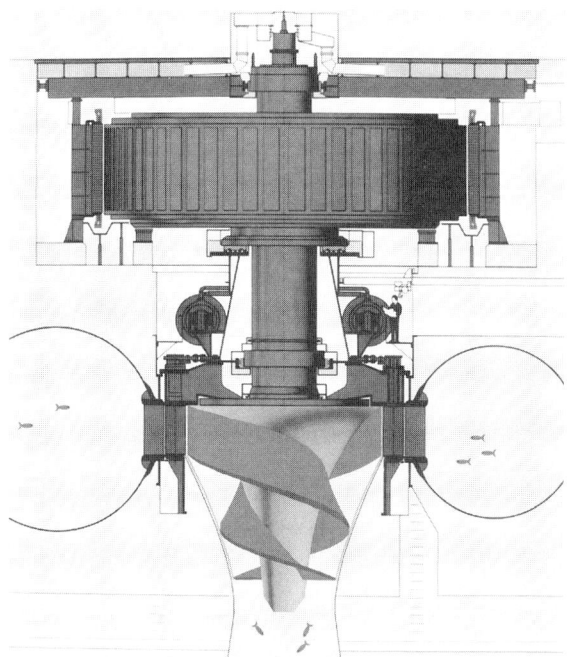

FIGURE 14 Cutaway of an ARL/NREC fishfriendly turbine. From the Alden Research Laboratory.

The ARL–NREC design proceeded into the proof-of-concept stage, where the biological and engineering tests were finalized. Part of the proof of concept was to verify the biological design assumptions and the issue of whether the results of biological testing of a smaller model can be scaled upward to a full-sized turbine.

Many turbine manufacturers have begun designing environmentally friendly turbines, based on a potential market not only in the United States but also worldwide. In 2002, DOE selected three projects where environmentally friendly turbines will be installed and tested. Two projects involve improving downstream fish passage, and one involves increasing dissolved oxygen in discharged water.

6.3 Other

Other federal agencies and utilities are conducting research looking at how hydro plants can be operated to improve fish habitat.

7. FUTURE OF R&D

In 1992, and again in 2001, HCI Publications conducted the Hydro R&D Forum. The purpose of the forum was to develop an agenda for hydro industry R&D activities worthy of pursuit during the years ahead. The forum brought together representatives from various parts of the hydro industry.

R&D needs were identified and prioritized by the following topic areas:

- Performance and production
- Environmental issues
- Innovative operations and maintenance (O&M)
- Multiple-purpose values
- Human resource issues

The top issues in each area were then prioritized. Following are the top 10 R&D needs identified in 2001:

1. *Quantify and communicate hydro's value compared with the power mix.* Needs include developing objective criteria for measuring and valuing hydropower's nonpower attributes. Criteria should be based on internationally recognized standards, permit comparisons across generation sources, recognize emissions as well as effects on aquatic and land resources, and take into account the entire life cycles of facilities associated with electricity production.

2. *Continue AHTS program development and deployment.* Achieving progress in the AHTS program, which aims to reduce mortality of fish passing through turbines, is a goal. Recommended new areas of emphasis include improving water quality without compromising energy generation efficiency.

3. *Develop business model for O&M decision support based on condition monitoring.* Condition monitoring equipment and systems are increasingly being used. However, the availability of new information poses the conundrum of how to use this information optimally for making both major and minor business decisions in support of facilities operation and maintenance.

4. *Improve methods to quantify costs and benefits from ancillary services.* Hydro plant owners and operators recognize that the ancillary services often provided by their facilities are valuable. However, market structures often fail to appropriately recognize these values. As a consequence, returns to owners and projects are often less than they would be if these values were acknowledged appropriately.

5. *Improve hydro education and outreach.* There is a need to ensure that constituencies are suitably informed about hydro issues. In this area, there are needs to identify effective educational methods and content, research the best ways in which to transfer knowledge, update and maintain factual content and methods, and develop communication networks to inform policymakers, nongovernmental organizations, and the public.

6/7 (tie). *Support green power acceptance for hydro.* In some regions, electricity is marketed as "green" (e.g., from sources that are environmentally beneficial compared with the alternatives). Projects in this area include determining how to gain acceptance for hydro within green marketing programs.

6/7 (tie). *Improve hydropower competency management program.* Workforce issues are increasing in priority, especially in light of loss and projected loss of personnel due to retirements. Projects in this area include focusing on inventorying the skills and competencies that are needed for hydro facilities operation and developing techniques and strategies for addressing identified needs.

8. *Improve hydro-related R&D technology transfer.* Much information is available that is, or may be, applicable to problems and areas of need. However, better tools and processes are needed for communicating and sharing information, both from research activities and from operational experience.

9. *Determine operating life effects due to more severe load operations.* During recent years, many hydro projects have begun to operate in ways that were not envisioned during their original designs. Work is needed to determine how equipment is affected as well as to identify mitigative actions and tools for predicting and preventing failures.

10. *Establish protocols for measuring mitigation effectiveness.* During recent years, hydro project owners have engaged in many instances of implementing environmental mitigation measures. However, it is generally not known whether the measures have been effective. That is, did they accomplish the desired results? Knowledge gained from research in this area could help in designing future mitigative measures and in avoiding ineffective mitigation.

SEE ALSO THE FOLLOWING ARTICLES

Hydropower Economics • *Hydropower, Environmental Impact of* • *Hydropower, History and Technology of* • *Hydropower Resettlement Projects, Socioeconomic Impacts of* • *Hydropower Resources*

Further Reading

Alward, R., *et al.* (1979). "Micro-Hydro Power: Reviewing an Old Concept." National Center for Appropriate Technologies, Butte, MT.

Armstrong, E. (1977). Selection of the type of dam. *In* "Handbook of Dam Engineering" (A. Golze, Ed.). Litton Educational Publishing, New York.

Chappell, J. (1984a). "Hydropower Hardware Descriptions." EG&G Idaho, Idaho Falls, ID.

Chappell, J. (1984b). "New Technology for Small Hydropower Installations." EG&G Idaho, Idaho Falls, ID.

Gulliver, J. and Arndt, R. (1991) *Hydropower Engineering Handbook*, McGraw Hill, Inc.

Idaho National Engineering Laboratory. (1997). "Hydropower Research and Development." Idaho National Engineering Laboratory, Idaho Falls, ID.

McKinney, J., *et al.* (1983). "Microhydropower Handbook." EG&G Idaho, Idaho Falls, ID.

Indoor Air Quality in Developing Nations

MAJID EZZATI
Harvard School of Public Health
Boston, Massachusetts, United States

1. Household Fuel Use Patterns in Developing Countries
2. Solid Fuel Smoke and Indoor Air Pollution
3. Human Exposure to Indoor Smoke from Solid Fuels
4. Health Effects of Exposure to Indoor Smoke
5. Health Benefits of Household Energy Transitions
6. Knowledge Gaps and Research Needs for Reducing Indoor Air Pollution
7. Emerging Indoor Air Quality Issues in Developing Countries

Glossary

acute lower respiratory infections (ALRI) Infections of the lower part of the respiratory tract. Although the division between lower and upper parts of the respiratory tract sometimes takes place at the level of the larynx, many infections affect multiple parts of the respiratory tract, especially where the affected areas are smaller. Further, infections of the bronchi (brochitis) and of the lungs (pneumonia) are often considerably more severe than those in other parts of the respiratory tract and these infections have more specific symptoms. For these reasons, many public health and medical protocols use ALRI to refer to bronchitis, pneumonia, and broncho-pneumonia and combine the infections of other parts of the respiratory tract in the category of acute upper respiratory infections (AURI).
acute respiratory infections (ARI) Infections of the respiratory tract, which consists of the lungs, the bronchi, the trachea, the larynx, the pharynx, the tonsillar glands, the eustachian tube, the nasal cavities, and the sinuses. Otitis media (middle ear infection) is also sometimes included in the definition of ARI.
PM_{10} Particulate matter less than 10 μm in aerodynamic diameter.

Indoor air quality (IAQ) research deals with the presence, levels, health effects, and control of physical, chemical, and biological factors in indoor environments, including homes, workplaces, and vehicles. IAQ research in industrialized countries has examined hundreds of specific factors (e.g., temperature, various chemicals, and mold), sources of pollution (e.g., environmental tobacco smoke, occupational factors, consumer cleaning products, and moisture), and control technologies (e.g., ventilation). In developing countries, however, the great majority of IAQ research, which is considerably less extensive compared to research in industrialized nations, has focused on the role of household energy use and indoor smoke from cooking and heating because this source of indoor air pollution has the largest public health effects.

1. HOUSEHOLD FUEL USE PATTERNS IN DEVELOPING COUNTRIES

Globally, almost 3 billion people rely on biomass (wood, charcoal, crop residues, and dung) and coal as their primary source of domestic energy. Biomass accounts for more than one-half of household energy in many developing countries and for as much as 95% in some lower income regions (Fig. 1). There is also evidence that in some countries the declining trend of household dependence on biomass has slowed, or even reversed, especially among poorer households. Slow economic growth, especially in sub-Saharan Africa, lack of energy infrastructure in remote rural areas for delivery of alternative energy sources, and uncertainty about the price of alternative fuels are among the likely causes of the persistence of biomass fuels.

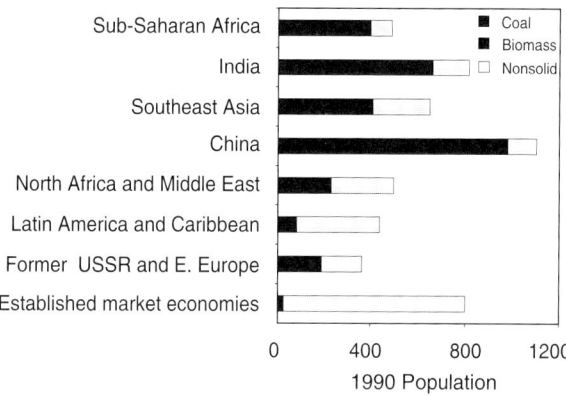

FIGURE 1 Sources of household energy by region. Data from the World Health Organization (1997).

FIGURE 2 The traditional "three-stone" open fire is used in many developing country households. Photograph by M. Ezzati.

2. SOLID FUEL SMOKE AND INDOOR AIR POLLUTION

Smoke from biomass and coal combustion contains a large number of pollutants with known health hazards, including particulate matter, carbon monoxide, nitrogen dioxide, sulfur oxides (mainly from coal), formaldehyde, and polycyclic organic compounds (e.g., carcinogens such as benzo[a]pyrene). The concentrations of each of these pollutants vary among the different forms of solid fuels, with animal dung and crop residues having some of the highest level emissions of particulate matter, one of the important indicator pollutants for health effects.

Pollution levels are especially high when burning is done using simple technologies such as open "three-stone" fires, which result in incomplete and inefficient combustion (Fig. 2). Monitoring of pollutants and personal exposures in biomass-burning households has shown concentrations many times higher than

FIGURE 3 Wood smoke rising from the roof of a house in Central Kenya on a typical day. Photograph by M. Ezzati.

those observed in industrialized countries (Fig. 3). The latest National Ambient Air Quality Standards of the U.S. Environmental Protection Agency, for instance, required that the daily average concentration of particles below 10 μm in diameter (PM_{10}) be less than 150 μg/m^3 (annual average below 50 μg/m^3). In contrast, typical 24-hour average concentrations of PM_{10} in homes using biofuels may range from 200 to 5000 μg/m^3 or more throughout the year, depending on the exact type of fuel, stove, and housing. Levels of carbon monoxide and other pollutants also often exceed international guidelines.

3. HUMAN EXPOSURE TO INDOOR SMOKE FROM SOLID FUELS

Exposure to air pollutants is very high in indoor environments in developing countries. Smith has estimated that at the aggregate level (i.e., without accounting for particle size, chemical composition, and source), approximately 80% of total global exposure to airborne particulate matter occurs indoors in developing nations. Details of exposure for various household members, and the roles of both pollution and behavior (e.g., location with respect to stove and activities), have been studied and evaluated using new tools and technology.

Biomass smoke is very episodic, with peaks occurring when fuel is added or moved, the stove is lit, cooking pots are placed on or removed from the fire, or food is stirred. These are also the times that those household members who cook are closest to the fire (Fig. 4). When these patterns of exposure, which incorporate both pollution and behavior, are accounted for, it can be seen that those household members who take part in cooking or are near the

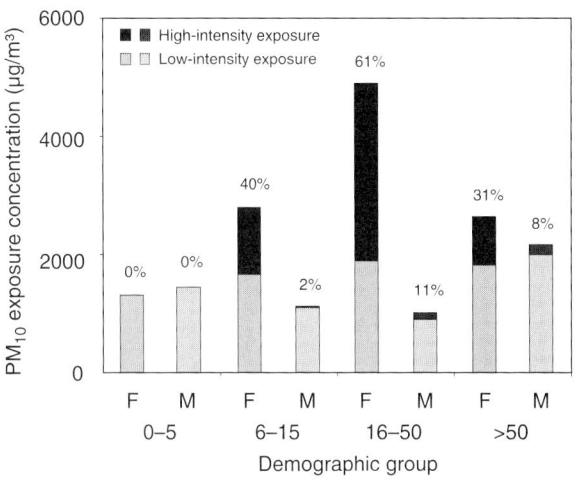

FIGURE 4 Household members involved in cooking are exposed to episodes of high levels of pollution when they work directly above the fire. Photograph by M. Edwards. From Ezzati and Kammen (2002), *Environmental Health Perspectives*.

FIGURE 5 Exposure to indoor air pollution from biomass fuels in rural households of central Kenya. For each demographic subgroup, the total height of the column is the group average exposure concentration divided into the average for high-intensity (darker shade) and low-intensity (lighter shade) components. The percentages indicate the share of total exposure from high-intensity exposure. The high-intensity component of exposure occurs in less than 1 hour, emphasizing the intensity of exposure in these episodes. From Ezzati, Saleh, and Kammen (2000), *Environmental Health Perspectives*.

cooking area (mostly women and children) are exposed to pollution levels that are tens of times higher than those in even the most polluted cities in the world (Fig. 5). Further, a substantial portion of this exposure occurs during short peak periods when emissions are highest and the cook is closest to the fire (referred to as high-intensity exposure in Fig. 5).

4. HEALTH EFFECTS OF EXPOSURE TO INDOOR SMOKE

Exposure to indoor air pollution from the combustion of solid fuels has been implicated, with varying degrees of evidence, as a causal agent of several diseases in developing countries. In a review of the epidemiological evidence for the health effects of indoor smoke from solid fuels, Bruce *et al.* concluded that, despite some methodological limitations, epidemiological studies provide compelling evidence of causality for acute respiratory infections (ARI) and chronic obstructive pulmonary disease (COPD). This is reinforced by experimental data, studies of pathogenesis, and indirect evidence from environmental tobacco smoke (ETS) and ambient air pollution studies. The relationship between coal smoke (but not biomass smoke) and lung cancer has also been consistently established in a number of epidemiological studies. A relationship between biomass smoke and lung cancer may also exist, because biomass smoke contains carcinogens, although at lower concentrations compared to coal smoke. Due to smaller risks and greater difficulties in exposure quantification, epidemiological studies on biomass smoke and lung cancer have so far been inconclusive. For other diseases, including asthma, upper aerodigestive tract cancer, interstitial lung disease, low birth weight and perinatal conditions, tuberculosis, and eye diseases, Bruce *et al.* classified the evidence as tentative.

Conservative estimates of global mortality due to indoor air pollution from solid fuels show that, in 2000, more than 1.6 million deaths were attributed to this risk factor. This accounts for approximately 4–5% of total mortality worldwide. Approximately 1 million of the deaths were due to childhood acute

TABLE I

Global Mortality Attributable to Indoor Smoke from Solid Fuels in 2000[a]

Country grouping	Number of deaths in children under 5 years	Number of deaths in adults[b]
High-mortality developing countries	808,000	232,000 (74%)
Lower mortality developing countries	89,000	468,000 (76%)
Demographically and economically developed countries	13,000	9,000 (79%)

[a] Data from World Health Organization (2002).
[b] Numbers in brackets show the fraction of adult deaths that are among women.

respiratory infections, with the remainder due to other causes, dominated by COPD and then lung cancer among adult women (Table I). The estimated health effects, currently limited to those diseases with strong epidemiological evidence, would of course be even larger if causality and quantitative relationships were established for other diseases. At the same time, even the diseases quantified so far show that the public health effects of indoor smoke in developing countries are enormous. As a result, coupled with the attention to details of exposure, health research has shifted to the "exposure–response" relationship for indoor air pollution and disease. Unlike early studies that divided the population into exposed and nonexposed, the exposure–response relationship, which quantifies the health effects along a continuum of exposure levels (or a larger number of categories), allows considering the benefits of partial exposure reduction for designing new interventions (Fig. 6).

5. HEALTH BENEFITS OF HOUSEHOLD ENERGY TRANSITIONS

Reducing exposure to indoor air pollution from household energy use can be achieved through interventions in of the following areas:

- Emissions source and energy technology (fuel–stove combination).
- Housing design and ventilation.
- Behavior and time–activity budget.

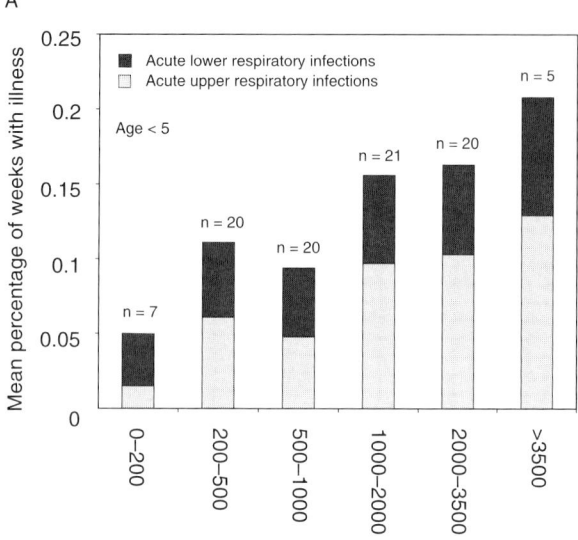

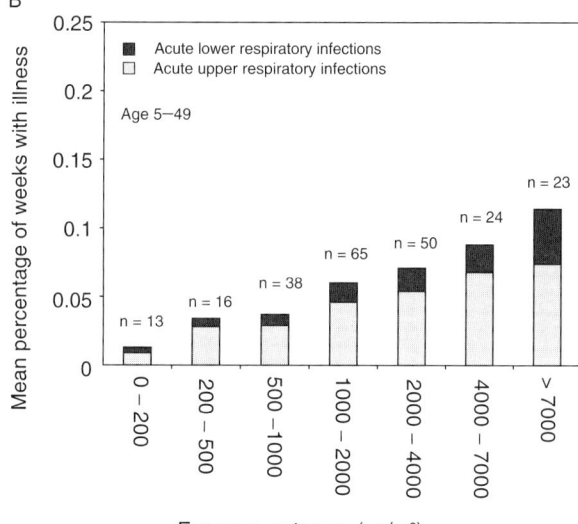

FIGURE 6 The relationship between exposure to indoor particulate matter and acute lower (more severe) and upper respiratory infections, for (A) children under the age of 5 years and for (B) young and middle-aged adults from a study in Central Kenya. As seen, disease increases with exposure level but the relationship is not linear; n is the number of people in each exposure group. Reproduced from Ezzati and Kammen (2001), with permission.

Interventions involving emissions sources and fuel and stove combinations (Fig. 7) include using cleaner fuels (e.g., nonsolid fuels such as kerosene, natural gas, and electricity, or even charcoal, which has lower emissions compared to wood), stoves with better combustion conditions and lower emissions, and enclosed stoves with a chimney that removes

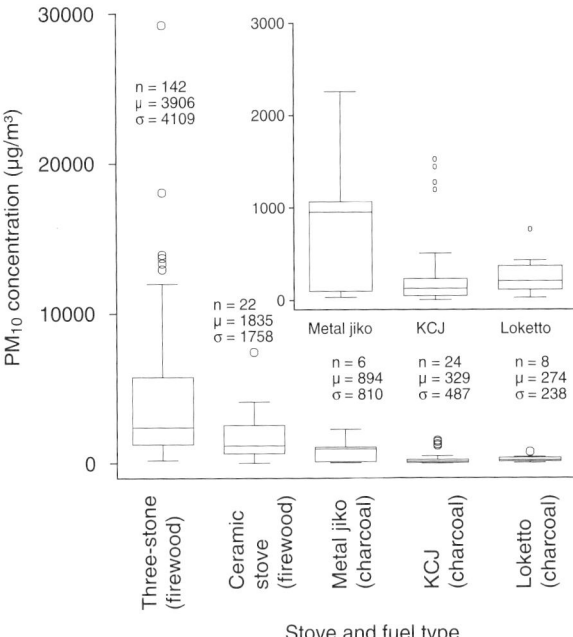

FIGURE 7 Daylong average of PM_{10} concentration for various stove–fuel combinations in Central Kenya (calculated over burning period). The insert in the upper right-hand corner is a more detailed version of the plot for the last three stoves; n is the number of days of measurement, μ is the sample mean, and σ is the standard deviation. The box-plots used here show a summary of the distribution of the variable. The lower and upper sides of each rectangle show the 25th and 75th percentiles and therefore enclose the middle half of the distribution. The middle line, which divides each rectangle into two, is the median. The large variability for different stove–fuel combinations shows that how a stove is used is as important in determining pollution as the technology. KCJ, Kenya ceramic jiko. Reproduced from Ezzati and Kammen (2002), *Environmental Health Perspectives*.

much of the emissions from indoor environments. Choices in housing and ventilation include houses with a separate cooking area or more windows. Modifications in behavior and in time–activity budgets include drying wood before use, having shorter burning periods, improving fire-handling practices to reduce the number and frequency of very smoky smoldering events, and keeping children away from the cooking area. The different intervention options are likely to have differential benefits for various household members. For example, cleaner fuels with lower emissions would benefit all household members regardless of age and gender. On the other hand, it has been estimated that relocating the stove to an outside cooking location without reducing its emissions provides larger relative benefits for male household members and children who do not cook.

Most current research in developing countries focuses on household energy technology (fuel–stove combinations) as the primary means for exposure reduction. Although it is fairly well established that transitions to cleaner fuels can provide large reductions in most pollutant levels, recent emphasis has been on "improved" (high-efficiency and/or low-emission) stoves and fuels, which may provide more affordable options in the near future compared to a complete shift to nonsolid fuels.

Attempts to reduce indoor air pollution through the design and introduction of new household energy technology should acknowledge the inherent interaction of technology and socioeconomic factors and include this interface in the research process. Initial research and development efforts to improve stove designs were often marked by a lack of detailed data on stove performances. Efficiencies and emissions, for example, were often measured in controlled environments, with technical experts using the stoves under conditions very dissimilar to those in the field. More recently, the attention of the research community has shifted from such ideal operating conditions to monitoring stove performance under actual conditions of use, taking into account the various social and physical factors that would limit the use of these stoves altogether or would result in "suboptimal" performance. These include factors such as food preferences or the ability to use the technology, the cost of the stove and fuel, the durability and ease of repair or replacement of the stove, uncertainties in fuel costs and access due to seasonality or political factors. Despite these complexities, a number of efforts have provided evidence that behavioral changes and the introduction of new energy technology can provide health benefits that are comparable in magnitude to other interventions (such as hospital-based case management for childhood ARI).

6. KNOWLEDGE GAPS AND RESEARCH NEEDS FOR REDUCING INDOOR AIR POLLUTION IN DEVELOPING COUNTRIES

Important research over the past few decades has illustrated that indoor smoke from solid fuel poses important health risks in developing countries. Unfortunately, affordable and effective interventions for reducing these risks are limited. This may be because in designing new interventions, the complexities of household energy use and exposure have been often overlooked, and there is a lack of infrastructure to support technological innovations, marketing and dissemination, and maintenance. Even less is known about combinations of technologies that may be used in a household and the factors that motivate the households to adopt them.

In broad terms, design and implementation of appropriate interventions and policies that will positively impact health effects of exposure to indoor smoke require answers to five research questions:

1. What factors determine the details of exposure and what are the relative contributions of each factor to personal exposure? As described earlier, these factors include the energy technology (stove–fuel combination), the housing characteristics and ventilation (such as the size of the house and the building materials, the number of windows, and arrangement of rooms), and behavioral factors (such as the amount of time spent indoors or near the cooking area). It is important to be aware that these factors often interact, requiring a more integrated approach to research. Attention to details of individual exposure has been a subject of more recent research.

2. What is the quantitative relationship between exposure to indoor air pollution and disease (i.e., the exposure–response relationship) along a continuum of exposure levels?

3. Which determinants of human exposure will be influenced, and to what extent, through any given intervention or group of interventions?

4. What are the impacts of any intervention on human exposure and on health outcomes, and how would these impacts persist or change over time?

5. What are the broader environmental effects of any intervention, its costs, and the social and economic institutions and infrastructure required for its success?

7. EMERGING INDOOR AIR QUALITY ISSUES IN DEVELOPING COUNTRIES

In terms of aggregate health effects, household solid fuel use is currently the most important source of indoor air pollution in developing countries. Although relatively equitable economic and rural development has contributed to reducing the use of solid fuels in some settings (e.g., in some newly industrialized Asian countries), indoor air pollution is likely to remain an important health risk in poorer developing countries in the absence of successful intervention programs. More broadly, indoor air quality issues in developing countries are dynamic phenomena that require dynamic research and policy responses.

One of the emerging issues in indoor air quality relates to the demographic, economic, and ecological changes of the past few decades. Some of the poorest households have been forced to use inferior sources of fuel (e.g., dung or crop residues instead of wood and charcoal). This is partially due to the rising price of fuels and because real income has remained unchanged or has even declined in some developing countries, especially in sub-Saharan Africa. The interactions of poverty, population, environment, and health, however, have complex characteristics that are dependent on local socioeconomic, geographical, and demographic factors. For example, if a stand of mature trees is cleared to open space for additional cultivation or grazing area, or for use in the lumber industry, the fallen trees are often burned *in situ* or processed into charcoal for sale in a distant town. Households that formerly relied on fallen limbs and dead wood from that stand of trees find they must travel farther to meet their fuelwood needs. If no mature stands with a sufficient stock of deadwood remain within a reasonable distance, poor households, unable to afford commercial fuels, may begin to cut smaller trees, which leads to a further loss of tree cover. If smaller trees are not an option, or prove insufficient to meet demand, then some households may turn to agricultural residues or animal manure. This shift has consequences that extend beyond the use of a lower quality and more polluting fuel. Crop residues are often used as fodder, and using traditional fodder as fuel can lower the value of a family's livestock or lower the quality of the animals' manure. When they are not used as fodder, crop residues are often left in the field as ground cover to protect topsoil between growing seasons. They may be ploughed back into the soil or burned on top of it before the next crop is planted; both activities return nutrients to the soil. Using these residues for fuel can leave topsoil unprotected between the harvest of one crop and the sowing of the next, leading to soil erosion and a loss of nutrients. Similarly, using animal manure as fuel takes away a valuable fertilizer, leading to lower yields or forcing the family to rely on expensive inorganic fertilizers. When coupled with increased malnutrition and exposure to other risks associated with poverty in these same households, the health effects of indoor air pollution would be greatly magnified. Therefore, while much of the literature has considered improvements in indoor air pollution as a result of economic development, the energy and health inequalities as a result of income inequality need particular attention.

Another emerging issue relates to urbanization and other forms of sociodemographic change, to changes in housing and neighborhood design, and to increased use of manufactured chemicals. The implications are that indoor air quality for some developing country households will be increasingly determined by pollution from sources other than their own energy use (nearby households, industries, transportation, insecticides, and chemical cleaners). Outdoor/indoor air quality interactions have been subject to research in industrialized countries. In developing countries, these issues should begin to appear on research and policy agendas, because many involve decisions that cannot be easily reversed.

Finally, smoking has been on the rise in developing countries over the past few decades. Environmental tobacco smoke, which has been subject to regulation in a number of industrialized countries, will increasingly affect the health of the many sectors of society in developing nations (e.g., workers in buildings where workplace smoking is allowed) and should receive early attention.

SEE ALSO THE FOLLOWING ARTICLES

Air Pollution from Energy Production and Use • Air Pollution, Health Effects of • Clean Air Markets • Climate Change and Public Health: Emerging Infectious Diseases • Development and Energy, Overview • Hazardous Waste from Fossil Fuels • Indoor Air Quality in Industrial Nations • Technology Innovation and Energy • Transitions in Energy Use • Women and Energy: Issues in Developing Nations • Wood in Household Energy Use

Further Reading

Agarwal, B. (1983). Diffusion of rural innovations: Some analytical issues and the case of wood-burning stoves. *World Dev.* **11**, 359–376.

Balakrishnan, K., Sankar, S., Parikh, J., Padmavathi, R., Srividya, K., Venugopal, V., Prasad, S., and Pandey, V. L. (2002). Daily average exposures to respirable particulate matter from combustion of biomass fuels in rural households of southern India. *Environ. Health Perspect.* **110**, 1069–1075.

Bradley, P. N., and Campbell, B. M. (1998). Who plugged the gap? Re-examining the woodfuel crisis in Zimbabwe. *Energy Environ.* **9**, 235–255.

Bruce, N., Perez-Padilla, R., and Albalak, R. (2000). Indoor air pollution in developing countries: A major environmental and public health challenge. *Bull. World Health Org.* **78**, 1078–1092.

Ezzati, M., and Kammen, D. M. (2001). Indoor air pollution from biomass combustion as a risk factor for acute respiratory infections in Kenya: An exposure-response study. *Lancet* **358**, 619–624.

Ezzati, M., and Kammen, D. M. (2002). Evaluating the health benefits of transitions in household energy technology in Kenya. *Energy Policy* **30**, 815–826.

Ezzati, M., and Kammen, D. M. (2002). The health impacts of exposure to indoor air pollution from solid fuels in developing countries: Knowledge, gaps, and data needs. *Environ. Health Perspect.* **110**, 1057–1068.

Ezzati, M., Mbinda, B. M., and Kammen, D. M. (2000). Comparison of emissions and residential exposure from traditional and improved biofuel stoves in rural Kenya. *Environ. Sci. Technol.* **34**, 578–583.

Ezzati, M., Saleh, H., and Kammen, D. M. (2000). The contributions of emissions and spatial microenvironments to exposure to indoor air pollution from biomass combustion in Kenya. *Environ. Health Perspect.* **108**, 833–839.

Kammen, D. M. (1995). Cookstoves for the developing world. *Sci. Am.* **273**, 63–67.

Leach, G., and Mearns, R. (1988). "Beyond the Woodfuel Crisis: People, Land, and Trees in Africa." Earthscan, London.

Manibog, F. R. (1984). Improved cooking stoves in developing countries: Problems and opportunities. *Annu. Rev. Energy* **9**, 199–227.

Reddy, A. K. N., Williams, R. H., and Johansson, T. B. (eds.). (1996). "Energy after Rio: Prospects and Challenges." United Nations Publications, New York.

Smith, K. R. (1987). "Biofuels, Air Pollution, and Health: A Global Review." Plenum Press, New York.

Smith, K. R. (1988). Air pollution: Assessing total exposure in developing countries. *Environment* **30**, 16–34.

Smith, K. R. (1993). Fuel combustion, air pollution exposure, and health: Situation in developing countries. *Annu. Rev. Energy Environ.* **18**, 529–566.

Smith, K. R., Metha, S., and Feuz, M. (2002). The global burden of disease from indoor air pollution: Results from comparative risk assessment. *In* "Indoor Air 2002: Proceedings of the 9th International Conference on Indoor Air Quality and Climate," pp. 10–19. Indoor Air 2002, Monterey, California.

Smith, K. R., Samet, J. M., Romieu, I., and Bruce, N. (2000). Indoor air pollution in developing countries and acute lower respiratory infections in children. *Thorax* **55**, 518–532.

Smith, K. R., Shuhua, G., Kun, H., and Daxiong, Q. (1993). One hundred million improved cookstoves in China: How was it done? *World Dev.* **21**, 941–961.

Spengler, J. D., and Chen, Q. (2000). Indoor air quality factors in designing a healthy building. *Annu. Rev. Energy Environ.* **25**, 567–600.

Spengler, J. D., Samet, J. M., and McCarthy, J. F. (eds.). (2001). "Indoor Air Quality Handbook." McGraw-Hill, New York.

Spengler, J. D., and Sexton, K. (1983). Indoor air pollution: A public health perspective. *Science* **221**, 9–17.

Westhoff, B., and Germann, D. (1995). "Stove Images: A Documentation of Improved and Traditional Stoves in Africa, Asia, and Latin America." Brandes & Aspel Verlag, Frankfurt, Germany.

World Health Organization (WHO). (1997). "Health and Environment in Sustainable Development." World Health Organization, Geneva.

World Health Organization (WHO). (2002). "World Health Report 2002: Reducing Risks, Promoting Healthy Life." World Health Organization, Geneva.

World Resources Institute (with UNEP/UNDP and World Bank). (1999). "World Resources 1998–1999: A Guide to the Global Environment." Oxford University Press, New York.

Indoor Air Quality in Industrial Nations

ADRIAN F. R. WATSON
Manchester Metropolitan University
Manchester, United Kingdom

1. Introduction
2. Sick Building Syndrome
3. Contribution of the Physical Environment to Indoor Air Quality
4. Indoor Chemical Pollutants
5. Indoor Biopollutants
6. Specific Microenvironments
7. Control of Indoor Air Pollution
8. Conclusions

Glossary

biopollutants A collective term to describe airborne microorganisms, including fungi, bacteria, viruses, inflammatory agents (endotoxins), and indoor allergens such as house dust mites, which are found commonly within all indoor environments.

building-related illness A discrete, identifiable disease or illness that can be traced to a specific pollutant or source within a building (Legionnaires' Disease, for example).

environmental tobacco smoke (ETS) The mixture of smoke from the burning end of a cigarette and the smoke exhaled by the smoker (also called passive smoking).

radon (Rn) A radioactive gas formed in the decay of uranium. The radon decay products (also called radon daughters) can be inhaled into the lungs, where they continue to release radiation as they decay further.

sick building syndrome (SBS) A collection of common symptoms that affect the health and productivity of a substantial number of workers within a building; these symptoms may diminish or go away during periods when occupants leave the building. The symptoms cannot be traced to specific pollutants or sources within the structure.

ventilation The replacement of stale air within a space by clean air through dilution of existing indoor air with either cleaned and filtered outdoor air or recirculated indoor air; in combination with the removal of indoor pollutants using exhaust fans and filtration.

volatile organic compounds (VOCs) The collective term for all vapor-phase chemicals that contain organic carbon. There are numerous indoor sources of VOCs, including cooking and heating activities, common household products, and building materials.

Good indoor air quality is important for occupant health, contributing to a favorable working or home environment and a sense of comfort, health, and well being; when indoor air pollutants are present, however, indoor air quality deteriorates. To date, much of the attention relating to air pollution has focused on the risk and health effects of outdoor air quality, despite the fact that indoor levels of air pollutants are often 2–5 (and occasionally more than 100) times higher than outdoor concentrations. Recently, though, comparative risk studies such as those performed by the U.S. Environmental Protection Agency (EPA) have consistently ranked indoor air pollution among the top five environmental risks to public health. Furthermore, in industrialized countries, people spend the majority of time indoors, thereby increasing their exposure to indoor air pollutants relative to outdoor air pollutants. A typical adult spends approximately 60% of their time at home; 25% is spent in the workplace; 5%, at public places, shops, leisure pursuits, etc.; 5%, within transport vehicles; and 5–10%, outdoors. Some populations, often those most susceptible to health effects, such as the elderly, infants, the sick, and their caregivers, spend even more time indoors, usually in the home. However, within the general public, there is a notable lack of appreciation of this risk; the level of education regarding indoor air quality issues is low and the indoor environment is considered a place of refuge and safety. Furthermore, unlike outdoor air quality, indoor air problems are often subtle and do not always produce easily recognized impacts.

1. INTRODUCTION

The main factors influencing indoor air quality include building design, construction materials, ventilation rates, and building maintenance; outdoor air quality; seepage of pollutants through the ground; occupant activities and preferences; and the occupants themselves. Figure 1 shows typical sources of indoor air problems in the home. The indoor air pollutants posing the greatest hazard include carbon monoxide, which is responsible for many accidental deaths every year; and radon and environmental tobacco smoke, both of which are implicated in the incidence of cancer. Other pollutants of concern are volatile organic compounds (VOCs), nitrogen dioxide, house dust mite allergens (all of which are implicated in the incidence of respiratory disease, especially in children), fungi, bacteria, and asbestos. A summary of the key indoor pollutants, sources, and effects is shown in Table I.

2. SICK BUILDING SYNDROME

Of all indoor air issues, sick building syndrome (SBS) has probably been the subject of most interest and study. It is a poorly defined syndrome that is described as involving a "building in which complaints of ill health (usually >20% of occupants) are more common than may be expected" and has been formally recognized as a disease by the World Health Organization (WHO) since 1986. SBS concerns are largely workplace based and relate to the health of individual workers and the productivity of the organizations employing them. A clear distinction can be made with building-related illnesses (BRIs), which are clinically diagnosed diseases such as humidifier fever or Legionnaires' disease, and are not considered to be SBS.

SBS is a relatively recent phenomenon; its incidence grew in the 1970s with the advent of more

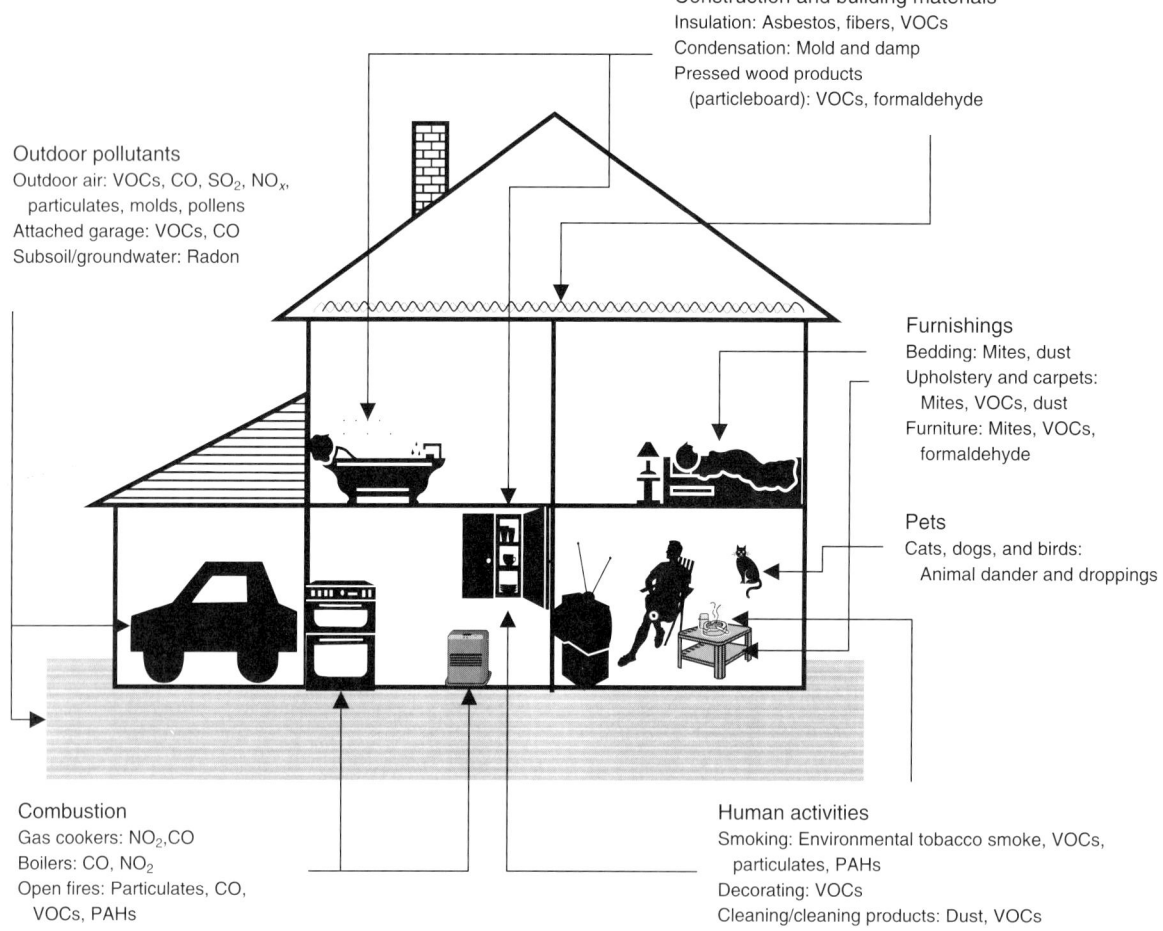

FIGURE 1 Sources of indoor air pollutants in the home. VOCs, Volatile organic compounds; PAHs, polycyclic aromatic hydrocarbons.

TABLE I

Indoor Air Pollutants, Sources, and Health Effects[a]

Indoor pollutant	Sources	Health effects
Carbon monoxide (CO)	Combustion appliances, environmental tobacco smoke, infiltrated exhaust from attached garages	Reduces capacity of blood to carry oxygen; increases frequency and severity of angina; headaches, decreased alertness, flulike symptoms in healthy adults; exacerbation of cardiopulmonary dysfunction in compromised patients; asphyxiation leading to death by suffocation
Radon	Rocks and soils containing naturally occurring radium and radon, well water, some building materials	Lung cancer
Environmental tobacco smoke (ETS)	Tobacco smoking (cigarettes, cigars, second-hand smoke)	Lung cancer, headaches and nausea, irritation to mucous membranes, chronic and acute pulmonary effects, cardiovascular effects
Biological contaminants (viruses, bacteria, fungi, molds, insect and arachnid excreta, endotoxins, pollen, animal and human dander)	Outdoors, humans, animals, damp building areas, improperly maintained heating ventilation and cooling systems, carpeting	Infectious diseases, allergic reactions (e.g., acute cases of asthma and skin irritations), toxic effects such as Legionnaires' disease
Volatile organic compounds (VOCs)	Paints, stains, adhesives, solvents, cleaners, pesticides, building construction materials, office equipment, upholstery fabrics, environmental tobacco smoke	Eye and respiratory irritation, neurotoxic effects, hepatotoxic effects, cancer
Formaldehyde (also a VOC)	Environmental tobacco smoke, foam insulation, resin in particle board, plywood paneling furnishings, carpets, upholstery fabrics	Headaches, dizziness, upper respiratory irritation, allergic responses (e.g., skin rashes), cancer
Polycyclic aromatic hydrocarbons (PAHs)	Environmental tobacco smoke, kerosene heaters, wood stoves	Cancer, cardiovascular effects, decreased immune function, atherosclerosis etiology
Pesticides	Pesticide application indoors and outdoors	Neurotoxicity, hepatotoxicity, reproductive effects
Asbestos	Asbestos cement, ceiling and floor tiles, insulation, other building materials	Easily inhaled to cause lung damage, asbestosis, mesothelioma, cancer
Nitrogen dioxide (NO_2)	Combustion appliances, environmental tobacco smoke	Eye and respiratory tract irritations, decreased pulmonary function (particularly in asthmatics and children)
Particulate matter	Combustion appliances, environmental tobacco smoke	Cancer (soot, polycyclic aromatic hydrocarbons adsorbed to particles), irritation of respiratory tissues and eyes, decreased lung function
Dust, sprays, cooking aerosols	Personal activity	Unknown (can range from irritation to cancer)

[a] Data from Goodish (2001).

thermally efficient buildings, brought about by the 1973–1974 energy crisis. It is most common in new and recently refurbished buildings that contain complex heating, ventilation, and air-conditioning (HVAC) units, although it can be found in naturally ventilated buildings. The common features of SBS buildings are that they are airtight, warm (isothermal), energy-efficient buildings with forced ventilation and internally there are often a lot of textile coverings and soft furnishings. However, the symptoms of SBS cannot usually be traced to specific pollutants or sources within the building.

SBS symptoms are common among the general population but are experienced more frequently in the workplace. Typical symptoms include dryness or irritation of the eye, nose, or throat; lethargy, headaches, and dizziness; dry skin, itching, or rashes; airway infections, coughs, and wheezing; and unpleasant odor and taste sensations. Occasionally, more serious symptoms, such as high blood pressure, have been reported. The symptoms are relieved when the worker leaves the workplace and recur following return to the affected building. SBS is poorly understood; because the etiology, symptoms, and cures often appear vague, it has been suggested that nonphysical issues within the workplace, such as organizational structure and work intensity, may contribute to an outbreak of SBS.

3. CONTRIBUTION OF THE PHYSICAL ENVIRONMENT TO INDOOR AIR QUALITY

The physical environment, characterized by temperature, humidity, and ventilation properties (and lighting), is the major determinant of comfort in the workplace, and departures from satisfactory conditions can have harmful health effects. Although workplace physical conditions tend to be controlled by local health and safety laws, there are no specific standards on the physical environment. However, two major professional groups, the Chartered Institute of Building Services Engineers (CIBSE) and the American Society of Heating, Refrigerating, and Air-Conditioning Engineers (ASHRAE) have been responsible for developing guidelines for workplace temperature, humidity, and ventilation.

3.1 Temperature and Air Movement

The temperature and air movement in the workplace are often the result of local workplace negotiation and Approved Codes of Practice (ACoP), rather than being enshrined in law, because the wide range of building functions requires a flexible approach. Cold temperatures can affect dexterity and mobility, and may increase physical and visual strain, with added problems for people with muscular pain, arthritis, and heart conditions. The design criteria of the CIBSE suggest a minimum temperature of 16°C in winter with a normal range of 21–23°C dry resultant temperature. In summer, the normal dry resultant temperature range is 22–24°C. The World Health Organization (WHO) recommends 24°C as the maximum temperature for working in comfort. Exceeding this temperature can lead to dehydration, heat stress, and ultimately heat stroke.

Air movement and drafts are also controlled in the workplace; health and safety laws merely stipulate that every enclosed workplace must have effective and suitable ventilation to provide a sufficient quantity of fresh or purified air, without specifying specific ventilation requirements. To assess fresh air, the CIBSE and ASHRAE usually identify carbon dioxide levels as a suitable indicator of human bioeffluents (odors) and a diagnostic tool in determining whether human comfort (odor) criteria are satisfied. The CIBSE quotes air exchange levels of 8 liters/sec per person (lsp) in no-smoking environments although design consultants increasingly use higher values (12–16 lsp). In the home, typical air movements are approximately 7 lsp. In many homes and workplaces, windows or other openings provide sufficient ventilation, although mechanical ventilation should be used if the openings are not suitable.

3.2 Humidity

Humidity is the amount of moisture in the air, and at saturation point, the relative humidity (RH) is 100%. To ensure occupant comfort, RH should be maintained between 40 and 70%. Health effects have been observed at the low levels of humidity (<30% RH), often seen during cold, dry winter months; desiccating effects seen in individuals can result in flu-like symptoms, such as stuffy or runny nose or sore eyes. However, indoor humidity should be maintained below 70% because high humidities can support the development of molds and can promote the growth of pathogenic or allergenic organisms such as *Legionella*.

4. INDOOR CHEMICAL POLLUTANTS

Indoor air quality is often compromised by a wide range of different chemical pollutants from a broad range of sources; the majority are gaseous (e.g., carbon monoxide), but some have particulate components such as environmental tobacco smoke and some are fibrous. The associated health risks should not be exaggerated because concentrations of indoor pollutants within homes and workplaces only usually affect morbidity through mild irritation, but occasionally some indoor chemical pollutants can lead to more extreme effects, such as lung cancers or accidental death. The indoor chemical pollutants that are of most current concern and interest are carbon monoxide, nitrogen oxides, volatile organic compounds, environmental tobacco smoke, radon, and fibers.

4.1 Carbon Monoxide

Carbon monoxide (CO) is a colorless, odorless gas produced by the incomplete combustion of carbon-containing substances. It is poisonous at high concentrations and is responsible for many accidental deaths worldwide (e.g., 600 per year in the United States). Outdoor levels can be determinants of indoor levels, particularly when a home has an attached garage or is in close proximity to heavily trafficked roads; however, the major source of indoor CO is incomplete fuel combustion (e.g., gas-fueled cooking and heating appliances, unvented paraffin space heaters, leaking chimneys and furnaces, and fireplaces), particularly where poor installation or ventilation or malfunctioning equipment reduces the oxygen concentration necessary for complete fuel combustion. Typical indoor concentrations range from 0.9 to 2.7 mg m^{-3}, although 15-minute peak concentrations have been measured at 180 mg m^{-3} in homes using gas cooker grills.

CO preferentially combines with hemoglobin in the blood to form carboxyhemoglobin (COHb), interfering with the distribution of oxygen throughout the body. At high concentrations, COHb can cause unconsciousness and death, whereas lower concentrations cause a range of flulike symptoms such as headaches, dizziness, weakness, nausea, fatigue even in healthy people, and episodes of increased chest pain in people with chronic heart disease. Fetuses, infants, elderly people, and individuals with anemia or with a history of heart (angina) or respiratory disease can be especially sensitive to CO exposures. Because CO pollution is undetectable by human sensory perceptions of smell or taste, when accidental CO poisoning occurs, the victim tends to lapse slowly into unconsciousness and death with no realization about what is occurring. For this reason, CO is often given emotive terms such as "the silent killer."

Control of CO is achieved through regular professional maintenance and careful use of fuel appliances, ensuring that heating appliances are vented outdoors. Vehicles should not be idled inside a garage that is attached to the house.

4.2 Nitrogen Oxides

Like CO, the main indoor sources of nitrogen oxides (NO$_x$) are poorly ventilated combustion appliances, although in the absence of indoor sources, outdoor concentrations are often higher. The most commonly observed nitrogen oxides are nitrous oxide (N$_2$O), nitrogen oxide (NO), and nitrogen dioxide (NO$_2$), which is of most interest from a health perspective.

In the United Kingdom, 2-week average NO$_2$ concentrations range from 25 to 70 μg m^{-3} with a gas cooker in the home and from 13 to 40 μg m^{-3} without a gas cooker, although 1-hour averages of over 1000 μg m^{-3} have been recorded in kitchens, which breaches most of the outdoor guideline levels (WHO guideline, 200 μg m^{-3}) for NO$_2$. Evidence from health studies of indoor NO$_2$ is inconclusive, but there is some evidence of an association with respiratory symptoms in children.

4.3 Volatile Organic Compounds

In indoor air quality studies, the term "volatile organic compounds" is typically used to describe all vapor-phase organic compounds. However, because VOCs exhibit a wide range of boiling points, they can be further subdivided into very volatile organic compounds (VVOCs; gaseous) with a boiling point ranging from <0 to 50–100°C, volatile organic compounds (50–100 to 240–260°C boiling point), and semivolatile organic compounds (SVOC; 240–260 to 380–400°C boiling point).

The VOC class of indoor air pollutants consists of more compounds than any other class, and hundreds of individual compounds have been identified at concentrations of 1 μg m^{-3} or more. Total VOC concentrations typically range from 50 to 1000 μg m^{-3} over long periods. However, some newly built offices and houses have higher source emissions, and over 130 individual compounds have been measured at a total VOC concentration of 13 mg m^{-3} in these buildings. VOC sources are ubiquitous in the office and home environment, and include evaporative surfaces (such as vinyl flooring, carpets, or soft furnishings), building products, pressed wood products, adhesives, sealants, cleaning agents, paint, ventilation systems, combustion sources, cooking and space heating activities, tobacco smoke, or the infiltration of outdoor air.

Emissions of VOCs are either continuous (long-term, constant-strength-release sources dependent on temperature, relative humidity, and air velocity) or discontinuous (time-dependent, variable-strength, short-term emissions). Consequently, the variety of sources and emission characteristics confers on VOCs the most dynamic behavior of all indoor pollutants. Table II summarizes the common indoor VOCs and their sources.

One of the most researched individual VOCs is formaldehyde, a low-molecular-weight aldehyde found at variable concentrations indoors, ranging

TABLE II

Sources of Volatile Organic Compounds

Compound	Source materials
Aromatic hydrocarbons (benzene, toluene, xylenes, napthalenes) and aliphatic hydrocarbons	Paints, adhesives, sealants, gasoline, combustion products, damp-proof membranes, wallpaper, carpets, vinyl floor covering, creosote-impregnated timbers, tobacco smoke
Styrene	Insulation foam, jointing, textiles, disinfectants, plastics, paints, carpets, rubber floor tiles
Hexane	Floor covering, wallpaper, insulation foam, tobacco smoke
Chloroform	Chlorinated water
Terpenes (limonene, α-pinene)	Scented deodorizers, polishes, fabrics, fabric softeners, cigarettes, food, beverages
Polycyclic aromatic hydrocarbons and polychlorinated biphenyls	Combustion products (tobacco smoke, burning wood, kerosene heaters, burning paint)
Acrylic acid esters, alcohols	Polymers, aerosols, window cleaners, paints, paint thinners, cosmetics, adhesives, vinyl flooring
Ketones (e.g., acetone)	Lacquers, varnishes, polish removers, adhesives
Ethers	Resin, paints, varnishes, lacquers, dyes, soaps, cosmetics
Esters	Plastics, resins, plasticizers, lacquer solvents, flavors, perfumes

from 0.01 to $1.2\,mg\,m^{-3}$ within homes (by comparison, outdoor levels in the United Kingdom range from 0.002 to $0.030\,mg\,m^{-3}$). Formaldehyde is commonly used in urea-formaldehyde- or phenol-formaldehyde-based resins, in foam insulation, as a binder in particleboard and wood paneling, and in many other building and household materials. Formaldehyde emitted from these materials into the building behaves as a common odorant at low concentrations (0.06–$1.2\,mg\,m^{-3}$) and as an eye and airway irritant at higher concentrations ($>0.5\,mg\,m^{-3}$). Smoking also contributes significantly to personal exposure, and as much as 25% of indoor formaldehyde can be accounted for by environmental tobacco smoke.

Industrial and occupational control of VOCs is achieved through health and safety legislation, but in the indoor environment no standards exist, although guidelines have been suggested in many industrialized countries (e.g., the WHO has recommended a guideline concentration of formaldehyde of $0.1\,mg\,m^{-3}$ as a 30-minute average). Control is most often achieved through good ventilation practice and by the regulated prohibition or restricted use of products that emit VOCs, such as insulation foams or particleboards.

4.4 Environmental Tobacco Smoke

Environmental tobacco smoke (ETS), sometimes called secondhand smoke, has been recognized in many countries, including the United States, Australia, and the United Kingdom, as a leading cause of acute and chronic health effects in both smokers and nonsmokers, contributing to lung cancer, ischemic heart disease, serious respiratory illness, and asthma. Active cigarette smoking is considered the major preventable cause of morbidity and mortality in the majority of industrialized countries.

Environmental tobacco smoke is a combination of the sidestream smoke released from the burning end of the cigarette and mainstream smoke exhaled by the active smoker. It is a complex mixture of over 3800 gaseous and particulate (the visible smoke) chemical species, with burning tobacco the sole source. Smoking indoors raises the concentrations of indoor pollutants such as respirable particulates, NO_2, CO, acrolein, nicotine, and other VOCs at a rate determined by the number of smokers, the rate of smoking activity, room size, and air exchange rates within the room. Because of the lower temperatures associated with the burning cone of the cigarette, most partial-pyrolysis products are enriched in sidestream smoke rather than in mainstream smoke, and consequently sidestream smoke is often considered to contain higher concentrations of toxic and carcinogenic compounds.

Many health effects of involuntary smoking have been demonstrated in fetuses and children. Perinatal health effects on the baby include growth effects such as decreased birth weight, prematurity, and congenital malformations. Postnatal health effects on the developing child may lead to increased risk of Sudden Infant Death Syndrome, reduced physical and cognitive development, and possibly longer term effects such as increased risk to childhood cancers (e.g., leukemia). During childhood, involuntary exposure to ETS has been linked to lower respiratory tract illnesses such as bronchitis and pneumonia, increased occurrence of asthma, and acute, persistent, or recurrent middle-ear disease.

The weight of scientific argument has led the International Agency for Research on Cancer (IARC) and many other regulatory authorities to consider ETS as a known (class A) carcinogen in adults. Other effects of involuntary ETS exposure in adults include irritation to eyes, nose, and upper and lower airways. There have been reports of increased coronary heart disease morbidity and mortality in active smokers, and acute respiratory morbidity in both active and passive smokers.

Control of ETS involuntary exposure has proved difficult; ventilation has not been shown to be fully effective and regulations specifying segregation of smokers into smoking areas or prohibition of smoking in locations such as the workplace and public buildings are often an emotional issue among the public.

4.5 Radon-222

Unlike other indoor chemical hazards, radon arises as a result of natural processes, not public or industrial activities. Radon is an inert, odorless, colorless, and tasteless radioactive gas, produced as part of the nuclear decay chain of naturally occurring ores bearing uranium-238, such as granite. Because the radioactive half-life of radon-222 is 3.8 days, it has sufficient time to diffuse through the soil, into buildings, where it can reach potentially dangerous concentrations, or into the outdoor atmosphere, where dilution and dispersion render radon safe.

Radon enters buildings though three main routes: migration through cracks in foundations, basements, and lower floors of houses; dissolved in groundwater, where it can enter wells and then homes; or via radon-contaminated building materials used for home construction. Radon decays with the emission of an α particle, a form of high-energy, high-mass ionizing radiation that is very damaging to biological tissue, and a series of daughter decay products. The radioactively active decay products include materials such as plutonium-218 and lead-214, which can form molecular clusters or attach to aerosols and then be subsequently deposited on the lung epithelium, where further radioactive decay (with the release of both α and β particles) will damage the surrounding tissue, leading to lung cancer. Worldwide, there has been concern regarding the environmental health risk of radon in areas with the appropriate geology. The U.S. EPA considers radon to be the second greatest risk factor for lung cancer after smoking, and the cause of an estimated 5–20,000 deaths annually in the United States. Evidence also suggests that the risk of lung cancer increases 10–20 times if there is a combination of exposure to radon and ETS, as opposed to each pollutant individually.

Unlike most indoor air pollutants, radon is subject to many international regulations. Regional environmental or radiological protection agencies are required to test for radon in areas where concentrations in dwellings approach the action level (e.g., in Japan and United States, ~ 150 becquerels (Bq) m^{-3}; United Kingdom, 200 Bq m^{-3}; Sweden, 400 Bq m^{-3}). If levels are considered actionable, control is relatively easy, usually by the identification and sealing of entry points, and/or increasing the level of ventilation in the home to increase the air exchange rate, thus reducing exposure.

4.6 Fibers (Asbestos)

Fibrous products are commonplace in modern buildings and are often employed for their thermal and acoustic insulation properties. Commonly used fibers are asbestos, a generic group term for several natural fibrous silicate minerals that were used widely until the 1980s, and vitreous and cellulose fibers, which are now commonly used as asbestos replacements.

Asbestos-containing materials (ACMs) have been used widely in building construction chiefly because of the fire prevention, chemical resistance, and insulation properties they offer relatively inexpensively. Example uses include roofing and pipe cladding; floor tiles, spray-applied fireproofing surfacing, and cement products. The low cost of ACMs has led to their widespread use in construction, particularly in public buildings such as schools, hospitals, and government buildings. Asbestos and asbestos-containing materials are among the most strongly regulated materials in the industrialized world and are classified as highly dangerous by most regulatory agencies, despite research that often questions this perceived level of risk, which is actually lower than that attributed to radon exposure. Many countries now require its removal from public buildings and have banned its use in current construction.

Asbestos is a particular hazard because the crystalline structure that underlies the larger fibers can cause fracture of asbestos-containing materials into thin, biologically active fibers, >5 μm long and <3 μm in diameter, which can become airborne. The small size of the fibers facilitates their easy passage into the alveolar region of the lungs, where they can cause great damage. The three principal important diseases linked with asbestos exposure are mesothelioma, asbestosis, and asbestos-related lung cancer. Mesothelioma, a malignant tumor that arises from the outer lining of the lung, tends cause mortality by compressing lung tissue. It is a rare condition associated with a long latency period (20–40 years) from first exposure and is almost always fatal despite treatment. Asbestosis, a disabling irreversible generalized scarification of the lung causing loss of pulmonary elasticity, can be fatal in severe cases. Asbestos-related lung cancer is a malignancy that arises within the lung when asbestosis is present. The development of these diseases is dose-related and the doses needed to induce them are substantial, although even very brief exposures to high levels of asbestos have been linked to asbestosis. Because the dose thresholds for all three diseases are orders of magnitude greater than exposures normally experienced by people who do not work with ACM, the risks are considered to be primarily through occupational exposure.

Vitreous and cellulose fibers are both irritants, but are considered safer than asbestos. Vitreous fibers, manufactured from sand or mineral wool, fracture into relatively larger fibers with a morphology different from that of asbestos, tending to settle out of the air more quickly, reducing exposure. Cellulose fiber is a recycled product made from materials such as wood dust or newsprint, with the addition of a fire retardant. It can act as a growth medium for microbiological growth in the presence of moisture, with potential respiratory tract irritation effects.

5. INDOOR BIOPOLLUTANTS

Despite being implicated in 35–50% of recent indoor air quality cases, indoor biopollutants, in contrast to chemical pollutants, have not been as widely studied. The wide range of biopollutants includes many airborne microorganisms; fungi, bacteria, and viruses, of which the fungi and bacteria are the most well studied; and inflammatory agents (endotoxins) and allergens from house dust mites, cockroaches, pets, and vermin. Like chemical pollutants, biopollutants can have indoor and outdoor origins. Pollens, a major outdoor group of biopollutants, are found indoors at highly variable levels, determined by outdoor levels and indoor/outdoor air exchange rates, but they are still believed to have a significant effect on morbidity. Biopollutants believed to be of most concern indoors are fungi and bacteria, especially the bacterium, *Legionella*, and allergens.

5.1 Fungi and Bacteria

Although many outdoor sources exist, particularly for fungi, many bacteria and fungi are found indoors associated with various forms of organic matter (e.g., surface coatings of walls, wood, and foodstuffs). Commonly seen bacteria include *Bacillus*, *Staphylococcus*, and *Micrococcus*, and common fungal species include yeasts, *Cladosporium*, and *Aspergillus*. Some species may be opportunistically pathogenic to humans, including the fungus *Aspergillus fumigatus* and the bacterium *Legionella*. The diversity and biomass of indoor fungi and bacteria vary with nutrient source, water availability, and temperature.

Mold and dampness are particularly associated with poor ventilation, and improved ventilation can be expected to reduce mold and fungi growth. Both bacterial and particularly fungal counts are influenced by season, with higher levels observed in the summer and autumn than in spring and winter. Several indoor infections have been attributed to the indoor fungi and bacteria, and a number of studies have indicated a relationship between home dampness (although not specific bacteria or fungi) and respiratory morbidity in children and adults. However, a lack of standardized sampling and analysis protocols for fungi mean that clear dose–response mechanisms in damp buildings have still to be established. A summary of reported health effects associated with exposure to bacteria and fungi can be seen in Table III.

5.2 Allergens (Endotoxins and House Dust Mites)

Endotoxins, proinflammatory substances present found in the outer membrane of gram-negative bacteria, have been associated with sick building syndrome. Levels in house dust above a median level of $1\,\text{ng}\,\text{mg}^{-1}$ have been shown to increase the severity of asthma, particularly in the presence of dust mite allergen. Sources of endotoxins are poorly defined but include contaminated humidifiers and areas that have suffered from severe water damage.

TABLE III

Diseases and Disease Symptoms Associated with Exposure to Bacteria and Fungi[a]

Disease syndrome	Examples of causal organisms
Asthma	Various *Aspergillus* and *Penicillium* spp., *Alternaria*, *Cladosporium*, mycelia sterilia, *Wallemia sebi*, *Stachybotrys*, *Serpula* (dry rot)
Atopic dermatitis	*Alternaria*, *Aspergillius*, *Cladosporium*
Extrinsic allergic alveolitis	*Cladosporium*, *Sporobolomyces*, *Aureobasidum*, *Acremonium*, *Rhodotorula*, *Trichosporon*, *Serpula* (dry rot), *Penicillium*, *Bacillus*
Humidifier fever	Gram-negative bacteria and their lipopolysaccharide endotoxins; actinomycetes and fungi
Rhinitis (and other upper respiratory symptoms)	*Alternaria*, *Cladosporium*, *Epicoccum*

[a]Data from Humfrey et al. (1996).

Other allergens include house dust mites (HDMs), cockroach allergens, and pet allergens. House dust mites are microscopic insects that grow rapidly in warm, damp conditions, thriving in carpets and bedding and feeding on dust, animal dander, and flakes of skin. HDM feces contain allergens that can trigger asthmalike reactions. Cockroach allergens are often found in kitchens and schools, and the common pet allergens from cats, dogs, and birds are found ubiquitously in urban indoor environments, either through pet ownership of because of transport of the allergen on contaminated pet owners to other locations.

5.3 Legionella (Legionnaires' Disease)

Legionnaires' disease is an uncommon form of pneumonia, first identified in 1976 in the United States after a large outbreak among a group of retired American service personnel who were attending a legion convention in a hotel in Philadelphia, Pennsylvania. Typical symptoms, including coughing, upper respiratory tract symptoms, and high fever, mainly affect susceptible populations (for example, the elderly and heavy smokers), but outbreaks can be fatal and the relatively high mortality rate means that the disease is associated with considerable public anxiety.

Infection is through the lungs, i.e., the route is airborne, and occurs as people breathe air containing fine water aerosols contaminated with *Legionella*. *Legionella* is a common bacterium found naturally in environmental water sources such as rivers, lakes, and reservoirs, usually in low numbers. However, if sufficiently high numbers of the bacteria get into water systems in buildings, they can present a risk to humans. Notable building sources of *Legionella* are air-conditioning equipment (cooling water systems, condensers), hot/cold water systems, humidifiers, nebulizers and medical humidifiers, showers, spray taps/mixer taps, water-based coolants (lathes, diamond wheels, and other rotating machines that generate aerosols), and whirlpool baths/spas. Wet cooling towers are considered a particular hazard, and have been identified as the cause of several outbreaks, because they generate large amounts of fine droplets, usually at rooftop level where there is potential for air currents to carry the droplets over a relatively large area.

Prevention and control of *Legionella* is through the careful design, maintenance, and operation of cooling towers (ideally, dry cooling towers) and water systems, removal of the conditions for bacterial growth, i.e., careful control of temperature of water systems, marinating water flow through pipes to avoid stagnation, and cleaning and disinfection (chlorination) of water systems to kill bacteria. There is no evidence that water systems in domestic homes present any risk from *Legionella*, but a significant proportion of public buildings, such as hotels, hospitals, offices, and factories, are known to be contaminated.

6. SPECIFIC MICROENVIRONMENTS

Although the majority of indoor air quality issues are general concerns relating to either the home or the office environment, there are certain locations where the indoor air quality has raised particular anxieties, either because there may be specific localized pollutant sources, or because of health and amenity issues requiring specialized control measures.

6.1 Hospitals

Usually the most important factor to consider with air pollution health effects is that a fraction of the population, usually the very young, the elderly, and the ill, will be affected more than the rest. The hospitalized population is largely drawn from just this fraction, and in most cases they spend 100% of their time exposed to an indoor environment. Within the hospital environment, microorganisms such as viruses, bacteria, and fungi are all potential killers, patients have much lower immunity levels due to their poor health and because prescribed drugs may further suppress their immune system. Unfortunately, hospitals tend to be environments in which pathogens are concentrated; airborne microorganisms found in hospitals are generated within the building by the staff, patients, and visitors, with only a minority of the microorganisms, typically fungal spores, originating outdoors. Generally, the higher the occupancy level, the greater the microbial burden in the air; also, general hospital tasks can also contribute to the abundance of microorganisms, which means that microbial concentrations are often quite variable.

Examples of airborne infection sources include violent sneezing and coughing, which release and disperse droplets; activities such as bed making and areas such as cleaning, laundry, and waste storage rooms, all of which can cause microorganisms to become airborne. Airborne transmission is known to be the route of infection for diseases such as tuberculosis and has also been implicated in some hospital outbreaks of diseases such as methicillin-resistant *Staphylococcus aureus* (MRSA) and Legionnaires' disease. Nevertheless, the general role that airborne pathogen transmission has to play in hospital-acquired infection is still poorly understood, but is likely to be a much greater factor than is currently recognized.

Control of bioaerosols to prevent the spread of infection is best achieved through use of appropriate hospital ventilation systems and isolation rooms. This is technically difficult and costly, because many of the airborne pathogens are extremely small (e.g., viruses can be as small as $0.001\,\mu m$, although they infect via transport on larger particles), therefore air requires cleaning using high-efficiency particulate air (HEPA) filtration, which can achieve 99.997% removal of particulates. However, the combination of high cost and hospital policy often limits mechanical ventilation to the principal medical treatment areas, such as operating theaters and associated rooms, and patient wards are typically naturally ventilated, even though it is common practice for treatment to occur on these wards.

6.2 Schools

Although sick building syndrome is widely known to affect office workers and their performance, there is comparatively little awareness of the effects of the classroom environment on children's educational performance. This is despite the facts that children are particularly sensitive to air quality; because they breathe higher volumes of air relative to their body weights and are actively growing, therefore they have greater susceptibility to environmental pollutants compared to adults, and they spend more time in the school environment than in any environment other outside the home.

Schools, relative to other kinds of buildings, are seen as particularly likely to have environmental deficiencies that could lead to poor indoor air quality; furthermore, chronic shortages of funding in schools contribute to inadequate operation and maintenance of facilities. In the United States, approximately 25% of public schools have reported that ventilation was unsatisfactory, and indoor air quality has been reported to be unsatisfactory in about 20% of schools. Classrooms also have a high occupancy density in comparison to offices (United Kingdom occupancy density values: schools, $1.8-2.4\,m^2$ per person; offices, $\sim 10\,m^2$ per person), which can be a particular problem for primary school children, who spend prolonged periods in one classroom.

Current evidence linking pollutant levels in classrooms and student performance obtained from studies worldwide is limited but persuasive. Poor indoor air quality can adversely affect the comfort and health of students and staff, causing short-term health problems such as fatigue and nausea, as well as long-term problems such as asthma, all of which can affect concentration, attendance, student performance, and absenteeism levels with potentially important immediate and lifelong effects for the student.

6.3 Museums

The main function of museums, art galleries, libraries, and archives is the storage and display of important artifacts over long periods of time. Unfortunately, indoor pollutants, such as VOCs (particularly formaldehyde and acetic acid), water vapor, ozone, and sulfur dioxide, are all known for their ability to damage cultural property. Although deterioration is usually slow and progressive, prolonged exposure of archival material to some pollutants can cause severe damage. The pollutants can have outdoor sources but can also be released from the archive objects or from archive enclosures (particularly wooden enclosures). Therefore, to assist the safeguarding of collections, it is important to assess the air quality characteristics of the objects' microenvironments, galleries, and building as a whole. The maintenance of a stable humidity, to avoid the destructive effects of natural humidity variation, is seen as being very important. Many collections are stored at "dry" humidity with typical values of 20–30% RH being maintained for corrodible collections, and 50–60% RH for organic/wooden collections. The use of and exposure of artifacts to high-VOC emission materials such as paints, adhesives, and cleaning products should also be tightly controlled.

Examples of archive deterioration include the action of sulfur oxides on photographic materials and collections of papers, causing brittleness and yellowing, and the emission of corrosive formic and acetic acids from sealed oak wood display cabinets, causing tarnishing or corrosion of metals (e.g., bronze or lead) and calcareous materials (e.g., egg shells or terra cotta).

7. CONTROL OF INDOOR AIR POLLUTION

In comparison to outdoor air quality, control measures for indoor pollutants are often considered poor. Currently, the main control of indoor air quality is usually assumed to be suitably maintained and operated ventilation and air-cleaning equipment; unfortunately, although this is effective for some parameters, these systems are not designed to maintain all aspects of air quality. Improvements, particularly in the short and medium term, can be gained through application and development of policy initiatives to control the sources of pollutants, such as smoking cessation in public places, restricting the sale of polluting products, imposing limits to activities that cause pollution, or developing standards for indoor pollutants. Longer term improvements may be likely if public awareness of indoor air quality is improved, such that eventually public and political pressure is applied to policymakers. This can be achieved using public health initiatives, providing better consumer product information, and improving the awareness of consumers' legal rights and the liabilities of landlords and manufacturers. The following sections focus on ventilation, using heating, ventilation, and air-conditioning systems, as the best current practice for indoor air quality control, and on legislation as the best hope for future short- and medium-term improvement.

7.1 Heating, Ventilation, and Air-Conditioning Systems

Although the functions and architecture of buildings are often taken for granted, they are highly complex dynamic systems, and the internal air quality within buildings depends on many interrelated factors. The principal engineering solution to providing good indoor air quality and thermal comfort is through ventilation and air-cleaning strategies. Ventilation is the replacement of stale air from a space by clean air through a combination of dilution of the existing indoor air, with either cleaned and filtered outdoor air or recirculated indoor air; and removal of indoor pollutants, using exhaust fans and filtration. To achieve this, a comprehensive heating, ventilation, and air-conditioning (HVAC) system is required, which must also maintain temperature and humidity at comfort levels for the building inhabitants. The main ventilation strategies to improve air quality are shown in Fig. 2.

Ventilation has a major impact on the energy use in buildings: it is estimated that more than 40% of primary energy in Organization for Economic Co-operation and Development (OECD) and European countries is consumed in buildings and half of that is dissipated through ventilation and air infiltration. The energy crisis of 1973–1974 acted as an impetus to an improvement in ventilation and building design, with an aim of conserving energy while creating an environment with good internal air quality. Therefore, ventilation systems that recirculate a large proportion of the indoor air have become more popular, owing to the large energy savings hat can be made by retaining heat within climate-controlled buildings. However, the high increase in reported indoor air quality problems, such as sick building syndrome, started in the 1980s, and the rise in symptoms and complaints appeared to correlate with energy conservation procedures. Consequently,

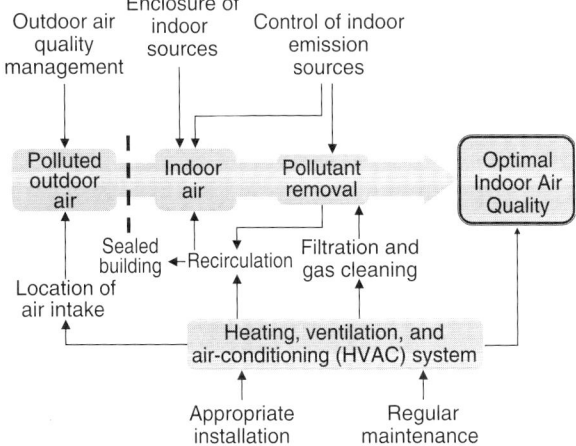

FIGURE 2 Ventilation strategies to achieve optimum indoor air quality.

careful design of HVAC systems is still required to provide a comfortable indoor environment for building inhabitants. To help achieve this, ventilation standards, such as standard 62 of the American Society of Heating, Refrigerating, and Air-Conditioning Engineers and guidelines of prepared by the Chartered Institute of Building Engineers have been developed to assist ventilation engineers.

The effectiveness of the HVAC system is dependent on a number of variables: initially, success depends on the proper design of the system relative to the building, then on installation and adherence to appropriate maintenance and operating procedures, the activity of people in the building, and the external air temperature, humidity, and air quality. Good outdoor quality air is a prerequisite for effective ventilation, because it provides the diluent for the internal stale air; however, in urbanized locations, constraints on outdoor air pollution such as traffic and industry mean that it is increasingly difficult to avoid outdoor pollutant entry across the building envelope. For this reason the location of the air intake duct is important, and close proximity to pollutant sources should be avoided.

Filtration is almost universally used in conjunction with mechanical ventilation and can be used to deal with both internal (recirculatory air filters) and external air pollutants. Gases are controlled using gas adsorption towers, and particle filters are available to trap particles of a wide range of different sizes. For example, in environments in which reduction of fine particle fractions of $<2.5\,\mu m$ is necessary, (e.g., hospitals), high-efficiency particle-arresting filters are available, although they are expensive to use because of the engineering and design constraints associated with such a high degree of filtration. Filtration is expensive and cannot be used in buildings that are naturally ventilated, or excessively leaky, because air will bypass the filtration units. Hence, for good ventilation, the building also has to be airtight, the tightness further contributing as a barrier to outdoor pollutants. Many energy-recovering ventilation systems cannot function unless the building is completely airtight.

Another feature of good ventilation design is the ability to contain polluting processes to specific locations and not disperse the pollutant around the building. This is achieved by negative-pressure regimes to keep pollutants in a space, and/or mechanical extraction to trap and remove the pollutants. Although HVAC systems can significantly improve air quality, if correct operation and maintenance procedures are not followed carefully indoor air quality problems can actually be created; for example, high humidities facilitate the growth of molds, and low humidity may be a source of discomfort.

7.2 Legislation

With the exception of radon and occupational environments, for which exposures are often covered by health and safety legislation, there are few indoor air pollution legislative measures or standards in industrialized countries. However, in 2000, the WHO issued statements and guidelines supporting the rights of individuals to healthy indoor air based on the basic principles of human rights, biomedical ethics, and ecological sustainability. The guidelines outlined the responsibility of both individual occupants and decision makers to create suitable indoor environments through public education programs, and other associated reduction or elimination strategies.

It is recognized that the creation of indoor air guidelines would be very difficult to implement for a number of reasons; for example, would guidelines be health or comfort based, should they be equivalent to outdoor standards for the same pollutants, would they apply to homes and offices, and who would be responsible for the compliance and monitoring of policies and guidelines? Despite these difficulties, it is apparent that there are benefits to establishing indoor air guidelines. Several countries, such as Germany, Norway, and Poland, have described target levels for pollutants, Australia uses indicators of good air quality rather than quantitative limits, and in some areas of the United States (California), VOC guidelines have been established for new building manufacturers. Other countries, such as the United Kingdom and South Africa, are also working to develop guidelines using typically observed concentrations and sources emission rates, which are both meaningful and contribute to public health.

8. CONCLUSIONS

As an environmental parameter, indoor air quality has the potential to affect health, comfort, and well being to an equal or greater extent than more well-known environmental issues such as outdoor air pollution; yet indoor air quality is still one of the more poorly understood environmental risks, by researchers, legislators, and the general public alike. Although the risks should not be overstated, it seems clear that there is a need for greater research into the indoor environment; for example, there is a need to study the roles of copollutants (such as a combined exposure to biological allergens and chemical irritants, which may exacerbate health responses) and to develop improvements in ventilation practices, to engineer improved air quality levels alongside energy efficiency requirements. Finally, as health-based standards exist for outdoor air pollutants, policymakers and scientists should collaborate to develop and apply guidelines or standards to the indoor environment as well, particularly because indoor air pollutant concentrations are usually higher.

SEE ALSO THE FOLLOWING ARTICLES

Air Pollution from Energy Production and Use • Air Pollution, Health Effects of • Climate Change and Public Health: Emerging Infectious Diseases • Hazardous Waste from Fossil Fuels • Indoor Air Quality in Developing Nations • Refrigeration and Air-Conditioning

Further Reading

Godish, T. (2001). "Indoor Environmental Quality." Lewis, Boca Raton, Florida.

Green, E., and Short, S. (1998). "IEH Assessment on Indoor Air Quality in the Home (2): Carbon Monoxide." Institute for Environment and Health, University of Leicester, UK.

Humfrey, C., Shuker, L., and Harrison, P. (1996). "IEH Assessment on Indoor Air Quality in the Home." Institute for Environment and Health, University of Leicester, UK.

Spengler, J. D., McCarthy, J. F., and Samet, J. M. (eds.). (2001). "Indoor Air Quality Handbook." McGraw-Hill, New York.

Industrial Agriculture, Energy Flows in

DAVID PIMENTEL
Cornell University
Ithaca, New York, United States

1. Energy Resources
2. Energy Inputs in Corn Production
3. Trends in Energy Use in U.S. Corn production
4. Conclusion

Glossary

british thermal unit (BTU) The thermal energy required to raise the temperature of 1 lb (avoirdupois) of water 1°F at or near 39.2° F, the temperature at which water is at maximum density; equal to about 0.252 calories.

calorie The amount of heat required at a pressure of 1 atm to raise the temperature of 1 g of water 1°C.

kilowatt-hour A unit of work or energy equal to that expended in 1 hour at a steady rate of 1 kW.

Energy is important in the development of land, water, biological, and human resources in crop production. In addition to human energy, sunlight and fossil energy are the primary energy resources utilized in agricultural production. Because all technologies employed in crop production require energy resources, the comparative measure of energy flow in crop production provides a sound indicator of the technological changes that have taken place in agriculture. A major advantage in assessing technological changes in crop production lies in the fact that energy values (kilocalories, kilowatt-hours, joules, and British thermal units) for various resources and activities remain constant, in contrast to economic values, which are continually changing depending on the relative supply and demand of the resources and services. Another advantage of using energy as a measure of change in crop production technology is that it can help assess the impact of substitution of different forms of energy for various practices, as well as the impact of substitution of land, water, biological, and labor resources for fossil energy. The objective here is to assess the changes that have occurred in intensive agricultural production in the United States during the past 55 years. To the agricultural ecologist, a historical analysis of the technological changes that have occurred in crop production over the past decades can provide valuable insight for future agricultural policies.

1. ENERGY RESOURCES

The use of fossil fuels in crop production is a relatively recent event; the use of tools and techniques requiring fossil fuels did not grow rapidly until after 1940 (Tables I–III). As recently as 1850, the primary energy resources for society were fuel wood and human labor. Today, the United States consumes nearly 100 quads (1 quad = 10^{15} Btu) of energy annually. On a per capita basis, this is the equivalent of about 8000 liters of oil per person per year. Approximately 17% of total U.S. energy is used in the food system, which represents about 1500 liters of fuel per person annually. This 17% includes 6% for agricultural crop and livestock production, 6% for processing and packaging of foods, and 5% for transport and preparation of food.

Since the oil crisis in 1973, the quantity of energy used in U.S. crop production has generally continued to grow. At the same time, the on-farm workforce has continued to decline. Some agriculturists proudly point to the fact that only about 2% of the U.S. workforce is involved in farming and thus are feeding the rest of the U.S. population. However, this is a misleading statistic because farmers hardly feed themselves entirely. They go to the same supermarket that everyone does to purchase food. Furthermore,

TABLE I

Quantities of Various Inputs to Produce Corn Using Mechanization, from 1945 to 1985[a]

Input	Year				
	1945	1954	1964	1975	1985
Labor (hours/ha)	57	42	27	17	10
Machinery (kg/ha)	22	35	49	50	55
Fuel (liters/ha)					
Gasoline	120	150	125	60	40
Diesel	20	30	65	70	75
Manure	3000	1000	1000	1000	1000
Nitrogen (kg/ha)	8	30	55	111	152
Phosphorus (kg/ha)	8	13	36	56	58
Potassium(k/ha)	6	20	28	62	75
Lime (kg/ha)	145	124	203	220	426
Seeds (kg/ha)	11	17	21	21	21
Insecticides (kg/ha)	0	0.2	0.4	0.5	0.6
Herbicides (kg/ha)	0	0.1	0.4	3	3.5
Irrigation (%/ha)	11	2	5	16	18
Drying (kg/ha)	43	77	725	2290	3800
Electricity ($\times 10^3$ kcal)	8	24	60	90	100
Transport (kg/ha)	170	262	325	298	322
Yield (kg/ha)	2132	2572	4265	5143	7400

[a] From Pimentel, Dazhong, and Giampietro (1990).

(Fig. 1). It was not until about 1945 that corn yields started to increase. At about this same time, hybrid corn varieties were introduced, commercial fertilizers and pesticides were applied, irrigation was used, and other new energy-intensive technologies in crop production were employed. From 1945 to 2000, corn yields increased about fourfold (Fig. 1). Concurrently, total energy input has increased also about fourfold (Tables I–III).

2.1 Labor Input

To raise corn by manpower alone requires about 1200 hours of labor per hectare during the growing season. Perhaps a few early settlers from Europe raised corn by hand, but by the time the United States gained independence, most corn culture involved the use of draft animal power, including oxen and horses. The use of 200 hours of oxen power to produce a hectare of corn reduces the manpower input, from 1200 hours for hand-produced corn, to about 400 hours. Thus, each hour of oxen input replaced 4 hours of manpower input. The estimate is that 120 hours of horsepower reduces the total labor input for corn production to about 120 hours, or 1 hour of horsepower replaces about 10 hours of manpower. Horsepower dominated U.S. corn production between 1910 and 1920.

Today, with the heavy mechanization of agriculture, the labor input in corn production has been reduced to about 6.2 hours per hectare (Table III). However, note that because of the technological changes, human labor is no longer required for direct power delivery. Thus, the labor input today is only 1/194th of the input required to produce corn by hand; however, this does not take into account, as mentioned, all the indirect labor input that goes into agricultural production, which includes the production of tractors and fuels. If the indirect labor input is taken into consideration, then current U.S. corn production uses about 1/40th of the labor input for hand-grown corn. This is still a dramatic reduction in the total amount of human activity required to produce corn over the past few decades.

2.2 Machinery and Power

Engine power has made a tremendous difference in U.S. society, as well as elsewhere in the world. This is illustrated by analyzing the manpower equivalent present in 3.78 liters (1 gallon) of fuel. This much fuel fed to a small gasoline engine will provide 20% of the heat energy produced in the form of power. Thus,

they depend on Detroit for their tractors and other farm machinery, depend on the oil industry to fuel their tractors and trucks, and depend on the petrochemical industry for fertilizers and pesticides. After crops and livestock are harvested on a farm, the farmer relies on the transport and food-processing sectors to move and process the foods. Eventually, the foods are shipped to wholesalers and are handled by food retailers. Thus, for every farm worker, it estimated there are three to four farm-support workers. If the total workforce in the food system is considered, then about 20% of the U.S. workforce is involved in supplying food. Interestingly, about 20% of the workforce was involved in the food system in the United States in the 1920s. The primary difference today is that only a small portion in the work force is located on the farm.

2. ENERGY INPUTS IN CORN PRODUCTION

Corn yields in the United States remained relatively static, at about 1880 kg/ha, from 1909 to 1940

TABLE II

Energy Input for Various Items Used in Corn Production from 1945 to 1985[a]

Item	Year				
	1945	1954	1964	1975	1985
Labor[b]	31	23	15	10	6
Machinery[c]	407	648	907	925	1018
Fuel[d]					
Gasoline	1200	1500	1250	600	400
Diesel	228	342	741	912	878
Nitrogen[e]	168	630	1555	2331	3192
Phosphorus[f]	50	82	227	353	365
Potassium[g]	15	50	70	155	187
Lime[h]	46	39	64	69	134
Seeds[i]	161	421	520	520	520
Insecticides[j]	ND	13	27	50	60
Herbicides[k]	ND	7	40	300	350
Irrigation[l]	125	250	625	2000	2250
Drying[m]	9	15	145	458	760
Electricity[n]	8	24	60	90	100
Transport[o]	44	67	89	82	89
Total	2492	4111	5935	8855	10,309
Ratio	3.4	2.5	2.9	2.3	2.9
Yield	8528	10,288	17,060	20,575	29,600

[a] In thousands of kilocalories/hectare. ND, No data exist, although the value is >0. From Pimentel, Dazhong, and Giampietro (1990).

[b] Food energy consumed per laborer per day was assumed to be 3110 kcal from 1700 to 1970, 3300 kcal in 1975, and 3500 kcal from 1980 to 1983.

[c] The energy input per kilogram of steel in tools and other machinery was 18,500 kcal.

[d] A liter of gasoline and a liter of diesel fuel were calculated to contain 10,000 and 11,400 kcal, respectively. These values include the energy input for mining and refining.

[e] Nitrogen = 21,000 kcal/kg.

[f] Phosphorus = 6300 kcal/kg.

[g] Potassium = 2500 kcal/kg.

[h] Limestone = 315 kcal/kg.

[i] When hybrid seed was used from 1945 to 1985, the cost was 24,750 kcal/kg.

[j] Chlorinated insecticides dominated use from 1945 to 1964, and the energy input was calculated to be 67,000 kcal; from 1970 to 1985, carbamate and phosphate dominated use and the energy input for these was calculated to be 100,000 kcal/kg.

[k] Phenoxyl herbicides dominated use from 1945 to 1959, and the energy input was calculated to be 67,000 kcal/kg; 1964 to 1985, other types of herbicides dominated use, and energy input for these are calculated to be 100,000 kcal/kg.

[l] Water used per irrigated hectare was assumed to be 37.5 cm from 1945 to 1970 and 45 cm from 1974 to 1983. Irrigation was calculated to be 300,000 kcal/cm.

[m] The quantity of corn per hectare that required drying is shown. The energy required per kilogram dried was 200 kcal.

[n] Includes energy input required to produce the electricity.

[o] For the goods transported to the farm, an input of 275 kcal/kg ws included.

from the 31,000 kcal in the 3.78 liters of fuel, about 7.2 kWh of power can be produced, which is the equivalent of 10 hp/hour or 100 manpower hours. Therefore, the 3.78 liters of gasoline provides about 2.5 weeks of human work equivalents. This is part of the reason for the dramatic reduction in labor inputs, from 1200 hours/ha to produce corn by hand, to only 6.2 hours/ha employing mechanization (Table III).

Interestingly, the total amount of fuel consumed has changed only slightly, from about 880 liters/ha in 1975 to about 850 liters/ha today (Tables II and III). The reasons for the improvements include more efficient engines, a change from gasoline to diesel fuel, and larger, more efficient farm machinery. The weight of the machinery per labor hour has grown from 2.2 kg/hour in 1945 to about 9 kg/hour in 2000.

TABLE III

Energy Inputs and Costs of Corn Production per Hectare in the United States in 2000

Inputs	Quantity	Energy (kcal × 1000)	Costs
Labor	6.2 hr[a]	250[b]	62.00[c]
Machinery	55 kg[d]	1414[e]	103.21[f]
Diesel	90 liters[g]	900[e]	23.40[h]
Gasoline	56 liters[g]	553[e]	14.60[h]
Nitrogen	148 kg[i]	2738[j]	81.40[k]
Phosphorus	53 kg[i]	219[l]	12.72[k]
Potassium	57 kg[i]	186[l]	17.67[k]
Lime	699 kg[i]	220[e]	14.00[m]
Seeds	21 kg[d]	520[e]	74.00[i]
Irrigation	8.1 cm[n]	941[o]	81.00[p]
Herbicides	2.1 kg[i]	210[e]	21.00[q]
Insecticides	0.15 kg[i]	15[e]	6.00[r]
Electricity	13.2 kWh[g]	34[e]	2.38[s]
Transportation	222 kg[t]	268[e]	66.60[u]
Total	—	8468	579.98
Yield, 8590 kg[v]	—	30,924	—
		Input: Output (kcal) = 1:3.80	

[a] National Agricultural Statistics Service (1999), Available on the Internet at http//usda.mannlib.cornell.edu. Accessed 3/8/2002.

[b] It is assumed that a person works 2000 hours/year and utilizes an average of 8100 liters of oil equivalents per year.

[c] It is assumed that farm labor is paid $10/hour.

[d] Pimentel and Pimentel (1996).

[e] Pimentel (1980).

[f] Hoffman et al. (1994).

[g] U.S. Department of Agriculture (1991), "Corn-State. Costs of Production." Stock #94018. USDA, Economic Research Service, Economics and Statistics System, Washington, D.C.

[h] Diesel and gasoline assumed to cost 26.5¢/liter.

[i] U.S. Department of Agriculture, (1997), "Census of Agriculture." Available on the Internet at http://www.ncfap.org. Accessed 8/28/2002.

[j] An average of energy inputs for production, packaging, and shipping per kilogram of nitrogen fertilizer; from Food and Agriculture Organization, United Nations (1999), "Agricultural Statistics," available on the Internet at http://apps.fao.org, accessed 11/22/1999; and Duffy, M. (2001), "Prices on the Rise: How will Higher Energy Costs Impact Farmers?" available at http://www.ag.iastate.edu, accessed 9/3/2002.

[k] Manitoba Agriculture and Food (2002), "Soil Fertility Guide." Available on the Internet at http://www.gov.mb.ca. Accessed 9/2/2002.

[l] Food and Agriculture Organization, United Nations (1999), "Agricultural Statistics." Available on the Internet at http://apps.fao.org. Accessed 11/22/1999.

[m] Assumed to be 2¢/kg; see Clary, G.M., and Haby, V.A. (2002), "Potential for Profits from Alfalfa in East Texas." Available on the Internet at http://ruralbusiness.tamu.edu. Accessed 9/2/2002.

[n] U.S. Department of Agriculture (1997), "Farm and Ranch Irrigation Survey (1998). 1997 Census of Agriculture," Volume 3, Special Studies, Part 1. USDA, Washington, D.C.

[o] Batty and Keller (1980).

[p] Irrigation for 100 cm of water per hectare costs $1000; see Larsen, K., Thompson, D., and Harn, A. (2002), "Limited and Full Irrigation Comparison for Corn and Grain Sorghum." Available on the Internet at http://www.Colostate.edu. Accessed 9/2/2002.

[q] It is assumed that herbicide prices are $10/kg.

[r] It is assumed that insecticide prices are $40/kg.

[s] Price of electricity is 7¢/kWh (USBC, 2002).

[t] Goods transported include machinery, fuels, and seeds that were shipped an estimated 1000 km.

[u] Transport was estimated to cost 30¢/kg.

[v] U.S. Department of Agriculture (2001), "Agricultural Statistics: USDA" I–1–XV–34p. USDA, Washington, D.C.

Hence, where liquid fuel inputs have declined, fossil energy inputs for machinery have risen. In corn production during the early years of mechanization, the principal fuel use in tractors was 120 liters of gasoline and only 20 liters of diesel fuel. Over time, there was a shift toward diesel fuel; today, at least

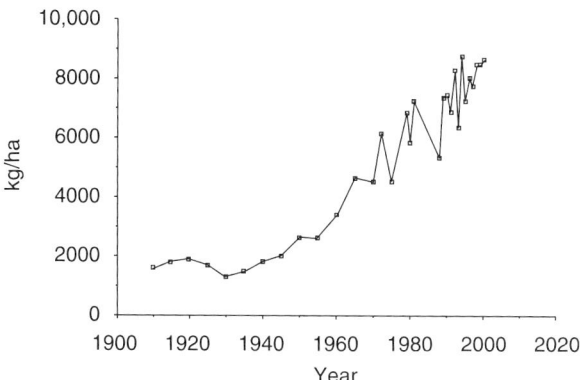

FIGURE 1 Corn grain yields from 1910 to 1999. From U.S. Department of Agriculture (1910–2000). "Agricultural Statistics." USDA, Washington, DC.

90 liters of tractor fuel is diesel and only 56 liters is gasoline (Tables I–III). Diesel fuel has advantages because it can provide 20–25% more power, or greater efficiency, per unit fuel energy, compared with gasoline. This is part of the reason for the improved corn yield per liter of fuel. Another reason for the improved fuel consumption per hectare has been the use of large farm equipment, which allows more operations to be performed during less time. The result is less fuel consumed per hectare of corn production.

2.3 Fertilizers

Early hand-produced corn and other crops depended on the nutrients that accumulated in the soil and derived from wild vegetation growing on uncultivated land during at least a 20-year fallow period. For example, in early slash/burn agriculture, vegetation was cut and burned to release the nutrients to the soil for crop production. Usually the land had to lay fallow for about 20 years so that nutrients would accumulate in the soil and vegetation. The land could then be tilled and planted for crops for approximately 2 years.

Early U.S. corn production was primarily organically based; that is, nutrients for corn production were provided mostly by livestock manure and green manure (legumes). In most cases, the farming system for corn required 2 ha of land to produce 1 ha of corn. For instance, 1 ha would be planted to a legume, such as clover or vetch, and the following year the legume would be plowed under and planted to corn. The legume would provide from 60 to 100 kg of nitrogen per hectare. As corn and other crops were harvested and removed from the land, the soils were being depleted of phosphorus, potassium, and calcium.

In 1945, when commercial fertilizers started to be used, only 8 kg/ha of nitrogen and phosphorus and 6 kg/ha of potassium were applied (Table I). By 2000, nitrogen application rates reached a high of 148 kg/ha, a nearly 20-fold increase. Note that the energy inputs for nitrogen alone in 1985 were greater than the total energy inputs for all items used in corn production in 1945 (Tables I and II). During the 38-year period, there was reported to be a 30% improvement in the efficiency of producing nitrogen fertilizer. Today the energy input for producing nitrogen for the farm, including production, packaging, and transport, is reported to be 18,590 kcal/kg.

Although the amounts of phosphorus and potassium applied per hectare rose significantly from 1945 to date, their use clearly did not grow as rapidly as occurred with nitrogen; both phosphorus and potassium require significantly less energy per kilogram to produce compared with nitrogen. The quantities of lime applied to land for growing corn also rose about threefold from 1945 to date, but lime is the least energy costly of the fertilizers that are used in corn production (Tables I–III).

2.4 Pesticides

Little or no pesticide was applied to U.S. corn crops in 1945. The quantity of insecticide used in corn production rose from 0.1 kg/ha in 1950 to 1.5 kg/ha in 2000. The insecticides applied to corn in the early 1950s and 1960s were primarily chlorinated insecticides such as dichlorodiphenyl trichloroethane (DDT) and dieldrin. Starting with the ban of DDT and other chlorinated insecticides in the early 1970s, there was a gradual shift from chlorinated insecticides to carbamate and phosphate insecticides. Although the quantity of insecticide applied per hectare increased only slightly with the newer insecticides, the energy required to produce a kilogram of insecticide increased and the toxicity increased 10- to 20-fold.

Changes in herbicide use also occurred in corn production starting in 1950. The first herbicide used in corn production was 2,4-dichlorophenoxyacetic acid (2,4-D), a phenoxy herbicide that was relatively efficient to produce in terms of energy input per kilogram. The newer triazines, such as atrazine, were applied during the 1960s to date and others were added later. The newer herbicides, although used at somewhat lower dosages, required more energy to produce compared with 2,4-D. The total energy input for chemical weed control increased about 30-fold from 1954 to date (Tables I–III).

2.5 Irrigation Water

Corn requires enormous amounts of water for production. For instance, to produce about 8000 kg/ha requires about 5 million liters of water for photosynthesis and other physiological activities in corn production. To pump this much water from a depth of only 100 m and apply it to corn requires 12 million kcal. Irrigation of corn acreage has increased from less than 1% in 1945 to 16% today. The energy required for irrigation correspondingly increased from 125,000 kcal/ha to 941,000 kcal/ha (Tables II and III). The energy input for irrigation water during the period from 1945 to 2000 rose more than 7.5 fold.

2.6 Corn Drying

Another major change in corn production technology during the past 55 years has been the way corn is harvested and stored. Initially, corn was harvested as corn on the cob. Gradually, more corn has been harvested as grain directly in the field, and most corn is currently harvested in this manner. Corn harvested directly as grain in the field contains 25–30% moisture. The corn must not contain more than 13–15% moisture before being placed in storage. About 200 kcal of energy is required to dry 1 kg of corn. Harvesting corn on the cob and then drying the corn in corn cribs has been calculated to use 33% less fossil energy compared to harvesting the corn as grain and then drying it using fossil energy. The corn on the cob in the crib is dried by wind or solar energy.

2.7 Electricity

Electricity is used for a variety of purposes on farms, including moving goods in and out of storage, running fans, and operating many pieces of equipment. The energy inputs for electricity given in the calculations in Tables I–III include the primary energy required to produce the electricity. About 3 kcal of coal, for example, is required to produce 1 kcal of electricity. Note that the electricity energy tally does not include electricity used by the farm family.

2.8 Transport

Transportation costs in Tables I–III are relatively small because only machinery, fuel, and nitrogen fertilizer transport costs are calculated; the transport costs for the other items were already included. Note that overall, energy inputs for transportation increased sixfold from 1945 to 2000, confirming the intensiveness of agricultural management over this period.

3. TRENDS IN ENERGY USE IN U.S. CORN PRODUCTION

Energy inputs in corn production have continued to increase from hand and draft animal power systems to highly mechanized systems of today. Only an estimated 716,000 kcal was used to produce 1 ha of corn by hand in the U.S. in 1700, whereas more than 8 million kcal is employed in U.S. corn production today. This represents more than a 10-fold increase in energy inputs (Tables I–III). Corn yields during this period grew slightly more than 4.5-fold.

Several improvements in resource conservation and in technology occurred starting in 1975 and continue to evolve to date. Most of these changes have occurred as the result of higher prices for inputs, such as fertilizers and pesticides, and for fuels (especially natural gas prices), and lower prices for corn. There was a move away from gasoline use, toward diesel fuel, with important energy conservation consequences. Diesel fuel improves efficiency in tractors and other vehicles because diesel fuel has more kilocalories per gallon compared to gasoline (8179 kcal/liter for gasoline versus 9235 kcal/liter for diesel fuel). At same time, machinery was getting larger and more appropriate for the larger agricultural fields that were managed. Farmers were also maintaining their farm machinery more efficiently and keeping the engines well tuned.

Fertilizers have also been utilized more effectively than previously. For instance, more nitrogen fertilizer is being applied today during the growing season, when the corn crop can use the nitrogen, rather than over the frozen ground during the winter months, as was the previous practice. During the late 1960s and early 1970s, nitrogen was often applied during the winter months when labor and machinery were relatively abundant. Some of the nitrogen applied during the winter months was wasted because it leached into groundwater or washed into streams and rivers. The nitrogen pollution problem continues today; nitrogen has become a serious pollutant in groundwater, rivers, and lakes.

How do solar energy inputs relate to the use of fossil energy use in corn production? The solar energy reaching 1 ha during the year averages about

14×10^9 kcal. During the 4-month growing season in the temperate region, 7×10^9 kcal reaches 1 ha. With a corn grain yield of 8590 kg/ha plus another 8590 kg/ha in stover (fodder) biomass, the total energy in the corn represents about 0.4% of the solar energy captured. For the corn grain, the percentage is only 0.2%, or an input of nearly 50 kcal of solar energy per kilocalorie of corn grain produced. During the winter months, there is no growth of corn or other vegetation in the temperate region.

4. CONCLUSION

During the past 55 years, there have been a great many changes in U.S. corn production technology. A major change has been the increased use of fossil energy for labor, cropland, irrigation, nutrients, and pest control, all of which has contributed to increased environmental degradation. Fossil energy use has increased more than 3.4-fold during the past 55 years; during the same period, corn yields have increased 4-fold. The increase in the intensity of energy use, especially concerning oil-based sources of energy, makes U.S. agriculture very vulnerable to fluctuations in world oil prices and supplies. An understanding of energy resource use can aid in the search for conservation measures and improved agricultural technologies.

SEE ALSO THE FOLLOWING ARTICLES

Aquaculture and Energy Use • *Diet, Energy, and Greenhouse Gas Emissions* • *Environmental Change and Energy* • *Fisheries and Energy Use* • *Food System, Energy Use in* • *Hunting and Gathering Societies, Energy Flows in* • *Industrial Ecology* • *Lifestyles and Energy* • *Livestock Production and Energy Use* • *Transitions in Energy Use*

Further Reading

Batty, J. C., and Keller, J. (1980). Energy requirements for irrigation. *In* "Handbook of Energy Utilization in Agriculture" (D. Pimentel, Ed.), pp. 35–44. CRC Press, Boca Raton.

Ferguson, A. (2003). Implications of the USDA 2002 update on ethanol from corn. *Optimum Population Trust* (April 2003), pp. 11–17.

Giampietro, M., Ulgiati, S., and Pimentel, D. (1997). Feasibility of large-scale biofuel production. *BioScience* 47(9), 587–600.

Hodge, C. (2002). Ethanol use in US gasoline should be banned, not expanded. *Oil Gas J.* (September 9), 20–30.

Hoffman, T. R., Warnock, W. D., and Hinman, H. R. (1994). "Crop Enterprise Budgets, Timothy-Legume and Alfalfa Hay, Sudan Grass, Sweet Corn and Spring Wheat under Rill Irrigation, Kittitas County, Washington. Farm Business Reports EB #1173." Washington State University, Pullman.

Lieberman, B. (2002). "The Ethanol Mistake: One Bad Mandate Replaced by Another." Competitive Enterprise Institute, September 17, 2002. Available on the Internet at http://www.nationalreview.com.

National Academy of Sciences NAS. (2003). "Frontiers in Agricultural Research: Food Health, Environment Communities." National Academy of Sciences, Washington, D.C.

Pimentel, D. (1980). "Handbook of Energy Utilization in Agriculture." CRC Press, Boca Raton.

Pimentel, D. (2001). The limitations of biomass energy. *In* "Encyclopedia of Physical Science and Technology," pp. 159–171. Academic Press, San Diego, CA.

Pimentel, D. (2003). Soil Erosion: A major environmental threat. *In*: "Making Development Work: A New Role for Science" (G. Leclerc, and C. A. S. Hall, Eds.). University of New Mexico Press, Albuquerque (in press).

Pimentel, D., and Pimentel, M. (1996). "Food, Energy and Society." Colorado University Press, Boulder.

Pimentel, D., Dazhong, W., and Giampietro, M. (1990). Technological changes in energy use in U.S. agricultural production. *In* "Agroecology" (S. R. Gleissman, Ed.), pp. 305–321. Springer-Verlag, New York.

Pimentel, D., Doughty, R., Carothers, C., Lamberson, S., Bora, N., and Lee, K. (2002). Energy inputs in crop production: Comparison of developed and developing countries. *In* "Food Security & Environmental Quality in the Developing World" (R. Lal, D. Hansen, N. Uphoff, and S. Slack, Eds.), pp. 129–151. CRC Press, Boca Raton.

Pimentel, D., Houser, J., Preiss, E., White, O., Fang, H., Mesnick, L., Barsky, T., Tariche, S., Schreck, J., and Alpert, S. (1997). Water resources: Agriculture, the environment, and society. *BioScience* 47(2), 97–106.

Pimentel, D., Bailey, O., Kim, P., Mullaney, E., Calabrese, J., Walman, L., Nelson, F., and Yao, X. (1999). Will the limits of the Earth's resources control human populations? *Environ. Dev. Sustain.* 1, 19–39.

United States Bureau of the Census (USBC). (2002). "Statistical Abstract of the United States 2002." USBC, U.S. Govt. Printing Office, Washington, D.C.

Wereko-Brobby, C., and Hagan, E. B. (1996). "Biomass Conversion and Technology." John Wiley & Sons, New York.

Youngquist, W. (1997). "GeoDestinies: The Inevitable Control of Earth Resources Over Nations and Individuals." National Book Company, Portland, Oregon.

Industrial Ecology

AMIT KAPUR and THOMAS E. GRAEDEL
School of Forestry and Environmental Studies, Yale University
New Haven, Connecticut, United States

1. Introduction to Industrial Ecology
2. Methods and Tools of Industrial Ecology
3. Industrial Ecology and Energy
4. Conclusion

Glossary

design for environment An engineering perspective in which environmentally related characteristics of a product, process, or facility design are optimized.

ecoefficiency A business strategy to produce goods with lower use of materials and energy, to realize economic benefits of environmental improvements.

industrial ecology An approach to the design of industrial products and processes that evaluates such activities through the dual perspectives of product competitiveness and environmental interactions.

industrial metabolism A concept to emulate flows of material and energy in industrial activities from a biological systems perspective.

industrial symbiosis A relationship within which at least two willing industrial facilities exchange materials, energy, or information in a mutually beneficial manner.

life cycle assessment A concept and a methodology to evaluate the environmental effects of a product or activity holistically, by analyzing the entire life cycle of a particular material, process, product, technology, service, or activity. The life cycle assessment consists of three complementary components: (1) goal and scope definition, (2) inventory analysis, and (3) impact analysis, together with an integrative procedure known as improvement analysis.

material flow analysis An analysis of flow of materials within and across the boundaries of a particular geographical region.

pollution prevention The design or operation of a process or item of equipment so as to minimize environmental impacts.

recycling The reclamation and reuse of output or discard material streams for application in products.

remanufacture The process of bringing large amounts of similar products together for purposes of disassembly, evaluation, renovation, and reuse.

In industrial ecology, the approach to understand industry–environment interactions is to move from contemporaneous thinking, or thinking about past mistakes, to forward thinking. The objective is to minimize or eliminate environmental impacts at the source rather than to rely on traditional end-of-pipe measures in a command and control regime. If properly implemented, industrial ecology promotes business competitiveness and product innovation. In addition, industrial ecology looks beyond the actions of single firms to those of groups of firms or to society as a whole.

1. INTRODUCTION TO INDUSTRIAL ECOLOGY

Industrial ecology is a nascent and challenging discipline for scientists, engineers, and policymakers. Often termed the "science of sustainability," the contemporary origins of industrial ecology are associated with an article titled "Strategies for Manufacturing," written by Frosch and Gallopoulos and published in 1989 in *Scientific American*. However, historically, indirect references to the concept of industrial ecology date back to the early 1970s. The multidisciplinary nature of industrial ecology makes it difficult to provide a consistent and universally accepted definition, but the following statement captures the essence of the topic:

Industrial ecology is the means by which humanity can deliberately and rationally approach and maintain sustainability, given continued economic, economic, cultural, and technological evolution. The concept requires that an industrial ecosystem be viewed not in isolation from its surrounding system, but in concert with them. It is a

systems view in which one seeks to optimize the total materials cycle from virgin material, to finished material, to component, to product, to obsolete product, and to ultimate disposal. Factors to be optimized are resources, energy and capital.
—Graedel and Allenby, 2002

Several core elements characterize the discipline: the biological analogy, the use of systems perspectives, the role of technological change, the role of companies, ecoefficiency and dematerialization, and forward-looking research and practice. Each of the themes offers a plethora of methods and tools for analysis. In following section, some of the most important aspects and tolls of the core elements, especially those particularly relevant to energy, are discussed.

2. METHODS AND TOOLS OF INDUSTRIAL ECOLOGY

Industrial ecology offers a realm of methods and tools to analyze environmental challenges at various levels—process, product, facility, national, and global—and then to come up with responses to facilitate better understanding and provide suitable remedies. Some of the important components in the industrial ecology toolbox are discussed in the following sections.

2.1 Life Cycle Assessment

A central tenet of industrial ecology is that of life cycle assessment (LCA). The essence of LCA is the examination, identification, and evaluation of the relevant environmental implications of a material, process, product, or system across its life span, from creation to disposal or, preferably, to recreation in the same or another useful form. The formal structure of LCA, contains three stages: goal and scope definition, inventory analysis, and impact analysis, each stage being followed by interpretation of results. The concept is illustrated in Fig. 1. First, the goal and scope of the LCA are defined. An inventory analysis and an impact analysis are then performed. The interpretation of results at each stage guides an analysis of potential improvements (which may feed back to influence any of the stages, so that the entire process is iterative). There is perhaps no more critical step in beginning an LCA evaluation than to define as precisely as possible the evaluation's scope: What materials, processes, or products are to

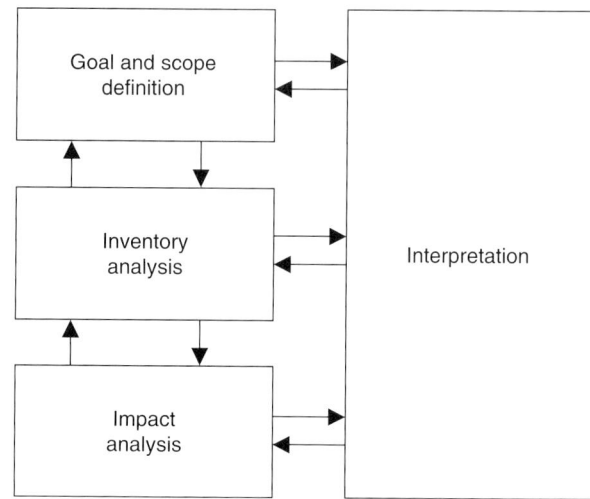

FIGURE 1 Stages in the life cycle assessment of a technological activity. The arrows indicate the basic flow of information. Reprinted from *Guidelines for Life-Cycle Assessment: A Code of Practice*, Society of Environmental Toxicology and Chemistry (1993), with permission.

be considered, and how broadly will alternatives be defined? To optimize utilization of resources in an LCA exercise, the depth of analysis should be keyed to the degree of freedom available to make meaningful choices among options, and to the importance of the environmental or technological issues leading to the evaluation.

The inventory analysis is by far the most well-developed component of LCA. It uses quantitative data to establish levels and types of energy and materials used in an industrial system and the environmental releases that result. The impact analysis involves relating the outputs of the system to the impact on the external world into which outputs flow, or, at least to the burdens being placed on the external world. The interpretation-of-results phase occurs when the findings from one or more of the three stages are used to draw conclusions and recommendations. The output from this activity is often the explication of needs and opportunities for reducing environmental impacts as a result of industrial activities being performed or contemplated.

A comprehensive LCA can be expensive and time consuming. As a consequence, more efficient approaches (streamlined LCAs, or SLCAs) have been developed with the intention of retaining the useful broad-scope analysis of the LCA while making the activity more tractable. In the case of either LCA or SLCA, the effort helps the analyst think beyond the boundaries of a particular facility or process to encompass the full measure of associated environmental implications.

2.2 Design for Environment

Product design engineers are always faced with the challenge of optimizing the multitude of attributes that determine the success or failure of the product. The paradigm for such design considerations, Design for X (DfX), states that X may be any of a number of attributes, such as assembly, compliance, disassembly, environment, manufacturability, reliability, safety, and serviceability. Design for Environment (DfE) is the DfX-related focus of industrial ecologists. The core theme of DfE philosophy is that it should improve the environmentally related attributes of a product while not comprising other design attributes such as performance, reliability, aesthetics, maintainability, cost, and time to market. DfE approaches systematically evaluate environmental concerns during the product life cycle stages of premanufacture, manufacture, delivery and packaging, use, and end of life, and accordingly set targets for continual improvements. The choice of materials during pre-manufacture and their efficiency of utilization during product manufacture, energy use during manufacturing and product use, and the environmentally friendly disposal or reincarnation of products at end of life are some of the prime considerations in DfE. DfE is also a "win-win" proposition in that it provides a corporation with a competitive edge in an ever-tightening regulatory environment, and promotes ongoing product innovation.

2.3 Industrial Symbiosis

The industrial ecologist views the economy as a closed system, similar to a natural system, in which the "residues" from one system are the "nutrients" for another. The concept known as industrial symbiosis is a current topic of research for industrial ecologists and environmentalists in identifying strategies to enable businesses to "close the loop." The objective is to create or encourage the formation of industrial production systems that function similarly to biological food chains. In either natural or industrial systems, symbiosis occurs when two or more organisms form an intimate association, either for the benefit of one of them (parasitic symbiosis) or for both or all of them (mutualistic symbiosis), such that there is high degree of synergy between input and output flows of resources. The most well-known industrial symbiosis example is the Kalundborg industrial system in Denmark. The qualitative material flows at Kalundborg are shown in

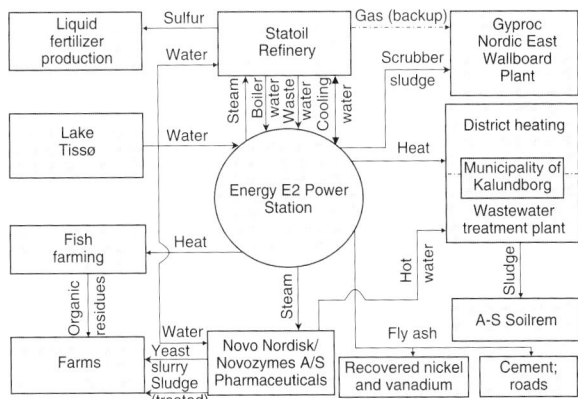

FIGURE 2 Material flows in the Kalundborg, Denmark industrial ecosystem. Reprinted from the research work of M. R. Chertow, Yale University, with permission.

Fig. 2. The heart of this industrial ecosystem is Denmark's largest power plant, which exchanges various residues (gypsum, steam, fly ash, etc.) with neighboring entities, i.e., the Statoil refinery, the Nova Nordisk pharmaceutical unit, the Gyproc plasterboard facility, and the town of Kalundborg.

Industrial symbiosis may occur opportunistically or can be planned. Coordinating industrial symbiosis appears to offer the promise of developing industrial ecosystems that are superior environmentally. Such a system would need to involve a broad sectoral and spatial distribution of participants, and to be flexible and innovative. The formation of ecologically balanced industrial systems results in numerous environmental and economic benefits. Economic benefits, which are shared by participating businesses, governments, and communities, are the primary driving force for setting up such industrial configurations. Entrepreneurs can gain appreciable cost savings from reduced waste management, reduced infrastructure costs, and improved process and product efficiency. There are opportunities for other cooperative ventures such as joint purchasing, combined waste recovery and treatment, employee training, environmental monitoring, and disaster response. The tangible environmental benefits include the reduction of greenhouse gas emissions and toxic air emissions, improving efficiency and conservation in the use of energy, materials and water, improving land use planning and green space development within the industrial complexes, and promotion of pollution prevention and recycling approaches.

2.4 Ecoefficiency, Dematerialization, and Decarbonization

The anthropogenic activities of production and consumption, both being integral parts of a robust economic world, are considered to be major factors in the sustainability debate. The concept of ecoefficiency evolved in the early 1990s, prior to the Earth Summit. Simply stated, ecoefficiency means doing more with less. The World Business Council for Sustainable Development defines ecoefficiency as being "attained by the delivery of competitively priced goods and services that satisfy human needs and bring quality of life, while progressively reducing ecological impacts and resource intensity throughout the life cycle, to a level at least in line with the earth's estimated carrying capacity." Businesses worldwide have begun to embrace ecoefficiency as a mutually beneficial strategy. Ideologically, ecoefficiency and dematerialization are synonymous. Dematerialization can be achieved by making products either smaller or lighter (e.g., personal computers today as compared to a decade ago), by replacing a product with an immaterial substitute (e.g., large-scale use of electronic mail rather than regular postal mail), or by reducing the use of material-intensive systems. Intensity of use of materials, defined as resource use per gross domestic product, is a commonly used metric to evaluate dematerialization trends. The key determinants that influence patterns of intensity of use are changes in material composition of products (material substitution and technological change) and product composition of income (structural changes in the economy and intrasectoral shifts). The variation of material intensity of use over the past century for some of the materials used extensively in the United States economy is shown in Fig. 3. There is an appreciable decline in the use of metals except for aluminum, whereas the use of paper and plastics continues to grow. These trends also indicate that the composition of materials in the United States economy changed over this period from dense to less dense, i.e., from iron and steel to light metals, plastics, and composites. At the product level, similar trends can also be observed; for example, the change in the weight composition of an average automobile in the United States is shown in Fig. 4. Plastics, composites, aluminum, and other specialty materials account for the decline in the use of conventional steel in automobiles over the 1978–2000 period.

In energy systems analysis, dematerialization is analogous to decarbonization, which refers either to moving away from carbon-intensive energy sources

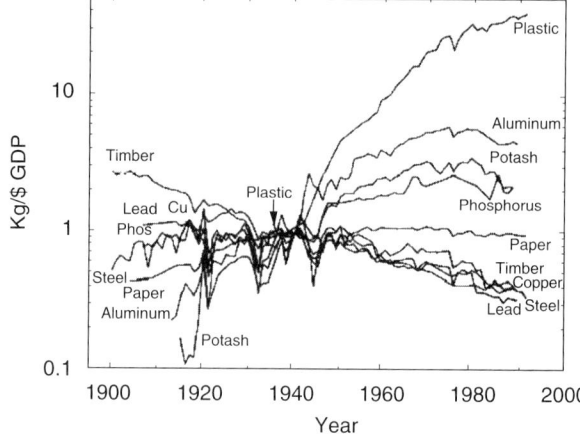

FIGURE 3 The intensity of use of materials in the United States, 1900–1990. The annual consumption data are divided by gross domestic product (GDP) in constant 1987 dollars and normalized to unity in the year 1940. Reprinted from Wernick (1996), with permission.

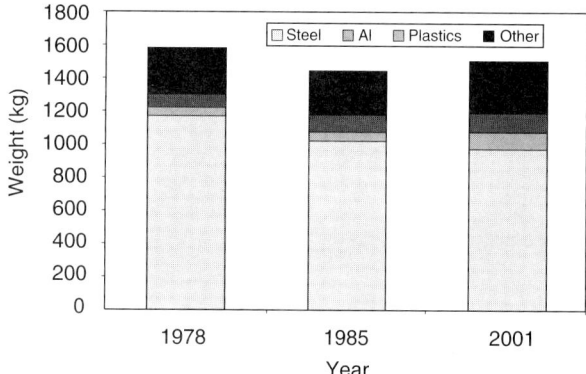

FIGURE 4 The average material composition of a domestic automobile in United States. Data from the U.S. Department of Energy (2001).

to completely carbon-free sources such as hydrogen or solar energy. Although evidence for any comprehensive dematerialization trend remains ambiguous, the situation is clearer with respect to energy. Historically, trends in fuel market share of global primary energy use indicate a transformation from carbon-intensive fossil fuels such as wood and coal to less intensive ones such as oil and natural gas, as shown in Fig. 5. The decline in carbon intensity has been very slow, approximately 0.3% per year over the 1850–1994 time period.

2.5 Industrial Metabolism

The extraction, production, and waste treatment of materials cause environmental problems that call for

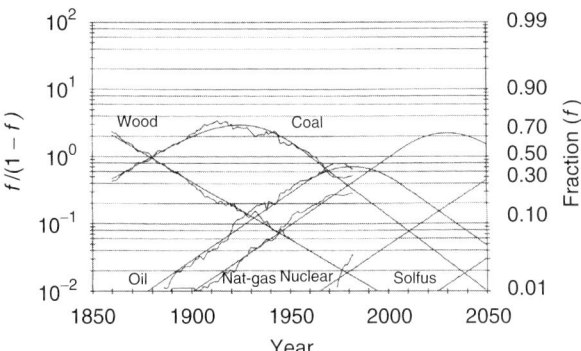

FIGURE 5 Global primary energy substitution, 1860–1980, and projections to 2050 (expressed in fractional market shares, f. Smooth lines represent model calculations and jagged lines are historical data. "Solfus" is a term employed to describe a major new energy technology (for example, solar or fusion). Reprinted from Nakićenović (1997), with permission.

an intervention by all the stakeholders: governments, corporations, and individuals. The approach to link society's management of materials to sustainable development and environmental quality is now well defined and understood. The concept of "industrial metabolism," first introduced by Robert Ayres, establishes an analogy between economy and environment on a material level. Societies mobilize material from Earth's crust to create "technomass," in analogy to nature's biomass. The ecosystems in the biosphere try to close the loop by cycling resources and wastes repeatedly. In contrast, modern society tends to use materials once and then discard them. Industrial ecologists have begun to assess the physical economy through the lens driven by biosphere guiding principles. The objective of such analytical studies from a systems perspective is to determine the anthropogenic contribution to natural material flows, the causal factors, and the spatial and temporal distribution of environmental problems.

Material balances, based on the universal law of mass conservation, have historically been used as an instrument to describe material flows through the economy. Within that framework, material flow analysis (MFA) and substance flow analysis (SFA) are two important approaches to assess the current and future state of resource flows and accumulation in the economy and environment. MFA usually tracks material intensities of national economies or sectors thereof, concentrating on bulk flows, whereas SFA is intended for flows of specific substances to identify specific environmental problems and propose remedial/prevention strategies.

An example of an SFA, a contemporary copper cycle for Germany, is shown in Fig. 6. The cycle for Germany represents a highly industrialized country without significant virgin copper resources. Unable to

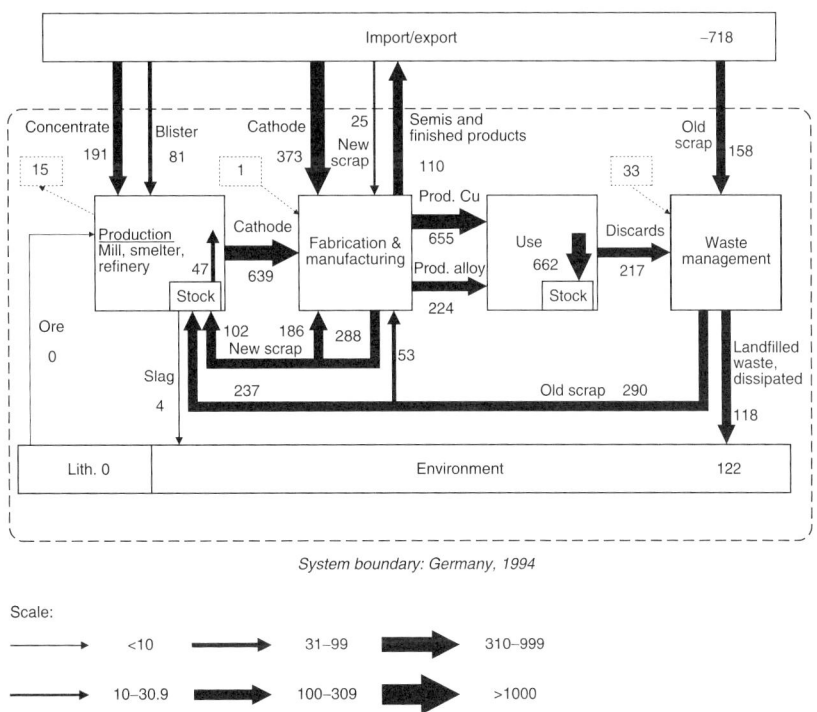

FIGURE 6 The contemporary copper cycle for Germany in 1994. All values are in Gg of Cu per year.

rely on an internal supply, Germany imports copper in different forms at every industrial life stage (production, fabrication and manufacturing, and waste management). Its reuse of discarded and imported scrap is so extensive that nonvirgin copper supplies represent about 45% of all inflows to its industrial facilities. Nonetheless, Germany landfills about 120 Gg Cu/year, more than any country in Europe. This substance diagram can be converted to an energy diagram if the flows are multiplied by the energy needed to effect a transfer of ore unit of copper from ore reservoir to another. The results make it obvious that the extraction of copper from ore is the most energy-intensive step in the entire cycle.

2.6 IPAT Equation

The environmental impact of materials use has often been conceptually formed by the "IPAT" equation:

$$I = P \times A \times T \quad (1)$$

in which the overall environmental impact I is expressed as the product of population P, the affluence A (expressed, for example, as gross domestic product [GDP] per person), and technology T (expressed, for example, as environmental impact per unit of per capita GDP). Historically proposed by Commoner, Ehrlich, and Holdren, the IPAT equation is now the "master equation" of industrial ecology. If the technology factor is expanded somewhat, the equation can be rewritten as

$$I = P \times A \times M \times D \times H \quad (2)$$

where M is the materials intensity, D is the dissipation factor, and H is the hazard factor, which depends on the chemical form of the material lost and the susceptibility of the receiving ecosystem. In words, Eq. (2) becomes

Environmental impact
$= $ (Population) $\times$ (GDP/person)
$\times$ (units of material/GDP)
$\times$ (units of pollution/unit of material)
$\times$ (impact/unit of pollution) $\quad (3)$

In the energy sector, the equivalent of the IPAT equation is the Kaya's identity, a mathematical expression for energy-related carbon emissions that can be written as

$$C = (P) \times (GDP/P) \times (E/GDP) \times (C/E) \quad (4)$$

Where the total energy-related carbon emissions C is expressed as a product of population P, GDP per capita, energy intensity E/GDP, and carbon intensity of energy use C/E. Although the master equation and Kaya's identity should be viewed as conceptual rather than mathematically rigorous, they can be used to suggest goals for technology and society. The technology-related terms M, E, and C offer the greatest hope for a transition to sustainable development, especially in the short term, and it is modifying these terms that is among the central tenets of industrial ecology.

3. INDUSTRIAL ECOLOGY AND ENERGY

Industrial ecology and energy use are inextricably linked because mobilization and utilization of materials in our present technological society are indispensable without the utilization of energy. However, global commons are threatened by environmental emissions from energy use. Simultaneous improvement in material and energy intensities of use is one of the desirable goals to preempt ecological damage across different scales and levels. In the following sections, the discussion focuses on energy considerations during different product life cycle stages.

3.1 Energy Considerations in Material Choice

Embodied energy, or "embedded energy," is a concept that includes the energy required to extract raw materials from nature, plus the energy utilized in the manufacturing activities. Inevitably, all products and goods have inherent embodied energy. The closer a material is to its natural state at the time of use, the lower is its embodied energy. Sand and gravel, for example, have lower embodied energy as compared to copper wire. It is necessary to include both renewable and nonrenewable sources of energy in an embodied energy analysis. The energy requirements for acquisition in usable form from virgin stocks of a number of common materials are shown in Fig. 7. From an industrial ecology perspective, a manufacturing sequence that uses both virgin and consumer-recycled material is usually less energy intensive than is primary production. The energy requirements for primary and secondary production of various metals are shown in Fig. 8. For one of the most commonly used industrial materials, aluminum, the energy requirement for secondary production of aluminum is approximately 90% less than the

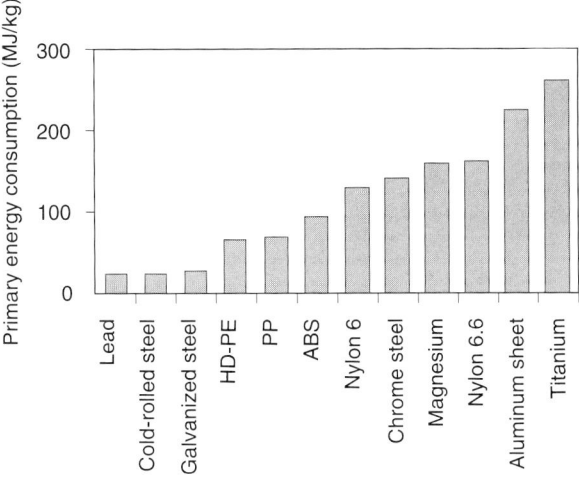

FIGURE 7 The primary energy consumption required to produce 1 kg of various materials. HD-PE, High-density polyethylene; PP, polypropylene; ABS, acrylonitrile butadiene styrene. Adapted from Schuckert et al. (1997).

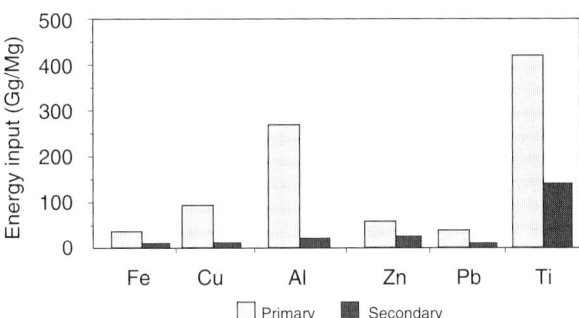

FIGURE 8 Energy requirements for primary and secondary production of metals. Data from Chapman and Roberts (1983).

requirement for primary production using virgin resource. Therefore, an efficient recycling operation can lead to potential savings in energy consumption and associated environmental damage.

3.2 Energy Considerations in Product Manufacture

Although the materials extraction and processing sectors have the highest energy intensity, these industries are suppliers to the intermediate processing industries, so there cannot be a plan to decrease industrial energy use solely by eliminating the extraction industries (which would not be possible in any case). The consumption of energy in selected manufacturing industries is shown in Fig. 9. Production petroleum and coal accounts for the largest energy use. Most of this energy use is attributable to

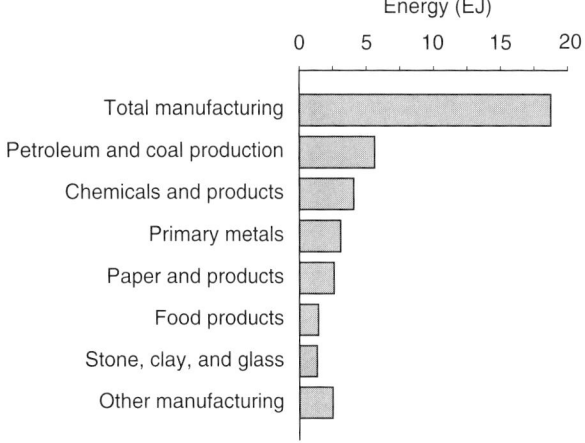

FIGURE 9 Consumption of energy in selected manufacturing industries. Data from the U.S. Department of Energy (1990).

the mining and processing of coal and the refining of petroleum. The trend toward desulfurization of crude oil and the production of high-octane gasoline without the use of metal-containing additives place ever-increasing energy demands on the refining operations. Refinery operations are generally subject to careful supervision and continuing engineering effort to improve efficiencies, but increased attention to cogeneration, heat exchange, and leak prevention is likely to offer opportunities for further improvements. Chemicals and chemical products rank among the industries depicted in Fig. 9 as well, although about a third of the amount shown represents petroleum and natural gas as feedstocks for products, rather than the fuel that is consumed to produce energy. Of the remaining two-thirds, a substantial amount is used in the generation or removal of process heat as a result of temperature differences between process streams and the heating and cooling streams. The production of compressed gases is another energy-intensive operation in this sector. Primary metals are the third industry listed in Fig. 9. Although the extraction of ore from the ground and its shipment are quite energy intensive, the bulk of the energy use is in crushing rock and recovering the target ore, and in generating the large amounts of process heat needed to extract metal from ore and to produce ingots and other purified products.

To maximize the efficiency with which water and energy are used in manufacturing and to minimize overall water and energy loss rates are major goals of pollution prevention. These actions are seen as complementary to those directed toward routine emissions to air, water, and soil, or to leaks or accidents. Energy audits for the different process operations and for the overall facility are always helpful to indicate opportunities to reduce energy use. A particularly successful energy conservation program to date was initiated by the Louisiana Division of the Dow Chemical Company in 1982. Many of the improvements embodied techniques useful industry-wide, such as installing insulation on pipes carrying hot fluids, cleaning heat exchanger surfaces frequently to improve heat transfer efficiency, and employing point-of-use fluid heaters when storage or long pipelines create the potential for heat loss. The Dow energy contest results are summarized in Table I. It is clear that energy conservation, either through good housekeeping, improved technology, or process change or modification, is always beneficial.

3.3 Energy Considerations during Product Use

Designing innovative products that provide maximum benefit and service to customers and that are also simultaneously environmentally responsible is a challenge and an opportunity for product designers. DfE guidelines, evolved through feedback from customer expectations and regulatory policies, offer a framework to identify, prioritize, and implement product design improvements.

On a life cycle basis, the "in-use" phase is often dominant in terms of energy use. For a modern jet engine, for example, an improvement of 0.1% in fuel-burning efficiency is a more important design change than would be difficult material choices, improved manufacturing efficiency, or end-of-life design, so far as the environment is concerned. Energy-efficient designs may sometimes involve new approaches; the result may not only lower cost operation but may also improve product positioning (from a sales standpoint), particularly in areas of the world that are energy poor.

TABLE I

Energy Conservation Projects at Dow Chemical Company, Louisiana Division

Project	1982	1984	1986	1988	1990	1992
Number	27	38	60	94	115	109
Average return on investment (%)	173	208	106	182	122	305

^aData from Nelson (1994).

3.4 Energy Considerations in Remanufacturing and Recycling

When products reach the end-of-life stage, they can revert into input materials if they are recovered and recycled. Recycling operations are usually viable if the quantities recovered are large enough and the discard stream is homogeneous and concentrated. Even with extensive recycling, however, a growing economy still has a need for virgin resources, and processing of any kind requires energy.

In remanufacturing, products are either refurbished or reconditioned to the same quality control as new products, and are subsequently returned to market. Remanufacturing is more energy efficient as compared to recycling, because no new materials have to be processed. Remanufacutring also saves on landfill space and costs. From a DfX perspective, design for disassembly is important in remanufacturing because modular design of products will facilitate disassembly with less effort and time. In the United States, a remanufacturing tax credit has been included in the Clean Energy Incentives Act to encourage businesses. However, the lack of product "take back" laws to make producers responsible for their products at the end of the product life span, and the typical consumer's psychological preference for new rather than refurbished products, provide barriers against an extensive and well-integrated remanufacturing industry.

A perspective for product designers, remanufacturers, and recyclers is shown in Fig. 10, and details are provided in Table II. Different types of products are plotted on the basis of their product lifetime and technology cycle. There are four product categories,

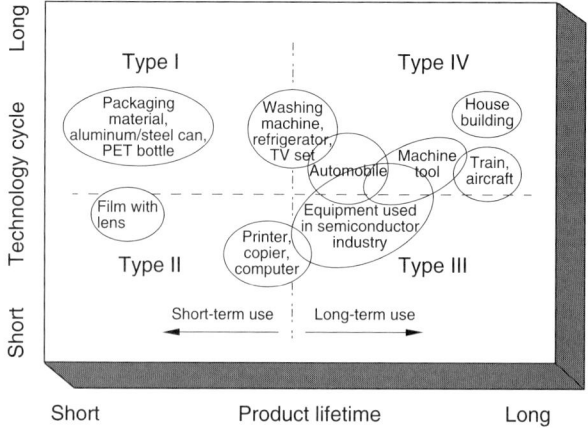

FIGURE 10 End-of-life plot for different product categories. PET, Polyethylene terephthalate. Reprinted from Masui (2002), with permission.

TABLE II

Guidelines for Product Designers and Recycling Technology Developers

Product category	End-of-life scenario	Product designer goal	Recycling technology developer goal
Type I	Material recovery	Ease separation of components for recycling high quality material	Develop separation technologies accounting for different physical properties of materials that cannot be sorted
Type II	Remanufacturing	Enhance reusability by using common parts and modular components in product family	Develop efficient cleaning and inspection technologies to reduce remanufacturing cost
Type III	Lengthen product life by upgrading	Extend product life by modular design of key devices that define value of product	Develop nondestructive techniques for removal of key components
Type IV	Lengthen product life by maintenance	Enhance ease of product disassembly for facilitating maintenance	Develop diagnostic technologies for maintenance

each of which offers a suite of environmental and nonenvironmental challenges to designers and recyclers. The complexity of the challenge increases in anticlockwise direction, moving from Type I products to Type IV in Fig. 10. Generally speaking, the energy consumption of these products also increases from Type I to Type IV. This characterization focuses on only two attributes and does not include functional and operational efficiency aspects, compatibility, number of parts, etc., but it does provide a framework for development of strategies to deal with products at their end of life.

4. CONCLUSION

To an extent not generally appreciated, the study of industrial ecology is simultaneously the study of energy—its methods of generation, its employment in driving the engines of industry, and its use by individuals everywhere as they employ the technology of the modern world. Consider the diagram in Fig. 11. The driver for all energy use, shown on the left, is the contribution of societal needs and cultural desires related to energy. These define the type and number of energy-using products required to satisfy those needs and desires. The requisite energy is produced by the energy industry, indicated on the right. In the center are the design, use, and recycling of products. It is these steps that are the province of industrial ecology, which aims to satisfy the needs and desires while minimizing the use of energy in product design and manufacture, the use of energy in product operation, the use of energy-intensive virgin materials, and the magnitude of energy-related

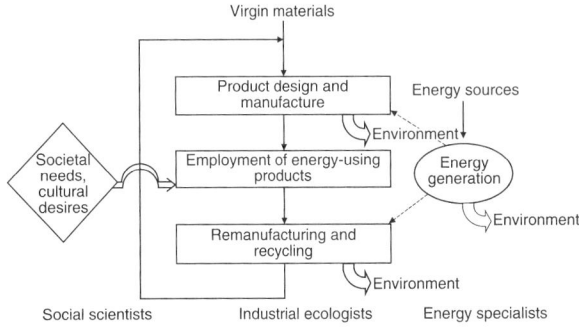

FIGURE 11 Satisfying the energy-related needs and desires of societies and cultures. The principal actors in each column of the diagram are shown at the bottom.

environmental impacts. The evolution of our society in the direction of increased sustainability will require not only that energy be widely available, but that it is used with maximum efficiency, and in such ways that its use produces minimal environmental impact. A close partnership between the energy industry and industrial ecologists will be required to make that vision a reality.

SEE ALSO THE FOLLOWING ARTICLES

Commercial Sector and Energy Use • *Ecological Footprints and Energy* • *Ecological Risk Assessment Applied to Energy Development* • *Goods and Services: Energy Costs* • *Industrial Agriculture, Energy Flows in* • *Industrial Energy Efficiency* •

Industrial Energy Use, Status and Trends • Industrial Symbiosis • Life Cycle Assessment and Energy Systems

Further Reading

Ayres, R. U. (1989). Industrial metabolism. *In* "Technology and Environment" (J. H. Ausubel and H. E. Sladovich, Eds.), pp. 23–49. National Academy Press, Washington, D.C.

Ayres, R. U., Norberg-Bohm, V., Prince, J., Stigliani, W. M., and Yanowitz, J. (1989). "Industrial Metabolism, the Environment and Application of Material Balance Principles for Selected Chemicals." International Institute for Applied Systems Analysis (IIASA) Report RR-89-11. IIASA, Laxenburg, Austria.

Ayres, R. U., and Simonis, U. E. (1994). "Industrial Metabolism, Restructuring for Sustainable Development." United Nations Press, Tokyo.

Ayres, R. U., and Ayres, L. W. (1996). "Industrial Ecology: Towards Closing the Materials Cycle." Edward Elgar Publishers, Cheltenham, United Kingdom.

Chapman, P. F., and Roberts, F. (1983). "Metal Resources and Energy." Butterworths, London.

Erkman, S. (2002). The recent history of industrial ecology. *In* "Handbook of Industrial Ecology" (R. U. Ayres and L. W. Ayres, Eds.), pp. 27–35. Edward Elgar, Northampton, Massachusetts.

Frosch, R. A., and Gallopoulos, N. E. (1989). Strategies for manufacturing. *Sci. Am.* **261**(3), 94–102.

Graedel, T. E. (1996). "Streamlined Life-Cycle Assessment." Prentice Hall, Upper Saddle River, New Jersey.

Graedel, T. E. (2000). Evolution of industrial ecology. *Environ. Sci. Technol.* **34**(1), 28A–31A.

Graedel, T. E., and Allenby, B. R. (2002). "Industrial Ecology," 2nd Ed. Prentice Hall, Upper Saddle River, New Jersey.

Graedel, T. E., van Beers, D., Bertram, M. B., Fuse, K., Gordon, R. B., Gritsinin, A., Kapur, A., Klee, R. J., Lifset, R. L., Memon, L. A., Rechberger, H., Spatari, S., and Vexler, D. (2004). The multilevel cycle of anthropogenic copper. Accepted for publication, *Environ. Sci. Technol.*

Grübler, A., and Nakićenović, N. (1996). Decarbonizing the global energy system. *Technol. Forecast. Social Change* **23**, 97–110.

Kaya, Y. (1990). "Impact of Carbon Dioxide Emission Control on GNP Growth: Interpretation of Proposed Scenarios." Paper presented to the International Panel on Climate Change (IPCC) Energy and Industry Subgroup, Response Strategies Working Group. IPCC, Paris, France [mimeo].

Kneese, A. V., Ayres, R. U., and d'Arge, R. C. (1970). "Economics and the Environment." Resources for the Future, Washington, D.C.

Lifset, R. L., and Graedel, T. E. (2002). Industrial ecology: Goals and definitions *In* "Handbook of Industrial Ecology" (R. U. Ayres and L. W. Ayres, Eds.), pp. 3–15. Edward Elgar, Northampton, Massachusetts.

Masui, K. (2002). "Life Cycle Strategies." Available on the Web site of the Agency of Industrial Science and Technology, Mechanical Engineering Laboratory at http://www.mel.go.jp (accessed on 11/16/02).

Nakićenović, N. (1997). Decarbonization as long-term energy strategy. *In* "Environment, Energy and Economy—Strategies for Sustainability" (Y. Kaya and K. Yokobori, Eds). United Nations University Press, Tokyo.

Nelson, K. E. (1994). Finding and implementing projects that reduce waste. *In* "Industrial Ecology and Global Change" (R. Socolow, C. Andrews, F. Berkhot, and V. Thomas, Eds.), pp. 371–382. Cambridge University Press, Cambridge, UK.

Schuckert, M., Beddies, H., Florin, H., Gediga, J., and Eyerey, P. (1997). Quality requirements for LCA of total automobiles and its effects on inventory analysis. *In* "Proceedings of the Third International Conference on Ecomaterials," pp. 325–329. Society of Non-Traditional Technology, Tokyo.

Society of Environmental Toxicology and Chemistry (SETAC). (1993). "Guidelines for Life-Cycle Assessment: A Code of Practice." SETAC, Pensacola.

United Nations Environment Program (UNEP). (2002). "The Industrial Symbiosis in Kalundborg, Denmark." Available on the United Nations Web site at http://www.uneptie.org (accessed on 11/16/02).

U.S. Department of Energy (DOE). (1990). "Manufacturing Energy Consumption Survey: Changes in Energy Efficiency 1980–1985." DOA/EIA-05169(85). Energy Information Administration, U.S. DOE, Washington, D.C.

U.S. Department of Energy (DOE). (2001). "Transportation Energy Data Book: Edition 21." Oak Ridge National Laboratory, U.S. DOE, Oak Ridge, Tennessee.

Wernick, I. K. (1996). Consuming materials: The American way. *Technol. Forecast. Social Change* **23**, 111–122.

Industrial Energy Efficiency

WOLFGANG EICHHAMMER
Fraunhofer Institute for Systems and Innovation Research
Karlsruhe, Germany

1. Definitions of Industrial Energy Efficiency
2. Industrial Energy Efficiency Trends
3. Energy Efficiency Benchmarking
4. Potentials for Industrial Energy Efficiency Improvement
5. Obstacles to Industrial Energy Efficiency Improvement and Transaction Costs
6. Long-Term Potential for Industrial Energy Efficiency Improvement

Glossary

barriers/obstacles to energy efficiency Factor that explains why energy efficiency measures are not carried out in the industrial sector despite the fact that they are economic according to company criteria.
energy efficiency benchmarking Comparison of the energetic performance of a company with the performance of other companies working in the same field.
energy efficiency improvement Decreasing the use of energy per unit service or activity without substantially affecting the level of these services.
energy intensity The amount of energy used per unit of activity or service.
exergy Ability of an energy form to perform work. Electricity has 100% exergy, while low-temperature heat has only a few percentage points.
transaction costs Costs occurring to change from one situation to another, such as from the current state to a more energy efficient state.

The industrial sector still represents in many countries a large fraction of energy consumption of typically 30 to 40% of the overall final energy consumption of a country. Although this sector has made continuous progress in energy efficiency over time (the article explains how to measure energy efficiency in the industrial sector), many examples from daily practice show that the economic energy efficiency potential in the industrial sector is far from being exhausted. Energy efficiency has many facets. Section 1 of this article introduces various concepts for energy efficiency in the industrial sector and explains why misunderstandings might arise from the nonexplicit use of different concepts. Section 2 uses the macroeconomic efficiency concept to describe the development of industrial energy efficiency, in particular the decoupling of economic growth and energy consumption, but also the pitfalls of this type of energy efficiency indicators. Section 3 shows how, at the microlevel, companies can use the benchmarking concept to improve their energy efficiency. Section 4 deals with the potentials for energy efficiency in the industrial sector (at a time horizon of 10 to 20 years) and how to realize them, while Section 5 deals with obstacles to industrial energy efficiency and transaction costs. Section 6 takes a look beyond the next two decades and points to the role of industrial energy efficiency research and development.

1. DEFINITIONS OF INDUSTRIAL ENERGY EFFICIENCY

When energy efficiency improvements are discussed for the industrial sector, quite often very different concepts are used for defining energy efficiency, which convey different messages that can even be contradicting. This is illustrated by the following tale.

In a public hearing in the year 2015, the (hypothetical) president of the European Commission was asked how much the energy efficiency of the iron and steel industry had improved between 2005 and 2015 in the framework of voluntary agreements on the reduction of CO_2 emissions. Naturally he did not know, but he had an expert to ask. The expert, after a few seconds of inquiry, came up with the following answer: "If you measure energy efficiency in the iron/steel industry in terms of energy consumed per value added, then efficiency has improved by 10%; if you measure energy efficiency in terms of energy consumed per ton of crude steel, then energy efficiency has got worse by 10%." The president was puzzled but decided to communicate only the first half of the sentence to the press.

The following analysis tries to provide an explanation to the contradictory answer of the expert to the hypothetical president of the European Commission. The figures used in the example are fictitious and chosen in such a way that the example is particularly striking: in the case of one indicator, the efficiency improves (monetary units); in the case of another, it deteriorates (physical units). The development of real indicators appears somewhat less spectacular because both types of indicators improve, but to a different degree (the energy efficiency improves more when measured in monetary units). A detailed analysis of the European iron/steel industry between 2005 and 2015 might show that the production of crude steel remained fairly stable as well as the energy consumed to produce 1 ton of crude steel. But what actually happened was that the iron/steel industry, in order to secure its future, discovered a new field of products (for example, the production of mobile phones), the production of which added, let's say, about 10% to the energy consumed for the production of the crude steel alone. Calculating the ratio between energy consumption and the constant crude steel production shows an increase of 10% of the specific energy consumption. At the same time, the value added created in the iron/steel industry increased due to the new activity by 20%. The energy intensity that links the 10% increased energy consumption to the 20% increased value added shows a decrease of roughly 10%.

This still does not answer the question whether the iron/steel industry has become more or less energy efficient. The contradiction is to be found in the different use of the term "energy efficiency" in the two parts of the sentence that the expert communicated to the president. When the term "energy efficiency" is used in conjunction with value added, it includes all kinds of structural changes within the branch considered—that is, energy efficiency is interpreted in a more economic sense: the structure is becoming more efficient; less energy is necessary to produce one unit of value. The correct answer to the president after a more refined analysis would have been: The energy efficiency of the core business has neither increased nor decreased. It has remained stable. The observed increase or decrease in "efficiency" has come about through the development of a new business with low energy-intensity (i.e., the sector has a different delimitation, which is not necessarily fully comparable with the previous state when looking at the overall branch). A further decomposition of the sector using physical units for the old crude steel production and value added for the new mobile phone business would reveal the correct answer that energy efficiency has not changed (assuming that the energy intensity for the production of the mobile phones was also fairly stable over time).

This example illustrates two important issues. First of all, when talking about improvements in industrial energy efficiency, it is important to mention which kind of energy efficiency definition is used (see Table I) and how it is operationalized in the form of energy efficiency indicators. The efficiency definitions in Table I start from very technical efficiency definitions at the level of processes: for example, efficiency in terms of energy output related to energy inputs in the case of a steam boiler. A second way would be by taking into account the different value or possibility to carry out work

TABLE I

Definitions for Industrial Energy Efficiency

Type of efficiency	Basic concept
First law efficiency	The energy transfer of a desired kind by a device or system divided by the energy supplied to the device or system to operate it.
Second law efficiency	The theoretical minimum available work needed for a given task divided by the actual available work used to perform the task.
Microeconomic I: Pareto efficiency	There are no feasible distributions of utility in which everyone is at least as well off, and at least one person is better off, than under the current distribution.
Microeconomic II: Cost-effectiveness	The rate of return on the investment equals or exceeds the hurdle rate of the investor.
Macroeconomic Energy efficiency I	The total consumption of energy over a relevant economic unit divided by the economic output of that unit.
Macroeconomic Energy efficiency II	The total consumption of energy over a relevant social unit divided by the current level of social welfare or level of sustainable income.
Institutional efficiency	Institutions are structured to minimize transaction costs.
Pareto efficiency with priced externalities	Standard microeconomic efficiency with the addition of monetary values assigned to externalities.
Distributional efficiency	Distributional objectives defined on moral and social grounds; prices set to achieve objectives cost-effectively.

(exergy) of different energy forms. For example, the exergetic efficiency of the same boiler would be much lower than its energetic efficiency. This can be understood in the following way: if the boiler could convert money coins, and would convert three Euro coins to three cent coins, the energetic efficiency would say three coins in, three coins out, so the efficiency is 100%, while the exergetic efficiency would consider what is written on the coins and would say three Euro in, three cents out (i.e., the efficiency is 3%). The third definition of efficiency is an economic definition at the microlevel stating that the world is always organized in such a way that an economic optimum at the microlevel is achieved. Other definitions occur at the level of macroeconomic units (e.g., whole countries) and relate energy inputs to the macroeconomic output of the units. The definitions continue to more general concepts, including externalities or institutional inefficiencies.

The second lesson is that in communicating results on energy efficiency improvements care has to be taken to avoid confusing messages by mixing up many different concepts. If one leaves the expert arena and goes to the political arena, where the indicators and concepts are to be applied, confusion is rapidly generated as the differences between the concepts are not recognized.

2. INDUSTRIAL ENERGY EFFICIENCY TRENDS

The objective of energy efficiency improvements in the industrial sector is the decoupling of the energy consumption and the corresponding environmental burdens (e.g., in the form of CO_2 emissions) from the economic growth of the sector. Figure 1 shows that for the European Union this aim was reached in the 1990s: while the industrial value added increased by 26% in that period of time, energy consumption remained stable at the level of 1990. CO_2 emissions from industrial activities decreased even more strongly. In other regions of the world, such as in developing countries or the United States, this decoupling is still far from being achieved. Also, electricity use in the industrial sector is still growing, even in the European Union, at a level close to the growth of the value added, but in some countries decoupling effects for electricity due to saturation start to become visible.

The most commonly indicator used to measure progress of energy efficiency in the industrial sector is the energy consumption per value added. Never-

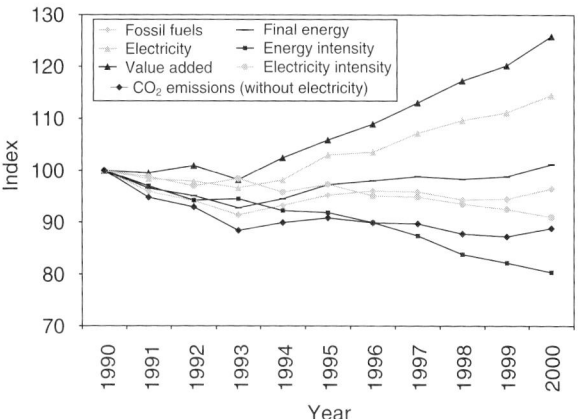

FIGURE 1 The decoupling of economic growth, energy consumption, and environmental burden: energy consumption, value added, energy intensity, and CO_2 emissions in EU industry, 1990–2000.

theless, this indicator has the disadvantage that it does not separate changes in energy intensities due to changes in the industrial sector structure from changes in energy efficiency in a more technical sense. Figure 2 illustrates this with the example of the countries forming the European Union. Not only the energy intensity decrease during the 1990s differs widely between the EU countries, but also the level of industrial energy intensity. This is mainly due to differences in the industrial structure of the countries (i.e., the respective share of more or less energy-intensive branches in total industrial production). These differences should be taken into account when comparisons of industrial energy intensities among countries are carried out. If the energy intensities are adjusted to an average European structure, the differences are considerably less pronounced than in the case of actual intensity (compare the dark arrow in Fig. 2 with the light gray arrow).

Why should one wish to separate different factors that influence the energy intensities? At the end it might not matter whether a reduction in energy consumption comes from changes in industrial structures or from energy efficiency measures. But, in fact, the point is that one might wish to influence energy efficient technologies by an adequate policy, while industrial structures might be dictated by considerations outside the energy field. Separation therefore allows to more adequately monitor factors of influence that one really wants to target by policy action.

As shown in Fig. 3 with the example of the European Union, the energy intensity of industry in a country or region tends to decrease more when the industrial value added increases strongly and to

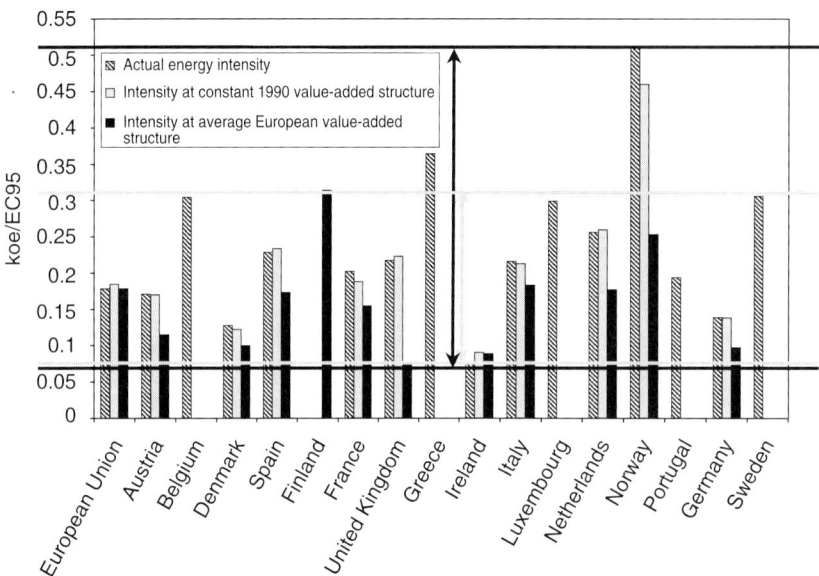

FIGURE 2 Manufacturing energy intensities in the EU countries 1999: actual, at constant 1990 value-added structure, at average European value-added structure.

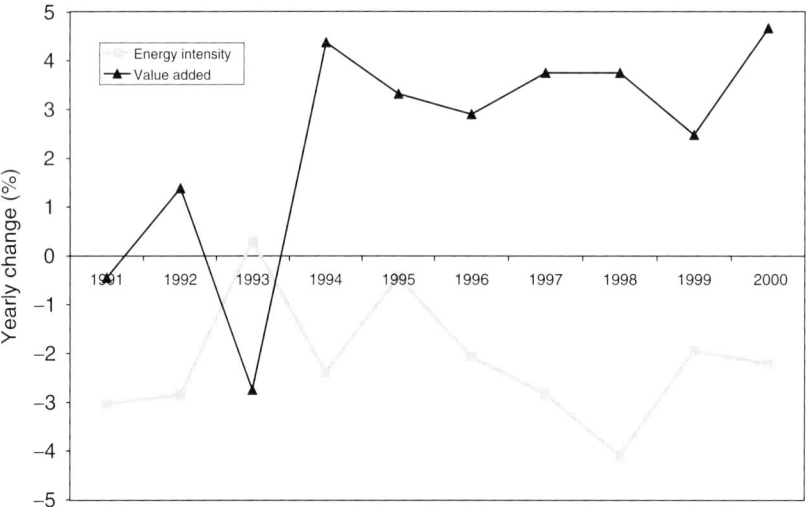

FIGURE 3 Annual growth rates of value added and energy intensity in EU industry, 1990–2000: The magnitude of the energy intensity variation is highly correlated to the industrial growth (the business cycles).

decrease less (or even increase) in the reverse situation (influence of the business cycle). The explanation is that there is one part of energy consumption that is rather independent from the production level (e.g., heating and lighting of premises) and another part that is more or less proportional to the production level (e.g., machinery, furnaces). In case of a production decrease, only the second part of consumption decreases, whereas the first one remains constant. As a result, the energy consumed per unit of production tends to increase.

3. ENERGY EFFICIENCY BENCHMARKING

At the level of companies, it appears also to be very useful to compare the own energy efficiency performance with other companies in the same field by using benchmarks. Benchmarks are widely used in many different areas of the society as well as in industrial practice to improve performances through competition and comparison with others. Energy efficiency benchmarking consists in the establishment of energy indicators that are related to the activity of an industrial company and allow in principle to compare the energetic performance of a single company with other companies in the same field of activity. Examples in practice have shown that differences in energy efficiency practice from simple to double are possible for the same type of activity (see Fig. 4).

Benchmarking, which can be found in many different areas outside the energy efficiency field, is an ongoing process of continuous improvement: finding out how the "best" companies meet these standards; setting targets for business activities according to the "best practice" that can be found; adapting and applying lessons learned from these approaches and ideas to meet and exceed the standards; identification of areas where improvement would make the greatest difference to the bottom line, to key areas of the business or to customer relationships, establishing what makes a difference in customers' perceptions between a run-of-the-mill and an excellent supplier. Why introduce energy efficiency benchmarking in a company? It significantly reduces waste, rework, and duplication. Benchmarking increases awareness of what you do and how well you are doing it. A better process understanding leads to a more effective management in both energy efficiency terms and in overall production efficiency terms. Benchmarking helps set credible energy efficiency targets and identifies what to change and why. It removes blinkers and attitudes.

Energy efficiency benchmarking is hampered in practice by a variety of different factors of influence such as production volume (see Fig. 5), use of production capacity over time, changes in the process mix of the company, differences in product type and quality, energy efficiency measures, non-production-elated energy use, and storage of energy carriers (solid and liquid fuels).

In a carefully designed benchmark process, it is possible to take into account a variety of these factors. Nevertheless, benchmarking appears as limited, especially when no detailed knowledge on energy consumption is available at the process level in the company. In this context, benchmarking can be seen as a prescreening step toward an energy audit. It gives a quick result and can be performed on a large number of companies. Benchmarking and energy audit are therefore complementary exercises. From an engineering point of view, very often critics is therefore manifested toward energy efficiency benchmarks by stating that "all companies are individuals and they cannot be compared with each other." On this basis, benchmarking is considered as useless. This argument can be debated, first of all, on more general grounds:

- We are well using benchmarks in many different occasions of daily life, starting from the grades received at school without taking care of the individuality of each person.
- We are using benchmarks in companies well in other fields, for example, to compare their economic performance without paying attention either to the individuality of companies.
- We underestimate the dynamics of improvement triggered by the comparison with others, which leads us to detailed questions about why we are different with respect to the energetic performance elsewhere observed and possible measures for improvement.

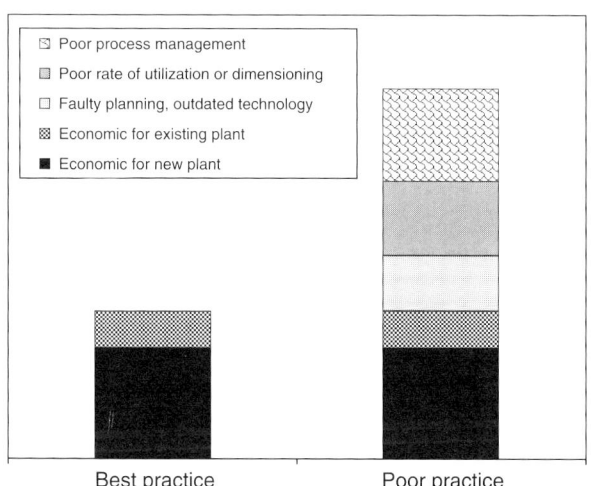

FIGURE 4 Why there are differences from simple to double in energy consumption for producing the same product.

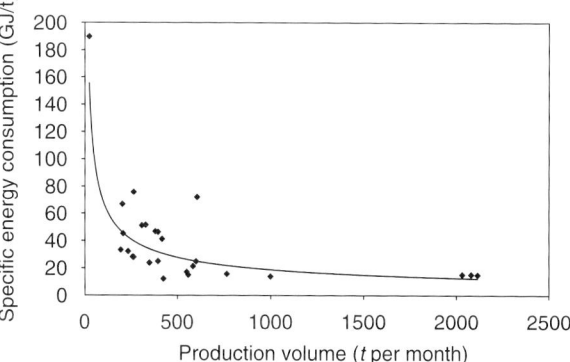

FIGURE 5 Scale effects of specific energy consumption (energy consumed per ton of product): yarn-producing companies in Thailand of different size.

In addition to these more general arguments in favor of benchmarking, there are increasingly also methodologies in development to design individually tailored benchmarks for each company. In fact, for this purpose the really observed energy consumption within the company is compared to a hypothetical company that has the same process mix but good practice processes. The advantage of this concept is that it is not necessary in particular to have detailed knowledge of the real energy consumption within the company at the process level. The disadvantage is, however, that extended knowledge has first to be provided to define and evaluate quantitatively process chains and basic processes for the comparison. This requires a lot of experience with benchmarking and a lot of knowledge accumulated with individual industrial processes and might be an obstacle at an early level of the introduction of a benchmarking process. Nevertheless, the so developed tailored benchmarks are possibly the most convincing ones from an engineering perspective.

4. POTENTIALS FOR INDUSTRIAL ENERGY EFFICIENCY IMPROVEMENT

Potentials for industrial energy efficiency improvement have been investigated on many occasions for different regions of the world. Results from investigations on European industries in the context of the contribution of energy efficiency to the reduction of global and local emissions provide the following wider conclusions, which are representative for many other studies and show the importance of energy efficiency for the general field of environment (see Table II):

- Cost-effective and currently available energy efficiency measures, which are technically proven, still offer the potential to make a significant contribution to reduction of pollution in the EU: an economic potential of 12 to 14% primary energy savings across the whole range of installations subject to the so-called Integrated Pollution Prevention and Control (IPPC) has been identified providing emissions saving of between 8 and 11%, mainly combustion related (CO_2, NO_x, SO_x).
- The calculate net cost benefit to the EU is 14 billion Euro at a total capital cost of 35 billion Euro for manufacturing industry.

TABLE II

Primary Energy Savings Calculated for European Union Industries in 2000 Covered by Integrated Pollution Prevention and Control IPPC (This Can Be Associated Roughly to Companies Entitled to Participate in Emission Trading Schemes)

Sector	IPPC coverage	Savings potential (PJ)	%
Food	31%	140–180	9–12
		(100–120)	(6–8)
Textiles	34%	70–80	12–14
Paper	98%	380–440	21–24
Chemicals	74%	460–790	16–27
		240–350	(8–12)
Nonmetallic minerals	86%	152	13
Steel	100%	387	15
Nonferrous metals	97%	147	21
Refineries	100%	770–980	25–32
Livestock	2%	2	10
Large combustion plants	98%	1830	6
Waste management	not quantified	40–90	

Note. It is assumed in this study that CHP replaces the average mix of thermal generation in the electricity supply industry. For some countries, it is more likely that CHP would replace gas turbines as these are the marginal technology. If this occurs, the primary energy savings would be much lower. The corresponding figures are shown in brackets in the table.

- Energy efficiency measures potentially save industry money, whereas end-of-pipe technologies to abate local pollution generally cost money.
- Low-cost measures such as energy management still offer significant scope for savings (around 8% of the total savings identified).
- Combined heat and power generation (CHP) offers a significant potential for primary energy savings (~36% of the identified savings) but involves relatively high capital costs. It is therefore likely that this potential does not represent a realistic short-term target for energy or emissions reductions.
- Energy savings from cost-effective improvements in cross-sector devices such as motors and drives, boilers, and compressed air plants are also available and are estimates to contribute 5% energy reduction.
- Process-specific energy savings represent a large proportion of the savings identified in many sectors (half of the total savings) and also benefit from other cost-savings factors such as throughput increases, quality, and labor savings.

This latter point illustrates that in many occasions energy efficiency measures do not only contribute to save energy in a company but while investigating these measures, often additional improvements for the production process are discovered at the same time, which improve considerably payback periods. Thus, energy efficiency, in addition to protecting the environment, contributes to render industrial processes more efficient and competitive (energy efficiency = production efficiency; see Fig. 6).

Energy efficiency improvements in the industrial sector may constitute large business opportunities. To illustrate this case, electricity saving measures in Western European industries on a time horizon 2015 are classified according to the following three categories:

- *Category I: Technical Add-on Improvements.* This category assembles measures that achieve electricity use reductions by modifying existing technologies, for example, with new or improved control devices or through improved process design.
- *Category II: Industrial Services.* This includes measures that help to improve the performance of existing technologies—that is, mainly by proper maintenance and operation. A major current trend is to outsource "housekeeping functions" that are not relevant for the core business of the company.
- *Category III: New Production Facilities.* This category represents the design and construction of new plants and production facilities (not necessarily new technologies).

Efficiency markets resulting from the implementation of efficiency measures that achieve electricity use reductions by modifying existing technologies (category I) or that help to improve the performance of existing technologies can be directly linked to the reduced energy costs. In total, the implementation of category I measures corresponds to a cumulated investment volume of 19,381 Mill € in 18 years for the Western European industrial sector (see Table III). The efficiency potential of category II measures results in an efficiency market volume of 7013 Mill. € for industry. All measures together result in a yearly efficiency market of 1466 Mill. € for industry in Western Europe.

Much higher market volumes arise from investments due to stock renewals and production growths of the analyzed technologies (category III measures). These investments bear a large potential for the uptake of energy efficient technologies. Especially the stock renewal of motors and motor-driven systems and the new construction of large office buildings bear large investment potentials.

5. OBSTACLES TO INDUSTRIAL ENERGY EFFICIENCY IMPROVEMENT AND TRANSACTION COSTS

There is a gap between the opportunities of the efficiency markets and the actual implementation of energy efficient measures in the industrial sector in practice. The reasons for this gap is commonly seen in a wide range of obstacles and market imperfections, which impede the widespread use of energy-efficient technologies. In theory, given all the benefits of energy efficiency at the microeconomic and macroeconomic levels, a perfect market would invest in, and allocate the rewards from, new energy-efficient technologies and strategies. But in practice, many obstacles and market imperfections prevent profitable energy efficiency from being fully realized:

- Improved energy efficiency is brought about by new technology, organizational changes, and minor changes in a known product. This implies that investors and energy users are able to get to know and understand the perceived benefits of the technical efficiency improvement. It also implies that investors and users have to be prepared to realize the improvement and to take time to absorb the new information and evaluate the innovation. But many users, especially small and medium-sized companies, do not have enough knowledge, technical skills, and

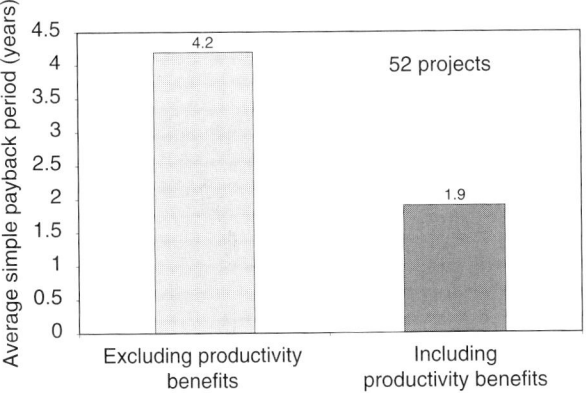

FIGURE 6 Energy efficiency = production efficiency.

TABLE III

Efficiency Markets in Western Europe for Electricity Savings in the Industrial Sector

	Electricity use, 1997 (in TWh)	Economic savings	Efficiency markets Western Europe			Markets per year
			Cat. I[a] Mill. €	Cat. II[b] Mill. €	Cats. I and II Mill. €	Mill. €/a
Industrial sector	966.8					
EAF steel	42.7	28%	187	152	339	
Primary aluminium	49.5	23%	454		454	
Chlorine	27.4	18%	1,017		1,017	
Air separation	13.3	6%	110		110	
Motors and drives	648.3	10%	11,663		11,663	
Compressed air	76.0	34%	706	3786	4492	
Pumps and fans	236.3	15%	3,515	3076	6,590	
Furnaces	61.8	28%	1,728		1,728	
Sum analysis	842.9		19,381	7,013	26,394	1,466

[a] Category I: Technical add-on improvements.
[b] Category II: Industrial services.

market information about possibilities for energy savings, particularly if energy converting off-sites are concerned (and not the production itself).

• The same energy consumers, even if they gain knowledge, often have trouble raising funds for energy-efficiency investments. Their capital may be limited, and additional credit may be expensive. Especially when interest rates are high, small firms tend to prefer to accept higher current costs and the risk of rising energy prices instead of taking a postponed energy credit.

• Large firms typically establish internal hurdle rates for energy efficiency investments that are higher than the cost of capital to the firm (e.g., by imposing payback requirements of 2 or 3 years for energy converting off-sites with technical life times of 15 years). This fact reflects the low priority that top managers place on increasing profits by raising energy productivity.

• In many companies, particularly small and medium-sized companies, all investments except for buildings and infrastructure are decided according to payback periods instead of internal interest rate calculations (see Table IV). If the lifetime of energy-saving investments is longer than that of existing production plants and machinery and if the payback period is expected to be even for both investments, entrepreneurs implicitly expect higher profits from energy-saving investments.

• Many firms (especially with the current shift toward lean firms) suffer from a shortage of trained technical staff because most personnel are busy maintaining production. Especially for small and medium-sized enterprises, installing new energy-efficient equipment is far more difficult than simply paying for.

• Insufficient maintenance of energy-converting systems and related control equipment causes substantial energy losses. Outsiders are not always welcome (external consultants, utilities), especially if proprietary processes are involved. Many companies cannot evaluate the risks connected with new equipment or control techniques in terms of their possible effects on product quality, process reliability, maintenance needs, or performance. Thus, firms are less likely to invest in new, commercially unproven technology.

• Cogeneration of heat and power has a considerable potential in a number of industrial branches, in particular the chemical industry, the pulp and paper industry, and the food sector. Yet the monopolistic structure of the electricity sector in many countries has led to high prices for maintenance and peak power and rather low buyback rates, as well as to dumping prices in the case of planning new cogeneration capacity. As a result, many auto producers restrict the capacity of the cogeneration plant to their minimum electricity and

TABLE IV

Internal Rate of Return in Percentage per Year for Industrial Energy Saving Measures (Continuous Energy Saving Is Assumed over the Whole Useful Life of the Plant)

Payback time (years)	Useful lifetime of a plant (in years)							
	3	4	5	6	7	10	12	15
2	24%	35%	41%	45%	47%	49%	49.5%	50%
3	0%	13%	20%	25%	27%	31%	32%	33%
4		0%	8%	13%	17%	22%	23%	24%
5			0%	6%	10%	16%	17%	18.5%
6	Unprofitable			0%	4%	10.5%	12.5%	14.5%
8						4.5%	7%	9%

Profitable investment possibilities (gray shaded area) eliminated by a 3-year payback time requirement.

heat needs, although they may wish to produce more heat by cogeneration. This obstacle is diminishing with increasing liberalization of the electricity markets opening up new opportunities for cogeneration and energy service companies.

It becomes obvious that manufacturers of energy-converting technologies and off-sites can surmount most of these obstacles by offering energy services in the form of contracting. In essence, many obstacles that are faced by small and medium-sized companies are business opportunities for large technology supplying companies. These opportunities are supported by the trend that companies in the industrial sector try to outsource their capital-intensive off-sites such as boilers, generation of technical gases, cooling, and logistics.

In the climate change debate, it is controversial to what extent measures should be undertaken to limit CO_2 emissions. Among the contentious aspects are the costs of these measures and, more specifically, the question whether there is a potential to reduce energy consumption at negative net costs, (i.e., a net benefit). According to numerous technoeconomic studies (i.e., studies that base their economic evaluations on detailed technology data), the investment costs of such measures will be recovered by the energy cost savings and other benefits they generate—even without counting their climate benefits. These so-called no-regret potentials are estimated for the year 2020 at 10 to 20% of global emissions of greenhouse gases. However, studies on the importance of this potential also affirm that multiple obstacles related to market imperfections or to juridical and administrative barriers impede their exploitation.

Other analyses with a more conventional economic background reject the possibility that profitable energy-saving measures may exist that are not undertaken autonomously by market forces. The fact that these measures are not implemented is interpreted as a rational response to real but hidden costs, which are supposedly not included in the evaluations of the no-regret potentials. In this context, the studies often refer to transaction costs as well as prohibitively high policy costs for the elimination of obstacles.

6. LONG-TERM POTENTIAL FOR INDUSTRIAL ENERGY EFFICIENCY IMPROVEMENT

Often the technical potentials of efficient energy use have a time horizon of only 15 to 20 years. Except for a few individual technologies, very little is known about the specific energy demand that industrial plant and machinery may have in, say, 40 or 60 years' time. This generally short time horizon of the (known) energy efficiency potentials gives rise to differing and inconsistent perceptions of the possibilities for technical and research-policy action on the energy supply and demand sides and therefore causes preference to be given to research on and development and application of new technologies in the energy conversion sector while devoting little funds to industrial energy-efficiency research and development (see Fig. 7).

Seen from a purely technological, natural science viewpoint, one can visualize potentials for eliminating more than 80% of present-day primary energy consumption in the industrialized countries.

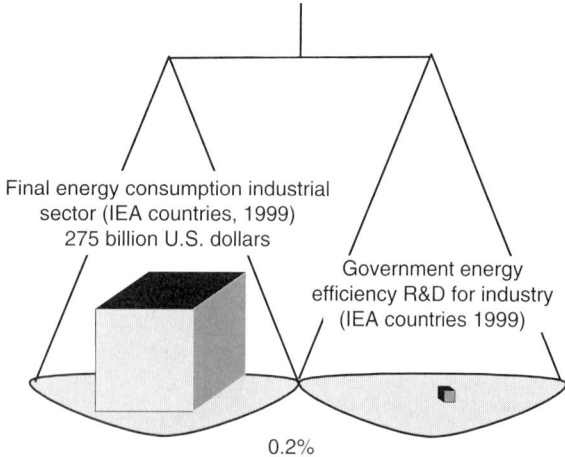

FIGURE 7 David against Goliath: research and devlopment spending for industrial energy efficiency in IEA countries versus the monetary value of industrial energy consumption, 1999.

Essentially, two basic principles can be adduced in support of this view: the better use of exergy of fossil fuels, heat, and electricity as well as the reduction of specific useful energy demand.

6.1 Better Use of Exergy of Fossil Fuels, Electricity, and Heat

The capability (or availability) of energy use (exergy) possessed by a given form of energy is often disregarded today when technical visions involving energy efficiency are made. The overall exergetic efficiency (i.e., the relation between useful energy and the availability of primary energy) is estimated for industrialized countries to be in the order of 10%. If this low efficiency is to be improved, it will be necessary to use energy with small losses of availability or only applying heat at small temperature differences, for example:

- Employ better use of available energy of burning fossil fuels via gas turbines, combustion engines, or fuel cells. In all these cases, electricity (of high exergetic value) is generated before the remaining energy in the form of heat is used.
- Try to avoid heat losses by using heat exchangers or vapor compressors. With increasing fuel prices in the long-term future, further technical progress, and additional economy-of-scale effects, heat recovery and energy cascading seem to have a bright future and an important role to improve energy efficiency in industry, the commercial sector, and the residential sector (as far as useful energy demand cannot be drastically reduced, discussed later).
- Try to increase temperature levels of waste heat in order to use the "upgraded" heat for useful purposes via heat pumps and heat transformers.

These technical possibilities open up a considerable energy-saving potential, in industry—for example, up to a theoretical maximum of 50%.

6.2 More Efficiency by Reduced Useful Energy Demand

The other major possibility for energy conservation is the reduction of specific useful energy demand, which is not a constant of nature but depends rather on the present level of technical know-how and energy price levels on the decision-making process and habits of consumption:

- The quality and quantity of thermal insulation used, for instance, reduce the heat demand in cooling equipment and production machinery. High-quality insulation material and vacuum solutions will be developed for heat transport, cooling equipment, and production machinery.
- Energy-intensive industrial processes may be substituted by less energy intensive processes or process chains (thin strip casting or even spraying of metals, of steel and aluminum in particular, introduction of the carbon technology in pig iron production, autothermal reforming in the production of ammonia and methanol, selective steam cradling in petrochemistry, fluidized bed cement production, direct route from salt to monomers of vinylchloride, etc.).
- Catalysts, enzymes biotechnology, and control techniques will render possible process substitutions and improvements in energy-intensive branches of industry to a very remarkable extent. High-energy demand of activation of chemical reactions, high pressure, and temperatures of processes may be rendered unnecessary by new catalysts or biotechnological processes. Traditional low- to medium-temperature processes may be substituted by biotechnological processes (e.g., dying of textiles or separation of intermediate products in the chemical or food industry).
- New applications of separation processes such as membrane processes, crystallization, extraction and absorption will only use a small percentage of the useful energy that is needed today in thermal separation processes in the basic products industry and the food industry.
- New materials (special plastics, for example, and foamed nonferrous metals) will reduce the

TABLE V

Long-Term Options to Improve Industrial Energy Efficiency by Levels of up to 80% beyond the Next 20 to 30 Years

Improve the use of the availability of energy (exergetic efficiency is only about 10% today) by	Reduce specific useful energy (it is not a constant of nature) by
Burning fuels: first via gas turbines, combustion engines, and fuel cells before heat is used	Insulation and air tightness
Using heat exchangers, vapor compressors, heat transformers, and heat pumps	Substitution of energy-intensive processes (catalysts, enzymes, membranes)
	New materials, lasers, inductive electric processes
	Recycling of energy-intensive materials
	Resubstitution by wood, natural fibres, new materials
	New concepts of vehicles and transportation systems
Theoretical energy saving potential: >50%	Theoretical energy saving potential: >60%
Total theoretical saving potential: >80%	

weight of moving parts and vehicles and, hence, reduce specific energy demand.

- Increasing recycling of energy-intensive materials (steel, aluminum, copper and other nonferrous metals, plastics, glass, paper/cardboard) can be anticipated, due to the necessity to reduce solid wastes that do not get societal acceptance for additional land fills and incineration plants.
- Substitution of energy-intensive materials by natural resources such as wood, natural fibers, and natural raw materials for chemicals is also a technical option, which will gain more attention in the long term because of its sustainability aspects.

The avoidance of heat and power losses, process substitution, and the closure of energy-intensive material cycles might have a theoretical technical energy-saving potential of at least 60% in industrialized countries. This estimate takes full account of the expectation that specific consumption of energy-intensive materials could be cut back by improved material properties and design changes.

Last, the useful service life of products made from energy-intensive materials could be lengthened by better corrosion protection and design for easier maintenance. This would reduce the cycling rate of energy-intensive materials, as well as the specific energy consumption of the respective energy services.

Adding together the above mentioned potentials, the theoretical long-term potential of efficient energy use undoubtedly lies above 80%. However, determined energy efficiency research and environmental protection policies are called for to maintain a high rate of efficiency increase over such a long period (see Table IV and V).

SEE ALSO THE FOLLOWING ARTICLES

Cogeneration • Commercial Sector and Energy Use • Discount Rates and Energy Efficiency Gap • Economics of Energy Efficiency • Energy Efficiency, Taxonomic Overview • Exergy Analysis of Energy Systems • Goods and Services: Energy Costs • Industrial Agriculture, Energy Flows in • Industrial Ecology • Industrial Energy Use, Status and Trends • Industrial Symbiosis

Further Reading

Bosseboeuf, D., Lapillonne, B., Eichhammer, W., and Landwehr, M. (1999). "Energy Efficiency Indicators: The European Experience." ADEME, Paris.

Golove, W. (1994). "A Preliminary Taxonomy of Efficiency Notions. International Comparisons of Energy Efficiency." Workshop Proceedings, Lawrence Berkeley Laboratory, Berkeley, CA.

Jochem, E., Mannsbart, W., Radgen, P., Schmid, Ch., Criqui, P., Worrell, E., and Price, L. (2001). "World-wide Electricity Demand and Regional Electricity Efficiency Markets with Particular Technological Focus on Western Europe." Research Report FhG-ISI, Karlsruhe.

Lapillonne, B., Schlomann, B., and Eichhammer, W. (2002). "Energy Efficiency in the European Union 1990–2000. SAVE-ODYSSEE." Project on Energy Efficiency Indicators (www.odyssee-indicators.org).

Ostertag, K. (2002). "'No-Regret' Potentials in Energy Conservation: An Analysis of Their Relevance, Size and Determinants." Physica Verlag, Heidelberg.

Industrial Energy Use, Status and Trends

ERNST WORRELL
Lawrence Berkeley National Laboratory
Berkeley, California, United States

1. Introduction
2. Global Manufacturing Energy Use
3. Energy Services and Energy Efficiency
4. Energy-Intensive Industries
5. Cross-Cutting Energy Services
6. Energy Intensity Trends
7. Potential for Energy Efficiency Improvement

Glossary

cross-cutting Energy technologies found throughout the industrial sectors and processes.
energy efficiency improvement Decreasing the use of energy per unit activity or service without substantially affecting the level of this activity or service.
energy intensity The amount of energy used per unit of activity or service.
energy-intensive industries Industries for which energy costs represent a large part (e.g., more than 10%) of total production costs.
energy service The activity or service supplied by the use of energy (e.g., lighting a room).
specific energy consumption The amount of energy used per unit of activity, expressed in physical terms (e.g., energy use per ton of steel).

Industry is a large energy user in nearly all countries. Approximately half of all industrial energy use is used in specific processes in the energy-intensive industries. On the other hand, various general energy conversion technologies and end uses can also be distinguished (e.g., steam production, motive power, lighting). Opportunities and potentials exist for energy savings through energy efficiency improvement in all sectors and countries. Technology development, as well as policies aimed at dissemination and implementation of these technologies, can help to realize the potential benefits. Technologies do not now, nor will they in the foreseeable future, provide a limitation on continuing energy efficiency improvements.

1. INTRODUCTION

Industrial production is the backbone of economic output in nearly all countries. Over the past decades, manufacturing industrial production has been growing in most economies. Industrial energy use can be broken down into that of the energy-intensive industries (e.g., primary metals, pulp and paper, primary chemicals, oil refining, building materials) and the non-energy-intensive industries (e.g., electronics, food). Energy use in the industrial sector is dominated by the production of a few major energy-intensive commodities such as steel, paper, cement, and chemicals. In any given country or region, production of these basic commodities follows the general development of the overall economy. Rapidly industrializing countries will have higher demands for infrastructure materials, and more mature markets will have declining or stable consumption levels. The regional differences in consumption patterns (expressed as consumption per capita) will fuel further growth of consumption in developing countries. In these "heavy" industries, energy is a very important production cost factor in addition to labor costs and raw material costs, driving a change toward higher energy efficiency (Table I).

Markets in the industrialized countries show a shift toward more service-oriented activities and, hence, non-energy-intensive industries. Still, energy-intensive industries will remain the largest energy consumers during the coming decades. Because of the great difference in energy intensity between energy-intensive industries and all others, changes in output

TABLE I
Energy Intensities and Energy Purchases in Selected U.S. Industries

Sector	1973		1985		1994	
	Energy intensity (primary energy)	Energy costs (percentage share of production costs)	Energy intensity (primary energy)	Energy costs (percentage share of production costs)	Energy intensity (primary energy)	Energy costs (percentage share of production costs)
Iron and steel (GJ/ton)	30.5	7	27.8	11	25.4	8
Pulp and paper (GJ/ton)	43.1	6	42.7	6	32.8	6
Cement (GJ/ton)	7.3	40	5.2	36	5.4	33
Primary aluminum (MWh/ton)	N/A	14	17.6	19	16.2	13
Petroleum refining (GJ/ton)	6.2	4	4.3	3	4.5	3

Source. Lawrence Berkeley National Laboratory.
Note. Energy intensity is expressed in primary energy, where the efficiency of electricity generation is assumed to be 33%. Energy intensity of primary aluminum production is given in MWh (1000 kWh).

shares of these industries can have a major impact on total industrial energy use. Many commodities (e.g., food, steel) are traded globally, and regional differences in supply and demand will influence total industrial energy use. Production trends also depend on regional availability of resources (e.g., scrap) and capital. Manufacturing energy use will also depend on the energy efficiency with which the economic activities are done.

This article discusses energy use patterns in industry and then assesses trends in industrial energy use and energy intensities. This is followed by a discussion of the potential for energy efficiency improvement and the future trends in industrial energy use.

2. GLOBAL MANUFACTURING ENERGY USE

In 1995, manufacturing industry accounted for 41% (131 exajoules [EJ], where $EJ = 10^{18}$ J) of global energy use. Between 1971 and 1995, industrial energy use grew at a rate of 1.7% per year, slightly less than the world total energy demand growth of 2.1% per year. This growth rate has slowed during recent years and was virtually flat between 1990 and 1995, primarily because of declines in industrial output in the transitional economies in Eastern Europe and the former Soviet Union. Energy use in the industrial sector is dominated by the industrialized countries, which accounted for 43% of world industrial energy use in 1995. Industrial energy consumption in these countries increased at an average rate of 0.6% per year between 1971 and 1990, from 49 to 54 EJ. The share of industrial sector energy consumption within the industrialized countries declined from 40% in 1971 to 33% in 1995. The decline partly reflects the transition toward a less energy–intensive manufacturing base, a shift toward a more service-oriented economy, and the continued growth in transportation demand resulting in large part from the rising importance of personal mobility in passenger transport use.

The industrial sector dominates in the economies in transition, accounting for more than 50% of total primary energy demand, the result of the emphasis on materials production, a long-term policy promoted under years of central planning. Average annual growth in industrial energy use in this region was 2.0% between 1971 and 1990 (from 26 to 38 EJ) but dropped by an average of 7.3% per year between 1990 and 1995. Even so, in 1995 industry accounted for 51% of primary energy use in these countries.

In the Asian developing countries, industrial energy use grew rapidly between 1971 and 1995, with an annual average growth rate of 5.9%, jumping from 9 to 35 EJ. It also accounted for the greatest share of primary energy consumption (i.e.,

58% in 1995). The fastest growth in this sector was seen in China and in other rapidly developing Asian countries. Growth in other developing countries was slightly lower.

The nature and evolution of the industrial sector varies considerably among developing countries. Some economies that are experiencing continued expansion in energy–intensive industry, such as those in China and India, show relatively unchanging shares of industrial energy use. In other countries, such as Thailand and Mexico, the share and/or growth of the transportation sector dominate. Many smaller countries have remained primarily agrarian societies with modest manufacturing infrastructures.

Future trends in industrial energy use will be affected by many different challenges. Globalizing trade patterns will affect industrial competition as well as technology transfer. Global environmental challenges such as climate change will lead to an increased focus on energy efficiency and changing energy consumption patterns and fuel choice. Regional and local environmental problems will also drive changes in industrial energy use in both industrialized and developing countries.

3. ENERGY SERVICES AND ENERGY EFFICIENCY

Energy is used to provide a service (e.g., producing a ton of steel, lighting a specified area). These services are called energy services. Energy consumers (including industry) are not interested in using energy; rather, they are interested in supplying the energy service in the most economic way. Energy efficiency improvement entails the provision of these services using less energy and without substantially affecting the quality of the service provided.

Approximately half of all industrial energy use is for specific processes in the energy-intensive industries. On the other hand, various general energy conversion technologies and end uses can also be distinguished (e.g., steam production, motive power, lighting). Hence, energy use in manufacturing industry can be broken down into various uses to provide a variety of services. A common breakdown distinguishes energy use for processes (called process specific) and energy use for buildings, utilities, and boilers (called cross-cutting). Note that boilers provide steam and hot water for processes as well as for cross-cutting energy services, as do motors.

Because of the wide variety in industrial processes, the discussion in the next section is limited to four energy-intensive sectors that are responsible for a large share of industrial energy use: iron and steel, pulp and paper, cement, and chemicals. This is followed in the subsequent section by a discussion of cross-cutting energy uses (i.e., boilers and motors).

4. ENERGY-INTENSIVE INDUSTRIES

4.1 Iron and Steel Industry

The first record of the use of iron goes back to 2500 to 2000 BC, and the first deliberate production of iron began around 1300 BC. Small furnaces using charcoal were used. High-temperature processes started to be introduced in Germany around 1300 AD. The design of these furnaces is essentially the same of that of modern blast furnaces. The furnaces still used charcoal, and the first reported use of coke was in 1718 in the United Kingdom. The higher strength of coke allowed larger furnaces to be built with increasing energy efficiency. By 1790, coke ironmaking contributed to 90% of the British iron production. The development of the modern blast furnace after World War II resulted in an annual reduction of energy intensity of 3 to 4%/year due to the use of improved raw materials, ore agglomeration, larger blast furnaces, and higher air temperature. Today, the blast furnace is the main process to make iron and provides the largest raw material stream in steelmaking.

Steel is produced by reducing the carbon content in the iron to levels below 2%. This reduces the brittleness of the material and makes it easier to shape. The first steelmaking process was invented in 1855 by Bessemer. In the Bessemer converter, air was blown through the hot iron, a process that oxidizes the carbon. This principle is still followed in modern steelmaking processes. In the United States, the last Bessemer converter was retired during the 1960s. During the late 19th century, the Siemens–Martin or open hearth furnace (OHF) was invented. The OHF process uses preheated air to oxidize the carbon and melt the steel. This process is currently found only in developing countries and in Eastern Europe. The United States was one of the industrialized countries that phased out the OHF at a very late stage. During the 1980s, the dominant process became the basic oxygen furnace (BOF), which uses pure oxygen instead of air. The BOF process was developed in Austria during the 1950s. The productivity of this process is much higher, as is the energy efficiency. An alternative process is the electric arc furnace (EAF). The EAF process is used mainly to melt scrap. Performance of EAFs has improved tremendously, starting to use fuel and oxygen in addition to electricity. It is expected that in the future, the BOF and EAF processes will follow similar developmental paths. Liquid steel is cast into ingot or slabs and is shaped in rolling mills to the final product. Although most energy use is concentrated in the iron- and steelmaking, reduced material losses and productivity gains in casting and shaping (e.g., continuous casting, thin slab casting) have contributed to dramatic increases in the energy efficiency of steelmaking.

Today, the global iron and steel industry produces approximately 840 million tons of steel. The industry is made up of integrated steel mills, which produce pig iron from raw materials (e.g., iron ore, coke) using a blast furnace and produce steel using a BOF, and secondary steel mills, which produce steel from scrap steel, pig iron, or direct reduced iron (DRI) using an EAF. The majority of steel produced is from integrated steel mills (62% in 2001), although the share of secondary steel mills (or "mini-mills") is increasing (35% in 2000).

Global energy consumption for steelmaking is estimated at 20 EJ. The worldwide average energy intensity of steelmaking is estimated at 24 gigajoules/ton (GJ/ton), although large variations occur between countries and plants. Today, the most energy-efficient process would use 19 GJ/ton for integrated steelmaking and 7 GJ/ton for making steel out of scrap. Analyses have shown that many technologies exist that could improve energy efficiency further. For example, in the United States, the potential for energy efficiency improvement is estimated at 18% using proven and cost-effective practices and technologies with a payback period of 3 years or less. New technologies that could considerably lower the energy intensity of steelmaking are under development. Smelt reduction in ironmaking would integrate the production of coke and agglomerated ore with that of ironmaking, leading to reductions in production costs and energy use. The development of direct casting techniques that abandon rolling would increase productivity while reducing energy use further. Combined, these technologies could reduce the energy intensity of primary steelmaking to 12.5 GJ/ton and could reduce that of secondary steelmaking to 3.5 GJ/ton, reductions of 34 and 50%, respectively. In the highly competitive and globalizing steel industry, manufacturers must continuously look for ways in which to lower their energy intensity and costs.

4.2 Pulp and Paper Industry

Paper consists of aligned cellulosic fibers. The fibers may be from wood or other crops or from recycled waste paper. Starting with wood fibers, the fibers need to be separated from the wood, a process that is done by pulping the wood. The separation can be done by chemicals, by heat, or by mechanical means. In the chemical pulping process, chemicals and hot water are used to separate the cellulosis from the ligno-cellulosis. The amount of pulp produced is about half the amount of wood used. Chemical pulping results in high-quality paper. In mechanical the wood is ground under pressure, separating the fibers from each other. In mechanical pulping, the ligno-cellulosis is not removed, resulting in a lower quality paper (e.g., paper used for newsprint) but also in a higher recovery ($\sim$90% of the used wood). In chemical pulping, a lot of steam is used to heat the water and concentrate the chemical by-products. Recovery of the by-products to be recycled in the process can actually produce sufficient steam for the whole paper mill. The mechanical process uses large quantities of electricity, whereas some processes can recover steam from the grinding process. Waste paper is pulped by mixing with water, after which ink is removed and the pulp is refined. Paper recycling reduces the energy needs of the pulping process. Waste paper use in the production of paper varies widely due to the different structures of the industry in different countries.

Although energy efficiency improvement options do exist in the pulping step, greater opportunities exist in the chemical recovery step. The most common pulping process in the United States is the Kraft pulping process. Black liquor is produced as a by-product. The chemicals are recovered in a recovery boiler, combusting the ligno-cellulosis. Because of the high water content, the recovery boiler is not very efficient, and the steam is used to generate electricity in a steam turbine and steam for the processes. Gasification of black liquor would allow the use of the generated gas at a higher level of efficiency. This would make a Kraft pulp mill an electricity exporter.

In papermaking, the pulp is diluted with water at about 1:100. This pulp is screened and refined. The solution with the refined fibers (or stock) is fed to the paper machine, where the water is removed. In the paper machine, the paper is formed into a sheet and water is removed by dispersing over a wire screen. At the end of the forming section, 80% of the water is removed. The rest of the water is removed in the pressing and drying section. Although only a small amount of water is removed in the drying section, most energy is used in the drying section. Hence, energy efficiency opportunities try to reduce the water content by increasing the water removal by pressing. In a long nip press, the pressing area is enlarged. The larger pressing area results in extra water removal. New technologies aiming to increase the drying efficiency considerably are under development. One technology, impulse drying, uses a heated roll, pressing out most of the water in the sheet that may reduce the steam consumption of the paper machine by 60 to 80%.

The pulp and paper industry uses approximately 6 to 8 EJ globally. Because energy consumption and intensity depend on the amount of wood pulped, the type of pulp produced, and the paper grades produced, there is a great range of energy intensities among the industrialized countries of the world. In Europe, energy use for papermaking varied between 16 and 30 GJ/ton of paper during the early 1990s. The Netherlands used the least energy per ton of paper, largely because most of the pulp was imported. Countries such as Sweden and the United States have higher energy intensities due to the larger amount of pulp produced. Sweden and other net exporters of pulp also tend to show higher energy intensities. Energy intensity is also influenced by the efficiency of the processes used. Many studies have shown considerable potentials for energy efficiency improvement with current technologies such as heat recovery and improved pressing technologies.

4.3 Cement Industry

Cement is an inorganic, nonmetallic substance with hydraulic binding properties. Mixed with water, it forms a paste that hardens due to formation of hydrates. After hardening, the cement retains its strength. There are numerous cement types due to the use of different sources of calcium and different additives to regulate properties. The exact composition of cement determines its properties.

In 1999, global cement production was estimated to be 1600 million tons. Because of the importance of cement as a construction material and the geographic abundance of the main raw materials, cement is produced in virtually all countries. The widespread production is also due to the relatively low price and high density of cement, and this in turn limits ground transportation due to high transport costs.

The most common raw materials used for cement production are limestone, chalk, and clay. The collected raw materials are selected, crushed, and ground so that the resulting mixture has the desired fineness and chemical composition for delivery to the pyro-processing systems. The grinding process differs depending on the pyro-processing process used. The feed to the kiln is called "raw meal."

Clinker is produced by pyro-processing the raw meal. The raw meal is burned at high temperatures, first by calcination of the materials and then by clinkerization to produce clinker. Various kiln types have been used historically or are used around the world. Besides the rotary kiln, the vertical shaft kiln is used mainly in developing countries. In industrialized countries, the ground raw materials are processed predominantly in rotary kilns. In processing without precalcination, the decomposition (calcination) of $CaCO_3$ to CaO and CO_2 takes place in the kiln. The clinker is cooled. The cooling air serves as combustion air. The largest part of the energy contained in the clinker is returned to the kiln in this way.

Grinding of cement clinker together with additives to control the properties of the cement is done in ball mills, roller mills, or roller presses. Coarse material is separated in a classifier to be returned for additional grinding. Power consumption for grinding depends strongly on the fineness required for the final product and the use of additives.

Cement production is a highly energy-intensive process. Cement making consists of three major process steps: raw material preparation, clinker making in the kiln, and cement making. Raw material preparation and cement making are the main electricity-consuming processes, whereas the clinker kiln uses nearly all of the fuel in a typical cement plant. Clinker production is the most energy-intensive production step, responsible for approximately 70 to 80% of the total energy consumed. Raw material preparation and finish grinding are electricity-intensive production steps. Energy consumption by the cement industry is estimated at 6 to 7 EJ or 2% of global primary energy consumption.

The theoretical energy consumption to produce cement can be calculated based on the enthalpy of formation of 1 kg of Portland cement clinker (which is ~ 1.76 MJ). In practice, energy consumption is higher. The kiln is the major energy user in the cement-making process. Energy use in the kiln depends basically on the moisture content of the raw meal. Most electricity is consumed in the grinding of the raw materials and finished cement. Power consumption for a rotary kiln is comparatively low.

4.4 Chemical Industry

The chemical industry produces many intermediate compounds that are used as the basis for many chemical products. The chemical industry produces more than 50,000 chemicals and formulations. For example, ethylene, one of the most important bulk chemicals from an energy point of view, is used to produce products varying from solvents to plastics. Also, many processes in the chemical industry produce different coproducts. Chemical industries consume fuels and electricity as energy and feedstock. This makes energy analysis of the chemical industry more complicated compared than that of other industries.

A small number of bulk chemicals are responsible for the largest part of the energy consumption in the chemical industry. These are the so-called basic chemicals that are used as building blocks for many chemicals down the production chain. The most important basic chemicals are the family of petrochemicals (ethylene, propylene, butadiene, and benzene) from the organic chemical industry as well as ammonia and chlorine/caustic soda from the inorganic chemical industry.

Ethylene and its derivatives are feedstocks for many plastics and resins as well as for fibers and detergents. Global ethylene production is estimated at more than 80 million tons and growing. The United States is the world's largest ethylene producer, accounting for less than 30% of the world capacity. Since 1974, ethylene production has grown by 3% annually, while propylene has grown by more than 4% annually. Propylene has grown more rapidly— 5% per year—during the past decade or so.

Ethylene and other coproducts are produced through cracking of hydrocarbon feedstocks. In the presence of steam, hydrocarbons are cracked into a mixture of shorter unsaturated compounds. In the cracking process, hydrocarbon feedstocks are preheated to 650°C (using fuel gas and waste heat), mixed with steam, and cracked at a temperature of approximately 850°C. A series of separation steps produce fractions consisting of ethylene, propylene, a C_4 fraction (e.g., butadiene), and pyrolysis gasoline (containing benzene, toluene, and xylenes). The gas mixture is rapidly cooled to 400°C (or quenched) to stop the reaction, during which process high-pressure steam is produced. Injection of water further decreases the temperature to approximately 40 to 50°C, and a condensate that is rich in aromatics is formed. The liquid fraction is extracted, while the gaseous fraction is fed to a series of low-temperature, high-pressure distillation columns. Feedstocks used are ethane, LPG, naphtha, gas oils (GOs), and (sometimes) coal-derived feedstocks. Many of the installations used today can handle different (if not all) types of feedstock.

The single most energy-consuming step in the petrochemical industry is the steam cracking of hydrocarbon feedstocks into a mixture of shorter unsaturated compounds. Recent estimates of global energy consumption for the production of ethylene and coproducts are not available. Global energy use is estimated at more than 1 EJ. Energy consumption for ethylene production can be separated in feedstock and energy use. Feedstock energy use is generally equivalent to the heating value of the product, that is, approximately 42 GJ/ton (lower heating value [LHV]). Specific energy consumption for the energy use of the process varies depending on the feedstock, technology, age of plant, capacity, and operating conditions. In general, it varies between 14 (ethane feedstock) and 30 GJ/ton of ethylene (gas oil feedstock), whereas older plants can use more energy.

The production of ammonia, a key component in the manufacture of nitrogenous fertilizers, is a highly energy-intensive process. Roughly 80% of ammonia production is used as fertilizer feedstock. Ammonia is produced through the high-pressure synthesis of gases. Ammonia production in 1999 was estimated at 130 million tons. Growth is found mainly in developing countries, where the production expansion is also concentrated.

Ammonia is produced by the reaction of nitrogen and hydrogen, the so-called Haber–Bosch process. The main hydrogen production processes used in the United States are steam reforming of natural gas and partial oxidation of oil residues. Hydrogen is produced by reforming the hydrocarbon feedstock and producing synthesis gas containing a mixture of carbon monoxide and hydrogen. The carbon monoxide is then reacted with steam in the water–gas–shift reaction to produce carbon dioxide and hydrogen. The carbon dioxide is removed from the main gas stream. The carbon dioxide is recovered for urea production or exported as a coproduct or vented. The hydrogen then reacts with nitrogen in the final synthesis loop to form ammonia. The ammonia is often processed to different fertilizers, including urea and ammonium nitrate, whereas ammonia is also used as an input to make other chemicals.

Energy intensity of ammonia making depends on the feedstock used and the age of the process. Natural gas is the preferred feedstock and is used for more than 80% of global ammonia production. Coal is still used in countries with limited natural gas supplies (e.g., China). The use of natural gas provides not only higher production efficiency but also lower capital costs. Part of the energy input is used as feedstock; that is, the hydrogen in natural gas ends up in the ammonia. This is generally equivalent to the thermodynamic minimum energy consumption for Haber–Bosch synthesis, corrected for losses. Hence, feedstock use is estimated at 19 to 21 GJ/ton of NH_3 (LHV). Energy use varies between 9 and 18 GJ/ton. The most efficient plants in the world consume approximately 28 GJ/ton of NH_3 (feedstock and energy use). In the United States, specific energy consumption is equal to approximately 37 GJ/ton (LHV). An analysis of energy intensities of ammonia manufacture in Europe showed that the average 1989 specific energy consumption (SEC) in the European Union was estimated at 35.5 GJ/ton (LHV).

Chlorine production capacity is estimated at 50 million tons at 650 sites. Approximately half of the chlorine is produced in Asia. The production of chlorine gas is an energy-intensive chemical process. In the process, a brine solution is converted into two coproducts, chlorine gas and sodium hydroxide (caustic soda), through electrolysis. The three main electrolysis cell types that are used to separate and produce the chlorine gas and caustic soda are the mercury flow, diaphragm, and ion-selective membrane. In the diaphragm and membrane cells, the caustic soda requires an additional step of concentrating the solution so that it can meet market specifications for most products.

Chlorine production is a main electricity-consuming process in the chemical industry, as is oxygen and nitrogen production. Energy use for chlorine production depends strongly on the cell type used for the electrolysis. Typically, power consumption varies between 2800 (membrane process) and 4300 kWh/ton (mercury process). In addition, steam may be used to concentrate the caustic soda in the membrane process. Membrane cells require the least amount of energy to operate. The membrane process is considered the state-of-the-art process. Countries in Asia and Europe have a higher share of membrane processes than do the United States and Canada. In the United States, chorine production uses 48 TWh, with an average intensity of 4380 kWh/ton.

5. CROSS-CUTTING ENERGY SERVICES

5.1 Steam Production and Use

Besides the energy-intensive industries, many smaller and less energy-intensive, or light, industries exist. Light industries can include food processing, metal engineering, and electronics industries. In light industries, energy is generally a small portion of the total production costs. There is a wide variety of processes used within these industries. In general, a large fraction of energy is used in space heating and cooling, motors (e.g., fans, compressed air), and boilers. Industrial boilers are used to produce steam or heat water for space and process heating and for the generation of mechanical power and electricity. In some cases, these boilers have a dual function such as the cogeneration of steam and electricity. The largest uses of industrial boilers by capacity are in paper, chemical, food production, and petroleum industry processes. Steam generated in the boiler may be used throughout a plant or site. Total installed boiler capacity (not for cogeneration) in the United States is estimated at nearly 880 million MW. Total energy consumption for boilers in the United States is estimated at 9.9 EJ.

A systems approach may substantially reduce the steam needs, reduce emissions of air pollutants and greenhouse gases, and reduce operating costs of the facility. A systems approach that assesses options throughout the steam system and incorporates a variety of measures and technologies can help to find low-cost options. Improved efficiency of steam use reduces steam needs and may reduce the capital layout for expansion, reducing emissions and permitting procedures at the same time. In specific cases, the steam boiler can be replaced nearly totally by a heat pump (or mechanical vapor recompression) to generate low-pressure steam. This replaces the fuel use for steam generation by electricity. Emission reductions will depend on the type and efficiency of power generation.

Another option to reduce energy use for the steam system is cogeneration of heat and power. Based on gas turbine technology, this is a way in which to substantially reduce the primary energy needs for steam making. Low- and medium-pressure steam can be generated in a waste heat boiler using the flue gases of a gas turbine. Classic cogeneration systems are based on the use of a steam boiler and a back pressure turbine. These systems have relatively low efficiency compared with a gas turbine system. Steam turbine systems generally have a power-to-heat ratio between 0.15 (40 kWh/GJ) and 0.23 (60 kWh/GJ). The power-to-heat ratio depends on the specific energy balance of the plant as well as on energy costs. A cogeneration plant is most often optimized to the steam load of the plant, exporting excess electricity to the grid. The costs of installing a steam turbine system depend strongly on the capacity of the installation. Gas turbine-based cogeneration plants are relatively cheap. In many countries (e.g., the Netherlands, Scandinavia), gas turbine cogeneration systems are standard in industry. The power-to-heat ratio is generally higher than that for steam turbine systems. Aero-derivative gas turbines may have a power-to-heat ratio of 70 to 80 kWh/GJ. Aeroderivative turbines are available at low capacities, but specific costs of gas turbines decrease sharply with larger capacities.

5.2 Motors and Motor Systems

Motor systems are used throughout any industrial operation. A motor system generally consists of an appliance, the drive train, and the motor. Motor systems consume more than 60% of industrial electricity use in the United States and nearly 70% in Europe. The share of motor electricity use varies by sector. For example, in the cement industry, as much as 90% of electricity is used in motors to grind raw materials and final products and to drive the kiln. Motors themselves are efficient, and well-designed and -maintained motors have conversion efficiencies of more than 90%. However, older and inefficient motors may have much lower efficiencies, and the efficiency losses are even higher in the total motor system.

Motors range from less than 1 kW to a few megawatts in size. The largest motors are found in kiln and mill drives in the cement industry or as compressors in the chemical industry. Smaller motors may be found in many locations and are often installed as a whole system (e.g., a packaged pump). Small motors account for more than half of the motors installed in U.S. industry but represent a much smaller share of electricity use because these motors are generally used for fewer hours, whereas the large motors may run continuously. The most important uses of motors are for pumps, compressed air, and fans. In the United States, nearly 25% of industrial motor electricity is used for pumping systems, 14% for fans, 16% for compressed air systems, and 23% for material processing. Other uses (e.g., material handling, refrigeration) represent more than 23% of motor electricity use.

There are many opportunities to improve efficiency in motor systems. Overall, the cost-effective potential for motor system efficiency measures is estimated at 15 to 25% of overall motor electricity use. The most important element of any motor efficiency effort is a systems approach, that is, to optimize the whole system and not focus on a single element (the motor). The most important opportunities are found in optimizing the system driven by the motor, managing the motor system through adjustable speed drives (ASDs), and improving the efficiency of the motor. Compressed air systems use approximately 16% of industrial motor electricity. However, the overall efficiency of compressed air systems is often less than 10%. There are large opportunities for reduction of energy use by reducing leaks, improving air intake, and abolishing inappropriate use of compressed air as an energy source (and replacing, e.g., by direct drives). ASDs have revolutionized motor systems by allowing for affordable and reliable speed control using rugged conventional induction motors. ASDs work by varying the frequency of the electricity supplied to the motor, thereby changing the motor's speed relative to its normal supply frequency. This is accomplished by rectifying supplied alternating current to direct current and then synthesizing an alternating current at another frequency. This is accomplished with an inverter, which is a solid-state device in modern ASDs.

6. ENERGY INTENSITY TRENDS

In aggregate terms, studies have shown that technical efficiency improvement of 1 to 2% per year has been observed in the industrial sector in the past. During and after the years of the oil shock, U.S. industrial energy intensity declined by 3.5% per year (between 1975 and 1983). Between 1984 and 1994, industrial energy intensity declined by less than 1% on average. Figure 1 gives an overview of energy intensity trends in the industrial sector in industrialized countries.

The trends demonstrate the capability of industry to improve energy efficiency when it has the incentive to do so. Energy requirements can be cut by improved energy management and new process development. In addition, the amount of raw materials demanded by a society tends to decline as the society reaches certain stages of industrial development, leading to a decrease in industrial energy use. At the same time, the mix of fuels used by industry changes over time, affecting energy intensity as well as emissions. Overall changes in the manufacturing fuel mix are driven by price changes, availability and reliability of supply, efficiency gains, and sector structure. For example, in the U.S. manufacturing industry, a general trend toward a reduced use of coal and oil since 1958 is obvious from Fig. 2. Although coal consumption remained fairly constant after the oil price shock of the early

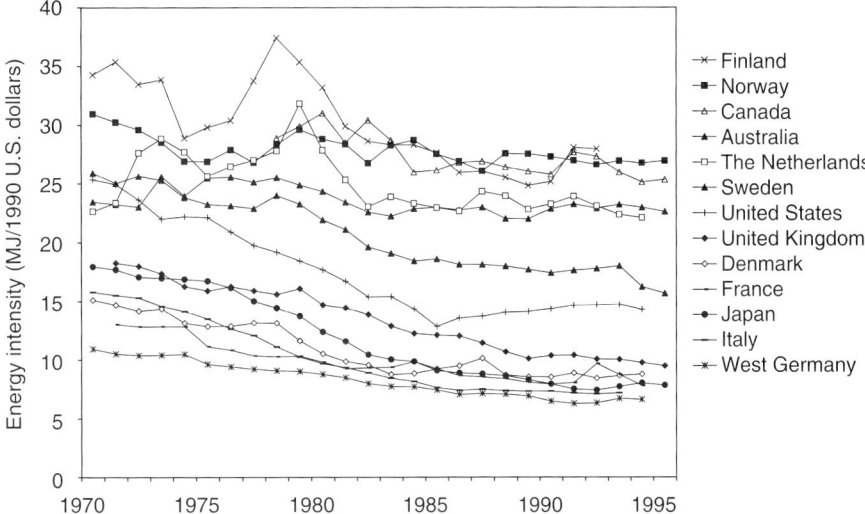

FIGURE 1 Industrial sector economic energy intensity trends in selected industrialized countries, 1970–1995. From the Lawrence Berkeley National Laboratory.

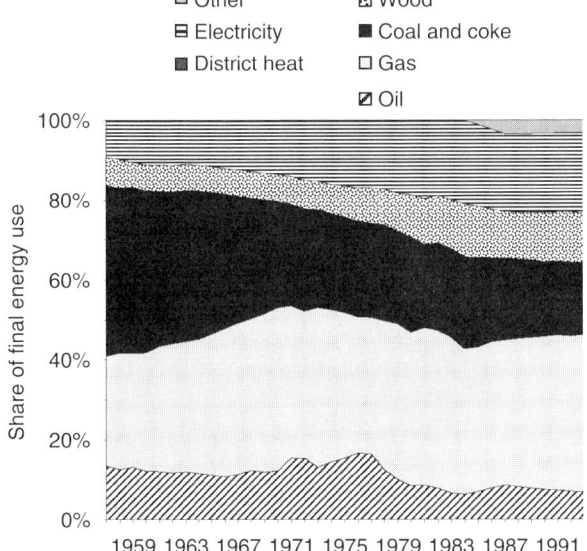

FIGURE 2 Relative fuel mix for U.S. manufacturing industry between 1958 and 1994. Energy is expressed by fuel as the share of the total final energy mix. From the Lawrence Berkeley National Laboratory.

1970s, it has declined further since. This change is partially a reflection of the decline of the energy-intensive industries in the U.S. economy. Natural gas, an easier and more efficient fuel to use, shows a dramatic increase in its share of the fuel mix over time. Also important is the increased use of electricity throughout industry, reflecting an overall trend in society of increased penetration of electric devices and equipment. Light industries tend to consume relatively more electricity than do energy-intensive industries.

The accounting of trends in structural shift, material intensity, and technical energy efficiency, as well as their interactions, can be extremely difficult. To understand trends in energy intensity, it is important to analyze the structure of the industrial sector. Reduction of energy intensity is linked closely to the definition of structure, structural change, and efficiency improvement. Decomposition analysis is used to distinguish the effects of structural change and efficiency improvement. Structural change can be broken down into intrasectoral (e.g., a shift toward more recycled steel) and intersectoral (e.g., a shift from steel to aluminum within the basic metals industry). A wide body of literature describes decomposition analyses and explains the trends in energy intensities and efficiency improvement. Decomposition analyses of the aggregate manufacturing sector exist mainly for industrialized countries but also for China, Taiwan, and selected countries such as those in Eastern Europe. The results show that different patterns exist for various countries, possibly due to specific conditions as well as differences in driving forces such as energy prices and other policies in these countries. More detailed analyses on the subsector level are needed to understand these trends better. Changes in energy intensities can also be disaggregated into structural changes and efficiency improvements at the subsector level. In the iron and steel industry, energy intensity is influenced by the

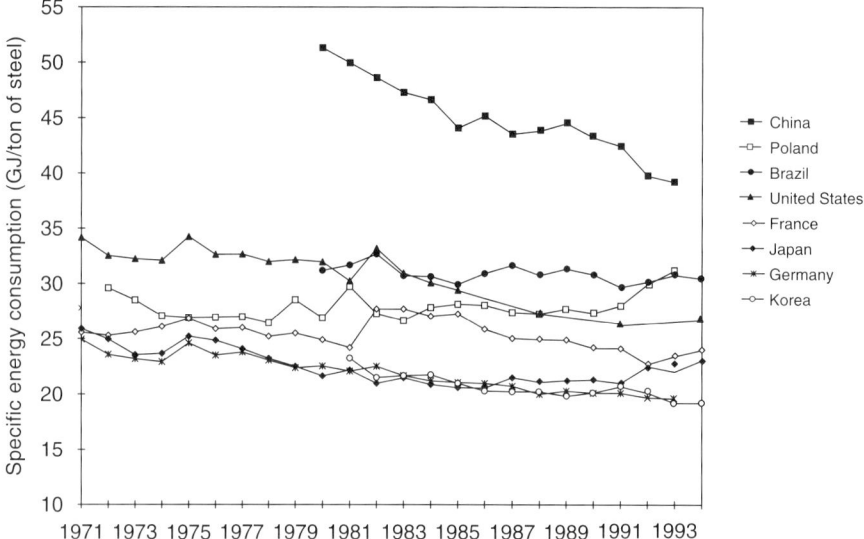

FIGURE 3 Trends in physical energy intensity (GJ/ton of crude steel) in seven countries between 1971 and 1994. From Worrell *et al.* (1997).

raw materials used (e.g., iron ore, scrap) and the products produced (e.g., slabs, thin rolled sheets). A study on the iron and steel industry examined trends in seven countries that together produced nearly half of the world's steel. Figure 3 shows the trends in physical energy intensity in these countries, expressed as primary energy used per ton of crude steel. The large differences in intensity among the countries are shown, as are the trends toward reduced intensity in most countries. Actual rates of energy efficiency improvement varied between 0 and 1.8% per year, whereas the energy intensity increased in the case of the restructuring economy of Poland.

7. POTENTIAL FOR ENERGY EFFICIENCY IMPROVEMENT

Much of the potential for improvement in technical energy efficiencies in industrial processes depends on how closely such processes have approached their thermodynamic limit. There are two types of energy efficiency measures: (1) more efficient use of existing equipment through improved operation, maintenance, or retrofit of equipment; and (2) use of more efficient new equipment through the introduction of more efficient processes and systems at the point of capital turnover or expansion of production. More efficient practices and (new) technologies exist for all industrial sectors. Table II outlines some examples of energy efficiency improvement techniques and practices.

A large number of energy-efficient technologies are available (Table II) in the steel industry, including continuous casting, energy recovery, and increased recycling. Large technical potentials exist in most countries, ranging from 25 to 50% even in industrialized countries. New technologies that are under development (e.g., smelt reduction, near net shape casting) will reduce energy consumption as well as environmental pollution and capital costs. In the chemical industry, potentials for energy savings in ammonia making are estimated at up to 35% in Europe and between 20 and 30% in Southeast Asia. Energy savings in petroleum refining are possible through improved process integration, cogeneration, energy recovery, and improved catalysts. Compared with state-of-the-art technology, the savings in industrialized countries are estimated at 15 to 20% and are higher for developing countries. Large potentials for energy savings exist in nearly all process stages of pulp and paper production (e.g., improved dewatering technologies, energy and waste heat recovery, new pulping technologies). Technical potentials are estimated at up to 40%, with higher long-term potentials. Energy savings in cement production are possible through increased use of additives (replacing the energy-intensive clinker), use of the dry process, and a large number of energy efficiency measures (e.g., reducing heat losses, using waste as fuel). Energy savings potentials of up to 50% exist in the cement industry in many countries through efficiency improvement and the use of wastes, such as blast furnace slags and fly ash, in cement making.

TABLE II

Examples of Energy Efficiency Improvement Measures in Energy-Intensive Industry

Iron and steel
- Heat recovery for steam generation, preheating combustion air, high-efficiency burners
- Adjustable speed drives, heat recovery coke oven gases, dry coke quenching
- Efficient hot blast stove operation, waste heat recovery for hot blast stove, top gas power recovery turbines, direct coal injection
- Recovery BOF–gas, heat recovery of sensible heat BOF–gas, closed BOF–gas system, optimized oxygen production, increased scrap use, efficient tundish preheating
- UHP process, oxy-fuel injection for EAF plants, and scrap preheating
- Heat recovery (steam generation), recovery of inert gases, efficient ladle preheating
- Use of continuous casting, "hot connection" or direct rolling, recuperative burners
- Heat recovery, efficient burners annealing and pickling line, continuous annealing operation

Chemicals
- Process management and thermal integration (e.g., optimization of steam networks, heat cascading, low and high-temperature heat recovery, heat transformers), mechanical vapor recompression
- New compressor types
- New catalysts
- Adjustable speed drives
- Selective steam cracking, membranes
- High-temperature cogeneration and heat pumps
- Autothermal reforming

Petroleum refining
- Reflux overhead vapor recompression, staged crude preheat, mechanical vacuum pumps
- Fluid coking to gasification, turbine power recovery train at the fluid catalytic cracker, hydraulic turbine power recovery, membrane hydrogen purification,
    unit-to-hydrocracker recycle loop
- Improved catalysts (reforming), hydraulic turbine power recovery
- Process management and integration

Pulp and paper
- Continuous digester, displacement heating/batch digesters, chemi–mechanical pulping
- Black liquor gasification/gas turbine cogeneration
- Oxygen predelignification, oxygen bleaching, displacement bleaching
- Tampella recovery system, falling film black liquid evaporation, lime kiln modifications
- Long nip press, impulse drying, other advanced paper machines
- Improved boiler design/operation (cogeneration), distributed control systems

Cement
- Improved grinding media and linings, roller mills, high-efficiency classifiers, wet process slurry
- Dewatering with filter presses
- Multistage preheating, precalciners, kiln combustion system improvements, enhancement of internal heat transfer in kiln, kiln shell loss reduction, optimize heat transfer in clinker cooler, use of waste fuels
- Blended cements, cogeneration
- Modified ball mill configuration, particle size distribution control, improved grinding media and linings, high-pressure roller press for clinker pregrinding, high-efficiency classifiers, roller mills

Source. Worrell, Levine, Price, *et al.* (1997).

In the United States, various studies have assessed the potential for energy efficiency improvement in industry. One study assessed the technologies for various sectors and found economic energy savings of 7 to 13% over the business-as-usual trends between 1990 and 2010. Technologies such as the ones described previously (Table II) are important in achieving these potentials.

However, barriers may partially block the uptake of those technologies. Barriers to efficiency improvement can include unwillingness to invest, lack of available and accessible information, economic disincentives, and organizational barriers. The degree to which a barrier limits efficiency improvement is strongly dependent on the situation of the actors (e.g., small companies, large industries). A range of policy instruments is available, and innovative approaches or combinations have been tried in some countries. Successful policy can contain regulation (e.g., product standards) and guidelines, economic instruments and incentives, voluntary agreements and actions, information, education and training, and research, development, and demonstration policies. Successful policies with proven track records in several sectors include technology development and utility/government programs and partnerships. Improved international cooperation to develop policy instruments and technologies to meet developing country needs will be necessary, especially in light of the large anticipated growth of the manufacturing industry in this region.

SEE ALSO THE FOLLOWING ARTICLES

Commercial Sector and Energy Use • *Decomposition Analysis Applied to Energy* • *Global Energy Use: Status and Trends* • *Goods and Services: Energy Costs* • *Industrial Agriculture, Energy Flows in* • *Industrial Ecology* • *Industrial Energy Efficiency* • *Industrial Symbiosis* • *Information Technology and Energy Use* • *International Comparisons of Energy End Use: Benefits and Risks*

Further Reading

Ang, B. W. (1995). Decomposition methodology in industrial energy demand analysis. *Energy* **20**, 1081–1096.

Brown, M. A., Levine, M. D., Romm, J. P., Rosenfeld, A. H., and Koomey, J. G. (1998). Engineering–economic studies of energy technologies to reduce greenhouse gas emissions: Opportunities and challenges. *Annu. Rev. Energy Envir.* **23**, 287–385.

De Beer, J., Worrell, E., and Blok, K. (1998). Future technologies for energy efficient iron and steel making. *Annu. Rev. Energy Envir.* **23**, 123–205.

Howarth, R. B., Schipper, L., Duerr, P. A., and Strom, S. (1991). Manufacturing energy use in eight OECD countries: Decomposing the impacts of changes in output, industry structure, and energy intensity. *Energy Econ.* **13**, 135–142.

International Energy Agency. (1997). "Indicators of Energy Use and Efficiency: Understanding the Link between Energy and Human Activity." IEA/OECD, Paris.

Nadel, S., Elliott, N., Shepard, M., Greenberg, S., Katz, G., and de Almeida, A. T. (2002). "Energy-Efficient Motor Systems: A Handbook on Technology, Program, and Policy Opportunities," 2nd Ed. American Council for an Energy Efficient Economy, Washington, DC.

Nilsson, L. J., Larson, E. D., Gilbreath, K. R., and Gupta, A. (1995). "Energy Efficiency and the Pulp and Paper Industry." American Council for an Energy Efficient Economy, Washington, DC.

Price, L., Michaelis, L., Worrell, E., and Khrushch, M. (1998). Sectoral trends and driving forces of global energy use and greenhouse gas emissions. *Mitigation Adaptation Strategies Global Change* **3**, 263–319.

Schipper, L., and Meyers, S. (1992). "Energy Efficiency and Human Activity: Past Trends, Future Prospects." Cambridge University Press, New York.

World Energy Council. (1995). "Energy Efficiency Utilizing High Technology: An Assessment of Energy Use in Industry and Buildings" (prepared by M. D. Levine, E. Worrell, N. Martin, and L. Price). WEC, London.

Worrell, E., Cuelenaere, R. F. A., Blok, K., and Turkenburg, W. C. (1994). Energy consumption by industrial processes in the European Union. *Energy* **19**, 1113–1129.

Worrell, E., Levine, M. D., Price, L. K., Martin, N. C., van den Broek, R., and Blok, K. (1997). "Potential and Policy Implications of Energy and Material Efficiency Improvement." UN Commission for Sustainable Development, New York.

Worrell, E., Martin, N., and Price, L. (1999). "Energy Efficiency and Carbon Emission Reduction Opportunities in the U.S. Iron and Steel Industry." Lawrence Berkeley National Laboratory, Berkeley, CA.

Worrell, E., Price, L., Martin, N., Farla, J., and Schaeffer, R. (1997). Energy intensity in the iron and steel industry: A comparison of physical and economic indicators. *Energy Policy* **25**, 727–744.

Industrial Symbiosis

MARIAN R. CHERTOW
Yale University
New Haven, Connecticut, United States

1. Introduction
2. Definition of Industrial Symbiosis and Related Terms
3. The Kalundborg Model
4. Elements and Tools of Industrial Symbiosis
5. Spatial Scale of Industrial Symbiosis
6. Technical, Regulatory, and Business Issues
7. Future Directions

Glossary

by-product exchange An organized means of sharing discarded products and materials among firms.
cascading of water and energy Repeated use in different applications of a resource, such as water or energy, that degrades over successive uses.
cogeneration The simultaneous production of electricity and heat, usually in the form of steam, from the same fuel-burning process.
eco-industrial park or estate A community of manufacturing and service businesses located on a common property seeking enhanced environmental, economic, and social performance through collaboration in managing environmental and resource issues.
industrial ecology The study of the flows of materials and energy in industrial and consumer activities, of the effects of these flows on the environment, and of the influences of economic, political, regulatory, and social factors on the flow, use, and transformation of resources.
industrial ecosystem The place where industrial symbiosis occurs, ideally for optimizing the consumption of materials and energy where the effluents of one industrial process serve as the raw materials for another process.
integrated biosystem Exchange in which at least two biological subsystems are part of an integrated process to reduce emissions and reuse agricultural by-products.
materials tracking A means of identifying and quantifying all significant material inputs and outputs of each firm in a given industrial system.
sustainable development Development that meets the needs of the present without compromising the ability of future generations to meet their own needs.
symbiosis A biological relationship in which at least two otherwise unrelated species exchange materials, energy, or information.
utility sharing Cooperative effort among proximate firms and other organizations to pool and assign water and energy resources that might otherwise be generated or supplied individually or through a large centralized authority.

Industrial symbiosis is part of a new field called industrial ecology. Industrial ecology is principally concerned with the flow of materials and energy through systems at different scales, from products to factories and up to national and global levels. Industrial symbiosis focuses on these flows through networks of businesses and other organizations in local and regional economies as a means of approaching ecologically sustainable industrial development. Industrial symbiosis engages traditionally separate industries in a collective approach to competitive advantage involving physical exchange of materials, energy, water, and/or by-products. The keys to industrial symbiosis are collaboration and the synergistic possibilities offered by geographic proximity.

1. INTRODUCTION

The term industrial symbiosis was coined in the small municipality of Kalundborg, Denmark, where a well-developed network of dense firm interactions was encountered. The primary partners in Kalundborg, including an oil refinery, a power station, a gypsum board facility, and a pharmaceutical company, share ground water, surface water, wastewater, steam, and fuel, and they also exchange a variety of by-products that become feedstocks in other processes. High levels of environmental and economic efficiency have been achieved, leading to many other less tangible benefits involving personnel, equipment, and information sharing. Many other examples of industrial symbiosis exist around the world and illustrate how the concept is applied.

This article describes the nature of the industrial symbiosis in Kalundborg. Then, it discusses the many elements of industrial symbiosis such as energy and water cascading, cogeneration, and materials exchange. It also examines tools such as input/output matching, stakeholder processes, and materials tracking. Then, the article discusses how industrial symbiosis is a useful umbrella term because it can describe exchanges across entities regardless of whether they are colocated, located near one another but not contiguous, or located within a broader spatial area such as regionally. It also examines technical and regulatory considerations that have come into play in various locations and that can facilitate or inhibit industrial symbiosis. Finally, it considers future directions with regard to industrial symbiosis based on historical and current experience.

2. DEFINITION OF INDUSTRIAL SYMBIOSIS AND RELATED TERMS

The term symbiosis builds on the notion of biological symbiotic relationships in nature where at least two otherwise unrelated species exchange materials, energy, or information in a mutually beneficial manner, with the specific type of symbiosis known as mutualism. So, too, industrial symbiosis consists of place-based exchanges among different entities that yield a collective benefit greater than the sum of individual benefits that could be achieved by acting alone. Such collaboration can also foster social values among the participants that can extend to surrounding neighborhoods. As described in what follows, the symbioses need not occur within the strict boundaries of a park, despite the popular use of the term eco-industrial park to describe organizations engaging in exchanges.

At the same time interest began to develop in industrial symbiosis and eco-industrial parks, a number of other parallel tracks advanced that might be construed, broadly, as green development. These include residential, commercial, industrial, and community development as captured in terms such as sustainable architecture, green buildings, sustainable communities, and smart growth. Eco-industrial development or sustainable industrial development narrows down the possibilities to refer predominantly to industrial and commercial activities. Cooperating businesses that include a materials/water/energy exchange component qualify the activity as industrial symbiosis. The materials exchange component has also been referred to as a by-product exchange, by-product synergy, or waste exchange and may also be referred to as an industrial recycling network.

As with the term industrial park, the term eco-industrial park refers to eco-industrial development on a particular plot of real estate. An eco-industrial park may include many ecologically desirable goals, including mechanisms to reduce overall environmental impact, conserve materials and energy, and foster cooperative approaches to resource efficiency and environmental management. The terms industrial estate and eco-industrial estate are more commonly used in Asia and can include communities of workers who live in or near the group of businesses constituting the industrial estate. Gunther Pauli popularized the notion of zero-emissions parks to emphasize the drive toward sustainable industrial development. The agricultural community has found that while traditional agriculture incorporated cyclical reuse of by-products, industrial agriculture became much more linear, consuming materials and disposing of wastes. Therefore, many zero-emissions researchers describe integrated biosystems in which at least two biological subsystems are part of an integrated process to reduce emissions and reuse agricultural by-products.

Some writers refer to eco-industrial networks to capture a broad range of environmental and economic activities among businesses. Just as economic clusters have come to mean a group of businesses that are sectorally related by the products they make and use, such as the furniture cluster in central North Carolina, the term eco-industrial clusters is sometimes used to describe environmental interactions among firms in the same or related industries. Industrial complexes of sectorally related firms have been successful for the past several decades in overall pollution reduction in industries such as pulp and paper, sugarcane, textiles, and plastics.

Although materials and energy exchanges have been a significant part of industry for centuries, focus on environmental attributes is much more recent. In a foundational 1989 article on industrial ecology, Frosch and Gallopoulos described the underlying notion of an "industrial ecosystem" in which "the consumption of energy and materials is optimized and the effluents of one process ... serve as the raw materials for another process." Others have extended the ecosystem metaphor to see related industrial activities as a food web and to interpret the roles of various scrap and remanufacturing businesses as the scavengers and decomposers of these systems.

In any multidisciplinary field such as industrial ecology, there are strands from many disciplines and paths of research that are the antecedents of current understanding. Industrial symbiosis was not popularized until the first articles and analyses of the industrial region in Kalundborg were published during the early 1990s. One origin of industrial symbiosis is from the chemical industry, which embeds an intrinsic value chain of materials as they are degraded. Other terms discussed previously originate in ecology, agricultural studies, engineering, and/or business economics. Cogeneration and utility sharing have been justified on engineering, environmental, and economic grounds. Indeed, private industry sees cost efficiency as a driving factor, whereas city planners, economic development experts, and real estate developers also emphasize land use, social and environmental aspects, and the synergies that can arise from colocation. Van Berkel summarized three types of synergies that help to characterize trends emerging in different types of symbiosis network: (1) synergies across the supply chain, (2) synergies from shared use of utilities, and (3) synergies from local use of by-products. As new projects are formed and old projects evolve, more common definitions will emerge.

3. THE KALUNDBORG MODEL

The model of industrial symbiosis was first fully realized in the industrial district at Kalundborg. Although it is continually evolving, there are currently some 20 exchanges occurring among the symbiosis participants involving water, energy, and a wide variety of residue materials that become feedstocks in other processes. The waste exchanges alone amount to some 2.9 million tons of materials per year. Water consumption has been reduced by a collective 25%, and the power station has experienced a 60% reduction in water use through cycling. Some 4500 homes receive district heat, which replaced 3500 small oil-fired units. A key coordinating role is played by the Kalundborg Center for Industrial Symbiosis, formed by the symbiosis partners under the auspices of the Industrial Development Council of the Kalundborg region. In addition to several companies that participate as recipients of materials or energy, many of which are beyond the local region, the specific industrial ecosystem in 2003 consisted of six main business partners located in Kalundborg as follows:

- *Energy E2 Asnæs Power Station:* the largest plant producing electricity in Denmark, a 1037-MW facility fired by orimulsion and coal
- *Statoil Refinery of Kalundborg:* Denmark's oldest and largest refinery, with a production capacity of approximately 5.5 million tons of crude oil annually
- *Gyproc A/S:* part of BPB Gyproc, producing plasterboards and plasterboard-based systems for the construction industry
- *Novo Nordisk A/S:* an international, biotechnology, and pharmaceutical company offering a wide range of insulin products to world markets
- *Novozymes A/S:* the largest producer of enzymes in the world, with its largest single factory in Kalundborg
- *A/S Bioteknisk Jordrens Soilrem:* specialist in remediation of soil contaminated by oil, chemicals, and heavy metals that cleans approximately 500,000 tons of contaminated soil annually

The municipality of Kalundborg and the inter-municipal waste treatment company Noveren I/S also participate actively in the symbiosis.

Figure 1 shows the interrelationships of the symbiosis participants. Each exchange was developed as an economically attractive business arrangement between participating firms through bilateral contracts. It is significant to mention that this symbiosis was not based on a planning process and that it continually evolves. In 1961, low levels of ground water locally led the refinery to pipe in surface water from a nearby lake, which now also supplies the power station and pharmaceutical plant. The gypsum board plant began receiving excess butane gas from the refinery beginning in 1972, and this was recently discontinued. Since the 1980s, the number of exchanges has accelerated. Regulation has played an indirect role over the years; for example, the national ban on placing organic waste streams into landfills caused the pharmaceutical company to seek arrangements to apply its sludges on agricultural lands. Social cohesion is regularly cited as a key element of success in the Kalundborg symbiosis.

Rather than a static system of locked-in firms and technologies, individual participants in the symbiosis have changed significantly over time, and the

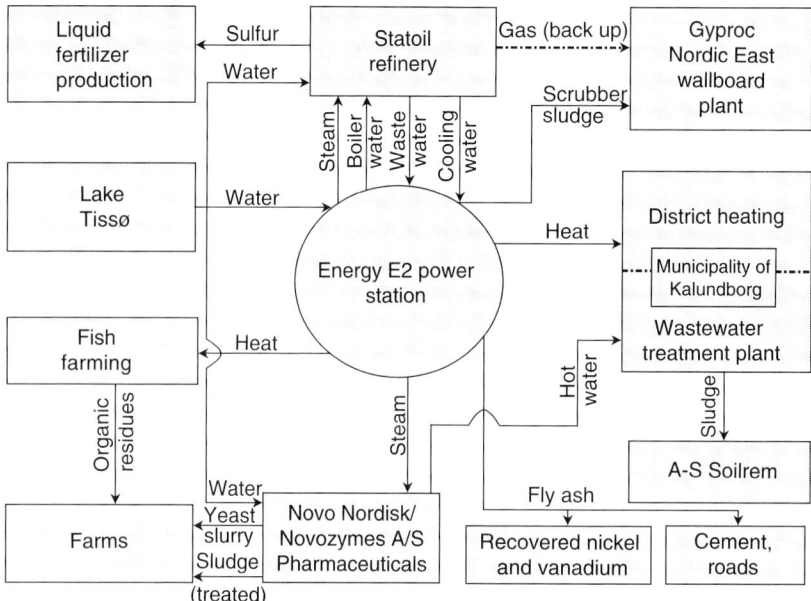

FIGURE 1   The industrial symbiosis at Kalundborg, Denmark.

ecosystem as a whole has adapted. Over the past several years, Kalundborg's Statoil Refinery doubled its capacity based on North Sea claims, the Asnæs Power Station switched from coal to orimulsion to comply with mandated carbon dioxide (CO_2) reduction, and the pharmaceutical plant split into two ventures, eliminated some product lines (including penicillin), and increased others. Although each individual business change alters the makeup of the industrial symbiosis, the changes collectively have not diminished the overall nature of the symbiosis. In fact, in the case of the power station's flue gas desulfurization program that creates calcium sulfate (or gypsum) used by the plasterboard plant, the change in fuel at the power station from coal to orimulsion led to the creation of even more gypsum, thereby increasing the benefits for the participants. Rather than tie themselves to a single supplier, the symbiosis participants try to insulate themselves from supplier interruptions by diversifying sources to reduce business risk, just as in traditional supplier–customer relationships.

4. ELEMENTS AND TOOLS OF INDUSTRIAL SYMBIOSIS

Drawing on industrial ecology, industrial symbiosis incorporates many elements that emphasize the cycling and reuse of materials in a broader systems perspective. A brief discussion of several elements is presented, followed by a discussion of useful analytical tools for industrial symbiosis. These elements include embedded energy and materials, life cycle perspective, cascading, loop closing, and tracking material flows.

4.1 Embedded Energy and Materials

To create a product, resources are used for extraction of materials, transportation, primary and secondary manufacturing, and distribution. The total energy and materials used is the amount embedded in that product. Reusing by-products in an industrial symbiosis preserves the embedded materials and energy for a longer period as part of the cycling emphasized in industrial ecology. Cogeneration is a specific means of cycling embedded energy by reusing waste heat to produce electricity or by using steam from electric power generation as a source of heat.

4.2 Life Cycle Perspective

As described by Graedel and Allenby in the first textbook in the field, industrial ecology is "a systems view in which one seeks to optimize the total materials cycle from virgin material, to finished material, to component, to product, to obsolete product, and to ultimate disposal." A life cycle perspective ensures a breadth of focus that is not limited to what happens within one facility or factory but rather considers the entire set of environmental impacts that occur at each stage of industrial development and use across entities. With respect to industrial symbiosis, a life cycle perspective is helpful in assessing symbiotic opportunities—the junctures in the product life cycle at which the by-product of concern can be considered for another use.

4.3 Cascading

Cascading occurs when a resource, such as water or energy, is used repeatedly in different applications. In successive uses, the resource is of lower quality, a lower level of refinement, and/or lower value. A cascade must include at least one use beyond the virgin use of the resource and is generally conceptualized as a downward step diagram. The cascade terminates when either a considerable amount of energy must be put in to recover value from the resource or the resource is discarded. Cascading is a common strategy for industrial symbiosis because the firm producing the used resource can save on treatment or disposal and possibly can earn compensation in exchange for the value of the resource. The environmental benefits of cascading are numerous, including the reduced use of virgin resources, the avoided impact of resource extraction, and the reduced deposition of waste into the environment.

4.4 Loop Closing

If cascading is conceptually stepwise, loop closing is more circular. A general name for many different variations of reuse and recycling of resources, loop closing occurs when a resource has a cyclical flow embedded in the industrial ecosystem and the resource, rather than becoming entirely degraded, reappears akin to its original form. Thus, glass bottles may be washed out directly and returned to use, or glass bottles may be collected and crushed into cullet before the glass cullet is melted to make new glass containers. Both methods close the loop and return the glass to a form similar to its previous form. The economic and environmental benefits of loop closing are similar to those of cascading.

4.5 Tracking Material Flows

Also key to industrial symbiosis studies is the tracking of material, water, and energy flows. This form of accounting captures instances of loop closing, cascading, and unidirectional flows. Over time, many applications of materials tracking at different scales have yielded specific tools, such as material flow analysis (MFA) and substance flow analysis (SFA), that formalize tracking practices. Materials tracking for symbiosis identifies and quantifies all significant material inputs and outputs of each firm in the subject industrial system. The results are analyzed to suggest opportunities for exchange of materials among firms as well as opportunities for more efficient resource use in the industrial ecosystem.

To the extent that new symbioses can be planned or existing exchanges can be augmented, several tools have proven to be useful in industrial symbiosis analysis. These tools are industrial inventories, input/output matching, stakeholder processes, and materials budgeting.

4.6 Industrial Inventories

Once a district has been identified as a candidate for industrial symbiosis, it is useful to begin with an inventory of local businesses and other resources, including utilities and relevant institutions. Because confidentiality is a critical aspect of dealing with private companies, data can be collected generically concerning the inputs and outputs of relevant industrial processes to achieve a base analysis from which to assess further goals.

4.7 Input/Output Matching

Key to symbiosis is the matching of inputs and outputs to make links across industries. There are various ways in which to collect these data in a systematic fashion, including written and oral surveys and literature review. The U.S. Environmental Protection Agency (EPA) commissioned three pieces of input/output matching software during the late 1990s to be used as a planning tool to allow communities to investigate what mix of specific types of industries might support industrial symbiosis. FaST (Facility Synergy Tool) was a database of industry profiles describing typical inputs and outputs of specific types of facilities. It had a data input screen and a search mechanism to identify possible input/output matches among facilities. DIET (Designing Industrial Ecosystems Tool) allowed scenario analysis of various combinations of facilities. It included a linear programming optimization model to enable planners to optimize environmental, economic, and/or employment objectives and to change the relative weights of each. REaLiTy (Regulatory, Economic, and Logistics Tool) assisted with sorting out regulatory hurdles likely to be confronted depending on the actual materials chosen for exchange.

A caveat on these and related models is that they can overemphasize idealized what-if scenarios with too little recognition of the time-consuming processes involved in attracting any business, let alone the optimal suite of symbiotic partners. It is also appropriate to add the cautionary note that most industrial products generally have not been designed for reuse, and so great care must be taken to see that the proposed match across systems is sound with respect to regulation and integration of best practices.

4.8 Stakeholder Processes

To the extent that industrial symbiosis involves different layers of unconnected participants, a broad array of community involvement techniques is warranted. Both the Londonderry Eco-Industrial Park in New Hampshire and the Sustainable Industry Park in Cape Charles, Virginia, assembled many diverse stakeholders and then conducted design charettes to seek input on what an eco-industrial project should look like in the local context. Whether and how to pursue specific covenants or conditions as a type of deed restriction is one topic of stakeholder meetings. Applied Sustainability gathered experience convening stakeholders from business and government in its efforts to create by-product synergy in Tampico, Mexico, and Alberta, Canada. Openness among participating companies and continued coordination by a stakeholder group such as an advisory council is important both to establish and to maintain the momentum of a symbiosis.

4.9 Materials Budgeting

Materials budgeting, a type of materials tracking discussed previously, can be used to map energy and material flows through a chosen system. Formally, in industrial ecology, materials budgeting embraces three concepts: (1) reservoirs, where a material is stored; (2) flux, which is the amount of material entering or leaving a reservoir per unit time; and (3) sources and sinks, which are rates of input and loss of specific materials entering or leaving a system. Because it helps to identify both stocks and flows, materials budgeting can be a basic building block of an industrial symbiosis analysis. By tracking material flows, Schwarz and Steininger determined the existence of a symbiotic network in Styria, Austria, larger than that at Kalundborg.

5. SPATIAL SCALE OF INDUSTRIAL SYMBIOSIS

In general, industrial symbiosis occurs locally or regionally across participating companies. Increasing the distance among firms lessens the breadth of exchange opportunities because it is not cost-effective to transport water and steam beyond regional boundaries, whereas by-products can often travel much farther. Observing numerous instances of industrial symbiosis, Chertow devised a taxonomy of materials exchange types to consider spatial and organizational elements. These include through waste exchanges (type 1); within a facility, firm, or organization (type 2); among firms colocated in a defined eco-industrial park (type 3); among local firms that are not colocated (type 4); and among firms organized "virtually" across a broader region (type 5).

5.1 Type 1: Through Waste Exchanges

Most often focused at the end-of-life stage of a product or process, examples of these exchanges would include contributions of used clothing for charity and collection of scrap metal or paper by scrap dealers or municipal recycling programs. Waste exchanges formalize trading opportunities by creating hard-copy or online lists of materials that one organization would like to dispose of and another organization might need. The scale of trades can be local, regional, national, or global. The exchanges accomplish various input/output savings on a trade-by-trade basis rather than continuously. They feature exchange of materials rather than of water or energy.

5.2 Type 2: Within a Facility, Firm, or Organization

Some kinds of materials exchange can occur primarily inside the boundaries of one organization rather than with a collection of outside parties. Large organizations often behave as if they are separate entities and may approximate a multifirm approach to industrial symbiosis. Significant gains can be made within one organization by considering the entire life cycle of products, processes, and services, including upstream operations such as purchasing and product design. In the context of state-owned enterprises popular in Asia, material exchanges can extend along the supply chain of a product still under single ownership. The Guitang Group, a state-owned enterprise in China, expanded from sugar refining to include alcohol and paper production as a means of deriving income from the use of its own by-products, specifically molasses and bagasse. The group has recently extended its exchange network into a community-wide one to receive by-products of other sugar producers.

5.3 Type 3: Among Firms Colocated in a Defined Eco-industrial Park

In this approach, businesses and other organizations that are contiguously located can exchange energy, water, and materials and can go further to share information and services such as permitting, transportation, and marketing. Type 3 exchanges occur primarily within the defined area of an industrial park or industrial estate, but it is also common to involve other partners over the fence. The areas can be new developments or retrofits of existing ones. The Londonderry Eco-Industrial Park was established in a green field adjacent to an industrial zone. The Industrial Estate Authority of Thailand has a plan to adopt sustainable practices at 28 industrial estates at various stages. In Canada, Burnside Industrial Park in Dartmouth, Nova Scotia, is an example of retrofitting an existing industrial park to improve environmental performance. The park is spread over 2500 acres, with more than 1200 businesses employing some 18,000 people. Côté reported that researchers from Dalhousie University have been devising principles and strategies to work within the park to encourage its transformation into an industrial ecosystem.

5.4 Type 4: Among Local Firms That Are Not Colocated

Partners in this type of exchange need not be sited adjacent to one another but rather are located within a small geographic area, as in Kalundborg, where the primary partners are within roughly a 2-mile radius of each other. Type 4 exchanges draw together existing businesses that can take advantage of already generated material, water, and energy streams and also provide the opportunity to fill in new businesses based on common service requirements and input/output matching. In the Kwinana industrial area south of Perth, Australia, heavy process industries were established beginning in the 1950s and now employ some 3600 people. Through their industrial council, the companies have resolved to develop regional synergies and have doubled the number of companies involved in exchanges over the past decade or so.

5.5 Type 5: Among Firms Organized Virtually across a Broader Region

Given the high cost of moving and other critical variables that enter into decisions about corporate location, very few businesses will relocate solely to be part of an industrial symbiosis. Type 5 exchanges depend on virtual linkages rather than colocation. Although still place-based enterprises, type 5 exchanges encompass a regional economic community in which the potential for the identification of by-product exchanges is greatly increased by the larger number of firms that can participate. An additional attractive feature is the potential to include small outlying agricultural and other businesses. Self-organized groups, such as the network of scrap metal dealers, agglomerators, and dismantlers that feed particular mills and subsystems (e.g., auto recycling), could be considered as type 5 systems. Specific projects organized in the Triangle J region of North Carolina, in Tampico, and in Alberta sought to identify relevant material and energy inputs and outputs and to match companies within their regions to each other when doing so is economically and environmentally efficient.

6. TECHNICAL, REGULATORY, AND BUSINESS ISSUES

Because industrial symbiosis requires interorganizational cooperation, including knowledge of physical flows, this creates both barriers and opportunities beyond those of more conventional development projects. These include technical issues, regulatory issues, and business issues.

6.1 Technical Issues

In general, symbiotic industrial facilities need to be in close proximity to avoid large transportation costs and energy degradation during transit. High-value by-products, such as pure sulfur from sour gas treatment, are exceptions. Industrial symbioses of types 3 and 4 usually incorporate at least one anchor tenant with a large, continuous by-product stream such as a power plant or refinery. Wastes that are largely organic in nature (e.g., the effluent from fermentation of pharmaceuticals or brewing), as well as raw agricultural or forestry by-products, can also be attractive. Use of organic streams from fermentation as feed or fertilizer requires assurance that toxic components or organisms are absent. Materials production, such as the manufacture of wallboard, is technically challenging and requires close matching of compositions. Supply security is important to the users of by-product streams, as would be the reliability of more conventional materials suppliers located farther away. The problem of achieving sufficient scale when aggregating by-products is significant because even if discarded materials are collected from numerous facilities, the total volume can fall far short of the raw materials necessary to support a new operation.

In Japan, scarcity of land for waste disposal has prompted significant cooperation and technological change in affected industries. The Japanese cement industry now consumes approximately 6% of the country's waste, including half of the fly ash produced by power plants. As a result, the Japanese cement industry reports the smallest energy consumption per ton of cement produced among developed countries. Taiheiyo Cement has been developing a new technology to convert plastic waste (including PVC) and other industrial wastes into raw materials and fuel. Another technology decomposes dioxin from fly ash through heating at high temperatures inside cement kilns. Heavy metals are extracted by wet-type refining technology to be recycled for the nonferrous metal industry.

6.2 Regulatory Issues

Industrial symbiosis is often at odds with environmental regulatory requirements, which may preclude by-product exchanges or at least serve as a very strong disincentive. For example, in the United States, the Resource Conservation and Recovery Act (RCRA) regulates the treatment, storage, and disposal of numerous wastes as a means of averting risks stemming from the improper management of hazardous waste. The law requires that by-products be matched to specific mandatory protocols through a very extensive set of rules that leaves little room for innovative schemes for by-product reuse as feedstocks elsewhere. This inflexibility is based in large part on a deep-seated fear of sham recycling, an undertaking where the generator of a waste product makes a show of reusing that by-product merely to escape treatment requirements. Industrial symbiosis must be distinguished from such efforts if it is to develop within the current regulatory system.

In many developing countries, governments lead the process for establishing industrial parks and estates. Lowe found that this often slows the development process. In addition, these projects are likely to rely on funding from development banks for which industrial symbiosis projects might seem unconventional and, thus, fall outside of development bank guidelines.

In some instances, regulatory actions encourage industrial symbiosis. Landfill bans in key European countries have driven symbiotic practices such as the reuse of organic wastes prohibited from land disposal in Denmark and The Netherlands. Very high tipping fees for waste disposal in Canada and climate change levies in the United Kingdom have been cited as stimulating innovation and action in by-product reuse.

6.3 Business Issues

Although private actors need not be the initiators, they clearly must be committed to the implementation of industrial symbiosis because, in most instances, the industrial symbiosis flows either belong to private actors or will be shared with them in the case of municipal wastewater linkages. Whether key private actors can appropriate sufficient benefit from environmental gains is a challenge to industrial symbiosis, especially given that the level of business benefit to various partners is not uniform. Businesses generally want to address nonproduct problems at the lowest cost and with the least use of resources. Such objectives might not include the time or inclination to work with others, especially concerning low-value wastes.

As a practical matter, all significant development projects take a long time, require substantial capital investment, and must constantly meet the test of whether investment in them will create sufficient returns. These issues are compounded with eco-industrial projects by the need for multiparty planning and coordination and the attendant transaction costs, including the risk that a proposed partner will relocate.

Industrial symbiosis raises the question of whether the desire to reuse waste streams comes at the expense of adhering to pollution prevention principles calling for the elimination of waste at the front end of the process. Some suspect that industrial symbiosis projects favor older dying industries and keep them going rather than fostering a new generation of clean technology. Overall, industrial symbiosis could potentially discourage companies from updating their systems, plant, and equipment, instead substituting the veil of interdependence.

At the first level of analysis, it is reasonable to assume that companies will do what is in their economic interest. If, through incremental improvements or through broader scale process redesign, a company can eliminate waste in a cost-effective manner, rational actors will do so. In this sense, pollution prevention comes first. However, it is plausible that the opportunity for symbiosis might make the proposed process improvement fall lower in priority in a company's capital outlay scheme, in which case the company's own economic decision making might favor the symbiosis over pollution prevention.

7. FUTURE DIRECTIONS

An interesting debate in the industrial symbiosis literature is the extent to which industrial ecosystems are, like their natural counterparts, self-organizing and, if so, the extent to which additional policy and planning would advance eco-industrial development more quickly. Coordinating activities through representative councils, however, has been an important step in advancing symbiosis. Conceivably, the private sector could grab hold of industrial symbiosis as a logical extension of resource productivity by groups such as the World Business Council for Sustainable Development and its affiliates. Governments could latch onto eco-industrial parks as another way in which to redevelop brownfields and structure new industrial developments. On the one hand, a public already nervous about genetic engineering may become more restive about material exchanges that bring one industry's residues to an altogether different industry. On the other hand, region-specific water shortages and the scarcity of some materials accelerate the trend toward cycling and reuse.

Variety and experimentation in industrial symbiosis developments around the world will shed light on what is and is not successful, what the largest risks are, how regulatory hurdles can be overcome, what can be financed, and what is most environmentally beneficial and technologically desirable. The form could splinter back toward specific industry clusters around key materials such as plastics or around wastes such as in recycling parks. Technology may embed its own reuse such as the cycling of steam back into the industrial process in combined cycle power plants. It is also possible that much broader visions, which combine industrial symbiosis with other types of urban and regional planning, will prevail. The Japanese Central Government and the Ministry of Trade and Industry have designated several eco-towns to promote environmentally friendly practices such as zero emissions and material exchanges. The quest to find the most beneficial forms of resource sharing is part of the journey toward sustainable industrial development.

SEE ALSO THE FOLLOWING ARTICLES

Cogeneration • *Earth's Energy Balance* • *Ecological Footprints and Energy* • *Ecological Risk Assessment Applied to Energy Development* • *Industrial Agriculture, Energy Flows in* • *Industrial Ecology* • *Industrial Energy Efficiency* • *Industrial Energy Use, Status and Trends* • *Sustainable Development: Basic Concepts and Application to Energy*

Further Reading

Chertow, M. (2000). Industrial symbiosis: Literature and taxonomy. *Annu. Rev. Energy Environ.* **25**, 313–337.

Cohen-Rosenthal, E., and Muskinow, J. (eds.). (2003). "Eco-Industrial Strategies: Unleashing Synergy between Economic Development and the Environment." Greenleaf Publishing, Sheffield, UK.

Côté, R. P. (2000). "A Primer on Industrial Ecosystems: A Strategy for Sustainable Industrial Development." www.mgmt.dal.ca/sres/pdfs/primer.pdf.

Ehrenfeld, J., and Chertow, M. (2002). Industrial symbiosis: The legacy of Kalundborg. *In* "Handbook of Industrial Ecology" (R. Ayres and L. Ayres, Eds.). Edward Elgar, Cheltenham, UK.

Frosch, R., and Gallopoulos, N. (1989). Strategies for manufacturing. *Sci. Am.* **261**(3), 144–152.

Gertler, N., and Ehrenfeld, J. R. (1996). A down-to-earth approach to clean production. *Technol. Rev.* **99**(2), 48–54.

Graedel, T., and Allenby, B. (2003). "Industrial Ecology." Prentice Hall, Englewood Cliffs, NJ.

Lambert, A., and Boons, F. (2002). Eco-industrial parks: Stimulating sustainable development in mixed industrial parks. *Technovation* **22**, 471–484.

Lowe, E. (2001). "Eco-industrial Park Handbook for Asian Developing Countries," report to Asian Development Bank. RPP International, Emeryville, CA.

Inflation and Energy Prices

PAULO S. ESTEVES
Banco de Portugal
Lisbon, Portugal

PEDRO D. NEVES
Banco de Portugal and Universidade Católica Portuguesa
Lisbon, Portugal

1. Evolution of Oil Prices and Inflation since the Early 1970s
2. Pathway from Oil Prices to Inflation
3. Impact of Oil Price Changes on Inflation
4. Oil Price Volatility

Glossary

energy consumption The expenditures made by households on energy products (fuels, electricity, gas, and heatenergy), which represent, in general, slightly less than 10% of total household expenditures.
energy intensity Energy consumption per unit of output.
first-round effects The impact of energy price changes on overall consumer prices, reflecting the reaction to energy prices (direct effect) and to energy content costs of other consumer goods and services (indirect effect).
inflation A generalized and sustainable increase in prices, usually measured using the rate of growth of the Consumer Price Index or the gross domestic product deflator.
oil price shock A high and abrupt increase (decrease) in oil prices.
second-round effects The feedback effects on consumer prices produced by other variables (typically wages) that were primarily influenced by the first-round effects of energy prices on inflation.

The remarkable energy price stability that characterized the period following World War II was interrupted in the early 1970s; the sharp increase in oil prices that occurred at that time followed the creation of the Organization of Petroleum Exporting Countries in 1973. Oil prices are, by far, the key factor underlying energy prices, even today, despite the growing role of alternative energy sources. Changes in energy prices, in turn, have an effect on inflation. The channels through which changes in oil prices affect inflation are extremely complex and are very likely to change over time. Using a study by the Organization for Economic Cooperation and Development, however, estimates of the impact of oil price increases on inflation can be made (despite the considerable uncertainty surrounding the estimates, they constitute very useful rules of thumb). A distinction can be made between first- and second-round effects. The expenditure share on oil and the taxation of energy products are the main factors influencing first-round effects. The credibility of monetary policy plays a key role in the magnitude of the second-round effects on prices. One of the most notable characteristics of oil price behavior is its extreme volatility. In particular, oil price volatility is an important source of changes in inflation; oil price futures, despite the very nice empirical properties that they exhibit, such as typically reverting to a mean price of $20–25 (U.S. dollars) per barrel, tend to produce considerable deviations in oil prices.

1. EVOLUTION OF OIL PRICES AND INFLATION SINCE THE EARLY 1970s

In the post-World War II period, until the beginning of the 1970s, oil price fluctuations were very small. From 1949 to 1970, average annual fluctuations of oil prices in U.S. dollars, as measured by the absolute value of year-to-year price changes, were of the order of 1%. Therefore, the real price (i.e., inflation adjusted) slightly declined throughout this period. This so-called Golden Age period was characterized by a remarkable price stability and very strong gross domestic product (GDP) growth in the main industrialized economies. The stability of oil prices was an important element behind the low inflation and strong economic growth.

In the 1970s, the price of oil increased dramatically as a result of the deliberate action of the Organization of Petroleum Exporting Countries (OPEC), by controlling the quantity supplied to the world market. In 1974, the oil price reached $11.50 per barrel, more than triple the price from the previous year and almost five times above the 1972 price (unless otherwise specified, the prices given herein are U.S. dollars). Some years later, in 1980 and 1981, oil prices reached values above $30 per barrel, duplicating levels achieved in the previous years.

The direct link between oil prices and inflation is illustrated by examining inflation rates; in Organization for Economic Cooperation and Development (OECD) countries, inflation reached maximum levels of about 15% in 1974 and 13% in 1980 (Fig. 1), as a result of rising oil prices. It is now widely accepted that monetary policy did not deal adequately with the first oil price shock in 1973, contributing to a situation that led to the so-called Great Inflation period. Indeed, ambivalent monetary policy, conflicted over price stability and short-term growth objectives, failed to provide an anchor to inflation expectations. It is also worth mentioning additional circumstances that were threatening price stability at the time: latent inflation pressures had emerged in OECD countries at the beginning of the 1970s, reflecting capacity constraints; powerful trade unions and the overall acceptance of wage indexation favored the existence of second-round effects; and the Bretton Woods institutional framework collapsed and floating exchange rates were adopted.

The rise in worldwide inflation after the oil price increase of 1979–1981 was much more short-lived than the one in the mid-1970s. The reaction of monetary policy to the oil price shock was clearly different, mainly orientated toward the objective of maintaining price stability. In this context, it is worth mentioning that a new strategic framework for monetary policy was adopted in some countries; the intermediate monetary targeting adopted by the Bundesbank deserves special reference.

In the first half of the 1980s, the oil cartel was not able to enforce the production quotas set for its members. The supply of oil increased when some countries produced more than their quotas and new producers entered the market, and oil prices started to decline. The price collapse of 1986, following serious disagreements within OPEC, marked a turning point from the previous upward trend in oil prices. Oil prices started to fluctuate within a narrower band (Fig. 1), while inflation recorded a noticeable downward trend from double-digit figures to figures close to 2%. However, oil price volatility remained a central feature of the world oil market. Oil prices decreased to around $10 per barrel in late 1998 and early 1999. After seriously misjudging the oil market during that period and even contributing to the collapse of prices, OPEC successfully pushed prices upward, overshooting its goal, and prices recovered vigorously, reaching slightly more than $30 per barrel in 2000. In 2003, oil prices were affected by the war in Iraq, remaining close to $30 per barrel in the first half of the year.

Despite all of these fluctuations in oil prices, inflation remained close to 2–3% in the main advanced economies, at least from the second half of the 1990s onward. Within the main advanced economies, the sensitivity to oil price fluctuations has declined over the past two decades. Moreover, during this period, an increase in the credibility of monetary policy decisively diminished the size of the second-round effects associated with oil price fluctuations. Central banks clearly signaled that they were focused on future inflation, rather than on current inflation, and therefore were not reacting to changes in headline-driven inflation created by transitory factors, unless if these were likely to induce changes in inflation expectations.

2. PATHWAY FROM OIL PRICES TO INFLATION

The channels through which oil price changes affect consumer prices are distinguished by first-round effects (including both direct and indirect effects)

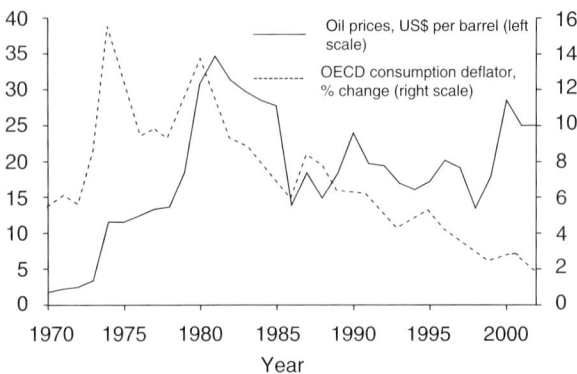

FIGURE 1 Oil prices and inflation, 1971–2002. Data from the OECD and Datastream.

and second-round effects. First-round effects reflect the fact that oil, as well as goods and services dependent on oil (transport, for instance), is included in the Consumer Price Index (CPI). The impacts on the energy components of the index are the so-called direct effects, and the impacts on the components of the index that have a content of energy are the so-called indirect effects. The size of first-round effects is affected by a large number of factors, including the share of energy included in total consumption expenditures, the energy intensity of production, the rate of taxation of energy, profit margins, and exchange rates.

The larger the share of energy (fuels, electricity, gas, and heat energy) in consumption expenditures, the larger the direct impact on the CPI of a given energy price change. The impact of a change in oil prices in the energy component of the price index is not instantaneous, however. Consumer fuel prices react almost immediately, but price adjustments in gas or heat energy can take somewhat longer to occur in normal circumstances. The magnitude of direct effects is affected by the importance and structure of tax rates, because oil is heavily taxed. If the tax rate has an *ad valorem* structure (i.e., the tax corresponds to a fixed proportion of the final price), an increase in the cost of energy of 1% will be transmitted to the consumer also as a 1% increase. However, if the tax corresponds to a specific tax (i.e., x cents by unit of energy), an increase in the cost of energy of 1% will be transmitted to the consumer as a somewhat smaller than 1% increase, at least before any adjustment of the tax, which is not likely to coincide in time with the usual fluctuations in oil prices in the international markets. Usually, tax systems are more complex than these very simple illustrations, reflecting the existence of both *ad valorem* and specific components. Therefore, the direct impact of oil price changes on the energy component of the consumer price index is affected by taxes. Figure 2 shows the sizable discrepancies of the tax component in unleaded gasoline prices for the G7 countries (G7 refers to the Group of Seven leading industrial nations that meet annually for economic and political discussions).

As previously stated, changes in energy prices will cause changes in the prices of goods and services that have been produced using energy resources. These changes are indirect effects. The most obvious case is the transport sector (air and surface), which is based on components that are highly energy intensive, but a large proportion of the price index is also likely to be affected. The magnitude of the effects on prices of goods and services depends, naturally, on the importance of energy as an input. The intensity of oil dependence can be measured by the consumption of oil per unit of GDP, as illustrated in Fig. 3. Given the strong decline in the use of oil per unit of GDP, the impact on prices of a given oil price shock is presently smaller than it was in the 1970s. This decline in oil intensity reflects the adjustment of advanced economies to a new era of higher and more unpredictable oil prices, through the use of alternative sources of energy and more energy-efficient technologies. Worldwide, the primary energy intensity has declined by more than 25% since 1970, slightly less so in the United States. The greatest decline in primary energy intensity has occurred in Japan and in the European Union countries, where the amount of oil used per unit of GDP has approximately fallen by half. The reduced sensitivity of the world economy to oil price changes resulting

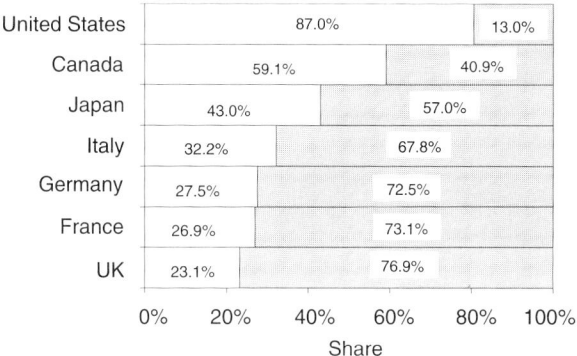

FIGURE 2 Gasoline prices and taxes (percent shares for unleaded gasoline, fourth quarter of 2002). Unshaded bars, before taxes; shaded bars, tax component. Data from the OECD.

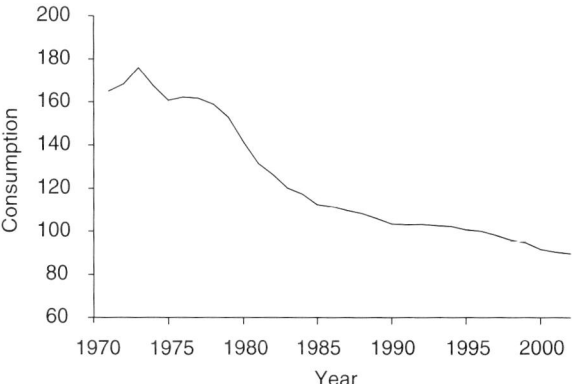

FIGURE 3 OECD consumption of oil per unit of GDP (in real terms; 1995 = 100). Data from the OECD.

from lower oil intensity has been reinforced by the decline of the oil prices in real terms (Fig. 4). Using the U.S. GDP deflator as the relevant price indicator, the relative oil price stood throughout the 1990s at levels that correspond to about half of its maximum levels reached in the early 1980s.

The size of the first-round effects is affected by additional factors, namely, the exchange rate and the profit margins. Oil prices are set in dollars in the international markets. Therefore, developments in oil prices denominated in an alternative currency (the euro, for example) reflect also the behavior of the exchange rate. Concerning profit margins, retailers always have the option to accommodate (partially or totally) the increases in a given input, such as energy, by contracting the profit component of the final price. The behavior of profit margins reflects a wide range of factors, including the cyclical position of the economy or the financial position of the retailers. Table I shows for the G7 countries the very strong correlation between annual rates of change of oil prices, based on national currencies, and the consumer energy components of the national CPIs. Table I also shows that the energy component of the CPI tends to react by slightly more than a ratio of 1:10 to the change in oil price. Combining this sensitivity with the present weight of energy in final household consumption, which tends to be slightly below 10%, it can be seen that inflation reacts by approximately 1:100 to the change in oil price. This rule of thumb gives a rough approximation of the magnitude of the direct first-round effects of an oil price change on inflation.

In addition, price developments are also affected by the so-called second-round effects. As shown in Table II, changes in consumer energy prices tend to produce contemporaneous but also lagging effects on the remaining components of CPI. This is the main reason why inflation measures that exclude the evolution of energy prices—usually referred to as core inflation indicators or measures of underlying inflation—tend to be lagging rather than leading indicators of price developments. Typically, second-round effects are associated with a circular wage–price causality. If workers manage to increase their nominal wages in line with the rise of consumer prices, rather than accepting lower real wages, additional inflation pressures arise, through a wage–price spiral. Therefore, the magnitude of the second-round effects clearly depends on the credibility of monetary policy, which is key to the

TABLE I

Correlations between Oil Prices and Energy Consumer Prices[a]

Country	Consumer energy prices[b]	
	Correlation coefficients	Elasticity
United States	0.756	0.141
Japan	0.765	0.127
Germany	0.661	0.097
France	0.811	0.137
Italy	0.797	0.154
United Kingdom	0.563	0.098
Canada	0.418	0.065
Average	0.682	0.117

[a] Annual rates of growth, 1971–2001. Data from the OECD and Eurostat.
[b] Prices based on national currencies.

TABLE II

Correlation between Energy Prices and Overall Consumer Prices Excluding Food and Energy in period $t + i^{ab}$

Country	$t-1$	t	$t+1$	$t+2$
United States	0.343	0.694	0.734	0.503
Japan	0.491	0.707	0.479	0.333
Germany	0.343	0.694	0.734	0.503
France	0.550	0.676	0.708	0.591
Italy	0.506	0.672	0.607	0.493
United Kingdom	0.642	0.877	0.747	0.379
Canada	0.642	0.717	0.658	0.452
Average	0.503	0.720	0.667	0.465

[a] Standard deviation of annual rates of growth, 1971–2001. Data from the OECD.
[b] $i = -1, 0, 1, 2$

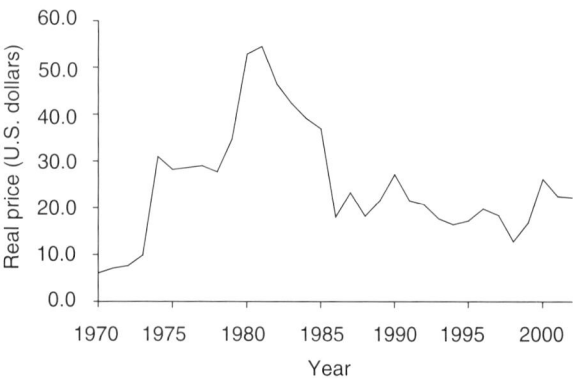

FIGURE 4 Oil prices in real terms (using the U.S. GDP deflator; 1995 = 1). Data from the OECD and Datastream.

formation of inflation expectations. For instance, second-round effects associated with the first oil price shock were larger than the ones associated with the second oil price shock, and this can to a large extent be explained by the different reactions of the monetary authorities to each of the shocks.

Considering the oil price shock of 1973–1974, there is strong evidence that substantial second-round effects took place in many industrial economies. This outcome reflects the combined effects of latent inflationary pressures (already present at the beginning of the 1970s), the large magnitude of the oil price shock, the relatively low flexibility of labor markets, and, last but not least, the accommodative stance of monetary policies in the more advanced economies. It is commonly accepted by economists that a tighter monetary policy was needed, as suggested by the fact that real interest rates became negative in 1974 and remained so until 1978, in a large number of countries. The lack of antiinflationary credibility contributed to unsustainable levels of real wages and thus to marked increases in the unemployment rate.

The experience with the second oil price shock was different. Despite the fact that inflation remained high in the period 1981–1982 (at double-digit figures), it fell significantly in subsequent years. A decisive factor for that evolution was the decline in real wages, in clear contrast to what had happened in the first oil price shock. In addition, the response of monetary policy, characterized by increases in the nominal interest rates, assured that real interest rates were positive and clearly above the levels observed in 1973–1974. Monetary policy was relatively successful in the second oil price shock, as measured by the subsequent reduction of inflation and the moderate evolution of inflation expectations.

3. IMPACT OF OIL PRICE CHANGES ON INFLATION

The focus of this section is on the quantitative assessment of the impact of oil price fluctuations on inflation. For that purpose, the empirical evidence provided by an OECD study is briefly described. The study considered a permanent oil price increase of about $10, vis-à-vis a baseline where oil prices were projected to remain at the $20–25 level (this shock corresponds to a price increase of about 40–50%). The OECD study presents estimates of the impact on GDP, world trade growth, and inflation, for the G7 countries. The estimates were obtained through the simulation of the OECD macroeconometric model Interlink, assuming fixed nominal exchange rates, real interest rates kept at baseline levels, and unchanged real government expenditures. The main results of this exercise, with respect to the impact on inflation, are reported in Table III.

Considering the average results for G7 countries, the impact of oil prices in the first year is slightly above the 1:100 rule of thumb, corresponding basically to the direct first-round effects, considering also some initial indirect effects. Effects in subsequent years are, naturally, smaller, corresponding to some lagging indirect first-round effects and to the second-round effects. As mentioned previously, the magnitude of the second-round effects depends on the assumptions on the reaction functions of the monetary authorities (constant real interest rates in this simulation).

Discrepancies in the estimates for different countries, Italy and Canada in particular, illustrate the difficulties associated with this type of exercise, owing to several factors. In first place, the transmission channels underlying the effects are extremely complex and are very likely to change over time. A macroeconometric model constitutes, in general, a very simplified representation of the real world and therefore does not capture precisely the functioning of the economy. Second, econometric simulations tend to replicate the average behavior of the economy since the early 1970s, despite the marked changes that took place in key variables, such as the oil intensity of the G7 economies or the weight of indirect taxation in final consumer prices. Finally, the estimates of the second-round effects associated with the change in oil price will depend on the different assumptions on the reaction of the monetary

TABLE III

Oil Price Effect on Inflation[a]

Country	Years after the shock		
	1	2	3
United States	0.5	0.2	0.1
Japan	0.4	0.0	0.0
Germany	0.4	−0.1	−0.2
France	0.6	0.1	0.1
Italy	1.0	0.2	0.0
United Kingdom	0.4	0.3	0.4
Canada	0.1	−0.1	0.2
Average	0.5	0.1	0.1

[a] A 50% rise. Data from Dalsgaard et al. (2001).

authorities, i.e., the credibility of monetary policy, and on how economic activity is affected. There is no doubt that there is an inverse relationship between oil prices and aggregate economic activity. Among the several channels through which oil price movements affect economic activity, the supply-side effect is probably the most common one. Rising oil prices, signaling reduced availability of a basic input to production, reduce potential output and the natural rate of unemployment, simultaneously explaining a negative relationship between oil price and economic activity and a positive relationship between oil price and inflation. However, a lot of uncertainty remains about the quantification of these effects. For instance, recent research indicates that there is a nonlinear relationship between oil price and real activity. In particular, the evidence indicates that oil price increases tend to produce stronger effects than do oil price decreases, and that increases after a period of stable prices tend to produce larger effects, as compared to increases that correct previous declines in prices. This type of result is also likely to hold when assessing the relationship between oil price changes and inflation.

4. OIL PRICE VOLATILITY

The volatility in oil prices since the early 1970s is a remarkable feature of energy economics. Annual fluctuations in the oil price level, as measured by the absolute value of year-to-year price changes, averaged only 1% in the 1949–1970 period; from 1970 to date, these fluctuations increased dramatically, reaching an order of magnitude of 30% per year. Even in the relatively stable period from 1986 to 1997, oil prices were more volatile than other primary commodities.

There is a very large set of reasons behind oil price volatility, including seasonal changes in demand; shifting demand conditions, in line with world GDP fluctuations; shocks in supply conditions that routinely affect, on a regular but rotating basis, some of the main oil producers (Iraq and Venezuela, in the recent past, for instance); and difficulties within OPEC in exercising its market power. The problems experienced by OPEC are due both to limitations of the knowledge of all the relevant aspects in the functioning of the market and to difficulty in maintaining cooperation between cartel members; these underlying problems have sometimes contributed to undesirable swings in the price of oil, as occurred in 1998–2000.

4.1 Oil Price Volatility as an Important Source of Inflation Changes

The contribution of energy prices to the volatility of the CPI is considerable. This point is illustrated in Table IV for the G7 countries. Despite the previously mentioned decline in oil prices in real terms, the average increase in consumer energy prices was slightly more than the overall increase in consumer prices. This evolution reflects an increase in taxation of energy products, which has been commonly justified for ecological reasons. At the same time, and this is the important aspect to stress at this stage, energy prices exhibited a volatility that was more than double the observed volatility in total consumer prices. Therefore, oil price changes are extremely important for the inflation level, because energy consumption exerts a weight slightly below 10% on total consumption, and even more important for fluctuations of inflation.

4.2 How to Forecast Oil Prices?

The evolution of oil prices is typically subject to a very high degree of uncertainty, given the extremely volatile nature of conditions that affect prices. Information on the future of oil prices is, however, extremely important for market operators as well as for central banks. Central banks, in particular, have a forward-looking perspective, attaching a very important role to inflation forecasts. This uncertainty tends to reinforce the importance of the relationship

TABLE IV

Oil Prices and Inflation[a]

Country	Average		Standard deviations	
	Total consumer prices	Consumer energy prices	Total consumer prices	Consumer energy prices
United States	5.1	5.8	3.1	9.4
Japan	3.8	3.3	5.0	9.3
Germany	3.3	4.5	2.0	7.4
France	5.8	6.7	4.3	9.5
Italy	8.9	9.0	6.0	11.1
United Kingdom	7.6	8.0	5.7	9.2
Canada	5.3	7.1	3.5	7.6
Average	5.7	6.3	4.2	9.1

[a]Using annual rates of growth, 1971–2001. Based on authors' computations using OECD data.

between oil prices and inflation, independently of estimates of the effects of a permanent change in oil prices on inflation. Two alternative technical assumptions on oil prices are worth mentioning:

1. The carryover assumption assumes that oil prices will remain constant; as in other markets (the exchange market provides a clear illustration), this hypothesis is justified by the fact that short-term changes in oil prices are statistically close to a pure mean zero random variable.

2. The futures market assumption assumes that the path of oil prices will follow the prices implicit in the agreements for future oil transactions.

Unfortunately, both techniques have produced considerable errors in past assessments, contributing to wrong evaluations of future inflation developments. On one hand, the carryover assumption reflects the idea of ignorance. If it is not possible to know if oil prices will increase or decrease, this assumption corresponds to putting the speculation somewhere in the middle, assuming that oil prices will remain constant. Obviously, this assumption will not account for the short-term volatility of oil prices and for the mean reversion process that has characterized the evolution of oil prices since the mid 1980s. On the other hand, futures markets tend to anticipate this mean reversion property, in the sense that they tend to converge, in a 1-year horizon, toward a range of $20–25 per barrel. However, the track record of futures on oil prices from 1995 onward has been rather poor (Fig. 5). In the short run, futures prices quite often missed the direction of change in oil prices. Additionally, the speed of adjustment and the mean level underlying the mean reversion process characterizing oil price behavior do not seem to be constant.

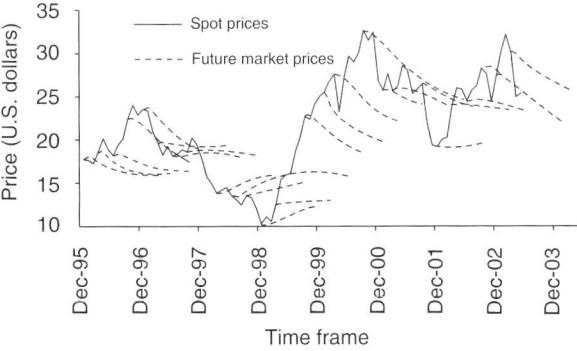

FIGURE 5 Spot prices versus future market prices (monthly figures, December 1995 through March 2003). Data from the International Petroleum Exchange.

An adequate measure of the degree of uncertainty attached to the evolution of oil prices in the near future is provided by the implicit volatility of the distribution of contracts. In that respect, it may also be interesting to analyze the role of speculators in energy futures markets. This has received a great deal of scrutiny in recent years, even though recent research suggests that the speculators, an important group among the commodity-fund managers, do not seem to be the driving force behind the volatility often observed in the crude oil market.

SEE ALSO THE FOLLOWING ARTICLES

Business Cycles and Energy Prices • *Economics of Energy Demand* • *Economics of Energy Supply* • *Energy Futures and Options* • *Innovation and Energy Prices* • *Oil Price Volatility* • *OPEC Market Behavior, 1973–2003* • *Prices of Energy, History of* • *Stock Markets and Energy Prices*

Further Reading

Bank of England. (2000). What do the recent movements in oil prices imply for world inflation? *Q. Bull.* (May 2000), 147–149.

Brown, S. P. A., and Yücel, M. K. (2002). Energy prices and aggregate economic activity: An interpretative survey. *Q. Rev. Econ. Finance* **42**, 193–208.

Bruno, M., and Sachs, J. (1985). "Economics of Worldwide Stagflation." Harvard University Press, Cambridge, Massachusetts.

Dalsgaard, T., André, C., and Richardson, P. (2001). "Standard Shocks in the OECD Interlink Model." Organization for Economic Cooperation and Development (OECD) Economics Department Working Papers, No. 306. OECD, Paris.

European Central Bank (ECB). (2000). *ECB Econ. Bull.* (November 2000), 21–26.

Hamilton, J. D. (2002). What is an oil shock? *J. Econometr.* **133**, 363–398.

Hooker, M. (1986). What happened to the oil price–macroeconomy relationship? *J. Monetary Econ.* **38**, 195–213.

Kohl, W. L. (2002). OPEC behavior, 1998–2001. *Q. Rev. Econ. Finance* **42**, 209–233.

Marques, C. R., Neves, P. D., and Silva, A. (2002). Why should central banks avoid the use of the underlying inflation indicator? *Econ. Lett.* **75**, 17–23.

Smith, J. L. (2002). Oil and the economy: Introduction. *Q. Rev. Econ. Finance* **42**, 163–168.

Stuber, G. (2001). The changing effects of energy-price shocks on economic activity and inflation. *Bank Can. Rev.* (Summer 2001), 3–14.

Weiner, R. J. (2002). Sheep in wolves' clothing? Speculators and price volatility in petroleum futures. *Q. Rev. Econ. Finance* **42**, 391–400.

Information Technology and Energy Use

KURT W. ROTH
TIAX LLC
Cambridge, Massachusetts, United States

1. The Direct Impact of Information Technology on Energy Consumption
2. Indirect Energy Consumption
3. Conclusions

Glossary

active power mode Usage mode when equipment is performing its primary function (e.g., copier printing).

computer network equipment Equipment central to the operation of computer networks, including server computers, uninterruptible power supplies (UPSs), local area network (LAN) switches, routers, hubs, and wide area network (WAN) switches.

indirect impacts Energy consumption influenced by, but not directly attributed to, information technology (IT) equipment (e.g., the energy required to manufacture a copier).

office equipment Equipment typically deployed in offices to enhance productivity (e.g., personal computers, monitors, copiers, printers, facsimile [fax] machines).

off mode Usage mode when equipment is turned off but still plugged in (e.g., a copier manually turned off).

power management-enabled (PM-enabled) rate The percentage of information technology (IT) equipment that enters lower power draw modes after periods of nonuse.

stand-by mode Usage mode when the equipment is ready to perform, but is not actually performing, its primary function (e.g., copier fuser roll warm and ready to print).

suspend mode Usage mode when the equipment has powered down and cannot directly transition to active mode (e.g., copier transitioned to low-power mode with cold fuser roll).

telecommunications network equipment Equipment central to the operation of telecommunications networks, including: wireless base stations, fiber-optic terminals, analogue phone network equipment, private branch exchanges (PBXs), and wireless phones.

Information technology (IT) equipment includes devices that produce and process information (e.g., office equipment) and that transmit information electronically (e.g., computer and telecommunications network equipment). IT equipment directly consumes energy by drawing power. Analyses estimate that IT equipment directly accounts for approximately 3% of annual U.S. electricity consumption or a bit more than 1% of annual U.S. primary energy consumption. To clarify, primary energy includes the energy consumed to generate, transmit, and distribute electricity to end-use sites. For example, the average kilowatt-hours (kWh) of site electricity in the United States in 2000, equal to 3413 Btu of energy, required 10,775 Btu of source energy. (This Energy Information Administration estimate takes into account the electricity production mix for the entire country.) The nation's hundreds of millions of devices also have several potential indirect impacts on energy consumption. IT equipment can influence the national energy intensity (i.e., energy consumption per dollar of gross domestic product [GDP]) due to productivity changes, changes in work patterns (e.g., telework and telecommuting, e-commerce), and shifts in the national economy. On a local level, IT equipment affects peak electricity demand, whereas on the building level, it alters the energy consumption of heating and cooling systems. The manufacture of IT equipment also consumes energy, as does the manufacture of the paper used by printers, copiers, and facsimile (fax) machines.

1. THE DIRECT IMPACT OF INFORMATION TECHNOLOGY ON ENERGY CONSUMPTION

Information technology (IT) equipment (Table I) annual energy consumption (AEC) depends on three

TABLE I

Major Energy-Consuming Information Technology Equipment

Office equipment	Network equipment
Copy machines	Analogue phone network
Facsimile machines	Fiber optic terminal
Monitors and displays	Local area network switches
Personal computers	Server computers
Printers (laser and inkjet)	Routers
	Server computers
	Uninterruptible power supplies
	Wireless telephony base stations

factors: equipment operating patterns, power draw by mode, and the installed base (Fig. 1). As a result, most IT equipment AEC estimates are bottom-up estimates that quantify each of the three factors. The operating pattern represents the number of hours spent per year in various modes of operation, each with a different power draw level. Typically, office equipment has at least three modes, as illustrated in Table II.

Unit energy consumption (UEC) for a given device equals the sum of the product of hours in each operating mode and the average power draw for each mode. Because the active power mode typically has much higher power draw levels than the other modes, the number of hours spent in active mode has a major impact on the UEC. Two factors, the night status and the power management-enabled (PM-enabled) rate, determine the number of hours spent in active mode for most kinds of office equipment. Weekend and night hours account for a sizable majority of potential operating hours, so night audits of commercial buildings are performed to determine the approximate operating patterns of office equipment during these periods. Many types of office equipment have PM capability, by which the equipment enters the low-power suspend mode after a prescribed period of inactivity. Clearly, the PM-enabled rate influences the daytime and night/weekend operating patterns of various equipment and is best derived from audits of commercial buildings. More accurate UEC estimates need to consider the actual (i.e., measured) power draw in each mode

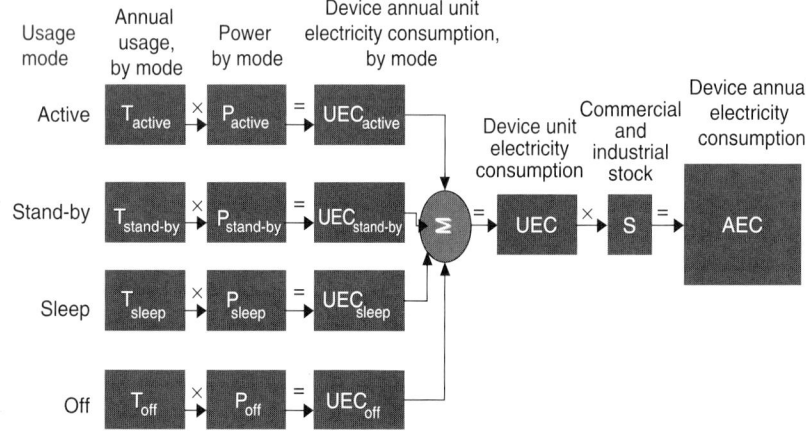

FIGURE 1 Bottom-up annual electricity consumption calculation methodology. From Roth *et al.* (2002).

TABLE II

Office Equipment Usage Modes

Mode type	Description	Examples
Active	Device carrying out intended operation	Monitor displays image; copier printing
Stand-by	Device ready to out intended operation (but not doing so)	Monitor displays screen saver; copier ready to print
Suspend	Device not ready to carry out intended operation (but on)	Monitor powered down but on; copier powered down but on
Off	Device not turned on (but plugged in)	Monitor off, plugged in; copier off, plugged in

instead of the nameplate power rating given that the active mode power draw for most IT devices is typically less than 50% of the nameplate value. The installed base of various types of IT equipment comes from either industry studies of equipment, market saturation data, or estimates based on sums of annual device sales over equipment lifetimes.

Circa 2000, nonresidential IT equipment energy consumption was roughly six times greater than residential IT equipment energy consumption. Consequently, this section focuses on nonresidential IT equipment and energy consumption.

1.1 Commercial and Industrial IT Equipment

In 2000, nonresidential IT equipment consumed just under 100 TWh Table III. Personal computers and their monitors represented just over 40% of the total AEC (~42 TWh site or 0.46 quads). The equipment forming the backbone of the Internet (server computers, computer networks, telephone networks, and uninterruptible power supplies [UPSs]) consumed approximately 30% of all nonresidential IT equipment electricity (~30 TWh site or 0.33 quads primary). The following subsections discuss the energy consumption of the major equipment types in further detail.

1.1.1 Personal Computers and Workstations

The installed base of approximately 71 million personal computers and 2.5 million workstations in 2000 consumed 17.4 and 1.8 TWh, respectively. In most desktop PCs, the chipset, central processing unit, memory, and AC/DC power supply draw the most power in active mode, with an average 2000 vintage desktop PC drawing 55 to 60 W in active mode. More recent PCs can draw much more power in active mode, sometimes in excess of 100 W, particularly when configured with powerful graphics cards. On the other hand, the steadily growing notebook (laptop) computer segment typically draws 15 to 25 W total (including the 14- or 15-inch LCD monitor) and employs more sophisticated PM schemes in a quest to increase "unplugged" running times. Large quantities of PCs left on at night (~50% based on limited night audits) and low PM-enabled rates (~25%) are responsible for the majority of PC annual energy consumption. The past incompatibilities between common PM schemes and operating systems that caused the systems to crash have been resolved, creating the potential for higher future PM-enabled rates. On the other hand, PC network cards remain a problem in that entering suspend mode can cause the PC to drop the network connection. To avoid the inconvenience of disconnection from the network, some users disable PM features.

Future PC energy consumption will depend on increases in microprocessor and graphics card active power draw, changes in PM-enabled rates, and transitions to notebook (and smaller) computing devices.

1.1.2 Monitors and Displays

Most of the nation's approximately 61 million monitors are associated with desktop PCs and workstations, whereas some 13 million displays rely on video input (as seen in airports). Together, they consumed approximately 22 TWh of electricity in 2000. That same year, cathode ray tube (CRT) monitors accounted for more than 95% of the display market due to their low cost, with liquid crystal display (LCD) devices accounting for most of the remaining market. A CRT converts electrical

TABLE III

U.S. Nonresidential IT Equipment Annual Electricity Consumption Summary

Device type	AEC (TWH)	Percentage of AEC
Communications networks	30	31
Server computers[a]	11.6	13
Telephone network equipment[b]	6.6	7
Computer network equipment[c]	6.4	7
UPSs	5.8	6
Monitors and displays	22	23
Monitors	18.8	19
General displays	3.4	4
PCs[d]	20	20
Imaging devices	15	16
Copiers	9.7	10
Printers	5.7	6
Other	10	10
Total	97	100

Source. Roth et al. (2002).
[a] Includes data storage.
[b] Includes cell site equipment, transmission (fiber-optic), PSTN, PBXS, wireless phones.
[c] Includes LAN switches, routers, hubs, WAN switches, Modems/Remote Access Server (RAS), cable modem termination systems (CMTS).
[d] Includes desktop personal computers, workstations, and laptop/notebook personal computers.

signals to the visual display seen on the screen, using an electron gun to emit a beam of electrons and project them onto a screen. Anodes accelerate the electrons, which are then steered by a varying electromagnetic field onto different parts of the screen, where they interact with a coating of phosphor compounds that convert the electron beam into an image. In essence, the CRT "paints" an image on the phosphor layer. In contrast to monochrome displays, which use a single electron gun, color monitors use three separate electron guns (red, blue, and green) to create color images.

Most CRT monitors (17 inches is the most common size) draw from 70 to 90 W in active mode, whereas similarly sized LCDs draw approximately 60% less power. Monitors have significantly higher PM-enabled rates than do PCs ($\sim 60\%$), resulting in monitors spending fewer hours in active mode. Recent growth in the market share of LCD monitors (to $\sim 20\%$ worldwide in 2002) and laptop computers ($\sim 25\%$ of global PC sales), as well as increases in PM-enabled rates, will tend to decrease net monitor energy consumption, whereas further increases in screen size will increase energy consumption.

1.1.3 Copy Machines

Commercial buildings house approximately 9 million copy machines that consumed nearly 10 TWh of electricity in 2000. Analogue and newer digital copy machines all function in the same basic manner. The machine captures the image, either via a light source (analogue) or via a scanner (digital), and transfers it to the hot toner drum, which then transfers it in the form of toner to paper sheets. Copy machines come in a wide range of bands (speeds), ranging from several copies per minute (cpm) for desktop units to approximately 100 cpm for high-performance units.

Copy machines have high UEC values due primarily to the high power draw during copying (typically in excess of 1000 W) and the high stand-by power levels (hundreds of watts) needed to maintain fuser rolls in a hot ready state. In fact, the stand-by (hot fuser roll) mode typically accounts for most copy machine energy consumption, whereas active copying mode accounts for only a fraction of the consumption. Copiers wait substantial time periods before powering down into suspend mode because reheating a cooled fuser roll can take several minutes, which is often an unacceptably long delay in the fast-paced office environment. Advances in copier fuser systems, including toner materials with lower melting temperatures, could substantially decrease copier energy consumption in active and stand-by modes by decreasing the fusing temperature, which would reduce the amount of energy needed to keep the fuser rolls hot. Lower fuser roll temperatures also would decrease the warm-up time, perhaps to only a few seconds, enabling copiers to spend more time in the low-power suspend mode. The gradual integration of copier and printer function into a single device will also affect future copier and printer energy consumption.

1.1.4 Printers

Three printer types—laser, inkjet, and impact—constitute most of the printer market and consume nearly 6 TWh of electricity per year, with laser printers accounting for approximately 80% of total printer AEC. Typically, a laser printer is a shared resource among several users in a computer network, whereas inkjet printers may serve as personal printers or the sole printers in smaller businesses. Impact printers are generally used in the "back office" as part of order fulfillment systems where it is necessary to print information onto multiple copies of forms via carbon copies.

Although inkjet printers dominate the residential market, laser printers remain the printer of choice for offices due to their high print quality and printing rates. Laser printers consume much more energy than do inkjet devices primarily because the fuser rolls must remain at high temperatures to bond the toner to the paper quickly and completely. During printing, the laser printer actively supplies resistance heat to ensure effective bonding. In addition, laser printers in stand-by mode require heat to avoid delays in response to a print request. Most laser printers have PM capabilities, but surveys suggest that only perhaps half of laser printers are PM enabled.

If color laser printers become faster and their prices continue to decrease, their presence in the office could increase significantly in the future. This would challenge the inkjet market to find another niche to fill. On the other hand, the development of higher speed and quality inkjet printers could supplant much of the laser printer market and lead to an overall reduction in printer energy consumption.

1.1.5 Server Computers

In today's usage, a server generally refers to a computer that is not directly associated with a specific human user. Instead, servers provide common functions to a group of users or perform back-end processing invoked on a scheduled basis or by other computers. Servers come in a wide range of sizes. At the upper end are the traditional mainframes, repositioned as high-end servers. The market segment formerly called minicomputers is now classified as midrange or workhorse servers. Server functions run the gamut of computing tasks, with the exception of those involving direct human interaction. High-end servers often carry out traditional batch processing applications (e.g., billing) or fulfill high-volume transaction processing applications (e.g., banking, airline reservations). Midrange servers typically handle database applications and can be the main back-end computers for medium-sized businesses. These larger systems feature superior reliability relative to PC-derived systems and often require large storage and input/output (I/O) handling capacity. Low-end servers typically take on tasks that require smaller storage and/or lower criticality (e.g., local area network [LAN] file and print management). Many perform as Web servers at large hosting centers (also known as server farms, data centers, or colocation facilities) operated by Internet service providers (ISPs). Other common roles for low-end servers include e-mail or lower volume specialized applications.

In general, CPUs and hard drives, as well as power supplies, account for the bulk of server power draw, with actual power draw depending significantly on the server class (Table IV) and configuration. Additional peripherals, primarily data I/O devices, play a key role in the functioning of server computers, particularly in high-end machines. There are two types of data storage: optical and disk systems. Optical storage is used for archiving data and involves only the writing of data to the optical storage disks. Disk storage typically interacts (reading and writing data) more frequently with outside systems and must be constantly accessible. Optical and disk data storage devices consume approximately 0.9 and 0.7 TWh per year, respectively.

1.1.6 Uninterruptible Power Supplies

Although UPSs are not IT equipment per se, they play an increasingly important role in ensuring its reliability. UPSs ensure the continuous flow of high-quality power to critical IT equipment. They provide power quality to eliminate transient spikes or sags in power that could adversely affect the performance or could damage electricity-consuming equipment. In case of power service failure, they also provide sufficient backup power to last through the outage, shut down the electronic device, and/or bring back up power generation online.

Battery chargers/rectifiers, stand-by batteries (usually lead acid), and inverters are the primary components of most UPSs. UPS applications relevant to IT equipment range from small (1 kVA) stand-by devices for individual PCs or workstations to very large (>100 kVA) online UPSs for data center applications. Overall, IT-related UPSs represent approximately 70% of UPS sales and consume nearly 6 TWh annually.

There are three major types of UPSs: stand-by, online, and line interactive. Stand-by and line-interactive systems have similar operating principles; relative to stand-by systems, line-interactive architectures typically offer superior PM because they have a live inverter. In standard operational mode, both architectures allow power to pass through the UPS and into the electronic device while providing varying degrees of power conditioning, typically surge suppression to counter voltage spikes and a filter to reduce unwanted harmonics. If the UPS battery has run down, the UPS battery charger will charge it. When power ceases flowing to the device, the UPS detects the lack of electric power and the transfer rapidly establishes (clamps) an electrical connection between the device and the battery power source to enable continued operation. Line-interactive and stand-by systems typically consume only a

TABLE IV

U.S. Server Computer Characteristics circa 2000

Attribute	Server class			
	Low end <$25,000	Workhorse $25,000–$100,000	Mid-range $100,000–$1 million	High end >$1 million
Installed base (thousands)	4,050	575	185	16.5
Typical active power draw (W)	125	650	1225	2500
AEC (TWh)	4.5	3.3	2.0	0.4

few percent of the electronic device load because in default operational mode they allow power to pass through to the load with minimal PM. The ferroresonant UPS is a notable exception in that it generates significant quantities of heat due to inefficiencies in the transformer.

In contrast, online UPSs accept AC power, which they condition and rectify (convert from AC to DC). Subsequently, the DC output charges the batteries and then passes into an inverter, which converts the DC power to well-conditioned AC power. If the input power fails, the battery continues to supply power to the inverter and provide high-quality power to the load without interruption because it has no transfer switch. Online systems offer high reliability but cost more than other UPS systems and are relatively inefficient, typically dissipating 10 to 15% of the power flow in heat when operating near the rated capacity. A new UPS architecture enables the input power to largely bypass the AC–DC–AC conversion process. Instead, the UPS primarily supplies the difference between the input and load powers, resulting in significantly higher efficiencies than is the case with double conversion systems (typically dissipating less than 5% of power flow near the rated capacity).

UPS efficiency also depends on the actual load seen by the UPS relative to the rated load. UPSs have a basic power demand (for fans, power supplies, etc.) that establishes a "base" load. However, the electric components are sized for the full UPS power rating and dissipate less heat at smaller loads. The net result is that the overall UPS efficiency curve is relatively flat when the load falls between 50 and 100% of the off rated load but often drops off below 50% and falls off precipitously under 20%. In general, UPSs are oversized, often designed for 70 to 80% load on larger systems and typically operating at 50% or less of design load. Several reasons for this practice exist, including general design conservatism, allowances for future expansion, greater redundancy, longer load runtimes after power failure, and parallel redundant systems for maintenance purposes. Circa 2002, many data centers had low occupancies, causing UPS loads to fall well short of expected loads. As a consequence, many UPS systems are likely see less than 25% of their rated loads and operate at low efficiencies.

1.1.7 Computer Network Equipment

Computer networks use a range of equipment that collectively connects the global Internet with millions of desktop and server computers. This equipment is generally divisible into two categories: LAN (within a building or campus) and wide area network (WAN, beyond the campus). LAN gear interconnects desktops and servers using private bandwidth, whereas WAN gear generally uses common carrier facilities to provide external access, either within a company (intranet), between companies (extranet), or to the global Internet.

In general, computer network equipment power draws vary little with data throughput rates, as devices always remain "hot" to respond to surges in traffic. In spite of round-the-clock operation, the total electricity consumed per year by all computer network equipment is only approximately 6 TWh per year (or $\sim 6\%$ of nonresidential IT energy consumption). LAN switches account for perhaps half of the total. Specialized processors and AC/DC power supplies are the major energy-consuming components of all computer network equipment except passive hubs.

1.1.7.1 LAN Gear: Hubs and Lan Switches In buildings, desktop computers are generally connected via short wiring to some kind of hub. Whether this is a simple hub or a LAN switch, it is often a rack-mounted device supporting anywhere from as few as 4 to a few hundred ports in a single rack. Functionally, hubs provide a physical connection between the local network and other devices, such as computers and printers, with most hubs providing signal conditioning in addition to connectivity (in contrast to the passive hubs of the past). LAN switches are more sophisticated devices that provide multiple paths between inputs and outputs, so that traffic between two nodes (e.g., one desktop and one server) does not affect traffic between other nodes. These higher function switches provide additional security features typically performed by routers, but they are scaled down to be cost-effective on the LAN. As the price approaches levels once occupied by simple hubs, the market is moving toward greater acceptance of higher layer switching.

1.1.7.2 Routers Routers have a basic set of primary functions, including examining, filtering, and routing incoming data packets. Routers examine each incoming packet and determine where to send it based on its address. Large high-capacity routers with high-speed interfaces (e.g., OC-48 or greater) serve major backbone routes, whereas subscriber connections tend to use large numbers of lower speed interfaces (e.g., DS-1 to DS-3). At subscriber locations, a small edge router usually provides the connectivity. Corporate users and smaller ISPs typically have mid-sized routers.

1.1.7.3 WAN Switches Like routers, WAN switches help to manage the flow of large amounts of data between locations. They differ in their use of connection-oriented telecommunications protocols, with some hybrid switch/router products providing both functions. WAN switches tend to be large devices used by phone service carriers, ISPs, and/or high-end enterprise customers.

1.1.8 Telecommunications Network Equipment
Telephone networks differ from computer networks in that they were designed primarily to carry voice information instead of data. Four primary categories of telephone switching equipment exist. The transmission network includes long-distance fiber-optic connections among major cities and locations handling long-distance telephone calls. The mobile telephone network includes the base stations (in towers) that connect mobile telephones to the larger phone network. The public switched telephone network (PSTN) denotes the established copper wire connections between telephone companies' central offices and residences and buildings. Lastly, the private branch exchange (PBX) is a private in-house phone system, typically found in larger office buildings or campus settings.

Although telephone systems were originally designed to carry voice traffic, the distinction between voice and data traffic has become increasingly blurred since the advent of the Internet. For example, local public telephone networks originating from central offices now provide Internet access via dial-up modems and digital subscriber lines (DSL), and the long-distance fiber-optic transmission networks originally installed to handle voice traffic now carry Internet data. Mobile phone networks have begun to offer basic Web access, notably in Japan and Europe.

In 2000, telephone network equipment consumed approximately 6.6 TWh of electricity (or just under 7% of all electricity consumed by nonresidential IT equipment).

1.1.8.1 Mobile Telephone Networks Wireless (also known as cellular) phones became commonplace during the 1990s. Cell sites contain base station radio transmission gear (analogue and/or digital) that transmits and receives signals to and from cellular telephones, with the number of transmitters and their power dependent on the cell size, location, and chosen radio technology. The transmission gear is located as close as possible to the antenna (e.g., in a hut at the foot of a tower). Cell sites are characterized by their size (i.e., broadcasting power and effective receiving and transmission radius), which ranges from large macro cells to small pico cells. Transmission gear power draws vary significantly with the amount of traffic passing through the site, although the equipment does require a base load to keep it hot (i.e., ready to transmit information). Peak power draws depend greatly on the size of the installation, ranging from close to 10 kW for macro cells to as little as a few hundred watts for pico sites.

The approximately 100,000 cell sites in the United States in 2000 consumed more than 2 TWh, representing the largest component of telephone network energy consumption (this does not include additional energy used to cool the sites). The rollout of third-generation wireless networks with higher power draw and the expected uptick in geographical coverage will likely increase the average power draw at the base station as well as the number of base stations, making it nearly certain that base station energy consumption will increase.

Battery-powered, compact, and designed to maximize time of operation without recharging, cellular phones consume energy at low levels (on the order of a watt). Roughly 28 million nonresidential cellular phones, representing approximately 30% of the total stock of cellular phones, consumed close to 0.5 TWh in 2000.

1.1.8.2 Fiber-Optic Terminals The original long-distance transmission networks used copper wire, with microwave towers entering long-distance service during the 1950s. Fiber conversion began during the 1980s, buoyed by the enhanced signal quality and much higher bandwidth of fiber service compared with microwave service. Most long-distance traffic now passes through fiber. The current U.S. fiber network consists of numerous trunk routes between major hubs in each state, with fiber beginning to supplement and replace copper wire to major facilities, such as office buildings, to take advantage of its higher voice and data bandwidth.

Fiber-optic communication systems transmit data by converting electronic signals, such as voice signals and data packets, into laser-generated light pulses. The light pulses pass through a glass fiber wrapped in plastic (multiple fibers combined together make a fiber-optic cable) to a receiver, where they are translated back into electronic signals. Light signals that pass through long distances require an amplifier to maintain signal quality and intensity. As the quality of fiber-optic devices has improved, the distance required between amplifiers has grown (to ~80 km in 2000).

Fiber terminals, which send, receive, and multiplex data (e.g., to a T1 connection), dominate the energy consumption of fiber-optic networks. Drawing roughly 200 W per terminal, the approximately 1 million terminals in 2000 consumed nearly 2 TWh of electricity. Continuing increases in network traffic, coupled with growth in fiber connectivity (to commercial buildings and potentially to homes), will likely increase fiber terminal energy consumption in the future.

1.1.8.3 Public Switched Telephone Network The PSTN refers to the established copper wire connections between telephone companies' central offices and residences and buildings. Analogue phone lines to residences and businesses dominate energy consumption of the PSTN, accounting for roughly 165 million of the more than 190 million phone lines installed in the United States in 2000. Central office switches typically power analogue phone lines, but the analogue line components that account for much of the switch power draw only draw power on establishing a connection between two phones (i.e., when a call is completed). Other switch elements that continuously draw power include common equipment (one administrative module per switch) and line equipment (e.g., concentration modules, trunk ports). The sheer variety of switching equipment in vintage, scale, and complexity makes it difficult to characterize the installed base. Furthermore, there is a dearth of publicly available power draw data. Together, these two factors complicate development of an accurate estimate of total energy consumption. Nonetheless, because each phone draws an average of approximately 3 W per active line, the energy consumed by analogue phone lines appears to be quite small (on the order of 1 TWh per year).

1.1.8.4 Private Branch Exchanges PBXs are private in-house phone systems, typically found in larger office buildings or campus settings. Commercial phone systems, such as those made by Lucent and Nortel, drive the energy consumption of PBXs, which operate round-the-clock and consume approximately 1 TWh per year.

1.1.9 Other Equipment Types
Other IT equipment accounts for approximately 10% of nonresidential IT energy consumption. Facsimile (fax) machines (3 TWh) and point-of-service terminals (1.5 TWh) consume the most energy. Other equipment types consume smaller quantities of energy: automated teller machines, desktop calculators (adding machines), dictation equipment, scanners, small handheld devices (e.g., personal digital assistants), supercomputers, typewriters, very small aperture terminals (VSATs), and voice mail systems.

1.2 Residential IT Equipment

As noted earlier, residential IT equipment energy consumption appears to be roughly six times less than nonresidential IT energy consumption, with desktop PCs and monitors accounting for the majority. The actual AEC of residential IT equipment has a very high uncertainty, reflecting poorly understood usage patterns. Nonetheless, there are several reasons to believe that residential IT energy consumption is significantly less than nonresidential IT energy consumption. Most residential IT equipment almost certainly has significantly lower usage (in active mode) due to the length of the business day relative to residential usage patterns. Furthermore, residences have much smaller installed bases of large energy-consuming devices, notably server computers, telephone network equipment, copiers, many types of computer network equipment, and laser printers. Finally, the more energy-intensive (and typically larger) IT equipment tends to be deployed in nonresidential settings, for example, faster copy machines and laser printers, larger UPSs, core routers, and LAN switches.

The potential exists for residential IT equipment energy consumption to grow rapidly in the future. PCs (and associated monitors and printers) have yet to achieve residential saturation equal to the level of the commercial sector, meaning that there is room for growth. Similarly, as the percentage of households connected to the Internet continues to increase from just over 50% circa 2001, increases in residential IT usage will follow. The fact that only approximately 10% of Internet subscribers in 2001 had broadband (primarily cable modems and DSL), which requires modems and tends to increase PC usage (to support "always-on" operation), represents another growth opportunity. Finally, the long-awaited arrival of widespread home networking and its associated infrastructure could increase future residential IT energy consumption.

2. INDIRECT ENERGY CONSUMPTION

2.1 Overview: Effect on National Energy Intensity

IT equipment indirectly affects energy consumption in several ways. For example, IT-enabled process controls enhance production efficiency, whereas IT-enabled telework and telecommuting (T&T) changes building usage and personal transportation patterns. The demand for IT equipment results in additional energy consumption to manufacture IT equipment. Together, these and many other facets of IT affect the macroeconomic energy intensity of the national economy (energy consumed per dollar of gross domestic product (GDP). Because of the number of factors at work and the sheer complexity of economic systems, quantifying the indirect energy impact of IT is significantly more difficult than understanding the direct energy impact. Nonetheless, evidence strongly suggests that the magnitude of the indirect impacts is comparable to—and likely exceeds—the magnitude of the direct impact.

Some of the most compelling evidence for this claim comes from energy intensity data from the late 1990s, when the Internet was experiencing widespread commercialization. During this period, the energy intensity of the U.S. economy began decreasing at a sharper rate, approaching decline rates characteristic of the early and mid-1980s, a time of high energy prices (Fig. 2). Notably, the sharp decrease in energy intensity, which coincided with a significant increase in economic productivity, occurred despite historically low (in real terms) energy prices, leading many to ascribe much of the sharp decrease in energy intensity to IT. Because the steeper energy intensity decrease began only recently, it is premature to draw a solid conclusion. If IT-based innovations do accelerate the long-term decrease in energy intensity, the impact of IT equipment on national energy consumption could be significantly larger than the 1% of energy directly consumed by such equipment.

IT equipment affects national energy consumption in at least three general ways: (1) it causes structural changes to the economy, (2) it changes the mechanics of commerce and work (e-commerce), and (3) it consumes energy during manufacture.

2.2 Structural Changes in the Economy

Rapid growth in the low-energy intensity IT sector may decrease the energy intensity of the entire economy by shifting a larger portion of the nation's economic output to lower energy intensity activity. Overall, the IT-related sectors of the economy directly consume less energy per dollar of GDP produced than does the economy as a whole. Consequently, if IT sectors grow faster than the economy as a whole (i.e., if they become a larger portion of the economy), the energy intensity of the entire economy will be decreased. From 1990 to 1997, IT-related sectors grew about five times faster (in dollar terms) than the U.S. economy as a whole, contributing more than one-third of real GDP growth over that period. This may have accounted for a portion of the decrease in U.S. energy intensity during the late 1990s.

2.3 E-Commerce

E-commerce, short for electronic commerce, denotes the myriad of ways in which electronic transmission of content can substitute for the exchange of physical goods (dematerialization). Online shopping and banking, electronic music files and books, automated business-to-business exchanges, auctions, and electronic purchasing systems all are manifestations of e-commerce. E-commerce has long existed in certain forms (banking and stock trading), but its use increased when the public began embracing the Internet on a larger scale during the mid-1990s. Telecommunications and computer networks, along with servers and data storage devices, form the backbone of e-commerce. Other devices, including computers, monitors, and printers, enable people to directly interact and exchange information with other people on these networks.

Although e-commerce may result in net national energy savings, many of its applications have an

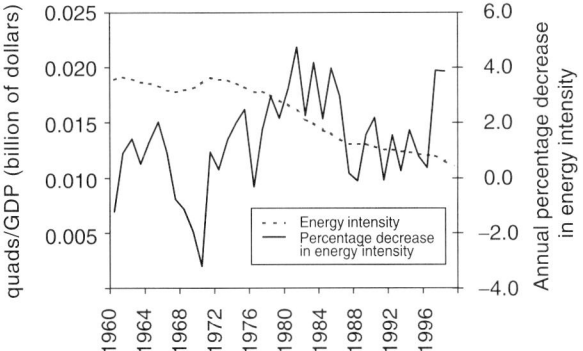

FIGURE 2 U.S. energy intensity and energy intensity decrease, 1960 to 1998. From Roth *et al.* (2002).

ambiguous net energy impact after considering all of the ramifications. For example, at first glance, it appears that a person ordering three books online reduces the energy consumed to acquire the books because the shipment to the front door via ubiquitous U.S. Post Office shipping (or by a private delivery company) replaces an automobile trip to a bookstore. However, many goods are shipped via air freight, which consumes nearly as much energy as does an auto trip. Furthermore, even if the books were bought online, the purchaser might have made the same automobile trip to buy something else at the same location as the bookstore or may have substituted a different trip for the trip to the bookstore. In either of these scenarios, the online order would show a net negative impact on energy consumption in the transportation sector. Online purchasing also requires significantly more packaging material, which consumes energy to manufacture and ship. Indeed, the net energy outcome is very sensitive to the specifics of the purchase: the transportation mode used and distance traveled for the store trip, the shipment mode(s) and distances traveled for e-commerce, the number of items purchased on the trip, the efficiency of the conventional book distribution network, the degree to which e-commerce reduces overproduction of books, and even the shipment packaging. "Rebound" effects (discussed in a subsequent subsection), such as any additional energy-consuming activities undertaken during the time saved (if any) by online ordering, also warrant consideration. All of these factors complicate this (or any) assessment of the net energy implication of e-commerce.

Many types of e-commerce require significant changes in personal and organizational behavior to achieve significant market presence, and these changes are typically difficult to realize and occur over long time periods. For example, shopping is not simply about the purchase; for many, it is a pleasurable activity that cannot be easily replaced by an online alternative. Similarly, the degree to which employees will adopt T&T is unclear. In some cases, T&T may be a favorable alternative for workers who spend a lot of time out of the office (e.g., sales personnel). On the other hand, a workplace fills important social needs and provides crucial opportunities for informal communication among workers. Security concerns about sharing personal information may also impede the development of e-commerce.

Ultimately, cost savings—not energy savings—drive companies to adopt e-commerce strategies, with the greatest savings expected to come from reduced procurement costs. In spite of optimistic projections for the growth of e-commerce during the current decade, e-commerce accounted for only a small fraction of U.S. and international commerce as recently as 2002. The net and potential impacts of e-commerce on national energy consumption will not become clear until businesses adopt e-commerce on a large scale.

The various manifestations of e-commerce have the potential to reduce energy consumption in the industrial (production), transportation, and building sectors.

2.3.1 Industrial Sector (Production)

Fundamentally, IT facilitates the acquisition and processing of data, increasing the flow of information about industrial processes and production. The information, intelligence, and control provided by IT can improve manufacturing and office productivity, reducing error rates and the time and energy expended to deal with errors. For example, in the back-office realm, a major manufacturer switching from a conventional ordering system to an online system could dramatically reduce its order rework rate. In one instance, such an improvement reduced order errors from 25 to 2%, saving the company $500 million while decreasing energy intensity.

2.3.2 Transportation Sector

Although not an aspect of e-commerce per se, IT equipment can also enhance the efficiency of industrial processes and product designs. IT equipment integrated with sensors could realize significant national savings through device improvement (e.g., an automobile) or enhanced system/process efficiency. For instance, a brewery installed a computer-based factory monitoring system integrated with a plant-wide intranet. The system improved process monitoring and production scheduling, resulting in a decrease in the level of waste bottles and cans from 5 to 0.1% of production.

IT can also enhance the effectiveness of product design by shortening the design cycle and improving the energy efficiency of design. To cite one example, aircraft companies routinely perform computational fluid dynamics (CFD) analyses on computers to model the flow around airplanes and calculate the drag. Relative to conventional wind-tunnel testing, CFD can be used to cost-effectively explore a wider range of design possibilities in a shorter period of time. This leads to improved airplane fuel efficiency, a key consideration in the purchase of commercial airliners.

E-commerce has the potential to reduce transportation energy consumption by reducing the number of trips and/or improving their efficiency. For example, roughly one-half of all trucks on the road carry no goods and are simply traveling to their next destination. An IT-enabled transportation exchange enables shippers to bid for unfilled space in truck fleets, increasing profits while also saving energy. Electronic distribution of music files and books precludes the need for transport of physical objects to the customer as well as the need for a trip to a store. In the case of T&T, electronic transfer of information (data and voice) makes trips to the office unnecessary. Similarly, video teleconferencing can replace long-distance travel.

2.3.3 Building Sector

E-commerce could reduce building energy consumption by reducing warehouse and retail building floor space. Improved supply chain management facilitated by IT should reduce inventories stored in warehouses, decreasing the square footage of floor space required and the energy consumed to condition and light that floor space. "Virtual storefronts" could replace a portion of the brick-and-mortar establishments. This could, in turn, greatly decrease (by more than 10-fold) the amount of energy required to operate businesses by reducing retail floor space (substituting low-energy warehouse-based operations for higher energy retail space) and increasing storage densities in warehouses relative to retail space. In both cases, reductions in warehouse and retail floor space would also displace a portion of projected new building construction, avoiding the energy consumed in the actual building construction and embodied in the building materials.

T&T can directly save energy by removing the need for office space. For example, assume that a major telecommunications company hopes to reduce office floor space per employee by approximately one-third by 2005 (relative to 2000) through vigorous promotion of telecommuting, which would lead to similar savings in energy consumption. In turn, reduced floor space demand would decrease the energy consumed in the construction sector, both directly (energy used to erect a building) and indirectly (energy consumed throughout the construction materials supply chains).

2.3.4 E-Commerce and the Rebound Effect

All purchasing decisions have ramifications well beyond the direct impact of a single purchase, making consideration of the energy impacts integrated over the entire economy very important. In energy terms, some rebound (the amount of additional energy consumed by other activities that occur due to the sale) is inevitable. Predicting the precise form of rebound experienced, as well as its magnitude, is often difficult. The rebound effect can operate on several levels relative to IT and may be intersector, meaning that an approach that reduces energy consumption in one sector tends to increase energy consumption in another.

Online bookstores clearly illustrate the importance of considering rebound effects. Lower costs for online book sales could force down the price of books, increasing the economy-wide demand for them as well as the national energy consumption to transport them. Net book production energy consumption would depend on the extent to which increased sales balance any decrease in book remainder rates. Moreover, if online sales do not actually displace physical bookstores, the energy used to support the infrastructure of online booksellers would not replace but rather supplement the energy already consumed in the retail sector. Rebound effects could also bring the savings of T&T into question. People who elect T&T may choose to live farther from the office, increasing automobile mileage for many trips and specifically during days when the employees do go to the office. If people move to less densely populated areas, the length of their other trips could increase. People working at home may also take advantage of their freedom to run more errands by car. On a somewhat larger scale, if T&T results in a net shift of the population to more remote locations, construction of the infrastructure (e.g., roads, electricity, sewage) to support the population shift will consume additional energy.

2.4 Manufacture of IT Equipment

2.4.1 IT Equipment

The manufacture of IT equipment consumes an appreciable amount of energy. A commodity I/O model of the U.S. economy based on 1997 U.S. Department of Commerce models and data enables life cycle analysis of the impact of the production of various equipment and materials, including IT equipment. It is important to note that the model includes not only the energy and resources directly consumed in device manufacture but also the resources consumed throughout the entire supply chain. The model indicates that the energy consumed throughout the supply chain to produce IT equipment in the United States in 1997 is on the same order as the direct energy consumption of all IT equipment in a single year. Presumably, manufacturing efficiencies have improved since 1997, while device shipment levels have grown in both unit and dollar terms over the same period. Consequently, the conclusion that the energy consumed throughout the supply chain to produce IT equipment is on the same order as direct IT equipment energy consumption still holds. When considering the energy impact of IT equipment manufacturing in a given country, however, it is important to consider that a portion of the equipment is manufactured in a different than where it directly consumes energy.

2.4.2 Paper

Copy machines, printers, and fax machines all consume paper. The direct energy consumption estimates for these devices take into account the energy consumed to print an image on the paper but do not reflect the energy used to manufacture the paper. A study during the late 1990s estimated that manufacturing a piece of paper from wood requires about 17 Wh and that manufacturing it from recycled paper requires 12 Wh, whereas electrographically producing an image consumes an average of roughly 1 Wh. As a result, the roughly 900 billion sheets of paper consumed in the United States in 2000 by inkjet and non-roll-fed laser printers, copiers, and fax machines required approximately 15 TWh to manufacture. In addition, very high-volume roll-fed laser printers consumed another 600 billion sheets. Assuming similar energy consumption values per sheet of paper, these 600 billion sheets would have consumed approximately 10 TWh to manufacture. In sum, the energy consumed to manufacture the paper consumed by copiers and printers in 1 year likely exceeds the total electrical energy consumed to operate all of the commercial copiers and printers for 1 year (∼16 TWh).

2.5 Building HVAC Energy Consumption

IT equipment located in conditioned areas of commercial buildings acts as an internal heat source, directly influencing heating, ventilation, and air-conditioning (HVAC) energy consumption. During cooling season, the electricity consumed by the equipment is equivalent to electric resistance heating, increasing the air-conditioning loads. The increase in cooling load also augments fan ventilation energy consumption by requiring the HVAC system to introduce additional cooled air into the conditioned space and remove it, and this in turn necessitates more cooling energy to mitigate the additional heat dissipated by the fans in the conditioned space. Conversely, during heating season, IT equipment displaces a portion of the heat that the building's heating system(s) would typically provide, generally 1 Btu of heating load per Btu dissipated by equipment. That is, the heat dissipated by the equipment reduces heating loads, supplanting the existing heating with inefficient resistant heating. Overall, IT equipment effectively lowers the outdoor balance temperature, above which the building requires air conditioning and below which the heating system operates. The precise impacts of the equipment on building loads vary greatly with the local climate as well as with the type of building in question, thereby complicating quantification of its impact. For example, the extremely high density of IT equipment in data centers requires year-round cooling in most cases, whereas IT equipment in office buildings displaces a portion of the heating load. The efficiency of the HVAC system also has a major impact on the energy expended to meet the modified loads. Taking all factors into account, the net national (U.S.) effect is likely to be a moderate increase in commercial building HVAC energy consumption.

HVAC system designers typically take IT equipment into account, modeling it as a heat source of from 0.4 to 1.1 W/ft^2, with an average value of approximately 0.8 W/ft^2. Naturally, some spaces, such as computer rooms, can generate much higher local loads, as do the portions of data centers that are densely packed with servers and computer and telecommunications network equipment (typically 20–60 W/ft^2, with levels projected to rise rapidly in the future).

2.6 Peak Electricity Demand Impact

The power draw of IT equipment during peak periods (typically mid- to late afternoon on hot summer days) makes three distinct contributions to demand. The first two, the direct power draw of the equipment and the power draw of the additional air conditioning required to cool the heat dissipated by the equipment, were described previously. The third, additional power demand resulting from equipment power factors, increases power demand at the power plant and building levels. Nonresidential IT equipment dominates the contributions to peak electricity demand due to operating patterns and the relative magnitude of its energy consumption.

Most nonresidential IT equipment will be in use or ready for use (in active or stand-by mode) during this period. This results in a ratio of peak to (annual) average power draw of approximately 1.2 to 1.5, which is less "peaky" than office buildings as a whole (ratio of ~ 1.8) due to dramatic increases in HVAC power draw during peak demand periods.

Power factors do not directly increase energy consumption per se but do increase power plant output. A study during the late 1990s found that most office equipment has power factors of approximately 0.6, primarily due to low-quality power supplies. In practical terms, this means that the power source must generate 1.67 times more power than is demanded by the equipment, and this increases the effective electricity demand as well as electricity transmission and distribution losses. On the other hand, server computers and network equipment tend to have higher power factors (>0.9). Taking into account all of these factors, on average, IT equipment appears to increase the total peak power demand in any given part of the United States by approximately 3%. The actual peak load impact for a given building depends on the actual density of IT equipment. For example, office buildings will face much larger increases in peak electricity demand from IT equipment than will retail establishments.

3. CONCLUSIONS

IT equipment directly consumed approximately 100 TWh of electricity in the United States in 2000, equal to just under 3% of all electricity produced, with monitors and desktop personal computers ranking as the two largest end uses (~ 40 TWh or 40% of the total). The infrastructure that powers the Internet—server computers, computer and telecommunications network equipment, and UPSs—consumed a total of approximately 30 TWh ($\sim 30\%$).

Indirectly, IT equipment influences national energy consumption in several ways that appear to be on the same order of magnitude as the direct impact, including the manufacture of IT equipment and several aspects of e-commerce: dematerialization, improved production efficiency, reduced inventories, and T&T.

National energy intensity (i.e., energy consumption per dollar of GDP) data reveal that the decrease in energy intensity sharpened during the late 1990s despite the low energy prices of that period. Because this coincided with the rise of the Internet in the United States, it suggests that IT may indeed reduce energy consumption relative to a business-as-usual forecast, slowing the growth of national energy consumption if not the absolute quantity of energy consumed. Closer consideration of e-commerce shows that careful consideration of changes in business paradigms and their interactions with multiple sectors of the economy, including rebound effects, is essential to developing a more accurate estimate of the magnitude and sign of the net energy impact of IT.

3.1 Future IT Energy Consumption

IT direct energy consumption will likely increase in the future due to growing equipment stocks to support the Internet infrastructure (server computers, telecommunications, and computer network equipment) and continued growth in the installed base and use of residential IT equipment. Both trends are driven by greater integration of society with IT. The future technology pathways of IT equipment will also play a major role, including the degree of substitution of portable computing for desktop computing, replacements for CRT monitors (e.g., with LCDs), and changes in PM-enabled rates. The deployment of ubiquitous sensors and controls in conjunction with communications could also increase direct energy consumption such as in home networks.

On a national scale, IT equipment has the potential to realize substantial reductions in future energy consumption relative to business as usual as e-commerce and other IT-related practices that dematerialize the economy and improve the efficiency of energy-consuming systems and equipment become more prevalent. The degree to which rebound effects will erode IT-enabled energy savings remains a critical issue warranting continued study.

SEE ALSO THE FOLLOWING ARTICLES

Electric Power Measurements and Variables • *Industrial Energy Use, Status and Trends* • *Microtechnology, Energy Applications of* • *Rebound Effect of Energy Conservation* • *Technology Innovation and Energy*

Further Reading

Matthews, H. S. (2001). "The Environmental Implications of the Growth of the Information and Communications Technology Sector." Organization for Economic Cooperation and Development, Paris.

Plepys, A. (2002). The grey side of ict. *Environ. Impact Assess. Rev.* **22,** 509–523.

Roth, K. W., Goldstein, F., and Kleinman, J. (2002). "Energy Consumption by Office and Telecommunications Equipment in Commercial Buildings—Vol. 1: Energy Consumption Baseline," final report by Arthur D. Little to the U.S. Department of Energy, Office of Building Technology, State and Community Programs, NTIS number PB2002-107657. www.tiax.biz/aboutus/pdfs/officeequipvol1.pdf.

Information Theory and Energy

SVEN ERIK JØRGENSEN
Danish University of Pharmaceutical Sciences
Copenhagen, Denmark

1. Introduction
2. Information and Entropy
3. Shannon's Information Theory and Kullbach's Measure of Information
4. Exergy and Information
5. Order, Information, Genes, and Proteins
6. Closing Remarks

Glossary

Boltzmann's constant A value (k) relating the average energy of a molecule to its absolute temperature; $k = 1.3803 \times 10^{-23}$ joules/molecules degree Kelvin.

detritus Dead organic matter; has an average free energy content of 18.7 kJ/g.

exergy The amount of work a system can perform when it is brought into thermodynamic equilibrium with its environment; entropy-free energy.

negentropy The difference between the entropy at thermodynamic equilibrium and the entropy of the present state.

The science of information theory has its root in the work of the theoretical physicists Ludwig Boltzmann and J. Willard Gibbs. At the beginning of the 20th century, Boltzmann and Gibbs related probability to entropy and free energy and thereby established the basis for the later development of information theory. The more modern formulation of information theory is attributed to Claude Shannon and Warren Weaver. Their approach was rooted in the problems of coding and decoding information during World War II. Shannon formulated an index of information known as Shannon's Information Index. This is sometimes referred to as "Shannon entropy," which has contributed to much confusion because of the unfortunate mixing of the two concepts, i.e., entropy and information. In the 1980s, biologists Daniel Brooks and E. O. Wiley, in their book *Evolution as Entropy*, added to this confusion. In addition, entropy has been applied by Ilya Prigogine and G. Nicolis to express the amount of information embodied in biological systems; the concept of negative entropy (negentropy) was defined by Erwin Schrödinger in his famous book *What is Life?*, and later also by Karl Popper. Jeffery Wicken should also be mentioned in this context, because he criticized the Brooks–Wiley attempt and introduced a more stringent interpretation of the relationship between entropy and information. Many researchers, including Myron Tribus, Werner Ebeling, Mikhail Volkenstein, and Rainer Feistel, have contributed to the clarification of the relationships between entropy and information, and although classical information theory is primarily discussed herein using Shannon's and Weaver's work and the later refinement by Leon Brillouin, the interpretations of the relationship between information and entropy of the many other contributors have been included.

1. INTRODUCTION

Thermodynamic information has found its widest application in biology due to the extremely high complexity of biological systems. This application of information theory will be presented here through an introduction to the concept of exergy. This thermodynamic concept was introduced in the 1950s, but was not applied in the ecological/biological contexts until the 1970s. These very complex concepts, involving the relationship between order and information as applied to biological systems, are presented here in the context of the genome and the proteom.

2. INFORMATION AND ENTROPY

In statistical mechanics, entropy is related to probability. A system can be characterized by averaging ensembles of microscopic states to yield the macrostate. If W is the number of microstates that will yield one particular macrostate, the probability P that this particular macrostate will occur as opposed to all other possible macrostates is proportional to W. It can further be shown that

$$S = k \ln W, \quad (1)$$

where k is Boltzmann's constant (1.3803×10^{-23} J/molecules degree Kelvin). Entropy is a logarithmic function of W and thus measures the total number of ways that a particular macrostate can be constituted microscopically. The entity S may be called the uncertainty or the required thermodynamic information, meaning the amount of information needed to describe the system, which must not be interpreted as the information actually available. The more microstates there are and the more disordered they are, the more information is required and the more difficult it will be to describe the system.

If an ecosystem is in thermodynamic equilibrium, the entropy, $S_{eq.}$, is higher than it is in an ecosystem in nonequilibrium. The excess entropy may be denoted as the thermodynamic information and is also defined as the negentropy, NE:

$$I = S_{eq} - S = \text{NE}. \quad (2)$$

In other words, a decrease in entropy, S, will imply an increase in information, I, and erosion or loss of information implies an increase in entropy. When $S_{eq} = S$ or $I = 0$, the system cannot be distinguished from its environment and the information about the system is thereby lost. Further, the principle of the second law of thermodynamics corresponds to a progressive decrease of the information content. An isolated system can evolve only by degrading its information.

Schrödinger formulated Boltzmann's Eq. (1) as follows:

$$S = k \ln D, \quad (3)$$

where S is the entropy, k is Boltzmann's constant, and D is the quantitative measure of the atomic disorder. The value D partly covers the heat motion and partly the random mixing of atoms and molecules. Furthermore, $1/D$ may be defined as order, Or. The equation may therefore be reformulated:

$$-S = k \ln(1/D) = k \ln(\text{Or}). \quad (4)$$

In other words, negative entropy is a measure of order.

That entropy is increased by dispersion, which creates more disorder, more possible microstates, and more randomness, can be shown by a simple model consisting of two chambers of equal volume, connected with a valve. One chamber contains 1 mol of a pure ideal gas (i.e., pressure and volume $pv = RT$, the ideal gas constants) and the second chamber is empty. If the valve between the two chambers is opened, and it is assumed that the change in energy $\Delta U = 0$ and time T is constant, an increase in entropy will be observed:

$$\Delta S = \int \partial Q/T = Q/T = W = R \ln V_2/V_1 = R \ln 2, \quad (5)$$

where ΔS is the increase in entropy, V_2 is the volume of gas in the chamber after the valve is opened, and V_1 is the volume of gas before the valve is opened. Thus, paradoxically, the greater the attempt to maintain order, the greater the requirement for energy and the greater the environmental stress (entropy), because all energy transformations from one form to another imply production of waste heat and entropy according to the second law of thermodynamics.

3. SHANNON'S INFORMATION THEORY AND KULLBACH'S MEASURE OF INFORMATION

Entropy of information is frequently applied to social and biological problems. Although entropy of information is analogous to thermodynamic entropy, it is not the same thing, because thermodynamic entropy is derived from physical laws only. Shannon and Weaver introduced a measure of information that is widely used as a diversity index by ecologists. Shannon's index states that

$$H = -K' \sum_{i=1}^{n} p_i \log_2(p_i), \quad (6)$$

where p_i is the probability distribution of a species and K' is a constant ($K' = 1$ when the equation is used to express biodiversity, for instance). Unfortunately, the term "Shannon's entropy" was introduced as an alternative name for Shannon's index. Several authors have attributed this to be due to John von Neumann, who is said to have used the argument that because nobody knows what entropy is, Shannon would have an advantage and would probably win any discussion. The symbol H is thus used here to avoid confusion and to stress that H and S to a certain extent are two different concepts. Both

S and H increase with an increasing number of possible (micro)states, but it is common to equate information with knowledge and entropy with lack of knowledge. The higher the Shannon's index is, the more knowledge is needed to cover, for instance, the entire representation of species in an ecosystem. By increasing diversity, H increases. If the information about this diversity is obtainable, which is assumed (for instance, the number of organisms for each of the species in an ecosystem must be counted to be able to discuss what the biodiversity is), the knowledge of the system has increased correspondingly. In contrast, S considers the uncertainty or the lack of knowledge that increases by increasing diversity (complexity). Entropy implicitly assumes that this knowledge is not available but is necessary to be able to describe the system fully. If the system is at thermodynamic equilibrium, whereby all of the components (molecules) are randomly distributed, the number of microstates is equal to the number of molecules. If, on the other hand, the system has a certain order, it will be possible to group some of the components (molecules) and the knowledge necessary to describe the system is correspondingly less. Notice also that Shannon's index uses $\log_2$ whereas entropy uses the natural logarithmic (ln; the number of microstates) measures, according to Eq. (1), for the complexity of the system.

In 1962, Brillouin formulated a rigid science of information theory, based on probability theory. His definition of information, I, is based on a comparison of the information content in two situations: (1) the *a posteriori* situation in which p_0 is the probability of an event or the presence of a component and the information is $I_0 = 0'$, and (2) the *a priori* situation in which the probability of the same event is p_1 and the information $I_1 > 0$. This can be found from Eq. (7), which is known as Kullbach's measure:

$$I_1 = kp_0 \ln p_0/p_1, \qquad (7)$$

where k is Boltzmann's constant. Equation (7) means that I expresses the amount of information that is gained as a result of the observations. In observing the system used for the computation of the entropy production by opening a valve between two connected chambers, it is expected that the molecules will be equally distributed in the two chambers, i.e., $p_0 = p_1$ is equal to 1/2. If, on the other hand, it is observed that all the molecules are in one chamber, $p_1 = 1$ and $p_0 = 0$. As can be seen, the entropy obtained by application of Kullbach's measure is the same as that obtained by the calculation of entropy,

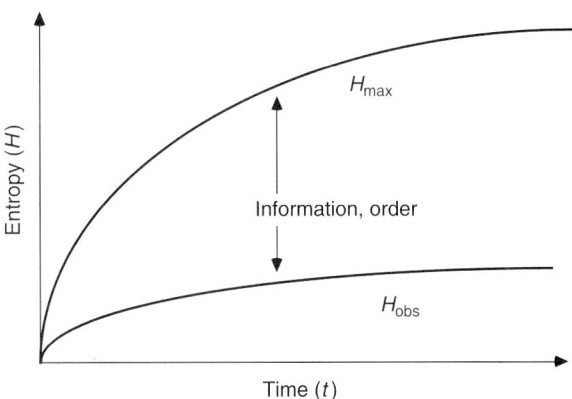

FIGURE 1 $H_{\max}$ and H_{obs} are plotted versus time. The difference between $H_{\max}$ and H_{obs} represents I, which increases with time, t. Data from Brooks and Wiley (1986).

because $R = kA$, where A is Avogadro's number, and there is proportionality to the number of molecules.

Figure 1 shows the relationship, as presented by Brooks and Wiley, in which $H_{\max}$ corresponds to the entropy of the ecosystem if it were at thermodynamic equilibrium, whereas H_{obs} is the actual entropy level of the system. The difference between $H_{\max}$ and H_{obs} covers the information or order. It means that

$$H_{\max} = \log W, \qquad (8)$$

where W is the total number of microstates available to the system. The value H_{obs} is defined according to the Eq. (9):

$$H_{\mathrm{obs}} = -\sum_{i=1}^{n} p_i \ln(p_i). \qquad (9)$$

Brooks and Wiley have interpreted this development of entropy in a variety of ways:

1. H_{obs} is interpreted as complexity: the higher the complexity, the more energy is needed for maintenance and therefore more is wasted as heat. The information in this case becomes the macroscopic information.

2. H_{obs} is translated to realization, whereas $H_{\max}$ becomes the total information capacity. Information may in this case be called constraints. Notice, however, that the strict thermodynamic interpretation of $H_{\max}$ is H at thermodynamic equilibrium, which does not develop (change) for an ecosystem on Earth.

3. H_{obs} represents the observed distribution of genotypes and $H_{\max}$ is any genotype equally likely to be found. The information becomes the organization of organisms over genotypes.

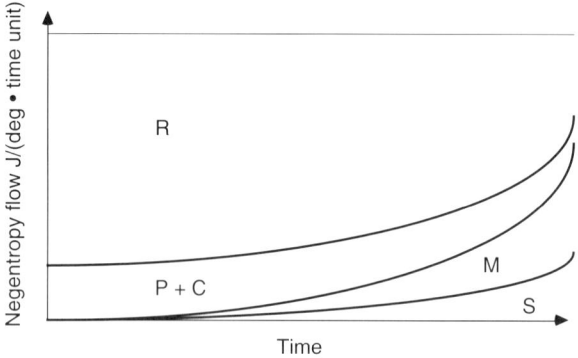

FIGURE 2 Due to biological evolution, there is tentative development in the application of the negentropy flow (solar radiation) on Earth versus time. Four cases are shown: the negentropy is not used, but is reflected (R); or is used by physical (P) and chemical (C) processes on Earth, P + C; or is used for maintenance (M) of the biological structure; or is used to construct biological structure (S).

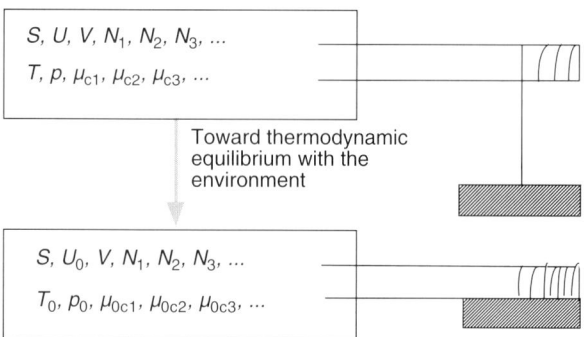

FIGURE 3 The definition of exergy.

Brooks and Wiley's theory seems inconsistent with the general perception of ecological development: order increases and entropy therefore decreases at the cost of entropy production of the environment. The misinterpretation probably lies in the translation of order to information. By increasing order, the amount of information needed decreases. Note that entropy in this context covers the amount of information needed.

Figure 2 attempts to give a different and tentative picture of the development as applied to the negentropy flow (i.e., the solar radiation) on Earth due to evolution. The negentropy flow is considered approximately constant, although the solar radiation has shown some (relatively minor) changes. Four different applications of negentropy are considered: unused negentropy (reflection of the radiation), negentropy used for physical and chemical processes, negentropy used for maintenance of life (respiration), and negentropy used for construction of biological structures. The latter two increase at the expense of reflection. It has been widely discussed whether ecosystems attempt to maximize H_{obs} (i.e., negentropy = information = $H_{max} - H_{obs}$). The difference between biological and thermodynamic entropy is also still the subject of discussion.

4. EXERGY AND INFORMATION

What is exergy? Definitions from a number of sources can be applied. Exergy is defined as the amount of work (entropy-free energy) a system can perform when it is brought into thermodynamic equilibrium with its environment. Figure 3 illustrates the definition. The system considered is characterized by the extensive state variables, S, U, V, N_1, N_2, N_3, ..., N_x, where S is the entropy, U is the energy, V is the volume, and N_1, N_2, N_3, ..., N_x are moles of various chemical compounds, and by the intensive state variables, T, p, μ_{c1}, μ_{c2}, μ_{c3}, ..., μ_{cx}. The system is coupled to a reservoir, a reference state, by a shaft. The system and the reservoir form a closed system. The reservoir (the environment) is characterized by the intensive state variables, T_0, p_0, μ_{0c1}, μ_{0c2}, μ_{0c3}, ..., μ_{0cx}, and because the system is small compared with the reservoir, the intensive state variables of the reservoir will not be changed by interactions between the system and the reservoir. The system develops toward thermodynamic equilibrium with the reservoir and is simultaneously able to release entropy-free energy to the reservoir. During this process, the volume of the system is constant because the entropy-free energy must be transferred through the shaft only. The entropy is also constant because the process is an entropy-free energy transfer from the system to the reservoir, but the intensive state variables of the system become equal to the values for the reservoir. The total transfer of entropy-free energy in this case is the exergy of the system. It is seen from this definition that exergy is dependent on the state of the total system (= system + reservoir) and is not entirely dependent on the state of the system. Exergy is therefore not a state variable. In accordance with the first law of thermodynamics, the increase of energy in the reservoir, ΔU, is

$$\Delta U = U - U_0, \quad (10)$$

where U_0 is the energy content of the system after the transfer of work to the reservoir has taken place.

According to the definition of exergy (Ex),

$$\text{Ex} = \Delta U = U - U_0. \qquad (11)$$

Because

$$U = TS - pV + \sum_c \mu_c N_i \qquad (12)$$

(see any textbook on thermodynamics), and

$$U_0 = T_0 S - p_0 V + \sum_c \mu_{0c} N_i, \qquad (13)$$

we get the following expression for exergy:

$$\text{Ex} = S(T - T_0) - V(p - p_0) + \sum_c (\mu_c - \mu_{0c}) N_i. \qquad (14)$$

Because the reservoir equals the reference state, it is possible, for instance, to select the same system but at thermodynamic equilibrium, i.e., all components are inorganic and at the highest oxidation state, if sufficient oxygen is present (nitrogen as nitrate, sulfur as sulfate, and so on). The reference state will in this case correspond to the ecosystem without life forms and with all chemical energy utilized or as an "inorganic soup." Usually, this implies considering that $T = T_0$ and $p = p_0$, which means that the exergy becomes equal to the difference between the Gibbs free energy of the system and the same system at thermodynamic equilibrium, or the chemical energy content includes the thermodynamic information of the system. Notice that Eq. (14) also emphasizes that exergy is dependent on the state of the environment (the reservoir = the reference state), because the exergy of the system is dependent on the intensive state variables of the reservoir.

Note that exergy is not conserved only if entropy-free energy is transferred, which implies that the transfer is reversible. All processes in reality are, however, irreversible, which means that exergy is lost (and entropy is produced). Loss of exergy and production of entropy are two different descriptions of the same reality, namely, that all processes are irreversible, and unfortunately there is always some loss of energy forms that can do work to energy forms that cannot do work (heat at the temperature of the environment). So, the formulation of the second law of thermodynamic by use of exergy is as follows: All real processes are irreversible, which implies that exergy inevitably is lost. Exergy is not conserved, whereas energy, of course, is conserved by all processes, according to the first law of thermodynamics. It is therefore wrong to discuss, as already mentioned briefly, the energy efficiency of an energy transfer, because it will always be 100%, whereas the exergy efficiency is of interest, because it will express the ratio of useful energy to total energy, which always is less than 100% for real processes, according to the second law of thermodynamics.

It is therefore of interest to set up for all environmental systems, in addition to an energy balance, also an exergy balance. The concern is loss of exergy, because it means that "first-class energy," which can do work, is lost as "second-class energy" (heat at the temperature of the environment), which cannot do work. So, the particular properties of heat and the fact that temperature is a measure of the movement of molecules create limitations in the possibilities to utilize all forms of energy to do work. Due to these limitations, a distinction must be made between exergy, which can do work, and anergy, which cannot do work, and thus all real processes imply inevitably a loss of exergy as anergy.

Exergy seems more useful than entropy to describe the irreversibility of real processes, because it has the same units as energy and is a form of energy, whereas entropy by definition is more difficult to relate to concepts associated with typical conceptions of reality. In addition, entropy is not clearly defined for "systems far from thermodynamic equilibrium," particularly for living systems. Moreover, it should be mentioned that the self-organizing abilities of systems are strongly dependent on temperature. Exergy takes temperature into consideration, but entropy does not. By definition, exergy at 0 K is 0 and is at a minimum. Negative entropy, as previously discussed, does not express the ability of a system to do work. Exergy therefore becomes, in contrast to entropy, a good measure of "creativity," which increases proportional to temperature. Furthermore, exergy facilitates the differentiation between low-entropy energy and high-entropy energy, because exergy is entropy-free energy. Finally, note that information contains exergy. Boltzmann showed that the free energy of the information at hand (in contrast to the information needed to describe the system) is $kT \ln I$, where I is the information known about the state of the system [for instance, that the configuration is 1 out of W possible (i.e., that $W = I$) and k is Boltzmann's constant ($= 1.3803 \times 10^{-23}$ J/molecules K)]. This implies that one bit of information has the exergy equal to $kT \ln 2$. Transference of information from one system to another is often almost an entropy-free energy transfer. If the two systems have different temperatures, the entropy lost by one system is not equal to the entropy gained by the other system, whereas the exergy lost by the first system is equal to the exergy transferred and is equal to the exergy gained by the other system, provided that the transference is not accompanied by any loss of exergy. In this case, it is obviously more convenient to apply exergy than entropy.

The exergy of a system measures contrast—it is the difference in free energy if there is no difference in pressure and temperature, as may be assumed for an ecosystem or an environmental system and its surroundings against the surrounding environment. If a system is in equilibrium with the surrounding environment, the exergy is of course zero. Because the only way to move systems away from equilibrium is to perform work on them, and because the available work in a system is a measure of ability, a distinction must be made between the system and its environment, or thermodynamic equilibrium alias (for instance, an inorganic soup). Therefore, it is reasonable to use the available work, i.e., the exergy, as a measure of the distance from thermodynamic equilibrium.

Because it is known that ecosystems, due to the through-flow of energy, have the tendency to move away from thermodynamic equilibrium, losing entropy or gaining exergy and information, the following proposal for the relevance for systems can be put forward: systems, receiving a through-flow of energy, attempt to move toward a higher level of exergy. This proposal for exergy development in ecosystems makes it pertinent to assess the exergy of ecosystems. It is not possible to measure exergy directly, but it is possible to compute it using Eq. (14). Assuming a reference environment that represents the system (ecosystem) at thermodynamic equilibrium, which means that all the components are inorganic, are at the highest possible oxidation states (if sufficient oxygen is present and as much free energy as possible is utilized to do work), and are homogeneously distributed in the system (no gradients), the situation illustrated in Fig. 4 is valid. The chemical energy embodied in the organic components and in the biological structures contributes by far the most to the exergy content of the system, thus there seems to be no reason to assume a (minor) temperature and pressure difference between the system and the reference environment. Under these circumstances, it is possible to calculate the exergy content of the system as coming entirely from the chemical energy: [$T = T_0$ and $p = p_0$ in Eq. (14)]:

$$\text{Ex} = \sum_c (\mu_c - \mu_{0c}) N_i. \quad (15)$$

Equation (15) represents the nonflowing chemical exergy, which is determined by the difference in

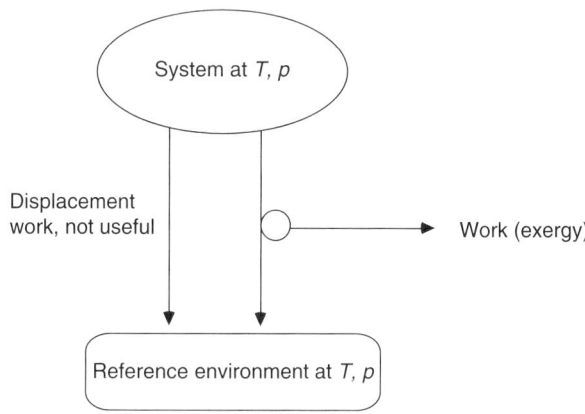

FIGURE 4 The exergy content of the system is calculated for the system relative to a reference environment of the same system at the same temperature (T) and pressure (p), but as an inorganic soup with no life, biological structure, information, or organic molecules.

chemical potential ($\mu_c - \mu_{0c}$) between the ecosystem and the same system at thermodynamic equilibrium. This difference is determined by the concentrations of the components in the system and in the reference state (thermodynamic equilibrium), as is the case for all chemical processes. The concentrations in the ecosystem can be measured, but the concentrations in the reference state (thermodynamic equilibrium) can be based on the usual chemical equilibrium constants. Given the process

Component A ⟷ inorganic decomposition products, (16)

there is a chemical equilibrium constant, K:

$$K = \frac{[\text{inorganic decomposition products}]}{[\text{component A}]}. \quad (17)$$

The concentration of component A at thermodynamic equilibrium is difficult to find, but can be based on the composition of A; the concentration of component A is found at thermodynamic equilibrium from the probability of forming A from the inorganic components. By these calculations, the exergy of the system, compared with the same system at the same temperature and pressure, but in the form of an inorganic soup without any life, biological structure, information, or organic molecules, is found. Because ($\mu_c - \mu_{0c}$) can be found from the definition of the chemical potential, replacing activities by concentrations, the following expression for the exergy is obtained:

$$\text{Ex} = RT \sum_{i=0}^{i=n} C_i \ln C_i/C_{i,0}, \quad (18)$$

where R is the gas constant, T is the temperature of the environment (and the system; see Fig. 4), and C_i is the concentration of the ith component expressed in a suitable unit (e.g., for phytoplankton in a lake, C_i could be expressed as milligrams/liter or as milligrams/liter of a local nutrient). Also, $C_{i,0}$ is the concentration of the ith component at thermodynamic equilibrium and n is the number of components. Of course, $C_{i,0}$ represents a very small concentration (except for $i = 0$, which is considered to cover the inorganic compounds), corresponding to a very low probability of forming complex organic compounds spontaneously in an inorganic soup at thermodynamic equilibrium. For the various organisms, $C_{i,0}$ is even lower, because the probability of forming the organisms is very low, due to their embodied information, which implies that the genetic code should be correct. By using this particular exergy, based on the same system at thermodynamic equilibrium as a reference, the exergy becomes dependent only on the chemical potential of the numerous biochemical components that are characteristic of life. This is consistent with Boltzmann's statement that life is struggle for free energy.

The total exergy of an ecosystem cannot be calculated exactly, because it is not possible to measure the concentrations of all components or to determine all possible contributions to exergy in an ecosystem. If the exergy of a fox is calculated, for instance, the calculations will reveal only the contributions coming from the biomass and from the information embodied in the genes of the fox, but not the contributions from blood pressure, sex hormones, and so on. These properties are at least partially covered by the genes, but is that the entire story? The contributions from the dominant components can be calculated (for instance, by the use of a model or measurements), which covers the most essential components for a focal problem. The difference in exergy by comparison of two different structures (species composition) is decisive here. Moreover, exergy computations always yield only relative values, because the exergy is calculated relative to the reference system.

It is possible to distinguish between the exergy of information and of biomass. The value p_i, defined as c_i/A, where

$$A = \sum_{i=1}^{n} c_i \quad (19)$$

is the total amount of matter in the system, is introduced as new variable

$$\text{Ex} = ART \sum_{i=1}^{n} p_i \ln p_i/p_{i0} + A \ln = A/A_0. \quad (20)$$

As $A \approx A_0$, exergy becomes a product of the total biomass A (multiplied by RT) and the Kullback measure:

$$K = \sum_{i=1}^{n} p_i \ln(p_i/p_{i0}), \quad (21)$$

where p_i and p_{i0} are probability distributions, *a posteriori* and *a priori* to an observation of the molecular detail of the system. Specific exergy is exergy relative to the biomass and for the ith component: Sp. $\text{Ex}_i = \text{Ex}_i/c_i$. This implies that the total specific exergy per unit of area or per unit of volume of the ecosystem is equal to RTK.

It is interesting in this context to draw a parallel with regard to development of entropy for the entire universe. Classical thermodynamic interpretations of the second law of thermodynamics predict that the universe will tend to "heat death," whereby the entire universe has the same temperature, no changes ever take place, and a final overall thermodynamic equilibrium results. This prediction is based on the steady increase of entropy according to the second law of thermodynamics: thermodynamic equilibrium is the attractor. It can, however, be shown that, due to expansion of the universe, the system is moving away from thermodynamic equilibrium at a high rate of speed. Due to incoming solar radiation energy, an ecosystem is able to move away from thermodynamic equilibrium (i.e., the system evolves, obtains more information, and undergoes organization). The ecosystem must produce entropy for maintenance, but the low-entropy energy flowing through the system may be able to more than compensate for this production of disorder, resulting in increased ecosystem order or information.

5. ORDER, INFORMATION, GENES, AND PROTEINS

Information theory has found widespread use in the analysis of biological macromolecules. Schrödinger, in his order principle, proposed that information about the organization of biological systems was laid down in the system. This was confirmed when Watson and Crick elucidated the structure of DNA, and a platform was provided that made it possible to apply information theory more comprehensively to biological systems.

Again, the expression for exergy per unit of volume is

$$\text{Ex} = RT \sum_{i=0}^{i=n} C_i \ln C_i/C_{i,0}, \quad (22)$$

where R is the gas constant, T is the temperature of the environment, and C_i is the concentration of the ith component expressed in a suitable unit (e.g., for phytoplankton in a lake, C_i could be expressed as milligrams/liter or as milligrams/liter of a local nutrient. The concentration of the ith component at thermodynamic equilibrium is $C_{i,0}$ and n is the number of components. This is, of course, a very small concentration, except for $i=0$, which is considered to cover the inorganic compounds; however, $C_{i,0}$ is not zero, but corresponds to a very low probability of forming complex organic compounds spontaneously in an inorganic soup at thermodynamic equilibrium.

It has been suggested that the exergy of structurally complicated material can be estimated on the basis of the elementary composition of the material. This, however, has the disadvantage that higher organisms and microorganisms with the same elementary composition will have the same exergy, which is in complete disagreement with the lower probability of forming more complex organisms, i.e., the lower concentration of $C_{i,0}$ in Eq. (22). The problem related to the assessment of $C_{i,0}$ has been discussed, and a possible solution was proposed by Jørgensen. For dead organic matter (detritus), which is given the index 1, the solution can be found from classical thermodynamics. Detritus has a free energy (chemical energy) of about 18.7 kJ/g. For the biological components, 2, 3, 4, ..., N, the probability, p_{i0}, consists at least of the probability of producing the organic matter (detritus), i.e., $p_{1,0}$, and the probability, $p_{i,a}$, of finding the correct composition of the enzymes (number and sequence, a, of amino acids) that determine the biochemical processes in the organisms. In living organisms, 20 different amino acids comprise the enzymes, and each gene determines, on average, the sequence of about 700 amino acids. For *Homo sapiens*, it has been found that as many as 38,000 amino acids are determined by one gene. The value for $p_{i,a}$ can be found from the number of permutations that represent the evolutionary selection of a particular organism based on its characteristic amino acid sequence. This means that

$$p_{i,a} = a^{-Ng_i}, \quad (23)$$

where a is the number of possible amino acids ($=20$), N is the number of amino acids determined by one gene ($=700$), and g_i is the number of non-nonsense genes. The following equations are available to calculate p_i:

$$p_{i,0} = p_{1,0}\, p_{i,a} = p_{1,0}\, a^{-Ng} \approx p_{1,0} \cdot 20^{-700g} \quad (24)$$

and the exergy contribution of the ith component (per unit of volume) can be found by combining these equations:

$$\begin{aligned}
\text{Ex} &= RTc_i \ln c_i/(p_{1,0} a^{-Ng_{c_{0,0}}}) \\
&= (\mu_1 - \mu_{1,0})c_i - c_i \ln p_{i,a} \\
&= (\mu_1 - \mu_{1,0})c_i - c_i \ln(a^{-Ng_i}) \\
&= 18.7 c_i + 700(\ln 20) c_i g_i.
\end{aligned} \quad (25)$$

The total exergy can be found by summing up the contributions originated from all components. The contribution by inorganic matter can be neglected because the contributions by detritus and (to a greater extent) by biological materials are much higher due to an extremely low concentration of these components in the reference system. The contribution by detritus (dead organic matter) is 18.7 kJ/g times the concentration (in grams/unit of volume), as previously indicated, and the exergy of living organisms consists of $\text{Ex}_{1,\text{chem}} = 18.7$ kJ/g times the concentration c_i (grams/unit of volume) and

$$\begin{aligned}
\text{Ex}_{i,\text{bio}} &= RT(700 \ln 20) c_i g_i \\
&= RT(2100) g_i c_i,
\end{aligned} \quad (26)$$

where $R = 8.34$ J/mol; if an average molecular weight of 10^5 is assumed for the enzymes (proteins), the following equation for $\text{Ex}_{i,\text{bio}}$ at 300 K is obtained: $\text{Ex}_{i,\text{bio}} = 0.0529 g_i c_i$, where the concentration now is expressed in grams/unit of volume and the exergy is in kilojoules/unit of volume. The exergy (per unit of volume) for the entire system, Ex_{total}, can be found:

$$\text{Ex}_{\text{total}} = 18.7 \sum_{i=1}^{N} c_i + 0.0529 \sum_{i=1}^{N} c_i g_i, \quad (27)$$

where g for detritus ($i=1$) of course is 0. Table I illustrates how each contribution depends on selected biological systems as well as on the detritus. A

TABLE I
Energy of Living Organisms

Organism	g_i	$\beta = (\text{Ex}_{i,\text{chem}} + \text{Ex}_{i,\text{bio}})/\text{Ex}_{i,\text{chem}}$	Exergy (kJ/g)[a]
Detritus	0	1	18.7
Minimal cell	470	2.3	43.8
Bacteria	600	2.7	50.5
Algae	850	3.4	64.2
Yeast	2000	5.8	108.5
Fungi	3000	9.5	178
Sponges	9000	26.7	499
Mold	9500	28.0	524
Plants, trees	10,000–30,000	29.6–86.8	554–1623
Worms	10,500	30.0	561
Insects	10,000–15,000	29.6–43.9	554–821
Jellyfish	10,000	29.6	554
Zooplankton	10,000–15,000	29.6–43.9	554–821
Fish	100,000–120,000	287–344	5367–6433
Birds	120,000	344	6433
Amphibians	120,000	344	6433
Reptiles	130,000	370	6919
Mammals	140,000	402	7517
Humans	250,000[b]	716	13389

[a] Based on the energy contained in detritus. On average, the exergy of 1 g of detritus is 18.7 kJ.
[b] See text for discussion.

weighting factor ß is introduced to be able to cover the exergy for various organisms in the unit detritus equivalent or chemical exergy equivalent:

$$\text{Ex}_{\text{total}} = \sum_{i=1}^{N} \beta_i c_i \quad \text{(as detritus equivalent)}. \quad (28)$$

Through the use of Eq. (28), the calculation of exergy accounts for the chemical energy in organic matter as well as for the (minimum) information embodied in living organisms. The latter contribution is measured as the extremely small probability of forming living components (for instance, algae, zooplankton, fish, mammals, and so on) spontaneously from inorganic matter. Weighting factors, defined as the exergy content relative to detritus (see Table I), may be considered quality factors that reflect how developed the various groups are and to what extent they contribute to the exergy due to their content of information, which is reflected in the computation. This is, according to Boltzmann, who gave the following relationship for the work, W, embodied in the thermodynamic information:

$$W = RT \ln N, \quad (29)$$

where N is the number of possible states, among which the information has been selected. For species, N is the inverse of the probability of obtaining the valid amino acid sequence spontaneously. The Kullback measure of information covers the gain in information, when the distribution is changed from p_{i0} to p_i. Note that K is a specific measure (per unit of matter). When multiplied by the total concentration, K yields the exergy [see Eqs (20) and (21)].

Exergy calculated by use of the equations given herein has some clear shortcomings:

1. Although minor, some approximations have been made.
2. The non-nonsense genes for all organisms are not known.
3. The exergy embodied in the proteins (enzymes) is calculated only in principle, and there are other components of importance for the life processes. These components contribute less to the exergy compared to the enzymes, and the information embodied in the enzymes controls the formation of these other components (for instance, hormones), but it cannot be excluded that these components contribute to the total exergy of the system.
4. The exergy of the ecological network is not included. If the exergy of models is calculated, the network will always be relatively simple and the contribution coming from the information content of the network is negligible.
5. A simplification (for instance, models or diagrams) of the ecosystem is always used. This implies that only the exergy contributions of the components included in the simplified image of the ecosystem are calculated. The real ecosystem will inevitably contain more components, which are not included in the calculations.

It is therefore proposed to consider the exergy found by these calculations as a relative minimum exergy index to indicate that there are other contributions to the total exergy of an ecosystem, although they may be of minor importance. In most cases, however, a relative index is sufficient to understand the reactions of ecosystems, because the absolute exergy content is irrelevant for the reactions. It is, in most cases, the change in exergy that is of importance to understand the ecological reactions.

The weighting factors presented in Table I have been applied successfully in several ecological models. The relatively good results obtained, despite the uncertainty of their assessment, seem to be explicable only by the robustness of the application of the factors in modeling and other quantifications. The differences between the factors (for microorganism, vertebrates, and invertebrates) are so obvious that it does not matter if the uncertainty of the factors is very high—the results are not influenced. On the other hand, it would be an important progress to get better weighting factors from a theoretical point of view, because it would enable modeling the competition between species that are closely related.

There is no doubt that the correct estimation of ß values should be based on the number of proteins that control the processes in the cells of various organisms. Information about all human genomes is available, and it has been found that the number of non-nonsense genes is not 250,000, as indicated in Table I, but rather is 40,000. On the other hand, it is also clear that the number of amino acids controlled by one gene is more than 700 for *Homo sapiens*. for some genes, it may be as high as 38,000. The weighting factor in Table I may therefore be approximately correct. The importance of proteins in biology and physiology is reflected in the intensive and enormous amount of analytical work that has been invested for many decades to find the composition of the human proteins. Current terminology for genetically determined proteins, "proteomes," emphasizes that these biological components are of particular importance. The great interest in proteoms is due to their roles in control of life processes. Many of these proteins may be the medicines of the future. The use of enzymes in industrial production is in its infancy, and there is enormous potential for their application in many more industrial processes.

Although the key to finding better β values is the proteom, knowledge about the number of proteoms in various organisms is very limited; more is known about the number of non-nonsense genes. It may be possible, however, to compare knowledge about non-nonsense genes, the overall DNA content, and the limited knowledge about the number of proteoms and evolutionary data, and see some pattern emerge. This could be used to give better (although still very approximate, at this stage) β values. For *Homo sapiens*, it is presumed that 200,000 different proteins are produced by the cells and that they contain about 15,000 amino acids on average. This would yield a biological exergy of $RT(\ln 20) \times 3 \times 10^9$ c_i; with a molecular weight of 100,000, the β value would be 12,275, or considerably higher than the value in Table I. The other values in the table may be similarly different. Current knowledge about the genome and the proteom is unfortunately very limited.

A live frog of 20 g will have an exergy content of $20 \times 6433 \, kJ \approx 0.1296 \, GJ$, whereas a dead frog will have an exergy content of only 374 kJ, although both frogs have the same chemical composition (at least within a few seconds after death). The difference is rooted in the information, or rather the difference in the useful information. The dead frog has the information a few seconds after its death (the amino acid composition has not yet been decomposed), but the difference between a live frog and a dead frog is the ability to utilize the enormous information stored in the genes and the proteom of the frog. The amount of information stored in a frog is surprisingly high. The number of amino acids required to exist in the right sequence is 84,000,000, which equates, for the correct selection of 1 of 20 possible amino acids, to $84,000,000^{20}$ possible microstates. This amount information is able to ensure reproduction and is transferred from generation to generation, which ensures that evolution can continue because what is already a favorable combination of properties is conserved through the genes. Because of the very high number of amino acids (84,000,000), it is not surprising that there will always be minor differences in the amino acid sequences from frog to frog. These may be the result of mutations or of minor mistakes in the copying process. Such variation is important because it provides possibilities to "test" which amino acid sequence gives the best result with respect to survival and growth. The best result, representing the most favorable combination of properties, will offer the highest probability of survival and will support the highest reproductive rate, and the corresponding genes will therefore prevail. Survival and growth mean more exergy and a bigger distance to thermodynamic equilibrium. Exergy can therefore be used as a thermodynamic function to quantify Darwin's theory. It is interesting in this context that exergy also represents the amount of energy needed to tear down a system. This means that the more exergy the system possesses, the more difficult it becomes to kill the system and the higher is the probability of survival.

6. CLOSING REMARKS

Information carries energy and also "negative" entropy in the sense that information about organization reduces randomness. Total randomness corresponds to maximum entropy, i.e., no information. All information contains energy, even first-class energy, which can do work, i.e., exergy.

A complete description of a dynamic system requires that all exchanges of mass (all elements and chemical compounds), energy, and information are known. Mass, energy, and information are the basic building blocks of all systems. Although information carries energy (exergy), it is necessary to specify all the information transfer processes, because not all energy is associated with information. For instance, heat energy has no or very little information content. Mass and energy are also convertible according to Einstein's famous equation: energy $= mc^2$. For complete information, it is necessary to know the transfer processes of both mass and energy. In this context, it is notable that information has a relatively low cost of energy [see; for instance; Eq. (29)]. Mass on Earth is limited, and the amount of energy is also limited to the diurnal inflow of solar radiation. Anthropogenic use of previously stored solar radiation in the form of fossil fuels is also limited due to environmental impacts. It is therefore not surprising that humans recognize the need to continue development by focusing on generation of information and a wider and better use of information. Success of the "information age" is based on the reality that information has (almost) no limitation, except, of course, for its low energy cost. The amount of information on Earth cannot, of course, increase by more than the inflow of solar radiation, unless, over a limited time, previously stored energy (e.g., fossil fuel) is used. The solar energy received on $1\,m^2$ of Earth is about 2 GJ/year. In accordance with Eq. (29), this would correspond to information on the order of about 10^w, where $w \approx 10^{27}$, provided, of course, that 100% of the energy could be transformed into information. So, there is a long way to go before solar radiation is exhausted by conversion to information. The exergy content of an average human being (75 kg) would, according to Eq. (28) and Table I, be around 750 GJ, or the solar radiation on $375\,m^2$/year, but it has taken 4 billion years to reach the concentration of information found in a human being. These figures are maybe the best way to get an idea about the relationships between information and energy.

SEE ALSO THE FOLLOWING ARTICLES

Complex Systems and Energy • Entropy and the Economic Process • Exergoeconomics • Exergy Analysis of Energy Systems • Exergy: Reference States and Balance Conditions • Thermodynamics, Laws of • Value Theory and Energy

Further Reading

Boltzmann, L. (1905). "The Second Law of Thermodynamics. Populare Schriften, Essay No. 3" (address to the Imperial Academy of Science in 1886). [Reprinted in English in "Theoretical Physics and Philosophical Problems, Selected Writings of L. Boltzmann." D. Reidel, Dordrecht.]
Brillouin, L. (1949). Thermodynamics and cybernetics. *Am. Sci.* **37**, 554–568.
Brooks, D. R., and Wiley, E. O. (1986). "Evolution as Entropy." University of Chicago Press, Chicago.
Jørgensen, S. E. (2002). "Integration of Ecosystem Theories: A Pattern." 3rd Ed. Kluwer Academic Publ., Dordrecht, Boston, and London.
Jørgensen, S. E., and Meyer, H. F. (1977). Ecological buffer capacity. *Ecol. Modelling* **3**, 39–61.
Jørgensen, S. E., and Mejer, H. F. (1979). A holistic approach to ecological modelling. *Ecol. Modelling* **7**, 169–189.
Jørgensen, S. E., Patten, B. C., and Straskraba, M. (2000). Ecosystem emerging IV: Growth. *Ecol. Modelling* **126**, 249–284.
Jørgensen, S. E., and Bendoricchio, G. (2001). "Fundamentals of Ecological Modelling." 3rd Ed. Elsevier, Oxford.
Jørgensen, S. E., and Marques, J. C. (2001). Thermodynamics and ecosystem theory, case studies from hydrobiology. *Hydrobiologia* **445**, 1–10.
Kay, J., and Schneider, E. D. (1992). Thermodynamics and measures of ecological integrity. *In* "Ecological Indicators." pp. 159–182. Elsevier, Amsterdam.
Prigogine, I. (1980). "From Being to Becoming: Time and Complexity in the Physical Sciences." Freeman, San Francisco.
Shannon, C., and Weaver, W. (1949). "The Mathematical Theory of Communication." University of Illinois Press, Chicago.
Ulanowicz, R. E. (1986). "Growth and Development. Ecosystem Phenomenology." Springer-Verlag, Berlin and New York.
Ulanowicz, R. E. (1997). "Ecology, the Ascendent Perspective." Columbia University Press, New York.
Wicken, J. S. (1987). "Evolution, Thermodynamics and Information. Extending the Darwinian Program." Oxford University Press, Oxford.

Innovation and Energy Prices

DAVID POPP
Syracuse University
Syracuse, New York, United States

1. The Concept of Technological Change
2. Macroeconomic Trends
3. Invention and Innovation
4. Diffusion

Glossary

diffusion The final step of technological change; the process during which new innovations are gradually adopted by new users.
factor substitution A shift in the proportion of inputs (or factors) of production used that occurs in reaction to a change in the relative prices of these inputs.
induced innovation The theory that the direction and magnitude of innovative activity is shaped by market forces such as prices.
innovation The second step of technological change, where a new concept is developed into a commercially viable product.
invention The first step of technological change, marking the initial creation of a new product.
research and development (R&D) Spending by individual actors on the creation and development of new inventions and innovations.
technological change The process by which new technologies are developed and brought into use; includes three steps: invention, innovation, and diffusion.
total factor productivity (TFP) Output per unit of total input, where increases in the level of output produced for a given level of inputs suggest that the productivity of these inputs has increased.

Technological change has potentially broad impact in the field of energy. Improvements in energy efficiency enable production to proceed with less energy consumption per unit over time. The potential development of alternative energy sources, such as fuel cells and solar energy, offers prospects for environmental improvements and helps to address concerns over exhaustibility of current energy resources. As such, it is important to understand the process by which such technologies evolve. This article focuses on the means by which incentives, through either prices or policy, influence technological change in the energy sector.

1. THE CONCEPT OF TECHNOLOGICAL CHANGE

It is useful to understand how economists decompose the reaction of individual actors to a price change. Consider, for example, the reaction of a firm to higher energy prices. Economists use a production function to represent the relationship between inputs and outputs. Typical notation is

$$Y = f(L, K, E, M, t),$$

where Y is output, L is labor, K is capital, E is energy, M is materials, and t is technology. A production function can be represented by an isoquant that illustrates all possible combinations of labor and energy that can produce a given level of output. Figure 1 provides an example of a production process using two inputs: labor and energy. Note that along any one isoquant, technology is held constant; that is, each isoquant represents production possibilities for a given technology. The straight line P_0P_0 in the figure is an isocost line. The isocost line represents all combinations of labor and energy that cost the same. Costs increase for isocost lines moving away from the origin. The firm chooses how to produce a given level of output by finding where these two lines are tangent. At that point, the costs of production are lowest. Graphically, this is the lowest possible isocost line that touches the isoquant.

When energy prices rise, we expect the firm to use fewer energy inputs. In Fig. 2, an increase in energy prices makes the isocost line steeper, as shown by the line segment $P'_0P'_0$. Intuitively, for the same level of expenditure, the firm now purchases fewer units of

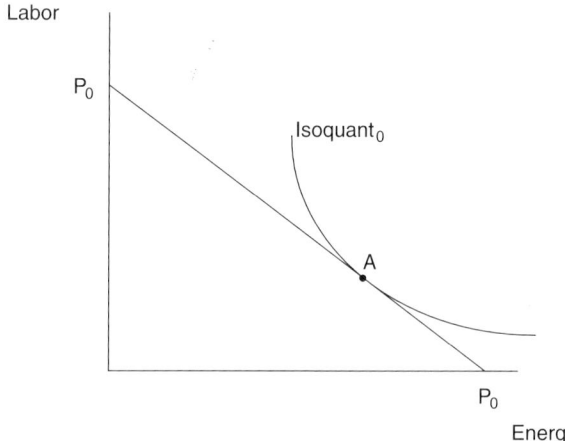

FIGURE 1 Diagram illustrating the use of an isoquant and isocost line to find the cost-minimizing choice of inputs. At time t, the production function is represented by isoquant$_0$. Along the isoquant, total output is constant. Initial prices are given by the isocost line P_0P_0. (Note: In this diagram, P does not represent actual prices but is simply a label for the relative price lines.) Initial production is at point A, where the two lines are tangent. This is the lowest possible isocost line that touches the isoquant. The interpretation is that this is the cheapest way possible to produce the level of output represented by the isoquant.

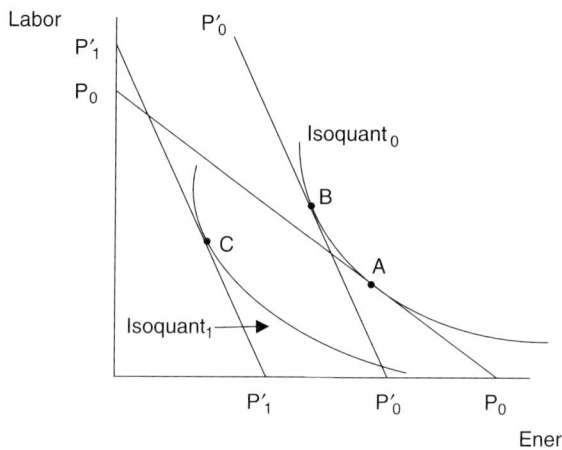

FIGURE 2 Diagram illustrating the distinction between factor substitution and technological change. At time t, the production function is represented by isoquant$_0$. Initial prices are given by the isocost line P_0P_0. Initial production is at point A. As energy prices rise, the isocost curve becomes steeper, as shown by the line $P'_0 P_0$. As a result of the price increase, the choice of inputs shifts to point B; that is, more labor and less energy are used. Because technology is constant along the isoquant, the shift from A to B represents factor substitution. During the next period, technological advances shift the isoquant to isoquant$_1$. If relative prices remain the same, the cost-minimizing combination of labor and energy is now at point C. The move from B to C is an example of technological change.

energy. As a result of the price increase, the cost-minimizing combination of labor and energy now involves more labor and less energy. An example of this would be a firm making less use of energy-intensive machinery and replacing the machines' capabilities with more manual labor. Such a change in reaction to a price change, illustrated by the shift from point A to point B on Fig. 2, is known as factor substitution. Note that the same production technology is being used but that different amounts of inputs are used in the process.

Of course, factor substitution is not the only reaction that a firm may have to higher energy prices. Over time, we would expect new machines that use energy more efficiently to be produced. New technology is represented by a new production function:

$$Y = f(L, K, E, M, t').$$

Because technology is constant along an isoquant, technological change results in an inward shift of the isoquant; the same level of output can now be produced using fewer inputs. In Fig. 2, technology shifts the isoquant from isoquant$_0$ to isoquant$_1$. Assuming that the ratio of prices for labor and energy remains the same, the choice of inputs moves from point B to point C. The movement from point B to point C on the figure is an example of technological change.

This process of technological change proceeds in stages. Writing in 1942, Schumpeter referred to the process as "creative destruction." First, an idea must be born. This stage is known as invention. New ideas are then developed into commercially viable products. This stage is referred to as innovation. Often, these two stages of technological change are studied together under the rubric of research and development (R&D). Finally, to have an effect on the economy, individuals must choose to make use of the new innovation. This adoption process is known as diffusion. At each stage, incentives, in the form of either prices or regulations, will affect the development and adoption of new technologies. This article begins with an overview of results pertaining to innovation and diffusion. A discussion of diffusion follows.

2. MACROECONOMIC TRENDS

To provide an introduction to the links between energy prices and innovation, it is helpful to look at trends in macroeconomic data. Figure 3 presents data on innovation, energy efficiency, and energy prices in the United States from 1972 to 1999.

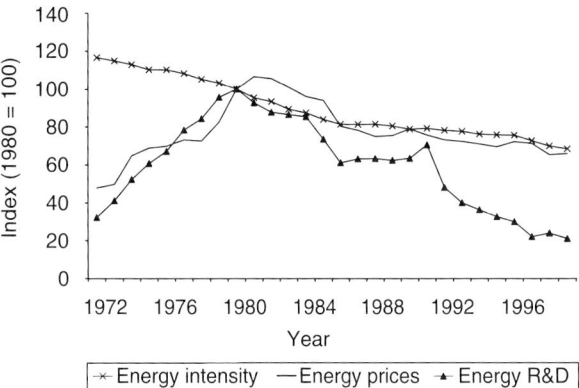

FIGURE 3 Diagram illustrating trends in energy prices, energy R&D, and energy intensity. All values are indexed so that $1980 = 100$. As one would expect, there is a positive relationship between energy R&D and energy prices but an inverse relationship between both prices and R&D and energy intensity. Note that energy R&D begins to fall earlier than do energy prices, providing potential evidence of diminishing returns to energy R&D. Also note that energy intensity continues to fall even as energy prices and energy R&D both decline, suggesting continued diffusion of new energy technologies.

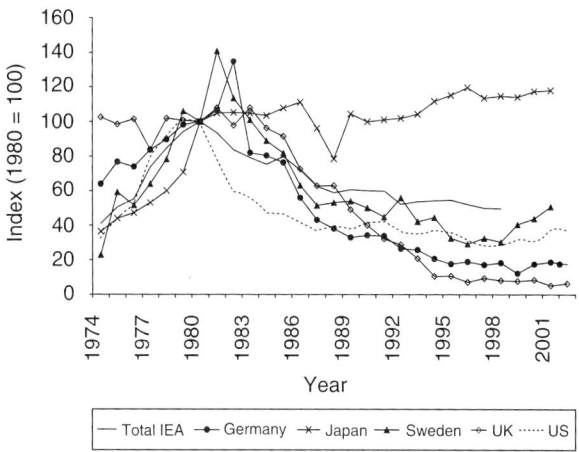

FIGURE 4 Diagram illustrating international trends in energy R&D spending. Values are indexed so that $1980 = 100$. To create the index, all R&D figures have been converted to 2002 U.S. dollars. Total IEA spending includes totals for 23 IEA member countries (data for Luxembourg were unavailable). (Source: author's own calculations, based on data from the IEA Energy Technology R&D Statistics Service.)

Efficiency is represented by energy intensity—that is, the ratio of total energy consumption to real gross domestic product (GDP). The energy price index combines prices on all energy sources, weighted by consumption. Finally, energy R&D data represent all energy R&D performed by industry, whether financed by industry or by government. In each case, the data are indexed so that 1978 values equal 100.

The data provide a few interesting trends. First, note the strong relationship between energy prices and energy R&D. Such trends are found not just in the United States but also in other industrialized countries. Figure 4 shows the same energy R&D data for selected International Energy Agency (IEA) member countries as well as for a total of 23 member countries. Only Japan has maintained high energy R&D levels since energy prices have fallen.

Second, looking at the relationship between energy intensity and energy prices in Fig. 3, note that energy intensity fell as energy prices were rising. Price-induced substitution away from energy certainly played a role in the decline of energy intensity, especially immediately after the first energy crisis. However, simple substitution does not tell the whole story. Even when energy prices returned to pre-crisis levels, energy intensity remained low. If the technologies being used had not changed during the 1970s, there would be no reason for firms to continue using fewer energy inputs as energy prices returned to low levels. Note also that energy intensity did not level off until several years after energy prices and energy R&D had fallen. This suggests that technology played a role and that the diffusion of these new energy-efficient technologies took time.

Empirical analysis of these trends supports the interpretation that technological change played a role in improving energy intensity. A 2001 study by Popp, looking at energy consumption and innovation in 13 energy-intensive industries, found that approximately two-thirds of the decrease in energy consumption in response to higher energy prices comes from factor substitution, with the remaining one-third being due to the adoption of new technologies resulting from innovation induced by the price shock. At the macro level, studies of the effect of technological change on energy consumption typically regress energy consumption on the prices of energy and other inputs and include a time trend to capture the effects of technological progress. Early studies of industrial energy consumption that focused on data from before the first energy crisis of 1973 found that technological change was energy using; that is, energy use per unit output increased over time. Of course, using data from this period would ignore energy-saving innovations developed after the energy crises of the 1970s. Indeed, more recent work that included post-1973 data has found technological progress since the energy crisis to be energy saving.

3. INVENTION AND INNOVATION

3.1 Incentives and Innovation

To better understand the links between energy prices and innovation, it is important to look beyond the macroeconomic trends to understand how individual firms and consumers react to changes in energy prices. This section begins by looking at the process of innovation. To consider the links between energy prices and innovation, we need to begin with a more general question: what factors influence inventive activity? In general, economists focus on the ways in which individual actors respond to incentives. For a firm deciding whether or not to undertake a potential research project, potential incentives to consider include the likely cost of performing the research, the likelihood that the project will end with a successful result, and the potential payoffs from a successful project.

The hypothesis that higher energy prices lead to the development of more energy-saving technologies is derived from demand-pull theories of innovation. Demand-pull theories note the relationship between market forces and potential returns to an innovation. For example, higher energy prices increase the potential payoffs to energy research. Higher energy prices make energy-efficient inventions more valuable, either because the dollar value of potential energy savings is larger or because the market for energy-efficient inventions will be larger.

Other theories of innovative activity focus on the role that existing scientific knowledge plays in determining the pace and direction of technological change. Such theories, referring to the supply of knowledge available, are known as technology-push theories. Research is an uncertain activity. Even when energy prices are high, rational investors will not invest research dollars into energy-efficient research if the prospects of success are not good. Technology-push theories of technological change focus on the importance of the existing base of scientific knowledge to new inventions. The base of knowledge on which inventors can build helps to determine the likelihood of success. Technological advances make new inventions possible, and it is these breakthroughs that influence the direction of technological change. For example, the invention of microcomputer chips led to the development of a generation of electronic equipment. Thus, a thorough treatment of incentives and innovation must look at both demand- and supply-side influences.

The idea that the creation of ideas through invention and innovation will be influenced by economic incentives is not new. The concept of induced innovation was first introduced in 1932 by John Hicks, who noted that changes in factor prices would lead to innovation to economize on use of the more expensive factor. In this framework, one would expect higher energy prices to lead to greater research efforts to develop new energy sources and to improve energy efficiency.

As measures of innovative activity, such as patents and R&D, have become more readily available, empirical economists have begun to estimate the effects that prices and environmental policies have on energy and environmentally friendly innovation. Such studies take one of two approaches. First, authors have compared measures of innovative activity to environmental regulatory stringency. Typically, data on pollution abatement spending are used as a proxy for regulatory stringency in such studies. These studies find correlations between innovation and regulation both across nations and across time. Other studies use energy prices and related regulations as the mechanism that induces innovation. Although the observed price changes might not be policy related, the results can also be applied to situations where policy affects prices, such as a carbon tax. In one recent study, Newell, Jaffe, and Stavins examined the effect of both energy prices and energy efficiency regulations on technological advances in energy efficiency for air conditioners and natural gas water heaters. They found that both lead to more innovation and that energy prices have the largest inducement effect.

Figure 5 presents an example of typical findings in this literature. The figure shows how patenting

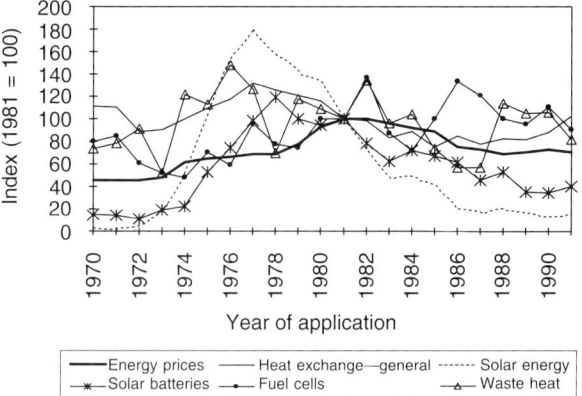

FIGURE 5 An index of patenting activity in five fields related to energy as well as an index of energy prices. Patenting activity is measured as successful U.S. patent applications per year. Only patents granted to U.S. inventors are included. From Popp (2002). Copyright 2002 American Economic Association; reproduced with permission of the *American Economic Review*.

activity for various energy technologies responded to changes in energy prices. Patents provide a detailed record of each invention. Classifications found on the front page of each patent can be used to identify its technological field. Economists have found that patents, sorted by their date of application, provide a good indicator of R&D activity. As a result, patent counts not only serve as a measure of innovative output but also are indicative of the level of innovative activity itself. Note that technologies such as solar energy experienced large jumps in patenting activity immediately following the first energy crisis. For example, there were just 10 solar energy patents in 1972. This figure jumped to 36 in 1973, 104 in 1974, and 218 in 1975. Thus, prices (or other regulations that increase the cost of using fossil fuels) can be expected to stimulate new research quickly.

Figure 5 also provides evidence of the importance of considering technology-push influences. Not only did energy R&D respond quickly to energy prices, but it also dropped off more quickly than did energy prices. Energy prices did not reach a peak until 1981. Nonetheless, patenting activity in most technologies peaked during the late 1970s. These data highlight the importance of the potential returns to R&D. Inventors "stand on the shoulders" of their predecessors. As a result, the quality of the knowledge stock available to an inventor is an important positive contributor to the level of innovative activity at any time. As innovation proceeds, it becomes more and more difficult to improve on the existing technology. In this fashion, doing research can be seen as analogous to drilling for oil. As more and more knowledge is drained from the "well," it becomes more difficult to achieve successful advances. The combination of standing on the shoulders of previous inventors and diminishing returns to research over time suggests that the level of induced R&D will fall over time.

3.2 The Returns to Research

As noted previously, in a market economy, researchers will not undertake a research project unless the expected returns justify the cost. Evidence from the returns to R&D comes from many sources such as case studies of successful innovation and econometric studies of firm productivity. Consistently, economists studying the returns to research have found that the returns to society as a whole are quite high. Most studies find rates of return of between 20 and 50%. Such rates are much higher than the rates of return on investment that firms typically receive.

These high rates of return would suggest that further investments in R&D are warranted. Given this, why aren't R&D investments higher?

Two main characteristics about R&D investment help to explain this phenomenon. One is that research is an uncertain activity. The returns to research are highly skewed. Figure 6 illustrates the distribution of returns to individual research projects typically found in the literature. Although many projects yield little value, successful projects can yield returns of billions of dollars. High returns for successful projects help to compensate firms for the risk of undertaking projects that yield little or no returns. Second, the knowledge created by R&D is a public good. Firms capture the fruits of their research through sales of new products or via cost savings in production. However, these are not the only benefits that new knowledge provides to society. Once new knowledge is publicly available, it can be applied repeatedly without decay and can serve as the building block for other improvements. Moreover, without public policy such as intellectual property rights, others cannot be excluded from making use of the new knowledge. As a result, the firm is unable to capture all of the benefits of new knowledge to society, so that the social returns to R&D are higher than the private returns to R&D. Typical findings are that the social rates of return are approximately four times higher than private rates of return.

The high social returns of research have two important implications. First is that government policy can play an important role in the creation of new knowledge. Government policies help either by enabling firms to capture more of the benefits of their research (e.g., by providing monopolies via patent protection) or by financing research that the private sector is unwilling to conduct on its own. The latter

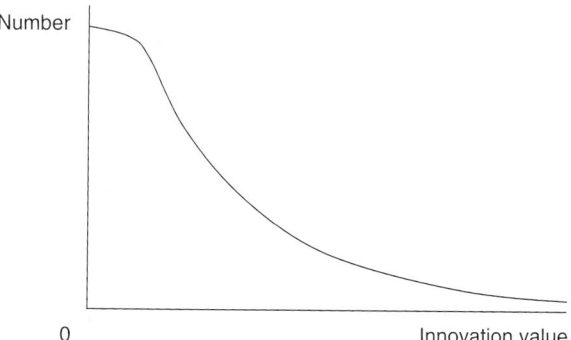

FIGURE 6 Diagram showing a typical distribution of the returns to R&D. The bulk of innovations have little or no commercial value. However, the right tail of the distribution includes innovations that are worth billions of dollars.

is particularly important for basic research that broadens general knowledge but does not have specific commercial applications. For example, basic advances in chemistry and physics are essential to developing new solar energy technologies. Because basic research serves as the building block for many other research efforts, and because the commercial benefits of basic research might not be readily apparent, firms are particularly reluctant to take on basic research projects. Indeed, basic research carried out by the government often provides the impetus for future research projects carried out by the private sector.

Second, high social returns suggest that the potential opportunity costs of doing R&D must be considered when looking at energy R&D policy. High social returns exist for all types of research, not just energy research. Given this, consider a simple economy in which output, Q_t, is devoted to consumption, C_t, investment in physical capital, I_t, or two types of research—energy, $R_{E,t}$, and other, $R_{O,t}$—so that

$$Q_t = C_t + I_t + R_{E,t} + R_{O,t}.$$

When energy R&D increases by $1, $1 less is available for the other activities. The opportunity cost of the first two is simply valued at $1. However, if the social rate of return is four times higher for R&D than for other investments, giving up $1 of other R&D activity has the same effect as giving up $4 of other investment.

This is important because empirical work suggests that at least some increases in energy R&D will come at the cost of other forms of R&D. Research activities are carried out by highly trained scientists and engineers. Because years of training are needed to enter the field, the supply of scientists and engineers available at any one time is inelastic; that is, it cannot increase quickly when new research needs arise. For example, a 1998 study by Goolsbee found that one of the chief beneficiaries of R&D tax subsidies are scientists and engineers, who receive larger wages when subsidies are increased.

Applying this to energy R&D, a simple regression of energy and total R&D spending in the United States suggests that approximately one-half of the energy R&D spending that took place during the 1970s and 1980s came at the expense of other R&D. Thus, at least part of the increased energy R&D of the 1970s and early 1980s came at the expense of other potentially productive R&D projects. At the same time, productivity of the economy as a whole fell during the energy crisis. Researchers are still trying to grapple with the macroeconomic effects of such a dramatic shift in research efforts. Much empirical work has focused on potential causes of decreased productivity during the 1970s, and the general conclusion is that increased energy prices are not enough to explain the entire decrease. Nonetheless, increased energy prices and the resulting shift in innovation toward energy efficiency played at least a partial role.

4. DIFFUSION

Of course, induced invention and innovation is only half the story. Technological advances are of little use unless society makes use of the innovation. Thus, diffusion is also important. Diffusion is likely to play a particularly important role for problems dealing with long-term consequences such as climate change. Most innovation takes place in highly industrialized countries. In 1998, 85% of all R&D in the Organization for Economic Cooperation and Development (OECD) countries was conducted in just seven countries. In fact, roughly 44% of all OECD R&D was done in the United States alone. Thus, as policymakers pay increased attention to potential increases in fossil fuel consumption for fast-growing countries such as China, it is important to pay attention to not only how policy will help to induce the development of new technologies but also how policy can help to encourage the adoption of these technologies in developing countries.

4.1 Diffusion within a Country

Studies of the diffusion of individual technologies consistently find that diffusion is a gradual process. Typically, the rate of diffusion can be represented by a sigmoid, or "S-shaped," curve over time; that is, the rate of adoption rises slowly at first, speeds up, and then levels off as market saturation approaches. Figure 7 illustrates such a curve, with market penetration plotted on the y axis and the passage of time plotted on the x axis.

Traditionally, researchers have used one of two approaches to modeling diffusion of a new technology. Each offers a slightly different explanation for the pattern observed in Fig. 7. The epidemic model of diffusion proposes that information is the primary factor limiting diffusion. Adoption is slow at first because few people (or firms) know about the technology. However, as more people adopt the technology, knowledge of the technology spreads

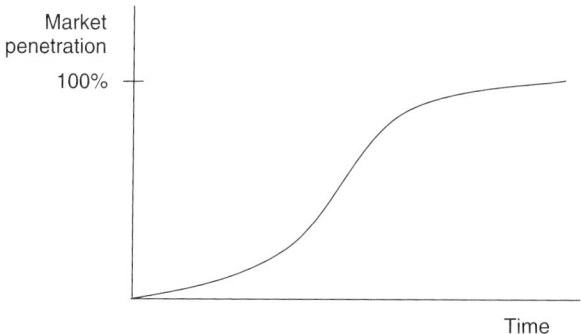

FIGURE 7 Diagram showing how the adoption of new technologies varies over time. Initially, only a few early adopters will choose to use a new innovation. Eventually, mass adoption takes place until only a few stragglers remain. As these stragglers eventually adopt the technology, adoption approaches 100% of the potential market.

quickly, leading to a period of rapid adoption. Economists often use the analogy of a contagious disease to describe this period of adoption; the more people who are "infected" by the technology, the more likely that others will also become "infected." Eventually, few potential adopters remain, as nearly everyone has adopted the technology, so that the rate of adoption levels off again. In a classic 1957 study using this framework, Griliches noted that the rate of diffusion is at least partially determined by economic factors such as the expected rate of return for adoption. Other work using the epidemic model typically focuses on firm characteristics such as firm size to explain variations in the rate of diffusion. These studies find that larger firms adopt new innovations first. Such firms are more likely to have access to capital necessary for new investments, and their size enables them to better bear the risks that come with the adoption of any new technology.

The second approach to studying diffusion focuses is the probit model. The probit model focuses on heterogeneity among firms. In this model, firm heterogeneity leads to a distribution of expected returns from adopting the new technology. Because adoption is costly, only firms above a threshold great enough to justify the costs of adoption will choose to adopt the technology at any given time. Over time, the technology gets cheaper and its quality improves, so that more firms cross the adoption threshold. In this framework, environmental policy and energy prices enter by increasing the expected returns from adopting new energy technologies, thus, increasing the likelihood that firms will cross the threshold. As would be expected, studies using this framework find that higher energy prices encourage the adoption of more energy-efficient technologies. Similarly, the presence of stronger environmental regulations also increases the probability of adopting such technologies. Nonetheless, the response still varies by firm characteristics, and again larger firms tend to adopt first.

Note that the two models suggest different interpretations of the gradual rate of diffusion. The epidemic model suggests that adoption generates positive externalities and, thus, is slower than would be optimal. In this framework, policies to encourage early adoption of new technologies, such as alternative energy sources, would be desired to ensure rapid deployment of new technologies. Conversely, the probit model suggests that gradual diffusion is optimal because differences in adoption decisions simply result from profit-maximizing decisions made by heterogeneous firms. Recent work combining these two models suggests that variations across firms are the main explanation for differences in adoption, as suggested by the probit model.

Turning to adoption studies that focus on energy-efficient technology, a common theme emerges: such technologies are underused. In many cases, individual firms fail to adopt technologies for which the present discounted value of energy savings is greater than the costs of the new technologies. Research suggests several possible explanations for this "energy efficiency paradox." One is that consumers lack information about the potential energy savings that new technologies offer. A second is that consumers place more importance on the upfront installation costs of new technologies than on potential future savings. This may be because they are uncertain about future savings (the upfront installation costs are certain, but the potential energy savings are uncertain given that energy prices may change), because they face credit constraints, and/or because they truly value future consumption at a low rate.

4.2 International Diffusion

The studies mentioned previously focus on the adoption of a single technology across individual actors within a single country. At the international level, there is a large body of research on technological diffusion generally but little work focusing on energy technologies specifically. One exception is a 1996 study by Lanjouw and Mody that looked at patenting activity for various environmental and energy-efficient technologies. This study found that the majority of environmental technology patents in developing countries came from foreign countries.

Moreover, it found that policies in one nation may affect innovation of technologies in another nation. For example, the authors noted that the majority of vehicle air emissions patents granted in the United States were from foreign nations, even though the United States was the first country to adopt strict emissions standards.

Given the importance of diffusion for bringing new energy technologies to developing countries, understanding the general lessons from empirical work on international diffusion is important. There are two potential avenues through which foreign knowledge can have an influence in the domestic economy. First, foreign inventions may be adopted directly by domestic firms. Second, rather than being adopted directly by domestic users, foreign knowledge may affect the productivity of domestic R&D. The "blueprints" represented by foreign patents may serve to inspire additional innovation by domestic inventors. Such productivity increases are knowledge spillovers in that the knowledge represented by the foreign patent creates a positive externality—knowledge from abroad that is borrowed by domestic inventors.

Because most R&D takes place in just a few countries, both types of diffusion are important. For direct adoption, many of the same lessons from diffusion studies within a country apply here as well. In particular, local conditions are an important determinant of the rate of diffusion across countries. Regarding knowledge spillovers, researchers have followed several strategies. At a macro level, several published articles estimate the effects of domestic and foreign R&D on total factor productivity (TFP) growth. Typically, accumulated levels of R&D are combined to create both domestic and foreign stocks of R&D effort. The foreign R&D stocks are weighted by international trade flows, so that these articles attempt to measure the effect of international trade on knowledge flows. These articles find a positive effect for foreign R&D, although this effect may vary by country. In general, the relative contribution of foreign knowledge is inversely related to economic size and level of development. At a more micro level, studies looking at the effect of foreign direct investment (FDI) and foreign knowledge spillovers have found mixed results. Whether the effect of FDI on growth is positive or negative depends on individual industry and country characteristics, with inward FDI being more successful for developed countries. These studies suggest that the ability of a country to successfully absorb new knowledge, such as having a well-educated workforce, is important for successful technology transfer.

SEE ALSO THE FOLLOWING ARTICLES

Business Cycles and Energy Prices • Depletion and Valuation of Energy Resources • Energy Futures and Options • Inflation and Energy Prices • Prices of Energy, History of • Stock Markets and Energy Prices • Technology Innovation and Energy • Transitions in Energy Use

Further Reading

Binswanger, H. P., and Ruttan, V. W. (eds.). (1978). "Induced Innovation: Technology, Institutions, and Development." Johns Hopkins University Press, Baltimore, MD.

Goolsbee, A. (1998). Does government R&D policy mainly benefit scientists and engineers? *Am. Econ. Rev.* **88**, 298–302.

Griliches, Z. (1957). Hybrid corn: An exploration of the economics of technological change. *Econometrica* **25**, 501–522.

Jaffe, A. B., Newell, R. G., and Stavins, R. N. (2003). Technological change and the environment. In "Handbook of Environmental Economics." (K. G. Mäler and J. Vincent, Eds.), Vol. 1, pp. 461–516. North-Holland, Amsterdam, Netherlands.

Lanjouw, J. O., and Mody, A. (1996). Innovation and the international diffusion of environmentally responsive technology. *Res. Policy* **25**, 549–571.

Newell, R. G., Jaffe, A. B., and Stavins, R. N. (1999). The induced innovation hypothesis and energy-saving technological change. *Q. J. Econ.* **114**, 941–975.

Popp, D. (2001). The effect of new technology on energy consumption. *Resource Energy Econ.* **23**, 215–239.

Popp, D. (2002). Induced innovation and energy prices. *Am. Econ. Rev.* **92**, 160–180.

Rosenberg, N. (1982). "Inside the Black Box: Technology and Economics." Cambridge University Press, New York.

Schumpeter, J. (1942). "Capitalism, Socialism, and Democracy." Harper, New York.

Stoneman, P. (ed.). (1995). "Handbook of the Economics of Innovation and Technological Change." Blackwell, Cambridge, MA.

Input–Output Analysis

STEPHEN D. CASLER
Allegheny College
Meadville, Pennsylvania, United States

1. Introduction
2. The Input-Output Model
3. Extensions of the Basic Model
4. Secondary Products
5. The Energy-Based Input-Output Model
6. Structural Decomposition Analysis: Evaluating Changes over Time
7. Conclusions

Glossary

energy intensity A measure of the total Btu's of energy needed both directly and indirectly to produce a dollar's worth of a product for sale to final demand.

fixed coefficients In production theory, a situation in which the required quantity of each input used is a constant multiple of the level of output.

intermediate product Output from an industry that is used as an input by another producing sector, as opposed to output sold to final demand.

inverse coefficient A measure of the total value of output needed both directly and indirectly to produce a dollar's worth of a product for sale to final demand.

open model In input-output analysis, a formulation of the model in which all elements of final demand for goods and services are assumed to be exogenously determined.

producer prices Prices that exclude the transportation costs and wholesale and retail trade margins that are included in the actual prices paid by purchasers.

secondary product A good or service produced by an establishment that differs in classification from the establishment's principal or primary product.

structural decomposition analysis The evaluation and measurement of causes of change in input-output relations through time.

technological/structural coefficient The ratio of the amount of an input used by a sector to the sector's total output, assumed to be constant under the fixed-coefficients assumption.

The input-output approach to analysis of economic activity provides a concise means of representing interrelationships between producing, final demand, and value-added components of the economy at any point in time. As its name implies, the range of inputs used by industries is accounted for, as is the sale of outputs by industries to other industries and to final demand. Because these supply and demand relationships are considered simultaneously for the full range of producing sectors, the input-output model provides a general equilibrium framework useful for analyzing all transactions within the economy.

1. INTRODUCTION

The input-output model's historical origins lie with the French Physiocrat Francois Quesnay, whose *Tableau Economique* portrayed the nature of exchange between farmers, landlords, and manufacturers in the mid-1700s. Fundamental theoretical and empirical work on the modern model was conducted by Wassily Leontief, who was awarded a Nobel Prize in economics for his contributions. The model lends itself to applications ranging from studies of local to national economies. Relative to other forms of analysis, the input-output model's main attributes are its consideration of the full range of production activities, the number of sectors it can accommodate, its ability to measure the mutual interdependence among sectors of the economy in meeting final demand requirements, and its reliance on data from only one time period. The model is also very flexible. With slight modification, it enables the study of issues related to energy use, employment, pollution emissions, and much more.

1.1 The National Income and Product Accounts and the Input-Output Model

The relations that underlie the input-output model are organized around the various measures of income and output found in the national income and product accounts. This consistency is seen in the equivalent measures of gross national product as the sum of final expenditure on goods and services, the sum of the factor payments, and the sum of the values added. The data used in the input-output model show inter-industry sales of intermediate products among the industries or sectors of the economy. Intermediate products are produced in one sector of the economy and then used as inputs by other sectors in the production of their output. The model also accounts for sales to final demand—goods and services purchased for personal consumption, investment, government activities, and net exports (exports minus imports). Finally, value added, the wages, interest, profit, and rent paid to factors of production during the production process by each industry, is included.

To illustrate how inter-industry sales, final demand, and value added are connected in the input-output framework, consider an economy composed of three producing sectors: agriculture, manufacturing, and services. Transactions among these sectors are shown in the first three rows and columns of Table I. The arrows pointing to the right indicate the sales of the goods and services produced by the sectors listed in the rows. For example, in row 1, agricultural output is shown as being sold for use as an input to the agriculture sector itself and to manufacturing and services. Its output is also sold to the various components of final demand, consumption, C, investment, I, government, G, and net exports, X. The sum of the sales of agricultural output across this row equals the total value of output produced by the agriculture sector. A similar interpretation holds for the rows of the table showing sales of output by manufacturing and services.

Within columns of the table, for any sector, the value of each of the inputs used by that sector is measured. Intermediate inputs are measured in the first three rows of each column. Further down the table, within the value-added row, are the total factor payments to labor, capital, entrepreneurs, and land for each producing sector. The sum of these factor payments equals total value added for each sector. The sum of the values added generated by all sectors equals gross national product, as does the sum of the values of products sold to final demand. Finally, the value of total output produced by any sector—sales of its output to producing sectors plus output sold to final demand—is equal to the sum of the values of intermediate inputs used by that sector plus that sector's value added. Corresponding row and column sums are equal in value.

In the United States, the input-output accounts are constructed by the U.S. Department of Commerce, Bureau of Economic Analysis. Benchmark accounts are based primarily on economic census data, which are collected by the Bureau of the Census at 5-year intervals. The 1992 benchmark accounts measure the inputs used and outputs sold by 498 industries. The inputs and outputs of the model are generally measured in dollars, but they can also be measured in physical units. The accounts published by the Bureau of Economic Analysis are dollar based, and goods and services are valued using producer prices.

2. THE INPUT-OUTPUT MODEL

There are two ways in which the output produced by any industry in the economy can be used: as an input by other producers and to satisfy final demand. For example, some of the corn produced in the agriculture sector is used in that sector to produce more corn, some is sold as feed to cattle producers, some is used in the production of corn-based sweeteners by manufacturers, and some is sold directly to consumers as part of final demand. Such transactions by all producing sectors are the foundation of the input-output model. For applications and manipulations of the model, the range of sales to producers and final demand is represented using linear equations. Using the standard notation and the convention that when two subscripts are present the first refers to input and the second output, X_{ij} represents the total amount of input i used to produce output j, X_i represents the total output of sector i, and Y_i represents the total final demand for output i.

2.1 The Fundamental Balance Equation

In equation form, the relation between the total production of output i and its various uses across the economy is represented as

$$X_i = X_{i1} + X_{i2} + \ldots + X_{in} + Y_i. \qquad (1)$$

That is, the total output of good i on the left side of the equation equals the sum of good i sold as an input to all n producing sectors of the economy, the X_{ij}'s, plus the final demand for good i. Equation (1) shows the fundamental balance between the output produced by a sector and all its potential uses.

TABLE I

Interindustry Transactions, Final Demand, and Value Added[a]

	Intermediate sales			Final demand	
	Agriculture	Manufacturing	Services	$C + I + G + X$	Total output
Agriculture	→ Agriculture input to agriculture	→ Agriculture input to manufacturing	→ Agriculture input to services	→ Agriculture sales to final demand	Total agriculture output
Manufacturing	→ Manufacturing input to agriculture	→ Manufacturing input to manufacturing	→ Manufacturing input to services	→ Manufacturing sales to final demand	Total manufacturing output
Services	→ Services input to agriculture	→ Services input to manufacturing	→ Services input to services	→ Services sales to final demand	Total services output
Value added	Value added by agriculture	Value added by manufacturing	Value added by services	Total value added = Total final demand = GNP	
Total output	The value of inputs to agriculture plus its value added equals the value of agriculture output.	The value of inputs to manufacturing plus its value added equals the value of manufacturing output	The value of inputs to services plus its value added equals the value of services output		

[a] The input-output model is organized around the sales of products used as inputs by industries and consumed by final demand. There is a balance between the value of a sector's total output and the value of its inputs plus value added.

2.2 Technological Coefficients

A key assumption of the input-output model is that output is related to input use through fixed coefficients. This assumption implies that to double the amount of output produced requires that all inputs be exactly doubled, which in turn implies constant returns to scale (but not vice versa). The fixed-coefficients assumption means that a sector's use of any input is a constant multiple of its output. In equation form, this relationship is expressed as

$$X_{ij} = a_{ij}X_j, \qquad (2)$$

where a_{ij} represents the ij^{th} technological or structural coefficient, the units of input i required to produce a single unit of output j. For example, suppose sector i produces eggs and sector j produces cakes. If it takes two eggs to produce one cake, then $a_{ij} = 2$; if it is desired to produce 500 cakes, then $X_{ij} = 2 \times 500$, or 1000 eggs. If input-output relations are measured as values rather than physical units, physical inputs and outputs are weighted by their respective prices in forming technological coefficients. Thus, if the price of an egg is 50 cents and a cake costs $3, the technological coefficient corresponding to two eggs per cake is 0.33, 33 cents of egg input is required per dollar of cake output. Production of $1500 of cake output requires $500 worth of egg input. Incorporating the definition shown in Eq. (2) into Eq. (1), the balance equation between sector i's total output and its various uses across the economy becomes

$$X_i = a_{i1}X_1 + a_{i2}X_2 + \ldots + a_{in}X_n + Y_i. \qquad (3)$$

2.3 Solving for Total Outputs

When all the sectors of the economy are considered, a system of simultaneous linear equations results. The endogenous variables are the outputs of each sector, the X_i's; the fixed technological coefficients or a_{ij}'s are parameters, and the Y_i's or final demands are exogenous variables. This system of equations for an economy of n sectors appears as

$$\begin{aligned} X_1 &= a_{11}X_1 + a_{12}X_2 + \ldots + a_{1n}X_n + Y_1 \\ X_2 &= a_{21}X_1 + a_{22}X_2 + \ldots + a_{2n}X_n + Y_2 \\ &\vdots \\ X_n &= a_{n1}X_1 + a_{n2}X_2 + \ldots + a_{nn}X_n + Y_n. \end{aligned} \qquad (4)$$

On the right-hand side of every equation, output levels X_1 through X_n are present, and each is multiplied by a unique technological coefficient, a_{ij}. All goods are not necessarily required as inputs to particular sectors; some of the technological coefficients can be zero.

Using matrix algebra, the various elements of the model are conveniently organized in terms of outputs, technological coefficients, and final demand. The equivalent matrix-algebra representation of the n equations shown in Eq. (4) is

$$\begin{bmatrix} X_1 \\ X_2 \\ \vdots \\ X_n \end{bmatrix} = \begin{bmatrix} a_{11} & a_{12} & \ldots & a_{1n} \\ a_{21} & a_{22} & \ldots & a_{2n} \\ & \vdots & & \\ a_{n1} & a_{n2} & \ldots & a_{nn} \end{bmatrix} \begin{bmatrix} X_1 \\ X_2 \\ \vdots \\ X_n \end{bmatrix} + \begin{bmatrix} Y_1 \\ Y_2 \\ \vdots \\ Y_n \end{bmatrix}. \quad (5)$$

With the letter X representing the column vector of total outputs, A the $(n \times n)$ matrix of technological coefficients, and Y the column vector of final demands, the entire system of equations is expressed simply as

$$X = AX + Y. \quad (6)$$

The outputs contained in the vector X are the unknowns in the model, and they can be solved for as follows. First, rearranging Eq. (6),

$$X - AX = Y$$

or

$$(I - A)X = Y.$$

Here, the letter I is the identity matrix, which serves many of the same functions in matrix algebra as the number 1 does in standard algebra. It consists of one's along its main diagonal with zeros elsewhere. These equations imply that final demand is equal to what remains when goods used as inputs by producing sectors are subtracted from total output. However, final demand is not a residual—the purpose of production is to satisfy a given level of final demand. The solution for the vector of total outputs that will do so is

$$X = (I - A)^{-1} Y. \quad (7)$$

The $n \times n$ matrix $(I-A)^{-1}$ is known as the Leontief inverse. Solving the system of equations in terms of the inverse, $(I-A)^{-1}$, is simply a more efficient means of finding a solution compared with solving the system through substitution. Its use makes it especially easy to solve for the new output levels that arise when there is a change in the final demand vector—rather than re-solving the entire system by substitution, it is only necessary to post multiply the existing Leontief inverse by the new final demand vector to find the new levels of outputs.

Elements of the Leontief inverse account for the full range of interrelationships between producing sectors. For example, a $1 rise in the final demand for a product simultaneously results in increased demand for all inputs used by the sector that produces it. But producing this first set of inputs implies an increased demand for the inputs required by all sectors that supply the initial set, and so on. Thus, inverse coefficients measure the direct and indirect requirements of a good that are necessary to produce $1 of some output for final demand.

2.4 Finding the Inverse and Total Output: Numerical Example

To illustrate the interpretation of the Leontief inverse and the mathematics involved in solving for the vector of sectoral outputs, consider a hypothetical economy consisting of two goods, coal and steel. It is assumed that the output of each good is measured in dollars. Production of either good is also assumed to require both coal and steel. Thus, to produce $1 of coal requires 10 cents of coal and 40 cents of steel. To produce $1 of steel requires 55 cents of coal and 20 cents of steel. In the notation of the input-output model, with sector 1 designated as coal and sector 2 as steel,

$$a_{11} = 0.10, a_{21} = 0.40, a_{12} = 0.55, a_{22} = 0.20.$$

2.4.1 The Technological Matrix

Using the hypothetical numerical values, the technological or structural A matrix from Eqs. (6) and (7) is

$$A = \begin{bmatrix} 0.10 & 0.55 \\ 0.40 & 0.20 \end{bmatrix}.$$

The rows of the technological matrix show how much of each particular input is required to produce $1 of output in the two producing sectors of this economy. Columns reveal the input requirements of each sector. For sector 1, coal, a total of 50 cents of intermediate inputs is required to produce $1 worth of coal output, 10 cents of coal and 40 cents of steel. The sum of technological coefficients in each column must be less than 1 or else the production process is not economically feasible—producing $1 of output should not cost more than a dollar. Aside from intermediate inputs, in producing a dollar's worth of output, the remaining value comes from value added. Hence, the remaining 50 cents of the $1's worth of coal output is comprised of coal's value added. Similarly, based on the second column of the technological matrix, which shows the steel sector's input requirements, a total of 75 cents of intermediate inputs is required to produce $1 of steel; therefore, to make up the rest of the dollar's worth of steel output, 25 cents of each dollar of steel is comprised of value added.

2.4.2 Solving for the Leontief Inverse

The solution for the economy's vector of outputs requires formation of the Leontief inverse. The foundation for this inverse is the matrix $(I - A)$, calculated as

$$(I - A) = \begin{bmatrix} 1 & 0 \\ 0 & 1 \end{bmatrix} - \begin{bmatrix} 0.10 & 0.55 \\ 0.40 & 0.20 \end{bmatrix}$$
$$= \begin{bmatrix} 0.90 & -0.55 \\ -0.40 & 0.80 \end{bmatrix}.$$

Based on the procedure used to find the inverse of a matrix,

$$(I - A)^{-1} = \begin{bmatrix} 1.6 & 1.1 \\ 0.8 & 1.8 \end{bmatrix}.$$

2.4.2.1 Interpreting the Inverse Coefficients

As previously noted, the values within the inverse represent the direct and indirect output necessary to supply $1 of output for final demand. Thus, to supply $1 of coal for final demand requires a total value of coal production of $1.60. One dollar of this amount represents the dollar's worth of coal that is required to satisfy final demand. The remaining 60 cents represents the value of the coal used to produce coal and well as the value of coal used to produce the steel that is necessary for coal production, and so on.

As another example, the direct and indirect coal required to produce $1 of steel equals $1.10. The interpretation and implicit derivation of this measure of direct and indirect coal use can be understood by considering the various *linkages* between the coal and steel sectors. From the technological matrix, A, when steel output increases by $1, the steel sector requires 55 cents of coal directly ($a_{12} = 0.55$). However, since it takes 10 cents of coal to produce $1 of coal ($a_{11} = 0.1$), an additional 5.5 cents of coal is needed *indirectly* to produce the 55 cents of coal used by the steel industry ($0.1 \times 0.55 = \$0.55$). In addition, production of a $1 of coal requires 40 cents of steel ($a_{21} = 0.4$). Since it takes 55 cents of coal to produce $1 of steel, to produce 40 cents of steel it takes 22 cents of coal ($0.55 \times 0.4 = \$0.22$). It also takes steel to produce steel ($a_{22} = 0.2$). But the production of the 40 cents worth of steel used by coal requires the use of 8 cents of steel ($0.2 \times 0.4 = \$0.08$), which again implies the use of more coal to produce it. When the direct requirement for coal by the steel sector is added to the infinite chain of all such indirect requirements, the inverse coefficient 1.1 is derived.

2.4.2.2 Output Multipliers

The sums of column entries within the Leontief inverse are known as output multipliers. Given a $1 increase in final demand for a particular good, the corresponding output multiplier provides a measure of the resulting change in total output in the economy. Output multipliers reveal where a dollar of spending on some component of final demand will have the largest impact on total output. For example, the output multiplier for the coal sector is 2.4—a $1 rise in final demand for coal will lead to a rise in overall production of $2.40, with $1.60 of output from the coal sector and $0.80 of output from the steel sector. Alternatively, if the final demand for steel rises by $1, the output multiplier is 2.9.

2.4.3 Finding Total Outputs

In addition to the measurement of the direct and indirect requirements per dollar of output, one of the most important uses of the input-output model is the determination of the total output necessary to satisfy some level of final demand. This determination involves a direct application of Eq. (7). For example, suppose final demand for coal and steel equals $100 and $200, respectively ($Y_1 = \$100$ and $Y_2 = \$200$). From Eq. (7), total output is calculated as

$$X = (I - A)^{-1} Y$$

or

$$\begin{bmatrix} X_1 \\ X_2 \end{bmatrix} = \begin{bmatrix} 1.6 & 1.1 \\ 0.8 & 1.8 \end{bmatrix} \begin{bmatrix} 100 \\ 200 \end{bmatrix}.$$

The equivalent algebraic expression is

$$X_1 = (1.6 \times 100) + (1.1 \times 200) = \$380$$
$$X_2 = (0.8 \times 100) + (1.8 \times 200) = \$440.$$

These equations imply that to satisfy a $100 final demand for coal and a $200 final demand for steel requires total coal production of $380 and steel production of $440. Alternatively, if final demand is *predicted* to rise to $120 for coal and $210 for steel, estimated coal and steel output equals $423 and $474, respectively.

2.5 The Model's Accuracy over Time

Economists understand that the technological coefficients used to form the Leontief inverse are only representative of input requirements during the time period for which data are collected, because relative price changes and technological change lead to changes in input requirements per unit of output. However, there is general agreement that the model accurately approximates the impact on output of changes in final demand for short periods of time around the date for which the table was formulated. For example, the Leontief model based on data for 1939 was able to correctly predict steel requirements at the end of World War II. Therefore, even though benchmark tables are not available on a yearly basis in the United States and are generally published several years after the benchmark date, the data they contain are still useful for analysis of the state of the economy in nearby years.

3. EXTENSIONS OF THE BASIC MODEL

A number of important extensions and applications of the input-output model are seen in empirical investigations. The most widely used include the price model, which makes it possible to estimate the prices of outputs that are consistent with physical technological coefficients and value added; the labor model, which enables the estimation of direct and indirect labor requirements and the employment levels necessary to meet a given level of final demand; the closed model, in which elements from final demand and value added are made endogenous; and the pollution-based model, which provides a means of linking production activities to the emission of pollutants.

3.1 The Leontief Price Model

The technological coefficient a_{ij} can be measured as the physical units of input i per unit of output j or the dollars of input i required to produce $1 of output j. In the Leontief price model, the a_{ij}'s are measured in physical units, for example, the tons of coal it takes to produce a ton of steel. Multiplication of the ij^{th} technological coefficient by the ith price, $P_i a_{ij}$, results in a measure of the dollars of input i required per physical unit of output j (dollars per unit of input i times units of i per unit of j equals dollars of input i per unit of output j). For example, if 2 tons of coal are required to produce 1 ton of steel and if the price of coal is $4 per ton, then $8 of coal are required to produce a ton of steel. The total value of all intermediate inputs required to produce one physical unit of output j is therefore equal to the sum of each of sector j's price-weighted input coefficients.

In addition to the per-unit value of all the intermediate inputs used in its production, the price of a good is comprised of the value added per unit of output that is generated through the use of labor and other factors of production. For example, if 40 cents of coal and $1.60 of steel are required as intermediate inputs to produce a ton of coal, and value added per ton of coal is $2, then the price of coal is $4. When all sectors of the economy are considered, the following system of price equations results:

$$P_1 = P_1 a_{11} + P_2 a_{21} + \ldots + P_n a_{n1} + va_1$$
$$P_2 = P_1 a_{12} + P_2 a_{22} + \ldots + P_n a_{n2} + va_2$$
$$\vdots$$
$$P_n = P_1 a_{1n} + P_2 a_{2n} + \ldots + P_n a_{nn} + va_n.$$

Each price on the left-hand-side of the equation is seen to equal the sum of the values of the given sector's intermediate inputs per unit of output plus the sector's value added per unit of output. In matrix algebra, this system of products and sums is

$$P = PA + va. \qquad (8)$$

where P is a $1 \times n$ vector of prices and A is the same matrix of technological coefficients found in the standard model, with coefficients measured in physical units of input per physical unit of output. Based on Eq. (8), the vector of prices is solved for as

$$P = va(I - A)^{-1}. \qquad (9)$$

Equation (9) shows that the row vector of prices is equal to the value added row vector postmultiplied by the Leontief inverse.

3.1.1 The Price Model: Numerical Example

To illustrate the calculation of prices, consider the two-sector economy comprised of coal and steel production. As in the previous example, coal is sector 1 and steel is sector 2. In physical units, it takes $a_{11} = 0.1$ tons of coal to produce a ton of coal, $a_{12} = 1.375$ tons of coal to produce a ton of steel, $a_{21} = 0.16$ tons of steel to produce a ton of coal, and $a_{22} = 0.2$ tons of steel to produce a ton of steel. The economy's matrix of physical-unit technological coefficients is therefore

$$A = \begin{bmatrix} 0.100 & 1.375 \\ 0.160 & 0.200 \end{bmatrix}.$$

The coefficients for coal's use of coal and steel's use of steel are exactly the same as in the value-based example—since the same price of coal applies to both the coal used as an input and the total amount of coal produced, the prices in the numerator and denominator cancel, with a similar effect for steel's use of steel. Subtracting the matrix of technological coefficients from the identity matrix and solving for the inverse yields

$$(I-A)^{-1} = \left(\begin{bmatrix} 1 & 0 \\ 0 & 1 \end{bmatrix} - \begin{bmatrix} 0.100 & 1.375 \\ 0.160 & 0.200 \end{bmatrix} \right)^{-1}$$

$$= \begin{bmatrix} 1.60 & 2.75 \\ 0.32 & 1.80 \end{bmatrix}.$$

Suppose value added per ton of coal is $2 and value added per ton of steel is $2.50. Based on Eq. (9), the solution for the economy's prices is

$$P = va(I-A)^{-1} = [2.0 \quad 2.5] \begin{bmatrix} 1.60 & 2.75 \\ 0.32 & 1.80 \end{bmatrix}$$

$$= [4 \quad 10].$$

The price of coal is $4 per ton and the price of steel is $10 per ton.

3.1.2 Prices and the Dollar-Based Model

Prices can also be calculated using the dollar-based technological matrix and the vector of dollar-based values added:

$$P = va(I-A)^{-1} = [0.5 \quad 0.25] \begin{bmatrix} 1.6 & 1.1 \\ 0.8 & 1.8 \end{bmatrix}$$

$$= [1 \quad 1].$$

Prices estimated using the dollar-based model always equal 1. This result follows from the fact that such prices represent the dollar value of output per *dollar* of output (versus dollars per *unit* of output for a standard price). The dollar value of output per dollar of output equals 1 by definition. For each good, such prices can be thought of as price indices, which equal 1 in the base year or, in this case, the year in which the input-output model is formulated. Using a dynamic version of the price model, the effects on prices of changes in elements of the technological matrix or the vector of values added can be analyzed relative to this base. Such changes are discussed in the section on structural decomposition analysis.

3.2 Direct and Indirect Labor Requirements

Another important and widely used application of the input-output model is the estimation of direct and indirect labor requirements—measures of the total labor required given a $1 increase in final demand for an output. These measures can also be used to estimate the economy's total employment given any level of final demand. Calculation of such labor intensities requires data on the direct labor input requirements per dollar of output for each sector of the economy. For example, to produce $1000 of coal might require five workers, while production of a $1000 of steel might require four workers. Estimation of the row vector of direct and indirect labor requirements, λ, simply involves postmultiplication of the $(1 \times n)$ vector of direct labor requirements per dollar of output, L, by the value-based Leontief inverse:

$$\lambda = L(I-A)^{-1}. \qquad (10)$$

3.2.1 Direct and Indirect Labor Requirements: Numerical Example

Using the value-based Leontief inverse for the two-sector economy comprised of coal and steel, if it takes five workers to produce $1000 of coal and four workers to produce $1000 of steel, the direct labor coefficients per dollar of output are 0.005 workers per dollar of coal and 0.004 workers per dollar of steel. The economy's direct and indirect labor requirements are then calculated as

$$\lambda = L(I-A)^{-1} = [0.005 \quad 0.004] \begin{bmatrix} 1.6 & 1.1 \\ 0.8 & 1.8 \end{bmatrix}$$

$$= [0.0112 \quad 0.0127].$$

To produce $1000 of coal requires a direct and indirect labor input of 11.2 workers, while production of $1000 of steel requires 12.7 workers from all sectors of the economy.

The logic underlying these measures can be seen for the coal sector as follows. Since it takes 0.005 workers to produce $1 of coal and $1.60 of coal directly and indirectly to produce $1 of coal for final demand, (0.005×1.6) or 0.008 workers are required everywhere in the economy to produce the $1.60 of direct and indirect coal output. Coal production also requires steel, and $1 of steel production requires 0.004 workers. Therefore, production of the $0.80 of steel required directly and indirectly to produce $1 of coal takes 0.0032 workers. The sum of these values equals the 0.0112 direct and indirect workers necessary to produce $1 of coal for final demand.

3.2.1.1 Measuring Total Labor Requirements

Finding the economy's total labor requirements to satisfy a given level of final demand simply involves postmultiplication of Eq. (10) by the final demand vector. If final demand for coal is $100 and final demand for steel is $200, total employment, E, is 3.66 workers:

$$E = L(I - A)^{-1}Y = [0.0112 \quad 0.0127] \begin{bmatrix} 100 \\ 200 \end{bmatrix}$$
$$= 3.66.$$

This result follows directly from the fact that $(I-A)^{-1}Y$ measures total output by sector, which for the simple economy used in the value-based example led to production of $380 of coal and $440 of steel. If it takes five workers to produce $1000 of coal, it takes 1.9 workers to produce $380; if four workers are needed to produce $1000 of steel, 1.76 workers are required to produce $440 of steel. Total employment in the economy is therefore, $(1.90 + 1.76)$, 3.66 workers.

3.3 Closing the Model with Respect to Labor

Rather than using direct employment coefficients to calculate direct and indirect labor requirements, it is possible to form similar measures by closing the input-output model with respect to households and the labor services they provide. In general, closing an input-output model refers to incorporating elements from final demand and value added directly into the matrix of technological coefficients. In the case of labor, some or all of the consumption goods used by households within final demand are assumed to be the inputs required to produce labor services (e.g., food is a necessary input for physical human effort). This consumption component is stripped out of final demand and included as an additional column of the inter-industry transactions table. Division of consumption expenditure for each good by total consumption expenditure results in the technological coefficients for the production of labor services. Similarly, payments for labor services by industries are stripped out of each sector's value added and included as an additional row of the transactions table. Division by the value of each sector's output results in technological coefficients that measure the value of labor input per dollar of output for each industry. Given these adjustments, the expanded technological matrix is used to find the Leontief inverse, which now includes dollar-based direct and indirect labor requirements.

3.4 Direct and Indirect Pollution Emissions

The estimation of direct and indirect labor requirements makes use of direct labor coefficients, measures of the direct labor input required to produce $1 of a sector's output. Such coefficients can be defined for any variable not included in the inter-industry transactions table and are generally defined as direct impact coefficients. For example, if there is an interest in measuring direct and indirect water use, the direct impact coefficients are gallons of water used per dollar of output for each sector, and direct and indirect water requirements can be solved for in the same way that direct and indirect labor requirements are found.

An important use of such coefficients is the estimation of direct and indirect pollution emissions. For example, suppose production of each $1000 of coal leads to 10 tons of carbon dioxide emissions, while production of a $1000 of steel results in 20 tons of emissions. The direct-impact coefficients are then 0.01 and 0.02. Analogous to calculation of direct and indirect labor requirements, direct and indirect carbon dioxide emissions are solved for as

$$\pi = E(I - A)^{-1} = [0.01 \quad 0.02] \begin{bmatrix} 1.6 & 1.1 \\ 0.8 & 1.8 \end{bmatrix}$$
$$= [0.032 \quad 0.047].$$

Here, E represents the row vector of carbon dioxide's direct emissions coefficients for each sector. The Leontief inverse, $(I - A)^{-1}$, is estimated using the value-based model. The measures that result from this calculation show that each $1000 of coal produced results in 32 tons of direct and indirect carbon dioxide emissions, while the total emissions everywhere in the economy from the production of $1000 of steel equals 47 tons.

Total carbon dioxide emissions for a given level of final demand are found by postmultiplying the row vector of direct and indirect emissions by the column vector of final demand. Based on the previous hypothetical values, total emissions are

$$ETOT = \pi Y = [0.032 \quad 0.047] \begin{bmatrix} 100 \\ 200 \end{bmatrix} = 12.6.$$

Since each dollar of coal production results in direct and indirect carbon dioxide emissions of 0.032 ton, production of $100 of coal for final demand leads to 3.2 tons. Added to the 9.4 tons of emissions induced by the final demand for steel, (0.047 × $200), satisfaction of the economy's total final demand leads to 12.6 tons of carbon dioxide emissions.

4. SECONDARY PRODUCTS

Complicating the estimation of input-output coefficients for real-world applications are secondary products, products that do not fall within the classification of goods that are primary to the industry that produced them. For example, steel is the primary product of the steel industry, but steel producers may produce aluminum as well. If secondary products are not accounted for, sales of output from the steel industry to other sectors will give misleading estimates of the amount of steel used in particular production processes—some buyers of output from the steel industry might really be using aluminum to produce their output.

To deal with the secondary product problem, a distinction is made between a commodity and an industry. A commodity is any well-defined good or service, consistent with the government's classification of products at a specified level of aggregation. An establishment can produce many different commodities but is classified into a particular industry based on the primary commodity produced—the commodity that makes up the bulk of the establishment's output.

One of two assumptions is generally used to reallocate industrial inputs to each distinct commodity that is produced. First is the industry-technology assumption, which treats secondary products as if their production required the same inputs as those used to produce total output in the industry where they are produced. For example, if the steel sector produces aluminum as a secondary product, and aluminum constitutes 20% of the steel industry's output, then 20% of each of the steel industry's inputs is assigned to the production of aluminum as a commodity. The other assumption used in estimating commodity input requirements is the commodity technology assumption, under which the composition of inputs needed to produce a unit of commodity output is the same regardless of the industry in which the commodity is produced. The industry technology assumption is most appropriate in cases where the mix of inputs used by primary producers is fundamentally the same whether secondary products are produced or not. Examples include chemicals that are secondary products of refining operations and the extraction of coal as a secondary product in metal ore-mining operations. Alternatively, the commodity technology assumption is most appropriate when only one input recipe is associated with the production of an output.

4.1 The Use and Make Tables

Under either the industry technology or commodity technology assumption, reallocation of inputs to the production of commodities involves two matrices: the use and make tables. The rows of the use table represent commodities, while its columns represent industries. Each cell in the use table shows the amount of the commodity listed on the row that is used as an input by the industry listed in the column. For example, the amount of the commodity coal used as an input by the steel industry is found at the intersection of the coal row and steel column. The sum of each row in the use table, which measures all intermediate commodity sales to industries, plus sales to final demand equals total commodity output.

As its name implies, the make table shows all the commodities that are produced by particular industries. Industries are listed along the rows, and the commodities they produce are designated in the columns. The main diagonal of the make table shows the values of commodity outputs produced by the industries to which they are primary. For example, the intersection of the coal row and column shows the total amount of the commodity coal produced by the coal industry. If some steel output is a secondary product of the coal industry, the value of this steel production is found at the intersection of the coal row and the steel column. Similarly, the intersection of the steel row and the coal column shows the coal produced as a secondary product by steel. The sum of values across any row of the make table equals total industry output for the industry listed on that row. The sum of values down any column of the make table accounts for all producers of a commodity and equals the total output of that commodity.

4.2 Estimating Commodity Output under the Industry Technology Assumption

Since it is used most frequently in estimating input-output relations when secondary products are present, the means by which the industry technology assumption leads to a commodity-based transactions table and inverse is illustrated. To produce their primary and secondary products $13.42 of the commodity coal is used by the coal industry, $266.58 of coal is used by the steel industry, $171.81 of the commodity steel is used by the coal industry, and $68.19 of the commodity steel is used by the steel industry. In the use matrix, these commodity sales to industries appear as

$$U = \begin{bmatrix} 13.42 & 266.58 \\ 171.81 & 68.19 \end{bmatrix}.$$

The first row of this table shows the use of the commodity coal by the coal and steel industries, respectively, while the second row shows the use of steel by these industries. If final demand for coal is $100 and final demand for steel is $200, the total commodity output of coal is equal to $380, the sum of elements in the first row of the use table plus final demand for coal. Similarly, the total commodity output of steel is $440.

Next, suppose that $338 of the commodity coal is produced by the coal industry—its primary product, along with $62 of steel—its secondary product. The steel industry produces $42 of coal as a secondary product and $378 of steel. The economy's make matrix is

$$V = \begin{bmatrix} 338 & 62 \\ 42 & 378 \end{bmatrix}.$$

Adding elements across rows of the make table equals total industry output. Therefore, the total value of output in the coal industry, comprised of $338 of coal and $62 of steel, is $400. Similarly, the total value of output produced by the steel industry is $420. Adding entries in the columns of the make matrix yields total commodity output, $380 of coal and $440 of steel, which are the same values found by adding across rows of the use matrix, plus final demand.

Through the industry technology assumption, commodity inputs to industries in the use table are allocated to the production of commodities in proportion to the fraction of industrial output they comprise. These proportions are calculated by pre-multiplying the make matrix by the matrix $\hat{g}^{-1}$, which is composed of the reciprocals of industry outputs along its main diagonal, with zeros elsewhere:

$$\hat{g}^{-1}V = \begin{bmatrix} 1/400 & 0 \\ 0 & 1/420 \end{bmatrix} \begin{bmatrix} 338 & 62 \\ 42 & 378 \end{bmatrix}$$
$$= \begin{bmatrix} 0.845 & 0.155 \\ 0.100 & 0.900 \end{bmatrix}.$$

The first row of the resulting matrix product shows that 84.5% of the output of the coal industry is coal, and 15.5% is steel. Similarly, 10% of the steel industry's output is coal and 90% is steel. Under the industry technology assumption, 84.5% of the commodity inputs to the coal industry found in the use matrix is used in the production of the commodity coal. This fraction of the coal industry's inputs is added to 10% of the inputs used by the steel industry to yield the total commodity input requirements of the commodity coal. Similarly, 15.5% of the coal industry's commodity inputs is assumed to be used in the production of the commodity steel, along with 90% of the inputs to the steel industry. The transformation of the use table into a table of commodity-by-commodity requirements is seen in the matrix product

$$U(\hat{g}^{-1}V) = \begin{bmatrix} 13.42 & 266.58 \\ 171.81 & 68.19 \end{bmatrix} \begin{bmatrix} 0.845 & 0.155 \\ 0.100 & 0.900 \end{bmatrix}$$
$$= \begin{bmatrix} 38 & 242 \\ 152 & 88 \end{bmatrix}.$$

The coal industry used $13.42 of coal to produce both coal and steel, but its coal output constituted 84.5% of total industry production. Hence, 84.5% of $13.42 ($11.34) from the coal industry is allocated to production of the commodity coal as an input requirement. The steel industry used $266.58 of coal to produce both steel and coal. Since 10% of steel's output is coal, 10% of $266.58 ($26.66) is assumed to be used for coal production. The total amount of the commodity coal used to produce coal by all industries in the economy is therefore $38. A similar explanation holds for the other elements in the final matrix product.

To form commodity-based technological coefficients, commodity inputs are divided by their respective total commodity outputs. This transformation is accomplished using the matrix $\hat{q}^{-1}$, which consists of the reciprocals of commodity outputs along its main diagonal, with zeros elsewhere. The resulting matrix is defined as BD. Using matrix algebra, the complete transformation of the use table into a commodity-based matrix of technological coefficients is seen as

$$BD = (U\hat{g}^{-1})(V\hat{q}^{-1}) = (U\hat{g}^{-1}V)\hat{q}^{-1}$$
$$= \begin{bmatrix} 38 & 242 \\ 152 & 88 \end{bmatrix} \begin{bmatrix} 1/380 & 0 \\ 0 & 1/440 \end{bmatrix}$$
$$= \begin{bmatrix} 0.10 & 0.55 \\ 0.40 & 0.20 \end{bmatrix}.$$

Finally, industry technology commodity-by-commodity total requirements are estimated as

$$(I - BD)^{-1} = \begin{bmatrix} 1.6 & 1.1 \\ 0.8 & 1.8 \end{bmatrix}.$$

Elements within this inverse are interpreted in the same way as elements within the usual Leontief inverse, $(I - A)^{-1}$, except that they relate to direct and indirect commodity input requirements (versus industry requirements) needed to produce $1 of commodity output for final demand. Postmultiplication of the inverse by the final demand vector yields the vector of total commodity outputs.

5. THE ENERGY-BASED INPUT-OUTPUT MODEL

The energy-based input-output model uses a mixed or hybrid-unit modification of the standard approach to measure direct and indirect energy input use. It traces its development to the work of Hannon, Herendeen, and Bullard. In the model, energy flows are measured in physical units, usually British thermal units or Btu's, while the output of nonenergy sectors is measured in dollars. This formulation enables analysts to determine which sectors of the economy are the most energy-intensive in terms of direct energy inputs, the energy it takes to produce energy, and in the use of energy to produce nonenergy inputs. Relative to the dollar-based model, it provides an accurate means of estimating the economy's physical energy requirements for a given level of final demand and of estimating changes in energy requirements given final demand changes. Measurement in physical units provides useful information regarding the depletion of finite energy products such as crude oil, natural gas, and coal. The model can be easily modified to measure the emissions of pollution that result from the combustion of fossil fuels. Finally, the energy-based model provides a foundation for evaluating direct and indirect changes in energy efficiency and understanding the factors responsible for such changes.

5.1 Elements of the Energy-Based Model

In the standard input-output model, the balance between the total output produced by some sector and its use as an input across sectors and final demand is

$$X_i = a_{i1}X_1 + a_{i2}X_2 + \ldots + a_{in}X_n + Y_i,$$

where a_{ij} represents the dollars of input i used to produce $1 of output j, X_j represents the total value of output j, and Y_i is the value of final demand for good i. In the energy input-output model, if a sector produces one of the energy products coal, crude oil and natural gas, refined petroleum products, gas utilities, or electricity, its energy use and sales are measured in Btu's. The matrix that shows the use of energy inputs by energy producers, nonenergy producers, and final demand is called the direct energy transactions matrix. For the energy sectors of the economy, it supplants the dollar-based flows of the standard input-output table.

With energy use and total energy output measured in Btu's and nonenergy input use and total output measured in dollars, there are four ways in which the units of the energy input-output model's technological and corresponding inverse coefficients appear. The technological coefficient a_{ij} measures Btu's of input per dollar of output when input i is an energy product and output j is a nonenergy product. It measures Btu's of input per Btu of output if a technological coefficient for energy use by an energy-producing sector is considered. For nonenergy inputs to energy sectors, the units of a_{ij} are dollars of input per Btu of output, since the outputs of energy producing sectors are measured in Btu's. Finally, if sales of nonenergy inputs to nonenergy producers are considered, the units of a_{ij} are the standard dollars of input per dollar of output.

In constructing an energy-based model, it is customary to rearrange the industrial classification system so that the energy inputs to producing sectors are shown in the top rows of the table, with nonenergy producing sectors below. With input-output sectors so arranged, Table II shows the units of technological and inverse coefficients for the various dollar and energy-based measures.

Based on the mixed-unit technological matrix, energy intensities are calculated using the equation

$$\varepsilon = e(I - A)^{-1}. \qquad (11)$$

TABLE II

Units of Technological and Inverse Coefficients in the Energy-Based Model[a]

	Energy producers	Non-energy producers
Energy inputs	Btu/Btu	Btu/$
Nonenergy inputs	$/Btu	$/$

[a] Using the mixed-unit formulation of energy-based input-output model, technological and inverse coefficients are measured in four possible ways.

Equation (11) shows that the economy's energy intensities are a simple function of the mixed-unit Leontief inverse. Specifically, the matrix e selects the energy rows from the inverse. In the general case of m energy products, ε is an $m \times n$ matrix of energy intensities and e is an $m \times n$ matrix composed of ones and zeroes, with a 1 appearing in each row in the location that corresponds to the energy-producing sector.

5.1.1 Primary Energy Intensities

A summary measure of the Btu's of all energy types used by a sector is known as a primary energy intensity—a weighted sum of energy intensities for each fuel, where, to avoid double counting, weights are based on the fraction of total output comprised of newly extracted energy produced by fuel sectors. Examples of primary energy include oil from oil wells, energy obtained from geothermal sources and water power, and energy released from nuclear reactors. In calculating the primary energy intensity for each sector, the weight for the coal and crude oil and natural gas intensities is 1, because 100% of the energy output of these sectors comes from the earth. Alternatively, the weight for refined petroleum products is about 0.05, because only 5% of the energy sold by that industry originates as a refined petroleum commodity. The rest comes from sales of crude oil input to refining operations.

5.1.2 The Energy Input-Output Model: Numerical Example

Using the two-sector model composed of coal and steel production, suppose the coal sector requires 950 thousand Btu's of coal and steel requires 6050 thousand Btu's of coal to produce their total output. Final demand for coal is 2500 thousand Btu's. Hence, the total output of coal is 9500 thousand Btu's. To produce coal, $152 of steel is required by the coal sector, and $88 of steel is required by the steel sector. Final demand for steel is $200. Total output of steel therefore equals $440. Dividing the inputs to each sector by total sectoral output leads to the formation of the hybrid-units technological matrix, A.

$$A = \begin{bmatrix} 950/9500 & 6050/440 \\ 152/9500 & 88/440 \end{bmatrix}$$
$$= \begin{bmatrix} 0.10 & 13.75 \\ 0.016 & 0.20 \end{bmatrix}.$$

Each Btu of coal production directly requires 0.1 Btu's of coal (the 1000s of Btu's in the numerator and denominator cancel). Production of 1000 Btu's of coal requires 1.6 cents of steel input. Each dollar of steel output requires 13.75 thousand Btu's of direct coal input and 20 cents of direct steel input.

Using Eq. (11), the solution for the economy's vector of energy intensities is

$$\varepsilon = e(I-A)^{-1} = \begin{bmatrix} 1 & 0 \end{bmatrix} \begin{bmatrix} 1.600 & 27.5 \\ 0.032 & 1.8 \end{bmatrix}$$
$$= \begin{bmatrix} 1.6 & 27.5 \end{bmatrix}.$$

To produce 1 Btu of coal for delivery to final demand requires a total of 1.6 Btu's of coal everywhere in the economy (i.e., the coal required to produce coal, and the coal required to produce the steel used by coal as an input, and so on). To produce $1 of steel output requires 27.5 thousand Btu's of coal, directly and indirectly.

Consistent with the energy conservation condition, the product of the vector of energy intensities and the mixed-unit final demand vector equals the Btu equivalent of total coal production:

$$\varepsilon Y = \begin{bmatrix} 1.6 & 27.5 \end{bmatrix} \begin{bmatrix} 2500 \\ 200 \end{bmatrix}$$
$$= 4,000 + 5,500 = 9,500.$$

Note that this relationship corresponds in principle to the identity in the all-dollar model that relates the economy's vector of total outputs, X, to final demand, Y, $X = (I-A)^{-1}Y$.

5.1.3 Energy Input-Output Models: The Accuracy of Alternative Techniques

Using the dollar-based table and direct energy coefficients (ratios of Btu's of direct energy use to the value of total output) to calculate energy intensities in a manner analogous to calculating labor intensities will lead to inaccurate results when the prices paid for energy products differ across using sectors, which is generally the case. Such price differences result from variations in the mix of fuels that make up energy input aggregates (bituminous versus lignite under the classification coal, for example) or price differences brought about by the volume of fuel purchased. The problem with energy intensities calculated using direct energy coefficients and the dollar-based inverse lies with the correct estimation of total physical energy use given changes in final demand. For example, if the price of coal used by the coal sector differs from the price of coal paid by steel, the direct and indirect dollars of coal used by coal and steel will be associated with different quantities of coal, depending on whether the coal use occurs in the coal or steel sector. The hybrid-units approach does not suffer from this problem—it will correctly estimate total energy requirements for any level of final demand.

6. STRUCTURAL DECOMPOSITION ANALYSIS: EVALUATING CHANGES OVER TIME

Applications of the input-output model can be extended to measure and explain sources of change in the economy through time. The evaluation of such changes is known as structural decomposition analysis, which traces its roots to Leontief, Anne P. Carter, and Hollis B. Chenery. The modern classification of studies falling under this heading is attributable to Rose and Miernyk. As a form of comparative static analysis, the object of most structural decompositions is to isolate the separate sources of changes in the variables that comprise the input-output model and to measure their importance. The foundation for such analysis begins with the fundamental input-output identity that relates the economy's vector of total output to final demand:

$$X = (I - A)^{-1} Y.$$

From one time period to the next, assuming that changes are infinitesimal, the change in the vector of outputs is

$$\Delta X = \Delta (I - A)^{-1} Y + (I - A)^{-1} \Delta Y. \quad (12)$$

Based on Eq. (12), there are two major sources of change. First, for a given final demand, changes in the vector of outputs are seen to result from changes in the Leontief inverse, $\Delta(I - A)^{-1}$, which are brought on by a change in the technological matrix, A. In turn, such change stems from variations in any of the n^2 structural coefficients within the A matrix. Second, changes in output will occur given a change in the final demand vector, ΔY, for a given Leontief inverse. These changes are caused by the overall growth of final demand and by variations in the demand for particular products within and across final demand categories. Underlying changes in both structural coefficients and final demand are the full range of economic forces, including relative price changes, economies of scale, and technological change.

6.1 Decomposing Changes in Final Demand

The input-output model makes it possible to estimate new output requirements given changes in final demand by simply postmultiplying the Leontief inverse by the new or predicted final demand; however, it is possible to extend the range of measured effects. For example, the final demand vector can be decomposed so that energy and nonenergy sectors are considered separately. The inverse matrix is assumed to be constant from one period to the next. Given n sectors in the economy, Y_t^e is an $n \times 1$ vector that shows the final demand in period t for energy products, with zeros elsewhere, and Y_t^{ne} is an $n \times 1$ vector showing final demand for nonenergy products in period t, with zeros in the energy locations. The sum of these vectors equals the final demand vector in period t:

$$Y_t = Y_t^e + Y_t^{ne}.$$

Given a change in final demand from period t to $t + 1$, the overall change in final demand is $\Delta Y = Y_{t+1} - Y_t$, which can be expressed in terms of energy and nonenergy inputs as

$$\Delta Y = \Delta Y^e + \Delta Y^{ne}.$$

Using these distinct changes in final demand, the overall change in output for all sectors of the economy can be broken down into that caused by changes in final demand for energy and that caused by changes in final demand for nonenergy products:

$$\Delta X = (I - A)^{-1} \Delta Y^e + (I - A)^{-1} \Delta Y^{ne}.$$

6.1.1 Structural Decomposition of Final Demand: Numerical Example

Using the two-sector dollar-based model, suppose that final demand for coal rises from $100 to $120, while final demand for steel rises from $200 to $240.

The overall change in final demand is

$$\Delta Y = \begin{bmatrix} 20 \\ 40 \end{bmatrix}.$$

Based on the standard output identity, the overall changes in coal and steel outputs are $76 and $70, respectively.

$$\begin{bmatrix} \Delta X_1 \\ \Delta X_2 \end{bmatrix} = (I - A)^{-1} \Delta Y = \begin{bmatrix} 1.6 & 1.1 \\ 0.8 & 1.8 \end{bmatrix} \begin{bmatrix} 20 \\ 40 \end{bmatrix} = \begin{bmatrix} 76 \\ 88 \end{bmatrix}.$$

In terms of energy and nonenergy effects, final demand change is decomposed as

$$\Delta Y = \Delta Y^e + \Delta Y^{ne} = \begin{bmatrix} 20 \\ 0 \end{bmatrix} + \begin{bmatrix} 0 \\ 40 \end{bmatrix}.$$

The estimated changes in both coal and steel output caused solely by the change in final demand for the energy product coal are $32 and $16, respectively:

$$\begin{bmatrix} \Delta X_1^e \\ \Delta X_2^e \end{bmatrix} = (I - A)^{-1} \Delta Y^e$$

$$= \begin{bmatrix} 1.6 & 1.1 \\ 0.8 & 1.8 \end{bmatrix} \begin{bmatrix} 20 \\ 0 \end{bmatrix} = \begin{bmatrix} 32 \\ 16 \end{bmatrix}.$$

The changes in output for both coal and steel due to the change in final demand for the nonenergy product steel equal $44 and $72, respectively:

$$\begin{bmatrix} \Delta X_1^{ne} \\ \Delta X_2^{ne} \end{bmatrix} = (I - A)^{-1} \Delta Y^{ne}$$

$$= \begin{bmatrix} 1.6 & 1.1 \\ 0.8 & 1.8 \end{bmatrix} \begin{bmatrix} 0 \\ 40 \end{bmatrix} = \begin{bmatrix} 44 \\ 72 \end{bmatrix}.$$

Aside from breaking final demand into mutually exclusive groups of rows that include changes for particular sectors in the economy (energy, mining, manufacturing, etc.), it is also possible to consider the effects of changes in the separate categories of final demand—consumption, investment, government spending, and net exports. In addition, it is possible to isolate the effects of changes in the mix and level of final demand on the level of output.

6.2 Decomposing Changes in the Inverse

Changes in the Leontief inverse are brought about through changes in the elements of the technological matrix, A—that is, the a_{ij}'s. One approach to separating and measuring the effects of such change begins with the definition of inverse change and leads to an additive set of partial inverse-change effects, whose sum will equal the true change in the inverse, when changes in technological coefficients are infinitesimal. To understand this approach to inverse decomposition, let $\gamma = (I - A)^{-1}$. It follows that

$$\gamma(I - A) = I \text{ or } \gamma - \gamma A = I.$$

Through the product rule of calculus, the overall change in the inverse is

$$\Delta\gamma - \Delta\gamma A - \gamma \Delta A = 0$$

or

$$\Delta\gamma = \gamma \Delta A \gamma. \quad (13)$$

From Eq. (13), the change in the inverse over two time periods is equal to the change in the structural matrix, pre- and postmultiplied by the base-period inverse. Using this equation, changes in the inverse induced by changes in individual technological coefficients or groups of coefficients can be evaluated. Possible groupings include input coefficients for each sector (the columns of the technological matrix) or the use of particular inputs by all sectors of the economy (the rows of the technological matrix). Using the two-sector model in which coal energy requirements are shown in row 1 and requirements for the nonenergy product steel are shown in row 2, the overall change in the technological matrix can be written

$$\Delta A = \Delta A^e + \Delta A^{ne}$$

$$= \begin{bmatrix} \Delta a_{11} & \Delta a_{12} \\ 0 & 0 \end{bmatrix} + \begin{bmatrix} 0 & 0 \\ \Delta a_{21} & \Delta a_{22} \end{bmatrix}.$$

Using this definition of the change in the structural matrix in Eq. (13) yields the distinct effects of inverse change brought about by changes in energy and nonenergy input requirements:

$$\Delta\gamma = \gamma(\Delta A^e + \Delta A^{ne})\gamma = \gamma(\Delta A^e)\gamma + \gamma(\Delta A^{ne})\gamma.$$

Postmultiplication by the final demand vector results in measures of the separate effects on total output brought on by the changes in the energy and nonenergy inputs.

6.2.1 Structural Decomposition of the Inverse: Numerical Example

With sector 1 producing coal and sector 2 producing steel, suppose the technological matrix and inverse in period t are

$$A_t = \begin{bmatrix} 0.10 & 0.55 \\ 0.40 & 0.20 \end{bmatrix} \text{ and }$$

$$\gamma = (I - A_t)^{-1} = \begin{bmatrix} 1.6 & 1.1 \\ 0.8 & 1.8 \end{bmatrix}.$$

In period $t+1$, suppose the technological coefficients change such that

$$A_{t+1} = \begin{bmatrix} 0.12 & 0.60 \\ 0.48 & 0.24 \end{bmatrix}.$$

The overall change in the technological matrix and the distinct effects of changes in coal and steel input requirements are

$$\Delta A = \begin{bmatrix} 0.02 & 0.05 \\ 0.08 & 0.04 \end{bmatrix}$$

$$= \begin{bmatrix} 0.02 & 0.05 \\ 0.00 & 0.00 \end{bmatrix} + \begin{bmatrix} 0.00 & 0.00 \\ 0.08 & 0.04 \end{bmatrix}.$$

Given a $100 final demand for coal and a $200 final demand for steel, an estimate of the change in output for both the coal and steel sectors caused solely by the changes in coal energy requirements is

$$\begin{bmatrix} \Delta X_1^e \\ \Delta X_2^e \end{bmatrix} = (\gamma \Delta A^e \gamma) Y$$

$$= \left(\begin{bmatrix} 1.6 & 1.1 \\ 0.8 & 1.8 \end{bmatrix} \begin{bmatrix} 0.02 & 0.05 \\ 0.00 & 0.00 \end{bmatrix} \begin{bmatrix} 1.6 & 1.1 \\ 0.8 & 1.8 \end{bmatrix} \right) \begin{bmatrix} 100 \\ 200 \end{bmatrix}$$

$$= \begin{bmatrix} 47.36 \\ 23.68 \end{bmatrix}.$$

Similarly, the estimated change in outputs caused solely by changes in technological coefficients associated with the use of the nonenergy-input steel is

$$\begin{bmatrix} \Delta X_1^{ne} \\ \Delta X_2^{ne} \end{bmatrix} = (\gamma \Delta A^{ne} \gamma) Y$$

$$= \left(\begin{bmatrix} 1.6 & 1.1 \\ 0.8 & 1.8 \end{bmatrix} \begin{bmatrix} 0.00 & 0.00 \\ 0.08 & 0.04 \end{bmatrix} \begin{bmatrix} 1.6 & 1.1 \\ 0.8 & 1.8 \end{bmatrix} \right) \begin{bmatrix} 100 \\ 200 \end{bmatrix}$$

$$= \begin{bmatrix} 52.80 \\ 86.40 \end{bmatrix}.$$

6.2.2 Fields of Influence and Inverse Important Coefficients

The process of inverse decomposition can be used to illustrate the concept of fields of influence and inverse important coefficients. Fields of influence refer to the degree and range of inverse coefficient change in response to a change in a single technological coefficient. Inverse important coefficients are those technological coefficients that are associated with the strongest fields of influence. They play the largest role in bringing about changes in output through their effect on the inverse. Knowing which coefficients are inverse important can be useful in identifying where technological improvements will have their greatest impact and in updating key coefficients to achieve the greatest level of accuracy when complete data for an input-output table in a new time period are unavailable.

6.3 Structural Decomposition: Problems in the Discrete Time Framework

The measurement of structural changes caused by inverse and final demand change is complicated in real-world applications where variations in technological coefficients and final demand occur simultaneously and in discrete rather than infinitesimal increments. In this environment, the change in total output is

$$\Delta X = \Delta(I-A)^{-1}Y + (I-A)^{-1}\Delta Y + \Delta(I-A)^{-1}\Delta Y.$$

The primary sources of output change caused by inverse and final demand change are still present, but there is an added effect. The final term to the right of the equality, $\Delta(I-A)^{-1}\Delta Y$, is an interaction term that involves simultaneous changes in both the inverse matrix and final demand vector. Its presence makes it difficult to isolate the effects of pure changes in the inverse and final demand.

In empirical applications, interaction terms are sometimes ignored, sometimes split in equal proportions and assigned to primary effects, and sometimes reported as separate effects along with the primary sources of change. While there is no generally agreed upon solution to the problem, one internally consistent means of separating interaction effects involves the use of weights related to the magnitudes of primary sources of change.

7. CONCLUSIONS

As this short overview shows, the input-output model provides a simple and convenient way to portray transactions among the various sectors of the economy. At any point in time, the model provides a snapshot of the inputs required by specific industries and the requirements of all industries for specific goods and services. Because these transactions are considered simultaneously, through the calculation of inverse coefficients, the model makes it possible to show how production in one sector of the economy affects production in all others. This overview barely scratches the surface of the wide range of input-output applications. It is an especially important tool in regional analysis, where the mutual interdependence of multiregional economic activity is investigated. The model has been used to study the interrelationship between the economy and the environment, where, for example, pollution output from industrial activity becomes an input to various ecological processes. Aside from these applications, there is also continuing interest in improving its accuracy. Because input-output tables are published infrequently, research continues to focus on the stability of coefficients through time, as well as on finding ways to create accurate updates to tables when complete data for new time periods are unavailable. Similarly, since benchmark tables are not available for all regions, there is ongoing interest in developing accurate regionalized versions of the national tables.

SEE ALSO THE FOLLOWING ARTICLES

Complex Systems and Energy • *Depletion and Valuation of Energy Resources* • *Economics of Energy Demand* • *Economics of Energy Supply* • *Energy Futures and Options* • *Goods and Services: Energy Costs* • *Modeling Energy Supply and Demand: A Comparison of Approaches* • *Multicriteria Analysis of Energy*

Further Reading

Bullard, C., and Herendeen, R. (1975). The energy cost of goods and services. *Energy Policy* **1**, 268–277.

Carter, A. P. (1970). "Structural Change in the American Economy." Harvard University Press, Cambridge, MA.

Casler, S. D. (2001). Interaction terms and structural decomposition: An application to the defense cost of oil. *In* "Input-Output Analysis: Frontiers and Extensions" (M. Lahr and E. Dietzenbacher, Eds.), pp. 143–160. Palgrave, New York.

Hannon, B., Blazeck, T., Kennedy, D., and Illyes, R. (1984). A comparison of energy intensities: 1963, 1967, and 1972. *Resources and Energy* **5**(1), 83–102.

Leontief, W. W. (1951). "The Structure of the American Economy." Oxford University Press, New York.

Miernyk, W. H. (1965). "The Elements of Input-Output Analysis." Random House, New York.

Miller, R. E., and Blair, P. D. (1985). "Input-Output Analysis: Foundations and Extensions." Prentice Hall, Englewood Cliffs, NJ.

Rose, A., and Casler, S. D. (1996). Input-output structural decomposition analysis: A critical appraisal. *Economic Systems Research* **8**(1), 33–62.

Rose, A., and Chen, C. Y. (1991). Sources of change in energy use in the U.S. Economy: A structural decomposition analysis. *Resources and Energy* **13**, 1–21.

Rose, A., and Miernyk, W. (1989). Input-output analysis: The first fifty years. *Economic Systems Research* **1**, 229–271.

Sonis, M., and Hewings, G. J. D. (1989). Error and sensitivity input-ouput analysis: A new approach. *In* "Frontiers of Input-Output Analysis" (A. Rose, K. Polenske, and R. Miller, Eds.), pp. 244–262. Oxford University Press, New York.

U.S. Department of Commerce, Bureau of Economic Analysis (1997). Benchmark Input-Output Accounts for the U.S. Economy, 1992. *In* "The Survey of Current Business," December.

Integration of Motor Vehicle and Distributed Energy Systems

TIMOTHY E. LIPMAN
University of California, Berkeley and Davis
Berkeley and Davis, California, United States

1. Introduction
2. Distributed Power Generation and Distributed Energy Resources: An Emerging Paradigm in Power Generation
3. Electric-Drive Vehicles: Battery, Hybrid, and Fuel Cell Vehicles to Improve Efficiency and Reduce Emissions
4. Vehicle-to-Grid (V2G) Power: Using Motor Vehicles to Complement Utility Grids
5. EVs for Remote and Backup Power
6. EVs for Power Generation and Ancillary Services
7. Hydrogen Energy Stations and Other Motor Vehicle and Distributed Generation Integration Schemes
8. Key Technical and Regulatory Issues for EVs Used as Distributed Generation
9. Conclusions

Glossary

battery electric vehicle An electric-drive vehicle that derives the power for its drive motor(s) from a battery pack.

distributed energy resources Power generation, storage, and metering/control systems that allow power to be used and managed in a distributed and small-scale manner, thereby siting generation close to load in order to minimize electricity transmission and maximize waste heat utilization.

fuel cell vehicle An electric-drive vehicle that derives the power for its drive motor(s) from a fuel cell system; hybrid fuel cell vehicles would also derive drive motor power from a supplemental battery or ultracapacitor.

hybrid electric vehicle An electric-drive vehicle that derives part of its propulsion power from an internal-combustion engine and part of its propulsion power from an electric motor, or that uses an internal combustion engine to power a generator to charge a battery that in turn powers one or more electrical drive motors.

vehicle-to-grid (V2G) power Electrical power that is produced or absorbed by a motor vehicle to provide real power, reactive power, or electrical grid ancillary services to local electrical loads and/or utility grids.

A new generation of advanced, small-scale power technologies is rapidly emerging based on the concept of distributed generation (DG) and, more broadly, distributed energy resources (DER). DER systems have the potential to allow for decentralized electricity production with reduced needs for transmission and distribution of power, thereby increasing the efficiency of end-user energy use, particularly when coupled with the utilization of waste heat for local heating or cooling needs (known as combined heat and power [CHP] or "cogeneration"). At the same time, DER can help to reduce the need for siting large power plants and transmission lines. Furthermore, DER systems can also offer environmental benefits through installation of clean technologies such as photovoltaic (PV), wind, and fuel cell systems, and by replacing or displacing the construction of relatively dirty peaker power plants.

1. INTRODUCTION

The motor vehicle industry is undergoing dramatic change with regard to the vehicle technologies that are being introduced and those that are being researched and demonstrated. After years of experimentation with different types of alternative-fuel vehicles (AFVs), the automobile industry is now commercializing hybrid electric vehicles (HEVs) with hopes that fuel cell vehicles (FCVs), now in a precommercial prototype development phase, can be commercialized by the second decade of the 21st century. Toyota Motor Company has led the way in HEV development, based on the successful launch of the Prius hybrid vehicle in the United States and Japan and the hybrid Estima minivan in Japan. Toyota has now sold more than 100,000 HEVs globally, and Honda has also been successful with the hybrid Civic, introduced in 2002, after limited success with the small Insight hybrid.

These electric-drive vehicles (EVs)—including battery electric vehicles (BEVs) as well as HEVs and FCVs—could have a variety of interactions with the stationary power sector and, with distributed energy resources (DERs) in particular, in ways that could provide value and reduce the costs of vehicle ownership. The only conventional connection would be recharging BEVs as simple load for the electric utilities (or DER systems), or similarly recharging plug-in HEVs or refueling hydrogen FCVs with hydrogen made from water using electricity to operate an electrolyzer. Of these options, recharging BEVs is occurring and will continue, although these vehicles have met with limited market success where they have been introduced. It is unclear if plug-in HEVs will be commercialized, as they have the advantage of being able to operate in a zero-tailpipe emission mode but will tend to be more expensive than other types of HEVs. Finally, refueling FCVs with grid-power electrolysis is not very efficient or cost-effective, and will likely only be done in the early years of vehicle introduction until better hydrogen refueling arrangements for FCVs are made.

However, all of these types of EVs could interact with utility grids, especially DER systems, in far more complex and interesting ways than simply as consumers of electricity. First, vehicles equipped with electrical power generating or storage capacity have the ability to supply needed ancillary services to utility grids. These services include grid voltage support, frequency regulation, spinning reserves to provide adequate power reserve margins, volt-amperes reactive (VARs) for power factor correction, and emergency backup power, among others. Furthermore, with regard to the ability to store or produce power, these vehicle types can act as buffer systems for the utility grid, taking up power off-peak when it is inexpensive (as many BEVs already do) and then delivering it on-peak when it is often much more highly valued. This could prevent the need to construct additional peaker power plants that are only needed during times of peak demand and that tend to be less tightly controlled for emissions than constantly-operating baseload plants. One need only consider the fact that an electrified vehicle fleet in the United States would have approximately 14 times as much power generating capacity as all of the stationary power plants in the country to realize that there is an enormous amount of power embodied in our fleets of motor vehicles!

This article reviews the potential linkages and synergies between emerging distributed power systems and EVs, particularly with regard to the potential for vehicle-to-grid (V2G) power from various vehicle types. The chapter is organized as follows. First, DG and DER systems are briefly described, along with a discussion of their advantages relative to other options. Second, electric-drive vehicle introduction is also briefly summarized. Third, various applications in which DER systems and EVs can potentially interact are discussed. Finally, emerging concepts and needs for future research are summarized, along with key points from the previous sections.

2. DISTRIBUTED POWER GENERATION AND DISTRIBUTED ENERGY RESOURCES: AN EMERGING PARADIGM IN POWER GENERATION

While DG is generally considered to include only power generation technologies—such as fuel cells, microturbines, small generator sets, PV systems, and so on—DER also includes energy storage and demand-side technologies and systems (for example, smart metering and thermostats, used in conjunction with real-time electricity pricing). These definitions vary, as does the upper-end definition of DG size above which power generation is considered central and not distributed—this can be 1 megawatt (MW), 10 MW, or even 100 MW. Regardless of the exact definition, however, the possibility of DER adoption on a significant scale offers interesting possibilities for improving the environmental quality of energy production while at the same time potentially working hand-in-hand with electricity market changes such as real-time pricing of electricity (to which DER could respond by turning on and off or ramping up or down) to complement the needs of the utility grid.

Table I shows some of the emerging DER technologies that are being developed and deployed, along with a few important characteristics. As the costs of these technologies fall through better design and economies of scale, and enabled by constraints on powerplant and transmission line siting, they are

TABLE I

Example Types of DER Technologies and Systems

System type	Primary use	Likely fuel/energy input	Commercialization status
Absorption cooling	CHP[a]	Waste heat	Early commercial
Energy storage systems	Energy storage	Electrical energy	Commercial
Fuel cells	Power generation	Natural gas, landfill gas, biomass, wind, others	Early commercial
Industrial turbines	Power generation	Natural gas	Commercial
Microturbines	Power generation	Natural gas, landfill gas, biomass, others	Commercial
Photovoltaics (PV)	Power generation	Solar radiation	Commercial
Real-time pricing systems	Load curtailment	Minimal electrical energy	Demonstration
Reciprocating engines	Power generation	Natural gas, landfill gas, biomass, others	Commercial
Waste heat recovery	CHP	Waste heat	Commercial
Wind turbines	Power generation	Wind energy	Commercial

[a]CHP is combined heat and power.

expected to compete more and more favorably with the traditional option of central power generation.

With regard to the challenge of developing and deploying DER systems, it seems clear that (1) DER penetration into electricity/natural gas markets is not only revolutionary technically but also controversial politically due to its potential economic impacts and (2) in order to meet growing energy needs while considering the environmental consequences of power generation, growth in the use of renewable energy, CHP, and other very clean energy production will be very important to meeting air quality and climate stabilization goals. With regard to this second point, in 1998 the former U.S. secretary of energy developed a goal for the United States, which is still official Department of Energy policy, of doubling the contribution of CHP to the nation's power supply by 2010 (46 gigawatts [GW] in 1998 to 92 GW). Furthermore, in California, the state assembly approved a renewable portfolio standard (RPS) that required 20% of California's electricity to be renewably generated by 2017, and the New York legislature passed a 25% RPS standard, following on the success of other RPS standards in Texas and other states. Programs such as these help to advance the renewables-based advanced power-generation programs in these states and also provide a general boost for DER technologies and efforts to work on grid interconnection issues.

The overall vision for the future that some electric industry practitioners and observers share—of a network of DER, wireless, and other information technologies that become integrated into the electricity system to facilitate real-time pricing of electricity and a significant amount of local power generation for local use—will surely take many years to become a reality. However, various states in the United States and countries around the world are well on their way toward pursuing this vision by developing the underlying technologies and complete systems and working to also develop the needed utility interconnection standards, address market and rate design issues, and redesign administrative protocols.

These efforts can be contentious with many different stakeholder groups involved, and they are also intimately connected with other broader issues that are also often controversial such as the overall concept of utility deregulation. The various roles played by state legislatures, public utility commissions, the Federal Electricity Regulatory Commission, electrical grid operators such as the California Independent System Operator, governors' offices, and interest groups are complicated and overlapping to some extent, and even making clear which groups have authority in certain areas has been far from straightforward. Nevertheless, there is growing interest in DER systems among a variety of industrial, commercial, governmental, and nongovernmental entities, and continued development and deployment of these systems will occur as the regulatory and administrative issues are slowly resolved.

While the politics involved make it difficult to predict the timing or exact nature of the manner in which DER systems will ultimately be integrated into existing utility grids (greenfield-type developments being a somewhat different matter), it is clear that the prospects of DER adoption also open many interesting side possibilities. These include several involving transportation sector linkages, such as (1) V2G power and grid ancillary services; (2) artful integration of DER in conjunction with such concepts as hydrogen energy stations (H$_2$E stations) that would co-produce electricity for local loads or the utility grid along with high-purity hydrogen to refuel fuel cell vehicles (FCVs), and (3) combination with information technologies such real-time pricing of electricity and intelligent transportation systems.

DER can also potentially interface with many other sectors, including the obvious potential benefits to end-user groups (commercial, industrial, ranches/farms, etc.), such as maritime, recreational, and military applications. Furthermore, technological innovation related to DER development for markets in developed nations may ultimately spill over into village power and remote/rural applications throughout the developing world, just as cell phone technology obviated the need for telephone cables and wires, thereby potentially affecting the lives of billions of people in the future.

3. ELECTRIC-DRIVE VEHICLES: BATTERY, HYBRID, AND FUEL CELL VEHICLES TO IMPROVE EFFICIENCY AND REDUCE EMISSIONS

Within the automotive industry, the development of electric-drive vehicles is about as revolutionary as the concept of DER relative to traditional power generation. After nearly 100 years of dominance by steel-bodied, gasoline-powered, internal combustion engine vehicles, we now are seeing introduction of HEVs and much research and development for even more advanced vehicles such as FCVs that would operate on hydrogen, or possibly methanol or other fuel. Toyota and Honda have already commercialized HEVs in the United States and Japan, and General Motors, Ford, and Nissan, among others are planning to follow in the next few years with HEV market introductions of their own.

BEVs, HEVs, and FCVs share many similarities, while being rather distinct in other respects. First, all of these vehicles are likely to incorporate battery systems for energy storage, but for different reasons. For BEVs, batteries are the sole energy storage device for the vehicle and provide all of the power to the electric drive motor (or motors). For HEVs, batteries power the electric motor (or motors), but additional power comes from the combustion engine. The main purpose of the batteries and motor is to allow the combustion engine to run better and to turn off when not needed, as well as to allow braking energy to be recaptured when the vehicle is decelerating as in BEVs (i.e., regenerative braking). For FCVs, regenerative braking is also a rationale for adding batteries, but additional considerations are the need for batteries to help start up the fuel cell system and the prospect of reducing costs by using battery power to reduce the size (and thus cost) of the fuel cell system.

Thus, in addition to battery and regenerative braking systems, these vehicle types all employ an electric motor to add propulsion power to the vehicle. For BEVs and FCVs, one or more electric motors provide all of the vehicle propulsion power. Most types of HEVs would also use a combustion engine to provide power to the wheels, but one type known as a series HEV would simply use the combustion engine to run a generator to help charge the battery and not use the engine to power the wheels.

The electric motors used in these vehicle types are of two different forms: (1) alternating current (AC) induction motors and (2) brushless permanent magnet (BPM) motors. To convert the direct current (DC) produced by the battery or fuel cell system to the proper current (and voltage) needed to operate these types of electric motors, a power inverter is required. These power inverters use insulated gate bipolar transistors (IGBTs) as the high-power switching devices to convert DC power to AC (or a similar synchronous output to operate BPM motors). Since these types of electric vehicles will all necessarily include a power inverter device that is capable of producing a high-quality AC waveform and power of at least 10 kilowatts (kW) and up to 100 kW (for typical BEV, HEV, and FCV designs), it is then a relatively minor matter to design in an interface to allow these vehicles to interact with utility grids so that they can both absorb and discharge real power and VARs, as well as acting as spinning reserves and backup power units.

One rationale within the industry for developing electric vehicles (particularly HEVs), other than the potential environmental and social benefits that they can provide, is that requirements for onboard power in motor vehicles are growing and beginning to stretch the limits of traditional 12-volt automotive electrical power systems. New motor vehicles are requiring this additional power to operate new types of accessories, such as vehicle navigation and auto alert safety systems; onboard televisions and DVD players, electrically actuated and controlled engine, exhaust, and auxiliary systems; and additional electric motors to control seats and window blinds. Other electrical gadgetry soon to be incorporated into some vehicle models includes electronic tire pressure monitoring/control, bumper-mounted side-view and rear-view cameras, and 120-volt AC accessory outlets. To meet these growing needs for onboard electrical power, a new 42-volt standard is being incorporated into future automobile designs, and this can easily be accommodated by BEVs, HEVs, and FCVs by using a DC-to-DC converter to step the full system voltage (typically 150 to 500 volts) down to the 42-volt level needed for the auxiliary systems.

4. VEHICLE-TO-GRID (V2G) POWER: USING MOTOR VEHICLES TO COMPLEMENT UTILITY GRIDS

As noted previously, EVs of all types can interact with utility grids in much more complex ways than simply as load for electric utilities. With the partial deregulation of the utility industry in several U.S. states, opportunities have arisen for the various sections of the electricity supply value chain to be supplied by different groups in competition with one another (for generation, operation of the grid, billing and customer services, etc.) rather than by a single vertically integrated utility as has historically been the case in most places. Within this context, of a competitive market for a range of services related to the provision of electricity, EVs can play interesting roles, either similar to or actually in conjunction with those of DG and DER systems.

There are several potential possibilities for using EVs as distributed generating resources. Various types of EVs could be used to do the following:

- Produce power to meet the demands of local loads
- Provide additional power to the grid in a net-metered or electricity buy-back scenario, helping to meet demands in times of capacity constraint
- Provide emergency backup power to residences, offices, hospitals, and municipal facilities
- Provide peak shaving for commercial sites, simultaneously reducing electricity energy charges ($/kWh) and demand charges ($/kW-peak per month)
- Provide ancillary services to the grid, such as spinning reserves, voltage and power quality support, and VARs
- Provide buffering and additional power for grid-independent systems that rely on intermittent renewable energy systems, or for microgrids that can benefit from additional generation or storage capacity

Although only some of these possibilities have been explored, research has shown that EVs can be used to provide a range of important services to utility electrical grids. In fact, the potential for producing electrical power from vehicles is enormous—the generating capacity of an electrified U.S. motor vehicle fleet would be approximately 14 times the entire capacity of all of the stationary power plants in the country. For example, in California a fleet of 100,000 FCVs could produce about 1.5 GW of power for the grid, assuming 30 kW net fuel cell output power per vehicle and 50% vehicle availability. This 1.5 GW of power represents approximately 5% of California's peak electrical power demand on a typical day. The following sections describe various potential uses of electric vehicles as DER systems, as well as other schemes for integrating motor vehicles into these small-scale energy systems.

5. EVs FOR REMOTE AND BACKUP POWER

One obvious use for EVs in a power generation role is the notion of using the vehicles to produce power for remote uses (such as construction sites, recreational uses such as camping and fishing, etc.). This is similar to the idea of using EVs for backup power in the event of a power loss to a residential or commercial building, except that under the U.S. national electrical code the latter is not permissible (due to concerns about line-worker safety) while the former is allowed. However, with adequate safeguards it should soon be possible to allow EVs and other grid-tied DG systems to be used in an islanded mode with adequate safeguards to provide emergency backup power, with minor changes in existing codes and regulations.

One prototype vehicle has already been built with this remote/backup power capability in mind. The Dodge contractor special is a HEV with the capability of producing approximately 2 kW of electrical power for sustained periods of time.

In fact, an emerging standard that may eventually be incorporated into nearly all future vehicles is based on an electrical system for vehicle auxiliary systems that operates at 42 volts versus the 14-volt standard for conventional systems. This standard has become necessary due to increasing desire to equip vehicles with electronic systems and the need to provide more power for these systems without increasing wire diameter and raising costs. New or planned electronic gadgetry on vehicles include navigation systems, extensive onboard communications, voice-actuated controls, exterior AC power supplies, computer controlled power-assisted active suspension, collision-avoidance systems, electric air conditioner compressors, drive-by-wire steering, side- and rear-view bumper cameras, electronic tire pressure control, and generally greater computer power for increasing control of various vehicle systems. The power demands from these new electrical systems are expected to push the power demands onboard vehicles from several hundred watts to a few thousand or perhaps even several thousand watts in the future.

At the same time that the 42-volt standard can support the addition of electrical devices in conventional vehicles, the increased power levels available from these systems in even conventional combustion-engine vehicles could enable the provision of a significant level of emergency or backup power—perhaps on the order of 3 to 5 kW—although local regulations may prevent backup power from being provided with the vehicle engine on. The duration of the power supply capability would then be limited to a few hours, by the capacity of the 42-volt battery system onboard the vehicle.

6. EVs FOR POWER GENERATION AND ANCILLARY SERVICES

With regard to using EVs to generate power and support utility grids, research has shown that EVs can be used to provide a range of important services in this regard. When not in use, BEVs could be used for emergency backup power or to buffer the utility grid by charging off-peak, when electricity is plentiful, and supplying it back during times of peak demand and capacity constraint. HEVs and FCVs could act as generators, producing electricity from a liquid or gaseous fuel to meet the demands of connected local loads or other nearby power needs. EVs could also provide certain types of support services to utility grids, such as spinning reserves and grid frequency regulation. While contracting for these services would likely require a service aggregator to bid the reserve or regulation function from EVs to local system operators, BEVs in particular appear to be well suited to these uses. For example, Alec Brooks of AC Propulsion has estimated that the frequency regulation of the California grid (a service that costs the California Independent System Operator millions of dollars per year) could be provided by 50,000 to 100,000 EVs, and these vehicles could provide this service as they were recharging their battery packs. AC Propulsion has demonstrated this capability with one of their prototype vehicles, and Fig. 1 shows the scheme by which they envision EVs receiving, by radio signal, the control error in the frequency of the utility grid and then responding as needed.

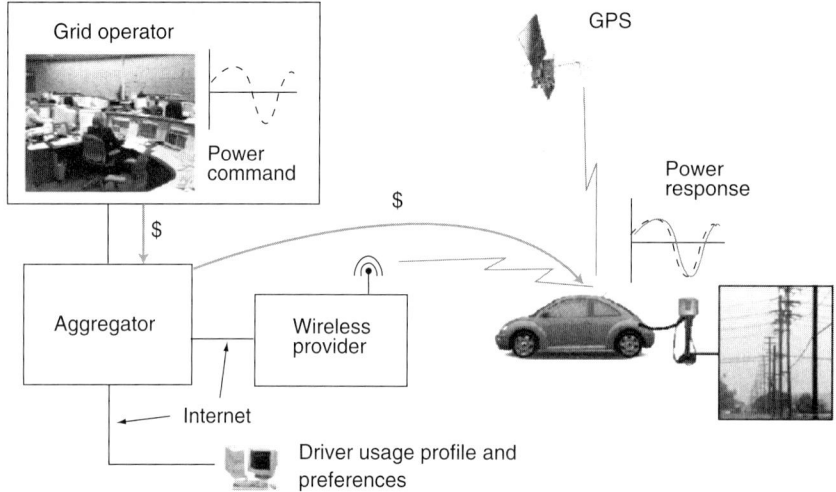

FIGURE 1 One concept of V2G Operation. BEVs acting as grid frequency regulation service providers. Data from Brooks (2002).

As noted earlier, the gross power generating capability of vehicles is enormous, and vehicles with significant onboard electrical generating capability allow that power to be generated efficiently for various uses, many of which have economic value that can translate into reduced costs of ownership for these new vehicle types. The most comprehensive published analysis of the potential for EVs and FCVs to act as DG resources was conducted by Kempton *et al.* This report was prepared for the California Air Resources Board by researchers at the University of Delaware, Green Mountain College, and the University of California. The analysis also involved collaboration with EV drivetrain manufacturer AC Propulsion, whose Generation 2 electric motor controller unit allows for bidirectional grid interface that can support V2G connections. The Kempton *et al.* effort examined the economic potential of using various types of EVs to produce power for buildings and the grid, as well as to provide grid ancillary services such as spinning reserves, nonspinning reserves, and grid frequency regulation. Table II presents some of the key findings of the Kempton *et al.* report, in terms of the range of annual values that might be expected for different EV types and for three different power-generation or ancillary services.

The particular case of using FCVs to produce power has drawn particular attention due to the potential for FCVs to become commercialized as mass replacements for conventional vehicles during the 2008 to 2012 timeframe. The significant power generation capability of these vehicles—from perhaps 20 kW for small hybrid FCVs (with additional battery power available) to 100 kW or more from larger FCVs with minimal battery or ultracapacitor— suggests that FCVs have the potential to become significant power generation resources in the future.

In the study discussed earlier, Kempton *et al.* concluded that FCVs could compete in the peak power market, with generating costs on the order of $0.18/kWh, but they could not compete with base-load power. Also, an earlier analysis of the potential for FCVs to provide power for buildings was conducted by Kissock. This analysis assumed that FCVs equipped with 25 kW nominal (37.5 kW peak) proton exchange membrane (PEM) fuel cell systems would be used to produce electricity, or electricity and heating/cooling in a cogeneration mode, for New Jersey and Texas residences, for a New Jersey hospital, and for a Texas office building. The analysis concluded that annual savings of up to $2500 per FCV docking station could be realized with the residential setting (with some cases showing negligible benefits), that annual savings of $1200 to $8800 were possible for each docking station at the hospital, and that annual savings of $2300 to $2900 were possible for the office building cases.

In a somewhat more detailed analysis, Lipman *et al.* have shown that while the costs of producing power from stationary and motor vehicle fuel cell systems are rather different in terms of the costs involved—for stationary systems much of the cost is capital cost, while for FCVs much of the cost is for periodically refurbishing the vehicle's fuel cell stack—the economics of producing power from these systems can be comparable, and potentially competitive with grid power in the future, if the systems are operated with similar overall efficiencies. However, results are highly sensitive to the compounding effects of variations in several key input variables such as natural gas and electricity prices, and fuel cell system costs, durability

TABLE II

Vehicle Owner's Annual Net Profit from V2G

	Peak power	Spinning reserves	Regulation services
Battery, full function	$267	$720	$3162
	(510–243)	(775–55)	(4479–1317)
Battery, city car	$75	$311	$2573
	(230–155)	(349–38)	(4479–1906)
Fuel cell, on-board H$_2$	$-50 (loss) to $1226	$2430 to $2685	$-2984 (loss) to $811
	(2200–974 to 2250)	(3342–657 to 912)	(2567–1756 to 5551)
Hybrid, gasoline	$322	$1581	$-759 (loss)
	(1500–1178)	(2279–698)	(2567–3326)

The figures represent $net and (revenue-cost). These are representative midrange figures extracted from full analysis in the report.
Source. Kempton *et al.* (2001).

levels, and fuel cell stack refurbishment costs. Where FCVs are used to produce power to meet very low load levels, such as at individual residences (e.g., and operate at very high turndown ratios), their efficiencies suffer and they become much less economically attractive at producing power. This suggests that net-metering policies—which would allow FCVs to be operated at higher power and efficiency levels and for excess power to be supplied to the grid for a credit—may be critical to improving the economics of FCVs used to produce power in this way.

Figures 2 and 3 present results from the Lipman *et al.* analysis, comparing the costs of producing power from FCVs and stationary PEM fuel cells at both California office building and residential locations, and with three different sets of economic assumptions. These results show that the costs of power production at residences from FCVs can be highly variable, but that the economics of producing power from FCVs at commercial locations are better due to lower commercial natural gas prices, better fuel cell system efficiencies in meeting higher loads, excess FCV power availability during the day when it is highly valued, and other factors.

7. HYDROGEN ENERGY STATIONS AND OTHER MOTOR VEHICLE AND DISTRIBUTED GENERATION INTEGRATION SCHEMES

In addition to using various types of vehicles with electricity-generation capability for standalone, backup, and power/ancillary services provision, there also are several other ways in which vehicles can interact with distributed energy systems. These includes opportunities for fueling motor vehicles in ways that are complementary to the development of distributed energy systems, and additional opportunities for EVs to act in conjunction with emerging schemes and plans for electrical power generation.

One interesting option relates to the need for developing hydrogen-refueling infrastructure for early FCVs and is the concept of the hydrogen energy station (or H_2E station). These H_2E stations would be either dedicated refueling facilities, or a key component of the energy production, use, and management portion of a commercial or industrial facility. The energy station component would consist of a natural gas reformer or other hydrogen generating appliance, a stationary fuel cell integrated into the building with the potential capability for combined heat and power (CHP) production, and a hydrogen compression, storage, and dispensing facility. In essence, H_2E stations would seek to capture synergies between producing hydrogen for a stationary fuel cell electricity generator that provides part or all of the power for the local building load (as well as the capability to supply excess electricity to the grid) and refueling FCVs with additional high-purity hydrogen that is produced through the same hydrogen-generation system.

In principle, many different H_2E station concepts and designs are possible, including the following:

- Service station–type designs that are primarily intended to produce hydrogen for FCV refueling

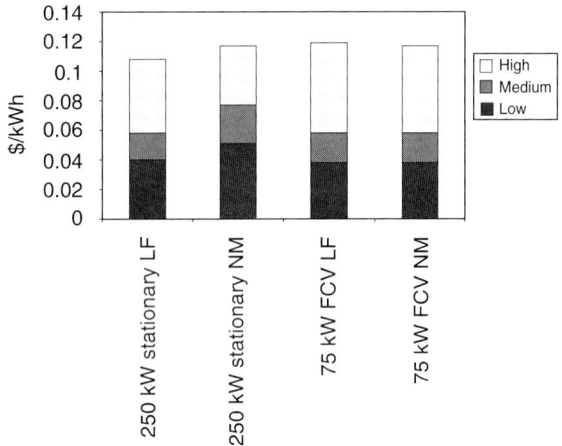

FIGURE 2 FCVs used as DG versus stationary PEM fuel cells at medium California office building setting. LF, load following; NM, net metering. Data from Lipman *et al.* (2002b).

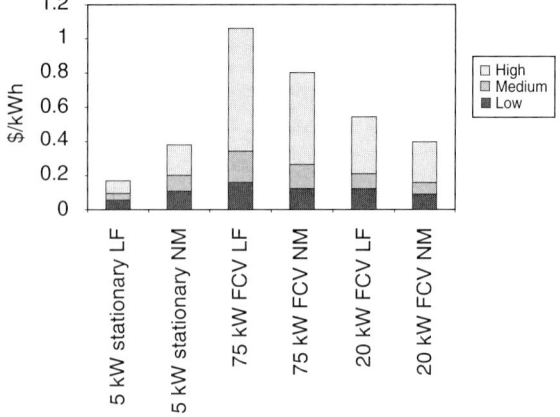

FIGURE 3 FCVs used as DG versus stationary PEM fuel cells at California residential setting. LF, load following; NM, net metering. Data from Lipman *et al.* (2002b).

but that also supply electricity to the service station
- Office building–based designs that primarily provide electricity and waste heat to the building but also include a small off-shoot for FCV refueling
- Distributed generation facilities that are primarily intended to supply excess electricity to the power grid but that also include some provision for FCV refueling

In addition, FCVs parked near the H_2E station for any sizable length of time could in principle supply electricity to the building or grid, since they would have access to a fuel supply.

Analysis of the economics of several potential H_2E station designs shows that for the California setting examined, where prevailing commercial and residential electricity rates are high, H_2E stations appear to be relatively attractive ways of supplying small numbers of FCVs with hydrogen. In general, H_2E stations tend to be more attractive economically than dedicated hydrogen refueling stations, especially for low numbers of vehicles supported per day, but the economics depend importantly on several variables including natural gas and electricity prices, captial equipment costs, the hydrogen sales price, and fuel cell maintenance and stack refurbishment costs. Figure 4 shows that for this set of the service station cases examined the H_2E station designs are more attractive economically than dedicated hydrogen refueling stations. None of the dedicated or H_22E stations are economically profitable with so few vehicles supported, but the H_2E stations offer a lower cost pathway to ultimately supporting greater numbers of vehicles profitably.

Figure 5 shows how after a transitional period with low numbers of vehicles supported, H_2E stations could begin to turn a profit with greater numbers of vehicles refueled per day and with reasonable hydrogen selling prices of around $20 per GJ. This figure shows that a 10% simple ROI target can be met with the service station type of H_2E station, but only with relatively high hydrogen sales prices of about $20 per GJ and only with about 50 or more vehicles per day refueled. At lower hydrogen sales prices of $10–15/GJ, the economics of this type of station do not look attractive, even with significant numbers of vehicles refueled. However, with lower natural gas prices than the $6/GJ assumed in this future high cost case, this picture would change somewhat, with hydrogen sales prices of $15/GJ potentially becoming profitable for large numbers of vehicles per day supported.

Additional options for integrating motor vehicles with distributed energy systems include incorporating EVs into designs for microgrids. Microgrids are small clusters of DER technologies that act together to supply power and heating/cooling to one or more adjacent buildings and that would connect with the

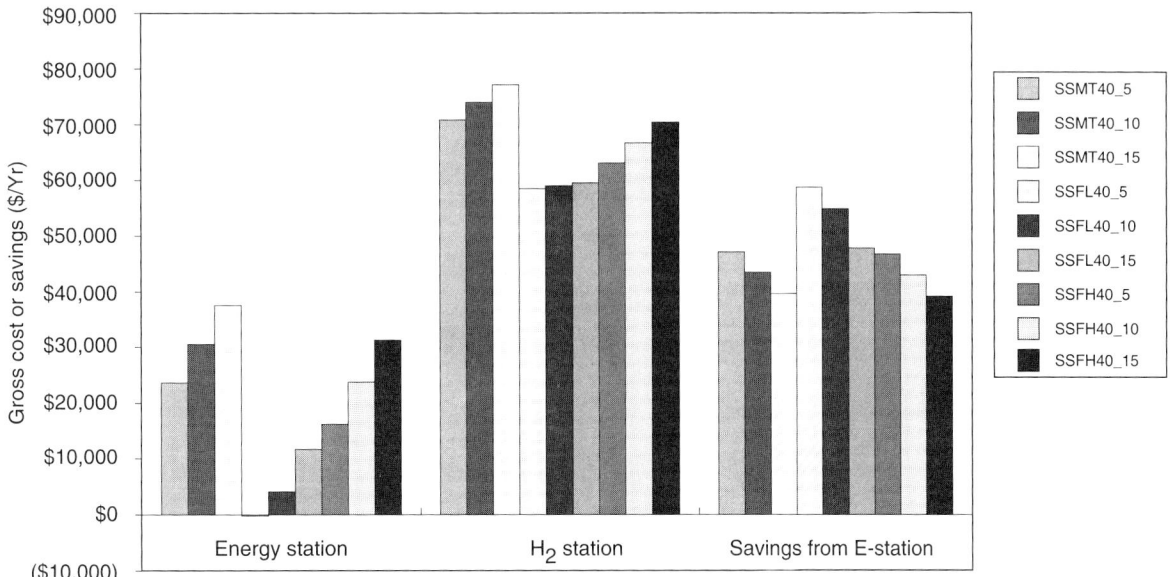

FIGURE 4 Estimated future annual costs of H_2E stations with 25 and 40 kW fuel cell and 5 to 15 FCVs refueled per day, compared with costs of dedicated H_2 stations (H_2 price of $10/GJ). FL, future low cost case; FH, future high cost case; MT, medium-term case; SS, service station; X_Y, fuel cell size in kW_# of FCVs per day refueled. Data from Lipman *et al.* (2002a).

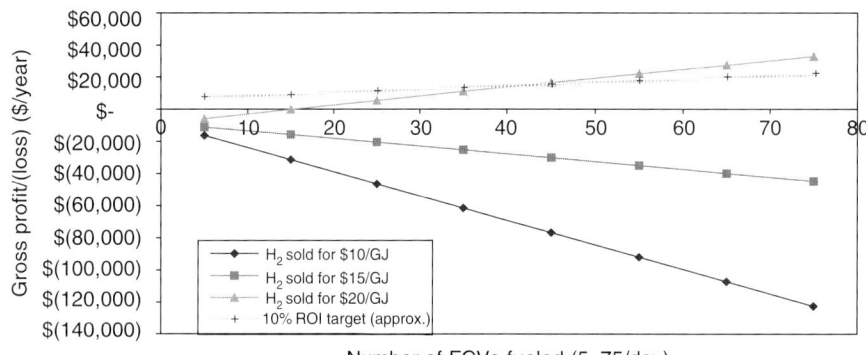

FIGURE 5 Estimated profit/loss from H$_2$E-service station with 40 kW fuel cell, 5 to 75 FCVs refueled per day, and future high costs (w/approx. 10% ROI target). Data from Lipman *et al.* (2002a).

main grid at one interface point. Microgrids could act as model citizens for the grid by supporting it in various ways, and they could benefit themselves from supplying local generation with grid power at various times (depending on electricity rate schedules and other variables). EVs could play interesting roles in microgrids by supplying electrical storage or power-generation capability, with the availability of the EV resources coinciding with the arrival of people at various commercial and industrial microgrid locations and their associated energy needs.

Similarly, EVs could help to provide additional power and buffering for small standalone renewable energy systems based on PV and wind power, along with a battery storage system. Small renewable systems that are off-grid because of their remote location are becoming increasingly common, but often are of relatively low power to keep costs low. Again, these systems could be bolstered by the use of EVs to provide additional power to operate high-power equipment for short periods of time, thereby allowing the size of the remote renewables system to be minimized.

Finally, another interesting opportunity for combining motor vehicles and distributed energy systems involves efforts to reduce emissions from heavy-duty diesel trucks. Approximately 30% of the fuel use and emissions associated with long-haul diesel trucks is due to the idling that these trucks do when they are stopped, primarily to support electronic devices in the truck cabin (refrigerator, microwave, television, climate control, etc.) by operating a generator with the main diesel engine. This is highly inefficient and polluting, and efforts are under way to examine opportunities for reducing these emissions. One such option would be to equip trucks with fuel cell auxiliary power units (APUs), running on pure hydrogen or possibly diesel fuel that is converted into a hydrogen-rich gas stream with an onboard fuel reformer. If operated on hydrogen, these heavy-duty truck APU systems could be refueled with hydrogen produced at truck stop hydrogen stations, possibly of a type of H$_2$E station design as discussed earlier, where hydrogen is used both for truck APU and FCV refueling, as well as to produce electrical power and heat for the truck stop using a stationary fuel cell. An alternative to using fuel cells as APUs onboard the trucks would simply be to have the trucks plug in to electrical outlets when near truck stops, and the electricity could be supplied either by the main utility grid or by a DG system at the truck stop. This would only reduce idling near truck stops, but would perhaps be easier to implement that the fuel cell option. Also note that each of these types of installations would be good candidate for integration of renewable PV or (if remote sites) wind power to further reduce environmental impacts.

8. KEY TECHNICAL AND REGULATORY ISSUES FOR EVs USED AS DISTRIBUTED GENERATION

Despite the potential benefits of using EVs to produce power and other grid services when the vehicles are parked, there are several key issues confronting the use of EVs in this manner. Some of these issues, and some potential solutions, are as follows.

First, with regard to grid interconnection issues for electric vehicles, reverse flow of electricity from EVs in event of loss of power in the grid (e.g., for emergency backup applications) is not permitted in the United States under the national electrical code. Vehicles could in principle interconnect with utility grids in a manner such that they were supplying net power, assuming they were to meet relevant DG interconnection standards, but this would only be allowed by a few states under existing net metering laws. Also, from a technical perspective, reverse flow of electricity into local distribution systems eventually will reach a limit due to the generally unidirectional character of the existing high-voltage transmission system. Also, utility on/off control of grid-connected EVs may be essential for lineworker safety, requiring a complex control system. Solutions to these issues include revising the national electrical code to allow reverse-flow from vehicles for backup generation, with appropriate safeguards. Public utility commissions and other utility industry groups should consider analysis of retrofitting substations to ensure that transformer tap changers and line-drop compensators are compatible with reverse flow into high voltage transmissions systems, as well as investigating utility-to-vehicle wireless communication technology to provide emergency shut-down, real-time load, and electricity price information for utility control centers and vehicle owners.

Second, with regard to fuel cell operation when using FCVs to generate power, nonhybrid fuel cell systems with 75 to 100 kW peak power (a typical power level for full-sized vehicles) will likely be power limited to 30 or 40 kW for continuous operation while the vehicle is at a standstill due to thermal management issues. Fuel cell systems connected to loads in grid-independent operation (e.g., where they cannot lean on the grid) may be subject to transient demands that would require a hydrogen fuel supply buffer or battery support system for adequate performance. Also, operating fuel cell systems in residential settings and meeting local loads only, which are often as low as 1 kW to 2 kW and rarely exceed 4 kW to 5 kW, will likely produce low fuel cell system operating efficiencies, particularly for stacks designed to operate at high pressure. Potential solutions to these problems are to operate fuel cell systems at 4 kW to 30 kW and sell excess power to grid through net metering or other buyback scenarios. Alternately, industry may develop off-board auxiliary blower air supply systems for fuel cell stacks that allow low pressure, higher efficiency operation at low load levels, or power-on-demand systems where only a part of the vehicle's fuel cell stack is used in a distributed power mode, similar to displacement-on-demand systems that are beginning to be incorporated into light-duty trucks.

Third, with regard to fuel supply for FCVs, vehicles designed to operate on pure hydrogen may not be able to use reformate (a mix of hydrogen and other gases produced from a liquid or gaseous hydrocarbon fuel) due to carbon monoxide and sulfur catalyst poisoning issues. To operate for longer than a few hours by using some of the hydrogen in the vehicle's tank, vehicles will require some sort of hydrogen production support system based on steam methane reforming, electrolysis, partial oxidation reforming, or auto thermal reforming. These issues can be addressed by potentially employing vehicle fuel cell stacks with platinum-ruthenium catalysts that can operate on either neat hydrogen or reformate (e.g., the Ballard Mark 900), or by producing purified hydrogen at small scales. In this regard, there are important research and development requirements for developing low-cost and small-scale hydrogen reformers, and particularly interesting would be the development of multifuel reformers for vehicles that can operate on pipeline natural gas as well as the vehicle's primary fuel, thus eliminating the need for an off-board fuel reformer for V2G power from FCVs.

9. CONCLUSIONS

The emergence of DER solutions for distributed power production and the commercialization of various type of EVs in the motor vehicle industry are in some ways similar, with both representing potential paradigm shifts in their respective sectors. As these evolutionary and perhaps ultimately revolutionary changes occur, whether slowly or quickly, a range of interesting opportunities will emerge for interaction between the two sectors and synergistic development of infrastructure and services. Some of the possibilities for integration of these types of systems include the following:

- Using various types of EVs for emergency backup or remote power
- Using various types of EVs for peak power or grid ancillary services such as grid frequency regulation and spinning reserves
- Developing H_2E stations that would coproduce electricity/heat/cooling for local loads and the grid as well as pure hydrogen to refuel FCVs
- Combining efforts to reduce diesel truck idling emissions with distributed energy systems for truck stops

Among these options and the various possibilities within them, particularly promising appear to be the prospects of using HEVs and FCVs for utility grid spinning reserves, FCVs for electricity peak shaving at commercial buildings, BEVs for utility grid frequency regulation, H_2E stations as a potential way to reduce the transition costs of developing hydrogen infrastructure for FCVs, and opportunities for reducing diesel truck idling through truck stop electrification or hydrogen refueling for fuel cell APUs. Initial investigations have suggested that all of these ideas are promising, but additional details such as how vehicles would bid these services into power markets and be compensated for the services they provide have yet to be fully considered. However, advances in information technologies such as wireless communications and real-time metering suggest that these logistical issues associated with V2G are unlikely to be major obstacles.

In conclusion, the tremendous amount of power available in the motor vehicle fleet and the prospects for electrifying the fleet in the future suggest a potentially large power resource that could be tapped to assist in the provision of future electrical power and electricity market services. With some careful planning and engineering, and in the context of the overall market development of distributed energy systems, the ideas discussed here may be just the beginning of the possibilities for integrating the motor vehicle and stationary power systems in creative and mutually beneficial ways.

SEE ALSO THE FOLLOWING ARTICLES

Alternative Transportation Fuels: Contemporary Case Studies • *Batteries, Transportation Applications* • *Biodiesel Fuels* • *Ethanol Fuel* • *Fuel Cell Vehicles* • *Fuel Cycle Analysis of Conventional and Alternative Fuel Vehicles* • *Fuel Economy Initiatives: International Comparisons* • *Hybrid Electric Vehicles* • *Hydrogen, End Uses and Economics* • *Internal Combustion Engine Vehicles* • *Transportation Fuel Alternatives for Highway Vehicles* • *Vehicles and Their Powerplants: Energy Use and Efficiency*

Further Reading

Brodrick, C-J., Lipman, T. E., Farschi, M., Lutsey, N., Dwyer, H. A., Sperling, D., Gouse, S. W., Harris, D. B., and King, F. G. (2002). Evaluation of fuel cell auxiliary power units for heavy-duty diesel trucks. *Transportation Research-D* 7(4), 303–315.

Brooks, A. N. (2001). "Electric Drive Vehicles: A Huge New Distributed Energy Resource." EVAA Electric Transportation Industry Conference, Sacramento, CA.

Brooks, A. N. (2002). "Vehicle-to-Grid Demonstration Project: Grid Regulation Ancillary Service with a Battery Electric Vehicle." Prepared for the California Air Resources Board, Contract number 01–313, December 10.

Kempton, W., and Letendre, S. (1997). Electric vehicles as a new power source for electric utilities. *Transportation Res.-D.* 2(3), 157–175.

Kempton, W., Tomic, J., Letendre, S., Brooks, A., and Lipman, T. (2001). "Vehicle-to-Grid Power: Battery, Hybrid, and Fuel Cell Vehicles as Resources for Distributed Electric Power in California," Inst. of Transportation Studies, Davis, UCD-ITS-RR-01–03, Prepared for California Air Resources Board (under contract #ARB00–612) and Los Angeles Department of Water and Power, June.

Kissock, J. K. (1998). "Combined Heat and Power for Buildings Using Fuel-Cell Cars." ASME International Solar Energy Conference, Albuquerque, NM.

Lipman, T. E., Edwards, J. L., and Kammen, D. M. (2002a). "Economic Analysis of Hydrogen Energy Station Concepts: Are 'H2E-Stations' a Key Link to a Hydrogen Fuel Cell Vehicle Infrastructure?" Energy Development and Technology Working Paper Series, EDT-003, University of California Energy Institute (UCEI), November.

Lipman, T. E., Edwards, J. L., and Kammen, D. M. (2002b). "Economic Implications of Net Metering for Stationary and Motor Vehicle Fuel Cell Systems in California." Program on Workable Energy Regulation (POWER) Paper Series, PWP-092, University of California Energy Institute (UCEI), February.

Intelligent Transportation Systems

SUSAN A. SHAHEEN
University of California, Berkeley and Davis
Berkeley and Davis, California, United States

RACHEL FINSON
University of California, Berkeley
Berkeley, California, United States

1. Introduction
2. Definition of ITS
3. ITS and Energy Impact Evaluation Tools
4. Predicted and Early Findings
5. System Integration and Conclusion

Glossary

bus rapid transit (BRT) Seeks to improve bus service by reducing travel time and providing enhanced rider information; exclusive rights-of-way, bus lanes, adjusting stop locations, wider doors, preboarding payment, and supportive land use policies contribute to BRT improvements.

commercial vehicle operations (CVO) The application of electronic and wireless intelligent transportation systems technologies to address a range of trucking industry concerns; CVO approaches include border crossing clearance and safety via electronic clearance and manifesting, automatic vehicle location, vehicle-to-fleet management center communications, on-board safety diagnostics to enable more effective roadside safety inspections, and hazardous materials incident response.

electronic toll collection (ETC) Enables the instant payment of highway tolls when a vehicle passes through a toll station via an electronic roadside antenna (or reader) and a pocket-sized tag containing a radio transponder (typically placed inside a vehicle's windshield); the toll tag transponder transmits radio frequencies to the toll reader, and the appropriate fare is deducted automatically.

incident management Consists of three key areas: traffic surveillance (incident detection and verification), clearance, and traveler information; also covered by this area are emergency management services, which coordinate local and regional incident response to traffic accidents, security threats, and hazardous material spills.

intelligent transportation systems (ITS) A wide range of technologies—including electronics, information processing, wireless communications, and controls—aimed at improving safety, efficiency, and convenience of the overall surface transportation network.

intelligent transportation systems (ITS) user services User services that categorize a wide range of ITS technologies from the user's perspective. User service areas currently address high-level problems and needs such as route guidance, electronic payment, and parking services.

ramp metering Traffic signals employed at freeway on-ramps to control the rate of vehicles entering the freeway; metering rates are set to optimize freeway flow and to minimize congestion.

traffic management and surveillance Includes incident management, ramp metering, traffic signal control, traveler information, and traffic surveillance; traffic surveillance tools provide the data needed to manage the roadways.

traffic signal control Can integrate freeway and surface street systems to improve traffic flow and vehicular and nonmotorized traveler safety and also provide priority services for transit or high-occupancy vehicles; traffic control technologies include traffic surveillance, ramp metering, lane control systems, and traffic signals.

transit management Consists of four key areas: (1) transit vehicle tracking, (2) fare payment, (3) traveler information, and (4) personalized public transportation alternatives; transit vehicle tracking includes communication between vehicles and transit centers.

transportation management center (TMC) The hub where transportation, operations, and control data are collected, combined, and distributed to manage the transportation network (including emergencies and incidents) and to generate traveler information; the TMC relies on various intelligent transportation systems tools to collect its data, including electronic toll collection, radar, closed circuit video equipment, and loop detectors.

traveler information Provides the traveling public with information regarding available modes, optimal routes, and costs in real time either pre-trip or en route via in-vehicle information and changeable message signs along roadsides or at transit stations.

vehicle control technologies Intelligent transportation systems technologies that can help to avoid collisions, prevent or lessen injuries when crashes do occur, and ultimately lead to full vehicle automation; some existing vehicle control technologies include adaptive cruise control, antilock brakes, and electronic system malfunction indicators.

Although world energy consumption has increased only marginally since 1998 ($\sim 0.5\%$), greater efficiencies are still required in transportation. Indeed, the Texas Transportation Institute estimated total congestion costs (delay plus wasted fuel) at $67.5 billion for 75 U.S. urban areas in the year 2000. This represents 3.6 billion in delay hours and 5.7 billion gallons of excess fuel consumed due to congestion-related delays. The goal of intelligent transportation systems (ITS) is to improve surface transportation safety, efficiency, and convenience.

1. INTRODUCTION

Intelligent transportation systems (ITS) technologies include state-of-the-art wireless, electronic, and automated technologies. Collectively, these technologies have the potential to integrate vehicles (transit, trucks, and personal vehicles), system users, and infrastructure (roads and transit). Automated and in-vehicle technologies include precision docking for buses, automated guideways, and collision avoidance systems. Many ITS technologies can help to optimize trips (route guidance), diminish unnecessary miles traveled, increase other mode use, reduce time spent in congestion, reduce dependence on foreign oil, and improve air quality. Furthermore, when ITS technologies are applied to system management (transit and highways) and vehicle design, they can reduce fuel consumption by

- facilitating optimal route planning and timing;
- smoothing accelerations/decelerations and stop-and-go driving;
- reducing congestion;
- enabling pricing and demand management strategies;
- increasing the attractiveness of public transportation mode use;
- adjusting vehicle transmission for varying road conditions and terrain;
- facilitating small platoons of closely spaced vehicles (i.e., safer vehicles could enable weight reduction without compromising occupant safety).

Although ITS technologies are still in the early phase of deployment, many have shown the potential to reduce energy use. During the past 10 years or so, fuel consumption impacts of the following ITS technologies have been studied: (1) traffic signal control, (2) traffic management and surveillance (e.g., ramp metering), (3) incident management, (4) electronic toll collection (ETC), (5) traveler information, (6) transit management, (7) commercial vehicle operations (CVO), and (8) vehicle control technologies. They are the focus of this article. Nevertheless, ITS impacts, including benefits, unintended consequences, and aggregate effects, are still not well understood. The field of intelligent transportation systems, energy consumption impacts, current measurement tools, and early findings are discussed in this article.

2. DEFINITION OF ITS

In 1991, the concept of ITS emerged when transportation professionals recognized that electronic technologies could begin to play a significant role in optimizing surface transportation, and the U.S. Congress legislated the national ITS program. Since then, computer, communication, and sensor technologies have improved dramatically, and ITS technologies have emerged in highway and transit jurisdictions worldwide.

ITS deployment can been categorized into three stages:

- Stage 1: Test and implement early ITS technologies (or building blocks).
- Stage 2: Link early ITS technologies.
- Stage 3: Develop an integrated system of ITS technologies.

The public sector has been the dominant driver of stage 1 ITS technology research and development. During this phase, early ITS technologies were applied to improve traditional operations. Key accomplishments include the deployment of

- traffic management centers in urban areas to monitor freeway traffic and early incident notification;
- traffic signal control and ramp metering to improve traffic flow and safety;
- improved traveler information;

- commercial vehicle screening and ETC;
- satellite-based dispatching systems in transit operations;
- ride-matching services via the Internet;
- in-vehicle navigation systems in private vehicles.

Whereas stage 1 of ITS will continue for years, stage 2—linking early ITS technologies—is also progressing, as is evident by efforts to mainstream ITS with conventional capital improvement projects (e.g., connecting traffic management centers with advanced traffic signal coordination, ramp metering adjusted in real time, traveler information systems coordinated with in-vehicle devices). During this phase, industry has already initiated work to understand the fundamental nature of driving and of driver/operator behavior in transportation models. Furthermore, the process has started to move from a heavy focus on stage 1 public sector investment toward customer-oriented operations. Examples of next steps—included in stage 2—are (1) public transportation system improvements, (2) coordination among various freight modes (e.g., trucks, rail), (3) improved system management through real-time data and performance tools, and (4) in-vehicle crash avoidance systems.

Ultimately, further integration is needed for the longer term goal of widespread ITS deployment to be realized—culminating in stage 3. An integrated system requires a network of technologies working together along critical corridors and urban centers. Nevertheless, the challenge still remains in getting from "here" (individual building blocks) to "there" (a comprehensive network of technologies managing and disseminating information).

2.1 ITS Categories and Deployment Planning

To guide the development and deployment of ITS technologies, the U.S. Department of Transportation released its National ITS Architecture (designed in conjunction with its external advisory committee—ITS America—and stakeholder input) during the mid-1990s. The National ITS Architecture reflects a mature product, developed and updated over a 9-year period. It provides a common framework and language for planning and implementing ITS so that systems could be integrated functionally and geographically and be interoperable from one location to another. The National ITS Architecture is composed of 32 ITS user services bundled into eight categories: (1) travel and transportation management, (2) public transportation options, (3) electronic payment, (4) CVO, (5) emergency management, (6) advanced vehicle control and safety systems, (7) information management, and (8) maintenance and construction. For an overview of the 32 ITS user services, see Table I.

The National ITS Architecture provides an implementation roadmap that starts with ITS building blocks (or early market packages). The building blocks are characterized as low risk, reflecting early market opportunities and tangible benefits. Early market packages include (1) surface street control, (2) freeway control, (3) ETC, (4) traveler information, (5) transit vehicle tracking, (6) transit operations, and (7) electronic clearance (e.g., tolling, commercial vehicles).

The National ITS Architecture is tied to a 10-year plan (released in 2002) that addresses goals, objectives, user service requirements, and expected benefits. The five main goals outlined in the 10-year plan are (1) safety, (2) security, (3) efficiency/economy, (4) mobility/access, and (5) energy/environment. Furthermore, the 10-year plan develops a series of programmatic and enabling themes to describe the opportunities, benefits, and challenges of future transportation systems. There are four programmatic themes: (1) an integrated network of transportation information, (2) advanced crash avoidance technologies (including in-vehicle electronics), (3) automatic incident detection, notification, and response, and (4) advanced transportation management (including traffic and transit).

Two of the four ITS programmatic themes (the second and fourth themes just listed) could result in significant energy consumption benefits in the future. The advanced crash avoidance technologies programmatic theme focuses primarily on reducing the number of vehicle crashes. Adaptive in-vehicle electronics, integral to crash avoidance, is forecasted to reduce fuel consumption by (1) smoothing accelerations and decelerations (particularly for commercial and transit vehicles), (2) responding automatically to stop-and-go driving, (3) anticipating and adjusting the throttle and transmission for varying road conditions and terrain, and (4) enabling the safe movement of platoons of tightly spaced trucks and transit and other vehicles. In the longer term, such safety devices could also permit the introduction of lighter weight vehicles for greater fuel economy. In addition, route guidance products that help drivers to plan optional routes—in case of an incident—may significantly reduce miles driven, saving fuel and helping to mitigate congestion.

TABLE I
ITS User Services

User services bundle	User services
Travel and transportation management	Pre-trip travel information
	En route driver information
	Route guidance
	Ride matching and reservation
	Traveler services information
	Traffic control
	Incident management
	Travel demand management
	Emissions testing and mitigation
	Highway rail intersections
Public transportation operations	Public transportation management
	En route transit information
	Personalized public transit
	Public travel security
Electronic payment	Electronic payment services
Commercial vehicle operations	Commercial vehicle electronic clearance
	Automated roadside safety inspection
	On-board safety monitoring
	Commercial vehicle administration processes
	Hazardous materials incident response
	Commercial fleet management
Emergency management	Emergency notification and personal security
	Emergency vehicle management
Advanced vehicle control and safety systems	Longitudinal collision avoidance
	Lateral collision avoidance
	Intersection collision avoidance
	Vision enhancement for crash avoidance
	Safety readiness
	Pre-crash restraint deployment
	Automated vehicle operation
Information management	Archived data function
Maintenance and construction management	Maintenance and construction operations

Source. From Iteris Inc. and Lockheed Martin (2002). "Executive Summary National ITS Architecture Version 4.0." U.S. Department of Transportation, Washington, DC.

The advanced transportation management programmatic theme is also predicted to have a considerable effect on energy use in the future. Tools included in this area aim to intelligently and adaptively manage vehicle flows within the physical infrastructure and often across multiple jurisdictions and modes. Advanced transportation management systems rely on area-wide surveillance and detection, rapid acquisition and evaluation of traffic flow data, and predictive capabilities.

ITS America forecasts that these programmatic areas—crash avoidance and transportation management systems—will be critical to achieving the 10-year plan's energy goal: saving a minimum of 1 billion gallons of gasoline per year. The remainder of this article focuses on tools (models, field operational tests, and deployments) used to evaluate ITS impacts on energy consumption and early findings.

3. ITS AND ENERGY IMPACT EVALUATION TOOLS

In 1998, the U.S. Environmental Protection Agency (EPA) released a technical report that examined methodologies and research efforts aimed at evaluating the energy and environmental impacts of ITS. The report concluded that developing an ITS fuel consumption and emission impact assessment is an exceptionally challenging exercise due to the complex relationship among ITS, travel behavior, and transportation system management. Traffic simulation and travel demand models can aid in this research; however, more work is needed.

A key theme of ITS is the integrated deployment of information networks to support travel, increase transportation infrastructure use, and better manage demand. However, this integration requires models that can simulate the effects of information on traffic flow at corridor levels and travel behavior at regional levels. Independently, traditional tools (microsimulation, regional travel demand, and emission and fuel consumption models) cannot adequately capture critical linkages among ITS technologies and various feedback loops.

For example, travel demand models are typically unresponsive to information-related improvements (e.g., road conditions) and intersection- and corridor-level changes. Such impacts are usually evaluated using traditional microsimulation tools, which lack the behavioral assumptions of travel behavior models. Thus, new approaches are needed to capture complex ITS interactions and impacts. These tools must be flexible enough to (1) reflect different scenarios (e.g., market penetration and behavioral assumptions), (2) capture ITS impacts on dynamic mode and route choice, and (3) simulate individual vehicle driving patterns.

Given the need for more complex tools, several advanced microsimulation models (e.g., INTEGRATION, DYNASMART) were developed to simulate the dynamic interactions among ITS technologies—including energy impacts. For example, INTEGRATION was designed with the purpose of simulating integrated networks, composed of freeways and arterial roads, for various ITS scenarios. INTEGRATION provides effectiveness measure estimates for individual vehicles, links, origin–destination trip pairs, and complete networks—including travel time, fuel consumption, and vehicle emissions. In addition, Traf–Nestim (another modeling tool) can be used to enhance the modal fuel consumption algorithms of INTEGRATION.

DYNASMART was developed to evaluate the dynamic effects of advanced traveler information and management systems on networks. This tool employs simple fuel consumption and emission models that reflect the relationship between uniform acceleration rates and speed. Consequently, this model cannot make adjustments for ambient temperatures, cold starts, and second-by-second driving behavior.

TRANSIMS, another ITS model initiative, stands for Transportation Analysis and Simulation System. This tool integrates a set of new travel forecasting and analysis procedures. TRANSIMS was developed to meet the travel forecasting and air quality analysis needs of metropolitan and state transportation and planning agencies. It is composed of six integrated modules: (1) population synthesis, (2) activity-based travel demand generation, (3) intermodal trip planning, (4) traffic microsimulation, (5) environmental analysis, and (6) feedback. The environmental module estimates air quality, energy consumption, and carbon dioxide emissions based on travel behavior outputs. As of 2003, TRANSIMS had been tested using data from the city of Dallas, Texas, and the city of Portland, Oregon. Developers anticipated that the fuel consumption module would be available for use in 2005.

Another common tool for evaluating ITS technologies is field operational tests (FOTs). Researchers and practitioners frequently conduct FOTs to demonstrate and evaluate ITS technologies in real-world settings. In some cases, there is an overlap between microsimulation tools and FOTs. Modeling is frequently used to simulate and predict real-world ITS deployment impacts. Furthermore, FOT data are often employed in modeling studies to simulate ITS impacts at different levels (e.g., corridor and higher market penetration). Early results, reported in the next section, reflect modeling, FOT, and deployment findings.

4. PREDICTED AND EARLY FINDINGS

Understanding of ITS fuel consumption impacts is still limited. A summary of predicted impacts for eight ITS categories in which fuel consumption has been studied over the past decade or so (at various levels) is presented in Table II. The eight categories are (1) traffic signal control, (2) traffic management and surveillance (e.g., ramp metering), (3) incident management, (4) electronic toll collection, (5) traveler information, (6) transit management, (7) commercial vehicle operations, and (8) vehicle control technologies. Energy impacts listed are both direct and indirect and are unsubstantiated by modeling or actual field test results.

Over the past decade or so, ITS technologies have been modeled, tested in limited FOTs and evaluated in full-scale deployment (e.g., electronic tolling, ramp metering, bus automatic vehicle location). Knowledge about ITS impacts is growing, as is an understanding of how to maximize benefits. Although fuel consumption is not the primary motivator of ITS (congestion relief and traffic and transit management are the key drivers), understanding of ITS energy impacts is increasing. Indeed, by reducing stop-and-go traffic and improving route guidance, ITS can have a positive impact on fuel consumption.

In the following subsections, early modeling, FOT, and deployment energy consumption results are quantified (when applicable). Although this analysis spans a wide range of ITS technologies, it does not represent an exhaustive review of ITS and fuel consumption impacts. Finally, there is a brief discussion regarding the potential effects and complexities of understanding latent or new demand that may result from ITS capacity enhancement strategies.

4.1 Traffic Signal Control

Traffic signals serve a variety of functions. They can be used to manage traffic speeds, vehicle merging and corridor crossings, and interactions between vehicles and low-speed or nonmotorized modes (e.g., bicycles, pedestrians, wheelchairs) at intersections. ITS

TABLE II
Predicted ITS Technology Impacts

ITS technology	Predicted impacts
Traffic signal control Traffic control technologies include traffic surveillance, ramp metering, lane control systems, and traffic signals	Reduced stop/idle delay times for vehicles traveling on main lanes and at intersections Better response to incidents and special events Some mode shifts to/from transit Reduced bus round-trip travel time, implying increased speed
Traffic management and surveillance ITS-based traffic management approaches include incident management, ramp metering, traffic signal control, traveler information, and traffic surveillance	Reduced freeway travel times due to ramp metering Increased ramp travel times Changes in acceleration (increased emissions) Dynamic changes in freeway exit points due to changeable message signs, thereby increasing freeway speeds Increases in high-occupancy vehicle travel resulting from high-occupancy vehicle bypass to ramp metering Increased travel times on arterials due to spillage of vehicles waiting to get on freeways Better use of existing freeway capacity as a result of lane control improvements
Incident management Incident management consist of three key areas: (1) traffic surveillance (incident detection and verification), (2) clearance, and (3) traveler information; in addition, this area includes emergency management services	Reduced incident response times Increased travel reliability Reduced delays, especially those due to incidents on highways and freeways (e.g., lane closures) Fewer secondary incidents caused by initial incidents Dynamic changes in destination, mode, and/or route choice
Electronic toll collection ETC technologies enable the instant payment of highway tolls when a vehicle passes through a toll station via an electronic roadside antenna (or reader) and a pocket-sized tag containing a radio transponder	Increase in toll lane capacity Fuel savings and a decrease in mobile emissions by reducing or eliminating waiting times
Traveler information Traveler information systems serve two primary functions, providing (1) information about transportation services and performance of the system to users and (2) an information linkage among various ITS components	Increased shifts in mode, destination, time of trip, and route choice over time
Transit management Transit management consists of four key areas: (1) transit vehicle tracking, (2) fare payment, (3) traveler information, and (4) personalized public transportation alternatives	Decreased boarding times Increased transit system reliability Improved transit fleet use Improved coordination between transit services, such as bus and rail transfers, promoting overall transit use Increased opportunity to provide more demand-responsive transit services (e.g., mode shifts)
Commercial vehicle operations CVO approaches include border crossing clearance and safety via electronic clearance and manifesting, automatic vehicle location, vehicle-to-fleet management center communications, on-board safety diagnostics to enable more effective roadside safety inspections, and hazardous materials incident response	Electronic clearance, safety, credentialing, and administrative processing save time by automating traditionally manual screening and inspection procedures Because of vehicle location technologies, emergency response teams can more easily locate and respond to HazMat accidents Reduced waiting times at international borders due to electronic clearance (inspectors can focus on noncompliant vehicles) On-board safety monitoring automatically alerts drivers of deficiencies in their vehicles' performance, ensuring greater roadway maintenance and safety

continues

Table II continued

ITS technology	Predicted impacts
Vehicle control technologies Advanced sensing, communication, and computing technologies represent the range of ITS technologies that can help to avoid collisions, prevent or lessen injuries when crashes do occur, and ultimately lead to full vehicle automation; some existing vehicle control technologies include adaptive cruise control, antilock brakes, and electronic system malfunction indicators	Reduce accidents resulting from unsafe headway, driver inattention, and errors in recognition and judgment Reduce antisocial driving behavior such as road rage Increased freeway capacity due to more closely spaced vehicle platoons Reduced fuel consumption and emissions due to traffic flow smoothing and vehicle platooning

microsimulation models enable researchers to assess the impacts of various traffic signal controls on emissions, travel times, and fuel consumption. Traffic signal controls can be set to optimize one or more desired goals (e.g., time savings, energy reduction). To maximize fuel efficiency, traffic signal controls can be fixed to reduce vehicle accelerations, decelerations, and idle times, all of which contribute to increased fuel consumption. Numerous—and sometime conflicting—goals must be balanced in traffic management. For instance, emission and fuel efficiency goals are typically not achieved using the same traffic control strategies employed to reduce travel times.

Using traffic signal controls to minimize energy consumption is not new, and studies to quantify energy impacts in this area predate ITS. For example, during the two energy crises of the 1970s (1973–1974 and 1979), researchers found that synchronizing traffic signal green lights to minimize stop-and-go traffic reduced energy consumption.

More recently, traffic signal control has become a predominant component of ITS, and its impacts have been evaluated in simulation and real-world settings. Overall, traffic signal control studies reveal fuel efficiency benefits ranging between 1.6 to 50.0%, with most results at less than 20%. Variability in estimated savings is a result of the following factors: quality of existing timing plans, network configuration, traffic patterns, and signal equipment.

4.2 Traffic Management and Surveillance

Ramp metering is one of several ITS technologies designed to manage traffic flow. The goal of ramp metering is to safely space vehicles merging onto a highway while minimizing speed disruptions to existing flows. Considerations include (1) public misunderstanding and system dislikes, (2) overflow of cars onto surface streets while waiting to enter ramps, and (3) driver use of arterial streets to avoid ramp meters. The most significant benefit of ramp metering is passenger time savings. Emission and fuel consumption impacts are mixed.

Ramp metering causes vehicles on ramps to stop and go, and this behavior consumes more fuel than does free-flow driving. Ramp metering also results in smoother vehicle flow on freeways because vehicles enter in a staggered and controlled manner, reducing bottlenecks that would otherwise impede traffic. This results in reduced fuel consumption. These two factors (increased stop-and-go traffic on on-ramps and decreased traffic flow disruption on highways) appear to negate each other. More detailed studies are necessary to understand how ramp metering effects interact and how they affect fuel consumption.

Whereas the larger category of traffic management and surveillance also includes incident management, traffic signal control, and traveler information, this subsection focuses on ramp metering primarily because each of the other categories is examined separately. Other traffic management tools include improved surveillance using loop detectors, closed circuit television, radar, lasers, video image processing, and vehicles equipped with toll tags or global positioning systems to determine travel times. Their impacts on fuel consumption relate directly to their use in incident management and traveler information.

4.3 Incident Management

ITS contributions to incident management include improved surveillance, verification, and dispatch to manage an incident. The use of changeable message signs and personal communications devices, such as cell phones and personal digital assistants (PDAs), can assist with early notification for upstream drivers, resulting in reduced incident-related congestion because drivers have more time to select an alternative route. Improved incident management can decrease fuel consumption by reducing the delay and congestion associated with the blocked traffic. Although incident delay reductions are limited, model calculations for a Maryland initiative (called CHART) have shown fuel savings of 4.1 million gallons per year.

4.4 Electronic Toll Collection

ETC allows for electronic payment of highway and bridge tolls as vehicles pass through a toll station. Vehicle-to-roadside communication technologies include electronic roadside antennas (or readers) and pocket-sized tags containing radio transponders (typically placed inside a vehicle's windshield).

Studies show that ETC saves time and reduces energy consumption by reducing the stop-and-go traffic associated with vehicle queues approaching toll plazas, stopping to pay a toll, and accelerating to rejoin regular traffic flow. Analyses also support that fuel savings offset operating costs. ETC systems have the capacity to move five times as many vehicles as do conventional toll lanes. One study along the New Jersey Turnpike found savings of 1.2 million gallons of fuel per year due to reduced delays at toll plazas employing ETC. Approximately three-fourths of the reported savings accrued to passenger cars and one-fourth to commercial vehicles.

4.5 Traveler Information

Effective traveler information requires the accurate collection and dissemination of real-time travel information to transportation managers and the public to aid them in making informed decisions about travel time, mode, and route. A wide array of ITS technologies assist with traveler information, including in-vehicle guidance, Web sites, cell phones, PDAs, and changeable message signs to distribute user information.

The actual impact of traveler information on fuel consumption depends on a number of factors. For example, if ITS technologies assist drivers with route selection and guidance, fuel consumption benefits will likely be greater the less familiar a driver is with an area. The timeliness and delivery medium of information will also influence the degree to which travelers use it and subsequent energy impacts. Fuel consumption benefits might include mode shifts (e.g., from a single-occupancy vehicle to transit or bicycle) and energy savings proportional to travel time reductions achieved by taking alternate routes.

4.6 Transit Management

Transit managers are already implementing ITS technologies to improve service, including automatic vehicle location, real-time bus arrival signage, traffic signal priority, and automated information announcements. The degree to which ITS technologies improve transit services and attract new riders determines the significance of fuel consumption impacts.

Bus rapid transit (BRT) encompasses the use of a series of ITS technologies, route planning, exclusive right-of-ways, and management to improve service—each of which can reduce travel times. Increases in bus ridership due to BRT implementation have been reported in the United States, Australia, and Europe. BRT improvements along certain routes attract new riders and encourage existing riders to use the service more frequently. If a mode shift occurs from single-occupancy vehicles to BRT, there is an efficiency benefit. If the previous mode was nonmotorized, such as walking or cycling, the impact on fuel efficiency is negative. If additional riders are attracted from another bus route, the impact on fuel efficiency is neutral. An estimated 30 to 50% of additional riders on a new BRT route are new transit riders. Furthermore, faster journey times and reduced acceleration, deceleration, and idle times—resulting from fewer stops and signal priority—have been shown to reduce fuel consumption. Indeed, signal priority modeling results indicate a 5% reduction in fuel consumption.

4.7 Commercial Vehicle Operations

ITS applications in commercial vehicle management primarily include automatic vehicle identification and weigh-in-motion. The purpose of automatic identification and weigh-in-motion technologies in CVO is to enable the weighing and cataloging of trucks without causing vehicles to stop and queue in line.

Simulation modeling and on-road testing reveal increased fuel efficiency due to weigh-in-motion technologies. Measured against static scales, high-speed weigh-in-motion systems demonstrate the greatest fuel benefits. Weigh-in-motion ramps that require trucks to slow—but not stop—also result in fuel savings, although not as significant as the savings with high-speed weigh-in-motion stations.

The purpose of automatic vehicle identification is to identify trucks, drivers, and loads as a companion function to weigh-in-motion.

In addition, CVO can result in fewer trucks being forced to bypass weigh stations due to full queues at static scales. Although commercial ITS applications demonstrate a clear fuel benefit, this value depends on the number and nature of stations passed.

4.8 Vehicle Control Technologies

ITS technologies that automate vehicle control systems aim to improve vehicle safety, efficiency, and comfort. These technologies include intelligent cruise control, antilock brakes, electronic system malfunction indicators, and automated highway systems (e.g., platooned vehicles). Simulation research indicates that some automated vehicle control technologies could have a positive impact on fuel consumption.

Intelligent cruise control refers to technologies that can identify the distance in front of a vehicle on a highway and correspondingly modify a car's controlled speed to accommodate lane merging and changes in the speed of vehicles ahead. This results in reduced fluctuations in the speed of controlled cars, resulting in a positive fuel efficiency effect. One simulation study showed fuel savings ranging from 8.5 to 28.5% when 10% of vehicles in a lane are equipped with intelligent cruise control.

Another group of vehicle control technologies is being tested for automated highway systems. The concept behind automated highways is to employ technologies that facilitate vehicle-to-vehicle and vehicle-to-roadside communication to improve safety and system efficiency. In this way, vehicles can operate in very close proximity to each other. Simulations indicate a 5 to 15% reduction in fuel consumption due to aerodynamic drafting effects.

4.9 Human Factors

As already demonstrated, many ITS technologies can have a positive impact on fuel consumption. Some of the energy benefits accrue from reduced congestion and stop-and-go driving, resulting in smoother traffic flows. In effect, ITS technologies can increase existing roadway capacity without increasing infrastructure. Latent (or additionally generated) demand is often at the forefront of infrastructure enhancement discussions. In regions with the greatest congestion, travel demand may be suppressed due to travel time costs. Under such conditions, adding or managing infrastructure to ease congestion may result in additional travel demand until delay costs rise enough to suppress travel again. Studies show that increased roadway infrastructure can also result in modal shifts from transit ridership to personal vehicles.

If ITS technologies are successful at reducing congestion and travel delay, roadway capacity will be increased without adding new infrastructure. An outstanding question regarding ITS capacity enhancement is whether or not added capacity will result in latent demand for highway use—similar to some demonstrated infrastructure expansion impacts (e.g., mode shifts from transit to auto use). If improved traffic throughput due to ITS results in increased highway demand and modal shifts away from transit, some of the systemwide fuel efficiency benefits may be negated. Because evidence to support latent demand is still not definitive, and final stage 3 ITS implementation has not yet been achieved, no conclusions can be drawn regarding the causal relationship between ITS capacity improvements and latent demand.

Other ITS technologies (e.g., real-time road pricing) may be used to offset increased travel demand to send signals to the traveling public regarding travel costs. By employing sophisticated surveillance technologies, vehicle identification, and electronic payment systems, ITS can assist in administering congestion pricing within incremental time blocks, thereby reducing congestion and energy use.

5. SYSTEM INTEGRATION AND CONCLUSION

At the current stage of ITS development and deployment, several early predictions regarding fuel efficiency impacts are being realized (e.g., electronic tolling, signal priority for buses). However, modeling and early ITS deployments are still occurring largely in discrete applications rather than across integrated regionwide networks. To understand the full effects of ITS on transportation systems (traffic and transit), technologies must be deployed in a comprehensive manner. Similarly, the full energy impacts of ITS cannot be known until technologies are integrated and complex dynamics, including human factors, are modeled and tested. Given the complex interaction of ITS technologies and human factors, development and use of suitable tools to measure environmental consequences, including fuel consumption, will remain important. Interrelationships among various ITS elements will determine the ultimate direction and degree of impacts. In addition, impacts will vary among regions with different traffic patterns and system use.

Overall, integration of individual ITS components can be expected to multiply benefits by providing the traveling public with a wider array of choices and real-time information. In the EPA's 1998 technical report on ITS, the agency identified four key dimensions to comprehensive ITS, requiring integration of

- information across key ITS elements such as transit management, freeway management, emergency and incident response, and traffic signal control;
- information across regions such as coordination between jurisdictions to ensure smoother traffic flow;
- transit management systems across regions to ensure integrated user services;
- incident management information across regions to provide the fastest possible response.

Ultimately, the integrated ITS vision should provide operational efficiencies and interjurisdictional coordination benefits that result in comprehensive and improved system management. The magnitude of these benefits will depend on factors such as market acceptance of available technologies to deliver information, user-perceived accuracy of the information provided, and level of personalized traveler information. Not surprisingly, the greatest travel time and energy benefits will come from traveler information persuading users to take public transportation or postpone their trips until congestion has cleared.

SEE ALSO THE FOLLOWING ARTICLES

Energy Efficiency, Taxonomic Overview • Passenger Demand for Travel and Energy Use • Transportation and Energy, Overview • Transportation and Energy Policy • Vehicles and Their Powerplants: Energy Use and Efficiency

Further Reading

Barth, M. (2000). An emission and energy comparison between a simulated automated highway system and current traffic conditions. *In* "2000 IEEE Intelligent Transportation Systems Conference Proceedings." Dearborn, MI.

Bose, A., and Ioannou, P. (2001). "Analysis of Traffic Flow with Mixed Manual and Intelligent Cruise Control Vehicles: Theory and Experiments" (California PATH Research Report UCB-ITS-PRR-2001-13). California PATH, Richmond, CA.

BP. (2002). BP statistical review of world energy 2002. www.bp.com/centres/energy2002/usenergy.asp.

Intelligent Transportation Society of America. (2002). Delivering the future of transportation: The National Intelligent Transportation Systems program plan—A ten-year vision. www.itsa.org.

Intelligent Transportation Systems Joint Program Office, U.S. Department of Transportation. (2003). ITS benefits and unit costs database. www.benefitscost.its.dot.gov.

New Technology and Research Program. (1996). "Advanced Transportation Systems Program Plan." California Department of Transportation, Sacramento.

Schrank, D., and Lomax, T. (2002). The 2002 urban mobility report. www.mobility.tamu.edu.

Shaheen, S. A., Young, T. M., Sperling, D., Jordan, D., and Horan, T. (1998). "Identification and Prioritization of Environmentally Beneficial Intelligent Transportation Technologies" (Research Report UCD-ITS-RR-98-1). Institute of Transportation Studies, University of California, Davis.

U.S. Department of Transportation (2002). The National ITS Architecture version 4.0. itsarch.iteris.com/itsarch.

U.S. Environmental Protection Agency. (1998). "Assessing the Emissions and Fuel Consumption Impacts of Intelligent Transportation Systems (ITS)" (EPA 231-R-98-007). EPA, Washington, DC.

Internal Combustion Engine Vehicles

K. G. DULEEP
Energy and Environmental Analysis, Inc.
Arlington, Virginia, United States

1. Introduction
2. Vehicle Energy Efficiency
3. Improving Efficiency by Reducing Tractive Energy Required
4. Improvement of Engine Efficiency
5. Increasing the Efficiency of Spark Ignition Engines
6. Increasing the Efficiency of Compression Ignition (Diesel) Engines
7. Intake Charge Boosting
8. Alternative Heat Engines

Glossary

engine efficiency Amount of energy produced by the engine per unit of fuel energy consumed.
fuel economy Vehicle distance traveled per unit volume of fuel consumed.
internal combustion engine vehicle Vehicle where primary motive power is derived from an engine that converts fuel energy to work using the air-fuel mixture as the working fluid.
light duty on highway vehicles Cars and light trucks with a fully loaded weight below 6000 k (13,200 lbs).
off-highway vehicle Vehicles designed to operate primarily on unpaved surfaces.

The vast majority of vehicles used in the world are powered by internal combustion engines (ICE). Other forms of propulsion such as electric motors or external combustion steam engines are used in specialized applications that account for a small fraction of the total vehicle fleet. Most vehicles are now powered by reciprocating piston engines that use the Otto cycle (also called the spark-ignition engine) or the diesel cycle (also called the compression-ignition engine). Gas turbines are used primarily in marine vessels and aircraft and are not discussed here. A small number of vehicles using the Wankel engine have also been sold. Internal combustion engine–powered vehicles typically account for one-quarter to one-third of total energy consumption in most countries, and their fuel consumption and fuel efficiency are issues of major concern.

1. INTRODUCTION

The on-highway fleet of vehicles accounts for over 95% of all vehicles in operation worldwide (which is in excess of a billion vehicles). The remainder is composed of off-highway vehicles, equipment such as forklifts or bulldozers, and motorcycles. Annual sales of on-highway vehicles exceeded 57.6 millions units worldwide in 2002 with 39.5 millions unit classified as passenger cars and 18.1 million units classified as trucks. The distinction between cars and trucks is not always clear (especially for light trucks) but trucks are usually used for cargo hauling or for carrying more than six passengers. Cars span the gross vehicle weight (GVW) range from 1 to 3 tons, while trucks typically span the GVW range of 2 to 40 tons. Off-road and specialized vehicles can be much heavier.

The majority of cars and light trucks (under five tons GVW) are powered by spark ignition engines, while most trucks that weigh more than 5 tons GVW are powered by diesel engines. Since the early 1990s, diesel engines have become more popular for cars and light trucks in the European Union (EU). The diesel engines' share in the new car market was over 50% in 2002 in countries such as France and Austria. In contrast, few diesel engine–powered cars and light trucks are sold in North America.

2. VEHICLE ENERGY EFFICIENCY

Vehicle energy efficiency is generally defined in terms of fuel economy measured in miles per gallon (mpg) or kilometers per liter of fuel. It can also be measured in terms of fuel consumption, which is the inverse of fuel economy, in units of liters per 100 km or gallons per 100 miles. The fuel economy of a vehicle is strongly dependent on the vehicle's overall weight, but is also dependent on the efficiency of the engine, as well as the matching of the engine characteristics to the vehicle's operational requirements. The fuel economy of a particular vehicle is dependent on the load carried, the driving cycle, the ambient temperature, and the characteristics of the road such as its gradient and surface roughness. Hence, the fuel economy of a specific vehicle can vary widely depending on how and where it is used.

In most developed countries, on-road light vehicles (cars and light trucks) are certified for emissions and fuel economy by the government. The fuel efficiency rating is measured in a laboratory-controlled environment and on a specified driving cycle. In the United States and Canada, for example, light vehicle fuel economy is measured on a "city cycle" with an average speed of about 20 mph and a "highway cycle" with an average speed of about 50 mph. All aspects of the fuel economy test, ranging from the ambient temperature to the specification of the fuel used, are tightly controlled and this results in a fuel economy measurement that is repeatable to within $\pm 2\%$, typically. While this measured fuel economy may differ significantly from the fuel economy for the same vehicle in any specific use, the measured value provides a comparative benchmark for vehicle fuel economy that is useful from a vehicle buyer's perspective and from an engineering perspective.

The sales-weighted average test fuel economy of all new vehicles sold in the United States is about 28 mpg for cars and 21 mpg for light trucks. Much of this difference between car and light-truck fuel economy is attributable to the fact that light trucks are larger and heavier than cars, but some of the difference is also attributable to the fact that cars utilize higher levels of efficiency enhancing technology. Fuel economy levels in Australia and Canada are similar to the U.S. levels, but cars in the EU have about 25% higher fuel economy, on average. The higher fuel economy in the EU is partly due to the smaller size and weight of cars sold and partly due to the higher penetration of diesel engines, which are more efficient than spark ignition engines.

Studies conducted by technical agencies have concluded that vehicle fuel economy can be increased substantially from average values without any reduction of attributes such as interior space of cargo carrying ability. The sources of energy loss and the technology available to reduce these losses are described later.

A simple model of energy consumption in conventional automobiles provides insight into the sources and nature of energy losses. In brief, the engine converts fuel energy to shaft work. This shaft work is used to overcome the tractive energy required by the vehicle to move forward, as well as to overcome driveline losses and supply accessory drive energy requirements. The tractive energy can be separated into the energy required to overcome aerodynamic drag force, rolling resistance and inertia force. It is useful to consider energy consumption on the U.S. city and highway test cycles, which are reference cycles for comparing fuel economy.

Denoting the average engine brake specific fuel consumption over the test cycle as bsfc, we have fuel consumption, FC, given by

$$FC = \frac{bsfc}{\eta_d}[E_R + E_A + E_K] + bsfc\, E_{AC} + G_i(t_i + t_b)$$

where η_d is the drive train efficiency, E_R is the energy to overcome rolling resistance, E_A is the energy to overcome aerodynamic drag, E_K is the energy to overcome inertia force, E_{AC} is the accessory energy consumption, G_i is idle fuel consumption per unit time, and t_i, t_b are the time spent at idle and braking.

The first term in the above equation represents the fuel consumed to overcome tractive forces. Since the Federal Test Procedure (FTP) specifies the city and highway test cycle in terms of speed versus time, E_R, E_A, and E_k can be readily calculated as function of the vehicle weight, the tire rolling resistance, body aerodynamic drag coefficient, and vehicle frontal area. Weight reduction reduces both inertia force and rolling resistance.

It should be noted that not all of the inertia force is lost to the brakes, as a vehicle will slow down without the use of brakes, at zero input power due to aerodynamic drag and rolling resistance. Braking energy *loss* is approximately 35% in the city cycle and 7% on highway cycle. The fuel energy is used not only to supply tractive energy requirements but also to overcome transmission losses, accounting for the transmission efficiency that is in the first term.

The second term in the equation is for the fuel consumed to run the accessories. Accessory power requirements are required to run the radiator cooling fan, alternator, water pump, oil pump, and power steering pump. Air conditioners also absorb power but are not reflected in official fuel economy estimates since they are not turned on during the FTP. Idle and braking fuel consumption are largely a function of engine size and idle RPM, while transmission losses are function of transmission type (manual or automatic) and design. The engine produces no power during idle and braking but consumes fuel, so that factor is accounted for by the third term. Table I shows the energy consumption (as a percentage) by all of these factors for a typical U.S. midsize car of mid-1990s vintage, with a 3-liter displacement s.i. engine, four-speed automatic transmission with lock-up, and power steering.

The values in Table I can be utilized to derive sensitivity coefficients for the reduction of various loads. For example, reducing the weight by 10% will reduce both rolling resistance and inertia weight forces, so that tractive energy is reduced by $(30.35 + 40.22) \times 0.1$ or 7.06% on the composite cycle. Fuel consumption will be reduced by $7.06\% \times 0.6544$, which is the fraction of fuel used by tractive energy, or 4.6%. This matches the common wisdom that reducing weight by 10% reduces fuel consumption by 4 to 5%. However, if the engine is also downsized by 10% to account for the weight loss, fuel consumption will be reduced by 5.8% since idle and braking fuel consumption will be reduced in proportion to engine size. In addition, there will be some reduction (0.5%) in transmission and drivetrain loss.

Fuel economy can be improved by two primary methods: (1) by reducing the power required to propel the vehicle and (2) by increasing the engine efficiency. To estimate the effects of different technology improvements that affect engine power required or the efficiency of the engine, it is useful to keep certain vehicle attributes constant. Vehicle attributes of interest to consumers are passenger room, cargo space or payload capability, acceleration performance, and vehicle comfort/convenience features. The impact of technology on fuel economy is typically measured while keeping these attributes constant.

Reducing the power required to propel the vehicle reduces engine load and can be accomplished by reduction of weight, aerodynamic drag, rolling resistance, or accessory loads. Engine efficiency increases can be accomplished not only by engine technologies but also by improved drivetrain technologies that improve the match between engine operating point and vehicle power requirements. Spark ignition engines convert only about 20 to 25% of fuel energy to useful work during typical driving so that a doubling of engine efficiency is theoretically possible without changing the basic Otto cycle.

3. IMPROVING EFFICIENCY BY REDUCING TRACTIVE ENERGY REQUIRED

Since vehicle weight is one of the most important variables determining fuel economy, weight reduction is an important method of improving fuel economy. The vehicle's weight is distributed between the body structure, the drivetrain, the vehicle's

TABLE I

Energy Consumption as a Percentage of Total Energy Requirements for a Typical Midsize Car[a]

	City	Highway	Composite[b]
Percentage of total tractive energy			
Rolling resistance	27.7	35.2	30.35
Aerodynamic drag	18.0	50.4	29.43
Inertia (Weight) force	54.3	14.4	40.22
Total	100	100	100
Percentage of total fuel consumed			
Tractive energy	57.5	80.0	65.44
Accessory energy	10.0	6.5	8.76
Idle + Braking consumption	15.0	2.0	10.41
Transmission + Driveline loss	17.5	11.5	15.39

[a] Midsize car of inertia weight = 1588 kg, $C_D = 0.33$, $A = 2.1 m^2$, $C_R = 0.011$, 3L OHV V-6, power steering, four-speed automatic transmission with lock-up, air conditioning.

[b] Highway fuel economy is 1.5 times city fuel economy, and composite figures are based on the U.S. EPA 55% city/45% highway fuel consumption weighting.

interior, and vehicle suspension/tires. The first two component groups account for over 75% of a vehicle's weight. Weight can be reduced in all four component groups by improved structural design as well as by the use of alternative materials.

Improved structural design and packaging has been made possible through advanced computer simulations of structural strength, so that material use and shape can be optimized for the loads encountered. Most modern cars feature unibody designs where the body panels carry the structural loads, but several older models as well as many light trucks continue to use a separate chassis to carry structural loads. Heavy trucks, however, almost always utilize a separate chassis on which body components are mounted. A new architecture called space frame designs have emerged where structural loads are carried on skeletal frame from which body panels are hung. Improved packaging by optimization of component placement, body layout, and drivetrain location can also yield weight benefits. The placement of the engine transversely between the front wheels and driving the front wheels provides significant packaging benefits over front engine, rear-wheel-drive packages for light-duty vehicles.

The use of alternative materials such as ultra-high-strength steel, aluminum, and plastic composites is another way to reduce weight. Because of its low cost, steel, and cast iron continue to be the material of choice for body structures. Aluminum is already widely used for engine blocks and cylinder heads, and it is also used in critical suspension components. Some luxury cars now feature all-aluminum bodies, which weigh 30 to 35% less than their steel counterparts. Plastic composites are also widely used in body closures such as fenders, hood, and decklid with weight savings of 20 to 25% relative to steel parts. Such composites also see wide usage for lightweight interiors in the vehicle dashboard, seats, and door panels. Specially constructed prototypes maximizing the use of lightweight alternative materials have shown that weight reduction of 25 to 30% (relative to a conventional average steel vehicle) is possible, although with higher cost and with manufacturability constraints. The use of alternative materials in heavy trucks may not reduce loaded weight but will permit a larger payload to be carried.

Aerodynamic drag can be reduced by styling the vehicle's exterior shape and guiding the vehicle's interior airflow. At the speeds experienced by a typical vehicle, low drag shapes are a result of careful attention to airflow at the front of the vehicle, rear wheel wells and outside mirrors, and at the end of the roof. A measure of the drag is the aerodynamic drag coefficient, C_D, which is defined as

$$C_D = \frac{\text{Drag force}}{\frac{1}{2}\rho v^2 A},$$

where ρ is the density of air, v is the velocity of airflow, and A is the vehicle frontal area.

In the early 1980s, cars had drag coefficients of 0.45 to 0.5. By 2000, the most aerodynamic cars had drag coefficient of 0.25 to 0.28. Trucks typically have higher drag coefficients because of their boxy shape and increased ground clearance, relative to cars. Prototype cars with drag coefficients as low as 0.15 have been built, but such designs typically involve reduction of vehicle attributes such as reduced rear passenger headroom, reduced rear visibility, or reduced cargo space. Nevertheless, drag reduction still offers opportunities to reduce fuel consumption.

The tires' rolling resistance is the third major contributor to overall load. The tire rolling resistance coefficient (C_R) is a measure of tire energy loss, and is defined as

$$C_R = \frac{T}{L \cdot R},$$

where T is the torque required at any speed, R the tire radius, and L the vertical load on the tire. Typically, most modern tires have a C_R in the range of 0.009 to 0.012.

The tire rolling resistance results from a combination of tire-to-road friction and hysteresis. As the tire deforms, heat is dissipated in the tire's sidewall and tread due to the visco-elastic nature of rubber. In comparison, a steel wheel riding on steel rails has about one-tenth the rolling resistance of a rubber tire. Tire rolling resistance can be reduced by improved design of the tire tread, shoulders, and belts. In addition, the tire material formulation can significantly reduce hysteresis loss. The use of silica compounds mixed with rubber has been found to reduce rolling resistance, without affecting other desirable properties such as braking and wet traction. Design improvements and changes in belt material are also capable of reducing C_R with limited or no reduction of desirable attributes. It appears possible to reduce C_R by 15 to 25% in the short term and by up to 40% over the long term (~25 years).

4. IMPROVEMENT OF ENGINE EFFICIENCY

Engine efficiency on the driving cycle is the most significant determinant of vehicle fuel economy for a vehicle of a specific weight. Heat engine efficiency can be stated in several ways. One intuitively appealing method is to express the useful energy produced by an engine as a percentage of the total heat energy that is theoretically liberated by combusting the fuel. This is sometimes referred to as "the first law" efficiency, implying that its basis is the first law of thermodynamics, the law of conservation of energy. Another potential, but less widely used, measure is based on the second law of thermodynamics, which governs how much of that heat can be converted to work. Given a maximum combustion temperature (usually limited by engine material considerations and by emission considerations), the second law postulates a maximum efficiency based on an idealized heat engine cycle called the Carnot cycle. The ratio of the first law efficiency to the Carnot cycle efficiency can be utilized as a measure of how efficiently a particular engine is operating with reference to the theoretical maximum based on the second law of thermodynamics. However, the most common measure of efficiency used by automotive engineers is termed brake specific fuel consumption (bsfc), which is the amount of fuel consumed per unit time per unit of power. In the United States, the bsfc of engines is usually stated in pounds of fuel per brake horsepower hour, whereas the more common metric system measurement unit is in grams per kilowatt-hour (g/kwh). The term brake here refers to a common method historically used to measure engine shaft power output. Of course, all three measures of efficiency are related to each other.

The efficiency of Otto and diesel cycle engines is not constant but depends on the operating point of the engine as specified by its torque output and shaft speed (revolutions per minute or RPM). Engine design considerations, frictional losses, and heat losses result in a single operating point where efficiency is highest. This maximum efficiency for an s.i. engine usually occurs at relatively high torque and at low to mid-RPM within the operating RPM range of the engine. At idle, the efficiency is zero since the engine is consuming fuel but not producing any useful work. When considering efficiency in a vehicle, the maximum efficiency need not, by itself, be an indicator of the average efficiency under normal driving conditions, since engine speed and torque vary widely under normal driving. The maximum efficiency of an engine is of interest to automotive engineers, but a more practical measure of efficiency is its average efficiency during "normal" driving or during the official city and highway fuel economy test.

4.1 Theoretical Maximum Engine Efficiency

The characteristic features common to all piston internal combustion engines are as follows:

1. Intake and compression of the air or air-fuel mixture
2. Raising the temperature (and hence, the pressure) of the compressed air by combustion of fuel
3. The extraction of work from the high-pressure products of combustion by expansion
4. Exhaust of the products of combustion

Combustion of the homogenous air-fuel mixture in a spark ignition engine takes place very quickly relative to piston motion and is represented in idealized calculations as an event occurring at constant volume. According to classical thermodynamic theory, the thermal efficiency, η, of an idealized Otto cycle, starting with intake air-fuel mixture drawn in at atmospheric pressure, is given by

$$\eta = 1 - 1/r^{n-1}, \qquad (1)$$

where r is the compression (and expansion) ratio and n is the ratio of specific heat at constant pressure to that at constant volume for the mixture. The equation shows that efficiency increases with increasing compression ratio.

Using an n value of 1.4 for air, the equation predicts an efficiency of 58.47% at a compression ratio of 9:1. A value of $n = 1.26$ is more correct for products of combustion of a stoichiometric mixture of air and gasoline. A stoichiometric mixture corresponds to an air-fuel ratio of 14.7:1, and this air-fuel ratio is typical for most spark ignition engines sold in the United States. At this air-fuel ratio, calculated efficiency is about 43.5%. Actual engines yield still lower efficiencies even in the absence of mechanical friction, due to heat transfer to the wall of the cylinder and the inaccuracy associated with assuming combustion to be instantaneous. Figure 1 shows the pressure-volume cycle of a typical spark ignition engine and its departure from the ideal relationship.

Compression ratios are limited by the octane number of gasoline, which is a measure of its resistance to preignition or knock. At high compression ratios, the heat of compression of the air-fuel mixture becomes high enough to induce spontaneous combustion of small pockets of the mixture, usually those in contact with the hottest parts of the combustion chamber. These spontaneous combustion events are like small explosions that can damage the engine and

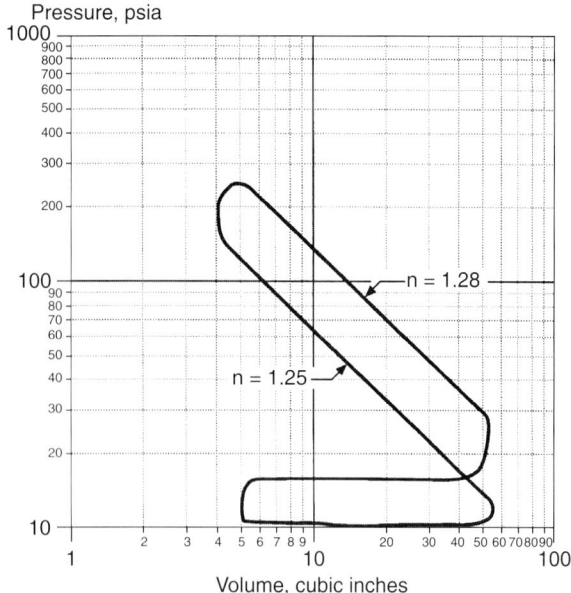

FIGURE 1 Pressure-volume diagram for a gasoline engine. Compression ratio = 8.7–11.

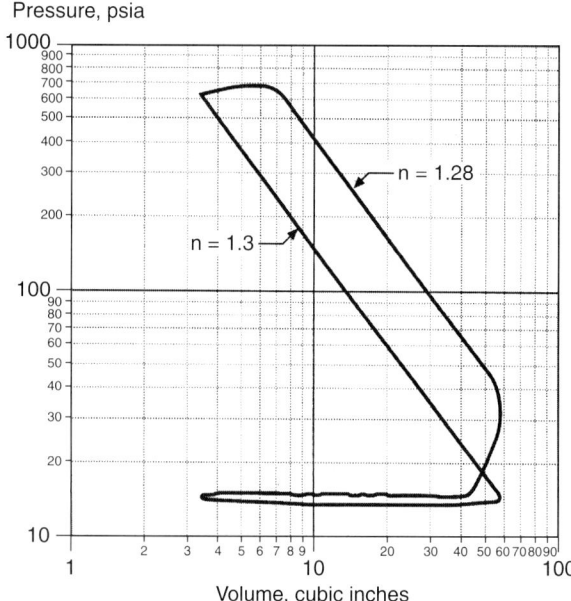

FIGURE 2 Pressure-volume diagram for a diesel engine. Compression ratio = 17–13.

can reduce efficiency depending on when they occur during the cycle. Higher octane number gasoline prevents these events, but also costs more and requires greater energy expenditure for manufacture at the refinery. The octane number is measured using two different procedures resulting in two different ratings for a given fuel, called motor octane and research octane number. Octane numbers displayed at the pump are an average of research and motor octane numbers, and most engines sold in the United States require regular gasoline with a pump octane number of 87. Typical compression rates for engines running on regular gasoline are in the range of 9:1 to 10:1.

The diesel, or compression ignition, engine differs from the spark ignition engine in that only air, rather than the air-fuel mixture, is compressed. The diesel fuel is sprayed into the combustion chamber at the end of compression in a fine mist of droplets and the diesel fuel ignites spontaneously upon contact with the compressed air due to the heat of compression. The sequence of processes (i.e., intake, compression, combustion, expansion, and exhaust) is similar to that of an Otto cycle engine. However, the combustion process occurs over a relatively long period and is represented in idealized calculations as an event occurring at constant pressure (i.e., combustion occurs as the piston moves downward to increase volume and decrease pressure at a rate offsetting the pressure rise due to heat release). Figure 2 shows the pressure-volume cycles for a typical diesel engine and its rela-

tionship to the ideal diesel cycle. If the ratio of volume at the end of the combustion period to the volume at the beginning of the period is r_c, or the cutoff-ratio, the thermodynamic efficiency of the idealized constant-pressure combustion cycle is given by

$$\eta = \frac{1}{r^{n-1}} \left[\frac{r_c^n - 1}{n(r_c - 1)} \right]. \qquad (2)$$

It can be seen that for $r_c = 1$, the combustion occurs at constant volume and the efficiency of the diesel and Otto cycle are equivalent.

The term r_c also measures the interval during which fuel is injected, and it increases as the power output is increased. The efficiency equation shows that as r_c is increased, efficiency falls so that the idealized diesel cycle is less efficient at high loads. The combustion process also is responsible for a major difference between diesel and Otto cycle engines. In an Otto cycle engine, intake is air throttled to control power while maintaining a near constant air-fuel ratio; in a diesel engine, power control is achieved by varying the amount of fuel injected while keeping the air mass inducted per cycle at near constant levels. In most operating modes, combustion occurs with considerable excess air in a c.i. engine, while combustion occurs at or near stoichiometric air-fuel ratios in a modern s.i. engine.

At the same compression ratio, the Otto cycle has the higher efficiency. However, diesel cycle engines normally operate at much higher compression ratios, since there are no octane limitations associated with this cycle. In fact, spontaneous combustion of the fuel is required in such engines, and the ease of spontaneous combustion is measured by a fuel property called cetane number. Most c.i. engines require diesel fuels with a cetane number over 40.

In practice, there are two kinds of c.i. engines, the direct injection type (DI) and the indirect injection type (IDI). The DI type utilizes a system where fuel is sprayed directly into the combustion chamber. The fuel spray is premixed and partially combusted with air in a prechamber in the IDI engine, before the complete burning of the fuel in the main combustion chamber occurs. DI engines generally operate at compression ratios of 15 to 20:1, while IDI engines operate at 18 to 23:1. The theoretical efficiency of a c.i. engine with a compression ratio of 20:1, operating at a cutoff ratio of 2, is about 54% (for combustion with excess air, n is approximately 1.3). In practice, these high efficiencies are not attained, for reasons similar to those outlined for s.i. engines.

4.2 Actual versus Theoretical Efficiency

Four major factors affect the efficiency of s.i. and c.i. engines. First, the ideal cycle cannot be replicated due to thermodynamic and kinetic limitations of the combustion process, and the heat transfer that occurs from the cylinder walls and combustion chamber. Second, mechanical friction associated with the motion of the piston, crankshaft, and valves consume a significant fraction of total power. Since friction is a stronger function of engine speed rather than torque, efficiency is degraded considerably at light load and high RPM conditions. Third, aerodynamic frictional losses associated with airflow through the air cleaner, intake manifold and valves, exhaust manifold, silencer, and catalyst are significant, especially at high airflow rates through the engine. Fourth, pumping losses associated with throttling the airflow to achieve part-load conditions in spark ignition engines are very high at light loads. Note that c.i. engines do not usually have throttling loss, and their part load efficiencies are superior to those of s.i. engines. Efficiency varies with both speed and load for both engine types.

Hence, production spark ignition or compression ignition engines do not attain the theoretical values of efficiency, even at their most efficient operating point. In general, for both types of engines, the maximum efficiency point occurs at an RPM that is intermediate to idle and maximum RPM and at a level that is 60 to 75% of maximum torque. On-road average efficiencies of engines used in cars and light trucks are much lower than peak efficiency, since the engines generally operate at very light loads during city driving and steady-state cruise on the highway. High power is utilized only during strong accelerations, at very high speeds, or when climbing steep gradients. The high load conditions are relatively infrequent, and the engine operates at light loads much of the time during normal driving.

During normal driving, the heat of fuel combustion is lost to a variety of sources and only a small fraction is converted to useful output, resulting in the low values for on-road efficiency. Figure 3 provides an example of the heat balance for a typical modern small car with a spark ignition engine under a low speed (25 mph or 40 mph) and a high-speed (62 mph or 100 mph) condition. At very low driving speeds typical of city driving, most of the heat energy is lost to the engine coolant. Losses associated with other waste heat include radiant and convection losses from the hot engine block and heat losses to the engine oil. A similar heat loss diagram for a diesel c.i. would indicate lower heat loss to the exhaust and

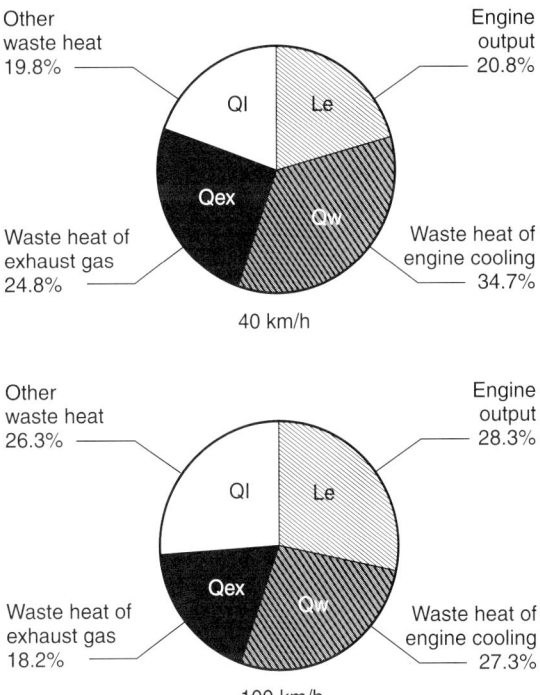

FIGURE 3 Heat balance of a passenger car equipped with 1500cc engine 17.

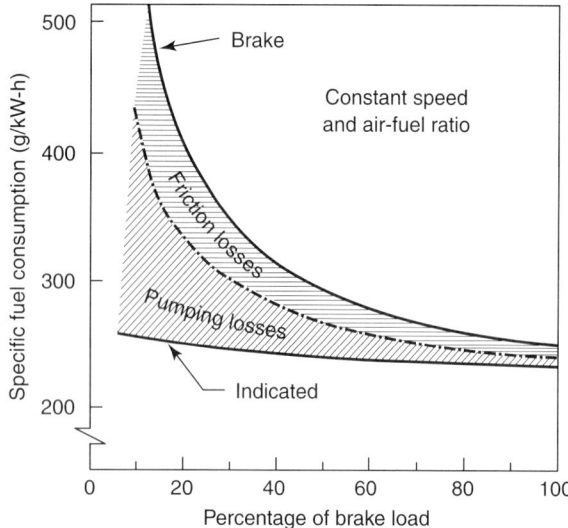

FIGURE 4   Specific fuel consumption versus engine load 18.

coolant and an increased fraction of heat converted to work, especially at the low speed condition. During stop-and-go driving conditions typical of city driving, efficiencies are even lower than those indicated in Fig. 3 because of the time spent at idle where efficiency is zero. Under the prescribed U.S. city cycle conditions, typical modern spark ignition engines have an efficiency of about 18%, modern IDI c.i. engines have an efficiency of about 21%, and modern DI diesel have an efficiency of about 23%.

Another method of examining the energy losses is by allocating the power losses starting from the power developed within the cylinder. The useful work corresponds to the area that falls between the compression and expansion curve depicted in Figs 1 and 2. The pumping work that is subtracted from this useful work, referred to as indicated work, is a function of how widely the throttle is open and, to a lesser extent, the speed of the engine. Figure 4 shows the dependence of specific fuel consumption (or fuel consumption per unit of work) with load, at constant (low) engine RPM. Pumping work represents only 5% of indicated work at full load, low RPM conditions, but increases to over 50% at light loads of less than two-tenths of maximum power.

Mechanical friction and accessory drive power, on the other hand, increase nonlinearly with engine speed but do not change much with the throttle setting. Figure 5 shows the contribution of the various engine components as well as the alternator, water pump and oil pump to total friction, expressed in terms of mean effective pressure, as a function of RPM. The brake mean effective pressure is

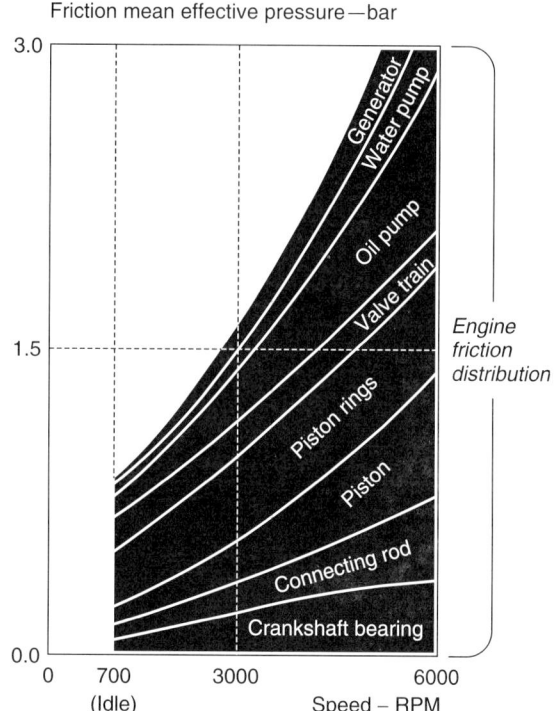

FIGURE 5   Friction distribution as a function of engine speed 19.

a measure of specific torque, or torque per unit of engine displacement; typical engine brake mean effective pressure (bmep) of spark ignition engines that are not supercharged range from 8.5 to 10 bar. Hence, friction accounts for about 25% of total indicated power at high RPM (~ 6000) but only for about 10% of indicated power at low RPM (~ 2000) in spark ignition engines. Friction in a c.i. engine is higher because of the need to maintain an effective pressure seal at high compression ratios, and the friction mean effective pressure is 30 to 40% higher than that for a dimensionally similar s.i. engine at the same RPM. Since the brake mean effective pressure of a diesel is also lower than that of a gasoline engine, friction accounts for 15 to 16% of indicated maximum power even at 2000 RPM. Typical bmep values for a naturally aspirated c.i. engine range from 6.5 to 7.5 bar.

5. INCREASING THE EFFICIENCY OF SPARK IGNITION ENGINES

5.1 Design Parameters

Engine valvetrain design is a widely used method to classify spark ignition engines. The first spark ignition engines were of the side-valve type, but such engines have not been used in automobiles for several decades, although some engines used in off-highway applications, such as lawn mowers or forklifts, continue to use this design. The overhead valve (OHV) design supplanted the side-valve engine by the early 1950s, and continues to be used in many U.S. engines in much improved form. The overhead cam engine (OHC) is the dominant design used in the rest of the developed world. The placement of the camshaft in the cylinder heads allows the use of simple, lighter valvetrain, and valves can be opened and closed more quickly as a result of the reduced inertia. This permits better flow of intake and exhaust gases, especially at high RPM, with the result that an OHC design typically can produce greater power at high RPM than an OHV design of the same displacement.

A more sophisticated version of the OHC engine is the double overhead cam (DOHC) engine where two separate camshafts are used to activate the intake and exhaust valves, respectively. The DOHC design permits a very light valvetrain as the camshaft can actuate the valves directly without any intervening mechanical linkages. The DOHC design also allows some layout simplification, especially in engines that feature two intake valves and two exhaust valves (4-valve). The 4-valve engine has become popular since the mid-1980s and Japanese manufacturers, in particular, have embraced the DOHC 4-valve design. The DOHC design permits higher specific output than an OHC design, with the 4-valve DOHC design achieving the highest specific output, in excess of 70 BHP/liter of displacement.

5.2 Thermodynamic Efficiency

Increases in thermodynamic efficiency within the limitations of the Otto cycle are obviously possible by increasing the compression ratio. However, compression ratio is also fuel octane limited, and increases in compression ratio depend on how the characteristics of the combustion chamber and the timing of the spark can be tailored to prevent knock while maximizing efficiency.

Spark timing is associated with the delay in initiating and propagating combustion of the air-fuel mixture. To complete combustion before the piston starts its expansion stroke, the spark must be initiated a few crank angle degrees (advance) before the piston reaches top dead center. For a particular combustion chamber, compression ratio, and air-fuel mixture, there is an optimum level of spark advance for maximizing combustion chamber pressure and, hence, fuel efficiency. This level of spark advance is called MBT for maximum for best torque. However, MBT spark advance can result in knock if fuel octane is insufficient to resist preignition at the high pressures achieved with this timing. Hence, there is an interplay between spark timing and compression ratio in determining the onset of knock. Retarding timing from MBT reduces the tendency to knock but decreases fuel efficiency. Emissions of hydrocarbons and oxides of nitrogen (NO_x) are also dependent on spark timing and compression ratio, so that emission constrained engines require careful analysis of the knock, fuel efficiency, and emission trade-offs before the appropriate value of compression ratio and spark advance can be selected.

Electronic control of spark timing has made it possible to set spark timing closer to MBT relative to engines with mechanical controls. Due to production variability and inherent timing errors in a mechanical ignition timing system, the average value of timing in mechanically controlled engines had to be retarded significantly from the MBT timing. This protects the fraction of engines with higher than average advance due to production variability from knock. The use of electronic controls coupled with magnetic or optical sensors of crankshaft position has reduced the variability of timing between production engines and also allowed better control during transient engine operation. Engines have been equipped with knock sensors, which are essentially vibration sensors tuned to the frequency of knock. These sensors allow for advancing ignition timing to the point where trace knock occurs, so that timing is optimal for each engine produced regardless of production variability.

High-swirl, fast-burn combustion chambers have been developed to reduce the time taken for the air-fuel mixture to be fully combusted. The shorter the burn time, the more closely the cycle approximates the theoretical Otto cycle with constant volume combustion and the greater the thermodynamic efficiency. Reduction in burn time can be achieved by having a turbulent vortex within the combustion chamber that promotes flame propagation and mixing. The circular motion of the air-fuel mixture is known as swirl, and turbulence is also enhanced by shaping the piston so that gases near the cylinder wall are pushed rapidly towards the center in a motion known as squish. Improvements in flow visualization and computational fluid dynamics have allowed the optimization of intake valve, inlet port, and combustion chamber geometry to achieve desired flow characteristics. Typically, these designs have resulted in a 2 to 3% improvement in thermodynamic efficiency and fuel economy. The high-swirl chambers also allow higher compression ratios and reduced spark advance at the same fuel octane number. The use of these types of combustion chambers has allowed compression ratio from about 8:1 in the early 1980s to 10:1 in the early 2000s, and further improvements are likely. In newer engines of the 4-valve DOHC type, the sparkplug is placed at the center of the combustion chamber, and the chamber can be made very compact by having a nearly hemispherical shape. Engines incorporating these designs have compression ratios of 10:1 while still allowing the use of regular 87 octane gasoline. Increases beyond 10:1 are expected to have diminishing benefits in efficiency and fuel economy and compression ratios beyond 12:1 are not likely to be beneficial unless fuel octane is raised simultaneously.

5.3 Reduction in Mechanical Friction

Mechanical friction losses are being reduced by converting sliding metal contacts to rolling contacts, reducing the weight of moving parts, reducing production tolerances to improve the fit between pistons and bore, and improving the lubrication between sliding or rolling parts.

Friction reduction has focused on the valvetrain, pistons, rings, crankshaft, crankpin bearings, and the oil pump. Valvetrain friction accounts for a larger fraction of total friction losses at low engine RPM than at high RPM. The sliding contract between the cam that activates the valve mechanism through a pushrod in an OHV design, or a rocker arm in an OHV design, can be substituted with a rolling contact by means of a roller cam follower. Roller cam followers have been found to reduce fuel consumption by 2 to 4% during city driving and 1 to 2% in highway driving. The use of lightweight valves made of ceramics or titanium is another possibility for the future. The lightweight valves reduce valve train inertia and also permit the use of lighter springs with lower tension. Titanium alloys are also being considered for valve springs which operate under heavy loads. There alloys have only half the shear modulus of steel and fewer coils are needed to obtain the same spring constant. A secondary benefit associated with lighter valves and springs is that the erratic valve motion at high RPM is reduced, allowing increased engine RPM range and power output.

The pistons and rings contribute to approximately half of total friction. The primary function of the rings is to minimize leakage of the air-fuel mixture from the combustion chamber to the crankcase, and oil leakage from the crankcase to the combustion chamber. The ring pack for most engines is composed of two compression rings and an oil ring. The rings have been shown to operate hydrodynamically over the cycle, but metal-to-metal contact occurs often at the top and bottom of the stroke. The outward radial force of the rings are as a result of installed ring tension and contribute to effective sealing as well as friction. Various low-tension ring designs were introduced in the 1980s, especially since the need to conform to axial diameter variations or bore distortions have been reduced by improved cylinder manufacturing techniques. Reduced tension rings have yielded friction reduction in the range of 5 to 10%, with fuel economy improvements of 1 to 2%. Elimination of one of the two compression rings has also been tried on some engines, and two-ring pistons may be the low friction concept for the future.

Pistons have also been redesigned to decrease friction. Prior to the 1980s, piston had large "skirts" to absorb side forces associated with side-to-side piston motion due to engine manufacturing inaccuracies. Pistons with reduced skirts diminish friction by having lower surface area in contact with the cylinder wall, but this effect is quite small. A larger effect is obtained for the mass reduction of a piston with smaller skirts and piston skirt size has seen continuous reduction since the 1980s. Reducing the reciprocating mass reduces the piston-to-bore loading. Secondary benefits include reduced engine weight and reduced vibration. Use of advanced materials also result in piston weight reduction. Lightweight pistons use hypereutectic aluminum alloys, while future pistons could use composite materials such as fiber-reinforced plastics. Advanced materials can also reduce the weight of the connecting rod, which also contributes to the side force on a piston.

The crankshaft bearings include the main bearings that support the crankshaft and the crankpin bearings and are of the journal bearing type. These bearings contribute to about 25% of total friction, while supporting the stresses transferred from the piston. The bearings run on a film of oil and detailed studies of lubrication requirements has led to optimization of bearing width and clearances to minimize engine friction. Studies on the use of roller bearings rather than journal bearings in this application has shown further reduction in friction is possible. Crankshaft roller bearings are used only in some two-stroke engines such as outboard motors for boat propulsion, but their durability in automotive applications has not been established.

Coatings of the piston and ring surfaces with materials to reduce wear also contribute to friction reduction. The top ring, for example, is normally coated with molybdenum, and new proprietary coating materials with lower friction are being introduced. Piston coatings of advanced high temperature plastics or resin are used widely and are claimed to reduce friction by 5% and fuel consumption by 1%.

The oil pumps generally used in most engines are of the gear pump type. Optimization of oil flow rates and reduction of the tolerances for the axial georotor clearance has led to improved efficiency, which translates to reduced drive power. Friction can be reduced by 2 to 3% with improved oil pump designs for a gain in fuel economy of about half of a percent.

Improvements to lubricants used in the engine also contribute to reduced friction and improved fuel economy. There is a relationship between oil viscosity, oil volatility, and engine oil consumption. Reduced viscosity oils traditionally resulted in increased oil consumption, but the development of viscosity index (VI) improvers had made it possible to tailor the viscosity with temperatures to formulate multigrade oils such as 10W-40 (these numbers refer to the range of viscosity covered by a multigrade oil). These multigrade oils act like low-viscosity oils during cold starts of the engine, reducing fuel consumption, but retain the lubricating properties of higher viscosity oils after the engine warms up to normal operating temperature. The development of 5W-20 oils and 5W-40 oils can contribute to a fuel economy improvement by further viscosity reduction. Friction modifiers containing molybdenum compounds have also reduced friction without affecting wear or oil consumption. Future synthetic oils combining reduced viscosity and friction modifiers can offer good wear protection, low oil consumption, and extended drain capability along with small improvements to fuel economy in the range of 1 to 3% over current oils.

5.4 Reduction in Pumping Loss

Reductions in flow pressure loss can be achieved by reducing the pressure drop that occurs in the flow of air (air-fuel mixture) into the cylinder and the combusted mixture through the exhaust system. However, the largest part of pumping loss during normal driving is due to throttling, and strategies to reduce throttling loss have included variable valve timing and lean-burn systems.

The pressure losses associated with the intake system and exhaust system have been typically defined in terms of volumetric efficiency, which is a ratio of the actual airflow through an engine to the airflow associated with filling the cylinder completely. The volumetric efficiency can be improved by making the intake airflow path as free of flow restrictions as possible through the air filters, intake manifolds, and valve ports. The shaping of valve ports to increase swirl in the combustion chamber can lead to reduced volumetric efficiency, leading to a trade-off between combustion and volumetric efficiency.

More important, the intake and exhaust processes are transient in nature as they occur only over approximately half a revolution of the crankshaft. The momentum effects of these flow oscillations can be exploited by keeping the valves open for durations greater than half a crankshaft revolution. During the intake stroke, the intake valve can be kept open beyond the end of the intake stroke, since the momentum of the intake flow results in a dynamic pressure that sustains the intake flow even when the piston begins the compression stroke. A similar effect is observed in the exhaust process, and the exhaust valve can be held open during the initial part of the intake stroke. These flow momentum effects depend on the velocity of the flow which is directly proportional to engine RPM. Increasing the valve opening duration helps volumetric efficiency at high RPM but hurts it at low RPM. Valve timing and overlap are selected to optimize the trade-off between high and low RPM performance characteristics.

Efficiency improvements can be realized by changing the valve overlap period to provide less overlap at idle and low engine speeds and greater overlap at high RPM. In DOHC engines, where separate crankshafts actuate the intake and exhaust valves, the valve overlap period can be changed by rotating the camshafts relative to each other. Such mechanisms have been commercialized engines show low RPM torque improvements of 7 to 10% with no sacrifice in maximum horsepower attained in the 5500 to 6000 RPM range. Variable valve overlap period is just one aspect of a more comprehensive variable valve timing system.

The oscillatory intake and exhaust flows can allow volumetric efficiency to be increased by exploiting resonance effects associated with pressure waves similar to those in organ pipes. The intake manifolds can be designed with pipe lengths that resonate, so that a high-pressure wave is generated at the intake valve as it is about to close, to cause a supercharging effect. Exhaust manifolds can be designed to resonate to achieve the opposite pressure effect to purge exhaust gases from this cylinder. For a given pipe length, resonance occurs only at a certain specific frequency and its integer multiples so that, historically, tuned intake and exhaust manifolds could help performance only in certain narrow RPM ranges. The incorporation of a resonance tanks using the Helmholtz resonator principle in addition to tuned length intake pipes has led to improved intake manifold design that provide benefits over broader RPM ranges. Variable resonance systems have been introduced, where the intake tube lengths are changed at different RPM by opening and closing switching valves to realize smooth and high torque across virtually the entire engine speed range. Typically, the volumetric efficiency improvement is in the range of 4 to 5% over fixed resonance systems.

Another method to increase efficiency is by increasing valve area. A 2-valve design is limited in valve size by the need to accommodate the valves and sparkplugs in the circle defined by the cylinder base. The active flow area is defined by the product of valve circumference and lift. Increasing the number of valves is an obvious way to increase total valve area and flow area, and the 4-valve system, which increases flow area by 25 to 30% over 2-valve layouts, has gained broad acceptance. The valves can be arranged around the cylinder bore and the sparkplug placed in the center of the bore to improve combustion. Analysis of additional valve layout designs that take into account the minimum required clearance between valve seats and the sparkplug location suggest that fivevalve designs (3 intake, 2 exhaust) can provide an additional 20% increase in flow area, at the expense of increased valvetrain complexity. Additional valves do not provide further increases in flow area either due to noncentral plug locations or valve-to-valve interference.

Under most normal driving conditions, the throttling loss is the single largest contributor to reduction in engine efficiency. In s.i. engines, the air is throttled ahead of the intake manifold by means of a butterfly valve that is connected to the accelerator pedal. The vehicle's driver demands a power level by depressing or releasing the accelerator pedal, which in turn opens or closes the butterfly valve. The presence of the butterfly valve in the intake air stream creates a vacuum in the intake manifold at part throttle conditions, and the intake stroke draws in air at reduced pressure, resulting in pumping losses. These losses are proportional to the intake vacuum and disappear at wide open throttle.

Measures to reduce throttling loss are varied. The horsepower demand by the driver can be satisfied by any combination of torque and RPM since

$$\text{Power} = \text{Torque} \times \text{RPM}.$$

The higher the torque, the lower the RPM to satisfy a given power demand. Higher torque implies less throttling, and the lower RPM also reduces friction loss so that the optimum theoretical fuel efficiency at a given level of horsepower demand occurs at the highest torque level the engine is capable of. In practice, the highest level is never chosen because of the need to maintain a large reserve of torque for immediate acceleration and also because engine vibrations are a problem at low RPM, especially near or below engine speeds referred to as lugging RPM. Nevertheless, this simple concept can be exploited to the maximum by using a small displacement high specific output engine in combination with a multispeed transmission with five or more forward gears. The larger number of gears allows selection of the highest torque/lowest RPM combination for fuel economy at any speed and load, while maintaining sufficient reserve torque for instantaneous changes in power demand. A specific torque increase of 10% can be utilized to provide a fuel economy benefit of 3 to 3.5% if the engine is downsized by 8 to 10%. In light vehicles, the number of forward gears has been increasing from 3 to 5, and six-speed transmissions are likely to be standard in the future. The continuously variable transmission is other development that allow continuous change of gear ratios over a specific range.

Lean-burn is another method to reduce pumping loss. Rather than throttling the air, the fuel flow is reduced so that the air-fuel ratio increases, or becomes leaner. (In this context, the c.i. engine is a lean-burn engine.) Most s.i. engines, however, do not run well at air: fuel ratios leaner than 18:1, as the combustion quality deteriorates under lean con;ditions. Engines constructed with high swirl and turbulence in the intake charge can run well at air: fuel ratios up to 21:1. In a vehicle, lean-burn engines are calibrated lean only at light loads to reduce throttling loss, but run at stoichiometric or rich air: fuel ratios at high loads to maximize power. The excess air combustion at light loads has the added advantage of having a favorable effect on the polytropic coefficient, n, in the efficiency equation. Modern lean-burn engines do not eliminate throttling loss, but the reduction is sufficient to improve vehicle fuel economy by 8 to 10%. The disadvantage of lean burn is that such engines cannot yet use catalytic controls to reduce emissions of oxide of nitrogen (NO_x), and the in-cylinder NO_x emission control from running lean is sometimes insufficient to meet stringent NO_x emissions standards. However, there are developments in lean NO_x catalysts that could allow lean-burn engines to meet the most stringent NO_x standards proposed in the future.

Another type of lean-burn s.i. engine is the stratified charge engine. Research is focused on direct-injection stratified charge (DISC) engines where the fuel is sprayed into the combustion chamber, rather than into or ahead of the intake valve. Typically, this enables the air:fuel ratio to vary axially or radially in the cylinder, with the richest air:fuel ratios present near the sparkplug or at the top of the cylinder. Stratification requires very careful design of the combustion chamber shape and intake swirl, as well as of the fuel injection system. Advanced direct injection systems have been able to maintain stable combustion at total air:fuel ratios as high as 40:1. Such engines have been commercialized in 2000 in Europe and Japan.

Variable valve timing is another method to reduce throttling loss. By closing the intake valve early, the intake process occurs over a smaller fraction of the cycle, resulting in a lower vacuum in the intake manifold. It is possible to completely eliminate the butterfly valve that throttles air and achieve all part load settings by varying the intake valve opening duration. However, at very light load, the intake valve is open for a very short duration, and this leads to weaker in-cylinder gas motion and reduced combustion stability. At high RPM, the throttling loss benefits are not realized fully. Throttling occurs at the valve when the valve closing time increases relative to the intake stroke duration at high speeds, due to the valvetrain inertia. Hence, throttling losses can be decreased by 80% at light load, low RPM conditions, but by only 40 to 50% at high RPM, even with fully variable valve timing.

Variable valve timing can also provide a number of other benefits, such as reduced valve overlap at light loads/low speeds (discussed earlier) and maximized output over the entire range of engine RPM. Fully variable valve timing can result in engine output levels of up to 100 BHP/liter at high RPM with little or no effect on low-speed torque. In comparison to an engine with fixed valve timing that offers equal performance, fuel efficiency improvements of 7 to 10% are possible. The principal drawback has historically been the lack of a durable and low-cost mechanism to implement valve timing changes. A number of new systems have been introduced that are ingenious mechanisms with the required durability.

6. INCREASING THE EFFICIENCY OF COMPRESSION IGNITION (DIESEL) ENGINES

Compression ignition engines, commonly referred to as diesel engines, are in widespread use. Most c.i. engines in light-duty vehicle applications are of the indirect injection type (IDI), while most c.i. engines in heavy-duty vehicles are of the direct injection type. In comparison to s.i. engines, c.i. engines operate at much lower brake mean effective pressures of (typically) about 7 to 8 bar at full load. Maximum power output of a c.i. engine is limited by the rate of mixing between the injected fuel spray and hot air. At high fueling levels, inadequate mixing leads to high black smoke, and the maximum horsepower is usually smoke limited for most c.i. engines. Naturally aspirated diesel engines for light-duty vehicle use have specific power outputs of 25 to 35 BHP per liter, which is about half the specific output of a modern s.i. engine. However, fuel consumption is significantly better, and c.i. engines are preferred over s.i. engines where fuel economy is important.

Due to the combustion process, as well as the high internal friction of a c.i. engine, maximum speed is typically limited to less than 4500 RPM, which partially explains the lower specific output of c.i. engine. In light-duty vehicle use, an IDI engine can display between 20 to 40% better fuel economy depending on whether the comparison is based on engines of equal displacement or of equal power output in the same RPM range. The improvement is largely due to the superior part load efficiency of the c.i. engine, as there is no throttling loss. At high vehicle speeds (>120 km/hr), the higher internal friction of the c.i. engine offsets the reduced throttling loss, and the fuel efficiency difference between s.i. and c.i. engine narrows considerably.

Most of the evolutionary improvements for compression ignition engines in friction and pumping loss reduction are conceptually similar to those described for s.i. engines, and this section focuses on the unique aspects of c.i. engine improvements.

6.1 Design Parameters

Note that c.i. engines have also adopted some of the same valvetrain designs as those found in s.i. engines. While most c.i. engines were of the OHV type, European c.i. engines for passenger car use are of the OHC type. The c.i. engine is not normally run at high RPM, so that the difference in specific output between an OHV and an OHC design is small. The OHC design does permit a simpler and lighter cylinder block casting, which is beneficial for overcoming some of the inherent weight liabilities. OHC designs also permits the camshaft to directly activate the fuel injector in unit injector designs, which are capable of high injection pressure and fine atomization of the fuel spray. Four-valve OHC or DOHC designs allow central placement of the fuel injector in the cylinder, which enhances uniform mixing of air and fuel.

6.2 Thermodynamic Efficiency

The peak efficiency of an IDI engine is comparable to or only slightly better than the peak efficiency of an s.i. engine, based on average values for engines in production. The contrast between theoretical and actual efficiency is notable, and part of the reason is that the prechamber in the IDI diesel is a source of energy loss. The design of the prechamber is optimized to promote swirl and mixing of the fuel spray with air, but the prechamber increases total combustion time. Its extra surface area also results in more heat transfer into the cylinder head.

Direct injection (DI) systems avoid the heat and flow losses from the prechamber by injecting the fuel into the combustion chamber. The combustion process in DI diesels consists of two phases. The first phase consists of an ignition delay period followed by spontaneous ignition of the fuel droplets. The second phase is characterized by diffusion burning of the droplets. The fuel injection system must be capable of injecting very little fuel during the first phase and provide highly atomized fuel and promote intensive mixing during the second. Historically, the mixing process has been aided by creating high swirl in the combustion chamber to promote turbulence. However, high swirl and turbulence also lead to flow losses and heat losses, thus reducing efficiency. The newest concept is the quiescent chamber where all of the mixing is achieved by injecting fuel at very high pressures to promote fine atomization and complete penetration of the air in the combustion chamber. New fuel injection systems using unit injectors can achieve pressures in excess of 1500 bar, twice as high as injection pressures utilized previously. Quiescent combustion chamber designs with high-pressure fuel injection systems have provided to be very fuel efficient and are coming into widespread use in heavy-duty truck engines. These systems have the added advantage of reducing particulate and smoke emissions.

DI engines have entered the light-duty vehicle market, but these engines still utilize swirl type combustion chambers. In combination with turbocharging (see Section 6), the new DI engines have attained peak efficiencies of over 41%. Fuel economy improvements in the composite cycle relative to IDI engines are in the 12 to 15% range, and are up to 40% higher than naturally aspirated s.i. engines with similar torque characteristics. It is not clear if quiescent combustion chambers will be ever used in DI engines for cars, since the size of the chamber is quite small and fuel impingement on cylinder walls is a concern.

Although the efficiency equation shows that increasing compression ratio has a positive effect on efficiency, practical limitations preclude any significant efficiency gain through this method. At high compression ratios, the size of the combustion chamber is reduced, and the regions of dead air trapped between the cylinder and piston edges and crevices became relatively large, leading to poor air utilization, reduced specific output, and, potentially, higher smoke. Moreover, the stresses on the engine increase with increasing compression ratio, making the engine heavy and bulky. The compression ratios are already somewhat higher than optimal to provide enough heat of compression so that a cold start at low ambient temperature is possible.

6.3 Friction and Pumping Loss

Most of the friction reducing technologies that can be adopted in s.i. engines are conceptually similar to those that can be adopted for diesels. There are limitations to the extent of reduction of ring tension and piston size due to the high compression ratio of c.i. engines, but roller cam followers, optimized crankshaft bearings, and multigrade lubricants have also been adopted for c.i. engine use. Since friction is a larger fraction of total loss, a 10% reduction in fraction in a c.i. engine can lead to a 3 to 4% improvement in fuel economy.

Pumping losses are not as significant a contributor to overall energy loss in a c.i. engine, but tuned intake manifolds and improved valve port shapes and valve designs have also improved volumetric efficiency of modern c.i. engines. Four-valve designs, in widespread use in the heavy truck market, have appeared in passenger cars, but their benefits are smaller in c.i. engine use due to the low maximum RPM relative to s.i. engines. Nevertheless, the 4-valve head with a centrally mounted injector is particularly useful in DI engines since it allows for symmetry in the fuel spray with resultant good air utilization.

Variable valve timing or any form of valve control holds little benefit for c.i. engines due to the lack of throttling loss, and lack of high RPM performance. Valve timing can be varied to reduce the effective compression ratio, so that a very high ratio can be used for cold starts, but a lower, more optimal, ratio for fully warmed up operation.

7. INTAKE CHARGE BOOSTING

Most c.i. and s.i. engines for light vehicle use intake air at atmospheric pressure. One method to increase maximum power at wide open throttle is to increase the density of air supplied to the intake by precompression. This permits a smaller displacement engine to be substituted without loss of power and acceleration performance. The use of a smaller displacement engine reduces pumping loss at part load and friction loss. However, intake charge compression has its own drawbacks. Its effect is similar to raising the compression ratio in terms of peak cylinder pressure, but maximum cylinder pressure is limited in s.i. engines by the fuel octane. Charge boosted engines generally require premium gasolines with higher octane number if the charge boost levels are more than 0.2 to 0.3 bar over atmospheric in vehicles for street use. Racing cars use boost levels up to 1.5 bar in conjunction with a very high octane fuel such as methanol. This limitation is not present in a c.i. engine, and charge boosting is much more common in c.i. engine applications. Most c.i. engines in heavy-duty truck applications use charge boosting.

Intake charge boosting is normally achieved by the use of turbochargers or superchargers. Turbochargers recover the wasted heat and pressure in the exhaust through a turbine, which in turn drives a compressor to boost intake pressure. Superchargers are generally driven by the engine itself and are theoretically less efficient than a turbocharger. Many engines that use either device also utilize an aftercooler that cools the compressed air as it exits from the supercharger or turbocharger before it enters the s.i. engine. The aftercooler increases engine specific power output by providing the engine with a denser intake charge, and the lower temperature also helps in preventing detonation, or knock. Charge boosting is useful only under wide open throttle conditions in s.i. engines, which occur rarely in normal driving, so that such devices are usually used in high-performance vehicles. In c.i. engines, charge boosting is effective at all speeds and levels.

7.1 Turbochargers

Turbochargers in automotive applications are of the radial flow turbine type. The turbine extracts pressure energy from the exhaust stream and drives a compressor that increases the pressure of the intake air. A number of issues affect the performance of turbomachinery, some of which are a result of natural laws governing the interrelationship between pressure, airflow, and turbocharger speed. Turbochargers do not function at light load because there is very little energy in the exhaust stream. At high load, the turbocharger's ability to provide boost is a nonlinear function of exhaust flow. At low engine speed and high load, the turbocharger provides little boost, but boost increases rapidly beyond a certain flow rate that is dependent on the turbocharger size. The turbocharger also has a maximum flow rate, and the matching of a turbocharger's flow characteristics to a piston engine's flow requirements involves a number of trade-offs. If the turbocharger is sized to provide adequate charge boost at moderate engine speeds, high RPM boost is limited and there is a sacrifice in maximum power. A larger turbocharger capable of maximizing power at high RPM sacrifices the ability to provide boost at normal driving conditions. At very low RPM (for example, when accelerating from a stopped condition), no practical design provides boost immediately. Moreover, the addition of turbocharger requires the engine compression ratio to be decreased by 1.5 to 2 points (or 1 to 1.5 with an aftercooler) to prevent detonation. The net result is that turbocharged engines have *lower* brake specific fuel efficiencies than engines of equal size, but can provide some efficiency benefit when compared to engines of equal mid range or top end power. During sudden acceleration, the turbocharger does not provide boost instantaneously due to its inertia, and turbocharged vehicles can have noticeably different acceleration characteristics than naturally aspirated vehicles. New variable geometry turbochargers have improved response and better boost characteristics over the operating range.

Turbochargers are much better suited to c.i. engines since these engines are unthrottled and the combustion process is not knock limited. Airflow at a given engine load/speed setting is always higher for a c.i. engine relative to an s.i. engine, and this provides a less restricted operating regime for the turbocharger. The lack of a knock limit also allows increased boost and removes the need to cap boost pressure under most operating conditions. Turbocharged c.i. engines offer up to 50% higher specific power and torque, and about 10% better fuel economy than naturally aspirated c.i. engines of approximately equal torque capability.

7.2 Superchargers

Most s.i. engine superchargers are driven off the crankshaft and are of the Roots blower or positive displacement pump type. In comparison to turbochargers, these superchargers are bulky and weigh considerably more. In addition, the superchargers are driven off the crankshaft, absorbing 3 to 5% of the engine power output depending on pressure boost and engine speed.

The supercharger, however, does not have the low RPM boost problems associated with turbochargers and can be designed to nearly eliminate any time lag in delivering the full boost level. As a result, superchargers are more acceptable to consumers from a driveability viewpoint. The need to reduce engine compression ratio and the supercharger's drive power requirement detract from overall efficiency. In automotive applications, a supercharged engine can replace a naturally aspirated engine that is 30 to 35% larger in displacement, with a net pumping loss reduction. Overall, fuel economy improves by about 8% or less, if the added weight effects are included.

Superchargers are less efficient in combination with c.i. engines, since these engines run lean even at full load, and the power required for compressing air is proportionally greater. Supercharged c.i. engines are not yet commercially available, since the turbocharger appears far more suitable in these application.

8. ALTERNATIVE HEAT ENGINES

A number of alternative engines types have been researched for use in passenger cars but have not yet proved successful in the market place. A brief discussion of the suitability of four engines for automotive power plants is provided next.

The Wankel engine is the most successful of the four engines in that it has been in commercial production in limited volume since the 1970s. The thermodynamic cycle is identical to that of a four-stroke engine, but the engine does not use a reciprocating piston in a cylinder. Rather a triangular rotor spins eccentrically inside a Fig. 8–shaped casing. The volume trapped between the two rotor edges and the casing varies with rotor position, so that the intake, comparison, expansion, and exhaust stroke occur as the rotor spins through one revolution. The engine is very compact relative to a piston s.i. engine of equal power, and the lack of reciprocating parts provides smooth operation. However, the friction associated with the rotor seals is high, and the engine also suffers from more heat losses than an s.i. engine. For these reasons, the Wankel engine's efficiency has always been below that of a modern s.i. piston engine.

The two-stroke engine is widely used in small motorcycles but was thought to be too inefficient and polluting for use in passenger cars. One development is the use of direct injection stratified charge (DISC) combustion with this type of engine. One of the major problems with the two-stroke engine is that the intake stroke overlaps with the exhaust stroke resulting in some intake mixture passing uncombusted into the exhaust. The use of a DISC design avoids this problem since only air is inducted during intake. Advanced fuel injection systems have been developed to provide a finely atomized mist of fuel just prior to spark initiation and to sustain combustion at light loads. The two-stroke engines of this type are thermodynamically less efficient than four-stroke DISC engines, but the internal friction loss and weight of two-stroke engine is much lower than a four-stroke engine of equal power. As a result, the engine may provide fuel economy equal or superior to that of a DISC (four-stroke) engine when installed in a vehicle. Experimental prototypes have achieved good results, but the durability and emissions performance of advanced two-stroke engines is still not established.

Gas turbine engines are widely used to power aircraft, and considerable research has been completed to assess its use in automobiles. Such engines use continuous combustion of fuel, which holds the potential for low emissions and multifuel capability. The efficiency of the engine is directly proportional to the combustion temperature of the fuel, which has been constrained to 1200°C by the metals used to fabricate turbine blades. The use of high-temperature ceramic materials for turbine blades coupled with the use of regenerative exhaust waste heat recovery were expected to increase the efficiency of gas turbine engines to levels significantly higher than the efficiency of s.i. engines.

In reality, such goals have not yet been attained partly because the gas turbine components become less aerodynamically efficient at the small engine sizes suitable for passenger car use. Part load efficiency is a major problem for gas turbines due to the nonlinear efficiency changes with airflow rates in turbomachinery. In addition, the inertia of the gas turbine makes it poorly suited to vehicle applications, where speed and load fluctuations are rapid in city driving. As a result, there is little optimism that the gas turbine powered vehicle will be a reality in the foreseeable future.

Stirling engines have held a particular fascination for researchers since the cycle closely approximates the ideal Carnot cycle, which extracts the maximum amount of work theoretically possible from a heat source. This engine is also a continuous combustion engine like the gas turbine engine. While the engine uses a piston to convert heat energy to work, the working fluid is enclosed and heat is conducted in and out of the working fluid by heat exchangers. To maximize efficiency, the working fluid is a gas of low molecular weight like hydrogen or helium. Prototype designs of the Stirling engine have not yet attained efficiency goals and have had other problems, such as the containment of the working fluid. The Stirling engine is, like the gas turbine, not well suited to applications where the load and speed change rapidly, and much of the interest in this engine has faded.

SEE ALSO THE FOLLOWING ARTICLES

Alternative Transportation Fuels: Contemporary Case Studies • *Combustion and Thermochemistry* • *Fuel Cycle Analysis of Conventional and Alternative Fuel Vehicles* • *Fuel Economy Initiatives: International Comparisons* • *Hybrid Electric Vehicles* • *Internal Combustion (Gasoline and Diesel) Engines* • *Transportation Fuel Alternatives for Highway Vehicles* • *Vehicles and Their Powerplants: Energy Use and Efficiency*

Further Reading

Amman, C. (1989). "The Automotive Engine—A Future Perspective." GM Research Publication GMR-6653.

Lichty, C. (1967). "Combustion Engine Processes." John Wiley & Sons.

National Academy of Sciences (2001). "Effectiveness and Impact of Corporate Average Fuel Economy Standards." National Academy Press.

Office of Technology Assessment (1995). "Advanced Automotive Technology—Visions of a Super—Efficient Family Car." Report to the U.S. Congress, OTA-ETI-638.

Weiss, M.A.(Ed.) (2000). "On the Road in 2020." Massachusetts Institute of Technology Report MIT-EC00-003.

Internal Combustion (Gasoline and Diesel) Engines

ROBERT N. BRADY
HiTech Consulting Ltd
Burnaby, British Columbia, Canada

1. Introduction to Basic Force/Motion Laws
2. Constant Volume versus Constant Pressure Cycles
3. Pressure–Volume Curve
4. Two-Stroke Cycle Gasoline Engine: Basic Operation
5. Four-Stroke Cycle Gasoline Engine: Basic Operation
6. Four-Stroke Cycle Polar Valve Timing
7. Two-Stroke Cycle Diesel Engine: Basic Operation
8. Two-Stroke Cycle Diesel Polar Valve Timing
9. Four-Stroke Cycle Diesel Engine: Basic Operation
10. Direct versus Indirect Diesel Engine Injection Concepts
11. Review of Two-Stroke versus Four-Stroke Cycles
12. Comparison of Two-Cycle versus Four-Cycle Engines
13. Horsepower Description
14. Torque Description
15. Diesel/Gasoline Engine Comparison
16. Summary: Gasoline versus Diesel Power

Glossary

blower A gear driven air pump usually consisting of two, two-lobe, or three-lobe rotors rotating within an aluminum housing to supply air flow to the engine cylinders at low pressure; it is typically used on two-stroke cycle heavy-duty diesel engines.

compression ratio A term used to describe the cylinder and combustion chamber swept volume when the piston is at bottom dead center (BDC) versus that when the piston is at top dead center (TDC); although the term compression ratio is used, in reality it is a volumetric ratio and not a pressure ratio.

naturally aspirated An non-supercharged internal combustion engine that relies solely on valve timing and piston movement to inhale or breathe the air/fuel mixture into the cylinder on a gasoline engine or the atmospheric air only on a diesel engine.

supercharged An engine that, through a combination of a gear-driven blower/pump and valve timing, contains a cylinder pressure greater than atmospheric at the start of the compression stroke; the degree of pressure retained determines the level of supercharging.

turbocharger An exhaust gas-driven device consisting of a shaft with two vaned fan-type wheels at each end; at the hot end, the turbine fan/wheel is driven by the hot pressurized exhaust gases, whereas at the opposite end, the compressor fan/wheel pressurizes the ambient air supply into the engine intake manifold.

In dealing with internal combustion (IC), reciprocating (back and forward motion) gasoline and diesel engine operating cycle theory, one needs to consider the basic laws of energy because the forces developed within the engine cylinder permit work to be developed. Both gasoline and diesel engines operate on the principle of IC, where chemical energy releases heat from the burning of gasoline in the former and diesel fuel in the latter. The heat is then converted into mechanical work.

1. INTRODUCTION TO BASIC FORCE/MOTION LAWS

In dealing with internal combustion (IC), reciprocating (back and forward motion) gasoline and diesel engine operating cycle theory, one needs to consider the basic laws of energy because the forces developed within the engine cylinder permit work to be developed. During the years 1665 and 1666, Isaac Newton developed three laws that describe all of the states of motion, rest, constant motion, and accelerated motion. These three laws explain how forces cause all of the states of motion:

1. An object at rest will remain at rest, and an object in motion will remain in motion at

constant velocity, unless acted on by an unbalanced force.
2. Newton's second law of motion shows how force, mass, and acceleration are related, where Force = Mass × Acceleration.
3. The third law of motion states that for every action, there is an equal and opposite reaction.

The gasoline engine has been in use for well over a century now and was patented by the German technician Nikolaus August Otto in 1876; hence, the term that is sometimes applied to this type of engine is the "Otto cycle." The sequence of events is based on four individual piston strokes, two up and two down, consisting of intake, compression, power, and exhaust, with one power stroke occurring every 720° or two full engine crankshaft rotations.

The diesel engine concept was named after the French-born German engineer Rudolf Christian Karl Diesel. Diesel's concept was also designed to operate on the four-stroke cycle principle, and a patent was granted to him in 1892. However, it was not until 1895, on his third design attempt, that he successfully proved his theory of compression ignition operation.

Both gasoline and diesel engines operate on the principle of IC, where chemical energy releases heat from the burning of gasoline in the former and diesel fuel in the latter. The heat is then converted into mechanical work. The combustion chamber is formed between the crown (top) of the piston and the underside of the cylinder head. The term "thermodynamic cycle" is often attached to the operation of the gas and diesel engine because the word "thermodynamic" describes the science concerned with the conversion of heat to mechanical energy, the control of temperature, and the like. "Thermo" refers to heat, whereas "dynamic" refers to motion or being full of force and energy. Because gas and diesel engine pistons reciprocate, the combustion heat is transferred from the piston, through a connecting rod attached to the piston at one end and the crankshaft of the engine at the opposite end, to create mechanical rotary motion. Figure 1 illustrates the major components of a four-stroke cycle, heavy-duty, electronically controlled, overhead camshaft diesel engine. A four-stroke cycle, electronically controlled, gasoline fuel-injected engine is very similar in its arrangement other than the fact that a spark plug is used. To create power, all IC chamber engines require the following:

- Air (oxygen) for combustion
- Fuel (something to burn)
- Ignition (a means of starting the fire)

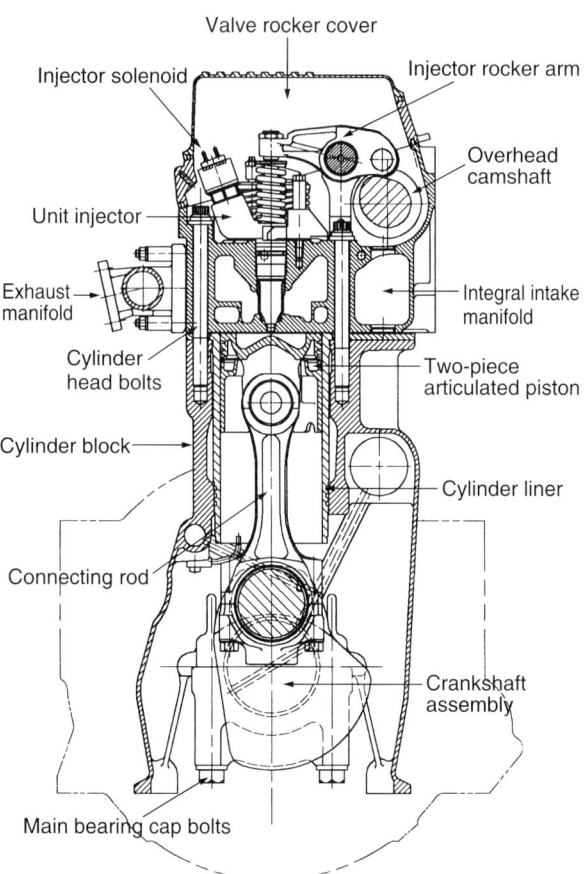

FIGURE 1 Design features of a 3406E Caterpillar heavy-duty truck engine that employs a turbocharger, air-to-air charge cooling (AACC), an overhead camshaft, and electronically controlled unit injectors and governing. Courtesy of Caterpillar Inc.

2. CONSTANT VOLUME VERSUS CONSTANT PRESSURE CYCLES

There are two major differences between a gasoline engine and a diesel engine. First, the gasoline engine operates on a theoretical air standard cycle commonly referred to as being a "constant volume cycle," where periodic combustion and generation of work take place. This term more fully indicates that combustion of the air/fuel charge is completed at the instant the piston is at top dead center (TDC) on its compression stroke, where pressure increases but the volume above the piston crown remains constant. Combustion requires a high-tension spark from an electrical/electronic ignition system distributed to a spark plug to initiate a flame front. Combustion time is primarily a function of the speed with which the flame propagates, typically in the range of 20 to 40 ms. In most engines, the air/fuel mixture is formed outside the combustion chamber. Therefore, it forms a largely homogenous (composed of parts of the same kind) charge at the time combustion is initiated. In gasoline direct injection (GDI) engines, the air/fuel mixture formation is introduced directly into the combustion chamber and is referred to as being heterogeneous (composed of parts of different kinds).

Today's automotive gasoline engines operate in the "stoichiometric" air/fuel ratio range. Simply put, it is the ratio (usually of mass) between air and flammable gas or vapor at which complete combustion or chemical combination takes place. Approximately 14.5 kg of air is required for complete combustion of 1 kg of gasoline. Therefore, one can say that the air/fuel mixture is approximately 14.5:1.0. This ratio permits the three-way catalytic converter to treat raw exhaust emissions with maximum effectiveness as per the U.S. Environmental Protection Agency (EPA) standards. This is achieved by the use of electronic engine controls/sensors in harmony with one or more exhaust gas oxygen feedback sensors to constantly advise the engine electronic control module (ECM) as to how far from stoichiometric after combustion the air/fuel ratio is at a given time. The ECM then varies the injector solenoid on/off time (duty cycle) either to lean out or to enrich the air/fuel mixture so as to maintain the desired stoichiometric setting.

Second, unlike the gasoline engine, the diesel engine does not operate on a stoichiometric air/fuel ratio. In a diesel, the air/fuel ratio is extremely lean at an idle speed (minimum fuel delivery) because the air entering the cylinders is unthrottled (not restricted as it is on a typical gasoline engine). At higher loads/speeds, the diesel air/fuel ratio will run richer because more fuel is injected to produce more power. Therefore, at an idle speed, the diesel engine can exhibit air/fuel ratios as high as 90:1 or 100:1, with some engines running even leaner. At full-load/high-speed operation, the air/fuel ratio can drop down as low as 25:1 or 30:1. In addition, a diesel engine operates on what is referred to as a theoretical "constant pressure cycle," where fuel is presumed to be supplied as the piston moves down the cylinder on its power stroke at a rate such that the cylinder pressure remains constant during the combustion process. The diesel is also said to operate with a heterogeneous charge of compressed air produced during the piston's upward-moving compression stroke, supported by a finely atomized spray of high-pressure liquid fuel injected prior to the piston reaching TDC. The generated heat created in the trapped air from the upward-moving piston compression stroke (hot high-pressure air only) causes the injected diesel fuel to vaporize. After a short time (ignition) delay, where the air and injected fuel mix, the self-ignition properties of the air/fuel mixture initiate combustion to establish a flame front within the combustion chamber. Therefore, additional fuel that is being injected has no ignition delay but burns instantly. The higher compression ratio used with current design, high-speed, heavy-duty diesel engines can develop peak cylinder pressures between 1800 and 2300 psi (12,411–15,858 kPa) and peak temperatures of 3500°F (1927°C). These high-pressure waves and their speed of propagation throughout the combustion chamber represent one of the items that creates the inherent noise attributable to a diesel engine when running.

(In reality, no IC engine, whether gasoline or diesel, operates on either the constant pressure or the constant volume combustion phase. Each requires a few degrees of crankshaft rotation for completion of combustion, and there is also a rise in cylinder pressure during the combustion process. Therefore, a dual cycle of the two that falls somewhere between the Otto and Diesel curves would more closely represent the theoretical curve for consideration with both the gasoline and diesel IC engine cycles. Because IC engines do not operate according to ideal cycles but rather operate with real gas during combustion, they are characterized by flow and pumping losses, thermodynamic losses, and mechanical losses due to friction.)

3. PRESSURE–VOLUME CURVE

To clarify the constant volume and constant pressure cycles described in the previous section, Fig. 2 illustrates what actually transpires during the two most important strokes within a four-stroke cycle engine, namely during the compression and the power strokes. A four-stroke cycle diesel has been chosen for this description, but the curve would be somewhat similar for a gasoline engine. This pressure–volume (PV) diagram represents the piston from a position corresponding to 90° before top dead center (BTDC) as it moves up the cylinder on its compression stroke to 90° after top dead center (ATDC) on its power stroke.

4. TWO-STROKE CYCLE GASOLINE ENGINE: BASIC OPERATION

The term "two-stroke cycle" is based on the fact that the piston moves through two individual strokes (one

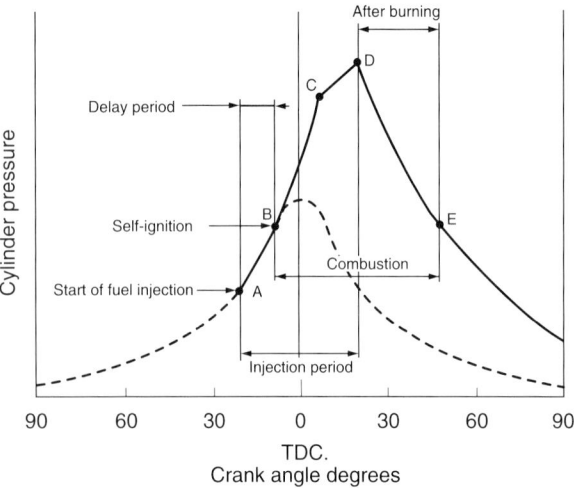

FIGURE 2 Graph illustrating the pressure–volume curve diesel engine combustion operating principle. Courtesy of ZEXEL USA, Technology and Information Division.

up and one down) to provide one power stroke for every 360° of crankshaft rotation. Two-cycle gasoline engines have been in use for many years in motorcycles, lawnmowers, outboard motors, and small industrial applications due to their simplicity of design, construction, lightness, and inexpensive mass production techniques. Although many models are air cooled, there are some that can be liquid cooled. However, these engines tend to emit a burning smell from the fuel/oil mix, are noisy, are fuel thirsty, emit higher exhaust emissions, and are more prone to failure compared with their four-stroke cycle counterparts.

The actions of both intake and exhaust occur between the compression and power strokes. In Fig. 3A, the power stroke occurs once the spark plug fires, the air/fuel mixture is ignited, and an immediate flame front is created. The heat and pressure created from the chemical energy of the burning air/fuel mixture forces the piston down the cylinder. The chemical energy is converted into mechanical energy by both the piston and the connecting rod to rotate the engine crankshaft.

In Fig. 3B, as the piston is driven down the cylinder, the specially shaped piston crown uncovers the exhaust port shown on the left-hand side and the hot burned gases start to escape from the combustion chamber. On the right-hand side of the diagram, as the piston continues down, the intake port is opened to permit a new air/fuel mixture charge from the engine crankcase to enter the cylinder. In Fig. 3C, a small reed valve opens to permit the air/fuel mixture to enter the crankcase.

This occurs on the piston upstroke (compression) due to a slight vacuum created in the crankcase. Atmospheric pressure through the carburetion system causes the air/fuel mixture to flow through the reed valve and enter the crankcase. Also shown in Fig. 3C, as the piston begins its upstroke and both cylinder ports are covered, the air/fuel mixture from the sequence in Fig. 3B is compressed into the combustion chamber. When the piston reaches TDC, the spark plug ignites the mixture and the power stroke is repeated.

5. FOUR-STROKE CYCLE GASOLINE ENGINE: BASIC OPERATION

Four-stroke cycle gasoline and diesel engines predominate as the global prime mover of choice. In gasoline engines, the introduction of electronic controls and GDI systems, in conjunction with dual overhead camshafts, distributorless ignition systems, variable geometry turbochargers, intercoolers, four-valve cylinder head designs, three-way exhaust catalytic converters, and the like, has greatly improved not only their thermal efficiency (fuel economy) but also the control of exhaust gas emissions.

In the four-stroke cycle, the piston moves through four individual strokes (two up and two down). The two upward-moving piston strokes are the compression and exhaust strokes, whereas the downward-moving piston strokes are the intake and power strokes within each cylinder. The individual strokes of the piston shown in Fig. 4 include the following:

- *The intake stroke where the piston moves down the cylinder.* The intake valve(s) is held open by a rotating camshaft to permit an air/fuel mixture to enter the cylinder. This stroke can typically range between 240 and 260° obtained by opening the intake valve BTDC and then closing the valve ABDC.
- The compression stroke where the piston moves up the cylinder to pressurize the air/fuel mixture while the intake and exhaust valves are held closed by valve springs.
- *The power stroke where the air/fuel mixture is ignited by a high-tension spark from the spark plug.* The length or duration of the power stroke is controlled by opening the exhaust valve(s) before bottom dead center (BBDC). Typical duration of the power stroke averages approximately 140° of crankshaft rotation. The number of engine cylinders will

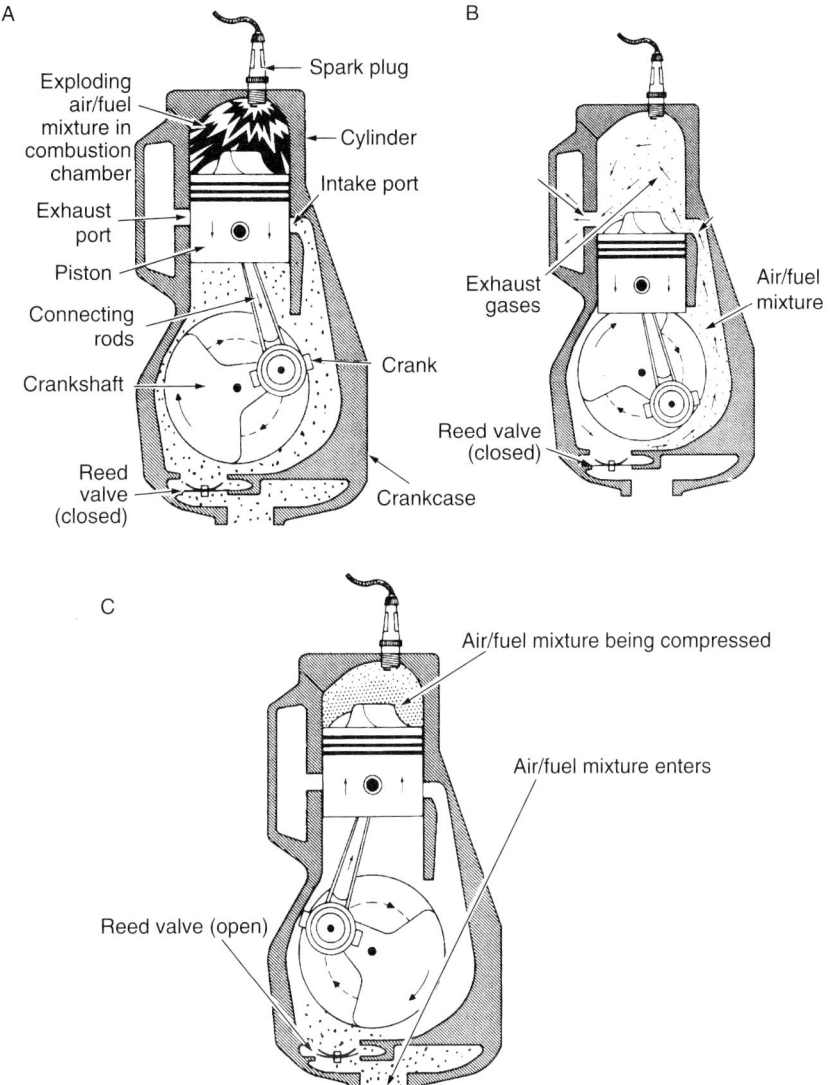

FIGURE 3 Two-stroke cycle gasoline engine operating cycles with crankcase scavenging. (A) Power or downstroke. (B) Exhaust and intake. (C) Compression or upstroke. Courtesy of Prentice Hall.

determine how many power strokes are delivered within the 720° of crankshaft rotation in the four-cycle engine. For example, in a four-cylinder engine (720° divided by four), one power stroke will be delivered for every 180° of crankshaft rotation, whereas in a six-cylinder engine (720° divided by 6), one power stroke will be delivered for every 120° of crankshaft rotation.

- The exhaust stroke, which begins when the exhaust valve(s) is opened before the piston reaches bottom dead center (BDC), typically between 30 and 40° of crankshaft rotation BBDC. The exhaust valves may remain open until 20° of crankshaft rotation after top dead center (ATDC) for a total duration of between 230 and 240° of crankshaft rotation.

6. FOUR-STROKE CYCLE POLAR VALVE TIMING

The actual valve opening and closing in a four-cycle gasoline or diesel engine can be considered similar. Figure 5 illustrates in graphical form the piston position and valve timing duration of the individual strokes.

Figure 6 illustrates one example of the opening and closing degrees for a four-stroke cycle,

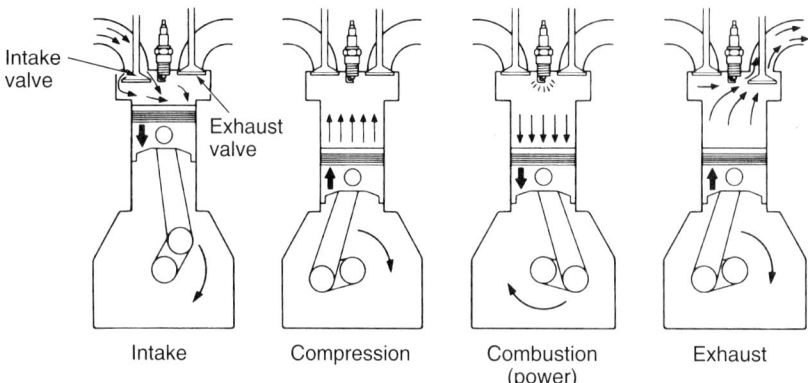

FIGURE 4 Schematic operation of a four-stroke cycle gasoline engine. Courtesy of Prentice Hall.

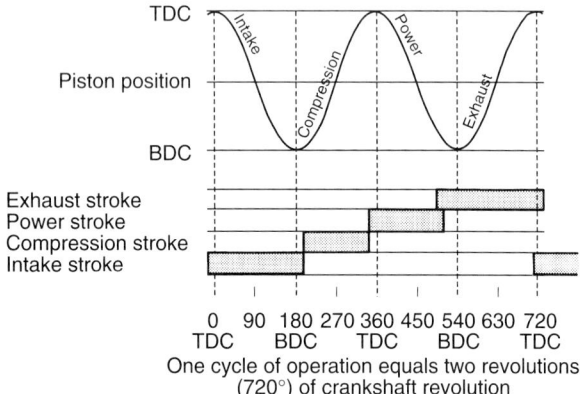

FIGURE 5 Graphical representation of the piston position and individual stroke duration for a four-stroke cycle gasoline engine. Courtesy of Prentice Hall.

high-speed diesel engine. This diagram is commonly referred to as a "polar valve timing diagram" because it shows the piston TDC at the north pole position and the piston BDC at the south pole position. For simplicity, the figure shows the four piston movements throughout the 720° of crankshaft rotation superimposed in a graphical 360° format. On the right-hand side of the diagram, the figure shows the piston downward movement that occurs during both the intake and power strokes, whereas on the left-hand side, it shows the upward-moving piston during both the compression and exhaust strokes. Whereas the individual strokes are illustrated in graphical form in Fig. 5 through a 720° rotation, Fig. 6 shows the number of degrees of valve opening in crankshaft degrees as being either BTDC, ATDC, BBDC, ABDC, TDC, or BDC.

In Fig. 6, the following sequence of events occurs:

1. *Intake stroke:* occurs when intake valves open at 16° BTDC and close at 36° ABDC (total duration is 232° of crankshaft rotation)
2. *Compression stroke:* occurs when the intake valves close at 36° ABDC until TDC (total duration is 144° of crankshaft rotation)
3. *Power stroke:* starts at TDC and continues until the exhaust valves open at 60° BBDC (total duration is 120° of crankshaft rotation)
4. *Exhaust stroke:* occurs when exhaust valves open at 60° BBDC and close at 16° ATDC (total duration is 256° of crankshaft rotation)

In this polar valve timing diagram, note that both the intake and exhaust valves are open for a period of 16° BTDC and for 16° ATDC. This design feature provides what is commonly referred to as a positive valve overlap condition, which in this example amounts to 32° of crankshaft rotation. Opening the intake valve(s) toward the end of the exhaust stroke (16° BTDC) permits the incoming fresh air to help fully scavenge any remaining exhaust gases from the cylinder through the still open exhaust valve(s) that is closed at 16° ATDC.

7. TWO-STROKE CYCLE DIESEL ENGINE: BASIC OPERATION

In the high-speed, heavy-duty, two-stroke cycle diesel engine design shown in Fig. 7, only camshaft-operated poppet exhaust valves are used. The intake of air is obtained through a series of ports machined around the circumference of the cylinder liner at its midsection. A gear-driven roots-type blower is employed to supply the air necessary for engine operation. Typically, the blower supplies air at a pressure between 4 and 7 psi (27.6–48.2 kPa) above atmospheric. High-speed, higher power output two-cycle engines can also employ an exhaust-driven turbocharger in series with the blower to supply a

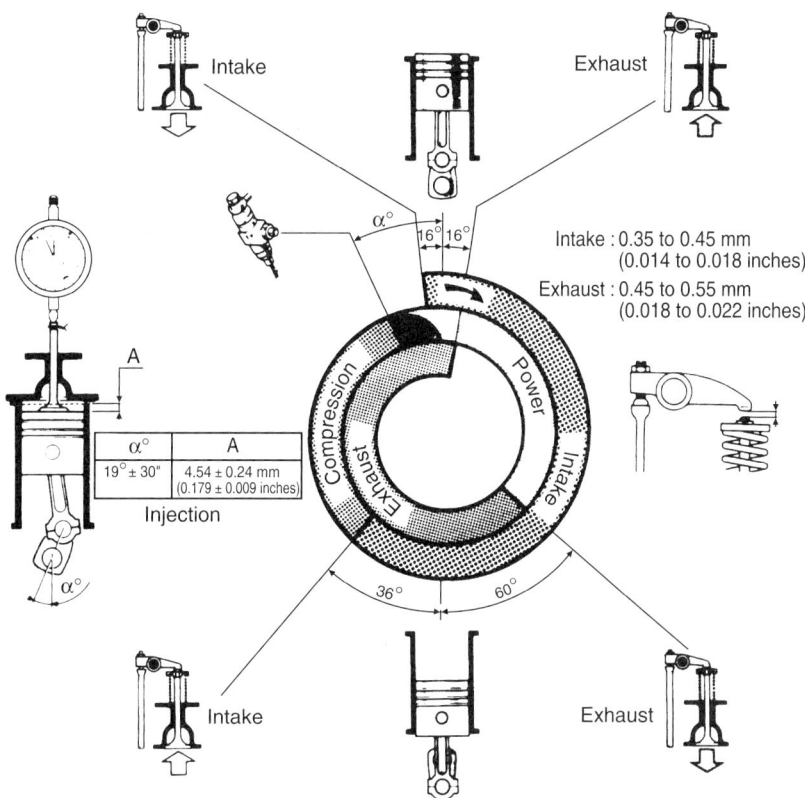

FIGURE 6 Polar valve timing diagram showing the individual stroke degrees for a four-cycle diesel engine. Courtesy of Volvo/Mack/RVI Inc.

higher boost air supply. This results in a boost under full-load rated speed conditions of between 50 and 60 inches Hg (24.5–29.5 psi or 169–203 kPa. In the two-cycle diesel, every upstroke of the piston is compression and every downstroke creates power. Therefore, only 360° of crankshaft rotation is necessary to create one power stroke versus the 720° of rotation required with the four-stroke cycle engine.

In the four-stroke cycle engine design, the pistons basically act as an air pump to permit air to enter the cylinder and to expel exhaust gases from the cylinder. In the two-stroke cycle design, the individual intake and exhaust strokes are basically eliminated. The pressurized air from the blower/turbo is stored in an air box that completely surrounds the ports of every cylinder liner. The air box acts somewhat like a plenum or storage chamber. Anytime the cylinder ports are uncovered, pressurized air can flow into and through the liner. If the exhaust valves are closed, air is trapped within the cylinder. If the exhaust valves are open, scavenging of exhaust gases will occur. The gear-driven blower, or the blower in conjunction with a direct mounted or upstream mounted exhaust-driven turbocharger, supplies the air flow necessary for several functions:

- Scavenging of exhaust gases from the cylinder
- Cooling of internal components such as the cylinder liner, the piston, and exhaust valves (approximately 30% of the engine cooling is achieved by air flow)
- Air required for combustion purposes
- Crankcase ventilation by controlled leakage of air past the oil control rings when the piston is at TDC

The four operational events that occur in the two-cycle engine are as follows:

1. When the piston is at TDC on its power stroke, it is pushed down the cylinder by the force of the high-pressure expanding gases. At approximately 98° ATDC, the exhaust valves are opened, permitting the pressurized exhaust gases to start leaving the cylinder via the opening exhaust valves.

2. At approximately 60° BBDC, the downward-moving piston begins to uncover the cylinder liner

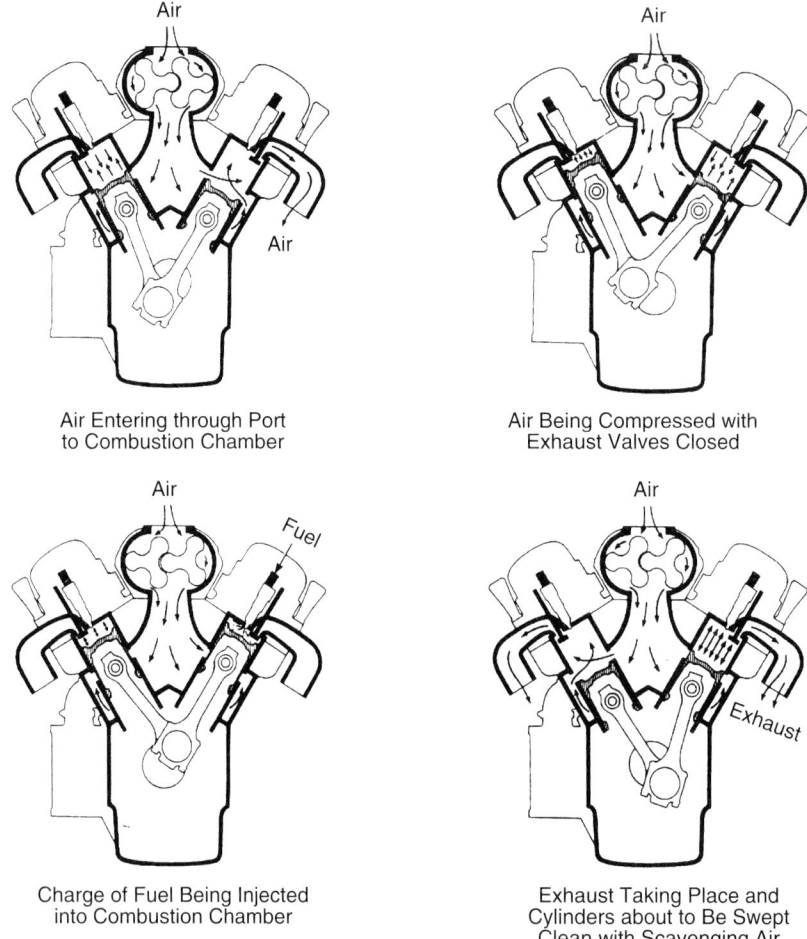

FIGURE 7 Two-stroke, heavy-duty, high-speed diesel principle of operation. Courtesy of Detroit Diesel Corporation, Daimler–Chrysler.

ports. At this point, the air box pressure is higher than the cylinder exhaust gas pressure, permitting this air to start entering the cylinder to assist in scavenging the burned exhaust gases from the cylinder. The liner ports are fully open when the piston reaches BDC for maximum air flow.

3. At approximately 60° ABDC, the upward-moving piston closes the liner ports. During the duration of 60° BBDC and 60° ABDC (120°), the air box pressure functions to provide exhaust scavenging, cooling, and fresh air for combustion purposes. It is during this 120° of liner port opening that exhaust and intake occurs.

4. The camshaft is timed to permit the exhaust valves to remain open for several degrees after the liner ports are covered by the upward-moving piston. This ensures that all burned exhaust gases have been fully scavenged from the cylinder and that a full charge of fresh air is retained in the cylinder. When the exhaust valves are closed, the compression stroke begins and the trapped air pressure and temperature increases. Based on engine speed and load, high-pressure finely atomized fuel is injected into the combustion chamber BTDC. The burning fuel reaches maximum pressure at TDC to force the piston back down the cylinder on its power stroke, and the sequence is repeated.

8. TWO-STROKE CYCLE DIESEL POLAR VALVE TIMING

Figure 8 illustrates one example of the actual degrees of valve timing used for a two-stroke cycle turbocharger and blower scavenged engine model.

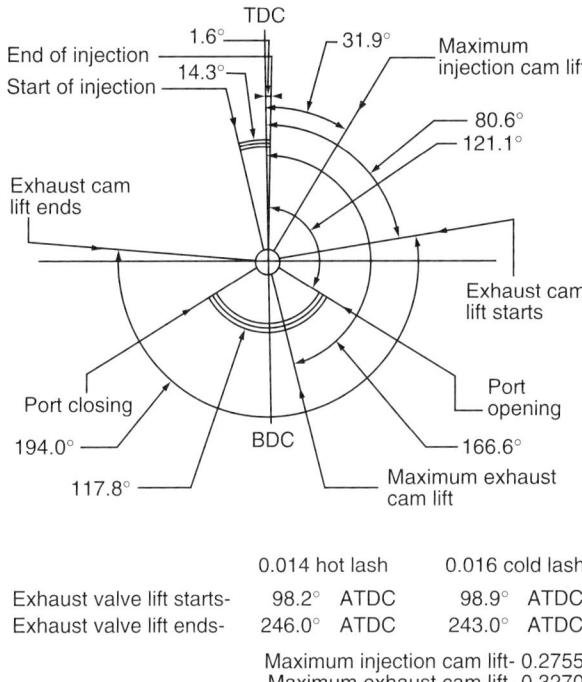

FIGURE 8 Polar valve timing diagram for a two-stroke cycle, heavy-duty, high-speed diesel engine. Courtesy of Detroit Diesel Corporation, Daimler–Chrysler.

9. FOUR-STROKE CYCLE DIESEL ENGINE: BASIC OPERATION

The operation of the four-stroke cycle diesel engine is illustrated in Fig. 9. On the intake stroke, only air (not an air/fuel mixture) is supplied into the cylinder, whether the engine is naturally aspirated or turbocharged. Air flow into the diesel engine cylinder is nonrestricted, meaning that the operator does not control a butterfly valve within the intake manifold as he or she does on a gasoline engine; rather, the throttle pedal or hand throttle is directly connected to the fuel injection pump control rack in mechanically governed fuel systems so as to directly control the fuel delivery rate. In an electronically controlled diesel engine, the position of the hand or foot throttle determines the voltage input signal to an ECM. The ECM then activates the electric solenoid on the injector (Fig. 1) to allow fuel delivery into the combustion chamber. Therefore, the start, duration, and end of fuel delivery (injection) is based on the duty cycle or injector solenoid activation period in crankshaft degrees.

Because of the much higher compression ratio used in the diesel, the heat of the compressed air toward the end of the compression stroke is sufficiently high to cause the injected liquid atomized diesel fuel droplets to convert to a vapor after a small ignition delay. Once the fuel vaporizes, ignition occurs to establish a flame front and the remaining injected diesel fuel burns immediately. By the time the piston reaches the end of the compression stroke, the burning fuel causes a further increase in the pressure above the piston.

As more fuel is injected and burns, the internal gases get hotter and the pressure increases further, pushing the piston and connecting rod downward to rotate the crankshaft during the power stroke. Once the exhaust valve(s) is opened, the pressurized gases force their way out of the cylinder and the exhaust stroke begins. When the piston starts its upward stroke again, it pushes the remaining exhaust gases from the cylinder. With a positive valve overlap condition, the intake valve(s) in the cylinder will be opened before the piston reaches TDC. Typical valve timing in a four-stroke cycle diesel is similar to that in a four-stroke cycle gasoline engine.

10. DIRECT VERSUS INDIRECT DIESEL ENGINE INJECTION CONCEPTS

Diesel engines can use either a direct injection (DI), an indirect injection (IDI), or what is referred to as a swirl chamber (SC) design. DI concepts are the design of choice in most diesel engines today due to their superior fuel consumption qualities, higher available injection pressures, and cleaner exhaust emissions. Figures 1, 7, and 9 show a DI design where the high-pressure atomized diesel fuel is injected directly into the combustion chamber that is formed by the shape of the piston crown and the underside of the cylinder head. Injection pressures ranging between 25,500 and 30,000 psi (176–207 Mpa) are obtained in today's electronically controlled, high-speed, heavy-duty engines. Various piston crown designs are in use with DI engines. The injector spray tip uses multiple holes/orifices to deliver the injected fuel throughout the compressed air mass of the combustion chamber, with the number, size, and spray-in angle of the orifices being determined by the piston crown shape.

The IDI design is also referred to as a precombustion chamber (PC) type, shown in Fig. 10A, where an electric glow plug facilitates ease of combustion, particularly in cold weather operation. In the PC or

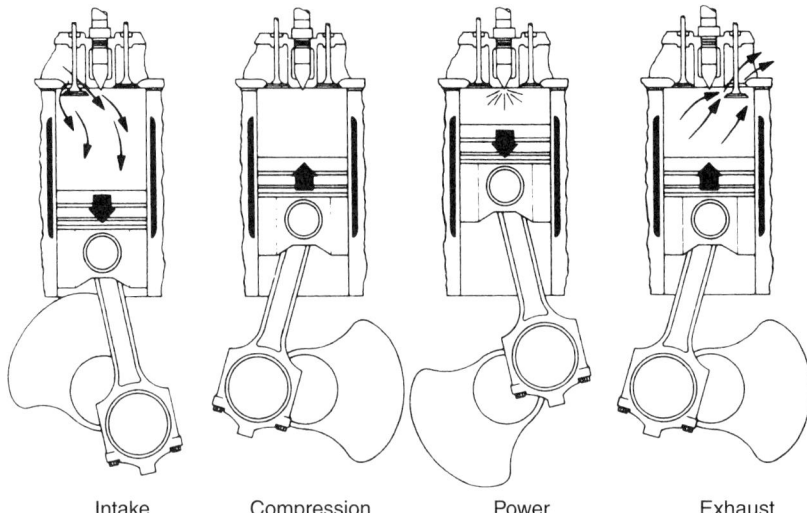

FIGURE 9 Sequence of individual piston and valve events for a four-stroke cycle diesel engine. Courtesy of Detroit Diesel Corporation, Daimler–Chrysler.

IDI design, a small auxiliary chamber, generally centrally located with respect to the main combustion chamber, supports the glow plug and the fuel injector/nozzle. The nozzle spray tip used generally consists of a single hole/orifice to deliver fuel. The small auxiliary chamber typically contains between 25 and 35% of the clearance volume with the piston at TDC. Once ignition occurs, the expanding high-pressure gases escape into the main chamber above the piston to force it down. Design variations exist between the PC and IDI concepts. Some engines also employ an SC design with a glow plug, shown in Fig. 10B, where the combustion chamber is usually located off to one side of the cylinder head. The high-pressure gases enter through a throat or passage onto the top of the piston to drive it down the cylinder on the power stroke. An SC design, although similar to the PC system, contains approximately 50 to 60% of the total clearance (compression) volume when the piston is at TDC. Both the SC and PC/IDI designs can be characterized as divided chamber engines where they are widely used in high-speed light industrial and automotive passenger cars. However, later engine models in these applications are now using more DI designs.

11. REVIEW OF TWO-STROKE VERSUS FOUR-STROKE CYCLES

These two cycles can be summarized by considering that the piston operation is divided into what are commonly referred to as "closed" and "open" periods. The closed period occurs during the power stroke, and the open period occurs during the time when the inlet and exhaust strokes are occurring. Consider the following sequence for the two-stroke cycle:

- Closed period:
  - a–b: compression of trapped air
  - b–c: heat created by the combustion process
  - c–d: expansion or power stroke
- Open period:
  - b–e: blow-down or escape of pressurized exhaust gases
  - e–f: scavenging of exhaust gases by the blower and/or blower–turbocharger combination
  - f–g: air supply for the next compression stroke.

All of the preceding events in a two-cycle diesel engine occur within 360° or one complete turn of the engine crankshaft/flywheel. Now consider the following sequence for the four-stroke cycle:

- Closed period:
  - a–b: compression of trapped air
  - b–c: heat created by the combustion process
  - c–d: expansion or power stroke
- Open period:
  - d–e: blow-down or escape of pressurized exhaust gases
  - e–f: exhaust stroke
  - f–g: inlet and exhaust valve overlap
  - g–h: induction stroke
  - h–i: compression stroke

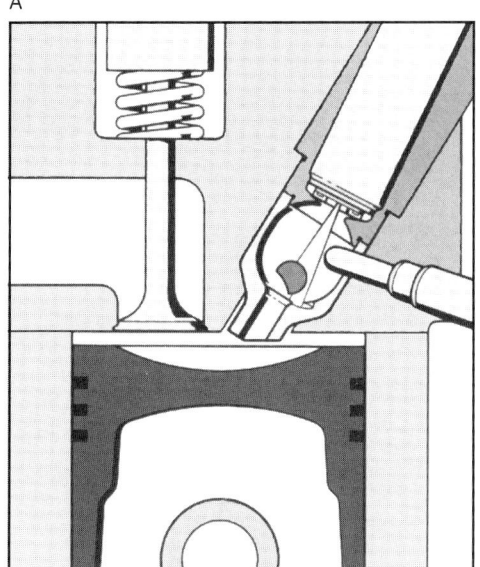

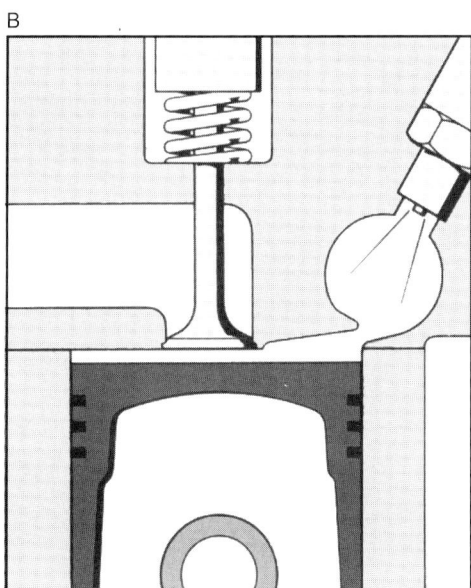

FIGURE 10 (A) Precombustion chamber design concept. (B) Swirl chamber design concept. Courtesy of Robert Bosch GmbH.

All of the preceding events require 720° of crankshaft/flywheel rotation, in contrast to the 360° required in the two-cycle engine model.

12. COMPARISON OF TWO-CYCLE VERSUS FOUR-CYCLE ENGINES

Consider the following specific operating differences between the two-stroke and four-stroke cycle engines, both gas and diesel. In the four-stroke cycle engine, there is a longer period available for the scavenging of exhaust gases and the separation of the exhaust and inlet strokes. In addition, with a shorter valve overlap period versus the port valve concept in the two-stroke engine, there tends to be a purer air charge at the start of the compression stroke in the four-cycle model than there is in a conventional blower air-supplied two-cycle model.

The four-stroke cycle engine pumping losses occur during the intake and the exhaust strokes when the pistons basically act as air pumps. The thermal (heat) loading on the piston, valves, cylinder head, and cylinder liner tend to be lower on a four-cycle engine because the power stroke occurs only once every two crankshaft revolutions versus once every revolution in the two-cycle engine. It is also easier to lubricate the pistons, rings, and liners in a four-stroke engine due to the absence of ports that are required around the midpoint of the two-stroke liner. Adding a turbocharger and the blower can increase the air flow substantially in the two-stroke engine, closer to the characteristics of the four-stroke model. Both four-stroke and two-stroke engine designs have pumping losses. In the two-cycle engine, the power losses required to drive the gear-driven blower reduce the net engine power output. In addition, the two-stroke engine requires a much larger air flow capacity to operate than does the equivalent displacement four-cycle engine.

Compare the pumping losses mentioned previously between the two-stroke and four-stroke cycle engines. In general, on a non-turbocharged two-cycle engine, the power required to drive the blower is less than the four-cycle pumping losses when the engines are operating at less than 50% of their rated speed/load. However, from 50% up to rated speed, the four-cycle engine's pumping losses tend to be roughly two-thirds those of the two-cycle model. On two-cycle models that also employ an exhaust gas-driven turbocharger along with a bypass blower feature, these pumping losses are reduced substantially. The two-cycle engine tends to have a slightly higher fuel consumption curve due to its double power stroke principle through the 720° for a four-stroke cycle model plus the higher overall pumping losses experienced in the two-cycle model above 50% of rated speed. The brake mean effective pressure (BMEP), which is the average pressure exerted on the piston crown during the power stroke, is generally lower on a two-cycle diesel. If one compares equivalent displacement and horsepower-rated engines at 450 to 475 hp (336–354 kW), the BMEP in the two-stroke engine model would be approximately 115 to 128 psi (793–883 kPa). On the four-cycle model, the BMEP would be approximately double the BMEP figures cited for the two-cycle model.

The brake-specific fuel consumption (BSFC) of a two-stroke engine tends to be higher than that for a comparably rated four-cycle engine. BSFC is simply the ratio of fuel burned to the actual horsepower produced.

To summarize the advantages and disadvantages of these two different cycles, we can cite the following:

Advantages—four cycle:

1. Better volumetric efficiency (VE) over the engine speed range (VE is a measure of the "breathability" of the engine or the extent to which the cylinder of an engine is completely filled by the incoming charge following an exhaust stroke. It is also important to understand that the VE is a ratio of masses, not of volumes.)

2. Low sensitivity to pressure losses in the exhaust system

3. Better control of the cylinder charging efficiency through control of valve timing and matching of the intake manifold system

Disadvantages—four cycle:

1. Valve control potentially more complex

2. Engine power density lower due to only one piston power stroke every 720° of crankshaft rotation

Advantages—two cycle:

1. Simple engine design, low weight, and low manufacturing costs

2. More favorable torsional forces (shorter number of crankshaft degrees between power strokes)

Disadvantages—two cycle:

1. Higher fuel consumption

2. Higher hydrocarbon exhaust emissions (lower scavenging efficiency

3. Lower BMEP due to lower VE

4. Higher thermal load because every upstroke is compression and every downstroke is power (no separate intake and exhaust strokes)

5. Can exhibit poorer idle quality due to the higher percentage of residual gases retained in the cylinder (particularly in gasoline engines).

13. HORSEPOWER DESCRIPTION

To clarify and appreciate the work produced in an IC engine, it is appropriate to understand the two main terms "horsepower/kilowatts (kW)" and "torque." Failure to understand these differences will result in confusion when discussing the operation of any IC engine.

The term "horsepower" is used in the nonmetric vernacular to describe how much work an IC engine can produce, with work being equal to Force × Distance. Power produced is a measure of the rate at which work is done during a given time period, with 1 hp equivalent to a rate of work equal to 33,000 foot-pounds/min. In the metric system of measurement, 1 hp is equal to 0.746 kW; therefore, 1 kW is equal to 1.341 hp. In an IC engine, the force of the hot high-pressure expanding gases acting on the piston crown during the power stroke forces the piston down the cylinder and transfers the reciprocating (back and forward) motion via a connecting rod to convert it into rotary motion at the engine crankshaft. Therefore, the rate of work performed within the engine cylinder will increase as the engine crankshaft revolutions per minute (rpm) rotates faster (greater number of power strokes delivered during a given time period). The product of the BMEP in the cylinder and the piston displacement produces the piston work, and the actual number of working cycles (power strokes) per unit of time is what determines the indicated horsepower (ihp), which is nonusable power at the crankshaft or internal power produced in the cylinder without taking frictional losses into consideration. For example, an engine specification that indicates the engine can produce 500 bhp (brake horsepower or usable power at the flywheel) at 5000 rpm will not produce this same power at 2000 rpm because the rate of doing work is much less at the lower speed. Also, one should bear in mind that horsepower cannot be multiplied.

14. TORQUE DESCRIPTION

The word "torque" means a twisting or turning force applied to a fulcrum point that tends to cause torsion or rotation. Torque is expressed in pound-feet or pound-inches in the English system of measurement, whereas it is expressed as Newton-meters in the metric system. In an IC engine, the power that drives the piston down the cylinder is transferred to torque through the connecting rod that is bolted to the rotating crankshaft. The connecting rod functions as a mechanical reaction lever; therefore, its length and the distance between the center line of the connecting rod and the center line of the rotating crankshaft will

TABLE I

Comparison of Pressures and Temperatures Created within a Gasoline Engine versus a Diesel Engine

	Gasoline	Diesel
Intake	Air and fuel	Air only
Compression	8.0:1 to 10.5:1 ratio	15:1 to 23:1 ratio
Typical average values	130–175 psi pressure (896–1206 kPa) 540–650°F temperature (282–343°C)	425–600 psi pressure (2930–4137 kPa) 1000–1200°F temperature (538–649°C)
Combustion	Spark ignition	Compression ignition
Power	Maximum pressure 450–600 psi (31103–41137 kPa)	Maximum pressure 1800–2300 psi (12411–15858 kPa)
Exhaust	Temperature of 1300–1800°F (704–982°C)	Temperature of 700–1100°F (371–593°C)
Thermal efficiency[a]	22–28% average	32–44% average

Note. At the end of the compression stroke in some gasoline engine models with a homogenous air/fuel charge, the pressure can be as high as 290–580 psi (20–40 bar), with temmperatures ranging between 752 and 1112°F (400–600°C).

[a] Percentage of return in work value for investment in fuel or percentage of return for every dollar of fuel consumed in the engine. Current fuel injected and electronically controlled gasoline engines can return thermal efficiency values in the 30% range. Waste heat recovery stationary diesel power plants can return total thermal efficiency values in the middle to high 50% range.

have a direct influence on the amount of torque generated at a given engine speed. The torque is actually what keeps the crankshaft rotating; therefore, it is responsible for the ability of the engine to handle a load and for the engine's capacity to do work. Maximum torque is usually generated at approximately 60 to 65% of the engine's rated revolutions per minute because it is a combination of force and time acting on the piston crown plus volumetric efficiency. Unlike horsepower, engine torque can be multiplied by directing its energy through clutches and/or torque converters, then through a series of gear ratios within the transmission, and then through the additional gearing in the axle. Therefore, if one increases the engine speed, torque will decrease but horsepower will increase, and vice versa.

15. DIESEL/GASOLINE ENGINE COMPARISON

Table I highlights the basic differences between a gasoline engine and a diesel engine during the four strokes of intake, compression, power and exhaust.

16. SUMMARY: GASOLINE VERSUS DIESEL POWER

The heavier construction, higher compression ratio, nonthrottled air supply, and greater heat value per pound of diesel fuel, as well as the ability to produce rated horsepower (kW) at a lower speed, develop substantially higher torque curves, return superior fuel economy (better thermal efficiency), and offer "life-to-overhaul" typically three to four times longer than that of a gasoline engine, are major attributes of the diesel engine. Today's heavy-duty truck electronically controlled diesel engines offer warranties of 500,000 miles (805,000 km) plus in heavy-truck applications, with accumulated mileage of between 800,000 and 1,000,000 miles (1,287,000–1,609,000 km) between major overhauls. Factored over the life of the engine/equipment, a substantial savings amounting to thousands of dollars per year in operating costs is realized by an owner/operator. In fleet use with hundreds or thousands of pieces of rolling stock, these features and advantages can amount to literally tens of millions of dollars.

SEE ALSO THE FOLLOWING ARTICLES

Biodiesel Fuels • Combustion and Thermochemistry • Internal Combustion Engine Vehicles • Mechanical Energy • Rocket Engines • Transportation Fuel Alternatives for Highway Vehicles • Vehicles and Their Powerplants: Energy Use and Efficiency

Further Reading

Brady, R. N. (1996). "Modern Diesel Technology." Prentice Hall, Englewood Cliffs, NJ.

Brady, R. N. (1997). "Heavy-Duty Trucks, Powertrains, Systems, and Service." Prentice Hall, Upper Saddle River, NJ.

Brady, R. N. (2001). "Automotive Electronics and Computer Systems." Prentice Hall, Upper Saddle River, NJ.

Brady, R. N., and Dagel, J. F. (2002). "Diesel Engine and Fuel System Repair." Prentice Hall, Upper Saddle River, NJ.

Lilly, L. R. C. (1984). "Diesel Engine Reference Book." Butterworth, London.

Robert Bosch Gmb, H. (1999). Bosch Automotive Handbook, 5th ed. Robert Bosch GmbH, Stuttgart, Germany. (Distributed by Society of Automotive Engineers, 400 Commonwealth Drive, Warrendale, PA 15096-0001.)

International Comparisons of Energy End Use: Benefits and Risks

LEE SCHIPPER
World Resources Institute
Washington, D.C., United States

1. What Are International Comparisons of Energy Use?
2. Approach and Data Sources
3. Sectoral Examples
4. Overall Comparisons of Energy Use
5. Lessons from International Comparisons

Glossary

activity (or output) The basic unit of accounting for which energy is used at the sectoral level.
degree day The difference between the average indoor and outdoor temperatures during a day, integrated over the entire heating season.
delivered energy (or final consumption) Energy supplied, for example, to a building, factory, or fuel tank and ultimately converted to heat, light, motion, or another energy service; transformation and distribution losses up to the final user are not included.
energy efficiency The ratio of energy converted into a desired form to energy put into the conversion; more commonly, it is a measure of energy converted to useful output divided by the energy "consumed" to produce this output.
energy intensity Energy "consumed" per unit of activity or output; it is usually presented at the most disaggregated subsectoral or end use level, but it can also be presented as energy use for an entire sector per unit of sectoral activity or output.
energy services Implies actual services for which energy is used (e.g., heating a given amount of space to a standard temperature for a period of time).
primary energy Delivered energy plus losses incurred in converting energy resources into purchased heat and electricity.
structure or subsectoral activity (or subsectoral output) The shares or modal mix (e.g., trucks, rail, and domestic shipping for freight; cars, two-wheelers, buses, rail, ship, and air for travel), ownership of using equipment by households (e.g., area per capita heated, refrigerators per capita), and manufacturing output by branch expressed as a share of total output.
useful energy Delivered energy minus losses assumed to occur in boilers, furnaces, water heaters, and other equipment in buildings; it is used for estimates of heat provided in space and water heating and cooking.

International comparisons of energy use entail detailed examination of the patterns of energy use over two or more countries. Comparisons are carried out in a disaggregated, bottom-up manner that permits comparing between or among countries measures of different economic and human activity and the energy uses associated with each activity.

1. WHAT ARE INTERNATIONAL COMPARISONS OF ENERGY USE?

International comparisons strive to understand how energy uses differ among countries of similar or varying geography, resource endowments, energy policies, incomes, and other differentiating characteristics. When undertaken carefully, international comparisons both reveal how energy uses differ and point to why these differences arise. Energy use patterns are complex, and comparisons among countries (or regions) at an aggregate level rarely reveal the nature of differences, the reasons for these differences, or the underlying policies that may contribute to these differences. The risks of international comparisons lie in the temptation to draw conclusions at too shallow a level of analysis. The benefits, which outweigh those risks, reveal themselves in insights as well as in pointers as to where to probe even deeper.

1.1 Why International Comparisons?

The most important reason for international comparisons lies in the need to understand how energy and development are linked. As economies grow worldwide, energy use increases, albeit at a slower rate. Analyses of how past development of wealthier countries affected energy use yield clues as to what is in store for developing countries as they grow. Still, over a time span of many decades, this growth in demand is worrisome due to pressures on world oil supplies, the environment, and global releases of carbon dioxide (CO_2) and other greenhouse gases (GHGs) associated with energy use. Energy use and its link to CO_2 emissions through fossil fuels plays a key role in negotiations and agreements related to climate. Understanding how wealthier countries use—and abuse—energy provides keys to less wealthy countries on how economic growth might diverge increasingly from energy use growth through careful management and clever policies.

Another reason for undertaking cross-country comparisons of energy use is to identify energy-efficient technologies. This is because politicians are wary of advocating energy use strategies that rely on more than just technologies. Unfortunately, a lack of sufficient data does not allow this to be done fully. One reason is that determining the true technological component of a particular energy intensity requires detailed study. For example, consider the case of space heating. The intensity of space heating in Sweden, measured as heat per unit of floor area and average winter outdoor temperature, is considerably lower than that in most other European countries. This suggests—but does not by itself prove—higher efficiency for heating in Sweden, that is, better insulation as well as efficient heating systems, heat distribution, and controls. New Zealand and Japan also have very low intensities of space heating. But how efficient are the space heaters, and what are the real thermal transfer properties of walls and windows, in New Zealand or Japanese homes vis-à-vis those in Sweden? Perhaps New Zealanders or Japanese simply do not heat very much, even in the cold regions. Available data suggest that this is the case, although adequate information for other countries is difficult to obtain at an aggregate level. Only a very detailed study of actual energy use and internal temperatures in hundreds of homes, as was done twice in Sweden, can verify that the real levels of thermal comfort in Sweden are very high. The conclusion is indeed that space heating in Sweden, with high indoor temperatures but low intensities, is very efficient. International comparisons alone point to an interesting possibility, but further work is needed to verify whether the superficial differences are really related to efficiency.

Comparisons guide policy formulation. Understanding how another country arrived at a particular structure may provide insights into how to steer one's own course. For example, careful study of the efficient state of housing in Sweden could lead to many ideas for promoting efficiency in the United States. Comparison of the effects of New Zealand's pricing of liquid petroleum gas (LPG), compressed natural gas (CNG), and diesel road fuels (whatever the reasons for the various levels of prices) with those for other countries illustrates the impacts of price differentials on fuel choices for new vehicles. Such international comparative studies can highlight important technologies that save energy as well as key policies that might promote or hinder enhanced energy efficiency.

In the end, international comparisons of energy use promote understanding on deep analytical levels, as well as on political levels, as to how and why energy use patterns differ. These differences explain much of national differences in energy and environmental strategies, particularly those aimed at restraining GHG emissions. Without a good understanding of each other's basic patterns, it is hard for leaders to agree on how to mold policies and measures to restrain carbon emissions. With that understanding, policymakers still need to look before they leap. Fortunately, international comparisons of energy use help to map out both known and unknown facets of energy use, enlightening policy formulation and execution for many purposes.

1.2 Aggregate Comparisons: Why They Tell Us So Little

Figure 1, from a series of national studies compiled by the International Energy Agency (IEA), shows a comparison of energy use and gross domestic product (GDP) among a number of countries and groups of countries. The division of energy use by GDP (or, alternatively, by population) is useful because the denominator normalizes energy use by an extensive quantity that is known to be a zero or even first-order driver of energy use. Without such a normalization, comparing any of the four Nordic

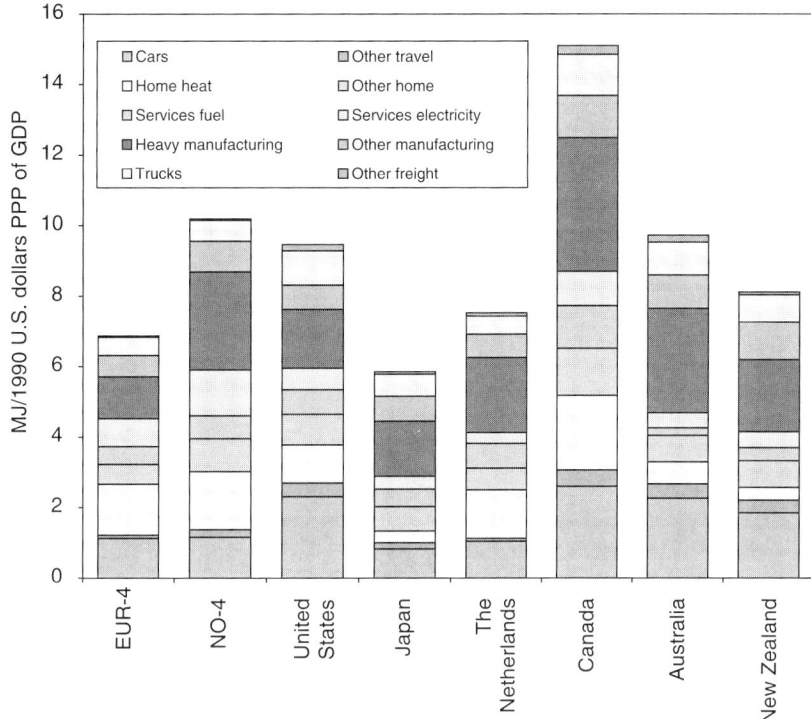

FIGURE 1 Energy use from 10 sectors/end uses relative to GDP. The EUR-4 consists of France, West Germany (or unified Germany from 1991 onward), Italy, and the United Kingdom. The NO-4 consists of Denmark, Finland, Norway, and Sweden. GDP is measured using purchasing power parity (PPP) to convert real local 1990 currency to U.S. dollars.

countries (NO-4) with the United States, which dwarfs them in geographical size, population, and total GDP, seems meaningless. Even comparing Canada, the United States, and Australia, three of the largest and wealthiest countries in the world, is hard to justify because their climates, industrial output mix, and housing structures differ considerably. Disaggregating energy uses into the sectors and activities shown, but keeping GDP as shown, is insightful but still cloaks most of the differences in how energy is used and, therefore, hides many of the reasons why energy use differs.

Instead of comparing energy use per sector against GDP, comparisons must start at a disaggregated level. Decomposition of changes in energy use can be summarized by the relation known as ASI. Put simply, it relates energy use E to three multiplicative terms on the right-hand side of an equation, with the sums over both indexes:

$$E = A_k * S_{k,j} * I_{k,j}.$$

In this decomposition, A represents overall sectoral activity (e.g., GDP in manufacturing). S represents sectoral structure (shares of output by two- or three-digit ISIC [International Standard Industrial Classification], branch of manufacturing or other industry, modal shares in transportation, and per capita equipment ownership in households). The k runs over the main sectors (here households, the service sector, manufacturing, other industries, travel, and freight). The j represents the subsectors, branches, or activities within each sector. The $I_{k,j}$ represents the energy intensities of each subsector j of activity k. The structure or relative output of each sector is represented by $S_{k,j}$. The exact units of each of these variables are defined in what follows. The product $A_k * S_{k,j}$ in sector k gives output or activity in absolute values (e.g., total steel production, total distance traveled). This is also known as energy services. Multiplying by the energy intensity $I_{k,j}$ gives total energy use for subsector j of sector k. A fourth term could be added to the right-hand side to differentiate among energy carriers given that this is important for judging the importance of electricity or calculating carbon emissions for a given energy use.

Divide both sides by GDP (G), suppressing the subsectors j for notational ease:

$$E/G = [(A * S_k)/G] * I_k.$$

Clearly, E/G can differ because the sectoral A_k/G differ, because the elements of S differ, or because the I differ. Differences in individual I values are related to, but not identical to, differences in energy efficiency. Differences in S values represent, within each sector in the structure, differences in energy use that can be related to technology, climate, industrial policy, natural resources endowment, land use, urban structure, and the like. Differences in A values, which aggregate sectoral activity into a few key indicators, also indicate broad differences in structure, output, and geography. The overall differences by sector (in A) or by subsector (in S) represent information that is as important to explaining differences in energy use/GDP as is the I. Because the I values are themselves only related to energy efficiencies, the overall ratio E/G cannot give an accurate picture of differences in energy efficiencies. Forming the ratio E/GDP lumps all of the differences in each country's structure into one parameter and then lumps all of the energy uses into a second parameter. These two aggregate numbers then represent neither the end uses of energy nor the structural features that drive many of these differences. This mixing leaves nothing to compare for meaningful international comparisons of energy use. Needless to say, the quotient of E/GDP itself cannot represent energy efficiency given that all of the activities and energy uses are aggregated first.

A particular problem is with GDP itself. Private vehicle uses, household activities related to energy consumption, and use of trucking fleets by firms for their own haulage account for 40% or more of national energy use but do not appear in GDP beyond the associated expenditures for energy because there are no transactions involved. The "outputs" of these activities do not show up as transactions anywhere.

This also means that Fig. 1 cannot tell us whether energy uses in any country or group of countries shown are efficient or inefficient. The figure only reveals how much fuel is used for cars relative to GDP, not relative to kilometers driven, which is the intensity related to fuel efficiency. However, energy use per kilometer only partly explains the obvious differences revealed in the figure. Total kilometers driven, relative to GDP, turns out to be just as important as fuel use per kilometer in explaining the total level of car fuel use/GDP shown in the figure. Knowing both of these vital elements—the S and the I—is crucial to making fair comparisons. Knowing enough about the characteristics of vehicles that shape the I, as well as the individual technological components, adds insights but becomes harder and harder unless only new vehicles (or homes, appliances, etc.) are compared due to enormous data availability problems. Hence, the minimum requirement for an international comparison is to be able to compare basic elements of A, S, and I between or among two or more countries.

Furthermore, even the use of the word efficient has to be scrutinized. It is dangerous to say that any country is energy efficient, even based on the disaggregated uses of energy shown in Fig. 1. First, countries themselves do not use energy; rather, individuals and public and private entities use energy. This is not a trivial point. For example, even in the centrally planned economy of the former Soviet Union, energy uses were affected by firms' and individuals' choices as to both how and how well to use energy. Although the ratio of energy use to GDP in the Soviet Union was far higher than in most other countries, the inefficiencies of energy uses explained only perhaps half of the differences. The rest was explained by very high production of energy-intensive raw materials, huge shipping distances, and cold climate. The size and cold climate of the Soviet Union certainly do not make that country inefficient, just as the lack of major space heating needs in Australia does not make that country efficient. Offsetting these factors were very small floor space for homes and commercial buildings, few household appliances, and very few cars (driven not very far) compared with Western countries. These latter factors did not make the Soviet Union any more efficient than the former ones made that country inefficient. Although choices of factory, transportation, and housing technologies may have been largely the results of central planning, these choices lead to explanations for only part of the high ratio of energy use to GDP. Those choices may be labeled inefficient, but the country per se may not be. For market-oriented nations of the Organization for Economic Cooperation and Development (OECD), it is even less obvious how to define national energy efficiency. Instead, international comparisons must be undertaken to explain how ratios of energy to GDP differ and then explain why the results occur.

Moreover, scholars attach energy efficiency to a particular energy-using process. When that process is taken by itself, such as the conversion of fuel to motion in a car motor (in reality, a collection of processes), its efficiency can be compared with that in other countries. But even at this level, automobiles differ in size and features. Thus, it turns out that automobiles in the United States actually use less energy per kilometer and per unit of weight than do those in nearly every other country. By that measure, U.S. automobiles are efficient. But they also use more energy per kilometer because they weigh more than cars elsewhere. Thus, one can say that the U.S. car fleet is more fuel intensive than fleets in any other advanced country. The reason is that American behavior "favors" large cars, largely due to low car prices and low fuel prices. Behavior and cultural choices, inhibited or reinforced by policies, affect how the efficiencies of technologies are transformed into the actual energy intensities of energy uses one can measure or infer. Thus, analysis should be very mindful of what is being labeled as efficient.

The example of space heating noted previously is also instructive. Are the Japanese or New Zealanders, who heat to lower indoor temperatures than do Americans and most Europeans, more efficient at space heating or simply frugal? Here, as with the example of cars, differences in revealed energy intensities may represent what is seen as economically efficient for each country's households. Because that implies a different meaning of efficiency than one often used by engineers, one must use the word efficient carefully. Indeed, it might be better to avoid misunderstandings—to reduce the risks of international comparisons—by avoiding the use of the word efficiency entirely.

What can we learn about how different societies use energy? Energy uses and activity levels for approximately three dozen energy uses or activities are known in more than a dozen IEA member countries, in most cases over a period of time from years to decades. Similar estimates have been made for former Soviet bloc countries and at least for a year for most of the largest developing countries. These comparisons are reasonably accurate portrayals of the structure of energy use and its corresponding energy intensities. But the comparisons rarely measure energy efficiencies. Measuring how well energy is used on an economy-wide basis takes far more data than are normally available. With this distinction firmly in mind, the benefits of comparisons are maximized and the risks of misunderstandings are minimized.

1.3 Early Work in International Comparisons

Much of the forgoing emerges from studying literature on energy uses that has emerged since the 1960s, when data illustrating disaggregated national energy use patterns became available. At the same time, the energy and economics literature is replete with analyses and comparisons of energy consumption and GDP. However, such aggregate comparisons should not be mistaken for comparisons of energy use. These comparisons flourished due to the lack of data. Although the manufacturing sectors of most countries were nominally covered by energy balances whose detail matched national accounts or other measures of output, few countries separated energy consumption into the residential, commercial/services, and agricultural sectors from a common aggregate, and very few countries attempted to split the transportation sectors into modes (e.g., automobiles, trucks). Understandably, then, most scholars turned to the energy/GDP comparisons, squeezing information like water from a stone.

Work during the 1970s by Strout and colleagues at the Massachusetts Institute of Technology (MIT), as well as work undertaken at the World Bank, began to relate aggregate energy use to important "structural" variables measuring what happens in an economy such as production of raw materials, vehicle ownership, and even winter climate. Although it lacked real energy use data for these variables, the work did show that, on an aggregate energy use level, structural variables explained part of intercountry differences or changes in use in a country over time. High production of energy-intensive raw materials (e.g., steel), cold winter temperatures, and large shipping distances help to explain high energy use relative to GDP; conversely, low values of these parameters yield lower values of E/GDP.

Adams and Miovic made an important early disaggregated comparison of energy use in the manufacturing sector. They found the reduction in energy use per unit manufacturing during the 1950s and 1960s hard to explain because the price of energy fell during that period. However, they discovered that increases in energy efficiency and shifts of fuel mix toward oil and gas (with fewer combustion losses than coal), all occurring as the scale of manufacturing facilities increased, naturally occurred even as energy became cheaper; that is, technology outpaced energy prices. This work was the first to reveal the strength of international comparisons at a disaggregated level.

Much of the lack of detail in comparisons beyond the manufacturing sector made during the 1960s and 1970s may have been due to the problems of obtaining data. A massive ad hoc data collection for more than a dozen countries, using 1972 as a base year and building on the ASI relationship shown previously, was done for the World Alternative Energy Strategies study at MIT. Systematic data began to appear only during the late 1970s, supported by national energy authorities in North America, much of Europe, and Japan. Experts reconstructed energy use patterns, often back to the 1960s, again with the ASI approach. The goal was nearly always to build a firm understanding of the past as a way of projecting future options in a disaggregated manner.

An early prominent binary comparison of energy uses was carried out by SRI International, comparing West Germany and the United States using data from 1972–1973. A more detailed binary comparison of energy uses, comparing the United States and Sweden, was published in 1976.

Both studies found that structural differences, which measured different levels of activity relative to GDP, accounted for significant differences in energy use relative to GDP. For example, Sweden and West Germany produced more steel, relative to GDP, than did the United States, raising energy use in the former two countries. But auto ownership and use in West Germany was lower than that in the United States relative to GDP, lowering energy use in the former country, whereas in Sweden this structural feature was close to that of the United States. On the other hand, both studies also identified differences in energy intensities, particularly automobile fuel use per kilometer, energy use per year for each major appliance, and energy use per tonne (metric ton) of various raw materials, suggesting that U.S. energy intensities were higher than those in the two European countries. For the first time, analysts could pinpoint key elements of the differences among sets of countries.

The first full comparison of a number of countries was carried out by Resources for the Future (RFF) in 1977. In "How Industrial Societies Use Energy," RFF used available data to quantify key differences in the structure and intensity of energy uses among the United States, Japan, Canada, and several European countries (the United Kingdom, France, Sweden, Italy, and The Netherlands). This epic investigation found that differences in both energy intensities and the structure of energy use contributed to differences in the overall ratio of energy use to GDP.

A detailed series of sectoral analyses of individual countries that included comparisons with up to 11 other countries was carried out at Lawrence Berkeley National Laboratory (LBNL) during the 1980s and 1990s and then at the IEA during the late 1990s. Drawing on 20 years of data (from 1973 to the mid-1990s) not available to the RFF study, and adding important countries such as Australia (with nearly no heating demand but a very energy-intensive industry structure) and Finland (the coldest of the IEA countries) to those studied, these analyses reinforced some of the key RFF conclusions but also quantified how the relative importance of structural and intensity differences evolved over that period. For one thing, structural difference among countries narrowed. The importance of manufacturing of heavy materials dropped in many countries, particularly Japan. Automobile ownership and use increased in Japan and Europe, closing the gap somewhat with the United States. Ownership of central heating in Europe and ownership of appliances in Europe and Japan gained on levels in the United States, closing that gap somewhat.

The IEA in 1997 developed energy indicators for more than a dozen countries based on these comparisons. (The United States, Canada, Japan, the four Nordic countries, West Germany, France, Italy, The Netherlands, the United Kingdom. Australia, and New Zealand were added in subsequent works.) The objective of this work was to describe the majority of IEA countries at the aforementioned level of detail, with the ultimate goal of introducing such data into the IEA databases for ongoing analysis and comparisons. An effort was made to standardize the approach and encourage other countries to develop data. Much of the effort was directed toward EUROSTAT as a way in which to ensure the highest level of quality of, and access to, data and indicators.

2. APPROACH AND DATA SOURCES

The landmark RFF comparison was the most disaggregated comparison of several countries ever undertaken through the late 1970s. Since the time of that work's base year (1972), many industrialized countries began to systematically improve their data. The IEA work, which has been the most exhaustive published work to date, described a number of key steps to define an approach to international comparisons of energy use, either across entire economies or in certain sectors. The same steps also enable disaggregated time-series analysis of energy uses in a given country. The key step is to be able to attach an activity measure to each energy use, that is, to make a disaggregated normalization.

- Disaggregation of each sector's energy uses for a given year (or over time) into subsectors (e.g., a limited number of branches of manufacturing), modes (e.g., passenger and freight transport), or end uses (e.g., households)—in all, approximately 32 to 36 energy uses or sectors (certain compromises may permit lowering the number of subsectors, as is noted in the chapter reviewing each sector)

- Identification and quantification of structural parameters—output or activity—for each of the subsectors or branches of energy use
- Quantification of energy intensities for each output or activity
- Comparison of structural parameters and energy intensities
- Synthesis of results

If time-series analysis is required, this can be done use by use:

- Comparison of time trends in energy uses, energy intensities, and structural measures
- Brief comparison of trends with changes in measures of income and, in some cases, energy prices
- Decomposition of total sectoral energy use trends into underlying components and comparisons of these results with similar analyses for other countries

Using index techniques, individual sectoral trends in structure or intensity can be reaggregated and compared among sectors within a country over time. If many countries are involved, these aggregates can be compressed further to look at how overall trends in activity, structure, or intensity affect total energy use or energy use relative to GDP.

A hybrid approach is the safest. A comparison across individual years is acceptable. However, comparing time series of individual parameters or indicators shows whether the values for any particular country are robust and stable over time or waver due to errors, uncertainties, or unusual events. An aviation strike in Australia caused an unusual drop in actual flying in 1990 as planes flew empty after the strike. Thus, the intensity of aviation for that year was not representative of any long-term trend. For this reason, many of the comparisons presented here will be shown over time.

The hybrid approach can also be expanded in two dimensions. For example, plots of a structural variable (e.g., GDP) against another variable (e.g., floor area, automobile use) can be shown. Both variables evolve over time, and this is implicit in the portrayal. But by coupling two variables together and eliminating time as an axis, one can portray an approximate causal relationship.

2.1 Sectors and Definitions

The sectors commonly studied are listed in Table I. These typically cover 85% of all the final or primary energy use consumed in most IEA countries. The end uses or subsectors account for all energy use in each sector except in transport. There, approximately 10% of the fuels normally consumed are ignored (e.g., off-road vehicles, military, private aviation). Some gaps in data commonly arise in other sectors; for example, parts of agriculture, mining, and construction, as well as street lighting and public utilities (e.g., pumping of water or district heat), are either lumped in with energy sector losses or aggregated with the commercial/services sector. Finally, bunkers of international shipping or aviation and, in some countries, international trucking can be estimated but are usually not included in comparisons due to lack of activity data.

The approach in most studies uses final energy to measure the amount of fuels, electricity, or heat consumed within a branch of manufacturing, by a certain kind of vehicle, or within a house for a specific end use. Trends in primary energy use are referred to occasionally. Primary energy assigns to electricity (and heat in some countries) the average ratio of primary energy resources consumed to produce electricity (or heat) made available to the economy during the year in question. There is a large literature on how to count primary electricity from hydro, geothermal, and nuclear sources, so important that the official Swedish energy balances give aggregate primary energy using three different accounting schemes. If carbon emissions are of interest, standard coefficients can be used to infer the carbon emissions from combustion of fossil fuels.

Good comparisons are carried out in a transparent way. Data sources are clear, and data are available to allow analysts to adopt different conventions for counting primary energy or for measuring key structural variables such as floor area and vehicle ownership and use. Some countries, notably the United States, make the original survey data available at nominal or no cost, whereas others keep the disaggregated data "secret." Although some kinds of raw data may indeed be commercially valuable or may need to remain confidential to protect information from individual consumers, an open debate about energy use policies requires open access to data on how energy is used. Unfortunately, some major countries still do not conduct surveys regularly to determine how energy is used, in part due to the expense of such surveys and in part due to the burden

TABLE I

Measures of Activity, Sectoral Structure, and Energy Intensities for International Comparisons

Sector (k)	Subsector (j)	Activity (A_k)	Structure ($S_{k,j}$)	Intensity ($I_{k,j} = E_{k,j}/S_{k,j}$)
Residential/household		Population		
	Space heat		Floor area per capita	Heat[a]/floor area
	Water heat		Person/household	Energy per capita[b]
	Cooking		Person/household	Energy per capita[b]
	Lighting		Floor area per capita	Electricity/floor area
	Appliances (5–7)		Ownership[c] per capita	Energy/appliance[c]
	Other household electricity uses		—	—
Passenger transport		Passenger-kilometers		
	Cars		Share of total passenger-kilometers	Energy/passenger-kilometer
	Bus		"	"
	Rail		"	"
	Inland air		"	"
Freight transport		Ton-kilometers		
	Trucks		Share of total ton-kilometers	Energy/ton-kilometer
	Rail		"	"
	Coastal shipping		"	"
Commercial/services			"	
	Services total	Floor area or GDP	(Not defined)	Energy/floor area or GDP
Manufacturing (with ISIC code)		Value added		
341	Paper and pulp		Share total value added	Energy/value added
351 352	Chemicals			
36	Nonmetallic minerals			
372	Iron and steel			
372	Nonferrous metals			
31	Food and beverages			
All other three-digit codes	Other manufacturing			
Other industry		Value added		
	Agriculture and fishing		Share total value added	Energy/value added
	Mining			
	Construction			

[a] Adjusted for annual winter climate variations and for changes in the share of homes with central heating systems.
[b] Adjusted for home occupancy (number of persons per household).
[c] Includes ownership and electricity use for six major appliance types.

of response. This lack of information also clouds the energy debate.

3. SECTORAL EXAMPLES

Table I gives a breakdown of six main energy use sectors used in international comparisons. This section reviews three of these in some detail and makes shorter remarks on two others.

3.1 The Residential/Household Sector

Population is the basic measure of activity in the residential/household sector. The residential/household sector "produces" space and water heat, cooking, cool temperatures for refrigeration and air conditioning, light, heat and mechanical action for washing and drying, and a number of other energy services. Lacking real measures of "output" (i.e., liters of water heated to a given temperature), some of the measures of structure are based on per capita ownership of appliances or characteristics of dwellings themselves such as home area per capita. Changes in these values cause the "structural changes" in this sector.

The residential/household sector consists of energy consumed in dwellings. Calculation of this can be complicated due to secondary dwellings, homes in nonresidential buildings, and the like. Surveys of private expenditures, which follow people and households rather than dwellings, offer a reliable way of counting households or families that can be related to the number of occupied dwellings in one-, two-, or multifamily buildings. Usually, detached (one-family), semidetached (two-family), and row houses and mobile homes are counted with one- or two-family dwellings. Homes in dormitories and larger industrial or commercial premises are usually included with multifamily dwellings. Because both household and housing surveys have been carried out for decades in IEA countries, all of these results are generally compatible among countries.

Energy consumption as reported by energy suppliers cannot be easily separated into the residential/household and commercial building sectors because some fuels, notably oil, are often provided to large buildings where the supplier cannot say to which sector the customer belongs. Even if a gas or electric subscription is labeled residential, there is no guarantee that the label is correct given that so many utilities classify customers by size of overall consumption, not by the nature of the consumers. Because most energy suppliers belong to worldwide trade associations, there is good agreement across countries over the definitions of the residential/household sector and general recognition of where the definitions and data are the weakest.

Estimates of total energy use in the residential/housing sector vary in accuracy. Typically, the estimates for electricity and gas are more accurate than those for oil, coal, and wood. Only a "bottom-up" household energy use or energy expenditure survey, as carried out regularly in a handful of IEA countries (e.g., Japan, United States, The Netherlands, Sweden, France), gives an accurate measure of how much energy is used in this sector. Remaining countries make estimates based on utility deliveries, estimates of distribution of fuels supplied to large buildings for central heating, and so on. Overall, counting of gas and electricity from suppliers is the most accurate method, whereas counting of liquid and solid fuels is somewhat less accurate, with errors as large as 10% occurring due to these problems of misidentification.

3.1.1 Key Trends and Indicators for Comparison

The overall position of the residential/household sector in the mid-1990s in a number of countries and regions is shown in Fig. 1. Final energy use in the household sector had grown relatively slowly in most IEA countries since the 1980s as the main energy uses of space heating and electric appliances became saturated. Water heating and cooking represent the other two major categories of energy use. These have grown slowly during the past two decades or so, and their relative importance in overall energy uses continues to decline. The share of electricity has been rising both because electricity has gradually captured space and water heating (and cooking) from fuels and due to the increased importance of appliances, particularly those for information and communication. The uneven penetration of electricity for these is cause for caution in making comparisons. With much fewer losses in conversion within the home, the use of electricity (or district heat) for a given purpose appears to be lower than that of fossil fuels, which suffer from combustion losses in furnaces, boilers, and stoves. In contrast, counting the fuel primary energy requirements of electricity may overstate energy use relative to fossil fuels. In extreme cases of high electricity penetration (e.g., Norway, Sweden, Canada) or high use of district heating (e.g., Sweden, Denmark, Finland), it is advantageous to try to separate electricity, district heat, and nonelectricity, and this requires more data and provides an even more disaggregated form of the ASI equation.

Space cooling, largely from electricity, is usually contained in electric appliances. It is significant (5% of final energy use) only in Japan and the United States, but it is recorded (if under 2%) in France, Italy, Spain, Greece, Australia, and New Zealand. In warm developing countries such as India, Thailand, China, and Brazil, space cooling is rising in importance among household energy uses. Because many of these countries have little heating supplied by fossil fuels, electricity tends to be the most important energy form supplied to the residential/household sector.

3.1.2 Space Heating

Energy use for space heating ranges from 4 to 5 GJ per capita in warmer countries to as much as 30 to 40 GJ per capita in the United States, Canada, and countries in Northern Europe. Space heating is driven by the number of homes, the indoor temperature relative to the outdoor temperature, and the number of hours per year the homes are actually heated. Although population and the average area per home are growing slowly in IEA countries, the number of homes continues to increase with the number of households at an annual rate as much as 1% faster than the population. Hence, per capita home area is increasing due to both factors. Higher GDP per capita both permits larger homes and allows more people to be able to afford to live apart. Thus, per capita area per unit of per capita GDP is a structural variable driving heating, and this is shown in Fig. 2. What can be seen is a strikingly large spread of per capita area at a given GDP.

To normalize for differences in climate, the number of degree-days is included. A degree-day represents the difference between the average indoor and outdoor temperatures during a day added up for the entire heating season; values (compared with a nominal 18°C indoor temperature) vary from as low as 900 degree-days in Australia to more than 4600 degree-days in Canada and Finland. These figures are averaged for the approximate distribution of population. An important conclusion is that both home area (relative to GDP) and winter climate differ enormously among the developed countries, contributing to significant differences in energy use.

The most useful intensity variable is formed by dividing total energy use for heating by total floor area and dividing further by degree-days. To portray the approximate differences between fuels, with combustion losses, and electricity or district heating, with no combustion losses in the home, "useful" energy is introduced. A nominal 66% efficiency for liquid and gaseous fuels and a 55% efficiency for solid fuels is taken, and the totals are added to the delivered values of heat and electricity. This avoids some of the problems of differences in fuel mix noted previously. The result gives a meaningful comparison of space heating intensity, as shown in Fig. 3.

What is striking about Fig. 3 is that, after normalizing for these important structural variables, large differences in space heating intensity still remain. The low values for Sweden, Denmark, and Norway reflect very good levels of insulation, as is somewhat the case for the United States. The low value for Japan reflects lower indoor temperatures and only partial heating of homes or rooms.

3.1.3 Electric Appliances

There are seven major appliances: white goods (refrigerator, refrigerator–freezer, and freezer), wet goods (dish and clothes washers, dryers), and air conditioners. The ownership (saturation or penetration, i.e., number/100 households) of major goods is known from private household surveys, censuses and household expenditure surveys, and household energy surveys. Important differences in size and characteristics are difficult to capture in a stock description but are known approximately. Such characteristics include freezer compartment size, whether washers use hot or cold intake water, and the like. Although these cannot be translated directly into energy use, they give qualitative explanations for differences in energy use.

Energy uses are more complicated to estimate. Only a few experiments have actually metered uses over enough households to make statistically valid estimates of annual energy use. Household ownership and total electricity consumption data can be analyzed using regression techniques to yield rough estimates of consumption per device. Indeed, in most

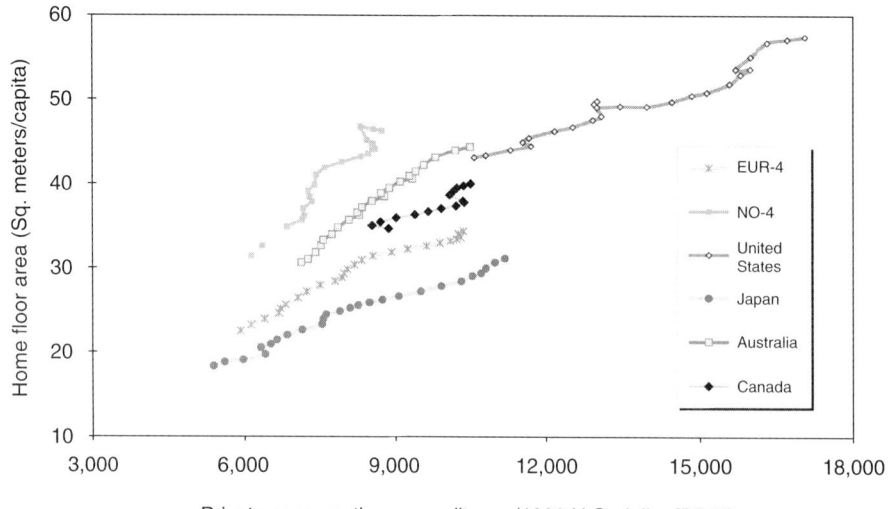

FIGURE 2 Home floor area and private consumption expenditures. See Fig. 1 for abbreviations.

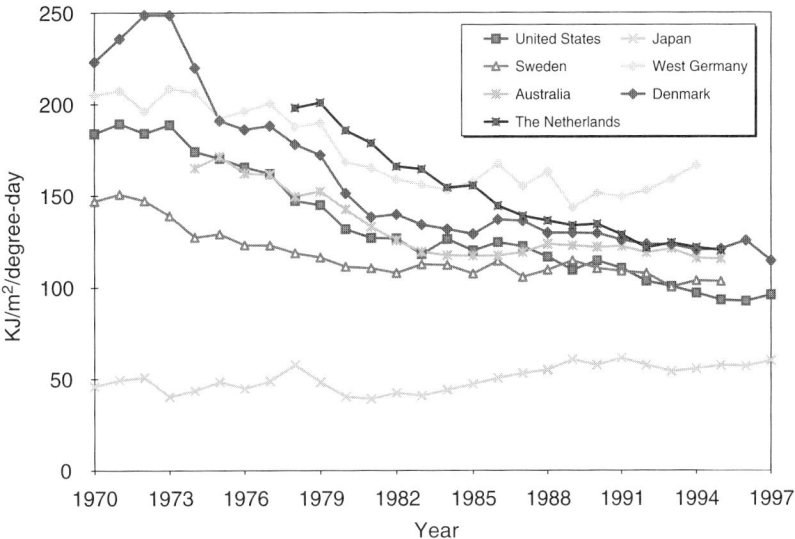

FIGURE 3 Space heating intensity: useful energy.

countries, utilities and their associations, researchers, and statistical bureaus often build models of ownership of each appliance by its age. Tests of new devices are more accurate measures of consumption or at least of given cycles (for wet goods).

Knowing the approximate energy use of new appliances of each type and year made, the model can estimate average energy for each appliance by integrating over the presumed age distribution of appliances. Survey or anecdotal information is used to estimate use of wet appliances (e.g., number of loads of washing or drying). The results can be compared with information on how residential electricity use varies for the entire sector by season and by hour of the day, and this also gives clues as to how much power each kind of appliance uses. The final results, in kilowatt-hours/appliance/year, give a rough indication of how much electricity is used for each kind of appliance. Although appliances in the United States have fallen in energy intensity since the imposition of efficiency standards during the late 1980s, they tend to use more energy than those in Europe and Japan due to differences in size/features and efficiency. But taken together, ownership, size, and features are still a greater source of differences in electricity use for appliances than is efficiency today.

3.1.4 Cooking and Water Heating
These end uses have become progressively less important in some countries but still account for 3 to 10 GJ per capita of delivered energy. By the 1980s, hot water penetrated into virtually every household in upper income countries. Some water is heated in dishwashers and clothes washers by electric elements and so should be transferred to "water heating."

Cooking has become a minor energy use. Typically, a gas stove might account for only 1 GJ per capita of home energy use, and an electric stove might account for barely half of that. Fewer and simpler meals eaten at home, more meals cooked with microwaves or other electric specialty appliances, and smaller household size all have reduced final cooking energy significantly from its values during the 1960s.

There is no obvious structural variable for either water heating or cooking. Hot water consumed (e.g., baths, loads of clothes) is not known to any precision, nor is the number (or kind) of meals cooked. Experience suggests that the amount of energy used for either purpose grows with the square root of household size. In comparing energy use for either purpose across countries, differences in household size can account for important differences in energy use per household. However, on a per capita basis, the effect is reversed; large households share energy uses across more members and, hence, tend to have lower per capita energy use. Thus, as household size shrinks, energy use per capita climbs. Either way, cooking and water heat use and habits explain the greater part of differences in per capita use.

There are modest differences in the efficiency of cooking appliances and water heaters. Imposition of electronic ignition for gas appliances, use of microwave or halogen cooking, and greater water tank insulation for all kinds of tanks explain small differences among countries.

3.1.5 Summary of Residential/Household Energy Uses

A comparison aggregating residential/household energy use by end use that accounts for key structural differences among countries can be devised. This permits separation of structural factors that are income driven, such as home area, from those that are endogenous, such as climate and household size. Such comparisons strip away differences not directly due to policies or technologies, thereby making comparisons more fruitful.

For example, Fig. 4 summarizes residential/household energy uses across a number of countries. (In this figure and subsequent graphs, Europe denotes the average of France, Italy, West Germany, and the United Kingdom.) Electricity is shown separately from fuels and district heat to illustrate its penetration. For comparison across a wide range of winter climates, space heating is scaled to the average number of degree-days for IEA countries ($\sim$2700 degree-days to base 18°C). This scaling alone shows how household energy use in Japan and Australia would increase while falling in Denmark and especially Canada.

The high use in the United States after normalization is explained by the largest homes and largest appliances as well as by somewhat higher than average intensities for heating and appliances. The low use in Japan, even after adjustment for winter climate, is explained by the small size of homes and low indoor temperatures (i.e., frugality). The normalization shows that Denmark's use is raised relative to that of somewhat warmer European countries, whereas that of Australia is lowered considerably by a very mild winter climate. Relatively low energy prices in Australia, the United States, and Canada boost energy use in those countries relative to the four European countries and especially Denmark and Japan.

3.2 The Commercial/Services Sector

As Fig. 1 shows, the commercial/services sector accounts for 10 to 15% of final energy use. The basic energy services include those of the household sector as well as ventilation. Power for computers, information, motors, and air conditioning in this sector is proportionally more important than that in homes, whereas energy use for water and space heating is less important. The basic measure of output is GDP arising in ISIC sectors 6 to 9. The basic measure of structure that is readily used is the total area within buildings. Some further detail of how that area is distributed among building types is known, and there are occasional surveys of energy use by fuel and building types in a handful of countries.

This sector is very heterogeneous and contains buildings that are not used primarily for homes,

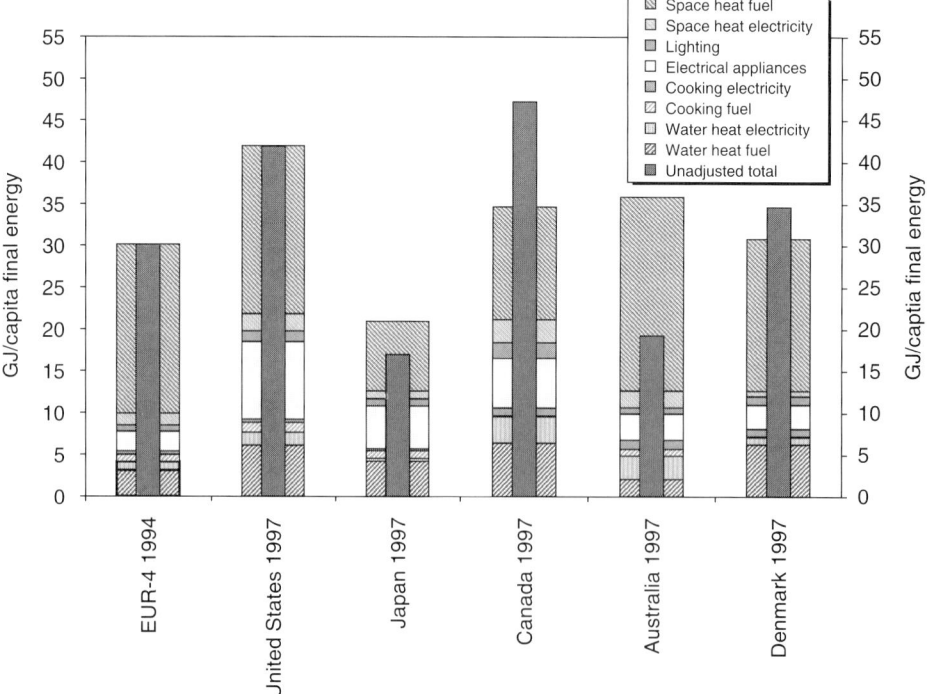

FIGURE 4 Energy use in homes, with space heating normalized to an average winter climate. See Fig. 1 for abbreviations.

industry, or farms. The economic sectors include warehousing and storage, retail and wholesale, transportation, private services (e.g., banking and insurance, hotels/restaurants), and public services (e.g., medical, educational, cultural, and sports facilities). The physical sectors by which buildings are surveyed directly might include "offices," "medical," "educational," "restaurants," "stadiums," and/ or even "parking." Economic data are reported by building owners or operators, whereas physical data have to be gathered by careful surveys of large samples of buildings. Both are useful for understanding how the sector uses energy so as to facilitate comparisons. Unfortunately, neither kind of definition overlaps well with the other; much cooking occurs in buildings other than restaurants, offices are found in all kinds of buildings, and so on. Hence, comparisons of individual building types and their energy uses must be viewed with caution, although such comparisons are probably fair at the national level (e.g., all hospitals, all schools).

Unlike the residential/household sector, less is known about end uses of energy in the commercial/ services sector. The reason is the enormous heterogeneity of the sector. As implied previously, laundry (and sanitary hot water, sterilization, etc.) is important in hospitals but is relatively unimportant in offices and schools, and space heating is important in some buildings but is increasingly provided indirectly in buildings, such as offices and stores, by heat from other end uses, particularly electronic machines. Comparisons of energy use based on metering studies and engineering estimates do exist, but they are very difficult to compare except among similar building types. Although a great deal is known about the efficiencies of individual motors, space heaters, lighting systems, and the like, it is virtually impossible to aggregate these into any meaningful figure of overall efficiency. Comparisons of intensities of fuel and electricity use, and of space heating use, are justified. However, beyond that, only very detailed examination of buildings through surveys yields enough information to say how efficient energy use is in any building type, not to mention across the entire sector or between countries.

3.2.1 Key Trends and Indicators for Comparison

The commercial sector's extent in square meters, and hence its energy use, has been driven by growth in the services component of GDP. Figure 5 shows this coupling clearly. Countries with large areas relative to sectoral GDP have a great deal of space (e.g., United States, Australia) or are relatively cold (e.g., Canada, Sweden, Denmark). Japan has little space because it is crowded, whereas area in Italy appears to be underreported. The large differences in floor use at a given GDP implied in the figure are clues to differences in overall sectoral energy use as suggested in Fig. 1.

Energy use in the services sector, in all but the warmest countries, was traditionally driven by the need for space heating, which was provided by fossil fuels and (in a few countries) by district heating supplied from central heat or heat-and-power stations. The same fuels generally provided water heating as well. The importance of heating is

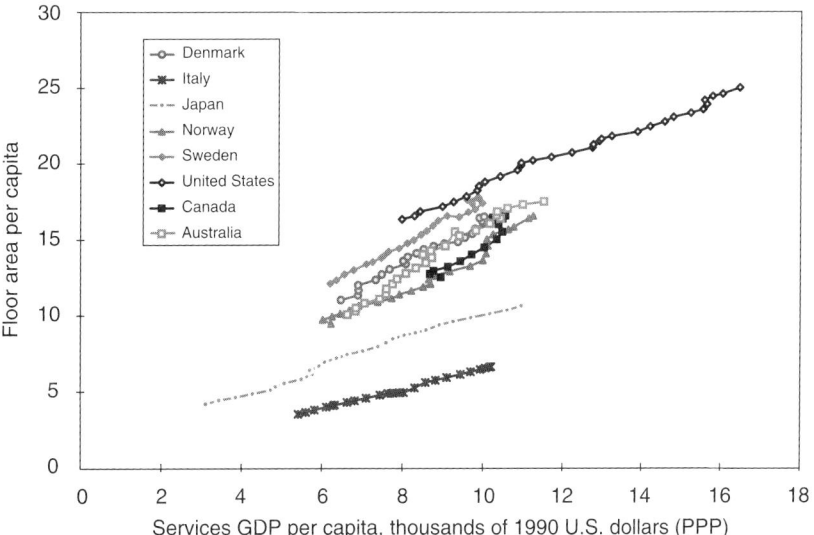

FIGURE 5 Commercial sector: floor area and GDP. PPP, purchasing power parity.

apparent from the high values of fuel/district heating per square meter for Finland and Canada from the 1970s and 1980s. The decline reflects both heat saving and switching to electric heating. In contrast, the very low value for Norway reflects the very high penetration of electricity for space heating even during the early 1970s as well as tight building shells. The waves of energy saving during the 1980s, and some substitution toward electricity, left fuel and heat use much lower, relative to area, as Fig. 6 shows. Although this indicator is inexact as a measure of space heating, detailed heating surveys by fuel and heating type that cover France, Denmark, the United States, and Sweden give approximately the same picture for each main heating fuel over time.

The decline in heating needs, coupled with growth in the demand for electricity services (e.g., lighting, information systems, ventilation, motors), has left electricity as the most important energy source, typically providing at least 40% of final energy (and much more if electric space heating is common). Figure 7 shows an interesting portrayal of electricity use in services as a function of service sector GDP. The high uses at a given GDP in Norway, Sweden, and Canada are largely a result of high space heating penetration. The sudden drop in Norway's value at

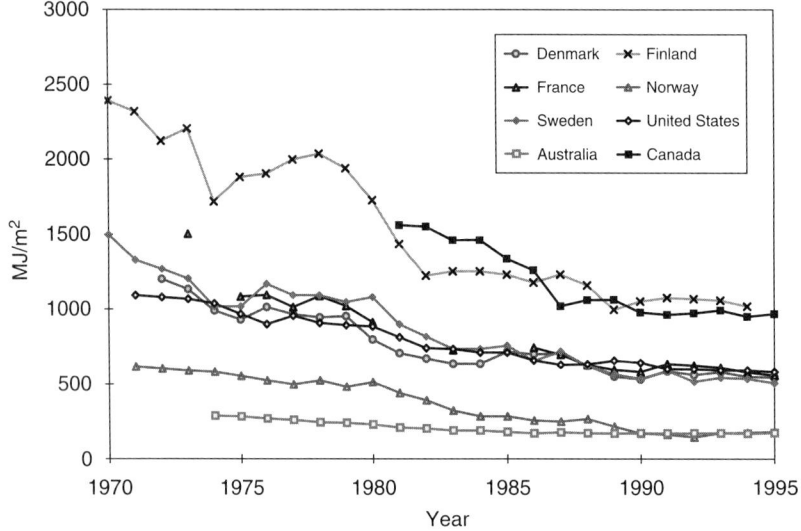

FIGURE 6 Fuel/district heat intensity in the services sector.

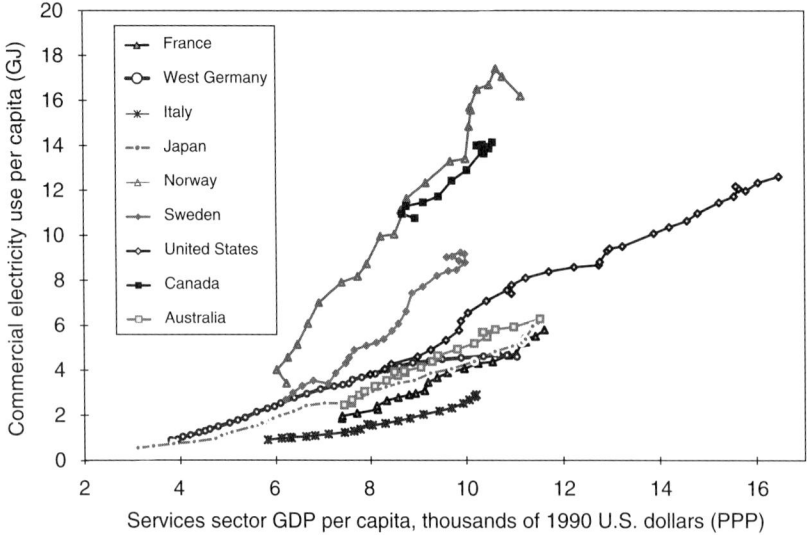

FIGURE 7 Services sector: electricity and GDP.

approximately U.S. $10,500 per capita (which occurred during the mid-1990s) arose when electricity use for heating was cut during a mild winter after electricity prices rose. No country appears to show a bulge of growth during recent years that might signal more electricity use for information and communication.

Although it is very difficult to estimate various end uses in this sector, results for space heating shown in Fig. 8 do at least suggest how space heating needs increase with increasing winter severity. The slope of this relationship should increase with colder temperatures. During mild periods, most of the heating needs of a large building can be met from waste heat from other uses. But during the coldest periods (or in the coldest countries), a disproportionate amount of heat comes from the heating systems. The unusually high position of Italy and Japan may be due to a significant undercounting of total heated area, also suggested in Fig. 5. Indeed, this is an example where international comparisons themselves point to fundamental problems with data, giving cause to reject results from some countries.

3.3 Manufacturing

Manufacturing accounts for 15 to 35% of energy use in IEA countries. Although it was traditionally the most important energy consumer before 1973, manufacturing's role in overall energy use has steadily eroded as the sector's overall output growth has slowed, energy has been saved, and other sectors have surged ahead.

3.3.1 Introduction

Manufacturing (ISIC 3) is a subset of the industry sector, which also includes agriculture (ISIC 1), mining (ISIC 2), and construction (ISIC 4) in most countries. Manufacturing typically accounts for 90% of energy use in industry unless a country has a large energy sector. Largely because these sectors are so poorly documented, most analysis is limited to manufacturing.

3.3.2 Methodology and Data

Manufacturing energy use is commonly disaggregated into six separate subsectors and a subsector that contains all remaining subsectors. In the formulation, output is measured in value added or GDP arising in each sector. The point where value is added corresponds approximately to the point where energy consumption is measured. The six subsectors are food and kindred products (ISIC 31), paper and pulp (ISIC 341), chemicals (ISICs 351 and 352), nonmetallic minerals (ISIC 36), iron and steel (ISIC 371), and nonferrous metals (ISIC 372). In a few countries with very specialized heavy industry (e.g., Norway, Sweden, The Netherlands, Australia, Finland), disaggregating chemicals, nonferrous metals, and/or paper and pulp gives additional insights into the very high energy intensities of those branches when disaggregated only this far. On the other hand, a more common aggregation— two digits at the ISIC

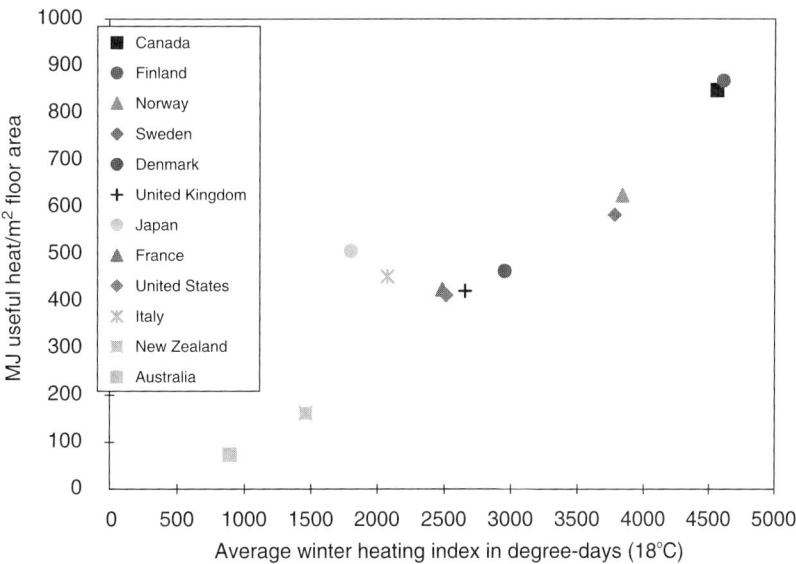

FIGURE 8 Estimated services sector space heating energy use and winter climate.

level—that combines energy-intensive branches, such as paper/pulp and heavy chemicals, with their respective less energy-intensive relatives, such as printing and rubber/plastics, obscures important features of manufacturing.

Alternatively, Phylipsen and colleagues at Utrecht University in The Netherlands and Price and colleagues at Lawrence Berkeley National Laboratory in the United States have carried out extensive comparisons and time-series analyses using data on physical output of materials and energy use for manufacturing those materials. Data on production and energy use typically cover paper, pulp, basic steel, primary and secondary aluminum, other nonferrous metals, cement, some basic chemicals, and refined petroleum. But these data are not available for all countries or for long time periods except in a few countries and industries. The economic approach is useful for explaining overall changes in national energy use, whereas the physical approach is more appropriate for understanding the energy efficiencies of basic processes. But ultimately, only the economic approach covers virtually all of manufacturing, and it is the one adopted in the long series of IEA studies.

3.3.3 Manufacturing Output and Structure
Manufacturing output, measured as aggregate value added or GDP arising, has generally risen slowly in most IEA countries, rising rapidly during economic boom times but declining sharply with recessions. As a share of total GDP across the IEA, manufacturing has fallen slightly due to the continued rise of services or, as in the United Kingdom and Norway, growth in the energy sector. However, manufacturing has not lost a significant share of GDP to services; instead, it is other sectors of industry that have suffered.

Within the manufacturing sector, important structural changes have taken place. These are measured by the share of output (as GDP or value added) in each branch, weighted by the energy intensity of that branch in a base year (rolling weights may also be chosen). In Japan, and to some extent West Germany and the United States, the energy-intensive industries have lost significant share of total output over the past three decades or so. In countries with particular niches for raw materials or cheap energy, such as Australia, Norway, and The Netherlands, the reverse has occurred. Not surprisingly, there is a wide variation of the share of output concentrated in the aforementioned energy-intensive industries, with Denmark at less than 20% and Finland, The Netherlands, Australia, and Norway at well above 30%.

These important movements and differences alone suggest that total manufacturing value added or GDP alone is not a good measure with which to compare countries. At a minimum, the shares in the several subsectors must be used for comparison. Overall differences in output per capita explain part of the difference in energy use per capita or GDP over all sectors. Differences within manufacturing—the share of energy-intensive raw materials in overall output—are nearly as important in explaining differences in total manufacturing or total energy use per capita.

3.3.4 Energy Use and Intensities
Figure 9 shows the energy intensities of the key sectors (also known as branches) of manufacturing in a number of IEA countries during the mid-1990s. The wide differences among sectors result from the intrinsic differences in the processes employed in each sector. But the large differences in intensities among countries within a sector arise for two reasons. Differences in energy efficiencies of processes do cause differences in sectoral energy intensities. But further differences in the composition of even a three-digit ISIC sector also lead to differences among countries. For example, aluminum production in Norway, Australia, and Canada is dominated by primary aluminum production, whereas in West Germany and Japan most is from recycled scrap and/or imported raw aluminum. Paper production in West Germany and Japan is fed largely by imported pulp, whereas "paper" production in Sweden, Norway, and Canada consists of a large surplus of energy-intensive, low value-added pulp that is exported. Similar differences arise in the chemicals and iron/steel sectors. Therefore, the results presented in Fig. 9 cannot be used to make widespread conclusions about energy efficiencies of processes. However, these intensities can be used to explain differences in aggregate energy use relative to output or changes over time. Again, within limits, differences in individual branch intensities can be very useful in explaining national differences in energy use for manufacturing.

3.3.5 Physical Intensities
With some care, physical intensities of production of raw materials can be derived from industry-wide data on production, energy use, and energy expenditures as well as from information on individual facilities. For example, Lawrence Berkeley National Laboratory and the University of Utrecht have carried out a series of comparisons of production and energy for steel, cement, paper and pulp, and

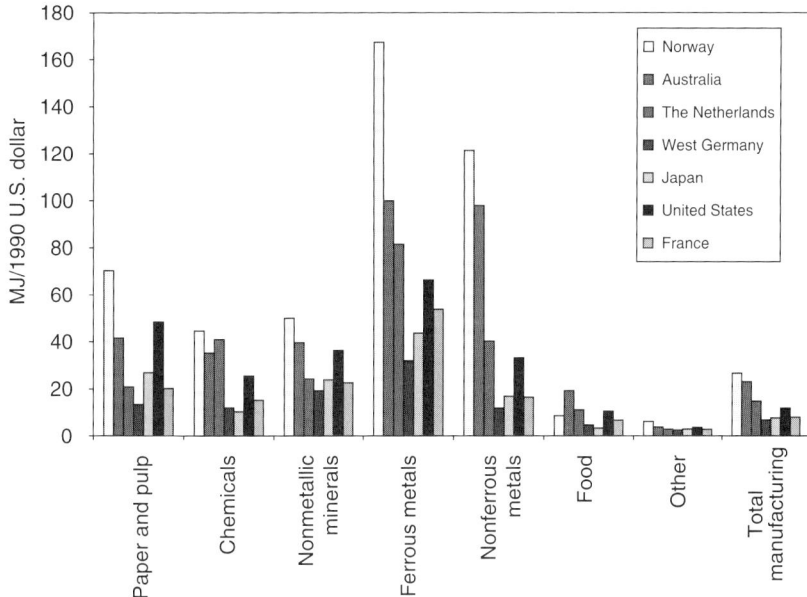

FIGURE 9 Energy intensities in manufacturing: 1995.

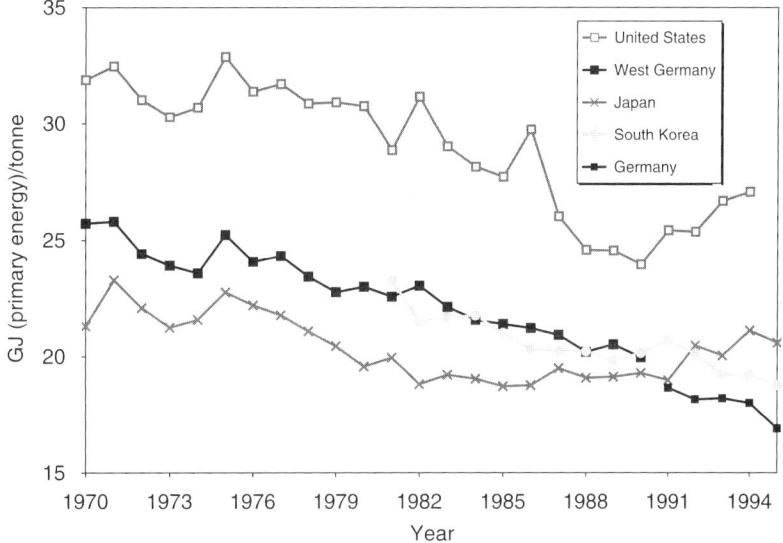

FIGURE 10 Physical intensities of steel production. Data from the Lawrence Berkeley National Laboratory and the University of Utrecht.

other important materials. The results for steel production are shown in Fig. 10.

Differences in national figures are explained by the actual efficiency of production, by the amount of recycled material used, by the nature of the process (e.g., basic oxygen vs blast furnace for steel, different pulping or cement-making processes), and by the actual nature of the final product (e.g., form of steel, type of paper). Efficiency depends on the age and size of the plant; newer and larger facilities are nearly always more energy efficient than older ones. And the level of housekeeping and vigilance in each factory also determines how much energy is used for each basic product. The differences over time are explained both by energy savings occurring within a certain kind of plant or process and by shifts toward processes or plants that are intrinsically less energy intensive and, to a great extent, more energy efficient.

An important caveat must be stated here. The fact that processes are physically more efficient in one country does not mean that they are the appropriate processes for another country, nor does it mean that one level of energy efficiency is "best." Factors such as plant size and vintage depend on many conditions, including energy prices. And, of course, energy prices also influence the level of housekeeping undertaken to save energy.

3.3.6 "Mine–Yours": A Summary Comparison

Summarizing overall differences in manufacturing energy use is tricky. Each part of the ASI equation exhibits significant differences among countries, and within a country the *I* values vary enormously from branch to branch of manufacturing. A comparison that has been developed to summarize these components is used to compare aggregate energy intensity of manufacturing (it was noted previously that output per capita itself varies by more than two to one among IEA countries). To account for differences in the mix of output, one calculates the aggregate manufacturing energy intensity that would have occurred for each country if all countries had the same shares of branch output that make up the average for 13 IEA countries but if each maintained its own branch energy intensity. The resulting aggregate intensity is shown in Fig. 11 ("IEA-13 structure and national intensity") next to the actual aggregate manufacturing intensity in 1994. Where the intensity increased in the alternative calculation (e.g., United States, United Kingdom, Italy, France), the country in question had an own-output structure that was less energy intensive than the average for all countries; where a decline occurred, the own-output structure was more energy intensive. For Australia, The Netherlands, Finland, Norway, and Canada, the difference is large, indicating the importance of the high shares of energy-intensive products. Unquestionably, the mix of branch output accounts for a significant variation in aggregate manufacturing energy intensity.

Differences in individual branch intensities can be summarized in a similar way. In Fig. 11, "IEA-13 intensity and national structure" displays the energy intensity that would have occurred in each country if it had the average energy intensity for the 13 countries in every manufacturing sector but its own sectoral output mix. In this case, an increasing energy intensity (e.g., France, West Germany, Japan, Italy, United Kingdom) compared with the actual value means that a country's energy intensity was lower than the average for all countries; a decline indicates that the country's intensity was higher than average. This is especially the case for the big producers of raw materials (e.g., Australia, Canada, The Netherlands, Norway, Finland). Note that the United States is the only country with higher energy intensity than average and, at the same time, with a (slightly) less energy-intensive structure than average within this group of countries. As with mix, differences in the

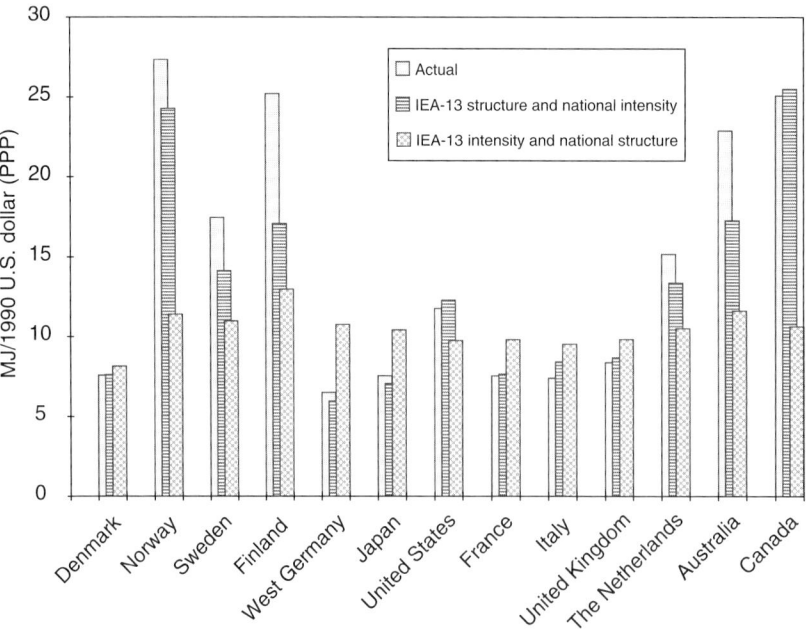

FIGURE 11 "Mine–Yours" comparison of manufacturing energy use. PPP, purchasing power parity.

energy intensities of branches also account for much of the differences in aggregate manufacturing energy intensities.

3.4 Passenger Transport

Passenger transport, or travel, measures human mobility. The most common measure is passenger-kilometers. These are measured by the distances that vehicles move multiplied by the number of people in the vehicles (e.g., 1.5 people/vehicle in private cars, 10–200 in buses, more in trains and aircraft). Roughly 80% of passenger transport is provided by people with their own cars, cycles, or feet. The rest is provided by transport for hire, most of which is air travel in the United States or intercity and commuter rail (including metros) and buses in Europe. The distribution of travel by mode measures the structure of the sector. Except for electricity used for rail and trolley buses and a small amount of CNG, liquid fuels dominate energy use in this sector.

Passenger cars dominate energy use in passenger transport. Their ownership and use grow with income, as Fig. 12 shows. Car use per car varies very little over time, so portraying car use captures mainly the effects of greater ownership. The growth in Japan at the highest income occurred because car sales spurred by tax changes continued to boom during the early 1990s even as economic growth slowed down, pushing up car use. Similarly, car use in Canada grew even when GDP per capita fell during the early 1990s. These twists illustrate the importance of looking at more than 1 year of data in an international comparison. They also show how the portrayal of a key variable against a driving factor such as income suggests important relationships that can be probed by formal statistical tests.

Portraying car use versus income shows the importance of income as a driving factor, but saturation (i.e., decreased growth in ownership or use) is clear for the countries in the upper right hand part of Fig. 12. Note that for every country shown, car use during the mid-1990s was higher than it was during the early 1970s, similar to the upward march of per capita house area shown in Fig. 2.

The wide variation at a given income between the United States, Canada, and Sweden, on the one hand, and Europe and Japan, on the other, is explained in part by fuel prices, incomes and car ownership policies (e.g., new car taxes, annual fees), and geography. Surprisingly, Sweden had relatively low fuel prices (for Europe) until the late 1980s and had very low taxation of company cars, which once represented nearly half of all new cars sold. This kind of practice probably accounts for Sweden reaching the level of car use/GDP seen by the United States, Canada, and Australia, that is, countries with even larger areas and lower fuel prices. This illustrates how a comparison points to, but does not prove, how a policy may have important impacts on energy use. A similar implication arises from the large areas of homes relative to private expenditures shown in Fig. 2.

When data on driving and the number of people in a car are combined, travel in passenger-kilometers can be developed. Together with estimates of travel from other commercial modes (including data from national travel surveys), the overall pattern of mobility can be established, as is shown in Fig. 13.

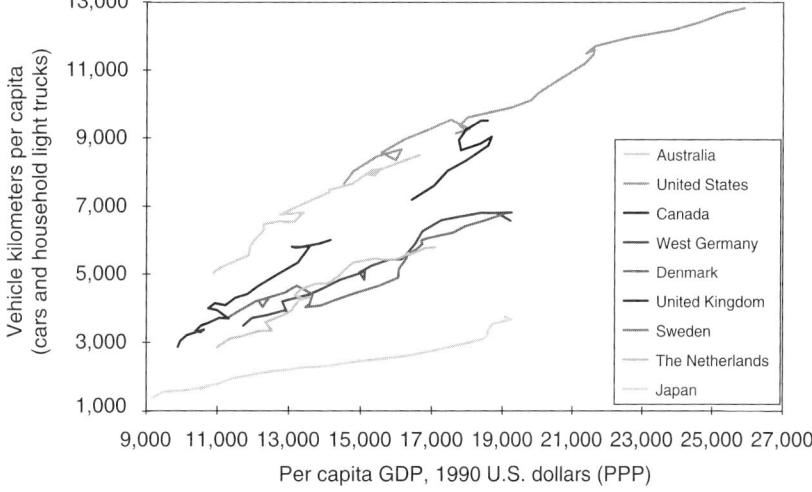

FIGURE 12 Car driving and per capita GDP: 1970 to 1998. PPP, purchasing power parity.

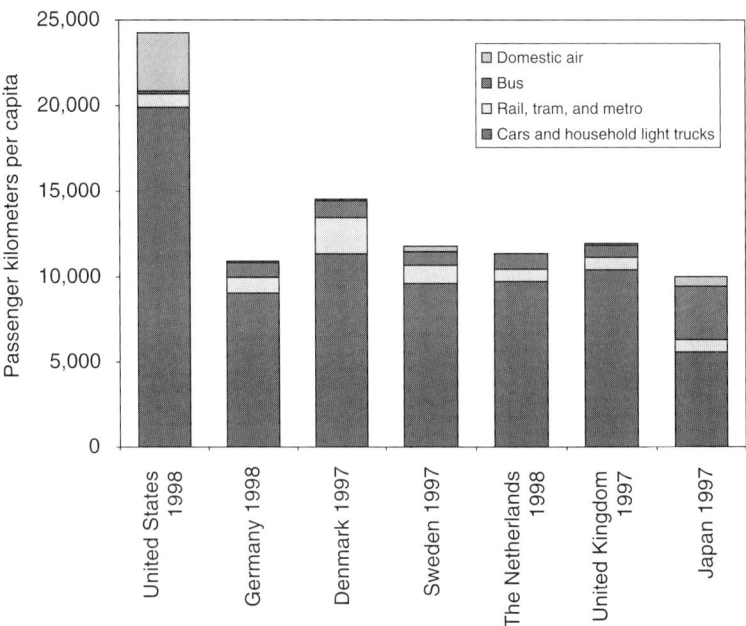

FIGURE 13 Per capita passenger transport by mode.

The high values for the United States follow both from higher car ownership relative to GDP and from higher annual driving distances. For every country shown, total travel was higher during the 1990s than during the 1970s and was increasingly dominated by cars. Indeed, the growth shown in Fig. 12, as well as some increases in air travel, accounted for virtually all of the increases in travel from the 1970s to the 1990s.

The most important energy intensity in the sector is automotive fuel economy or fuel consumed per unit of distance. Figure 14 shows that the United States has the most fuel-intensive cars, followed closely by Australia, Canada, and Japan. The latter might seem surprising, but poor traffic in Japan raises fuel intensities.

When the average loads are included, the energy intensities of car travel can be tabulated in megajoules per passenger-kilometer. These can be compared with the (lower) values for rail and bus as well as with the values for air travel, which are higher than those for car travel in Europe but lower than those for car travel in the United States. Multiplying each modal intensity by travel per capita in that mode gives energy use per capita by mode, which is shown in Fig. 15.

Although each mode serves a different purpose and has different characteristics, the difference in intensities among the modes, combined with the shifts in modes toward car and air travel, explains much of the change—usually an increase—in overall average energy use per passenger-kilometer of travel. Combined with higher per capita travel, this meant that energy use increased over the past three decades. Note that the only country in which per capita energy fell over the period shown was the United States, a result of the strong decline in the intensities of both automobile and air travel. Canada, which was affected by the same trends as the United States, had similar declines in intensities but a greater increase in car ownership and use; hence, per capita energy for travel was slightly higher in 1995 than in the earlier period. Note that by decomposing total change in energy into activity, structural, and intensity components, the analysis is able to show which caused energy use to increase and which caused it to decrease. This decomposition gives credit for policies, technologies, or efforts that led to saving energy but whose effects may have been obscured by the impact of other forces on overall energy use.

Using the data on modal structure and modal intensities, it is possible to compare transportation energy use among countries in an innovative "mine/yours" way that brings out the relative importance of these three components in much the same way as the same approach was used to explain differences in aggregate manufacturing energy intensities. However, this time the activity term is included explicitly, so the variable is per capita energy use and the results of imputing other countries' activity, modal structure, and modal intensities are compared with actual per capita energy use. Figure 16 shows first actual per capita use for a number of countries. Next,

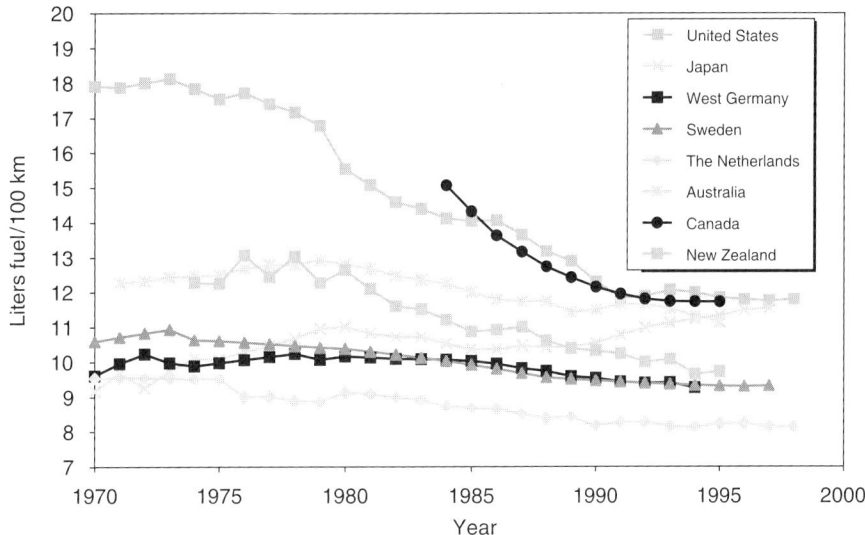

FIGURE 14 Automobile/personal light-truck fuel intensities: 1970s to 1990s.

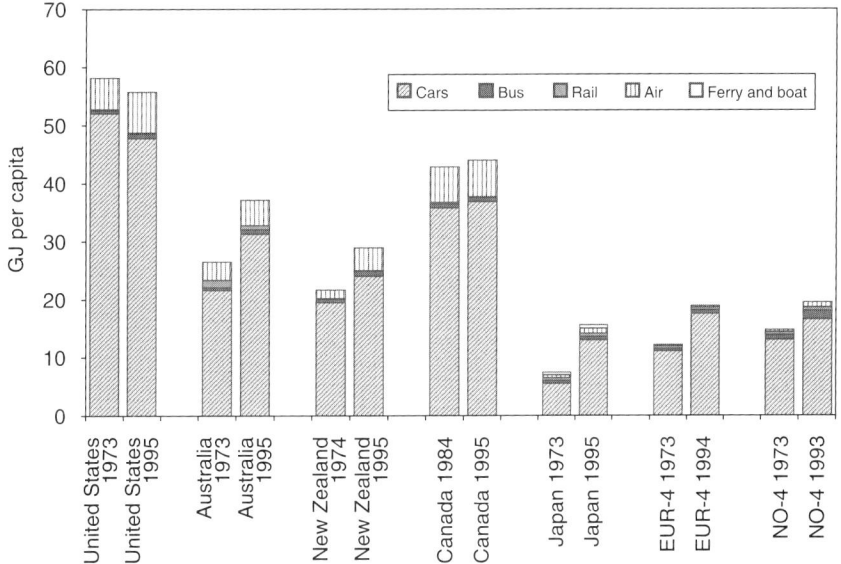

FIGURE 15 Per capita energy use for travel by mode. See Fig. 1 for abbreviations.

energy for each country is calculated as if total travel corresponded to the average value for the 13 countries considered. Then, the same calculation is carried out using the average modal mix and activity. Finally, the calculation is done for actual activity and modal mix but with the average intensities.

When a country's "total" is reduced by substitution of one set of average parameters, that is an indication that the country's own parameters boost energy use. The comparison shows that total travel and modal intensities explain most of the differences among countries. Thus, one can see for the United States, Australia, and Canada that activity, modal structure, and intensities all lead to higher energy use than the average. Conversely, giving Japan or Europe higher activity or a modal split more weighted toward cars would raise energy use in those countries. Note that with the exception of Japan, the impact of changing modal mix alone is small. This is because in every country except Japan, travel is dominated by cars (Fig. 14).

3.5 Domestic Freight Transport

Freight accounts for roughly 15% of a nation's energy use (Fig. 1), and its importance has been

550 International Comparisons of Energy End Use: Benefits and Risks

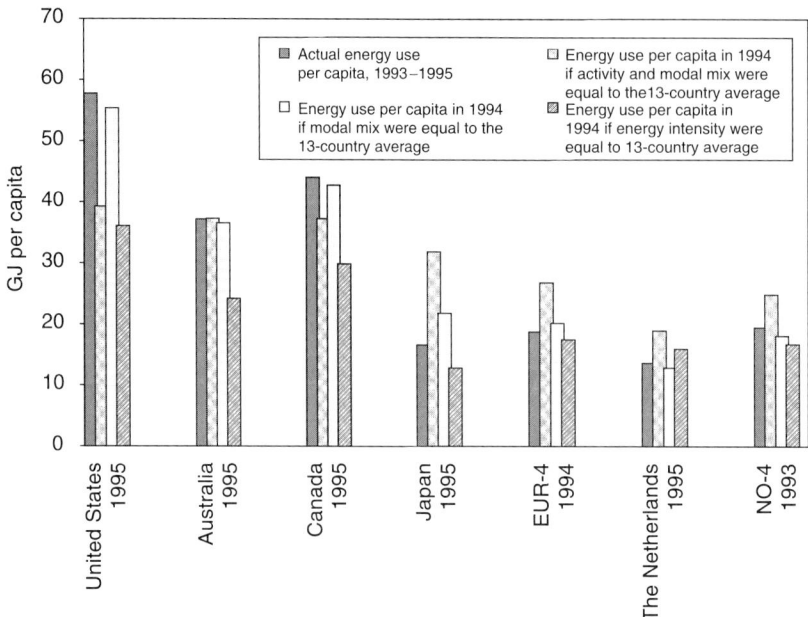

FIGURE 16 "Mine–Yours" comparison of energy use for passenger transport (travel). See Fig. 1 for abbreviations.

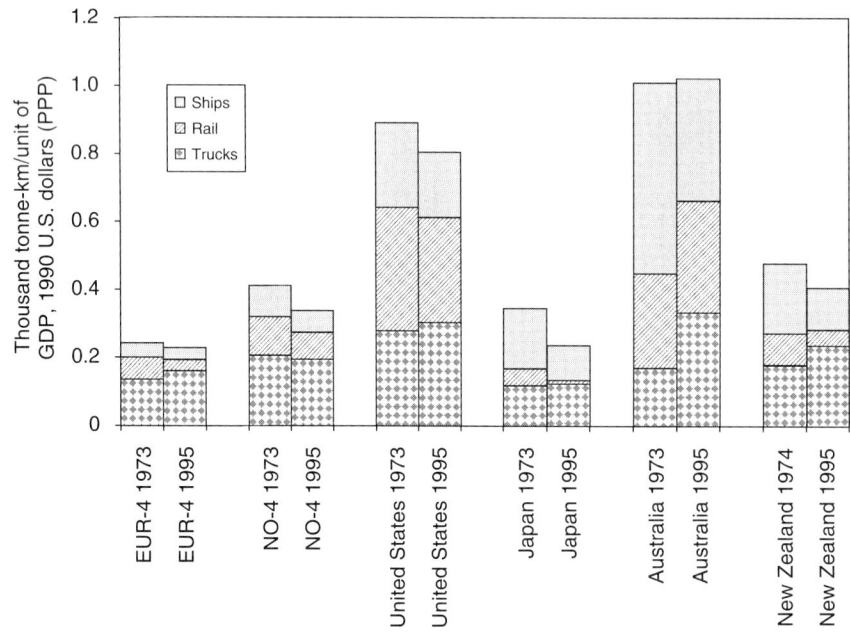

FIGURE 17 Domestic freight activity relative to GDP. See Fig. 1 for abbreviations.

rising. Output is measured in ton-kilometers. A modal mix of overall ton-kilometers—trucking, rail, and domestic shipping (coastal and river activity)—makes up structure. Freight hauled by pipeline, air freight, and freight sent overseas are excluded. The key driver for freight is GDP, which makes a convenient normalization, in addition to population, for aggregate comparison. Most energy used is for trucking, which also dominates total freight hauled in Japan and Europe.

Figure 17 gives a glimpse of domestic freight hauled in 1995 by mode relative to GDP. That normalization removes the most gross differences among countries. Clearly, the high levels of freight relative to GDP in the largest countries (e.g., United States, Australia) are related to their size. It is also clear from further analysis that these countries produce more bulk raw materials than do the smaller countries portrayed. These larger countries also haul a much larger share of their freight on rail or domestic shipping (river barge and ship, coastal shipping). These differences explain much of the overall differences in the ratio of freight energy use to GDP implicit in Fig. 1.

Energy use for trucking has to be estimated carefully from surveys to separate it from energy use for cars and other road vehicles. Although diesel fuel dominates trucking, diesel is also important in a number of European countries for cars, so it cannot simply be assigned to trucks. On the other hand, many light trucks used exclusively for freight or commercial purposes run on gasoline, so this must also be added to energy use. For rail, it is difficult to separate fuels and electricity into that used for freight and that used for travel, although in principle national railroads could do this. Energy use for domestic shipping is well identified because marine fuels are so identified and because fuels used for foreign or overseas shipping (bunkers) are counted separately for tax purposes. Overall energy use is so dominated by trucking that if this mode is portrayed accurately, comparisons are relatively straightforward.

Trucking is the most energy-intensive mode after air freight, which still accounts for less than 3% of all freight. Trucking energy intensity depends mostly on the type and size of vehicle, use of each vehicle's capacity, and traffic conditions. In contrast, actual efficiencies of trucking (in terms of fuel used per unit of horsepower or load pulled) vary little from vehicle to vehicle due to widespread competition to sell trucks. This means that comparisons of trucking energy use either per kilometer driven or per ton-kilometer hauled are not representative of the efficiencies of individual types of vehicles. Unfortunately, few data exist on energy use per kilometer of trucks of a given size, so little can really be said about truck efficiency on the basis of international comparisons. Thus, the energy intensity of trucking is analogous to that of a manufacturing branch or space heating—related to, but not identical with, the inverse of energy efficiency.

Over time, the share of trucking freight haulage has increased, but trucking has become somewhat less energy intensive. Because trucking is so much more energy intensive than other modes, this shift raised energy use nearly enough to offset the impact of lower energy intensities. Still, trucking dominates energy use in all of the countries studied, as Fig. 18 shows.

Notice that the differences in energy use/GDP for freight among countries are smaller than the differences in levels of freight/GDP. This is because the most freight-intensive countries rely the least on energy-intensive trucking and because trucking tends to be less energy intensive along wide-open roadways than on the congested highways of the smaller countries in Europe and Japan. Thus, averaging in the 65% of U.S. domestic freight that is hauled by ship or rail to a relatively non-energy-intensive

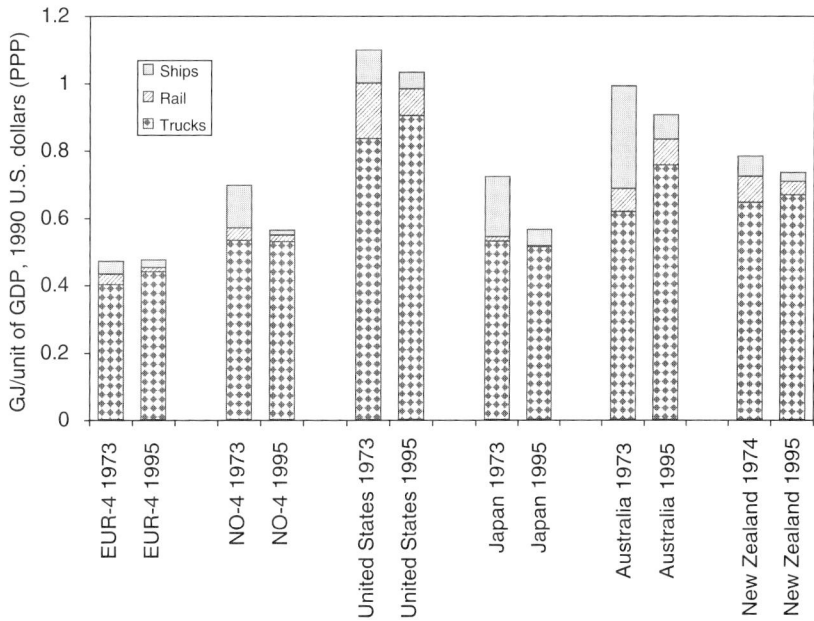

FIGURE 18 Energy use for domestic freight relative to GDP. See Fig. 1 for abbreviations.

trucking sector gives the United States a relatively low aggregate freight energy intensity and overall freight energy use/GDP that is only $2\frac{1}{2}$ times that in Europe. This is an important example often found in international comparisons: a feature that raises energy use (in this case, country size) is associated with lower energy use due to modal structure (share of trucking) and energy intensity.

4. OVERALL COMPARISONS OF ENERGY USE

It was argued early in this article that the ratio of energy use to GDP is a misleading way in which to compare overall energy use among countries. Still, it is important to tie together all of the information presented here in a meaningful way. The results can and should be related to GDP as well because wealthier nations generally do use more energy. This section summarizes a method based on the mine–yours formulation shown for travel and manufacturing that decomposes overall differences in energy/GDP ratios into structural and intensity components.

Figure 19 shows the results expressed in terms of E/GDP. The first column shows the actual ratio of final energy to GDP in 1995. The second column shows what that ratio would be if the average set of structural parameters for the 13 countries were used in calculating E/GDP. The final column uses the average energy intensities throughout the economy to repeat the calculation. (It is worth noting that in this calculation, each country is held out of the calculation of "others." Otherwise, the United States and, to some extent, Japan show little variation because these two countries dominate the averages.)

The results are not startling given the sectoral differences shown previously. Overall energy use in the United States would be significantly lower if it had the structure of other countries' energy use, due mainly to transportation, households, and services. This is important because the public perception—that manufacturing contributes to the United States' higher energy use—is actually the reverse. Both the size and structure of manufacturing restrain U.S. energy use relative to the average of the other countries. Sweden's energy use would also be lower for similar reasons because average structure would mean warmer winters and less energy-intensive manufacturing. Australia's energy use would be higher because the impact of its very energy-intensive manufacturing sector is more than offset by the lack of demand for space heating. The energy use of Japan and France would also rise due to imposing colder winters, greater car use, larger homes, and greater service sector areas. Thus, the comparisons show that structural differences are significant in explaining differences in energy use among countries.

Energy intensities are nearly as important. For Canada, the imposition of other countries' intensities would have a very large downward impact, but this is also true, to a lesser extent, for the United States,

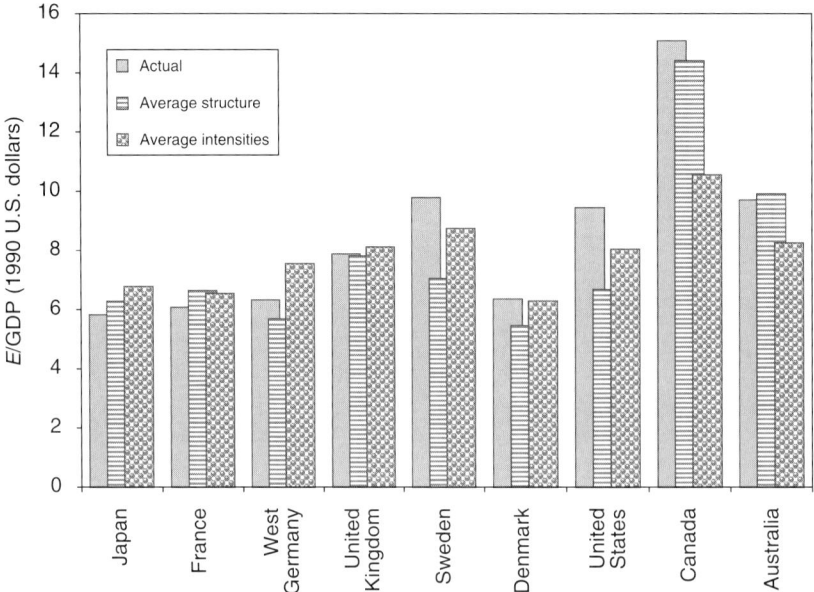

FIGURE 19 Summary comparison across all sectors and uses: mine–yours.

Australia, and Sweden. Thus, we can say that these countries have the highest energy intensities. Japan, France, and West Germany have the lowest energy intensities.

Overall, this approach illustrates how differences in the structure of energy use, as well as differences in individual energy intensities, explain differences in the ratio of energy use to GDP. Interestingly, the United Kingdom emerges as the "average" country that is affected the least by this procedure for aggregating differences.

5. LESSONS FROM INTERNATIONAL COMPARISONS

The most important lesson from international comparisons is that structural differences explain about as much of the differences in energy use relative to GDP as do intensity differences. This is not surprising given the huge variation in key structural parameters listed in Table II. Factors that drive these differences—incomes, housing and vehicle taxation policies, natural resource endowments, subsidies to one industry or another—may in fact be as important as energy policies per se and energy prices in explaining the differences in energy uses.

The second lesson is that energy intensities make acceptable, if imperfect, indicators of energy efficiencies. Considerable work is required before the efficiency of processes and devices can be evaluated. It was noted that the fuel intensity of cars is related both to characteristics consumers choose (e.g., weight, power) and to the real technological efficiency of cars (e.g., energy/weight or energy/power plus kilometers driven). Similarly, the energy intensity of space heating depends both on indoor heating habits and on the insulation of the structure or efficiency of the heating systems. And the intensity of appliances depends on their size and features, their frequency of use, and the efficiency of energy conversion to heat, cooling, motor power, and the like. Usually, but not always, intensities make good indicators of efficiencies. Thus, the comparisons of efficiencies implied by intensities must be made with caution.

Can these comparisons be extended to developing and transitional countries? The answer is a cautious yes. Data for Mexico, India, Brazil, China, and a number of transitional countries (e.g., Poland) have been presented for analysis in a number of forums. For example, Fig. 20 shows such data for Mexico. The data for the household sector come from

TABLE II

Key Differences in Economic Structure and Energy Prices: Late 1990s

- Heating degree-days (using a base indoor winter temperature of 18°C) vary from 900 in Australia and 1800 in Japan to more than 4500 in Canada and Finland. The population-weighted average for 14 countries is 2800
- Automobile use per capita varies from a high of 13,500 km/year in the United States to a low of 3900 km/year in Japan. Trucking ton-kilometers per capita vary by a factor of two, and the ratio of freight hauled/GDP varies by more than a factor of two, due to geography and other considerations.
- Home size varies from more than 155 m^2 per capita (or roughly 60 m^2 per capita in the United States, at the high end to less than 90 m^2 and closer to 30 m^2 per capita in Japan at the low end.
- Per capita manufacturing output varies by two to one. The manufacturing share of GDP varies from less than 20% in Australia and Denmark to nearly 30% in other countries, and the share of the energy-intensive raw materials sectors varies from more than 30% in Australia, The Netherlands, Finland, and Sweden to only 15% in Denmark.
- Similar considerations in the service sector reveal nearly a two-to-one ratio of built-up area to population or to service sector GDP.
- Real road fuel prices vary by a factor of three when national purchasing power is taken into account. Home heating fuel prices vary by a factor of two, and electricity prices in any sector vary by as much as a factor of three. GDP per capita varies by 1.75 to 1 across the countries considered, again when compared using purchasing power parity. Some countries permit full deduction of mortgage interest from incomes (e.g., United States), whereas others permit almost none at all. New care taxation varies from very low (e.g., United States) to effectively 200% (e.g., Denmark). Company car schemes accounted for as much as 50% of sales in Sweden and Great Britain during many periods between 1970 and 1995.

surveys, whereas those for the productive sectors except transport come from detailed energy balances. For these sectors, structural data are also available. The difficult sectors are travel and freight. The problem for most developing counties is the lack of accurate data on the number of vehicles actually in circulation and the distances they are driven as well as a breakdown of how gasoline and diesel (and in a few cases LPG or CNG) are divided among two- or three-wheelers, cars, light trucks, heavy trucks, buses, and other vehicles.

The final lesson is the most difficult one. Collecting and processing data for these comparisons is not expensive, but it is not free. The United States, which pioneered surveys in the stationary sectors (e.g., households, manufacturing, services), has reduced its effort in both coverage and time and has abandoned most of its effort to accurately measure both activity

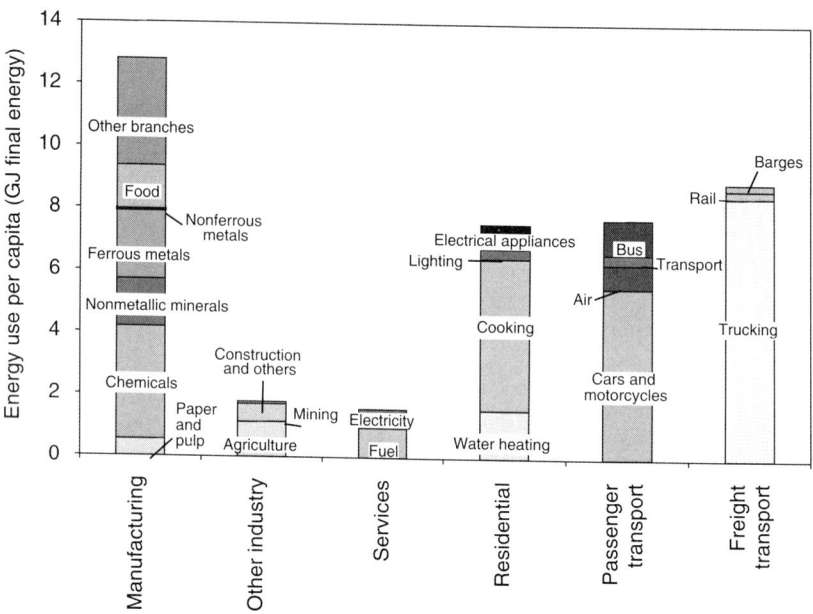

FIGURE 20 Energy uses in Mexico: 1994. From Claudia Sheinbaum, Universidad Autonoma de Mexico.

and fuel use in highway vehicles. Yet fairly accurate measures of energy use for other sectors and uses portrayed in this article are still available and even used to produce detailed inventories of carbon emissions. Other countries, notably Canada and Australia, have stepped up their efforts to institutionalize both data collection and data analysis. The IEA efforts noted previously have also stimulated a number of IEA member governments to take careful inventory of their own energy data and how they are analyzed and published.

These improvements should not stop and, in fact, should be extended to developing countries. The reason was given early in this article: a fair deal on restraining carbon emissions depends on all parties understanding the basis of each other's emissions, which for most countries are dominated by carbon arising in the combustion of fossil fuels.

Thus, the benefits of international comparisons of energy use are a better understanding of one's own consumption patterns as well as those of one's neighbors, partners, and even adversaries. The costs of those comparisons, in terms of data-gathering efforts, are relatively small given that many of the structural data are collected anyway. Even with sectoral surveys every few years, the price tag is less than one-thousandth of a country's monetary energy budget. The risks of misuse of information, particularly the confusion of intensities with real efficiencies and the assignment of "targets" for energy use based on other countries' experience, are real. But the risks of not understanding even one's own energy use patterns, and hence of missing opportunities for applying more efficient uses or for applying (or removing) policies that lead to inefficiencies or waste, may be even larger.

SEE ALSO THE FOLLOWING ARTICLES

Commercial Sector and Energy Use • Conservation Measures for Energy, History of • Development and Energy, Overview • Economic Geography of Energy • Energy Ladder in Developing Nations • Fuel Economy Initiatives: International Comparisons • Global Energy Use: Status and Trends • Industrial Ecology • Industrial Energy Efficiency • Industrial Energy Use, Status and Trends • Sustainable Development: Basic Concepts and Application to Energy

Further Reading

Adams, F. G., and Miovic, P. (1968). On relative fuel efficiency and the output elasticity of energy consumption in Western Europe. *J. Industr. Econ.* 17, 41–56.

Darmstadter, J., Dunkerley, J., and Alterman, J. (1978). "How Industrial Societies Use Energy." Resources for the Future, Washington, DC.

Krackeler, T., Schipper, L., and Sezgen, O. (1998). Carbon dioxide emissions in OECD service sectors: The critical role of electricity use. *Energy Policy* 26, 1137–1152.

Milstein, J., and White, R. (1975). "Comparison of Energy Use in the U.S. and West Germany." Stanford Research Institute International, Menlo Park, CA.

Phylipsen, G. J. M., Blok, C., and Worrell, E. (1998). "Handbook on International Comparisons of Energy Efficiency in the Manufacturing Industry." Department of Science, Technology, & Society, Utrecht University, Netherlands.

Price, L., Phylipsen, D., and Worrell, E. (2001). "Energy Use and Carbon Dioxide Emissions in the Steel Sector in Key Developing Countries." Lawrence Berkeley National Laboratory, Oakland, CA.

Schipper, L. (1997). "Indicators of Energy Use and Efficiency." International Energy Agency, Paris.

Schipper, L. (1997). "The Link between Energy and Human Activity." International Energy Agency, Paris.

Schipper, L., Johnson, F., Howarth, R., Andersson, B., Andersson, B. G., and Price, L. (1993). "Energy Use in Sweden: An International Perspective." Lawrence Berkeley National Laboratory, Oakland, CA.

Schipper, L., and Lichtenberg, A. (1976). Energy use and well being: The Swedish example. *Science* **194**, 1001–1013.

Schipper, L., and Marie-Lilliu, C. (1999). "Carbon-Dioxide Emissions from Transport in IEA Countries: Recent Lessons and Long-Term Challenges." Swedish Board of Transportation and Communications Research, Stockholm.

Schipper, L., Murtishaw, S., and Unander, F. (2001). International comparisons of sectoral carbon dioxide emissions using a cross-country decomposition technique. *Energy J.* **22**(2), 35–75.

Schipper, L. J., Ting, M., Khrushch, M., and Golove, W. (1997). The evolution of carbon dioxide emissions from energy use in industrialized countries: An end-use analysis. *Energy Policy* **25**, 651–672.

Schipper, L., Unander, F., Marie-Lilliu, C., and Walker, I. (2000). "Energy Use in New Zealand in an International Perspective: Comparison of Trends through the Mid-1990s," prepared for the Energy Efficiency and Conservation Authority of New Zealand. International Energy Agency, Paris.

Schipper, L., Unander, F., Marie-Lilliu, C., Walker, I., and Gorham, R. (2000). "Energy Use in Australia in an International Perspective: Comparison of Trends through the Mid-1990s." International Energy Agency, Paris.

Schipper, L., Unander, F., and Ting, M. (1996). "Energy Use and Efficiency in Denmark in an International Perspective." Energistyrelsen, Copenhagen, Denmark.

Unander, F., Karbuz, S., Schipper, L., Khrushch, M., and Ting, M. (1999). Manufacturing energy use in OECD countries: Decomposition of long-term trends. *Energy Policy* **27**, 769–778.

Unander, F., and Schipper, L. (2000). "Trends in Norwegian Stationary Energy Use: An International Perspective." International Energy Agency, Paris.

International Energy Law and Policy

THOMAS W. WÄLDE
Center for Energy, Petroleum and Mineral Law and Policy, University of Dundee
Dundee, United Kingdom

1. Introduction
2. Organization of Petroleum Exporting Countries
3. International Energy Agency
4. Energy Charter Conference and Secretariat
5. International Atomic Energy Agency
6. United Nations
7. OECD Nuclear Energy Agency
8. European Union
9. Conclusions

Glossary

Energy Charter Conference and Secretariat The mission of the Energy Charter is to promote open, efficient, sustainable, and secure energy markets and a constructive climate conducive to energy interdependence on the basis of trust between nations.

International Atomic Energy Agency (IAEA) An independent intergovernmental science- and technology-based organization, in the United Nations family, that serves as the global focal point for nuclear cooperation.

International Energy Agency (IEA) The energy forum for 26 countries whose member governments are committed to taking joint measures to meet oil supply emergencies. The IEA collects and publishes a wide array of energy and environmental data and research reports.

Nuclear Energy Agency (NEA) A specialized agency within the Organization for Economic Cooperation and Development, an intergovernmental organization of industrialized countries. The mission of the NEA is to assist its member countries in maintaining and further developing, through international cooperation, the scientific, technological, and legal bases required for the safe, environmentally friendly, and economical use of nuclear energy for peaceful purposes.

Organization of Petroleum Exporting Countries (OPEC) Eleven-nation organization, the members of which supply a major share of global oil; members collectively agree on global oil production and/or price targets. Members include Algeria, Indonesia, Iran, Iraq, Kuwait, Libya, Nigeria, Qatar, Saudi Arabia, United Arab Emirates, and Venezuela.

World energy laws and policies are the responsibility of numerous international agencies, including the Organization of Petroleum Exporting Countries, the International Energy Agency, the Energy Charter Conference and Secretariat, the International Atomic Energy Agency, the United Nations and its agencies, the Organization of Economic Cooperation and Development's Nuclear Energy Agency, and the European Union. Understanding the development and operation of these organizations is integral to understanding the role and effectiveness of each organization in promoting sustainable development.

1. INTRODUCTION

The concept of sustainable development now dominates the natural resources, energy, and environmental discourse with its accommodating notion of developing resources now, without compromising the future. Sustainable development suffers from an imbalance: rhetoric overwhelms action, affirmation of moral values overwhelms implementation, and good intentions prevail over getting good results. Results can be achieved only if the "game" is properly understood: this requires understanding the context of an action, including the economic, social, and cultural elements of the relevant environment; the concepts, which are usually culturally formed in some relationship with the often historical and possibly obsolete economic and technical contexts; and the roles, parameters of action, interests, and dominant perceptions of the main players.

Although sustainable development can perhaps be described as ideological input, activities such as technical and financial assistance and work with rules (formulating, implementing, revising) are the output of international agencies. This output is one of the factors that determines if, how, and to what extent and in what shape and form sustainable development leaves the realm of morality and ideology and enters the realm of socioeconomic and organizational reality. Very little practical research seems to have been done on how international agencies participate in the emergence of rule systems in the field of energy. The analyses here are meant to sketch out observations on the role of selected key international agencies in the field of energy law (policies and regulation). The outcome should help to understand somewhat better than before how sustainable development can be achieved more effectively through the activities of international agencies in the energy sector. The emphasis is not only on the formal mandates, a popular approach in the examination of international organizations by public international lawyers, but also and perhaps with greater emphasis on the "real-life" operations and strategies pursued by such agencies.

International organizations play a central role in the emergence of international energy law. They represent a response to the globalization of the regulatory challenges. As national regulation loses its grip, international agencies emerge to mirror the global scope of action of relevant actors (multinational companies, banks, nonstate actors, and criminal groups), but also the global scope of problems, i.e., transboundary environmental impact, and the need to create more level playing fields for competition in global trade, development, and administration of a system of global legal disciplines to ensure that the wealth-creating machine of the global economy can function properly, including its social safeguard. International agencies in this area are therefore more a prototype of true international regulation; they help member states to regulate better by providing model guidelines, collective and better regulatory intelligence, and a forum of dialogue and collegiality. As we shall see, they are primarily intergovernmental collegial networks. This, in turn, helps national regulation to be more aligned. National administrations usually establish formal and informal channels for collaboration as part of their participation in international agency activities (conferences, workshops, elaboration of model codes).

There is an explicit, visible role of international agencies in the regulatory process—servicing the negotiation and administration of treaties, formal elaboration of standards and guidelines, and, though rarely, direct regulatory power. Elaborating technical standards, usually in collaboration with experts in governments, companies, and industry associations, is a latent, but key, direct regulatory role. Such standards may be directly binding, but in most cases they have an indirect legal effect: as standards of due diligence for defining civil responsibility, by explicit incorporation into contracts or by reference in national or international (e.g., General Agreements on Tariffs and Trade/World Trade Organization-based) law. They also work by providing national governments, domestic experts, and social forces with a ready-made set of usually well-researched, tested, and often elsewhere-applied rules.

But international organizations also have an implicit, informal, and less visible role: Serving as a forum for discussion, i.e., an actor (often with some autonomy and independent dynamics) with agenda-setting power, international agencies also provide legitimacy and a benchmark for national regulatory action. There is a feedback process between the national level, the nongovernmental organization (NGO) and business level, the authoritative media (foremost the *Financial Times*), and the international level. Innovations and new concepts jump from the domestic level to the international agency level. They are here processed, refined, and negotiated into a more universally accepted format. They then leapfrog by way of an agency endorsement to the level of national regulation. At times, international agencies acquire true international regulatory competencies. The European Union (EU) is the best example. The new EU energy law emerging out of the 1996/1998 electricity and gas directives and its current amendment proposals illustrate the balance between true EU Commission powers and national regulatory powers to be exercised within the EU energy law framework. The more technical the subject and the more homogeneous the membership, the more it is likely that international agencies obtain true and direct regulatory powers, often exercised jointly with national enforcement authorities. The International Maritime Organisation (IMO), the World Intellectual Property Organization (WIPO), and other specialized agencies illustrate this angle. Here, there are superior technical and efficiency reasons for transferring direct regulatory powers, and less "sovereignty" reasons related to more politicized issues for keeping such power at home.

International agencies serve as vehicle by transnational coalitions (e.g., national ministries, international business associations, and international agency officials and their expert networks, or, with a similar structure, national environment agencies, international NGOs, and the environmental agency network) to impose or influence by persuasion of national regulation, usually to counter the stronger influence of domestic pressure groups (protectionist business or labor lobbies; ideological pressure groups). The agency, with its experts, funds, and machinery for consultancy work, meetings, and publications, serves as forum for the elaboration of common rules by such specialized communities. The national representatives then return to their domestic regulatory process with the added legitimacy and technical and informational underpinning gained in the international agency process. The international agency, so to speak, is the nest to which and from where "birds of the same feather" fly into their domestic political and regulatory context. This leads to another nonconventional observation.

One generally ignored quality of international agencies is their scapegoat role for domestic politics. There is nothing more convenient than putting the blame and political responsibility for necessary action, which is strongly contested by vociferous interest groups, on international agencies and international obligations. The EU is the perfect example of national civil servants negotiating a regulatory agenda with the EU Commission staff. National politicians then blame an external and therefore intrinsically unpopular organization for the result they have all designed jointly. The current politicization of the most visible international agencies [World Bank, International Monetary Fund (IMF), World Trade Organization (WTO), Organization for Economic Cooperation and Development (OECD)], resulting in agitation by the inchoate and disparate forces of the antiglobalization movements, is a consequence of governments redirecting protest to faceless international agencies.

The primary and traditional function of international agencies is to provide the secretariat services for conferences, out of which emerge regulatory instruments such as multilateral treaties. The WTO services the General Agreements on Tariffs and Trade (GATT)-related treaties (and subsidiary instruments such as protocols and understandings), the Energy Charter Secretariat, the protocols and decisions of the Energy Charter Conference, and the IMO for various offshore pollution and dumping protocols. International agencies always proclaim that they are mere "servants" of the governments meeting in conference. This is an essential element of the official self-presentation. They will use such defense in particular when attacked by outside forces (the press, NGOs) for disliked treaties, treaty administration, and their mandate-based activities. This is formally correct: Intergovernmental organizations consist of a "secretariat" providing faceless "clerical" services to their governmental masters: preparing studies, conference services, support of negotiations, and reports on implementation. Formal decision powers, treaty-making, and the making of subsidiary law under treaties rest, as a rule, with the conferences or governing bodies composed of government representatives—economic ministry delegates in the case of the OECD, finance ministry delegates in the case of the World Bank (here acting through the World Bank executive directors), environmental ministry delegates in the United Nations Environment Program (UNEP), and so on.

This view of the role of the organization is usually described in the charters and other legislative documents setting up the organization. But such defensive escape into anonymity, though formally correct, also intentionally ignores the reality. An international organization, while professing to be only the servant of its governmental masters, develops within a short while a life, interests, external constituencies, and an organizational dynamic of its own. Pure formal descriptions focusing on founding instruments and organizational self-representation are therefore unable to capture the reality of an international agency and its role in the evolution of energy law. International agencies foremost develop a corporate self-interest in perpetuating themselves, even when their original purpose has evaporated. Because they are usually not set up with explicit "sunset" clauses, i.e., a finite end, like a task force, they will continue to exist even when the original purpose no longer exists nor requires the organization's continued existence. International agencies will therefore work hard at pretending that the original purposes still exist and are valid and will justify the considerable expense, usually by membership fees, to maintain them. They will, in addition, adapt to respond to issues they see as suitable for their survival. The fact that international agencies leap like lemmings, usually all together, after the most recent fad, reflects this survival instinct. A study of the self-justifications of agencies would lead to the identification of a "life cycle" of fashionable terms, with innovation, reluctance, gradual acceptance, enthusiastic pursuit, widespread use, and finally obsolescence and quiet disappearance.

International agencies will develop constituencies (in more fashionable terms, "stakeholders," e.g., environmental ministry delegates, academic experts, consultants, or environmental NGOs) that will support the continuous existence of the agencies and the need to fund them. Because few if any agencies are subject to an independent assessment to gauge the continuing need for them, they are very hard to eradicate once established. Most informed assessment is partisan, and a nonpartisan assessment is usually uninformed, because it is without access and resources for an in-depth assessment. To develop a proper understanding of the role of international agencies is therefore quite difficult, if not impossible. It is understandable why most of the international law literature does little more than regurgitate statutory instruments and organizational propaganda rather than attempting an independent examination with the aim of at least coming closer to an image of reality.

The work of international agencies is closely related to the functioning of efficient regional and global energy markets. Only the market mechanism can unleash the human forces of innovation and efficiency. The market mechanism has, if left to its own devices, shortcomings (externalities, failure to provide desirable public goods, excessive volatility). The task of regulation, and, therefore, for regional and global markets, of the protoregulatory functions of international agencies is to deal with these deficiencies, i.e., to strengthen the effective functioning of international energy markets, to internalize externalities, to take care of truly "public goods," and to smooth out, to the extent feasible, excessive volatility. There is an ever-present risk of overregulation: "state failure" is the other side of market failure. Overregulation imposes excessive transaction costs, prevents innovation, and thus damages the capacity of society to create both economic and technological resources to deal with the challenges of sustainable development. In international organization, the risk of looking to state-led solutions is greater than in a national regulator: the international agency is twice removed from democratic legitimation, almost fully removed from the pressure of competitive markets, and interacts mainly with likeminded constituencies, i.e., governmental delegates, NGOs, and agency-selected experts with a natural proagency and usually proregulatory bias. This inherent weakness of international organizations is to some extent mitigated by the reduced effectiveness of their action, in particularly of the mainly "talk-shop" organizations.

In the remainder of this discussion, the relevant international agencies are reviewed, with comments on their roles and functions. Due to space constraints, only the international agencies of most significance for energy are discussed.

2. ORGANIZATION OF PETROLEUM EXPORTING COUNTRIES

In attempting to deal with truly "international" energy law, the Organization of Petroleum Exporting Countries (OPEC) is the international organization with the greatest impact on the oil sector; OPEC also has the greatest influence on energy and energy-related environmental questions, and not only on production and trade, but also on investment. OPEC countries currently control about 75% of the world's oil reserves and 40% of oil production. What is more, most low-cost oil is produced by OPEC countries. This means that a prolonged slump in prices will tend to enhance these market shares. [Low oil prices will also tend to increase the efficiency in high-cost areas (e.g., the North Sea). Significant results in terms of lowering production cost have been achieved by the Cost Reduction Initiative in the New Era (CRINE) incentive in the United Kingdom, for example. This incentive for efficiency is not present in the Middle Eastern low-cost producers. But the difference between, say, North Sea and Saudi production cost is still very large.] High-cost producers also tend to deplete their reserves much more rapidly. This means that the lower the price, the greater the subsequent enhancement of the future OPEC's market share.

The suppression of the existence of OPEC found in the Western international law literature reflects a Euro- or U.S.-centric perspective. OPEC is an actor that does not fit well into a traditional perspective given that it has been seen, since the 1970s, as at least coresponsible for high oil prices, nor into the modern "green" perspective, because it pushes, naturally, for hydrocarbon consumption. The fact that a high oil price has as much an energy-efficiency impact as most other measures undertaken by Western government does not help—it is a high oil price imposed by the interest of producing countries and not on, as the EU states, large gasoline excise taxes, for, at least ostensibly, environmental reasons. It does also not fit into the mind-set of Western NGOs: OPEC does not conform with the dominant view of developing countries as victims and passive beneficiaries in need of high-minded Western support and guidance, but it is rather an independent actor outside Western control. It is also barely, if at all, the object of NGO attention or exposed to NGO pressure. NGOs play virtually no role in the political systems of almost all OPEC countries.

OPEC was founded in the 1960s through a Venezuelan–Saudi initiative to reduce dependence on the international oil companies. At the root of OPEC's formation were pricing disputes with the companies and United States import restrictions. The main objective was to establish a common front vis-à-vis the oil companies with respect to several fiscal issues ("expensing royalties" in lieu of crediting them against tax; calculation of the proper reference price for oil taxes and royalties). Its objectives evolved in the 1960s in response to U.S. government and oil company action moving gradually from a focus on tax matters to oil pricing. First put forward as a developing country producers' association to develop links, information, and discussion, it supported the efforts of major producing countries in the 1960s and 1970s (Iran, Gulf countries, and Venezuela) to take over foreign-owned operations and to set prices. OPEC Resolution XVI.90 of 1968 formulates a common agenda and the legal position of producing countries. This crucial, now quite dated, resolution also expresses explicitly the otherwise implicit OPEC policies restricting acreage access by the international oil companies; the principle is still alive in the Middle East, but has been disregarded in Indonesia, Venezuela, and Nigeria. In terms of impact on international law, that was perhaps the period of the major impact: OPEC's position was fully aligned with the then-prevailing "Third World" claim for a "new international economic order (NIEO)" with "permanent sovereignty over natural resources" and exclusive jurisdiction over foreign investment, including the right to nationalize. OPEC's success then fed directly into the formulation of the NIEO resolutions of the United Nations General Assembly, and those in turn provided and amplified legitimacy to OPEC actions, without much attention paid to the fact that the increase in the oil prices then affected developing countries most.

With most of the upstream oil and gas reserves and production in the hands of producing countries' state enterprises by 1980, OPEC became the forum for dialogue, and sometimes collective action, by its members. The very early days of OPEC focused on royalty/tax questions, but by the 1970s, the focus was price setting. This was later replaced by the use of production quotas for each country. The intention then was that through the change of the production quota in response to market conditions, prices within desired price bands could be achieved. With oil revenues dramatically increased after 1973 and again 1981, the OPEC Special Fund for development aid was set up.

OPEC is an international organization headquartered in Vienna; there is a staff of about 120 (with 25 senior professionals from member countries). The secretariat does not, unlike other international agencies, seem to have developed a major autonomous role. It is engaged in research, monitoring of market trends, and technical preparation of the meetings of ministers. Attempts to acquire a larger role of its own seem to be regularly rebuffed by members. The International Energy Agency (IEA), for example, born as a reaction to the formation of OPEC, has a much larger, much more serious, and even more autonomous role and outreach. It has built up a reputation for high-level competence, probably due to the autonomy given by its members and the emphasis on professional competence. An important part of OPEC's mandate is presenting OPEC's view to the public, and, in particular, developing informed response to the criticism of OPEC, mainly in Western countries—a function of bundling the expertise in and for OPEC countries and providing an "OPEC view" to the world. But even so, it seems that the major countries (in particular, Saudi Arabia, able to rely on ARAMCO, the world's largest oil company) do not fully trust the OPEC secretariat, either with respect to sharing intelligence with it or using its work, but instead operate their own oil market intelligence machinery. The use of OPEC by its member states is quite limited. Perhaps given the great differences among them (with nothing uniting them other than oil dependence), OPEC member states seem to use OPEC much less for jointly building strategic expertise and negotiating capacity and for developing a professional network for intelligence, standard-setting, and policy formulation than, for example, the much more homogeneous industrialized countries grouped in the IEA. The studies that are presently being carried out by the World Bank on petroleum revenue management, or work on petroleum law, contracts, taxation, safety regulation, and environmental protection, would constitute the natural mandate for an organization of oil producers. But member states seem to prefer to limit their solidarity to the essential core, i.e., production and pricing policies, rather than build up OPEC as a common producing countries' research, think-tank, and standard-setting organization, as is the function of international agencies such as the OECD, IEA, or Nuclear Energy Agency (NEA). It seems that the organization, including its system of decision-making, its mandate, its staffing, and its budget level, is much in need of modernization to adapt to post-2000 globalization. This is a world that is very different from the confrontation with the "Seven Sisters" in 1959, which was at the root of OPEC's foundation. OPEC (or better, its member states) would benefit from carrying systematic benchmarking of the structure, organization, recruitment, and funding mechanisms in place in organizations such as the IEA.

The posts of the Secretary-General (mainly 3 years) and of the Chair of the board of governors (1 year, on alphabetic rotation) are filled in relatively short-term rotation. The principal organ is the Ministerial Conference; management is under the authority of the board of governors. Special issues (in particular, monitoring of the oil price and production quotas) is handled by specialized ministerial committees. The Economic Commission reviews reports and is responsible for technical recommendations for the ministerial council. Its budget (1995) was about $15 million (U.S. dollars). From all accounts, the organization seems to have little influence on the oil policy per se. There is little evidence that the secretariat brokers compromise in the typically very fractious dialogue. OPEC is also hampered by the fact that more significant secretariat positions can be filled only by nationals from OPEC member states. Because OPEC has, in comparison to the IEA, for example, a much smaller number of member states, this is one constraint (there are others) for its staff work. The pool of qualified professionals is therefore much smaller, because often in international agencies, member-state influences play a large role in recruitment. Member-state control over OPEC activities (including recruitment) reportedly goes much further than in other international organizations. This may be due, *inter alia*, to the small number of member states and the sensitivity of oil production to those states, but also to the prevailing culture of caution in the most influential member states. The fact that OPEC members are, apart from being major oil producers, much less culturally homogeneous than, say, the OECD and IEA members, may also be a hindrance to the full development of the inherent potential of OPEC's secretariat. It seems that OPEC does not grant to its staff the status of independent civil servants. Though this principle is, in the other international organizations established, not always taken fully seriously in practice, it does help to establish a professional culture. Long-term, contractually tenured, international civil servants tend to develop an "esprit de corps" and identify with the organization rather than with their country of origin. Civil servants, on the other hand, who are appointed on relatively short- and fixed-term contracts on the recommendation of their sponsoring home governments, will naturally have their prime loyalty to the home government.

In spite of such constraints, OPEC has developed considerable expertise in the operation of petroleum markets, probably much through links with the very large state oil companies in its member countries. It publishes regularly the *OPEC Bulletin*, the *OPEC Review*, and the *Monthly Oil Market Report*; these bring together know-how, with very much a focus on authors from its member states. OPEC, as an international organization, provides and services a forum for discussion and determination of concerted action by the member states. Such action consists, in the main, as of 2002, of production quotas per member state plus their adjustment in response to oil prices. Currently, OPEC pursues a policy of increasing production levels when prices over a period exceed a specified price band, and lowering production levels when prices decline below the specified price band (the farewell open letter by the then outgoing Secretary General Ali Rodriguez Araque, available from the OPEC Web site, is of interest in this regard).

OPEC's current role in the evolution of international energy law is marked by two key issues. First, the organization was founded on the basis of the producing countries' natural interest to increase and stabilize revenue, i.e., "mineral rent" from its control over oil and gas resources. That is still its raison d'etre. But there are tensions between short-term maximization through price versus long-term strategies centered on market share for OPEC oil and gas as against non-OPEC competitors and nonhydrocarbon alternatives. Here, there is a conflict between the high excise taxes of Western (in particular, EU) governments and OPEC policy. [OPEC estimates that the Group of Seven (G7) nations in 1996 obtained oil tax incomes totaling U.S. $270 billion, while OPEC petroleum export revenues were U.S. $160 billion.] High excise taxes (up to four times or more the price in gasoline in the United Kingdom, for example) have an environmental justification. They internalize external costs to the environment, and by road traffic. But they are also a convenient cover for large tax income to compensate for the more visible lowering of income tax rates. The OPEC-consuming country conflict is not about a higher price for petroleum-based energy, but rather about who gets most of it. The EU and the Unite States have tried to deflect political blame for high gasoline prices onto OPEC. They have, however, not been ready to accept the OPEC interest in stabilized oil prices and a "fair" OPEC share of the mineral rent on a formal negotiating agenda. Second, OPEC is naturally disinclined to view with favor the use of Western (very much currently EU) government policy to use heavy pressure to move away from hydrocarbons in favor of renewable energy sources, because this would devalue the OPEC reserves. But such policies could go hand in hand with a price- and production-based supply restriction.

Climate change is not a mid-term threat to OPEC countries if production (based on investment) is kept in balance with demand. The OPEC policy of implicitly keeping controls on investment and explicitly on production is quite compatible with the more extreme antihydrocarbon positions (exiting from hydrocarbons by restricting supply) taken by NGOs such as Greenpeace. OPEC policy can be interpreted not only as a price stabilization (increase) policy, but also as a conservation policy, in the sense of Article XX of GATT. But current Western thinking is not favorable to the use of regulatory instruments (trade, investment, and pricing rules) to smooth pricing volatility. The history of the largely failed commodity stabilization instruments of the 1970s and 1980s does not encourage new tinkering with similar instruments. Country-based income stabilization is another matter. In a volatile industry, it makes sense to skim off surplus in rich years and add to invested funds to add income in lean years. Many, if not all, OPEC and Western producing countries (Alaska, Norway; Kuwait; Abu Dhabi, and Venezuela) have developed different types of oil income funds. These are now proposed or established for new developing-country producers. In essence, income is stored away and made more difficult to access except in cases of emergency or severe budget pressures due to historically low oil prices. Such income stabilization may not make oil prices less volatile, but would make low price periods easier to suffer.

The role of OPEC is also likely to come under scrutiny from WTO law; several OPEC members are now in the process of accession. GATT/WTO obligations do not apply to OPEC, which is not a WTO member, but to its member states. Production quotas such as the ones currently used are "export quotas" under GATT Article XI. One justification may be under Article XI(2)(b), i.e., measures "necessary for international marketing of commodities"; the next defense could be found in GATT Articles XX and XXI. The question is justifiability under Article XX, mainly XX(g), conservation of exhaustible natural resources, or XX(h), pursuance of obligations under any intergovernmental commodity agreement that conforms to criteria submitted to the WTO parties and not disapproved by them. Acceptance as a legitimate measure under an international commodity agreement might be one way, but it is unlikely that governments, unless a comprehensive deal is struck with OPEC, would approve such arrangements at present. OPEC quotas are intended primarily to maintain and increase price levels. Do they have a conservation function? It is at present not a primary rationale for these measures, but it can be seen as a secondary justification. Conservation is certainly the effect of a higher price and government-induced limitation on production. But the condition is that such conservation measures must be applicable equally to domestic production. That this is done currently is questionable, but needs more in-depth investigation. OPEC countries would also rely on the national security exception (GATT Article XXI); acceptance of this is far from certain, but the concept has been interpreted by major trading countries (United States; EU) very widely. The dependency of OPEC countries on oil production, not comparable to the role of oil in other countries, would be an argument. GATT does not include any formal reference to "permanent sovereignty over natural resources" (UN General Assembly Resolution 1801 of 1962) or "energy sovereignty" [Energy Charter Treaty (ECT) Article 18]. But this principle could be seen as controlling or at least influencing the interpretation of the national security and conservation exceptions (GATT Article XX), either directly as customary international law or indirectly as a result of GATT interpretation for maximum compatibility with customary international law. Accession negotiations and conditions could carve out an exception for participation in OPEC export quota schemes (UAE, Nigeria, Qatar, Venezuela, Indonesia, and Kuwait are WTO members; Algeria and Saudi Arabia are in accession negotiations; Iran and Libya's applications for accession are explicitly being blocked by the United States; Iraq is not a member and is not involved in accession discussions). However, accession to the WTO is now subject to increasingly restrictive conditions; getting early into the club means having to live with less such constraints. The WTO may have been, in 1948, primarily about access to manufacturing goods, with little interest in energy security. But this has changed for the influential blocks in the WTO. There is and will be more and more of an effort to extract concessions favoring U.S. and EU energy security concerns from the resource-owning countries requesting membership. For example, in the case of Russia, the dual energy price (i.e., higher export price, lower domestic prices for both energy exports, and pipeline tariffs) is currently a stumbling block [as is the prohibition on the WTO Trade Related Investment Measures (TRIMs) rules]. Nevertheless, the legal instruments for a "deal" are available, but political will and some creativity are required to identify the contours of a deal that will improve the situation of both sides.

Possible future membership of the OPEC countries in the Energy Charter Treaty would raise the same questions as raised for the WTO, because the ECT provides for non-GATT members the application of GATT provisions, with some qualifications. Different from the GATT, however, the ECT, in Article 18, explicitly recognizes "Energy Sovereignty" and "the optimization of (resource) recovery and the rate at which they may be depleted or otherwise exploited" [Article 18(3)]. Arguably, ECT membership therefore poses fewer problems for OPEC countries than does the GATT (though the GATT/WTO include several OPEC members and others in accession discussions, whereas the ECT does not). There may, however, be soft-law disciplines under the ECT for "export taxes" on oil [Article 29(4): states shall "endeavor" not to increase export levies], and under Article 6 (competition law) as well as Article 5, prohibition on TRIMs. Some of these issues could be solved through "understandings" negotiated by countries requesting accession (or, in the case of Russia, before ratification). Such understandings could include a limitation or long transition process for the TRIMs obligation (Article 5), a recognition of OPEC production control (limiting any argument about the competition law, Article 5), or an understanding that the ECT does not affect issues that are controversial between the EU and energy exporters (e.g., Russia and Algeria) such as destination clauses and initial access to new infrastructure such as pipelines.

To sum up, OPEC, as an international organization and forum that facilitates concertation among the major oil- and gas-producing countries, is now increasingly pulled into the institutional structure of the global economy. There will have to be a give-and-take on both sides to conclude such integration successfully. OPEC fulfills, different from more hostile sentiments in the 1970s, silently important functions for both domestic producers and international oil companies by being the organization most keen (and most potent) to help stabilize prices by helping producers to manage production. Whether this ability, which was not evident in the 1980s and 1990s, is maintained or, as happens with most cartels, will fade again is outside our ability to forecast. Sustainable development requires greater application of energy efficiency, minimization of emissions that are harmful for the global (and localized) climate, and possibly restrictions on the supply and use of hydrocarbons. Such policies, eagerly pursued by NGOs and the EU, for example, are unlikely to succeed if proper account is not taken of OPEC, the major international agency of the major oil-producing countries. This analysis suggests that there may be more compatibility than meets the eye or that is intuitively implicit in the conventional reference to the "OPEC cartel." An overall deal is possible, but requires a more active and creative effort at identifying communities of interest and much stronger leadership in pursuing and negotiating them on both sides. An arrangement could require some concessions by OPEC in terms of managing the oil price as a contribution to a stabilizing world monetary policy (e.g., lower prices in a recession, higher in a boom). It would require better guarantees of security of supply to concerned parties (e.g., the United States, EU, and China). Oil prices could also be linked to import prices for the producing countries. In exchange, there could be some examination of the very high excise taxes on gasoline and some other developed-country policies affecting the producer states. A higher price for oil together with a discipline on supply could be in the interest of both the OPEC countries and the environmentalist community, and in the long-term interest of consumer countries in a stable and secure oil supply. An unfettered global oil market is probably not in the interest of anybody—contrary to recurrent allegations, in particular in the United States. An unfettered oil market without political influence never existed. It would drive down oil prices to very low levels, would close down most non-OPEC production (including in the United States), countercurrent to Kyoto and energy efficiency objectives, would discourage development of renewable energy, and would very likely result in extreme (and therefore, bad for the global economy) swings of the oil price.

3. INTERNATIONAL ENERGY AGENCY

The IEA in Paris is the Western response to OPEC, though to mention this reverse-mirror role explicitly seems to be taboo. The IEA is the main international organization dealing with energy, though its mandate, membership, and operations are very limited; essentially, it is an intergovernmental energy policy institute for Western countries and also manages an emergency sharing system. Its importance also derives from the fact that there is no global intergovernmental energy agency. Quite likely, the division between the Western (IEA) and producer country world (OPEC) is the reason that there has not been enough political interest and effort to create a World Energy Agency (WEA). Existing international agencies with an energy mandate (UN, UN Conference on Trade and Development (UNCTAD), UNEP, UN International Development Organization (UNIDO), and World Bank) would also see their turf threatened if a World Energy Agency were to be created and would seek to take on such mandates.

The IEA was founded in 1974 on a suggestion by Henry Kissinger made in 1973, i.e., at the height of the first oil crisis, when there were sudden increases in oil prices and takeovers of foreign-owned oil production by the producing countries. OPEC, whose existence was disregarded in the 1960s, came to be seen as the main instigator and as a powerful cartel threatening the oil supply of western countries–in particular, the supply of the United States. The IEA was, and is, explicitly limited to western (that is, most OECD) countries. [Members, at present, include the EU countries, plus Hungary and Czech Republic, Turkey, the United States, Japan, Canada, Australia, New Zealand, and South Korea. The EU participates, but is not a member (though it could accede). Norway is an associate of the IEA, but not of the emergency system, based on a special agreement with the IEA.] South Korea, Hungary, and the Czech Republic have recently joined the IEA. There have been discussions with Mexico (which originally applied for membership), but in the meantime, Mexico has withdrawn its application. The IEA was not meant to be or to become a universal energy organization. Like with the EU, there have been intermittent calls for a UN Energy Organization, but these have so far come to nothing. The basic purpose of the IEA was to develop a system of collective energy security mirroring the collective producer power embodied, and then at its height, in OPEC. Such collective energy security operates through the continuing emergency-sharing system administered by the IEA. This system was never put into operation, though that may have been close at the time of the first Gulf war in 1991. [At the time of writing this article, a second Gulf war looked likely. The IEA was again preparing plans to ready its emergency-sharing system. But ultimately, the IEA desisted from implementing the emergency-sharing system because Saudi Arabia committed to use its massive production reserve to increase production. The Saudi "emergency production potential" is clearly superior to the IEA emergency sharing, though its availability in the future rests on the solidity of the U.S.–Saudi relationship, which, at present (summer 2003), is in jeopardy.] The most significant supply disruptions in the EU came about not because of OPEC and Middle Eastern conflict, but because of resistance to another round in gasoline tax increases by the British and other EU governments in October 2000.

The IEA's raison d'être has therefore diversified away from its earlier focus—Western solidarity in the face of OPEC-faced threats—toward energy-focused research and market intelligence, i.e., a type of work that is as well carried out by academic institutions, consultancy, and investment firms. The IEA advantage or difference is that it is publicly and internationally funded, and therefore with much more historical continuity, financial stability, and less dependence on markets, clients, and national budgeting processes, as compared to private or nationally based public energy research institutes. Though still in charge of the OECD countries' energy-sharing mechanism, it now fulfills a function of centralized research and intelligence, quite similar to the role of national energy institutes or the pooled research and intelligence function of international business associations. Focused on something that is still seen as strategic, and insulated from the questioning of national agencies of that type by its character as an international organization, it may have to justify its continuing usefulness by defining more closely the "public good" it delivers and its distinctive comparative cost and quality advantage over private, more market-driven organizations.

The International Energy Program (IEP) of 1992 is the treaty constituting the IEA. Membership is limited to OECD countries, but there is no automatic membership of OECD countries. Accession therefore requires a double hurdle—to the OECD and then to the IEA. The close-to-membership association agreement with Norway might, though, be a relatively simple way of protomembership for non-IEA countries (if this was considered desirable). An IEA path toward a universal energy agency is therefore in all likelihood blocked; given the common economic and political orientation, a link-up with the Energy Charter Conference could be advocated, if the United States were to join (*vide infra*), but a full merger seems not to be in the cards as long as Russia and the other former Soviet countries are not OECD members. The IEA is open to membership by the European Community (Article 72 of the IEP), but this has not happened as yet. The EU Commission participates in all meetings without voting right. There seems to be a low-profile competition with the EU Commission. Decision-making by its governing board is mainly by majority voting, not by consensus as in the OECD. This should make the IEA in theory able to act more rapidly; presumably, in view of energy disruptions, that was the intention of a voting arrangement different from the OECD arrangement. Votes are weighted by pro-rata oil consumption, highlighting the economic weight and role in emergency sharing.

The IEA is in a complex relationship with the OECD as an "autonomous agency," with its own Executive Director and its own budget (though formally integrated into the OECD budget) and governance process, with some integration into and use of OECD administrative services. Funding is by membership contribution based on the OECD scale, plus voluntary, project-related contributions. The IEA has a separate division on "nonmember countries," which is used for an energy policy dialogue, economic studies, and joint research projects. A modest "global energy window" is thus open to the IEA via its nonmember activities and structures. Different from national or international agencies (e.g., World Bank, UN, UNIDO, UNCTAD, IMF, and IMO), the IEA seems not to have carried out technical assistance (policy advice; legislative, tax, and institutional reform; training; assessment of energy projects and programs; privatization; investment promotion) in member or nonmember countries. That is regrettable, given that the IEA has a large core of expertise, a comparative edge in its specialty field, and an energy policy link with all the IEA and many non-IEA countries. Energy assistance, when provided by other agencies (including the EU Commission), is never the central focus of expertise and interest and therefore is often provided with rather uneven know-how. Member states must have preferred energy policy assistance to come, rather, through their bilateral aid agencies, perhaps with the idea to gain a competitive advantage for their own energy industries and consulting firms, rather than provide regular program funding for technical assistance to the IEA. Also, the IEA may have intentionally avoided vigorously seeking voluntary contributions that would have placed it into competition with most other international agencies. The rough-and-ready tumble of international aid competition does usually not sit easily with well-funded international organizations staffed by Western civil servants on secondment. The dual character—rarefied and rare emergency sharing here, intergovernmental research institute there—may also not sit easily with the competitive vigor required to develop a sizable role in policy advice.

The traditional core of the IEA is its emergency oil-sharing program, now rather described better with the wider term as "emergency response measures." This consists, first, of measures by member states to reduce demand and to maintain oil stocks at 90 days of net imports. If emergency situations for the whole IEA group occur (two levels of group shortage: a 7% and a 12% shortfall, to be determined by the IEA Executive Director), a rationing plan is triggered that also requires surplus countries to provide for imports into deficit countries. IEA-based oil companies may have to be directed by member states to reorder supplies. There is a regular coordination mechanism between the IEA and such oil companies. The trigger is an actual shortage, rather than a sudden price explosion, so that the mechanism is rather more akin to a wartime international rationing plan than to a market intervention mechanism. The distinction between the two is sometimes questioned. Price spikes can both indicate and cause a disruption of physical supplies, and market intervention can help to manage scarcity situations. The United States, for example, has been more interventionist in deploying their strategic reserve in high-price situations, e.g., in 2000/2001. With two more recent governing board decisions in 1984 and 1995, the flexibility of the IEA and the member states to respond in a coordinated fashion to preemergency situations was increased, basically by a consultation procedure leading to the use ("drawdown") of the oil stocks. None of these mechanisms has ever been activated, though the beginning of the Gulf war in 1991 led to calls to start at least with some use of oil stocks to calm markets.

Otherwise, and in view of the largely dormant character of the emergency program, the IEA has become very much a collective study organization for its members. There is regular reporting to enhance the transparency of oil markets; in-depth review of country energy policies (with recommendations and preparation of "outlooks" that provide national and commercial actors with some idea about the future of energy demand and supply), and a formulation of "shared goals," which at present comprise liberalization of markets and environmental issues. The IEA is unlikely to take a position on implementation of the Kyoto Protocol to the UN Framework Convention on Climate Change as long as there is a strong divergence between the United States (and most of its companies) and the EU (and its companies). Different from other more visible international agencies (e.g., the OECD, with its failed effort to negotiate a multilateral agreement on investment; the WTO, World Bank, and IMF, with their exposure to the anti-globalization movements), the IEA has so far escaped from much public attention. This relative immunity, though, is also likely to make it less aware of nongovernmental views and demands and less able to engage in a dialogue with them. Finally, the IEA provides an institutional vehicle for research in energy technology through, so far, over 40 collaboration agreements it sponsors, including with nonmember countries. There should be a potential here to engage not only scientific and technological, but also more applied and policy-oriented, technical assistance.

The IEA is in a curious situation. A child of the oil agitation of the 1970s, it may find itself in quest for justification of its existence. With the realization that the oil-exporting states are much more dependent on export than are the importing states, the fact that the emergency program has so far never been activated can be more easily explained. With oil declining in the energy mix, substituted in part by gas, and now, under the signs of Kyoto change, the new push toward renewable energy resources, the oil focus of the IEA risks obsolescence. Energy security is now no longer exclusively a matter of oil supply, but also of gas, coal, uranium, and electricity supply. Energy security for the IEA member countries, in particular the EU, means a favorable investment situation in producing countries and favorable legal and institutional conditions for transport and transit of energy resources (and electricity) and physical infrastructure (plus a regulatory framework maximizing its use) such as pipelines, interconnectors, and storage facilities. Though the IEA has dealt with such issues in various studies, it has no operational, or policy, or policy advisory role. Also, its character as a strictly Western, OECD-type of organization may be in question as globalization and the forces now triggered call rather for universal organizations, with an ability to conduct a global dialogue with all relevant stakeholders, conduct globally focused research, and prepare globally relevant policy studies corresponding in coverage with the globalization of energy markets. This was brought home starkly to the OECD when it tried to negotiate (in its club atmosphere) a multilateral investment code that was mainly relevant and intended to be ultimately applied to non-OECD countries.

There does seem to be a need for a truly universal energy agency, because energy continues to be the mainstay of the global economy. One way would be to maintain the emergency-sharing system of oil, but to expand the organization's focus on all energy sources, expand membership (perhaps in associate form) to all countries wishing to join, and put more emphasis on developing energy-related technical assistance. The IEA's nonmember countries and its research contract areas would seem to provide a nucleus for such expansion. An IEA consisting of two components—an emergency-sharing, OECD-based side and a universal, all-energy-based side—might be envisaged. This could be developed gradually, and might emerge over time anyway, but it would also make sense to consider amending the IEA agreement to provide for formal associate membership to non-OECD countries. In terms of influence on international energy law, the IEA's contribution has been mainly the emergency-sharing system as a free-standing element. IEA studies, generally of high quality and often prepared by seconded government and industry staff, will have an influence on policymaking by accentuating policy shifts (as presently toward liberalization and environmental mechanisms), but there is little direct influence on national energy law reform, as, for example, through bilateral or World Bank and UN technical assistance, nor has the IEA so far played a visible role in the evolving WTO-, North American Free Trade Agreement (NAFTA)-, or EU-based trade law of energy nor in the many environmental treaties, protocols, and guidelines now emerging.

With or without the IEA, the policy recommendation of the author is for a truly global World Energy Agency (WEA). The energy industries are coalescing into a truly globalized industry. This means there is a need for the research institute-type work of the IEA, the market-stabilizing influence of OPEC, and the protoregulatory work of the OECD, but involving, on an equal level, all stakeholders, i.e., governments, companies (for example, in the way they are involved in the International Labor Organization procedures), other international agencies (IEA, OPEC, OECD, World Bank, UN agencies), and nonstate actors such as industry, professional associations, and NGOs. As the EU Commission recently noted, energy in developing countries is "an orphan without a parent international organization."

4. ENERGY CHARTER CONFERENCE AND SECRETARIAT

The Energy Charter Conference, served by its Secretariat, is the most recent addition to specialized, energy-focused international organizations. It is based on the Energy Charter Treaty (with a headquarters agreement with Belgium), and has the formal status of an international organization. The Secretariat services the negotiations for secondary and follow-up instruments (protocols) and supports treaty implementation. The ECT is an energy-focused treaty with all European countries, the states of the former Soviet Union, plus Australia and Japan as members. (The United States and Canada have signed the 1991 precursor European Energy Charter, but not the 1994 Energy Charter Treaty. China, Nigeria, Algeria, Saudi Arabia, and other countries have acquired observer status; several Mediterranean countries are considering the possibility of accession. Russia and Norway have signed, but have not as yet ratified. They are obliged to apply the Treaty provisionally, though.) The treaty deals mainly with investment protection (in the style of modern bilateral investment treaties) and with trade (adopting WTO rules for energy trade between states where at least one state is not a WTO member (now mainly Russia, Ukraine, and the Asian countries of the former Soviet Union). The ECT also deals with transit in a novel way.

The main activities of the Conference/Secretariat are at this time the elaboration of a transit protocol providing more specifics to the more general Article 7 of the ECT. A supplementary treaty to provide legally binding and specific rules for access for foreign investors ("preinvestment") was supposed to be negotiated when the ECT was signed in 1994. Negotiations took place between 1995 and 1998. Their final conclusion depends on the resolution of a number of outstanding political issues. The political interest in the supplementary treaty seems to have expired. There was also resistance, reportedly mainly from France and the EU. The interest in expanding investment arbitration (available under Article 26 of the ECT) may also have waned as the full potential of investment arbitration, in particular under the NAFTA, became clear.

The trade part of the Treaty is losing significance, because most East European countries are about to join the WTO. It is, for example, still relevant for the issue of energy import restrictions now being imposed by EU countries (in particular, Germany and Austria) on nuclear-based electricity, to the extent that such restrictions would affect non-WTO members (e.g., Russia or Ukraine).

The ECT, different from the quite similar (content-wise) Multilateral Agreement on Investment (MAI; *vide supra*), has not been politically very visible. As a result, the treaty and the ECT organization have been spared NGO onslaught. As in the IEA case, that has also compelled them, less so than the large international organizations, to engage in a dialogue with nongovernmental voices. The ECT's future lies mainly in securing its current membership through ratification by Russia (which seems possible), extending its membership to other Asian and Mediterranean countries, and serving at least as a model for regional energy relations in Asia (Asia–Pacific Economic Cooperation, or APEC), Latin America, and Africa. The strength of the treaty and the justification for a permanent Secretariat will also depend on the respect for the treaty's investment protection, and here mainly in the higher risk former Soviet Union countries. But the role of the Secretariat in getting such respect is limited, though the Secretary-General has a role under Article 7, in particular, for appointing a conciliator with provisional decision-making powers in case of transit disputes.

The staff of the ECT is quite small (under 30) and it has at this time no substantial budget for extracurricular activities, e.g., studies and country reports, like the IEA style, or technical assistance, like the UN and World Bank style. EU technical assistance has, though, at times supported ECT implementation projects. The Energy Charter Secretariat sits not too comfortably between the EU Commission (both are located in Brussels) and the IEA in Paris. Formally, the EU Commission has no particular role in the EC Secretariat; the EC is one of the 52 members. But politically and financially, the EU in its entirety is the main sponsor of the ECT. The absence of the United States reinforces this role.

The ECT is one of several privileged dialogue facilities between Europe and the former Soviet

Union, though for East European countries, the ECT will decline in significance because accession to the EU, and thereby full adoption of the energy "acquis communautaire," has acquired priority. But for the EU and the Commission in particular, it does not seem clear whether they wish to use the ECT machinery or its internal instruments for creating a legal and institutional framework facilitating trade and investment with its major energy partners. Several EC initiatives (e.g., the Inogate energy transit project) seem to overlap, duplicate, or compete with the ECT transit protocol effort. Similarly, instruments of EU external energy policy, such as association, partnership, and cooperation agreements and the Lome/now Cotonou agreements, could substitute for work through the ECT.

In terms of influencing non-OECD countries (transitional and developing economies) toward market-economy models for organizing the energy sector, the EC Secretariat/Conference overlaps with much larger organizations such as the EU and the World Bank, and the nonmember activities of the IEA. If the United States were part of the ECT, an argument could be made for much more collaboration, up to merger, between the EC Secretariat/Conference and the IEA. If, on the other hand, Russia were to ratify the ECT formally and other countries were to accede, the ECT could become either a major channel of EU dialogue with energy producers around the world (including OPEC countries) or, alternatively, a jointly "owned" and therefore more equal organization for regulating their economic relationship, rather than the more one-sided reliance on EC agreements.

The ECT and its organizational contribution to international energy law is currently mainly through the service of the ECT, one of the very few multilateral treaties exclusively devoted to energy, and thus a key element of international energy law. By support of negotiations for an energy transit protocol, the Energy Charter organization is also directly involved in the emergence of new and very relevant international energy law. As energy markets integrate regionally and globally, the role of transport, transit, and interconnectors becomes much more important than it was in a period when most energy industries were segregated into national areas. By the end of 2003, the negotiations for a transit protocol came to a halt, due to so far irreconcilable differences mainly between Russia (strongly influenced by Gazprom) and the EU.

5. INTERNATIONAL ATOMIC ENERGY AGENCY

The International Atomic Energy Agency (IAEA) is the one universal agency dealing with the peaceful use of nuclear energy. It has played a vital role in international nuclear security and a minor role, if at all, in the development and application of nuclear power. [The IAEA has, over the past 15 years, been relatively starved for cash, part of the general policy of Western countries to curtail funding to the universal international agencies. This has, arguably, not been a very wise policy, because the IAEA's core functions—nuclear safety, controls over trade in nuclear materials, and disciplines on nonproliferation—have become, even before September 11, 2001, more acute than ever, in the context of the collapse of the former Soviet Union nuclear industries, acquisition of nuclear power (for peace and for war) by developing countries, and the threat of nuclear terrorism. It is assumed that the Al-Qaeda attack on New York in 2001 will reverse that trend.] The IAEA was established by treaty, effective in 1957, one of the few instances of a successful East–West collaboration during the Cold War. Seated in Vienna, with a staff of over 2000, a budget of over U.S. $300 million, and the usual setup (general conference, board of governors, and headed by a Director-General), it is not much known outside the specialized nuclear community, and has, as yet, not been exposed to the antinuclear movement in any significant way. [This is arguably so, because the security functions of the Agency (setting of security standards, control of nuclear materials trade, and nonproliferation) have been largely supported by the antinuclear movement. The IAEA has also never taken a strong position in favor of establishing more nuclear power capacity.] As a universal organization (related to the UN as such, but not a specialized UN agency in the narrow sense), it is used by Western countries for purposes of controlling nuclear risk in the developing, and now former Communist, countries, but it has to coexist (with some underlying competitiveness) with the specialized nuclear agencies of the OECD countries (OECD Nuclear Energy Agency; *vide infra*) and Euratom, the EU's special nuclear agency. (The Euratom treaty, essentially governing nuclear materials transfers within and into the EU, was concluded in 1957.)

Among other things, the IAEA original mandate included (1) research on the peaceful use of nuclear energy, including scientific and technical information exchange and training, (2) a safeguard system to ensure that nuclear materials are not diverted to military purposes, and (3) setting of safety standards. Over the years, the main functions have been standard-setting and the safeguards system. Its main task now is to safeguard the use of nuclear materials and facilities in member countries under the Non-Proliferation Treaty (NPT; Article III deals with IAEA verification), but also in non-NPT countries (e.g., India, Pakistan, and Israel, although most of their facilities are kept outside the reach of IAEA safeguards). The nuclear powers emerging in the Cold War encouraged nonnuclear powers into the IAEA and NPT treaty system. Based on special bilateral agreements with the state, the IAEA carries out monitoring of facilities to ensure safety in terms of operational standards, but also in terms of nonproliferation. The nonproliferation issue has become relevant not only in relation to states (e.g., the Iraqi efforts to build secretly a nuclear arms industry), but also in relation to illegal trade in nuclear materials and weapons from state to state and possibly to terrorist groups. In 2002, with the security threat by terrorists to nuclear installations seen in a much more acute light, the future high-priority IAEA activities are most likely to focus on nuclear security, both with respect to defense against attacks on nuclear installations and with respect to terrorist threats to build and use small-scale nuclear explosives against civilian targets.

Apart from nonproliferation, which now has a significance for "rogue states" and smaller states keen on nuclear weaponry (i.e., not countries that are too big for international pressuring, such as China and India), the IAEA is to develop and help apply technical standards and guidelines for nuclear plant safety, waste disposal, and decommissioning of nuclear plants. The IAEA has here, like most specialized agencies, no direct regulatory powers, but has a major, if not the most dominant, influence on regulation by national agencies with direct regulatory powers. For those agencies, in particular from smaller countries, it would be inefficient to replicate the amount of effort at information-gathering, consultation, and best-practices definition that goes into IAEA standard-setting. In a material, though not formal, way, the IAEA therefore acts as a global nuclear regulatory agency. The standard-setting activities are the most relevant for nuclear operators. Standard-setting is preceded by extensive information-gathering, including the International Nuclear Information System (INIS) for scientific literature. Reactor safety protection and radiation protection are implemented through Operational Safety Review Teams assessing specific nuclear power plants; the Operational Safety Indicators Program, providing for plant-specific safety indicator; and the Incident Reporting System, which reports on incidents in nuclear power plants.

The safeguarding activities are the third major, and currently very relevant, function. The IAEA carries out a large number of inspections of individual nuclear facilities and materials (over 2000 usually per annum, depending on budgetary constraints). Several treaties (the nonproliferation treaty of 1968; Tlatelolco, for Latin America; and Rarotonga of 1985, for the South Pacific) oblige member states to submit to IAEA inspections. Bilateral safeguards agreements, now with most states, govern the details of the safeguarding inspections. It is hard to deny in these situations that the IAEA has acquired not only material, but also formal, regulatory powers in such situations by delegation via treaty. Following the revelations of Iraqi efforts to build an atom bomb, the safeguards system was enhanced. Under the Additional Protocol (INFCIRC/540) to the NPT safeguards agreement (INFCIRC/153), the IAEA acquired greater and more immediate access to sites in NPT countries, and state parties were obliged to disclose more information about their nuclear activities and installations. Ratification, though, has been relatively slow, much due to the U.S. government's depreciation of multilateral arms control. With nuclear terrorism now threatening the United States, this is likely to change. As other developing countries embark on major nuclear power programs (e.g., Iran, China, and Pakistan), they will be encouraged to accede to the Additional Protocol as a sign of goodwill and to avoid "blacklisting" and other forms of sanctions. The IAEA has also an important role in nuclear trade control. Under the Nuclear Suppliers Guidelines, to which nearly all supplier countries subscribe, nuclear materials exported must be placed under permanent IAEA safeguards.

The IAEA provides technical assistance in matters of nuclear safety (out of voluntary contributions), but it is here in some competition with other agencies, e.g., the European Bank for Reconstruction and Development (EBRD) program on nuclear safety in Eastern Europe or the OECD/NEA. The EBRD is involved only in the Ukraine (Chernobyl Shelter Fund), Lithuania (Ignalina nuclear power plant closure), Bulgaria (Kozloduy plant upgrade/closure), and Russia (Rovno Unit 4 and Khmelnitsky Unit 2 plants) under its Nuclear Safety Account. The OECD/NEA is involved only in some advisory activities on emergency testing, technical aspects of power plants, and a project in Russia (the TASPLAV project, concerning a Russian experimental facility where the reactor core material can be melted).

The IAEA has been a significant contributor to international nuclear law, in particular in the area of nonproliferation and safety of materials and installations, by administering relevant treaties, servicing the negotiation of new multilateral treaties and protocols, and designing the periodically reformed technical standards. The IAEA also publishes regularly, in its *Legal Series*, updated information and analysis on nuclear law (the *Nuclear Law Bulletin*, on the other hand, is published by the OECD/NEA). The IAEA was the moving force in creating the Vienna Convention on Civil Liability, and also the Brussels Convention on the Liability of Operators of Nuclear Ships, the Convention on Civil Liability for Maritime Carriage of Nuclear Materials, the Convention on Physical Protection of Nuclear Materials, the Convention on Early Notification of Nuclear Accidents, the Convention on Assistance in Case of Nuclear Accidents, and the revised Vienna Convention on Civil Liability for Nuclear Damage. The Vienna Convention served as a model for the subsequently negotiated Paris Convention. Other agreements sponsored by the IAEA relate to radioactive waste, nuclear safety, radiation, emergency planning, safeguards, and the nuclear liability agreements mentioned (with some involvement of the OECD/NEA) in the Paris and Brussels nuclear liability conventions.

6. UNITED NATIONS

The UN system consists of the United Nations proper with its various departments and other units and specialized agencies. Some of those other groups, such as the World Bank or the IMO, are for practical purposes completely independent. The survey here cannot do justice to the panoply of activities by the UN system, its main and secondary organs, and many secretariat groups and specialized agencies. What follows is therefore a selection with comments, rather than a systematic survey.

6.1 Climate Change Secretariat

For the energy industries, in particular the oil and gas and coal industry, the one UN activity with most relevance is the UN Framework Convention on Climate Change, with its Secretariat in Bonn The fate of the Kyoto Protocol is not clear, with its specified caps on CO_2 and other relevant greenhouse gas emissions in industrialized (including post-Soviet) countries, unspecific goodwill obligations on developing countries, introduction of emission trading and other emission reduction measures [Clean Development Mechanism (CDM) and Joint Implementation (JI)], and absence of the United States, the largest CO_2 emitter. (The target for post-Soviet Union socialist countries is currently largely irrelevant because it was based on the much higher CO_2 output during Communist times. This was the political price to get these countries on board the Kyoto protocol, but it also means that they do not have to undertake any meaningful efforts at present.) But the negotiations held by the Committee of the Parties (COP) for implementing the Kyoto Protocol are likely to put pressure on governments, in particular within the EU and the EU accession countries, to favor renewable (and possibly later nuclear) electricity generation and reduce coal, and possibly later oil and then gas-based power generation and consumption in transportation. With U.S. absence, there will also be a trade issue to the extent that the implementation of the Kyoto mechanisms is likely to develop intracorporate, national, and international trade in emission rights, but also financing of joint implementation and clean development mechanisms; it is hard to see how U.S. companies, operating outside the Kyoto membership, can be full beneficiaries of the emerging trade in emission rights and equipment/services for climate-change management when there is a tension between free trade under the WTO agreements and restricted trade among the Kyoto member states. [It is necessary to examine here the implications of GATT Article XXI(g) and analogies to the role of regional economic integration organizations and the noninclusion of international organizations in the WTO membership and obligation system.]

6.2 Compensation Commission

The political arm of the UN has had an involvement in oil and gas affairs through the UN Compensation Commission (UNCC), instituted after the Gulf War to administer Iraqi liability for war damage, in particular large-scale environmental and other damage to the oil production facilities in Kuwait by Security Council resolutions 687 and 705 (1991) and 986 (1995). The UNCC was created in 1991 as a subsidiary organ of the UN Security Council. Its mandate is to process claims and pay compensation for losses and damage suffered as a direct result of Iraq's unlawful invasion and occupation of Kuwait. A specified percentage of the revenue (in 2001, 25%) from authorized Iraqi oil exports is earmarked for the compensation of damages resulting directly from the invasion of Kuwait by Iraq. Such damages are to include, according to the Security Council resolution, commercial losses, "environmental damage," and depletion of natural resources. [There is no doubt that Iraq by far exceeded the right of extraction of a belligerent occupant. It seems to have carried out a large-scale destruction of the Kuwaiti oil industry installations, according to the UN Compensation Commission Governing Council (Document S/AC.26/2001/16 of 22 June 2001, in particular p. 65, re Kuwait). The issue is, if the belligerent occupant, with the right of usufruct, can continue to extract the "normal" amount of (technically depletable) hydrocarbon (or water) resources.- With the occupation of Iraq in 2003, the priority allocation of Iraqi oil revenues to compensation under the UNCC will change, for the benefit of Iraqi occupation purposes and to the detriment of 1991 Gulf War compensation.]

The UNCC is in form and name not a tribunal, but an administrative process set up to expedite claims. (See the explicit statement on the UNCC Web site on "claims processing": "The Commission is thus neither a court nor a tribunal with an elaborate adversarial process. Rather, the Commission was created as a claim resolution facility that could make determinations on a large number of claims in a reasonable time. As such, the Commission operates more in an administrative manner than in a litigation format. The Commission's claims processing procedures were prescribed by the Security Council and were further elaborated by the Governing Council.") The Commission has received over 2.3 million claims. Practical and expedient justice is therefore the primary aim, different from the U.S.–Iran Claims Tribunal, wherein claims are still being litigated in depth. Most claims have been settled as of 2002. There have been decisions awarding very large amounts to Kuwait for damage to its oil installations, for oil extracted, and for environmental damage caused by oil spills. These have been subject to criticism of overvaluation, an issue that may be more problematic in view of the legal and financial resources available to Iraq to put its own position effectively. In practice, however, two things have transpired that, in the opinion of many, may vitiate this concern. First, at the Panels' direction, the Commission staff has taken an aggressive role in verifying the claims. As a result, the Commission sought and received significant budget increases to permit the legal and valuation staff to conduct thorough investigations of the circumstances of many of the commercial and environmental claims. Some claimants objected that the process was more intrusive than they would have experienced under most normal adversarial processes. Budget figures and the success rates for category E claims bear this out. Second, Iraq has had far more access than the Commission's designers anticipated. Iraq receives all claims submitted to the E1 (oil sector) Panel and is permitted to make its own submissions. In the larger claims, the E1 Panel has also held oral proceedings and has permitted Iraq to appear through counsel and argue the issues raised for decision. The panels only issue reports and recommendations—the final decision (not always the same) is made by the Governing Council. Reportedly, various governments, including Russia and France, used experts to review the E1 (oil-related) awards. On the other hand, participants of the process, both inside and outside the UNCC, have also communicated to this author that the Governing Council's decisions were intensely political, with only marginal adjustment of the procedures to afford to Iraq more than a mere formal opportunity to argue against, in particular, the valuations proposed by the UNCC's consultants.

The UN Compensation Commission practice should lead to international precedent for valuation of damage to the oil industry and oil-related environmental damages. The activity of the UNCC in the field of oil industry-related liabilities would merit deeper examination. It is regrettable that this significant precedent impact is somewhat weakened by the absence of full "due process" to the de facto defendant, Iraq. The panels of the UNCC have reportedly, in response of such criticism, made considerable efforts to stretch the existing rules to provide as much of a hearing to Iraq as possible. Future litigants will cite the actions as examples of legitimate ways to address problems such as valuation of oil and gas losses. This is particularly so because there is not much direct precedent. The remaining work of the UNCC will undoubtedly be overshadowed by the reconstruction of Iraq, which, by the end of 2003, was still not very clear. The U.S.-led effort to obtain a general and far-reaching debt reduction from all relevant debtor countries may therefore also impact on the UNCC, which, after all, manages Iraqi debt toward, *inter alia* but primarily for Kuwait.

6.3 United Nations Development Program

The UN Development Program (UNDP) is the main UN development funding program, fed by voluntary contributions. Though the main funding is allocated to each country according to a population/poverty factor and spent according to national priorities, UNDP also runs several energy-related programs relating to small-scale energy development, development of renewable energies [mainly through the Global Environmental Facility (GEF), jointly with the World Bank], and implementation of the Montreal Protocol by assisting developing countries to eliminate activities that contribute toward depletion of the ozone layer. It is difficult to discern any appreciable effect on energy law in these activities; there is occasional funding by UNDP, if it fits into country priorities, of technical assistance for legislative reform in the oil and gas or energy sector.

The UN/UNDP in technical and financial assistance in the energy sector is, apart from the inevitable special attention for Kyoto and Montreal Protocol issues (renewable energy; ozone layer), not in any particular way focused on energy and certainly not on energy law. The significant developments in the energy industry over the past decade—privatization, liberalization, and postprivatization economic regulation—seem to have passed the UN system almost unnoticed. This is probably because the World Bank (for transition economies in some competition with EBRD) has taken leadership and "ownership" of these issues, but also because the UN system has had trouble modernizing its internal culture and outlook (and staff), all mired like old generals in the philosophy of the NIEO, with little substantial contact with industry, banks, or modern market-oriented thinking. Multilateral development funding has also been declining, and UNDP projects have shifted from the "harder" topics of energy to the topics that are more fashionable in the UN discourse (that is, poverty eradication, human rights, advancement of women, and sustainable development). In competition with the World Bank, the UN system has not been able to capitalize on its competitive advantage—greater sympathy for and trust by developing countries and (somewhat) greater independence from the United States—to develop concepts that are both in tune with the modernization of formerly state-oriented economies and less ideological than the philosophies that the Western-dominated institutions (World Bank, EBRD, and OECD) have imposed with rigor and purity, but also with less realism, practicality, and critical judgment, on developing and post-Soviet Union countries in the 1990s. It is both easy to speculate on the role of the UN in the energy sector and difficult to prove anything, because there is no independent assessment. As with all other international agencies, information and competence for critical assessment are divorced and self-assessment generally amounts to a mixture of propaganda and paraphrase of formal remits and mostly fallacious, if not even sometimes fraudulent, reports on own successes. From the accounts available, the UN activity appears diluted, ad hoc, not subject to systematic independent assessment in terms of cost-effectiveness, and with little, if any, lasting impact. This may be due partly to the organization's mode of operation and heavy bureaucratic processes, but also due to the fact, relevant for any provider of technical assistance, that aid may not be the least effective method to upgrade economic and energy competencies in countries that are seriously underdeveloped, in institutional, structural, governance, and cultural terms. Aid, to put it directly, does in principle not work where there is no absorptive capacity and culture.

6.4 United Nations Environment Program

The UN Environment Program in Nairobi is mandated to develop a global approach to environmental issues of sustainable development. As do all UN organizations, UNEP organizes training workshops for disseminating state-of-the-art know-how to developing countries, conferences to identify key issues and develop policy recommendations, and technical assistance to help developing countries to adopt modern policies. There is some focus on incorporating environmental considerations into energy planning. More of interest for international energy law is UNEP's work providing administrative support, including for the negotiation for subsequent protocols, to international treaties, such as the Vienna Convention for the protection of the ozone layer and its 1987 Montreal Protocol, the Basel Convention on hazardous waste transport, and the United Nations Framework Convention on Climate Change (UNFCCC). These, though not directly "energy laws," have at times a tangential impact on, in particular, the oil industry.

UNEP, though, has never been fully accepted as the lead agency on global environmental challenges and the global community's policy response. As part of the UN system, it suffers from the lack of political and technical credibility throughout Western countries and international companies; major activities in its field are carried out by the Global Environment Facility rather than by UNEP, or the UNFCCC Secretariat in Bonn. The "greening" of the World

Bank under its president, James Wolfensohn, has simultaneously reduced the need of Western countries for a UN institution in the environment field. Different from some of the accepted specialized agencies, seen in practical terms as fully independent and professionally very competent, UNEP has not achieved such status. It is rather covered by the negative view from which most of the standard UN activities (conferences, constant organizational restructuring, and making of ever more pious resolutions) suffer. A significant contribution to energy law, though not international or national environmental law, has, so far, been absent. That may differ, though, if UNEP's current efforts to promote guidelines on best practices with respect to environmental disclosure in the oil industry develop. Here, UNEP has stepped into the middle of nonconventional international energy law evolution by authoritative international and internal corporate guidelines.

In summary, energy is not treated well in the UN system. Its political appeal has been overshadowed by the great popularity of environmental and now human rights activities, which offer the chance to develop mutual benefits for both NGOs and the UN system to help each other to more political legitimacy. Energy is not only one of the most significant nuts-and-bolts issues of economic development, intercountry trade, but also a core element of sustainable development and climate change. With energy, after liberalization and privatization, developing from a mostly mere country issue to a great opportunity for mutual benefits from transnational trade, developing countries seem to gain less from this potential, because their institutional framework and intercountry politics are in most cases discouraging energy trade.

There is a case for a specialized UN agency dealing with energy matters, both in terms of monitoring world developments, linking with other issues (environment, transport, nuclear, shipping, and climate change), developing policy proposals for a global negotiating agenda (where really needed), relating with industry, and providing technical assistance in technical, institutional, and regulatory areas. Such an agency would work satisfactorily only if it were not part of the UN system as such, but rather a professionally competent specialized agency. It should be organized with a considerable input from both industry and competent NGOs and professional associations, i.e., in the way that the IMO, IAEA, IEA, and WTO are set up, rather than as a general UN department, with its inevitable slack and wastage.

7. OECD NUCLEAR ENERGY AGENCY

The NEA is a specialized agency integrated (much more closely than the IEA) into the OECD, though with some internal operational autonomy and directly under the Nuclear Energy Steering Committee, a subcommittee of the OECD Council. The 28 members comprise the OECD countries, with some exceptions (Poland and New Zealand); they account for 85% of the world's installed nuclear energy capacity. The main functions of the NEA relate to research, data collection, and the information exchange relating to the peaceful use of nuclear power, in particular, safety of operations, transport of nuclear materials, workers' protection, and waste management. In the past, the NEA managed nuclear operations directly. It works in the areas of nuclear safety and regulation; nuclear energy development; radioactive waste management; radiological protection and public health, nuclear law, and liability; nuclear science; and data collection related to the nuclear industry.

The NEA also has a protoregulatory role by present decisions of the OECD to its member states in the area of operational safety of nuclear plants, severe nuclear accidents, and radioactive waste disposal. Its own guidelines and standards are recommendatory, i.e., not legally binding. It works here through the Committee on Nuclear Regulatory Activities, the Committee on the Safety of Nuclear Installations, and the Radioactive Waste Management Committee; these bring together the national nuclear authorities. Different from the IAEA, it does not carry out safeguard inspections. Its relations are mainly with the nuclear authorities in member countries. As in all other international agencies, there is a connection between its primary network function and the identification of "best practices" arising out of the technical dialogue and regulatory comparison. The NEA also functions as a channel to transpose radiation protection norms developed by the International Commission on Radiological Protection into OECD decisions. A process of revision of the 1990s radiological protection recommendations is currently underway. In 1998, there was an evaluation of the NEA. The principal recommendations have been to incorporate sustainable development into its conceptual framework, to integrate better with the broader energy policy perspectives of the IEA, to develop a better collaboration with the IAEA based on complementarity and through an agreement, and to accept new members (in particular, with major nuclear operations), but also to avoid duplication (in particular, in the area of technical assistance). It is not certain that the inherent organizational logic and self-interest of the NEA will allow such cooperative strategies to be implemented (interagency cooperation is usually recommended by external advisers and agencies pay lip-service to it); the reality is usually that there is interagency competition for interesting projects, funding, public profile, and organizational mandates, with often large-scale duplication.

One of the NEA functions is to develop and disseminate information on nuclear law. The objective is greater harmonization. It has carried out technical assistance on nuclear law reform (including nuclear liability in case of accidents) in Eastern Europe and Asia, usually in collaboration with the EU and the IAEA. It publishes the authoritative *Nuclear Law Bulletin* and has compiled several analytical/comparative studies on nuclear law in its member countries and Eastern Europe. It also runs a professional training program on nuclear law with the University of Montpellier. Finally, the NEA services the Paris Convention on third-party liability in the field of nuclear energy and the Brussels Convention complementing it. (The NEA Steering Committee may recommend the exclusion of nuclear installations or materials from the operation of the Paris Convention.) The NEA also collaborates with nonmember countries (particularly in the former Soviet Union) in the area of nuclear law.

Different from the IAEA (which has real regulatory powers), or the EU (which has real money for technical assistance), the NEA should be considered rather in the club model of the OECD: a forum for exchange and therefore dissemination and improvement of best nuclear regulatory practices. From the antinuclear perspective, the NEA is not necessarily an adversary: Its emphasis on nuclear safety, including its work on decommissioning of nuclear plants, fits as well with an antinuclear perspective. There is no record that "civil society" so far has included the NEA in its group of top evil forces driving globalization, such as the World Bank, the IMF, and the WTO. The NEA has in the past not taken much of a position in the debate over the continued justification of nuclear power. But with its most recent study on the link between greenhouse gases, climate change, and nuclear power, it has identified in detail the contribution that nuclear power makes, and can make, to achieving the Kyoto targets. It demonstrates that nuclear power is responsible for virtually zero CO_2 emissions, quite different from the large to very large amounts of CO_2 emissions from coal, oil, and gas consumption in power plants. Given the current debate about nuclear power, with opposition from traditionally antinuclear NGOs, and support based on its pro-Kyoto impact, the NEA should be expected to continue to be a significant, if not vital, function in the very specialized field of nuclear law and regulation.

8. *EUROPEAN UNION*

The EU is not an international agency, but rather, under international law, a persona that is situated between a supranational organization and a federation of states. Each of the European Communities (EC; Euratom) has its own legal personality. The European Union combines the European Communities plus, as established by the Treaty of EU (the Amsterdam Treaty), the foreign and security policy and justice and home affairs "pillars." The Council can authorize the Presidency to negotiate agreements with third parties, binding the EU. The EU's formal legal status is, therefore, as with many EU matters, unclear; perhaps it could be qualified as a "partial and budding international legal person." The "energy law" it produces is therefore both part of international energy law and the internal, domestic energy law of both the Union and (through direct effect and implementation by national law) the member states. The EU has, among all international organizations, been the most active producer of energy law over the past 15 years, primarily in the design and implementation of its target, an integrated energy market. This development is now in full motion, including implementation in member states, but far from completion. The EU is also a most interesting case to watch: its energy law is a pilot exercise for creating integrated energy markets in other regions (e.g., North America, South America and the Americas, Asia-APEC, and around Russia). It is also the dominant model in spheres of intensified economic cooperation of the EU [accession countries, Eastern Europe and Commonwealth of Independent States (CIS), and Mediterranean] where there is now a legal obligation and a de facto pressure to adopt the single-market instruments. Finally, the energy integration methods and experience of the EU provide an example of energy (and wider economic) integration in the global economy. It is the world's laboratory for ways to create integration benefit out of opening up national, hitherto largely segregated, markets for cross-border investment and trade; for identification of the obstacles that have been overcome (and that have not yet been overcome, including the current cultural, institutional, and political limits to integration); and for the new challenges (in particular, environment and climate change) and ways to deal with them. The EU is therefore at this time the most relevant precedent case, not necessarily for instant copying, but for identifying challenges, issues, and policy instruments and their likely impact (including resistance to them). The importance of the EU as the global economy's laboratory for modern, postprivatization energy law as an instrument of economic and environmental regulation in emerging integrated energy markets cannot be overestimated.

In terms of organizational structure, the European Commission is both the conventional Secretariat servicing the treaties [Treaty of Rome and Treaty of Amsterdam, Euratom, and European Coal and Steel Community (ECSC)] and also an independent actor with colegislation, regulatory, and enforcement powers. In terms of political weight, it is the driver of integration policies, setting the agenda, organizing the process of information, consultation, and coalition building. Energy competence is located mainly in the Directorate General Transport and Energy (DG TREN, formerly DG XVII; handling the single energy market, Synergie), but energy-related competencies are also exercised by the DG Competition and the DG Environment, and the directorates [in particular DG Relex (ex-DG I), handling international assistance and foreign affairs—e.g., the TACIS and PHARE programs]. The Council represents the member states and the intergovernmental facet of the EU.

The European Court of Justice (ECJ) is the most influential and active international court. It has been instrumental in developing key notions of EU law going beyond the intergovernmental character of international treaties and developing the EC/EU into something between a supranational organization and a federal state, mainly through the concepts of supremacy and direct effect of EU law, and by a mostly integration-oriented and policy-based, rather than letter-based, interpretation of EU law. It has, though, been reticent to decide on matters that would require a large-scale industrial restructuring and establishment of a regulatory system, as in the 1997 case involving various EU member states' energy import/export monopoly. In these cases, the Court has observed "regulatory restraint" and has waited for the negotiations for new energy directives led by the Commission to achieve their result.

The new EU energy law cannot be compared to conventional international public law, which primarily deals with, first, division of proprietary and regulatory jurisdiction, and with, second, the existing quite general, unspecific, and indirect impact of many, often not yet ratified, international conventions, on the energy industry. EU energy law has a quite different goal: it is aimed primarily at restructuring the institutional and legal foundation of national energy industries so that a truly EU-wide energy industry can gradually emerge, while at the same time introducing on the Community-level and member state-level a more even playing field, with respect to environmental regulation and initiatives toward reducing greenhouse gas emissions. It is part of the global paradigm of privatization and liberalization. This entails dismantling of existing monopolies and barriers to cross-border trade, providing the legal basis for competitive markets, and, finally, nudging the industries toward real competition. It is the legal form of a proactive, rather than reactive, industrial restructuring in the EU energy industries.

EU energy law consists primarily of the EU treaty and its key general provisions (Articles 28–31, 49, 50, 81, 82, 86, and 87) for freedom of movement and controls on anticompetitive conduct as applied specifically to energy situations, on one hand, and, on the other, on a series of directives, chief among them being the 1996 Electricity and the 1998 Gas Directives (and their amendments in 2003/2004). There is no separate chapter or policy in the EU Treaty on energy (though it has often been advocated), but only some references [Articles 3(u), 154, and 174; to natural resources and environment, 175) of marginal significance. Because energy is one of the most strategic industries, it can be argued that an energy chapter is desirable, but also that energy is automatically covered and included as a key component in any reference to industry, commercial, and economic affairs. The energy directives are being implemented by the member states (and by the East European accession states based on "Europe agreements"), with perhaps still a too large diversity. Primary law could have been used to dismantle the export and import monopolies and provide third-party nondiscriminatory access to the "essential facilities" of electricity and gas transport, storage, and distributions systems owned mainly through monopoly, but there was not enough political will and power in the Commission, too much resistance from member states and the energy monopolies, and no green light from the European Court of Justice. In its October 1997 judgment on export/import monopolies, the ECJ essentially told the Commission it had to restructure the existing regime by negotiated and agreed-upon specific directives, and not by ad hoc actions focusing on specific issues in the much more complex sequence of energy operations endorsed by the Court. The Court in essence declined political responsibility for such restructuring and mandated the Commission to seek a negotiated (and thereby accepted and easier to implement) solution.

Liberalization initiatives have been accompanied by emerging measures for compliance with the EU's Kyoto targets for greenhouse gas emissions by promotion of energy efficiency and renewable energy sources (RESs). These will, in turn, require compatibility with the EU's rules on state aids (which are responsive to renewable energy and other truly environmental goals), transparent and nondiscriminatory procurement, and EU-wide trade. Other significant measures have been the obligation of state and private energy utilities and oil and gas licensees to procure in a transparent and nondiscriminatory way [i.e., to abstain from formal or informal protectionism (Utilities Directive)] a duty on member states to issue oil and gas exploration and development licenses in a similar transparent and nondiscriminatory way [i.e., exclude preference for domestic companies or companies with a desired domestic procurement record (Licensing Directive)].

The EU, however, not only produces "internal" energy law, but it also participates actively, though in a still inchoate form, in "general" international energy law. The gradually coalescing "federal" character, with its tension between necessary, unitary negotiating and deal-making competence toward the outside world, and the maintenance of such elements of national sovereignty by member states make the EU a particularly awkward, indecisive, and contradictory international actor. It is now recognized, based on several authoritative ECJ decisions, that the EU has exclusive competence in trade matters (e.g., WTO negotiations), but only "mixed" competence in "investment matters" [e.g., General Agreement on Trade in Services (GATS), Trade Related Aspects of General Property Rights (TRIPS), ECT, and MAI]. There is no explicit competence for the EC to enter into treaties relating to energy matters in the EC Treaty, thus trade, investment, and other powers are relied on, and usually in the form of joint EC/member states accession. The result is that the EU is very inflexible and slow to react and is therefore at a disadvantage in international negotiations, which require decisiveness and clarity. The EU is very dependent on import of primary energy sources, in particular oil (apart from the United Kingdom), gas (apart from The Netherlands, Denmark, and the United Kingdom). This important dependence is coupled with the fact that most energy sources are in volatile regions close to the EU (Russia, Central Asia, Caucasus, Algeria, and other Mediterranean countries; Middle East/Gulf countries; Angola and Nigeria), with political insecurity in these countries spilling over to the EU in terms of security of supply and domestic political disruptions (terrorism, volatile situations involving ethnic minorities from these countries, and the unwilling but inevitable implication of the EU in the U.S./Israel and Arab conflict). But a unitary and focused EU action is here impeded by numerous special interests of its member states: the United Kingdom and French relationships with former colonies; the German sensitivity to anything endangering peaceful relations with its Eastern neighbors; the Norwegian special status as non-EU member, but more or less subject to EU energy law via the European Economic Area (EEA) agreement (and a major, and stable, oil and gas supplier); the United Kingdom's "special relationship" with the United States; and the French "special" competitiveness with the United States in political and cultural affairs. EU actions are also influenced by the opaque relationship with the United States. The EU needs the United States as senior partner (in particular, for security measures, because the EU usually cannot act decisively nor employ effective security forces), but there is also an underlying tension, out of economic competition, some resentment (in particular, France) at U.S. hegemony, and much greater linkage of the EU to various countries and social, political, and religious forces in its greater neighborhood.

The most visible success of the EU is the Energy Charter Treaty; here, the EU has managed its probably most visible tangible success in being essential in moving a +52-country EC investment, trade, and transit treaty to legal effectiveness and implementation. The ECT is not legally "owned" by the EU, but the EU is its major financial and political sponsor. The treaty's transit Article 7 is now being developed into a specific energy transit protocol that reflects the EU's interest in facilitating diversified supply of energy (oil, gas, and perhaps electricity), in particular from the prospective oil and gas countries now making up the former Soviet Union. The Treaty, though, has not been followed, as was the original negotiating mandate, by a "supplementary agreement" dealing with privatization and access for investors (though a full-text draft exists). The Treaty is now of increasing interest to countries outside its original sphere (East and West Europe); for OPEC countries, it would be a multilateral treaty whereby they would not have to face U.S. obstruction.

Other noted instruments have been the "Europe agreements" with the East European accession states; these oblige the accession states to transpose EU law (including energy law and the new directives), within a time span, into their national system. There is a focus on developing "trans-European energy networks" (i.e., pipelines, transmission grids, and interconnectors) and other current priorities of the EU in the energy sector: promotion of renewable energy, compliance with Kyoto obligations, and nuclear safety of the problematic safety standards of nuclear reactors in Eastern Europe. The accession countries are under considerable pressure, as weaker, EU-entry-seeking parties, to adopt the EU "acquis communautaire," i.e. the current state of EU energy law. On the other hand, though they are being pressured to adopt this "acquis," the making of which they have had no influence, they are not likely to benefit from the liberalization of the EU if powerful domestic interests of domestic member states are affected. There is, for example, considerable reluctance (and search for legitimate reasons, mainly based on environmental pretext) to provide for free energy imports into the EU from Eastern Europe. But this is the nature of the relative bargaining power—seekers to join a club have to accept its rules, without being certain that their presence is universally welcome.

The major direction of current EU involvement in international energy law is through the Kyoto process. The EU has become the main promoter and sponsor of the Kyoto Protocol, setting targets on greenhouse emissions, and this role has increased with the exit of the United States. It is not dissimilar from the way the Energy Charter Treaty has moved to completion, with the EU pushing and the United States exiting. The strong position taken in the Kyoto process reflects the EU's comparatively greater domestic interest in environmental matters, as compared to the United States. The reasoning is that, first, the European Parliament has a sizable representation of "green party" members and that, second, the EU Commission, as an international agency searching for ways of making up for its low level of political legitimacy, is more responsive to pressures from environmental NGOs. Parallel to Kyoto, the EU has also been promoting tighter measures against marine pollution, mainly by using port state leverage, influencing the Oslo/Paris Atlantic (OSPAR) marine environment convention committee toward full prohibition of offshore decommissioning of oil platforms, participating in all relevant international conventions with an at least tangential impact on energy industry operations, issuing a number of directives mandating environmental impact assessment for power plants, and developing systems of ecolabel measures (e.g., energy efficiency of appliances).

8.1 Euratom

Euratom, a separate treaty, but fully integrated into the Community institutions and managed by the Commission, was the EC's early response to the need to develop and monitor safety systems for nuclear energy materials. It was originally premised on the 1950s enthusiasm for the "peaceful use of nuclear energy," fears over security of supply of oil, and a supply-sharing system for uranium. None of these factors present at the origin of Euratom is present today, except to some extent the security-of-supply aspect. Euratom's original objectives were frozen during the early decades by conflicts with France, which wanted to see Euratom deployed to develop an exclusively European (i.e., not U.S.-linked, and largely French-influenced) nuclear industry.

The nuclear industry in the EU (particularly outside France and the southern member states) is an embattled industry. Austria has closed down its nuclear industry and Sweden and Germany are engaged in a slow process toward closing and decommissioning. The future of these processes can, however, not be predicted with any certainty because the enthusiasm for the Kyoto Protocol and the opposition to nuclear industry are difficult to square. If nuclear industry is on the wane, then the future task of Euratom will be decommissioning and safe storage of used nuclear materials. If nuclear industry reemerges, then safety rules, rehabilitation of East European nuclear power plants, and management of nuclear waste will be priorities. The new threat from terrorism will pervade as well in European nuclear law. There will be a tightening of rules of nuclear safety (installations, materials). The implications of a successful or almost successful terrorist attack on a European nuclear installation are hard to predict; the consequence could be an acceleration of the trend toward closure or a much greater investment in security.

Euratom develops safety norms, supports research, cooperates closely with industry, scientists, and other international organizations (in particular, the IAEA and the OECD/NEA), and can invest in projects, including the right to raise loans for that purpose. Nuclear research constitutes one of the major components of Commission-funded research. The current nuclear-related EU Commission research budget is over 1.2 billion euros, from 2002 to 2006; loans available to fund rehabilitation of East European nuclear reactors should exceed 2 billion euros. Euratom produces, through the Community institutions, directives, regulations, and administrative decisions. It is given the right to own nuclear materials within the EU. It runs a system of safeguarding for nuclear materials (excepting those for defense purposes) within the EU. Like its sister organizations (the IAEA and the OECD/NEA), Euratom has managed to keep largely out of the limelight of public opposition to nuclear power development. Its specialized, technical character, its generally endorsed focus on nuclear safety, and its absence from public debate over nuclear energy may have helped. Such positioning may have also helped the nuclear activities of the European Commission to have a more comfortable life, but it is not certain that retreat from challenge and debate has served these organizations or the issue of nuclear energy. As organizations with special knowledge of nuclear industry and the most extensive networks, the Euratom service of the Commission (as the IAEA and OECD/NEA) should have developed a position to be able to make informed arguments about nuclear energy and highlight in the public eye both the risks and the particularly newly emerging benefits in terms of climate change. At present, the Commission's nuclear services have taken a more active role in the debate on nuclear industry: They stress the climate change/Kyoto target and security-of-supply advantages of nuclear energy; they currently propose a much deeper engagement with the safety issues of problematic nuclear reactors in Eastern Europe and improved safety standards and safeguard procedures.

The function of Euratom in safeguarding nuclear materials will only increase as international terrorism develops increasing sophistication and inevitably tries to utilize nuclear materials. Here, though, the main risk of access to nuclear materials, equipment, and expertise is likely to be in countries with rundown nuclear establishments and weak systems of governance. The Euratom service of the Commission as well as the IAEA and OECD/NEA are likely to gain more prominence because they will have to serve as international instruments to deny access to nuclear power to terrorist organizations.

In summary, the EU is currently, in spite of its many and deep institutional weaknesses, the one supranational organization with the most visible impact on national energy laws (mainly in accession, transition, and many developing countries), the main laboratory for economic regulation of cross-border energy trade and investment under the current paradigm of environment-friendly postprivatization liberalization, and the main sponsor of major international energy initiatives, mainly the Kyoto Protocol and the Energy Charter Treaty. International energy law can no longer be seriously studied without understanding both the internal and the external dimensions of EU energy policy. EU energy policy is likely to progress at a tortoise pace, pushed away from conventional energy (coal, oil, and gas) by the new emphasis on renewable energy sources, but, on the other hand still, and for all the foreseeable future, dependent on strategic oil and gas imports from highly insecure producing countries. It has not yet been able to grapple with its major energy dilemma: A large part of the EU's energy demands is covered by nuclear energy. Nuclear electricity is currently and for the foreseeable future, together with large-scale hydropower, the only substantial energy source that is not responsible for greenhouse gas emissions. The EU's 2000 Security of Supply Green Paper and subsequent reports make this clear. Simultaneously, there is no political leeway for the EU to promote maintenance and expansion of nuclear power until the green movements in the EU have made a choice between their traditional opposition to nuclear energy, based on the risk of accidents and disposal of nuclear waste, and the indubitable fact that nuclear energy, among the currently relevant energy sources, is the most Kyoto-friendly choice. Should the reorientation of subsidy and support lead to a true upsurge of nonnuclear renewable energies, then this dilemma might fade, but if it does not, then there seems to be no choice—either embracing, vigorously and expansively, nuclear again or forgetting about control of greenhouse gases.

9. CONCLUSIONS

All international agencies have at their core a "cross-border network" function: They bring "colleagues" together, typically from a ministry with a similar regulatory and operational task. There is a Secretariat, which acts to organize the network and its typically very formalized gatherings and provides permanent expertise (mainly of a comparative nature). This function is not often perceived so clearly, but it may be the vital function of the organization. The network now increasingly includes, often relegated to a lesser rank in communication priority, nonstate actors; these are nongovernment, commercial, and professional organizations and consultants. If the agency does not have direct regulatory powers (they rarely do), the "network" nevertheless exercises considerable influence over national and international energy law: it identifies best or at least prevailing and current practices. These provide a persuasive and practical blueprint for national regulation. Experiences are discussed and models are taken home or brought to the meetings. Copying other people's work and working off legal precedent has always constituted the practical core of the legal profession.

The much more formal and directly "legal" focus of international agency work revolves around their respective treaties. The Secretariats service the treaties and the governmental delegates negotiate (with low-profile Secretariat influence) such treaties and then "implement" them at home. It is only in the EU, a hybrid between an international organization and a federated country, that the international authorities (Commission, Council, and Court) have some direct regulatory powers.

Different from human beings, who grow largely along the lines of a genetic program, international organizations mutate: they come with a fixed mandate, the organizational constitution, which seems at the beginning to be written in stone. But all agencies here surveyed have developed quite differently; some parts of their original mandate have become obsolete and dead text, other parts have been developed and new mandates have been effectively acquired, often based on bold reinterpretations of the original terms of reference. The OECD and its predecessor started out to manage the Marshal Fund; it is now the major intergovernmental policy think tank on any subjects that move governments. The World Bank started off as a financing agency for reconstruction in the countries that were U.S. allies in Europe after World War II; the World Bank sees itself now as the "premier development institution." The IAEA started off to encourage nuclear power; it is now mainly an institution to develop and enforce disciplines related to nuclear risk (by accident), mismanagement, or terrorism. What the IAEA lacks is a natural death. It is very hard to make an obsolete international agency disappear; the agency and its leaders, staff, clients, and constituency will cling to life, sometimes with (self-) deception over its continued usefulness, often with a desperate attempt to latch on to the currently fashionable paradigm so that criticism against the agency can be presented as criticism of the current high-ground moral values. International agencies seem to adapt more easily in form than in substance. They rebaptise readily what they have always done and will continue to do in the fashions of the day; they are much better at redesigning their public relations than what they do and how they do it. Probably every organization examined is in some way or other seriously out of date, mainly with respect to its internal structure and organization. All should be modernized toward more professional competence, but also toward inclusion of nonstate actors in their formal decision processes, although that is unlikely given the highly conservative and anti-change bias built into their organizational structure and culture. Their funding, almost exclusively by government contribution (e.g., national taxes), is also in need of review. More dependence on the value of services rendered in a more competitive situation would make them more modern, responsive, and efficient. Some of them (e.g., OPEC or the IEA) were created in response to a particular crisis that no longer exists in the same form; here, adaptation has been particularly difficult and modernization particularly pressing. But as intergovernmental organizations, they are mainly controlled by diplomats, who are removed from elections, politics, competition, and markets, so that the need for modernization is delayed by a double wall of insulation.

There is a considerable difference between two types of international agencies. The general ones (primarily the UN, also to some extent the OECD) are mainly "talking-shops." They do not have a very specific mandate and certainly very little or no regulatory or operational focus, but they serve to accommodate the public concerns and themes of the day, to process them into organizational language of some (though in reality much less than is claimed) authority. This is an important function because they conduct a public dialogue, albeit in very stilted form, about global politics. The weaker the nation state, and the greater pressure for public participation, now expressed on the international level mainly by the NGOs, the more is there a need for a "parliament" in the original sense, i.e., a "talking-shop." Not surprisingly, general international agencies have found it least problematic to accommodate the pressure of NGOs for involvement. Both actors look for legitimacy, and reciprocal recognition of legitimacy helps both. They operate politically in symbiosis. The problematic legitimacy of both is less of a problem because the operational impact is minimal; even international treaties emerging are typically neither legally effective (because not ratified) nor (even if ratified) specific enough to make a difference.

Both the international agencies and the manifold actors within "civil society" need to open up to each other, in a professional way. There is a good reason for "civil society" to be present in the talk-shop forums. But there is also the temptation that NGOs and such agencies continuously run like lemmings into the same, regularly changing, fashionable direction, scoring successes that do not count. There is a good reason for the specialized, and practically influential, international organizations to open up, not just formally, and not only by being more transparent, but by effectively allowing formal representation of nonstate actors on their governing boards. Multinational companies sometimes now have representatives from public-interest organizations and academia on their boards. Intergovernmental organizations are a child of the club of nation states. Such adjustment to a world in which governments do not play exclusively and in which many nonstate actors are now significant actors is difficult for international organizations and their nation-state governors. Even if nonstate actors now compete with governmental actors on the global scene, it is difficult for the places where governments still call the tune to yield a part of their formal power. Such an opening of the decision-making structures would only accommodate the 1940-type of intergovernmental structure to the changing reality of global society where non-state actors are now operating next to the conventional sovereign states. How to do it is much more difficult. NGOs, for example, are mainly self-appointed, with weak governance, transparency, and accountability. Commercial companies, and their associations and professional groupings, on the other hand, are better structured and in a formal sense more easily identifiable. In principle, international organizations should gradually co-opt those who represent power (commercial, financial, political, and public opinion) into their governing structures. Similarly, treaty negotiations should, as the OECD debacle with the MAI has shown, incorporate in a much more active sense those who have power and a voice in the field. There are inchoate precedents: the ILO tripartite system of decision-making (itself a child of the 1950s corporatist worldview) and the meek OECD Business Industry Advisory Council (BIAC) and Trade Union Advisory Council (TUAC) (largely, it seems, lunch opportunities for retired functionaries). But these need to be developed, experimented with, and, in the end, formalized. There is no reason why those with a powerful voice (International Union for the Conservation of Nature [IUCN], Shell, the International Chamber of Commerce, or the International Bar Association) should not be able to be part of the directorium of international organizations and part of the treaty negotiators whose consensus is necessary. For me, the unresolved issue is if the not fully responsible "activist" NGOS, such as Greenpeace *et al.*, should be in the governing boards of international organizations. As in domestic law for political parties, there should probably be a set of quality standards reflecting transparency, governance, and accountability. From "civil society" this would require a sea change in attitude: from a merely critical, destructive, and political-campaigning attitude to a position whereby constructive alternatives have to be designed and bargained for, and whereby positive responsibility and accountability for results (not just for criticism) have to be accepted.

Finally, this survey of the role of international agencies, with a focus on energy law and policy, also posits that there is an ongoing process of formalizing power relationships over weakly governed, underdeveloped countries reminiscent of (though certainly not identical with) the colonialism of the 19th century. Then, the power of the state, the wealth of the investors, and the values of the missionary movements propelled European states to control a large part of the world. Today, the formal trappings of statehood and sovereignty are everywhere. But there is exercise of "soft" and "structural" power by the rich societies through their NGOs (reborn 19th-century missionaries), through their multinational companies, through their controlled and funded international agencies, and through the legal, financial, educational, and cultural tools of exercising influence and co-opting elites, rather than outright formal ownership, as in the past. The treaties and their subsidiary tools now being used to impose "good governance" on less civilized societies all tend to be formal, on their face, but asymmetric in substance and actual practice. They are all directed toward telling underdeveloped countries and their governments what to do, but rarely, in practice, do they target the rich countries.

Acknowledgments

This article is based on a more extensive and in-depth study in OGEL 4 (2003), available from http://www.gasandoil.com/ogel;

also from Wälde, T. (2003). The role of selected international agencies in the formation of international energy law and policy towards sustainable development. In "Energy Law and Sustainable Development" (A. Bradbrook and R. Ottinger, Eds.), pp. 171–203. IUCN Environmental Law Programme, Paper No. 47.

SEE ALSO THE FOLLOWING ARTICLES

European Union Energy Policy • *Fuel Economy Initiatives: International Comparisons* • *Global Energy Use: Status and Trends* • *International Comparisons of Energy End Use: Benefits and Risks* • *Labels and Standards for Energy* • *Nongovernmental Organizations (NGOs) and Energy* • *OPEC, History of* • *Renewable Energy Policies and Barriers* • *Sustainable Development: Basic Concepts and Application to Energy* • *United Nations Energy Agreements* • *World Environment Summits: The Role of Energy*

Further Reading

Bunter, M. (1998). "Promotion and Licensing of Petroleum Prospective Acreage." Kluwer, Amsterdam.

Bradbrook, A., and Ottinger, R. (2003). "Energy Law and Sustainable Development." IUCN Environmental Law Programme, Paper No. 47. International Union for Conservation of Nature and Natural Resources, Bonn, Germany.

CEPMLP Internet Journal (available on the Internet at http://www.cepmlp.org/journal).

International Energy Law and Taxation Review (published by Sweet & Maxwell).

Journal for Energy and Natural Resources (published by the International Bar Association).

Maniruzzaman, A. (2002). Towards regional energy cooperation in the Asia Pacific, some lessons from the energy charter treaty. *J. World Invest.* 3, 1061.

Oil, Gas Energy Law Intelligence (available on the Internet at www.gasandoil.com).

Redgwell, C. (2001/2002). International regulation of energy activities. In "EU Energy Law," (M. Roggenkamp, Ed.), pp. 13–96. Oxford Univ. Press, Oxford.

Smith, E., *et al.* (2001). "International Petroleum Transactions." 2nd ed. Rocky Mountain Mineral Law Foundation, Denver.

Tomain, J., Hickey, J., and Hollis, S. (1989). "Energy Law and Policy." Anderson, Cincinnati.

Waelde, T. W. (1996). "The Energy Charter Treaty, Gateway for East–West Investment Trade." Kluwer, Amsterdam.

Zedalis, R. J. (2000). "International Energy Law—Rules Governing Future Exploration, Exploitation and Use of Renewable Resources." Ashgate, Aldershot.

Investment in Fossil Fuels Industries

MARK E. FISCHER
Fischer–Seitz Capital Partners, LLC
Lyons, Colorado, United States

1. Breadth of Investment Opportunities in Fossil Fuels Industries
2. Crude Oil and Natural Gas Exploration, Development, and Production
3. Coal Development and Production
4. Crude Oil Refining and Marketing
5. Natural Gas Processing, Transmission, and Distribution
6. Electricity Generation, Transmission, and Distribution

Glossary

barrel of oil-equivalent A unit measure of oil and natural gas that assumes that 6000 (United States) or 10,000 cubic feet (Canada) of natural gas is equivalent to one barrel of oil; the United States focuses on average energy equivalency, whereas Canada works off average historical market value.

discretionary cash flow Usually defined as the sum of net income, depletion, depreciation, amortization, exploration expense (or dry hole costs), and deferred income taxes.

finding, development, and acquisition costs (FD&A) The capital costs of exploring for, developing, and acquiring reserves, usually expressed in a currency per proved barrel of oil-equivalent added during the period of spending.

hedging The practice by producers of avoiding spot pricing for some of their oil or natural gas volumes through the use of forward, future, swap, and option contracts.

price realization The unit price of oil or natural gas realized by a producer for certain volumes; individual price realizations typically differ from benchmark pricing due to differences in energy content, impurities, distance from major markets, and fixed-price contracts.

proved reserves The estimated quantities of crude oil, natural gas, and natural gas liquids that geological and engineering data demonstrate with reasonable certainty to be recoverable during future years from known reservoirs under existing economic and operating conditions; the standards that reserves must meet to be defined as "proved" are set by the Securities and Exchange Commission in the United States and by similar regulatory authorities in other countries.

standardized measure of discounted future net cash flows (SEC PV10) A measure of the net present value of a set of proved reserves established by the U.S. Securities and Exchange Commission; the measure projects the future net cash flows that would be generated from the liquidation of a set of reserves under certain pricing and cost assumptions and then uses a standard 10% discount rate to calculate the net present value of the projected stream.

successful efforts accounting An optional accounting practice whereby only capital spending that results in the booking of proved reserves is capitalized as property, plant, and equipment and the rest is written off immediately as exploration expense; the alternative practice is full-cost accounting, whereby all capital spending for exploration and development is immediately capitalized regardless of the results.

While the fossil fuels industry is mature, it offers multiple investment opportunities driven by technological innovation, its cyclicality, an ever-changing competitive landscape, increasing environmental requirements, and a declining resource base. Successful investment in the fossil fuels industry requires a solid understanding of the supply and demand dynamics of the highly cyclical oil and natural gas markets as well as of the unique assets of hydrocarbon producers and how they are accounted for financially and accurately valued. Investment in the regulated subsectors of the fossil fuels industry further necessitates a firm grounding in government regulation of natural monopolies. This article covers the key issues of which an investor in the fossil fuels industry should be aware.

1. BREADTH OF INVESTMENT OPPORTUNITIES IN FOSSIL FUELS INDUSTRIES

Although fossil fuels industries are typically considered mature, they actually offer a wide variety of investment opportunities, from aggressive high-tech growth to ultra-safe income. Growth in global consumption of fossil fuels is slower than overall economic growth because energy efficiency continues to rise, but above-average growth opportunities are offered by companies able to increase their production of low-cost energy and, thus, steal market share from companies producing high-cost energy. Service companies that develop new technologies to reduce fossil fuels production costs also offer growth opportunities. As protection of the environment becomes increasingly important, the definition of "cost" expands from consumption of capital and labor resources to include the creation of pollution. Thus, today's energy companies with above-average growth prospects are not only oil and gas exploration and production operations that can find and produce hydrocarbons at below-average costs but also companies developing cleaner coal-burning technologies and companies trying to commercialize innovative fuel cell technologies.

Although prospects for industry-wide above-average unit growth are dim, profit growth could accelerate if technology begins to lose the battle against the ever-declining resource base and real energy prices rise. Until now, real prices for fossil fuels, as for nearly all natural resources, have been flat or down over time, as improved extraction technologies have offset ever scarcer resources to exploit. However, it is conceivable that the decline in the resource base could overwhelm technological advances as global demand continues to climb and that real fossil fuels prices would have to rise to cover the higher production costs. Although rising costs would fully offset the benefits of higher prices in the very long term, in the nearer term producers with low sunk capital costs would be net beneficiaries. Growth opportunities would also arise for producers of renewable energy as well as for oilfield services and equipment firms, which would experience greater demand for their products and services per unit of energy produced. Energy industry analysts hotly debate how close the world is to experiencing a secular increase in real fossil fuels prices. Less highlighted is the possibility that an unforeseen technological breakthrough will result in lower real fossil fuels prices in the future, perhaps because technological breakthroughs are by definition difficult to foresee.

The lack of above-average growth opportunities in fossil fuels industries may limit the long-term appreciation potential of many energy investments, but it in no way restricts the cyclicality and volatility of most investments in the unregulated, capital-intensive sectors of the industry. These sectors typically experience large swings in capital investment that result in alternating shortages and surpluses. When energy is short and prices rise, energy investments perform well, but when overinvestment leads to surpluses, energy investments perform poorly.

The exception to this rule is investments in energy industries with government-regulated returns. For example, in the transmission and especially the distribution of energy, in most cases it would be inefficient to foster the construction of two or more competing transportation systems, so the government allows a single company to maintain a monopoly on transportation of gas or electricity in a given market. However, the government sets transportation rates for the company on a cost-plus basis targeting a "fair" return for the company's shareholders. Profitability of these heavily regulated companies typically depends on management's ability to keep costs in line with, or lower than, those projected for the regulatory agency's return calculation. Profitability can shift sharply if costs drift widely from projected levels and the regulatory agency suddenly adjusts pricing up or down to restore returns to "fair" levels.

2. CRUDE OIL AND NATURAL GAS EXPLORATION, DEVELOPMENT, AND PRODUCTION

With crude oil and natural gas providing approximately 62% of the world's energy supply, the oil and gas exploration and production industry offers significant investment opportunities. That said, many of the world's oil and gas reserves are owned by the governments of the countries in which they are located and may not be invested in by outsiders. The oil and gas reserves available for investment typically are owned by large so-called integrated oil companies or smaller so-called independent oil and gas producers. Integrated oil companies ("integrateds") not only explore for, develop, and produce oil and gas but also at least refine the oil, and most of these companies transport it and market it as well. In addition, some own and operate petrochemical plants. Investments in integrateds offer exposure to the exploration and production business, but they are not so-called pure plays. They also provide exposure to the refining, transportation, and marketing businesses.

In contrast, independent oil and gas producers ("independents") mainly explore for, develop, and produce oil and gas, and they rarely engage in other businesses. Although nearly all integrateds are public companies and have equity market capitalizations of $20 billion to $200 billion U.S., many independents are small private companies and some have capitalizations of less than $1 million U.S. The public independents typically have equity market capitalizations of $10 million to $20 billion U.S. Independents offer a pure play on the oil and gas exploration and production business.

Exposure to the oil and gas exploration and production business is also available through investments in so-called oilfield services and equipment companies ("oil service companies"), which are firms that supply specialized services and equipment to integrateds and independents to help them find, develop, and produce oil and gas. For example, oil service companies plan, perform, and analyze seismic surveys, rent onshore and offshore drilling rigs with crews, run sensors down wells to detect oil and gas (so-called wireline logging), fracture reservoirs with acid or water pressure to maximize production, and cement casing and tubing in wells to prepare them for production. The upper end of the range of equity market capitalizations of oil service firms is a little higher than that of independents.

2.1 Key Drivers of Value Addition for Oil and Gas Producers

For the oil and gas exploration and production operations of integrateds and independents, the key drivers of value addition are unit growth, pricing, and costs. As in any business, companies that add value for shareholders are those whose return on capital employed exceeds their weighted cost of capital. However, because the lead time between the discovery of new reserves and first production is often years, sophisticated analysts often include in their calculation of annual returns not only net income but also the estimated increase in the net present value (NPV) of a company's reserves.

2.1.1 Unit Growth

Oil and gas investors monitor both production and reserves growth. Typically, to grow profits, an oil and gas producer must increase production. To do this, oil and gas companies drill new production wells, boost volumes from existing producing wells, or purchase producing wells from other companies. New well-stimulation and reservoir-fracturing technologies continue to increase the rates at which oil and gas fields can be drained. To experience net growth, companies must add more production than they lose to the natural declines on their existing producing wells. Depending on the rate at which hydrocarbons flow out of a reservoir and underground pressure on the hydrocarbons drops, production from a producing well can drop 5 to 70% per year. Typically, the faster a reservoir produces, the faster its production declines. The worldwide average annual natural decline rate for crude oil production is estimated to be 10 to 15%, but for North American natural gas production the estimate is 20 to 30%. Although companies are eager to monetize their oil and gas reserves as quickly as possible to maximize their return on capital, from a practical standpoint, public companies find it difficult to manage high decline rates. The timing of reinvestment opportunities does not always match a high-production, high-decline rate company's strong cash inflows, which can lead to volatile production rates and profitability. Just as public market investors accord premium valuations to companies with high stable growth rates, they impose discount valuations on companies with volatile returns, even if the long-term growth rate is above average. A private oil and gas producer with a limited need for external capital is freer to pursue high-return, high-decline rate investments because the associated profit volatility affects its cost of capital much less.

To be able to drill new production wells, an oil and gas company must discover new oil and gas reservoirs through exploration activities or it must purchase so-called undeveloped acreage that contains discovered hydrocarbons that have not yet been fully brought into production or even brought into production at all. Larger, well-capitalized oil and gas companies that can afford the often long lead times between discoveries and first production tend to favor exploration to replenish their development opportunities, whereas most smaller firms adopt a so-called acquisition-and-exploitation strategy. They mostly buy into existing discoveries that are not fully developed and try to complete the development and produce the reserves with a more favorable cost structure than those of the larger companies typically selling the properties. Nearly all oil and gas producers maintain a portfolio of assets ranging from so-called unproven acreage, where no discoveries have yet been made, to mature producing properties on the verge of generating negative cash flow. The weightings of higher risk, higher potential early-stage assets and lower risk, lower potential mature properties in a company's mix define its risk/reward profile.

For production growth to be sustainable, an oil and gas company must replenish not only the production it is losing to natural declines but also the underlying reserves that it is producing. Theoretically, an oil and gas company could engage exclusively in exploration and/or reserves trading activities, leaving production from the reservoirs to those companies to which it sells reserves. In reality, the handful of independents that have come closest to implementing this strategy have encountered cash flow problems and financing difficulties. As with production, to grow reserves, a company must add more reserves than it produces. Unlike production, reserves are more difficult to define. The accounting systems in the countries where stocks of companies with hydrocarbon assets trade all define reserves differently. However, there are three basic categories of reserves: possible, probable, and proved. In addition, there are three subcategories of proved reserves: undeveloped, proved developed nonproducing, and proved developed producing.

2.1.1.1 Possible These reserves are also called "unrisked potential reserves." Reservoir engineers estimate this quantity by assuming that all of the potential reservoirs visible from seismic surveys of a given set of acreage are full of hydrocarbons and that the distribution of reservoirs across the acreage where seismic has not been acquired is similar to that found across the acreage where seismic has been acquired. The financial statements of oil and gas companies rarely refer to possible reserves because they are such an approximation.

2.1.1.2 Probable These are reserves that analysis of drilling, geological, geophysical, and engineering data does not demonstrate to be proved under current technology and existing economic conditions but where such analysis suggests the likelihood of their existence and future recovery. Although the U.S. Securities and Exchange Commission (SEC) does not require oil and gas companies listed on U.S. exchanges to report estimates of probable reserves, these are reported by Canadian and U.K. oil and gas companies. As the name suggests, probable reserves are more likely to be there than not, but odds of an upward or downward revision of the estimate are nearly even. Supposedly, the estimate has been risked for a betting person, and theoretically, this is the number that investors should care about most. If an investor can find a set of probable reserves that are undervalued, he or she should purchase them because an upward revision is slightly more likely than a downward revision and a return to fair value is likely over time. In reality, many investors fear that probable reserves are estimated too optimistically.

2.1.1.3 Proved These are the estimated quantities of crude oil, natural gas, and natural gas liquids that geological and engineering data demonstrate with reasonable certainty to be recoverable during future years from known reservoirs under existing economic and operating conditions. Most investors focus on proved reserves as defined by the SEC. Nearly all major oil and gas companies in the world now have their stocks trading on U.S. exchanges, either as ordinary shares or as so-called American depositary receipts (ADRs) or shares (ADSs), so U.S. oil and gas accounting has become the de facto global standard. All companies with oil and gas reserves whose stocks trade on U.S. exchanges must report their year-end reserves according to U.S. generally accepted accounting principles (GAAP) in a Form 10-K (for ordinary shares) or 20-F (for ADRs or ADSs).

Unlike probable reserves, proved reserves are defined conservatively to make downward revisions much less likely than upward revisions. In other words, on average, oil and gas companies have more reserves than their reported proved reserves. This partially explains why oil and gas companies typically trade at premiums to reasonable estimates of the liquidation value of their proved reserves. Investors correctly assume that the companies' actual reserves are probably larger than the reported proved amounts. The other explanation for the typical premium is growth potential. Unless an oil and gas company is a so-called royalty trust or is poorly managed, investors assume that it is not a static entity in liquidation but rather a profitable growing enterprise.

There are three types of proved reserves: undeveloped, proved developed nonproducing, and proved developed producing:

- *Undeveloped.* These are reserves that are expected to be recovered from new wells on developed acreage where the subject reserves cannot be recovered without drilling additional wells. These reserves have been "proved" by a successful exploration well, but they have not been readied for production. No proved reserves of any kind are typically booked according to seismic interpretation alone. After the first successful exploration well, incremental proved reserves are booked as additional successful wells are drilled and define the size of the discovery. Most oil and gas companies are reluctant to book any proved reserves until a development plan for the discovery has been approved by the partners in the field as well as by the host government. Given the high-cost and time-consuming nature of capital-intensive development work (e.g., drilling wells, building production platforms and facilities), proved undeveloped reserves are usually worth only 15 to 30% of proved developed producing reserves. In other words, after reserves are discovered, 70 to 85% of the capital needed to bring them into production still must be spent. Given the often multiple-year lead times between discovery and initial production, the time value of money has a much more negative impact on the NPV of proved undeveloped reserves than on that of proved developed producing reserves.

- *Developed nonproducing.* These are proved reserves that can be expected to be recovered from existing wells with existing equipment and operating methods. These proved reserves have been readied for production, but production has not commenced. Oil and gas wells often encounter reservoirs at various depths, and for technical reasons it does not always make sense to commingle production from several different reservoirs at various depths in the same well. The reserves in the reservoirs that are "in line" awaiting their turn to be produced are proved developed nonproducing.

- *Developed producing.* These reserves have the highest NPV of all the categories because they are the only reserves generating cash flow. In addition, they are reserves for which all of the exploration and development capital already has been spent.

Understanding growth in proved reserves is much more complicated than understanding production growth. For simplicity's sake, many investors focus on growth in total proved reserves, but this can be misleading. For example, a company could add significant value by spending capital to transfer proved reserves to the developed category from the undeveloped category without any growth in total proved reserves. Conversely, another company could boost proved undeveloped reserves significantly with a large discovery in a frontier area but add limited value due to a long lead time until first production and heavy capital requirements to develop the discovery. Based on growth of total proved reserves, the latter company would appear to be the faster grower, but it could easily be adding value more slowly than is the former company. In other words, the NPV of proved reserves per unit varies widely depending on their stage of development. In addition, NPV per unit is affected by the quality of the oil or gas (which determines price received), the distance from markets, and the remaining unit development costs.

The key unit of measure for crude oil is the barrel, which is equal to 42 gallons. For natural gas, it is the Mcf or unit of 1000 cubic feet. Although different grades of oil and gas have different levels of energy content, on average, 1 barrel of oil contains the same amount of energy as does 6 Mcf of natural gas or approximately 6 million British thermal units (BTUs). In the United States, investors use this approximate energy equivalency to calculate so-called barrels of oil-equivalent or cubic feet of natural gas-equivalent. For example, a company with 1000 barrels (MB) of oil reserves and 6 million cubic feet (MMcf) of natural gas reserves is said to have total reserves of 2000 barrels of oil-equivalent (MBoe) or 12 million cubic feet of natural gas-equivalent (MMcfe). Somewhat differently, in Canada, an equivalency of 10 Mcf of natural gas per barrel of oil is standard. The Canadian ratio reflects the long-term historical pricing relationship between the two forms of hydrocarbon that was driven by the premium price per BTU accorded to oil because it can be used as a transportation fuel and not only as a heating fuel. However, during recent years, the price ratio has been narrowing for global crude oil and North American natural gas because North American natural gas production is maturing faster than global crude oil production (i.e., unit costs are rising faster for gas), and increasingly stringent environmental requirements are generating a new price premium for more environmentally friendly natural gas.

2.1.2 Pricing

Oil and gas are commodities, and in a perfectly competitive market all of the producers would be price takers. Although this model accurately describes the North American natural gas market, it is too simple for the global crude oil market. Approximately 41% of global crude oil production is controlled by the nations that belong to the Organization of Petroleum Exporting Countries (OPEC), whose goals are to maximize the global price of crude oil without losing market share and to stabilize the price of oil. OPEC controls much of the world's lowest-cost oil production, and in a free market its market share would be expected to rise. However, OPEC fears that any increase in its market share would be more than offset by a decline in prices. Indeed, if OPEC increased low-cost production faster than total oil consumption, the world would need less oil from the high-cost end of the curve and the cost of the marginal high-cost production that sets the price would fall. OPEC could boost prices by actually relinquishing market share and forcing consumers up the supply curve to even higher cost marginal high-cost production, but as OPEC discovered during the early 1980s, such a strategy is unsustainable. In 1986, Saudi Arabia lost interest in artificially maintaining the price of oil above $25 per barrel by cutting back its production because it realized that its production would soon need to approach zero to achieve its goal.

OPEC also tries to stabilize crude oil pricing by cutting back its production temporarily during times of weak demand and by moving its production up toward maximum capacity when demand is cyclically strong. Fluctuating levels of cohesion among OPEC members and economic uncertainty make the process highly imperfect.

Although OPEC has an undeniable impact on the global crude oil market, it has not been able to eliminate the cyclicality of the industry. Like the freer North American natural gas market, the oil market is characterized by alternating periods of under- and overinvestment driven by pricing. Low prices cause underinvestment, which leads to shortage, which in turn raises prices and results in ensuing overinvestment, which makes prices fall again. Investors as a group typically overreact to pricing of the moment, making oil and gas investments more attractive when hydrocarbon prices are low and less appealing when oil and gas prices are high.

Because oil and gas prices are highly volatile and have such a dramatic impact on the ability of oil and gas producers to add value, investors make tremendous efforts to forecast fossil fuels prices. However, because of the large number of factors affecting oil and gas prices, they are extremely difficult to forecast with any accuracy over any period of time. In the past, the best assumption for crude oil for the long term has been reversion to mean historical real pricing, but as was highlighted earlier, some analysts believe that a secular upward move in real crude oil prices is under way, perhaps dating back to the early 1970s. For the 100 years through 1970, the real price of crude oil (in 2001 U.S. dollars) fluctuated mostly between $10 and $20 per barrel, but since 1970 the typical range has increased to $20 to $30 per barrel. Real North American natural gas prices may have also already entered a secular upward trend based on diminishing returns to drilling activity, but for now this trend may be muted by a loss of gas demand to cheaper competing fuels such as coal and residual fuel oil.

Although investors and the press often refer to benchmark prices, such as the near-month futures contracts for West Texas Intermediate (WTI) crude oil, Brent U.K. North Sea crude oil, and Henry Hub Louisiana natural gas, the actual prices realized by each oil and gas producer from each field in each region can vary widely from the benchmarks. So-called price realizations are affected by the energy content of the oil or gas, the quantity of impurities to remove (e.g., sulfur, carbon dioxide), the quantity of valuable by-products to extract (e.g., natural gas liquids), the need for refining, and the transportation costs to major markets. In addition, oil and gas producers are increasingly using futures and options to hedge the price realizations for a portion of their future production (so-called hedging). Their main goal is to add stability to their operating cash flows so as to be able to maintain more consistent capital spending programs in the face of oil and gas price volatility. Some companies also try to lock in pricing that is above historical averages when they can. One of the results of more widespread hedging is that the financial results of oil and gas operations no longer necessarily reflect spot market pricing of oil and gas during a given period.

2.1.3 Costs

Perhaps the aspect of an oil and gas operation that is most important to investors is its ability to maintain low costs and further reduce them. Essentially, even volume growth is a function of low costs. Companies that can add reserves and production at low unit costs are able to add more units with less capital. The price umbrella provided by the marginal high-cost producers provides strong returns to the industry's low-cost producers.

Accountants divide costs into several categories: lease operating expense, production taxes, exploration expense, depletion, depreciation and amortization, general and administrative expense, interest expense, income taxes, and dividends.

2.1.3.1 Lease Operating Expense
These are the costs incurred to maintain and run the facilities that

produce the oil and gas. Low lease operating expenses are typically driven by high flow rates per well and widespread automation. These costs usually range from less than $1 per barrel of oil-equivalent at prolific world-class oil and gas fields up to more than $8 at high-cost marginal operations. Average lease operating costs are $3 to $4 per barrel of oil-equivalent for diversified public companies.

2.1.3.2 Production Taxes These are severance, ad valorem, and excise taxes on the revenues generated from sales of oil and gas produced. They vary widely from one fiscal regime to another. Investors typically analyze them as a percentage of revenues. The percentage usually ranges from 0 to 15%.

2.1.3.3 Exploration Expense Most oil and gas companies use so-called successful efforts accounting as opposed to full-cost accounting. In successful efforts accounting, all expenditures for exploration activities are immediately expensed during the period in which they are recognized to have been unsuccessful. Only exploration spending that results in booked proved reserves is capitalized as property, plant, and equipment. In contrast, in full-cost accounting, all exploration spending is capitalized regardless of the outcome. Exploration expense is the periodic charge off of the exploration spending that has been unsuccessful at "successful efforts" companies. For oil and gas companies with significant exploration programs, the number can be large at times and typically fluctuates widely. In contrast, companies adopting an acquisition-and-exploitation strategy generally have low and stable exploration expense. For "full-cost" companies, these expenditures pass through the income statement as depletion of capitalized reserve value over time and occasionally in a single large charge called a ceiling test writedown. Every quarter, full-cost companies are required by the SEC to compare the book value of their oil and gas reserves to an estimate of the NPV of their reserves assuming quarter-end spot oil and gas prices. When oil and gas prices dip, full-cost companies with a lot of capitalized unsuccessful exploration spending usually must take ceiling test writedowns.

2.1.3.4 Depletion When oil and gas companies spend money to explore for and develop reserves, most of it is capitalized as property, plant, and equipment. As the reserves are produced, companies take a noncash charge for every barrel of oil and Mcf of gas produced. This allows the companies to match the revenues from selling production during a given period with the costs incurred to generate that production, even if some of those costs were incurred many years earlier. Although there are different accounting methods, essentially the per unit charge is calculated by dividing the net capitalized costs associated with the proved developed reserves by the number of reserves. Unit depletion swings in a range very similar to lease operating expense: from less than $1 to more than $9 (U.S.) per barrel of oil-equivalent. The average for public companies is $4.50 to $5.50.

Over time, unit depletion should be similar to so-called unit finding, development, and acquisition costs. For oil and gas companies, investors monitor the annual cost of adding proved reserves. They divide the total annual spending for exploration, development, and acquisitions by the number of proved reserves added. Although this is an imprecise calculation on its own without reference to the NPV of the reserves added and not accounting for the number of proved reserves moved to the developed category from the undeveloped category, investors look for unit finding, development, and acquisition costs of diversified oil and gas companies to run less than $6 per barrel of oil-equivalent over 3- to 5-year periods. Note that for successful efforts companies and full-cost companies that have taken significant ceiling test writedowns, unit depletion is typically less than unit finding, development, and acquisition costs because a portion of these costs has been expensed through exploration expense or ceiling test writedowns.

2.1.3.5 Depreciation and Amortization With most assets of oil and gas operations in depleting reserves and not in depreciating factories or amortizing goodwill, this is not usually a significant expense.

2.1.3.6 General and Administrative Expenses Although these are not generally large for capital-intensive oil and gas companies, low unit general and administrative expenses can be a sign of corporate efficiency. However, it is usually not possible to calculate unit general and administrative expenses for integrateds because general and administrative expenses are not broken out for the exploration and production division but instead are reported for the whole company.

2.1.3.7 Interest Expense This is the interest on a company's debt as well as fees to maintain credit lines. It is important to know how much debt is at fixed interest rates and how much is at variable rates. Variable rate debt is usually less expensive but subject to the risk of interest rate increases. For oil and gas companies, investors are typically comfortable with ratios of debt to total book capitalization of 45% or less and ratios of debt to operating cash flow of two times or less. Of course, operating cash flow fluctuates with oil and gas prices, so investors' comfort with a given debt level may change with the oil and gas price outlook.

2.1.3.8 Income Taxes Like production taxes, these vary from one fiscal regime to another. In general, the lower the operating costs of extracting hydrocarbons, the higher the taxes imposed by the government with jurisdiction over the reserves. Essentially, governments try to "levelize" the risk-adjusted returns available to oil and gas companies in the global marketplace for reserves. If a country is fortunate enough to hold low-cost oil and gas reserves, its government tries to keep the excess return for itself, allowing the oil and gas company extracting the reserves a return only marginally high enough to be competitive. That said, actual after-tax returns vary widely because the marketplace demands upside potential in exchange for assuming high risk, and when high-risk projects work out the payoff can still be large.

2.1.3.9 Dividends Some oil and gas companies include preferred stock in their capital structure and pay cash dividends on the preferred stock. These dividends, unlike interest on debt, are typically not deductible from income taxes. In general, the larger an oil and gas company, the more likely it pays a cash dividend to common shareholders and the larger the dividend itself. Larger oil and gas companies usually cannot find good reinvestment opportunities for all of their operating cash flow and so return some of it to common shareholders in the form of cash dividends. Although cash dividends are income tax-inefficient because common shareholders are taxed twice on the company's profits (once as taxable income to the corporation and then again as dividend income to the shareholder), they continue to be paid largely as a way of signaling stability to investors. An oil and gas company that can afford to pay a high dividend is viewed as more stable than a firm that does not pay a dividend. Greater stability results in a higher valuation, all other things being equal, and this may reduce a company's cost of capital enough to offset the higher tax costs of paying a dividend.

2.2 Valuing Oil and Gas Producers

The inventories of reserves held by oil and gas producers make valuing these operations more complex, but also potentially more precise, than valuation of other industrial concerns. Essentially, all investments are worth the NPV of future expected cash flows. But because the further in the future, the more uncertain the cash flows, investors typically focus on expected cash flow for the next 12 months, which is difficult enough to forecast. Once this has been projected, an estimate of maintenance capital spending (usually just depletion, depreciation, and amortization) is subtracted to project free cash flow (for which accounting net income or earnings is a proxy). A target multiple is then assigned to this free cash flow (or net earnings) projection based on the investor's assessment of the accuracy of the projection as well as growth prospects and risks beyond the next 12 months.

An investor could value oil and gas assets in this way, but it would be difficult to quantify the impact on growth prospects and risks of companies' differing inventories of reserves. For example, if two companies both hold a 50% interest in the same producing oilfield but one also has a nonproducing interest in another oilfield (to be brought into production in 3 years), both would be expected to generate identical free cash flow over the next 12 months, but the growth prospects of the company with the second nonproducing interest would be superior to those of the company with only the producing interest. The question is how much higher a multiple of next 12 months projected free cash flow should be assigned to the company with both the producing and nonproducing interests.

2.2.1 Asset Values: Vice to Virtue

The fact is that oil and gas producers' large inventories of proved reserves make their future cash flows more certain than those of many other types of commercial and industrial operations; thus, they can be turned into a valuation advantage. Making key assumptions about natural production decline rates, future oil and gas prices, future development costs, and the growth in future operating costs and taxes, investors can calculate a relatively precise estimate of the liquidation value of a set of oil and gas reserves for a given discount rate. The SEC recognized this fact when it decided to require all U.S.-traded companies with oil and gas assets to calculate annually its so-called Standardized Measure of Discounted Future Net Cash Flows, referred to as the "SEC PV10" by investors because it estimates the NPV of the net cash flows expected using a standardized 10% discount rate. For the other key assumptions, companies go to their reservoir engineers and accountants, although the SEC requires that all companies use year-end oil and gas price realizations flat into the future. Companies that want to add credibility to their SEC PV10 have it audited, along with their proved reserves, by an independent reservoir engineering firm. The SEC PV10 is published is each company's 10-K or 20-F.

The SEC PV10 contains valuable information that is underused by many investors. The key analytical problem with the SEC PV10 is that each year a different set of price assumptions is used according to year-end pricing. For example, if an investor believes that the real price of oil will be approximately $20 per barrel in the future, but because the year-end spot price dipped to $14 per barrel the SEC PV10 shows him or her the liquidation value of a given set of reserves at $14 per barrel, the investor might not find the SEC PV10 to be a useful valuation tool. Unfortunately, this investor might end up ignoring valuable information about decline rates and future costs embedded in the SEC PV10 because of an unpalatable pricing assumption. However, the SEC PV10 offers the sophisticated investor enough information to broadly replicate the underlying discounted cash flow analysis with the pricing assumptions used, allowing him or her to solve for estimated decline rates and future costs. The sophisticated investor can then replace the year-end pricing assumptions mandated by the U.S. SEC with those to his or her liking and calculate a valuable estimate of the NPV of the expected future cash flows from a given set of reserves. To get from here to a value for a company's equity, the investor need only subtract its net debt.

Although sophisticated investors can squeeze much incremental value out of the SEC PV10 with in-depth analysis, even just as reported, the SEC PV10 can be helpful. In a given year, SEC PV10s for different companies have the virtue of having been calculated using the same year-end oil and gas price assumptions. Investors trying to understand the true value of a company's proved reserves can divide its SEC PV10 by its number of proved reserves to calculate its unit SEC PV10 (expressed in dollars as SEC PV10 per barrel of oil-equivalent). In comparing this value to the SEC PV10s per barrel of oil-equivalent of other oil and gas assets, investors can discover by how much the NPV of the company's reserves is at a premium or discount to the industry average. Investors can then compare the premium or discount of the company's market value (market equity plus net debt) per barrel of oil-equivalent relative to the industry average to its unit SEC PV10 premium or discount to assess the accuracy of its market valuation relative to that of peers. Note that this valuation analysis is less effective at the end of years in which the relationship of the price of oil to that of natural gas is atypical, and particularly in analyzing oil and gas companies with an oil and gas mix very different from the industry average. The closer the mix to the industry average, the more effective the analysis.

In practice, unit SEC PV10s range from X/Boe to about $3X$, where X depends on pricing assumptions. As highlighted previously, reserves are more valuable, the more developed, the faster the production rate, the higher the quality of the oil and gas, the closer to major markets, the lower the development and operating costs, and the more benign the fiscal regime where they are located. Although some investors assign a fixed asset value per barrel of oil and per Mcf of natural gas to all reserves, analysis of unit SEC PV10s demonstrates how needlessly imprecise this form of analysis is.

2.2.2 Cash Flow: Simple but Potentially Misleading

Many investors value oil and gas assets based on operating cash flow or earnings before interest, taxes, depletion, depreciation, amortization, and exploration expense (EBITDAX). Like the free cash flow (or net earnings) analysis mentioned previously, this is not a formal discounted cash flow analysis but rather an analysis of next 12 months expected operating cash flow or EBITDAX and the appropriate multiple to apply to it. Investors typically focus on expected operating cash flow before volatile and difficult to forecast working capital changes, which is essentially the sum of net income, depletion, depreciation, amortization, exploration expense, and deferred income taxes. This is also called discretionary cash flow because it is cash flow that management has the discretion to use in multiple ways (i.e., exploration, development, acquisitions, debt reduction, stock repurchases, dividend increases).

When it comes to oil and gas assets, investors do not like free cash flow or net earnings because the volatility in oil and gas prices makes these residual metrics even more volatile. Multiples of free cash flow and net earnings are virtually unanalyzable much of the time. In contrast, operating cash flow and EBITDAX have the virtue of being less "residual" and, thus, less volatile, but they have the defect of offering less useful information. If an earnings-like valuation metric does not account for maintenance capital spending (as operating cash flow and EBITDAX do not), it has little useful information about earning power. In fact, a company that has analyzable operating cash flow, but whose maintenance capital spending is higher than its operating cash flow, is effectively in liquidation. There is no way in which to gauge this from operating cash flow or EBITDAX alone.

Operating cash flow and EBITDAX are actually more proxies of asset value than of free cash flow. They can help investors to compare oil and gas operations with each other so long as the operations have similar ratios of reserves to production. However, when one operation has a significantly higher ratio of reserves to production than does another, the analysis breaks down. Essentially, using operating cash flow or EBITDAX as a proxy for asset value means assuming a fixed relationship between cash flow and asset value, and this does not hold when one company has much more asset value (or reserves) per dollar of cash flow (or barrel of oil-equivalent of production) than does another. Even when reserves-to-production (R/P) ratios are similar, asset value per dollar of cash flow can vary for the same reasons that unit SEC PV10s differ from company to company, but varying R/P ratios show the pitfalls of cash flow valuation even more clearly.

Although reported operating cash flow and EBITDAX are less volatile than free cash flow and net earnings, they still fluctuate too widely with oil and gas prices to be solid anchors of value. Sophisticated investors often recalculate historical and projected cash flow using so-called normalized pricing, just as they calculate asset values using projections of average long-term real pricing for oil and gas. Normalized pricing makes operating cash flow and EBITDAX better proxies for asset value, although it does not solve the other problems highlighted previously.

Using multiples of market value (defined as market equity plus net debt) to EBITDAX to value oil and gas assets carries all the problems listed previously, but multiples of market equity alone to operating cash flow have yet another defect. Operating cash flow moves with financial leverage. When an oil and gas producer purchases oil and gas properties using incremental debt, its operating cash flow nearly always rises even if there is no increase in its net asset value. In fact, operating cash flow typically increases even if net asset value falls. This occurs because without reinvestment, operating cash flow from oil and gas properties falls with natural production declines while the interest on the incremental debt is constant (assuming fixed-rate debt). Even if the NPV of the expected stream of operating cash flows from the oil and gas properties is below the NPV of the stream of expected interest payments (i.e., shareholder value has been destroyed in the purchase), the first 12 months of operating cash flow nearly always exceeds the first 12 months of interest, resulting in a net increase in operating cash flow for the acquiring oil and gas producer. (Of course, this does not occur if the properties acquired are relatively underdeveloped and production is low relative to reserves.)

The analysis is slightly different, but the conclusion is identical, if one looks at the acquired properties as a going concern rather than as a simple liquidating entity. If they are a going concern, a large portion of the operating cash flow needs to be reinvested to maintain and grow production. The expected future net cash flows from the properties become the flat or rising operating cash flow minus the capital required to maintain or grow the production. Even if the NPV of this expected cash flow stream were lower than the NPV of the interest on the incremental debt used to fund the acquisition (i.e., shareholder value is being destroyed), expected operating cash flow for the next 12 months for the acquiring company would rise because the acquired properties' operating cash flow would exceed the interest on the debt assumed to fund the purchase.

The valuation problem arises when investors decide that the equity of a certain oil and gas producer is worth a given multiple of its operating cash flow and that producer increases its operating cash flow through a debt-financed acquisition that adds no net asset value. These investors might mistakenly conclude that the value of the equity has risen with the operating cash flow. Instead, the value of the equity in this example is unchanged, and the appropriate multiple of the market equity to operating cash flow has actually declined. In fact, high-debt oil and gas independents typically see their equity trade at steep discounts to the peer average relative to operating cash flow, and this is not only because of concerns over the risks of high leverage. These same high-debt companies usually see their total market value (market equity plus net debt) trade at much smaller discounts, if any, to the peer average relative to EBITDAX. This unleveraged multiple corrects for the distortions created by the timing differences between oil and gas property operating cash flow and interest payments.

2.3 Oilfield Services and Equipment Companies: Less Would Be More

Investors who believe that rising hydrocarbon demand and the declining hydrocarbon resource base are about to overwhelm the pace of technological improvements like oilfield services and equipment ("oil service") investments. The less oil and gas the industry gets out of every square kilometer shot with seismic and every meter drilled into the earth, the more demand for oilfield services and equipment. The oil service industry has leverage to diminishing returns to extraction activity, and unlike the producing industry, it should experience a minimal cost squeeze. Although oil and gas producers must purchase many industry-specific services and equipment, oil service companies purchase many industry-nonspecific inputs (e.g., steel, electronic components, construction services) that they render oil and gas industry specific.

Of course, leverage cuts both ways. When oil and gas prices decline, producer capital spending, which is tied to producer cash flows, typically falls more, and producer capital spending is oil service companies' revenues. The equities of many oil service companies are among the worst performers during oil and gas industry downturns. And if an unexpected technological breakthrough were to result in lower real oil and gas prices and less demand for oilfield services and equipment on a secular basis, oil service companies would probably be the worst performers in the oil and gas industry.

Fortunately, the financial analysis of oil service companies is less complex than that of oil and gas producers. Oil service companies are industrial concerns that purchase capital and labor inputs and transform them into services and equipment that are sold to oil and gas producers. They do not have large inventories of hydrocarbon reserves at various stages of development to value. However, operationally, they can be more complex to analyze. Some produce proprietary patented products and services, for which they can extract monopolistic or oligopolistic pricing until competitors develop even better products or services. Investors need to understand the market for each key product or service that a given oil service company offers. For each product or service, is the market growing or contracting, and at what rate? Is this company gaining market share with superior offerings in terms of quality or price, or is it losing share? Is pricing in each market rising due to high demand or underinvestment, or is it falling due to low demand or overinvestment? Are this company's costs increasing or decreasing, and what is the trend for competitors' costs? These all are questions that investors in oil service companies must address, and they are similar to the issues facing investors in any industrial concern.

Although financial theory would dictate that oil service companies be valued based on free cash flow or its accounting proxy, net earnings, many investors also focus on operating cash flow and earnings before interest, taxes, depreciation, and amortization (EBITDA). As with oil and gas producers, investors find reported net earnings difficult to analyze because of the broad cyclicality of the industry. Reported net earnings of oil service companies are sometimes unanalyzable for multiple quarters during lengthy industry downturns. Some investors use estimates of so-called normalized earnings, but others fear that these types of estimates might not be sufficiently grounded in reality. So, many investors fall back on operating cash flow and EBITDA because they are less volatile and analyzable quantities are reported quarterly. Investors analyze historical cycles in the oil service industry and observe the multiples of operating cash flow and EBITDA at cyclical peaks and troughs as well as at mid-cycle. They then compare the current multiples with the historical ones to value the companies. Larger, more diversified oil service companies operating in the less cyclical sectors of the industry typically trade at higher multiples, as do smaller companies operating in fast-growing niche sectors.

3. COAL DEVELOPMENT AND PRODUCTION

Although coal is broadly criticized as an energy source because it creates the most pollution, it remains a very important fuel, supplying approximately 25% of the world's energy needs. Investing in coal-producing operations raises many of the same issues as does investing in oil and gas operations, with a few key differences. Unlike oil and gas, for which there is active exploration, coal is already so abundant that few companies are searching aggressively for more reserves. Instead, coal-producing operations focus their efforts on reducing extraction and transportation costs.

Like oil and gas, coal is found with different characteristics in different deposits. The value of coal rises with its energy content and vicinity to markets, and it falls with the quantity of associated sulfur and other pollutants.

Most governments are trying to reduce their countries' use of coal because of the associated pollution. However, global energy needs may continue to force up the use of coal. At the same time, the power generation industry is trying to develop technologies that will allow cleaner burning of coal so that this low-cost abundant fuel may continue to be used.

With coal so abundant and yet so unpopular, investors typically do not expect any secular increase in real pricing anytime soon. The key to value creation for coal-producing operations is to reduce extraction and transportation costs and increase their production of low-cost, high-quality coal, crowding out of the market higher cost, lower quality competing coal.

Although there are several public companies whose principal business is coal production, many coal-producing operations belong to private companies and publicly traded conglomerates. There are a limited number of publicly traded pure plays on coal production.

4. CRUDE OIL REFINING AND MARKETING

The crude oil refining business is well known to investors for its difficulty in generating attractive returns. Although the business is cyclical, profit margins typically swing from poor to mediocre and then back to poor without ever getting good. Essentially, the refining industry suffers from chronic overinvestment. Ironically, few new refineries have been built in developed countries during recent decades. However, refining capacity has continued to grow as engineers upgrade and de-bottleneck existing plants and effectively increase their capacity. Increasing environmental requirements have forced companies to invest in their refining assets to reduce emissions, and this type of work often boosts capacity with limited incremental investment. In developing countries, refineries continue to be built despite weak returns because the plants create employment and allow crude oil-producing countries to export more finished product.

Although the ability to transport refined products (e.g., gasoline, jet and diesel fuels, heating oil, residual fuel oil) around the world inexpensively creates a global market for these products, marketing profitability is highly regionalized. In truly competitive markets, it is difficult to generate a competitive advantage in marketing refined products. However, for gasoline and diesel fuel, the geographic location of service stations can be a competitive advantage. Many buyers focus on convenience more than on price, so highly trafficked locations can be advantageous. For most other refined products, as well as for gasoline and diesel fuel, it takes a monopolistic or oligopolistic market for marketers to be able to generate above-average returns. A further problem facing gasoline and diesel fuel marketers has been the emergence of so-called hypermarkets that sell transportation fuels at very thin margins to attract motorists to the other retail outlets in their shopping centers. Although the hypermarkets have the potential to more than offset their thin margins in gasoline and diesel fuel marketing with higher profits from other stores, nondiversified marketers in the same market must suffer below-average profitability to remain price competitive.

Like coal-producing assets, refining and marketing operations typically are part of conglomerates, mostly part of so-called integrateds. As we noted previously, integrateds are usually very large companies with oil and gas exploration and production operations, oil refining and marketing assets, and (often) petrochemical operations. However, there are a small number of publicly traded companies that are pure plays on refining and marketing. Investors in refining and marketing assets usually try to take advantage of the cyclicality in the industry by buying assets when margins are depressed and trading out of them when margins recover. Some investors are expecting growing global demand for refined products to outstrip supply capacity increases, which may be constrained by more and more stringent environmental requirements, but so far this has not occurred.

5. NATURAL GAS PROCESSING, TRANSMISSION, AND DISTRIBUTION

5.1 Processing

Once natural gas has begun to flow from the wellhead, it must be treated and transported to end users. The degree to which it must be processed depends on both its natural gas liquids (NGLs) content and its impurities. Natural gas often contains NGLs, which comprise up to 15% of the volume. These include ethane, propane, butanes, and pentanes. Approximately 60% of NGLs produced go into the refinery and petrochemical markets, whereas the other 40% are used for space heating. Because

NGLs usually have approximately 65% of the energy content of crude oil, they normally trade at a composite 65% of crude prices. When they occasionally trade below that level, the lighter NGLs (ethanes and some propanes) are kept in the gas stream and sold for their heating value. The impurities include brine, silt, and water (so-called BS&W), and these are filtered out with the help of a separator. In several producing basins, gas is produced with high proportions of sulfur (hydrogen sulfide) and carbon dioxide. The hydrogen sulfide and carbon dioxide are removed before the gas enters the pipelines that take it to end users.

Until the energy industry consolidation of the 1990s, there were public companies for which natural gas processing was a dominant business. However, today nearly all natural gas processing operations are part of much larger energy conglomerates, so there are no more pure plays on natural gas processing. That said, the key profit driver of natural gas processing operations is the spread (or gross margin) between composite NGL prices and the "wet" gas feedstock, which is known as the fractionation (or "frac") spread. The typical NGL barrel contains about 37% ethanes, 33% butanes, 10% normal butanes, and 13% pentanes (or natural gasoline). The frac spread in the United States has averaged approximately $6/barrel (or $0.14/gallon) for the past decade or so, but it has been volatile due to the volatility of both crude oil and natural gas prices.

Another determinant of processing profitability is the type of gas purchase contract used by the NGL processors. There are three types of contracts: keep whole, percentage of proceeds (POPS), and fee based. Keep-whole contracts were the norm historically but have become obsolete due to the increased volatility of gas prices. With this type of contract, the NGL processor buys so-called wet gas at the plant gate, strips out the NGLs, and resells them and the now "dry" gas (just methane) back into the market. The processor takes all of the price risk. Today, approximately 35% of NGL volumes are still processed under keep-whole contracts. In contrast, processors with POPS contracts agree to pay their wet gas suppliers a fixed percentage of their proceeds from reselling the NGLs stripped out and the residue dry gas, so their margins are less volatile. POPS contracts cover approximately 40% of the NGL market. The remaining 25% of NGL volumes are processed under fee-based contracts. With this type of contract, the NGL processor simply strips out the NGLs for a fee. These contracts are most popular with Rocky Mountain processors because gas price volatility is so high in this region of North America.

5.2 Transmission

Natural gas transmission refers to the transportation of natural gas from a producing area to a consuming area, whereas distribution indicates the transportation of natural gas from so-called city gates on the transmission lines to the very locations where it is burned (e.g., residences, office buildings, factories, power plants). For distances of up to 4000 to 5000 miles across land or shallow water, natural gas is typically transported via pipelines. For longer hauls, and especially across deep bodies of water, natural gas is usually liquefied at low temperatures (–260° Fahrenheit) and transported to markets in seafaring liquefied natural gas (LNG) tankers. The methane is then warmed up and re-gasified at LNG terminals near end-user markets. Historically, the high capital and operating costs associated with the construction and operation of natural gas liquefaction facilities and LNG tankers have made LNG most attractive to energy-importing countries more interested in diversification and security of energy supply than in lowest-cost energy. Liquefaction facilities cost more than $1 billion, LNG tankers cost approximately $160 million, and re-gasification plants cost approximately $300 million. Japan, South Korea, and Taiwan, which rely on imports to meet most of their energy needs, have chosen to purchase LNG from Southeast Asia to reduce their reliance on oil from the Middle East.

Because natural gas pipelines are costly to build, it is often most economical for there to be only one or two lines from a producing region to a given consuming one. This market structure would give the pipeline owners a monopoly, so governments typically intervene to protect consumers from monopoly pricing of transmission services. Governments set the prices to be charged so that the owners of a pipeline can earn a "fair" return on their investment. In the United States, these returns are normally calculated as a 12% return on stockholders' equity, assuming a 40 to 60% debt-to-equity capital structure. Periodically, U.S. pipeline owners have the right to file for new tariffs via a so-called general rate case. However, as many as 15 of the 30 transmission companies in the United States have not filed rate cases during the past decade or so. Because of the way in which rates are designed and the lack of any rate case requirement, U.S. interstate pipelines typically offer among the steadiest streams of earnings in the energy industry.

Until April 8, 1992, U.S. natural gas pipeline companies were gas merchants, buying the commodity from producers and reselling it to local distribution companies (LDCs), power plants, and industrial users. The Federal Energy Regulatory Commission's Order 636 reregulated the pipelines as so-called common carriers. Pipelines now sell space to owners of natural gas that want it transported. Owners may include marketing arms of diversified energy companies as well as natural gas producers or consumers. Pipeline customers can choose so-called firm transportation, which guarantees them space in a pipeline over a set time period (often 5–10 years), or they can buy often-discounted "interruptible transportation," which provides them with transportation services on a space-available basis. Typically, industrial customers and power generators that can switch between natural gas and residual fuel oil use interruptible transportation to minimize costs. In contrast, LDCs that have commitments to provide gas to heat residences in winter must use more expensive firm transportation.

Natural gas pipelines are owned mostly by publicly traded so-called diversified energy companies, which may also own unregulated oil and gas exploration and production assets, gas processing plants, and natural gas trading and marketing operations as well as other regulated utilities such as local gas distribution companies and even power generation, transmission, and distribution assets. The profits of regulated natural gas pipeline companies are driven by their ability to find new capital projects that regulators will approve and to maintain as wide a spread as possible between their regulated revenues and their costs. Earnings are also affected by regulators' changing definitions of a "fair" return on assets. Regulators' assessments of fair returns can depend on interest rates and perceived equity risk premiums. Natural gas pipeline companies in North America typically have limited expansion opportunities, generate significant free cash flow, and pay common dividends above stock market averages.

A corollary business to natural gas transmission is gas storage. In the colder regions of the world, a large portion of natural gas consumption is for space heating, which creates wide seasonal fluctuations in demand. In these countries, gas storage is critical to meeting peak seasonal demand requirements without having to build up excess productive capacity that would remain idle much of the year. Because gas is expensive to transport over large distances, unlike crude oil, it is not economical to move gas hemisphere to hemisphere to satisfy seasonal fluctuations in consumption. Technology has been developed that allows natural gas to be reinjected in emptied oil and gas reservoirs. Gas storage caverns are also owned mostly by diversified energy companies, which rent space to electric and gas utilities, wholesale merchants, and other commodity players.

5.3 Distribution

LDCs in North America are both natural gas merchants and common carriers, but some gas marketing companies would like to see all of them become common carriers so that they could sell gas directly to residential and commercial consumers. But so far, most so-called retail unbundling programs have not been successful. Supporters of current regulation question the benefit to consumers of being allowed to choose from which company to buy a commodity such as natural gas, especially given that in many instances LDCs do not charge a premium over their cost of gas and just pass it through. Whether local distribution companies remain merchants or become common carriers like transmission companies, their monopolies on gas distribution in a given region will almost certainly continue to be regulated to protect consumers from monopoly pricing.

Like transmission companies, LDCs charge prices that are set by regulators to give the companies a "fair" return on their investment. Profitability is driven by factors similar to those that affect the earnings of regulated transmission companies. LDCs are typically either publicly traded or owned by publicly traded diversified energy companies. LDCs usually have limited expansion opportunities, generate significant free cash flow, and pay common dividends above stock market averages.

6. ELECTRICITY GENERATION, TRANSMISSION, AND DISTRIBUTION

A lot of the elements reviewed in natural gas transmission and distribution are applicable to the electricity market. Historically, electricity generation, transmission, and distribution assets have been owned by regulated integrated electric utility companies. Like natural gas transportation and distribution companies, electric utilities have held monopolies in their markets and had their prices set by government regulators to protect consumers from monopoly pricing and to offer the utilities a "fair" return on investment. However, during recent years, there has been significant political pressure to deregulate (or "unbundle") the electricity generation and transmission industry. The movement started in Chile and the United Kingdom, where each part of the business (distribution, transmission, and generation) was separated. As with the natural gas industry, transmission and distribution remain a natural monopoly and, thus, will continue to be regulated. Generation is clearly the best candidate for competitive forces.

The genesis of competition in North America and the United Kingdom stems from the development of an unregulated generation market. Early on, so-called independent power producers (IPPs) and, in some cases, so-called cogenerators (which produce steam/heat and electricity) introduced competitive power and superior technology in markets where inefficiencies and old technology dominated. Although electricity is generated by hydropower, nuclear power, and the burning of oil, natural gas, and coal, most of the innovation in generation has been associated with natural gas, which is appealing because it is environmentally friendly and plant construction costs are relatively low. The emergence of highly efficient, so-called combined-cycle gas turbines (CCGTs) has been an important driver.

There are fundamentally three different turbine technologies in the market today. The oldest is the so-called steam turbine system, in which a plant generates power by boiling water using a fuel (e.g., coal, oil, gas, nuclear fission, waste) to produce steam. The pressure from the steam turns a turbine, which in turn creates electricity. The steam is recycled back into water in the boiler, creating a loop. The second turbine technology is actually based on a technology as old as civilization in that it involves wind, hydropower, and other generation technologies that use a physical force to turn a turbine as a substitute for burning fuel. The third is the gas turbine technology, which uses the equivalent of a big jet engine to produce power; the turbine is directly injected with natural gas, which burns and turns the turbine in the process. This is also called a "simple cycle" gas turbine. A CCGT plant combines the first and third technologies. Natural gas is fed into a gas turbine, which produces both electricity and heat. The heat is then applied to a boiler, which produces steam to turn a second turbine, thereby creating two outputs or cycles.

With the emergence of IPPs and cogenerators and their new, cleaner CCGT plants, many power markets seemed ready for competition during the 1990s. However, so far it has turned out that the level of capacity use in a market has been a key determinant of the success or failure of deregulation. One of the main reasons why deregulation in the United Kingdom has worked so well is that there is substantial overcapacity in its generation market. Where supply exceeds demand by a comfortable margin, price is driven by marginal high-cost production. However, in underserved markets, price can temporarily far exceed marginal high cost as scarce power is rationed by price to the end users most desperate for it. In natural gas markets, storage and competing residual fuel oil serve as natural ceilings for prices, but electricity is difficult to store and substitute. As a result, unregulated power prices can be very volatile. In fact, over the past few years, North America has experienced periods of electricity shortage when prices, normally about $30 per megawatt-hour (MWh), traded higher than $7000/MWh in some regions.

Beyond IPPs and cogenerators, deregulation has created another new entrant on the power scene: the unregulated trading and marketing company, also known as a merchant company. Merchant companies, which have appeared mostly in North America, purchase power from generators and resell it to distribution companies and industrial customers. Merchant players try to profit from arbitrage opportunities between regions, closing out these inefficiencies. When a merchant's parent company owns power plants as well, the parent may even temporarily shut down a plant if it can make more money by selling the fuel originally committed to the plant to another user for a higher price through its merchant arm.

Although the goal of deregulation has been to increase efficiency and reduce costs to consumers, it recently resulted in apparent market manipulation in California, and there is now political pressure to reregulate electricity generation in some states of the United States. Power marketing companies continue to argue that consumers would benefit if they were permitted to choose from which company they would like to purchase electricity. The U.S. Federal Energy Regulatory Commission has proposed a so-called standard market design, which would open up the power transmission grid to the wholesale marketplace. Currently, only about 30% of the power generation market is unbundled. All power lines would become common carriers, from which marketing companies and others could purchase space on transmission lines.

Many traditional electric utilities and new independent power producers are publicly traded or are owned by publicly traded companies. Like natural gas transportation companies, electric utilities typically have limited expansion opportunities, generate significant free cash flow, and pay common dividends above stock market averages. During the 1990s, looming shortages of power generation capacity in North America made growth companies of some independent power producers and merchant participants with plans to build new power plants. However, the slower economic growth of recent years has resulted in many of these projects being cancelled or deferred.

SEE ALSO THE FOLLOWING ARTICLES

Electric Power Generation: Fossil Fuel • *Energy Futures and Options* • *Hazardous Waste from Fossil Fuels* • *Natural Gas Resources, Global Distribution of* • *Oil and Natural Gas Liquids: Global Magnitude and Distribution* • *Petroleum Property Valuation* • *Resource Curse and Investment in Energy Industries* • *Trade in Energy and Energy Services*

Further Reading

Duarte, J. (2002). "Successful Energy Sector Investing: Every Investor's Complete Guide." Prima Publishing, Roseville, CA.

Jansen, R. A. (1998). "Profits from Natural Resources: How to Make Big Money Investing in Metals, Food, and Energy." John Wiley, New York.

McCormack, J., Smith, A. L., and Fan, A. C. (2001). One way to close the oil industry's credibility gap: Hedge PDP production and buy back stock. *EVAluation* **3**(6).

McCormack, J.L.,Vytheeswaran, J. (1998). How to use EVA in the oil and gas industry. *J. Appl. Corp. Finance* **11**(3).

Labels and Standards for Energy

LLOYD HARRINGTON
Energy Efficient Strategies
Warragul, Australia

PAUL WAIDE
PW Consulting
Manchester, United Kingdom

1. Why Labels and Standards?
2. Energy: The Invisible Attribute
3. When to Label and When to Use MEPS
4. Types of Labels
5. Regrading of Labels
6. Setting MEPS Levels
7. Review and Update of MEPS Levels
8. Efficiency Targets
9. Related Program Measures
10. Enforcement and Implementation Issues
11. Test Procedures
12. Tested Energy Consumption versus In-Use Energy
13. Performance versus Energy Consumption
14. Evaluation
15. Impacts of Efficiency Standards and Labeling Programs

Glossary

categorical labels Where products are allocated into predetermined energy efficiency categories or classes, usually defined as a function of capacity and energy consumption; categories remain fixed, irrespective of the range of products on the market at any particular time.

check test A part or full test to verify the energy and performance claims or compliance claims, minimum energy performance standards (MEPS), or other performance requirements.

comparative labels Allow consumers to form a comparative judgment about the energy efficiency (or energy consumption) and relative ranking of all products that carry a label; comparative labels usually carry information such as size and/or capacity, energy consumption, and an energy efficiency rating scale.

continuous scale labels Where the energy consumption of the best and worst products on the market are shown, together with the relative energy consumption of the product that is labeled, allowing the consumer to assess the position of the product under consideration in comparison with the best and worst products; continuous scale labels can effectively compare products only of a similar type and size.

endorsement labels Indicate that products belongs to the "most energy efficient" class of products or that they meet a predetermined standard or eligibility criteria; these labels are essentially "seals of approval" that the products meet these requirements.

minimum energy performance standards (MEPS) A policy tool used to force low-efficiency products out of the market; MEPS either ban the sale of products that fail to meet specified minimum energy efficiency levels or restrict the availability of low-efficiency products.

Energy labeling is a policy tool that informs consumers about the energy performance of appliances and thereby encourages them to purchase appliances that provide the services they need with less energy consumption. In providing information to consumers about equipment energy consumption and operating costs that would otherwise be invisible or unavailable, energy labeling enables consumers to make more balanced and rational purchasing decisions. Energy labels can also help consumers to identify the most efficient products on the market. In effect, energy labeling attempts to provide a market "pull" for more energy-efficient products while simultaneously presenting information that might discourage the purchase of less efficient products. Energy labeling programs can be voluntary or mandatory and can vary considerably in style and type. Minimum energy performance standards (MEPS), also called minimum energy efficiency standards, are a policy tool used to force low-efficiency products out of the market. MEPS either ban the sale of products that fail to meet specified minimum energy efficiency levels or restrict the availability of low-efficiency products. MEPS are used to "push" the energy efficiency of the market. To be effective, most MEPS programs need to be mandatory, although there are examples of voluntary standards that have been effective, usually in the form of efficiency targets.

1. WHY LABELS AND STANDARDS?

Energy consumed by equipment and appliances in the industrial, commercial, and residential sectors of the developed world is a major source of global greenhouse gas emissions as well as of other pollutants such as NO_x and SO_x. Energy efficiency programs such as energy labeling and minimum energy performance standards (MEPS), where legislation and regulation are used to improve product energy efficiency, are among the most cost-effective and widely used measures employed to reduce these emissions (Fig. 1). Improved energy efficiency also has many economic benefits. Because of this, labels and MEPS are now a couple of the most common energy policy tools in use today. Nearly 60 countries (with the expansion of the European Union) now have labeling and/or MEPS programs in operation, and together these cover roughly 40 product types. Most Organization for Economic Cooperation and Development (OECD) countries now have labeling and MEPS programs in force for a range of products.

2. ENERGY: THE INVISIBLE ATTRIBUTE

Many people argue that consumers are not interested in energy per se but rather are interested in the services that energy can provide such as cold beer, hot showers, conditioned air, lighting, and operation of machinery and computers. Although this is undoubtedly true (what can one do with a raw kilowatt-hour or megajoule?), energy consumption is the key determinant of operating cost of most appliances and equipment and, therefore, is of significant concern to consumers. However, in places where energy prices are low (or where users are not charged directly for energy), energy consumption and efficiency are of less interest.

Some product attributes are readily observable on inspection by consumers or are readily available in

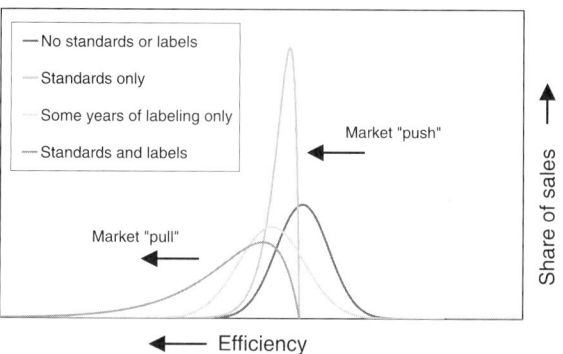

FIGURE 1 Conceptual impact of energy labeling and MEPS.

product literature, whereas other attributes are not at all obvious. For example, in the case of a refrigerator, the volume available for the storage of food is clearly visible to consumers when the unit is inspected in a showroom. Product information also gives storage volumes and external dimensions (this information is essential for installation). Conversely, inspection of an appliance such as a refrigerator does not reveal its energy consumption, energy efficiency, or standard of performance. In fact, the determination of these performance attributes requires careful testing and the use of sophisticated equipment, which is not readily accessible to consumers.

Information on nonobservable attributes can be obtained by consumers only where the manufacturer chooses to provide such information or where there is a requirement for the information to be disclosed such as through an energy labeling program. Where such nonobservable attributes are declared (and hence become visible to consumers), and where even a small market segment responds to these attributes (so-called informed consumers), manufacturers tend to concentrate on improving the rating and performance of those observable and visible attributes at the expense of other nonobservable attributes.

3. WHEN TO LABEL AND WHEN TO USE MEPS

It is important to remember that labeling and MEPS programs aim to influence the selection of products by consumers at the point of sale by making higher efficiency units more attractive (through labeling) or by making less efficient ones unavailable (through MEPS). Labeling or MEPS cannot be expected to have any significant ongoing influence on consumers' use of products once they have been purchased and installed.

Many of the necessary prerequisites for a successful labeling or MEPS program are common to both; that is, these programs are most applicable

- to products that use a significant amount of energy;
- when there is a large market (in terms of sales volume) for the product or where rapid growth is forecast;
- when there is already a range of energy efficiencies on that market or where significant improvements in energy efficiency are technically possible and economically feasible.

The main point to consider when selecting either energy labeling or MEPS is the purchasing process and payment of energy bills. Energy labeling is likely to be most effective where consumers inspect products in a retail outlet, select the products, and are also responsible for the payment of the resulting energy bills. This would typically apply to the purchase of major household appliances. Energy labeling is likely to be less effective in cases where

- a third party (e.g., a contractor/tradesperson) is responsible for the selection of the product (e.g., a water heater may be selected and installed by a plumber);
- products are not generally on retail display to allow direct comparisons between products (e.g., products are generally selected from catalogs or lists rather than on display in a showroom);
- the purchaser of the product is not responsible for the payment of the energy bills (e.g., a landlord purchasing an appliance for a tenant, capital purchases being separated from operational budgets such as energy bills in a large company or department)—the so-called "split incentives" market failure.

In these cases, purchasers are likely to be "label resistant" in that they either do not see the label or have no economic interest in selecting a more efficient product at the time of purchase to reduce operating costs. Therefore, MEPS is often a more effective policy option in these cases; removal of the least efficient models means that purchasers have no choice but to acquire a more efficient unit.

MEPS is invisible to purchasers, whereas labeling attempts to engage consumers to consider energy efficiency and use this as a selection criterion in their purchasing decisions. Labeling is an attempt to improve the operation of the market by bringing to the attention of consumers the operating cost at the time of purchase. From this perspective, even those who oppose regulatory energy policy intervention are usually in favor of energy labeling because this is seen as an essential part of smooth rational market operation. MEPS, on the other hand, is an attempt to overcome market failures such as split incentives.

An energy label works

- by providing consumers with data on which to base informed choices (to select the most efficient and suitable product available);
- by encouraging manufacturers to improve the energy performance of their models;
- by encouraging distributors and retailers to stock and display efficient products (i.e., those with good energy ratings).

In many ways, the most influential of these points is likely to be the second. If a manufacturer withdraws an inefficient product from the market and replaces it with a more efficient one, consumers are no longer able to purchase the less efficient product. For MEPS programs in isolation, the first point is not relevant, whereas manufacturers and retailers are forced (rather than encouraged) to stock more efficient products.

Often, products are most suited to either labeling or MEPS. However, some products, particularly those that are ubiquitous and that use substantial amounts of energy, may justify simultaneous use of both energy labeling and MEPS. A common product that is subject to both energy labeling and MEPS is household refrigerators. Energy labeling may provide an effective pull at the more efficient end of the market, whereas MEPS can redress those lower efficiency elements of the market that are "label resistant." Another strategy that is often employed is to introduce energy labeling initially and to assess the market impact over a period of, say, 5 years. If the rate of progress in terms of energy efficiency is judged to be too slow, MEPS may be introduced in parallel to accelerate progress. The advantage of this approach is that labeling can be used to gather a good deal of data in terms of products on the market and their energy efficiency, making subsequent setting of MEPS levels and implementation more straightforward.

4. TYPES OF LABELS

Energy labels fall broadly into two main types: comparison labels and endorsement labels.

4.1 Comparison Labels

Comparison labels are where key information on the product energy consumption and/or performance compared with that of products of an equivalent type is shown for comparative purposes. Additional measures of energy efficiency (e.g., a star, numerical, or other efficiency rating) may also be shown. This type of system works best when it is mandatory for all products to carry a label (so that poor performers can be more readily identified and avoided by consumers). Experience has shown that where labeling is not mandatory or where mandatory provisions are not enforced, energy labels on appliances with lower ratings are actively removed by retailers to improve their chances of selling these products (retailers and manufacturers actively dislike having products with poor energy ratings on display). Comparison labels tend to be sponsored by governments, although there are exceptions.

4.2 Endorsement Labels

Endorsement labels indicate that the products carrying the labels belong to the "most energy-efficient" class of products or meet a predetermined standard or eligibility criteria. These labels generally consist of a logo or mark that indicates the products have met the standard and generally contains little or no comparative energy efficiency information (although this may be available through lists of endorsed products). These labels merely inform consumers that the products meet the required standard. Criteria for some schemes are updated on a regular basis; therefore, these labels may carry the year of qualification. Endorsement labeling programs are mostly of a voluntary nature. An endorsement label may be specifically for energy efficiency or may be a broader "eco" label. Eco label programs endorse products that have low impact across a wide range of environmental factors; even though energy consumption levels often have a high priority within the rating system. It is a system that works best when only a limited proportion of the market is eligible for endorsement, typically the top 10 to 40% of performers. Endorsement labels can be sponsored by governments, but sponsorship by utilities, industry, and environmental groups is also very common.

Endorsement and comparative labels can coexist and do so in many countries and for many products. They tend to be complementary rather than competitive.

The most commonly used comparison labels employ a scale with absolutely defined energy efficiency categories (so-called categorical labels). These labels allow consumers to easily assess the efficiency of a product in relation to an absolute scale by means of a simple numerical or ranking system. The concept is that it is much easier for consumers to remember and compare a simple efficiency ranking scale (e.g., 1, 2, 3; 1 star, 2 star, 3 star; A, B, C) than to remember and compare energy consumption values.

The other type of comparison label design is continuous scale labels, where the best and worst energy consumption of products on the market is shown together with the relative energy consumption of the products that are labeled, allowing consumers to assess the position of the products under consideration in comparison with the best and worst products available.

In general, the visual designs of comparison labels in use around the world (Fig. 2) can be grouped into three basic types:

- *Dial label.* This type of label has a "dial" or gauge, with greater efficiency linked to advancement along the gauge (more efficient represented by a clockwise arc). This type of label is used in Australia, Thailand, and Korea. The number of stars or the "grading" numeral on the scale depends on the highest preset threshold for energy performance that the model is able to meet.
- *Bar label.* This type of label uses a bar chart with a grading from best to worst. All grade bars are visible on every label, with a marker next to the appropriate bar indicating the grade of the model. This approach is used in Europe, Russia, and Mexico as well as in some South American, African, and Middle Eastern countries.
- *Linear label.* This label has a continuous linear scale indicating the highest and lowest energy uses of models on the market, locating the specific model within that scale. Because energy (rather than efficiency) is used as the comparison indicator, it is necessary to group models into similar size categories for comparison. This model is used in the United States and Canada.

There are also many other comparative energy labels that have no graphic concept to support the indication of energy efficiency. These generally rely on text to explain the efficiency or some numeric indicator of efficiency (e.g., EER for air conditioners). Some labels show only energy consumption.

The majority of endorsement labeling schemes use a wide range of criteria to determine eligibility. Although energy often plays a prominent role in

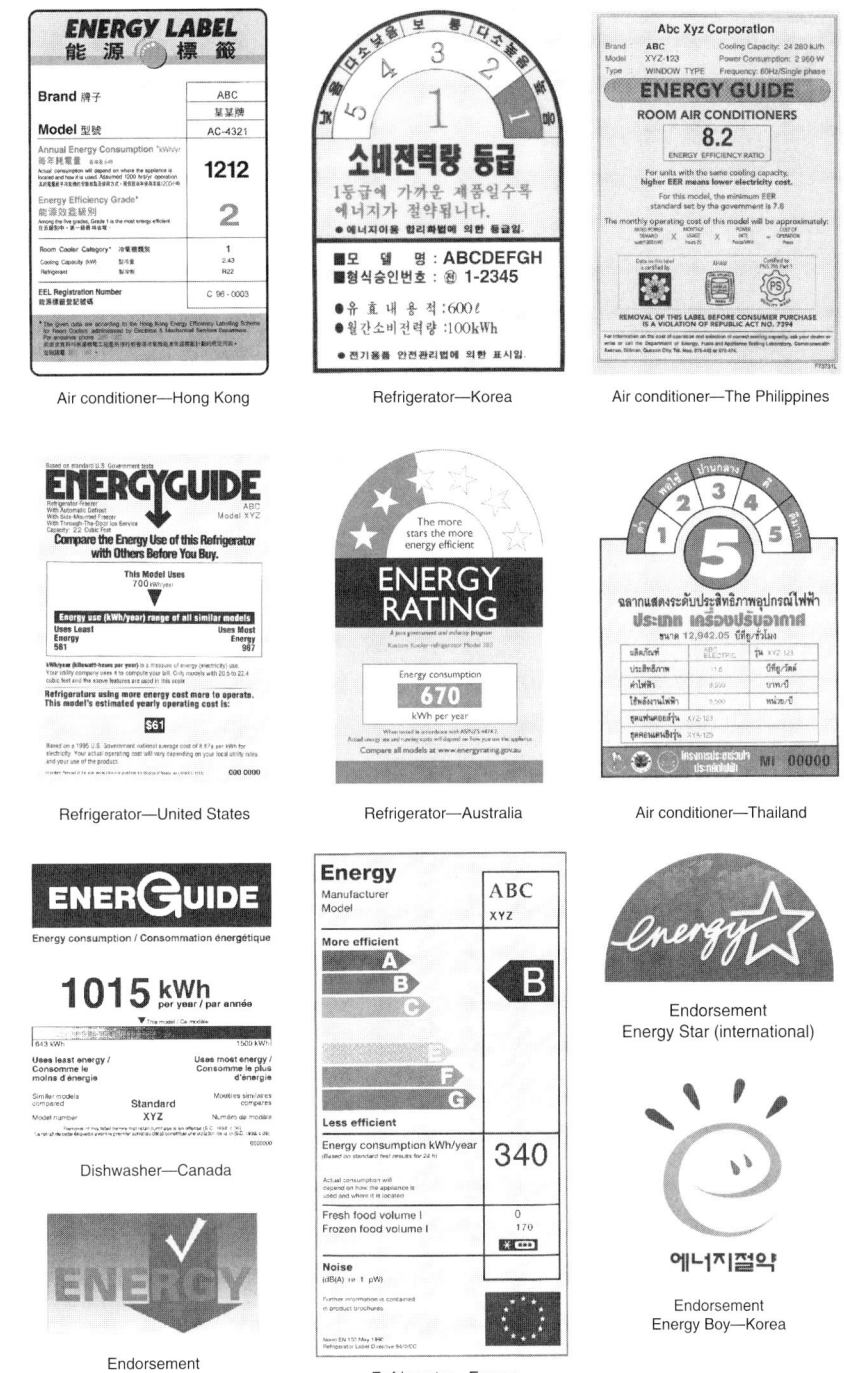

FIGURE 2 Examples of energy labels from around the world. Labels are presented for illustrative purposes only, and are not to scale or the correct color.

many of these programs, it is by no means the predominant criteria in most schemes. However, there are a few endorsement labeling schemes in place that focus exclusively on energy and energy efficiency. These are the U.S. Environmental Protection Agency (EPA) Energy Star label (used as an international endorsement label for information technology equipment and consumer electronics, with the emphasis on standby power consumption for these products, and domestically in North America for a wider range of products), the Korean Energy Boy label (essentially covering international products and criteria covered by Energy Star), and the European Group for Efficient Appliances label for consumer electronics (emphasizing standby power consumption for these products).

Labeling programs are not necessarily restricted within country borders. For example, the European Union (EU) label not only extends to EU member states but also has been adopted by several other European and non-European countries. The U.S. Energy Star program for office equipment has been adopted in many countries around the world. In the South and Latin American region, attempts have been under way to design and implement a unified energy labeling program.

The purpose of an energy label is to convey key information to consumers to assist them in a purchase decision. Therefore, it stands to reason that there are likely to be key differences in the type of information that it is necessary to convey to consumers, depending on their socioeconomic, physical, and cultural contexts. Differences in language and, to a lesser extent, literacy are also key factors that will influence the design and presentation of an energy label within a particular region and culture. If an energy label is to be effective, it has to be understood, liked, and used by consumers.

5. REGRADING OF LABELS

If an energy label is doing its task, the efficiency of products available on the market will improve gradually over time. This means that adjustments will need to be made to energy labels to keep them relevant.

In the case of continuous scale labels, where the best and worst products currently on the market are shown, it is necessary to update the end points of the labels on a regular basis, typically annually. Although this keeps the labels relevant, it creates administrative difficulties in that new end points need to be regularly defined, notified, and distributed. Sometimes, labels with older and newer end points can end up in the same store and there is nothing to identify these differences. This system creates problems if a new product that lies outside the current range comes onto the market between updates (this is difficult to convey to consumers in a nonambiguous fashion). The other issue is that continually changing scale end points acts as a disincentive for manufacturers to improve the efficiency of their products in that they have no idea where their products will lie relative to end points that are continuously changing. There is also evidence that continuous scale labels are more difficult for consumers to use in an effective manner.

For categorical labels, the efficiency category definitions usually remain fixed for a considerable time, typically 5 but up to 10 years (or longer if the energy efficiency progress in the market is slow). From manufacturers' perspective, this tends to encourage the development of more efficient products because they know during the development phase how their products are likely to rate on the categorical efficiency scale for some time to come and will make some effort to reach the next highest rating category if this can be done in a cost-effective manner. As products improve in their efficiency over time, there tends to be efficiency category creep; that is, the proportion of high-efficiency ratings increases and that of low-efficiency ratings decreases (Fig. 1). The disadvantage of the categorical system is that when a rating system has to be changed or regraded (and it will have to be revised eventually), there is likely to be some disruption to the market during the transition. Model efficiency will appear to fall as categories are regraded, and the appearance of labels with two different rating scales in retailers at the same time both need to be handled carefully to avoid consumer confusion. The other element of difficulty is that manufacturers tend to be strongly resistant to a downgrading of the current efficiency ratings for their products (they dislike supplying products that have very low efficiency ratings; however, there is no point in having a categorical label scale where the bottom two or three scales are empty, with all models bunched into the higher efficiency categories). So, an orderly and well-planned transition is essential.

6. SETTING MEPS LEVELS

MEPS can be set at a range of levels, from gentle to aggressive. At the gentle end of the scale, MEPS levels would eliminate only a modest part of the current market, would eliminate only the least efficient models currently offered for sale, or would even set as a backstop that does not eliminate any current models but would stop any lower efficiency models from entering the market. At the most aggressive level, MEPS can be set at levels that eliminate all current models from the market and/or mandate the adoption of new technology and designs that are at the current leading edge to meet the prescribed efficiency levels. Of course, there is a continuum between these levels, and the precise level selected will depend on a range of factors such as capability of local manufacturers (and whether there are any), manufacturing costs, local energy prices, and incomes. For modest MEPS levels, the implementation period can be relatively short (1 year or less), but for more stringent levels, the implementation period tends to be longer (up to 3 years in more extreme cases). The balance of local manufacture and imports for a particular product may also influence the levels and timing of MEPS.

There are two traditional methods for setting MEPS levels: the so-called statistical approach and the engineering approach. In the statistical approach, efficiency levels are set so that the market impact is manageable in terms of the number of models eliminated (typically 20–50% of models would be eliminated from the market, with at least some models left in all major categories and types of products). In this approach, the MEPS levels are defined in terms of the range and characteristics of products currently on the market. This approach is practical only where there is some significant variation in the efficiency of products already on the market.

The engineering approach is where a technical analysis of product efficiency is used to assess a range of MEPS levels. Typically, a range of measures and their incremental cost are assessed against a base case market product. The cost-effectiveness of each additional measure is assessed to determine the optimum efficiency level. Other factors that are typically considered when setting MEPS using an engineering approach are the capability and financial position of local product manufacturers, energy prices, and likely impact on retail product prices after MEPS.

An alternative hybrid approach to setting MEPS levels is to review the levels already in use by a country's major trading partners and to adopt one of these as a local MEPS level. This approach has the merit of avoiding questions regarding the technical feasibility of the proposed MEPS level and also ensures that suitable products would be available in the market after the implementation of MEPS. Harmonization of MEPS levels, although of secondary importance, would also reduce restrictions to free trade of products. This approach is feasible only where test procedures and product types in each of the relevant market are similar or related.

MEPS as a policy option has an advantage over energy labeling in that faster, larger, and more certain energy savings can usually be achieved through implementation of aggressive MEPS levels. Although this may be justified in terms of cost-effectiveness, implementation of aggressive MEPS levels can create problems and difficulties among manufacturers, especially where they are forced to invest large amounts of capital to meet MEPS requirements. In this regard, enforcement of MEPS, especially where the levels are aggressive, is critical to protect the investments of those manufacturers that comply with the requirements.

7. REVIEW AND UPDATE OF MEPS LEVELS

As with energy labels, improvement of efficiency levels within the market may warrant the reassessment of MEPS levels from time to time. The approaches used to review MEPS levels are the same as setting initial levels and depend on the market structure, mix of products, and level of stringency proposed. Where overseas levels are adopted locally, a periodical review of changes in overseas MEPS levels may be warranted.

Strategically, there is merit in setting initial MEPS levels at a moderate level (even with a short implementation time frame) to allow the regulatory system to be initiated and proved in a framework where the stakes for the main players are not too severe. Once the system operation is settled and manufacturers are accustomed to the concept and processes involved, the scene is set for implementation of more stringent or aggressive MEPS levels.

One issue to bear in mind is that when more stringent MEPS levels are implemented for products that also have energy labels, it may be necessary to regrade the energy label categories to take account of the new MEPS levels because all low-efficiency categories may be eliminated from the market. In this respect, close coordination between MEPS revisions and energy labeling revisions need to be maintained. In addition, the implementation of very aggressive MEPS levels may reduce or even eliminate energy efficiency differences among products, obviating the benefits of energy labeling, at least in the short term. The viability of energy labeling in conjunction with very aggressive MEPS levels (especially where these are upgraded regularly) would need to be reviewed from time to time.

8. EFFICIENCY TARGETS

As an alternative to regulated mandatory MEPS programs, it is possible to have program variations that achieve comparable outcomes yet are implemented on a voluntary basis. These usually take the form of voluntary agreements between governments and industry to achieve prescribed efficiency targets. Efficiency targets may be sales-weighted averages within each manufacturer's sales or product range or may be the voluntary elimination of products that do not meet prescribed minimum efficiency levels. These arrangements work best where strong industry associations (whose members dominate the supply of products to the market) are a party to the arrangements. Examples of these types of arrangements can be found in Europe for products such as televisions, videocassette recorders, clothes washers, dishwashers, water heaters, and external power supplies. A related approach is used in Japan (the "Top Runner" program), where the government sets voluntary efficiency targets that are based on the most efficient products on the market today. Manufacturers are asked to ensure that their sales-weighted average efficiencies meet these targets by the prescribed date (typically 5 years). Although there is no regulatory framework to enforce the voluntary targets, manufacturers clearly understand that those who fail to meet these will be publicly named and humiliated by the government, making the system quasi-mandatory in effect.

9. RELATED PROGRAM MEASURES

As an adjunct to energy labeling and MEPS, there are a number of related measures that can complement these programs. These include retailer support, integrated promotion, efficiency awards, and lists of current models on the market.

9.1 Retailer Support

Retailers play a key role in the effectiveness (or lack thereof) of an energy labeling program. A hostile retailer can neutralize the impact of an energy label by discrediting the information or the label's validity or message. Retailer education and supply of point-of-sale information can help to build the effectiveness of the program.

9.2 Integrated Promotion

Where a number of products are included in a regulatory program, there is an opportunity to market the program in a coordinated fashion and to deliver to the marketplace complementary messages that reinforce the importance of energy efficiency. This can extend to other products, such as homes, cars, and windows, that have a similar rating system.

9.3 Efficiency Awards

An awards scheme for the most efficient products will encourage manufacturers to seek these awards and will highlight the most efficient products to consumers. A range of options are possible, including annual or ongoing awards for products that meet specified criteria and/or the development of an endorsement label that is used in conjunction with a comparison label. Lists of the most efficient products (e.g., those eligible for awards), either at the point of sale or on the Internet, will make these products more attractive and accessible.

9.4 Lists of Current Models on the Market

This is a complementary tool that is primarily for products that carry energy labels but that can also be useful for products covered by MEPS or efficiency targets. A complete lists of products can be made available in the form of brochures or, more commonly now, through product lists on the Internet. These are usually sorted so that the most efficient products are at the top to encourage consumers to consider these products first. Of course, interactive product lists on the Internet can be sorted by whatever criteria are desired by users.

10. ENFORCEMENT AND IMPLEMENTATION ISSUES

A range of issues and options need to be considered when implementing an energy labeling or MEPS program. Most of these relate to regulatory options.

10.1 Registration of Products

Some programs make registration of all regulated products mandatory, whereas others require manufacturers to hold details only of the products that are submitted to market. Registration certainly makes compliance and enforcement more straightforward, but this involves some organization and additional work for all parties, especially where there are a large number of products on the market. It is very difficult to keep track of products on the market without some sort of registration system.

10.2 Supply of Test Reports

Manufacturers may be required to supply test reports with their registrations to confirm energy labeling or MEPS claims. Alternatively, they may be required to retain such reports for a specified period.

10.3 Self-Testing

In some cases, manufacturers are allowed to test their own products in their own laboratories, whereas other programs require independent third-party testing.

10.4 Accreditation

Tests may or may not have to be undertaken in an accredited or approved laboratory.

10.5 Certification

Manufacturers may be allowed to certify that their own products conform to the requirements/claims made. Alternatively, they may be required to use an independent third-party certification system.

10.6 Check Tests

It is critical that claims of energy efficiency be verified through a program of targeted compliance checks. These might not be random; in fact, they are most cost-effective if all available market intelligence is used to identify products that are most likely to not comply (e.g., excessive or unlikely claims of efficiency, brands with poorer compliance records, third-party referrals from competitors, products with high model turnovers). This requires a system of credible response to third-party claims of noncompliance and expedient enforcement of cases where noncompliance is proved.

10.7 Correct Display of Labels

An important aspect of a mandatory labeling program is checking retailers to ensure that labels are correctly displayed on appliances at the point of sale. An expedient way of dealing with minor infringements (e.g., spot fines) may be the most effective way in which to ensure high levels of compliance.

11. TEST PROCEDURES

Test procedures (often also called "test standards") are the method of testing used to determine appliance performance, energy consumption, and hence energy efficiency. They are critical in that they allow the comparison of products on a fair basis.

The test procedures used to determine the energy consumption (and, where relevant, the performance) of an appliance or a product can also have a large influence on the measured energy consumption. Factors such as ambient temperatures (for refrigerators and air conditioners), minimum wash temperatures (for dishwashers and clothes washers), and initial moisture content (for clothes dryers) all are critical. Although some of these parameters are specified in international standards, these do not always suit regional or national requirements for energy and performance testing. Often, a national standard will contain test conditions that are specific to and reflective of local climate and/or consumer patterns of use.

For appliances and equipment, the energy test procedure is the foundation for MEPS, energy labels, and other related energy programs. It provides manufacturers, regulatory authorities, and consumers a way in which to consistently evaluate energy use and savings across various appliance models. A well-designed test procedure services the needs of its users economically and with an acceptable level of accuracy and correspondence to real conditions of use. On the other hand, a poorly designed energy test procedure can undermine the effectiveness of everything built upon it.

Therefore, energy test procedures are a critical element underpinning all energy programs that seek to measure and improve the energy efficiency of appliances and equipment.

Test procedures cover all aspects of testing of the product such as the following:

- Ambient temperature
- Water quality and temperature
- Test loads
- Instrumentation and equipment
- Special materials and methods
- Duty cycles and/or loading patterns

Ideally, a good test procedure should have the following characteristics:

- Repeatability (producing the same result each time the same model is tested in the same test laboratory)

- Reproducibility (producing the same result each time the same model is tested in different test laboratories)
- Reflective of consumer use
- Simple but effective
- Covering existing products as well as new and forthcoming technologies
- Able to represent cultural and environmental influences and user patterns

Unfortunately, it is rarely possible for a test procedure to meet all of these requirements simultaneously. Clearly, the requirements for repeatability and reproducibility are paramount in any test procedure that is used to regulate products or judge manufacturer claims. These are often regarded as key and fundamental requirements for any test procedure and tend to take precedence over other aspects.

The ability to be reflective of real consumer use is also a key requirement. The problem is that consumers usually behave and act in different ways, so more often than not, there is a distribution of patterns of use rather than a single pattern. Variation in the frequency of use (e.g., of a clothes washer) is reasonably easy to determine, and it may even be possible to reflect this to some degree in an energy label or advisory brochure. However, consumer selection of various wash programs is more difficult to represent, as is variation in, say, washing temperatures. Being able to represent (or simulate) climate and temperature impacts is also an important consideration.

The issue of new and emerging technologies is a critical one. Increasingly, there appear product configurations and types that were never envisaged by the people who developed the original test procedure for a product. The presence of load-sensing devices (in products such as clothes washers), automatic dryness sensors (in clothes dryers), adaptive and smart controllers, and fuzzy logic (adjusting and optimizing operation and performance on the basis of the previous day/week/month/year of use) makes it increasingly difficult to test some product types, especially with outmoded procedures. Many of these devices and controls will actually save energy in actual use. However, some savings will be fake; that is, the savings will appear only in the test procedures. The key issue is how to test products without being tricked into thinking that they are more or less efficient than they are likely to be in actual use.

The alignment of international test procedures will

- facilitate international trade;
- decrease testing and approval costs for manufacturers;
- allow the free movement of the most efficient products (note that products with low energy efficiencies may still be barred if they do not meet local MEPS levels);
- facilitate international comparisons;
- assist in the diffusion of advanced energy-saving technologies.

12. TESTED ENERGY CONSUMPTION VERSUS IN-USE ENERGY

A key (but by no means obvious) difference between energy labeling programs is the assumptions that lie behind the calculation of the energy consumption and related performance data shown. Many energy labels have built into them assumptions about the frequency and duration of use for the calculation of energy consumption and related parameters. For example, in the case of clothes washers, the data shown on some energy labels is based on an assumed number of washing loads per year. Such estimates are usually based on averages from surveys and data collected from the country or region where the energy label is to be used. However, it is important to remember that no consumer or user is typical; there is always likely to be a wide distribution of patterns of use and consumer related behavior for all products.

Some analysts argue that the energy label will, or at least should, be of most economic value and of most interest to customers who use their appliances most intensively, so it may be appropriate for values on the label to be calculated for a higher than average frequency of use (or perhaps median use). In terms of values shown on an energy label, large "annual" figures for energy consumption and cost will have a greater influence on consumer decisions than will small numbers such as energy per usage cycle or per day. For example, 10-year running cost data (a proxy for expected minimum appliance life for many products) shown on energy labeling brochures would often be comparable to the appliance purchase cost. In this respect, there is a potential contradiction, albeit a minor one, between energy labels as a consumer information program and as a policy measure for increasing energy efficiency.

The economics of energy efficiency is a key area of consumer interest with respect to energy labeling. Most consumers express interest in the cost of energy used to operate an appliance. However, conveying this information through an energy label has many problems, including variations in energy tariffs within a country and by time of day. In addition, consumers can easily confuse cost information shown on energy labels. In fact, it has been found to be unclear to some consumers whether the cost figures shown on a label relate to the cost of energy used to operate the appliance or to the savings in energy operating cost. Some consumers also mistakenly believe that energy cost values on an energy label are related to the appliance purchase price.

Climatic considerations are critical for some products (especially air conditioners and refrigerators and, to some extent, water heaters), and historically this has been poorly handled in international and regional test procedures for these product types. These products, which typically have widely varying temperature performance coefficients for different models, are usually tested under a single static temperature condition, which often neither is representative of nor facilitates the estimation of performance under other conditions (including real use). Of course, "real use" and a "single representative test point" can never be reconciled given that many countries have climate zones ranging from cool temperate to humid tropical; a single test condition can never be representative of such a range. For such products, a complex conversion algorithm (i.e., a computer simulation model) is probably the only feasible long-term option by which in-use energy consumption can be estimated more accurately from results recorded under standard test conditions.

13. PERFORMANCE VERSUS ENERGY CONSUMPTION

The measurement of appliance performance in conjunction with energy consumption is a challenging issue. Within a particular appliance type, the energy consumption and performance are related (e.g., making food colder in a particular refrigerator requires more energy, increasing the wash temperature of a clothes washer increases both the wash performance and the energy consumption). Measurement of the performance of products is widely accepted as part of most product test procedures. Things such as internal temperature control of a refrigerator, cooling or heating capacity of an air conditioner, shaft power output of an electric motor, and volume of hot water delivered by a water heater all are measures of performance. In fact, consumers and businesses are not generally as interested in the energy that is consumed by a product as they are in the energy service that the product performs for them. Nearly all test procedures measure the performance in some way.

Performance is critical as the energy service and energy consumption combine to provide a measure of the energy efficiency of the product, that is, the "energy service delivered per unit of energy consumption." In general, statements of energy efficiency are meaningless unless the level of performance is either specified or declared.

If there is no performance measurement in the test procedure (or no realistic task for the product to perform), it is often necessary to prescribe attributes such as capacity and configuration. Some advanced products are able to trick a nonperformance test procedure. For example, dishwashers or clothes washers that have soil sensors could terminate the washing operation prematurely if a clean load were used. Such machines would operate only with very short program cycles and would achieve inordinately high energy ratings under the test procedure, whereas in actual use the energy would be much higher. Hence, comparative advice based on such tests would mislead consumers.

14. EVALUATION

It is critical that any energy program be subjected to ongoing evaluation and assessments. This will ensure that the program is operating smoothly, that problems are identified and resolved, and that an accurate estimate of program impacts is obtained. There are two main types of evaluation of labeling and MEPS programs: process evaluation and impact evaluation.

14.1 Process Evaluation

Process evaluation is usually qualitative and is a measure of the operation of the program. Unfortunately, this is sometimes seen as relatively less important by policymakers. In reality, process elements are especially critical to the successful implementation of a labeling program (in contrast to a MEPS program) because they cover aspects such as the following:

- consumer priorities in purchasing an appliance;
- consumer awareness levels of labels;
- correct display of labels in retailers;

- correct indication of energy performance on labels;
- the operation of the regulatory system such as ease of product registration.

Elements such as check testing and evaluation of administrative efficiency (e.g., registration times and processes) may be relevant to both labeling and MEPS programs.

14.2 Impact Evaluation

Impact evaluation is an attempt to quantify the energy and related environmental (especially greenhouse gas) effects arising from the implementation of a labeling or MEPS program. Impact elements include tracking of product efficiency trends and quantification of energy, greenhouse gas, and demand savings resulting from the implementation of energy labeling or MEPS. Impacts can be very difficult to determine accurately in some cases, especially where it is problematic to benchmark the efficiency of the market prior to program implementation (e.g., due to lack of data). Another problem is that once an energy program has been in place for a period of time, it becomes increasingly difficult to determine a "base case" against which to compare the program impact. (There is only "one reality"; for example, after 20 years of energy labeling, a "no label" scenario can hardly be considered as a base case for comparison.) Even where the overall impact can be determined with reasonable accuracy, it can be difficult to accurately allocate the share of the efficiency improvement where a number of influences have been working in combination, for example, where labeling, MEPS, and the introduction of new technology (which may be unrelated to labeling or MEPS) all have had an influence on the energy efficiency of products over a defined period. Advanced techniques can be used to partly separate the impact of these various influences, but this is not always precise.

It is critical that planning and data collection for evaluation be incorporated into the program design and implementation. Collection of relevant data needed to establish a baseline after the program has been implemented is usually difficult or impossible.

15. IMPACTS OF EFFICIENCY STANDARDS AND LABELING PROGRAMS

Notwithstanding the difficulties of evaluating the impacts of appliance energy efficiency programs, various efforts have been made to quantify their impact to date as well as their potential impact if they were to be strengthened. A publication from the International Energy Agency (IEA) in 2003 pulled together much of the available information among OECD countries and produced a new synthesis of the probable impacts of the MEPS and labeling programs that are currently in place. At the time, residential appliances and equipment used 30% of all electricity generated in OECD countries and accounted for 12% of all energy-related carbon dioxide (CO_2) emissions. They were the second largest consumer of electricity and the third largest emitter of greenhouse gases in the OECD. Since 1973, primary energy demand in the OECD residential sector has grown faster than that in all other sectors (other than transport), and in terms of electricity demand growth, it has outstripped all but primary energy demand in the commercial buildings sector over this same period.

Figure 3 shows aggregate (projected) residential electricity demand in OECD countries from 1990 to 2030 for three scenarios:

- The "no policies" scenario that projects what would have happened if no MEPS, targets, or labeling programs had been implemented in OECD countries
- The "current policies" scenario that projects electricity demand with the existing set of policies

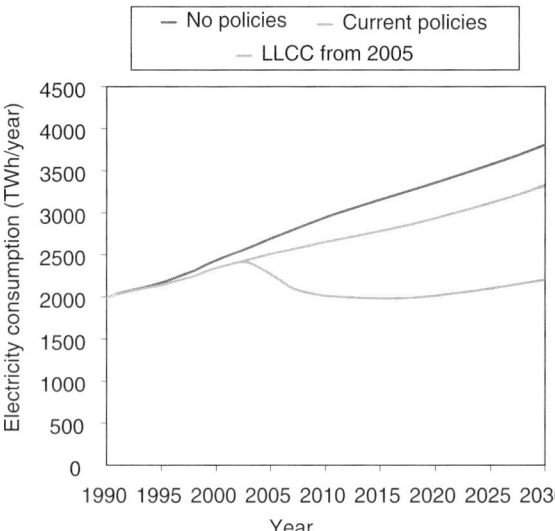

FIGURE 3 Aggregate residential electricity demand in the OECD for three appliance efficiency scenarios. LLCC, least life cycle cost.

- The "LLCC from 2005" scenario that projects electricity demand if the efficiency of residential electrical equipment sold from 2005 onward were to correspond to the least life cycle cost efficiency level from consumers' perspective, that is, if the efficiency of the products sold from 2005 onward were to be at a level by which the sum of the purchase prices and the discounted costs of operating the appliances over their lifetimes are minimized

The IEA results show that the appliance policies enacted since 1990 reduced OECD residential electricity consumption by 3.8% in 2000 and will go on to reduce it by 12.5% in 2020 compared with what would have happened if the policies had not been introduced. These savings resulted in a reduction of some 37 million tonnes (Mt) of CO_2 emissions in 2000 and avoided the need for at least 20 gas-fired power stations. Even without further strengthening, these same policies will go on to reduce emissions by 126 Mt CO_2/year by 2010 as more efficient equipment replaces less efficient equipment in the stock.

However, the IEA analysis also shows that a very significant cost-effective potential still exists to strengthen the impact of appliance efficiency programs. By using efficiency policy to target the most cost-effective level of efficiency (the least life cycle cost) for appliances from 2005 onward, IEA member countries could save more than 642 TWh electricity/year, or some 322 Mt CO_2/year, by 2010 compared with what they will save under existing policy settings. In terms of greenhouse gas emissions, this would be equivalent to taking more than 100 million cars off IEA roads or doing without nearly 200 gas-fired power stations. The IEA estimated that targeting the least life cycle cost for residential appliances could achieve up to 30% of OECD member countries' targets under the Kyoto Protocol on climate change. By 2030, a policy of targeting the least life cycle cost for residential appliances (from 2005) would avoid more than 1110 TWh/year of final electricity demand, or 572 Mt CO_2/year, equivalent to taking more than 200 million cars off OECD roads.

Because the efficiency level associated with the least life cycle cost necessarily corresponds to the lowest net equipment purchase and operating costs, the pursuit of these policies is highly cost-effective even if climate change targets are not a concern. The IEA estimated that each tonne of CO_2 saved through 2010 by current policies in OECD member countries in North America will be attained at a net cost saving of U.S. $78; that is, the net cost of CO_2 abatement is U.S. −$78/tonne CO_2. For OECD member countries in Europe, the equivalent value is Euro 241/tonne CO_2. Under the least life cycle cost from 2005 scenario, the net cost of CO_2 abatement is U.S. $65.5/tonne CO_2 in OECD member countries in North America and Euro 168.9/tonne CO_2 in OECD member countries in Europe by 2020. These are slightly less cost-effective per tonne saved than are the existing policies, but they result in much greater total cost savings.

SEE ALSO THE FOLLOWING ARTICLES

Consumption, Energy, and the Environment • Economics of Energy Efficiency • Energy Efficiency, Taxonomic Overview • Industrial Energy Efficiency • International Energy Law and Policy • Obstacles to Energy Efficiency • Vehicles and Their Powerplants: Energy Use and Efficiency

Further Reading

Bertoldi, P., Ricci, A., and de Almeida, A. (Eds.) (2001). "Energy Efficiency in Household Appliances and Lighting," proceedings of the Second International Conference on Energy Efficiency in Household Appliances and Lighting. Springer-Verlag, Berlin.

Bertoldi, P., Ricci, A., and Wajer, B. (Eds.) (1998). "Energy Efficiency in Household Appliances," Proceedings of the First International Conference on Energy Efficiency in Household Appliances and Lighting. Springer-Verlag, Berlin.

Egan, K., and du Pont, P. (1998). "Asia's New Standard for Success: Energy Efficiency Standards and Labeling Programs in 12 Asian Countries." International Institute for Energy Conservation, Washington, DC.

Energy Efficiency Strategies, et al. (1999). "Review of Energy Efficiency Test Standards and Regulations in APEC Member Economies," Report APEC 99-RE-01.5, prepared by Energy Efficiency Strategies and others for Asia-Pacific Economic Cooperation (APEC) secretariat, Singapore.

Harrington, L., and Damnics, M. (2001). "Energy Labelling and Standards Programs throughout the World," report commissioned and published by the National Appliance and Equipment Energy Efficiency Committee, Australia. www.energyrating.gov.au.

International Energy Agency. (2000). "Energy Labels and Standards." IEA, Paris. www.iea.org.

International Energy Agency. (2003). "Cool Appliances: Policy Strategies for Energy Efficient Homes, April 2003." IEA, Paris. www.iea.org.

Wiel, S., and McMahon, J. (eds.). (2000). "Energy Efficiency Labels and Standards: A Guidebook for Appliances, Equipment, and Lighting." Collaborative Labelling and Appliance Standards Program (CLASP), Washington, DC. www.clasponline.org.

Land Requirements of Energy Systems

VACLAV SMIL
University of Manitoba
Winnipeg, Manitoba, Canada

1. Biomass Energies
2. Direct Solar Radiation
3. Wind and Water Energy
4. Fossil Fuels and Thermal Electricity Generation
5. Aggregate Demand and Energy Consumption
6. Energy Transitions

Glossary

insolation The average share of solar (electromagnetic) energy absorbed by the earth's surfaces that amounts to only about 12% of the solar constant.

phytomass Living or dead organic matter (biomass) of plants; in energy literature, it refers particularly to that of trees and crop residues used for fuel.

power density Average long-term power flux per unit of land area, usually expressed in watts per squared meter (W/m^2).

solar constant Total flux of solar (electromagnetic) energy that falls on 1 square meter of the extraterrestrial area above the earth's atmosphere at a vertical angle; its value is approximately $1.37 \, kW/m^2$.

thermal electricity generation Generation where either the combustion of one of the fossil fuels or nuclear fission is used to produce steam to drive turbogenerators.

Land requirements of energy production depend primarily on the nature of harnessed resources. The two sets of activities that define the modern high-energy civilization—extraction of fossil fuels and thermal generation of electricity—require a relatively limited amount of land. In contrast, all forms of renewable energy conversions are highly land intensive, but the differences among individual techniques range over three orders of magnitude. Production of phytomass fuels, the principal source of heat in preindustrial economies, makes the largest relative claims on land. In contrast, direct conversions of solar radiation, whose contribution to the world's supply of commercial energy remains minuscule, need one or even two orders of magnitude less land than does either the natural growth of forests or the cultivation of phytomass for biofuels. Hydroelectric generation can be as land demanding as phytomass production, and although it provides less than 3% of the world's commercial primary energy supply, the aggregate area of reservoirs makes up approximately 60% of all land occupied by the world's energy infrastructures. Socioeconomic and environmental implications of this pronounced spatial dichotomy will become increasingly important during the 21st century as the world's energy supply, now dominated by fossil fuels and thermal electricity, shifts gradually to a system that will be much more dependent on conversions of renewable energies. By far, the most revealing way of comparing land requirements of various energy systems is to calculate their power densities, that is, their energy flux per unit area of the earth's surface. The standard unit of power density is expressed in watts per square meter (W/m^2), and this fundamental measure is applicable to all energy flows, be they parts of production or consumption systems, natural or anthropogenic processes, or renewable or fossil fuel conversions. There are immutable natural limits to all commercial power densities and movable technical barriers that can raise the rates through higher efficiencies. Natural limits for fossil fuel-based techniques are given by the accumulations of coals and hydrocarbons, and unless we resort to space-based capture of solar energy, all renewable energy conversions will be limited to a fraction of solar radiation that reaches the earth's surface.

1. BIOMASS ENERGIES

Because of the inherently low efficiency of the photosynthetic process, no form of energy supply has such low power densities, and hence such high land demands, as does the production of phytomass. Recent estimates of the global terrestrial net primary productivity (NPP) average approximately 55 billion metric tons (tonnes) (Gt) of carbon per year, that is, approximately 120 Gt of dry biomass that contains (assuming the mean energy density of 15 billion joules per metric ton [GJ/tonne]) some 1800×10^{18} joules (EJ). This productivity prorates to less than 0.5 watts per square meter (W/m^2) of ice-free land. Rates for forests are naturally higher, but they do not surpass 1.1 W/m^2 even in the richest tropical ecosystems. In all natural ecosystems, a large part of the NPP is consumed by heterotrophs (ranging in size from bacteria to megaherbivores); hence, the phytomass that is actually available for energy conversion is only a fraction of the originally produced biomass.

These inherently low power densities mattered little so long as the woody phytomass was required in relatively small quantities, but they led to considerable logistic problems for traditional societies whose growing cities and expanding industries were entirely dependent on phytomass for their thermal energy needs. Clearing of a primeval forest in preindustrial Europe could produce several hundred tonnes of wood per hectare, but this was obviously a historical singularity. Once that rich growth was gone, the wood was commonly cut in 10- to 20-year rotations from coppicing hardwoods whose annual increment was between 5 and 10 metric tons per hectare (tonnes/ha) or 0.25 to 0.5 W/m^2. Converting wood to more energy-dense and less-polluting charcoal greatly increased the primary demand because primitive methods of charcoaling needed as much as 10 kg (and commonly at least 5 kg) of wood per kilogram of charcoal and resulted in production power densities of no more than 0.1 W/m^2. Spatial implications of these low power densities became burdensome for all large preindustrial cities and even more so for the highly energy-intensive iron smelting.

For example, in 1810, when the United States produced less than 50,000 tonnes of pig iron, approximately 2600 km^2 of forest, an area of roughly 50×50 km, had to be cut down every year to produce the requisite charcoal. A century later, the annual output of 25 million tonnes per year would have required, even with much lower charging rates, approximately 170,000 km^2 of forest every year, a square with the side equivalent to the distance between Philadelphia and Boston. This single example makes it obvious that only the substitution of charcoal by coal-derived coke made it possible to expand the modern pig iron production to more than a half-billion tonnes per year.

Although modern intensively cultivated field crops and fast-growing trees are grown with power densities that match or surpass the photosynthetic rates in the world's most productive tropical ecosystems, land requirements of biofuel production remain well below 1 W/m^2 (Fig. 1). Only with fast-growing trees (most commonly poplars, willows, pines, and eucalypti) that are intensively cultivated as short-rotation crops and produce exceptionally high harvests of 20 tonnes/ha of dry matter, will the above-ground phytomass yield more than 1 W/m^2. Even the best possible conversions—with efficiencies greater than 50% for cogeneration of electricity and heat and 55 to 60% for conversion to methanol or ethanol—result in power densities of just 0.5 to 0.6 W/m^2 for the final energy supply. These rates are one to two orders of magnitude below the power densities of electricity generation using water, wind, or solar radiation. These low power densities will limit the share of the total primary energy that could be derived from biofuels.

In fact, any large-scale schemes of biofuel production would be questionable even if they were to have much better net energy returns than has been the case so far. This is because we are already appropriating approximately 40% of all terrestrial NPP through our field and forest harvests, animal grazing, land clearing, and forest and grassland fires. Any future large-scale reliance on cultivated biofuels would have to increase this appropriation by a large margin, a shift that not only would preclude food production and eliminate wood harvests for timber and paper from areas devoted to biofuels but also would destroy many irreplaceable environmental services that are now provided by mature natural ecosystems or by only partially disturbed forests.

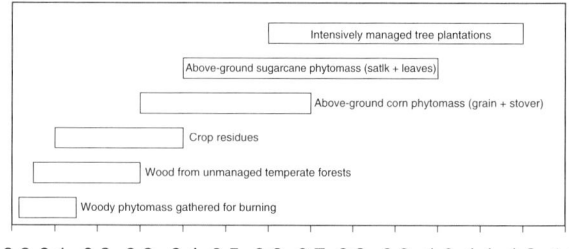

FIGURE 1 Power densities of various forms of biomass production.

2. DIRECT SOLAR RADIATION

Global mean of the total insolation—the solar radiation absorbed by the earth's surfaces—averages 168 W/m^2, only approximately 12% of the extraterrestrial solar constant. Annual maxima surpass 250 W/m^2 in the cloudless, unpolluted, high-pressure belts of subtropical deserts, where the daily peaks are approximately 1.2 kW/m^2. Except for passively heated structures (every house, no matter how poorly built and unsuitably oriented, is a solar house), preindustrial societies had no technical means to convert solar radiation into useful energy. Moreover, traditional architectures were much better for devising solutions for passive cooling in warm climates than for capturing solar radiation in cold locations. Modern conversions of solar radiation have followed two distinct routes: central solar power techniques, which concentrate the insolation by means of troughs, mirrors, or dishes, and photovoltaics (PV), which converts radiation directly into electricity.

Peak solar-to-electric conversion efficiencies are just over 20% for troughs, 23% for power towers, and 29% for dishes, but annual rates are much lower, that is, 10 to 18% for troughs, 8 to 19% for towers, and 16 to 28% for dishes. Thus, troughs and towers located at the sunniest sites and operating with the best efficiencies could have peak power densities of approximately 60 W/m^2 of collecting surface, and in less sunny locations the densities would go down to less than 40 W/m^2. Spaces between collectors, service roads, and buildings as well as exclusion zones will reduce those rates by at least 30%.

Maximum theoretical efficiency of PV cells is limited by the range of photon energies, whereas the peak practical rates are reduced above all through reflection from cell surfaces and the leakage of currents. Large differences remain when comparing theoretical, laboratory, and field performances as well as when contrasting the best performances of high-purity single crystals, polycrystalline cells, and thin films. Theoretical single-crystal efficiencies are 25 to 30%, and lenses and reflectors can be used to focus direct sunlight onto a small area of cells and boost conversion efficiencies to more than 30%.

Stacking cells sensitive to different parts of the spectrum could push the theoretical peak to 50% and to 30 to 35% in laboratories. Actual efficiencies of commercial single-crystal modules are now between 12 and 14%, and after a time in the field they may drop to below 10%. Thin-film cells can convert 11 to 17% in laboratories, but modules in the field convert as little as 3 to 7% after several months in operation. Multijunction amorphous Si cells do better, converting at least 8% and as much as 11% in large-area modules. The latter performance translates to no more than 15 to 18 W/m^2 for most locations, whereas with the eventual 17% field efficiency PV cells would be averaging approximately 30 W/m^2.

3. WIND AND WATER ENERGY

Wind energy is distributed very unevenly in both space and time. Many sites around the world have mean annual wind speeds of 7.0 to 7.5 m/s, enough to produce power densities of 400 to 500 W/m^2 of vertical area swept by rotating blades 50 m above ground. Translating these rates into horizontal power densities requires a number of sequential assumptions. Sufficient spacing between the units is needed to eliminate upstream wakes as well as to replenish kinetic energy. Spacing equal to 5 rotor diameters is enough to avoid excessive wake interference, but at least twice that distance is needed for wind energy replenishment in large installations. In addition, no more than 16/27 (59.3%) of wind's kinetic energy can be extracted by a rotating horizontal axis generator (the rate known as the Betz limit), and the actual capture will be still lower.

Consequently, machines with 50-m hub height and with 10 × 5 diameters spacing would have to be 500 m apart, and in locations with wind power density averaging 450 W/m^2 (the mean value common in the Dakotas, northern Texas, western Oklahoma, and coastal Oregon), they would intercept approximately 7 W/m^2. However, the average 25% conversion efficiency and 25% power loss caused by wakes and blade soiling would reduce the actual power output to approximately 1.3 W/m^2. Power density of 700 W/m^2, turbine efficiency of 35%, and power loss of 10% would more than double that rate to approximately 3.5 W/m^2. Actual rates are highly site specific. Rated performance of California's early wind projects was almost exactly 2 W/m^2, and the most densely packed wind farms rate up to 15 W/m^2. In contrast, more spread-out installations have power densities mostly between 5 and 7 W/m^2.

Unlike in the case of phytomass, the low power density of wind is not in itself an insurmountable obstacle to harnessing this large resource because all but a small share of areas occupied by wind facilities (commonly less than 5% of the total) can still be used for farming or ranching. Moreover, in large countries, windy land is abundant. A recent U.S. study that used a variety of land exclusion scenarios (due to many environmental and land use reasons) concluded that the country's windy land adds up to approximately 460,000 km^2, or about 6% of all land in the contiguous United States, and that it could support generating capacity of approximately 500 GW.

Hydroelectricity generation operates with generally low power densities, and reservoirs have inundated large areas of natural ecosystems, including many unique and highly biodiverse forests and wetlands. Reservoirs behind the world's large dams (those higher than 30 m) now cover nearly 600,000 km^2 (an area nearly twice as large as Italy), and those used solely or largely for electricity generation add up to approximately 175,000 km^2. This prorates to approximately 4 W/m^2 in terms of installed capacity and to 1.7 W/m^2 in terms of actual generation, but power densities of individual projects span three orders of magnitude. Dams on the lower courses of large rivers impound huge volumes of water in relatively shallow reservoirs. The combined surface of the world's seven largest reservoirs is as large as the total area of the Netherlands, and the top two impoundments—Ghana's Akosombo on the Volta River (8730 km^2) and Russia's Kuybyshev on the Volga River (6500 km^2)—approach the size of small countries such as Lebanon and Cyprus. Power densities of these projects are less than 1 W/m^2 (Fig. 2).

In contrast, dams on the middle and upper courses of rivers with steeper gradients have much smaller and deeper reservoirs, and their power densities are on the order of 10^1 W/m^2. Itaipu, the world's largest operating hydro project so far (12.6 GW), has a power density of 9.3 W/m^2. Grand Coulee (6.48 GW) rates nearly 20 W/m^2. China's Three Gorges (Sanxia, which is to become the world's largest hydro project, with 17.68 GW of installed capacity, when finished in 2008) will have approximately 28 W/m^2. In addition, some Alpine stations surpass 100 W/m^2 (Fig. 2). The world's record holder will be the planned Nepali Arun project, whose 42-ha reservoir and 210-MW installed capacity translate to approximately 500 W/m^2. Because load factors of most hydro stations are typically less than 50%, and because many projects are built to supply electricity during the hours of peak demand and hence have even shorter operating hours, effective power densities are only 25 to 50% of the just-cited theoretical rates.

4. FOSSIL FUELS AND THERMAL ELECTRICITY GENERATION

Energy density of recoverable coal deposits varies greatly with the thickness of seams and heat content of the fuel. Typical energy densities of large coal basins range from no more than 10 GJ/m^2 for poor lignites in thin seams to about 50 GJ/m^2 for good bituminous coals in multiple seams. The best sites have densities one or nearly two orders of magnitude higher: Arizona's Black Mesa (∼200 GJ/m^2); Montana's Ashland (415 GJ/m^2); Fortuna, West Germany's largest brown coal mine in Rheinland (430 GJ/m^2); parts of Victoria's Latrobe Valley, with brown coal seams up to 100 m (>800 GJ/m^2); and Queensland's Blair Athol, with 30 m of bituminous coal (∼900 GJ/m^2).

Power densities of coal extraction are the lowest (less than 500 W/m^2) in poorly designed surface mines producing low-quality lignites or sub-bituminous coals (with the best example being that of large brown coal mines in the former East Germany), whereas open cast mines working thick, high-quality seams of bituminous coal can extract several kilowatts per square meter (Australia has a number of such operations). Most coal mines, be they surface (with proper land reclamation) or underground (including all above-ground facilities), produce the fuel with power densities between 1 and 10 kW/m^2 (Fig. 3).

Prorating the most likely global estimate of recoverable hydrocarbons, an equivalent of nearly 750 Gt of crude oil according to the latest U.S.

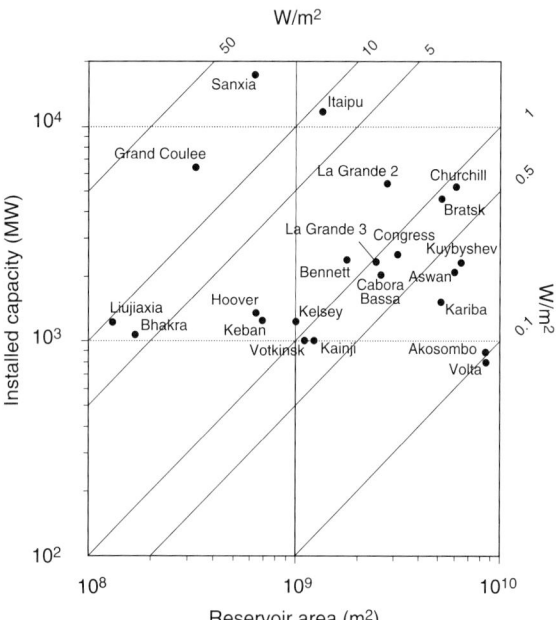

FIGURE 2 Power densities of large reservoirs used for hydroelectricity generation.

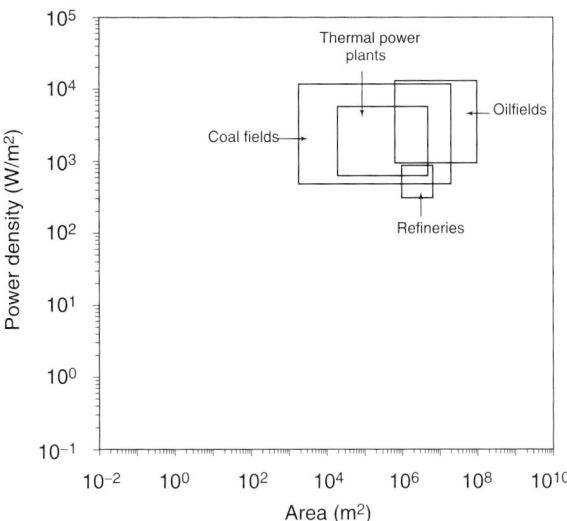

FIGURE 3 Typical power densities of fossil fuel extraction and processing.

Geological Survey study, more than roughly 70 million km^2 of known and prospective sedimentary basins gives a median energy density of approximately 450 million joules (MJ)/m^2. But the richest basins contain 1 to 10 GJ/m^2, and many giant hydrocarbon fields go well above that. Alaskan Prudhoe Bay rates about 25 GJ/m^2; Saudi al-Ghawar, the world's largest oilfield with roughly one-seventh of the world's oil reserves, contains close to 100 GJ/m^2; and Kuwaiti al-Burqan, with the world's second largest oil reserves, contains approximately 1000 GJ/m^2. Recovery of hydrocarbons has always required much less land per unit of extracted energy than does coal mining, and recent advances in directional and horizontal drilling mean that even extreme reaches of an underground oil or gas reservoir can be accessed from a single borehole, further minimizing any surface disruption.

Because usually less than 1% of the area overlying an oilfield is actually taken up by surface structures, even unremarkable hydrocarbon fields, such as the old U.S. ones dominated by stripper wells, can produce oil with power densities surpassing 2 kW/m^2 of land affected by the extraction. This includes not just exploratory and production wells but also mud and water ponds, storage yards, buildings and sheds, gathering pipelines, and access roads (Fig. 3). Rates for the world's richest Middle Eastern oilfields are commonly between 10 and 20 kW/m^2. Typical power densities of natural gas extraction are as high as 10 to 15 kW/m^2, but this rate is reduced by up to an order of magnitude once the rights-of-way for gathering pipelines and field facilities for gas processing are added to well enclosures.

The following generalizations summarize the land requirements of fossil fuel production and processing. Coals and hydrocarbons are now produced with power densities unmatched by any other fuel or source of primary electricity; 400 to 4000 W/m^2 might be a good conservative range, including the majority of coal mining and oil and gas extraction rates. In comparison with their extraction, processing of fossil fuels requires relatively little space. Coal preparation and cleaning have throughput densities between 8 and 10 kW/m^2. Crude oil refining requires more space due to safety precautions in siting high-pressure operations and storage facilities and in preventing spills, and its throughput power densities are generally lower (3–4 kW/m^2).

Pipelines usually need 25 to 30 m for construction, after which only access strips of up to 10 m wide may be necessary. Compressor stations take up to 20,000 m^2 at 80- to 120-km intervals, and pumping stations take up to 10 times as much every 130 to 160 km. Tanker terminals and liquefied natural gas facilities need relatively small areas but require extensive security zones. Railroads that were built exclusively to move coal by using unit trains over distances of 500 to 1500 km to large power plants have annual throughput densities between 100 and 400 W/m^2.

Only a small part of land required by modern thermal electricity generating plants is devoted to structures housing boilers and turbogenerators. Combustion of fossil fuels in large boilers proceeds at power densities often surpassing 5 million watts (MW)/m^2, and modern turbogenerators have outputs higher than 1 MW/m^2. In contrast, auxiliary structures (including switchyards, storage sheds, and office buildings), water-cooling facilities, and air pollution controls take up much more space, and even more land is needed when substantial on-site fuel storage is necessary (Fig. 4). Once-through cooling (returning warmed water into streams or bays) has minimal spatial claims, but spray ponds require approximately 400 m^2/MW and ordinary cooling ponds require between 4500 and 5000 m^2/MW. Cooling towers cut this demand to between 30 and 60 m^2/MW for both wet and dry natural draft structures, whereas mechanical draft units need as little as 10 m^2/MW.

Electrostatic precipitators removing more than 99% of all fly ash are now the norm on coal-fired stations, but because they are required to handle huge volumes of hot gas, they take up about as much space as do boilers. Desulfurization units can be more compact, but the disposal of captured fly ash

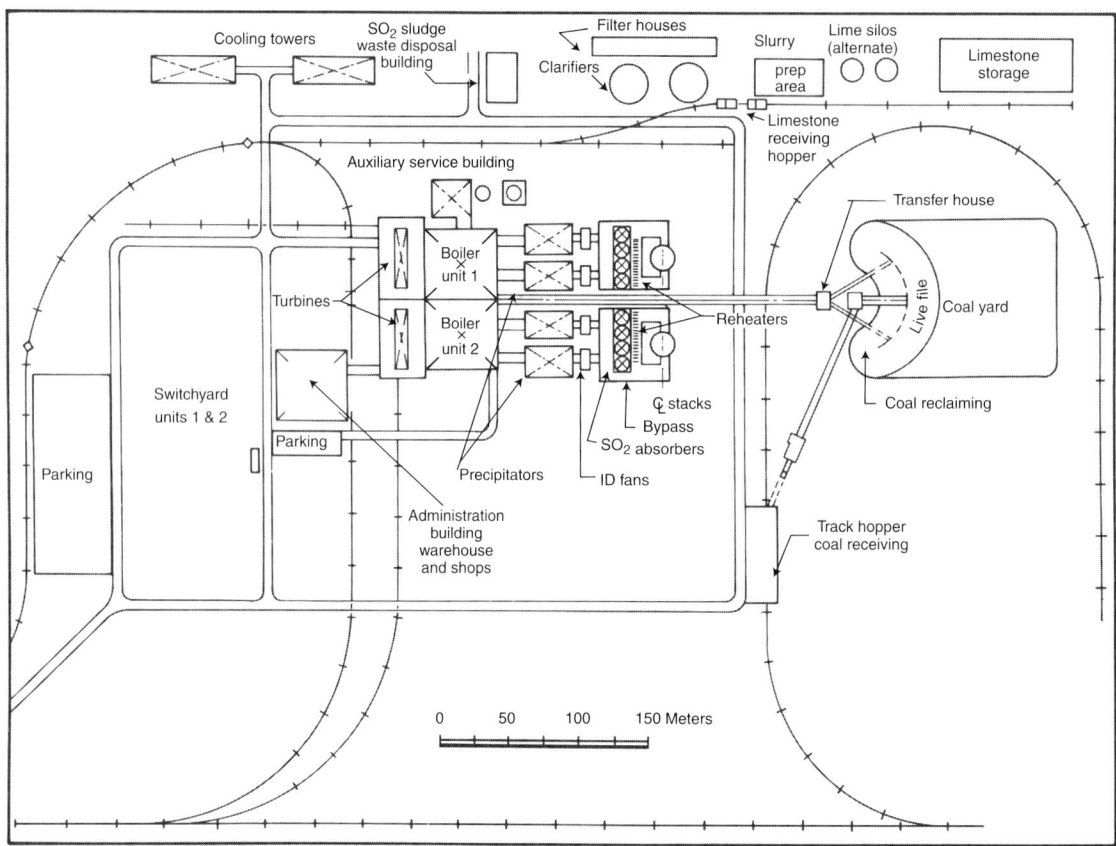

FIGURE 4 Land requirements of a coal-fired electricity-generating plant with two units rated at 600 MW$_e$ each.

and ponding of the sulfate sludge require relatively large areas of adjacent land or costly arrangements for trucking or railway transport to dump sites. Long-term storage of fly ash captured by electrostatic precipitators needs 20 to 40 m^2/MW, and during a plant's lifetime of 35 to 50 years between 700 and 2000 m^2 of ash disposal space will be needed per installed million watts (assuming that coal contains about 10% of ash). Flue gas desulfurization facilities claim between 400 and 600 m^2/MW, and the lifetime disposal of sludge requires between 200 and 600 m^2/MW of installed capacity.

Hydrocarbon-fueled plants supplied by pipelines will have only small storage requirements, but coal piles needed for 60 to 90 days of operation (in case the railway deliveries were interrupted by a strike) are a different matter; they have storage densities of 25 to 100 thousand watts (kW)/m^2, similar to those of switchyards. Buffer zones between plants and the nearest settlements and land kept in reserve for future expansion may easily double or even triple the total land claim.

Given the variety of fuels, water-cooling and air pollution control arrangements, and fuel storage needs, it would be misleading to cite a single typical figure, but most coal-fired power plants will have overall power densities (including all land within a protective fence) of less than 1 kW/m^2, and the power density of the complete fuel cycle (extraction, transportation, and generation) will be less than 300 W/m^2. Analogical rates, for the plant alone and for the total fuel cycle, are approximately 2 kW/m^2 and less than 500 W/m^2 for oil-fired stations and are 3 and 1 kW/m^2 for the natural gas-fired ones. Land claims of nuclear plants that require no extensive fuel storage and no air pollution control facilities are dominated by exclusion and buffer zones, and with rates commonly approximating 1.5 kW/m^2, they are less than those of oil-fired stations but more than those of coal-fired ones.

5. AGGREGATE DEMAND AND ENERGY CONSUMPTION

Any global figure of the total area of land claimed and affected by infrastructures devoted to the extraction and processing of fossil fuels and to generation of electricity must be only a useful approximation rather than an accurate summation. The author calculated that in the year 2000 the aggregate land claim of the global energy infrastructure reached approximately 290,000 km^2, equal to an area slightly smaller than Italy. Although the absolute figure is certainly large, it is perhaps more revealing to see the total in a relevant global perspective and to compare it with other anthropogenic conversions of land. The area of 290,000 km^2 amounts to less than 2% of the total area of natural ecosystems (mostly forests and grasslands) that has been converted to crop, pasture, industrial, transportation, and urban land during the past 250 or so years.

Less than 2% of the nearly 300,000 km^2 was claimed by the production and processing of coals and hydrocarbons, implying an average power density of approximately 1.75 kW/m^2 for the world's fossil fuel extraction (Fig. 5). Rights-of-way of hydrocarbon pipelines claimed roughly one-tenth of the total, and those of high-voltage transmission lines slightly claimed more than a quarter of the total. Naturally, claims by pipelines and transmission lines are a form of land use that is not directly comparable to the first category (extraction and generation) given that rights-of-way, particularly underneath high-voltage lines, can be used for crop farming, animal grazing, or shrub or small tree plantations. By far the largest share, approximately three-fifths of the total, was occupied by reservoirs storing water for the generation of hydroelectricity.

A more accurate estimate for the United States shows only approximately 1500 km^2 occupied by fossil fuel extraction and processing and by thermal electricity generation but nearly 20,000 km^2 under water reservoirs. A similar amount of land has been devoted to rights-of-way by hydrocarbon pipelines and high-voltage transmission lines. In addition, there is more than 10,000 km^2 of land that was disturbed long ago by coal mining and that has yet to be properly reclaimed. All of these claims added up to approximately 50,000 km^2 during the mid-1990s, implying average power densities of approximately 1.9 kW/m^2 for the fossil fuel extraction and 45 W/m^2 for the entire U.S. energy infrastructure. These averages are just slightly higher than the global means of 1.75 kW/m^2 and 40 W/m^2, respectively.

But the actual effects of energy infrastructures on land use and land cover cannot be measured simply by adding up the areas whose use has been affected. Many energy facilities require land near large cities and, hence, often take over highly productive alluvial soils, whereas others are located in extraordinarily biodiverse and relatively fragile environments. Good examples include oil and gas production in the deltas of the Niger and Mississippi rivers and on Alaska's North Slope. Moreover, transportation and transmission corridors needed to develop energy resources may contribute to the impoverishment of surviving ecosystems because they increase their fragmentation, a process that is well known to reduce biodiversity of many species.

Although power densities of energy production range over five orders of magnitude, final energy uses in modern high-energy societies are confined mostly to between 10^1 and 10^2 W/m^2. Depending on the climate, construction, and the kind of converters inside the building, typical rates for single-family houses in temperate zones range from 20 to 100 W/m^2 of a foundation. Similar rates apply to low-rise office buildings and to structures that house industrial production of low-energy intensity, particularly assembly line manufacturing. Supermarkets and medium-height multiple-story office buildings use 200 to 400 W/m^2.

However, energy-efficient design and state-of-the-art converters can reduce these prevailing rates by large margins. For example, the average energy requirement of U.S. office buildings was approximately 33 W/m^2 of floor space during the late 1990s, whereas the Centex building in Dallas (the most efficient structure in its category as of 2002) used only 10 W/m^2 or only 90 W/m^2 of its foundations. Energy-intensive industrial enterprises (steel mills, smelters of nonferrous metals, refineries, and many chemical syntheses are the most prominent examples

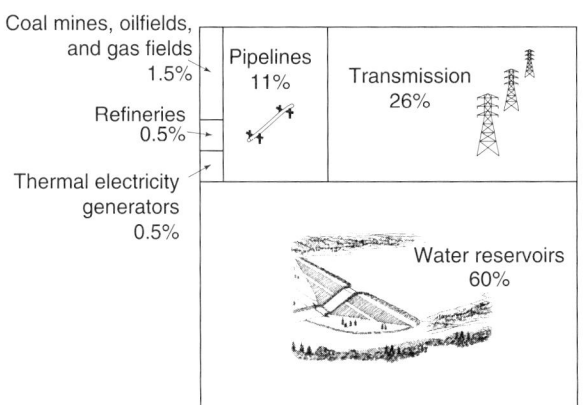

FIGURE 5 Approximate distribution of land claimed by the world's energy infrastructures during the late 1990s.

in this category) require 300 to 900 W/m², and high-rise buildings require up to 3 kW/m².

Comparison of production and consumption power densities shows clearly that our overwhelmingly fossil-fueled civilization extracts fuels and generates thermal electricity with power densities that are commonly one to two (and sometimes even three) orders of magnitude higher than the common power densities of final energy use in buildings, factories, and cities (Fig. 6). These concentrated energy flows are then diffused through pipelines, railways, and high-voltage transmission lines, and that is why space taken up by extraction and conversion of fuels is relatively small in comparison with rights-of-way required to deliver fuels and electricity to consumers. For example, in the United States, the total area occupied by extraction of fossil fuels is on the order of a mere 0.01% of the country's area.

6. ENERGY TRANSITIONS

If a future overwhelmingly solar society were to inherit the existing urban and industrial infrastructure, it would have to rely on an energy-harnessing strategy that is an exact opposite of today's arrangements, that is, on concentration of diffuse energy flows that it would convert to useful forms of energy at rates ranging mostly between 0.3 and 30 W/m² (Fig. 7). Consequently, it would need several hundred, or even several thousand, times more space than does today's fossil fuel extraction to produce the same amount of energy. But it would also need extensive transmission networks to deliver electricity from regions endowed with high intensities of solar radiation or with strong winds.

A gradual increase of average conversion efficiencies (higher power densities of renewable generation) and reduced energy demand (lower power density), most notably for heating and lighting energy-efficient houses, should result in much more closer matches of power densities between production and use and, hence, lead to a much wider adoption of the distributed generation of electricity by individual households or small settlements. But to energize its cities and industrial areas, any society would have to concentrate diffuse flows to bridge power density gaps of several orders of magnitude.

By the year 2030, approximately 60% of the world's population will live in cities, and nearly a quarter of that total will live in "megacities" (with more than 10 million people) that will be dominated by clustered high-rises; parts of today's Hong Kong are a good preview of that future. Supplying these megacities by locally generated renewable energies will be impossible because the power density mismatch between the production and demand is simply too large. The same is true about manufacturing and even more so about often highly energy-intensive processing, especially smelting and chemical syntheses that are being increasingly concentrated in industrial parks.

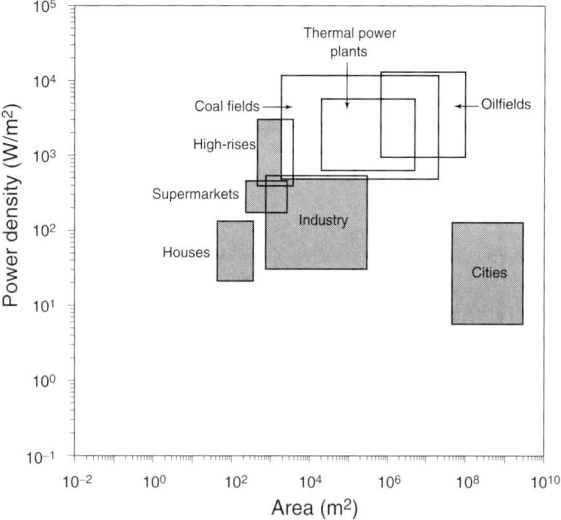

FIGURE 6 Comparison of typical power densities of fossil fuel production and modern urban consumption. The illustration shows how our current arrangements are based on the diffusion of concentrated primary energy supplies.

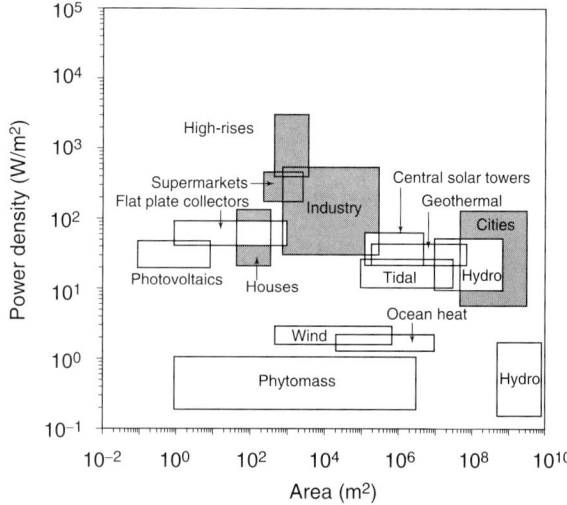

FIGURE 7 Comparison of typical power densities of renewable energy conversions production and modern urban consumption. The illustration shows the opposite process, that is, the need to concentrate diffuse sources of direct and indirect solar and geothermal flows.

In addition, a high-energy society relying on renewable flows would lose most of the flexibility now available in locating electricity-generating stations that can be sited either near large load centers (to minimize transmission distances) or far away from densely populated areas (to minimize air pollution levels). In contrast to these flexible arrangements, high-capacity concentrations of wind turbines and PV cells (or geothermal stations or land-based sea wave-converting machines) must coincide with the highest power densities of the requisite renewable flows. Unfortunately, nearly every contemplated instance of such a development would call for new high-capacity long-distance transmission links as well as for extensive upgrading of the existing lines.

For example, the United States now has only a rudimentary capacity to move electricity from the Southwest, the region of the most advantageous central solar and PV locations, to the Northeast. Similarly, there are no high-capacity transmission links from the windy Northern Plains to California. And the country is poorly prepared for even shorter electricity transfers. Although the U.S. transmission capacity nearly doubled between 1975 and 2000, when it is divided by peak load, the resulting ratio, which represents the distance to which a given plant can expect to sell energy during the conditions of peak demand, has been falling since 1984, and the trade area of a given generator is now only approximately 70% of the mean during the mid-1980s.

Complete absence or inadequate capacities of transmission links between regions endowed with high power densities of direct and indirect solar flows and densely populated load centers can be found in every large country that is contemplating higher reliance of renewable conversions. For example, China's windiest and sunniest areas are the northern grasslands in Inner Mongolia and the arid unpopulated western interior of Xinjiang and Tibet, respectively, whereas its largest load centers are along the eastern and southern coast. Linking the two would require spanning a distance of more than 3000 km and crossing some of Asia's most forbidding terrain.

Stochasticity of all solar energy flows would further increase the spatial impact of renewable conversion due to the need for either extensive energy storage or backup peak demand systems. Unfortunately, building pumped storages and using off-peak electricity generated by nuclear or fossil-fueled power plants to lift water into nearby reservoirs remains the only practical, but obviously a highly land-intensive, way of storing large quantities of almost instantly deployable energy. In spite of more than a century of diligent efforts to develop other effective forms of storage, all other options remain either inefficient or inadequate.

Even greater challenges would arise with any large-scale use of biofuels. For example, replacing the world's coal consumption (nearly 90 EJ in 2000) by woody biomass would require the cultivation of high-yielding trees on approximately 330 million ha, an area larger than all of the forests in the European Union and United States combined. And running the entire fleet of U.S. vehicles registered in the year 2000 on corn-derived ethanol would require planting the crop (with grain yields averaging 7.5 tonnes/ha and ethanol yields of 0.3 kg of alcohol/kg of grain) on an area 20% larger than the country's entire cultivated cropland. Such schemes are obviously impossible, but even their partial realization would have enormous environmental impacts.

These realities do not make large-scale reliance on renewables impossible, but they make it considerably more expensive and challenging to accomplish. Mismatch between low power densities of renewable energy flows and relatively high power densities of many modern final energy uses means that any large-scale adoption of renewable energy conversions by societies dominated by megacities and concentrated industrial production will require a profound spatial restructuring of the existing energy infrastructure, a process that will have many major environmental and socioeconomic consequences.

SEE ALSO THE FOLLOWING ARTICLES

Biomass Resource Assessment • *City Planning and Energy Use* • *Ecological Footprints and Energy* • *Electric Power Generation: Fossil Fuel* • *Global Energy Use: Status and Trends* • *Global Material Cycles and Energy* • *Heat Islands and Energy* • *Hydropower Resources* • *Solar Thermal Power Generation* • *Transitions in Energy Use* • *Wind Resource Base*

Further Reading

Dracker, R., and De Laquill, P. (1996). Progress commercializing solar-electric power systems. *Annu. Rev. Energy Envir.* **21**, 371–402.

Goetzberger, A., Knobloch, J., and Voss, B. (1998). "Crystalline Silicon Solar Cells." John Wiley, Chichester, UK.

Markvart, T. (ed.). (2000). "Solar Electricity." John Wiley, New York.

McGowan, J. G., and Connors, S. R. (2000). Windpower: A turn of the century review. *Annu. Rev. Energy Envir.* **25**, 147–197.

Perlin, J. (1999). "From Space to Earth: The Story of Solar Electricity." Aatec, Ann Arbor, MI.

Rojstaczer, S., Sterling, S. M., and Moore, N. J. (2001). Human appropriation of photosynthesis products. *Science* **294**, 2549–2551.

Smil, V. (1991). "General Energetics: Energy in the Biosphere and Civilization." John Wiley, New York.

Smil, V. (1994). "Energy in World History." Westview, Boulder, CO.

Smil, V. (2002). "The Earth's Biosphere: Evolution, Dynamics, and Change." MIT Press, Cambridge, MA.

Smil, V. (2003). "Energy at the Crossroads: Global Perspectives and Uncertainties." MIT Press, Cambridge, MA.

Leisure, Energy Costs of

SUSANNE BECKEN
Lincoln University
Lincoln, Canterbury, New Zealand

1. Background
2. Concept of Leisure, Recreation, and Tourism
3. Upstream/Downstream
4. Recreational Activities
5. Tourist Accommodation
6. Tourist Transport
7. Conclusion

Glossary

ecolabel (tourism) Certification recognizing good practice that enhances the environment or minimizes environmental impacts and aims to influence the purchasing behavior of tourists.

ecotourism Ecologically sustainable tourism with a primary focus on experiencing natural areas that fosters environmental and cultural understanding, appreciation, and conservation.

leisure Freely chosen activities pursued in free time that are often intrinsically satisfactory.

recreation Leisure activities occurring close to home.

sustainable tourism Tourism that meets the needs of current tourists and host regions while protecting and enhancing opportunities for the future.

tourism Leisure that involves traveling away from home for more than 1 night but less than 1 year.

Leisure constitutes an integral part of Western societies, which have increasing free time, higher disposable incomes, and an increasing demand for leisure activities. These trends are fortified by demographic trends such as an increasing number of senior citizens who engage in recreational activities or travel. Little is known about leisure in developing countries, and this may be explained to some extent by the (possibly perceived) limited economic significance of leisure and an absence of statistics on leisure behavior and trends. It seems highly likely that leisure activities in developing countries are less energy intensive than those in developed countries. However, on a global scale, it may be that particularly large countries, such as China and India, contribute significantly to energy consumption from leisure. Already, China is seen as one of the most promising markets for tourism development and also as an important source of tourists in the future.

1. BACKGROUND

Given the lack of knowledge on developing countries, this article focuses on Western countries, particularly those where leisure participation is increasingly frequent and diverse. In countries such as the United States, Canada, Japan, Germany, Australia, and New Zealand, the leisure industry continuously develops new products, provides an increasing variety of leisure experiences, and seeks to turn citizens of all ages into leisure consumers. This manifests in the increasing "commercialization" of leisure that encourages providers to compete in the market by providing more and more "thrills" and excitement. Examples of these developments include amusement and theme parks, indoor skiing centers, and entertainment complexes. Many of these offer light amusement that does not require active involvement by the leisure participant but involves considerable inputs of energy to sustain the amusement.

At the same time, in these societies there is a trend toward outdoor activities that provide opportunities for connection with the natural environment. Much of this outdoor recreation is energy intensive, not only through the activity itself (e.g., skiing, diving, boating) but also as a result of transport used to get to the natural amenity where the activity takes place.

Growing mobility manifests not only in travel to recreational activity sites but also in popular travel to foreign places either within the country of residence or abroad. This form of leisure, namely tourism, is a widely acknowledged and researched phenomenon of modern society now constituting the third largest industry globally (after the automobile and petroleum industries). Tourism is an important economic driver contributing between 10 and 12% to the world's gross domestic product.

Although nearly 80% of the world's tourism activity occurs within national borders (domestic tourism), international tourism is the faster growing segment. International visitor arrivals reached 699 million worldwide in 2000. This number is forecast to grow to 1.56 billion tourists by 2020. The top five countries in terms of visitor arrivals are France, Spain, the United States, Italy, and China; however, the market is becoming increasingly diverse, with many new destinations entering the scene. As a result of technological development, a highly competitive market with decreasing airfares, and increasing travel experience and expectations on the part of tourists, long-haul air travel is growing continuously. Tourists travel more frequently for shorter holidays to ever further destinations.

In many developing countries, tourism evolved as a promising tool to stimulate economic development while at the same time offering the potential to support efforts to protect natural and cultural heritage. Sustainable tourism, and more specifically ecotourism, has been discussed widely in this respect.

The trends just outlined result in an increasing demand for energy, mainly for transport but also for energy-intensive infrastructure and facilities. It has been recognized that recreation and tourism contribute considerably to resource use and environmental impacts such as climate change. For this reason, the UN Division for Sustainable Development identifies the indicator "time spent on leisure, paid and unpaid work, and traveling" as 1 of 17 indicators for the provisional core set of "indicators for changing consumption and production patterns."

2. CONCEPT OF LEISURE, RECREATION, AND TOURISM

Leisure has been defined from various perspectives. Most often, leisure is understood as time free of obligations to work, family, or community or to meeting other essential personal needs. Leisure further includes experiences, which are a source of intrinsic satisfaction and which are commonly associated with self-chosen behaviors—passive or active.

Recreation and tourism are leisure activities that have to be seen within the context of leisure. Recreation typically occurs spatially and temporally within reach of home, whereas tourism is leisure away from home. It can be argued that the same factors that encourage tourism (e.g., cheaper travel) also encourage people to extend their spatial and temporal expectations with regard to recreational activities. For example, a kayaking club might not be satisfied with local river conditions and may organize trips away from home to experience more demanding conditions. In this sense, recreation, travel, and tourism may be combined.

However, global statistics draw a line between tourists and recreationists by defining a tourist as any person being away for more than 24 h but less than 12 months from his or her usual environment. By this definition, tourists engage in travel and require accommodation; therefore, they rely on a broader range of services and products to meet their needs than do recreationists.

Visitors who travel for business or to visit friends or relatives are usually included in tourism statistics because they do not differ from "leisure tourists" in their infrastructure needs despite their differing travel motivations. Research on the emerging segment of MICE (meetings, incentives, conventions, and exhibitions) tourists has shown that these travelers engage in recreational activities to a substantial degree during their trips.

Much leisure takes place in the home. Watching television, reading, and listening to music are generally the most frequently undertaken activities. A compilation of studies of leisure participation in Western countries by Cushman, Veal, and Zuzanek provides country-specific information on leisure behavior. For example, Australians walk for pleasure and go swimming, in addition to engaging in home-based entertainment and visiting friends or relatives, in their free time. The most popular activities in Japan are eating out, driving, domestic sightseeing, karaoke, watching videotapes, and listening to music. For the American market, it was found that 94.5% of all U.S. citizens take part in outdoor recreation. Walking (66.7%), visiting a beach or waterside (62.1%), and family gatherings (61.8%) were the most popular activities. More energy-intensive forms of outdoor recreation, such as boating (30.0%) and off-road driving (13.9%), were also undertaken by large numbers of U.S. citizens.

The range of activities just listed indicates varying degrees of energy use associated with both recreation and tourism. Energy is demanded by leisure participants (e.g., consumption of electricity or petroleum) or by the industry. The leisure industry includes those involved in providing goods, services, and facilities to people in their leisure time. Conceptually, it is useful to distinguish energy demand for upstream activities (e.g., construction or production of goods) and downstream activities (e.g., waste disposal) from energy use resulting from the operation of a recreational facility or activity.

2.1 The Leisure Industry

This article discusses the energy costs of leisure, starting with recreational activities and continuing with accommodation and transport. Recreational facilities or locations (this includes home) are used and visited by both recreationists and tourists. In the context of tourism, these are often called tourist attractions (built structures or natural attractions such as beaches and national parks). Tourist attractions constitute the core of the tourism product because they are the very reason why tourists travel. However, to support leisure activities, a much larger infrastructure and industry are required. Leisure participants get to their activities through various means of transport (walking and cycling are considered transport modes as well). Furthermore, they rely on (often activity-specific) infrastructure and equipment provided by ancillary industries. Tourists require transport and accommodation, and sometimes they also need the services of tour operators. International tourists make further arrangements for travel to their destinations. The various levels of leisure activities and associated services demanded by recreationists and both domestic and international tourists are illustrated in Fig. 1.

Ancillary services and tour operators are not included in the discussion. Although these industries are an important part of the whole leisure industry and are also likely to exert a substantial degree of influence through their purchasing behavior (cf. the tour operators initiative by the UN Environmental Program [UNEP]), they are generally small contributors in terms of energy use. The food and beverage industry is another important industry that provides services to leisure participants. However, little information is available, and this industry is not discussed separately.

3. UPSTREAM/DOWNSTREAM

Upstream and downstream activities constitute the indirect component of energy use such as the construction of recreational facilities and infrastruc-

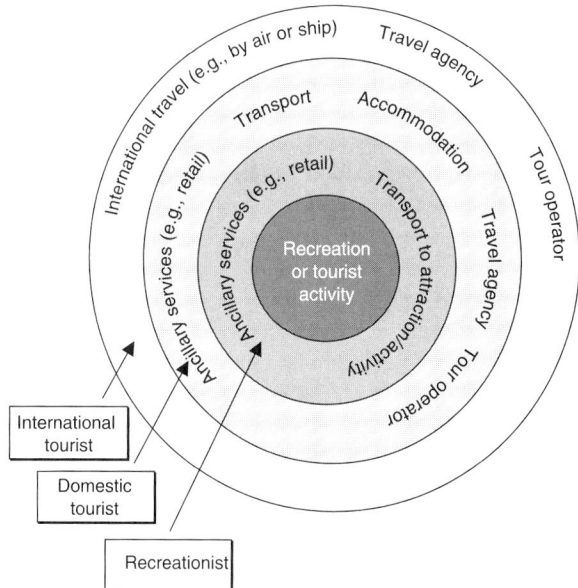

FIGURE 1 Overview of leisure components for recreationists, domestic tourists, and international tourists.

ture and the production of activity-specific equipment. For example, the theater industry uses direct energy for heating and lighting and uses indirect energy via cleaning services, shops that sell the cleaning agents, transport operators who deliver the cleaning agents, and so forth. Energy involved with these processes is called "embodied" or indirect energy. The increasing commodification of leisure has contributed to an increased direct and indirect input of energy and material.

For some activities, energy use associated with upstream activities is considerable such as in motor sports and air activities (e.g., [para]gliding, skydiving). Skiing and snowboarding are other popular activities that require large energy inputs outside of the actual activity. In a recent study in Switzerland, Stettler found that the energy input for infrastructure and equipment amounts to 200 MJ per skier-day compared with 55 MJ for tennis and 15 MJ for soccer. In these examples, the indirect energy use is more than twice the direct energy use. In the case of snow sports, considerable energy costs result from artificial snowmaking, which can be considered an upstream activity because it is often undertaken to enable ski fields to open early in the season. A study on environmental impacts resulting from leisure in Japan found that the indirect energy use from leisure is about seven times larger than the direct energy use, whereas a study undertaken in New Zealand indicated that indirect energy costs of tourism are about the same as direct costs. In both cases, the consideration of flow-on effects adds a substantial amount of energy to the direct energy costs.

Little specific information is available on downstream activities such as the disposal or recycling of equipment. However, it is believed that considerable energy use is involved with the disposal of sometimes highly technical sports equipment. Similarly, waste management associated with tourist accommodation is likely to consume considerable amounts of energy such as for transportation but also for waste (both solid and liquid) treatment.

Information on indirect energy use of leisure is scarce; therefore, this article focuses on direct energy use of recreational and tourist activities. It is acknowledged that total (both direct and indirect) energy use exceeds the direct energy use discussed in the remainder of this article.

4. RECREATIONAL ACTIVITIES

The discussion of recreational activities in this section focuses on active recreation away from home (often tourism related) because it is believed that energy use of such activities is larger than that of passive entertainment at home. It is acknowledged that home-based activities consume some energy, both directly (e.g., electricity) and indirectly (e.g., embodied energy in equipment or appliances). Recreational activities away from home include both visits to attractions as a permanent establishment or point of interest and "activities" in the true sense of the word. Activities include walking, swimming, biking, taking photographs, water sports, mountaineering, and taking guided tours.

4.1 Operation of Attractions and Activities

Many people engage in leisure on an individual basis (e.g., going for a walk, reading a book). Many of these activities require little direct energy input. However, an increasing number of activities are offered through institutions or businesses that consume energy for their buildings, operations, and transport.

The energy use of various recreational attractions or activities is strongly dependent on visitor volumes. The largest amusement park worldwide is Disneyland Tokyo, attracting more than 16 million visitors per year. On the other hand, there are many small (tourism) businesses operating for an exclusive market of several hundred visitors per season.

In a topical report on tourism, the European Environment Agency stated that recreation facilities can constitute a big source of energy demand in the local context. Ski fields in France consume between 2.1 and 2.6 PJ for 4000 ski lifts. Similarly, leisure facilities in urban environments are often main consumers of energy. The opera theater in Zurich, Switzerland, consumes approximately 1190 GJ per year, and the ice skating stadium in Berne, Germany, consumes approximately 5950 GJ per year. The very popular Japanese activity of playing Pachinko (similar to pinball) consumed a total of 49 PJ in 1995, just behind the energy consumption of eating and drinking places (59 PJ) and well ahead of other leisure activities (e.g., sports facilities, gardens, and amusement parks [9.3 PJ]).

In New Zealand, tourist attractions (e.g., museums, parks, historic sites, geothermal sites) consume on average 400 GJ per year, and entertainment facilities (e.g., shopping malls, theaters, amusement parks) consume approximately 1600 GJ per year. In contrast, businesses that offer recreational activities (e.g., scenic boat cruises, guided walks, adventure tour operators) are comparatively small energy users, using approximately 500 GJ per business per year. When comparing these attractions and activities on a per capita basis, the attractions become energy efficient, with an energy use of approximately 20 MJ per visit, compared with activities that require energy inputs of at least 50 MJ and up to approximately 2000 MJ per visit (e.g., scenic flights). Activity packages are energy intensive because they often offer an individual style and service orientation that requires energy at various stages as compared with (mass tourism) attractions. Furthermore, leisure activities often build on the use of motorized vehicles, either to get to the attraction or activity or for the activity itself (e.g., snowmobiling). In general, the input of petroleum fuels is characteristic of "activities," whereas "attractions" require electricity for their built environment.

An example of a typical tourist activity package in New Zealand is guided (alpine) walks (Fig. 2). Energy is consumed in operating the building where customers book the tour and receive their equipment. A shuttle bus takes customers to the glacier, where they start the guided glacier walk or go to the helicopter pad in the case of heli-hikes. After the activity, tourists are taken back to the main building, where they are served hot drinks. The total energy use per tourist is 88 MJ on average. Energy use per visit or per participation in an activity is a useful indicator for comparing various leisure activities (Table I).

The degree to which an activity depends on motorized vehicles largely determines its overall energy intensity. Sports traffic is an interesting example of the use of transport for recreation. For example, the average travel distance to sports activities in Switzerland was estimated to be 25 km return, resulting in an energy use of 60 MJ per sportsperson (considering specific transport mixes). Travel distances differ considerably for sports activities; the average one-way travel distance for auto sports is approximately 710 km, for skiing 87 km, for golf 32 km, and for jogging (the lowest) 4.6 km. Most sports travel is based on private motorized vehicles (78%), whereas only 18% are by public transport and 4% are by bike or on foot. Overall, sports traffic makes up more than 10% of primary energy demand in Switzerland; in 1995, 24 PJ was used by sports participants, 1.8 PJ by trainers and officials, and another 3.7 PJ by spectators.

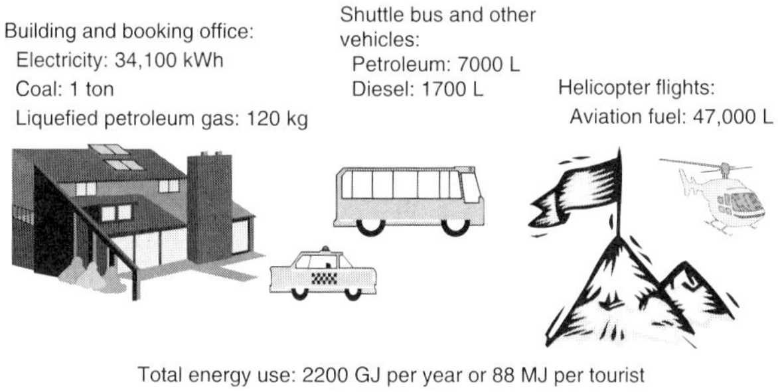

The relatively new, yet very popular, "ecotourism" and "ecotourism activities" seem to be responsible and nature-based forms of tourism that seek to contribute to nature conservation, yet they are rarely discussed in the light of energy use. However, many ecotourism operators provide "ecotours" that take visitors to natural assets with various types of motorized vehicles, often four-wheel drives or boats. The environmental impact resulting from the energy use embodied in these tours has often been neglected, with researchers focusing on local environmental impacts. Clearly, against this background, there is a strong contradiction within the concept of ecotourism.

4.2 Best Practice

Many initiatives have been reported to minimize the energy use of leisure activities. For example, the Australian National Maritime Museum in Sydney seeks to increase energy efficiency by improving the saltwater-cooled air-conditioning system, reducing fan speed, installing lighting sensors, and using energy-efficient office equipment. These initiatives resulted in considerable cost savings. Another example is SeeQuest Adventures (Canada), which has renounced motorized vehicles and has clients walk or paddle to campgrounds on its guided tours. The replacement of gasoline motors with electric ones by Costa Rica Expeditions helped to decrease the use of

Building and booking office:
Electricity: 34,100 kWh
Coal: 1 ton
Liquefied petroleum gas: 120 kg

Shuttle bus and other vehicles:
Petroleum: 7000 L
Diesel: 1700 L

Helicopter flights:
Aviation fuel: 47,000 L

Total energy use: 2200 GJ per year or 88 MJ per tourist

FIGURE 2 Direct energy use per year of a guided alpine walk operator in New Zealand.

TABLE I

Examples of Energy Intensities for Various Leisure Activities

Activity	Direct energy use per visit (MJ)	Source	Country of reference
Watching television (3.8 h/day)	1.8	Müller	Switzerland
Experience centers	29	Becken and Simmons	New Zealand
Museums, art galleries	10	Becken and Simmons	New Zealand
Gondola ride to Pilatus (Luzern)	8.5[a]	Geisel	Switzerland
Restaurant meal	18.0	Müller	Switzerland
Swimming in public pool	46.8	Müller	Switzerland
Swimming in public pool	119[a]	Motiva	Finland
Indoor ice skating	28.8	Müller	Switzerland
Skiing	90[a]	Motiva	Finland
Golf	12	Becken and Simmons	New Zealand
Rafting	36	Becken and Simmons	New Zealand
Heli-skiing	1300	Becken and Simmons	New Zealand
Scenic boat cruises	165	Becken and Simmons	New Zealand

[a] Primary energy, otherwise secondary energy (measured as heat content).

fossil fuels, whereas Odyssey Safaris reduces fuel consumption by training drivers and maintaining vehicles (e.g., tire pressure). Tasmania's West Coast Charters (Australia) achieved energy savings worth $6000 (Australian) per year by using wind-powered yachts instead of motorized boats.

Australia has implemented the first Nature and Ecotourism Accreditation Programme to ensure that an accredited company is committed to best environmental practice and provides quality experience. Energy efficiency is one key criterion in this program, specified as minimizing energy use in natural tourism and ecotourism facilities (including a biannual energy audit) and minimizing energy use for transport.

It can be concluded that the energy requirements of recreational activities range from minimal to extremely large. When attractions or activities are operated by businesses, the determining factors for varying energy efficiency include the visitation levels, management style and commitment, service level (e.g., shuttle transport), technical equipment, fuel mix, and vehicle use (including type of vehicles, occupancies, and travel distances).

5. TOURIST ACCOMMODATION

5.1 Description of the Accommodation Sector

The accommodation sector is extremely diverse. Not only are there very different types of accommodation within specific regions, but there is also a large diversity worldwide as a result of different cultural backgrounds. "Indigenous tourism" often involves staying at traditional housing forms such as long houses in Indonesia, maraes in Fiji, and yurts in Mongolia. In Western countries, commercial accommodation categories include mostly hotels (often rated through quality systems such as stars), resorts, motels or motor lodges, apartments, youth hostels and other budget accommodation, camping grounds (often with cabins or caravans), and partially commercial forms such as homestays, farmstays, and bed & breakfasts.

The energy use while staying in the private homes of friends or relatives is another important aspect of tourism accommodation. For example, U.S. domestic travelers spend 46% of their nights during travel at the homes of friends or relatives. In New Zealand, the share is even bigger, with domestic tourists staying more than half of all visitor nights at private homes.

Commercial accommodation offers various services in addition to a sleeping facility. These include washing and cooking facilities, eating and drinking places, entertainment services such as television lounges and casinos, and other amenities such as swimming pools, tennis courts, and fitness studios. At the very top end of services provided are "all-inclusive" resorts, where all tourist activities are provided in a packaged form through the accommodation business. This form of accommodation is widely established in the Caribbean and is now developing in places such as Mexico, Gambia, Malaysia, and the Mediterranean. In the case of all-inclusive holidays, the accommodation business often becomes a tourist destination in its own right, making the geographic location largely irrelevant. This high service level clearly has implications for the total energy use given the high energy inputs required to sustain such megaresorts. However, parts of this may be outweighed by the decreased transport needs of all-inclusive tourists.

5.2 Energy Use by Various Accommodation Categories

In strongly developed tourism destinations, the accommodation industry may contribute substantially to total energy demand. For this reason, a "roundtable" on ecoefficiency in the tourism sector was held in 1997 in the Caribbean. A large portion of electricity used on the Caribbean islands is to satisfy the large energy demand of the numerous tourist resorts (consumption of 6.1 PJ in 1994). This energy consumption carries a large "rucksack" given that electricity is usually generated from oil shipped to the islands. Total energy use for accommodation amounts to 29.2 PJ per year in Japan (0.1% of total national energy demand of $\sim$25,000 PJ), whereas accommodation in New Zealand consumes approximately 4 PJ per year ($\sim$1% of total national energy demand of 450 PJ).

For a single accommodation business, the energy use depends on a number of factors such as climate, size and type of building, technological standard, staff awareness, and visitation levels. For example, in the case of hotels, operational energy use ranges between 2000 GJ per year (e.g., New Zealand) and 3000 GJ per year (e.g., North America). Most energy savings could probably be achieved through the design, construction, or refurbishment of places by considering aspects of location and building design.

An important factor for energy use is the level of service provided. Accommodation can be split into two broad categories: "comfort oriented" and "purpose oriented." Hotels, luxury lodges, and resorts clearly fall into the first category, whereas youth hostels, huts, and camping facilities are purpose oriented. This distinction makes it easier to establish benchmarks or codes of practice for accommodation businesses, which otherwise would ignore very different consumer needs (e.g., a business traveler vs a rucksack tourist) and, hence, different energy requirements to meet these needs.

The energy use per visitor night differs considerably between comfort-oriented and basic forms (Table II). Hotels and bed & breakfasts consume about three to five times the energy of motels, backpacker hostels, and campgrounds.

Energy costs constitute approximately 3 to 10% of business costs, which is the third highest cost after staff and food. Space heating or cooling, water heating, and refrigeration account for most of the energy use of accommodation businesses (Fig. 3). In warmer climates, space heating is redundant, but hotels use air conditioning to keep temperatures comfortable. In the Caribbean, air conditioning accounts for 44% of the energy use in a large hotel (more than 100 rooms).

The energy mix determines energy costs and environmental impacts. Most accommodation businesses rely mainly on electricity. For example, in New Zealand, 94% of the energy use for motels is supplied by electricity, compared with 74% for hotels and 44% for campgrounds. When generated from fossil fuels, electricity has a large amount of embodied energy. On the other hand, when based on renewable sources, particularly at remote locations (e.g., ecotourism lodges), equipment based on electricity (e.g., for water heating) has the potential to be sustainable.

5.3 Best Practice

Examples of best practice include building designs allowing for maximum use of sunlight, cross-ventilation, and weather protection; solar panels; a quantum hot water system; low-watt fluorescent lights and other energy-efficient equipment; and the installation of guest room occupancy sensors for heating, cooling, and lighting.

There are various environmental practice codes that support businesses in decreasing energy use. *Case Studies on Environmental Good Practice in Hotels* by UNEP and publications by the International Hotels Environment Initiative are two examples. There are also location-specific codes such as the energy code for energy-efficient buildings in Hawaiian hotels developed specifically for the unique conditions of Hawaii.

Single businesses engage in best practice to save costs, to decrease environmental impacts, and to gain a competitive advantage. For example, Turtle Island, Fiji, uses solar power for hot water and wind

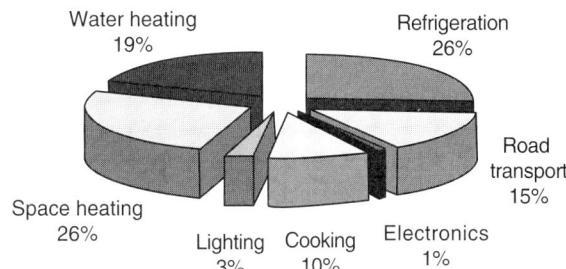

FIGURE 3 Energy end use in the New Zealand hotel sector. (Source: Energy Efficiency and Conservation Authority. (1996). Hotel sector energy use: Highlights. *Energy-Wise Monitoring Q.*, No. 4 [Wellington, New Zealand]).

TABLE II

Examples of Energy Intensities for Various Accommodation Categories

Accommodation type	Direct energy use per visitor-night (MJ)	Source	Country of reference
Hotel	151	Müller	Switzerland
Hotel	200	Brunotte	Germany
Hotel	circa 260	Gössling	Zansibar
Commercial accomodation	94	Bundesministerium für Umwelt, Naturschutz und Reaktorsicherheit	Germany
Hotel	155	Becken *et al.*	New Zealand
Bed & breakfast	110	Becken *et al.*	New Zealand
Motel	32	Becken *et al.*	New Zealand
Backpacker	39	Becken *et al.*	New Zealand
Campground	25	Becken *et al.*	New Zealand

generators for other electricity use. The Binna Burra Mountain Lodge in Queensland, Australia, is a Green Globe 21-certified company that reduced electricity use by 20% through energy-efficient lights. Total energy use was reduced by 15,000 kWh (189 metric tons CO_2) per year as a result of weatherproofing buildings, replacing diesel with gas for water heating, and using gravity for water supply. Transport energy use can also be reduced by transporting customers and/or staff in mini-buses to the accommodation businesses (e.g., Freycinet Experience, Tasmania, Australia), by replacing four-wheel drives with two-wheel drives (e.g., Australis, Australia), and by encouraging cycling. The need for travel can also be reduced by increasing the length of stay at a single location (e.g., Crystal Creek Rainforest Retreat, Queensland, Australia).

6. TOURIST TRANSPORT

6.1 Importance of Transport

Travel for leisure purposes is different from other travel in that it is chosen largely in terms of destination, length of travel, and transport mode. It has been estimated that leisure travel constitutes more than half of all travel in Western countries. Transport for recreation has already been discussed in the context of energy use associated with recreational activities or attractions. In this section, transport is examined in a broader tourism-related context.

Mobility is an essential prerequisite for tourism. Research on tourism systems has long emphasized that transportation connects the tourist generating and the destination region via transit routes and that transportation also ensures mobility within a destination, either to visit tourist attractions or to travel along recreational routes. Often, transport is more than a means to get from one place to another; it also constitutes an attraction in itself, as in the case of cruising trips, scenic flights, and railway journeys.

In the case of single-destination trips (often resort based), transport is used mainly to travel to the destination or to undertake multiple-day trips from the base location. An increasing number of trips are to more than one destination; this includes "touring tourists," who stay in different places within one country, and tourists who visit different countries on one trip such as round-the-world travelers. New Zealand is a typical site for both types of multiple-destination travel. Approximately 75% of all international visitors travel to more than one place within New Zealand, and 36% visit another country on their way to or from New Zealand.

The development of various forms of transport, such as private vehicles, high-speed trains, and airplanes, makes it possible for a large number of tourists to engage in travel. This manifests in mass tourism phenomena such as organized coach tours and chartered flights to resort holiday destinations. Today, one in five German air travelers is going on a package holiday by charter flight.

6.2 Transport Modes Used by Tourists

Leisure travel is based largely on private vehicles and airplanes. A study on travel in the United States in 1995 showed that a large majority of trips (81%) were made in personal-use vehicles and that these trips accounted for more than half of all person-miles. Air travel, although undertaken by only 15% of all travelers, accounted for most of the other person-miles. Other transport modes played a minor role for U.S. citizens (Table III). The situation is similar in New Zealand, where 77% of all domestic tourist trips are car based, with another 13% being airline flights. In the European Union, the share of other transport modes is higher. Although most travel is still by road (61%) and by (21%), train (15%) and ship (3%) are also frequently used modes of travel. The use of public transport for leisure is generally low because it is often perceived unreliable, expensive, time-consuming, and inconvenient.

TABLE III

Travel (>100 Miles) in the United States

Transport mode	Person-trips (thousands)[a]	Percentage	Person-miles (millions)	Percentage
Personal-use vehicle[b]	813,858	81.3	451,590	54.6
Commercial airplane	155,936	15.6	347,934	42.1
Intercity bus	3,244	0.3	2,723	0.3
Chartered tour bus	14,247	1.4	9,363	1.1
Train	4,994	0.5	4,356	0.5
Ship, boat, or ferry	614	0.1	1,834	0.2

Source. 1995 American Travel Survey.
[a] Of these trips, 23% are business trips, 33% are for visiting friends or relatives, and 30% are leisure trips.
[b] Private car, truck, or camper van.

The distance traveled per trip is considerable. For example, in the European Union, tourists travel on average 1800 km per trip, whereas in New Zealand, domestic tourists travel 623 km per trip (with 4.4 trips taken per person per year) and international tourists travel 1950 km during their stays in New Zealand. In 1995, the average trip length of Americans in a personal-use vehicle was 590 km and in the case of air travel was 1732 km.

6.3 Energy Use of Transport

In most countries, transport is an important source of energy demand. Worldwide transportation contributes 20 to 25% to all fossil fuel consumption. Given the dominant role of leisure in transport demand, there is a clear need to integrate leisure traffic into transport management plans and strategies to reduce energy use.

In Japan, total energy use associated with leisure transport amounted to 260 PJ in 1995, with the largest single contributor being rental car and taxi services (98 PJ). Tourism transport alone (excluding recreation) makes up 4% of New Zealand's national energy consumption. A recent report on tourism and the environment by the German government states that inbound holiday travel (excluding business and visiting friends or relatives) makes up 7% of all passenger travel's greenhouse gas emissions.

The most energy-intensive forms of travel are by air and sea. On a per passenger-kilometer (pkm) basis, air travel consumes between 2 MJ for international flights (e.g., 1.9 MJ/pkm for Lufthansa, 2.2 MJ/pkm for British Airways, 2.5 MJ/pkm for Scandinavian Airlines) and 3 MJ for domestic flights (e.g., 2.75 MJ/pkm for Air New Zealand). Ferry travel is comparably energy intensive (e.g., 3.5 MJ/pkm for Sydney Harbor Ferries, 2.0 MJ/pkm for Brisbane Ferries). In comparison, private cars consume between 3 and 4 MJ per vehicle kilometer; however, the per capita energy use can be decreased substantially by high load factors. The energy intensity of leisure-specific transport modes in the New Zealand context is presented in Table IV.

Vehicles that are designed particularly for recreational purposes, such as jet boats, motorboats, snowmobiles, and off-road vehicles, are often comparatively energy intensive. This may be explained by the fact that power and speed, rather than efficiency and economy, are features of the leisure experience.

On the other hand, a range of nonmotorized recreational transport options exist. These include cycling, walking, horseback riding, kayaking, sailing, and cross-country skiing. It is likely that these nonmotorized recreational transport options appeal to only a small sector of recreational consumers. Cycling has the largest potential to be accepted as an alternative transport mode, whereas the other examples are likely to be considered as mere recreational activities. Many countries recognize the integration of cycling into transport planning as an important tool for achieving sustainable transport. The Netherlands, Denmark, Germany, and Great Britain already have extensive cycle routes for recreational biking and networks that support long-distance cycle tourism. The conversion of former rail tracks into multipurpose user trails for bikers and walkers in North America has proven to be very popular.

TABLE IV

Energy Intensities of Transport Modes Used by Leisure Travelers in New Zealand

Transport mode	Direct energy use per passenger-kilometer (MJ/pkm)
Domestic air	2.75
Private car	1.03
Rental car/Taxi	0.94
Organized coach travel	1.01
Scheduled (intercity) bus	0.75
Backpacker bus	0.58
Camper van	2.06
Train	1.44
Cook Strait ferries	2.40
Yacht/Boat	1.75
Other ferries	3.53
Motorcycle	0.87
Helicopter	4.68

Source. Becken, S. (2002). "Tourism and Transport in New Zealand: Implications for Energy Use" (TRREC report 54). Lincoln University, New Zealand.

6.4 International Travel

International travel, which includes both air and sea travel, plays a special role because it is responsible for the largest share of energy use. Furthermore, it is not accounted for in national energy or greenhouse gas accounts. Because air travel in particular is growing consistently at rates of approximately 9% per annum, paralleling tourism growth rates, energy use by international travel is becoming increasingly important.

Aviation contributes approximately 2 to 3% to global energy demand, and at least half of this energy use is due to tourist travel. An increasing number of countries, often developing or remote ones, rely on international tourism as a major economic activity and foreign exchange earner. However, associated with this is the dependency of these countries on fossil fuels. External costs of aviation are not accounted for at this stage in international agreements and airfares, but once they are implemented (e.g., through the Kyoto Protocol during the second commitment period), they could severely affect these destinations.

In the New Zealand case, the total energy use involved in transporting tourists to the country and back home by air amounts to 55.6 PJ per year. If this were added to the national energy account, the total would increase by more than 12% and render compliance with international agreements (e.g., the Kyoto Protocol) difficult.

Similar considerations apply to cruise ship tourism, although the total tourist volumes are much smaller than those for air travel. In 1994, there were approximately 5 million cruise passengers worldwide who undertook some 37 million cruising-passenger days. Many of these tourists buy "fly and cruise packages." Energy use by these tourists has rarely been analyzed for several reasons. First, cruising companies are large international firms that are reluctant to reveal commercially sensitive energy consumption figures. Second, fuel consumption varies considerably among different types of cruising ships. Third, the fuel is consumed for transport and electricity generation, so it includes on-board hospitality functions. Finally, cruise ship tourists consume additional energy on their often frequent land trips, and this renders accounting for the total energy requirements of cruise tourists even more complex. On the basis of shipping companies advertising on the World Wide Web, it is estimated that a cruising ship consumes between 1500 and 4000 MJ per passenger-day, and this is large compared with other transport or accommodation options.

6.5 Best Practice

A large number of initiatives have been made by the tourism industry to make transportation more sustainable. These are represented in business actions and initiatives by tourist destinations.

Business initiatives exist at different levels, ranging from small and locally operating companies to large international organizations. The largest reductions in energy use are expected from large organizations. For example, Budget Rent-a-Car rents electric cars in France and operates an electric shuttle bus at the Los Angeles airport (LAX). It has established a national bike hire service and distributes fuel-saving tips to customers. Similarly, Avis increased the percentage of small cars in its fleet and offers renting schemes based on mileage instead of days to provide an incentive to travel less.

British Airways is well known for its environmental actions, for example, by reducing the energy use of land-based operations, by providing mileage allowances for business travel by bicycle, and by encouraging the use of public transport. Both KLM Royal Dutch Airlines and Lufthansa developed "Air–Rail Projects" where connecting flights are substituted by rail travel. This initiative is supported by the German Touristik Union International (TUI). Furthermore, Lufthansa seeks to continuously modernize its fleet, improve air traffic management (navigation), and decrease ground energy use by employing cogeneration to heat water, leasing bicycles to staff, optimizing vehicle travel, and using flight simulators. The latter alone saves 450,000 tons of fuel per year. The East Japan Railway Company aims to decrease energy consumption by reducing the weight of train cars and by using regenerative brakes. Electric power plant generators will be replaced by combined steam and gas turbine cycle generators.

Tourist destinations have developed a variety of strategies to decrease transport, partly because the attractiveness of many tourist resorts is reduced by noise, pollution, and congestion. The Gemeinschaft Autofreier Schweizerischer Tourismuorte (GAST) is one example, where nine communities joined in 1988 to promote car-free tourist resorts. Although the use of motorized vehicles in these nine resorts is prohibited, most tourists (71% during summer, 69% during winter) still travel by car to and from the resorts. Another program in the European Alps, Sanfte Mobilitaet, explored the potential of cards for travel on local public transport (Bodenmais, Germany), cycle routes (Oberstdorf, Germany), and a "Talerbus" that connects several neighboring valleys (Lungau, Austria). The Talerbus proved to be very successful by attracting a large number of visitors. However, problems arose when the short-term funding provided by the European Union expired and the future existence of the bus system became uncertain.

The numerous actions by operators and destinations to decrease traffic are promising. However, it is the tourists or recreationists themselves who make travel decisions and who exert substantial influence through their consumer behavior. In this sense, it is critical to incorporate travel behavior into current transport management.

7. CONCLUSION

Tourism and recreation, especially outdoor recreation, depend on natural resources and a healthy environment. For this reason, tourism is often seen as a potential ally of nature conservation. However, the large energy consumption linked to leisure makes this association tenuous. The total leisure-related energy use in Switzerland was estimated to make up approximately 43% of total national energy use; 16% of total energy is consumed in the living environment (home), for example, 11% in leisure transport, and 6% each in recreational activities and tourism In Japan, the total energy use for leisure contributes about 12% to national energy demand. Tourism alone makes up approximately 6% of energy use in New Zealand, not including energy use of international travel to and from New Zealand.

Recreational facilities (e.g., parks, sports grounds) and activities (e.g., walking, other physical activities) often constitute the core of leisure. However, energy use and resulting impacts are mostly local and relatively small compared with other components of the leisure industry. The accommodation industry and other ancillary services are more important in terms of energy consumption. However, the dominant energy-consuming sector is transportation. Energy use for leisure transport is considerable and is expected to continue growing, particularly due to the heavy use of private vehicles and the increasing popularity of air travel. Transport is also the dominant contributor to an individual traveler's total energy use. International tourists in New Zealand consume approximately 4000 MJ per trip within New Zealand, of which 69% is for transport. If the energy use of international return flights by overseas visitors were included, transport would constitute 97% of a tourist's total energy use. Similar relations were found for holiday travel within Germany (excluding business and visiting friends or relatives); fully 63% of total energy use was for transport, 28% for accommodation, and 9% for recreational activities. Clearly, tourism is a highly energy-dependent industry that would be severely affected by economic measures to decrease energy use such as fuel or emissions taxes.

The leisure, recreation, and tourism industries have been proactive in reducing resource use and mitigating environmental impacts. Numerous business actions illustrate a shift in practice, and the introduction of ecolabels for the tourism industry is another step in this direction. There are approximately 100 different certification programs, ranging from regional to global schemes. Green Globe 21 is an example of a global certification program that was launched by the World Travel and Tourism Council but is now an independent company. Green Globe 21 seeks to increase awareness among tourism businesses and provides tools to implement environmentally sound practices (e.g., benchmarking). Reducing energy costs is a key element of the Green Globe 21 system.

To decrease energy use associated with leisure, and more specifically energy use associated with tourism, in an effective manner, concerted action by industry and consumers will be necessary. For example, the utility of ecolabels is uncertain because it is not yet clear whether business efforts and possibly increased costs in the short term are compensated by gaining a competitive advantage in the market. It seems particularly difficult to change the transport behavior of leisure participants who expect a high level of comfort and promptness of transport during leisure time. The balance between tourism as an important industry in many countries (and as a mechanism for development in some others) and as a significant contributor to greenhouse gas emissions (mainly as a result of long-distance flights) remains.

Acknowledgments

I am grateful for the useful comments on an earlier draft by Bob Gidlow and Chris Frampton.

Further Reading

Becken, S. (2002). Analysing international tourist flows to estimate energy use associated with air travel. *J. Sustainable Tourism* **10**, 114–131.

Becken, S., Frampton, C., and Simmons, D. (2001). Energy consumption patterns in the accommodation sector: The New Zealand case. *Ecol. Econ.* **39**, 371–386.

Becken, S., and Simmons, D. (2002). Understanding energy consumption patterns of tourist attractions and activities in New Zealand. *Tourism Mgmt.* **23**, 343–354.

Brunotte, M. (1993). "Energiekennzahlen für den Kleinverbrauch: Studie im Auftrag des Öko-Instituts." Freiburg, Germany.

Bundesministerium für Umwelt, Naturschutz, und Reaktorsicherheit. (2002). "Tourismus und Umwelt." www.bmu.de/fset800.php.

Cushman, G., Veal, A. J., and Zuzanek, J. (1996). "World Leisure Participation: Free Time in the Global Village." CAB International, Wallingford, UK.

Ecotourism Association of Australia. (2002). "NEAP: Nature and Ecotourism Accreditation Programme," 2nd ed. www.ecotourism.org.au/neap.pdf.

Font, X., Buckley, R. C. (eds.) (2001). "Tourism Ecolabelling: Certification and Promotion of Sustainable Management." CAB International, Wallingford, UK.

Geisel, J. (1997). "Ökologische Aspekte zum Reisebustourismus in Luzern und Ausflugstourismus zum Pilatus: Diplomarbeit im Studiengang Geooekologie." http://cobra.hta-bi.bfh.ch/home/gsj/downloads/uni_ka/da_goek.pdf.

Gielen, D. J., Kurihara, R., and Moriguchi, Y. (2001). The environmental impacts of Japanese leisure and tourism: A preliminary analysis. www.resourcemodels.org/page8.html.

Gössling, S. (2000). Sustainable tourism development in developing countries: Some aspects of energy use. *J. Sustainable Tourism* **8**, 410–425.

Holding, D. M (2001). The Sanfte Mobilitaet project: Achieving reduced car-dependence in European resort areas. *Tourism Mgmt.* **22**, 411–417.

Høyer, K. G. (2000). Sustainable tourism or sustainable mobility? The Norwegian case. *J. Sustainable Tourism* **8**, 147–160.

Motiva. (eds.) (1999). "Enduser's Energy Guidebook for Schools." www. motiva.fi.

Müller, H. R. (1999). "ETH-Pilotprojekt "2000 Watt Gesellschaft": Arbeitsgruppe "Freizeit und Energie." Bern, Switzerland.

Life Cycle Analysis of Power Generation Systems

JOULE BERGERSON and LESTER LAVE
Carnegie Mellon Electricity Industry Center
Pittsburgh, Pennsylvania, United States

1. Introduction
2. Methods for Life Cycle Analysis
3. Brief Historical Review
4. Coal
5. Natural Gas
6. Hydro
7. Oil
8. Nuclear
9. Biomass
10. Wind
11. Solar
12. Conclusions

Glossary

discharges Discharge of liquid effluent or chemical emissions into the air from a facility through designated venting mechanisms.

economic input-output life cycle analysis (EIOLCA) A tool used to assess the environmental impact of a particular product or service by linking all the various economic transactions, resource requirements, and environmental emissions required in its manufacture.

SETAC/EPA life cycle analysis Holistically analyzing the cradle-to-grave environmental impact of products, packages, processes, and activities, from raw material acquisition to final disposition.

To make informed decisions about electricity generation options, companies, concerned citizens, and government officials need good information about the environmental implications of each fuel and generation technology. To be most helpful, an analysis must examine the life cycle of each fuel/technology, from extraction of the materials to disposal of residuals. In this article, we review studies examining the life cycle environmental and health implications of each fuel and technology. We begin with an examination of the methods of life cycle analysis, then present a brief historical overview of the research studies in this area. Finally, we review and critique the alternative methods used for life cycle analysis. Our focus is on the recent studies of the health and environmental implications of each tehnology.

1. INTRODUCTION

Increases in electricity demand and the retirement of old generating plants necessitate investment in new generation. Increasingly stringent environmental regulations, together with other regulatory requirements and uncertainty over future fuel prices, complicate the choice of appropriate fuels and technologies.

Electricity generation, a major source of CO_2, SO_2, NO_x, and particulate matter, also produces large quantities of solid waste and contributes to water pollution. To make informed decisions about refurbishing old plants or investing in new ones, companies, concerned citizens, and government officials need good information about the environmental implications of each fuel and generation technology. New issues have surfaced recently, such as discharges of mercury and total greenhouse gas emissions. Since other potential issues loom (e.g. other heavy metals), an environmental analysis must examine the life cycle of each fuel/technology, from extraction of the materials to disposal of residuals.

This article reviews studies examining the life cycle environmental implications of each fuel and technology. It focuses on the coal fuel cycle since (1) it accounts for more than half of the electricity generated in the United States; (2) historically, the coal fuel cycle has been highly damaging to the environment and to health; (3) there are huge coal reserves in the United States, China, and Russia; and (4) the fuel is inexpensive to mine and likely to be used in large quantities in the future. The article begins with an examination of the methods of life cycle analysis. Recent advances in life cycle analysis offer large improvements over the methods of three decades ago and should help in choosing among fuels and technologies as well as modifying designs and practices to lower the health and environmental costs. Then a brief historical overview of the research studies in this area is presented. Finally, the alternative methods used for life cycle analysis are reviewed and critiqued. The article focuses on recent studies of the health and environmental implications of each technology. The studies agree that coal mining, transport, and combustion pose the greatest health and environmental costs. Among fossil fuel fired generators, natural gas power turbines are the most benign technology. Light water nuclear reactors received a great deal of attention in the early literature but are neglected in recent U.S. studies. The earlier studies found that the health and environmental costs of light water reactors were low, at least for the portions of the fuel cycle that were evaluated. The studies did not evaluate the disposal of spent fuel and thus are incomplete.

2. METHODS FOR LIFE CYCLE ANALYSIS

Life cycle analysis (LCA) is needed for informed decisions about alternative fuels and technologies. Modern LCA is divided into (1) scoping, (2) discharge inventory, (3) impacts, and (4) improvement. Since a comprehensive analysis is impossible, each analyst must decide, explicitly or implicitly what will be considered in the analysis.

The Environmental Protection Agency (EPA) and the Society of Environmental Toxicologists and Chemists (SETAC) developed and formalized methods for conducting LCAs in the 1990s. The basic steps are conducting mass and energy balances of each relevant process. Thus, the analyses tend to be time consuming and expensive; the first step, scoping, refers to reducing the number of processes to be studied to a feasible level. An LCA can be quicker if the results from previous analyses are used for each process. However, using old data lowers accuracy. Unfortunately, the quantitative estimates are generally uncertain and often controversial. For example, the "best" technology can change as the scope of the study changes. Holdren's 1978 criticism that electricity life cycle analyses exclude important aspects and take insufficient care still applies.

A new approach to LCA was developed using the Department of Commerce's 485×485 commodity input-output model of the United States economy and publicly available environmental data by Lave and Hendrickson. This approach is quick and inexpensive. The disadvantage is that it is at an aggregate level.

3. BRIEF HISTORICAL REVIEW

In the early 1970s, Lave and Freeburg as well as Sagan performed the first comprehensive analyses on the effects of power plants. They found that coal posed significant environmental risks, from mining, transport, and generation. Both studies found that oil and natural gas have much smaller environmental and health costs. Finally, both found that light water reactors have an even lower health burden, although neither could assess the environmental and health burdens of dealing with spent fuel, decommissioning old reactors, or of potentially catastrophic events.

Numerous additional studies were conducted throughout the 1970s and early 1980s that also primarily focused on coal and also found major problems. More recent papers have examined new technologies and newer data. In 2002, Pacca and Horvath as well as Meier conducted studies using a new life cycle analysis tool. Some recent studies focus on environmental burdens and impacts. Some compare several fuel cycles using CO_2 payback times (the time required to "payback" the CO_2 emitted from constructing the power plant), while others evaluate the global warming potential (GWP) of a fuel/technology. Pacca and Horvath proposed a metric, global warming effect (GWE) to compare technologies based on global impact by accounting for the time dependency of GWPs. Table I shows Pacca and Horvath's estimates of the total GWE for different fuels/technologies (coal, hydro, PV, wind farm, and natural gas) after 20 years of operation.

Renewable technologies have received increasing attention in recent studies. Many studies have concluded that renewable technologies, such as wind, hydro, solar-thermal and photovoltaic, are less attractive environmentally when evaluated using an LCA.

In 2002, Gagnon et al. reviewed previous LCAs. This paper focused on hydropower (run-of-river and

TABLE I

Global Warming Effect of Five Fuels/Technologies[a]

	Hydroelectric	Photovoltaic	Wind farm	Coal	Natural gas
Output (TWh)	5.55	5.55	5.55	5.55	5.55
	Emissions (MT CO_2 equiv.)				
CO_2 ($\times 10^6$)	5	1	0.8	90	50
CH_4 ($\times 10^6$)	0.01	0.008	0.0005	0.4	0.5
N_2O ($\times 10^6$)	0.09	0.09	0.007	2	4
GWE ($\times 10^6$)	5	1	0.8	90	50

[a] From Pacca and Horvath (2002).

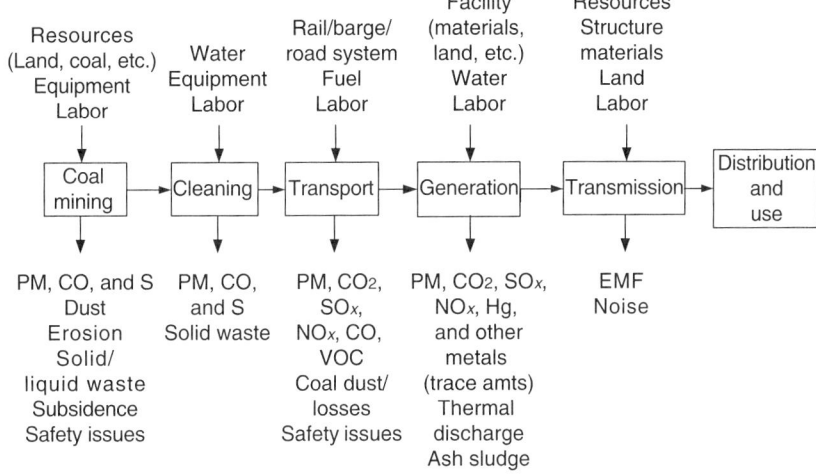

FIGURE 1 Example of life cycle of electricity (from coal).

with reservoir), nuclear energy, and wind power. They concluded that, although many studies have demonstrated technological innovation that promises to reduce emissions in one phase of the life cycle, it often increases emissions in other phases. They also suggest that a fair comparison requires including the backup power needed to achieve the same level of reliability. The environmental issues associated with management and storage of spent nuclear fuel as well as the potential for catastrophic events is only mentioned briefly while a considerable portion of the paper discusses the excellent performance of nuclear energy. This paper also suggests that estimates of land use should include the land that is damaged due to climate change and acid rain.

4. COAL

Figure 1 shows the fuel cycle of coal. We focus on comprehensive studies conducted by research teams at the Oak Ridge National Laboratory-Resources for the Future (ORNL-RFF), Argonne National Laboratory (ANL), and National Renewable Energy Laboratory (NREL). The ORNL-RFF study focused on valuing the externalities, NREL focused on a complete inventory of discharges of the main processes (including the construction and decommissioning of the plant), and ANL focused on the design of the plant and the quantification of impact on the environment.

The 1994 ORNL-RFF study, the most detailed of the three, was part of a series that also looked at natural gas, hydro, biomass, oil, and nuclear. Each study examined one plant in the southeast United States and one in the southwest United States. The plants were selected on the basis of easily available data, but are not representative of most plants in the United States. The 1999 NREL study assessed the environmental impacts of three pulverized coal boiler systems: a currently operating plant, a plant that meets the New Source Performance Standards

(NSPS), and a low-emission boiler system (LEBS) based on the design of a potential future plant. The 2001 ANL study focused on an advanced technology, an integrated gasification combined-cycle (IGCC) plant design based on the Shell entrained-flow gasifier; this plant was used to produce either electricity or both electricity and hydrogen. The assumptions made for the three studies are summarized in Table II.

Environmental standards for coal combustion have tightened considerably since 1970. For example, the current New Source Performance Standards (NSPS) are 0.60, 0.60, and 0.03 lb per million BTU of energy from coal for NO_x, SO_2, and particulate matter, respectively. A low-emissions boiler has emissions standards one-sixth of the NSPS for NO_x and SO_2 and one-third of the NSPS for particulate matter. Similarly, standards for underground mine safety, acid mine drainage, and restoration of strip mined land have become more stringent over time.

4.1 Mining

Problems from coal mining include injuries and chronic lung disease in miners, acid mine drainage, unrestored mining sites, dumping hilltops into neighboring valleys, air pollution, erosion, mining waste, subsidence, and disruption in underground water flows and storage. The environmental aspects of mining have received little analysis.

U.S. coals vary in moisture content (2 to 40%), sulfur content (0.2 to 8%) and ash content (5 to 40%). The energy content varies from lignite to subbituminous to bituminous coal. The ORNL-RFF study looks at two levels of sulfur content (0.7% and 2.1%). The NREL and ANL studies focused on high-sulfur coal (Illinois #6), providing no information about the range of coals currently used in the United States.

The ORNL-RFF study assumed that the coal came from surface mining. The NREL study examined the impacts of underground (longwall) and surface (strip) mining but concluded that the results were not significantly different. The ANL study assumed underground mining but did not conduct a full analysis of the impacts of mining.

4.2 Transportation

Coal is transported by rail, barge, truck, and pipeline. The environmental impacts and injuries vary considerably across modes. The ORNL-RFF estimates for rail injuries and deaths are shown in Table III. Transporting coal causes nearly 400 deaths annually in the United States, almost all to the general public.

4.3 Electricity Generation

Resources required to build the power plant were considered, but some studies gave little detail.

The ORNL-RFF study examined steel, concrete, land, and water. The coal feed requirements were taken into account in order to estimate the mining impacts (e.g., accidents), transportation impacts (e.g., road damage), and generation impacts (e.g., NO_x emissions). The study did not consider the resources or environmental issues associated with opening a new mine or constructing a new transport system, since they assumed that coal came from existing mines and transport systems. NREL evaluated concrete, steel, aluminum and iron, analyzing the resources used in the mining and transportation phases of the fuel cycle, including the transportation vehicles, land reclamation activities, and mining equipment.

4.4 Transmission of Electricity

Transmission has been neglected in most studies, based on the argument that all power plants require transmission. However, energy resources are located in different places, requiring different amounts of transmission. A few studies evaluate the environmental impacts of transmitting electricity. The impacts appear to be small, unless there are important health consequences from exposure to 60 hertz electromagnetic fields. In 1996, Knoepful developed and tested a framework for comparing the environmental impacts associated with various methods of shipping energy in Europe. This study concluded that for coal, generating electricity early in the fuel cycle and shipping the energy through high-voltage transmission lines can lead to significant impact reductions when compared to coal transport by barge and train. The results for oil and gas were not as clearly beneficial but have potential for environmental improvements.

4.5 Environmental Discharges and Impacts

The studies considered a wide range of discharges and impacts, from greenhouse gas discharges to water pollution discharges, and from human health impacts to recreational fishing impacts.

TABLE II

Summary of Assumptions for Three Coal Studies

	ORNL-REF SE and SW Ref	NREL			ANL	
		National average	NSPS	LEBS	Base	Co-product
Date of study	1994	1999			2001	
Plant size (MWe)	500	360	425	404	413	76/423
Technology	PC	PC	PC	PC	IGCC	H2/CC
Efficiency	35%	32%	35%	42%		
Capacity	75%	60%	60%	60%		
Type of plant	Based on two reference sites	National average	Average attaining NSPS	LEBS	Design simulation	Design simulation
Other plant details	Wet lime/limestone scrubber (90% efficient), electrostatic precipitator (99.2% efficient), Low nox burners, can meet NSPS	Baghouse filter, FGC system, heat recovery steam generator, steam turbine	Same as national average except higher FGC efficiencies	Low NO_x system with advanced burners, air staging, wet ash slagging system, enhanced FGC (CuO)	Conventional IGCC plant releasing CO_2 by combustion of the synthesis gas in a gas turbine	Production of electricity and H2 as energy carriers, 90% of CO_2 is recovered for disposal in geological storage
Coal type	Kentucky no. 9 and Navajo	Illinois no. 6	Illinois no. 6	Illinois no. 6	Illinois no. 6	Illinois no. 6
Extraction and processing	Surface mining only (strip and contour)	Raw material extraction, equipment manufacture, coal mining (surface-strip and underground mining-longwall), surface coal mining reclamation requirements, coal preparation and cleaning–jig washing (electricity and water required, refuse landfilled)			Underground mining, coarse cleaning at mine mouth (assumed refuse returned to mine)	
Transportation	Rail and truck	Transport and chemicals/materials to the mine site and power plant as well as the transport of coal—railcar, railcar, and barge, mine mouth			By rail only—coal losses from train considered, diesel fuel use, open rail cars loaded with crushed coal, did not include manufacturing diesel fuel or manufacture and maintenance of railcars	
Generation	Operation only	Operation, construction, and demolition, ash treated and landfilled or alternate use			Operation of power plant as well as construction and demolition of power plant, CO_2 and hydrogen pipelines	
Transmission	Not considered	Not considered			Not considered	
Data sources	Coal Technology—DOE 1988 coal technology, water and solid emissions—Meridian 1989, EPRI 1989, DOE 1989 material requirement emissions—Meridian 1989	Power generation—TEAM database, FETC (Utility Data Institute, 1993, 1993, and 1996; Keeth et al., 1983; U.S. EPA., 1985; Schultx, and Kitto, 1992; Combustion Engineering, 1981; Ladino et al., 1982; Walas, 1988; Collins, 1994; Larinoff, 1994; and Darguzas et al., 1997); plant details—average and NSPS plant—(Utility Data Institute, 1996), TEAM LEBS Plant—Ruth, 1997, Surface and Underground Mining—Bureau of Mines—Sidney et al., 1976 (surface), and Duda and Hemingway, 1976 (underground)			LCA analysis; LCAdvantage, process design; ASPEN simulation	

NSPS, New Source Performance Standards; LEBS, low emission boilers; PC, pulverized coal; IGCC, integrated coal gasification combined cycle; H2/CC, electricity and hydrogen as energy carriers with carbon capture. From ANL (2001); NREL (1999); and ORNL-RFF (1994).

TABLE III

Number of Deaths/Injuries from Southeast Reference Site/Year by Rail[a]

		Public	Occupational				Total
			Trans.	Maint.	Other	Total	
Injuries	Low	0.7	2.1	3.7	0.37	6.2	6.9
	Mid	0.83	2.5	4.2	0.37	7	7.8
	High	0.97	2.8	4.6	0.43	7.8	8.7
Fatalities	Low	0.28	0.0052	0.0047	0.0013	0.011	0.29
	Mid	0.34	0.0062	0.0053	0.00155	0.013	0.35
	High	0.39	0.0071	0.0058	0.0017	0.015	0.41

[a] From ORNL-RFF (1994).

TABLE IV

Comparison of Emissions from Coal Studies[a]

Emissions (ton/GWh)	ORNL-RFF		NREL			ANL		Pacca and Horvath
	Southeast ref site	Southwest ref site	Average	NSPS	LEBS	Base case	Coproduct case	
CO_2	1100	1200	1100	1000	820	940	120	900
SO_2	1.8	0.87	7.4	2.8	0.79			
NO_x	3.0	2.3	3.7	2.6	0.60			
Particulate matter (PM)	1.6	1.6	10	11	0.12			
CO	0.27	0.27	0.23	0.28	0.21			
HC	0.099	0.13	0.23	0.22	0.21			
Trace metals								
As ($\times 10^{-4}$)	2.0	2.0	0.54					
Cd ($\times 10^{-6}$)	3.0	3.0	4.5					
Mn ($\times 10^{-4}$)	1.3	1.3	0.47					
Pb ($\times 10^{-5}$)	9.0	9.0	3.3					
Se ($\times 10^{-4}$)	0.50	0.50	4.5					

[a] Adapted from ANL (2001); NREL (1999); ORNL-RFF (1994); Pacca and Horvath (2002).

The ORNL-RFF study estimated monetary values for some impacts and characterized rather than measured or calculated additional impacts. NREL distinguished between human health and ecological health when discussing the impacts associated with the production of electricity. There was no attempt to estimate the dollar loss or magnitude of these impacts. ANL considered (1) natural environment impacts, such as acidification, eutrophication, smog, global climate changes, and ecotoxicological impacts (aquatic and terrestrial toxicity); (2) human health impacts, such as toxicological impacts, PM_{10} inhalation effects, and carcinogenic impacts; and (3) natural resources impacts (the depletion of fuels and consumption of water).

ANL made an attempt to calculate the relative impact of the plant designs studied. Table IV shows that modern technology can lower the adverse discharges from a coal-fired generation plant, due both to greater efficiency and better processes. SO_2 and NO_x emissions can be lowered by almost a factor of ten and particulate matter emissions by a factor of 100. Since little or no attention has been given to CO and HC, the new technologies have little effect.

4.6 Other Considerations

4.6.1 Abandoned Coal Mine Problems

Abandoned coal mines are ubiquitous, as shown in the Office of Surface Mining's (OSM) map (Fig. 2).

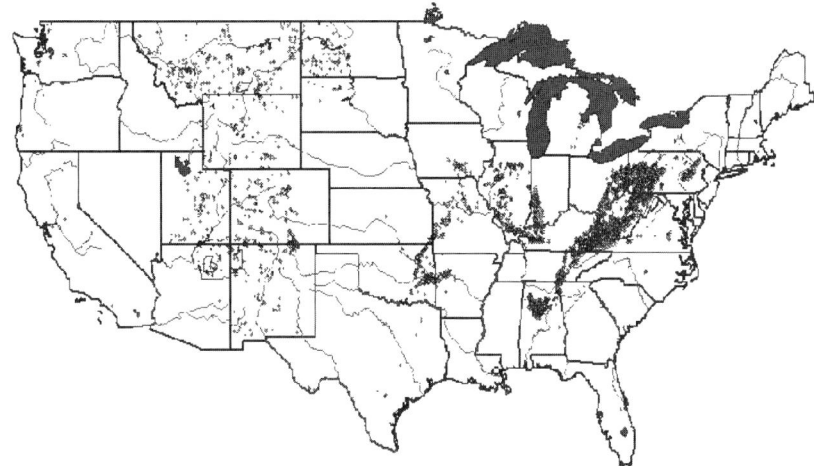

FIGURE 2 Locations of abandoned mine land problems eligible for funding from the U.S. Office of Surface Mining (OSM). From http://www.osmre.gov/aml/intro/zintro2.htm.

TABLE V

Office of Surface Mining—Unreclaimed Public Health and Safety Coal Related Problems by Problem Type[a]

Problem description	Measured as	Units unreclaimed	Cost of reclaiming problems (000$)
Clogged streams	Miles	1,700	48,000
Clogged stream lands	Acres	25,000	190,000
Dangerous highwalls	Feet	4,200,000	600,000
Dangerous impoundments	Count	760	14,000
Dangerous piles and embankments	Acres	8,800	250,000
Dangerous slides	Acres	4,700	72,000
Gases: Hazardous/explosive	Count	94	2,800
Hazardous equipment and facilities	Count	2,600	26,000
Hazardous water body	Count	970	54,000
Industrial/residential waste	Acres	400	10,000
Portals	Count	5,600	21,000
Polluted water: Agriculture and industry	Count	540	100,000
Polluted water: Human consumption	Count	4,200	3,70,000
Subsidence	Acres	8,500	480,000
Surface burning	Acres	440	17,000
Underground mine fire	Acres	4,200	860,000
Vertical opening	Count	2,400	37,000
Total			6,500,000

[a] From OSM—Abandoned Mine Land Inventory—Current Database.

Funding for site restoration is small. For example, Pennsylvania is estimated to need $15 billion worth of restoration work, while the tax collected from coal mining operations available to Pennsylvania is only about $21 million per year. The "high priority" problems monitored by the OSM are listed in Table V.

4.6.2 Acid Mine Drainage

Acid mine drainage is the main cause of polluted water in the United States with devastating effects on biological activity in many streams. In 1995, 2400 of 54,000 miles of streams in Pennsylvania were polluted by acid mine drainage from old mining operations. The summary data from OSM show that there are still a total of 4688 miles of waterways that are affected by acid mine drainage; the cost of reclaiming those waterways is approximately $3.8 billion. Other organizations estimate that the total costs to reclaim waterways from acid mine drainage are much higher ($5 billion in Pennsylvania alone). While measures have been put in place to minimize the effects of acid mine drainage, it still occurs in abandoned mines as well a small percentage of new mines.

4.6.3 Coal Mine Fires

Coal mines contain hazardous and explosive gases, and there is a potential for long-lasting fires. The OSM estimates that there are currently 4163 acres burning, including 94 sites where hazardous or explosive gas is being emitted from underground mine fires, which can have an effect on humans in the vicinity of the site. The estimated cost of extinguishing these fires is $860 million. The most extreme case in the United States is in Centralia, Pennsylvania, where an underground fire has been burning for longer than 30 years. Attempts to extinguish it have failed, leading the government to buy all the property at a cost of $42 million as well as costs associated with the attempts to fight the fire.

5. NATURAL GAS

Nearly all of the new generation in the United States in the past 5 years has been fueled by natural gas. NREL estimates that 22% of energy consumed in the United States is natural gas; the Department of Energy predicts that by 2020, 33% of the electricity will be generated from natural gas. The life cycle stages of natural gas are construction and decommissioning of the power plant, construction of the natural gas pipeline, natural gas production and distribution, ammonia production and distribution, NO_x removal, and power plant operation.

The ORNL-RFF study concluded that the major sources of damage from the gas fuel cycle are emissions of particulate matter and ozone. The gas fuel cycle has lower net emissions of CO_2 than other fossil fuel cycles but still has greater discharges than renewable energy sources. Other important environmental consequences of this fuel cycle are the consumption of water, land impacts from drilling and exploration, the potential for pipeline fires and explosions.

TABLE VI

Comparison of Three Natural Gas Studies

	Natural gas		
	ORNL-RFF	NREL	Pacca and Horvath
Technology	CCGT	NGCC	CCGT
Capacity (MW)	500	505	1000
Efficiency	45-47%		
Output	3.2 TWh		5.5 TWh
Emissions (ton/GWh)			
CO_2	640	440	500
SO_2	neg.	0.32	
NO_x	0.50	0.57	
Particulate matter (PM)	9.9	0.13	

From NREL (1999); ORNL-RFF (1994); and Pacca and Horvath (2002).

A summary of the results obtained from LCAs of natural gas are shown in Table VI. There is good agreement concerning CO_2 emissions. The studies don't make common assumptions about the plants.

6. HYDRO

Until recently, hydroelectric power was considered the most environmentally benign form of electricity. In recent years, however, many people have concluded that hydro may be one of the worst fuel cycles in terms of environmental damage. Most major waterways that have the potential to be used as hydroelectric generators in the United States have already been developed. However, projects involving retrofitting current dams as well as smaller scale diversion structures are possible. The ORNL-RFF study states that more than 77,000 dams in the United States have the potential for hydroelectric power development. The discharges from the hydroelectric fuel cycle are shown in Table VII. These values represent the discharges experienced in producing the materials to construct the hydroelectric power plant.

7. OIL

Since the mid-1970s, the share of oil has declined. In 1999, petroleum generated only 3.3% of U.S.

TABLE VII
Comparison of Emissions from Two Hydro Studies

Emissions (ton/GWh)	Hydroelectric	
	ORNL-RFF	Pacca and Horvath
	2° Emissions from manufacture	Upgrade of existing dam (5.5 TWh/yr)
CO_2	8.7	5
SO_2	0.027	
NO_x	0.074	
PM	0.0052	

From ORNL-RFF (1994) and Pacca and Horvath (2002).

electricity. The ORNL-RFF study investigated a hypothetical plant to be built in 1990, assuming effective pollution abatement technologies.

8. NUCLEAR

In 1987, 107 nuclear power plant facilities were operating in the United States. No new permits to construct nuclear power plants have been issued in three decades, and no applications are imminent in the United States. A major advantage to nuclear power is that generation does not release the pollutants that are a problem with fossil fuels. However, there are major concerns about the treatment and risks associated with the generation and storage of radioactive wastes and the possibility of a catastrophic release of radioactive material, as occurred at Chernobyl. Since the 1970s, little assessment has been done in the United States, in contrast to studies in other nations. An ORNL-RFF study investigates two hypothetical plants using pressurized water reactors. This is not state-of-the-art technology but reflects typical plants in the United States today. However, the study was conducted as if a new plant were being built with this technology.

9. BIOMASS

Biomass is a renewable fuel that could substitute for much of the coal currently used. In 1999, Rafaschiari *et al.* compared an integrated gasification combined cycle plant fired by dedicated energy crops (poplar short rotation forestry) to a conventional power plant. They found that biomass had less environmental impact than coal in almost all of the ecoindicators and normalized effects considered in this study. The most significant environmental effects from this fuel cycle are caused by the use of chemicals and fertilizers.

Several recent studies have concluded that the use of crops to generate electricity is preferred to their use as transport fuels from both an ecological and socioeconomical criteria. However, financial incentives are required to make these crops competitive fuels for electricity generation.

Some studies focus on the greenhouse gas emissions of this fuel cycle while others insist that considering ecological and socioeconomic sustainability of biomass crops is essential to gaining a clear perspective of this fuel cycle. A comparison of co-combustion between different biofuels and hard coal for electricity production from hard coal alone has also been conducted.

The ORNL-RFF study investigated two hypothetical plants. This study concluded that there are significant differences in damages, and thus externalities, among different sites (for example, benefits from erosion reduction differ by a factor of three) and for different biomass technologies. The use of advanced biomass conversion technologies could reduce NO_x emissions significantly compared to conventional wood burners. This biomass fuel cycle has near-zero emissions of CO_2. This study concluded that a biomass fuel cycle has less impact on global climate change than the fossil fuel cycles.

10. WIND

Wind power has been used by humans for thousands of years. It has been used to generate electricity (on a small scale) since the early to mid 1900s. By 1995, it was estimated that there were 17,000 commercial wind turbines in the United States. The main advantage of wind is that the generation phase does not emit environmentally harmful pollutants. However, there are several major issues to consider. In evaluating wind, it is important to account for the environmental impacts associated with the manufacture of the wind turbines as well as the land used for the wind turbine. The amount of energy that can be extracted from wind goes up with the cube of wind speed, making location important. Good wind locations often are located far from electricity demand centers and therefore require additional transmission to deliver the electricity. Finally, due to the intermittency of this technology, backup power is required to supply electricity when the wind is not blowing.

In 2002, Lenzen and Munksgaard reviewed studies concerning the environmental impact of wind turbines. They suggest using an input-output based hybrid technique in order to minimize the uncertainties as well as using a standardized method of assessment. In 2000, McGowan *et al.* conducted a general review of the technology, design, trends, and their subsequent environmental impact.

11. SOLAR

The sun is the earth's greatest source of energy and the source of most renewable energy. The sun radiates energy (approx. 2.1×10^{15} kWh per day on earth) in the form of electromagnetic radiation. Although biomass, hydro, and waves are indirect forms of solar energy, solar thermal or photovoltaic are the direct ways of using solar energy to generate electricity. Solar thermal technology uses the radiation directly to heat water. Photovoltaic technology converts the sun's rays directly to electrical energy. One of the advantages of solar radiation is that the conversion of electromagnetic radiation to electricity occurs without environmentally harmful discharges. However, other stages of the fuel cycle do contribute to environmental damage. One of the major environmental issues with this fuel cycle is the manufacture and disposal of solar cells and other equipment required to capture the radiation before it is transformed into electricity.

The renewable technologies, except for wind, are not used widely anywhere in the world because of their cost. Because the technology will not be disseminated widely until its costs are competitive, we assess promising prototypes. Many studies evaluate the environmental implications of fuel cycles in terms of their contribution to global warming. This is only one aspect of the life cycle and may mislead readers.

In 1996, Mirasgedis *et al.* estimated the level of atmospheric pollutants emitted during the manufacturing process of solar water heating systems. They found that the LCA gaseous pollutant emissions from the production of solar water heating systems are much less than that resulting from generating electricity through conventional means in Greece.

In 2001, Greijer *et al.* evaluated the environmental life cycle implications of a nanocrystalline dye sensitized solar cell and compared this to a natural gas combined cycle power plant. This evaluation focused on CO_2 and SO_2 emissions per kWh. They found that the gas power plant emitted about 10 times the CO_2 emissions of the solar cell. The largest impact from the solar cell was the process energy required to produce it.

In 2000, Koner *et al.* looked at a photovoltaic generator and used the life cycle energy cost analysis to compare it to fuel generators (kerosene and diesel). They found that, at current market prices, the photovoltaic generators were comparable or less expensive than the fuel generators.

The toxic and flammable/explosive gases of concern in photovoltaic power systems are silane, phosphine, and germane as well as cadmium. Recycling the cell materials is possible, but the environmental consequences must be considered. Depletion of rare materials is also a concern. Energy use in the manufacturing stage is the largest contributor to emissions. An LCA of solar systems should consider the system integration aspects such as energy storage and the treatment of imports and exports.

12. CONCLUSIONS

A substantial amount of research has explored the life cycle implications of generating electricity using a range of fuels and technologies. This work has developed the framework and life cycle method as well as the implications of each fuel/technology. Most of the U.S. research has focused on coal, since it is the fuel for more than half of the electricity that is generated. The early technologies for generating electricity from coal produced many deaths and injuries from mining and transport as well as highly polluted air and water due to acid mine drainage and burning the coal. Increasingly stringent regulatory pressure has lowered both the injuries and environmental pollution from the life cycle of using coal to generate electricity. Large remaining problems are underground mining, transport of the coal, and CO_2 emissions from burning the coal. Although the technology exists to solve the remaining environmental problems, little of that technology has been implemented. Additional incentives will be needed to solve these problems.

Nuclear-powered turbines are perhaps the most benign fuel/technology, with only relatively small amounts of injury and environmental discharges. Concerns about disposing of radioactive waste and protecting plants against mishaps or terrorist attacks are not fully resolved. The next most benign is likely to be natural gas. Biomass offers a solution to the CO_2 emissions problem, but this fuel is more expensive and may not be less polluting. Petroleum has been phased out of the U.S. electricity fuel market over the past quarter century; it is no longer important. On a life cycle basis, the renewable fuels have much higher environmental costs than might be suspected from an examination of a single part of the fuel cycle.

LCA has a major contribution to make in choosing among fuels and generating technologies, as well as in finding the parts of the fuel cycle of each that are most important to fix.

SEE ALSO THE FOLLOWING ARTICLES

Clean Coal Technology • Coal Mine Reclamation and Remediation • Electricity, Environmental Impacts of • Electric Power Generation: Valuation of Environmental Costs • Greenhouse Gas Emissions from Energy Systems, Comparison and Overview • Hydropower, Environmental Impact of • Life Cycle Assessment and Energy Systems • Nuclear Waste • Turbines, Gas • Wind Energy Technology, Environmental Impacts of

Further Reading

Argonne National Laboratory (ANL). (2001). "Life-Cycle Analysis of a Shell Gasification-Based Multi-Product Systems with CO_2 Recovery." The First National Conference on Carbon Sequestration. May 15–17. Washington, DC.

Carnegie Mellon Green Design Initiative, Economic Input-Output Life Cycle Assessment. (2003). http://www.eiolca.net/.

Gagnon, L., Belanger, C., and Ychiyama, Y. (2002). Life-cycle assessment of electricity generation options: The status of research in year 2001. *Energy Policy* **30**(14), 1267–1278.

Hendrickson, C., Horvath, A., Joshi, S., and Lave, L. B. (1998). Economic input-output models for environmental life cycle analysis. *Environ. Sci. Technol.* **32**(7), 184A–191A.

Holdren, J. P. (1978). Coal in context: Its role in the national energy future. *Houston Law Review* **15**(5), 1089–1109.

Knoepfel, I. H. (1996). A framework for environmental impact assessment of long-distance energy transport systems. *Energy* **21**(7/8), 693–702.

Lave, L. B., and Freeburg, L. C. (1973). Health effects of electricity generation form coal, oil and nuclear fuel. *Nuclear Safety* **14**, 409–428.

Lenzen, M., and Munksgaard J. (2002). Energy and CO_2 life-cycle analyses of wind turbines–Review and applications. *Renewable Energy* **26**(3), 339–362.

Oak Ridge National Laboratory and Resources for the Future (ORNL-RFF). (1992–1996). "Estimating Externalities of Fuel Cycles." Utility Data Institute and the Integrated Resource Planning Report. United States.

Office of Surface Mining. Abandoned Mine Land Reclamation. http://www.osmre.gov/osmaml.htm.

Pacca, S., and Horvath, A. (2002). Greenhouse gas emissions from building and operating electric power plants. *Environ. Sci. Technol. ACS* **36**, 15.

Rafaschieri, A., Rapaccini, M., and Manfrida, G. (1999). Life cycle assessment of electricity production from poplar energy crops compared with conventional fossil fuels. *Energy Conversion Manage.* **40**, 1477–1493.

Sagan, L. A. (1974). Health costs associated with the mining, transport, and combustion of coal in the steam-electric industry. *Nature* **250**, 107–111.

Society of Environmental Toxicology and Chemistry (SETAC). http://www.setac.org/lca.html.

Spath, P. L., Mann, M. K., and Kerr, D. R. (1999). "Life Cycle Assessment of Coal-Fired Power Production." National Renewable Energy Laboratory. NREL/TP-570-27715. Golden, CO.

U. S. Environmental Protection Agency. Life Cycle Assessment–LCAccess. http://www.epa.gov/ORD/NRMRL/lcaccess/index.htm.

Life Cycle Assessment and Energy Systems

EVERT NIEUWLAAR
Copernicus Institute, Utrecht University
Utrecht, The Netherlands

1. Definition and Role of Life Cycle Assessment
2. Goal Definition and Scoping
3. Inventory Analysis
4. Impact Assessment
5. Interpretation
6. Energy Issues in Life Cycle Assessment
7. Life Cycle Assessment of Energy Systems

Glossary

functional unit Quantified performance of a product system for use as a reference unit in a life cycle assessment study.
life cycle Consecutive and interlinked stages of a product system, from raw material acquisition or generation of natural resources to the final disposal.
life cycle assessment (LCA) Compilation and evaluation of the inputs, outputs, and potential environmental impacts of a product system throughout its life cycle.
life cycle impact assessment (LCIA) Phase of life cycle assessment aimed at understanding and evaluating the magnitude and significance of the potential environmental impacts of a product system.
life cycle interpretation Phase of life cycle assessment in which the findings of the inventory analysis, the findings of the impact assessment, or both are combined, consistent with the defined goal and scope, to reach conclusions and recommendations.
life cycle inventory (LCI) analysis Phase of life cycle assessment involving the compilation and quantification of inputs and outputs for a given product system throughout its life cycle.

The environmental impact of energy systems has a diverse origin. The combustion of fuels leads directly to emissions and potential environmental harm, whereas the use of electricity does not lead directly to environmental impacts. However, the production of electricity from fuels leads to emissions of, for example, carbon dioxide and nitrogen oxides, which can lead to climate change and acidification, respectively. If wind turbines or photovoltaic cells produce the electricity, these emissions are avoided. However, the material-intensive production of wind turbines and photovoltaic cells is associated with environmental releases. Therefore, assessing the environmental impacts of energy carriers involves not only the process of using the energy carrier but also all related processes: from extracting primary energy from nature, its conversion into secondary energy carriers, its use, and any processing of waste flows at the end. Environmental life cycle assessment is the "cradle-to-grave" approach for assessing the environmental impacts of products such as electricity and petroleum. The cradle-to-grave approach involves all steps between extracting materials and fuels from the environment until the point where all materials are returned to the environment. The methodology of life cycle assessment and its application to energy systems are the subject of this article.

1. DEFINITION AND ROLE OF LIFE CYCLE ASSESSMENT

1.1 Definition and Characteristics

A life cycle refers to the life span of a product, from resource extraction, to manufacture, to use, to final disposal. The cycle aspect reflects that materials extracted from the environment are followed until they are ultimately returned to the environment.

Life cycle assessment (LCA) refers to the analysis and evaluation of product life cycles. In general, the assessment is limited to environmental issues only, although LCA could also imply the assessment of other issues (e.g., social, economic). Therefore, LCA is shorthand for environmental life cycle assessment. Economic tools such as life cycle costing do exist but are not considered in this context as part of LCA.

LCA addresses environmental aspects of products—actually product systems, as is explained later. This is a major distinction of LCA from other environmental tools such as environmental impact assessment (EIA) and substance flow analysis (SFA). The product-oriented approach in LCA means that environmental impacts related to a product are examined. Products can be physical items such as pencils, potatoes, and cars. The term "product" also includes services such as transport and health care. Actually, it is not the physical product or service that has the focus in an LCA; it is the function that is fulfilled by the product that is central. Products are provided by product systems that fulfill a certain function. For example, milk cartons and milk bottles are provided by packaging systems that provide the packaging function. For energy systems, the products are delivered energy (e.g., in the form of petroleum, electricity, or heat) delivered by energy supply systems from primary energy carriers. Ultimately, the delivered energy is used to provide energy functions (e.g., lighting, mechanical work, production of materials).

Another important characteristic of LCA is that it addresses complete chains. In principle, all processes from resource extraction through production, consumption, and ultimately waste management are included in the assessment (cradle-to-grave analysis). In most cases, this means that the activities in the life cycle are not controlled by a single company or consumer and that companies involving themselves with LCA must face the fact that the environmental profile of their products also depends on the environmental performance of other companies.

LCA is not only comprehensive with respect to the number of processes included in the assessment; it is also comprehensive with respect to the number of environmental impacts. In principle, all environmental impacts related to the product system are addressed. The environmental impacts assessed include, but are not limited to, the environmental issues most often related to energy conversion systems: acidification, climate change, depletion of primary energy carriers, human toxicity, and ecotoxicity.

Where possible, quantitative results are presented in an LCA. These can range from listing all emissions amounts, aggregated values within environmental themes, to single-value indicators such as ecopoints or ecoindicators. However, single-value indicators reflecting all environmental impacts are value laden because this would involve weighting of different types of environmental damage.

Because of its nature, LCA is not location specific or time specific; that is, it cannot directly tell what the actual environmental impact of the product will be at a certain time or place. All phases of the life cycle (e.g., manufacture, use, waste management) are usually not concentrated on one location, and the time between manufacture of the product and waste management after discarding the product can be several years.

1.2 Energy and LCA

In many LCAs, energy plays an important role. The use of energy by itself is not always considered to be an environmental issue. The depletion of primary energy carriers and the strong link between energy use and certain other environmental issues (most notably acidification and climate change) have often led to the use of energy as an environmental indicator. Energy supply systems are also a major contributor to environmental themes.

1.3 Methodology Outline

LCA methodology was developed during the 1990s and is still under further development. These developments have led to standardization by the International Organization for Standardization (ISO) in the ISO-14040 series. In these standards, the following steps are distinguished (Fig. 1).

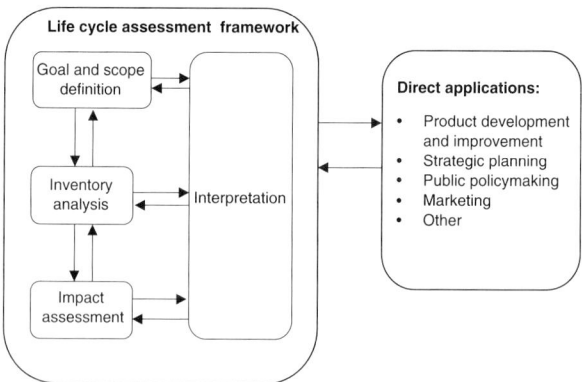

FIGURE 1 Phases of a life cycle assessment. From ISO (1997). Reproduced with the permission of the International Organization for Standardization, ISO. These standards can be obtained from any ISO member and from www.iso.org. Copyright remains with ISO.

1. *Goal definition and scoping.* The application and type of LCA are described, the product systems to be evaluated are defined, and the geographical and temporal scope are defined as well. This step also includes the definition of the functional unit, which will act as the reference for the subsequent steps.

2. *Inventory analysis.* The environmental releases (e.g., emissions, resource extractions) for the product systems and the functional unit are determined.

3. *Impact assessment.* The (potential) environmental impacts caused by the environmental releases analyzed in the previous step are determined.

4. *Interpretation.* The results of the inventory analysis and impact assessment are discussed, conclusions are drawn, and recommendations are made.

These four steps in an LCA are further discussed in subsequent sections of this article.

2. GOAL DEFINITION AND SCOPING

In the goal definition and scoping phase of LCA, the purpose and method of including environmental information in decision making are defined. The primary goal of an LCA is the selection of the best product alternative in terms of having the least effect on human health and the environment. This almost always means that comparisons are made: the product is compared with a competing product, or improved products are compared with the original or a design target. Also, the LCA can be used to identify life cycle stages where opportunities for reduction in resource use and emissions might be found. LCA can also play a role in guiding the development of new products. More specific goals can be defined; for example, how can the product system be changed such that a specific environmental impact (e.g., acid rain) can be reduced without the introduction of adverse effects in other environmental impact categories?

The goal definition and scoping phase determines the amount, type, and quality of the data requirements and, therefore, also the time and resources needed for the LCA study. For instance, during early stages of product development, many design choices exist. Limited amounts of rough data and a focus on the most important steps may be considered sufficient in such cases, especially when the results are used internally. In the case of comparing competing products, the justification of environmental claims may require comprehensive data acquisition and, therefore, more time and resources for the LCA study.

Depending on the goal, the LCA practitioner may decide not to include all life cycle phases in the scope of the research. This may be justified in cases where such phases are known to have negligible impact or where such phases are equal for all product alternatives considered.

LCA data and results will be expressed in terms of a functional unit that describes the function of the product system studied and also gives a quantification of this function. For electric power production, the function can be defined as "to deliver electric power to the electric grid." The functional unit is then 1 kilowatt-hour (kWh) electric power delivered to the electric grid. Comparisons between product alternatives (e.g., between coal-fired power stations and gas-fired power stations) must be made based on the same functional unit for all product alternatives. This means that 1 kWh of electricity from a coal-fired power station must be functionally equivalent to 1 kWh of electricity from a gas-fired power station.

3. INVENTORY ANALYSIS

3.1 Definition

In the life cycle inventory (LCI) analysis, the life cycle is drawn up and all energy and material requirements; emissions to air, water, and soil; and other environmental releases are quantified. The steps involved in performing an LCI are the development of a flow diagram, data collection, multi-output processes, and reporting.

3.2 Development of a Flow Diagram

The flow diagram reflects all of the processes that make up the product system and the inputs and outputs to each of the processes within the product system. From the goal and scope definition, it follows which processes will be included within the system boundary. Figure 2 gives an example of a product system for LCI analysis. It shows the processes that are included in the system boundary as well as product flows to or from other systems that are not included. Two types of processes are modeled here in a more generic way: transports and energy supply. This represents the fact that in nearly all cases some form of transport is needed between processes and the fact that energy is needed in many operations. All processes are connected with flows of (intermediate)

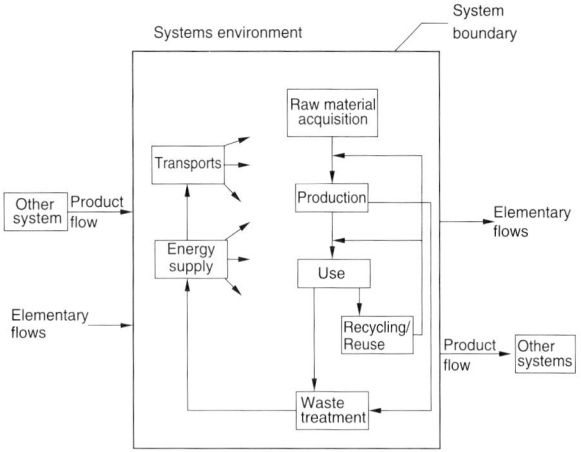

FIGURE 2 Example of a product system for life cycle inventory analysis. From ISO (1998). Reproduced with the permission of the International Organization for Standardization, ISO. These standards can be obtained from any ISO member and from www.iso.org. Copyright remains with ISO.

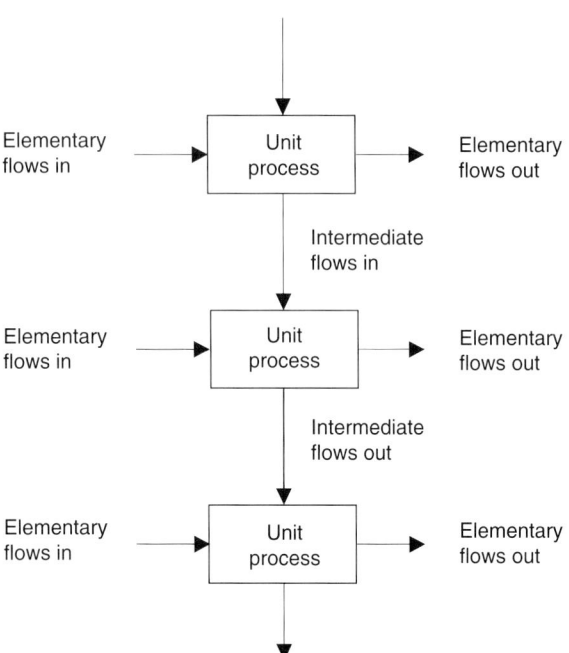

FIGURE 3 Example of a set of unit processes within a product system. From ISO (1998). Reproduced with the permission of the International Organization for Standardization, ISO. These standards can be obtained from any ISO member and from www.iso.org. Copyright remains with ISO.

products. Furthermore, elementary flows to and from the environment are shown. Elementary flows are either material or energy flows entering or leaving the system that have been drawn from the environment or discarded to the environment without previous or subsequent human transformation, respectively. In other words, elementary flows originate directly from the environment (e.g., energy and material resources, land use) or are discarded directly to the environment (e.g., emissions, heat, radiation, sound). Elementary flows originate from the processes within the system boundary, as is shown in Fig. 3.

3.3 Data Collection

Data collection is probably the most time-consuming step. From the goal definition and scoping step, it must be clear which requirements for data accuracy must be fulfilled. Most often, this will be a mix of site-specific data (obtained through measurements and calculations from measurements) and more generic, non-site-specific data that can be obtained from available databases or expert estimates.

3.4 Multi-Output Processes

In many cases, processes are encountered that do not have a single useful output. In such multi-output processes, often only one of the outputs is used in the life cycle. Other outputs are used in other life cycles that are not under investigation. The material and energy inputs and the elementary flows of these processes have to be divided (allocated) among the useful outputs in a "proper" way. Various possibilities exist for the allocation in multi-output processes. A well-known multi-output process is the combined production of heat and power (CHP). The fuel input in CHP plants is to be divided between heat and electricity. One possibility is system extension by introducing a reference technology for one of the products. In the case of heat production, the fuel input for heat is calculated as the fuel required for generating the heat in a reference boiler. The fuel input for electricity from the CHP plant is then obtained by subtracting the fuel input for heat production from the CHP plant fuel requirement. Of course, a reference electricity plant could be introduced instead of a reference boiler. Another approach is the use of a physical property of the products (e.g., mass or energy content) as an allocation factor. For CHP, the use of mass as an allocation factor is meaningless. Energy content (either on a heat/enthalpy basis or on an exergy basis) is more meaningful. A third approach is the use of economic value of the individual products as the proper allocation factor. All three approaches have been used in LCAs.

3.5 Reporting

The outcome of the LCI analysis is a list of all elementary flows to or from the environment for each product system investigated, resulting from the functional unit provided by the product systems. For the purpose of analyzing these results, they can be organized according to life cycle stage and/or media (air, water, or land). It is possible to use these results for further analysis such as comparison of product alternatives and identification of life cycle stages where significant environmental releases take place. However, these releases do not express the potential environmental impacts that may be caused. The translation of environmental releases to potential environmental impacts is the subject of the next section.

4. IMPACT ASSESSMENT

4.1 Definition

Life cycle impact assessment (LCIA) is the phase of an LCA where the evaluation takes place of the potential environmental impacts stemming from the elementary flows (environmental resources and releases) obtained in the LCI. LCIA consists of the following steps:

1. Selecting the relevant impact categories
2. *Classification*: assigning the elementary flows to the impact categories
3. *Characterization*: modeling potential impacts using conversion factors obtaining an indicator for the impact category
4. *Normalization (optional)*: expressing potential impacts relative to a reference
5. *Grouping*: sorting or ranking the impact indicators
6. *Weighting*: relative weighting of impact categories; and evaluation and reporting

4.2 Impact Categories

Environmental impacts are the consequences caused by the elementary flows on human health, plants, and animals (ecological health) or by the future availability of natural resources (resource depletion). A number of impact categories are listed in the first column of Table I.

In the classification step, the LCI results are organized and combined into the impact categories selected in the previous step. An example is formed by the greenhouse gases that are classified in the global warming category (third column of Table I).

In the characterization step, characterization factors are used (where possible) in each impact category to combine all of the elementary flows to one indicator for the impact category. For greenhouse gases, the global warming potential for each greenhouse gas is used to combine all greenhouse gases into one indicator (global warming equivalent), expressed in carbon dioxide equivalents. The final two columns of Table I show the characterization factors and a description of the procedure. For some impact categories (e.g., global warming, ozone depletion), there is a consensus on acceptable characterization factors. For other impact categories, methods have been developed but further consensus regarding their application has to be established. Furthermore, characterization factors for regional environmental issues (e.g., acidification) may sometimes be defined such that they are relevant only for the region for which they were derived. Use of these factors in other regions has to be investigated.

In the (optional) normalization step, the indicator scores obtained in the characterization step are normalized by dividing by a reference value. Reference values can be the total emissions or resource uses for a given area (global, regional, or local) or the total emissions/resource uses for a given area per capita. The results do not imply a certain weighting among impact categories; they merely give an indication how much the environmental profile of the product contributes to the total environmental profile of, for example, the region.

4.3 Grouping and Weighting

In the grouping step, the indicators are sorted by characteristics such as emissions (e.g., air, water) and location (e.g., local, regional, global). Other forms of grouping, such as using a ranking system (high, low, or medium priority), also exist. In the last case, value choices regarding what is to be considered a high or low score play an important role, and transparency regarding the procedure used is essential.

The weighting (or valuation) step assigns relative values to the various impact categories based on perceived importance. Although scientific procedures may be used (e.g., multiple-criteria analysis), the weighting factors obtained are not strictly scientific. At best, they correctly reflect the preferences of the stakeholders considered in the (scientific) procedure. Because such preferences can change in time and among stakeholders, it can also be considered to use only the characterization (or normalization) results.

TABLE I
Life Cycle Impact Categories

Impact category	Scale	Relevant LCI data (i.e., classification)	Characterization factor	Description of characterization factor
Climate change	Global	Carbon dioxide (CO_2) Dinitrogenmonoxide (N_2O) Methane (CH_4) Chlorofluorocarbons (CFCs) Hydrochlorofluorocarbons (HCFCs)	Global warming potential (GWP)	Converts LCI data to carbon dioxide (CO_2) equivalents (Note: global warming potentials can be 50-, 100- or 500-year potentials)
Stratospheric ozone depletion	Global	Chlorofluorocarbons (CFCs) Hydrochlorofluorocarbons (HCFCs) Halons Methyl bromide (CH_3Br)	Ozone depletion potential (ODP)	Converts LCI data to trichlorofluoromethane (CFC-11) equivalents
Acidification	Regional local	Sulfur oxides (SO_x) Nitrogen oxides (NO_x) Hydrochloric acid (HCl) Hydrofluoric acid (HF) Ammonia (NH_3)	Acidification potential (AP)	Converts LCI data to sulfur dioxide (SO_2) equivalents
Eutrophication	Local	Phosphate (PO_4) Nitrogen monoxide (NO) Nitrogen dioxide (NO_2) Nitrates (NO_3) Ammonia (NH_3)	Eutrophication potential (EP)	Converts LCI data to phosphate (PO_4) equivalents
Photochemical smog	Local	Non-methane volatile organic compounds (NMVOC)	Photochemical ozone creation Potential (POCP)	Converts LCI data to ethylene (C_2H_4) equivalents
Ecotoxicity	Global continental	Releases to air, water, and soil	Fresh water aquatic ecotoxicity potential (FAETP) Marine aquatic ecotoxicity potential (MAETP) Fresh water sediment ecotoxicity potential (FSETP) Marine sediment ecotoxicity, potential (MSETP) Terrestrial sediment ecotoxicity potential (TETP)	Converts LCI data to 1,4-dichlorobenzene equivalent; separate factors for fresh water aquatic/sediment Marine aquatic/sediment, and terrestrial ecotoxicity (Note: different factors for global/continental scale and infinite or 20-, 100-, or 500-, year time horizon
Human toxicity	Global continental	Releases to air, water, and soil	Human toxicity potential (HTP)	Converts LCI data to 1,4-dichlorobenzene equivalents (Note: different factors for global/continental scale and infinite or 20-, 100-, or 500-year time horizon.)
Resource depletion, abiotic	Global	Quantities of minerals/fossil fuels used	Abiotic depletion factor (ADP)	Converts LCI data to antimony equivalents
Land use	Global regional local	Land occupation	Increase of land competition	Converts LCI data to cumulative land use expressed in square meters/year

Sources. U.S. Environmental Protection Agency (2001), with corrections; Guinée (2002). See Guinée (2002) for a more complete listing, guidelines, and values for the characterization factors.

In fact, the ISO 14040 standard does not allow the use of weighting methods in comparative assertions that are to be disclosed to the public.

5. INTERPRETATION

Life cycle interpretation is the procedure during which the results of the LCI and LCIA are identified, checked, and evaluated. Life cycle interpretation consists of the following steps:

1. Identify significant issues.
2. Evaluate the completeness, sensitivity, and consistency of the data.
3. Draw conclusions and make recommendations.

Significant issues are the data elements that contribute the most to the results of the LCI and LCIA. These can be certain inventory parameters (e.g., energy use), impact categories (e.g., acidification), or certain life cycle stages (e.g., manufacturing).

In the second step, checks are performed regarding completeness, sensitivity of the significant data elements, and consistency (with regard to the goal definition and scope of the study). Such checks are required to reach consistent and reliable conclusions. In most cases, a clear answer cannot be given to the question of which product alternative is better from an environmental point of view. This can be caused by the uncertainty in the results or by differing outcomes in the impact categories that were considered relevant. The LCA process can then still give useful results in terms of better understanding of the environmental impacts of each product alternative (e.g., which impacts, where these impacts occur).

When LCA results are to be used to support local decision making, additional analysis addressing local aspects is required. For this, various forms of environmental impact assessment and risk analysis can be used.

6. ENERGY ISSUES IN LIFE CYCLE ASSESSMENT

Because energy is used in nearly every life cycle step, energy production often plays a key role in the LCA of any product system. Therefore, a consistent and accurate treatment of energy in LCA is required.

In a strict sense, energy is not a separate issue in LCA. The inputs and outputs of energy carriers are recorded individually in the inventory phase, but aggregate input/output is not. Similarly, the use of energy is not seen as an impact category during the impact assessment phase. The use of energy carriers leads to depletion of biotic and abiotic resources and can also lead to the emissions of various pollutants. The impacts in the related categories (e.g., [a]biotic depletion, climate change, acidification) are of concern in environmental assessment. On the other hand, an account of primary energy use in a product life cycle can be made relatively easy when the LCA data are available. Such data can be used to show overall energy performance and may also be used as a limited indicator of environmental performance.

The use of electricity in an LCA asks for careful treatment. Electricity is generated in various power plants within a certain country or region, each having its own environmental profile. It can be considered to take average values of energy use and emissions during a specific year by averaging over all power plants in the region. However, this might not reflect what is actually happening. For instance, if incandescent lamps are replaced by compact fluorescent lamps, a marginal approach is more adequate than an average approach. In the short-term marginal approach, the electricity savings caused by the substitution of lamps do not stem from the average of all plants (e.g., based mainly on coal) but rather stem from the plant that will not produce the electricity anymore. This could be an efficient natural gas-fired power plant or an older, less efficient power plant that can now be taken out of operation. For product choices that reduce electricity use in the future, present average and short-term marginal values may be inadequate. In such cases, long-term marginal values are needed. In many regions of the world, the dominant technology for new electric power plants is the natural gas-fired combined cycle power plant. Environmental data from this power plant are more adequate than present or future averages.

7. LIFE CYCLE ASSESSMENT OF ENERGY SYSTEMS

LCA methods can be used to analyze energy systems. In principle, the scope for such analyses can range from the analysis of individual energy production technologies to, ultimately, the analysis of complete national energy systems.

A convenient functional unit for energy supply systems is an amount of energy produced. For electric power systems, the kilowatt-hour of electricity produced is an example. However, an account

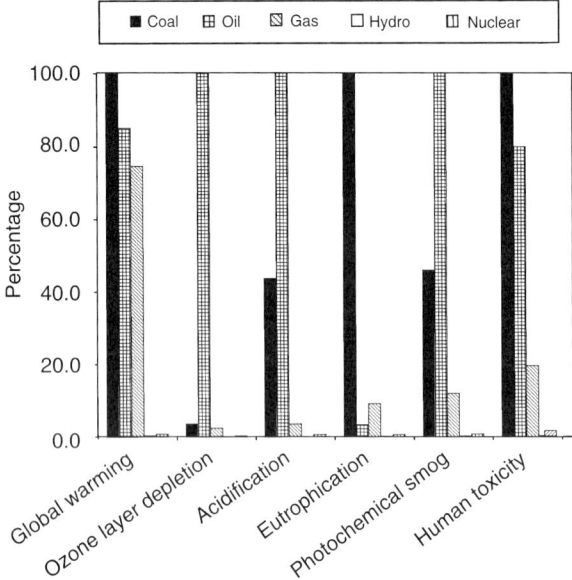

FIGURE 4 Comparison of various electricity generation systems with respect to various impact categories. For every impact class, the most polluting plant type is set to 100% (no normalization is applied here). From inventory data from Frischknecht [1996]; impact assessment characterization factors from Guinée [2002].

must be made of various factors such as operation times (load factor), time patterns in the supply of energy (especially for solar and wind energy), and capacities. These factors make that kilowatt-hour of electricity produced from one source (e.g., coal, natural gas, wind, solar, nuclear) not fully comparable to a kilowatt-hour produced from another source. So long as the total power generation mix satisfies certain demands regarding, for example, reliability, comparisons on a kilowatt-hour basis can be made. When high levels of intermittent supply sources (e.g., solar, wind) require the installation of storage options, these should be included in the LCA.

For fossil fuel-based energy systems, important environmental impacts will result from the energy conversion processes (mostly combustion) and the waste and flue gas treatment directly related to these conversion processes. In cases such as nuclear power, solar power, and wind power, the environmental profile of the materials used will largely determine the total environmental profile.

As an example, Fig. 4 shows the results for typical electricity-generating systems in Europe for six impact categories. It shows that coal is most polluting in the impact categories of global warming (climate change), eutrophication, and human toxicity and that oil is most polluting with respect to ozone layer depletion, acidification, and photochemical smog. Results are sensitive to the quality of the fuels and the specific cleaning technologies used.

SEE ALSO THE FOLLOWING ARTICLES

Climate Protection and Energy Policy • Ecological Risk Assessment Applied to Energy Development • Electric Power Generation: Valuation of Environmental Costs • Green Accounting and Energy • Industrial Ecology • Life Cycle Analysis of Power Generation Systems • Multicriteria Analysis of Energy • Net Energy Analysis: Concepts and Methods

Further Reading

Frischknecht, R. (1996). "Okoinventare für Energiesysteme." Bundesamt für Energiewirtschaft, Zurich, Switzerland.

Guinée, J. B. (ed.). (2002). Handbook on Life Cycle Assessment: Operational Guide to the ISO Standards. Kluwer Academic, Dordrecht, Netherlands.

International Organization for Standardization. (1997). "Environmental Management: Life Cycle Assessment—Principles and Framework" (ISO 14040). ISO, Geneva, Switzerland.

International Organization for Standardization. (1998). "Environmental Management: Life Cycle Assessment—Goal and Scope Definition and Inventory Analysis" (ISO 14041). ISO, Geneva, Switzerland.

International Organization for Standardization. (2000). "Environmental Management: Life Cycle Assessment—Life Cycle Impact Assessment" (ISO 14042). ISO, Geneva, Switzerland.

International Organization for Standardization. (2000). "Environmental Management: Life Cycle Assessment—Life Cycle Interpretation" (ISO 14043). ISO, Geneva, Switzerland.

U.S. Environmental Protection Agency and Science Applications International Corporation. (2001). "LCAccess: LCA 101." www.epa.gov/ord/nrmrl/lcaccess/lca101.htm.

Lifestyles and Energy

KORNELIS BLOK
Utrecht University
Utrecht, The Netherlands

1. Introduction
2. What Are Lifestyles?
3. The Structure of Household Energy Requirement
4. Trends in Household Energy Requirement
5. What Can a Consumer Do?
6. How Can Consumption Patterns Be Influenced through Policies?

Glossary

consumption patterns The total of household consumption defined as the package of goods and services that a household consumes (generally measured in terms of household purchases).

direct energy requirement The primary energy required to deliver energy carriers (e.g., electricity, natural gas, and gasoline) to households.

energy intensity of household consumption The primary energy requirement for a consumption item divided by the expenditure on this item. The energy intensity can be determined for individual goods and services but also for categories or for the complete consumption of households in a country.

indirect energy requirement The primary energy required to deliver goods and services other than energy carriers to households.

lifestyles A set of basic attitudes, values, and patterns of behavior that are common to a social group. In this article, lifestyles are operationalized in terms of consumption patterns.

primary energy requirement of a product The amount of primary energy needed to deliver a product to a consumer, including the energy needed for the production, use, and disposal of the product.

Through a range of studies, a good overview exists of how consumption patterns determine the primary energy requirements of households. Income is an important determinant. Household consumption is likely to increase in monetary terms and so may the primary energy requirement because there is not much tendency toward less energy-intensive consumption patterns. Changes in consumption patterns that substantially reduce household energy requirement are possible. To date, no clear policies can be defined that could actually lead to such changes.

1. INTRODUCTION

The concept of lifestyles is not a very well defined one. In all kinds of discussions about how to solve environmental problems, it is often suggested that people especially in industrialized countries should "change their lifestyles." Before we can discuss the relation between lifestyles and energy, we must better define the concept. This article first discusses how consumption patterns of households determine energy use. It distinguishes direct energy requirement (associated with, e.g., in the form of electricity or gasoline) and indirect energy requirement (needed to produce goods and services consumed by households). Then, attention is paid to the ways households can change their demand on primary energy carriers by changing their consumption patterns. On this basis, conclusions can be drawn on the extent to which energy use can be influenced by changing our lifestyles.

2. WHAT ARE LIFESTYLES?

In many policy documents and also in the scientific literature, lifestyles are often quoted. In one of the assessment reports of the Intergovernmental Panel on Climate Change, lifestyle is defined following the sociologist Max Weber as "a set of basic attitudes, values, and patterns of behavior that are common to a social group, including patterns of consumption and anticonsumption." As a concrete example of a lifestyle change, the report mentions the change of people's preference from car to bicycle transport.

The concept is sometimes used in social science, but there are also scholars that doubt the usefulness of the concept. Lifestyles are associated with social status groups. According to many scientists, in modern societies stable status groups no longer exist. The limited use of the lifestyle concept is also demonstrated by the fact that lifestyle research has not led to a uniform and widely used classification. Nevertheless, the lifestyle concept is often used in marketing circles, although classifications vary widely.

This article focuses on people in households. Households carry out a wide range of activities, and most of them are associated with the purchase of goods and services. Some of these goods are energy carriers (electricity, heating fuels, and gasoline); however, all other goods and services also require input of energy over the life cycle to produce the good or service. We therefore make the concept of lifestyles operational in terms of consumption patterns: The lifestyle is defined by the package of goods and services that households consume. The energy impact of a consumption pattern is the total amount of energy associated with the production, use, and disposal of these goods and services.

3. THE STRUCTURE OF HOUSEHOLD ENERGY REQUIREMENT

A detailed study to establish the relation between household consumption patterns and associated energy use was carried out for The Netherlands and in somewhat less detail for the main European Union member states (Fig. 1A). In another study, the induced CO_2 emissions were established (Fig. 1C). It shows that the well-known categories, such as space heating and transport, are fairly important. However, food, an often neglected category in energy studies, contributes substantially. The direct energy use of households (heating fuels, electricity, and gasoline) comprises approximately half of the total energy use. Also, data are available for India (Fig. 1B). Here, the category food is even more important. Note that total energy requirement per capita in India is approximately a factor of 10 smaller than in the European Union countries.

The main determinant for the energy impact of household consumption is income. In the 1970s, it was established through input–output analysis that the income elasticity of household energy requirement is approximately 0.7 or 0.8 (i.e., with every

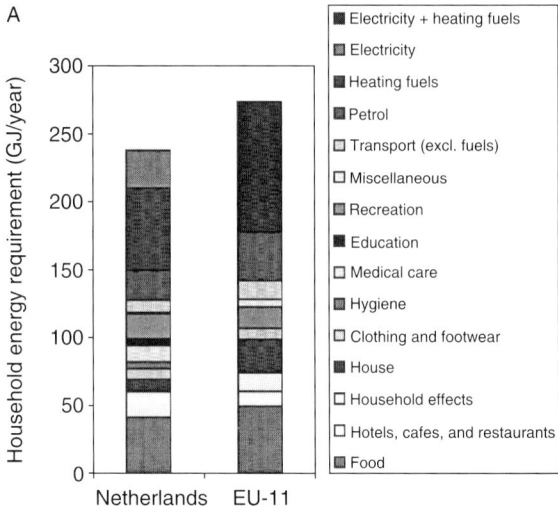

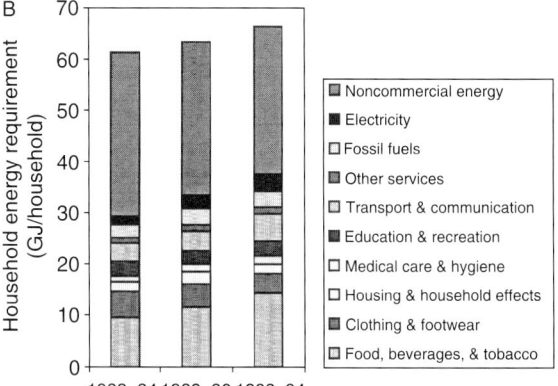

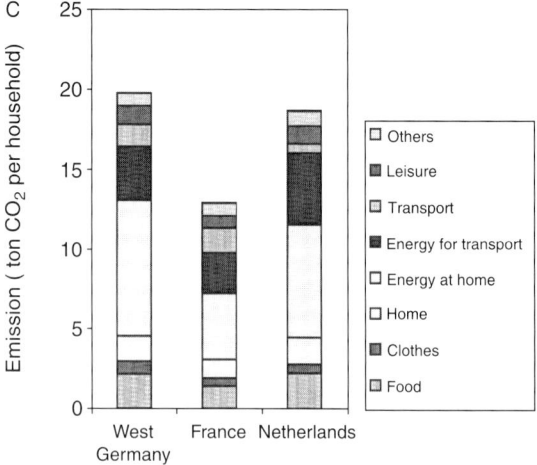

FIGURE 1 The breakdown of total energy requirement for households (A) in The Netherlands (from Vringer and Blok, 1995) and in 11 European Union member states (from Reinders et al., 2003), (B) in India for various years (from Pachauri and Spreng, 2002), and (C) total induced CO_2 emissions in some countries (from Weber and Perrels, 2000).

percent of growth of household income, the household energy requirement increases by 0.7 or 0.8%). Recent studies have derived lower values (e.g., a value of 0.63 was found for The Netherlands and 0.67 was found for India, the latter actually indicated expenditure elasticity). The strong relation between income and household energy requirement is not only found within a country but also in a comparison between European countries (Fig. 2).

The composition of the consumption package changes with income (Fig. 3). Some categories only increase slowly with income (e.g., food and space heating fuels), whereas others are relatively unimportant at low incomes but increase strongly with higher incomes (e.g., electricity, transportation, and recreation).

Other factors, such as the size of the household, the age of the oldest household member, the life cycle phase of the household, the degree of urbanization, and the level of education, are less important in explaining household energy requirement. The degree to which they explain differences in household energy requirement varies between studies, but income, or expenditure, is always the most important variable. However, there are strong correlations between household income and some of the other factors (e.g., the number of persons per household and education level).

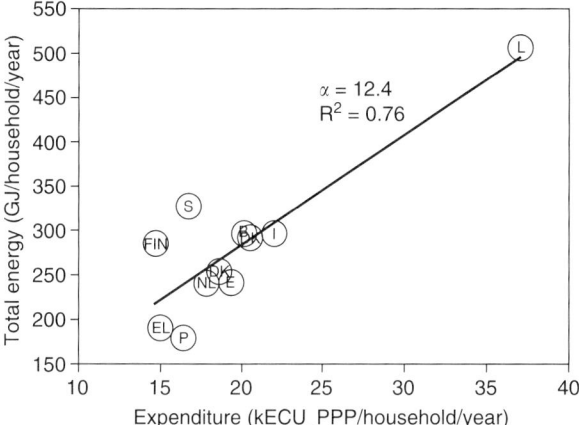

FIGURE 2 The relation between average household expenditure and average household energy requirement for European Union member states. B, Belgium; DK, Denmark; E, Spain; EL, Greece; FIN, Finland; I, Italy; L, Luxembourg; NL, Netherlands; P, Portugal; S, Sweden; UK, United Kingdom. From Reinders et al. (2003).

4. TRENDS IN HOUSEHOLD ENERGY REQUIREMENT

An important question regards how household energy requirement develops in time. We distinguish again direct and indirect energy requirement. Most information is available on direct energy requirement. Table I gives an overview of growth rates of direct parts of energy use over time. In Organization for Economic Cooperation and Development (OECD) countries, energy use for transportation and electricity services is growing at a rate comparable to that of gross domestic product, whereas

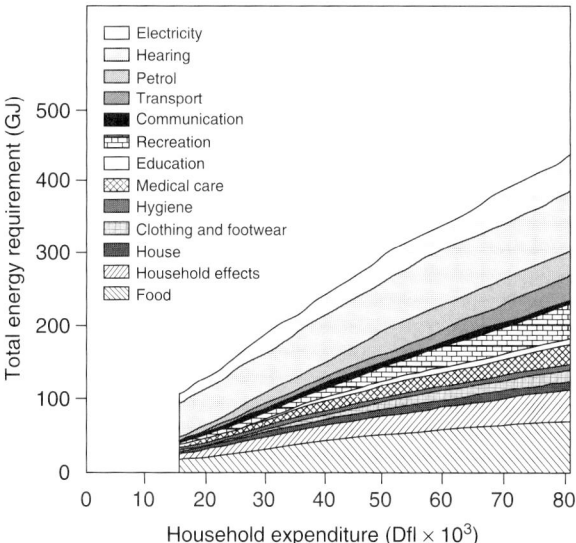

FIGURE 3 The development of total energy requirement for the various consumption categories with income. From Vringer and Blok (1995).

TABLE I

Overview of Growth Rates of Energy Use by Households Compared to Other Parameters[a]

	Average change during the period 1971–1995 (% per year)	Average change during the period 1985–1995 (% per year)
Household fuel use for space heating	0.0	0.8
Household electricity use	3.7	2.7
Fuel use for transportation	2.4	2.6
For comparison		
Population	1.5	1.6
GDP (ppp corrected)	0.9	0.8
Total energy use	2.7	2.4

[a] Source. "IEA/OECD Energy Balances." International Energy Agency, Paris, France.

energy use for space heating is lagging. There are several relevant factors.

For space heating, the development is the result of several factors: an increase in population (0.8% per year) and a decrease in the number of persons per dwelling (0.8–1.2% per year). These factors combined lead to an increase in the number of dwellings. In addition, an increase in the average size of the dwelling can be observed in most countries (0–0.8% per year).

For electric appliances, the increasing number of households plays a role. In addition, there is an increase in appliance ownership per household (typically by 1–3% per year during the past few decades). Only for a few large appliances is there a saturation of ownership rate. Apart from the ownership rate, other aspects change. For instance, there is a tendency to purchase larger refrigerators and to wash clothes more often.

For passenger cars, an increase in car ownership has occurred in all industrialized countries in the past decades; relatedly, total car mileage has increased. There is also an increase in car size and performance. Other transportation modes only contribute modestly to total passenger transportation performance. There are no substantial shifts between the various modes, except for the strong growth of the most energy-intensive mode—air transport.

All these factors can be considered as lifestyle factors, although some are strongly related to growth of population and economy. Most of the trends are likely to continue. All lead to an increase in primary energy requirement.

Adoption of improved technology (which can also be considered lifestyle related but is not discussed here in detail) partly or completely offsets the other tendencies. In the case of space heating, in most OECD countries the substantial effort to insulate buildings and improve boilers has led to a net decrease in energy use for this application. For the other areas, the energy efficiency efforts definitely had effects but were too dispersed to offset the growth drivers completely.

Also, in the case of the indirect energy requirement, there is an ongoing increase. For The Netherlands, an increase of 2.2% per year was observed during the period 1948–1988. Growth was smallest in the categories of food and clothes and highest in the categories of leisure, the house, and household effects. India, had a comparable growth rate, but especially food and transport and communication are growing.

We may ask whether the growth in household energy requirement is higher or lower than the growth in household expenditures. To answer this question, the concept of energy intensity of household consumption is introduced. This is defined by the energy requirement for a consumption item divided by the expenditure on this item. There are substantial differences in energy intensity of various consumer goods. A typical value for the average energy intensity of consumption is 7 MJ/Euro if direct energy use is excluded; if direct energy use is included, this increases to 12 MJ/Euro. Some items that have high energy intensities (20–30 MJ/Euro) are frozen food, eggs, sugar, refuse bags, toilet paper, aluminum foil, and flowers grown in greenhouses (the energy intensity of natural gas and electricity is more than 100 MJ/Euro; for gasoline, this value is lower if taxes are taken into account). Categories that have low energy intensities (<3 MJ/Euro) are music lessons, dance classes, maintenance and repair of electric appliances and central heating systems, use of taxis, telephone use, and expenditures on beauty parlors. The question now is did the average energy intensity change due to changes in household consumption patterns? Hardly. During a period of 50 years, the average decrease of energy intensity through changes in consumption patterns was only approximately 0.1% per year (Fig. 4). Note that the effect of energy efficiency improvements in the supplying sectors is likely to be much higher.

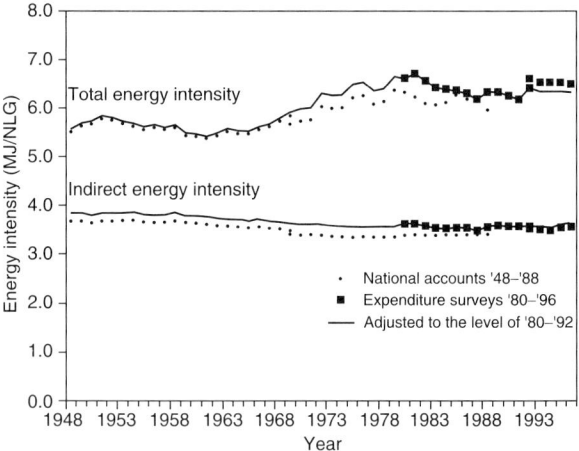

FIGURE 4 The development of energy intensity of household consumption in The Netherlands. The energy intensity of individual consumption items is kept constant at the 1990 level, so this figure only reflects the effect of changes in consumption patterns. The change in the average energy intensity of the indirect energy requirement is small. The total energy intensity shows changes caused by changes in the share of direct energy use: first an increase, mainly caused by the expansion of natural gas-based central heating systems, and later a decrease, mainly caused by insulation programs and improved boilers. 1 NLG = 0.45 Euro. From Vringer and Blok (2000).

For the indirect energy requirement, changes in consumer preferences in no way offset the increase in energy requirement related to increased income and total expenditures.

What are the mechanisms for the increase in primary energy requirement? As discussed by Røpke (1999), an important one is the increase in industrial productivity (forced by competition) that makes it possible to produce more goods, given the amount of hours worked by the total population. A variable is the total amount of time worked per person. Although there are tendencies to limit the number of working hours per week, there is only limited exchange of income for leisure time. Elements that play a role are the "hours-invariant" component of labor costs and the tendency to relate career making to full-time employment. Note that more leisure time could also lead to more energy use if the extra time is used for energy-intensive activities (e.g., those involving a lot of driving); however, the limitation of production and consumption that is ultimately related to shorter working time should lead to a limitation of energy use.

Psychological mechanisms on the consumer side are important. Although traditionally the role of producers in generating consumption (e.g., through advertising) is stressed, it is also important to recognize that the possession of goods is a way for people to position themselves in social relations.

One might further ask why there is not a stronger shift from material consumption to services consumption. Often, the idea is expressed that when basic material needs are fulfilled, consumption patterns will shift to nonmaterial items. One of the explanations is that productivity increases are larger for the production of material goods than for the production of services. This leads to lower relative prices for material goods than for services, and this limits the shift to buying more services.

These mechanisms together may explain the increasing consumption and the limited shift of consumption in a less energy-intensive direction. These are all strong mechanisms and they are not easily changed.

5. WHAT CAN A CONSUMER DO?

A consumer can change the energy impact of his or her behavior by making changes in direct energy consumption. Concrete actions include insulating the house, purchasing efficient appliances, and purchasing efficient cars. All these actions involve improving the energy efficiency of energy use within the household; households generally can save substantially on the direct energy requirement (typically 20–50%).

Although most energy can be saved with the choice of specific equipment, some savings are also achievable in the use of equipment. Consumers can save energy by switching off a light when a room is not in use, driving a car conservatively, lowering the thermostat; etc. Typical savings through good housekeeping range from 5 to 20%.

In this section, we focus on changes in the packaging of goods and services that are purchased by households with the aim of limiting the energy requirement of households. We can distinguish the following categories of relevant changes:

A. Households may choose an energy-saving variant within a product category. The functionality of the consumption essentially does not change. These variants sometimes require more effort or cost more money. Examples include the following:

- Increase the lifetime of products: This can be realized through the purchase of products with improved quality (e.g., shoes, furniture, and tools) or by having products repaired more often (e.g., household equipment and shoes).
- Product or material substitution: There are several examples in the area of food (e.g., the replacement of animal-based food by vegetable food, selection of food of the season, or selection of food produced locally) but also, for example, in the material choice for clothing and floor coverings.
- Share the use of products with other households (e.g., cars, tools, and newspapers).

B. Households may choose a more luxury variant within a product group. The functionality of the product remains the same in physical terms but is much more attractive for the household. The price per functional unit is substantially higher, which leads to a lower energy intensity. Examples include the following:

- The purchase of products with individualized design: Examples are more high-end or tailor-made clothes and furniture.
- Contract out certain activities, such as household work, clothes washing, or maintenance of the house.
- Eat outdoors or purchase luxury food (including ecological food).

C. Households may refrain from certain activities and replace them by other less energy-intensive ones. An example is a household that does not go on a long-distance vacation but chooses to buy a piano and music lessons with the money saved.

A quantitative assessment of all these options identified a potential shift in consumption patterns in a period of approximately two decades that can lead to a decrease in indirect energy requirement by 10% and a reduction in energy intensity by 30% (Table II). This means that in principle it is possible for household expenditures to keep increasing while at the same time there is a decrease in indirect energy requirement.

Following such analysis, an interesting experiment was carried out in The Netherlands. Fourteen households were selected to test whether it is feasible to combine higher "future" financial budgets with lower energy budgets. All households received a bonus of 20% of their net salary. Before the experiment started, the total (direct + indirect) energy requirement amounted to 260 GJ per household. They were given a maximum energy requirement target of 207 GJ, which they had to achieve in 2 years. They achieved this target and at the end of the experiment used 182 GJ. They reduced direct energy requirement from 60 to 50 GJ (the selected households already had an energy-efficient house), and indirect energy requirement from 200 to 132 GJ (all data are averages for the 14 households). Households reduced expenditures on energy-extensive products such as meat and flowers. They only slightly reduced their vacation expenditures but managed to reduce energy use by half through shorter distances and less use of air transport. In the area of transport, energy requirement was reduced by more than 20%. However, expenditures decreased less due to the increased use of taxis and first-class public transport.

Most households were satisfied with the low-energy lifestyle they had to develop. In the course of 2 years, the households developed an appetite for low-energy expenditures, including quality products, delivery services, delicious or ecological food, exclusive clothes, antique furniture, and fancy parties (i.e., something that could be considered a "light" lifestyle). The most problematic aspects were the constraints perceived on car transport and vacation destinations.

The experiment showed that a low-energy lifestyle can be developed. However, it must be noted that the participating households were very well informed, had computer-aided support, were coached on an individual basis, and were motivated by the income bonus. Nevertheless, this shows that the estimated potential is more than just theory.

A more systematic study of how consumers prefer to change their consumption patterns is presented in Table III. Technical measures (improving the energy efficiency of the home and equipment) are preferred over behavioral measures, especially shifts in consumption. Furthermore, home energy-saving measures were more acceptable than transport energy-saving measures.

TABLE II

Effect on the Energy Intensity of Household Consumption through Various Measures[a]

Category of measures[b]	Effect on indirect energy requirement (%)	Effect on expenditures (%)	Effect on energy intensity of goods and services (excluding direct energy use) (%)
A	−17	−4	−14
B	+3	+23	−17
C	+3	+8	−6
Total	−10	+26	−29

[a] *Source.* Vringer, K., Potting, J., Blok, K., and Kok, R. (1995). A reduction of the indirect energy requirement of households. *In* "Proceedings of the 1995 Summer Study: Sustainability and the Reinvention of the Government—A Challenge for Energy Efficiency" (A. Persson, Ed.). European Council for an Energy Efficient Economy, Stockholm.

[b] See description in text.

6. HOW CAN CONSUMPTION PATTERNS BE INFLUENCED THROUGH POLICIES?

The Intergovernmental Panel on Climate Change (IPCC) presents the following list of policy options:

1. Environmental education: Although public awareness is already fairly high, environmental education could be more effective if it is recognized that it is linked to lifestyle and self-awareness.

2. Decreasing marginal satisfaction with increasing private material consumption: People might prefer other forms of satisfaction than ever-increasing consumption of purchased commodities. A suitable index of welfare and the stimulation of public dialogue on the goal of economic action could raise awareness of this issue.

3. New emphasis on immaterial and common goods: In general, these goods have lower energy and environmental impacts.

TABLE III

Acceptability of 23 Energy-Saving Measures and Their Characteristics[a]

Energy-saving measure	Domain	Strategy	Amount	Mean[b]	SD[b]
Switching off lights in unused rooms	H	2	+	4.6	0.68
Appliances not on stand-by	H	2	+	4.4	0.87
Energy-efficient heating system	H	1	+ +	4.3	0.94
Walking or cycling short distances (<2.5 km)	T	2	+ +	4.3	1.00
House insulation	H	1	+ +	4.3	0.89
Line drying of laundry	H	2	+ +	4.0	1.13
Compact fluorescent light bulbs	H	1	+	4.0	0.94
Applying radiator insulation (foil)	H	1	+	4.0	0.99
Energy-efficient refrigerator	H	1	+	3.9	1.01
Shorter showers	H	2	+ +	3.6	1.10
Walking or cycling short distances (<5 km)	T	2	+ +	3.6	1.18
Econometer in car	T	1	+	3.6	1.16
Energy-efficient car	T	1	+ +	3.4	1.15
Car–pooling	T	2	+	3.4	1.33
Drive at most 100 km/h on the highway	T	2	+	3.3	1.35
Using public transport	T	2	+ +	3.2	1.42
No greenhouse vegetables	H	3	+	3.1	1.14
Thermostat maximally 18°C	H	2	+ +	3.1	1.27
Energy-extensive presents (no flowers)	H	3	+	2.9	1.19
Rinsing the dishes with cold water	H	2	+	2.9	1.31
Vacation by train	T	3	+	2.8	1.32
Altering food pattern	H	3	+	2.7	1.26
Hiring a housekeeper	H	3	+ +	2.7	1.31

[a] Abbreviations used: H; home measures; T; transport measures; 1; increasing technical energy efficiency; 2; a different use of products; 3; shifts in consumption; +; small energy saving; + +; large energy saving.
[b] Mean and standard deviation (SD) on a scale from 1("unacceptable") to 5 ("very acceptable"). Data obtained from a survey of 455 households in The Netherlands. From Poortinga *et al.* (2003).

4. New deals in collective action: By improving communication before people have to decide in social dilemmas, people might be more willing to cooperate in the common interest.

5. Environmental legislation: Governments could have people pay incrementally higher prices for environmentally more benign goods.

6. Creative democracy: For example, the local implementation of Agenda 21 could be considered; these actions are directed not only at concrete action but also at promoting understanding.

The IPCC concludes that lifestyle research is neglected compared to technology research.

To what extent will these policy proposals actually lead to a change in behavior and an actual impact on the development of energy use and greenhouse gas emissions? In policy analysis, three basic mechanisms to influence behavior are distinguished: communication, regulation, and economic incentives.

The list of options put forward by the IPCC mentioned previously relies heavily on various forms of communication. However, communication as a stand-alone instrument is weak; only a part (generally <10%) of the population changes its behavior. This portion often declines to 1% when substantial costs are involved; these costs can either have a monetary character or be a burden.

Regulation (command-and-control) makes certain behavior mandatory or prohibited. Given the wide variety of options, regulation would only have a substantial effect if an extended web of rules were implemented. Moreover, most options represent incremental changes in behavior (e.g., eating less meat), which are not very suitable for regulation. Nevertheless, for specific situations (e.g., forbidding car traffic in certain areas, such as inner cities and nature reserves) regulation might be an option.

Economic incentives include both subsidies, which stimulate desired behavior, and taxes, which discourage undesired behavior. Taxes and subsidies for individual options have the same problem as regulation: There are simply too many options to design specific taxes or subsidies for each of them. Generic energy or carbon taxes will discourage all options with a high energy or carbon dioxide emissions impact. However, the effect on behavior is probably minor due to dilution of the taxation in the product chain. For example, a carbon tax of 50 Euro/ton of CO_2 can be considered high (only present in some countries and only for small consumers). Such a tax will lead to a price increase of approximately 10% for a product with high energy intensity of 25 MJ/Euro. Such a price increase will probably have only a limited effect on the consumption of these products. Nevertheless, taxes or subsidies may be useful for a limited number of specific products with a high energy impact (e.g., meat consumption and air travel).

The discussion of the three basic mechanisms focuses on the adoption of individual options. It would be much more effective if complete lifestyle changes could be accomplished. However, how can this be stimulated? An intriguing idea is the rationing of individual carbon quota. Everyone receives a specified amount of carbon points, which can be traded. For every purchase, not only is the monetary price paid but also the carbon points are paid from one's carbon credit card. The practicality of such a system needs to be proven.

In conclusion, there is no golden key to the stimulation of consumption patterns with low energy impacts and greenhouse gas emissions.

SEE ALSO THE FOLLOWING ARTICLES

Consumption, Energy, and the Environment • Diet, Energy, and Greenhouse Gas Emissions • Global Energy Use: Status and Trends • Goods and Services: Energy Costs • International Comparisons of Energy End Use: Benefits and Risks • Leisure, Energy Costs of • Suburbanization and Energy • Transportation and Energy Policy • Urbanization and Energy

Further Reading

Anonymous (1997), "Indicators of Energy Use and Energy Efficiency." International Energy Agency, Paris.

Carlsson-Kanyama, A., Ekström, M. P., and Shanahan, H. (2003). Food and life cycle energy inputs: Consequences of diet and ways to increase efficiency. *Ecol. Econ.* **44**, 293–307.

Cohen, C. A. M. J., Schaeffer, R., and Lenzen, M. (2004). Energy requirements of households in Brazil. *Energy Policy*, in press.

Lenzen, M. (1998). Energy and greenhouse gas cost of living for Australia during 1993/94. *Energy* **23**, 497–516.

Noorman, K. J., and Schoot Uiterkamp, A. J. M. (Eds.) (1997). "Green Households? Domestic Consumers, Environment and Sustainability." Earthscan, London.

Pachauri, S. (2004). An analysis of cross-sectional variations in total household energy requirements in India using micro survey data. *Energy Policy*, in press.

Pachauri, S., and Spreng, D. (2002). Direct and indirect energy requirement of households in India. *Energy Policy* **30**, 511–523.

Poortinga, W., Steg, L., Vlek, C., and Wiersma, G. (2003). Household preferences for energy-saving measures: A conjoint analysis. *Econ. Psychol.* **24**, 49–64.

Reinders, A. H. M. E., Vringer, K., and Blok, K. (2003). The direct and indirect energy requirement of households in the European Union. *Energy Policy* **31**, 139–154.

Røpke, I. (1999). The dynamics of willingness to consume. *Ecol. Econ.* **28**, 399–420.

Schipper, L. (1996). Life-styles and the environment—the case of energy. *Daedalus* **25**, 113–138.

Toth, F., and Mwandosya, M. (convening lead authors) (2001). Decision making frameworks. *In* "Climate Change 2001—Mitigation" (B. Metz, O. Davidson, R. Swart, and J. Pan, Eds.), Report of Working Group III of the Intergovernmental Panel on Climate Change's Third Assessment Report. Cambridge Univ. Press, Cambridge, UK.

Vringer, K., and Blok, K. (1995). The direct and indirect energy requirement of households in The Netherlands. *Energy Policy* **23**, 893–910.

Vringer, K., and Blok, K. (2000). Long-term trends in direct and indirect household energy intensities: A factor in dematerialisation? *Energy Policy* **28**, 713–727.

Weber, C., and Perrels, A. (2000). Modelling lifestyle effects on energy demand and related emissions. *Energy Policy* **28**, 549–566.

Lithosphere, Energy Flows in

SETH STEIN
Northwestern University
Evanston, Illinois, United States

CAROL A. STEIN
University of Illinois at Chicago
Chicago, Illinois, United States

1. Plate Tectonics and Mantle Convection
2. Heat Flow
3. Heat Flow in Oceanic Lithosphere
4. Heat Flow in Continental Lithosphere
5. Global Heat Flow
6. Thermal Evolution of the Earth
7. Societal Implications

Glossary

asthenosphere The weaker shell of material underlying the lithosphere.
differentiation The process by which material within planets became compositionally segregated during their evolution.
geotherm The temperature as a function of depth in the earth.
heat flow The outward flow of heat through a portion of the earth's surface.
heat loss The outward flow of heat integrated over the earth's surface.
lithosphere The strong outer shell of a planet, approximately 100 km thick and composed of rigid plates on Earth.
plate Portion of Earth's lithosphere that moves as a coherent entity.
spreading center Divergent plate boundary in the oceans where hot material upwells to form new portions of the two plates.
subduction zone Convergent plate boundary where cooling oceanic lithosphere descends into the mantle.
thermal conduction Transfer of heat by molecular collisions.
thermal convection Transfer of heat by material motion due to temperature-induced density variations.
thermal history The temperature as a function of position and time on the earth or another planet.

It is said that heat is the geological lifeblood of planets. Planets are great heat engines whose nature and history govern their thermal, mechanical, and chemical evolution. On Earth, the heat engine is manifested by the plate tectonic cycle, whereby hot material upwells at spreading centers, also known as mid-ocean ridges, and then cools. Because the strength of rock decreases with temperature, the cooling material forms strong plates of the thick outer shell known as the lithosphere. The cooling oceanic lithosphere moves away from the ridges and eventually reaches subduction zones or trenches, where it descends in downgoing slabs, reheating as it goes. Thus, Earth's surface topography (e.g., mountain chains) and geological activity (e.g., volcanoes, geysers, earthquakes) reflect the planetary heat engine. As a result, measurements of heat flow at the earth's surface provide crucial constraints on how the heat engine works today and has evolved through time.

1. PLATE TECTONICS AND MANTLE CONVECTION

Plate tectonics is conceptually simple. It treats Earth's outer shell as made up of approximately 15 rigid plates, about 100 km thick, that move relative to each other at speeds of a few centimeters per year (roughly the speed at which fingernails grow). The plates are rigid in the sense that little deformation (and ideally no deformation) occurs within them, so they move as coherent entities and deformation occurs at their boundaries, causing earthquakes, mountain building, volcanism, and other spectacular phenomena. The direction of the relative motion between two plates at a point on their common boundary determines the nature of the boundary.

At spreading centers, such as the Mid-Atlantic Ridge, both plates move away from the boundary, whereas at subduction zones, such as the Aleutian Trench, the subducting plate moves toward the boundary. At the third boundary type, transform faults such as the San Andreas Fault, relative plate motion is parallel to the boundary (Fig. 1).

These strong plates form Earth's lithosphere and move over the weaker asthenosphere below. The lithosphere and asthenosphere are mechanical units defined by their strength and the way in which they deform. The lithosphere includes two chemically distinct layers: the crust and part of the underlying upper mantle.

Evidence from seismic waves show that Earth is made up of a series of concentric shells with varying properties. From the outside, these are the crust (which extends to about 6 km depth beneath the oceans and 35–50 km beneath the continents), the underlying rocky mantle (which extends to 2900 km depth), the fluid iron outer core (which extends to 5150 km depth), and the solid iron inner core. This layered structure provides the primary evidence for the process of differentiation by which material within planets became compositionally segregated during their evolution. The differentiation process is thought to be largely controlled by Earth's thermal history—how temperature within the planet evolved with time.

The key to Earth's thermal evolution is the thermal convection system involving the mantle and core that removes heat from Earth's interior. Rocks within the mantle are sufficiently hot that they flow slowly over geological time, causing heat to be transported by this motion (i.e., thermal convection). Although much remains to be learned about this convective system, especially in the lower mantle and core, there is general agreement that at shallow depths the warm, and hence less dense, material rising below spreading centers forms upwelling limbs, whereas the relatively cold, and hence dense, subducting slabs form downwelling limbs. The lithosphere is a very thin layer compared with the rest of the mantle (100 km is ∼3.4% of the mantle's radius), but it is where the greatest temperature change occurs, from more than 1400°C at a depth of 100 km to approximately 0°C at the surface. For this reason, the lithosphere is called a thermal boundary layer. Because of this temperature change, the lithosphere is much stronger than the underlying rock and so is also a mechanical boundary layer. This strong boundary layer is thought to be a primary reason why plate tectonics is much more complicated than simple models of convecting fluid. Moreover, the lithosphere, which contains the crust, is also a chemical boundary layer distinct from the remainder of the mantle. An important point is that the continental and oceanic lithospheres are distinct, although individual plates can contain both. The former is made of granitic rock, which is less dense than the basaltic rock composing the latter and so does not subduct. The oceanic lithosphere is continuously subducted and reformed at ridges and so never gets older than approximately 200 million years, whereas some continental lithosphere can be up to billions of years old.

2. HEAT FLOW

The challenge in understanding Earth's heat engine is that nearly all of our observations are made currently at the surface, whereas we seek to understand how it operates at depth over time. Our primary observation is the flow of heat at the surface today. Mathematically, heat flow q at the surface ($z = 0$) is the product of the vertical derivative of the temperature T with respect to depth z times the thermal conductivity:

$$q = k \frac{dT}{dz} \text{ at } z = 0.$$

(Normally, this equation requires a minus sign because heat flows from hot objects to cold ones. Without this sign, hot objects would get hotter. There is none here due to customary but inconsistent geophysical definitions; heat flow is measured upward, whereas depth is measured downward.) Hence, measuring the temperature at several depths

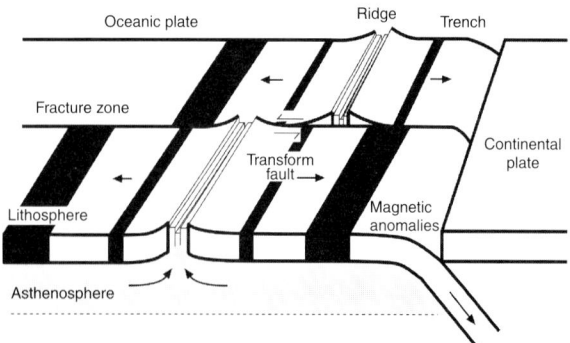

FIGURE 1 Plate tectonics at its simplest. Oceanic lithosphere is formed at ridges and subducted at trenches. At transform faults, plate motion is parallel to the boundaries.

and the thermal conductivity gives the heat flow. The heat flow is the energy per unit time (power) per unit area flowing out from the earth and is reported in milliwatts per meter^2 or in "heat flow units" (HFU), where 1 HFU is approximately 42 mW m^{-2}.

Such measurements began in 1939 on land and in 1952 at sea. By now, heat flow has been measured at more than 30,000 sites worldwide. This is done at sea by using a probe that penetrates into the soft sediment on the sea floor and on land by using bore holes that have been drilled for oil or other purposes.

To relate surface heat flow to the temperature at depth as a function of time, we use the heat equation that describes the temperature $T(z, t)$ as a function of depth z and time t. A simple form, assuming that heat is transferred only by conduction, is

$$\rho C \frac{\partial T}{\partial t} = k \frac{\partial^2 T(z,t)}{\partial z^2} + \rho H,$$

where C is the specific heat per unit mass that shows how much material will heat up for a given temperature change, ρ is the density (mass/unit volume), and H is the rate at which heat is produced (by radioactivity or chemical reactions) per unit mass.

3. HEAT FLOW IN OCEANIC LITHOSPHERE

Oceanic heat flow data (Fig. 2, center right) show a clear variation with the age of the lithosphere. Values greater than 100 mW m^{-2} occur near the ridges and decrease smoothly to approximately 50 mW m^{-2} in the oldest oceanic lithosphere. Similarly, ocean depth is approximately 2500 m at the ridges and increases to 5600 m for the oldest sea floor.

These variations are the primary data used to constrain models of the thermal evolution of oceanic lithosphere and, hence, of the forces driving plate tectonics. They can be described using a simple but powerful model for the formation of the lithosphere by hot material at the ridge, which cools as the plate moves away. In this model, material at the ridge at a mantle temperature T_m (1300–1400°C) is brought to the ocean floor, which has a temperature T_s. The material then moves away while its upper surface remains at T_s. Because the plate moves away from the ridge faster than heat is conducted horizontally, we can consider only vertical heat conduction. Mathematically, this is the same as the cooling of a half-space originally at temperature T_m, whose surface is suddenly cooled to T_s at time $t = 0$.

The solution of the heat equation for this case can be viewed in terms of isotherms, or lines of constant temperature, in the plate. The depth to a given temperature increases as the square root of lithospheric age (Fig. 2, upper left). Such square root of time behavior occurs for any process described by a diffusion equation (e.g., the heat equation). For example, after a lava flow erupts, it cools as the square root of time.

The cooling of the lithosphere predicts both the depth and heat flow variations. As the rock cools, it gets denser and contracts, so ocean depth should vary with the square root of age, as observed. Similarly, the heat flow should decrease as the square root of age, as observed. Because ocean depth seems to "flatten" at approximately 70 million years, we often use a modification called a plate model, which assumes that the lithosphere evolves toward a finite plate thickness with a fixed basal temperature. The flattening reflects the fact that heat is being added from below, so the predicted sea floor depth and heat flow behave for young ages like in the half-space model but evolve asymptotically toward constant values for old ages.

Comparison with depth, heat flow, and geoid (gravity) data shows that the plate thermal model is a good, but not perfect, fit to the average data because processes other than this simple cooling also occur. For example, water flow in the crust transports some of the heat for ages less than approximately 50 million years, as shown by the spectacular hot springs at mid-ocean ridges. Thus, the observed heat flow is lower than the model's predictions, which assume that all heat is transferred by conduction. Some topographic effects, including the spectacular volcanic oceanic plateaus, result from crustal thickness variations. Because these and other effects vary from place to place, the data vary about their average values for a given age.

The cooling of the oceanic lithosphere as it moves away from ridges is part of the plate tectonic cycle, as is the cooled lithosphere heating up again as it subducts. Both portions give rise to thermal buoyancy forces, which drive plate motions, due to the density contrast resulting from the temperature difference between the plate and its surroundings. Often, the first is called ridge push, which is a confusing term because it is zero at the ridge and increases linearly with plate age. Similarly, the force due to cold, dense, subducting slab is called slab pull. Although it is useful to think of the two forces separately, both are parts of the net buoyancy force due to mantle convection.

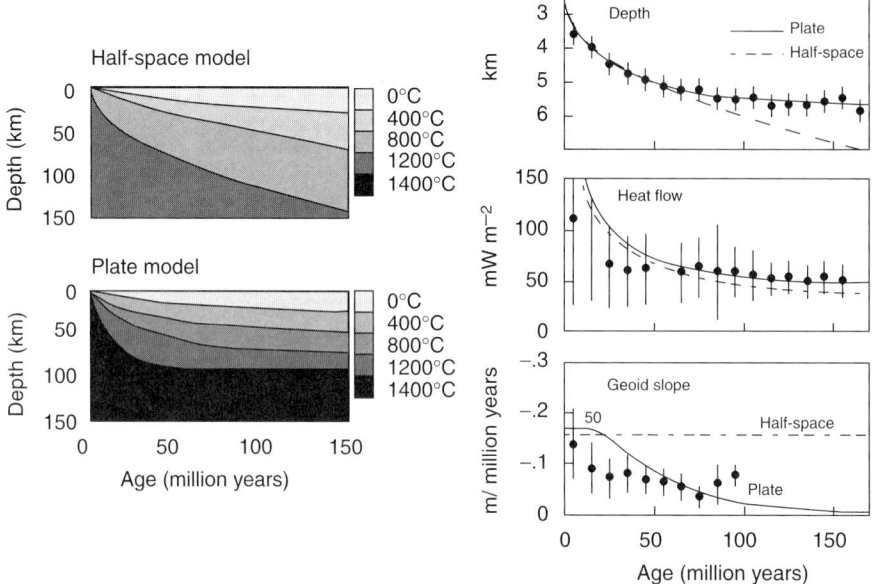

FIGURE 2 Models and data for thermal evolution of the oceanic lithosphere. Isotherms for thermal models are on the left-hand side. The lithosphere continues cooling for all ages in a half-space model and equilibrates for 70 million years lithosphere in a plate model with a 95-km thick thermal lithosphere. The plate model shown has a higher basal temperature than the half-space model. Comparisons of thermal model predictions to various data are on the right-hand side. All show a lithospheric cooling signal and are better (but far from perfectly) fit by a plate than by a half-space model. Reprinted from Stein and Wysession (2003), with permission.

4. HEAT FLOW IN CONTINENTAL LITHOSPHERE

Continental heat flow differs from oceanic heat flow because continental crust is not subducted. However, continental heat flow also varies with geological age, from approximately $80\,\text{mW}\,\text{m}^{-2}$ in the youngest rocks to $40\,\text{mW}\,\text{m}^{-2}$ in the oldest ones (Fig. 3).

Using the heat flow to infer the temperature with depth beneath continents is more complicated than doing so with depth beneath the oceans because the rocks of the continental crust contain much more of the radioactive heat-producing elements uranium, thorium, and potassium than does the oceanic crust. Hence, the geotherm, or temperature as a function of depth, is found near the earth's surface by making the simplifying assumption that the temperature is not changing with time. In this case, the heat equation can be solved to give

$$T(z) = T_s + \frac{q}{k}z - \frac{\rho H}{2k}z^2.$$

This solution (Fig. 4) has two interesting consequences. First, because of the heat production, the temperature at depth is less than it would be without heat production. Moreover, because the geotherm is less steep at depth, the heat flow at depth is less than

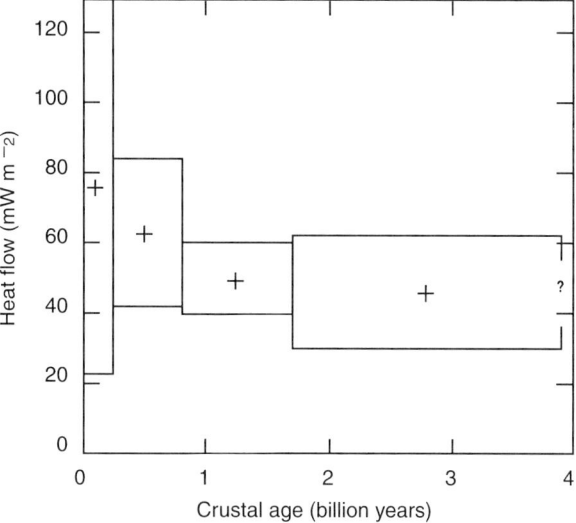

FIGURE 3 Heat flow versus age for continental lithosphere. Data from Sclater et al. (1980).

that at the surface. Physically, this is because some of the heat is generated in the crust.

A second consequence is that at relatively shallow depth, or approximately 100 km, the geotherm intersects the solidus, where rock begins to melt, and then the liquidus, where all of the rock has melted. This cannot be the case because seismic shear

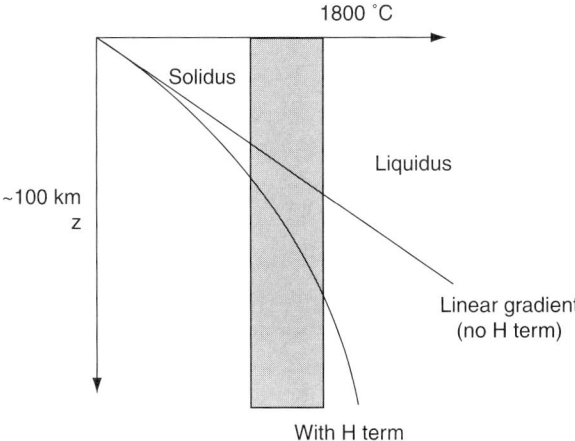

FIGURE 4 Schematic comparison of continental geotherms with and without crustal heat production (H term). In reality, the liquidus and solidus temperatures increase with depth, and convective effects need to be included.

waves travel through the entire mantle, but molten rock does not transmit them. The explanation for this paradox is that this geotherm is computed assuming that heat is transferred entirely by conduction, whereas at these high temperatures rock flows, and so heat, are transferred by convection. Hence, the lithosphere is the strong outer layer of the mantle convection system, within which heat transfer occurs by conduction, and is underlain by weaker rock, within which heat is transferred by convection that is much more efficient.

Intriguingly, the heat flow in old continental and old oceanic lithospheres are approximately equal. This is surprising given that the former includes the effect of crustal heat production, whereas the latter reflects only mantle heat flow. Whether this equality is simply a coincidence or reflects underlying physics is unclear.

5. GLOBAL HEAT FLOW

For many applications, we use the global average or total heat flow. To find these, we take the mean oceanic and continental heat flows of approximately 101 and 65 mW m^{-2} and weight them by their areas. This gives a global mean of about 87 mW m^{-2} or a total global heat loss of 44×10^{12} W. Of this loss, 70% is in the oceans and 30% is in the continents, reflecting both the higher heat flow and larger area of the oceans. The term "loss" indicates that the solid earth is losing heat. Heat loss is measured in units of energy per unit time (power).

6. THERMAL EVOLUTION OF THE EARTH

Heat flow provides a crucial constraint on the thermal evolution of the earth. A common approach is to consider conservation of energy for the whole earth. In this case, the change in the average temperature T as a function of time t is given by the balance between heat produced and that lost at the surface:

$$MC \frac{\partial T}{\partial t} = MH - Aq,$$

where M is the mass of the earth, A is its surface area, C is specific heat, q is the average heat flow, and H is the average rate of radioactive heat production. This equation can be applied to both the core and mantle or to the mantle alone. The heat flow used is an average of that coming from the mantle, or approximately 72 mW m^{-2}, which is estimated by removing the 17% or so thought to be produced by radioactivity in the continental crust.

The balance between surface heat flow and heat production is given by the Urey ratio

$$Ur = \frac{MH}{Aq},$$

so the temperature evolves with time according to

$$\frac{\partial T}{\partial t} = \frac{Aq}{MC}(Ur - 1).$$

The Urey ratio is thought to be between 0.5 and 0.8, so heat is lost faster than it is produced and the earth has been cooling with time. Thermal history models are derived by assuming an initial temperature and distribution of radioactive elements, both of which decay with time. The temperature is thought to have been high, owing to the release of gravitational potential energy as the planet accreted, gravitational potential energy that was released as iron sank to form the earth's core, and radioactivity. Analytic parameterized models are used to estimate how convection transferred heat from the interior to the surface over time. An important aspect of these models is the strong dependence of mantle viscosity on temperature. Early in the earth's history, the planet was hotter, making mantle viscosity lower and heat transfer by convection more efficient. With time (Fig. 5), the earth cooled and the viscosity increased. This idea seems plausible because rocks formed roughly 3 billion years ago solidified at temperatures approximately 300°C higher than those of comparable rocks today.

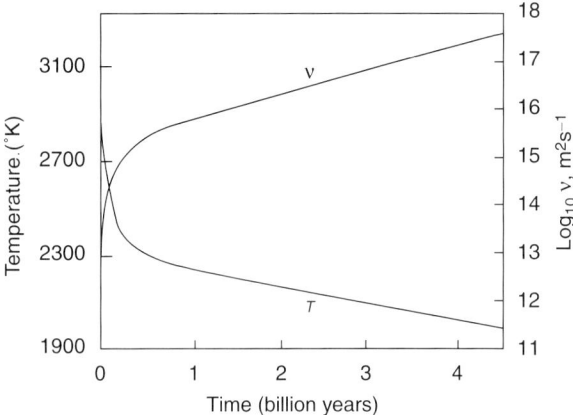

FIGURE 5 Average mantle temperature and viscosity history for a parameterized convection model of Earth's evolution. Reprinted from Schubert *et al.* (1980).

Hence, the plate tectonic cycle is evidence that the earth's cooling continues today as heat is transferred from the planet's interior. As cooling continues, we expect the strong lithosphere to thicken with time. Similarly, cooling causes the earth's solid inner core to grow with time at the expense of the liquid outer core. Because convection involves heat transfer by the motion of material, the thermal evolution of the earth gives rise to the chemical differentiation that produced the earth's current structure.

This approach gives insight into the differences among the terrestrial or inner planets of the solar system. What we know about the earth and our more limited knowledge of the moon and other planets suggests that although there are differences among the inner planets that reflect their initial compositions, there are also similarities in their evolution. As shown in Fig. 6, planets may follow a similar life cycle with phases including their formation, early convection and core formation, plate tectonics, terminal volcanism, and quiescence. This evolution is driven by the available energy sources and reflects the planets' cooling with time. Thus, even though the planets formed at about the same time, they are at different stages in their life cycles (as are a human and a dog born on the same date).

The earth is in its middle age, characterized by a relatively thin lithosphere and active plate tectonics. Conversely, it appears that the moon now has a thick lithosphere and is tectonically inactive. Thus, it seems to have lost much of its heat, presumably due to its small size, which favors rapid heat loss. In general, we would expect the heat available from the gravitational energy of accretion and radioactivity to increase with the planet's volume, whereas the rate of heat loss through the surface should depend on its surface area. Hence, the remaining heat should vary as

$$\text{remaining heat} \rightarrow = \frac{\text{available}}{\text{loss}} = \frac{(4/3)\pi r^3}{4\pi r^2} = \frac{r}{3},$$

so larger planets would retain more heat and be more active.

From such arguments, we might expect Mercury and Mars, which are larger than the moon but smaller than the earth, to have also reached their old age with little further active tectonics. Venus, which is comparable in size to the earth, might still be active but with episodic, rather than continuous, plate tectonics.

7. SOCIETAL IMPLICATIONS

One might not expect the flow of heat from the earth to be of great significance for society. This flow typically seems small because the ground is not hot when we touch it, except in geothermal areas. The average heat flow is much less than the solar constant, which measures the average power delivered per unit area at the top of the atmosphere, given that their ratio is $87 \text{ mW m}^{-2}/1372 \text{ W m}^{-2} = 6 \times 10^{-5}$. Even so, heat from the earth can have important consequences, both good and bad. For example, volcanic eruptions can be very destructive. However, islands such as Hawaii result. Moreover, the steady supply of volcanic gases are crucial to maintaining the atmosphere. Earthquakes are a destructive consequence of tectonic processes that shape topography that we enjoy and use.

Energy from the earth is also crucial for society's energy needs. Direct use of geothermal energy, although only a minor contributor to global energy supply, is important in a few areas, such as Iceland and New Zealand, where high temperature rock is close to the surface. Other energy sources ultimately require energy from the earth. Hydroelectric power depends on topography, raised by tectonic processes, to provide potential energy. Fossil fuels, such as coal, oil, and natural gas, result because organic matter has been buried to the depths required for the earth's heat to convert it to useful fuels. Nuclear power relies on the same radioactive elements that supply much of the earth's heat.

Finally, at a fundamental level, humanity evolved and survives in large part because Earth loses its heat by plate tectonics. Plate tectonics is crucial for the evolution of Earth's ocean and atmosphere because it

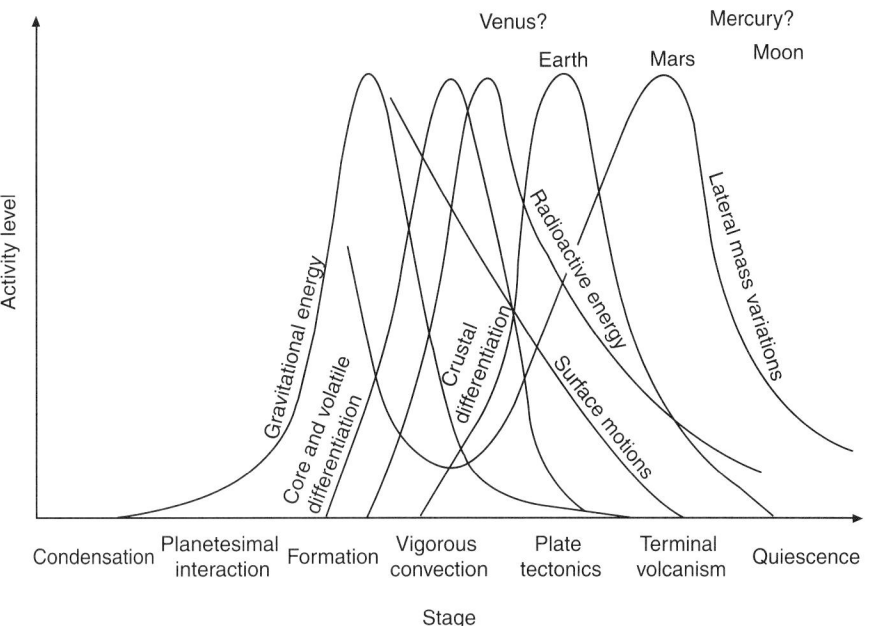

FIGURE 6 A model for the evolution of terrestrial planets showing the energy sources at each stage. Presented in text and verse by Kaula (1975).

involves many of the primary means (including volcanism, hydrothermal circulation through cooling oceanic lithosphere, and the cycle of uplift and erosion) by which the solid earth interacts with the ocean and atmosphere. The chemistry of the oceans and atmosphere depend in large part on plate tectonic processes, and many long-term features of climate are influenced by mountains that are uplifted by plate convergence and the positions of continents that control ocean circulation. Plate tectonics may explain how life evolved on Earth (at mid-ocean ridge hot springs) and is crucial for its survival (the atmosphere is maintained by plate boundary volcanism, and plate tectonics raises the continents above sea level).

SEE ALSO THE FOLLOWING ARTICLES

Earth's Energy Balance • *Geothermal Direct Use* • *Lunar–Solar Power System* • *Ocean, Energy Flows in* • *Ocean Thermal Energy* • *Sun, Energy from* • *Thermal Energy Storage*

Further Reading

Kaula, W. M. (1975). The seven ages of a planet. *Icarus* 26, 1–15.
Pollack, H. N., Hurter, S. J., and Johnston, J. R. (1993). Heat loss from the earth's interior: Analysis of the global data set. *Rev. Geophys.* 31, 267–280.
Schubert, G., Stevenson, D., and Cassen, P. (1980). Whole planet cooling and the radiogenic heat source contents of the earth and moon. *J. Geophys. Res.* 85, 2511–2518.
Schubert, G., Turcotte, D., and Olson, P. (2001). "Mantle Convection in the Earth and Planets." Cambridge University Press, Cambridge, UK.
Sclater, J. G., Jaupart, C., and Galson, D. (1980). The heat flow through oceanic and continental crust and the heat loss of the earth. *Rev. Geophys. Space Phys.* 18, 269–311.
Sleep, N. H. (1979). Thermal history and degassing of the earth: Some simple calculations. *J. Geol.* 87, 671–686.
Stacey, F. D. (1992). "Physics of the Earth," 3rd ed. John Wiley, New York.
Stein, C. A., and Stein, S. (1992). A model for the global variation in oceanic depth and heat flow with lithospheric age. *Nature* 359, 123–129.
Stein, C. A., and Stein, S. (1994). Constraints on hydrothermal heat flux through the oceanic lithosphere from global heat flow. *J. Geophys. Res.* 99, 3081–3095.
Stein, S., and Wysession, M. (2003). "Introduction to Seismology, Earthquakes, and Earth Structure." Blackwell, Oxford, UK.
Verhoogen, J. (1980). "Energetics of the Earth." National Academy of Sciences, Washington, DC.

Livestock Production and Energy Use

DAVID PIMENTEL
Cornell University
Ithaca, New York, United States

1. Animal Products Consumed in the U.S. Diet
2. Energy Inputs in Animal Product Production
3. Land Resources
4. Water Resources
5. World Food Needs
6. Conclusion

Glossary

cellulose Any of several fibrous substances constituting the chief part of the cell walls of plants.
erosion The slow breakdown of rock or the movement and transport of soil from one location to another. Soil erosion in crop and livestock production is considered serious worldwide.
forage Vegetable food (as hay, grain) for domestic animals.
fossil energy Energy from sources extracted from the earth such as oil, coal, and natural gas.
hectare (ha) A metric unit of area equal to 10,000 square meters or 2.47 acres.
monoculture The cultivation of a single crop or product to the exclusion of other possible uses of the land.
per capita By or for each person.
salinization The process by which salts accumulate in soil.
waterlogging The saturation of soil with water causing the water table to rise high enough to expel normal soil gases and interfere with plant growth or cultivation.

Worldwide an estimated 2 billion people live primarily on a meat-based diet, while an estimated 4 billion people live primarily on a plant-based diet. The shortages of cropland, freshwater, and energy resources require most of the 4 billion people to live primarily on a plant-based diet; however, there are serious food shortages worldwide. For instance, the World Health Organization recently reported that more than 3 billion people are malnourished in the world. This is the largest number and proportion of malnourished people ever recorded in history. In large measure, the food shortage and malnourishment problem are primarily related to rapid population growth in the world plus the declining per capita availability of land, water, and energy resources required for food production. Meat, milk, and eggs contribute valuable nutrients to the human diet in the United States and the rest of the world. To produce animal protein successfully requires the expenditure of human and fossil energy to supply livestock forage and grain. The land, devoted to grain or forage for livestock production, is exposed to soil erosion, which slowly diminishes the fertility of the soil and its productivity. Additionally, animal production requires large inputs of water for the grain and forage crops and, to a lesser extent, directly for animal consumption. All of these factors interact to determine the ultimate success of animal production systems. This article includes analyses of the quantities of animal products produced; energy, land, and water resource inputs in livestock production; and meat, milk, and egg production.

1. ANIMAL PRODUCTS CONSUMED IN THE U.S. DIET

In the United States, more than 8 billion livestock are maintained to supply the animal protein consumed annually. In addition to the large amount of cultivated forage, the livestock population consumes about seven times as much grain as is consumed directly by the entire U.S. population.

From the livestock population of more than 8 billion, approximately 7.5 million tons (metric) of animal protein is produced each year (Table I). If distributed equally, it would be sufficient to supply

TABLE I

Number of Livestock in the United States

Livestock	Number × 10⁶
Sheep	7
Dairy	13
Swine	60
Beef cattle	74
Turkeys	273
Broilers	8000
Eggs	77,000

TABLE II

Grain and Forage Inputs per Kilogram of Animal Product Produced, and Fossil Energy Inputs (kcal) Required to Produce, 1 kcal of Animal Protein

Livestock	Grain (kg)[a]	Forage (kg)[b,c]	kcal input/kcal protein
Lamb	21	30	57:1
Beef cattle	13	30	40:1
Eggs	11	—	39:1
Beef cattle	—	200	20:1
Swine	5.9	—	14:1
Dairy (milk)	0.7	1	14:1
Turkeys	3.8	—	10:1
Broilers	2.3	—	4:1

From [a]USDA (2001). "Agricultural Statistics." U.S. Department of Agriculture, Washington, DC; [b]Heischmidt, R. K., Short, R. E., and Grings, E. E. (1996). Ecosystems, sustainability and animal agriculture. *Animal Sci.* 74(6), 1395–1405; [c]Morrison, F. B. (1956). "Feeds and Feeling." Morrison Publishing Co., Ithaca, NY.

about 75 g of animal protein daily per American. With the addition of 37 g of available plant protein, a total of 112 g of protein is available per capita. In contrast, the recommended daily allowance (RDA) per adult per day is 56 g of protein for a mixed diet for an adult. Therefore, based on these data, each American is consuming about twice the RDA for protein per day. About 144 kg of meat, including fish, are eaten per American per year. In addition, 271 kg of milk and eggs are consumed per capita in the United States per year.

2. ENERGY INPUTS IN ANIMAL PRODUCT PRODUCTION

Each year an estimated 45 million tons of plant protein are fed to U.S. livestock to produce approximately 7.5 million tons of animal protein for human consumption. To produce this animal protein, about 28 million tons of plant protein from grain and 17 million tons of plant protein from forage are fed to the animals (Table II). Thus, for every kilogram of high-quality animal protein, livestock are fed nearly 6 kg of plant protein. In the conversion of plant protein into animal protein, there are two principal costs: (1) the direct costs of production of the harvested animal including the grain and forage and (2) the indirect costs for maintaining the breeding animals (mother and father).

The major fossil energy inputs for grain and forage production include fertilizers, farm machinery, fuel, irrigation, and pesticides. The energy inputs vary according to the particular crop and forage being grown. When these inputs are balanced against their energy and protein content, grains and some legumes like soybeans are produced more efficiently in terms of energy inputs than fruits, vegetables, and animal products. In the United States, the average protein yield of the five major grains (plus soybeans) fed to livestock is about 700 kg/ha. To produce a kilogram of plant protein requires about 10 kcal of fossil energy.

Forage can be fed to ruminant animals, like cattle and sheep, because they can convert forage cellulose into usable nutrients through microbial fermentation. The total plant protein produced on good U.S. pasture and fed to ruminants is 60% of the amount produced by grains (Table II). The current yield of beef protein from productive pastures is about 66 kg/ha, while the energy input per kilogram of animal protein produced is 3500 kcal. Therefore, animal protein production on good pastures is less expensive in terms of fossil energy inputs than grain protein production (Table II).

Of the livestock systems evaluated in this investigation, chicken-broiler production is the most efficient with an input of 4 kcal of fossil energy per 1 kcal of broiler protein produced (Table II). Broilers are a grain-only system. Turkey production, also a grain only system, is next in efficiency with a ratio of 10:1. Milk production based on a mixture of grain and forage also is relatively efficient with a ratio of 14:1 (Table II). Nearly all the feed protein consumed by broilers is grain, whereas for milk production about two-thirds is grain (Table II). Of course, 100% of milk production could be produced on forage. Both pork and egg production also depend on grain (see Table IV). Pork has a 14:1 ratio, whereas egg production is relatively more costly in terms of feed energy requiring a 39:1 ratio (Table II).

TABLE III

Calorie, Water, and Protein Availability per Kilogram of Animal Product

Livestock	Energy (kcal)	Water (%)	Protein (g)
Lamb	2521	47	220
Beef	2565	49	186
Turkey	1193	55	123
Egg	1469	74	116
Pork	2342	57	134
Dairy	647	87	34
Broiler	1357	71	238

From Pimentel, D. (1997). "Livestock Production: Energy Inputs and the Environment." Canadian Society of Animal Science, Proceedings. Canadian Society of Animal Science, Montreal, Quebec.

TABLE IV

Estimated Liters of Water Required to Produce 1 kg of Food and Forage Crops

Crop	Liters/kg
Potatoes	500
Wheat	900
Alfalfa	900
Sorghum	1100
Corn	1400
Rice	1900
Soybeans	2000
Broiler	3500
Beef	43,000

From Pimentel, D., Houser, J., Preiss, E., White, O., Fang, H., Mesnick, L., Barsky, T., Tariche, S., Schreck, J., and Alpert, J. (1997). Water resources: Agriculture, the environment, and society. *BioScience* 47(2), 97–106.

The two livestock systems depending most heavily on forage, but still using significant amounts of grain, are the beef and lamb production systems The lamb system with a ratio of 57:1 and the beef system with a ratio of 40:1 are the two highest (Table II). If these animals were fed only on good-quality forage, the energy inputs could be reduced by about half depending on the conditions of the pasture forage as well as the management practices. Note that beef fed 200 kg of forage and no grain had an energy input per kcal protein output ratio of 20:1 (Table II). Rainfall is critical for all productive pasture systems.

Per kilogram of animal product foods, broiler chicken flesh has the largest percentage of protein and milk the lowest (Table III). Beef has the highest calorie content because of its high fat content and relatively low water content. Of the all the animal products, milk has the highest water content with 87%.

The average fossil energy input for all animal protein production systems studied is about 25 kcal of fossil energy input per kcal of animal protein produced (Tale II). This energy input is more than 10 times greater than the average input to output ratio for grain protein production, which was about 2.5 cal per kcal of protein produced. As a food for humans, however, animal protein has about 1.4 times the biological value as a food compared with grain protein.

3. LAND RESOURCES

Livestock production requires a large number of hectares to supply the grains, forages, and pastures for animal feeds. In fact, nearly 300 million hectares of land are devoted to producing the feed for the U.S. livestock population. Of this, 262 million hectares are pasture and about 30 million hectares are for cultivated grains. In addition to the large amount of forages and grass that are unsuitable for human consumption and are fed to animals, about 323 million tons of grains—or about 816 kg per American in the United States are fed to livestock to provide meat, milk, and eggs.

More than 99.2% of U.S. food is produced on the land, while less than 0.8% comes from oceans and other aquatic ecosystems. The continued use and productivity of the land is a growing concern because of the rapid rate of soil erosion and degradation that is taking place throughout the United States and indeed throughout the world. Each year about 90% of U.S. cropland is losing soil at an average rate 13 times above the sustainable rate of 1 t/ha/year. On croplands where most grain is produced, soil loss averages more than 13 t/ha/year from the combined effects of water and wind erosion. Also, our rangelands are losing soil an average of 13 t/ha/year. About 60% of United States rangeland is being overgrazed and is subject to accelerated erosion.

The concern about high rates of soil erosion in the United States and the rest of the world is evident when it is understood that it takes approximately 500 years to replace 25 mm (1 in.) of lost soil. Clearly a farmer cannot wait for the replacement of 25 mm of soil. Commercial fertilizers can replace some nutrient loss resulting from soil erosion, but this requires large inputs of fossil energy.

The future of all agricultural production that requires land, including that targeted for livestock, will feel the effects of land degradation, particularly when fossil fuel supplies decline and prices increase. Soil erosion losses, compounded by salinization and waterlogging, are causing the abandonment of nearly 1 million hectares of U.S. agricultural land per year. Some of the abandoned, degraded cropland may find use as either pasture or forest.

The costs of soil erosion are well illustrated by the loss of rich U.S. soils. Iowa, which has some of the best soils in the world, has lost more than one-half of its topsoil after only 150 years of farming. Iowa continues to lose topsoil at an alarming rate of about 30 t/ha/year, which is about 30 times faster than the rate of soil formation. The rich Palouse soils of the northwestern United States have similarly lost about 40% of their topsoil in the past century.

Despite the efforts of the USDA Soil Conservation Service, erosion rates in the United States have decreased only slightly during the past 50 years. This is the result of major changes in agricultural production, such as an emphasis on commodity price-support programs; widespread planting of crop monocultures; crop specialization; abandonment of crop rotations; the removal of tree shelter-belts; leaving the soil without protective biomass cover; and the use of heavy farm machinery. Concurrently these changes have been accompanied by the creation of fewer and larger farms where increased mechanization is a necessity.

Although modern farming practices are contributing to the soil erosion problem, the failure of farmers and governments to recognize and address the soil erosion problem is equally important if soil depletion is to be halted. Erosion often goes unnoticed by some farmers because soil loss is difficult to measure visually. For instance, one night's wind- or rainstorm could erode 15 t of soil per hectare as a sheet, which would be only 1 mm of soil; the next morning, the farmer might not even notice. This soil loss continues slowly, quietly, year after year, until the land is no longer productive. In addition, governments tend to ignore erosion because of its insidious nature and because it does not seem to be a major environmental crisis such as floods or tornadoes.

4. WATER RESOURCES

Agricultural production, including livestock production, consumes more fresh water than any other human activity. Western U.S. agriculture accounts for about 81% of the fresh water consumed after being withdrawn. Growing plants render all water nonrecoverable through evaporation and transpiration. In the United States, about 62% of the water used in agricultural irrigation comes from surface sources and 38% ground water sources.

The transfer of water to the atmosphere from the terrestrial environment by transpiration through vegetation is estimated to range between 38% and 65% of the rainfall depending on the terrestrial ecosystem. The vital photosynthetic processes and temperature control necessitates that the plants consume enormous amounts of water.

The water required to produce various food and forage crops range from 500 liters to 2000 liters of water per kilogram of plant biomass produced (Table IV). For example, a hectare of U.S. corn producing about 8000 kg per year transpires about 5 million liters of water during the growing season. Approximately 1000 mm (10 million liters per hectare) of rainfall or other sources of water are needed during the growing season for corn production. Even with 800 to 1000 mm of annual rainfall in the corn-belt region, corn usually suffers from some lack of water during the summer growing season. Producing 1 kg of beef requires about 43 times more water than it takes to produce 1 kilogram of grain. Livestock directly use only 1.3% of the total water used in agriculture. However, when the water required for forage and grain production is included, this dramatically increases the water requirement for livestock production. Producing 1 kg of fresh beef requires about 13 kg of grain and 30 kg of forage (Table II). This much grain and forage requires a total of 43,000 liters of water. On rangeland where the animal consumes about 200 kg of forage to produce 1 kg of beef, about 200,000 liters of water are need to produce the 1 kg of beef. With forage and some cereal crops, livestock can be produced in areas with low rainfall ranging from 150 to 200 mm per year. However, crop production and yields are low under such conditions.

Animals vary in the amounts of water required for their production. In contrast to beef, 1 kg of broiler chicken can be produced with about 2.6 kg of grain requiring approximately 3500 liters of water (Table IV).

Water shortages are already severe in the western and southern United States. The situation grows worse as the U.S. population and its requirements for water, including for agriculture, rapidly increase.

5. WORLD FOOD NEEDS

Worldwide human food needs are rising and will continue to rise with the world population. Currently, more than 3 billion people are malnourished based on calories, protein, vital minerals, and vitamin shortages in their diets. There are currently 6.2 billion people on Earth, and it is projected that the world population will double to more than 12 billion in less than 50 years, based on the current growth rate. The U.S. population is also increasing rapidly. The U.S. population is currently at 285 million and is expected to double to 570 million in about 70 years. Food security becomes at risk as more and more people need food, while the required resources of land, water, and energy decline per person.

Food consumption patterns in the United States and most other developed nations include generous amounts of animal products. More than half of U.S. grain and nearly 40% of world grain is being fed to livestock rather than being consumed directly by humans. Grains provide 80% of the world's food supply. Although grain production is increasing in total, the per capita supply has been decreasing for nearly two decades. Clearly, there is reason for concern in the future.

If all the 323 million tons of grain currently being fed to livestock was consumed directly by people, the number of people who could be fed would be approximately 1 billion. Also, if this much grain were exported, it would provide approximately $80 billion each year in income—this is sufficient income to pay for our current oil bill of $75 billion per year. Of course, exporting all the grain currently fed to livestock would reduce the average protein consumption of Americans from 112 gram per day to approximately 73 grams per day. Yet this intake would still be greater than the 56 gram of protein suggested by the RDA.

Exporting all U.S. grain that is now fed to livestock assumes that livestock production would change to a grass-fed livestock production system. Animal protein in the diet would then decrease from the current level of 75 grams to 36 grams per day, or about one-half. Again, the diet for the average American would be more than adequate in terms of protein consumption, provided that there was no change in the current level of plant protein consumed. In fact, consuming less meat, milk, and eggs and eating more grains and vegetables would improve the diet of the average American.

6. CONCLUSION

Meat, milk, and egg production in the United States relies on significant quantities of fossil energy, land, and water resources. Grain-fed livestock systems use large quantities of energy because grain crops are cultivated; in contrast, cattle grazed on pastures use considerably less energy than grain-fed cattle. An average of 25 kcal of fossil energy is required to produce 1 kcal of animal protein and requires approximately 10 times the energy expended to produce 1 kcal of plant protein. However, it should be noted that animal protein is 1.4 times more nutritious for humans than plant protein.

Nearly one-third of the U.S. land area is devoted to livestock production. Of this, about 10% is devoted to grain production and the remainder is used for forage and rangeland production. The pastureland and rangeland is marginal in terms of productivity because there is too little rainfall for crop production.

Livestock production is also a major consumer of water because grains and forage consumed by livestock require significant amounts of water for growth and production. To produce 1 kg of grain requires about 1000 liters of water. Based on grain and forage consumption, about 43,000 liters of water are required to produce 1 kg of beef. In regions where water is already in short supply and where aquifers are currently being mined faster than they can be recharged, major decisions will have to be made concerning all agricultural production, including grain and forage crops for livestock.

As human food needs escalate along with population numbers, serious consideration must be given to the conservation of fossil energy, land, and water resources. The careful stewardship of these resources is vital if livestock production, and indeed agriculture, will be sustainable for future generations. In the end, population growth must be reduced, in the United States and in throughout the world, if we are to achieve a quality life for ourselves and for our grandchildren.

SEE ALSO THE FOLLOWING ARTICLES

Aquaculture and Energy Use • Conversion of Energy: People and Animals • Diet, Energy, and Greenhouse Gas Emissions • Fisheries and Energy Use • Food System, Energy Use in • Hunting and Gathering Societies, Energy Flows in • Industrial Agriculture,

Energy Flows in • Land Requirements of Energy Systems • Lifestyles and Energy

Further Reading

FAO (1998). Food Balance Sheet. http://apps.fao.org (November 16, 2000).

Pimentel, D. (2004). Soil erosion: A major environmental threat. *In* "Making Development Work: A New Role for Science" (G. Leclerc and C. A. S. Hall, Eds.). University of New Mexico Press, Albuquerque, New Mexico (in press).

Pimentel, D., Doughty, R., Carothers, C., Lamberson, S., Bora, N., and Lee, K. (2002). Energy inputs in crop production: comparison of developed and developing countries. *In* "Food Security & Environmental Quality in the Developing World" (L. Lal, D. Hansen, N. Uphoff, and S. Slack, Eds.), pp. 129–151. CRC Press, Boca Raton, FL.

Pimentel, D., and Pimentel, M. (1996). "Food, Energy and Society." Colorado University Press, Niwot, CO.

Pimentel, D., and Pimentel, M. (2002). Sustainability of meat-based and plant-based diets and the environment. *Am. J. Clinical Nutrition* **78** (Suppl.), 660–663.

Thomas, G. W. (1987). Water: Critical and evasive resource on semiarid lands. *In* "Water and Water Policy in World Food Supplies" (W. R. Jordan, Ed.), pp. 83–90. College Texas A & M, University Press, Station, TX.

Troeh, F. R., Hobbs, J. A., and Donahue, R. L. (1999). "Soil and Water Conservation." Prentice Hall, Englewood Cliffs, NJ.

USBC (2002). "Statistical Abstract of the United States 2002." Bureau of the Census, U. S. Government Printing Office, Washington, DC.

WHO (2000). Malnutrition Worldwide. http://www.who.int (July 27, 2000).

Lunar–Solar Power System

DAVID R. CRISWELL
University of Houston
Houston, Texas, United States

1. Isolated Spaceship Earth and the Human Race to Sustainability
2. The Immediate Global Energy Challenge
3. Lunar–Solar Power System
4. Lunar–Solar Power Demonstration Base
5. Scale and Cost for 20 TWe Lunar–Solar Power System
6. Conclusion

Glossary

astronomical unit (AU) The radius of a Keplerian circular orbit of a point mass having an orbital period of $2 \times \pi/k$ days (where k is the Gaussian gravitational constant) or 149,597,870.691 km.

bootstrapping Use of local materials to manufacture a significant fraction of the tools to make additional manufacturing facilities.

energy The capacity to do work, E, which is formerly determined as the scalar product of the vector quantities force and distance ($E = \mathbf{F} \cdot \mathbf{s}$); it is measured scientifically in joules (J), and large quantities are often quoted in pentajoules (1×10^{15} J) and exajoules (1×10^{18} J).

power Work per unit time and formerly measured in watts (1 W = 1 J/s), kilowatts (1 kW = 10^3 W) characteristic of the household level of use, megawatts (1 MW = 10^6 W) for a community or small industrial activity, gigawatts (1 GW = 10^9 W) for use by a large city, terawatts (1 TW = 10^{12} W) for use at the national or world level, and so on.

rectennas Many specialized radio antennas and diodes that are mechanically and electrically linked together into a large plane of rectifying antennas that receive microwave power and convert the incident microwave power into electrical power.

stand-alone power system Able to deliver a predictable and dependable flow of power without major ancillary equipment or alternative fuels or power sources.

Earth's inhabitants need, within 50 years, a new source of sustainable commercial power that provides at least 2 kilowatts of electricity per person or 20 terawatts globally. The new power system must increase Earth's resources and be independent of the biosphere, and the use of the power must not change the biosphere. Facilities built on the moon from lunar materials can capture a fraction of the 13,000 terawatts of solar power incident on the moon. This lunar–solar power system can deliver more than 20 terawatts of affordable electric power to Earth.

1. ISOLATED SPACESHIP EARTH AND THE HUMAN RACE TO SUSTAINABILITY

It is nearly impossible to comprehend the physical isolation of our tiny spaceship Earth and its cargo of humans and the other life within Earth's thin biosphere. To gain some sense of it, consider the Voyager 1 spacecraft, launched from Earth in September 1977. It is now traveling in a northward direction away from our solar system. Each year, it is 3.6 astronomical units (AU) farther away. On February 14, 1990, when Voyager 1 was approximately 11 AU away from our sun, it turned a long-range camera toward us and photographed our Earth, Venus, and our sun (Fig. 1A). The pale blue dot of Earth filled only 10% of one pixel of the camera (Fig. 1B).

Every living entity on Earth is totally dependent on sources of meager flows of geothermal power from within Earth—less than 30 terawatts of thermal power (TWt) and a tiny fraction of the approximately 175,000 terawatts of solar power (TWs) that intersects Earth. Life on Earth is totally subject to unpredictable major violent events. These include volcanic eruptions, sudden changes in regional and global climates, and impacts by asteroids and comets. Major volcanic eruptions can devastate large portions of a continent and quickly affect the atmospheric greenhouse. Earth's biosphere is affected over millions of years by changes in the orbit of Earth that change the solar power Earth receives. Its orbit around our sun is affected by the changing gravitational forces exerted on Earth by the sun, Jupiter, Saturn, and the other smaller planets. Earth was formed 4.6 billion years ago by the in-falling gas, dust, comets, asteroids, and primitive moons. The infall continues. Several metric tonnes of meteors are vaporized each day in Earth's upper atmosphere. Approximately once a year, a larger meteor releases the energy of a small nuclear bomb. It is projected that, on average, regional-scale disasters due to impacts of multi-hundred-meter-diameter objects will occur every 10,000 years, civilization-ending impacts by multi-kilometer-diameter objects will occur every 1 million years, and global cataclysms will occur every 100 million years. Such a collision 64 million years ago ended the age of dinosaurs and enabled small mammals to proliferate. One lineage evolved into humans.

The lifeless moon is our closest neighbor (Fig. 2). Surface temperatures near the equator of the airless moon vary from a daytime high of approximately 380°Kelvin (K) (107°C) to a predawn low of 120°K. A few craters near the lunar poles contain cold surfaces (~80°K) that have never seen the sun. Earth appears just above the horizon of the moon. Unlike the moon, Earth was large enough to gravitationally retain the water and gases that form its oceans and atmosphere. Earth's atmosphere provides a protective blanket, like that of a greenhouse, that recirculates about Earth a fraction of the solar power that would otherwise be reradiated quickly and directly back to space, as occurs from the lunar surface. This recirculation of solar heat provides Earth with a global mean temperature of approximately 295°K (22°C) within which water is liquid and water-based life can exist and evolve. Earth's atmosphere and

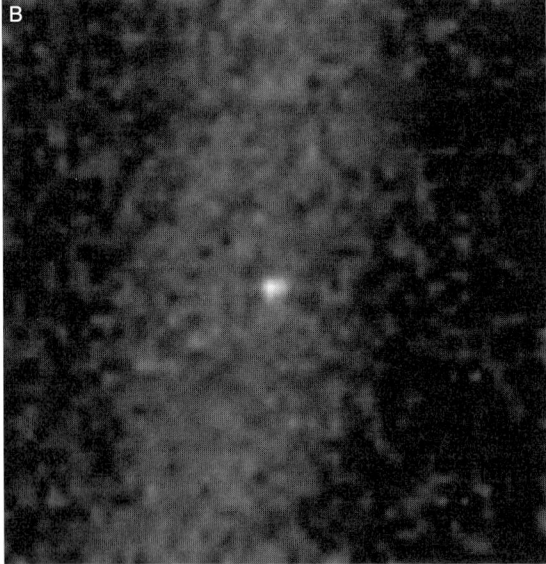

FIGURE 1 (A) Composite photograph of Earth (artifact-lined insert on left), Venus (middle artifact-lined insert), and our sun by the Voyager 1 spacecraft on February 14, 1990, at a distance of approximately 11 AU from the sun. (B) Image of Earth magnified to the greatest possible extent. It is a single pixel located at approximately the middle of the third "artifact" line from the top in the left-hand insert in panel A. Photographs extracted from the Voyager spacecraft file PIA00452 of the Jet Propulsion Laboratory.

FIGURE 2 Photograph of Earth, moon, and Apollo 11 lunar module returning from the moon to the command service module in 1969. From NASA, ID AS11-44-6642.

oceans operate as a heat engine, driven by the sun, to transport this recirculated solar heat about the planet. This process evens out the temperature differences between night and day and among the equator, mid-latitude deserts (extreme of $\leq 331°K$), and poles (extreme of $\geq 184°K$). Days- to weeks- to years-long variations of the recirculating atmosphere at any one point on Earth are termed "weather." Decades- and centuries-long patterns in the recirculating atmosphere and oceans are termed "climate."

Carbon dioxide is the minor noncondensing and relatively long-lived component of the atmosphere that controls the greenhouse over centuries. The effectiveness of Earth's greenhouse varies over shorter time scales due to the concentrations of water vapor, methane, and other minor constituents that vary quickly in time, latitude, longitude, and altitude. Over the life of Earth, its greenhouse has adjusted to approximately a 30% increase in the output of solar power and changes in Earth's orbit. However, the greenhouse is not perfect. Major and "minor" ice ages occur. The external and internal forces that drive climate change are not fully understood. Interactions between the drivers are even less understood. However, one thing is clear: human industry is now changing the face of the planet. Human commercial power systems now output the equivalent of approximately 14 TWt of thermal power and are rapidly consuming the relatively small sources of fossil energy accumulated over a period of more than 100 million years from the ancient biosphere. Also, human power systems and industry are degrading, in many ways, the ability of the living biosphere to support us. In particular, our commercial power systems are increasing the concentration of greenhouse gases in the atmosphere at a faster rate than have any previous natural processes. The inhabitants of the uncontrolled spaceship Earth (note Figs. 1B and 2) need a new source of sustainable abundant power that is independent of the biosphere and whose use does not fundamentally change the biosphere. This new source of power must be started, at industrial scale, within the next two decades and come to maturity within 50 years.

2. THE IMMEDIATE GLOBAL ENERGY CHALLENGE

The World Energy Council Statement 2000 issued this energy challenge:

Slightly more than 1 billion people in the industrialized countries (about 20% of the world's population) consume nearly 60% of the total energy supply, whereas just under 5 billion people in developing countries consume the other 40% of the total energy supply. ... The 2 billion poorest people ($1000 annual income per capita or less), a small but growing share of whom live in shantytowns with most still scattered in rural areas, use only 0.2 toe of energy [tonnes of oil-equivalent of thermal energy] per capita annually, whereas the billion richest people ($22,000 annual income per capita or more) use nearly 25 times more at 5 toe per capita annually [5 toe/person-year ~ 6.7 kilowatts of thermal power [kWt]/person].

In 2002 the UN World Summit on Sustainable Development challenged the world to "diversify energy supply and substantially increase the global share of renewable energy sources in order to increase its contribution to total energy supply."

Over the past century, the economic output of a unit of commercial energy has increased at approximately 1% per year. If this trend continues, the current 6- to 7-kWt/person power usage by Western Europeans and Japanese might decline by approximately 50% for the same level of economic output. Thus, many argue that per capita global energy use will decline significantly over the 21st century. Unfortunately, this is not a realistic prospect for a sustainable global economy. Both the developed and developing economies obtain significant quantities of energy embedded as fresh air and water and energy-rich biomass. As world population increases, the per capita quantities of these biospheric resources decline. The current economies obtain necessary chemicals from mines in which economically important chemicals have been concentrated by ancient geological processes. In 1976, Goeller and Weinberg estimated the per capita power necessary, assuming the use of non-fossil fuels systems such as breeder reactors, to indefinitely sustain the 1960s U.S. and global economies. This advanced sustainable economy will obtain its fresh water from desalination. All of the agricultural, process, and industrial chemicals will be obtained at their average abundance in the crust of the earth. Goeller and Weinberg projected that a sustainable 1968-style U.S. economy will require approximately 15 kWt/person and that a 1960s global economy will require 7.5 kWt/person. Global power consumption for 10 billion people of 75 to 150 TWt is implied. Ayres and colleagues demonstrated that economic growth of the U.S. economy, from 1900 to 1998, was proportional to the total "useful work" accomplished. Useful work, approximately a product of total input exergy (i.e., available energy) and effectiveness of its use, is now the dominant factor in U.S. economic growth. National economies are only now beginning to face the challenges of providing new levels of useful work to provide new goods and services that are essential to sustainable development. For the following discussion, it is assumed that ingenuity and economic adaptability will enable humankind to provide sustainable existence using only 6 to 7 kWt/person. However, it is likely that human needs and desires will require a greater level of per capita power.

The new power must be clean and significantly less expensive than it is now. The developed countries expend approximately 10% of their gross domestic product (GDP) on all phases of commercial energy. They now have a per capita GDP of approximately $22,000/person-year. By analogy, if the developing countries spend approximately 10% of their GDP on all phases of providing 6.7 kWt/person, the new source of power must initially supply thermal energy for less than 0.4¢/kWt-h. The new primary energy source and the new power system must be adequate for centuries and should have significant capacity for growth. Finally, it is practical to assume that the new commercial power system will use physical resources and technologies that are relatively well understood at this time.

Together, the developed and developing countries now consume the equivalent of approximately 14 TWt of thermal energy. To provide 10 billion people with 5 toe/person-year implies the global production of 67 TWt of commercial thermal power by approximately 2050. This corresponds to consuming the equivalent of approximately 900 billion barrels of oil per day or 11 times current global production. An energy-rich 10 billion people will consume 6700 TWt-years of energy per century. Commercial power is used to provide goods (e.g., steel, concrete) and services (e.g., fuel for transportation or heating). In 1880, only thermal power was available. However, by 1900, commercial electric power became available. By 2000, electricity powered 25% of the final production of goods and delivery of services. Electricity is projected to rise to approximately 100% of final power by 2050. This is because electricity, where it can be applied, is approximately three times more productive than thermal power. By 2050, approximately 20 to 30 terawatts of commercial electric power (TWe) will likely be as economically productive as 60 to 67 TWt of commercial thermal power.

The first column of Table I lists 25 options for generating global commercial power. The options are divided into five major categories: traditional fuels from the living and fossil biosphere, nuclear fuels, geothermal power, solar-derived power on Earth, and solar power supplied from facilities in space. The second column lists the equivalent total thermal energy, measured in terawatt-years, that can be provided by each option. The third column lists the estimated maximum electrical power, in terawatts, that can be provided by each option by 2050. The fourth column lists the anticipated trend in pollution production to 2050 for each power option. The final column lists the anticipated cost of electric energy in 2050 versus 1¢/kWe-h. Criswell provided considerable details on the analyses of these options and references.

Power flow through the living biosphere (row 1 in Table I) is simply too small to support a 20 TWe or 60 TWt commercial power system. Peat (row 2) and oil and gas (row 4) do not provide sufficient fuels. Coal (row 3) can potentially provide the order of 4500 TWt-year of fuel. However, a prosperous global economy would consume that fuel in approximately 70 years. It is estimated that natural gas hydrates (row 5), buried in the ocean silt along the margins of the continents, may contain the order of 10,000 TWt-year of thermal energy. However, the hydrate deposits are thinly spread over vast areas. Mining them likely will be expensive and disrupt the sea floor.

Conventional nuclear reactors that use uranium mined from the continents (row 6 in Table I) are projected to supply less than 430 TWt-year of commercial thermal energy. Breeder reactors can increase the total energy output by approximately a factor of 60. However, fission breeder reactors (rows 7 and 8) appear to be unacceptable politically and very expensive. Continuous net energy output from controlled nuclear fusion of deuterium (D) and tritium (T) is not yet demonstrated. It will be decades before the engineering and commercial characteristics of this "least challenging" fusion technology can be specified (rows 9 and 10). Fusion burning of helium-3 (^{3}He) with D (row 11) requires much higher temperatures and pressures. Again, it likely will be decades before commercial D–^{3}He fusion power plants can be engineered. In addition, the lunar soils are the only potential source of ^{3}He near Earth. The total quantity of ^{3}He that is potentially available is not known, and the economic efficiency with which it can be mined will require extensive demonstrations on the moon. Huge areas of the lunar surface would be totally disrupted by the mining operations.

Geothermal power (row 12 in Table I) is driven primarily by the slow radioactive decay of uranium (U) and thorium (T) within Earth. The original

TABLE I

Global Power System Options for 2050

Power system option	Fuel resource TWt-year thermal	2050 output electric power (TWe)	2050 pollution vs now	2050 cost vs 1 ¢/kWe-h
Biosphere and fossil				
1. Bioresources	<230	<0.2	More	>
2. Peat	<60	~0	More	>
3. Coal	<4,500	<4	Large	>
4. Oil and gas	<1,300	<8	Large	>
5. Natural gas hydrates	TBD >10,000	TBD	Large	Likely >
Nuclear				
6. Fission (no breeder, high-grade uranium ores)	<430	<1.5	Large	>
7. Fission (breed 238U/T)	<33,000	in #6	Large	>
8. Fission (breed ocean U)	~6,000,000	in #6	Large	>
9. Fusion (D–T/U-Th)	$<6 \times 10^9$	in #6	Large	Likely >
10. Fusion (D–T)	$\gg 1 \times 10^9$	0 likely	More	Likely >
11. Fusion (D–^{3}He lunar)	$? \sim 100$ to 10^5	0 likely	More	Likely >
Geothermal				
12. Nuclear decay and accretional energy (0–7 km depth)	~9,000,000	<0.5	Low	>
Solar within the biosphere				
13. Hydroelectric	<14	<1.6	Low	>
14. Salinity gradient to sea	~1,700	<0.3	TBD	Likely $\gg$
15. Salinity gradient to brine	~24,000	<0.3	TBD	Likely $\gg$
16. Tides	0	<0.02	Low	>
17. Ocean thermal	~200,000	<0.1	Large	$\gg$
18. Wind (not stand-alone)	0	<6	Low	>
19. Terrestrial thermal solar	0	<3	TBD	$\gg$
20. Terrestrial solar photovoltaics	0	<3	TBD	$\gg$
Mixed conventional system				
21. (1, 3, 4, 6, 12, 13, 16, 18, 19, 20)	≤3,200	11 (~33 TWt)	Large	>
Solar power from space				
22. GEO solar power satellites (deployed from Earth)	0	<1	Low	$\gg$
23. LEO solar power satellites (deployed from Earth)	0	<0.1	Low	$\gg$
24. SPS beyond GEO (made of nonterrestrial materials)	0	<1	Reduce	Likely ≥
25. Lunar–solar power system (made on the moon)	0	≥20	Reduce	Likely ≤

Note. t, thermal; e, electric; 3 Wt ~ 1 We in utility; TBD, to be determined; GEO, geosynchronous earth orbit; LEO, low earth orbit; SPS, solar power satellite.

materials that formed Earth also contributed a portion of their gravitational potential energy. The entire Earth releases only approximately 30 TWt of thermal energy. Most is released from the inaccessible sea floor and along the spreading boundaries between plates. Only a fraction of a terawatt is released as high-quality steam at continental sites such as the geysers in California. Large-scale geothermal power systems must mine the heat content of rocks between the surface of Earth and to a depth of less than 7 km. This shell of continental rocks contains less than 9 million TWt-years of thermal energy. However, it will be expensive to mine this relatively low-quality heat. Maintaining large-scale circulation of water through mined rocks is not demonstrated. Power installations on the surface will release considerable waste heat into the environment. The geological consequences of long-term mining of more than 600 TWt from this outer shell of Earth have not been considered.

Our sun delivers approximately 175,000 TWs of solar power to the disk of Earth. Unfortunately, the terrestrial renewable systems listed in Table I, such as wind (row 18) and solar (rows 19 and 20), are not stand-alone and must be supplemented by other power systems of similar capacity. This drives up the cost of the delivered energy significantly. Also, very large-scale renewable systems can directly affect the biosphere by changing the natural flows of power, water (row 13), and wind (row 18) and by modifying the reflective and emissive properties of large areas of Earth.

Expansion of Earth's existing commercial mixed power system in row 21 of Table I (a combination of coal, oil, natural gas, biomass, nuclear, hydroelectric, and miscellaneous renewables) can likely provide less than 3200 TWt-years of commercial energy due to economic and environmental constraints and a maximum output by 2050 of less than 11 TWe. In this mixed system, coal (row 3) and conventional fission (row 6) systems would exhaust their estimated fuels a few decades into the 22nd century. Even assuming increasing efficiency in economic output of a given unit of energy, the expensive mixed power system would not enable global prosperity by 2050. It must be noted that, under the modeling conducted by the International Institute for Applied Systems Analysis, decreasing the use of fossil fuels and increasing the use of renewable power systems resulted in decreased economic growth and less available per capita power.

Sustainable global energy prosperity requires low-cost, dependable, and direct access to solar power. Our sun, in Fig. 1A, is the dominant source of power and energy for Earth. It outputs 3.9×10^{14} TWs of solar power. It is powered by the nuclear fusion of hydrogen into helium and heavier elements, and it now converts approximately 1.4×10^{14} metric tonnes/year of hydrogen and helium into energy. Over the next 5 billion years or so of stable operation, the sun will release at least 2×10^{24} TWs-year of energy.

Commercial-scale solar power satellites, deployed from Earth (rows 22 and 23 in Table I) or the moon (row 24), will likely not be demonstrated until well into the 21st century. A huge fleet of extremely large satellites in orbit around Earth will be unacceptable due to concerns over safety and disruption of the night sky. Orders of magnitude improvements are required in materials, systems performance, assembly and maintenance, and transportation to space before such systems could provide commercially competitive electric energy to Earth. It appears that 24 of the power options cannot provide sustainable, pollution-free, and affordable global power by 2050.

The answer to supplying adequate, sustainable, and affordable commercial power is the moon. Earth's moon (Fig. 2) is dependably illuminated by 13,000 TWs of solar power. It is argued in the following sections that a reasonable fraction of the solar power incident on the moon can be sustainably delivered to Earth as commercial electric power at significantly less than 1 ¢/kWe-h.

3. LUNAR–SOLAR POWER SYSTEM

Figure 3 illustrates the essential features of the lunar–solar power (LSP) system: the sun, the moon and power bases (the bright spot on the left side and the dim spot on the right side of the moon), microwave power beams from each power base on the moon, relay satellites in orbit around Earth (not shown but discussed later), and a microwave receiver (i.e., rectenna) on Earth. Figure 4 shows the moon after the construction of an LSP system scaled to supply more than 20 TWe to Earth. There are 10 pairs of power bases on opposing limbs of the moon. The power bases receive sunlight, convert it to electricity, and then convert the electric power into microwave power beams. Each base transmits multiple microwave power beams directly to the rectennas on Earth when the rectennas can view the moon. Power beams between Earth and the moon are not esoteric. The Arecibo Radio Telescope in Puerto Rico has routinely beamed microwaves from Earth to the moon. Its beam intensity is approximately 20 W/m^2 going upward through the atmosphere. This is 10% of the maximum intensity ($\leq 230 \text{ W/m}^2$) proposed for transmission of commercial power.

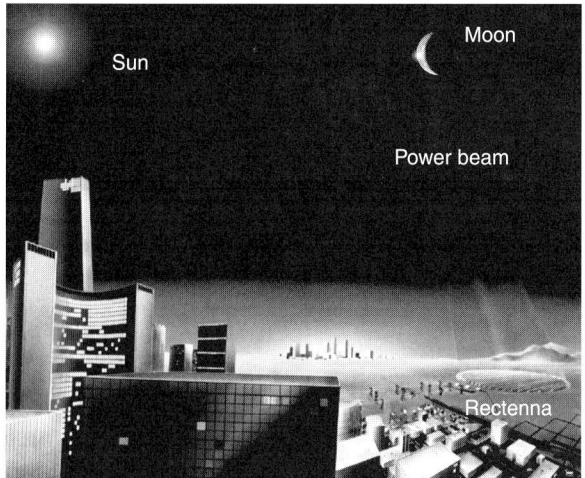

FIGURE 3 Sun, moon, beam, and rectenna.

FIGURE 4 Ten pairs of LSP bases.

Bases on the earthward side of the moon can be augmented by fields of photoconverters just across the far side of the moon from Earth. Power lines connect each earthward base and the extra arrays of photoconverters on the far side of the moon. One or the other of the two bases in a pair will receive sunlight over the course of a lunar month. Thus, a pair of augmented power bases can beam power toward Earth over the entire cycle of the lunar day and night.

The rectennas on Earth that receive the power are simply specialized types of centimeter-scale television antennas and electric rectifiers. A rectenna is illustrated in the lower right of Fig. 3. A rectenna converts the microwave beam it receives at more than 85% efficiency into electricity and outputs the pollution-free power to local electric distribution systems and regional grids. Most long-distance power lines will not be necessary. Rectennas are the major cost element of the reference version of the LSP system.

Microwave power beams, approximately 12 cm wavelength or 2.45 GHz, pass from the moon through Earth's atmosphere, clouds, fog, snow, smog, smoke, and normal rain with 1% or less of attenuation. An extremely heavy (≥ 25 cm/h) rain will attenuate the beam by less than 30%. This level of rain is very rare in most regions and lasts for only a few tens of minutes. Beam intensity can be increased to maintain the electric output of the rectenna. Other frequencies can be used but will not be as efficient. A rectenna that receives a load-following power beam of less than 230 W/m^2 will, over the course of a year, output approximately 200 W/m^2 of electric power. The rectenna will output this power whether it is located in a desert, the tropics, or a polar region. In contrast, stand-alone solar arrays on Earth output much less average power per unit of area. A stand-alone solar array is one that feeds power directly to a grid and also to a storage system so that power can be provided during the night or when the sky is obscured. A stand-alone solar array on Earth will have an annually averaged output of less than 3 W/m^2 if it uses 1980s technology and of less than 20 W/m^2 if it uses advanced technologies. Solar arrays on Earth are captives of the biosphere, season, and weather, whereas the power output of the rectenna is independent of these limitations.

Rectennas on Earth can view the moon and receive power only approximately 8 h each day. Earth-orbiting satellites can redirect beams to rectennas that cannot view the moon. The redirectors enable load-following power to rectennas located anywhere on Earth. Rectennas on Earth and the lunar transmitters can be sized to permit the use of Earth-orbiting redirectors that are 200 to 1000 m in diameter. Redirector satellites can be reflectors. Alternatively, a relay satellite can receive a power beam from the moon and then retransmit several new beams to different rectennas on Earth. Unmanned and manned spacecraft have demonstrated the transmission of beams with commercial-level intensity in low Earth orbit. Demonstration-scale reflectors and retransmission technologies have been, and are now, operating in space.

Power beams that will be 1 to 20 times more intense than the maximum that is recommended for continuous exposure by the general population are proposed. Each tightly focused beam will be directed to rectennas that are industrially zoned to exclude the general population. Microwave intensity under the rectennas will be reduced to far less than is permitted for continuous exposure of the general population. Secondary electrical shielding can reduce the microwave intensity to a negligible level. The power beams do not pose a hazard to insects or birds. Aircraft can simply fly around the beams. In addition, passengers will be shielded by the metal skin of their aircraft. Beams can be turned off in a few seconds or decreased in intensity to accommodate unusual conditions.

Earth can be supplied with 20 TWe by several thousand rectennas whose individual areas total approximately 100,000 km^2. Rectennas placed over agricultural land and industrially zoned facilities enable multiple uses of the same land. Rectennas can be as large in area as is necessary to produce larger electric power output. Conventional electric power systems use far larger total areas. In the cases of strip-mined land or power line rights-of-way, energy production degrades the land for several years to decades and/or precludes multiple uses. Rectennas could be placed over such land and can be located relatively close to major power users and, thereby, minimize the need for long-distance power transmission lines. A rectenna as small as approximately 0.5 km in diameter, with an output of 40 MWe, is feasible.

Unlike conventional power systems, the LSP system supplies net new power to Earth. Its net new electricity can be used to power the production and recycling of goods and the provision of services and to restore the biosphere. LSP energy replaces the tens of billions of tonnes of oil, coal, and natural gas that are now burned each year and the hundreds of billions of tonnes of wind and flowing fresh water that would otherwise be diverted from their natural courses to produce a much smaller output of electricity. The LSP system requires very little continuous processing of mass on Earth to receive and distribute the electric power. The fundamental waste product of lunar–solar power on Earth is waste heat that is eventually converted into useful work and then into infrared photons and radiated back to space.

Figure 4 depicts 20 power bases, 10 on each limb or visible edge of this full moon. They are adequate, assuming 1980s technology, to provide 20 TWe of electric power to Earth. More advanced technology will enable bases that are so small they will be unobservable to the naked eye. The moon receives 13,000 TWs, so the LSP has considerable growth potential.

Approximately 14 days after the view in Fig. 4, the opposite or far side of the moon will be fully illuminated by the sun. Additional fields of solar converters located approximately 500 km across the limb of the moon from each base in Fig. 4 will receive full sunlight. These cross-limb bases of solar converters can be connected by power lines to the bases on the earthward side of the moon. Approximately once a year, the moon will be totally eclipsed by Earth for up to 3 h. It is reasonable to precisely plan for the amount of additional LSP energy that must be provided to Earth for use during a lunar eclipse.

Continuous power can be provided on Earth by storing solar-derived energy on the moon and transmitting it to Earth during the eclipse. Alternatively, predictable amounts of energy can also be stored on Earth and released during the short lunar eclipse. Eclipse power can also be produced on Earth from conventional systems such as natural gas-fired turbines. Finally, mirrors, similar to proposed solar sails, could be placed in orbits high above the moon and reflect solar power to the bases during the eclipse. Such mirrors eliminate the need for expensive power storage.

4. LUNAR–SOLAR POWER DEMONSTRATION BASE

Figure 5 illustrates a demonstration power base. A power base is a fully segmented, multibeam, phased-array radar powered by solar energy. This demonstration power base consists of tens to

FIGURE 5 LSP demonstration base: multiple power plots [arrays of solar converters (#1), microwave reflector (#2), and microwave transmitter (#3)], set of mobile factory (#4) and assembly units (#5), and habitat/manufacturing facility (#6).

hundreds of thousands of independent power plots. A power plot is depicted in the middle to lower right portion of Fig. 5. Each power plot emits multiple sub-beams.

A power plot consists of four elements. There are arrays of solar converters (#1 in Fig. 5), shown here as north/south-aligned rows of photovoltaics. Solar electric power is collected by a buried network of wires and is delivered to the microwave transmitters. Power plots can use many different types of solar converters and many different types of electric-to-microwave converters. In this example, the microwave transmitters (#3) are buried under the mound of lunar soil at the earthward end of the power plot. Each transmitter illuminates the microwave reflector (#2) located at the anti-earthward end of its power plot.

Earth is fixed in the sky above the power base. The many small reflectors (#2 in Fig. 5) can be arranged in an area on the limb of the moon so that, when viewed from Earth, they appear to form a single large aperture as depicted by one of the power bases in Fig. 4. This enables a given power base to form very narrow and well-defined beams directed toward Earth. The moon has no atmosphere and is mechanically stable. There are no moonquakes. Thus, it is reasonable to construct the large lens from many small units. Individually controllable sub-beams send their microwave sub-beams toward each small reflector. The sub-beams are correlated to combine coherently on their way toward Earth to form one power beam. A mature power base can have hundreds to a few thousand sets of correlated microwave transmitters illuminating each reflector. The multiple reflectors will likely include additional sub-reflectors or lenses in front of each main reflector.

To achieve low unit cost of energy, the lunar portions of the LSP system are made primarily of lunar-derived components. Both the fixed (#6 in Fig. 5) and mobile (#4 and #5) factories are transported from Earth to the moon. Their high output of LSP components made on the moon from lunar materials greatly reduces the impact of the high cost of transporting the factories from Earth to the moon. Lunar factories will produce hundreds to thousands of times their own mass in LSP components.

Table II characterizes an LSP demonstration base. A period of 10 years is required to plan a demonstration base and establish it on the moon. Three sizes of base are modeled. Absolute costs are less important than the trend of cost versus the total power put in place after 10 years of operations. The machinery of the smallest base (column 1) installs 1 GWe of power that is delivered to Earth

TABLE II

Cost (1990 Dollars) of LSP Demonstration Base

	1 (smallest)	2	3 (largest)
GWe installed after 10 years on the moon	1	10	100
GWe-years of energy to Earth over 10 years	5	50	500
Gross revenue (billions of dollars) ($0.1/kWe-h)	4.4	44	438
Net revenue (billions of dollars)	−56	−47	195
Total costs (billons of dollars) (sum 1 + 2 + 3)	60	91	243
1. Research and development (sum a + b + c + d)	42	51	86
a. LPS hardware (billions of dollars)	11	11	11
b. Construction system (billions of dollars)	1	3	11
c. Facilities and equipment (billions of dollars)	5	10	30
d. Transport (billions of dollars)	26	27	35
2. Space & Operations (billions of dollars)	17	34	103
3. 1970s reference rectennas on Earth (billions of dollars)	0.6	6	55
Dollars/kWe-h	1.4	0.2	0.06
Moon (tonnes)	2300	6200	22,000
Space (tonnes)	970	2700	9700
People (moon, LLO, and LEO)	30	85	300

LLO, low lunar orbit; LEO, low earth orbit.

(1×10^9 We). A total cost of $60 billion is predicted, as measured in 1990 U.S. dollars. Cost of the LSP production equipment is $12 billion (1.a + 1.b). This base is estimated to have a mass of 2300 tonnes and requires 30 people on the moon. Electricity sold on Earth at $0.1/kWe-h generates $4.4 billion of revenue. The largest base (column 3) is scaled to install 100 GWe received on Earth. Total cost increases by a factor of 4. However, the sales of electric energy increase by a factor of 100. The largest base pays for itself. Cost of LSP production equipment, the mobile units in Fig. 4, increases by only a factor of 2. The production process will continue after the 10-year demonstration. Total power capacity (energy delivered to Earth) will grow, and net profit will continue to grow. Energy cost will drop steadily. The demonstration base can be established in less than 10 years. The size of the base and the production process can be steadily increased in size to enable 20 TWe delivered to Earth by 2050.

5. SCALE AND COST FOR 20 TWE LUNAR–SOLAR POWER SYSTEM

A full-scale LSP system, based on 1980s technology, would occupy approximately 15% of the surface area of the moon. This is illustrated in Fig. 4. The total area of all the bases in Fig. 4 appears to be small compared with this disk of the moon. However, this is an illusion. The area of each base curves around the spherical surface of the moon near the lunar limb as seen from Earth. Each base is six times longer in the east–west direction than it is wide in the north–south direction. As illustrated in Fig. 5, approximately 80% of the area would be the empty ground between the solar arrays. Using solar collectors on the far side of the moon and eliminating the solar mirrors in orbit around the moon increases the area of the 1980s-style bases to approximately 25% of the lunar surface. Technologies anticipated to be available in 2020, all demonstrated by advanced systems of today, and close packing of the solar cells on the surface of the moon reduce the area occupied by an advanced LSP system to less than 0.2% of the lunar surface.

Table III presents estimates of the tonnage of equipment that would be deployed from Earth over 70 years to deliver 20 TWe of electric power on Earth and the number of people required in space and on the moon. The estimates assume that a period of 10 years is required to place the production facilities on the moon. Next, there is 30 years of full-scale production that requires 53,000 tonnes/year of components and supplies launched to the moon. The model includes the additional equipment and materials needed to maintain the bases and rebuild 50% of the power collectors and transmitters between 2050 and 2070. "LSP REF" assumes that all production equipment, people, and supplies are taken from Earth to the moon. "10 × LSP REF" assumes that the equipment taken to the moon and used to make the power plots are 10 times more massive than, or one-tenth as productive as, the equipment in the LSP REF column. The "BOOT 90%" column assumes that 90% of the mass of the production equipment is made on the moon of lunar materials.

In the LSP REF model (column 3 in Table III), over the 70-year life cycle, the micromanufacturing

TABLE III

Life Cycle Cost for Heavy LSP, Reference LSP, and Bootstrapped LSP

	10 × LSP REF	LSP REF	BOOT 90%
Tonnes of lunar equipment			
Micro manufacturing	3,205,323	250,605	24,361
Hot forming	1,031,271	103,127	10,313
Beneficiation	321,229	32,123	3212
Habitats, shops, and mobile units	75,057	23,341	22,085
Chemical refining	246,945	24,941	2469
Gather and eject to orbit	43,185	4383	438
Excavation	83,115	831	80
Cold assembly	2776	278	28
Total (tonnes)	5,012,039	442,630	62,915
Number of people			
Moon	55,915	4717	436
Lunar orbit	4986	468	59
Earth orbit	5010	443	63
Cost of equipment and people (1977 dollars) for 20 TWe and 1000 TWe-year to Earth	$19.5 trillion	$1.75 trillion	$0.26 trillion
Energy engineering cost (dollars/kWe-h)			
Reference rectenna ($7.9 trillion)	0.0106	0.0037	0.0037
Reflective rectenna ($0.82 trillion)	0.0078	0.001	0.0004

of solar converters and similar products requires approximately 57% of the equipment mass or 250,000 tonnes. Approximately 23,000 tonnes of habitats, shops, and other support facilities equipment is required to support 4700 people needed during full production of 0.6 TWe/year of new power capacity. The intense and variable solar and cosmic radiation makes it impossible for people to work on the surface of the moon for long periods. A covering of approximately 3 m of lunar soil is required to protect people. Most surface operations are automated and/or controlled by people in habitats, as shown on the left side of Fig. 5.

The engineering cost in Table III is presented in 1977 U.S. dollars and does not include financing cost. The engineering cost of $1.75 trillion includes all lunar- and space-related activities and launch and manufacturing on Earth. The National Aeronautics and Space Administration (NASA) and U.S. Department of Energy (DOE) studies conducted during the 1970s were used to calculate the engineering cost of 20 TWe of rectennas at $7.9 trillion. Rectenna construction and operation is five times greater than the cost of all the lunar-related operations. A 20-TWe LSP system that delivers 1000 TWe-year of energy over a life cycle of 70 years is projected to achieve an average cost of $0.0037/kWe-h (0.37¢/kWe-h). Assuming a factor of three for inflation from 1977 to today, LSP REF electricity would be competitive with the lowest cost existing conventional power systems that provide electricity at a wholesale or busbar cost of $0.02 to $0.04/kWe-h. Power bases can be maintained indefinitely at far lower cost than is required to develop and construct them. Rectennas on Earth that use reflective concentrators will reduce the number of expensive rectifiers and also be less subject to mechanical loads from wind, rain, and ice. The projected cost of 20 TWe of reflector rectennas is approximately $0.8 trillion. The engineering cost of energy is reduced to $0.001/kWe-h. Over time, lower costs are possible. Thus, the LSP REF system can potentially provide electric energy at 10% or less of the cost of existing systems.

Table III assumes that the cost of launch from Earth to orbit is $470/kg (in 1977 U.S. dollars). Increasing the cost of Earth-to-orbit transportation to $5000/kg increases the upfront cost significantly, and life cycle engineering cost increases to approximately $5.2 trillion. However, the life cycle engineering cost of electricity from the LSP REF system increases by only approximately 20%. The system model assumes the extensive use of lunar-derived propellants. Solar electric propulsion is used for the transport of cargo from orbit around Earth to orbit around the moon.

The "10 × LSP REF" in Table III assumes that all of the production machinery must be 10 times more massive than that for LSP REF to achieve the same output of LSP components. Less productive manufacturing equipment (column 2), with lower output, will increase the size and cost of the lunar and space operations. Total tonnage of production machinery on the moon increases by a factor of 11. Engineering life cycle cost increases to approximately $20 trillion over the 70 years. However, the engineering cost of the electric power on Earth increases by only a factor of 3, to $0.0106/kWe-h, for the reference rectenna and to $0.0078/kWe-h for the reflective rectenna. The "10 × LSP REF" is still competitive with electricity from existing suppliers.

P. Glaser originated the concept of using large satellites in geosynchronous orbit around Earth to capture sunlight, change it into microwaves, and beam the power to Earth. During the 1970s and 1990s, NASA and DOE studied the deployment of large solar power satellites (SPS) from Earth to geosynchronous orbit. For delivery of energy to Earth, the LSP REF is projected to be at least 1000 times more efficient in its use of mass deployed from Earth than is deployment of an equal mass as an SPS.

On Earth, we make all of our tools for manufacturing and production from materials on Earth. This is called "local manufacturing" or "bootstrapping." The industrial materials obtained from lunar soils (e.g., glasses, ceramics, iron, aluminum, titanium, sulfur, silicon) can also be used on the moon to manufacture, as shown in Fig. 5, a large fraction of the mobile factories (#4 and #5), shops (#6), and habitats (#6). Of course, the production machinery in Fig. 5 must be designed for optimal use of parts manufactured on the moon. The bootstrapping equipment and operations will be fully refined during the demonstration phase.

The final column of Table III assumes that 90% of the mass of production equipment (#4, #5, and #6 in Fig. 5) is "booted" from lunar materials. The model is adjusted for the additional people needed to conduct this extra level of manufacturing. The model is also adjusted for a higher level of support from Earth in the form of remote monitoring and teleoperation. Only 63,000 tonnes of equipment is shipped to the moon over the period of 70 years. If components and supplies can be made entirely of lunar materials, then BOOT 90% enables the delivery of 1000 TWe-year to Earth at a specific energy of 63 tonnes/TWe-year. Engineering cost (in

1977 U.S. dollars) of electric energy could decrease to $0.0004/kWe-h (0.04¢/kWe-h). The production and logistics systems are assumed to make extensive use of tele-operation from Earth. The mass of facilities in orbit around Earth and the moon and the number of people are decreased significantly.

Major funding is required to build the LSP system, but it is reasonable when compared with global expenditures on energy and space to date. Today, the world expends approximately $4 trillion/year on all aspects of the 14 TWt global power system. Petroleum companies expend approximately $130 billion/year exploring and bringing to production new oil and gas fields. This continuing investment provides less than 6 TWt of commercial thermal power derived from petroleum. Global oil and natural gas production will likely peak within 10 years. The United States now expends more than $20 billion/year on civilian and military space activities. The current infrastructure of NASA, which can be adapted to enable the demonstration LSP, results from a total investment more than $600 billion since 1958.

Bootstrapping enables the exponential growth of the LSP power bases on the moon. Bootstrapping facilities are brought to the moon until sufficient manufacturing capacity is established to provide 20 TWe by 2050. Once full-scale production of power is achieved, many of the bootstrapping facilities can be directed to providing a wider range of new lunar goods and to enabling new services.

The Apollo- and Soviet-era lunar programs and the worldwide post-Apollo lunar research used well over $1 billion to provide the essential knowledge of lunar resources and the lunar environment. Post-Apollo satellites and deep space missions have mapped the major mineral and chemical composition of most of the moon.

The International Space Station program provides the international level of cooperation, technology, and operating systems that enables a quick return to the moon. A permanent international lunar base, established to conduct industrial research and development, can accelerate economic development of the moon. This can be done within a modest expansion of existing cash flow of the major civilian space programs. Automation and remote operations are a normal component of terrestrial industry. Major offshore oil and gas platforms use advanced capabilities, such as underwater robotics and complex operations and logistics, that can be employed on the moon and in space to build and maintain the lunar–solar power bases. Much work can be done by tele-operation and supervisory control from Earth.

6. CONCLUSION

The conventional energy resources of our isolated spaceship Earth (Figs. 1B and 2) are extremely limited and will be quickly consumed by a prosperous human race. The sun (Fig. 1A) is the only reasonable power source for a prosperous world. It contains adequate fuel, is a functioning fusion reactor that does not have to be built and operated, and retains its own ashes. The challenge is to build the transducers that can extract this power and deliver it to consumers on Earth at a reasonable cost. The moon receives 13,000 TWs of dependable solar power. The LSP system, built on the moon from lunar soils and rocks, is the transducer. The LSP system can deliver net new power to consumers on Earth that is independent of the biosphere and clean. Developing countries can afford LSP electricity. LSP electricity can accelerate the economic growth of all nations. Net new LSP electric power enables all nations to produce and recycle their goods and consumables independent of the biosphere. Transportation and services can be powered without consuming or affecting the biosphere. The challenges of the World Energy Council and the United Nations can be achieved by the year 2050.

SEE ALSO THE FOLLOWING ARTICLES

Earth's Energy Balance • *Global Energy Use: Status and Trends* • *Lithosphere, Energy Flows in* • *Sun, Energy from* • *Sustainable Development: Basic Concepts and Application to Energy*

Further Reading

Ayres, R. U., Ayres, L. W., and Ware, B. (2003). Exergy, power, and work in the U.S. economy, 1900–1998. *ENERGY: The Int. J.* **28**, 219–273.

Criswell, D. R. (1985). Solar system industrialization: Implications for interstellar migration. *In* "Interstellar Migration and the Human Experience" (B. R. Finney and E. M. Jones, Eds.), pp. 50–87. University of California Press, Berkeley.

Criswell, D. R. (2002). Energy prosperity within the 21st century and beyond: Options and the unique roles of the sun and the moon. *In* "Innovative Energy Strategies for CO_2 Stabilization" (R. Watts, Ed.), pp. 345–410. Cambridge University Press, Cambridge, UK.

Criswell, D. R. (2002). Solar power via the moon. *The Industrial Physicist*, April/May, pp. 12–15. (See letters to the editor and reprise in June/July, August/September, and October/November issues or at www.tipmagazine.com.)

Criswell, D. R., and Waldron, R. D. (1993). International lunar base and lunar-based power system to supply Earth with electric power. *Acta Astronautica* **29**, 469–480.

Clarke, A., and Trinnaman, J. (eds.). (1998). "Survey of Energy Resources 1998." World Energy Council, London.

Goeller, H. E., and Weinberg, A. M. (1976). The age of substitutability. *Science* **191**, 683–689.

Heiken, G. H., Vaniman, D. T., and French, B. M. (eds.). (1995). "LUNAR Sourcebook: A User's Guide to the Moon." Cambridge University Press, Cambridge, UK.

Murdin, P. (Ed.-in-Chief) (2001). "Encyclopedia of Astronomy and Astrophysics" Vols. 1–4. Nature Publishing Group, New York.

Nakicenovic, N., Grubler, A., and McDonald, A. (eds.). (1998). "Global Energy Perspectives." Cambridge University Press, Cambridge, UK.

Shirley, J. H., and Fairbridge, R. W. (eds.). (1997). "Encyclopedia of Planetary Sciences" (1st ed.). Chapman & Hall, London.

Weissman, P. J., McFadden, L-A., and Johnson, T. V. (eds.). (1999). "Encyclopedia of the Solar System." Academic Press, San Diego.

World Energy Council. (2000). "Energy for Tomorrow's World: Acting Now!" Atalink Projects, London.

Magnetic Levitation

DONALD M. ROTE
Argonne National Laboratory
Argonne, Illinois, United States

1. Introduction
2. Characteristics of Attractive-Force-Based Suspension Systems
3. Characteristics of Repulsive-Force-Based Suspension Systems
4. Maglev Development History
5. Advantages of Maglev Technology for High-Speed Transportation

Glossary

critical temperature The phase transition temperature of a superconductor.
ferromagnetic materials Technically, materials having very high magnetic permeability that depends on the magnetizing force. "Ferro" refers to iron, typical of the group. Physically, these are materials strongly attracted to magnets and are used in permanent magnets, including nickel and cobalt, and various ceramic compounds, notably those made of neodymium, iron, and boron.
magnetic levitation The noncontact support of an object with magnetic forces. In the context of transportation applications, Howard Coffey coined the term "Maglev."
magnetic permeability The ratio of change in magnetic flux density in a material to the change in magnetizing force applied to it.
reluctance The resistance of a material or combination of materials to magnetization.

This description of magnetic levitation focuses on applications of this technology to both low- and high-speed mass transportation systems. Methods of providing the three main functions of levitation, guidance, and propulsion are described in detail for the two basic types of magnetic suspension systems, namely, the attractive-force and the repulsive-force systems. Various magnetic field source design options, including the use of conventional electromagnets, permanent magnets, and superconducting magnets to accomplish these three functions, are discussed and compared. Alternative choices for propulsion system configurations, including the use of long- versus short-stator linear induction and synchronous motors, are also examined. This is followed by a brief history and update of the development of the technology and a discussion of the advantages and cost of maglev versus conventional high-speed trains.

1. INTRODUCTION

The term magnetic levitation has come to be used in a wide variety of different contexts ranging from suspending a small laboratory-scale stationary object so that it is isolated from vibrations of its surroundings (an isolation platform) to large-scale mobile applications such as maglev vehicles capable of carrying people and materials up to speeds of several hundred miles per hour or the proposed assisting in the launch of space vehicles. Depending on the nature of the application, some degree of physical contact may be required. However, if physical contact is to be completely eliminated, as in the case of very high-speed vehicles, then in addition to suspension, the functions of lateral guidance, propulsion, braking, energy transfer, and system control must be provided by noncontact means alone. Faced with a wide variety of options, maglev system designers must decide how the system should be configured and which components should be placed on board the vehicle and which should be mounted on the guideway. The decision depends on a variety of both technical and economic issues.

For maglev vehicle applications, two basic maglev suspension system designs have been widely used. The first employs the attractive force between magnets and ferromagnetic metals and is referred

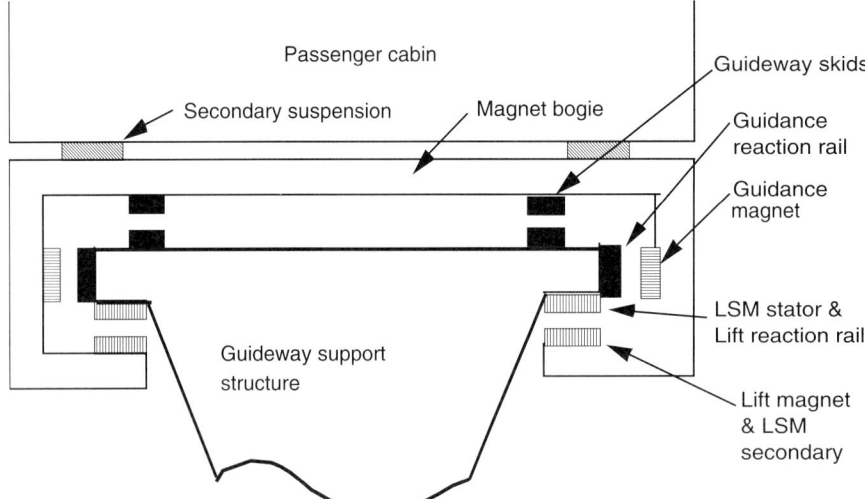

FIGURE 1 Schematic diagram of a long-stator EMS system. The magnet bogie, which supports the lift and guidance magnets, is connected to the passenger cabin via a secondary suspension system.

to as an electromagnetic suspension (EMS) system. The other uses the repulsive force generated by magnets moving relative to electrical conductors and is referred to as an electrodynamic suspension (EDS) system. Examples are illustrated schematically in Figs. 1 and 2.

2. CHARACTERISTICS OF ATTRACTIVE-FORCE-BASED SUSPENSION SYSTEMS

The task of trying to suspend a ferromagnetic object below a permanent or electromagnet magnet with an air gap between the magnet and the object is thwarted by an intrinsic instability. An air gap length exists where the force of magnetic attraction exactly balances the force of gravity. However, if the air gap length decreases slightly, the attractive force increases and the object is clamped onto the magnet. If the air gap increases slightly, the force is weakened and the object falls down. This inherent instability can be overcome by constantly changing the current supplied to an electromagnet's windings in response to signals being fed back to the magnet's power supply from a sensor that continuously monitors the air gap. More complex feedback control systems use additional control parameters, such as the absolute position of the object, its velocity, and its acceleration, to achieve certain suspension characteristics.

All EMS-based systems employ some type of feedback control mechanism to maintain stable levitation. (An interesting exception to the above rule, which has been demonstrated in the laboratory, consists of a permanent magnet suspended below a block of high-temperature superconducting material held below its critical temperature. Stable suspension results from a property of the superconducting state that immobilizes magnetic field lines in the superconducting material.)

A feature of EMS systems that is particularly attractive for maglev vehicles is that the suspension force is essentially independent of speed. Consequently, stable magnetic suspension can be achieved at rest as well as at any other speeds within the system's design limits. Typically, for vehicle applications, in order to compensate for the force of gravity with an attractive magnetic force, the vehicle and guideway components must be configured so that the onboard magnets are drawn upward toward the undersides of the guideway rails, as illustrated in Fig. 1.

The design of the lateral guidance and propulsion systems must, of course, be compatible with the lift system design. A variety of methods for producing lateral guidance in EMS systems have been developed. The German Transrapid high-speed EMS system utilizes separate magnets interacting with guideway-mounted side rails, as illustrated in Fig. 1. In that system, the guidance magnets are excited with feedback-controlled DC power supplies in essentially the same manner as the lift magnets. Several low-speed EMS systems, including the Japanese high-speed surface transport (HSST) and systems under development in Korea and the United States, use a suspension system in which the same magnets are

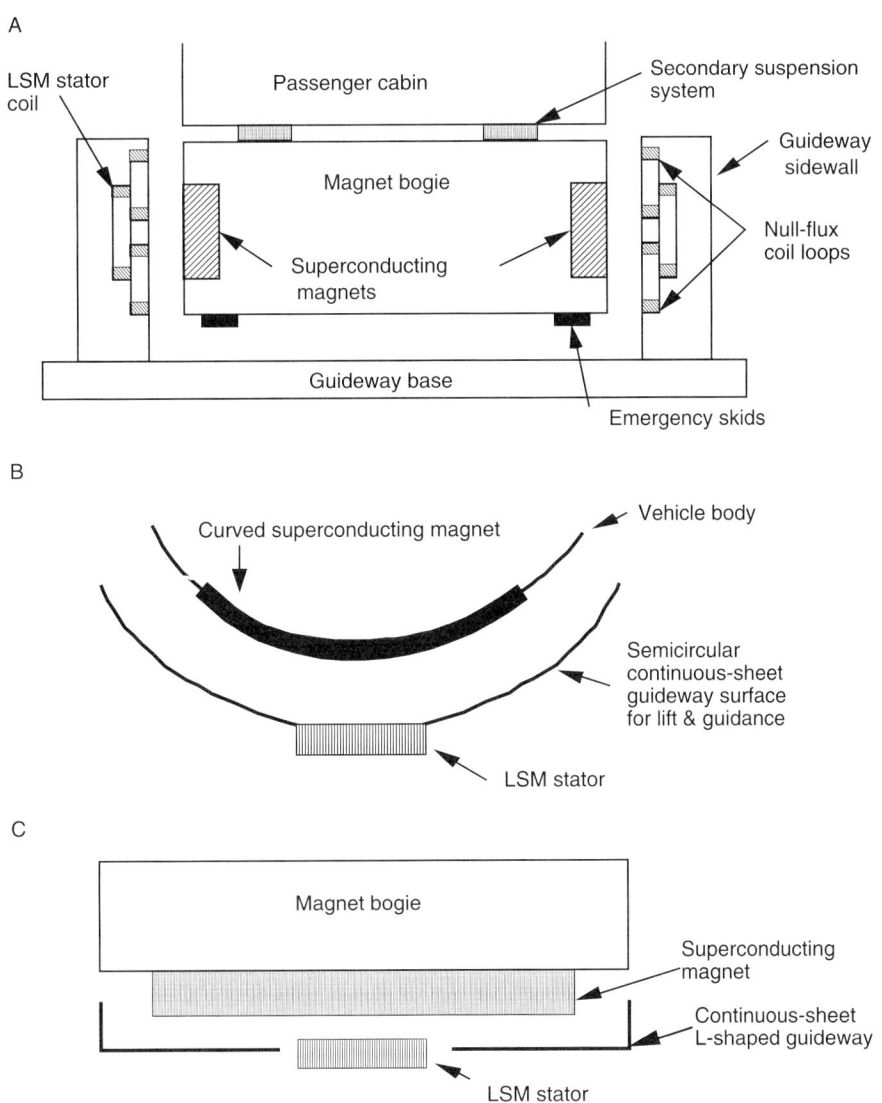

FIGURE 2 Examples of EDS systems using superconducting magnets. (A) Null-flux lift and guidance with air-core LSMs in a "U"-shaped guideway. (B) Continuous-sheet, semicircular guideway. (C) Continuous-sheet, "L"-shaped guideway.

used for both lift and lateral guidance, as illustrated in Fig. 3. The latter method utilizes the natural tendency of a ferrous metal object to orient itself relative to magnetic poles so as to minimize the reluctance of magnetic flux path to magnetization. In Fig. 3, the minimum reluctance position of the ferrous rail is centered between the two magnet poles. To increase lateral stiffness, the vehicle-borne magnets are alternately offset from the centerline. Departures from the lateral equilibrium position result in a strong centering force to be exerted on the rail. Another method proposed for combining lift and guidance functions in EMS systems is to place the reaction rails and magnets at an angle so that the magnetic attraction force has both a horizontal component and a vertical component.

In principle, any type of noncontact propulsion system could be used to propel a maglev vehicle. However, in practice, the choice depends on a number of important and often conflicting considerations, including compatibility with the lift and guidance system, power requirements, thrust requirements, operating speed, weight penalties, cost, and environmental constraints. With the exception of some special-purpose applications, linear induction motors (LIMs) and linear synchronous motors (LSMs) have generally been the propulsion means of choice. Both consist of a primary part that generates a traveling

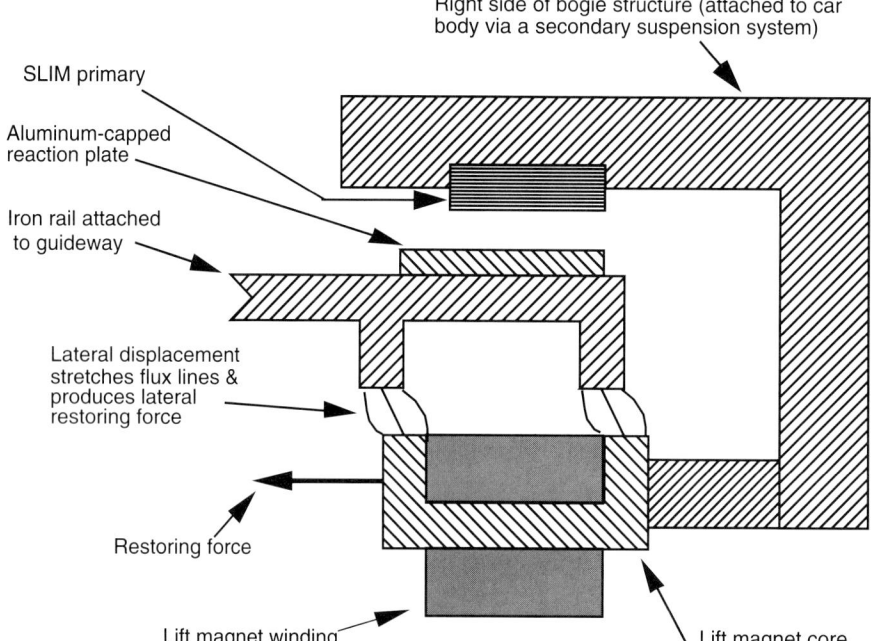

FIGURE 3 An EMS short-stator system used in several low-speed maglev designs.

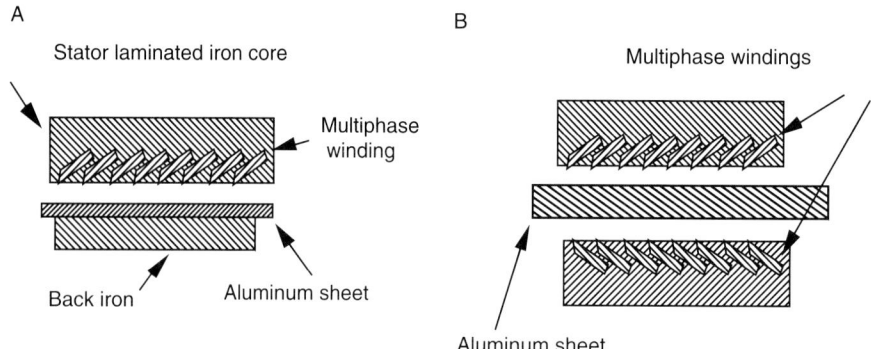

FIGURE 4 (A) Single-sided LIM. (B) Double-sided LIM.

magnetic wave, which in turn interacts with a secondary motor part to produce thrust. Air gaps for both types of linear motors are generally approximately 0.5 in. or less in EMS systems. A third type of linear motor uses electronic switching technology to turn electric currents on and off at precisely the right times to create thrust between the primary and secondary motor parts. The latter type, called sequentially excited linear motors, includes locally commutated linear synchronous motors and pulsed linear induction motors.

LIMs consist of an energized primary part, the stator, and a passive secondary part, the reaction rail (Fig. 4). The stator generally consists of a laminated iron core with transverse slots spaced along its length. Windings are placed in the slots to form a series of coils. In the most common form, three separate windings are placed alternately in the core slots so that when excited with three-phase electric current, a traveling magnetic wave is created. The speed of the traveling wave, called the synchronous speed, is given by the relation, $v_s = 2pf$, where p is the distance between stator magnetic poles (the pole pitch) and f is the frequency of excitation in Hertz. The primary power supply is called a variable voltage variable frequency (VVVF) power supply.

When the reaction rail is placed adjacent to the stator, the traveling magnetic wave exposes the rail to time-varying magnetic fields, which in turn cause eddy currents to flow in the rail. Those eddy currents react with the traveling wave and produce a force that has two components: one in the direction of motion, i.e., a thrust force, and another that acts normal to that direction, i.e., a repulsive force. As the speed between the traveling wave and rail (called the slip speed) is increased, the thrust initially increases, reaches a peak value, and then diminishes. The repulsive normal force increases with slip speed.

Depending on the motor configuration and its application, the reaction rail may be composed of ferrous or nonferrous metal or a combination of both. Single-sided LIMs (Fig. 4A) generally use iron or aluminum-capped iron rails for low-speed applications. Examples include the Japanese HSST and the maglev systems under development in the United States and Korea. The iron provides a return pathway for the magnetic flux emanating from the stator pole faces. This concentrates the magnetic flux in the air gap and improves motor performance in some cases. However, with iron reaction rails, the ferromagnetic effect produces a strong attractive force that acts normal to the direction of motion. As the slip speed increases, this attractive normal force is partially compensated for by the repulsive force between the stator magnetic pole faces and eddy currents induced in the rail as mentioned above. Since the iron is a relatively poor conductor of electricity, the repulsive normal force component tends to be small. When an aluminum cap is placed over the iron rail, the induced eddy currents are larger and the net normal force can be adjusted from attractive at low slip speeds (dominated by ferromagnetic attraction) to repulsive at high slip speeds (dominated by eddy-current repulsion). For high-speed applications, so-called end effects, particularly at the entry end of the motor, become very important and can considerably degrade motor performance. Consequently, to achieve satisfactory performance at high speed, they must be compensated for. Several techniques have been proposed, including the use of compensating stator windings and wound secondaries. Pulsed linear induction motors that eliminate the end effects have also been designed.

For LIM applications, a crucial question is whether to place the stator on board the vehicle and the reaction rail on the guideway (called the short stator option) or vice versa. The short-stator option adds substantially to the vehicle's weight, which puts an extra burden on the suspension system. It also requires that the propulsion power be generated on board or transferred to the vehicle from the guideway. On-board power generation requires the storage and conversion of chemical fuel to electricity, which produces pollutants and adds more weight to the vehicle. Use of fuel cells could reduce such impacts. Transfer of propulsion power to the vehicle from the guideway using sliding contacts or brushes has proven practical for low-speed systems only. The principle advantage of the short-stator system is low cost. With most of the technology carried on board, the guideway is relatively simple, and since the cost per mile of guideway must be multiplied by the guideway length, the total cost savings can be very substantial. However, several factors have limited the short-stator option to low-speed applications. First, the difficulty of transferring power to the vehicle increases with speed. Second, greater speed requires larger and heavier motors and accompanying power-conditioning equipment, which, in turn, places an ever-increasing burden on the suspension system, which, in turn, becomes larger and heavier and requires more power. A third issue is the degradation of performance of LIMs at high speed due to motor end effects. As noted earlier, the end effects must be either compensated for or eliminated by special LIM designs.

The long stator option, on the other hand, places the primary on the guideway and the reaction rails on the vehicle, thus reducing vehicle weight and eliminating the need to transfer propulsion power to the vehicle, which results in a lighter, simpler vehicle, but a more complex and expensive guideway. In principle, one could also achieve a better performing system, with greater acceleration and maximum speed. However, continuous generation of eddy currents in the on-board reaction rails is accompanied by ohmic heating, which poses large cooling requirements on board the vehicle.

Whereas low-speed maglev systems have generally used single-sided LIMs, double-sided LIMs using aluminum or copper sheets or shorted-turn coils (coils whose ends are connected together) have been used in applications requiring high acceleration, such as mass drivers and rail guns. Since this type of motor configuration has energized magnetic poles on both sides of the reaction rail, the net normal force is near zero and a good magnetic flux path is provided, leading to a high thrust capability.

LSMs also have a primary part that is energized by a VVVF power supply so as to produce a traveling magnetic wave. However, the secondary part differs in that it consists of a separate set of discrete magnetic poles (Fig. 5). The latter may be a series of permanent or electromagnets or ferrous

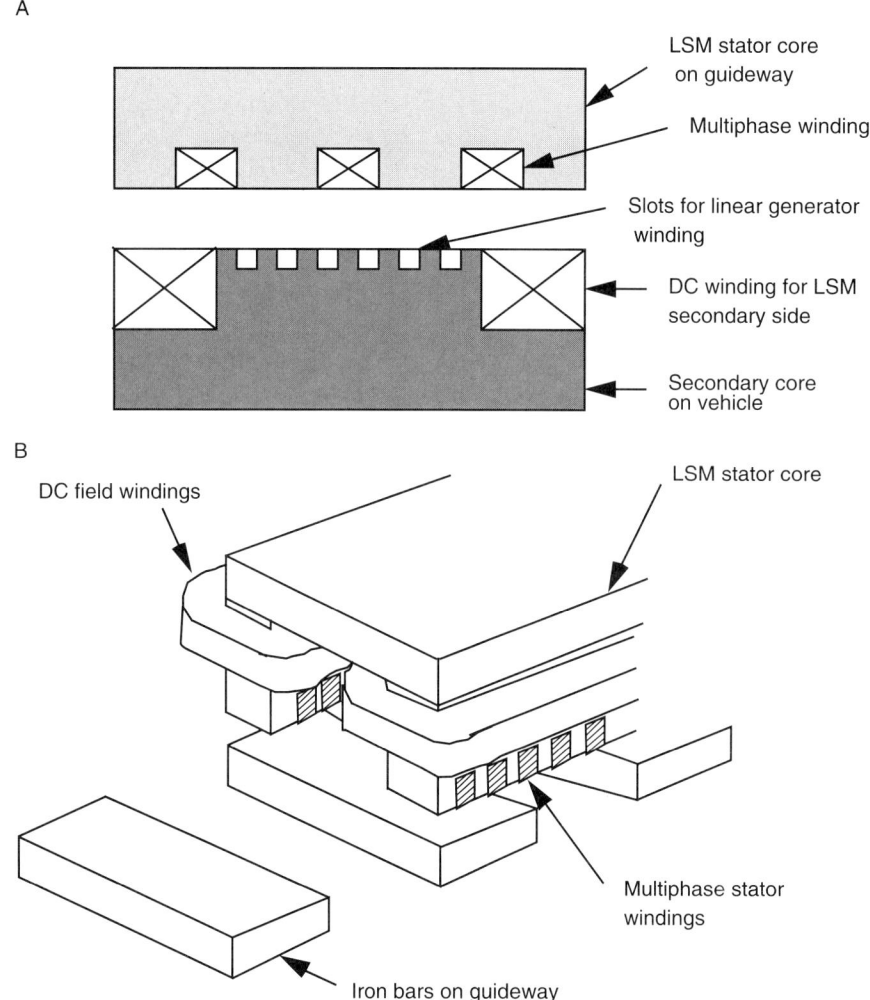

FIGURE 5 (A) Longitudinal cross section of a conventional iron-core LSM. (B) Isometric sketch of a homopolar LSM in which both the AC and DC windings are on the stator core.

metal bars that are magnetized by separate DC energized windings placed on the stator core. Thrust is produced by the interaction between the traveling magnetic wave and the secondary magnetic poles. In contrast to the LIM, the LSM secondary part always moves at the synchronous speed. Position sensors continuously monitor the position of the secondary poles relative to the poles of the traveling wave and their signals are used to maintain synchronization of the secondary part with the traveling wave. The speed is controlled by the frequency of the primary excitation and the thrust by the applied voltage.

The LSM configuration of choice for both EMS and EDS high-speed applications places the primary side on the guideway and the secondary on board the vehicle. This long-stator option has several advantages. It enables the same on-board magnets to be used for lift and as the secondary side of the LSMs. In the case of the German Transrapid system, the laminated iron core of the stator also serves as the reaction rail for the on-board lift magnets. Furthermore, it avoids the problem of transferring propulsion power from the guideway to the vehicle at high speed. It also reduces the weight burden on the vehicle suspension system and places vehicle control at the wayside, eliminating the need for an onboard engineer or driver. To improve efficiency, the primary side of the LSM is normally divided into blocks ranging in length from a few tenths of a mile to several miles. The wayside power control system monitors the position of each vehicle in the system and turns on a block as the vehicle approaches and turns it off after it leaves. The system is designed so that only one vehicle (which may consist of one or more cars linked together as in a train) can occupy a block at a time. This allows short headways while ensuring positive vehicle separation.

3. CHARACTERISTICS OF REPULSIVE-FORCE-BASED SUSPENSION SYSTEMS

In contrast to EMS systems, EDS systems can provide stable suspension without a feedback system. Imagine a magnet held above a moving flat sheet of aluminum or copper. Following Ampere's law of induction, an electric field is induced in the moving metal surface that causes eddy currents to flow in closed loops near its surface. The eddy currents in turn set up their own magnetic field, whose polarity, in accordance with Lenz's law, opposes that of the magnet's field. Consequently, the magnet is repelled from the moving metal surface, countering the force of gravity. If the magnet is pushed down toward the moving metal surface, the induced currents and the resultant repulsive force increase, restoring the equilibrium position automatically. Conversely, if the magnet is moved upward, the levitating force decreases. Consequently, the system is said to be inherently stable. (Of course, the magnet must be held in position because the induced eddy currents also cause an electromagnetic drag force that tends to pull the magnet in the longitudinal direction along with the moving metal sheet.)

In practice, the stability of a magnetically suspended object is more complicated because any disturbance from the equilibrium position in the repulsive force case results in an oscillation about the equilibrium position. The oscillation may decrease or increase with time depending on whether the net damping force present is positive or negative, respectively. Both active and passive damping mechanisms may be employed to provide net positive damping and ensure stable suspension. In the attractive force case, the feedback system mentioned above provides an effective positive damping force that attenuates any oscillations.

The configuration of choice for most EDS systems places the magnets on board and the electrical conductors on the guideway. One disadvantage of the EDS system is that, because the lift force depends on the speed, mechanical support must be used at rest and at low speeds until the lift-off speed where the magnetic suspension force exceeds the force of gravity is reached. Pneumatic tires mounted on retractable landing gears can provide suitable mechanical support. Support at rest or at low speed can also be achieved with repulsive magnetic forces. This can be accomplished using AC-excited coils interacting with eddy currents induced in reaction rails or coils. However, this method tends to be rather energy-intensive.

A variety of combinations of vehicle-borne magnets and guideway-mounted electrical conductor configurations have been devised for EDS systems. These include the use of continuous-sheet and discrete-coil guideways. Three configurations are illustrated in Fig. 2. In all cases, the guideway-mounted conductors are completely passive. The null-flux coil configurations have high lift efficiency and can produce forces in two opposing directions (up and down or left and right).

Both permanent and superconducting magnets (SCMs) can be used as the magnetic field sources. Designs utilizing permanent magnets have been investigated by both American Maglev Technology, Inc. and by Lawrence Livermore National Laboratory. The latter group has incorporated permanent magnets in Halbach arrays, which provide greater field strength at the pole faces per unit weight and substantially reduced stray fields behind the arrays.

The principal advantage of SCMs is greater force per unit weight. This is due to the strong magnetic fields that can be produced over large volumes of space and because the SCMs do not use iron cores. Air gaps on the order of 10 to 20 cm are possible with SCMs. Such large air gaps help ensure against contacting the guideway surfaces when guideway irregularities or disturbances including wind gusts and seismic activity are encountered. However, in practical designs, component parts tend to take up at least some of that space, reducing the actual clearance between the outer vehicle and guideway surfaces to 6–8 cm. The major disadvantage is that the superconducting windings must be kept below their critical temperatures to maintain their superconducting states. This requires the use of specially designed cryostats and cryogenic refrigeration systems. Superconducting magnets used in transportation applications are of the "low-temperature" type, which are operated at liquid-helium temperatures. The Japanese have successfully developed efficient cryogenic systems for use with SCMs on their high-speed "linear motor cars." The newer "high-temperature" superconductors, which can be cooled with liquid nitrogen, are not yet ready for such applications.

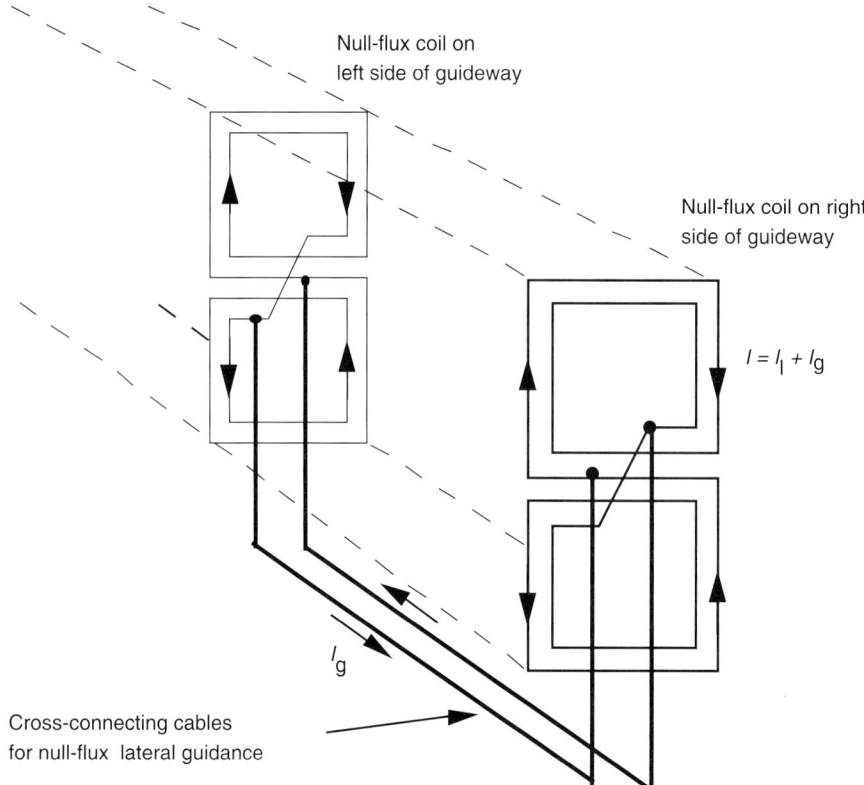

FIGURE 6 Connections of null-flux coil system for null-flux lift and guidance.

EDS systems use eddy current repulsion for guidance as well as lift. Depending on the type of lift and propulsion system being used, and the guideway configuration, either separate magnets or the same magnets can be used for lift and guidance. In the Japanese linear motor car system, the same onboard SCMs provide the magnetic field for lift and lateral guidance and serve as the field sources for the secondary side of the LSMs. Eddy currents induced in the null-flux coils mounted on the guideway sidewalls in this system interact with the magnetic fields of the SCMs to produce both lift and guidance forces. The guidance forces are enhanced by cross-connecting the null-flux coils on opposite sides of the guideway (Fig. 6). Null-flux lift and guidance rely on the same operating principle. That is, the two loops of a null-flux coil are wound in opposite directions so that when a magnet is in the neutral position, an equal amount of magnetic flux links both loops so that no net current flows in the two connected loops and hence, no force is generated. When the magnet is displaced relative to the neutral position, then one loop links more flux than the other, a net current flows in both loops, and a restoring force is produced.

4. MAGLEV DEVELOPMENT HISTORY

One of the earliest known mentions of the concept of high-speed ground transportation using maglev technology appeared in a fictional story written by Robert Goddard, father of the rocket, in 1907. By 1911, two patents were awarded to Graeminger for attractive force suspension systems. He proposed several methods, involving mechanical contact between the electromagnets and the reaction rail, for overcoming the inherent instability in attractive force systems. In the October 18, 1912 edition of *The Engineer*, the article "Foucault and Eddy Currents Put to Service" began with the line: "There was demonstrated some months ago at Mount Vernon, which is really a suburb of New York City, a levitating transmitting apparatus invented by Mr. Emile Bachelet." Bachelet actually conducted a series of experiments in which he levitated plates of various metals above a coil excited with 60 Hz single-phase current. The eddy currents induced in the plates by the time-varying magnetic field reacted against that field to produce the resultant lift force. He designed a test vehicle consisting of aluminum reaction plates for lift and iron framing pieces that served as the reaction rails for the sequentially excited coils of a simple type of LIM. The energy requirements, though large, were minimized by the clever use of limit switches that turned the levitating and propelling coils on and off as the vehicle passed. Little further development was reported until the 1930s, when Hermann Kemper, a German inventor, demonstrated the first noncontact method for stably suspending an object with attractive magnetic force. His inventions, for which he was awarded patents in 1934 and 1935, were the basis for modern attractive-force-based systems.

Following a long hiatus, widespread interest in maglev technology suitable for intercity and urban applications was renewed in the late 1960s and early 1970s. After a decade of experimentation and analysis in Germany, it was decided in 1978 to use EMS technology with separate lift and guidance electromagnets and long stator LSMs with iron cores for propulsion (see Fig. 1). By 1985, the first 20 km section of the Emsland test track, which included a closed loop at the north end, was completed and the first high-speed prototype commercial vehicle (the TR06) reached a speed of 355 km/h. A second 10 km long closed loop was added to the south end, permitting continuous running in 1987. Continued technical refinements resulted in the TR07 200-passenger vehicle prototype, which reached speeds in excess of 450 km/h in the early 1990s. Visitors could purchase rides, for 20 Deutsch Marks, on the TR07, which reaches speeds near 450 km/h. Construction of the first commercial route is completed in China and is expected to begin revenue service in early 2004.

At the same time that the Germans were developing the high-speed EMS maglev system, several countries including Germany, Japan, and the United Kingdom undertook the development of EMS technology for low-speed applications. Japan Airlines initiated a long-term program in the early 1970s to develop and commercialize an EMS maglev system using the same "U"-shaped magnets and reaction rails for both lift and guidance and short-stator LIMs for propulsion (see Fig. 3). Commercialization was taken over by the Chubu HSST Development Corporation in 1991. Various versions of the HSST family of vehicles have been demonstrated in public venues since the early 1980s, but thus far no systems have been put in revenue service. Developments continued in the 1990s on a 1.6 km test track in Nagoya. The M-Bahn system, which used permanent magnets stabilized by a mechanical contact system and propelled by a LSM, was developed in Germany in the early 1980s and placed in demonstration service in Berlin in the late 1980s but was never operated in revenue service. The first commercial EMS maglev system in the world was installed at the Birmingham Airport in the United Kingdom in 1984. Based on experiments at the British Rail Research and Development Division, lift and guidance were provided by the same controlled DC electromagnets reacting against laminated iron rails in the longitudinal flux configuration and a LIM provided the propulsion. The system shuttled passengers from the air terminal to an adjacent railway station until it was retired in 1995. Short-stator EMS systems are also under development in other countries including Korea and the United States. A consortium of organizations, including American Maglev Technology Corporation, is building a people-mover system at the Old Dominion University in Virginia. Additional efforts at urban maglev technology development, under Federal Transit Administration sponsorship, are in progress at Sandia National Laboratory and at General Atomics Corporation.

Development of the EDS system technology commenced with U.S. scientists Powell and Danby, who received a patent in 1969 for the design of the first SCM-based suspension and stabilization system. It utilized the same set of vehicle-borne SCMs for null-flux lift and guidance. The use of SCMs greatly reduced the energy requirements that stymied further development of repulsive-force, eddy-current-based suspension concepts first demonstrated by Bachelet in 1912. They also allowed large air gaps on the order of 10 to 20 cm. In the early 1970s, researchers around the world, including those in the United States, Canada, the United Kingdom, Germany, and Japan, subsequently picked up on some of Powell and Danby's ideas and began developing their own versions of EDS systems using SCMs. North American participants included researchers at Massachusetts Institute of Technology, Stanford Research Institute, Ford Motor Co., and the Canadian Institute of Guided Ground Transport at Queen's University. Government support for maglev research in the United States stopped in the mid-1970s but work continued in other countries, including Canada, the United Kingdom, and, most notably, in Japan, where development has proceeded to the stage of commercial readiness on the high-speed Japanese linear motor car.

Following several years of basic development and testing of their superconducting maglev concept at the Japanese Railway Technical Research Institute, construction of the 7.1 km long test track at Miyazaki on Kyushu Island was initiated in 1977. The guideway was of the inverted "T" type with levitation coils (called ground coils) placed horizontally on the floor and LSM stator windings mounted vertically on either side of the stem of the inverted "T." In 1979, the unmanned MLU-500 test vehicle incorporating SCMs reached a speed of 517 km/h. Shortly thereafter, the superstructure of the guideway was converted to a "U" shape and a long program of developmental and testing of the MLU family of vehicles commenced. In the late 1980s, the ground coils were replaced with null-flux lift coils mounted on the vertical sidewalls of the guideway. Up to that period, separate coils were used on the guideway for lift and combined propulsion and null-flux guidance. In the late 1980s and early 1990s, sections of the guideway coils were replaced with a new set of coils that served all three functions of lift, guidance, and propulsion. Meanwhile, the construction of a new 42.8 km test track began in the Yamanashi Prefecture approximately 70 km west of Tokyo. Originally planned so that it could be extended to form a commercial route from Tokyo to Osaka, the first phase consisted of an 18.4 km "priority section" that was completed in 1995. Due to the mountainous terrain, approximately 87% of its length is in tunnels. A 12.8 km portion of the priority section was double-tracked to permit tests of passing vehicles. The first test vehicle delivered to the test track in July 1995 consisted of a full-scale prototype commercial three-car train set referred to as the MLX01. Following preliminary trials, the test speed was gradually increased to 550 km/h in March 1998. After a second prototype vehicle arrived, tests were conducted with trains passing each other at relative speeds as high as 1000 km/h. A new system under development in Switzerland is the SwissMetro. It is a high-speed EDS system designed for operation in partially evacuated tunnels. Such a system is motivated by the mountainous terrain and the desire to reduce the propulsion power required to overcome the large aerodynamic drag force present at high speed in tunnels.

In the United States, the National Maglev Deployment Program, under the auspices of the Transportation Equity Act for the 21st Century, is in its second phase. Two consortia are competing for final approval of their proposed high-speed maglev projects. One connects Baltimore to Washington, DC and the other connects the Pittsburgh Airport to the city and its eastern suburbs. Another project, under the sponsorship of the state of Florida, and the direction of Powell and Danby, will link the Titusville Airport to the Kennedy Space Center and the cruise terminals at Port Canaveral. In addition, the development of low-speed maglev systems for urban mass transit applications is being sponsored by the Federal Transit Administration.

5. ADVANTAGES OF MAGLEV TECHNOLOGY FOR HIGH-SPEED TRANSPORTATION

High-speed maglev technology offers four main advantages: noncontact operation; low-mass vehicles (on a per seat basis, maglev vehicles weigh approximately one-third to three-quarters as much as high-speed trains); high speed; and wayside system control and self-propelled vehicles (each car contains its own secondary part of a LSM).

Noncontact operation means that vehicle traction does not depend on adhesion between contact surfaces, e.g., wheels and rails. Combined with lower mass, vehicles can accelerate and brake faster and can climb steeper grades independent of whether the surfaces are wet or dry. Noncontact also means lower noise levels than for conventional trains in urbanized areas, permitting higher urban and suburban speeds or, alternatively, reduced or eliminated noise abatement costs. In addition, maintenance costs associated with routine wheel, rail, and bearing wear are eliminated. (However, if the specific maglev system design requires a landing gear for low-speed operation, some maintenance costs will be incurred for landing gear wear.) The combination of noncontact operation and reduced vehicle weight translates into reduced stress on guideway support structures and construction costs. At grade, guideway costs have been estimated to be more for maglev systems than for conventional trains, but approximately the same as or even less than that for elevated guideways, depending on the specific technology. A technology assessment conducted by the U.S. Army Corps of Engineers in the mid-1990s estimated that the technology cost (omitting profits and land acquisition costs) for elevated maglev system concepts would range from $18 to $28 million/double-track mile and at grade systems from approximately $11 to $25 million/double-track mile (all in terms of 1993 U.S. dollars). Estimates place the total cost for the elevated German Transrapid system at approximately $35 million/double-track mile. Total costs for existing high-speed rail systems range widely depending on country, terrain, and percentage of elevated track and tunnels used. In terms of 1993 U.S. dollars, costs range from $7.3 to $21.5 million/double-track mile in France and from $32 to $50 million/double-track mile in Germany.

High speed translates into reduced trip times. Existing maglev systems have demonstrated operating speeds of passenger-carrying vehicles in the range of 450 to 550 km/h. Higher speeds are possible in open air but become increasingly energy-intensive due to the rapidly increasing aerodynamic drag force with speed. However, higher speeds are practical in partially evacuated tunnels. Figure 7 shows a comparison of line haul times versus trip length for commercial jet aircraft and high-speed ground transportation (HSGT) modes. Line haul time refers to time spent in the vehicle from terminal to terminal and does not include terminal access or egress times or times spent in the terminals. Time spent on the ground by aircraft in major airports is clearly a disadvantage that is only compensated for by higher speeds at flight altitude at flight distances greater than approximately 400 to 500 miles. Figure 8 shows a more realistic comparison of the total trip times, including access and egress time and times spent in the origin and destination terminals for maglev and jet aircraft. The maglev makes two additional urban stops to pick up and discharge passengers (reducing access and egress times); the aircraft makes no

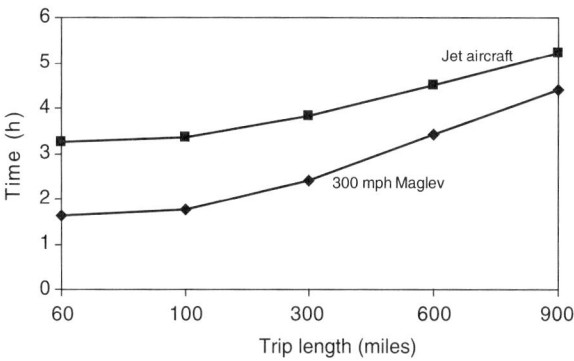

FIGURE 8 Total trip times versus trip length. Maglev makes two urban stops. Jet aircraft makes no stops.

additional stops. The maglev is the clear winner even for distances greater than 900 miles.

Since maglev cars are self-propelled, they can be operated as individual cars (each carrying 50 to 100 passengers) or as linked cars, depending on travel demand. Furthermore, since each car or set of linked cars is controlled by the wayside control system, headways (times between vehicles) can be reduced to less than 30 s, if necessary, to accommodate travel demand. In contrast, most high-speed trains are operated as standard train sets consisting of one or two locomotives and 8 or 10 coaches. Consequently, the technology provides considerable flexibility to meet a wide range of passenger demand and still maintain a very high-frequency service.

Two public benefits of maglev vehicle systems are reduced dependence on petroleum-based fuels and reduced emissions of air pollutants. The energy used by a maglev system comes from the electric utility grid. Consequently, the emissions produced and energy consumed are strongly dependent on the mix of electricity-generating technologies that supply the grid. To the extent that renewable energy sources, such as hydroelectric, wind, solar, and nuclear technologies, are used, the emissions can be greatly reduced. Furthermore, new fossil-fuel-generating technologies, including combined-cycle gas turbines and fuel cells, that are substantially more efficient than traditional technologies are coming into the mix. In addition, nationally, less than 4% of the electricity is generated by combustion of petroleum-based fuels. Hence, to the extent that trips are diverted from conventional modes to maglev vehicles, the United States moves closer to energy independence and a cleaner environment. The reduction in emissions and energy use comes about for a variety of reasons including the high efficiency

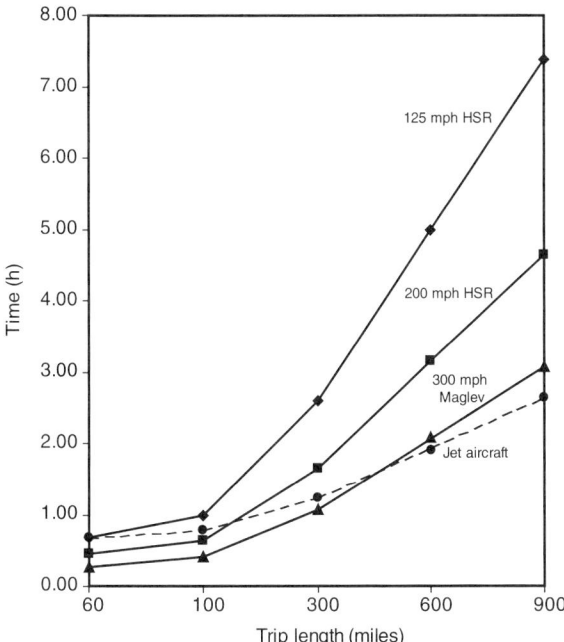

FIGURE 7 Line-haul trip times versus trip length for various high-speed modes.

of the maglev technology itself, the efficiency of the generating plants, and the manner in which the maglev vehicles are operated. The efficiencies are somewhat design-specific, e.g., systems incorporating SCM technology are more efficient than those that use conventional electromagnets. Generally, SCMs are operated in the "persistent current" mode, whereby once energized, the winding is disconnected from the power source and the current in the winding continues to flow without loss. Power is then required only to keep the SCM below its critical temperature. Propulsion energy is used only when the vehicles are moving (there is no need for idling). Maglev vehicle systems have the added advantage that they utilize three-phase power, which provides a more balanced load to the utility grid than catenary-based systems, which can use only single-phase power. In addition, the use of smaller coaches with shorter headways tends to distribute the load to the utility system more evenly over time. Finally, since maglev guideways are connected together electrically, there is a greater opportunity to convert the power derived from regenerative braking of some vehicles into propulsion power of other vehicles.

Figure 9 shows a comparison of energy used per seat·mile for jet aircraft and HSGT modes as a

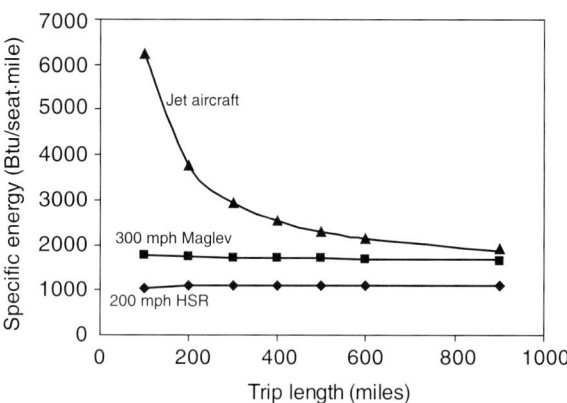

FIGURE 10 Same as for Fig. 9, but includes energy for two urban in-route stops and one in-route stop for HSGT and one in-route stop for jet aircraft.

function of trip length. Aircraft consume the greatest amount of energy except for distances greater than approximately 700 miles. Figure 10 shows a similar comparison for which the HSGT modes make two additional urban stops and one in-route stop, whereas the aircraft makes only one in-route stop. In this case, the aircraft uses more energy for distances further than 900 miles.

SEE ALSO THE FOLLOWING ARTICLES

Intelligent Transportation Systems • Magnetohydrodynamics • Transportation and Energy, Overview • Transportation and Energy Policy • Ultralight Rail and Energy Use

Further Reading

German Transrapid World Wide Web Site. Available at http://www.transrapid.de/en/index.html.

He, J. L., Rote, D. M., and Coffey, H. T. (1992). "Survey of Foreign Maglev Systems," Argonne National Laboratory Report No. ANL/ESD-17. Argonne National Laboratory, Argonne, IL.

Japanese Maglev World Wide Web Site. Available at http://www.rtri.or.jp/index.html.

Laithwaite, E. R. (1987). "A History of Linear Electric Motors." Macmillan Education Ltd., London, UK.

Lever, J. H. (ed.). (1998). "Technical Assessment of Maglev System Concepts," Final Report by the Government Maglev System Assessment Team, U.S. Army Corps of Engineers, Cold Regions Research and Engineering Laboratory, Special Report 98-12. National Technical Information Service, Washington, DC.

Moon, F. C. (1994). "Superconducting Levitation." Wiley-Interscience, New York.

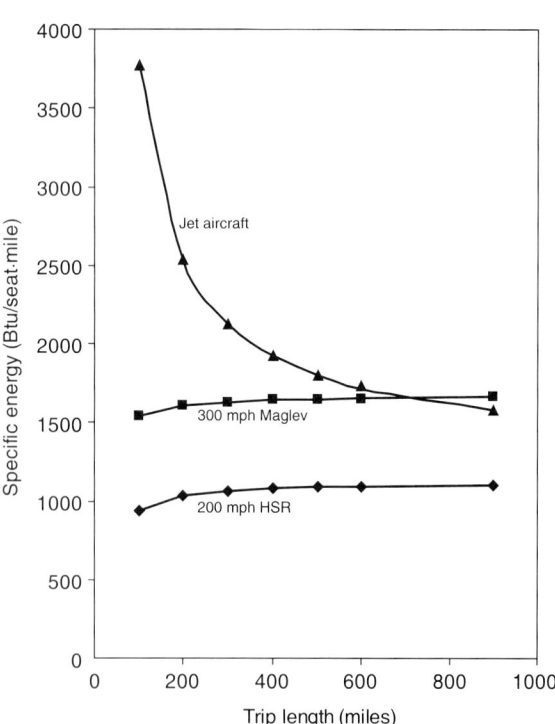

FIGURE 9 Energy per seat·mile used by HSGT and short-haul jet aircraft (no stops).

NMI Office. (1993). "Final Report on The National Maglev Initiative." FRA, U.S. Army Corps of Engineers, U.S. Department of Energy, Washington, DC.

Rhodes, R. G., and Mulhall, B. E. (1981). "Magnetic Levitation for Rail Transport." Clarendon, Oxford, UK.

Rote, D. M. (2001). "Guidelines for Estimating Trip Times, Energy Use, and Emissions for HSGT Technologies," Argonne National Laboratory Report No. ANL/ES/RP106522. Argonne National Laboratory, Argonne, IL.

Magnetohydrodynamics

PAUL H. ROBERTS
University of California, Los Angeles
Los Angeles, California, United States

PATRICK H. DIAMOND
University of California, San Diego
San Diego, California, United States

1. Introduction
2. Pre-Maxwell Approximation
3. Ideal Magnetohydrodynamics and Magnetostatics
4. Nonideal Magnetohydrodynamics
5. Stability of Magnetically Confined Plasmas

Glossary

Alfvén wave A wave that propagates along magnetic lines of force because of the tension of the field lines; also called a hydromagnetic wave.

Ampère's law The law that, in pre-Maxwell theory, determines the magnetic field created when electric current flows.

beta The name given to the ratio of the gas pressure to the magnetic pressure in a conducting fluid; plasma β is one of the figures of merit awarded to a magnetostatic equilibrium.

dynamo The process whereby a magnetofluid self-excites and amplifies a magnetic field.

electromotive force The electric field created by a changing magnetic field. Motive refers to its ability to move charge in a conductor (i.e., generate current).

Faraday's law The law that determines the electromotive force created by a changing magnetic field.

field line A curve that is everywhere parallel to the direction of the prevailing magnetic field; also called a line of (magnetic) force, it is part of a family of such lines, one of which passes through each point of space.

field line tension Part of the stress that a magnetic field $\mathbf{B}$ exerts on a medium and interpreted as a tension in a flux tube of B^2/μ_0, per unit cross-sectional area of the tube; μ_0 is the permeability of the medium.

figure of merit A number that quantifies the susceptibility of a plasma equilibrium to a certain type of magnetohydrodynamics instability.

flux tube See magnetic flux surface.

force-free field The field in a particular type of magnetostatic equilibrium in which the Lorentz force vanishes and in which field and current are parallel.

gradient A term used to indicate how rapidly a scalar or vector field changes with position. In the case of a scalar field, Φ, it is used to describe the vector field that is parallel to the direction in which Φ increases most rapidly; it is equal to the spatial rate of change of the scalar in that direction and is denoted by $\nabla\Phi$.

interchange-ballooning mode An instability driven by the combined effects of pressure gradient and average, or locally, unfavorable curvature of the field lines.

International Thermonuclear Experimental Reactor (ITER) a large-scale experiment, run by an international consortium, designed to study the physics of burning plasma.

kink safety factor A figure of merit quantifying the susceptibility of a plasma equilibrium to kink instability.

kink-tearing mode An instability driven by current gradients.

line of force A curve that is everywhere parallel to the direction of the prevailing magnetic field; also called a (magnetic) field line, it is part of a family of such lines, one of which passes through each point of space.

Lorentz force The force per unit volume on a fluid conductor that carries an electric current (density $\mathbf{J}$) and lying in a magnetic field $\mathbf{B}$; mathematically, it is the vector product of $\mathbf{J}$ and $\mathbf{B}$ and can also be represented by magnetic stresses.

magnetic helicity A property of a magnetic field that is preserved in the motion of a fluid that is a perfect electrical conductor.

magnetic pressure Part of the stress that a magnetic field $\mathbf{B}$ exerts on a medium and interpreted as an isotropic pressure $B^2/2\mu_0$, where μ_0 is the permeability of the medium.

magnetic reconnection The process by which a magnetofluid forms thin current layers that change the field line topology and dissipate magnetic energy.

magnetic relaxation The process by which a magnetofluid relaxes to its minimum energy state subject to certain constraints.

magnetic shear The rate at which the direction of a magnetic field changes with position **x**; a special case is used in plasma physics as a figure of merit.

magnetic stresses Stresses that are equivalent in their dynamical effect to the Lorentz force.

magnetic surface, also called a flux tube or a magnetic flux surface A surface composed of field lines so that the magnetic field, **B**, is everywhere tangential to it. The flux of **B** contained within the surface is the same everywhere along its length.

Ohm's law The law that states that the electric current density, **J**, produced in a motionless conducting medium by an electric field **E** is proportional to **E**. The constant of proportionality is the electrical conductivity of the medium.

pre-Maxwell theory The approximate form of electromagnetic theory that existed prior to Maxwell's discovery of displacement currents.

scalar product A convenient mathematical abbreviation used to describe a quantity derived from two vectors, **F** and **G**, and in magnitude is equal to the product of their magnitudes, |**F**| and |**G**|, and the cosine of the angle between them; it is written as **F** · **G**.

Taylor relaxation The process proposed by J. B. Taylor whereby a magnetofluid relaxes to its minimum energy state subject to the constraint of constant global magnetic helicity.

toroidal Alfvén eigenmode (TAE) A particular type of Alfvén wave appropriate to toroidal geometry.

vector product A convenient mathematical abbreviation used to describe a vector that is perpendicular to two other vectors, **F** and **G**, and that in magnitude is equal to the product of their magnitudes, |**F**| and |**G**|, and the sine of the angle between them; it is written **F** × **G**, where **F**, **G**, and **F** × **G** form a right-handed triad.

Magnetohydrodynamics (*MHD*) is the study of the movement of electrically conducting fluids in the presence of magnetic fields. The magnetic field influences the fluid motion through the Lorentz force, which is proportional and perpendicular to the magnetic field and the electric current flowing through the conductor: The magnetic field is affected by the electric current created by an electromotive force which is proportional and perpendicular to the magnetic field and the fluid velocity. It is this duality between magnetic field and fluid flow that defines the subject of MHD and explains much of its fascination (and complexity).

1. INTRODUCTION

Magnetohydrodynamics (MHD) is the marriage of hydrodynamics to electromagnetism. Its most famous offspring is the Alfvén wave, a phenomenon absent from the two subjects separately. In fact, many consider the discovery of this wave by Alfvén in 1942 to mark the birth of MHD. Initially, MHD was often known as hydromagnetics, but this term has largely fallen into disuse. Like MHD, it conveys the unfortunate impression that the working fluid is water. In reality, the electrical conductivity of water is so small that MHD effects are essentially absent. Moreover, many fluids used in MHD experiments are antipathetical to water. Even as fluid mechanics is now more widely employed than hydrodynamics the terms magnetofluid mechanics or magnetofluid dynamics, which are already sometimes employed, may ultimately displace MHD. Magnetofluid is already widely used in MHD contexts.

Since electric and magnetic fields are on an equal footing in electromagnetism (EM), it may seem strange that the acronym EMHD was not preferred over MHD. There are two reasons for this. First, to invoke EM theory in its full, unapproximated form would, in most contexts, add complexity without compensating enlightenment. It usually suffices to apply the form of EM theory that existed in the 19th century before Maxwell, by introducing displacement currents, cast the theory into its present-day form. In this pre-Maxwell theory of EM, there are no displacement currents and the electric and magnetic fields are not on an equal footing; the magnetic field is the master and the electric field the slave. Consequently, MHD is an appropriate acronym, but EMHD is not. Situations in which this is untrue and in which full EM theory is needed involve relativistically moving fluids and are too seldom encountered to be described here; in this article, the pre-Maxwell approximation is used throughout. The second reason why EMHD cannot be used is that the acronym is generally understood to mean electron MHD, which studies the MHD of the electron fluid in a two-fluid description of a plasma, the ions forming the other fluid. This topic is also outside the scope of this article.

A significant branch of MHD is the study of Magnetostatic equilibria (MSE). This subject is the MHD analogue of hydrostatics, the branch of fluid mechanics that deals with fluids at rest, with the pressure gradient in the fluid balancing external forces such as gravity. Similarly, in MSE the fluid is motionless and the pressure gradient balances the Lorentz force (and any other forces present).

The two main applications of MHD are technological—to liquid metals and to plasmas. There is little doubt that the former has had the greater impact on society. It includes the casting and stirring of liquid metals, levitation melting, vacuum-arc remelting, induction furnaces, electromagnetic valves, and aluminum reduction cells. Another application, the flow of a liquid metal in the blanket surrounding a thermonuclear reaction chamber, touches on the other main area: plasma MHD. The reactor contains a rarefied plasma of deuterium/tritium (DT) that is raised to a high enough temperature for these nuclei to fuse and release energy. The economic promise of such a device in generating magnetic fusion energy (MFE) has provided a powerful incentive for studying plasma MHD and has led to significant new insights, particularly into the structure and stability of MSE. In addition to these practical applications, the elucidation of a wide variety of magnetic phenomena in nature depends on an understanding of MHD. Astrophysics and geophysics provide abundant examples, including the magnetism of the earth, planets, and satellites, that of the sun and other stars, and that of galaxies.

This article describes the simplest form of MHD theory, in which the working fluid is a homogeneous, continuous medium. This may or may not (depending on context) be a satisfactory description of a plasma. In general, a plasma and its dynamics are described by kinetic equations that govern distribution functions for electrons, ions, and (if the plasma is incompletely ionized) neutral atoms. The distribution function, $f(\mathbf{x}, \mathbf{v}, t)$, of any of these species is proportional to the probablility that a particle of that species at position $\mathbf{x}$ is moving with velocity $\mathbf{v}$ at time t. In certain cases, such a description can be greatly simplified by replacing the kinetic equations for the functions f by a closed set of fluid equations. This happens when the dependence of each f on $\mathbf{v}$ is close to what it would be if the plasma were in local thermodynamic equilibrium, as seen by an observer moving with the mean velocity, $\mathbf{V}(\mathbf{x}, t)$, of the species at $\mathbf{x}$ and t. Such a distribution is termed Maxwellian. The distribution functions are then replaced by ion, electron, and neutral densities, n_i, n_e, and n_n, respectively, that give the total numbers of each species per unit volume at $\mathbf{x}$ and t, with each fluid having its own (mean) velocity. Although this may seem (and is) very complicated, it is far less forbidding than the alternative (i.e., solving for the evolution of the distribution functions).

The method of closure described previously necessarily involves an approximation of restricted validity. Single-fluid MHD also assumes that the characteristic length scale, $\mathscr{L}$, is large compared with the Debye length, $\mathscr{L}_{\text{debye}}$. This length quantifies the distance over which n_e and n_i can differ substantially from one another. When $\mathscr{L} \gg \mathscr{L}_{\text{debye}}$, these densities are closely equal and the plasma is very nearly electrically neutral. A second requirement for simple one-fluid MHD to hold is that $\mathscr{L}$ be large compared with the mean free path between particle collisions in the plasma and in comparison with the gyroradii. (A charged particle in a magnetic field would, in the absence of collisions with other particles, orbit about a field line in a spiral having a certain gyroradius that depends on its mass and charge and on the field strength.) These demands are more easily satisfied in many astrophysical contexts than in a thermonuclear reactor chamber. Nevertheless, one-fluid MHD can provide useful qualitative insights into the structure and stability of plasmas relevant to MFE.

To simplify the following discussion, the abbreviations shown in Table I will usually be employed. In particular, velocity means fluid velocity, density stands for the mass density of the fluid, field means magnetic field, current is short for electric current density, conductor means conductor of electricity, and potential means electric potential. Script letters—$\mathscr{L}$, $\mathscr{T}$, $\mathscr{V}$, $\mathscr{B}$, $\mathscr{J}$, $\mathscr{E}$, etc.—are used to indicate typical magnitudes of length, time, velocity, field, current, electric field, etc.

The aim of this article is to provide as simple an account of MHD as possible—one that is slanted toward the MFE field but one that involves a minimum of mathematics. Some familiarity with vector fields is necessary, however.

TABLE I

Abbreviations

Quantity	Symbol	Magnitude	SI unit
Time	t	$\mathscr{T}$	Seconds (s)
Position	$\mathbf{x}$	$\mathscr{L}$	Meters (m)
Velocity	$\mathbf{V}$	$\mathscr{V}$	m/s
Density	ρ	—	Kilogram/m^3 (kg/m^3)
Pressure	p	—	Newton/m^2 (N/m^2)
Temperature	T	—	Kelvin (K)
Field	$\mathbf{B}$	$\mathscr{B}$	Tesla (T)
Current	$\mathbf{J}$	$\mathscr{J}$	Amp/m^2 (A/m^2)
Charge density	υ	—	Coulomb/m^3 (C/m^3)
Potential	Φ	—	Volt (V)
Electric field	$\mathbf{E}$	$\mathscr{E}$	V/m
Conductivity	σ	—	Siemens/m (S/m)

2. PRE-MAXWELL APPROXIMATION

Some readers may value a brief review of the relevant concepts in EM theory that underpin MHD; others may prefer to skim this section and move ahead to Section 3.

Under discussion are systems having characteristic length, time, and velocity scales, $\mathscr{L}$, $\mathscr{T}$, and $\mathscr{V}$, respectively, for which $\mathscr{L} \ll c\mathscr{T}$ and $\mathscr{V} \ll c$, where c is the speed of light. In these circumstances, EM phenomena are well described by an approximation that was in use before Maxwell introduced displacement currents and his famous equations into EM theory. This approximation, which is also called the nonrelativistic approximation (since it implies that c is infinite), is summarized.

Electrostatics is the study of static electricity. Each positive (negative) electric charge experiences an attraction toward every negative (positive) charge and a repulsion from every other positive (negative) charge. The sum, $\mathbf{f}$, of all these forces on a charge e is proportional to e and is conveniently represented by an electric field, $\mathbf{E} = \mathbf{f}/e$. Suppose that, in steady conditions, a test charge is carried around a closed curve Γ, where by test charge is meant a charge so small that it does not have any effect on the other charges. As the charge describes Γ, the agency that moves it will sometimes do work against the force $\mathbf{f} = e\mathbf{E}$ and will sometimes receive energy from it but, when the circuit is complete the net gain or loss of energy is zero. Such forces are termed conservative and this fact is expressed mathematically by the statement that $\mathbf{E} = -\nabla\Phi$, where Φ is the electric potential. (Here, $\nabla\Phi$ denotes the gradient of Φ, a vector that has Cartesian components $\partial\Phi/\partial x$, $\partial\Phi/\partial y$, $\partial\Phi/\partial z$.) If positive, $\Phi_1 - \Phi_2$ is the energy received from the electric field when a unit positive charge moves from a location where the potential is Φ_1 to one where the potential is Φ_2; if negative, it is the work that must be done to move the charge from the first location to the second. The force $\mathbf{f} = -e\nabla\Phi$ acts to move a positive charge in the direction of $\mathbf{E}$ from a region of higher potential to one of lower potential.

If a conductor carries a net charge, the mutual repulsion between charges quickly drives them to the surface of the conductor, where they arrange themselves with a surface charge density such that the surface is at a uniform potential Φ. The electric field is then normal to the surface and cannot move charges along the surface. The charge distribution is therefore static. Beneath the surface charge layer, $\mathbf{E}$ is identically zero, as is the volumetric charge density ϑ. If different areas of the surface are held at different potentials, as at the two ends of a straight wire, $\mathbf{E}$ is no longer zero in the conductor and a charge e experiences the force $\mathbf{f} = e\mathbf{E}$. If $e > 0$ (< 0) this moves it from (to) the area at the higher potential to (from) the area at the lower potential. The resulting flow of charge is called an electric current. Its density, $\mathbf{J}$, is proportional to the force $\mathbf{f}$ and, according to Ohm's law, it is given by $\mathbf{J} = \sigma\mathbf{E}$, where σ is the electrical conductivity of the conductor in siemens/meter (S/m). To relate this to the more familiar form of Ohm's law, consider again the example of the straight wire. If its length is L meters, a potential difference of Φ volts between its ends creates an electric field of strength Φ/L (V/m) and therefore a current of density $J = \sigma\Phi/L$ A/m^2. If the cross-sectional area of the wire is A, the total current is $I = JA = \sigma A\Phi/L = \Phi/R$, where $R = L/\sigma A$ is the electrical resistance of the wire in ohms (Ω). This result, $I = \Phi/R$, is the form of Ohm's law encountered in elementary texts on electricity and magnetism.

In pre-Maxwell theory, the flow of charge resembles the flow of a constant density fluid, for which mass conservation requires that the net flow of mass into any closed volume is equal to the net flow out of that volume. In the same way, charge conservation requires that the inward flow of electric current into the volume balances its outward flow.

Whenever current flows, a magnetic field, $\mathbf{B}$, attends it. This is shown in Fig. 1, again for a straight wire C having a circular cross section of radius a. In this case, $\mathbf{B}$ is in the θ direction, where θ is the azimuthal angle around the wire. Some lines of (magnetic) force are shown. A line of force, also called a field line, is actually not a line but a continuous curve whose tangent is everywhere

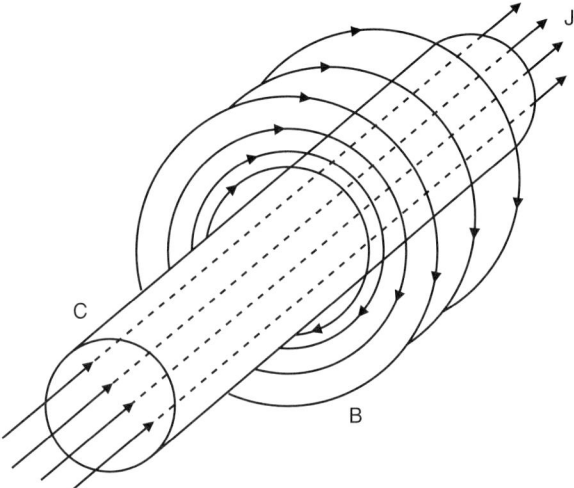

FIGURE 1 Lines of force created by a current-carrying wire C.

parallel to **B**. Although lines of force pass through every point in space, sketches such as Fig. 1 can only show a finite number. The density of lines indicates the strength of the field. Where lines bunch together (spread apart) the field is stronger (weaker). Magnetic charges, analogous to electric charges, do not exist. This means that a line of force may close on itself or cover a magnetic surface ergodically (i.e., without closing; Fig. 2), or it may continue to infinity, but it cannot terminate. It also means that (as for **J**) the net flow of **B** into any closed volume is equal to the net flow of **B** emerging from the volume. A bundle of lines of force contained in a magnetic surface form a flux tube (Fig. 3). The net flow of **B** into one end of a volume of the tube bounded by two of its cross sections is the same as the net flow that emerges from the other end. This is called the strength of the tube.

To determine **B** from **J** in a situation such as that sketched in Fig. 1, Ampère's law is required. This states that the component of **B** along any closed curve Γ is, when integrated around that curve, proportional to the net flux, I, of current through Γ. In a uniform medium, the constant of proportionality is its permeability, which in this article is always assumed to be the permeability of free space, $\mu_0 = 4\pi \times 10^{-7}$ H/m. (The fact that this value is exact is a quirk of the SI system of units, which makes use of one more quantity than is strictly necessary.) Taking the closed curve to be the (circular) line of force of radius r surrounding the wire, the required integral is $2\pi r B_\theta$, which by Ampère's law is also $\mu_0 I$, so that $B_\theta = \mu_0 I/2\pi r$; the field decreases with distance r from the wire as $1/r$. The field outside the wire does not depend on whether the current is uniform across the cross section of the wire ($J = I/\pi a^2$) or whether, as in a case considered in Section 3, it flows only along the surface of the wire with density $I/2\pi a$.

Faraday's law provides the final link between **B** and the electric field. This law applies to any closed curve Γ that is the periphery of a surface S. It states that the component of **E** along Γ, when integrated around Γ, is equal to the rate of decrease of the flux of **B** through S. This statement depends on a sign convention that can be understood from Fig. 4. If **E** is integrated around Γ in the direction of the arrows, then the flux of **B** through S is taken into the plane of the paper. (This right-handed rule also applies to Ampère's law; see Fig. 1.) Faraday's law shows that in unsteady EM states, **E** is not conservative (i.e., it cannot be expressed as $-\nabla \Phi$).

Faraday's law applies even when Γ is moving and changing shape, but in such cases **E** must be reinterpreted. To see why this is necessary, consider again Fig. 4. Suppose that **E** is zero and that **B** is uniform, constant in time and directed out of the plane of the paper. Now suppose that the sides ad and bc start to lengthen at speed U so that the area

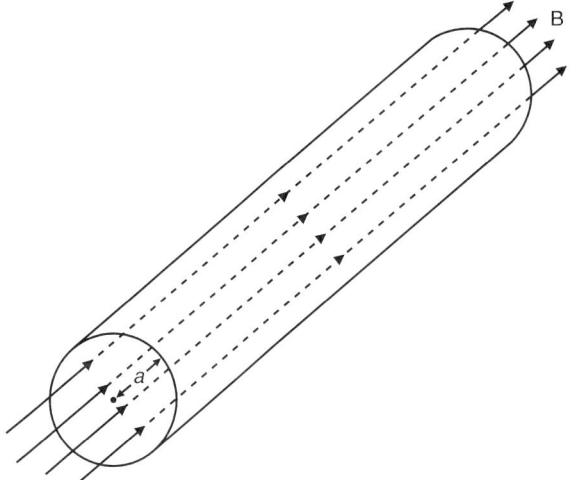

FIGURE 3 A straight flux tube. Its surface is composed of field lines.

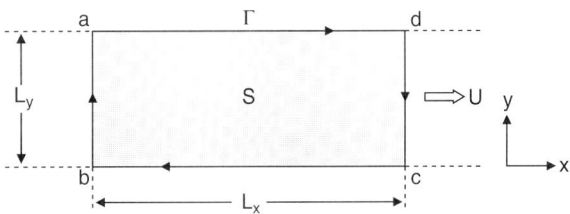

FIGURE 4 Application of Faraday's law to a growing rectangle S with periphery Γ. The sides da, ab, and bc are fixed. The side cd moves with speed U in the x direction. The sides ad and bc lengthen at the same rate.

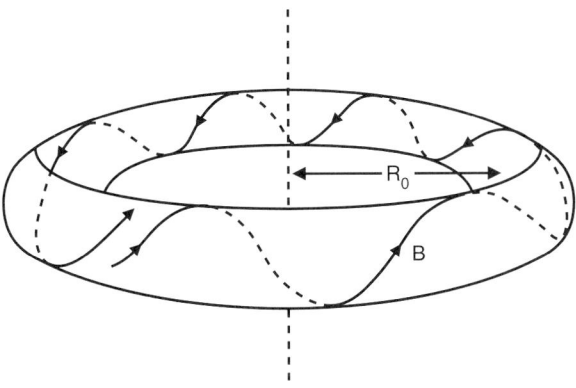

FIGURE 2 An axisymmetric magnetic surface filled ergodically by one line of force that does not close. The dashed line is the symmetry axis of the "donut"; it is also called the magnetic axis.

S of the rectangle increases at the rate UL_y. The flux of $\mathbf{B}$ out of the plane of the paper increases at the rate BUL_y; in the positive sense defined previously, it decreases at the rate BUL_y. By Faraday's law, this must also be the integral of the component of the electric field along the sides of the integral. However, $\mathbf{E}$ is clearly zero on all sides of Γ, particularly on the stationary sides da, ab, and bc. One must conclude that when integrating the electric field around Γ, one is not integrating E_y on the moving side cd but some other electric field, E_y'. The required integral is then $E_y' L_y$ and Faraday's law gives $E_y' = -UB$.

This argument has uncovered a very significant fact: Although $\mathbf{B}$ (and $\mathbf{J}$) are the same in all reference frames, the electric field is not. In the current example, $E_y' = -UB$ in the reference frame moving with speed U in the x direction. It is convenient here to make use of a shorthand notation: $\mathbf{U} \times \mathbf{B}$ is called the vector product of $\mathbf{U}$ and $\mathbf{B}$. It is proportional and perpendicular to $\mathbf{U}$ and $\mathbf{B}$; its Cartesian components are $U_y B_z - U_z B_y$, $U_z B_x - U_x B_z$, and $U_x B_y - U_y B_x$. In this notation, the electric field in the frame moving with the side cd is $\mathbf{E}' = \mathbf{U} \times \mathbf{B}$. More generally, when the electric field in one reference frame $\mathscr{F}$ is $\mathbf{E}$, then it is $\mathbf{E}' = \mathbf{E} + \mathbf{U} \times \mathbf{B}$ in the reference frame $\mathscr{F}'$ moving with velocity $\mathbf{U}$ relative to $\mathscr{F}$. Returning to Faraday's law in the case in which Γ moves or changes shape, it is $\mathbf{E}'$ and not $\mathbf{E}$ that is integrated around Γ, where $\mathbf{E}'$ is evaluated at each point P of Γ using the velocity $\mathbf{U}$ at P.

The force on a charge e that is stationary in $\mathscr{F}'$ is $\mathbf{f} = e\mathbf{E}'$, which in $\mathscr{F}$ is more conveniently written as $\mathbf{f} = e(\mathbf{E} + \mathbf{U} \times \mathbf{B})$. Analogously, Ohm's law in a solid conductor moving with velocity $\mathbf{U}$ is $\mathbf{J} = \sigma(\mathbf{E} + \mathbf{U} \times \mathbf{B})$. To illustrate this modification of Ohm's law, suppose that the rectangle sketched in Fig. 4 is electrically conducting, with ad and bc being parallel rails along which cd slides, the electrical contact between cd and the rails being perfect. In general, a current I will flow around Γ in the direction indicated by the arrows. This current is driven through $cb + ba + ad$ by a potential difference Φ created by charges that accumulate at the sliding contacts c and d. At one extreme, when the resistance R_1 of $cb + ba + ad$ is large, $I = \Phi/R_1$ is small. By charge conservation, I is also the current flowing in cd and it is small because the electric field $E_y = (\Phi_c - \Phi_d)/L_y$ in cd created by the charges at c and d cancels out almost completely the electric field $-UB$ induced by the motion of cd; i.e., $|E_y'|$ is small. At the other extreme, when R_1 is small, the path $cd \to ba \to ad$ is almost a short-circuit, the contact charges at c and d are small, and these points are at nearly the same potential; E_y is small, $E_y' \approx -UB$, and the current $I = UBL_y/R_2$ flows around Γ in the direction indicated, where R_2 is the resistance of cd. In the general case between these two extremes, $I = UBL_y/R$, where $R = R_1 + R_2$ is the total resistance of Γ.

The modified Ohm's law also applies when the conductor is a fluid, but now $\mathbf{U}$ is replaced by the fluid velocity $\mathbf{V}$, which is usually not the same everywhere. At a point at which the fluid velocity is $\mathbf{V}$, the current density is $\mathbf{J} = \sigma(\mathbf{E} + \mathbf{V} \times \mathbf{B})$. By Ampère's law, this current will influence $\mathbf{B}$. This fact exposes one half of the relationship between $\mathbf{V}$ and $\mathbf{B}$, alluded to in Section 1, that characterizes the subject of MHD: Flow alters field in a fluid conductor. The other half of the relationship is a consequence of a force that a conductor experiences when $\mathbf{J}$ and $\mathbf{B}$ are nonzero. The force is proportional to, and perpendicular to, both $\mathbf{J}$ and $\mathbf{B}$. This is the Lorentz force, $\mathbf{J} \times \mathbf{B}$ per unit volume. Through this force, field alters the flow of a fluid conductor.

To illustrate the action of the Lorentz force, consider the conducting rectangle Γ sketched in Fig. 4. The sides ab and cd experience equal and opposite forces in the x direction, IBL_y and $-IBL_y$ respectively. These oppose the continual expansion of the rectangle (an example of Lenz's law). To maintain the expansion, forces must be applied to the sides ab and cd, and these do work at the rate $UIBL_y = (UBL_y)^2/R = I^2/R$, which is precisely the rate at which electrical energy is converted into heat in the circuit Γ by its electrical resistance R. In this way, mechanical energy is converted into electrical energy and then "ohmically" into heat.

To state this result more generally, it is convenient to introduce a second shorthand notation: $\mathbf{F} \cdot \mathbf{G}$ is the scalar product of two vectors $\mathbf{F}$ and $\mathbf{G}$. In terms of their Cartesian components, it is $F_x G_x + F_y G_y + F_z G_z$. If $-\mathbf{V} \cdot (\mathbf{J} \times \mathbf{B})$ is positive at a point P, it quantifies the rate at which fluid loses its kinetic energy at P. This energy goes partially to increasing the magnetic energy density at P but some is radiated away from P through a "Poynting" energy flux $\mathbf{E} \times \mathbf{B}/\mu_0$. The remainder offsets the ohmic losses, J^2/σ per unit volume, at P. When $-\mathbf{V} \cdot (\mathbf{J} \times \mathbf{B}) > 0$ at P, the loss of kinetic energy provides a brake on the motion, which would cease even in the absence of viscous friction, unless maintained by some other force. When $-\mathbf{V} \cdot (\mathbf{J} \times \mathbf{B}) < 0$ at P, the reverse process occurs; field passes energy to motion, as in an electric motor.

The volumetric charge density ϑ in a moving fluid conductor is generally nonzero so that the fluid experiences a body force $\vartheta \mathbf{E}$ per unit volume. In the pre-Maxwell approximation, this is negligible in comparison with $\mathbf{J} \times \mathbf{B}$. These forces can be reinterpreted in terms of stresses on the fluid and, not surprisingly, the electric stresses are negligible in comparison with the magnetic stresses. It is also found that the energy density of the electric field is insignificant compared with the energy density, $B^2/2\mu_0$, of the magnetic field. This highlights the unimportance of **E** relative to **B** in pre-Maxwell theory.

The magnetic stresses alluded to previously have two parts. One is an isotropic magnetic pressure, $p_M = B^2/2\mu_0$; the other is nonisotropic and can be interpreted as a field line tension of B^2/μ_0 per unit area, or AB^2/μ_0 for a flux tube of cross-sectional area A. Both parts increase quadratically with B. The magnetic pressure, which is already approximately 4 atm for a field of 1 T, is 10,000 times greater for $B = 100$ T.

The action of the field on the flow is recognized by including the Lorentz force $\mathbf{J} \times \mathbf{B}$ in the equation of motion that governs the fluid velocity **V**. The remaining forces affecting the motion include the gradient of the (kinetic) pressure p and a term, proportional to the (kinematic) viscosity ν, that describes frictional effects. When the Lorentz force is equivalently represented by magnetic stresses, the magnetic pressure p_M joins p in a "total" pressure $P = p + p_M$. The ratio $\beta = p/p_M = 2\mu_0 p/B^2$ is usually called beta. It is a significant quantity in plasma physics, where it is a figure of merit—one of several used to quantify the excellence or otherwise of a plasma containment device.

3. IDEAL MAGNETOHYDRODYNAMICS AND MAGNETOSTATICS

A fluid is termed ideal if it has zero viscosity and zero thermal conductivity but infinite electrical conductivity. Since **J** must be finite even though $\sigma = \infty$, Ohm's law $\mathbf{J} = \sigma(\mathbf{E} + \mathbf{V} \times \mathbf{B})$ implies that $\mathbf{E} = -\mathbf{V} \times \mathbf{B}$. Consequently, Alfvén's theorem holds: Lines of force move with the fluid as though frozen to it.

To establish this important result, let Γ be the periphery of a surface element S that moves with the fluid. Since $\mathbf{E}'$ vanishes everywhere on Γ, Faraday's law shows that the flux of **B** through S is unchanging. Suppose S is part of the curved surface C of a flux tube (Fig. 5A). The flux of **B** through S is zero because **B** is tangential to it; by Faraday's law, it remains zero as the contents of the flux tube are carried to a new location by the fluid motion **V**. Since this applies to every such S drawn on C, **B** remains tangential to C everywhere. Thus, it is still a flux tube at its new location. On shrinking the cross section of the tube to a single line of force, it is seen that this line of force moves with the fluid, as though frozen to it. A stronger statement can be made. Let Γ encircle the tube so that S becomes a cross section of the tube (Fig. 5B). The flux of **B** through S is the strength of the tube, as defined in Section 2. By Faraday's law, it is the same at its new location. This applies to every cross section. The fluid motion therefore preserves the strength of flux tubes and the integrity of the magnetic surfaces that contain them.

Alfvén's theorem provides a useful way of envisioning MHD processes in highly conducting fluids. For example, consider the eruption of a flux tube from the relatively dense region below the solar photosphere and into the more tenuous solar atmosphere (Fig. 6). Gravity drains the material in the ejected flux loop back to the solar surface, leaving the flux loop at low density and pressure, where it is in approximate equilibrium with its new surroundings. Such processes are responsible for expelling from the sun the field created within it.

Next, consider the way in which fluid motions can exchange energy with the field. Imagine a straight flux tube, initially of cross-sectional area A_0, lying in

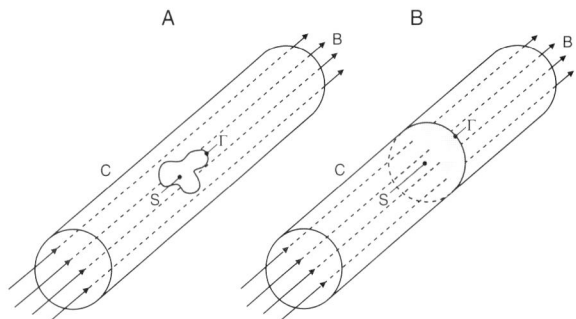

FIGURE 5 (A) An element S of surface area lying on the surface of a flux tube with periphery Γ. (B) A cross-section S of the flux tube with periphery Γ.

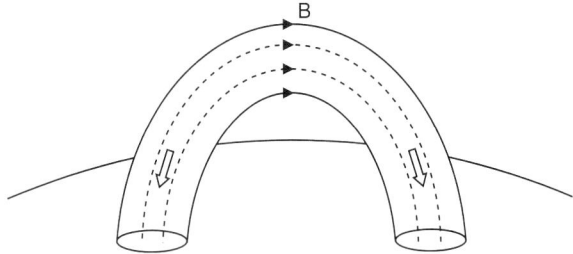

FIGURE 6 Fluid pulled by gravity down a flux tube that has erupted from the surface of the sun. Field directions are indicated by single arrows and fluid motion by arrowheads.

a compressible fluid. Suppose a motion compresses the tube uniformly so that, while remaining straight, its cross-sectional area is reduced to $A(<A_0)$. The density of the fluid it contains increases from its initial value of ρ_0 to $\rho = \rho_0(A_0/A)$. Since (by assumption) the fluid does not conduct heat, the tube compresses adiabatically, and its internal energy (i.e., the energy of the random molecular motions within it) increases so that its temperature rises. Because the strength of the tube cannot change, the field B_0 initially within the tube increases to $B = B_0 A_0/A$ (i.e., by the same factor as ρ_0); thus, B/ρ remains at its initial value, B_0/ρ_0. The magnetic energy per unit length contained by the tube, however, increases from $A_0(B_0^2/2\mu_0)$ to $A(B^2/2\mu_0) = [A_0(B_0^2/2\mu_0)](A_0/A)$. In short, the kinetic energy of the compressing motion has been transformed into internal energy and magnetic energy. The reverse happens if $A > A_0$; the flux trapped in an intense tube tends to expand into surroundings where the field is weaker. For these reasons, sound traveling across a field moves faster (and much faster when β is small) than it would in the absence of the field; the magnetic pressure created by the field intensifies the restoring force in the compressions and rarefactions of the wave.

The transformation of kinetic energy into magnetic energy also occurs in an incompressible fluid when motions stretch field lines against their tension. The process may be likened to the storing of elastic energy in a stretched rubber band or to the energy transmitted to a violin string by plucking it. If the flux tube of cross-sectional area A_0 containing field B_0 considered previously is lengthened from L_0 to L, its cross section will decrease by the same factor ($A = A_0 L_0/L$) and the field within it will increase by the same amount ($B = B_0 L/L_0$). The magnetic energy it contains, which is proportional to B^2, is enhanced by a factor of $(L/L_0)^2$ from $(B_0^2/2\mu_0)L_0 A_0$ to $(B^2/2\mu_0)LA = (B^2/2\mu_0)L_0 A_0 = [(B_0^2/2\mu_0)L_0 A_0](L/L_0)^2$. If $L = L_0 + \delta$, where $\delta \ll L_0$, the increase in magnetic energy is $(B_0^2/\mu_0)A_0\delta$. This is the work that the applied force had to do against the magnetic tension $(B_0^2/\mu_0)A_0$ of the field lines in stretching the tube by δ.

Tension has other important effects. Consider again the system shown in Fig. 1, but suppose now that the cylinder C is a compressible fluid or plasma surrounded by a vacuum. Suppose that the current I flows along the surface of C so that $I/2\pi a$ is the surface current density. The current I produces a field of strength $B_\theta = \mu_0 I/2\pi a$ on C. Because of their curvature, the field lines exert a "hoop stress" on C of $B_\theta^2/2\mu_0$ per unit area and directed radially inward. This compresses the fluid in C. Since **B** is zero in C, only (kinetic) pressure p can oppose the hoop stress. If $p = B_\theta^2/2\mu_0$, the forces are in balance, and the configuration is in *MSE*.

One may imagine that this MSE is brought about in the following way. Initially, A is large and p is small. After the current I and its associated field B have been set up, the forces on C are not in balance and the hoop stresses exerted by the field lines encircling C pinch the plasma column in much the same way as a stretched elastic band squeezes what it encircles. As these lines of force shorten and A becomes smaller, magnetic energy is converted into internal energy. The interior of C becomes hotter and its pressure p increases to oppose the further contraction of the field lines. Plausibly, the final result is an MSE of the kind envisaged previously, in which a plasma column has been strongly heated and prevented from expanding by the encircling field. Hot plasma is then confined within a magnetic bottle and away from solid walls. The possibility of this kind of magnetic confinement is of considerable interest in the MFE field. A hot reacting plasma must be prevented from leaching impurities from the solid walls of the reactor because these would greatly enhance the radiative losses, cooling the plasma and quenching the reactions.

In an MSE, the Lorentz force $\mathbf{J} \times \mathbf{B}$ balances the pressure gradient ∇p. This means that **B**, being perpendicular to ∇p, is tangential to the constant-p surfaces, which are therefore magnetic surfaces, as defined in Section 2, and can be labeled by their value of p. Current lines (curves drawn parallel to **J**) also lie on these surfaces and, like the field lines, they may close or traverse the surfaces ergodically. The total pressure P is continuous across the special magnetic surface that is the boundary of a plasma; if the exterior is a vacuum, P must balance the magnetic pressure, $B^2/2\mu_0$, in the vacuum.

If a region exists in which $\mathbf{J} \times \mathbf{B} = 0$, the field within it is termed force-free. Since $\nabla p = 0$, the pressure p is constant in the region, and p is no longer available to label surfaces on which the field and current lines are constrained to lie. Since **J** and **B** are parallel in a force-free field, $\mathbf{J} = \lambda \mathbf{B}$, where λ is constant on each field line, this being the component $J_\parallel = \mathbf{J} \cdot \mathbf{B}/B$ of **J** parallel to **B** on that line. The parameter λ may act as a surrogate for p in labeling the magnetic surfaces and constraining the field and current lines. If, however, λ is also constant, the field and current lines become unconstrained. A single field line and a single current line may fill the region. These are termed stochastic lines. They magnetically connect every part of the region to every other part. Particular cases of this are presented in Sections 4.6 and 5.3.2.

The MSE just described is called the Z pinch because the current flows parallel to the axis of C, a direction usually labeled by z. The Z pinch is shown in Fig. 7A, together with two other MSEs. In Fig. 7B, the θ pinch, the current flows around the plasma column, a direction often denoted by θ. Between these extremes is the screw pinch (Fig. 7C). Here, the current flows in helices around C. The pitch $q_J = J_z/J_\theta$ of the helices depends on r, the radius of the cylindrical magnetic surface on which they lie. The field lines are also helices with a different pitch, q. The way that q varies with r is a significant factor for the stability of the equilibrium, and dq/dr is another figure of merit, called magnetic shear (see Section 5.3). It quantifies how rapidly the direction of the magnetic field changes as a function of r.

As applied to plasma devices, an obvious shortcoming of the linear or one-dimensional pinches shown in Fig. 7 is that, in practice, they must have ends, and these are sources of contamination. Greater interest therefore centers on toroidal or two-dimensional pinches. One may visualize a toroidal MSE as a linear pinch whose ends are joined together by bending it around on itself to form a donut of radius R_0, similar to that sketched in Fig. 2; R_0 is called the major radius and a the minor radius of the torus. The axis of symmetry is called the magnetic axis and distance from it is denoted by R. The z coordinate of the linear pinch is replaced by the angle ϕ around the magnetic axis, and increasing ϕ is called the toroidal direction. The coordinates r and θ of the linear pinch are approximately the same for both the linear and toroidal systems, r now being the distance from the axis running through the center of the donut and θ being the angle around that axis. Increasing θ is now called the poloidal direction. There is no toroidal MSE analogous to the θ pinch; in the absence of current flow in the toroidal direction, the Lorentz force cannot be balanced by a pressure gradient. Main interest centers on the toroidal screw pinch. The quantity $q = rB_\phi/R_0 B_\theta$ is another figure of merit called the kink safety factor (see Section 5.3.2).

In addition to the one- and two-dimensional MSEs, there are also interesting three-dimensional configurations, so-called because they have no symmetry. Of fundamental concern in all these MSEs is the stability of the equilibria. This topic is discussed in Section 5.

Alfvén's theorem leads to a new phenomenon, the Alfvén wave. Consider again the straight flux tube of cross-sectional area A. Suppose first that the fluid is incompressible, and imagine that a transverse displacement bends the tube, carrying its contents with it, in obedience to the theorem (Fig. 8). The tension $\tau = AB^2/\mu_0$ of the tube acts, as in a stretched string, to shorten the tube. A wave results that moves in each direction along the tube with speed $V_A = \sqrt{(\tau/\Sigma)}$, where $\Sigma = \rho A$ is the mass per unit length of the string, with ρ being the fluid density. The wave velocity is therefore $\mathbf{V}_A = \mathbf{B}/(\mu_0\rho)^{1/2}$, which is usually called the Alfvén velocity. It is also the velocity with which energy can be transmitted along the field lines. In a compressible fluid, the situation is more complex because the presence of the field and its associated Lorentz force make sound propagation anisotropic. For example, the speed s of sound waves traveling parallel or antiparallel to the field $\mathbf{B}$ is unaffected by it: $s = \sqrt{\gamma p/\rho}$, where γ is the ratio of specific heats. For sound traveling perpendicular to the field, $s = \sqrt{(\gamma p + B^2/\mu_0)/\rho}$. For small β this is approximately V_A.

The MHD or Alfvénic timescale, $\tau_A = \mathscr{L}/V_A$, is very significant in many MHD contexts. It provides an estimate of how quickly a system responds to changes in its state. Such changes generate acoustic and Alfvén waves, the former crossing the system in a time of order $\tau_s = \mathscr{L}/s$, which for $\beta \ll 1$ is indistinguishable from τ_A. Thus, τ_A is the time required for the initial change to permeate the system. In particular, it is the dynamic timescale on which an MSE responds to a perturbation, and it is usually the timescale on which that equilibrium will, if unstable, be disrupted. It should be noted, however, that even if an equilibrium is stable according to ideal MHD, it

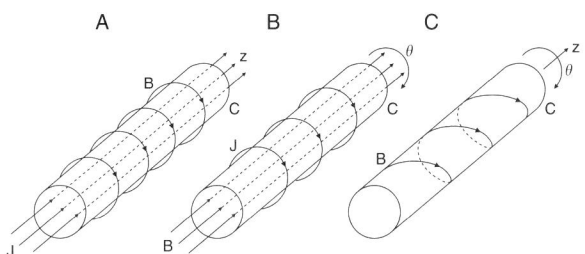

FIGURE 7 (A) Z pinch; (B) θ pinch; (C) screw pinch.

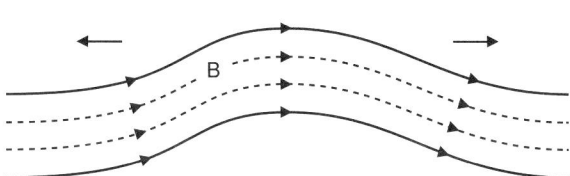

FIGURE 8 Alfvén wave propagation down a straight flux tube. The arrows indicate the directions of wave propagation.

4. NONIDEAL MAGNETOHYDRODYNAMICS

4.1 Quantifying Deviations from the Frozen in Law

Alfvén's theorem poses a paradox: If field and fluid are inseparable, how did the fluid acquire its field in the first place? The answer is by diffusion, by which is meant, by diffusion of magnetic field. This happens because every real fluid has a finite conductivity.

In the reference frame $\mathscr{F}'$ moving with a particular point P of the fluid, Ohm's law is $\mathscr{J} = \sigma \mathscr{E}'$, in order of magnitude. Ampère's law shows that $\mathscr{J} = \mathscr{B}/\mu_0 \mathscr{L}$, whereas Faraday's law implies that $\mathscr{E}' = \mathscr{B}\mathscr{L}/\mathscr{T}$. The timescale $\mathscr{T}$ on which **B** at P can change as P moves around is therefore the magnetic diffusion time, defined as $\tau_\eta = \mu_0 \sigma \mathscr{L}^2 = \mathscr{L}^2/\eta$, where $\eta = 1/\mu_0 \sigma$ is the magnetic diffusivity. On timescales $\mathscr{T}$ short compared with τ_η, the field is frozen to the fluid; for $\mathscr{T} \gg \tau_\eta$ it is not. Unless a field of scale $\mathscr{L}$ at P is maintained in some way, its lifetime is τ_η; also, a field of this scale cannot be implanted at P in a lesser time. One may consider these phenomenon as a type of EM memory that sets a minimum timescale for change, in the frame moving with the conductor.

Consider again the Z pinch described previously. After the potential difference has been applied between the ends of C at time $t = 0$, the current it creates initially flows only over the surface of C because, over short times, the plasma behaves as an ideal conductor in which **B** in the bulk of C cannot change from its initial zero value. Thereafter, the current sheet penetrates into C by diffusion, its thickness being of order $\delta_\eta = \sqrt{(\eta t)}$ at time t. After a magnetic diffusion time of $\tau_\eta = a^2/\eta$, the current will be distributed almost uniformly across the cross section of the column. As a digression, in practice the process may take rather longer since the diffusion of field into C is hampered by the increasing conductivity of the plasma as its temperature T rises through its adiabatic compression or heating. Similarly, the z-directed field initially trapped in a screw pinch grows as the column C collapses, whereas B_z in the surrounding vacuum does not. The tendency for the flux in C to diffuse out of the column is hampered by the diminishment of η that accompanies the compressional increase in T.

As another example, suppose that a plasma column C, such as the screw pinch shown in Fig. 7C, is contained within, but is not in contact with, a surrounding metallic wall. The field created outside C after the pinch has been initiated at time $t = 0$ cannot at first penetrate the walls because, over short times, they behave as perfect conductors in which **B** cannot change. The field is excluded from the walls by currents that flow in a thin skin on their surfaces, and these create their own field, which (when the walls are planar) is the mirror image of that of the plasma column (Fig. 9). The plasma column is repelled by its image (another example of Lenz's law), thus, preventing it from striking the wall. This stabilizing effect weakens in time because the thickness $\delta_\eta = \sqrt{(\eta t)}$ of the skin increases with t until ultimately the field completely penetrates the walls.

In a moving fluid, the advective timescale, $\tau_v = \mathscr{L}/\mathscr{V}$, quantifies the time taken (according to Alfvén's theorem) for a field of scale $\mathscr{L}$ to be carried over that distance. However, electrical resistance diffuses the field on a timescale of τ_η. If the magnetic Reynolds number, $Rm = \tau_\eta/\tau_v = \mathscr{L}\mathscr{V}/\eta$, is large, advection is more rapid than diffusion and Alfvén's theorem is useful in visualizing MHD processes. When $Rm \ll 1$, as is usually the case in technological applications of MHD involving liquid metals, the concept of frozen fields is not very useful. The prevailing field then provides an anisotropic friction that attempts to damp out motions perpendicular to itself on a timescale of $\tau_d = \rho/\sigma V_A^2 = \eta/V_A^2$, often called the magnetic damping time or the Joule damping time.

Energy can also be transmitted in a fluid by waves, and particularly significant for MHD systems is

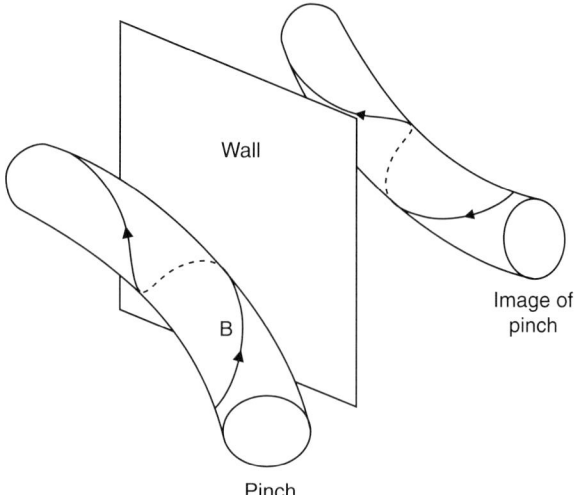

FIGURE 9 Image of a screw pinch in a conducting wall.

Alfvén radiation. This introduces a second way of quantifying diffusive effects: the Lundquist number, Lu. Alfvén waves are damped both by electrical conduction and by the viscosity v of the fluid. The timescale over which this occurs is a combination $\tau_c = \mathscr{L}^2/(\eta + v)$ of τ_η and the viscous diffusion time, $\tau_v = \mathscr{L}^2/v$. For a liquid metal ($v \ll \eta$), $\tau_c \approx \tau_\eta$, and we shall now use τ_η in place of τ_c. During a time τ_η, the Alfvén waves travel a distance, $V_A \tau_\eta$. If the Lundquist number, $Lu = V_A \tau_\eta / \mathscr{L} = V_A \mathscr{L}/\eta$, is large, this distance greatly exceeds the characteristic length scale $\mathscr{L}$, and Alfvén radiation is significant. If $Lu \ll 1$, this is not the case since the time $\tau_d (= \eta/V_A^2)$ taken for the wave to damp out is small compared with the time $\tau_A (= \mathscr{L}/V_A)$ that the Alfvén wave requires to traverse the distance $\mathscr{L}$. The Lundquist number can be thought of as a magnetic Reynolds number, with the wave speed V_A replacing the advection speed $\mathscr{V}$.

4.2 Magnetic Reconnection

According to Alfvén's theorem, the topology of field lines cannot change in ideal MHD, but in a finitely conducting fluid ($\eta > 0$), field lines can sever and reconnect. This can happen fastest where scales are smallest (i.e., where the gradients of **B** are greatest). Magnetic reconnection is the process that forms thin current layers that change field topology and dissipate magnetic energy, even though the resistivity is modest and smoothly varying. Indeed, it is very important to distinguish magnetic reconnection, a process in which very thin (nearly singular) current sheets form in otherwise smoothly varying systems, from simple diffusive dissipation of magnetic energy when sharp profiles are built in *ab initio*. In magnetic reconnection, resistive diffusion and the magneto-fluid dynamics combine to release energy stored in the magnetic fields and to reconfigure the magnetic topology of the system. Reconnection phenomena are usually subdivided into two classes, namely driven magnetic reconnection (discussed here) and spontaneous magnetic reconnection (described in Section 5 in the context of the tearing instability).

Perhaps the simplest and most fundamental model of magnetic reconnection is that of Sweet and Parker, hereafter referred to as the Sweet–Parker reconnection (SPR) model. This is a two-dimensional machine that steadily reconnects field lines. A cross section is sketched at successive times in Fig. 10. Resistivity acts everywhere but is most effective near a segment of the x axis. To simplify the argument, it is supposed that $\eta = 0$ everywhere except in reconnection region (shown

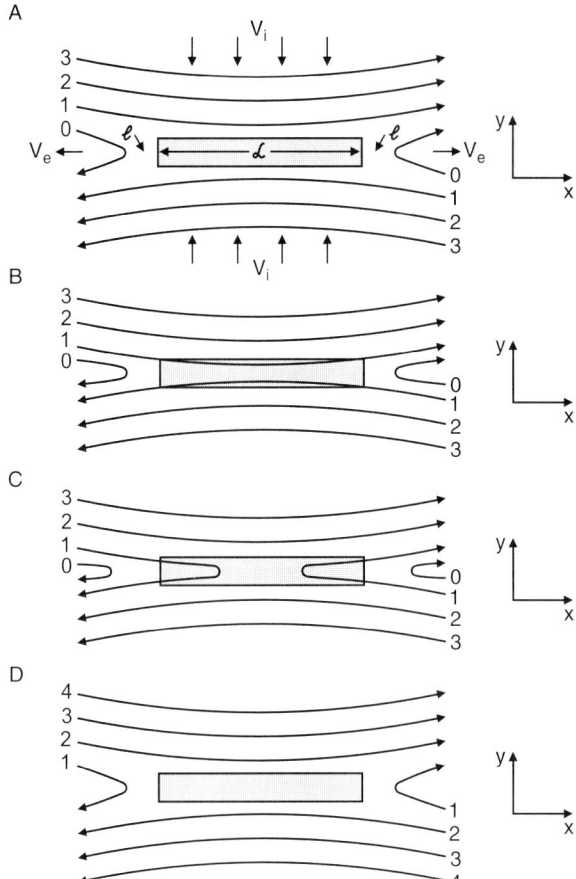

FIGURE 10 The Sweet–Parker reconnection mechanism. (A) Reconnection occurs when fluid motion brings slabs of oppositely directed magnetic fields together. The slabs have length $\mathscr{L}$ and velocity V_i. Pushing the slabs of field together creates a current sheet of thickness ℓ and ejects fluid at velocity V_e. The field lines labeled 1–3 approach the current sheet and field line 0 leaves it. (B) Field lines 1 enter the current sheet. (C) Field lines 1 are reconnected in the current sheet. (D) Field lines 1 leave the current sheet and move away in the $\pm x$ directions with velocity V_e. The configuration of field lines is now identical to that in A, but field lines n take the place of field lines n–1. Reconnection has changed the field line topology and dissipated magnetic energy.

shaded) of x width $\mathscr{L}$ and y width ℓ; it is assumed that ℓ is so much smaller than $\mathscr{L}$ that it is almost a current sheet. A steady inflow, V_i, in the $\pm y$ directions forces together two slabs of oppositely directed field $\pm B$ (Fig. 10A). The fluid is incompressible so that fluid must be ejected, with velocity V_e in the $\pm x$ directions. Figure 10B shows the field lines, labeled 1, moving into the reconnection region, and Fig. 10C shows them moving out after they have been reconnected. Their curvature implies a tension that assists the ejection of fluid in the $\pm x$ directions.

The object now is to determine V_i, V_e, and ℓ, assuming that the configuration is in a steady state and that the Lundquist number $Lu = V_A \mathcal{L}/\eta$ is large. The argument rests on three simple balance laws:

1. Balance of fluid mass flowing into the reconnection layer with that flowing out: This gives $V\mathcal{L} = V_e \ell$.
2. Balance of fluid momentum in the reconnection layer. Since the flow in the reconnection layer is approximately one dimensional, it is governed by Bernoulli's law, which states that the sum of the total pressure P and the dynamic pressure $\frac{1}{2}\rho V^2$ is constant. Furthermore, the total pressure P is, in turn, the sum of the thermal pressure p and the magnetic pressure $p_M = \mathbf{B}^2/2\mu_0$. The sum, $S \equiv p + \mathbf{B}^2/2\mu_0 + \frac{1}{2}\rho V^2$, is therefore constant throughout the layer. By symmetry, the center of the configuration is a stagnation point ($\mathbf{V} = 0$), and at the midpoints on the long sides of the reconnection layer $\mathbf{V}$ is negligibly small so that $S \approx p + B^2/2\mu_0$. Outside the layer (i.e., for $|x| \gg \mathcal{L}$), the magnetic field is small so that at the ends of the layer where fluid is ejected in the $\pm x$ directions, $S \approx p + \frac{1}{2}\rho V_e^2$. Equating these two expressions for S shows that the ejection velocity is simply the Alfvén velocity: $V_e = V_A$.
3. Balance of magnetic energy: In the steady state, the influx of magnetic energy to the layer must equal the rate of ohmic dissipation in the layer. The influx rate is simply $(2V\mathcal{L})B^2/2\mu_0$ since the efflux from the narrow side edges of the reconnection region is negligible. The rate of ohmic dissipation is $(\mathcal{L}\ell)J^2/\sigma$, where by Ampère's law, $J = B/\mu_0 \ell$. Since $\eta = 1/\mu_0 \sigma$, the inflow velocity is related to the layer thickness by $V = \eta/\ell$.

Taken together, these three demands imply that

1. Since $Lu \gg 1$, the reconnection velocity, $V = V_A/\sqrt{Lu}$, is much smaller than V_A. It is proportional to a fractional power of the resistivity of the system; the growth rate of the tearing mode (discussed in Section 5) shows a similar fractional dependence.
2. The aspect ratio, $\ell/\mathcal{L}$, of the reconnection region is small, of order $Lu^{-1/2} \ll 1$. The direction of the merging fields imposes this anisotropy. It is consistent with the idea that magnetic energy is dissipated in a thin current sheet. Of course, the simultaneous smallness of V/V_A and the aspect ratio $\ell/\mathcal{L}$ is a consequence of their equality, by the balance law, and of the assumption $Lu \gg 1$.

Although conceptually simple and plausible, SPR tells a rather pessimistic story, namely that reconnection is a relatively slow process (i.e., $V = V_A/\sqrt{Lu} \ll V_A$, where $Lu \gg 1$ in cases of greatest interest). Many physical phenomena appear to require, for their explanation, fast reconnection (i.e., reconnection on ideal timescales or, equivalently, at Alfvénic velocities $V \sim V_A$). This is especially true of many solar and astrophysical phenomena, such as solar flares. The physics of fast reconnection is an active area of research in MHD and plasma physics.

Two main routes to fast reconnection have been proposed. Within the framework of MHD, one originally proposed by Petschek suggests that shock waves carry energy away from the reconnection layer. This rapid process of energy removal is thought to increase both the aspect ratio of the reconnection region and the reconnection rate. The Petschek mechanism predicts a reconnection velocity of order V_A, with only a weak logarithmic dependence on Lu [i.e., $V \sim V_A/\ln(Lu)$]. The alternate route to fast reconnection appeals to small-scale, non ideal, or kinetic processes that cannot be studied using the one-fluid model of MHD. All models of fast reconnection are controversial and they will evolve considerably in the years ahead.

4.3 MHD Turbulence

Many fluid systems are turbulent. This fact presents formidable obstacles to theorists and is likely to embarrass them for the foreseeable future. The motions and fields of a turbulent fluid have many length and timescales. In the classical picture of turbulence, these range from macroscales, $\mathcal{L}_{macro}$ and $\mathcal{T}_{macro}$, corresponding to scales on which the fluid is stirred, to dissipation scales, $\mathcal{L}_{diss}$ and $\mathcal{T}_{diss}$, which are dominated by viscosity. It is argued that $\mathcal{L}_{macro}/\mathcal{L}_{diss} \sim (Re)^{3/4}$, where $Re = \mathcal{V}_{macro} \mathcal{L}_{macro}/\nu$ is the Reynolds number. This is typically greater than 10^4, so the range of scales spans many decades and cannot be adequately resolved even by modern, high-speed computers.

In classic turbulence, interscale interaction occurs via a process of cascade in which larger eddies (i.e., fluctuations that have the appearance of vortical whorls) fragment into smaller eddies, thus producing a range of fluctuations on scales from $\mathcal{L}_{macro}$ to $\mathcal{L}_{diss}$. This process is encapsulated by a well-known parody by Louis Fry Richardson of a verse of Jonathan Swift:

Big whorls have little whorls, that feed on their velocity.
And little whorls have lesser whorls, and so on to viscosity.

In contrast to ordinary fluid turbulence, which can be thought of as a soup of eddies, MHD turbulence consists of a mixture of eddies and small-scale Alfvén waves and is controlled by viscosity and resistivity. Thus, although the idea of a cascade remains useful for describing MHD turbulence, the interaction processes in MHD are considerably more complex than in nonmagnetic fluids. Perhaps the most important difference is that, since MHD turbulence involves Alfvén waves, it necessarily retains an element of memory or reversibility. Here, Alfvénic memory refers to the tendency of even small-scale magnetic fields (which are always present when Rm is large) to convert the energy in eddies (which cascades irreversibly) to reversible nondiffusive Alfvén wave motion. This is a consequence of the fact that wave motion is necessarily periodic and involves a restoring force.

Despite the difficulties discussed previously, there is considerable effort in the research community directed at numerical simulation of MHD turbulence. The large scales can be resolved numerically in so-called large eddy simulations (LES), but only if the effects of the unresolvable or subgrid scales (SGS) on the large scales are parameterized in some way. One popular, although controversial, expedient originated from an idea by Osborne Reynolds in the 19th century. He drew an analogy between the diffusive effects of the small-scale eddies and collisional diffusion processes at the molecular level. This suggested that the SGS similarly spread out the large-scale motions and fields diffusively, but on timescales $\tau_{v/\text{turb}} = \mathcal{L}^2/v_{\text{turb}}$ and $\tau_{\eta/\text{turb}} = \mathcal{L}^2/\eta_{\text{turb}}$ determined by turbulent diffusivities, v_{turb} and η_{turb}, of order $\mathcal{L}_{\text{eddy}} \mathcal{V}_{\text{eddy}}$ that might be very large compared with the corresponding molecular values, v_{mol} and η_{mol}.

Reynolds's idea is not totally satisfactory but is qualitatively useful. For example, the fact that $\eta_{\text{turb}} \gg \eta_{\text{mol}}$ in the solar convection zone explains, in a rough and ready way, why the timescale $\mathcal{T}$ of magnetic activity on the sun, as estimated by the period of the solar cycle, is on the order of a decade, even though $\tau_{\eta/\text{mol}}$ exceeds the age of the sun: $\mathcal{T}$ is determined by η_{turb} and not η_{mol} so that $\mathcal{T} \approx \tau_{\eta/\text{turb}}$.

Reynolds's idea also correctly implies that turbulence enhances viscous and ohmic dissipation. Although the magnetic Reynolds number Rm may be large when defined from $\mathcal{L}_{\text{macro}}$ and $\mathcal{V}_{\text{macro}}$, it may be small when $\mathcal{L}_{\text{eddy}}$ and $\mathcal{V}_{\text{eddy}}$ are used instead. Even if the characteristic macroscale field, $\mathcal{B}_{\text{macro}}$, is large compared with the typical eddy field, $\mathcal{B}_{\text{eddy}}$, so that the magnetic energy resides mostly in the macroscales, the magnetic energy may be dissipated mainly by the eddies. This is because $\mathcal{L}_{\text{eddy}} \ll \mathcal{L}_{\text{macro}}$ so that $\mathcal{J}_{\text{macro}} = O(\mathcal{B}_{\text{macro}}/\mu_0 \mathcal{L}_{\text{macro}})$ may be smaller than $\mathcal{J}_{\text{eddy}} = O(\mathcal{B}_{\text{eddy}}/\mu_0 \mathcal{L}_{\text{eddy}})$, even though $\mathcal{B}_{\text{eddy}} \ll \mathcal{B}_{\text{macro}}$. In many situations however, $\mathcal{B}_{\text{eddy}} \gg \mathcal{B}_{\text{macro}}$ so that the eddies contain most of the magnetic energy and are mainly responsible for its dissipation. Similarly, they may contain the majority of the kinetic energy and be the main cause of its viscous dissipation. The reader is cautioned, however, that the expressions for v_{turb} and η_{turb} given here are estimates only. In particular, it has been appreciated that the Alfvénic memory intrinsic to MHD turbulence may cause significant reductions in v_{turb} and η_{turb} in comparison with the estimates given here. The self-consistent calculations of v_{turb} and η_{turb} are active topics of current research in MHD.

MHD propulsion and drag reduction are topics of considerable interest and usually concern systems for which $Rm \gg 1$ and in which MHD turbulence occurs. An extensive discussion of these topics is beyond the scope of this article.

4.4 Buoyancy in MHD

A real (nonideal) fluid has a nonzero thermal conductivity, K. The appropriate measure of the resulting conduction of heat is the thermal diffusivity, $\kappa = K/\rho C_p$, where C_p is the specific heat at constant pressure; the thermal diffusion time is $\tau_\kappa = \mathcal{L}^2/\kappa$. The finiteness of κ gives rise to the magnetic buoyancy of flux tubes. Such tubes are created by, for example, the turbulent motions in a stellar convection zone, such as that of the sun. These cause the prevailing magnetic field to become intermittent, with regions of strong field (i.e., flux ropes) being surrounded by regions of comparatively weak field. During a time $t \ll \tau_\eta$ after its formation, flux loss from a tube can be ignored. Radiative transport of heat, however, is very effective in a star and $\tau_\kappa \ll \tau_\eta$. Any difference in temperature between the rope and its surroundings diffuses away in a time of order τ_κ. It is reasonable to suppose that in the time interval $\tau_\kappa \ll t \ll \tau_\eta$, the temperature of the rope is the same as its surroundings but that the field B in the interior of the rope differs from the field $B_0(<B)$ outside it. Magnetostatic balance requires that the total pressure exerted by the rope on its surroundings is approximately the same as the total pressure the surroundings exert on it: $p + B^2/2\mu_0 = p_0 + B_0^2/2\mu_0$. This means that the kinetic pressure p of the gas in the tube is less than that of its surroundings, p_0. However, since its temperature is the same, its density must be less. The tube is therefore buoyant and rises toward the stellar surface.

When the tube breaks through the stellar surface, dark regions occur that are called starspots (or sunspots in the case of the sun). That these spots are dark is another illustration of the constraining effects of the field. Near the stellar surface, heat is carried outward mainly by convective overturning, which within a tube is coupled to Alfvén waves that radiate energy and therefore tend to suppress the convective motions. The emerging heat therefore has to find a route that goes around the spots rather than through them. The spots are thus cooler and darker than their surroundings.

4.5 Amplification of Magnetic Fields in MHD

A magnetic field threading a sphere of solid iron of radius 1 m would disappear in approximately 1 s unless maintained by external current-carrying coils. Since $\tau_\eta \propto \mathscr{L}^2$, larger bodies can retain their fields longer, but many celestial bodies (planets, stars, galaxies, etc.) are magnetic and are believed to have been magnetic for times long compared with their magnetic diffusion times. The age of the geomagnetic field, for instance, exceeds 3×10^9 years, even though τ_η is less than 10^5 years. A process must exist that replenishes these fields as they diffuse from the conductor. This process is the same as that operating in commercial power stations: self-excited dynamo action. From a field $\mathbf{B}$, fluid motion creates the electromotive force (emf) $\mathbf{V} \times \mathbf{B}$ that generates the electric currents $\mathbf{J}$ necessary to create, by Ampère's law, $\mathbf{B}$.

Although it was stated previously that in turbulent flow eddies may be responsible for greatly enhancing diffusion of the large-scale part, $\mathbf{B}_{macro}$, of $\mathbf{B}$, these small-scale motions can, if they lack mirror symmetry, also act to replace the lost flux. The simplest manifestations of such symmetry breaking are helical motions. Helicity is defined as $\mathbf{V} \cdot \boldsymbol{\omega}$ where $\boldsymbol{\omega}$ is the vorticity of the flow, a vector that has Cartesian components $\partial V_z/\partial y - \partial V_y/\partial z$, $\partial V_x/\partial z - \partial V_z/\partial x$, $\partial V_y/\partial x - \partial V_x/\partial y$ and that is often written as curl$\mathbf{V}$ or $\nabla \times \mathbf{V}$. Helicity is a measure of how mirror symmetric the flow is; like a common carpenter's screw, a helical flow does not look the same when viewed in a mirror. Helicity is created naturally by the rotation of large systems such as the solar convection zone, where the rising and falling motions created by buoyancy acquire vertical vorticity through the action of the Coriolis force. Small-scale helical motions create a large-scale emf that, in the simplest case, is proportional to $\mathbf{B}_{macro}$ and is conventionally written $\alpha \mathbf{B}_{macro}$. This α effect may suffice to maintain $\mathbf{B}_{macro}$ and hence $\mathbf{B}_{turb}$ as well, thus creating a turbulent dynamo. The α effect is the cornerstone of a subject now called mean field electrodynamics. (Confusingly, this is often known by the acronym MFE, which we reserved as an abbreviation for magnetic fusion energy.)

The α effect illustrates the danger, mentioned previously, of adopting the Reynolds ansatz too uncritically; through their anisotropy and lack of mirror symmetry, small-scale motions can create large-scale emfs that are not recognized when η_{mol} is merely replaced by η_{turb}. It should also be mentioned that the issue of Alfvénic memory appears once again in the context of the dynamo and the α effect. Recent research has suggested that because of memory effects and constraints on the rate of magnetic reconnection, the α effect at large Rm may be much weaker than it would be were the dynamical effects of the small-scale eddies ignored, as was done in the early days of mean field theory. Further developments in this important research area may be expected in the near future.

The interested reader can get an intuitive, hands-on perspective on a particular type of dynamo process by the following home demonstration, which also illustrates several of the processes described in this section and the last. Obtain an elastic (rubber) band. This loosely corresponds to the flux tube in Fig. 11A, its tension being the sum of the tensions in the magnetic field lines it contains (i.e., it is proportional to the energy of the field within it).

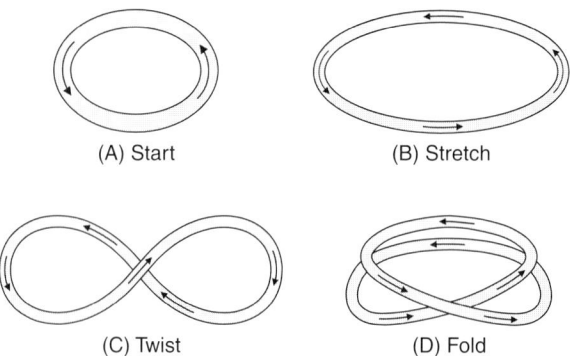

FIGURE 11 The stretch–twist–fold dynamo. (A) The initial undeformed flux tube; (B) the result of stretching it to twice its length; (C) the stretched tube is twisted into the shape of an infinity sign ∞; and (D) the result of folding one loop of the ∞ over onto the other. Reconnection in D can produce two loops of the same size as the initial tube.

The tension in the band may be systematically increased by a three-step, stretch–twist–fold process:

1. Stretch the rubber band to double its length.
2. Twist it into the shape of an infinity sign, ∞.
3. Fold the loops on top of one another to form two linked loops of the same size as the original band.

These steps are illustrated in Fig. 11B–D. In a similar way to the rubber band, the energy within and the tension of the flux tube are increased by a factor of 4. This is the essence of a dynamo, the creation of magnetic energy by helical motion.

In principle, these steps can be repeated over and over again, with the sense of twist in step 2 being always the same. This gives the band the sense of handedness of a screw motion and the broken reflection symmetry and helical motion that are crucial to the success of the dynamo. The tension in the band (analogous to the magnetic tension) increases progressively as the steps are repeated, and at some stage the reader's hands will tire, making it impossible to increase the tension further. This illustrates the end of the kinematic regime (in which fluid motion is prescribed *a priori* and the field grows) and the beginning of the dynamic regime (in which the field quenches its own further growth). If the reader lets go of the band at any stage, it will immediately relax back to its initial state. This illustrates the crucial importance of magnetic reconnection as a means of locking in the amplified magnetic field. In other words, magnetic reconnection provides the crucial element of irreversibility in the dynamo. The analogue of magnetic reconnection in the case of the rubber band could be achieved by fusing the two loops together after the stretch–twist–fold process.

In recent years, much effort has been expended on the study of fast dynamos, which are defined as processes of field amplification that operates on timescales independent of τ_η in systems with $Rm \gg 1$. It should be apparent to the reader that a fast dynamo necessarily requires fast reconnection in order to lock in the dynamo-generated field. Thus, the problems of fast reconnection and those of the fast dynamo are intimately linked.

4.6 Magnetic Helicity and Relaxation in MHD

No doubt the reader has noticed that the past three sections, which discuss magnetic reconnection, turbulent transport of magnetic fields, and the dynamo process, all concern the following broader question: Given certain initial conditions, what magnetic configuration does the magnetofluid ultimately adopt? This question seeks to identify the final state when some very complex dynamical processes, involving the dissipative phenomena mentioned previously (as well others, some unknown), are at work.

Apart from a very few simple, exactly soluble cases (that nearly always have an unrealistically high degree of symmetry built in), there are two main ways of finding the final state.

The first is to integrate the MHD equations directly using a computer. This approach forces the researcher to confront all the thorny issues of MHD turbulence, SGS modeling, etc. described previously. Moreover, such brute force tactics are often expensive and inefficient. One is naturally motivated to seek other approaches that are bolder but simpler. Foremost among these is constrained relaxation. This variational method seeks to identify the final state as one that minimizes the magnetic energy, subject to certain constraints. The central issue then becomes the identification and inclusion of the most important constraint or constraints.

A natural candidate for consideration as a constraint is the magnetic helicity. This is the analogue of the (kinetic) helicity, which was previously defined as $\mathbf{V} \cdot \boldsymbol{\omega}$, where $\boldsymbol{\omega} = \text{curl}\mathbf{V}$. Conversely, $\mathbf{V}$ can be derived from $\boldsymbol{\omega}$ by integration in an uncurling operation. Likewise, a vector $\mathbf{A}$ can be obtained from $\mathbf{B}$ by a similar uncurling operation. This defines a vector potential, $\mathbf{A}$. The scalar product $K = \mathbf{A} \cdot \mathbf{B}$ is the magnetic helicity. It is conserved in the motion of an ideal fluid, and this is why it is significant in MHD. Physically, magnetic helicity quantifies the self-linkage or knottedness of the magnetic field lines and is thus a measure of the topological complexity of the field. Indeed, the invariance of field line topology in ideal MHD is the origin of the conservation of magnetic helicity.

If an MSE is analyzed using ideal MHD, it is found that magnetic helicity is conserved in detail in the sense that if we regard the MSE as an assembly of infinitesimally thin flux tubes of volume Δv and labeled by parameters α and β, their individual magnetic helicities, $K_{\alpha,\beta} = (\mathbf{A} \cdot \mathbf{B})_{\alpha,\beta} \Delta v$, are conserved, although they are not the same for each tube. If instead nonideal MHD is used in analyzing the MSE, the configuration will relax toward its final state on a timescale determined by the resistivity and other slow diffusion mechanisms. In this process, individual flux tubes lose their identity through magnetic diffusion and turbulent cascade. Thus, the local magnetic helicity, $K_{\alpha,\beta}$, associated with a given thin flux tube, is *not* conserved. However, J. B. Taylor made the insightful conjecture that the global magnetic helicity (i.e., the sum of $K_{\alpha,\beta}$ over the system) would be approximately conserved on the timescale of the relaxation.

Global helicity is the integral of K over the entire system. It is, in some sense, a coarse-grained topological invariant that is also related to the self-inductance of the plasma. Taylor relaxation theory proposes to determine relaxed states by minimizing magnetic energy subject to the constraint of constant global magnetic helicity. The resulting Taylor states have the property that the current is linearly proportional to the field (i.e., $\mathbf{J} = \lambda \mathbf{B}$, where λ is a constant). This, is a force-free field of the special type described in Section 3 for which the component, $J_\parallel$, of $\mathbf{J}$ parallel to $\mathbf{B}$ is everywhere the same. In Section 5.3.2, we discuss kink-tearing instabilities of toroidal MSE. These instabilities are current driven (i.e., they exist only because $\nabla J_\parallel$ is nonzero). Since $\nabla J_\parallel = 0$ in the Taylor relaxed state, it is stable to kink-tearing modes.

Taylor relaxation theory has been very successful in explaining the relaxed mean field profiles of low β confined plasmas, especially the reversed field pinch (RFP) and spheromak. Of particular note is Taylor's successful prediction of the F–Θ curve for the RFP. The F–Θ curve relates the degree of field reversal to the poloidal field at the edge of the RFP and is an experimentally derived signature of the final RFP state. In relation to Section 4.4, it is worth noting that the relaxed state is maintained by a cyclic dynamo process in which dissipative excursions from the Taylor state trigger instabilities that drive the plasma back to relaxed state.

The main mystery of Taylor relaxation theory is why it works so well. In other words, why is the global helicity the most important constraint on turbulent relaxation? There are at least three plausible answers:

1. *Enhanced dissipation:* Turbulence and dissipation drive magnetic reconnection, which destroys domains of local magnetic helicity on all but the largest scale. Thus, as the system evolves, the global magnetic helicity is the only surviving topological invariant.

2. *Field line stochasticity:* In the initial state of the configuration, the field lines lie on magnetic surfaces, each labeled by its hydrostatic pressure p. However, as the configuration relaxes to the force-free state of constant p, the identity of the surfaces evanesces, and field lines become increasingly stochastic. Ultimately, as the turbulent relaxation proceeds, the entire MSE is threaded by a single field line. Since magnetic helicity is calculated by volume integration over a region enclosed by a magnetic surface, and all other magnetic surfaces are destroyed, only global magnetic helicity is relevant to a stochastic state.

3. *Selective decay:* In three-dimensional MHD turbulence, energy cascades to small scales. It can be shown, however, that magnetic helicity cascades inversely to large scales. As a result, magnetic helicity accumulates on the largest scales of the system, with minimal coupling to dissipation. Thus, magnetic helicity is dynamically rugged, whereas energy decays.

It should be noted that these are only plausibility arguments, and that a rigorous justification of the Taylor hypothesis remains an elusive goal of current research.

5. STABILITY OF MAGNETICALLY CONFINED PLASMAS

5.1 Generic Issues

Configurations of MHD fluids are analyzed using the concepts of equilibrium and stability. These are used to classify a variety of possible configurations according to their intrinsic interest, potential utility, importance for science and technology, and so forth. They are vital in the context of magnetic plasma confinement for the success of controlled fusion (hereafter referred to as MFE). This is the only application considered here, but there are many others in astrophysics, geophysics, space plasma dynamics, and MHD power generation.

MHD equilibria are configurations in which the plasma pressure gradient, ∇p, balances the $\mathbf{J} \times \mathbf{B}$ force (Lorentz force) and any other body forces acting on the magnetofluid. An equilibrium in which the balance is dominated by the pressure gradient and the Lorentz force is referred to as a *MSE*. All equilibria of interest in the context of MFE are MSEs. In general, the magnetic fields in MSEs are produced in part by external coils and in part by plasma currents. They are characterized by the pressure and current profiles $p(r, \theta, \phi)$ and $\mathbf{J}(r, \theta, \phi)$. In toroidal MSEs, these are independent of ϕ, the angle around the torus in the toroidal direction, but still depend on θ, the angle that increases in the so-called poloidal direction.

Interesting MSEs almost always have closed magnetic flux surfaces (i.e., surfaces that enclose a specified amount of magnetic flux). These are labeled by an effective radius r, and because they are so significant, we generally suppress the dependence of p and $\mathbf{J}$ on θ and ϕ. Charged particles in a plasma cannot readily cross a flux surface and are therefore confined. For this reason, an MSE is sometimes called a magnetic bottle.

MSEs have a high degree of symmetry, corresponding to closed toroidal, spheroidal, or helical configurations. Toroidal MSEs of interest in the context of MFE include the tokamak (based on a Russian acronym for maximum current) and the RFP. Together, these two configurations are referred to as toroidal pinches since plasma is confined by a toroidal current that produces a magnetic field, which in turn acts to pinch the plasma column. A tokamak has a strong, externally applied toroidal magnetic field, whereas a RFP plasma generates its own toroidal field by a process of constrained relaxation, whereby some of the toroidal current is dynamically twisted into the poloidal direction. This relaxation process is closely related to that of the dynamo described previously. RFPs are intrinsically dynamic (i.e., not quiescent) in nature. Spheroidal MSEs include the spheromak and the spherical torus, whereas helical MSEs include the heliac, a type of stellarator. Stellarator MSEs do not rely on inductively driven currents as do toroidal pinches, but they sacrifice the virtues of toroidal symmetry. In this section, the discussion is limited for reasons of brevity and clarity to toroidal pinches in general, and tokamaks in particular. The MSE of a tokamak is particularly simple in that the pressure and toroidal current profiles, $p = p(R)$ and $J_\phi = J_\phi(R)$, can be parameterized by a single coordinate, R, that may be thought of as a (generalized) distance from the symmetry axis.

Once the existence of a closed MSE is established, the next question is whether it is magnetohydrodynamically stable—that is, whether initial, small perturbations that break the symmetry of the equilibrium tend to reinforce themselves and grow. This growth is due to the release of potential energy stored in the initial MSE and is accompanied by a relaxation of the pressure and/or current profiles $p(r)$ and $J_\phi(r)$, which tends to lower the initial potential energy and degrade the quality of the nascent MSE. Instability is a linear concept and is relevant for only a few growth times (at most), until the dynamics enter the nonlinear regime, characterized by finite (i.e., not small) amplitude perturbations. Instabilities saturate (i.e., cease growing) either when the plasma arrives at a new secondary equilibrium, which is a MSE with lower potential energy than the initial equilibrium, in which the symmetry is usually broken as well, or when the kinetic energy of the plasma motion is lost through dissipation (e.g., through viscosity). The latter often occurs as a consequence of the cascade to small scales characteristic of MHD turbulence.

MHD instabilities are usually classified as either pressure driven [when they are associated with relaxation of $p(r)$] or current driven [when associated with relaxation of $J_\phi(r)$]. They are either ideal or resistive. Ideal instabilities occur without a violation of Alfvén's frozen flux constraint, whereas resistive instabilities require some decoupling of field and fluid, usually by collisional resistivity, to trigger the energy-release process. MHD instabilities are of great interest in the context of MFE, because they limit the class of viable MSEs and thus severely constrain the performance of magnetic fusion devices.

Entire books have been devoted to the MHD stability of confinement devices for MFE. There is space here to describe only a few of the basic concepts and their application to MFE. One way of investigating dynamical stability makes use of the fact that ideal MHD conserves the total energy E of the system. This is the sum of the kinetic energy E_K, the magnetic energy E_M, and the internal energy of the plasma, E_I. Before perturbation, $\mathbf{V} = 0$ and therefore $E_K = 0$ so that $E = E_M + E_I$. The perturbation creates a small initial E_K and this may grow at the expense of $E_M + E_I$, indicating that the initial state is unstable; if the initial state is linearly stable, E_K remains small.

The situation may be described conceptually by analogy with a ball rolling under gravity on a surface S. Then E_K is the kinetic energy of the ball and its height z above some reference level plays the role of $E_M + E_I$. A magnetostatic equilibrium is analogous to the ball being at rest at some point P where the surface is level (i.e., where the tangent plane at P to S is horizontal, as at the points indicated in Fig. 12). In Fig. 12A, the ball rests at a global minimum of z, and the system is not only linearly stable but also globally stable (i.e., stable to all perturbations no matter how large they are or in which horizontal direction they occur). In Fig. 12B, the ball is at an unstable equilibrium that is destroyed if perturbed in any horizontal direction. Figure 12C also illustrates an unstable state; although it is true that some

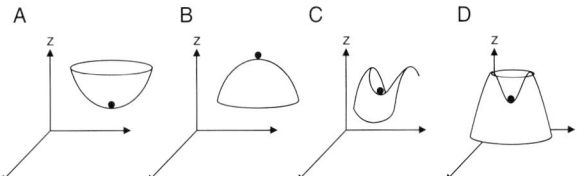

FIGURE 12 Equilibrium points (•) of a ball rolling on a surface. (A) global stability; (B) instability at a maximum of z; (C) instability at a saddle point of z; (D) stability for small perturbations but not large.

perturbations increase z (suggesting stability), other directions of perturbation exist that will cause the ball to descend permanently from the saddle point. In Fig. 12D, a sufficiently large perturbation will take the ball over an adjacent hill, after which it descends permanently to a state remote from its starting point (i.e., the initial state is linearly stable but unstable to finite amplitude perturbations).

Similarly, if $E_M + E_I$ increases for every small perturbation of a MSE, it is linearly stable, as in Figs 12A and 12D. The MSE is unstable if any perturbation exists that reduces $E_M + E_I$, as in Figs 12B and 12C. Figure 12C is the better analogue for a MSE since many directions of perturbation do not threaten a MSE; perturbations that bend field lines increase E_M and those that compress the plasma increase both E_M and E_I. Other directions of perturbation, however, may exist that reduce $E_M + E_I$, and these are the perturbations that should be understood and, as far as possible, eliminated. These directions are usually associated with perturbations that relax the current or pressure gradients.

5.2 Magnetically Confined Plasmas and MFE

The hope of controlling the release of thermonuclear energy by the fusion of light nuclei provides a very powerful incentive for the study of plasma physics and MHD. To make light nuclei react and release energy, they must be forced together against their mutual electrostatic repulsion. This happens naturally at the enormous pressures characteristic of the interiors of stars, such as the sun. The aim of MFE, of course, is to create high plasma pressure MSEs using magnetic fields. The efficiency of the MSE is determined (in part) by the dimensionless parameter β, (which is the ratio of the plasma pressure p to the magnetic pressure $B^2/2\mu_0$. For fusion to occur, high pressure must be sustained long enough for the ignited plasma to replace its energy, via the fusion burn, faster than it loses it through leakage of energy across the confining fields. Thus, both high plasma pressure and a sufficiently long plasma energy confinement time, τ_E, are necessary for the success of MFE. The Lawson criteria for ignition of a fusion burn are that the ion temperature T_i exceeds 4.5×10^7 K and that the product of density and energy confinement time $n\tau_E \geq 10^{20}$ m^{-3} s. These two conditions can be combined into a single criterion often referred to as the Lawson triple product. This requires that $nT\tau_E > 4.5 \times 10^{27}$ m^{-3} K s. This product is proportional to $\beta B^2 \tau_E$, indicating that high beta and large fields are desirable for fusion, assuming that τ_E does not degrade rapidly with either. In addition, high β is intrinsically desirable for economic reasons.

The required Lawson number is lower for a DT plasma than for a D plasma. Tritium, which is virtually nonexistent in nature, may be bred in a lithium blanket surrounding the ignited plasma in which alpha particles escaping from the burning DT plasma react to create T, which can subsequently be extracted to fuel the fusion reactions. Of course, the commercial viability of MFE demands even better performance. The fusion reactions occurring during confinement must recoup the cost of creating and heating the plasma and all other overheads of the system, such as collecting T from the lithium blanket and extracting the spent fuel from the reaction chamber. This necessarily raises the required Lawson triple product in comparison with the value needed for ignition. In practice, Lawson triple products in excess of 2.4×10^{28} m^{-3} K s are thought to be required for a fusion power plant producing 1 GW of power.

The performance of a magnetic confinement device is determined by

1. *The magnetic configuration*: the external magnetic fields and plasma currents that together define the geometry of the MSE and the strength of the magnetic bottle. Today's tokamaks rely heavily on optimization of the magnetic geometry by shaping of the cross section to achieve peak performance.

2. *Heating power and method*: the means by which the plasma is heated. These include ohmic heating, neutral beam injection, radiofrequency heating, and self-heating by the slowing down of fusion-generated alpha particles. Most scenarios for achieving ignition require auxiliary (i.e., non-ohmic) heating, usually via injection of radiofrequency waves that resonate with a characteristic frequency (known as the cyclotron frequency) of a class of plasma particles.

3. *Fueling*: the means by which fuel is injected. These include gas-puffing, pellet injection, and beam fueling.

4. *Boundary control*: the means by which the deleterious effects of impurity accumulation and other undesirable consequences of plasma–wall interaction are minimized by control of the plasma boundary. This is usually accomplished by a special magnetic configuration called a divertor. This diverts plasma outside the confinement region to a separate chamber, where interaction with walls occurs.

5. *Momentum input*: the means, usually via neutral beam injection, whereby plasma rotation is generated. Toroidal rotation is desirable for optimizing energy confinement and controlling certain types of MHD instabilities.

The principal limitations on fusion plasma performance are

1. *Collisional transport*: There is an irreducible minimum to the cross-field losses, that is, for a given magnetic geometry, *some* losses through diffusion are due to collisions between charged particles (analogous to, but dynamically different from, molecular diffusion in fluids) and these are unavoidable.

2. *Radiative losses*, especially due to *impurities* in the plasma.

3. *Macroscopic MHD instabilities*: MHD instabilities are large-scale perturbations that constrain possible pressure and current profiles. Robust ideal instabilities set hard limits on $p(r)$, $\mathbf{J}(r)$, β, etc. that place an upper bound on the possible Lawson number. Resistive instabilities may set soft limits (i.e., bounds that can be exceeded but only at unacceptably high cost in, for example, heating power). Both ideal and resistive instabilities can lead to disruptions, which are catastrophic events that result in termination of the plasma discharge and may also damage the confinement vessel.

4. *Microinstabilities*: These are small-scale instabilities, driven by local temperature and density gradients, leading to small-scale turbulence that degrades energy and particle confinement through turbulent transport. Microinstabilities typically require a description in terms of two (or more) fluids or one using kinetic equations. Discussion of these is beyond the scope of this article. Microinstabilities typically produce cross-field leakage of energy and particles that far exceeds that due to collisions. They essentially control τ_E, which enters the Lawson triple product.

Instabilities are the key players that limit achievable Lawson numbers. MHD phenomena typically determine β and the plasma pressure p, whereas microinstabilities determine τ_E. Thus, the interplay of macro- and microinstabilities is a recurring theme in the design of MFE devices.

MHD instabilities have played a prominent role in the history of MFE, especially in elucidating the behavior of tokamaks. The early predictions of rapid progress in MFE relied on naive expectations based only on considerations of collisional and radiative losses. Fusion researchers soon learned, however, that an evil genie named instabilities lurked in the magnetic bottle and was only too anxious to escape. The instabilities it created were difficult to control. Thus, the hopes of the MFE pioneers were dashed by the struggle against unexpected, premature termination of discharges by disruptions created by MHD instabilities. The genie had to be persuaded to remain in the bottle. The instabilities had to be understood and minimized or eliminated.

Much, indeed most, research on tokamaks in the late 1960s and 1970s focused on current-driven instabilities called kink-tearing modes, which can disrupt the current profile and thus the discharge. The output of this phase of fusion research was a greatly improved understanding of the parameter space of viable tokamak operations, particularly with respect to current profile and magnetic field configurations. The emphasis of tokamak research in the 1980s and early to mid-1990s shifted to microinstabilities and the turbulent transport associated with them. The aim here, simply stated, was to understand and predict the parameter scaling of τ_E. Great progress was made.

A watershed in this line of research was reached with the discovery of spontaneously formed transport barriers at the edge of, and within, the plasma. A transport barrier is a localized region in which turbulence due to microinstabilities is greatly reduced or extinguished so that cross-field leakage declines to very low, collisional levels. This, in turn, results in the formation of regions of steep pressure gradient. Transport barriers usually form via a spontaneous transition to a state of strongly sheared poloidal and/or toroidal flow, which tears apart the eddies driven by microinstabilites before they can cause significant leakage. Present-day transport barriers can sustain temperature gradients that seem almost incredibly large, in excess of 1 million degrees per inch. Profiles with and without transport barriers are shown in Fig. 13.

The discovery and exploitation of transport barriers stimulated a renewed interest in MHD stability in the late 1990s, a trend that continues to

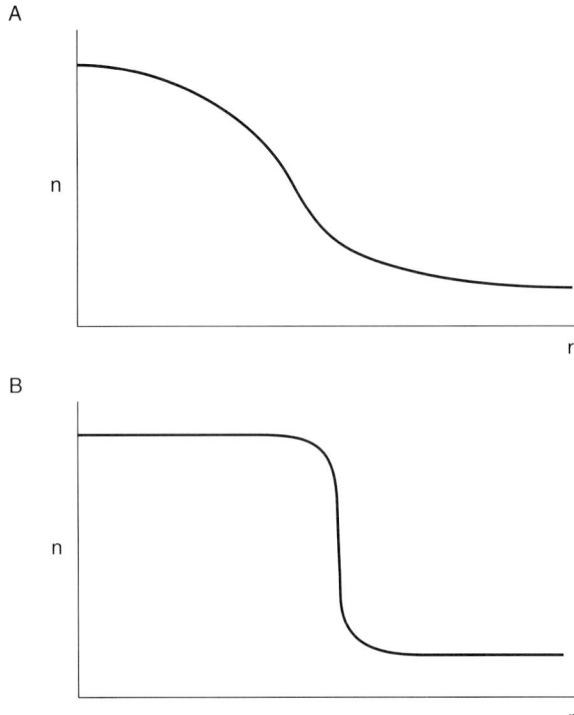

FIGURE 13 Profiles with and without transport barrier. (A) Normal profile; low confinement. (B) Profile with transport barrier; high confinement. Particle density is denoted by n.

this day. In confining the plasma better, a transport barrier also creates a larger pressure gradient in the plasma column that brings it closer to the verge of ideal MHD instability. In the jargon of MFE, enhanced confinement is necessary to reach the β limit. Thus, much of the current research on MFE is devoted to avoiding or mitigating the effects of pressure-driven ideal MHD instabilities via cross-section shaping, active feedback techniques, and profile control. Paradoxically, after 40 years of research on how to exterminate microinstabilities, effort is now being expended on finding ways to stimulate them at opportune times. The aim is to reap the full benefits of transport barriers by avoiding disruptions while at the same time controlling impurity accumulation.

5.3 Examples of MHD Instabilities in MFE Plasmas

As mentioned previously, the MHD instability of MFE plasmas is an exceptionally complex topic but is the subject of several excellent monographs. Here, we give only an introduction to the major pressure gradient and current gradient instability

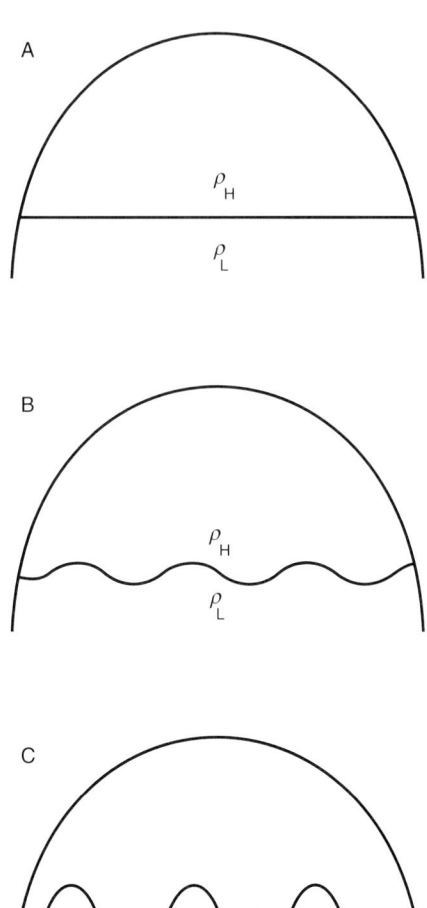

FIGURE 14 Development of Rayleigh–Taylor instability. (A) Initial equilibrium, with heavy fluid (density ρ_H) on top of light fluid (density ρ_L). (B) Ripples form on the interface. (C) Ripples are amplified by Rayleigh–Taylor instability.

mechanisms; interchange-ballooning modes and kink-tearing modes, respectively.

5.3.1 Interchange-Ballooning Modes

These modes, like all ideal MHD instabilities driven by the pressure gradient, are related to Rayleigh–Taylor instability, familiar from classical (nonmagnetic) fluid dynamics. Rayleigh–Taylor instability occurs when ripples are excited on the interface between a heavy fluid (e.g., water) sitting atop a lighter fluid (e.g., air) in a gravitational field, as shown in Fig. 14. Everyone knows what happens: The water falls out of the glass. What is not so apparent is that the initial configuration of the water and air is in equilibrium, but that the equilibrium is unstable to the growth of ripple perturbations. The experimentally inclined reader (with inexpensive carpeting) may easily convince himself or herself of this by filling a glass to the brim, placing a strong piece of cardboard (preferably two-ply) over the surface of the water and in contact with the rim of the glass, and then inverting the glass. Because the presence of the cardboard prevents the formation of surface ripples, the water will not fall, so long as the integrity of the cardboard is maintained. Of course, the free energy source for this instability is simply the gravitational potential energy stored in the initial configuration (i.e., the elevated heavy fluid); the gradient that relaxes is simply the density gradient.

The analogue of the Rayleigh–Taylor instability in the MHD of magnetically confined plasmas is the interchange instability. Interchange instability relaxes the gradient in density, or more generally in the pressure gradient, by lowering the effective gravitational potential energy of the system. It does this by interchanging a tube of high-pressure fluid with a tube of low-pressure fluid, as shown in Fig. 15. In a magnetically confined plasma, the role of gravity is played by the centrifugal force exerted on the plasma particles as they traverse curved field lines. This results in a net body force that resembles a gravitational force. Thus, if field lines curve or sag away from regions of higher pressure, the system is interchange unstable, whereas if they curve or sag toward regions of higher pressure, the system is interchange stable (Fig. 16). Equivalently, an interchange stable system is said to have favorable curvature, whereas one that is unstable is said to have unfavorable curvature. Also, it is important to realize that the conceptual image of an interchange of two plasma tubes is motivated by the fact that potential energy release will be maximal when the perturbation does not spend any portion of its energy budget on bending field lines. In other words, a pure interchange instability does not couple to Alfvén waves. In reality, the physical appearance of an interchange instability resembles that of a convection roll (Fig. 17).

In magnetic confinement devices of practical interest (including tokamaks), the magnetic curvature is not constant but, rather, varies along the field lines. Interchange stability is then determined by an average of the magnetic curvature over the extent of the field line. One key virtue of the tokamak is that its configuration has favorable average curvature and thus is interchange stable. However, as one follows a field line around the tokamak, one traverses regions of locally unfavorable curvature on the outboard side of the torus and locally favorable curvature on the inboard side. Thus, if perturbations are larger and stronger in regions of locally unfavorable curvature

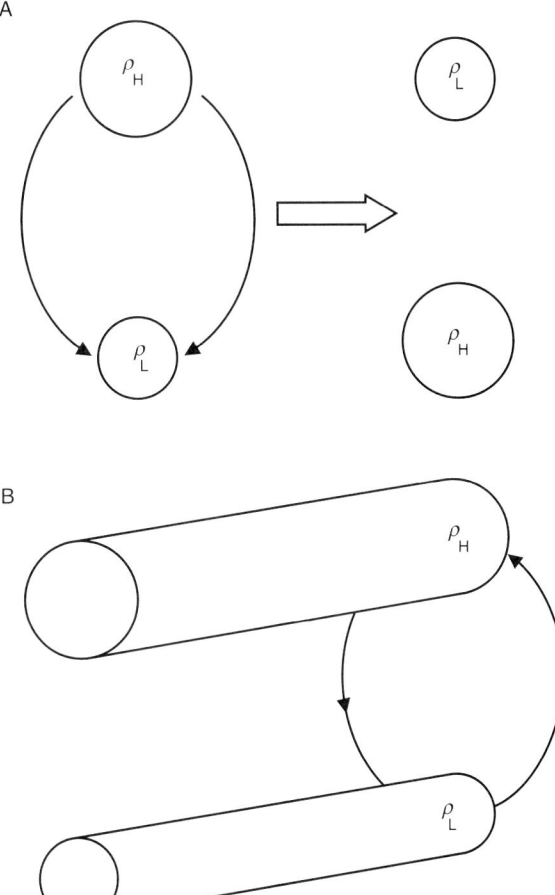

FIGURE 15 Interchange instability. (A) Interchange instability switches tubes of heavy and light fluid (densities ρ_H and ρ_L, respectively. (B) Oblique view of tube interchange; note that tubes, which are aligned with field lines, are not bent.

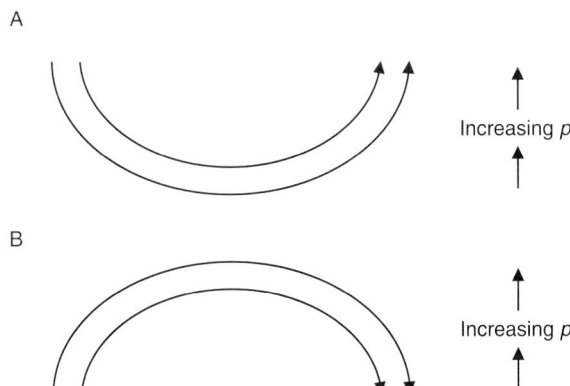

FIGURE 16 Favorable and unfavorable curvature. (A) Unfavorable curvature: Field lines sag away from the region of high pressure p. (B) Favorable curvature: Field lines curve toward the high-pressure region.

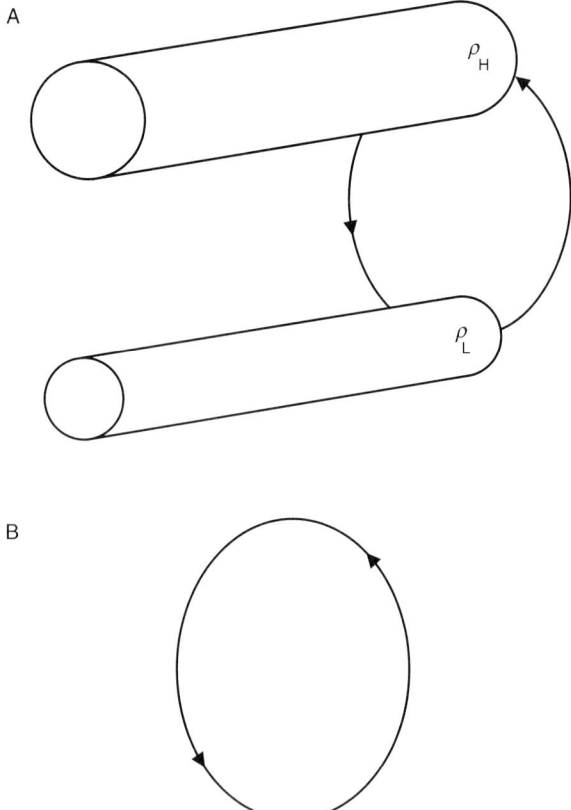

FIGURE 17 Depiction of an interchange convection roll; see legend to Fig. 15.

than they are in the locally favorable regions, instability is possible. These are called ballooning instabilities because the perturbation in the plasma pressure balloons the plasma outward in regions of unfavorable curvature. A physical picture of a ballooning instability may be gained by thinking of what happens when one overinflates a bicycle tire. Of course, the tire eventually ruptures. If there is no flaw in the tire material, the rupture will be likely to occur on the outboard portion of the wheel because the inboard portion is supported by the structural frame of the wheel and is thus mechanically more resilient. Indeed, sometimes an aneurysm will form in the tire. Finite-amplitude ballooning may be thought of as a kind of aneurysm in the magnetic surface that is driven by high plasma pressure interior to that surface. Finite-amplitude ballooning-kink instabilities are shown in Figs 18 and 19. Figure 18 shows development of the instability in a poloidal cross-section. Notice how fingers form on the outboard side of the torus. Figure 19 shows the magnetic flux surfaces. Notice how the ballooning instability produces crenelations in these.

Although ballooning instabilities are closely related to interchange instabilities, they differ crucially in that the perturbations vary along the field lines and therefore must bend the field lines. Thus, ballooning instabilities couple localized interchange motions to Alfvén waves. For ballooning instability to occur, the energy released by interchange motions in the unfavorable regions must exceed the energy expended in bending the magnetic field lines. Thus, there is a minimal or critical pressure gradient required for ballooning instability to occur. This is in sharp contrast to the interchange mode (in a system with a curvature that is on average unfavorable), which, in ideal MHD, can occur for any pressure gradient. The critical pressure gradient for ballooning instability plays a central role in the ultimate determination of the maximum achievable β for a specific configuration (i.e., its beta limit). In practice, the magnetic geometry and magnetic field strength determine the critical pressure gradient for ballooning instability. One important feature of magnetic geometry is the magnetic shear, which parameterizes how rapidly the direction of field lines changes as a function of radius r. Magnetic shear plays an important and subtle role in the detailed dynamics of ballooning instabilities. A complete discussion of this is beyond the scope of this article.

5.3.2 Kink-Tearing Modes

Kink-tearing modes are all driven by the current gradient, although the symptom of instability is often related to the pitch of the magnetic field lines. The physical nature of the modes is best illustrated by a sequential discussion of sausage instabilities, kink instabilities, and tearing instabilities.

Consider the Z pinch (Fig. 7A). Figure 20 illustrates a type of perturbation, sometimes called a sausage mode, which is axisymmetric around the axis of the pinch. The existence of the sausage mode of instability can be explained rather simply. As discussed in Section 2, the field B_θ outside the plasma column, created by a current I flowing down it, is azimuthal in direction and of strength $\mu_0 I/2\pi r$. On the surface of the column, the field strength is therefore $\mu_0 I/2\pi a$, where a is the local radius of the column, which varies when the tube is pinched. The pinched column then resembles a tube of sausage meat that has been compressed to create a chain of links. The surface field and the associated field line tension are greater where the column necks and smaller where it bulges. This is the field that confines the plasma column, and since I is constant, the tension it exerts on the necks of the chain further

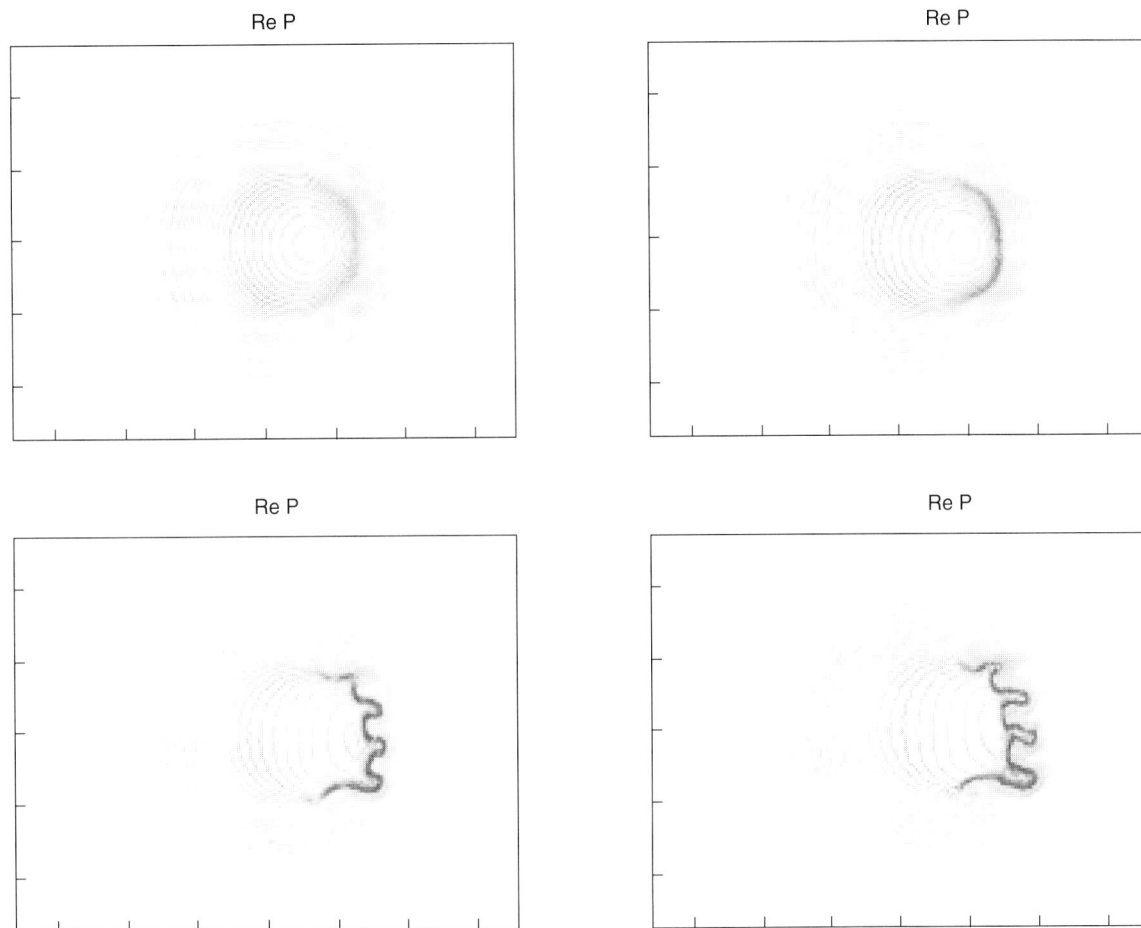

FIGURE 18 Development of a ballooning mode as seen in a poloidal cross section of the torus. The magnetic axis is on the left of each panel. The sequence of panels is as follows: top left, top right, bottom left, and bottom right. Note that perturbations are larger on the outboard midplane regions (on the right). Also note that as the perturbation grows, it evolves to form fingers or spikes. Courtesy of S. C. Jardin and the Center for Extended MHD Modeling, Princeton Plasma Physics Laboratory.

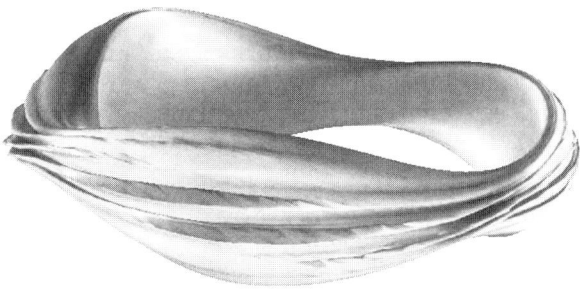

FIGURE 19 Development of a ballooning-kink mode in a torus. Note how a side view of the finger formation reveals the tendency of the ballooning mode to crenelate the plasma and break it into ribbons. Courtesy of S. C. Jardin and the Center for Extended MHD Modeling, Princeton Plasma Physics Laboratory.

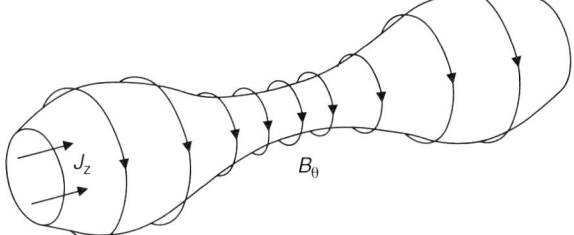

FIGURE 20 The sausage mode of instability of a Z pinch; lines of force encircling a current-carrying plasma column are shown. J_z indicates the cross section through which the current enters.

increases the initial deformation of the column. This self-reinforcing, positive feedback loop is indicative of an instability process that tends to break up the column into a chain of plasmoids (i.e., plasma droplets). For obvious reasons, the instability is called the sausage instability. The mechanism of sausage instability is closely related to that causing a thin stream of water from a faucet to break up into a line of spheroidal droplets. In this case, surface tension plays the role of field line tension.

The sausage instability may be stabilized by an axial field B_z of sufficient strength within the column, which provides a tension that opposes the formation of necks. The critical strength of B_z necessary for stabilization can be estimated by comparing the field energy released by necking of the column with the energetic costs of deforming the axial field B_z. Because necking motions must conserve axial magnetic flux, the change, $\Delta(B_z^2/2\mu_0)$, in energy density of the axial field is $-B_z^2 \Delta a/\mu_0 a$, where $\Delta a/a < 0$. The corresponding change, $\Delta(B_\theta^2/2\mu_0)$, in the azimuthal field energy density is $B_\theta^2 \Delta a/2\mu_0 a$. It follows that the net change in magnetic energy is positive if $B_z^2 > \frac{1}{2}B_\theta^2$. This gives the required axial field strength necessary to stabilize the instability, as depicted in Fig. 21. Note that when the plasma tube is bent into a torus, one arrives at a simple model of the tokamak configuration.

The dual presence of both a poloidal field and a strong, vacuum toroidal field greatly improves the MHD stability of a toroidal pinch by eliminating both the sausage instability and the related bending instability shown in Fig. 22 and by introducing magnetic shear, which is useful in controlling pressure-driven instabilities. Indeed, the tokamak

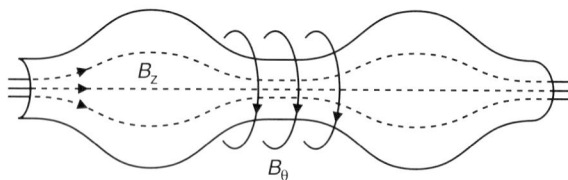

FIGURE 21 Stiffening of the Z pinch against sausage modes of instability by the addition of a field B_z along the plasma column (creating a screw pinch).

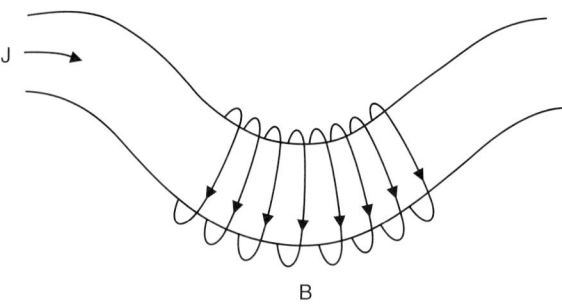

FIGURE 22 The bending mode of instability of the Z pinch. Note how the lines of force are crowded together on the inside of the bend and are moved apart on the outside of the bend; the concomitant difference in magnetic pressure acts to reenforce the displacement.

magnetic configuration is de facto defined by the pitch of its magnetic field lines, which is given by $q(r) = rB_\phi/R_0 B_\theta$. With the (possible) exception of a finite region around the magnetic axis, the $q(r)$ of standard tokamaks exceeds 1 and increases with r so that the magnetic shear is positive. The maximum $q(r)$ in most standard tokamaks is between 2 and 4. The intrinsic dual periodicity of toroidal configurations means that perturbations also have their own effective pitch. This is a rational number, m/n, where m and n are positive integers, with m being the number of times the perturbation circles the torus in the poloidal direction while completing n turns in the toroidal direction. Thus, it is possible for the field line pitch $q(r)$ to equal (or resonate with) the perturbation pitch m/n at certain radii, $r_{m,n}$, where $q(r_{m,n}) = m/n$. These special radii define magnetic surfaces that are called rational or resonant surfaces.

Resonant surfaces, and more generally field line pitch, are crucial to an important class of instabilities called kink-tearing instabilities. The onset of these instabilities is determined by $q(r)$ and by the radial gradient of the current profile. Since $q(r)$ is determined primarily by $B_\theta(r)$, kink-tearing instabilities are also conveniently described as current-driven instabilities. When the perturbation pitch m/n exceeds $q(a)$ for $r < a$, the instability does not involve magnetic reconnection and is called an external kink. It may then be described by ideal MHD. The effects of boundaries, particularly conducting walls, are vitally important to external kink dynamics and stability (see Section 4 and Fig. 9). When the perturbation is resonant [i.e., when $m/n = q(r)$ for $r < a$], reconnection is usually involved and the instability is called a tearing mode since magnetic field lines tear and reconnect. Tearing modes involve the formation of current filaments. In certain cases, the instability dynamics of a resonant kink (usually with $m = 1$) can be described, at least in its initial phases, by ideal MHD. Such instabilities are called ideal internal kinks.

Ideal kink instabilities are current driven, but it is really the pitch of the magnetic field lines that signals the onset of the instability. Kink instabilities have long wavelength in the toroidal direction and cause the plasma column to snake helically around its equilibrium position in the manner shown in Fig. 23. Before the column is perturbed, the current flows parallel to the magnetic field. After perturbation, it follows the helical path shown and produces a magnetic field resembling that shown in Fig. 23. This is only part of the total field; to obtain the total field, one must imagine augmenting the **B** shown in the

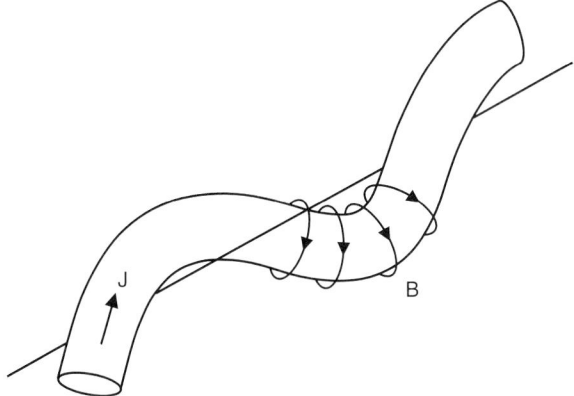

FIGURE 23 Kink instability of a toroidal equilibrium. In the initial state, the current flows along field lines. In the perturbed state (shown here), the current follows a helical path and creates a field, **B**, of the type shown. The remainder of the field (not shown) is along the helically distorted column. The net result is a field that has a component perpendicular to **J** that creates a Lorentz force enhancing the perturbation.

figure by the field along the column. The net result is a field, **B**, that has a component orthogonal to **J**. This creates a $\mathbf{J} \times \mathbf{B}$ force that reenforces the initial perturbation, thus promoting instability.

Kink instabilities are similar to the helical deformations of an elastic tube or band that occur when it is twisted beyond a critical value. An example of such kinks is sometimes seen on a wire connecting a telephone handset to its cradle. The twist of the band is clearly analogous to the initial pitch (or twist) of field lines, as given by $q(r)$. Kink instabilities reflect the tendency of the plasma to lower its energy by assuming a secondary, helical equilibrium that replaces the initial configuration with one of helical symmetry. The ultimate radial extent of the helical deformation is determined by several factors, including the amount of plasma current, the toroidal field strength, and the size of the gap between the plasma boundary and the conducting walls of the containment vessel. Since the plasma current channel expands during kink instabilities, these constitute a simple but very direct route to an undesirable disruption of the discharge, should the current column come into contact with a wall of the containment vessel.

In practice, external kink instabilities can be avoided by not operating the discharge with $q(a)$ close to a low-order rational number, such as, $\frac{3}{2}$, 2, or 3, and/or by having a conducting wall sufficiently close to the plasma column. Kink instabilities are also predicted to occur when $q(r) < 1$. For this reason, internal kinks are ubiquitous in tokamaks and are thought to be related to the sawtooth phenomenon.

This refers to quasi-periodic relaxation oscillations of the central temperature of the tokamak that, in diagnostic recordings, produce a jagged trace reminiscent of the teeth on a saw.

Since internal kink perturbations usually do not pierce the plasma boundary or bring the current channel into contact with the vessel wall, they are usually not directly associated with disruptions. Internal kinks do, however, affect the performance of tokamaks by limiting the central plasma pressure that can be achieved.

The dynamics of energy release in resonant kinks or tearing modes are determined by nonideal effects (i.e., by effects that break the Alfvén frozen flux constraint). Typical of such nonideal effects is plasma resistivity η. However, it is important to keep in mind that the free energy source for tearing modes is the same as that for kinks, namely the current gradient. Nonideal effects, such as resistivity, are only the triggers of the instability and are significant only in a narrow diffusion layer, surrounding the resonant surface, in which reconnection occurs. Just as kinks tend to form helical secondary equilibria, so do tearing modes. However, since tearing modes are resonant, reconnection occurs, resulting in the formation of magnetic islands in the region near the resonant surface, as shown in Fig. 24. These islands may be thought of as consequences of current filamentation and concomitant reconnection. Magnetic islands are MHD analogues of Kelvin cat's eyes or nonlinear critical layers, which form as a consequence of certain hydrodynamic shear flow instabilities. When magnetic islands grow to a size at which they touch the wall, or when magnetic islands at neighboring resonant surfaces overlap (resulting in stochastic field lines in a region that pierces the surface of the plasma), MHD turbulence is generated, resulting in turbulent transport, rapid heat loss, and a sudden expansion of the current channel. When the current strikes the wall, a large impurity influx commences, and this raises the effective

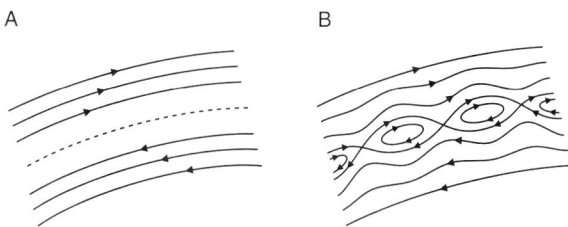

FIGURE 24 The formation of magnetic islands. (A) Initial state; the neutral line is shown dashed. (B) Field configuration after magnetic islands have formed around the neutral line.

plasma resistivity, which in turn quenches the current, creating a disruption. For this reason, tearing mode dynamics and stability are of great interest to MFE.

A final comment on tearing instabilities concerns their rate of growth. Ideal MHD instabilities grow exponentially, on Alfvénic (e.g., kink) or sonic (e.g., ballooning) timescales. Tearing modes initially grow at a rate proportional to $(\tau_A^{-2}\tau_\eta^{-3})^{1/5}$, where τ_A and τ_η are the Alfvén and (global) resistive diffusion times defined in Sections 3 and 4. Since $\tau_A \ll \tau_\eta$, tearing modes grow more slowly than ideal MHD modes. However, linear growth rates are of little relevance to the dynamics of the tearing mode since magnetic islands rapidly enter a nonlinear phase of algebraic growth. This occurs when the width of the magnetic island exceeds the width of the (narrow) reconnection layer of the linear theory. A major current research topic in tokamak MHD concerns the nonlinear growth and saturation of magnetic islands.

5.4 Interplay of MHD Instabilities and Tokamak Operation

Considerations of turbulent transport and MHD stability define the possible viable operating space of present-day tokamaks. The MFE program has sought to exploit understanding of MHD stability limits, etc. in order to expand and optimize the tokamak operating space. The tactics and tricks used include

1. *Shaping the cross section* to increase β limits: The goal here is to design a shape that increases the relative fraction of a magnetic field line that lies in the region of favorable curvature.

2. *Programming current and heating evolution* in order to avoid conditions that are prone to the onset of disruptions (i.e., in order to improve ideal stability limits).

3. *Controlling current profiles* by driving currents locally through the injection of radiofrequency waves.

4. *Using active feedback*, including current drive, to mitigate or retard the growth of undesirable perturbations.

5. *Driving plasma rotation* by neutral beam injection in order to stabilize certain MHD instabilities.

A recurrent theme in fusion research is the often rather diabolical ways in which various instabilities conspire to limit the operating space and tokamak performance. As mentioned previously, eliminating microinstabilities using transport barriers enhances confinement times but opens the door to pressure-driven instabilities (i.e., interchange ballooning modes). Operating at high current is good for energy confinement and helps limit pressure-driven ballooning modes, but it increases the possibility of disruption due to kink-tearing modes. Employing low currents improves stability but often degrades confinement and fails to achieve maximal machine performance. The list is long and continues to grow. These numerous and accumulating questions and concerns ensure that both the theoretical and experimental MHD of tokamaks and other MFE devices will remain vital and vibrant research areas well into the 21st century.

5.5 The Future of MHD in MFE

There is little doubt that in the near future the focus of the world MFE program will be the International Thermonuclear Experimental Reactor (ITER), a tokamak shown schematically in Fig. 25. The ITER program is an international project involving the collaboration of many countries. Its primary mission is to study plasma dynamics under conditions of a fusion burn, in which the main source of plasma heating is from collisional friction between the alpha particles produced by the thermonuclear reactions and the electrons.

A burning plasma is different from the standard electron–ion plasma discussed so far in (at least) two very important respects. First, a burning plasma contains a very substantial population of high-energy alpha particles. This situation is inconsistent with the foundations of MHD, which treats both ions and electrons together as a single fluid. The absolutely minimal viable theory of the dynamics of a burning plasma is a two-component description involving an MHD-like model of the bulk plasma and a kinetic equation for the energetic alphas; these have a velocity distribution that is far from Maxwellian and cannot therefore be adequately modeled by fluid equations. The two components of the model couple to each other electrodynamically.

The second important distinction between a burning plasma and a standard plasma is that high-energy alpha particles can resonate with plasma waves of higher frequency than can bulk particles, thus encouraging a new class of instabilities. Of particular importance among these are toroidal Alfvén eigenmodes (TAEs), which can be destabilized by resonance with alphas and so relax the alpha particle density and temperature gradients. TAEs are

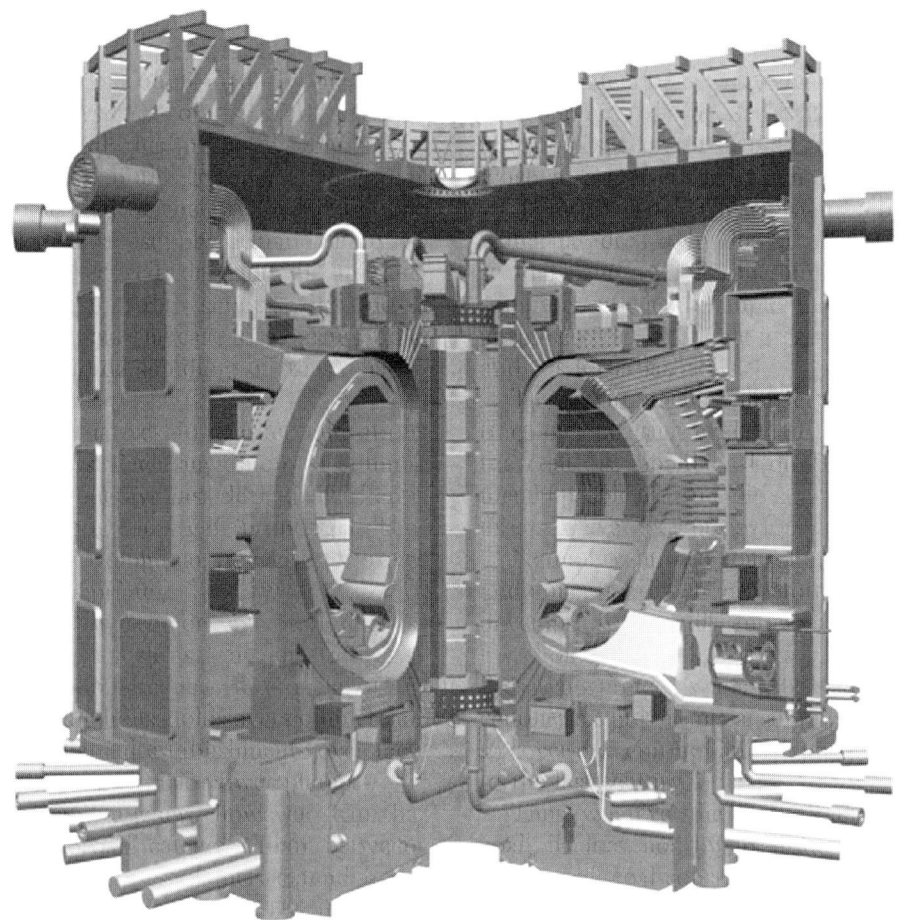

FIGURE 25 The tokamak ITER. To appreciate the scale of the device, see the figure of a person near the bottom.

critical to MFE since excessive instability-induced alpha losses can prevent, or prematurely quench, ignition. TAEs and other energetic particle-driven instabilities have been a topic of great interest in theoretical, computational, and experimental MFE research, with beams of externally injected high-energy particles used to simulate the effects of the alphas. Clearly, research in alpha particle-driven instabilities will be a major part of the ITER program.

Traditional MHD problems will be major foci of the ITER program as well. A key topic will certainly be feedback stabilization of instabilities, with the aim of avoiding or mitigating disruptions. Note that the disruption of a burning plasma could cause very significant damage to the ITER confinement vessel. Also, any successful fusion power plant must be effectively disruption free. Because the ITER plasma will be very hot, with concomitantly infrequent collisions between particles in the plasma, the traditional MHD model will need extension in order to include dynamical effects relevant to the ITER plasma. Such effects include possible nonideal, kinetic modifications to MHD as well as transport coefficients that describe dissipation due to microturbulence rather than collisions. Theoretical and computational research on such extended MHD models is being pursued vigorously throughout the world today. In addition, it is anticipated that subgrid scale models of turbulent transport will be an integral component of extended MHD models in the future.

SEE ALSO THE FOLLOWING ARTICLES

Electrical Energy and Power • *Electricity Use, History of* • *Electric Motors* • *Electromagnetism* • *Magnetic Levitation*

Further Reading

Bateman, G. (1978). "MHD Instabilities." MIT Press, Cambridge, MA.

Biskamp, D. (2003). "Magnetohydrodynamic Turbulence." Cambridge Univ. Press, Cambridge, UK.

Biskamp, D. (1997). "Nonlinear Magnetohydrodynamics." Cambridge Univ. Press, Cambridge, UK.

Chandrasekhar, S. (1961). "Hydrodynamic and Hydromagnetic Stability." Cambridge Univ. Press, Cambridge, UK.

Childress, S., and Gilbert, A. D. (1995). "Stretch, Twist, Fold: The Fast Dynamo." Springer, Heidelberg.

Davidson, P. A. (2001). "An Introduction to Magnetohydrodynamics." Cambridge Univ. Press, Cambridge, UK.

Drazin, P. G., and Reid, W. (1981). "Hydrodynamic Stability." Cambridge Univ. Press, Cambridge, UK.

Freidberg, J. P. (1987). "Ideal Magnetohydrodynamics." Plenum, New York.

Itoh, K., Itoh, S.-I., and Fukuyama, A. (1999). "Transport and Structure Formation in Plasmas." Institute of Physics Press, Bristol, UK.

Kadomtsev, B. B. (1992). "Tokamak Plasma: A Complex Physical System." Institute of Physics Press, Bristol, UK.

Krause, F., and Rädler, K.-H. (1980). "Mean-Field Magnetohydrodynamics and Dynamo Theory." Pergamon, Oxford, UK.

Mestel, L. (1999). "Stellar Magnetism." Oxford Univ. Press, Oxford, UK.

Parker, E. N. (1979). "Cosmic Magnetic Fields." Oxford Univ. Press, Oxford, UK.

Taylor, J. B. (1986). Relaxation and magnetic reconnection in plasmas. *Rev. Modern Phys.* **58**, 741–763.

Tillack, M.S., Morley, N.B. (1999). Magnetohydrodynamics. *In* "14th Standard Handbook for Electrical Engineer" (D. G. Fink and H. W. Beaty, Eds.), pp. 11-109–11-144. McGraw-Hill, New York.

Wesson, J. (1997). "Tokamaks." 2nd Ed. Oxford Univ. Press, Oxford, UK.

White, R. B. (1989). "Theory of Tokamak Plasmas." North-Holland, Amsterdam.

Manufactured Gas, History of

JOEL A. TARR
Carnegie Mellon University
Pittsburgh, Pennsylvania, United States

1. Origins and Characteristics of the Industry
2. The Period of Coal Gas Domination, 1816–1875
3. Innovations and Challenges from Competing Fuels, 1875–1927
4. The Transition to Natural Gas, 1927–1954
5. Environmental Impacts of the Manufactured Gas Industry
6. Conclusions

Glossary

carbureted water gas Gas produced by spraying blue gas (water gas) with liquid hydrocarbons and thermally cracking it to form carbureted water gas and tars.

coal gas A method of heating bituminous coal in a retort to destructively distill volatiles from the coal in order to produce fuel gas for illuminating and heating purposes.

water gas A fuel gas composed primarily of hydrogen and carbon monoxide produced through the interaction of steam with incandescent carbon, usually from anthracite coal or coke.

Welsbach mantle A mantle made of cotton fabric impregnated with rare earth that becomes incandescent when exposed to a gas flame.

This article examines the development of the manufactured gas industry in the United States from its origins early in the 19th century to its demise in the post-World War II period. It discusses the production technologies of coal gas, water gas, and oil gas and their methods of distribution. It also discusses industry innovations, such as the development of carbureted water gas and of the Welsbach mantle, and their effect on regulatory standards. It explores the shift from the use of manufactured gas to natural gas in the period from the 1920s through the 1950s, with the consequent demise of the industry. Finally, the article examines the environmental effects of the industry and the issues of the cleanup of former MGP sites.

1. ORIGINS AND CHARACTERISTICS OF THE INDUSTRY

Manufactured gas was one of the most critical energy sources and fuels that provided American cities, as well as other cities throughout the world, with light and energy during much of the century from 1850 to 1950. Manufactured gas is produced by heating coal or other organic substances such as wood, pitch, or petroleum, driving off a flammable gas made primarily of carbon monoxide and hydrogen as well as other gases. The discovery that flammable gas could be produced in this manner was first made in Great Britain by John Clayton in the late seventeenth century followed by the work of Frenchmen Jean Pierre Minckelers and Philippe Lebon in the late 18th and early 19th centuries. At the turn of the century, the British engineer William Murdock developed methods of gas lighting that were used in cotton mills.

Demonstrations and applications of gas lighting began in the United States in 1802, based on European developments. Gasworks were installed to provide better illumination for cotton mills and lighthouses during the first decade of the century. In 1817, the Baltimore municipality granted the Gas Light Company a franchise to lay gas pipe in the city streets and to light them. Others cities quickly followed: New York, 1825; Boston, 1829; New Orleans, 1835; Philadelphia, the first municipally owned plant, 1836; and San Francisco, the first western plant, 1854. The U.S. Census reported 30 manufactured gas plants in 1850, 221 in 1859, 390 in 1869, 742 in 1889, 877 in 1904, and 1296 in 1909, but the actual number was probably much larger.

The manufactured gas utilities that emerged ranged in size from very small plants in small towns (cf. 5000 population) that served a limited number of consumers (200–300) to very large plants with extensive distribution systems that served thousands of customers with gas for home lighting, cooking, and heating, served municipalities by illuminating city streets, and provided gas for a number of industries. Urban gas utilities were systems that generated fuel for their customers from a central plant (or plants in the case of the very large utilities) and distributed it through a piped system. By far, it was most widely used for lighting purposes. Individual gas production units were also manufactured and were used for isolated factories and homes (spirit gas lamps). From the late 19th century onward, various industries in need of low-Btu (British thermal unit) gas frequently installed coal-using producer gas units in their factories.

Municipal and state governments regarded gas manufacturing plants as public utilities and required that they acquire franchises to use the city streets for their distribution lines. In order to profitably survive, gas manufacturing utilities required long rather than short franchises—the exact term, however, produced many heated political battles. Franchises were originally local but frequently became the concern of state authorities because the natural monopoly characteristics of the utilities were resistant to market forces. Battles over issues such as price, luminosity and energy ratings, and technology, as well as price and duration, frequently took place between utility firms and municipal and state authorities. Although some municipally owned gas works existed, the industry remained largely private.

2. THE PERIOD OF COAL GAS DOMINATION, 1816–1875

In the period from 1816 to 1875, the number of centralized gas works grew to between 300 and 400, primarily using technology developed in Great Britain. The most important innovations during these years concerned fuel, distillation technology, distribution, and metering technology.

Although most firms eventually used coal as their feedstock for gas production, other substances were also initially used. The Baltimore Gas-Light Company, for instance, used pine tar as the feedstock for its initial gas production, the New York Gas Works used rosin, the Brooklyn Gas Works consumed British coals, and the Wilmington, North Carolina Gas Works used wood from the Carolina Pine Barrens. Several small eastern cities used distillate of turpentine to manufacture rosin gas and Philadelphia experimented with wood as late as the 1850s.

Utility managers originally avoided coal because of its impurities, but in the period from 1815 to 1820, British gas manufacturers developed methods to remove such impurities. In 1822, Baltimore became the first American city to consume coal to produce gas, using bituminous coal from Virginia mines for its feedstock.

Since bituminous coal mines were not easily accessible for most cities, the cost of the fuel kept prices relatively high or forced the use of organic materials such as wood. Expensive fuel, therefore, provided a reverse salient (a concept originated by historian of technology Thomas Hughes), in terms of extending the size of the market for improved lighting. Whereas bituminous coal, considered "the best in the world for gas light," existed in western Pennsylvania and was used in Pittsburgh for gas production and for home heating, transportation costs over the Allegheny Mountains were extremely high, limiting its use in eastern markets.

The major transportation breakthroughs came in the late 1840s and 1850s with the slack watering of the Pittsburgh area rivers and the completion of railroad connections from Philadelphia to Pittsburgh in 1852. These transport improvements facilitated the shipping of high-quality bituminous gas coal from western Pennsylvania to gas works in eastern cities at greatly reduced transportation costs. By the late 1850s, for instance, the Westmoreland Coal Company, a large western Pennsylvania producer of gas coal, was supplying 58 gas works.

Responding to the lower fuel costs and the anticipation of a larger market for cheaper light, a number of gas works were founded in the 1850s and 1860s. Between 1850 and 1860, the U.S. Census reported that the number of manufactured gas establishments jumped from 30 to 221, a rise of over 600%—an increase much greater than that of urban population growth. By 1870, the total number of manufactured gas plants reported was 390, another large jump.

The manufacture of coal gas consisted of three operations: distillation, condensation, and purification. Although there were many different processes patented for the production of coal gas, by the 1850s those most widely used operated by heating bituminous coal in a retort to destructively distill volatiles from the coal. The leading types of coal-carbonization retorts were the horizontal, inclined, and vertical. Horizontal retorts were most commonly used; until approximately 1850, they were made of cast iron, when superior clay refractories were substituted. A typical horizontal bench consisted of six retorts and a producer gas furnace that used coke or coal for heating the retorts. Coal tar condensed from the gas was also used as a retort fuel. Wet and dry lime processes were used in purification methods, as was iron oxide, beginning in 1870.

Proper modes of metering, gas storage, and distribution were critical to the successful operation of a gas firm. Gas holders, or gasometers, as they were called, were essentially inverted iron bells in a tub of water. The early gas holders were usually made of masonry and often set in the ground, with iron plates for the bell. A number of firms initially used wooden mains for distribution, but in ∼1820, cast iron pipes manufactured in British foundries were introduced into the United States; however, some wooden mains of tamarack logs lasted into the late 19th century. By the 1830s, American foundries were producing their own cast iron pipe, reducing dependence on the British. From ∼1820 through 1850, pipe was made in horizontally cast sections (4–5 ft in length). These were subject to considerable core sagging and slag accumulations but in 1850 foundries shifted to casting in vertical molds, producing a much stronger and longer (12 ft) pipe product. Nineteenth century cast iron gas mains often lasted 50 to 75 years.

The distribution system was another major cost for the gas companies. Firms occasionally used smaller diameter mains than was ideal, causing problems and poor service as the system expanded. In Philadelphia, for instance, the city-owned utility used pumps to force gas through mains too small to accommodate the demand, often resulting in customers receiving gas under excess pressure. Though the gas utilities seldom extended conduits to undeveloped streets, they quickly installed pipe on streets in growing areas in order to benefit from rises in real estate prices.

Cities initially used gas primarily for street illumination but cost reductions plus innovations such as improved burners and fixtures and metering led to a broader market. Gas companies originally charged flat rates per burner, leading to expansion limits for street lights and domestic sales. Meters were essential for price reduction for both municipalities and domestic consumers but were fought in many city councils. The original meters were "wet" meters, using water, but in the 1850s firms began replacing them with more accurate dry meters.

The manufacture of coal gas provided the potential for firms to recover and sell by-products such as coke, tar, light oil, and ammonia. However, unlike Great Britain, there was no market for such products in the United States (aside from the coke) before the late 19th century. The United States lacked a coal-based chemical industry and was almost completely dependent on imports for dyestuffs, drugs, and explosives and other products of coal tar and light oil origin. Ammonia was regarded as a nuisance until after the Civil War, when ammonium sulfate was used for fertilizer. Gasworks themselves burned by-product coke under their retorts and sold some on the open market, but it had limited appeal.

By the 1870s, over 400 manufactured gas utilities existed and the industry had spread throughout the country. Over half of the firms, with over 50% of the labor and capital, were located in just four states: New York (71), Massachusetts (52), Pennsylvania (43), and Ohio (32). Price, however, still limited home gas service to a relatively affluent segment of the market. Gas companies had originally hoped to make large profits from street lights, but municipal franchises applied stringent regulations as to the numbers of public street lights and rates paid, and many firms defaulted on this part of the contract or were continually petitioning city councils for rate increases. Because costs were high and margins low, gas companies were reluctant to expand to lower income markets, putting the industry at risk from lower priced fuels. In the 1850s, coal oil lamps provided much of the lighting needs of those who could not afford gas, the fuel supplied from refineries concentrated in Pittsburgh and Ohio. These lamps furnished up to 14 cp and were quite flexible to use. In the post-Civil War period, however, two new competitors for manufactured gas emerged to challenge the industry for the lighting market—kerosene and electricity.

3. INNOVATIONS AND CHALLENGES FROM COMPETING FUELS, 1875–1927

Substantial competition for urban lighting markets developed in the half century from 1875 to 1927. Municipalities and states also subjected the industry to increased regulation. The first competitor for the lighting market was the kerosene lamp, an outcome of the discovery of oil in Pennsylvania in 1859, which largely captured the low end of the illumination market. Improvements in petroleum refining and transportation techniques and the discovery of new sources of supply resulted in reduced oil prices. By the 1870s and 1880s, most of the urban working class, as well as a substantial part of the middle class, used kerosene lamps for domestic illumination.

Electric arc lamps began competing with manufactured gas for the street lighting market in the 1880s. Since gas companies found these lighting contracts unprofitable, it was a limited loss. By the 1880s, domestic lighting accounted for 90% of gas revenues and arc lamps posed no threat in this market because their intensity made them unusable in the home. In 1881, however, Thomas Edison, the inventor of the incandescent lamp, developed a central power station in New York City that supplied power for home lighting that directly competed with interior gas light. Edison had wisely based his distribution system on that of the gas industry, constructing an "electrical analogue to the gas system."

The manufactured gas industry responded to this challenge with the development of carbureted water gas. The concept behind water gas, which was first observed in 1780 and patented in France in 1834, involved the action of steam on incandescent carbon, usually anthracite coal or coke. When the steam, composed of oxygen and hydrogen, came into contact with the incandescent carbon, it produced a fuel gas composed of hydrogen and carbon monoxide, as well as other products. The gas had a heating value of approximately 300 Btu/ft^3, burned with a blue flame, and had little illuminating power. The last characteristic diminished the utility of water gas for commercial use, because the market for manufactured gas in most of the 19th century was almost entirely in lighting.

In 1875, however, Thaddeous S. C. Lowe, a Pennsylvania inventor, patented a process to enrich the blue gas by spraying it with liquid hydrocarbons and thermally cracking the gas to form carbureted water gas and tars. The Lowe process was acquired by the United Gas Improvement Company (UGI) of Philadelphia, formed in 1882. Before this time, utilities had been slow to adopt the water gas process and the states of Massachusetts and New Jersey actually outlawed water gas because of gas poisoning incidents. UGI, however, aggressively pushed the technology and the rapid growth of the petroleum industry made oil available at a relatively cheap price. The gas industry proceeded to shift a substantial amount of its manufacturing processes from the production of coal gas to carbureted water gas. Originally carbureted water gas manufacturers used naphtha as the process oil because it was relatively inexpensive, but after 1895, as rising demand caused the price of naphtha to rise, gas manufacturers switched to the petroleum fraction known as gas oil. The invention and adoption of the automobile increased the demand for the higher petroleum fractions even further and in $\sim$1930 the carbureted water gas industry switched to a yet cruder grade of petroleum.

The use of oil rather than coal as the feedstock for gas production was another major step by the manufactured gas industry. Oil gas was made by the thermal decomposition or cracking of petroleum. Whale oil had been used in several gas plants early in the 19th century, but it was not until 1889 that a modern refractory process to use petroleum was invented. L. P. Lowe, the son of the discoverer of carbureted water gas, was the inventor. The first oil gas plant was constructed in California and in 1902 an oil gas plant was erected in Oakland, California for lighting purposes. The so-called Pacific Coast oil gas process was used mostly in Pacific coast cities—an area that was poor in coal but rich in petroleum.

The number of carbureted water gas plants in the United States grew steadily. In 1890, the census reported returns from 742 gas manufacturing firms, probably approximately 75% of the total. In this census, coal gas represented 46.7% of the total product, whereas water gas made up 38% of the total. Between 1890 and 1904, coal gas production dropped precipitously to 11.2% of total gas produced, straight water gas rose to 48.6%, and a new category, mixed coal and water gas, absorbed 36.4% of the market. Between 1904 and 1914, coal gas declined even further, from 11.2 to 5.2% of total gas production; carbureted water gas also declined from 48.6 to 44.2% and mixed gas advanced from 36.4 to 42.4%. Because Great Britain (the first water gas plant was built in 1890) and other European nations did not have access to inexpensive petroleum, they did not adopt carbureted water gas to the same extent as did the United States. The U.S. gas industry, therefore, became increasingly dependent on oil, whereas the European industry remained largely based on coal.

Carbureted water gas had a number of advantages over coal gas. Its production required less space, less capital, and less labor, as well as a smaller storage capacity, because plants could be started up in a relatively short time. Many coal gas plants installed auxiliary carbureted water gas machines to increase their flexibility in regard to load, mixing the two gases. Without the auxiliary water gas equipment, firms would have had to add additional storage or construct additional retorts, which would have been idle for many hours in the day. Carbureted water gas also had a higher illuminating value and a flexible heating value; a plant could produce gas varying in heating value from 300 to 800 Btu/ft^3. Utility operators could produce carbureted water gas with widely varying Btu contents by changing the amount of oil cracked into the blue gas. Higher Btu carbureted water gas could be mixed with gas with a lower Btu grade from blue gas or coal gas retorts, thereby increasing the available gas volume without the necessity of increasing the number of water gas sets.

The addition of petroleum hydrocarbons to water gas provided a gas that had illumination values equal to those of coal gas but at a lower cost. Most municipal franchises to gas companies required illuminating power of between 15 and 17 cp. The manufactured gas burners most commonly used before 1886 were the bat-wing, the fish-tail, and the Argand burners. These burners depended on the luminosity of the flame in order to fulfill legal candlepower requirements. Therefore, when carbureted water gas was burned as a fuel, the burners produced an equal or better flame at a lower cost. The lamps, however, still depended on the incandescence of the carbon particles in the flame itself for their illuminating power.

In 1885, Carl Auer von Welsbach of Germany, working in the laboratories of Robert Wilhelm von Bunsen, produced another major innovation in the consumption of manufactured gas. Welsbach patented a gas mantle made of cotton fabric impregnated with rare earth. When the mantle was placed near a gas flame from a Bunsen burner, it "became incandescent," producing a steady white light of 30–50 cp. This increased the light that could be obtained from a given quantity of gas approximately six times, utilizing the principle of incandescence rather than the direct illuminating properties of the gas. Thus, it mirrored the innovation of the gas industry's chief competitor, the incandescent electrical lamp. The Welsbach mantle provided an inexpensive and superior incandescent lamp to meet the competition of the electric light and helped modernize an industry whose existence was under severe competitive threat.

Although carbureted water gas was necessary for the older burners in order to meet existing legal candlepower and calorific requirements, it did not have to be used with the Welsbach mantle, which produced an equally brilliant light with cheaper blue water gas. Yet, legal candlepower requirements for manufactured gas required that higher priced carbureted water gas rather than blue gas had to be used with the Welsbach mantle. At the time of the mantle's introduction, municipal franchises usually stipulated a candlepower standard for gas ranging from approximately 14 to 16 for coal gas and 18 to 36 for water gas. The illuminating power of the Welsbach Mantle, however, which was not dependent on a higher candlepower gas rating, made these standards obsolete.

Statutory change to accommodate the new technology occurred slowly and some utilities actually opposed shifting to a different standard because they feared a possible profit decline. Beginning with the Wisconsin Public Utilities Commission in 1908, however, states and municipalities began requiring that gas be measured in Btu or heat units per cubic foot of gas (as was done in Great Britain) rather than lighting units. Of the 1284 gas firms reporting in the 1914 Manufacturing Census, 1055 reported candlepower standards and 811 reported standards expressed in Btu heating values. By 1919, there had been a substantial change. Of 1022 establishments reporting, only 389 reported candlepower standards, whereas 737 reported Btu heating value standards.

The use of gas solely for lighting resulted in poor load factors for gasworks. Although they needed generating, storage, and distribution equipment capable of handling demand during peak lighting periods, their capital equipment was often utilized for only 10–12 h per day. The use of larger gasholders (gasometers) was one way utilities could reduce investment in gas-producing equipment. Smaller plants frequently ran their equipment during part of the day and relied on their gasholders to supply gas at night. Larger companies could not risk supply cutoffs and had to run separate units of gas producers, even though they too used gas holders for reserve. From the 1870s through the 1890s, several important improvements were made in gasholders, including the use of iron and steel tanks and the installation of the first three-lift holder tank. Because of limited capacity, however, gas holders were not a satisfactory solution to the load factor.

Competition from electricity, despite the success of the Welsbach mantle, consistently ate into the gas industry's market share. Utilities could attempt to spread their loads as a way to meet competition and to use their technology more efficiently. The gas industry moved increasingly into markets outside of lighting, such as cooking stoves, water heating, and house heating, although firm aggressiveness varied from place to place. In both Baltimore and Philadelphia, for instance, both large markets, the gas companies only gradually moved away from lighting as their only product. In other cities, however, such as Denver and Los Angeles, utilities introduced gas-consuming appliances more rapidly. Two technical innovations aided in the expansion of these markets: the Bunsen burner (1885), which premixed gas with air, and thermostatic controls, initially applied to water heating in 1899. This movement away from light and into household appliances reflected the recognition that manufactured gas was a fuel or energy source. It was also a "technical mimicry" of electricity and an attempt by an older technology to survive by copying a more recent technology, in this case electricity.

Even though the gas industry developed a substantial market share in the home utility domain, it still faced excess productive capacity. In ~1900, gas firms began to try to capture industrial customers to absorb this capacity. Efforts were made to market gas engines as well as to sell gas for heating purposes. Potential industrial customers, however, often found it more economical to use low-Btu gas to meet their energy requirements rather than high-Btu carbureted water gas. Frequently they used producer gas, a low-grade but inexpensive gas made on the firm's site from coal. Gas companies also found it difficult to enter industrial markets because of state requirements for a fixed number of Btu per cubic foot (usually 450–550 Btu) in their gas, which raised gas prices. This Btu rating was higher than many industries desired or would pay for, limiting the number of industrial customers. The gas industry, therefore, began to push for the sale of gas with Btu as a standard and sold by the therm (100,000 Btu), as was done in Great Britain, rather than by the cubic foot. Industries could then obtain less costly low-Btu gas to supply the heat they needed in their processes. Changes in this regard were reported in the 1920 Census of Manufacturers. The census noted that the prevailing Btu standard was lower in 1919 than it had been in 1914, as the number of gas firms providing high-Btu gas dropped sharply. Between 1919 and 1927, the industrial use of manufactured gas increased from 70.4 billion to 136.4 billion ft^3.

Gas utilities could also compete with electricity by investing in electrical generating equipment or by acquiring electrical companies. In 1887, approximately 40 gas companies supplied electric lighting, usually arc lights; by 1889, the number was 266, or approximately 25% of U.S. gas companies, and in 1899, 362 or nearly 40% of gas companies provided electric lighting. A counter trend was for electrical utilities to acquire gas utilities, and large electric firms or holding companies made acquisitions of small gas companies increasingly in the 20th century.

Water gas possessed another disadvantage compared to coal gas in terms of potential profits—the character of its by-products. Markets had developed in the late 19th century for coal gas by-products such as tar for creosote in railway tie preservation, pitch for roofing and waterproofing, and refined tar for road surfacing. The carbureting process produced gaseous, liquid, and solid by-products, but they were not as salable as those from coal gas. The liquid was "water-gas tar," which was lighter and contained less valuable constituents than coal tar. The solid was coke, the sale of which could substantially affect the profit position of plants. Water gas coke, however, was lower in thermal efficiency than coal gas coke. In addition, water gas firms, rather than entering the coke markets, consumed the coke they produced as water gas fuel. Finally, unlike coal gas, the water gas process did not produce ammonia, cyanides, or phenolic compounds as by-products. The ammonia was especially valuable and could constitute 6–10% of the cost of the coal used. Before the invention of the Haber process for the synthetic production of ammonia, coal carbonization was the main source of fixed nitrogen, and coal carbonization processes found increasing favor in the 1920s among some gas producers.

Between 1914 and 1929, the amount of manufactured gas distributed expanded from approximately 204 billion to over 450 billion ft^3. A little more than 50% of the gas produced in 1929 was water gas, approximately 40% was classified as coal gas, although the sum also included gas from by-product ovens, and oil gas was 6.2%. The most significant new addition to the census categories was gas from by-product coke operations.

Between approximately 1916 and 1929, a revolution occurred in the technology used by the iron and steel industry to make coke. From 1850 through 1910, most of the coke manufactured for use in blast furnaces was carbonized in beehive coke ovens. These ovens produced a high-quality coke from bituminous coal, but wasted all of the coal by-products. The by-product coke oven had originally been developed in Europe, and beginning in the 1890s several by-product coke plants were constructed in the United States. The real expansion of the United States industry, however, occurred just before World War I and in the following decade. The by-product coke oven captured the valuable coal chemicals for use and also produced a gas surplus. That is, by-product ovens could be heated by 35 to 40% of the coal gas produced in the oven, leaving approximately 60% to be marketed. Coke oven gas had a heating value of approximately 560 Btu and could be easily sold to other customers. By 1932, by-product coke firms produced 25% of the manufactured gas distributed by utilities to domestic, commercial, and industrial customers in the United States.

During the 1920s, the manufactured gas utilities themselves moved into the by-product industry. In order to meet peak period requirements, utilities with carbureted water gas technology had to manufacture large amounts of coke. By utilizing by-product ovens rather than retorts, utilities could combine gas manufacture with coke manufacture to meet peak load requirements and to produce a higher quality of coke for sale than that produced in retorts. During the 1920s, construction firms such as Koppers Company and Semet-Solvay built coke oven batteries for a number of large utilities, including Brooklyn Union Gas, Consolidated Gas Company of New York, New Jersey Public Service Gas & Electric, Rochester Gas & Electric Company, and People's Gas & Coke Company of Chicago. In addition, entrepreneurs built a number of merchant plants that had as their chief products foundry coke, domestic coke, and coke for making water gas, as well as producing gas to sell under contract to public utilities. These plants had the advantage of flexibility in meeting demand peaks for either gas or for coke as well as lowering the cost of gas generation. By the end of the decade, a clear trend existed for manufactured gas utilities to adopt carbonization equipment that produced both high-quality gas and high-quality coke, and by 1932, 18.7% of the manufactured gas produced by gas utilities was coke oven gas. Thus, as a 1926 text on fuels noted, "the gas and the coke industry gradually are merging as far as equipment requirements are concerned, and...the tendency is toward the same type of plant for both branches."

4. THE TRANSITION TO NATURAL GAS, 1927–1954

Natural gas is a substance that occurs in nature under three conditions: in porous domes that form a gas field (nonassociated gas), in association with petroleum deposits (casinghead or associated gas), and mixed with the oil so that it must be trapped out of it. Chemically, there is no difference between natural gas and manufactured gas—each is composed of hydrocarbons—but they differ in heating value. Manufactured gas usually averages between 550 and 600 Btu, whereas natural gas is approximately 1000 Btu. Natural gas is not distributed evenly throughout the environment. In the late 19th and early 20th centuries, there were four regions where natural gas was piped in for use in cities and industries: western Pennsylvania and West Virginia, northern and central Indiana, locations around Los Angeles including the San Joaquin Valley, and eastern Kansas. Towns and cities close to these fields utilized the natural gas until it was exhausted.

In the first decades of the 20th century, additional natural gas discoveries were made in the Southwest, but gas from these fields was initially wasted because of the absence of nearby markets and because the technology to produce pipelines long enough to reach markets was nonexistent. In the 1920s, however, significant breakthroughs occurred in the making and laying of welded pipe technology. In addition, the mathematical understanding of the role of tensile steel strength, pipe thickness and diameter, pipeline capacity, gas pressure, and compressor station spacing to pipeline capacity greatly increased. Pipeline construction passed beyond the craft stage into the stage where decisions were made by mathematical analysis.

From 1925 to 1935, pipeline companies and utilities constructed a number of pipelines ranging from 200 to over 1000 miles in length, connecting southwestern and California gas fields to urban markets. Cities such as Houston and Beaumont (1925), Wichita (1927), Denver (1928), St. Louis (1928), Atlanta (1929), San Francisco (1929), Minneapolis (1930), Chicago (1931) and Washington, DC (1931) were supplied with natural gas, driving out manufactured gas. The depression and the war slowed pipeline construction, but in 1944 the Tennessee Gas Transmission Company completed a 1265-mile pipeline from the Texas Gulf Coast to West Virginia as part of a wartime push to replace diminished Appalachian supplies.

The last major regions to be linked by pipeline with natural gas fields were the Middle Atlantic States and New England. New York City and Philadelphia were considered especially prized markets, since New York City alone consumed 40% of the nation's manufactured gas. These cities were supplied with natural gas in 1947 through the conversion of Big Inch and Little Inch, wartime government-constructed petroleum pipelines, to natural gas and the sale of these pipelines to private enterprise. New England was the last major region to receive natural gas, with Boston being the most distant urban market to draw from the southwestern gas fields. The postwar period also saw a boom in the construction of pipelines serving California and Florida, as well as new major pipelines to the midwest.

In these new markets for natural gas, the logistics of conversion of manufactured gas distribution systems into natural gas distribution systems was a major undertaking. Manufactured gas utilities did not always view conversion with favor, even though there seemed to be many advantages to the change. Some utility managers worried that natural gas supplies would prove insufficient to maintain their service commitments. Corporations were also reluctant to abandon their substantial investments in gas manufacturing plants and equipment.

In the conversion to natural gas as a fuel, the gas could be circulated in the same distribution system and consumed by the same appliances used by manufactured gas. However, because of the different Btu ratings of the gases, conversion required gas main and appliance adjustments. In addition, engineers and managers had to decide whether to use natural gas or natural gas mixed with manufactured gas after conversion. Occasionally regulatory commissions overturned the decisions made by utility managers.

The 1946 Federal Natural Gas Investigating Committee recommended that natural gas be used primarily for enrichment purposes. This strategy would expand the production capacity of existing plants, provide backup in case of supply interruptions, and furnish peak load shaving abilities. A number of firms followed the recommendation for mixed gas. In 1945, for instance, 301.4 million therms of manufactured gas and 988.5 millions therms of natural gas were used to produce mixed gas; by 1954, this total had reached 936 million therms of manufactured gas and 2012.7 million therms of natural gas. Other utilities, especially after World War II, went directly to natural gas, usually to avoid the costly necessity of adjusting gas appliances twice. In many cases, utilities kept their gas-making equipment for emergencies and for peak shaving purposes.

Once the decision concerning the type of gas to be used had been made, utilities faced major organizational tasks of conversion and adjustment, regardless of whether mixed or natural gas had been chosen. In 1931, an American Gas Association Subcommittee issued a report on "The Study of Factors Involved in Change-over from Manufactured to Natural Gas." The committee made four major recommendations: (1) that cities be divided into districts for conversion purposes and that each district be converted to natural gas separately; (2) that those handling the conversion task be thoroughly trained and that the work force be large enough so that the largest district could be covered in 1 week; (3) that customers be kept fully informed at all times of the conversion process; and (4) that a universal adapter be used to adjust orifices. The committee also warned of the drying-out effects of the natural gas on joints and valves in a system that had formerly only carried wet manufactured gas, with resultant leaks. The committee concluded, however, that compared to manufactured gas, natural gas presented fewer hazards. In approaching the conversion process, utilities communicated extensively with one another regarding the pitfalls to be avoided.

The largest single task facing a utility converting to natural gas was adjusting domestic and commercial appliances to the new gas. All gas-using appliances had to be adjusted at the orifice because of natural gas' higher Btu rating and its slower rate of flame propagation compared to 550–600 Btu for manufactured gas. Management of conversion could involve thousands of specially trained workers. In order to carry out the conversion tasks, the larger utilities hired engineering management firms. Some utilities, however, managed their own transitions rather than hiring outside contractors, forming schools and training men specifically for the job and relying on experience from other utilities.

With the transformation in the gas industry, the fuel flow now originated from a distant source rather than traveling a relatively short distance from gasworks to customer. From a regulatory point of view, this shifted responsibility to the federal government because pipelines often crossed state boundaries. Pipeline companies became subject, after the Natural Gas Act of 1938, to federal control, which had a constraining and shaping effect on the industry.

5. ENVIRONMENTAL IMPACTS OF THE MANUFACTURED GAS INDUSTRY

The various by-products produced by the manufactured gas industry, if not captured and disposed of in a safe manner or sold, could become environmental externalities, damaging surface and groundwater quality, poisoning soils with toxic substances, and producing odors and smoke. British gas engineers were aware during the industry's early years of the potential for environmental pollution. In London, for instance, in 1821, effluents from gasworks killed fish and eels in the Thames. In 1841, in his *Practical*

Treatise on the Manufacture of Gas, published in London, engineer Thomas Peckston warned of the possible escape of "tar and ammonical liquor" from the "tar cistern" because it could "percolate through the ground" and render water in springs and wells "unfit for use."

Americans also became aware of the injurious environmental effects of manufactured gas effluents on the environment as the industry expanded. New York State, for instance, passed a statute in 1845 restricting the disposal of gas house wastes in streams in the New York City area; Pittsburgh, Pennsylvania passed a similar statute in 1869 and New Jersey approved legislation limiting the disposal of gas house wastes in streams in 1903 and 1911. Plaintiffs filed numerous nuisance suits against manufactured gas companies in the late 19th and early 20th centuries for offenses such as odors, pollution of wells, damage to trees, and the killing of finfish and shellfish. Various city councils and state bodies undertook investigations of the environmental effects of manufactured gas wastes as well as their pricing policies beginning in the middle of the 19th century. Warnings about the pollution potential of gas house wastes appeared with increasing frequency in the professional literature during the first decades of the 20th century.

Because of the potential costs of nuisance suits as well as the damage to the industry's reputation, in 1919 the American Gas Association established a committee on the "Disposal of Waste from Gas Plants." The 1920 report of the committee noted that gas plant wastes could drive "away fish and damage oyster beds," create "damage to paint on pleasure boats," cause "objectionable odors," result in "pollution of wells [by] contaminating ground water," create "deposits in sewerage systems," and cause "pollution of drinking water supplies where water is chlorinated." The committee provided numerous recommendations to help manufactured gas plant operators to deal with pollution disposal questions and even volunteered to visit sites with problems. The committee continued to meet throughout the 1920s and to explore these issues.

The gradual phasing out of the manufactured gas industry in the 1950s and 1960s because of the conversion to natural gas resulted in the elimination of many plants and a gradual loss of institutional memory concerning their operations. The renewed interest in land and groundwater pollution that developed after Love Canal in 1978 and the subsequent passage of the Superfund Acts, however, focused renewed attention on the sites of former manufactured gas plants (MGPs), many of which had been stripped of gas-making technology and structures. Investigation of these sites by state and federal environmental officials identified a number of toxic by-products remaining in the soil and in groundwater from past MGP operations. Some of the MGPs had operated for a century or so and most sites contained accumulated wastes from the gas-making processes. Thus, an industry largely of the past that had operated approximately 2000 plants in towns and cities throughout the nation and provided it with fuel for light and heat for over a century was discovered to have left a heavy environmental burden for the future.

6. CONCLUSIONS

This article has focused on the evolution of the manufactured gas industry in the United States and the transition to natural gas as well as its environmental legacy. Manufactured gas systems were urban public utilities that, because they utilized city streets for their distribution systems, required first municipal and later state approval for their operations and rates.

Manufactured gas systems shared a number of characteristics with other urban networks, such as water supply, sewerage, and district heating. These were "grid-based systems" that required a special physical network devoted to supplying users. They were also centralized rather than decentralized and had natural monopoly characteristics, a factor that affected their rate-making and consumer relations and, at times, resulted in their becoming deeply involved in politics.

The substitution of natural gas for manufactured gas resulted in the gradual death of the manufactured gas industry as an entity that produced a product rather than merely distributing it. However, much of the original industry's grid remained in the form of inner-city distribution systems. The substitution of natural gas for manufactured gas did not require network reconfiguration but only adjustments because the delivery system and appliances from the first system were retained for use with the natural fuel.

The industry had large environmental effects, both during its period of operation and after, because of wastes discarded during its processing of coal and oil for purposes of producing a fuel for illumination and energy. As an industry it furnished both illumination and energy to American cities for many decades, but its significance today is measured primarily in terms of the environmental burdens it has left to contemporary society.

SEE ALSO THE FOLLOWING ARTICLES

Coal, Chemical and Physical Properties • Coal, Fuel and Non-Fuel Uses • Coal Industry, History of • Electricity Use, History of • Hydrogen, History of • Hydropower, History and Technology of • Natural Gas, History of • Nuclear Power, History of • Oil Industry, History of • Solar Energy, History of • Transitions in Energy Use

Further Reading

American Gas Association. (1956). "Historical Statistics of the Gas Industry." American Gas Association, New York.
Anderson, N., Jr., and DeLawyer, M. W. (1995). "Chemicals, Metals and Men: Gas, Chemicals and Coke: A Bird's-Eye View of the Materials That Make the World Go Around." Vantage Press, New York.
Binder, F. M. (1974). "Coal Age Empire: Pennsylvania Coal and Its Utilization to 1860." Pennsylvania Historical and Museum Commission, Harrisburg, PA.
Blake, A. E. (1922). Water gas. *In* "American Fuels" (R. F. Bacon and W. A. Hamor, Eds.), Vol. 2, pp. 995–1015. McGraw-Hill, New York.
Block, E. B. (1966). "Above the Civil War: The Story of Thaddeus Lowe, Balloonist, Inventor, Railway Builder." Howell-Nort Books, Berkeley, CA.
Castaneda, C. J. (1993). "Regulated Enterprise—Natural Gas Pipelines and Northeastern Markets, 1938–1954." Ohio State University Press, Columbus, OH.
Castaneda, C. J. (1999). "Invisible Fuel: Manufactured and Natural Gas in America, 1800–2000." Twayne Publishers, New York.
Castaneda, C. J., and Pratt, J. A. (1993). "From Texas to the East: A Strategic History of Texas Eastern Corporation." Texas A&M University Press, College Station, TX.
Collins, F. L. (1934). "Consolidated Gas Company of New York." The Consolidated Gas Company of New York, New York.
Davis, R. E. (1935). Natural gas pipe line development during past ten years. *Natural Gas* **16**, 3–4.
Eavenson, H. N. (1942). "The First Century and a Quarter of American Coal Industry." Privately Printed, Pittsburgh, PA.
Elton, A. (1958). Gas for light and heat. *In* "A History of Technology." (C. Singer, *et al.*, Eds.), Vol. IV, pp. 258–276. Oxford University Press, New York.
Harkins, S. M., *et al.* (1984). "U. S. Production of Manufactured Gases: Assessment of Past Disposal Practices." Environmental Protection Agency, Research Triangle Park, NC.
Hasiam, R. T., and Russell, R. P. (1926). "Fuels and Their Combustion." McGraw-Hill, New York.
Hughes, T. P. (1983). "Networks of Power: Electrification in Western Society, 1880–1930." Johns Hopkins University Press, Baltimore, MD.
Hunt, A. L. (1900). Gas, manufactured. *In* "U. S. Bureau of the Census. Twelfth Census of Manufacturers (1900), Part IV: Reports on Selected Industries," p. 19. U.S. Government Printing Office, Washington, DC.
Hyldtoft, O. (1995). Making gas: The establishment of the Nordic gas systems, 1800–1870. *In* "Nordic Energy Systems, Historical Perspectives and Current Issues" (A. Kaijser and M. Hedin, Eds.), pp. 75–100. Science History Publications, Canton, OH.
Jacobson, C. D., and Tarr, J. A. (1998). No single path: Ownership and financing of infrastructure in the 19th and 20th centuries. *In* "Infrastructure Delivery: Private Initiative and the Public Good" (A. Mody, Ed.), pp. 1–36. The World Bank, Washington, DC.
Kaijser, A. (1990). City lights: The establishment of the first Swedish gasworks. *FLUX* **1**, 77–84.
Kaijser, A. (1993). Fighting for lighting and cooking: Competing energy systems in Sweden, 1880–1960. *In* "Technological Competitiveness: Contemporary and Historical Perspectives on the Electrical, Electronics, and Computer Industries" (W. Aspray, Ed.), pp. 195–207. IEEE Press, New York.
Kaijser, A. (1999). Striking bonanza: The establishment of a natural gas regime in the Netherlands. *In* "The Governance of Large Technical Systems" (Olivier Coutard, Ed.), pp. 38–57. Routledge, London, UK.
King, T. (1950). "Consolidated of Baltimore 1816–1950." The Company, Baltimore, MD.
Leinroth, J. P. (1928). Industrial gas in the United States—Growth and future trends. *In* "Transactions of the Fuel Conference, Vol. II, The Carbonisation Industry Utilization of Fuels," pp. 1211–1217. Percy Lund, Humphries & Co., London, UK.
Matthews, D. (1987). The technical transformation of the late nineteenth-century gas industry. *J. Econ. History* **47**, 970–980.
Passer, H. C. (1953). "The Electrical Manufacturers 1875–1900." Harvard University Press, Cambridge, MA.
Passer, H. C. (1967). The electric light and the gas light: Innovation and continuity in economic history. *In* "Explorations in Enterprise" (H. G. J. Aitken, Ed.). Harvard University Press, Cambridge, MA.
Platt, H. L. (1991). "The Electric City: Energy and the Growth of the Chicago Area, 1880–1930." University of Chicago Press, Chicago, IL.
Rose, M. H. (1995). "Cities of Light and Heat: Domesticating Gas and Electricity in Urban America." Pennsylvania State University Press, State College, PA.
Sanders, M. E. (1981). "The Regulation of Natural Gas Policy and Politics, 1938–1978." Temple University Press, Philadelphia, PA.
Schivelbusch, W. (1988). "Disenchanted Night: The Industrialization of Light in the Nineteenth Century." University of California Press, Berkeley, CA.
Stotz, L., and Jamieson, A. (1938). "History of the Gas Industry." Stettiner Bros, New York.
Tarr, J. A. (1998). Transforming an energy system: The evolution of the manufactured gas industry and the transition to natural gas in the United States (1807–1954). *In* "The Governance of Large Technical Systems" (O. Coutard, Ed.), pp. 19–37. Routledge, London, UK.
Tarr, J. A., and Dupuy, G. (eds.). (1988). "Technology and the Rise of the Networked City in Europe and America." Temple University Press, Philadelphia, PA.
Thoenen, E. D. (1964). "History of the Oil and Gas Industry in West Virginia." Education Foundation, Charleston, WV.
Thorsheim, P. (2002). The paradox of smokeless fuels: Gas, coke and the environment in Britain, 1813–1949. *Environ History* **8**, 381–401.
Troesken, W. (1996). "Why Regulate Utilities: The New Institutional Economics and the Chicago Gas Industry, 1849–1924." University of Michigan Press, Ann Arbor, MI.

Tussing, A. R., and Barlow, C. C. (1984). "The Natural Gas Industry: Evolution, Structure, and Economics." Ballinger, Cambridge, UK.

Weber, F. C. (1922). The future of the artificial gas industry. *In* "American Fuels" (R. F. Bacon and W. A. Hamor, Eds.), Vol. 1, pp. 1095–1106. McGraw-Hill, New York.

Williams, T. I. (1981). "A History of the British Gas Industry." Oxford University Press, Oxford, UK.

Williamson, H. F., and Daum, A. R. (1959). "The American Petroleum Industry: The Age of Illumination 1859–1899." Northwestern University Press, Evanston, IL.

Marine Transportation and Energy Use

JAMES J. CORBETT
University of Delaware
Newark, Delaware, United States

1. Introduction
2. Overview of Marine Transportation
3. Energy Choice and Marine Propulsion Development
4. Trends, Opportunities, and Challenges

Glossary

bunker Marine fuel, typically for international shipping. The term is adapted from the days of coal-fired steam boilers; fuel was stored aboard ship in coal bunkers and was shoveled into the boiler; when oil replaced coal, the bunkers became fuel tanks, but the name stuck.

deadweight ton (dwt) The weight of cargo, fuel, and supplies needed to make a vessel fully loaded; in other words, it is the weight difference between a completely empty and a fully loaded vessel.

gross registered ton (grt) The capacity on a ship, in cubic feet, of all spaces within the hull (body of the ship) and the enclosed spaces above the deck available for passengers, crew, and cargo, divided by 100; simply, 100 cubic feet of capacity is equal to 1 gross registered ton; the equivalent term, gross ton (gt), is also used.

marine distillate oil (MDO) A group of distillate oils that can be burned in auto-ignited engines (typically called diesel fuels). MDO can be specified with a range of properties such as density, flashpoint, and maximum sulfur content. Sulfur contents may not exceed 2% by weight. This fuel requires very little pretreatment for proper combustion in marine engines.

marine gas oil (MGO) A group of distillate oils that are typically clean and clear, used for emergency diesel engines, lifeboat diesel engines, etc. Marine gas oil is comparable to No. 2 fuel grade (onroad diesel fuel) used commercially ashore. MGO can be specified with a range of properties, such as density, flashpoint, and maximum sulfur content.

residual fuel (heavy fuel oil) (HFO) A group of residual fuels that can be burned in auto-ignited engines (typically called diesel fuels). HFO can be specified with a range of properties such as density, flashpoint, and maximum sulfur content. This fuel typically requires heating and filtering for proper combustion in marine engines. The maximum sulfur content allowed is 5% by weight, but the world average is much lower (~2.7% sulfur).

tonne-kilometer (tonne-km) The transport of 1 tonne of cargo over a distance of 1 km.

More than any other transportation mode, most modern shipping activity is designed to minimize fuel consumption because it helps minimize operating costs. As a result, waterborne commerce generally is considered an energy-conserving mode of transportation. On average, ships and barge tows accomplish the work of moving cargo on oceans and inland rivers at much less energy per tonne-kilometer, compared to trucking, rail, or air modes. This is especially true for cargoes shipped in bulk, such as oil or grain (see Table I). Published data suggest that there are important exceptions to this effort: (1) naval ships are designed for power and performance for military readiness and warfare and (2) some passenger ships (e.g., fast ferries) need to meet high-speed service requirements as a condition of high-value passenger service. The focus in this article is on the characteristics of the marine transportation system, the role of energy conservation in marine propulsion development, and trends in energy use for marine transportation as ships meet increased demand for cargo trade worldwide.

1. INTRODUCTION

Conservation of energy in marine transportation has been important since before the classical times of the Greeks and Romans. Shipping has been an important

TABLE I
Comparison of Modal Energy Intensities[a]

Mode	National averages of energy intensity[b] (MJ/tonne-km)	Modeled ranges of energy intensity[c] (MJ/tonne-km)
Average road freight	1.8–4.5	
Heavy-duty trucks	0.60–1.0	
Rail (freight trains)	0.40–1.0	0.25–0.50
Marine freight (average)	0.10–0.40	
General cargo		0.20–0.50
Container ship		0.20–0.30
Bulk carrier		0.10–0.25
Oil tanker		0.10–0.20
Air freight	7–15	

[a] As the data show, marine transportation is less energy intensive, on average, compared to other transport modes. Specific energy intensities depend strongly on individual vehicle load factors and patterns of use.
[b] Data from United Nations Framework Convention on Climate Change Working Paper No. 1, Appendix A, Table 21.
[c] Data from International Maritime Study of Greenhouse Gas Emissions from Ships, Appendix A4, Figure 4.9.

human activity throughout history, particularly where prosperity depended primarily on commerce with colonies and interregional trade with other nations. From ancient times, when ships sailed mostly in sight of land using wind-driven sails or slave-driven oars, to today's oceangoing fleet of steel ships powered by the world's largest internal combustion engines, the history of shipping includes many examples of technical innovations designed to increase ship performance while conserving energy, capital, and labor. Whether the primary energy for ship propulsion has been wind and sail or engines and petroleum, all major innovations in marine transportation throughout history have involved balancing these factors. Since the 1800s, when industrialization made combustion power available to shipping, marine propulsion technology development has focused on ways to move a ship using less fuel and smaller crews (See Table I).

2. OVERVIEW OF MARINE TRANSPORTATION

The role of marine transportation in today's commerce and society must be considered to understand energy use practices and trends. Marine transportation in the 21st century is an integral, if sometimes less publicly visible, part of the global economy. The marine transportation system is a network of specialized vessels, the ports they visit, and transportation infrastructure from factories to terminals to distribution centers to markets. On a worldwide basis, some 44,000 oceangoing vessels move cargo more than 33 billion tonne-km annually (see Table II). In the European Union, marine transportation moves more than 70% (by volume) of all cargo traded with the rest of the world; in the United States, more than 95% of imports and exports is carried by ships. This work is accomplished by ships using 2 to 4% of the world's fossil fuels.

Maritime transportation is a necessary complement to and occasional substitute for other modes of freight transportation (see Fig. 1). For many commodities and trade routes, there is no direct substitute for waterborne commerce. (Air transportation has replaced most ocean liner passenger transportation, but carries only a small fraction of the highest value and lightest cargoes.) On other routes, such as some coastwise or short-sea shipping or within-inland river systems, marine transportation may provide a substitute for roads and rail, depending on cost, time, and infrastructure constraints. Other important marine transportation activities include passenger transportation (ferries and cruise ships), national defense (naval vessels), fishing and resource extraction, and navigational service (vessel-assist tugs, harbor maintenance vessels, etc.).

2.1 Marine Transportation System Complexity

The marine transportation system is a complex element of global trade. From an energy consumption perspective, the most important elements are the vessels; complexity of the marine transportation may be seen most easily by examining the complex mix of ships and boats that comprise the system's vehicles.

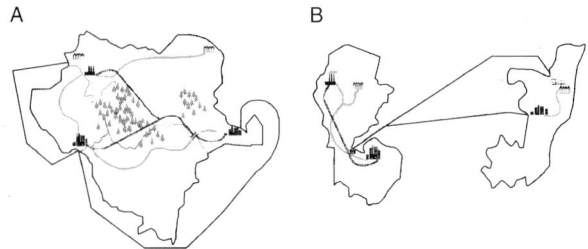

FIGURE 1 The marine transportation system as (A) a substitute for and (B) a complement to land modes.

TABLE II

Profile of World Commercial Fleet, Number of Main Engines, and Main Engine Power

Ship type	Number of ships	Percent of fleet	Number of main engines	Percent of main engines	Installed power (MW)	Percent of total power	Percent of energy demand[a]
Cargo fleet							
Container vessels	2662	2	2755	2	43,764	10	13
General cargo vessels	23,739	22	31,331	21	72,314	16	22
Tankers	9098	8	10,258	7	48,386	11	15
Bulk/combined carriers	8353	8	8781	6	51,251	11	16
Noncargo fleet							
Passenger	8370	8	15,646	10	19,523	4	6
Fishing vessels	23,371	22	24,009	16	18,474	4	6
Tugboats	9348	9	16,000	11	16,116	4	5
Other (research, supply)	3719	3	7500	5	10,265	2	3
Registered fleet total	88,660	82	116,280	77	280,093	62	86
Military vessels[b]	19,646	18	34,633	23	172,478	38	14
World fleet total	108,306	100	150,913	100	452,571	100	100

[a] Percent of energy demand is not directly proportional to installed power because military vessels typically use much less than their installed power, except during battle. The average military deployment rate is 50 underway time per year; studies indicate that when underway, naval vessels operate below 50% power for 90% of the time. Therefore, energy demand was adjusted in this table to reflect these facts.

[b] The data on which military vessel power was estimated specified the number of engines aboard ship associated with propulsion systems.

Ships, barges, ferries, towboats, and tugboats are among the most recognized and romanticized parts of the marine transportation system, perhaps second only to the water. Unlike automobiles, of which many thousands of each model are built and sold each year and a popular car design may remain largely the same for years, each ship design is unique to only one or a few vessels.

Because of different service requirements, ship vary greatly in terms of hull designs, power plants, cargo loading/discharge equipment, and other features. Inland-river and coastal towboats range in power from 0.5 to 4 megawatts (MW), or 600 to ~5000 horsepower (hp), and can push between 1 and more than 30 barges in a tow. Passenger ferries often have capacity to carry 150 to 300 passengers and range in power from 2 to 4 MW (~2600 to ~5000 hp), depending on speed requirements; some ferries, such as the Staten Island ferries in New York, have capacity for 6000 passengers. Roll-on, roll-off ships (or RoRos, so called because vehicles and other rolling cargo are driven onto the vessel rather than loaded by crane) can carry between 200 and 600 automobiles and have installed power between 15 and 25 MW (~20,000 and ~33,000 hp). Tankers and dry bulk carriers can carry more than 250,000 tons of cargo, with installed power often in the range of 25 to 35 MW (~33,000 to ~46,000 hp). Although container ships are not often as large as tankers, they have much larger power plants to accommodate greater vessel speeds. Average container ships carrying between 1750 and 2000 standard shipping containers have installed power of 20 to 25 MW (~26,000 to ~33,000 hp); larger container ships with capacity for more than 4000 containers can have installed power of 35 to 45 MW (46,000 to 60,000 hp), and the largest container ships can carry more than 6000 standard containers with engines rated as high as 65 MW (86,000 hp).

To accommodate complexity in geography, cargoes, and vessels, ports and terminals also vary greatly. The primary common goals at ports and terminals are to achieve rapid vessel loading and unloading and to maximize the rate of cargo transport through the port facilities. Many ports are designed to facilitate container movements between ship, truck, and rail. Transportation planners and engineers have been working to speed this port transfer with on-dock rail, larger cranes, and improved queuing schedules for trucks. Some ports are megahubs that attract much of the world trade (by value and volume), whereas other ports are networked ports that handle much of the niche cargoes or serve as feeders for megaports. Each of the world's ports is unique by location (proximity to market), waterway configuration, and ecosystem, making general characterizations difficult. (Although each of these developments affects the overall energy consumption required to move cargo, the land-side energy demands of industrial port terminals are not addressed herein.)

In terms of the work performed, measured in tonne-kilometers or ton-miles, the marine transportation system rivals the other modes. The relative share of cargo moved by water, compared with truck and rail modes, varies from year to year, but U.S. waterborne commerce represents between 22 and 24% of the total tonne-kilometer of cargo movements in recent years. Truck and rail modes in the United States each account for about 25–29% of annual cargo tonne-kilometer. In Europe, the European Commission reports that waterborne commerce between nations (called shortsea shipping) accounts for nearly 41% of the goods transport market, compared to 44% for trucking and only 8% for rail and 4% for inland rivers.

One of the reasons that marine transportation moves so many tonne-kilometers of cargo is cost efficiency. Choice of mode is a function of cost (shipping is usually cheaper), time (shipping is often slower), and other quality-of-service factors. Only pipelines move goods at lower average cost than shipping, because of their fixed capital and ability to move fluids (liquids and gases), at very low energy requirements. Bulk shipment of liquid and dry goods such as oil and grain costs about two to three times as much per tonne-kilometer as pipelines, averaging between 2.5 and 3.5¢/tonne-km according to U.S. national transportation statistics. These average costs are generally less than but similar to rail, whereas marine transportation of bulk goods can cost much less than trucking. Except for high-value containerized goods that are shipped intermodally, the cost per tonne-kilometer for trucking can be more than an order of magnitude greater than the cost of marine transportation.

In energy conservation terms, shipping is not only among the least costly modes of transportation but efforts to keep down costs focus operators directly on fuel savings. Fuel costs can be the first or second largest cost of operating a vessel (not considering vessel capitalization); labor is generally the other largest cost and may vary with crew nationality. Fuel costs can range between 20 and 60% of operating costs, depending on vessel design and route factors. In order to be competitive, shipping companies try to reduce costs and/or add value to their service. These economic motivators have been very strong in the shipping industry and motivated much of the progress in energy conservation on ships.

2.2 International Fleet Propulsion Profile and Vessel Energy Use

The world's ships are primarily powered by diesel engine systems, consisting of one or more diesel engines in one of three typical propulsion configurations. Diesel direct-drive engines are typically large, slow-speed diesel engines directly connected to the propeller; they rotate at exactly the same revolutions per minute (rpm) as the propeller (0–130 rpm), except when connected to controllable-pitch propellers. Diesel gear engines drive the propeller indirectly through a set of gears, allowing the engine speed to be selected independently of the propeller. Diesel electric engines drive generator systems that turn the propeller; these engines typically operate at a constant speed, with electric power providing variable power to the propeller shaft. Diesel electric and diesel gear engines can operate at speeds greater than propeller revolutions per minute; in the international fleet, these are generally medium-speed engine designs operating between 130 and 2000 rpm. A fourth major category of engine, steam turbine systems, relies on conventional steam generation to drive a turbine geared to the propeller shaft; these marine systems typically have thermal and fuel efficiencies lower than modern marine diesel efficiencies and are rarely installed on new vessels. Other propulsion systems, not in wide use today, include combustion turbines (gas turbines) and older or nontraditional designs such as wind-powered vessels, reciprocating steam pistons, and nuclear power (for some military vessels). The evolution of engine technologies is discussed further in Section 3.

Most of the energy required by marine transportation is used on cargo or passenger transport vessels (ships that move cargo or passengers from one place to another in trade). A profile of the internationally registered fleet of ships greater than 100 gross tons is shown in Table II. Transport vessels account for almost 60% of the ships and nearly 80% of the energy demand of the internationally registered fleet (not including military ships); considered along with military ships, cargo ships account for 40% of the world fleet and 66% of world fleet fuel use. Cargo ships are analogous to on-road vehicles because they generally navigate well-defined (if unmarked) trade routes, similar to a highway network. Other vessels are primarily engaged in extraction of resources (e.g., fishing, oil, or other minerals), or are primarily engaged as support vessels (vessel-assist tugs, supply vessels). Fishing vessels are the largest category of nontransport vessels and account for more than one-quarter of the total fleet. Fishing vessels and other nontransport ships are more analogous to off-road vehicles, in that they do not generally operate along the waterway network of trade routes. Rather, they sail to fishing regions and operate within that region, often at low power, to extract the ocean resources. As a result, fishing vessels require much less energy.

The exception to energy conservation designs in shipping would be the nearly 20,000 military vessels worldwide, which are sized for power and performance by their military mission, not for plant fuel economy. The naval ship is designed for sustained speeds in excess of their endurance (cruise) speed; as a result, the installed power in the world Navy is nearly 40% of the total installed power of the entire commercial fleet. However, the world's Navy ships use less energy than their installed power would suggest, because most of the time they operate in peaceful conditions at endurance speeds, partly to achieve fuel/cost savings. Average engine power characteristics for the U.S. Navy show that military ships operate below 50% power for 90% of the time that they are underway. Moreover, military vessels spend greater time in port than do modern commercial ships. Whereas commercial ships cannot effectively earn profits unless underway, Navy studies claim that military ships typically spend as much as 60–70% of their time in port; this is confirmed by public records of Naval ship activity, which show that typical deployment rates range between 40 and 55% of the fleet. This means that, in practice, military ships would demand some 14% of total energy required by the world fleet (commercial plus military fleets), much less than their installed power implies (Table II).

Fuel consumption is generally proportional to installed power on commercial vessels, and this power is concentrated in the larger commercial ships. About two-thirds of commercial ships are powered by four-stroke compression-ignition engines (operating on the compression-ignition, or diesel, cycle, and therefore referred to as diesel engines). More than one-quarter are powered by two-stroke diesel engines. Recent fleet statistics report that 6% of the ships have "unknown" diesel engines (i.e., either two- or four-stroke engines) and only 1% are turbine-driven. Most turbine-driven vessels (80%) are steam turbines with oil-fired boilers; the number of aero-derivative gas turbine engines in the commercial fleet is very low.

When estimating fuel consumption from commercial ships, the installed engine power rather than the number of engines or vessel size provides the most direct indication of energy requirements. The registered fleet has approximately 84,000 four-stroke engines with total installed power of 109,000 MW and some 27,000 two-stroke engines with total installed power of 164,000 MW. Engines with "unknown" cycle types and "turbines" together make up only about 2.5% of total installed power for main engines. This suggests that 27,000 two-stroke marine prime movers account for almost 60% of the commercial fleet's total energy output and fuel consumption, and more than half of the energy demanded by the world oceangoing fleet (including military ships). A majority of these engines are large-bore, low-speed diesel engines with above 10 MW rated output. Two-stroke engines are the main consumers of bunker fuel and therefore are the major sources of oceangoing ship emissions, followed by four-stroke engines.

The fuel types used in marine transportation are different from most transportation fuels. Marine fuels, or bunkers, used in today's fleet, can be generally classified into two categories: residual fuels and other fuels. Residual fuels, also known as heavy fuel oil (HFO), are a blend of various oils obtained from the highly viscous residue of distillation or cracking after the lighter (and more valuable) hydrocarbon fractions have been removed. Since the 1973 fuel crisis, refineries have used secondary refining technologies (known as thermal cracking) to extract the maximum quantity of refined products (distillates) from crude oil. As a consequence, the concentration of contaminants such as sulfur, ash, asphaltenes, and metals has increased in residual fuels. Residual fuel has been described by industry professionals as "black, thick as molasses, and has an odor that's pungent, to put it politely." By contrast, "other" bunkers used in international shipping include marine distillate oils of higher grade; however, these fuels often consist of a blend of heavier fractions (residuals) with distillate oils and are not to be confused with the more highly refined fuels used in motor vehicles. Domestically, ferry vessels and harbor craft may use a more familiar diesel fuel that currently meets less stringent standards for maximum contaminant concentrations; however, in the United States, at least much of the marine fuel used by inland towboats, workboats, and ferries is rebranded on-road diesel. Few commercial ships, but many recreational vessels (not addressed here), use spark-ignited engines and marine gasoline oils (MGOs).

To reduce fuel costs, marine engines have been designed to burn the least costly of petroleum products. Residual fuels are preferred if ship engines can accommodate its poorer quality, unless there are other reasons (such as environmental compliance) to use more expensive fuels. Of the two-stroke, low-speed engines, 95% use heavy fuel oil and 5% are powered by marine distillate oil (MDO). Fuel consumed by 70% of the four-stroke, medium-speed engines is HFO, with the remainder burning either MDO or marine gas oil. Four-stroke, high-speed engines all operate on MDO or MGO. The remaining engine types are small, high-speed diesel engines all operating on MDO or MGO, steam turbines powered by boilers fueled by HFO, or gas turbines powered by MGO.

2.3 National and International Energy Statistics for Ships

The world's fleet operates internationally, but individual ships are registered under national authorities. Understanding which nations have maritime and trade interests is important to understanding which nations provide fuel for these ships. In terms of fuel sales, the nations providing the most fuel for ships are typically nations with strong interests in the cargoes or services those ships provide. Table III presents a summary of the top nations selling international marine fuels during the 1990s, according to the World Energy Database. The United States provides most of the world's marine fuels by far, and together the top 20 nations selling international marine fuels (shown in Table III) account for more than 80% of total marine fuel sales.

According to the Energy Information Administration's World Energy Database, some 140 million tonnes of fuel has been recorded as sold for international marine fuel consumption in recent years. At least 70% of these fuel sales has been identified as residual fuel, although recent work by Jos Olivier and co-workers at the National Institute for Public Health and the Environment in The Netherlands [Rijksinstituut voor Volksgezondheid & Milieu (RIVM)] suggests that residual fuel probably accounts for more than 80% or more of oceangoing fleet energy. (Typically, statistics for international "other" bunkers include international fuel sold to aircraft, which are excluded from the quantities discussed here.) This lack of clarity in the definitions makes the statistic somewhat uncertain, but the general trends are well documented.

Organization for Economic Cooperation and Development (OECD) nations account for roughly half of these fuel sales and provide a good illustration of historical consumption trends in the overall fleet, according to the International Energy Agency. Detailed fuel statistics for OECD nations over the past several decades (Fig. 2) show that marine fuel consumption has increased gradually, along with a growth in global trade and transportation. Over the entire period shown, the increase in marine fuel sales

TABLE III

International Marine Fuel Sales by Nation

Percent of total sales (1990–1999, inclusive)	Nations selling residual bunkers	Nations selling other bunkers
>15%	United States (18%)	United States (22%)
6–15%	Singapore (9%), Russia (9%), Netherlands (8%), United Arab Emirates (8%)	Saudi Arabia (12%)
2–5%	Japan, Saudi Arabia, Belgium, South Korea, Spain, South Africa, Greece	Hong Kong, Singapore, Netherlands, South Korea, United Kingdom, Spain, Russia
~1%	France, Taiwan, China, Italy, Egypt, Netherlands Antilles, Hong Kong, United Kingdom, Germany	Thailand, Greece, India, Belgium, Italy, Brazil, Indonesia, Denmark, Egypt, Venezuela, Germany
<1%	47 other countries	92 other countries

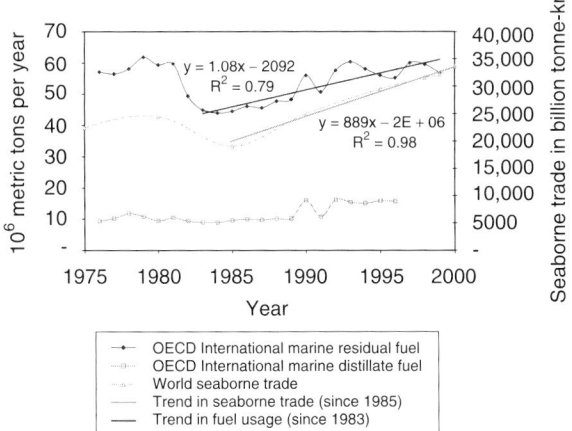

FIGURE 2 International marine fuel sales for Organization of Economic Development and Cooperation (OECD) nations and world seaborne trade, according to the International Energy Agency, Energy Information Agency, and United Nations data. The regression equations represent a simple linear statistical annual growth relationship. For residual fuels, the slope (1.08) suggests that, on average, an additional 1.08 million metric tons of residual marine fuel is sold each year by OECD nations; this represents an annual growth rate of about 2% for all international marine residual fuel sold. Note that there was significant variability in residual fuel sales during the 1990s and that the trend for international marine distillate furl is much flatter. For seaborne trade, the slope suggests that each year, ships move cargo an additional ~900 tonne-km; this represents an annual growth rate of about 3.5% in seaborne trade.

averages only about 0.2%/year, but since 1983 (i.e., after the recession in the early 1980s), these data show about a 2% annual average growth in residual fuel use. This growth in fuel consumption is about the same as recent annual growth rates in the fleet of ships, according to United Nations statistics. In other words, the fuel economy per ship may be roughly constant. However, over the long term, the fleet may be performing more useful work (i.e., transporting more cargo over longer distances) with a less-than-proportionate increase in energy demand. During a similar period (1985–2000), as shown in Fig. 2, seaborne trade increased slightly faster (~3.5%/year on average) than fuel sales; this could indicate that economies of scale and/or improved cargo logistics may work to reduce the energy consumed per cargo tonne-kilometer of work performed by the fleet.

An important caveat to consider is that these international fuel statistics do not represent the total fuel consumed by the international fleet of ships. Even making simplistic and conservative assumptions about the number of hours that vessels operate main engines and applying general fuel consumption rates for marine diesel engines, the international fleet of ships presented in Table II would require more fuel than reported as international marine bunkers. Recent estimates based on ship activity and installed engine power suggest that the world fleet of ships (including cargo, noncargo, and military vessels) consumes some 284 million metric tonnes of fuel per year, with more than 200 million metric tonnes required for cargo ships alone. This represents nearly 4% of all fossil fuel use.

Clearly, internationally registered vessels must be consuming domestic supplies of marine fuels. However, fuel statistics are generally well understood, especially for the OECD nations that sell most of the world's marine fuels. There are two reasons for the discrepancy. First, not all statistical sources for marine fuels have defined international marine fuels the same way. Original International Energy Agency (IEA) definitions have been reworded to be more consistent with reporting guidance under the *Revised 1996 IPCC Guidelines for National Greenhouse Gases*. The IEA defines "international marine bunkers (fuel) [to] cover those quantities delivered to seagoing ships of all flags, including warships. Consumption by ships engaged in transport in inland and coastal waters is not included." The IEA defines national navigation to be "internal and coastal navigation (including small craft and coastal vessels not purchasing their bunker requirements under international marine bunker contracts). Fuel used for ocean, coastal and inland fishing should be included in agriculture." Second, domestic fuel consumption inventories include fuel provided to oceangoing ships, at least during voyages that do not meet the definition criteria for international marine fuels.

Reconciling fleet fuel consumption between international and domestic fuel supplies is not simple. To get agreement with international fuel statistics and with the definitions, about 31% of the fuel consumed by transport vessels must occur while operating in shortsea or coastwise service using domestic fuels, and all of the nontransport ships (fishing, research, tugs, etc.) operate on fuel included in domestic inventories. This appears to be the case. The U.S. Energy Information Administration publishes state-by-state consumption statistics by fuel type and by end-use sector for the United States, the world's largest provider of marine fuels over the past 10 years at least (accounting for some 21% of marine residual sold by OECD nations). These Combined State Energy Data System (CSEDS) statistics identify residual fuel consumption in the transportation sector in the United States domestic inventory, which represents marine residual fuels primarily because no other mobile source category consumes residual fuel.

A comparison of marine residual fuels on a state-by-state basis, alongside state-by-state waterborne commerce statistics, shows good agreement. First, total residual fuel consumption reported in the CSEDS agrees exactly with domestic residual consumption statistics in the World Energy Database, providing a consistency check. Second, the 10-year average residual consumption by marine transportation in the United States equals more than 40% of total residual consumption, more than enough to accommodate the difference between fleet energy requirements and international residual fuel statistics. Third, the top states for waterborne commerce (measured by total cargo volume) are also the top states for marine fuel sales. This suggests that most of domestic marine fuel consumption is going to cargo transport ships, and to vessels that directly support commerce. In terms of energy conservation, it is also worth noting that, although domestic land-based consumption of residual fuel has decreased significantly over the past two decades, marine transportation fuel consumption has not changed much, either internationally (Fig. 3) or domestically (Fig. 4).

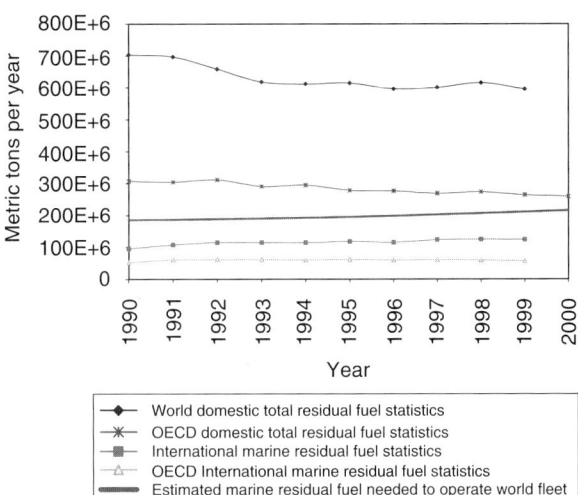

FIGURE 3 Comparison of domestic residual fuel consumption and international marine residual fuels worldwide and by Organization for Economic Cooperation and Development (OECD) members.

3. ENERGY CHOICE AND MARINE PROPULSION DEVELOPMENT

Aside from the shift of human labor (oars) to wind-driven sails, the first modern energy conversion in marine transportation was the shift from sails to combustion. Two primary motivators for energy

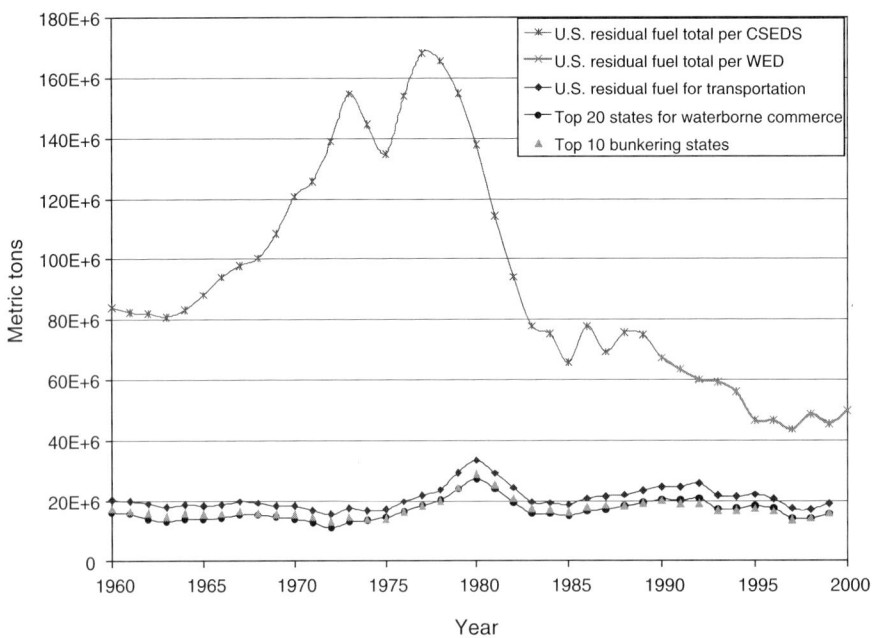

FIGURE 4 United States domestic residual fuel statistics, including residual fuel for transportation. Nearly 100% of residual fuel used in transportation is used by oceangoing marine vessels; nearly all inland vessels, harbor craft, and small commercial workboats use diesel fuel. CSEDS, Combined State Energy Data System; WED, World Energy Database.

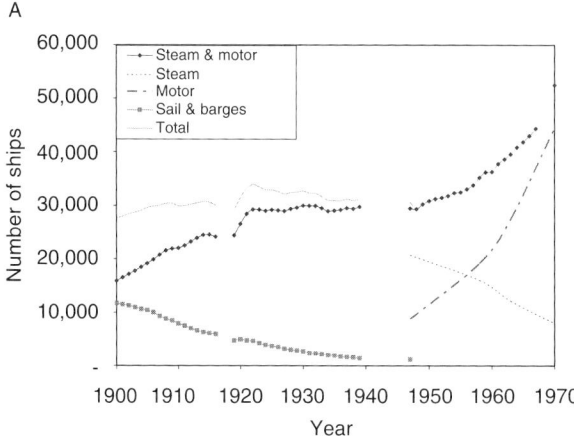

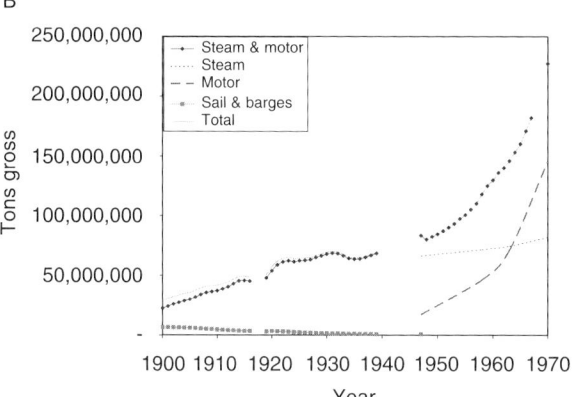

FIGURE 5 Conversion of ship propulsion, from sail to combustion propulsion, for internationally registered ships greater than 100 gross tons. (A) The conversion from sail as measured by number of ships took approximately 50 years to complete. (B) The effective fleet conversion of vessel tonnage to higher performing combustion systems took less than 20 years. During the shift from steam to motor, after 1948, the number of steamships decreased steadily. Economies of scale meant that ships with steam plants were still being built (mostly tankers) and that they were among the largest ships in the fleet. Note that ship registry data were not published during major world wars. Data from Lloyd's Register of Shipping Statistical Tables for various years.

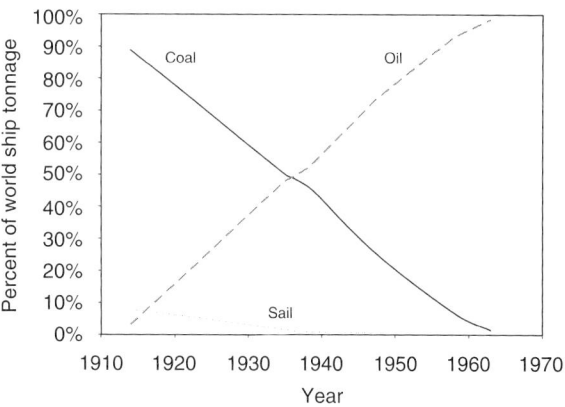

FIGURE 6 First major shift of world shipping to alternative fuel (coal to oil). Note that due to sparse data, particularly during world wars, these curves may misrepresent how smooth the transition was; the switch to oil was likely accelerated during periods of heavy shipbuilding (e.g., during the war periods and times of rapid economic growth). Data from Lloyd's Register of Shipping Statistical Tables for 1914, 1935–1939, 1947, 1958, and 1963.

technology innovation—greater performance at lower cost—caused this conversion. Figure 5 illustrates how this shift was completed during the first half of the 20th century, using data from Lloyd's Register Merchant Shipping Return for various years. Essentially, newer and larger ships adopted combustion technologies as part of an economy of scale. These technologies enabled trade routes to emerge, regardless of the latitudes, without consistent winds (referred to as the doldrums) supporting both international industrialization and modern political superpower expansion. As shown in Fig. 5, the conversion of fleet tonnage to the preferred technology was achieved much more rapidly than was the phaseout of smaller ships using the outdated technology; this lead in conversion by tonnage was because the new technology was installed on the larger and newer vessels. Initially, these ships were powered by coal-fired boilers that provided steam first to reciprocating steam engines and later to high-speed steam turbines that drove the propellers. Later, the introduction of the industry's first alternative fuel, petroleum oil, enabled the introduction of modern marine engines. This pattern is repeated in many technology changes for marine transportation: some ship operators continue to use long-lived vessels purchased on the secondhand market while industry leaders replace their fleets to achieve new markets or realize economies of scale.

The switch from coal to oil was clearly motivated by a desire to conserve fuel. Switching fuels was both a direct means of conservation and a technology enabler during the introduction of internal combustion engines that ultimately replaced steam turbines and boilers. Of course, the reasons were economic: According to British Admiral Fisher's remarks to Winston Churchill in 1911 (quoted in Yergin's 1991 book, *The Prize*, page 155), a cargo steamer could "save 78 percent in fuel and gain 30 percent in cargo space by the adoption of the internal combustion propulsion and practically get rid of stokers and engineers." Essentially, the commercial sector was converting to oil-fired boilers powering marine turbines and introducing oil-fueled internal-combustion, compression-ignition engines, commonly known as the diesel engine.

Figure 6 illustrates how international shipping switched from coal to oil over five decades. It is worth emphasizing that energy conservation was not the primary purpose for this conversion to "alternative fuel" in the early 1900s. Oil-powered commercial ships required fewer crew members and enjoyed a greater range of operations between fueling. This was not only of commercial interest; military vessels appreciated these advantages and the fact that refueling at sea could be accomplished more quickly and easily. Oil-powered ships also accelerated more quickly than did coal-powered systems, and could achieve higher speeds. When Churchill converted the British Navy to oil, despite the fact that this conversion made the British fleet dependent on a fuel not found within the British islands, the industry conversion was irreversible. After the fuel conversion was implemented, the next big shift was to more fuel-efficient marine diesel engines. A historical overview by Joe Evangelista in the American Bureau of Shipping's *Surveyor* describes how marine diesels came to dominate marine propulsion:

> *The first oceangoing diesel ship, the 370-ft Selandia, was delivered to the East Asiatic Company in 1912 by Burmeister & Wain (B&W) shipyard, Copenhagen, Denmark. In the 20 years since the first commercial diesel motor was proven, through the collaboration of Rudolph Diesel and Maschinenfabrik Augsburg-Nürnberg (MAN), the new type engine had been the subject of intense development and speculation. It promised industry a revolutionary new power source, and in so doing threatened the world's coal producers. So Selandia's sailing was one of the biggest news items of the time. Winston Churchill, Britain's First Lord of the Admiralty, visited the vessel and hailed it as an achievement of world proportion.*
>
> *Selandia was powered by twin B&W eight cylinder, four-stroke diesels, each developing 1,250 hp. Within the year, competitors like MAN, Sulzer, Vickers and Krupp were working to develop two-stroke engines with an output near 2,000 hp per cylinder—beginning a power competition that has not paused since. Observing these early efforts, Rudolph Diesel commented, "If, as seems probable, these tests give satisfactory results, the era of very large diesel engines has come."*
>
> *The king of propulsion for the biggest commercial vessels—tankers, bulkers, and containerships—is the large, slow-speed, two-stroke diesel engine. It became a market force during the late 1920s, as growing ship size and power requirements began slipping beyond the capabilities of four-stroke engines. By making the intake and exhaust phases occupy a fraction of a stroke, instead of a full stroke each, the two-stroke conferred several advantages. It doubled the power output per unit size over four-strokes, and made higher power outputs accessible through bigger piston sizes and longer strokes. In addition, the low speed two-stroke allowed direct coupling to the propeller. Over time, two-stroke manufacturers would also come to claim—as they do today—higher efficiency and greater overall reliability for these engines.*
>
> *The two-stroke diesel's 50-year battle against steam for supremacy of the seas was settled amid the skyrocketing fuel prices of the 1970s. In 1975, for example, of 614 engines delivered for the biggest commercial ships, 78 percent (482) were low-speed diesels, and the remaining 123 were steam turbines. In 2000, of the 1,059 main engines delivered to vessels over 2,000 dwt, 60 percent were low-speed two-strokes. Most of the remainder was medium-speed four-strokes.*
>
> *Four-stroke marine diesels are applied everywhere from racing boats and motor yachts to cruise ships, (freight and passenger) vessels and shuttle tankers (and in a large number of stationary power plants). Their higher speed requires some intermediary between engine and propeller (reduction gearing or diesel-electric drive), a setup offering advantages of quieter operation, lower vibration, and highly flexible engine room layout and location. Competition for this variety of business is divided among a large field of manufacturers, led by Wärtsilä (51 percent of the market in 2000) and MAN B&W Diesel.*
>
> *In their competition for higher performance [and fuel economy], diesel designers focused most of their efforts on raising power while controlling engine weight. To gage their success, consider just the power increase between two 420 mm-bore, 12-cylinder two-stroke engines, separated by 30 years. In 1968, B&W's 1242-VTBF-90 delivered a total output of 6,600 bhp on a stroke of 900 mm at a speed of 220 rpm. In 2001, MAN B&W's 12S42-MC delivered 17,640 bhp on a stroke of 1,764 mm at a speed of 136 rpm—a power increase of about 2.5 times within nearly the same engine silhouette. Such power increases were realized across the diesel spectrum. Top engine power jumped in that time from some 40,000 to 100,000 bhp. Average piston speeds increased from about 6.6 m/s to about 8.5 m/s.*
>
> *The tremendous increase in internal forces and pressures that accompanied this power rise was handled not by doubling the physical size of the engine, but through advanced materials, forging, and structural technologies. There have also been great gains in thermal efficiency, the measure of an engine's ability to get mechanical work out of the energy potential of the fuel. Efficiency has been raised from roughly 40 percent in 1975 to 50 percent today, with hybrid systems pushing 55 percent. In the orderly, high-tech workshops of today's diesel designers, the massive components of even the biggest engines—like piston crowns nearly a meter in diameter and fuel injectors 600 mm long—are hand-machined to watchworks precision in order to coax every fraction of power from the combustion process. New methods of fuel treatment, computerized injection and cylinder controls, and new exhaust handling technologies optimize power while lowering the pollutant content of engine emissions. And now, the last bastion of steam propulsion, the LNG carrier, is under assault from compact, attractively-priced diesels that burn a combination of oil and natural gas.*
>
> —Joe Evangelista, in the American Bureau of Shipping's *Surveyor* (Spring 2002), pp. 14–20

Although the introduction of motor ships dominated new construction, the phaseout of existing sail and steam vessels still took many decades. For example, in 1948, steam ships accounted for 68% of the ships in the fleet and 79% of the fleet tonnage, whereas motor ships accounted for 29% of ships and only 20% of the tonnage; sail still powered 4% of vessels but only 1% of registered ship tonnage. By 1959, motor ships accounted for 52% of vessels and 39% of registered tonnage in the fleet, and in 1963, motor ships represented 69% of vessels and 49% of registered tonnage. By 1970, motor ships dominated the fleet both in terms of ships and cargo tonnage, with 85 and 64%, respectively.

For maritime nations, this rapid modernization of the world fleet was anything but shared. In fact, a nation's fleet modernizes according to economic and political motivators within each country. This means that some large merchant fleets simply age while some of the latest technologies go to emerging maritime nations. For example, after World War II, the United States had the second largest fleet by number and the largest fleet by tonnage capacity; its fleet was populated by ships built during the war to carry supplies. These ships were nearly all steam turbine powered, and they would last at least two decades. As a result, the U.S. merchant fleet of ships did not invest significantly in newer motor-driven vessels; moreover, despite stated intentions to promote a modern and competitive fleet, policymakers continued a practice of industry subsidy and protection that discouraged domestic shipbuilding and subsidized older, costlier ships that burned more fuel and required larger crews. Even as late as the 1990s, when the world merchant fleet was more than 97% diesel powered, the U.S. oceangoing fleet was more than 50% steam-powered.

4. TRENDS, OPPORTUNITIES, AND CHALLENGES

The marine transportation system has clearly considered the importance of energy conservation in its technological and economic development. However, whether these achievements continue will depend on how marine energy systems respond to growth in trade (increased demand to perform work), emerging environmental regulations (competing constraints on power system design), and increased attention to CO_2 reductions to mitigate climate change (higher expectations for energy conservation improvements). Consideration of each of these in terms of the global fleet of ships that will continue to provide a primary transportation and trade service worldwide is the focus of this final section.

4.1 Increasing Trade Volumes

Global trade has increased dramatically over the past century, and projections continue to forecast a doubling of trade volume for major nations, and as much as tripling in cargo volumes to major world ports. Carriage of these increasing trade volumes will by necessity include water transportation, and energy conservation will be a factor. There exist three possible ways for the marine transportation system to accommodate these forecasted cargoes. First, the existing fleet might simply carry more through better logistics and more fully loaded ships to maximize efficiency. Second, additional ships may be added to carry the additional cargoes, resulting in an increase in ship traffic overall. Third, the fleet might transition to larger vessels without increasing the overall number of ships in service in order to provide greater capacity within the same network of routes. In practice, each of these alternatives may be applied in some combination.

First, on certain routes and for certain cargoes, it may be possible to carry more cargo with the current fleet. However, there is an upper bound to this alternative. Most shipping routes are practically unidirectional, meaning that most of the cargo flows in one direction. For example, oil is primarily imported into the United States, and grain is primarily exported. Bulk carriers typically transit empty (or under ballast) in one direction and return with a load of cargo; this results in a practical upper bound of 50% capacity overall on routes for bulk carriers (e.g., averaging 100% one way and 0% the other, or more realistically, averaging 80–90% one way and 10–20% on the return). For container and general cargoes, some routes are able to secure a backhaul; however, there is limited ability for ships to realize this idealized goal. Where a strong backhaul market can be established, the backhaul freight rates usually are much lower and may not directly cover the costs of the voyage; here, as for bulk carriers, the freight rate for "inbound" cargoes must cover most of the cost of the round-trip voyage. For inland-river and shortsea shipping, backhauls are more practical, because vessel capacities are smaller. In fact, the industry appears to behave as though there is an upper bound. Most ships carry loads that average 50–65% capacity or less. When cargo capacities exceed 70%, it can be an indication that too few ships are available for the route; under these conditions, ships are added to the route. Where possible, this may help reduce growth in energy consumption by the fleet.

Second, adding ships to increase route capacity is easier than it might be for on-road vehicles in the current transportation system. This is primarily because ocean routes are "unconstrained" highways. For most route distances, few modifications will be needed to enable more vessel traffic. However, in and near port regions, this may not hold. The number of berths at terminals and cargo handling rates provide at least two important limits on how many vessels a port can accommodate in a given period. No consistent metric to evaluate port capacity exists, but clearly, the capacity of the marine transportation system will be limited if ships have to wait for a berth at a port. This may involve increased energy demand for waterborne transportation, and unless these added ships divert cargoes from land-based modes, the overall energy consumption will likely increase.

The third alternative is to continue increasing economies of scale for shipping with larger ships, better cargo stowage technologies, and improved cargo logistics planning. Examples of this include the 6000-tonne equivalent unit (TEU) containerized "megaships" and supertankers. Very large ships are being employed on some routes where the cargo volumes are strong, and this practice is projected to grow. However, trade logistics may limit the cost-effectiveness of applying this option across the fleet. There will continue to be important niche routes and specialty cargoes that require vessels smaller than the megaships. This alternative may allow for even greater growth in trade while minimizing increases in energy consumption.

In general, the challenge of decoupling energy requirements from a growing global trade will be significant. The cost motivation of carriers to reduce fuel consumption will continue to promote conservation research and innovation; however, cheaper fossil-based fuels may limit the conservation progress in shipping. This is an important consideration, because many of the alternative energy sources are much more expensive than residual fuels and will require matching engine technology to the fuel.

4.2 Environmental Constraints and Continuing Conservation Efforts

Environmental impacts from shipping have been the focus of increasing attention over the past decades, primarily focused on waterborne discharges and spills of oil, chemical, and sewage pollution. A number of international treaties and national laws have been adopted, along with industry best practices, to prevent these pollution releases through accident or substandard operation. However, the shipping industry is seen as an important hybrid between land-based transportation and large stationary power systems. On the one hand, these are non-point-source vehicles operating on a transportation network of waterways; on the other hand, these vehicles operate ship systems that compare to small power-generating plants or factories. As industry leaders and policymakers at state, federal, and international levels recognize these facts, environmental and energy performance is being measured and regulated. Recent efforts to mitigate environmental impacts from shipping (for example, invasive species in ballast water, toxic hull coatings, and air emissions) are relatively new for the industry and will take decades to address.

Environmental regulation of the shipping industry typically lags behind regulation of similar industrial processes (engines, etc.) on land. At the international level, the International Maritime Organization has effectively developed a set of treaties addressing much of these ship pollution issues. Under the International Convention for the Prevention of Pollution from Ships, these regulations (MARPOL 73/78, a combination of two treaties from 1973 and 1978) apply global standards to environmental protection practices aboard ship. However, national and state jurisdictions can apply stricter regulations to certain vessel operations and to the ports that receive these ships. For example, the U.S. Environmental Protection Agency recently regulated shipping under mandated authority in the Clean Air Act. Another example includes efforts by port states in Sweden and Norway, and by individual ports in California, to apply market-based or voluntary regulations to reduce air pollution from ships entering their ports.

Environmental regulation may make some energy conservation efforts more difficult, by introducing environmental trade-offs. For example, hull coatings are specialized for marine conditions to prevent hull fouling; some of these coatings have been shown to affect the marine environment, particularly in ports and harbors, by releasing toxic chemicals or metals. These compounds slowly "leach" into the seawater, killing barnacles and other marine life that have attached to the ship. But the studies have shown that these compounds persist in the water, killing sea life, harming the environment and possibly entering the food chain. Toxic hull coatings are being banned by national and international regulations, and this may threaten to increase energy consumption by shipping. Replacing these coatings with less toxic alternatives may result in increased rates of hull fouling and increased fuel requirements.

4.3 Future Alternative Fuels and Energy Conservation

Future energy conservation efforts in shipping are increasingly considering factors other than cost. Certainly, energy independence and fossil fuel resource availability continue to play a factor. Traditional pollution mitigation also plays a role. And most recently, climate change concerns are motivating reductions in CO_2 emissions that are directly proportional to fossil fuel combustion. Marine fuels, like all petroleum products, are about 86% carbon. If achieving greater reductions in CO_2 from marine transportation is to be achieved through means other than cost-driven fuel economy, then alternative fuels will likely be introduced.

It is worth noting that alternative fuels for energy conservation were the focus of significant attention during the energy crisis in the 1970s. In fact, the National Research Council conducted a large study, Alternative Fuels for Maritime Use, motivated explicitly by President Carter's national energy policy. That report proposed a 20-year plan to conduct research and development toward conversion of fleet to alternative energy sources; however, very little marine transportation research was pursued because energy independence descended from the top of the public policy agenda after the crisis. To a large extent, this work is being updated by current policy efforts to achieve CO_2 reductions to mitigate climate change impacts. Whether the political attention will sustain such a long-term research and development effort remains to be seen; after all, it took nearly five decades to switch the industry from coal to oil even with the performance advantages.

The current focus on alternative fuels in maritime applications involves various types of fuel, including biofuels such as biodiesel, clean diesel products, natural gas, and hydrogen. Solar, wind, and other noncombustion sources (e.g., nuclear) are sometimes identified but do not appear to be particularly suited for modern ships or global seaborne trade. Hydrogen is attracting increasing attention as a long-term goal among alternative fuels; for that reason it will be the main example for this discussion, although the issues discussed generally apply to other fuels as well.

There are challenges to the successful introduction of alternative fuel in any transportation mode. Five factors that may favor marine transportation as a cost-minimizing mode for hydrogen (or other alternative fuel) introduction have been discussed by Farrell et al. in a 2003 *Energy Policy* article. In general, converting to hydrogen fuel in freight modes first may offer certain cost-saving advantages over passenger-only modes of transportation. Some factors may apply equally or better to nonmarine modes as well, but here they are discussed only in context with marine transportation.

First, vehicle design and performance factors are important constraints on less dense fuels such as hydrogen. When the fuel-storage:payload ratio is high, switching to hydrogen will involve greater cost trade-offs. For smaller ferries, this may mean carrying significantly fewer passengers; for large cargo ships, the loss of cargo capacity may be less. Also, if the alternative fuel (hydrogen) has less energy density, then there may be impacts on acceleration, which would matter less to slower, oceangoing vessels than to high-speed harbor craft.

Second, ship operations and management may be well suited to adopting alternative fuels. Ship crews are generally well trained and licensed, and requirements for shipboard qualifications are becoming stricter. This means that commercial vessels with professional crews may be able to adopt innovative technologies more safely and at lower cost than the general public can accomplish. Moreover, if there is an increased risk of accident from the alternative fuel (which may not be strictly true for hydrogen versus petroleum fuels), then introducing a new fuel into marine cargo transportation may be accomplished through crew training that minimizes exposure to the general public. Also, fueling operations for marine vessels occur at commercial docks or by fuel barges in a commercial port harbor. Furthermore, because most commercial vessels are operated around the clock and throughout the year, investments to convert to alternative fuel may be spread across a greater period of operation than, say, is the case for passenger vehicles, which are typically driven only a few hours each day.

Third, one of the most important factors in alternative fuels is the cost of energy infrastructure. Larger, more centralized refueling locations tend to reduce the cost of delivering the alternative fuel. In the case of marine transportation, about 30 major

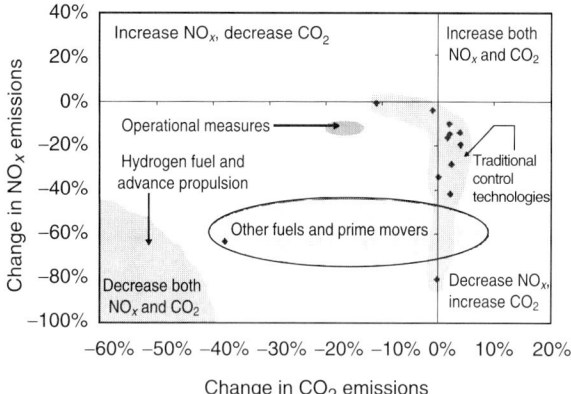

FIGURE 7 Potential trade-off between technologies to achieve reductions in traditional pollutants (NO_x) and fuel economy (in terms of CO_2).

fuel providers (refiners) in the United States supply most of the marine fuel, and most of that fuel is delivered in the top 20 ports for waterborne commerce.

Fourth, reduced air emissions are one of the benefits offered by many alternative fuels, and improved environmental performance may be greater in less regulated industries such as international and domestic shipping, compared to passenger cars, for which emissions standards have been applied increasingly for more than three decades. Converting ships to hydrogen will reduce emissions of traditional pollutants significantly, depending on the prime mover used. Figure 7 illustrates the potential trade-offs between different technologies and fuels aimed at reducing ship air emissions versus achieving greater energy conservation. It appears likely that any of the trade-offs will not be motivated by a reduction in vessel operating cost.

Fifth, shipbuilding and marine engine production rates will also be important to any fuel conversion. On the one hand, the facts that ships use tailored designs with generally similar power systems and that they are built one at a time make the process of innovation more effective than for massed-produced vehicles such as passenger cars. On the other hand, the few number of ships that are built annually and their long operating lifetimes make economies of scale for new propulsion systems difficult to achieve. However, marine engines are currently available that operate on gaseous fuels, and the long tradition of engine designs that can handle a wide range of fuel properties may contribute to successful demonstration efforts.

SEE ALSO THE FOLLOWING ARTICLES

Aircraft and Energy Use • *Fisheries and Energy Use* • *Internal Combustion (Gasoline and Diesel) Engines* • *Tanker Transportation* • *Transportation and Energy, Overview* • *Transportation and Energy Policy* • *Transportation Fuel Alternatives for Highway Vehicles*

Further Reading

Corbett, J. J., and Fischbeck, P. S. (1997). Emissions from ships. *Science* 278(5339), 823–824.

Corbett, J. J., and Fischbeck, P. S. (2000). Emissions from waterborne commerce in United States continental and inland waters. *Environ. Sci. Technol.* 34(15), 3254–3260.

Corbett, J. J., and Koehler, H. (2003). Improving the accuracy of ship emissions inventories. *JGR Atmospheres* 28 (D20), 4650–4666.

Energy Information Administration (EIA). (2001). "World Energy Database and International Energy Annual 2001." EIA, Washington, D.C.

Evangelista, J. (2002). Big diesel balancing act [quarterly magazine of the American Bureau of Shipping]. *Surveyor* (Spring 2002), 14–20.

Farrell, A. E., Keith, D. W., et al. (2003). A strategy for introducing hydrogen into transportation. *Energy Policy* 31(13), 1357–1367.

International Energy Agency (IEA). (1977–1997). "Quarterly Oil Statistics" [various issues]. Organization for Economic Cooperation and Development, Paris, France.

International Organization for Standardization (ISO). (1987). "International Standard, Petroleum Products—Fuels (Class F)—Specifications of Marine Fuels." ISO, Geneva, Switzerland.

Kendall, L. C., and Buckley, J. J. (1994). "The Business of Shipping." Cornell Maritime Press, Centreville, Maryland.

Lloyd's Maritime Information System (LMIS). (2002). "The Lloyds Maritime Database." Lloyd's Register, Fairplay Ltd., London.

Lloyd's Register of Shipping. (1970–1994). "Merchant Shipbuilding Return." Lloyd's Register of Shipping, London.

National Research Council (U.S.) Commission on Sociotechnical Systems. (MTRB). (1980). "Alternative Fuels for Maritime Use." National Academy of Sciences, Washington, D.C.

Skjølsvik, K. O., Andersen, A. B., et al. (2000). "Study of Greenhouse Gas Emissions from Ships (MEPC 45/8 Report to International Maritime Organization on the outcome of the IMO Study on Greenhouse Gas Emissions from Ships)." MARINTEK Sintef Group, Trondheim, Norway; Center for Economic Analysis, Carnegie Mellon University, Pittsburgh; and Det Norske Veritas, Oslo, Norway.

United Nations Conference on Trade and Development (UNCTAD). (2002). "Review of Maritime Transport 2002." United Nations, Geneva.

United States Army Corps of Engineers (USACE). (1999). "Navigation Data Center (NDC) Publications and U.S. Waterway Data." Available on compact disk from Water Resources Support Center, NDC, Alexandria, Virginia.

Yergin, D. (1991). "The Prize." Touchstone, New York.

Market-Based Instruments, Overview

STEPHEN FARBER
University of Pittsburgh
Pittsburgh, Pennsylvania, United States

1. Types of Instruments
2. Principles and Applications to Energy-Related Environmental Management

Glossary

command-and-control A regulatory method by which acceptable activities are tightly prescribed, such as designating the permissible technologies for pollution control.

double dividend The claim that environmental taxes not only benefit the environment, but the collected revenues can be beneficially used to reduce other tax distortions.

emissions allowances Certifiable permits to emit pollutants.

emissions reduction credits Certifiable credits received for reducing emissions below allowance levels or beyond required reductions.

externalities Inadvertent effects on persons, such as adverse health effects of pollution, from activities of production or consumption.

marginal control cost Additional costs of controlling emissions by one more unit (ton, pound, etc.).

market-based instruments Regulatory tools that use market-like forces to provide flexibility and incentives to meet regulatory objectives.

performance bonds A financial obligation to pay when performance criteria are not met.

pigovian taxes Levies that reflect the social costs of activities, such as pollution, that are not typically accounted for in traditional markets.

tradable permits Quantified permits to engage in an action, such as emissions, whose ownership can be transferred among parties.

Market-based instruments (MBI) are proposed as one of several types of regulatory tools for attaining environmental, ecological, and resource management objectives. MBI mimic processes found in traditional markets, such as price signals, trading, and information. They allow behavioral flexibility by entities acting in their own self-interest. MBI include the following specific instruments: prices, marketable rights, subsidies, liability, and information. They contrast with command-and-control (CAC) instruments, which stipulate with varying degrees of specificity what actors must do and how they must do it, for example, establishing a technology standard for pollution control or specifying allowable emissions. MBI are also referred to as incentive-based instruments, although the latter term is a bit misleading as CAC instruments with fines for violations create incentives. Each of these MBI create circumstances where some activities critical to regulatory objectives are not fully prescribed, leaving varying degrees of discretion open to the regulated parties. Interest in MBI reflects concerns for cost-effective management and the need to persuade regulated parties to buy into regulation. They provide more flexibility in the achievement of environmental goals than CAC. Whereas CAC instruments have been successful in moving toward these goals, their typical "one size fits all" rigidity and centralized specification of behavior standards have led to concerns of overly costly regulations, as actors are allowed little flexibility and have minimal incentives for developing new control technologies. Cost savings obtained by shifting from CAC to MBI may be a prerequisite to the attainment of higher, more costly, environmental standards.

1. TYPES OF INSTRUMENTS

1.1 Prices

The function of prices as a MBI is to force actors to explicitly consider the environmental, ecological, or resource implications of their activities. By placing a price on an activity, the actor can then consider the trade-offs involved in avoiding that activity. For example, a price placed on each ton of emissions should induce actors to consider the cost of reducing emissions by 1 ton compared to the price they must pay if they do not reduce. This is illustrated in Fig. 1. Emissions reductions become progressively more expensive, as the Marginal Control Cost illustrates. These control costs are attributable to any options the actor wishes to employ to avoid emissions, including end-of-pipe controls, internal process changes, input selection, and reduced product output. The actor would have complete flexibility in selecting options and would presumably select those that are least costly. If a price is levied for emissions, an actor seeking to keep emissions-related costs low will find it cheaper to reduce emissions up to level ER* than to pay the price of not reducing emissions. The total costs of reducing emissions by an amount ER* in Fig. 1 are represented by area B and the net cost savings by avoiding the price levy are represented by area A. The actor will not reduce emissions beyond ER*, as it costs more to do so than would be saved by avoiding the levy.

Prices may be levied directly on the activity of concern to the regulator, such as emissions, or on related activities, such as the sulfur content of fuels. Where the levy is placed determines its effectiveness. Administrative concerns may determine where and how the price is levied. For example, it is administratively simpler to place a price levy on permitted activities, such as emissions volumes, than on other less readily observable activities. Legal and political concerns may determine whether the levy can be a tax or a fee, where the latter is reserved for levies that compensate the regulator for services rendered, such as a permit fee for financing the permitting process.

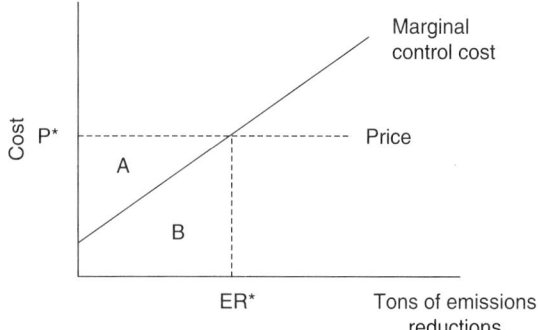

FIGURE 1 Emissions pricing and control.

The terms fines and penalties are reserved for charges based on violations of stipulated activities and are typically associated with CAC instruments.

Financial assurances, such as performance bonds, are a potentially useful MBI that place an explicit price on undesirable events. Typically the actor must post some financial assurance, such as a bond or letter of financial assurance, that can be drawn upon if some undesirable event, such as a spill or incomplete mining remediation, occurs. A price is placed on the event and the actor must pay it when the event occurs. The anticipation of this price obligation guides the actor's behavior. These assurances both manage undesirable activities through a price signal and ensure equity by holding the actor financially accountable, rather than leaving others with unwanted liabilities.

Pricing has many advantages from a social perspective. A single price sends the same signal to all actors, inducing each of them to control behavior to the same level of marginal control costs, which means that there are not some actors whose control costs far exceed the costs of others. It appears simple to administer, although it requires verification through monitoring, which may be more expensive than the cost savings from control efficiencies. Actors have incentives to consider new technologies that reduce their control costs, but only when the prices remain the same and are not lowered as all actors find cheaper methods of control. But actors may not be supportive of pricing, as they must pay for control plus the levies on remaining activities or emissions. Of course, levies may be placed on activities above a certain level to counteract potential opposition to this "double cost."

The benefits of pricing socially adverse environmental or ecological impacts of activities are reductions in welfare losses from overuse of the environment and overuse of resources in activities that harm the environment. In addition, some have claimed that using these environmental prices for public revenues allows the reduction in perverse or resource distorting taxes, such as income taxes, that diminish welfare. The benefit to welfare from environmental enhancement and reductions in otherwise distorting taxes is referred to as the "double dividend" or environmental tax reform. There is controversy over whether the second dividend is large or even positive. Environmental taxes may not raise significant revenues if the environmental activity can be easily reduced when taxes are imposed. The welfare costs of preexisting noncompetitive market structures may only be increased with such taxes. And revenue-neutral shifting of taxation to undesirable environmental activities and away from desirable activities, such as work, would increase commodity prices to such an extent that real wages would fall, discouraging labor participation.

1.2 Marketable Rights

Endowing the actor with rights to some activity and allowing the actor to market those rights provide flexibility in achieving regulatory objectives. For example, a power plant can be allowed to emit a given volume of SO_2. But if it is allowed to buy or sell these permitted emissions allowances, they become marketable rights or tradable permits. The prudent plant will then buy or sell them based on the market prices of these allowances compared to the cost of emissions control. Plants that find it more costly to control emissions to levels permitted under their initial allowances than the market price of allowances will seek to buy allowances in the market and emit more than their initial allocation and vice versa. The result is that high-cost plants can avoid excessive control costs and low-cost plants can control to higher levels than their initial allowances. These marketable rights shift control responsibilities across sources so that a total emissions objective can be attained at lower costs than under the initial allocation of responsibility and they provide individual plants with complete flexibility in determining how much and how they will control emissions.

The rights can be for allowable activities, such as emissions allowances, or for reductions in activities, such as emissions reduction credits. The principle of trading is the same; responsibility is shifted from those who find it more costly to meet objectives to those who find it less expensive. For example, in Fig. 1, if an actor is required to reduce emissions more than ER* and the price of emissions reduction credits is P*, the actor will buy emissions reduction credits, thus allowing emissions to be reduced to a lesser degree. The actor would end up reducing emissions by ER*, as it would be less expensive to buy reduction credits than incur the greater costs of control. Similarly, an actor with an initially assigned responsibility of less than ER* will find it profitable to reduce emissions more, as more can be received for the sale of these excess reductions than it costs to create them.

The prices established through marketable rights trading guide the behavior of the actors, just as direct prices set by regulators. So there is a strong parallel between the two sets of instruments. They both have the potential to accomplish regulatory objectives at lower costs and with more flexibility than CAC instruments.

1.3 Subsidies

Subsidies are the opposite of the pricing instrument outlined in Section 2.1. Whereas a price is a payment extracted from actors enabling them to do something, a subsidy is a payment to an actor to do something. An example would be a payment to a power plant for each ton of SO_2 removed from the waste stream. In Fig. 1, if the subsidy is P* for each ton removed, the plant would remove ER*, as the reward for removal exceeds the cost of emissions reduction up to that point. So the subsidy and price instruments can achieve the same regulatory objectives and associated flexibility and cost-effectiveness. The major difference is in the wealth position of the actors under the two instruments. With a pricing system, the actor's wealth after regulation is the original wealth minus control costs (B); in a subsidy system, the actor's postregulatory wealth is the original wealth plus the net of subsidy payments over control costs (A). Clearly, the actor's postregulatory wealth is greater under the subsidy system.

The subsidy could be directly on the behavior the regulator wished to encourage or it could be indirectly related to it. For example, the subsidy could be directly on emissions reductions or a payment for installing a scrubber. The subsidy could be an explicit payment, or an indirect benefit, such as a special tax privilege.

A deposit–refund system is a hybrid price-subsidy system. For example, an electric battery purchase may require a deposit, which is a price. But if the battery is returned, the actor receives a refund, which is a subsidy for its return. The magnitude of the deposit and the differential between the deposit and the refund guide both the original purchase and the disposal of the battery.

1.4 Liability

Legal principles, such as common laws of trespass and negligence or statutorily designated liabilities, place implicit costs on actors for undesirable activities. Actors then face potential liabilities for undesirable and harmful activities. These are liabilities beyond regulatory obligations such as CAC fines and penalties. In a sense they are prices, but with a higher degree of uncertainty over magnitude and incidence than explicit regulatory pricing instruments. The potential obligation for an undesirable activity is probabilistic, requiring a range of proofs from responsibility to magnitude of harm.

A variant of liability MBI is third-party insurance coverage. The purpose of insurance is to shift financial responsibility from an actor to the insurer. The insurer then has an incentive to establish premiums in accordance with risk and to monitor the behavior of the insured. Insurance coverage costs then reflect the insurer's assessment of risks, providing a signal of potential liabilities and the insurer becomes a quasi-regulator through monitoring. Insurance may not protect actors against liabilities due to negligence or illegal behavior.

1.5 Information

Information provided to the public about actors' behaviors may change environmental management activities. For example, public knowledge about a firm's emissions, or the "greenness" of its production processes, may galvanize public attitudes and induce firms to become good "corporate citizens." This moral suasion may extend to actual impacts on firms' profitabilities, as demands for their products are diminished or it becomes more difficult and costly to operate in the community.

2. PRINCIPLES AND APPLICATIONS TO ENERGY-RELATED ENVIRONMENTAL MANAGEMENT

The MBI are based on their ability to meet regulatory objectives in a manner that increases actor flexibility and reduces the costs of meeting objectives. The designs of these MBI are based on some principles of application. These principles suggest that some MBI are more appropriate for some circumstances and management objectives than others. These principles are discussed below, along with observations on strengths and weaknesses. Applications to energy-related issues are also presented for each MBI.

2.1 Prices

2.1.1 What to Price?
The regulator has control over actor behavior through the price. Prices may be placed on the behavior most immediately related to the regulatory objective. If the objective is to control emissions, the price can be placed directly on those emissions. The actor then has a host of activities that can be employed to avoid paying the price. However, the further the priced activity is from the regulatory objective, the less perfectly and the more costly it is for the regulator to manage for that objective. For example, SO_2 emissions from power plants may be regulated by placing a price on those discharges. Alternatively, levies based on the sulfur content of fuels would induce plants to shift to lower sulfur fuels but not induce them to change internal production processes, alter production levels, install scrubbers, etc. The pricing of fuel content does not induce the plant to engage in a whole host of activities downstream from the fuel purchase that could also reduce emissions, perhaps more cheaply.

Pricing the regulatory target may not be feasible, either economically or technologically. For example, nonpoint source run-off of nutrients from agriculture would be a candidate for targeted pricing if it were not so difficult to measure that run-off. Although it may not be impossible to measure, either directly or through nutrient transport modeling, it may be more feasible to price the nutrient inputs through charges on fertilizer. The total costs of meeting regulatory objectives must consider both the costs to agricultural production and the administrative costs. Pricing at points only indirectly related to the regulatory targets may be most cost-effective when considering both of these costs.

2.1.2 The Price Level
The level of the prices depends on the regulatory objective. A general economic principle is to set prices for activities at levels that reflect the cost to society of those activities. When traditional markets do not fully reflect these full costs, as would be the case for externalities such as pollution, a tax can be levied to correct for these price distortions. This tax is referred to as a "Pigovian Tax," named after the British economist who proposed taxing activities according to the damages they imposed on society but which were not accounted for in typical market decisions. This levy forces actors to "internalize" the adverse consequences of their actions. Such a levy would assure society that all benefits and costs of activities would become accounted for in market supply and demand conditions. For example, if it is known that a ton of emissions will change ambient concentrations by a known amount and that an estimable number of persons will become ill, one could set the emissions charge at a level reflecting the costs of those illnesses. For example, suppose that each ton of emissions harms society by an amount P^*

in Fig. 1. A levy of P* would induce actors to consider those emissions and they would reduce emissions as long as the levy exceeded the costs of emissions reduction. In Fig. 1, they would control to point ER*.

But regulators are not likely to know these social costs with much exactness or to a level of certainty to avoid challenges of "arbitrary and capricious." The alternative is to establish charges related to the known costs of control. For example, in Fig. 1, if the regulator knows the marginal control costs over a wide range of emissions reductions, it can set the price to induce the actor to reduce emissions by ER*. In fact, setting the price and observing actor behavior will indicate what the marginal cost function is, if the actor engages in a rational comparison of control costs to prices.

When price levels are designed to reflect the social costs of activities, such as pollutant damages, non-uniformly mixed pollutants must bear different prices. Suppose the same emissions from two plants result in different health impacts, where 1 ton from plant A has half the impact as 1 ton from plant B. The levy on plant A should then be half that of plant B.

Pigovian taxation is designed to induce actors to behave in the best interests of society by guiding their activities and resources so that the full social costs are considered. When prices on activities do not reflect the potential social costs of those activities, the guidance advantages of MBI are lost. For example, some emissions pricing, such as permit fees that vary with permitted volumes, is designed to collect sufficient revenues to fund the regulatory program. Such prices may be greater or less than the damages they impose on society. They are not Pigovian taxes, but simply fees for services rendered.

2.1.3 Pricing with Uncertainty

In principle, when a price is placed on activities, the actor will engage in control of those activities to the point where the cost of control equals the price, as shown in Fig. 1. When control costs are not known, and a small change in price can result in large changes in control behavior, it may not be prudent to use price as a MBI when the adverse consequences of the activity can be severe. For example, polychlorinated biphenyls have such serious health impacts that it would be better to directly regulate them through quantitative restrictions or CAC regulations than to depend on pricing to ensure safe and acceptable levels. Clearly, there is a trade-off between the cost of being wrong in setting inappropriate prices and the cost of foregoing the incentive and flexibility benefits of price MBI. When the cost of being wrong can be too great, price MBI are not adequate.

2.1.4 Pricing Applications

Pricing has been used as a means of managing energy-related resource activities and environmental impacts in a variety of ways. Severance taxes and leasing payments, although they have the potential of a MBI, are typically levied as sources of revenue and payment for transfer of ownership from the public to private actors rather than a resource management instrument. Superfund taxes in the United States on inputs to chemical and refinery production, which were ended in 1995, would not qualify as MBI, as they are designed for revenue-raising and not to create incentives for behavioral changes. Fuel-use taxes are also primarily revenue-generating rather than behavior-inducing. Many U.S. states charge water effluent or air emissions permit fees, but the levels of these fees are based on the revenue needs of the regulatory programs. Such taxes and fees are not Pigovian taxes.

France introduced an industrial emissions levy on large power plants and incinerators in 1985. The charge was on sulfur oxide emissions, but the purpose was to raise revenues to finance air monitoring and subsidize installation of pollution control equipment. Japan has implemented an emissions charge for sulfur oxides also, although the charge and revenues are based on needs to compensate applicants for the health damages compensation fund. The Netherlands has implemented a manure charge system, whereby farmers pay for phosphate loadings on their land above what they are allowed to apply. Sweden has levied a charge on nitrogen oxide emissions from large furnaces. Several European countries find it more convenient to place levies on products rather than on environmental impacts, per se. For example, Norway and Sweden have imposed levies on pesticides and fertilizers. These countries and Finland have also levied carbon taxes on motor fuels, but the primary purpose is to raise revenues.

The state of Maine has a complex air-emissions permit fee system, where fees per ton of several pollutants are based on volumes, with increasing per-ton fees for larger volumes and more toxic emissions. In the South Coast Air Quality Management District of California, facilities exceeding allowable emissions must pay excess emissions fees. This district also levies fees on both criteria pollutants and air toxics, and toxics fees are considerably higher than criteria fees. Many U.S. states levy charges on the generation and management of hazardous wastes. Taxes on landfill and incinerator disposal of hazardous waste have been high enough to reduce disposal by these methods, although it is unclear whether they have reduced the generation of such wastes or shifted disposal across state lines or to illegal disposal. The United States imposes gas guzzler taxes on automobiles, presumably to increase fuel efficiencies. The United States imposed taxes on ozone-depleting chemicals, based on their depletion factors, until such chemicals were completely phased out in the United States in 1996. Most U.S. states impose fertilizer fees, but they are based on revenues for fertilizer inspection. Texas introduced a clean fuel incentive surcharge on boiler fuel oil. Used oils and fuels derived from hazardous wastes are exempt. This surcharge is designed to shift fuel use rather than raise revenue.

2.2 Tradable Permits

The basis of tradable permits as a useful regulatory tool stems from the notion that all parties will benefit from free and voluntary trades. Markets are institutions for transfers of property rights. Ronald Coase, in his famous Coase Theorem, proposed that in the absence of transactions costs of trade, these property rights would be voluntarily redistributed so that the social value is maximized. It did not matter to whom the original property rights were assigned; the outcome would be the same and coincide with what society deemed best. By implication, if a fixed number of permits to do something were made available and trading was allowed, those permits would go to the entities most capable of profiting from them. For example, a fixed number of emissions permits, distributed across polluters, would be traded so that actors with low control costs would sell their permits to high-cost controllers; hence, low-cost controllers would do more control. The result was low social costs of attaining an emissions objective, a socially desirable goal.

2.2.1 What to Trade?

A wide variety of activities are amenable to the establishment of rights and trading. In fact, any activity that is identifiable and to which an enforceable property right can be feasibly established is ripe for tradable permitting. Emissions allowances establish a volume of permitted emissions, give an enterprise permits for that volume, and define the terms of trading. Enterprises may be given emissions reduction credits when they reduce emissions below allowable levels. They trade these reduction credits with others who will not reduce their emissions to allowable levels. Rules would be established that define the property rights implied by the permits, such as how long they are active, whether they can be banked and used later, whether some spatial trades are not allowed, and trading ratios across locations or types of activities. Tradable water rights, leaded gasoline, land development, and timber harvesting illustrate the broad range of tradable rights or permits. Some have even suggested biodiversity trading.

2.2.2 Problems with Trading

Permit and rights trading can have adverse consequences when the regulatory objective is to ensure an acceptable ambient concentration; for example, unrestricted trading of nonuniformly mixed pollutants may create "hot spots" where ambient conditions exceed objectives. This suggests trading rules in which restrictions or trading ratios are based on ambient concentration effects. A buyer of emission reduction credits may be restricted in how many they can purchase or trading among greenhouse gases may be based on CO_2 equivalents of different gases.

Since permits are required for activities, the level of permits one has will determine business viability. Actors may seek to hoard permits or employ the permitting process to enhance their market position. When tradable permitting results in market monopolization, the possible increased consumer costs must be weighed against any social gains from the tradable permit program.

A small number of trading partners limits the viability of the trading market in several ways. Transactions costs associated with finding and making trades may be higher than when the market has a large number of active participants. Also, manipulation of the market is more likely the smaller the number of traders. Any manipulation defeats the purpose of the trading and eliminates social gains. Monopolization of the trading market has the same disadvantages as monopolization of any market.

Limiting the time during which purchased permits or rights remain valid also reduces the viability of the trading market, as the property rights diminish over time. This is especially a problem when the regulator wants to reduce levels of the activity in the future. Allowing existing permits to depreciate just reduces the demand for them and reduces the potential gains from trading.

Although Coase's Theorem suggests that the initial assignment of rights or permits does not matter to the final outcomes, this may not be true when the initial assignment affects the wealth of actors. A permit or right endows wealth. This wealth can be employed in a number of ways. Otherwise, failing enterprises may be able to obtain access to capital that allows them to endure or this wealth may be used to alter political processes in an actor's favor. These wealth effects suggest that the initial assignment does matter to the final outcome.

2.2.3 Applications

Tradable permits have been used in a variety of environmental regulatory applications. The first U.S. application was in air nonattainment areas in the early 1970s, where new sources had to obtain emissions reductions credits, or offsets, from existing sources. Those who generated reductions credits were allowed to bank them for future use. In the course of phasing out lead in gasoline, U.S. refineries were allowed to trade lead content. It is instructive that U.S. Environmental Protection Agency enforcement of trading relied on reports and random gasoline samples. There were fraudulent trades discovered, but only in less than 1% of the cases. The United States allowed chlorofluorocarbon producers to trade during the phase-out of ozone-depleting chemicals during the early 1990s.

U.S. power plants have participated since the mid-1990s in an active sulfur dioxide trading program under the Clean Air Act's Acid Rain Program. Emissions allowances are calculated based on allowable emissions per million British thermal units (Btu) times average Btu consumption. Plants with emissions less than allowables can trade with plants that exceed their allowances. The U.S. Environmental Protection Agency holds some allowances and offers them at annual auctions. This trading program has been estimated to save approximately $2.5 billion per year compared to costs without trading. Eight northeastern U.S. states also participate in a NO_x trading program.

The South Coast Air Quality Management District in southern California established a Regional Clean Air Incentives Market (RECLAIM) in the early 1990s. Emissions allowances for NO_x and SO_x were established and sources were allowed to trade emissions reduction credits, where trading ratios were set for trades between geographic areas. The district estimates that compliance costs were reduced by nearly 50% as a result of trading. Several other states, including Illinois, Michigan, New Jersey, and Texas, have trading programs similar to RECLAIM.

Trading within and between point and nonpoint sources has been applied in several cases in the United States. The state of Washington has a grass burning tradable permit system. Water effluent trading has had limited applications in the United States. Biological oxygen demand trading in the Fox River of Wisconsin had limited success because of limitations the state placed on trades and the fact that since the Clean Water Act does not explicitly allow trading, minimum technology-based standards placed on effluents limited source control flexibility. Phosphorous discharge trading in the Dillon Reservoir in Colorado has also had limited success. Trading between point (wastewater treatment plants) and nonpoint (development) was allowed. As treatment plants found cheaper means of source control, their demand for permits diminished. As a result, the market has not been active. The ability of point sources to reduce nutrient effluents inexpensively has limited the application of tradable nutrient permits in the Tar Pamlico Basin in North Carolina.

The Kyoto Protocol allows greenhouse gas trading through several provisions. The Joint Implementation provision allows Annex I countries, larger developed countries, to acquire emissions reduction units from other Annex I countries. These reduction units can be based on projects that either reduce emissions or increase sinks for greenhouse gases. The Clean Development Mechanism allows Annex I countries to invest in emissions reduction or sink projects in non-Annex I countries in order to obtain emissions reduction units. International Emissions Trading provisions then allow Annex I countries to trade with one another to meet greenhouse gas targets.

Trading programs designed to protect ecosystems and resources have also been established. Wetlands mitigation banking in the United States allows persons who engage in activities that destroy or impair wetlands to purchase credits from a wetlands bank, where persons who have created or improved wetlands place deposit credits. Rules of trading often restrict trades to geographic areas and employ trading ratios that presumably ensure no net loss of wetlands. Credits can be based on simple acreage or on quality-based measures. Transferable development rights in the United States have been used to shift residential and commercial developments to more appropriate urban areas in order to avoid urban sprawl. Landowners in protected areas are provided development rights to alternative areas in lieu of development on their property and can sell these rights to others. This market has had limited success in the United States, partially due to municipalities' unwillingness or inability to restrict property rights to development and to the fact that effective sprawl-controlling trades may be between jurisdictions. Historic allocations of water rights in California to agriculture have created circumstances where water has not gone to its highest and best use, especially municipal and wetlands needs. The introduction of tradable permits has facilitated the movement of water rights from agriculture to these higher valued needs.

2.3 Subsidies

Subsidies are a means of rewarding actors for activities they may not have undertaken otherwise. Subsidies may be particularly useful when the actor is not legally obligated to undertake an activity or where restrictions on activities are not feasible for economic, political, or social reasons. For example, enforcement of nonpoint pollution restrictions in agriculture is costly and imperfect and farmers may be incapable, financially or politically, of controls. Pollution reduction may be achieved by subsidizing best management practices or the use of environmentally friendly fuels may be encouraged through subsidies to production or demand. Subsidies may be direct payments, such as price supports for crops used in ethanol production, or favored tax treatment, such as reductions in ethanol fuel taxes.

Although subsidies violate the "polluter pays" principle, by which social costs of activities are borne by the polluter, they may be the only means of achieving regulatory objectives. They face public resistance in cases where the undesirable activities are determined to be socially unacceptable and the public balks at paying someone for doing something they should have done anyway. Also, the perverse effect of subsidies may be to encourage the unacceptable activities in order to enhance the subsidy levels. For example, farmers may claim they would have drained more wetlands if they are subsidized for not farming wetlands.

The United States employs a variety of subsidy instruments for environmental and energy management. Favorable tax treatment in accelerated depreciation allowances for oil and gas drilling and production activities lowers energy costs. Reduced federal and state taxes on ethanol fuels, public procurement of recycled items, and favorable tax treatment of pollution control investments are further examples of subsidies impacting the demands for environmentally related goods and services. The RECLAIM program in southern California bought back old, high-polluting automobiles. An interesting, but short-lived, tax-based incentive program was the Louisiana Environmental Scorecard. Enterprises applying for industrial tax relief received exemptions based on their environmental performance, including violations, recent behavior, and Toxic Release Inventory (TRI) emissions per employee. A scoring system resulted in tax relief ranging from 0 to 100% of potential relief.

Of course, there are also many types of subsidies that harm the environment, such as energy development subsidies, below-cost timber sales, and grazing fees. Favorable tax treatment in timber, such as allowing enterprises to expense management and reforestation costs rather than capitalizing them, lowers timber costs and encourages timber development. Price support programs, such as those for sugar, have encouraged farming in sensitive ecosystems, such as the Florida Everglades. These agricultural price support programs also encourage the increased use of fertilizers.

2.4 Liabilities

The use of liabilities relies on the actor's willingness and ability to foresee the financial implications of its activities. Depending on the actor's foresight and time horizon, liabilities may be an effective means of attaining regulatory objectives. But there are many factors that may limit their effectiveness. The actor must foresee the implications of activities. There must be a high probability that adverse actions will eventually be penalized. But this depends on the effectiveness of the legal system and its enabling laws. Proofs must be made and penalties must ensue. Each element of the legal process is probabilistic. For example, can long-term health effects of a pollutant be proven? Can it be proven that a particular actor was responsible for the emissions? Will harmed parties pursue retribution? Will juries or judges convict? Will they impose punitive damages beyond proven damages? And to make liabilities an even less perfect enforcement tool, actors will use varying discount rates in assessing implications for current actions. They may accept bankruptcy as a better alternative to control.

Burdens of proof alter the liability landscape. It is one thing to have to prove that harm was done before halting an activity. It is another if actors must prove that no harm will be done before they can engage in an activity. Performance bonds may be one MBI that would hold society harmless while actors are allowed activities that may eventually prove harmless. Actors would post them while engaging in the activity and bond values would be based on worst-case damages scenarios. Required bonding levels would be diminished as the actions are proven to be less harmful than the worst-case basis. The bonding makes the burden of proof of no harm fall on the actor. This would be appropriate when activities may have tragic consequences, such as loss of life or irreversible changes in ecosystem conditions. Bonding is a means of allowing potentially harmful activities when it is uncertain what their impacts will be. Actors have an incentive to discover their true implications.

Environmental performance bonds are used in the mining industry to ensure proper remediation of surface mines. Mining bonds in Pennsylvania have been set at values only one-tenth the average reclamation costs, leaving the state holding the financial responsibility for complete restoration. Bonding for oil and gas mining activities could similarly be applied to exploration, development, and production activities, especially in ecologically sensitive areas. Bonding of farmers for nutrient runoff, erosions control, and other adverse effects of agriculture could be used, although the ability of farmers to finance such bonds may be limited.

In the United States, the Comprehensive Environmental Response, Compensation, and Liability Act (CERCLA) applies strict liability and joint and several liability to hazardous substance releases. Strict liability means that guilty actors are liable regardless of whether the act is purposeful or a result of negligence. Joint and several liability means that any party to a release can be held liable for the entire cost, regardless of its identifiable responsibility. Damages to natural resources from oil spills are included in CERCLA and require compensation to the public for lost resources.

Personal injuries from toxic agents can be litigated under U.S. tort law. However, a successful suit requires considerable demonstration of causality by plaintiffs. The willingness of courts to accept statistical analyses is limited, making proof of causation more difficult.

2.5 Information

Information provided to the public regarding actors' activities can have dramatic repercussions on the market conditions of the actors. Boycotts related to the environment can have impacts on product demand. Community opposition can increase costs of operations, such as legal battles over zoning and property assessments, and actors may just feel some social responsibility to be good citizens for the environment. Firms may be interested in the production conditions of suppliers and better environmental records may enhance labor supply conditions. Capital markets may look favorably at firms with better environmental records, as a reflection of potential liabilities.

Information may be casual and less than perfect, based on perceptions rather than facts. But it may also be well-established and certifiable, such as appliance and vehicle energy efficiencies or toxic releases. Good examples of effective information systems in the United States are the "right-to-know" federal and state statutes. The federal Emergency Planning and Community Right-To-Know Act introduced the publicly available TRI reporting system. Over 600 toxic chemicals have been reported by larger manufacturing facilities. Generation, treatment, and disposal information on the chemicals is available to the public. The TRI program has earned wide acclaim for its effectiveness. One study has shown that firms whose stock prices fell after releasing the TRI data were more likely to reduce TRI emissions in the future. California enacted Proposition 65, which requires firms to post warnings to the public of potential exposures to listed carcinogenic or toxic chemicals. Warnings include product labels and placards in business establishments.

Product labels, such as the Green Seal program in the United States, inform consumers about the environmental friendliness of products and their inputs and production processes. Manufacturers pay the Green Seal organization for an evaluation. The U.S. Federal Trade Commission has been developing "Green Guides" for recyclable, degradable, and environmentally friendly products. Scientific Certification Systems is a U.S. for-profit business that scores timber companies for their harvesting and management practices. The company also prepares internal environmental audits for firms. Several states and nonprofit foundations sponsor environmental awards.

SEE ALSO THE FOLLOWING ARTICLES

Clean Air Markets • *Economic Growth, Liberalization, and the Environment* • *Industrial Ecology* • *Market Failures in Energy Markets* • *Markets for Biofuels* • *Markets for Coal* • *Markets for Natural Gas* • *Markets for Petroleum* • *Modeling Energy*

Markets and Climate Change Policy • Stock Markets and Energy Prices • Subsidies to Energy Industries

Further Reading

Andersen, M. S., and Sprenger, R. (eds.). (2000). "Market-Based Instruments for Environmental Management: Politics and Institutions." Edward Elgar, Cheltenham, UK.

Costanza, R., and Perrings, C. (1990). A flexible assurance bonding system for improved environmental management. *Ecol. Econ.* **2**, 57–77.

Field, B. C. (1997). "Environmental Economics: An Introduction." McGraw-Hill, New York.

Hanley, N., Shogren, J. F., and White, B. (1997). "Environmental Economics in Theory and Practice." Oxford University Press, Oxford, UK.

OECD (1993). "Taxation and the Environment: Complementary Policies." OECD, Paris, France.

O'Riordan, T. (ed.). (1997). "Ecotaxation." St. Martin's Press, New York.

Rosenzweig, R., Varilek, M., and Janssen, J. (2002). "The Emerging International Greenhouse Gas Market." Pew Center on Global Climate Change, Arlington, VA.

Shogren, J. F., Herriges, J. A., and Govindasamy, R. (1993). Limits to environmental bonds. *Ecol. Econ.* **8**, 109–133.

Stavins, R. N. (Ed.). (1991). "Project 88—Round II: Incentive for Action, Designing Market-Based Environmental Strategies," A Public Policy Study Sponsored by Senator Timothy E. Wirth, Colorado, and Senator John Heinz, Pennsylvania, Washington, DC.

Tietenberg, T. (2000). "Environmental and Natural Resource Economics." Addison-Wesley, New York.

Tietenberg, T., Button, K., and Nijkamp, P. (eds.). (1999). "Environmental Instruments and Institutions." Edward Elgar, Cheltenham, UK.

Market Failures in Energy Markets

STEPHEN J. DECANIO
University of California, Santa Barbara
Santa Barbara, California, United States

1. "Classical" Market Failures
2. Is Optimization the Correct Model of Behavior?
3. Additional Considerations

Glossary

consumers' surplus A measure of the benefit to consumers, net of the cost of the good, of their being able to purchase it at a particular price.

cost–benefit analysis The formal comparison of the economic costs and benefits of some particular action or policy.

general equilibrium model A model designed to represent the utility- or profit-maximizing behavior of all the agents in the economy, the solution of which is a set of prices and goods allocations that satisfy the maximization conditions of the agents.

Herfindahl index A measure of industry concentration equal to the sum of the squares of the market shares of the firms in the industry.

hurdle rate A rate of return that investment projects must achieve in order to be undertaken.

inelastic demand (or supply) A situation in which large changes in the price of a good lead to only small changes in the quantity demanded (or supplied).

marginal cost The additional per-unit cost of production as output increases.

price taker An economic agent that is not able to influence, by its own actions, the price at which any good is transacted.

producers' surplus A measure of the benefit to producers, net of the cost of production, of being able to sell their output at a particular price.

risk-adjusted cost of capital The price a firm must pay to borrow funds for an investment project of a particular risk.

traveling salesman problem Given a finite number of cities along with the cost of travel between each pair of them, find the cheapest way of visiting all the cities and returning to the starting point.

Market failure can be understood only as the opposite of some concept of market success. But what is market success? Defining it as a general improvement of the conditions of life is too vague; requiring achievement of the best of all possible worlds is too utopian. For economists, market success is ordinarily thought of as achieving Pareto optimality, a situation in which no one's material well-being can be improved without reducing the well-being of at least one other person. If the economic decisions of all the individuals in the society (that is, their activities as they participate in markets of all types) lead to a Pareto optimum, the criterion of market success has been met. Anything that falls short constitutes market failure. One of the achievements of modern economic theory has been to demonstrate that there is a set of ideal conditions under which decentralized decision-making leads to a Pareto-optimal social outcome. This remarkable result is embodied in the so-called First Welfare Theorem, which states that if every relevant good is traded in a market at publicly known prices (i.e., if there is a complete set of markets), and if households and firms act perfectly competitively (i.e., as price takers), then the market outcome without any government regulation or intervention will be Pareto-optimal. This result sets a scientifically meaningful standard—one with empirical content that is not a tautology—against which real-world situations can be compared. Policies cannot improve on the economic efficiency of a Pareto optimum, although it should be noted that many different Pareto optima having different income distributions are possible. Closely related to the notion of Pareto optimality is the concept of the Kaldor-Hicks test. According to the Kaldor-Hicks criterion (first introduced in the 1930s), situation B is socially preferable to situation A if those who would gain from moving from A to B could compensate the losers and still have something left over. This type of move will be referred to subsequently as one that is cost-beneficial, drawing on the terminology of cost–benefit analysis. Market failure also occurs if the economy exhibits opportunities for cost-beneficial actions (investments, policy implementations, etc.) that are not undertaken. It should be noted that to be cost-beneficial, it is only necessary that a policy or private action generate enough additional output to make it *possible* for the gainers to compensate the losers. In the real world, such compensation is rarely paid even when cost-beneficial moves are undertaken. This is a feature of the political system, however, that is not entailed in the definition of market success or failure.

1. "CLASSICAL" MARKET FAILURES

In energy markets, the requirements of the First Welfare Theorem (FWT) are systematically violated, leading to a wide variety of market failures. Of course, the conditions for the FWT to hold are quite strong and are unlikely to be perfectly satisfied anywhere, but in the case of energy markets the failures are conspicuous.

1.1 Monopoly

Perhaps the most commonly recognized breakdown of the necessary conditions for market success is that energy markets are frequently subject to problems of monopoly. Some of the key agents do not act as price takers but would, in the absence of legal constraints, be able to control prices and quantities so as to improve their own position at the expense of general economic efficiency. The foremost source of such market power is the existence of economies of scale in production. Economies of scale imply that the unit cost of producing the output is declining as the scale of operations increases. If this decline in unit costs prevails over the entire range of market demand, the industry is said to constitute a natural monopoly. In energy markets, perhaps the clearest case of this is in electricity distribution, which historically has been characterized by very large fixed-cost investments in the grid.

Consider a simplified graphical representation of the electricity delivery market. Building the "backbone" grid requires a large initial investment, but once it is in place, the additional cost of connecting a new customer (that is, building a line from the grid to the customer) is small and roughly constant. In a

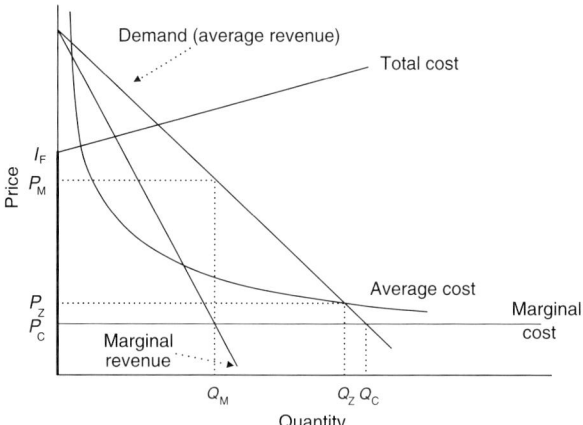

FIGURE 1 Prices and quantities under monopoly, zero profit, and perfect competition.

stylized representation, this gives rise to Total Cost, Average Cost, and Marginal Cost curves, as depicted in Fig. 1. Total Cost consists of two parts: a fixed initial investment (I_F) that must be incurred before the first unit of output can be sold, plus an amount that increases linearly with the quantity of electricity delivered. This type of Total Cost curve implies an Average Cost curve that is falling over its entire range and a Marginal Cost curve that is horizontal. Demand is given by the Average Revenue curve, which slopes downward over the relevant range of potential output. The Marginal Revenue curve (defined as the additional unit of revenue per unit of additional output sold) is also a downward-sloping line with a slope steeper than the demand curve. Linearity of the demand and marginal cost curves is not essential to the argument, but simplifies Fig. 1 and the exposition.

The electricity distribution utility's economic profit (after accounting for payments to all factors of production including management and the return to capital) will be zero when average revenue equals average cost; in contrast, the perfectly competitive equilibrium will be the point at which price equals marginal cost, whereas an unregulated monopoly would maximize its profit by delivering an amount of product such that the marginal revenue equals marginal cost. The prices in these three cases would be P_M for the monopoly, P_Z for the point of zero economic profit, and P_C for the perfectly competitive equilibrium.

Market failure occurs because the unregulated monopolist charges too high a price for the electricity and delivers too little of it. If the fixed costs are high enough, no potential competitor could enter the market and make a profit. Even if the entrant's fixed cost were low, no entrant would be able to make any positive profit so long as the entrant's average cost for whatever market share it could obtain were greater than the price prevailing in the market after the newcomer's entry. The monopolist collects "supernormal" profits, that is, profits over and above all the costs of production. Elimination of these monopoly profits requires that the price be no higher than average cost and the minimum output for which this is possible is (Q_Z, P_Z). This zero-profit point is often the target aimed for by utility regulators—the price the utility is allowed to charge is only enough to cover all costs including a "fair" return on capital.

Yet even at the zero-profit point (Q_Z, P_Z), the market failure is not entirely corrected. Although it is true that at this point there is no incentive for entry into the market and the existing firm is economically viable, the zero-profit point is not Pareto-optimal because it is possible to design taxes on consumers and subsidies to the utility that would induce the utility to expand its output (and lower its price) to (Q_C, P_C). At (Q_C, P_C) every member of society is potentially better off than at (Q_Z, P_Z). This can be seen by comparing the consumers' gain in moving from (Q_Z, P_Z) to (Q_C, P_C) to the subsidy that would be required to bring the utility back to the break-even point. Alternatively, it can be shown that the total surplus (producers' plus consumers') increases in moving from (Q_Z, P_Z) to (Q_C, P_C). Society gains on net in moving to (Q_C, P_C).

The example illustrated by Fig. 1 is a simple case in which economies of scale are the cause of the monopoly problem. In the real world, monopoly power is often more difficult to detect and may be very hard to control. Markets are rarely occupied by a monopolist producer, and although measures of concentration (such as the Herfindahl index) based on market shares are often used to provide a rough indication of potential market power, the absence of a general theory of oligopoly pricing makes unambiguous inferences of market power from data on market shares impossible. To complicate matters further, temporary market power can emerge in almost any dynamic setting. It often involves a time-sensitive component. Entering a market can take time—to build parallel infrastructure, establish a sales network, or obtain the required regulatory permissions, for example. Thus, it may be possible for a producer to enjoy a monopoly position in the short run that cannot be maintained over the long term. Much actual business practice involves seeking temporary monopoly advantages—being the first to offer a new product or service or developing a new process or technique that is protected by patent rights or as a trade secret. In such cases, the monopoly return may be viewed as a way to compensate firms for the expense of research and development and provides an incentive to innovate. The kind of temporary monopoly advantage that leads to technological progress can be an indicator of dynamism in the market economy. In the perfectly competitive equilibrium in which no one is earning any supernormal profits, there is no incentive for anyone's behavior to change. In the real world, the search for profit opportunities must have some positive probability of a payoff, or innovation would wither.

In many other circumstances, however, monopoly power not based on economies of scale has no social benefit. Taking advantage of temporary shortages, or collusion to create a shortage, is rightly condemned as a market failure. Energy markets are prone to problems of this type because of complex technical interdependencies or inelastic supply and demand curves (as in electricity markets). Controlling the opportunism that such conditions engender is one of the reasons firms exist and is also a legitimate role for government intervention. The debacle of "deregulated" electricity markets in California in 2000–2001 is an example of the kinds of things that can go wrong when such opportunities for the exercise of market power are not held in check. When electricity supply became tight in the summer of 2000, even producers with relatively small market shares could exercise market power because of the extreme inelasticity of short-term demand.

1.2 Externalities

A second source of market failure in energy markets stems from the existence of externalities. A negative externality arises when production or consumption decisions of some agents directly and adversely affect the happiness of other agents and the affected agents have no way of obtaining compensation for the harm done to them. (A similar inefficiency occurs if the externality is positive in that it provides benefits to some members of society that they do not have to pay for.) In the case of energy production and consumption, externalities are the rule rather than the exception. Examples include the local air pollution of coal-fired power plants (which produce harmful particulates and oxides of nitrogen and sulfur as a natural result of the combustion process), the congestion of streets and highways that results when individuals make uncoordinated personal transportation decisions, and the climate-changing effect of the carbon dioxide that is emitted whenever fossil fuels are burned.

Essentially, the market failure from externalities comes about because some markets simply do not exist. There is no way that individuals can buy or trade "climate stability," for example. At the root of the externality is the absence of well-defined and enforceable property rights. Clear property rights of a suitable type would enable individuals to obtain compensation for the adverse effect on their well-being of the externality-producing activities of others. When such rights do not exist, people have no way of expressing their preferences through their willingness to exchange goods they own. In the case of monopoly, market failure occurs because not all participants in the market are price takers; with externalities, markets fail because they are not complete. The First Welfare Theorem does not hold because not every good is traded.

A number of strategies can eliminate the market failures caused by externalities, but all involve some kind of policy decision or collective action. Taxes (or subsidies in the case of positive externalities) can be imposed that make the private costs associated with the externality-producing activities (such as the burning of fossil fuels) correspond to what producers would have to pay those harmed by the externality if the appropriate property rights existed. Alternatively, the government can create new property rights and assign them to individuals so that the requisite market transactions can occur with low transactions costs. This enabling of new markets is exemplified by the creation of SO_2 allowances under the 1990 Clean Air Act (CAA) Amendments in the United States. Under the CAA Amendments, SO_2 emissions are curtailed in the aggregate (thereby creating a cost for SO_2 emissions and correcting the externality) and trading of allowances among producers makes it possible for the net aggregate reduction in SO_2 emissions to be achieved a minimum cost. Provided the participants in the newly created markets behave as price takers and not as monopolists or oligopolists, this kind of policy can eliminate the externality that arises from nonexistent markets.

It should be noted here that the distribution of wealth and income will be affected by how the property rights are assigned or tax revenues are collected and spent. Different assignments of rights will give rise to different Pareto-optimal equilibria and those different outcomes need to be judged according to a criterion of justice rather than economic efficiency (because all Pareto-optimal outcomes are efficient). According to the definition being used here, market failure pertains to the failure to achieve an efficient outcome, not to the intrinsic desirability or fairness of the resulting pattern of prices, incomes, and consumption.

1.3 Public Goods

A third source of market failure in energy markets is that energy production and consumption entail the production or diminishment of public goods such as macroeconomic stability or the scientific knowledge derived from basic research. A public good is a commodity or service such that if it is supplied to any of the members of society, it can be enjoyed by all at no extra cost. The consumers of the public good are not able to exclude others from its consumption. The classic examples are national defense and public order. A military establishment that is successfully able to deter potential aggressors and defend the nation's territory against attack necessarily provides this service for all citizens alike. This "nonexcludability" of public goods gives rise to the problem of free riders—if national defense protects citizens equally regardless of their contribution, individuals have no incentive to pay for the military. If left to the unregulated market, not enough public goods will be supplied because of this free rider problem. This is the fundamental reason why taxes that are used to purchase public goods must be compulsory.

It might be thought that fuels and energy services are commodities like any other. In their direct benefits this is true: the owner of an automobile can exclude others from driving it and a consumer's electricity bill reflects only his or her own consumption of electricity. However, because the use of energy is ubiquitous, and because the means of converting fuels into useful work requires complex and expensive capital equipment (power plants, motor vehicles, pipelines, etc.), the commodity-like nature of energy products gives rise to the problem of macroeconomic stability. Fuel prices can change much more rapidly than the capital stock can be adjusted and therefore the volatility of fuel prices creates situations in which capital that was profitable at previous fuel price levels becomes uneconomic at current fuel price levels. Uneconomic capital equipment is taken out of production and this in turn leads to job losses and unemployment. Thus, movements in energy prices can have a ripple effect throughout the economy, resulting in fluctuations in employment and output that affect the well-being of all. This is precisely what happened during the oil price shocks of the 1970s. Unanticipated sudden increases in oil prices associated with market and political events had macroeconomic consequences around the world. Policy initiatives such as the Strategic Petroleum Reserve (SPR) are intended to stabilize the price of oil and thereby ameliorate macroeconomic disruptions caused by oil price volatility. The SPR provides a valuable buffer against the potential macroeconomic consequences of unexpected oil price movements just as discretionary monetary policy is an appropriate tool for responding to changes in aggregate demand.

Macroeconomic vulnerability to energy price shocks is sometimes characterized as a national security problem. Given that modern military vehicles require sophisticated power plants (whether fossil-fuel driven or nuclear), it is certainly true that the availability of suitable fuels is necessary for military operations. For some countries, such as the United States, domestic fuel sources are sufficient to satisfy military requirements in wartime so direct military needs are not part of the national security problem for those countries. Also, the relative proportions of foreign and domestic oil consumed are no indication of macroeconomic vulnerability. Because oil is a commodity traded in world markets, a supply disruption anywhere has an effect on the world oil price. So, for example, even if U.S. oil demand were supplied 100% domestically, a world oil price shock would have adverse macroeconomic effects in the United States as the domestic price followed the world price. Thus, the national security problem associated with energy has to do primarily with the flexibility of a nation's capital stock in the face of energy price fluctuations. In addition, foreign oil purchases may constitute a national security problem if the supply comes from a hostile state or states whose military capability is increased because of the wealth gained from their petroleum sales.

A second public goods aspect of energy markets is the role of research and development (R&D). Scientific knowledge is very much a public good, because once new knowledge is published it is available to all without the possibility of exclusion. The producer of the knowledge cannot hope to recover more than a fraction of the benefits resulting from the knowledge. Useful knowledge created through the research and development process has spillovers to firms or other organizations and individuals that did not carry out the research. As a result, there will be too little research and development in the aggregate unless the government intervenes. If the optimal level of research were being undertaken, the rate of return to R&D would be equal to the rate of return on other forms of investment. In fact, empirical studies consistently find social rates of return to R&D that substantially exceed the market return to capital. This discrepancy provides a rationale for government support of R&D, whether through universities, government research laboratories, tax incentives, or direct subsidies.

A related potential market failure comes about because of the tendency for unit costs to decline as producers and consumers gain experience with a product or technology—the phenomenon of "learning by doing." This dynamic pattern of cost decline has been observed across many technologies, including energy technologies such as wind power generators and solar voltaic panels. Unless private investors correctly anticipate the timing and extent of the cost reductions due to learning by doing, they may fail to invest sufficiently in promising emerging technologies. Consumers may be unwilling to pay the high price for new technologies in the early stages of their development, even though the technologies would be quite attractive if they could be utilized in sufficient volume to realize cost reductions from learning by doing and economies of scale. This is sometimes referred to as the "chicken and egg" problem. In some cases, market forces will lead to the emergence of the best technologies, but not always. A long history of the gradual improvement of an inferior technology (along with the growth of infrastructure associated with it) may lead to the lock-in of that technology and barriers to the adoption of superior alternatives.

2. IS OPTIMIZATION THE CORRECT MODEL OF BEHAVIOR?

The market failures described in Section 1 are commonplace staples of textbooks. There is another class of energy market failures about which there is less consensus. Prominent among these is the energy efficiency gap: the notion that even though sizable improvements in energy efficiency could be achieved by making investments that would be profitable at current prices, these investments are not carried out. Sometimes this phenomenon takes the form that the hurdle rates required for energy-efficiency investments are considerably higher than the cost of capital to the firm for projects of comparable risk. The existence of this type of market failure is controversial, because if profitable opportunities are available, why do profit-maximizing firms fail to undertake them? (By definition, a profitable investment is cost-beneficial because the resources devoted to it create more wealth than if the investment were not made.) A very large body of empirical literature attests to the existence of these opportunities, yet the systematic failure to maximize profits conflicts with standard practice in economic modeling. It would seem that either the assumption of profit maximization does not hold or the studies documenting the possibility for energy efficiency improvements have missed essential elements of full cost accounting.

To see why the resolution of this issue has implications for market success or failure, consider the two situations depicted in Fig. 2. In both Fig. 2a and Fig. 2b, the axes represent desirable outputs from different activities that might be undertaken by the firm. The horizontal axis shows the reduction in pollution that would follow from enhancing energy efficiency (by suitable investments, changes in operating procedures, etc.) and the vertical axis shows the economic output from the other activities of the firm. The curves PP' in the two diagrams represent the theoretical "production-possibilities frontier" of the firm. This frontier is the set of possible combinations of the two kinds of goods that the firm could produce from a given set of resources if it were fully optimized.

The empirical question is whether the firm is always operating on its production-possibilities frontier. Figures 2a and 2b represent alternate answers to this question. In both cases, the points A and A' represent the initial situation of the firm and points B and B' show the position of the firm after taking some action such as increasing its energy efficiency. If the firm is fully optimized, it is not possible to achieve reductions in pollution without sacrificing some other profitable activity. This case is shown in Fig. 2a. In contrast, Fig. 2b shows a situation in which it is possible for the firm to realize gains simultaneously from investments in energy efficiency and other activities. This case corresponds to the possibility that there are profitable energy-saving investments available to the firm. The increase in energy efficiency accounts for the pollution reduction along the horizontal axis and the additional profit earned from these investments allows the firm to expand its output in other directions. If the firm is not organized optimally to begin with, improvements are possible along multiple dimensions of its performance.

The situation displayed in Fig. 2b is commonly referred to as a market failure, even though it is perhaps more appropriate to describe it as an organizational failure or institutional failure. The point remains that in Fig. 2b it is possible to take actions that would be cost-beneficial. To assess how common or realistic is the situation shown in Fig. 2b compared to Fig. 2a, it is necessary to examine what kinds of factors might give rise to a firm's operating inside its production-possibilities frontier.

2.1 Nonunitary Organizations

Primary among the causes of organizational failure is the fact that firms are made up of distinct individuals having different interests. The firm has no mind or will of its own; it acts according to a set of decision-making procedures and there is no guarantee that the collective action of the individuals making up the firm will lead to fulfillment of the firm's formal objective (the maximization of profit). Even though

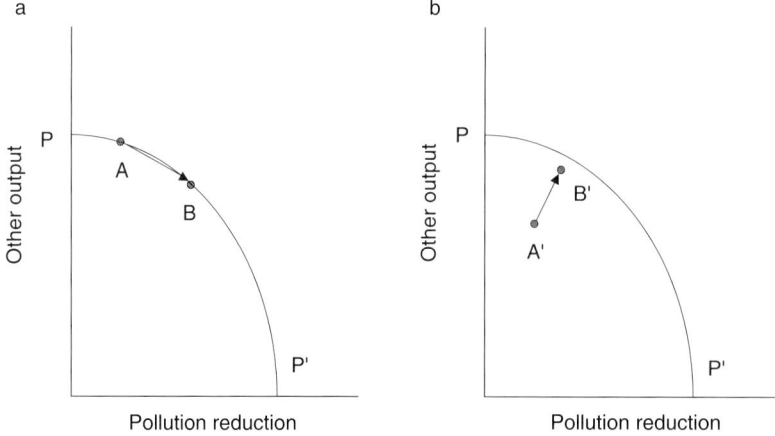

FIGURE 2 An energy efficiency gap? (a) Firm that is fully optimized. (b) Firm that is not fully optimized.

there are gains from cooperation (firms would not exist otherwise), the diverging objectives of the agents making up the firm are almost certain to result in a failure of complete optimization. The type of conflict arising from the nonunitary nature of firms and other productive organizations has historically been referred to as the principal–agent problem. The principal (in the case of firm, this would be the shareholders whose interest is that the firm maximize profits) cannot guarantee that the agents (the managers and employees of the firm) will act to further the principal's objective rather than improving the conditions of the workplace, creating perquisites of employment, or otherwise diverting profits into additional benefits for the agents. Because profit maximization by firms is required for Pareto optimality in a market economy, any deviation from that standard caused by managers' and employees' self-interested behavior is a cause of market failure.

The various manifestations of the principal–agent problem have been developed by scholars specializing in the theory of the firm. Adolph Berle and Gardiner Means in the 1930s identified how the separation of ownership and control in the modern corporation enables managers to advance their own interests at the expense of shareholders, but an awareness of the general problem can be traced back to Adam Smith and earlier. Managers typically have access to detailed inside information not available to the principals. To prevent managers from using this information to their benefit, principals sometimes have recourse to operating procedures that are second-best in terms of profit maximization, but that control the principal–agent conflict. This explanation has been proposed for the excessively high hurdle rates required for approval of investment projects in some firms. Principals cannot know the actual rates of return on all potential projects. If managers are interested in maximizing "organizational slack" (which includes managerial perquisites and inessential expenditures), the second-best solution for the principals involves setting a hurdle rate greater than the risk-adjusted cost of capital. The high hurdle rate serves as a substitute for expensive postproject auditing. The benefit to the principals comes at the cost of foregoing some profitable projects (for example, low-risk energy-saving investments in lighting retrofits or heating, ventilation, and air conditioning upgrades) having rates of return that are lower than the hurdle rate.

Managers also can exhibit myopia, risk aversion, and bias in project selection that make the managers' lives easier at the expense of the profit maximization objective of the owners of the firm. It is impossible for the owners to determine from the outside just what contribution each employee of the firm makes to its success; hence, the market rates of compensation for the agents cannot be known. Many of the incentive compensation plans that are observed in firms (stock options, performance bonuses, etc.) are compromises intended (on the part of the principals, at least) to bring the interests of owners and their agents into closer alignment. All of these sources of deviation from perfect profit maximization apply to the general operations of the firm, including the firm's energy use decisions.

2.2 Bounded Rationality and Agent-Based Models

The principal–agent problem undermines full optimization when all agents are entirely rational in the strong sense required by neoclassical economic theory. That is, the agents are assumed to have unlimited capacity to carry out whatever calculations are required to maximize their own utilities. This assumption of unlimited computational capacity is questionable, however. The alternative notion of "bounded rationality," first proposed by Herbert Simon, is based on the realization that the calculations necessary to achieve full optimization in economic activities are effectively beyond the reach of real people and firms.

Since the time of Simon's initial work, advances in mathematics and computer science have formalized the concept of computational complexity. It is widely believed (but not yet proven) that a wide class of practical problems such as the Traveling Salesman Problem require computation time that grows "faster than polynomially" with the size of the problem. John Rust and others have pointed out that many calculations in economics are subject to formal computational complexity constraints, and Michael Garey and David Johnson's long list of intractable problems (of the same class as the Traveling Salesman Problem) includes many standard management problems such as network design, storage and retrieval, database management, and sequencing and scheduling. Achievement of full maximization (including its energy efficiency component) involves calculations of this type in many cases. It is an empirical question how large the deviation from optimality might be in any particular instance, but the existence of computational limits constitutes a formidable barrier to complete optimization by firms.

Aside from these mathematical foundations of bounded rationality, as a practical matter limitations on management time and attention mean that cost-saving energy efficiency investments not central to firms' strategic priorities are often given short shrift. Even this commonsense observation indicates a departure from Pareto optimality, because optimally efficient firms would take advantage of every profitable opportunity. Bounded rationality provides one justification for policies that simplify business decision-making by setting standards. In addition, the success of voluntary pollution prevention programs in the United States and other Organization for Economic Cooperation and Development countries testifies to the possibility of policy-driven cost-beneficial improvements if the government provides high-quality information and support services to boundedly rational firms. Michael E. Porter has even proposed the bold hypothesis that environmental regulation may, in some circumstances, stimulate firms to better performance than they would have achieved in the absence of regulation.

Some economists have begun to embody the concept of bounded rationality in models in which the entities (or agents) follow relatively simple rules of behavior, rather than being presumed to maximize an objective function. Because the rules are simple, it is possible to set up numerical simulations of systems containing large numbers of such agents and observe the development of such systems over time in a laboratory-like setting. Models of this type have the potential for reproducing features of the population of firms, such as the size distribution or the birth and death rates of companies, that do not emerge naturally in conventional maximization models. If the performance of the firms in agent-based models is compared to the ideal of full optimization, room for improvement will be seen. If the economic system is in fact made up of agents following relatively simple rules, then cost-beneficial gains are possible.

2.3 Behavioral Economics

Both principal–agent problems and bounded rationality take for granted that at least the agents in the economy are trying to maximize utility as it is conceived by neoclassical economic theory. Both within and outside of economics, however, there is recognition that people do not necessarily behave that way. Psychologists have noted that decisions about energy-saving investments frequently do not conform to the strict requirements of the economic paradigm. "Organizational Behavior" and "Management Science" are recognized as fields of inquiry distinct from economics. Indeed, it would be difficult to understand the considerable demand for management training and expertise (Master of Business Administration and executive education programs, for example) were it not for the fact that management activities, such as motivating employees, coordinating complex operations, and visualizing the uncertain future, require skills and talents of the highest order. The intricacies and subtleties of human and organizational behavior are far from being reducible to a simple economic calculus.

Some economists have begun to explore the consequences of behavior outside the conventional maximizing framework. A Special Section in the March 2000 edition of the *Journal of Economic Behavior and Organization* was devoted to "Psychological Aspects of Economic Behavior" and a session of the year 2000 annual meeting of the American Economic Association included papers titled "Emotions in Economic Theory and Economic Behavior" and "Thinking and Feeling." If irrational exuberance can account for stock market bubbles measured in trillions of dollars, it is perhaps naive to think that all decisions regarding energy technologies and investments are explicable as purely rational investment decisions. Market pressures surely push firms and individuals in the direction of rational choice, but social forces that individuals are barely conscious of, such as the imperatives of maintaining prestige and self-image and even considerations of style and fashion, are also powerful influences on behavior.

2.4 Evolutionary Economics

The inadequacy of the neoclassical model of the firm ("maximization of profits subject to the constraint of current technology and resources") has led some economists and other social scientists to explore a variety of alternative characterizations of production. Among the most promising is evolutionary economics—the idea that firms and the economy can be described and analyzed in terms analogous to biological populations and ecosystems. In this view, the market acts in the same way as natural selection in evolutionary biology to shape the characteristics of the population of firms. If a firm is not sufficiently fit (i.e., profitable), it dies (goes bankrupt or is acquired by a healthier firm), whereas the more profitable firms expand and propagate.

What are the implications of the evolutionary model for market failure as it has been defined here? First, it is important to realize that the biological analogy does not carry over perfectly. In biological evolution, species evolve over time through mutation and natural selection, whereas in the economy, firms are capable of consciously changing or adapting to new circumstances. Humans can learn, so the "inheritance of acquired characteristics" is to be expected in economic evolution, unlike in biology. Despite this difference, the crucial similarity between economic and biological evolution is that in both the evolutionary pressures of natural selection lead to improvements in performance over time. However, it is a common mistake to confuse market pressures that reward profitability with a mechanism that somehow leads to the *maximization* of profits (or full optimization by firms). According to such a view, market selection results in firms acting "as if" they maximize profits, regardless of what the managers think they are doing or limitations on their access to information or computational capabilities. It was pointed out by Armen Alchian in 1950 that though market competition favors firms that realize positive profits, this by no means implies that surviving firms maximize profits.

In neither economics nor biology can the evolutionary process be expected to achieve optimization of the fitness of all members of the population. Evolution takes existing populations as its raw material and works through incremental change. The consequence is that evolutionary processes are path dependent. Small differences in initial conditions can lead to different paths of technological progress or institutional development or, alternatively, to different ways of accomplishing the same function through different institutional arrangements or technologies. There is no reason to think that the particular path "selected" by the evolutionary process arrives at an optimal solution. Similarly, in the evolutionary context, optimality can be defined only in terms of the environment in which the population's members operate. If that environment is changing rapidly, the evolutionary process will lead to improvements that never entirely catch up to the changing conditions. In addition, some degree of diversity in the population (of organisms or firms) is likely to provide resilience against sudden shifts in the environment. This indicates that not all firms will be equally adept at dealing with any particular environment that exists at a moment in time. Finally, it is a fallacy to think that the population in existence today is optimally evolved, for if that were so, why was it not true also of the population that existed a year ago, a decade ago, or a century ago?

These departures from complete optimization apply to the behavior of firms and economic agents in general and are not specific to the energy sector or to energy markets. Nevertheless, the ubiquity of departures from optimality certainly indicates that market failure in energy markets (and in the energy decisions made by firms) will also be widespread. There is no reason to think that incomplete optimization is more severe with respect to energy decisions than it is with respect to other important decisions taken by firms, say with regard to employee benefit packages, advertising, or organizational structure. Nevertheless, the energy sector and the energy decisions of firms have been scrutinized extensively in an effort to discover empirical examples of failures to optimize. This is partly an historical accident because of the intense interest in energy issues that followed the oil price shocks of the 1970s. From a scientific standpoint, this focus on energy has been highly productive, for it led to the large literature documenting the energy efficiency gap and the noneconomic factors that influence energy decision-making. This empirical literature does not support the presumption of complete optimization.

3. ADDITIONAL CONSIDERATIONS

The discussion thus far has been couched almost entirely in economic terms. As such, it has given no consideration to the very real possibility that markets can fail even if Pareto optimality is achieved. A perfectly functioning market system may be a failure from a higher social or philosophical perspective. Ordinary economic theory ignores the question of where individuals' preferences come from and how they are shaped by the social context. In the same way, economic analysis typically takes the pattern of ownership of property rights as given, even though different assignments of those rights (and creation by the State of rights distinct from the set currently in existence) can have a major impact on the justice of social arrangements. Even if energy markets were not subject to market failure as outlined above, it might still be the case that the prices and consumption patterns emerging from them would not appropriately reflect the interests of future generations or that they might violate standards of fairness that transcend economics.

In addition, there is no guarantee even given a particular set of endowments of property rights that the Pareto-optimal competitive equilibrium will be *unique*. Without any problems of monopoly, externality, etc., it is easy to set up simple models that exhibit multiple equilibria. Each of these equilibria is a Pareto optimum, but they can be quite different in their equilibrium prices, allocations, and distributions of income. There is no consensus on how society might go about choosing among such equilibria from a welfare perspective or about what dynamic processes might lead an economy to one or another of the equilibrium points (or from one to another if a particular equilibrium is disturbed). This lack of a theory of equilibrium choice is a gap in economic knowledge that would need to be filled before too much reliance is placed on Pareto optimality as a goal of policy.

It should also be noted that the Pareto criterion for market success makes operational sense only if individuals' utilities are not interdependent. If some individuals' utilities depend not only on their own consumption of goods but also on the consumption of others, then the First Welfare Theorem no longer holds. Striving after positional goods (that derive at least part of their value as a result of comparison to others rather than because of the goods' intrinsic characteristics) also leads to inefficiency. If organizations of one type (such as economics departments or universities) all seek to increase their academic rankings, a large part of their effort is bound to be wasted because not all of them can succeed in moving up. Market failures of this type are no less important than more conventional kinds of deviations from Pareto optimality nor should they be ignored in formulating policy.

At the same time, the ubiquity of market failures in the energy sector does not mean that markets should cavalierly be replaced by political decisions or central planning. Even where markets fail, there is no guarantee that their replacement by some other allocative process would perform better. The political process is also vulnerable to failure, and sometimes to more spectacular failure, than markets. Political systems may restrict competition, inhibit the generation and transmission of accurate information, and replace satisfaction of the wants and needs of the general population with implementation of the preferences of the decision-making elite. It is unrealistic to substitute the assumption that the government always acts in the best interests of society as a whole for the assumption that markets always conform to the requirements of the FWT. Neither assumption is justified. Nonetheless, an awareness of the possibilities of market failure should caution against embracing the market fundamentalism that rejects every attempt to make improvements in the functioning of the economic system and dismisses all forms of regulation as unwarranted government intrusion on the free functioning of the Invisible Hand. In energy markets, the balance between decentralization and regulation is a delicate one, but that only makes it more important to maintain.

SEE ALSO THE FOLLOWING ARTICLES

Economic Growth, Liberalization, and the Environment • *Economics of Energy Demand* • *Economics of Energy Supply* • *Energy Efficiency, Taxonomic Overview* • *Market-Based Instruments, Overview* • *Markets for Biofuels* • *Markets for Coal* • *Markets for Natural Gas* • *Markets for Petroleum* • *OPEC Market Behavior, 1973–2003* • *Stock Markets and Energy Prices*

Further Reading

Antle, R., and Eppen, G. D. (1985). Capital rationing and organizational slack in capital budgeting. *Manage. Sci.* **31**, 163–174.

Borenstein, S. (2002). The trouble with electricity markets: Understanding California's restructuring disaster. *J. Econ. Perspect.* **16**, 191–211.

Brekke, K. A., and Howarth, R. B. (2002). "Status, Growth and the Environment." Edward Elgar, Cheltenham, UK.

Conlisk, J. (1996). Why bounded rationality? *J. Econ. Lit.* **34**, 669–700.

DeCanio, S. J. (1993). Barriers within firms to energy-efficient investments. *Energy Policy* **21**, 906–914.

DeCanio, S. J. (2003). "Economic Models of Climate Change: A Critique." Palgrave-Macmillan, Houndmills, UK.

Dennis, M. L., Soderstrom, E. J., Koncinski, W. S. Jr, and Cavanaugh, B. (1990). Effective dissemination of energy-related information. *Am. Psychologist* **45**, 1109–1117.

Epstein, J. M., and Axtell, R. L. (1996). "Growing Artificial Societies: Social Science from the Bottom Up." M.I.T. Press, Cambridge MA.

Howarth, R. B., and Sanstad, A. H. (1994). Normal markets, market imperfections and energy efficiency. *Energy Policy* **22**, 811–818.

Interlaboratory Working Group. (2000). "Scenarios for a Clean Energy Future." Interlaboratory Working Group on Energy-Efficient and Clean-Energy Technologies, Oak Ridge National Laboratory and Lawrence Berkeley National Laboratory. ORNL/CON-476 and LBNL-44029, Oak Ridge, TN and Berkeley, CA.

Kehoe, T. J. (1998). Uniqueness and stability. *In* "Elements of General Equilibrium Analysis" (A. Kirman, Ed.), pp. 38–87. Blackwell, Oxford, UK.

Laitner, J. A., DeCanio, S. J., and Peters, I. (2001). Incorporating behavioural, social, and organizational phenomena in the assessment of climate change mitigation options. *In* "Society, Behaviour, and Climate Change Mitigation" (E. Jochem,

J. Sathaye, and D. Bouille, Eds.), pp. 1–64. Kluwer Academic, Dordrecht, The Netherlands.

Loewenstein, G. (2000). Emotions in economic theory and economic behavior. *Am. Econ. Rev. Papers Proc.* **90**, 426–432.

Mas-Colell, A., Whinston, M. D., and Green, J. (1995). "Microeconomic Theory." Oxford University Press, New York.

Nelson, R. R., and Winter, S. G. (1982). "An Evolutionary Theory of Economic Change." Belknap Press of Harvard University Press, Cambridge MA.

Nelson, R. R. (1995). Recent evolutionary theorizing about economic change. *J. Econ. Lit.* **33**, 48–90.

Porter, M. E., and van der Linde, C. (1995). Breaking the stalemate *Harvard Bus. Rev.* **73**, 120–134.

Romer, P. M. (2000). Thinking and feeling. *Am. Econ. Rev. Papers Proc.* **90**, 439–443.

Rust, J. (1997). "Dealing with the Complexity of Economic Calculations." Economics Working Paper Archive at Washington University, St. Louis. Available at http://econwpa.wustl.edu:8089/eps/comp/papers/9610/9610002.pdf (accessed 5/26/03).

Sanstad, A. H., DeCanio, S. J., Boyd, G. A., and Koomey, J. G. (2001). Estimating bounds on the economy-wide effects of the CEF policy scenarios. *Energy Policy* **29**, 1299–1311.

Markets for Biofuels

BENGT HILLRING
Swedish University of Agricultural Sciences
Uppsala, Sweden

1. Introduction
2. Experiences from Sweden
3. Experiences from Finland
4. Experiences from Denmark
5. Experiences from Austria
6. European Experiences
7. Conclusion

Glossary

bioenergy Super ordinate to the terms "biofuel" and "wood fuel."
biofuels Fuels produced directly or indirectly from biomass.
biomass Material of biological origin, excluding material embedded in geological formations and transformed to fossil.
forest fuel Wood fuel that has not been used previously.
upgraded wood fuels Wood fuel briquette with a diameter equal to or less than 25 mm; wood fuel pellet with a diameter less than 25 mm; wood fuel powder with a main fraction less than 1 mm.
wood fuel All types of biofuels originating directly or indirectly from woody biomass.
woody biomass Biomass from trees, bushes, and shrubs.

The use of biofuels is widespread over various countries and regions all over the world. However, the reasons for use of biofuels are different. Historically, it has been a very important resource for what we today call the industrialized countries. There are several examples of successful industrial use in mining industries, ore industries, and the like. For example, charcoal still is used in ironworking industries in Brazil. Biofuels may be used for cooking, heating, industrial applications such as steam production, electricity production, and fuel for vehicles as well as in the ironworking industry. Modern use of biofuels, as described in this article, includes industrial production in district heating networks and production of electricity. This development has increased trade of biofuels and in some European regions has created markets for various qualities of biofuels.

1. INTRODUCTION

The use of bioenergy is a strategic resource used in the work to fulfill the Kyoto Protocol to replace fossil fuels on a large scale. Bioenergy and biomass generally could be used to mitigate greenhouse gas emissions and the global climate change. In some countries, the use of biofuels already holds a significant share of the energy supply to industrial processes, heat production, and (in some cases) electricity production. The Nordic countries are examples of this. Large-scale production and use of biofuels has to be done in an environmentally sound way. New technology and system solutions give biofuels possibilities to compete in new markets.

Traditionally, biofuels have been used in the same geographical regions in which they were produced. During more recent years, this pattern has been changed in Northern Europe by industrial and large-scale use of various forms of biomass for district heating such as a vast supply of recycled wood and forest residues. The trade situation has come about as a result of means of control on waste and energy. Sea shipments allow bulk transports of biofuels over long distances at low cost.

Climate change due to anthropogenic emissions of greenhouse gases is one of the greatest environmental challenges today, and greenhouse gas concentrations in the atmosphere will continue to rise unless there are major reductions in greenhouse gas emissions. Carbon dioxide (CO_2) is the most important greenhouse gas, and increasing the use of biomass for energy is an important option for reducing CO_2 emissions.

The European Union (EU) has no common energy policy, but there are different common goals mainly given in the European Commission's (EC) "White Paper for a Community Strategy," which sets out a strategy to double the share of renewable energy in gross domestic energy use in the EU by 2010 (from the recent 6–12%, with some 85% of the renewables being bioenergy). The strategy includes a timetable of actions to achieve this objective in the form of double the share of renewable energy in gross domestic energy use in the EU by 2010, including a timetable of actions to achieve this objective in the form of an action plan. The EC's green paper in November 2000, "Towards a European Strategy for the Security of Energy Supply," introduced the objective of substituting 20% of traditional fuels by alternative fuels in the road transport sector by 2020. A proposal for a directive "on the promotion of the use of biofuels for transport," adopted by the EC in November 2001, requires that an increasing proportion of all diesel and gasoline sold in the member states be biofuels, starting with 2% in 2005 and increasing progressively so as to reach a minimum of 5.75% of fuels sold in 2010.

The EU member countries are jointly decreasing the emissions of six different greenhouse gases with by 8% on average during the period starting in 1990 and ending in 2008–2012 (depending on type of action and type of gas). An example of this policy is the EU-funded research programs, where the focus is on nonfossil fuels and nonfossil techniques to target the goal of reducing emissions of greenhouse gases.

The Scandinavian countries and Austria have been the pioneering countries in the modern use of bioenergy. Other countries, such as Germany, the United Kingdom, The Netherlands, and Spain, are expected to increase their use of biofuels significantly. There are variable forest resources and variable uses of biofuels in Europe, and these are the key factors behind the trade that has grown rather quickly during the past years. Because of the incentives to increase the use of biofuels, so the prediction for the short and medium term is an increase in trade biofuels in Europe.

2. EXPERIENCES FROM SWEDEN

Sweden is one of the leading European countries in use of biomass based mainly on the large forest resources. Biofuels are used in the forest products industry, in the district heating sector, and for traditional heating mainly of houses in the countryside. Industrial use is 184 PJ per year of mainly wood fuel and black liquor, with the majority of this being internal use. The residential sector uses 38 PJ per year, also mainly internal use, on farms or self-supply among house owners.

Swedish experience of a commercial biofuel market is based on the development of the district heating sector over a period of at least 20 years, from a low level during the 1970s to a substantial market reaching 96 PJ in 2000. Wood fuel counts for 55% of the biofuel supplied to the district heating sector, reaching 54 PJ in 2000. The average growth rate of supplied wood fuel between 1990 and 2000 was approximately 10% per year. Wood fuels compete on the market with other untaxed biofuels. The highly taxed fossil fuels are not competitively priced for heat production.

Prices for biofuels, peat, and the like are reported by the Swedish National Energy Administration.

2.1 Imports

During the early 1990s, the import of biofuels was established in the Swedish market.

Many different assortments and qualities of biofuels were traded. Various projects were carried out to adapt fuels to existing technology. In some areas, cofiring between various biofuels and biofuel and coal is established.

Later results from the imports indicate much large activity and import figures reaching as much as one-third of the supply of biofuels to the district heating sector. No reliable figures for later than 1997 were available at the time of this writing, but the estimated level of imports today is a stabilization of the levels from the mid-1990s. One reason is the weaker Swedish currency, which made it more expensive to buy from other countries. An example of this is the strong U.S. dollar, which made it too expensive to continue with the import of wood pellets from North America.

Swedish currency is stronger and prices have increased due to a larger market. This gives good ground conditions for an increased import (Table I).

Importers are the energy companies themselves or specialized fuel companies. The imported fuels show a spectrum of various assortments and qualities. Sometimes, imports are double counting, and this gives a range of the imports (Table II).

2.2 Exports

During recent years, the acceptance of the Kyoto Protocol has started activities in many European

TABLE I
Imports of Untaxed Fuels (Biofuels) to Sweden in 1992, 1995, and 1997 (in Pentajoules)

	1992	1995	1997
Biofuels	2–4	11–15	29–36
Approximate share of wood fuels	0.2 (estimated)	0.33–0.45	0.5–0.62

Sources. NUTEK (1993) and Vinterbäck and Hillring (2000).

TABLE II
Imports to Sweden of Untaxed Fuels in 1997: Combined Results from Telephone Survey to Fuel Users and Fuel Dealers

Assortment (quality)	Pentajoules	Share provided by importers (percentage)
Tall oil	8.73	59
Wood pellets	4.35	74
Recovered wood chips	3.65	80
Green wood chips	2.70	58
Peat	1.48	0
Rubber and tires	1.29	48
Railway sleepers	0.58	100
Round wood	0.21	100
Bark	0.19	100
Waste wood	0.17	100
Recycled paper	0.14	0
Sawdust	0.07	100
Total 1	23.56	(No double counting)
Total 2	14.70	(Only importers, i.e., maximum double counting)

Source. Vinterbäck and Hillring (2000).
Note. Figures in the survey did not represent the total population of possible users and importers. To get an estimation of the total, import figures were multiplied by 1.36.

countries, as mentioned earlier. Interest in exporting projects has started due to the fact that Sweden and the other Nordic countries, as well as the Russian Federation, have large forest resources. Importing countries in this future scenario are the densely populated European countries and some of the Mediterranean countries. No regular trade is carried out yet, but there have been projects in this field.

3. EXPERIENCES FROM FINLAND

In Finland, the annual use of solid wood fuels is approximately 127 PJ. During a recent 5-year period, more than 100 district heating plants and 500 MW_e of new additional capacity for electricity production from wood-based fuels were commissioned in Finland (total capacity 2000 MW_e). Each year, Finland exports approximately 1 PJ of wood pellets to Denmark, Sweden, and The Netherlands as well as peat and wood chips to Sweden. Each year, the Finnish forest industry imports approximately 13 million solid m^3 of wood for raw material, mostly from the Russian Karelia. Out of this, 11% is bark, which is used for energy production (11 PJ). This import of wood fuel as a separate fraction is equal to 1.3 PJ.

The Finnish national solid biofuel trade is carried out by various organizations. There are big diversified companies working on energy, wood processing, and biofuels business. These companies can exploit industrial wood residues from their own mills for production of wood chips and/or pellets. They can integrate harvesting of logging residues into timber or pulp wood harvesting and, in so doing, can keep fuel prices competitive. In small-scale heating of municipal buildings such as schools and small district heating plants, a special heating entrepreneurship model has been used in fuel procurement as well as in the operation and maintenance of the boiler plants. Usually, these entrepreneurs are cooperative, harvest small-sized wood of their own woodlots, or purchase industrial wood chips or cutter shavings from the local wood processing industry. The number of such entrepreneurs in Finland has increased to more than 130. In large-scale wood fuel procurement, various modern information technologies, such as mobile telephones and the Internet, are used in positioning fuel storage and logging sites, in ordering chippers, and so on.

In Finland and Austria, several electronic firewood marketplaces are established on the Internet (e-trade) for local, national, and even international consumers to select log and pellet suppliers.

4. EXPERIENCES FROM DENMARK

Denmark is a net importer of biofuels and imports trunks as well as wood chips and wood pellets. The import of wood chips covers approximately 25% of the annual Danish consumption of roughly 2 to 3 PJ. The Baltic states and Germany represent nearly 90%

of these imports, with the remainder coming from Poland. These imports consist mainly of wood qualities originating from clear-cuttings and trunks that are not chipped until after their arrival at a Danish harbor. Selling to the Danish market is competitive because it may be difficult to sell wood chips in Germany; furthermore, the prices in Denmark are a little higher than in Germany (Table I). In addition, an unknown amount of moist industrial by-products (e.g., bark, sawdust, chips) is imported.

The international market for wood fuels can be compared with the trading of straw for energy purposes. Straw is a local commodity, and in Denmark there is a geographic fluctuation in the prices due mainly to the bridge toll (Store Bælt) or ferry crossing, which make the straw transport very expensive. Long-distance transport of straw bales is expensive but is not an environmental problem given that a truck loaded with 12 tonnes of straw must drive 17,000 km with the straw before its emissions have compensated for the straw's CO_2-displacing effect in a heating plant.

5. EXPERIENCES FROM AUSTRIA

In Austria, biofuels have a 10.5% share of the annual consumption of approximately 105 PJ. Fossil fuels provide a significantly lower percentage of the primary energy in Austria compared with the European average. Austria has committed itself to an annual 13% reduction in greenhouse emissions. This is a highly ambitious goal that is all the more difficult and costly to achieve given that the per capita share of greenhouse gas emissions in Austria is already relatively low due to the fact that Austria meets a high percentage of its energy demand from hydropower, biomass, and natural gas.

The country is rich in both forest resources and agricultural land, giving Austria the opportunity to use bioenergy. Policy to promote the use of various biofuels in district heating is rather small scale compared with that in the Scandinavian countries.

Trade of wood fuels occurs with neighboring countries such as Germany, Slovenia, and Italy; however, such trade occurs at rather low levels.

6. EUROPEAN EXPERIENCES

Trade between countries has been established in various parts of Europe. In most countries, the customs statistics do not record trade in enough detail to identify international trade of different biomass types. Today, solid biofuels such as wood residues, pellets, and wood chips are already traded in Europe. In some countries, there is a growing interest in international trade because it can provide biofuels at lower prices, larger quantities, and better quality than can domestic alternatives.

6.1 European Trade

Solid biofuels such as wood residues, pellets, and wood chips are now traded in Europe and reached a level of nearly 50 PJ in 1999. Trade among European countries has resulted in a growing interest in biofuel trade because international trade can provide fuels at lower prices. In several cases, the national biomass market is not yet developed well enough for organized international trade. On the other hand, projects may benefit in countries with unexploited biomass resources when fuels are available on an international market. Although there may be cross-border trade of, for example, domestic firewood between neighboring countries, this trade is more or less occasional and beyond official statistics.

The largest volumes of biofuel are traded from the Baltic countries (Estonia, Latvia, and Lithuania) to the Nordic countries (especially Sweden and Denmark but also Finland). Some volumes are also traded from Finland to other Nordic countries as well as among neighboring countries in Central Europe, especially The Netherlands, Germany, Austria, Slovenia, and Italy. The traded biofuels most often include refined wood fuels (e.g., pellets, briquettes) and industrial by-products (e.g., sawdust, chips) as well as (in Central Europe) wood waste. The annual production of wood pellets in Europe is estimated to be 1.2 to 1.3 million tonnes.

Some biofuels are also traded intercontinentally. Sweden has imported biofuels from Canada, and Italy imports firewood from Northern Africa. In addition, Germany exports some firewood to the Middle East and Far East.

Scandinavian biofuel markets have increased, with national energy policies contributing strongly to this trend. Taxes on energy with a clear environmental profile were introduced in Scandinavian countries during the early 1990s. Fossil fuels are heavily taxed in heat production, whereas biofuels are untaxed. In electricity production, all fuels are untaxed, whereas consumers pay a tax. In Finland and Sweden, the investment supports called for growth in the

TABLE III

Minimum, Maximum, and Average Fuel Prices (Including Taxes) in 18 Selected European Countries: 1999 (in GJ)

Fuel	Minimum country	Maximum country	Average
Forest residues	1.02, Germany	8.33, Italy	3.42
By-products, forest products industry	0.58, Romania	9.07, Poland	2.38
Firewood	1.01, Slovakia	14.00, United Kingdom	5.26
Wood waste	−4.00, Ireland	3.31, Poland	0.97
Refined wood fuels	3.24, Latvia	18.22, Germany	8.37
Other biomass	0.83, Slovakia	12.00, Poland	4.68
Peat	2.10, Finland and Latvia	3.75, Ireland	2.83
Heavy fuel oil	1.40, Slovakia	12.00, Ireland	6.74
Light fuel oil	3.10, Slovakia	14.30, Denmark	6.74
Natural gas	1.10, Slovakia	16.21, Italy	5.80
Coal	1.19, Poland	12.78, Germany	4.53

Sources. Vesterinen and Alakangas (2001) and Hakkila *et al.* (2001).

TABLE IV

European Prices of Wood Fuels for Combustion Plants of the Size 1 to 5 MW_{th}: 1999 (in €/GJ, Including Energy Taxes and Excluding Value-Added Taxes)

Country	Bark, sawdust, and chips	Wood chips	Wood pellets
Denmark	4.2	4.5	5.0
Finland	1.6	3.0	7.5
Germany	3.1	3.7	6.1
Sweden	2.9	3.4	4.8
France	1.1	4.0	10.6
Latvia	0.8	1.6	3.3

Sources. Vesterinen and Alokongas (2001) and *STEM* (periodical).

capacities and also contributed to the demand for biofuels.

6.2 Fuel Prices

Prices of various fuels were collected in a 1999 survey. Since then, oil prices have risen significantly, and this also might have influenced the prices for other fuels (Table III).

6.3 European Price Level

The increase in trade among European countries has established international prices for wood fuels. Table IV gives examples of prices. Of course, there is a range in prices set by the market situation, production cost, and cost of competing fuels. By-products were traded at approximately 3 to 4 €/GJ, wood chips at 3.5 to 4.5 €/GJ, and wood pellets at 5 to 6 €/GJ.

7. CONCLUSION

There is a growing international trade in biofuels, and the largest volumes are traded from the Baltic countries (Estonia, Latvia, and Lithuania) to the Nordic countries (especially Sweden and Denmark but also Finland). Some volumes are also traded from Finland to other Nordic countries as well as among neighboring countries in Central Europe, especially The Netherlands, Germany, Austria, Slovenia, and Italy. The traded biofuels are most often refined wood fuels (e.g., pellets, briquettes) and industrial by-products (e.g., sawdust, chips) as well as (in Central Europe) wood waste.

There are international market prices established in Europe; however, volumes are still limited. Both the use and trade of wood fuels are expected to increase, at least in Europe, in the future.

SEE ALSO THE FOLLOWING ARTICLES

Biodiesel Fuels • *Biomass for Renewable Energy and Fuels* • *Biomass Resource Assessment* • *European Union Energy Policy* • *Forest Products and Energy* • *Market-Based Instruments, Overview* • *Market Failures in Energy Markets* • *Markets for Coal* • *Markets for Natural Gas* • *Markets for Petroleum* • *Wood in Household Energy Use*

Further Reading

Alakangas, E., Hillring, B., and Nikolaisen, L. S. (2002). Trade of solid biofuels and fuel prices in Europe. In "12th European Biomass Conference: Biomass for Energy, Industry, and Climate Protection" (W. Palz et al., Eds.). Amsterdam, Netherlands.

Hakkila, P., Nousiainen, I., and Kalaja, H. (2001). "Use and Prices of Forest Chips in Finland in 1999." VTT, Jyväskylä, Finland.

Hillring, B. (2002). Incentives for co-firing in bio-fuelled industrial steam, heat, and power production. *Renewable Energy* **28**, 843–848.

Nikolaisen, L. S. (2001). Trade with solid biomass. In "Proceedings of the Conference, Bioenergy 2001," pp. 77–80. BioPress, Risskov, Denmark.

NUTEK. (1993). "Forecast for Biofuel Trade in Europe: The Swedish Market in 2000." Swedish National Board for Industrial and Technical Development, Stockholm.

Swedish National Energy Administration. (2001). "Energy in Sweden: Facts and Figures 2001." SNEA, Stockholm.

Vesterinen, P., and Alakangas, E. (2001). "Export–Import Possibilities and Fuel Prices in 20 European Countries." VTT, Jyväskylä, Finland.

Vinterbäck, J., and Hillring, B. (2000). Aufbau eines Europäishen Handels mit Holz-Brennstoffen. *Holzforschung Holzverwertung* **6**, 98–102.

Markets for Coal

RICHARD L. GORDON
Pennsylvania State University, University Park
Pennsylvania, United States

1. Patterns of Coal Consumption
2. End Uses
3. The Basis of Change: Coal in "Energy"
4. The Mythology of Energy Shortages
5. Innovations in Mining, Transportation, and Use

Glossary

coal The solid fossil fuel ranging from hard anthracite to the softer subbituminous coals.
coking coal Coal converted into coke for pig iron manufacture and other uses.
energy Petroleum, natural gas, coal, nuclear power, waterpower, and alternatives such as geothermal and solar.
steam coal Coal used for heating in boilers.
synthetic fuels Liquid fuels and gases made from other sources such as coal.
unit trains A train dedicated entirely to hauling coal between a specific origin (almost always a coal mine) to a specific destination (either a consumer such as an electric power plant or a transfer terminal).

In many discussions of coal, its abundance is extolled with the implication that coal consumption can be and should be massively expanded. In practice, coal use became increasing specialized. Electricity generation is the major market. Although this is not true in every major consuming center, it is increasingly true for most. This article seeks to delineate and explain this pattern. Discussion begins by indicating patterns of coal use by country and end use. The situation in the early 21st century and differences from earlier years are examined.

1. PATTERNS OF COAL CONSUMPTION

Coal consumption is even more unevenly distributed around the world than total energy use. The amount of coal consumed is affected greatly by the level of economic activity in the country. The relationship, however, is not direct. Wide differences prevail among countries in "energy intensity" as measured by the amount of energy consumed per unit of gross domestic product (GDP) and in the relative role of coal in energy use. Were this not complicated enough, further issues arise about the proper ways to measure economic activity and energy use. Table I summarizes the critical data for the leading coal-consuming countries.

The most striking aspect of the tabulation is the existence of countries with high GDP and energy use and low coal use. France, Italy, Brazil, and Mexico are the main examples. In the French case, the critical influence was a decision to allow a slow death of the domestic coal industry and rely heavily on nuclear power. (Over 80% of French electricity has been nuclear generated since the middle 1980s.) The other three countries simply display a tendency to low coal use characteristic of countries lacking extensive domestic coal supplies.

Higher coal roles necessarily mean that a country will be more important in coal use than in total energy. Several important cases arise. Every country getting the majority of its energy from coal is among the leaders in coal tonnage consumed. China and India are the primary examples. Russia, the Ukraine, and Kazakhstan are notable for unusually high energy/GDP ratios.

Another consideration is that coal is a much less mobile fuel than oil. The basic problem is that solids are more expensive to move than liquids. Moreover, coal has 70% or less heat content per ton than oil.

The vast majority of coal is consumed in the country of production. Again, however, substantial international differences prevail (see Table II). In particular, the four leading coal-consuming countries are all substantial producers, and international trade in coal is less than 10% of each of these countries' output. In contrast, other major producers export

TABLE I

Economic Activity, Energy Use, and Coal Use

	GDP 1998[a] (billion US$ at 1990 prices and PPP[b]s)	Total Energy/GDP[a] (tonnes of oil equivalent per thousand 1990 US$ PPP)	Total energy 2001[c] (million tons oil equivalent)	Coal 2001[c] (million tons oil equivalent)	Coal share[c]	Electricity share in coal[a]
World	33,653.523	0.284	9124.8	2255.1	24.7%	64.6%
United States	7,043.638	0.310	2237.3	555.7	24.8%	95.1%
China	4,154.508	0.248	839.7	520.6	62.0%	44.9%
India	1,487.012	0.320	314.7	173.5	55.1%	66.7%
Russia	664.945	0.875	643.0	114.6	17.8%	64.4%
Japan	2,581.473	0.198	514.5	103.0	20.0%	47.2%
Germany	1,483.293	0.232	335.2	84.4	25.2%	87.7%
South Africa	195.894	0.567	107.0	80.6	75.3%	59.2%
Other Asia–Pacific Islands			95.8	58.7	61.3%	15.7%
Poland	257.313	0.375	87.7	57.5	65.6%	62.7%
Australia	362.814	0.289	109.9	47.6	43.3%	85.1%
Korea	539.059	0.303	195.9	45.7	23.3%	66.1%
United Kingdom	1,077.407	0.216	224.0	40.3	18.0%	70.3%
Ukraine	75.978	1.881	131.1	39.0	29.7%	69.7%
Taiwan	398.142	0.193	85.4	30.9	36.2%	66.8%
Canada	610.413	0.384	274.6	28.9	10.5%	84.8%
Kazakhstan	52.116	0.749	43.1	24.7	57.3%	60.9%
Czech Republic	92.227	0.445	41.6	21.3	51.2%	66.7%
Turkey	461.224	0.157	70.2	20.4	29.1%	46.5%
Spain	557.744	0.202	134.6	19.5	14.5%	81.3%
Indonesia	665.131	0.185	97.9	16.7	17.1%	69.1%
Brazil	907.057	0.193	173.6	14.0	8.1%	15.5%
Italy	1,037.380	0.162	177.2	13.9	7.8%	53.7%
France	1,138.864	0.225	256.4	10.9	4.3%	53.4%
Other Europe			41.1	10.6	25.8%	
Greece	118.096	0.228	31.1	9.5	30.5%	88.7%
Thailand	360.047	0.192	63.0	8.8	14.0%	57.6%
Netherlands	305.165	0.244	88.6	8.4	9.5%	69.3%
Romania	64.645	0.613	38.2	7.7	20.2%	61.2%
Other Middle East			91.9	7.2	7.8%	100.0%
Belgium and Luxembourg	209.057	0.298	63.9	7.1	11.1%	43.0%
Other Africa			95.8	6.5	6.8%	
Mexico	757.984	0.195	127.7	6.3	4.9%	62.9%

[a] *Source. Energy Balances of OECD Countries* and *Energy Balances of Non-OECD Countries* (disk versions), The International Energy Agency.

[b] PPP, purchasing power parities, which are calculations that compare currencies in terms of relative domestic purchasing power over all commodities.

[c] *Source. BP Statistical Review of World Energy*, June 2002.

higher proportions of their output and many major consumers are heavily import dependent (see Table III).

The United States alone of the long-time major coal producers remains cost competitive with oil, natural gas, and coal from other countries. The

TABLE II

2001 Consumption and Production of Coal[a]

	Coal consumption	Coal production	Consumption as percentage of production
United States	555.7	590.7	94.1%
China	520.6	548.5	94.9%
India	173.5	161.1	107.7%
Russian Federation	114.6	120.8	94.9%
Japan	103.0	1.8	5722.2%
Germany	84.4	54.2	155.7%
South Africa	80.6	126.7	63.6%
Other Asia Pacific	58.7	68.1	86.2%
Poland	57.5	72.5	79.3%
Australia	47.6	168.1	28.3%
South Korea	45.7	1.7	2688.2%
United Kingdom	40.3	19.6	205.6%
Ukraine	39.0	43.6	89.4%
Taiwan	30.9		
Canada	28.9	37.6	76.9%
Kazakhstan	24.7	40.6	60.8%
Czech Republic	21.3	25.8	82.6%
Turkey	20.4	12.9	158.1%
Spain	19.5	8.0	243.8%
Indonesia	16.7	56.9	29.3%
Brazil	14.0	2.1	666.7%
Italy	13.9		
France	10.9	1.5	726.7%
Other Europe	10.6	12.1	87.6%
Greece	9.5	9.0	105.6%
Thailand	8.8		
Netherlands	8.4		
Romania	7.7	7.3	105.5%
Other Middle East	7.2		
Belgium and Luxembourg	7.1		

[a] Source. BP Statistical Review of World Energy, June 2002.

TABLE III

Coal Production and Exports 2001[a]

	Production	Exports	Exports as percentage of production
China	940,000	91	9.68%
United States	927,868	44	4.74%
Russia	330,000	36	10.91%
India	315,000		
Australia	257,288	194	75.40%
South Africa	221,273	69	31.18%
Poland	103,000	23	22.33%
Canada	32,450	30	92.45%

[a] Source. Zahlen Zur Kohlenwirtschaft, Statistik der Kohlenwirtschaft, E.V. (June 2002).

United Kingdom, Germany, Japan, and France are countries with substantially reduced coal production and shifts to imported coal. In every case, depletion has eliminated resources that can compete with alternative fuels or coal from elsewhere.

New more competitive sources of coal have emerged. Many of these—Australia, Indonesia, Colombia, and Venezuela—export well over half their coal; another case, South Africa, approximately 30%. Japan, France, and South Korea are particularly striking cases where domestic production is a small fraction of consumption. The role of domestic supplies is greater in the United Kingdom and Germany but contrasts sharply from the prior tendency to be a net exporter. Korea and Taiwan were never significant coal producers.

2. END USES

The history of coal use involves greatly increased specialization of use. In particular, consumption to generate electricity became the dominant use. The United States is the prototype for this development. Data are available on U.S. coal use since 1920 and show a massive shift in coal-use patterns (Table IV). In 1920, coal was used throughout the economy. Over one-half was employed as a general-purpose fuel for "industry" (which as reported apparently included residential and commercial use). Over one-quarter of use was to fuel the railroads. The mix of use persisted to 1945 (and the level of use was only 10% higher in 1945 than in 1920).

A precipitous decline of coal use prevailed from 1945 to 1961. The adoption of the diesel-electric engine eliminated the railroad consumption of coal. Sales for general industrial use and retail deliveries declined. Use in coking fluctuated without showing any persistent trend. However, growth in electric power use was maintained.

The post-1961 history involved continuation of both gains in sales to electric utilities and declines elsewhere. Indeed, coking joined other industry and retail deliveries as a declining sector. Since rail use

TABLE IV

U.S. Coal Use by End Use[a]

	Electric utilities	Railroads	Coking	Industry	Retail	Bunkers	Total
1920	30,099	135,414	76,191	255,959		10,486	508,148
1929	39,729	131,100	86,787	257,214		4,287	519,117
1945	71,603	125,120	95,349	148,198	119,297		559,567
1960	176,685	3,046	81,385	96,017	40,948		398,081
1970	320,182	298	96,481	90,156	16,114		523,231
1980	569,274		66,657	60,347	6,452		702,730
1990	773,549		38,877	76,330	6,724		902,893
2001	968,990		26,075	63,361	4,127		1,062,554
1920	5.9%	26.6%	15.0%	50.4%	0.0%	2.1%	100.0%
1929	7.7%	25.3%	16.7%	49.5%	0.0%	0.8%	100.0%
1945	12.8%	22.4%	17.0%	26.5%	21.3%		100.0%
1960	44.4%	0.8%	20.4%	24.1%	10.3%		100.0%
1970	61.2%	0.1%	18.4%	17.2%	3.1%		100.0%
1980	81.0%		9.5%	8.6%	0.9%		100.0%
1990	85.7%		4.3%	8.5%	0.7%		100.0%
2001	91.2%		2.5%	6.0%	0.4%		100.0%
2001/1960	448.4%	−100.0%	−68.0%	−34.0%	−89.9%		166.9%

[a] *Sources.* U.S. Bureau of Mines Reports (for years up to 1945) and U.S. Department of Energy, Energy Information Administration, *Annual Energy Review*, Table 7.3 (for later years).

had almost disappeared, further losses in transportation were minute. Electric power use gains, however, began to outweigh the losses and coal use began rising. These trends have persisted. The other sectors shrink while electric power use rises. By 2001, electric power absorbed 91% of coal use. General industrial use was 6% and coking use was 2%.

The tendency to increased importance of electric power as a coal consumer prevails in most of the main coal-consuming countries. The details again differ but most countries and the world as a whole use more than half of their coal for electricity production. The main exceptions are China and Japan. In both cases, electricity is becoming an increasing portion of coal use. The Chinese electricity use already exceeds that for any other end use. Japanese electricity use is nearing parity with coking use.

Germany uses approximately 88% of its coal in electricity. Here the pattern of steady electricity growth and declines elsewhere produced steadily declining total coal use.

The below average degree of electricity use in South Africa reflects its unique venture into synthetic fuels from coal. South Africa under white rule undertook an extensive effort to produce liquid fuels, gas, and chemicals from coal. This generated at most approximately one-third of the inputs for producing petroleum products. Oil procurement problems were not as dire as feared. Even at the height of the United Nations boycott, major European oil companies such as Shell, British Petroleum (BP), and AGIP openly operated in South Africa. The majority of coal use in Poland is for electricity. The use has declined sharply since the fall of communism.

3. THE BASIS OF CHANGE: COAL IN "ENERGY"

The central proposition about coal is that it is a problem-plagued source of "energy." To see this, the basic propositions about energy need review. Energy as usually treated is defined to include only inanimate energy sources and usually not all of them. Five types of sources—coals, oils, gases, uranium, and waterpower—are invariably covered. Depending on their importance in a country and the interests of the observer, such substitutes as wood, animal wastes, other sources of biological origin, wind, and solar energy also are treated. (Coverage of these sources has expanded in large part reflecting interest in increasing the use of "alternative fuels." The data show that present utilization is heavily concentrated in poor countries. China and India consume the largest amounts of such fuels and together account for approximately 38% of reported world totals. However, in both cases, the majority of reported fuel use comes from the five usually reported fuels. Smaller, even more impoverished countries such as all sub-Saharan Africa other than South Africa obtain the vast majority of their fuel from alternative sources. In short, these fuels are widely considered the last resorts for those not integrated into the world economy.) Thus, energy consists of things that are burned to secure work or close substitutes for such material.

Moreover, all three fossil-fuel sources actually are secured from heterogeneous occurrences. They differ within as well as among deposits in such key characteristics as amount and form of the fuel contained, the extent of contamination by other material, and the ease of extraction. In particular, solid fuels range from the very hard (anthracite) to the boggy (peat). Although classifications differ among countries, coals extend at most to the softer, more ash-laden forms called lignites or brown coals.

Outside the United States, lignite is treated as being distinct from the harder coals—anthracite, bituminous, and subbituminous—and even subdivided into better and worse qualities. Thus, the Germans distinguish between braunkohle and pechkohle. In contrast, the U.S. reporting practice was established long before lignite became a significant part of output. U.S. data still combine lignite with harder coals. (Conversely, the procedures arose when the United States had substantial anthracite production. Anthracite was reported separately long after it ceased to be important. Under the new methodology, anthracite is reported with bituminous and lignite with inclusion of separate anthracite subtotals.)

Those hard coals used to make coke for iron-making are termed metallurgical or coking coals. All others are termed steam coals. The coals are not necessarily different; often the only difference is the use made. Considerable evidence exists that producers shift between the two sectors as market conditions warrant. (In particular, Queensland long has reported the disposition of coal from each of its mines, and shifts between steam and coking sales can be observed.) This may or may not alter what is mined and how it is processed. The flexibility, in any case, links the steam coal and coking coal sectors of the market.

The work done by these fuels, in turn, involves primarily providing heating, lighting, and cooling to residential, commercial, and industrial buildings, powering industrial processes, and fueling automobiles and other forms of transportation. With the rise of electric power in the late 19th century, a tendency has arisen increasingly to consume fuels to produce electricity.

The physical characteristics of each fuel are a major influence on their production, transportation, processing, and utilization economics and on the information that is known about them. Coals are solid fuels that are more readily found than oil, gas, or uranium. Being solid, however, is otherwise an economic drawback. Solids by definition lack the fluidity that makes oil and gas so attractive and, as noted, have less heat content per ton than oil. As a result, extraction, transportation, storage, handling, and use are more difficult than is true for oil and gas. With transportation, storage, and handling, more material must be handled by less supple methods. Coal must be pushed through every stage. For example, a coal-using facility requires larger, more complex, more expensive boilers than those needed to obtain the same amount of heat from oil or gas.

In particular, although oil can be fairly cheaply transformed physically and chemically, coals are expensive to transform. The expense has been such that only an economy such as South Africa with a strong desire for self-sufficiency heavily engages in coal transformation.

Among the drawbacks of coal are the environmental impacts. The original problem was of the high waste content that emerged as ash, much of which was discharged during combustion. This difficulty was alleviated at a cost by the development and adoption of techniques to collect the ashes before they would have been discharged. The next major concern was the sulfur contained. This was, in fact, also a problem for natural gas and oil but removal or shifts to naturally less-sulfur-containing sources were more economic than with coal. The coal options are naturally lower sulfur coals such as in the U.S. Rocky Mountain States or adoption of postcombustion clean-up technologies. Further concerns arise from other contaminants in coal.

The newest concern is that coal emits more of the carbon dioxide believed to be a critical greenhouse gas in the global warming threat. Thus, adoption of vigorous measures to limit greenhouse gas emission could produce major new pressures to reduce coal use.

During the 20th century, coal changed from a dominant fuel to one often beleaguered by competition from other fuels. With the prevailing scarcity pattern, the role of coal is to fill niches in which its drawbacks are less. Coal is attractive when location and size of operations justify use of coal. Energy use on a larger scale produces economies. Just where they are exhausted is unclear. It is at a level much smaller than the electric power market but apparently much larger than that maintained by most manufacturing plants and commercial facilities. These scale economies narrow, but do not eliminate, the cost disadvantages of using coal. If large cheap-enough coal supplies exist, savings on fuel costs will justify endurance of higher other costs and lead to coal use by electric utilities. Large boilers predominantly in electricity generation are a prime and indeed steadily growing market.

Coal as the input for the coke used in pig iron manufacture is a niche that has shrunk. The remainder of coal use is scattered, predominantly in manufacturing plants.

Steel is an old, beleaguered industry that has undergone remarkable changes. Coal use has been squeezed by adverse developments in this market. The alterations involve the location of raw material suppliers and steel producers, the technologies used in different stages of steelmaking, and the product mix. These have been good for consumers but bad for coal. Historically, coke-based iron-making provided most of the inputs for steelmaking. In what is termed the hot-metal approach, pig iron is refined into steel. Some scrap also can be used. Steelmaking technology has changed so that less coal is needed.

First, steel output has fluctuated without much net gain. New technology lessens processing losses and thus the gross input of steel per ton of finished steel output. The hot-metal route, moreover, is losing ground to the alternative of producing steel entirely from scrap (using electricity). This occurred despite introduction of the oxygen furnace that reduced costs of the ore-based route. A secondary consideration is the ability to supplement coke with other fuels such as oil and uncoked coal. Thus, coke has many direct and indirect rivals.

More innovations that would further reduce coal use are under consideration. Certain products such as the thin sheets used for automobile bodies could not be made satisfactorily from steel from scrap, but technology has emerged to overcome this limit. For decades, steel companies have experimented with technologies to make iron without coke. Application still is limited to countries in which coking coal is expensive to attain, but hope of change persists.

The growing dependence of the coal industry on electric-power customers has many important technical and economic implications. Coal suppliers are dealing with large entities participating in an industry with far more flexibility in fuel choice than any other fuel users. This means substantial substitution possibilities and widespread knowledge of these possibilities.

The components of an electric power network include generating units of different fuel-using capabilities in many different locations. A unit consists of a generator and whatever causes it to operate. A plant has the usual meaning of a total facility. Although one-unit plants exist, it is common to have several units. These sometimes consist of several identical units but often differ from one another in various degrees. Some simply have units that differ only in age and size, but many cases arise of major differences among units. One U.S. plant has a coal unit and a nuclear unit. Many have both boiler-powered plants and combustion turbine units (see below). Often only some of the boilers are designed to burn coal.

Nuclear units using technologies that rigidly determine fuel choice are one pole of the spectrum of flexibility. At the other extreme are plants on the U.S. East Coast physically able to burn coal, oil, and natural gas and located so that they can be economically supplied by all three fuels. Much lies in between. A boiler designed for coal can be (and often has been) converted to oil and gas use. In the 1960s era of cheap oil, utilities located at the center of interfuel competition on the U.S. East Coast built triple-firing capability and shifted among coal, oil, and gas as prices changed. Such plants were readily served by tankers carrying fuel oil and close to coalfields. Gas use was attractive at least during the warm months in which home heating use ceased. When natural gas appeared likely to remain a low-cost fuel, utilities in gas-producing states (most notably Texas and California) built plants designed to burn only gas.

When cheap oil seemed likely in the late 1960s, plants that burned only oil were built. This occurred mainly on the East Coast. Prevailing public policy then freed East Coast power plants from import controls that discouraged oil use elsewhere. A few inland (Illinois and Michigan) plants also arose in the expectation that Canadian oil sources would be an attractive alternative.

Finally, the demand for electricity fluctuates, with some of the variation being unpredictable; equipment, moreover, is not continuously available. Thus, the need exists to maintain spare capacity to meet demand surges given equipment availability. In these circumstances, great short-run flexibility exists. Output mixes can and do shift in response to changing fuel prices. Plants already using cheaper fuel sources are used more; plants that can easily shift to cheaper fuels do so and also are utilized more.

In the longer run, the mix of available plants can be altered to allow still greater use of cheaper fuels. Thus, the enthusiasm for nuclear power in the early 1970s inspired a flurry of orders that was (partially) implemented through the rest of the decade and throughout the 1980s.

Similarly, as gas became more expensive, gas-using utilities started building coal-fired units. This process started in West North Central and Mountain states and later spread to the large oil-producing states such as Texas, with California choosing for various reasons to rely on coal-generated electricity transmitted from plants in Arizona, New Mexico, Nevada, and Utah.

4. THE MYTHOLOGY OF ENERGY SHORTAGES

In contrast to the limited prospects just sketched, restoration of coal to a role of eminence often was advocated. Tendencies exist to exaggerate both the advantages and disadvantages of coal use. Coal is regularly proclaimed a highly abundant fuel. The implication often is that a great potential has not been realized and coal should be restored to its historical role as the leading source of energy for general use. These assertions, however, ignore or at least mute debate of the formidable barriers to such a massive return. The correct view is that the industry has done extremely well in a hostile environment.

At least immediately after the rapid oil price increases of 1973–1974, discussions of energy routinely argued that a massive return to coal is essential. (Those with longer memories know this idea often has risen and fallen since oil competition emerged in the middle of the 19th century and that the coal industry perennially reiterates the contention.)

The advocacy of a return to coal is based on the presumption that the economics will change radically within a few decades. Some add that since so much oil comes from a small group of politically unstable countries engaged in efforts to cartelize supplies, the supply problems will emerge sooner. Coal then will become competitive again.

This then inspires an unending debate in which both sides remain impervious. Both views have some truth on their side, albeit not to an equal extent. Nothing lasts forever, but oil and gas supplies are greater than those expecting an immediate demise realize. The disparity between expectation and outcome arose from misunderstanding of the basic economic principles affecting coal. Unfortunately, many people believe in the desirability of more expensive energy. Therefore, clearly premature predictions of vanishing oil and gas supplies frequently appear and error fails to discredit the advocates.

The available data suggest that the physical supply of coal is much greater than that of oil and gas. Since oil and gas, as noted, are harder to find than coal, the appearance may be incorrect. The data are biased because coal outcrops more than oil seeps. Thus, more can be and is readily learned about physical endowments of coal than about oil and gas. Indeed, coal resource data are based almost entirely on physical availability. In contrast, oil and gas reserves cover only the well-delineated resources in fields developed for production. Thus, whereas coal estimates change little, oil and gas resources are steadily augmented.

Ignorance remains about the true state of future energy supplies because acquiring the knowledge involves costs from which investors see inadequate benefit. Those who might use the knowledge to assist investment in developing energy supplies are aware that they can safely delay their decisions until possible use is nearer at hand.

Even if the assertions about physical abundance are true, relative physical availability does not dictate a massive swing to coal. Physical endowment is not economic supply and not even a good proxy for economic supply. Supply relates to the cost of extraction and use. The best is used first. The available physical stocks of oil and gas contain a larger portion of the cheapest resources to produce, process, and use than do the available coal resources. This is why oil and gas attained a larger energy role than coal.

The impacts of physical availability will occur so far in the future that research and development may have succeeded in making competitive energy sources that have even greater physical availability. It can readily be seen that at least another half-century's worth of extensive oil and gas supplies remains. Oil and gas prices may remain at 2002 levels for many more decades. The balance among coal, oil, and gas, therefore, will not change radically at least until the middle of the 21st century.

No guarantee exists that the coal industry is ensured an ultimate victory. The interim could permit the development of many alternative sources of energy. Instead of being the only option, coal may be one of several. Over such a time span, more will be learned about the availability of oil, gas, and coal and if profitable, even more ample energy supplies (e.g., uranium—given supply-augmenting techniques called breeding, nuclear fusion, and solar energy) will be developed as will new ways to use them.

For example, nuclear fission involves the release of energy by splitting atoms. Such fission occurs only with unstable materials such as uranium 235 and plutonium. However, as demonstration projects proved, it is possible to build reactors in which the splitting produces both energy and the conversion of uranium 238 into a fissionable material. Since more fissionable atoms are added than used, the process is termed breeding. Plans to build a larger demonstration project were terminated by the Carter administration. Although it stressed environmental concerns, another problem was that the effort was another example of reaction to premature warnings of resource exhaustion.

Indeed, in the years since 1973, coal has become increasingly déclassé. Resource pessimists are stressing a future in which conservation and alternative energy rather than coal (or even worse for them nuclear power) replace oil and natural gas.

Thus, many outcomes are possible. It would be a serious mistake to commit the world to any option for this distant future. The need to act has not arisen and it is still not known what the best approach will be when the time comes.

The enthusiasts of government assistance to promote rapid changes in energy utilization believe that oil and gas soon will become very expensive and that energy producers and consumers are not acting rapidly enough to prepare.

Two (questionable) reasons exist for arguing that public policy has a major role in ensuring a more efficient transition. The first relates to the externalities in research on energy possibilities. The core issue here is whether government assistance will be more efficient than forming consortiums of private companies. A second issue is the perennial debate about whether markets or governments take better care of the future. Governments possess greater resources and thus potentially are better able to hedge against risks. However, the need to win elections puts stress on preserving the status quo and concentrating on quick payoffs. Thus, controversy persists over how much of a role should be played by governments.

Contentions about the abundance of coal often lead to proposals for increasing coal use. The coal industry thought (probably incorrectly) that President Nixon was committed to doubling coal output; President Carter made a clear commitment to such a doubling. Neither president did much to promote the outcome.

In the Nixon case, optimism was killed by analysis. A massive effort was undertaken to appraise the feasibility of Nixon's call for attaining energy independence. Those studying coal-production prospects tentatively suggested that, with effort, coal output could be doubled. The work actually grossly understated the ability to expand coal output. It posited barriers that failed to matter.

However, this proved irrelevant. Those producing the computer analyses that synthesized all the studies into estimates of possible market clearing production–consumption patterns could not devise a plausible combination of market conditions that would lead to utilizing all the physical possibilities for expanding coal output and thus reflected economic realities that limit the ability to use coal.

Extensive evidence of the ability of energy markets to adapt to changing circumstances is available. Energy consumption growth slowed radically with the rises in oil prices of the 1970s. The pattern of energy use also changed. A substantial part of the alteration involved shifts by U.S. electric utilities to coal and nuclear power. U.S. refineries long produced a higher proportion of gasoline and less of fuels used in industry than other industrialized countries. The U.S. pattern reflected the availability of low-cost alternatives, such as natural gas and coal, that were formidable competition for oil products in industrial markets. Changing oil prices combined with rising natural gas, nuclear energy, and foreign coal supplies produced a European move to a similar pattern.

Coal possesses a niche that has expanded. Efforts to alter this must carefully appraise the consequences of reducing coal use.

As attention turned to the environmental side effects of energy production, processing, distribution, and use, it appeared that more serious effects were associated with coal than with oil and natural gas. This further reduced the competitiveness of coal.

Policy makers deal with these contradictions by professing to want more coal consumption while imposing barriers to its production and use. Whatever the realities, the practical implication has been that generally greater stress is placed on taming the environmental impacts of coal than on promoting its use.

Contrary to statements often made by alarmists about energy, the existence of "special" uses is another irrelevance. All the proposed nonmarket criteria prove defective. Both economic theory and energy practice indicate deficiencies with fascination about critical uses. Basic supply–demand analysis shows that what is served depends on market conditions. The greater the supply, the more demands that can be met. The case for consumption control involves misunderstanding of these basic economic principles.

The proposed alternatives to market allocation have serious drawbacks. Each suggestion for administratively allocating energy has clear defects and conflicts prevail among different proposals. The proposed rules for overriding market decisions prove neither clear nor appropriate.

One of several widely used concepts of urgency is satisfying some end use such as transportation and chemical production. Some concentrate on favoring types of users—favoring households over commercial establishments, commercial establishments over factories, and factories over electric power plants.

The differences in urgency, however, can be greater within any broad category than between categories. The chemical uses of fuels extend from life-saving drugs to fancy wrappings. The marginal chemical uses of fuels, such as packaging, are less urgent than some of the applications simply to provide heat. An apparently superior use may not persist as with coke for iron-making. A new basis, moreover, may conflict with the old. The widespread political preference for direct household use over use in producing goods for households is inconsistent with a stress on chemicals.

More basically, favoring direct consumption is clearly harmful to such final consumers. Indirect uses are undertaken to provide goods for final consumers. Such uses can outbid households for energy because the industrial use has more value to consumers than direct consumption of energy.

However, the price system is preferable to political earmarking in determining preferable uses of energy resources. Markets decide on the basis of willingness to pay; this approach responds to the differences within broad categories ignored by policies giving blanket favoritism. No convincing evidence exists of market failures to anticipate future demands. In addition, the preferred ways of meeting a need are not even the only ones already known and new modes could emerge. Thus, even if best uses could be identified, saving fuels for them may not prove the best way to serve them.

The contentions perpetrate one of the oldest errors in economics (dating back at least to John Law, a notorious speculator of the early 18th century) over when it is efficient to extend use beyond the most vital ones. The debate is epitomized by the diamond–water paradox—why does a decorative good sell for so much more than something vital for life? When supply is sufficiently ample, it proves good economics to serve less valuable uses—water for car washes and gas for a barbeque.

The environmental impacts associated with extracting and using coal justify regulation. However, the controls should deal directly with the environmental problems. An indirect indicator, such as energy use, can be a poor proxy for pollution impacts.

Nonintervention avoids both tying up resources and unnecessarily acting before relevant information develops. Too rapidly anticipating demand can lead to choice of technologies that prove obsolete. Preserving reserves may be as onerous for this century as preserving a supply of horses sufficient to cover the return of animal power would have been to the last century.

Thus, whatever the truth about the availability of coal, it does not by itself determine the optimal pattern of coal-consumption evolution. Even if a shift back to coal were ultimately to occur, this does not necessarily require that oil and gas be hoarded and more critically indicate that special government policies are needed to promote that hoarding.

5. INNOVATIONS IN MINING, TRANSPORTATION, AND USE

Coal and its producers have a reputation for stodginess that is belied by the record. Over the years since 1960, a major building program has taken place. Substantial expansion occurred in the western United States. Other important developments included those in Australia, South Africa, western Canada, Colombia, Indonesia, and Venezuela.

This has involved many innovations in extraction and distribution technology. Mechanization of underground mining has increased considerably. Better machines have been developed to cut the coal and transfer it to the intramine transportation system. Methods to mechanize the installation of mine supports have been adopted. An earlier simpler step was to develop a technology for placing bolts to hold up roofs. Traditionally, the ability to avoid expensive systems of supporting the roofs with elaborate props was an advantage of U.S. mining. Extensive European efforts to mechanize prop installation proved so attractive that they became widely adopted in U.S. mining. A technology that failed to reverse the unprofitability of deep, thin-seamed European mines even more greatly benefited shallower mines in thicker seams. Similarly, heavy equipment builders developed large machines called draglines to remove the masses of material that cover coal extracted by surface mining.

Other innovations have occurred in transportation. Railroads around the world run dedicated (unit) trains between mines and their customers (or a transfer point to water-borne shipment). Various methods have been devised to speed loading and unloading. One approach is to turn each car upside down and to empty it into a receiving pit; another is to empty the coal from the bottom of the cars as they slowly move over the receiving pit. Conveyers can move the coal to storage areas and then into ships or power plants. Bulk shipping terminals are used to transfer coal from trains to ships.

The most important changes in utilization were improvements in the efficiency of electric power plants. Such improvements clearly proceeded for much of the 20th century. Concerns are widespread, however, that the innovation process had hit serious barriers by at least the 1970s.

However, the intrinsic difficulties have thwarted the desire to widen use. Considerable attention has been devoted for many decades to a combination of transformation and utilization technologies that would facilitate coal use. Coal can be converted by long-known technologies into both gases and liquids of various qualities. Thus, the gas might be as good as naturally occurring gas (methane) or have a much lower energy content. The liquid might be gasoline or a thicker fuel suitable only for large boilers. Various alternative utilization technologies have been proposed. However, widespread commercial adoption of these technologies has been difficult to attain. The costs remain difficult to reduce to competitive levels.

A key element of coal promotion rhetoric is claiming that such new technologies will soon emerge. For example, in 1948, the Bituminous Coal Institute (an industry trade association that was subsumed into the National Mining Association) proclaimed in a pamphlet extolling the virtues of coal that "...coal and lignite comprise over 98% of the U.S. mineral fuel energy reserves...." (p. 12). The discussion suggested that coal use would expand through synthesis of liquids and gases from coal, greater use of electricity, and better methods of direct coal burning, including utilization in gas turbines (an adaptation of the jet engine for stationary use, called a combustion turbine). Only the electricity prediction came true. The gas turbine was developed but still cannot directly use coal. Attention shifted to coupling the turbine to a coal gasification plant.

Two main realms of research arise: improved combustion technologies and synthetic fuel approaches. In both realms, a range of options exists. At least three alternative combustion technologies have been explored. The range of synthetics extends from better solids to efforts to produce gasoline and methane (the principal component of natural gas).

Hybrids have been suggested. For example, the combined-cycle approach involves attaching a conventional boiler to a combustion turbine. The boiler utilizes the waste heat from the turbine. Efforts have been made to develop a technology to produce synthetic gas from coal with a heat content above that from 19th century gasification technologies but less than that of methane. Building such a gasifier next to a combined cycle could allow plant coal use. However, none of the technologies proposed has proved to be economically feasible.

These concepts have regularly been presented by various U.S. government agencies as new salvations for coal. U.S. government agencies have gone from the Office of Coal Research in the 1960s to the Energy Research and Development Administration of the 1970s to President Carter's synthetic fuels program to the Department of Energy's clean coal technologies in the 1990s. Throughout, it has been difficult to produce competitive new technologies. This necessarily engenders skepticism about current claims.

SEE ALSO THE FOLLOWING ARTICLES

Clean Coal Technology • *Coal, Chemical and Physical Properties* • *Coal Conversion* • *Coal, Fuel and Non-Fuel Uses* • *Coal Industry, Energy Policy in* • *Coal Industry, History of* • *Coal Mining, Design and Methods of* • *Coal Preparation* • *Coal Storage and Transportation* • *Market-Based Instruments, Overview* • *Market Failures in Energy Markets* • *Markets for Biofuels* • *Markets for Natural Gas* • *Markets for Petroleum.*

Further Reading

Adelman, M. A. (1995). "The Genie Out of the Bottle: World Oil since 1970." MIT Press, Cambridge, MA.

Adelman, M. A., Houghton, J. C., Kaufman, G., and Zimmerman, M. B. (1983). "Energy Resources in an Uncertain Future, Coal, Gas, Oil, and Uranium Supply Forecasting." Ballinger, Cambridge, MA.

Barnett, H. J., and Chandler, M. (1963). "Scarcity and Growth: The Economics of Natural Resource Availability." Johns Hopkins Press for Resources for the Future, Baltimore, MD.

Gordon, R. L. (1987). "World Coal: Economics, Policies and Prospects." Cambridge University Press, Cambridge, UK.

Simon, J. L. (1996). "The Ultimate Resource 2." Princeton University Press, Princeton, NJ.

Markets for Natural Gas

AAD F. CORRELJÉ
Delft University of Technology
Delft, The Netherlands

1. Introduction
2. Historical Development of Markets for Natural Gas
3. The Liberalization of Gas Markets
4. The Future of Natural Gas: New Patterns of Supply and Demand?

Glossary

distribution of gas Operations necessary to deliver natural gas to the end users, including low-pressure pipeline transportation, supply of natural gas, metering, and marketing activities vis-à-vis the several types of customers.

liberalization Reduction of public authorities' intervention to coordinate the behavior of actors in (gas) markets.

market for gas System of institutional, economic, and physical elements and determinants through which gas is produced, transported, and sold by suppliers to the consumers.

natural gas Mixture of methane and other gases, as it is found in gas fields in geological layers of the earth crust, that may be either associated or not with crude oil. Natural gas is not a fraction of distilled crude oil.

production Exploration, drilling, production, and the collection of gas from the fields' wellheads.

regulation of gas market Intervention by public authorities to coordinate gas businesses' and consumers' behavior through sets of rules, standards, and prescriptions.

transmission Long-distance, high-pressure pipeline transport of gas from the producers to the distribution systems in the consumer markets.

A continuous, steady growth of gas consumption in domestic households, industry, and power plants in North America, Europe, Asia, and Latin America has gradually turned natural gas into a major source of energy during the past century. Main drivers in this development are the technical and economic advantages of natural gas as a clean, versatile, and easily controllable fuel. A major challenge, however, has been the establishment and management of the markets for gas. This article examines how gas markets have been constituted in the past as a function of economic, technical, and institutional components and how they have been altered in recent times. In particular, it illustrates how the setup and later restructuring of these markets have been influenced by the characteristics of the complex production, transmission, and distribution systems through which the gas is transported from increasingly remote on- and offshore fields to a wide range of users.

1. INTRODUCTION

Natural gas supplies approximately one-fourth of the world's total commercially traded primary energy requirements. A continuous, steady growth in gas consumption in domestic households, industry, and power plants has gradually turned natural gas into a major source of energy. Main drivers in this development are the technical and economic advantages of natural gas. It is a clean, versatile, and easily controllable fuel for which no on-site storage is needed. Further growth in gas consumption is expected as a consequence of its relatively low carbon content compared to that of coal and oil products. On this basis, gas is often considered the form of energy that will be the "bridging fuel" to a sustainable energy system, sometime after 2050.

Unlike other main sources of energy, such as oil and coal, gas is not traded on an actual world market. This is because gas is made available to its consumers by means of complex production and transport systems, through which it is moved often from remote fields to its users. The geographical reach of these pipeline transport and distribution systems is essential for supply and demand to develop.

Traditionally, the development and exploitation of these systems posed a great challenge because of the great risks and uncertainties involved. Huge investments must be committed to facilities that, once constructed, have only one purpose and destination and no alternatives. In addition, producers, transporters, and consumers are tied into a relationship of mutual dependency. In response to these characteristics and the specific local setting of systems, a variety of contractual relationships and organizational structures were established to reduce the risk involved and to establish the terms of trade for a longer time period so that the producers' as well as the consumers' investments would not be jeopardized.

This has resulted in the development of truly regional gas markets in different areas of the United States and later in continental Europe, the United Kingdom, Japan, the Soviet Union, and Latin America. Each market has its own market structure, characteristic institutional framework and role for governments and local authorities, and specific outcome in terms of the economics of supply and demand.

During the 1980s, a gradual shift in economic thought began to take shape in which the stabilizing role of the state and the need to control markets in general were questioned. It was argued that the state would never be able to coordinate the economy more efficiently than the market. The state could never acquire and process the necessary information to do so, whereas government failures would undermine the economy's efficiency. Moreover, in the process of planning, the government ran a serious risk of being affected by interest groups or by political deadlock. The arguments for restructuring were reinforced by a plea to integrate national and regional markets for goods because international trade theory argued that economic welfare would be enhanced by allowing production to take place in the most efficient location or country. Because countries differed widely in their energy resource endowments, national (energy) markets needed to integrate to the extent that the process of producing and trading energy would no longer be confined to national territories. To achieve this, national trade regimes had to remove existing barriers to trade, whereas the physical infrastructure to efficiently transport energy between and within countries, such as pipelines, ports, and railways, had to be developed.

Gradually, in the several gas-consuming regions processes of structural and regulatory change were undertaken. Again, the evolution of these processes was influenced by local, economic, and (geo)political circumstances. This is reflected in the timing of these processes, the speed with which they evolved, and the structural models chosen to reform the gas supply "systems" into actual gas "markets." This article provides some illustrations of the manner in which some of the main gas markets have been organized in the past and how they are currently being restructured. Due to space limitations, it does not provide a wide overview of such developments. Neither does it present a quantitative overview of the actual development of gas consumption and supply in the countries and areas discussed.

2. HISTORICAL DEVELOPMENT OF MARKETS FOR NATURAL GAS

The use of natural gas on a large scale is a relatively recent phenomenon, occurring in the 20th century. Much earlier, however, gas was used in local systems as so-called "town gas" produced in coke works, sometimes associated with steel mills, or in specific gas manufacturing plants. In part, however, it involved natural gas, from small-scale underground gas reservoirs, supplying nearby villages and cities. Thus, in the United States the first recorded major use of natural gas was in 1821 in Fredonia, New York. In Russia, the first use of natural gas was associated with the production of oil in the Baku in 1871.

It was not until the 1920s that developments in welding technology facilitated the construction of steel pipes capable of handling higher pressures. This allowed the development of much larger systems over extended areas, through which actual systems of wells could be connected with consumers attached to local distribution systems, and in which depleted fields could be replaced by newly found deposits farther away. In the United States, the evolution of long-distance gas began in 1925. Elsewhere, it would take until the end of the 1950s until these systems developed. In Europe, major gas discoveries in The Netherlands in 1959 marked the start of the development of a natural gas system during the 1960s, stretching to Germany, Belgium, France, Switzerland, and Italy. The United Kingdom followed in 1967 after the discovery of significant fields in the North Sea. In the Soviet Union, large-scale use of gas began in the 1950s.

The systems that have developed since then normally involve the following: the production of gas, pipeline transportation, distribution, trading, and supply. The production segment involves exploration, drilling, production, and the collection of gas from the fields' wellheads to move it to the transmission pipelines. Gas transmission involves the long-distance, high-pressure pipeline transport of gas from the producers to the consumer markets. Trading refers to the resale of natural gas in the wholesale market and retail market. Natural gas distribution consists of the operations necessary to deliver natural gas to the end users, including low-pressure pipeline transportation, supply of natural gas, metering, and marketing activities vis-à-vis the several types of customers.

Traditionally, a major challenge has been the establishment and management of these complex production, transport, and distribution systems. There is great risk involved. Huge, specific investments have to be made into facilities that, once constructed, have only one purpose, namely to produce gas and transport that gas from a specific field to the consumers attached to the system. If either the consumers decide to turn to alternative fuels or the gas supplier stops producing, the system becomes useless and the capital invested worthless. Thus, the system ties producers and the consumers in a heavily dependent relationship. Each side has to face the risk that the other side will drop out for whatever reason or that prices will increase or decrease to unacceptable levels. For producers, the former implies a volume risk in the sense that sunk investments in exploration and production become worthless because they will not be able to sell the gas they produce; for consumers, it may mean that their gas-fired appliances and equipment will fail to perform. The price risk implies that because of the tight relationship between them, either producers or consumers are able to put pressure on the other side to sell the gas at either too high or too low prices.

In a growing market, a key challenge over time is to maximize capacity utilization of the system, as well as revenue, to cover high fixed costs, including an acceptable level of profit. The main instrument to achieve this goal is a supply portfolio that dynamically balances the sales to these several markets as a function of the relative prices that these markets will bear and the capacity available in the system. In the energy market, gas has to compete with other fuels. Depending on the sector, fuels such as coal, city gas, fuel oil, and heating oil are substitutes for natural gas.

Main sectors with specific characteristics are the domestic household sector, the commercial market (also involving public use in schools, hospitals, etc.), the large industry sector, and the electricity production sector. In each of these sectors, the use of gas for purposes of space or process heating, hot water supply, or as a feedstock involves particular patterns of demand with respect to the load factor and daily and seasonal swing factors. Gas for domestic and other space heating purposes is subject to an enormous seasonal load variance. The electricity sector is a dynamic market, which relatively easily switches from one source of energy to another. A cyclical pattern develops over time. The highly profitable market for domestic household hot water supply and cooking is normally associated with sales in the space heating segment, driven by the cold season. Sometimes short-term sales to industry may be used to fill the free capacity during the summer.

This cyclical pattern is very significant for gas marketing in systems in which normally a constant, close-to-supply capacity production of gas is required for geological and technological/economic reasons. It is also significant for the planning of transmission capacity, distribution, and storage systems used to transport the gas from the wellhead to areas of consumption and to individual consumers. The pattern imposes very stringent requirements for coordination of marketing and planning and investment in production, transport, and storage capacity. Depending on specific regional circumstances, these aspects may be more or less relevant, but in general they induce a great degree of risk and, thus, affect suppliers' and consumers' willingness to invest.

A crucial spatial element in the evolution of the supply systems during the second half of the 20th century was the phenomenon that, by and large, the distance over which gas was transported from the production location to the centers of consumption increased. As a principle, gas exploration, the development of reserves, and eventual production must take place as close to the location of gas consumption as possible. Transport is expensive and it reduces the flexibility of gas supply to meet the daily and seasonal variation in demand. Therefore, the development of gas resources should evolve in a pattern of concentric circles around the center(s) of gravity as regards consumption.

This pattern, of course, is not completely consistent with what is actually observed. Old fields maintain production because they are huge, because technological development facilitates a "second life," or because they are conserved for reasons of national resources policy. Nevertheless, most of the "greenfield" gas production will take place increasingly farther away, requiring longer pipelines and more compressors to move the gas through the pipelines. It will also involve the pipelines passing through more countries. Moreover, in many places large volumes of gas are produced in joint production, dissolved in crude oil. The characteristics of the joint production of gas and oil traditionally resulted in major difficulties in coordinating the output and marketing of both types of output. Often, the commercialization of these volumes of gas was not economical because oil production took place far away from potential customers. The cost of transporting the gas by pipeline was prohibitive, so it was flared. However, expected shortages in gas in some markets such as the United States, the declining cost of liquefied natural gas (LNG) transport and handling, and environmental considerations may change this practice. Of course, if oceans are to be crossed, the whole concept of supply through pipelines must be abandoned and LNG is the only option. Thus, geographical patterns of gas supply may experience radical changes in the near future.

In response to these characteristics, a variety of contractual relationships were established between gas producers, transporters, and consumers. In these contracts, the volume and the price risk were reduced by fixing prices and establishing specific terms of trade for a longer time period so that the producers' as well as the consumers' investments would not be jeopardized. Often, it involved a role for public entities, taking ownership and management over parts of these systems. These factors have resulted in the development of truly regional gas markets, each with its own market structure, characteristic institutional framework and role for governments and local authorities, and specific outcome in terms of the economics of supply and demand.

Traditionally, in the U.S. system, strict state and federal regulation stabilized a market with predominantly private firms. In Europe, a variety of national arrangements were unified within an international system of trade and supply. This Europe-wide system was highly coordinated under the aegis of a few governments of gas-producing countries and a limited number of gas-producing oil companies. In the United Kingdom, a single state-owned distribution company, British Gas, purchased all gas from private and public offshore producers and supplied it to the consumers. In Japan, LNG was imported under long-term contracts between buyers consortiums of Japanese gas and power companies and foreign suppliers, often state oil and gas companies from producer countries or the international majors. In the Soviet Union, the planning bureaucracy stabilized the market. In Latin America, stabilizing regulatory arrangements were established in the countries in which gas developed as a significant source of energy, namely Argentine, Chile, and Venezuela. The following sections illustrate the specific patterns of market coordination in the United States, Europe, and the Asian Pacific area (Figs. 1 and 2).

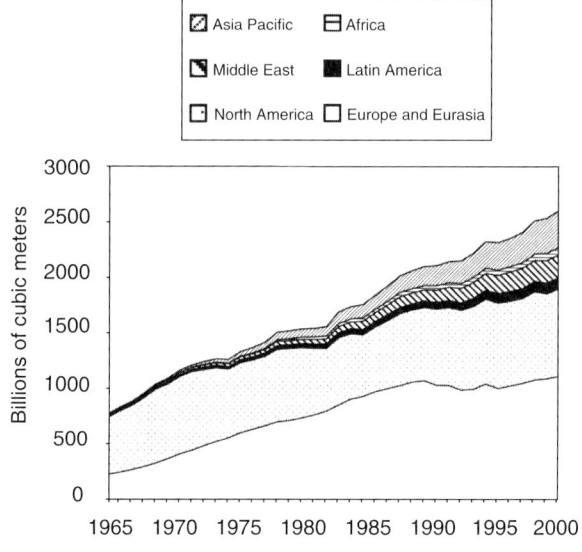

FIGURE 1 Natural gas consumption by region.

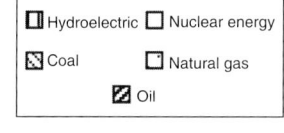

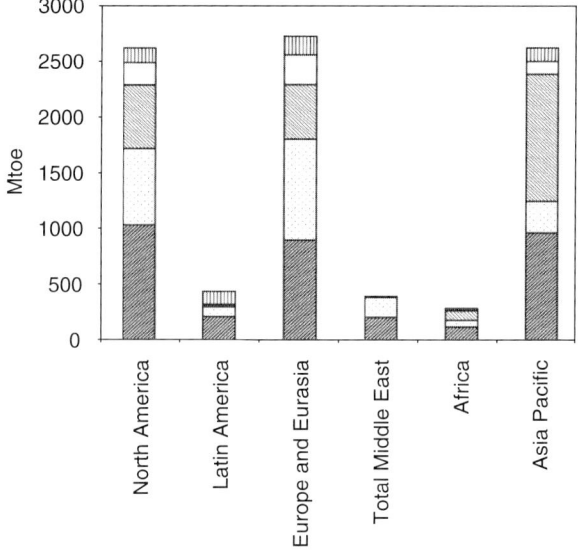

FIGURE 2 Primary energy consumption by source and region.

2.1 The United States

The U.S. natural gas industry is the largest gas industry in the world and it comprises a huge number of fields of various sizes. Initially, the development of gas use was haphazard and locally confined. After a period of limited oversight, with mainly municipality involvement, the Natural Gas Act of 1938 established a basis for the regulation of gas prices and the activities of companies. The following period saw a rapid expansion of the industry and the industry's regulation. Interstate transactions came under regulation by the Federal Energy Regulatory Commission (FERC). Intrastate transactions came to be regulated by the state public utility commissions. The industry was vertically separated into production, pipeline transportation, and distribution. However, all transactions were regulated and took place under long-term contracts. Thus, the industry was vertically integrated. Most of the gas produced was sold by the producers to pipeline companies under long-term, take-or-pay contracts. These pipeline companies supplied the distribution companies also under long-term contracts.

Attempts to regulate wellhead prices at a low level, however, discouraged gas exploration and production. Also, the transportation and distribution systems became private monopolies, imposing high costs on consumers. This delayed the growth of the industry considerably in a period in which enhanced indigenous energy supply was seen as crucial. The costly gas shortages in the 1970s prompted Congress to embark on a reform of the gas industry. In 1978, Congress adopted the Natural Gas Policy Act, through which the FERC was enabled to liberalize interstate natural gas markets.

2.2 Europe

Crucial to the development of the European system—and different from conditions elsewhere—was the ability to produce flexible load curves from the Groningen field in The Netherlands. Because of its geological structure and because of large investments in wellhead production capacity, the field had a capacity of approximately 100 billion m^3 annually. Daily capacity was sufficient to supply the required peak demand for the whole northwest European gas system in the early 1970s. On this basis, the European supply system was developed. Initially, The Netherlands supplied the lion's share of all requirements, to be sold to Germany, Belgium, and France.

However, during the 1970s, new foreign suppliers appeared, attracted by the high revenues from gas sales in Europe. The producers in these countries undertook the exploration and production activities and sold the gas produced to the state-owned wholesale buyer and exporters, such as Gasunie in The Netherlands, British Gas, GFU in Norway, Gazprom in Russia, and Sonatrach in Algeria. These firms coordinated the production and sales both nationally and abroad with other suppliers. They sold the gas at the borders to transmission companies and/or traders that transported the gas throughout Europe and supplied it to the national and local gas distribution companies and large industrial customers. With the exception of Germany, every country in Europe had one transmission company, operating the whole national high-pressure transport system. The local, municipal, or regional distribution companies operated and owned the local, low-pressure network, through which the gas was transported to small domestic users and businesses.

Trade between the several parties in the gas chain was arranged through medium- or long-term contracts. The producers delivered the gas via long-term contracts, (15–20 years) to transmission companies. These companies had medium-term contracts (1–5 years) with the local distribution companies and large-scale users. The small users had exclusive supply agreements with their local distribution companies. Throughout Europe, the transmission companies and local distribution companies had been given a local monopoly for pipeline operation and distribution.

By means of concessions, public ownership, indicative planning, and regulated prices, the producers and distributors were able to coordinate purchase and sales volumes of gas and prices. Through these contractual provisions, the risk involved in the funding of the expensive production and transport facilities over the longer term was reduced. This, of course, stimulated enormous investment in the system, as long as gas could be produced and transported at such a cost that the end-use price allowed for a reasonable profit. An important element of the contracts was the so-called take-or-pay clause, obliging the distributor or consumer to pay even when the whole volume of gas agreed upon had not been taken from the system, for example, because of lagging demand.

The only competition to gas emerged from the threat that consumers would switch to other energy sources, of which oil was the most relevant. However, the end-use gas prices were always slightly less than local prices for substitutable oil products. Importantly, the contracts contained destination clauses through which the volumes of gas were committed to a specific area and could not be sold elsewhere at a higher profit or undercut official sales. The revenues to the producers were thus dependent on the retail prices for oil products to the several types of end users (the netback principle). The transporters and distributors were remunerated with a fixed fee, which was dependent on the cost of the pipeline and the negotiation position of its operator (normally having a monopoly position).

2.3 Asia and the Pacific

Unlike the United States and the European natural markets, the Asian Pacific gas markets are essentially markets for LNG. Japan and other main LNG users, such as Korea and Taiwan, import most of their LNG under long-term contracts. Most of the gas is used in power generation, and relatively smaller shares, compared to the United States and the European Union (EU), are used in industry directly. The shares of domestic household use are comparable.

Traditionally, pipelines have been a difficult issue in Asian markets as a consequence of geopolitical tensions between supplier, transit, and consumer countries, sometimes in combination with problems of scale with regard to the size of the markets and relatively low prices for other fuels. As a consequence, the whole region consists of a number of national submarkets with their own dynamics, economics, and institutional frameworks.

The national markets in Japan, India, China, Taiwan, and Korea are heavily regulated. In some countries, natural gas is supported and subsidized, whereas in others gas is priced at its full cost. Both situations may hamper further development: The former because it may impose an additional burden on the subsidizing states, and the latter because it may scare away the (potential) consumers that have to make a choice for natural gas. Regarding the Asian natural market, there is a huge potential for growth in gas consumption in most countries, as well as in gas supply from a few of them. There are, however, major problems specific to each country in bringing about the expansion of the gas systems, not to mention a market.

3. THE LIBERALIZATION OF GAS MARKETS

As it is generally stated, liberalization of the gas industry envisages to reduce the costs of supply, enhance quality, and enhance the overall efficiency of the gas supply system. By providing the consumers, or traders, with a choice with respect to their suppliers and the type of contract, the customer is in a position to select the supplier offering the most attractive conditions. It is assumed that suppliers will try to gain, or protect, their market share by improving supply and price conditions. The traditional wholesale traders and distribution companies will have to develop new strategies and adjust their organizations because they will lose both their secured market, while the integration and coordination between their several activities in the supply chain disappears. These firms will also adjust their structure and operations—through mergers and acquisitions—so that their size and activities fit the requirements of the newly emerging competitive market.

Competition in the gas market can be achieved by altering the structure of the market, as suggested by economic theory. Whereas the traditional perspective denied the feasibility of competition in the gas industry, structural change started from the hypothesis that the introduction of competition would be possible in some segments and that this would improve the performance of the whole system. Actually, only the pipeline transportation and distribution segments of the industry were accepted as being a natural monopoly because of economies of scale and scope, high fixed costs of pipeline construction, and relatively low variable costs. The other segments, production and wholesale and retail trade, were considered to be potentially competitive markets. Therefore, by changing the ownership structure of the sector and by dividing the competitive segments into a number of different firms, competition could be introduced into the sector. A number of models can be distinguished to break up the industry.

One model separates production from the rest of the industry and introduces competition among producers. Producers sell natural gas to a gas utility, which then resells it to the end users. The transactions between the producers and the utility lead to the development of a wholesale natural gas market, in which natural gas is traded for further resale. Regulation is needed to restrict the market power of the gas utility relative to both the consumers and the producers. End-user prices are regulated. The price of gas sold by producers to the utility is determined through competitive bidding for a supply contract with the gas utility.

A more ambitious model introduces open access in pipeline transportation to third parties that either sell or buy gas from each other. The gas utility thus supplies gas to the captive, small consumers and it provides transportation services to large consumers that purchase gas independently in the wholesale market. A gas utility may also be further separated into a pipeline company and several distribution utilities, all providing access to their networks. Open access promotes competition in the wholesale and end-use market. Producers benefit because of the increase in the number of potential buyers, whereas distribution utilities or large end users benefit from a greater choice in gas suppliers and conditions. The high transaction costs involved in buying gas create room for natural gas traders, which aggregate demand and supply for a number of smaller market participants by purchasing natural gas and transportation services on their behalf. In this model, the end users have to be protected from the monopoly power of gas utilities. The price of transport and other services is another crucial factor in achieving competition in the wholesale market.

A third model separates natural gas supply from pipeline transportation and distribution and introduces full competition into these markets. Unbundling creates a level playing field for all participants in the natural gas market and creates a large number of supply companies that purchase natural gas in the wholesale market and resell it downstream using the transportation systems of pipeline and distribution companies. As competition among suppliers reduces excess margins and forces the savings on to the end users, the need for stringent price regulation at the wholesale and retail level will be reduced. In this model, new flexible short-term trading and contractual arrangements will be provided to balance supply and demand and give market participants the flexibility they need. Liquid spot markets are expected to emerge, yielding prices that continuously reflect the market value of natural gas at a specific location.

The advantages of a more efficient market may be offset by the transaction costs, high risks, and high volatility in volume and price. It increases the uncertainty of demand for gas and transportation services, exceeding or falling below available capacity. In part, this may be resolved by contracts that offer secure or interruptible transportation services, storage and conversion contracts, and other techniques to moderate changes in supply and demand. However, it is too early to effectively evaluate the outcome of processes of structural change. Only in a very few countries, such as the United Kingdom, has the third model been implemented to its full extent.

Other European countries are contemplating this step in several forms, and the United States is also moving in this direction.

It has become clear that a structural change of the gas market to introduce competition requires complex organizations and administrative systems to carry out and control all these transactions. Strict and independent regulatory oversight is required to keep owners of distribution networks or other parties from abusing their dominant economic positions. Indeed, such transport systems remain natural monopolies, of which the owners and operators are in the position to jeopardize others' interests.

3.1 The United States

In 1978, Congress adopted the Natural Gas Policy Act authorizing FERC to liberalize interstate natural gas markets. Since the major producing and consuming regions in are in different states, this had a major impact on the operation and efficiency of the natural gas industry. During the 1980s and 1990s, FERC gradually established a framework for the actions of market forces in the natural gas industry.

By Order No. 436, FERC instituted an open-access regime for interstate pipeline transportation. This regime enabled local distribution utilities and large end users to bypass pipeline companies' gas sales and purchase natural gas directly from producers. Although the open-access regime was voluntary for pipeline companies, it was widely accepted because it enabled them to increase the utilization of pipelines. However, large-scale implementation occurred only after FERC resolved the issue of how the costs of the transition to open access were to be distributed. Because customers were allowed to exit long-term supply contracts, the pipeline companies were left with large take-or-pay obligations to producers. FERC Order No. 500 allowed them to pass a share of the transition costs to procurers, distribution utilities, and end users. Subsequently, the Wellhead Decontrol Act of 1989 deregulated the wholesale price of natural gas in all interstate transactions, promoting competition in the wholesale natural gas market.

Order No. 636 required pipeline companies to unbundle natural gas sale operations from pipeline transportation activities and set up separate transportation and trading affiliates. This attracted many new companies into marketing and promoted fierce competition among marketing firms. Order No. 636 also reformed the regulation of interstate pipeline transportation. It allowed resale of transportation contracts by shippers, creating a secondary transportation market in which shippers can purchase pipeline capacity from other shippers that have temporary or permanent spare capacity. This enhances an efficient allocation of transportation contracts among shippers and high utilization of natural gas pipelines.

A subsequent series of measures was designed to promote transparency and flexibility trading practices. It facilitated short-term capacity resale, shippers' choice of delivery locations on interstate pipeline systems, and the standardization of contracts and pipeline system operation. Additional rules guided the development of a short-term transportation market in which capacity and interruptible contracts can be traded among pipeline companies and shippers.

However, by the end of the 20th century questions began to arise about the future of the U.S. gas and energy markets. Fear began to arise that the supply of new, additional volumes of gas, in response to the continuous increase in demand, would not materialize. Low prices in the context of specific market circumstances, fiscal regulations, disputes about pipeline construction, and difficulties in the establishment of storage facilities all contributed to an awkward situation with regard to gas supply. The difficulties in the operation of the market mechanism seem to be a consequence of the fact that excess production, transport, and storage capacities for gas and substitute fuels have been eroded, as illustrated by a number of local and regional energy crises. The workings of the market seem to have reduced the incentives for private industry to maintain sufficient spare capacity. It remains questionable whether this phenomenon is part of a learning experience, or whether the energy is structurally unable to establish long-term balanced behavior.

3.2 Europe

The situation in Europe has not reached that of the United States. There are no signs of immediate capacity shortages in the gas industry, although the situation in the United States is raising questions. The main issue, however, seems to be the progress of the liberalization of the EU gas market and the lagging implementation in several member countries. During the 1980s, the perception that the economies of the EU countries would be better off if linked, through integration and/or internationalization, was gaining widespread recognition as European countries were increasingly confronted with large internationally operating firms from the United States and Japan selling goods and services in their relatively small and often somewhat protected home markets. In 1985, Jacques Delors, supported by powerful European industrial leaders, launched the Single European Market initiative. The aim was to achieve a free market for the whole community by dismantling impediments to the free movement of goods, services, labor, and capital.

The EU gas directive (98/30/EC) mandated that member states of the EU had to adjust the organization and regulation of their national gas sectors in such a way that consumers would receive the right to contract with national or foreign gas suppliers of their choice. It had fundamental consequences for the structure of the existing gas market and the way in which the gas was produced and moved to the final consumer. First, the member states had to open up a first tranche of 20% of the national market and after 10 years 35% of the market had to be open for supplies by third parties.

To facilitate supplier and trader access to consumers, two main changes were necessary. First, the EU member states had to adjust their rules so that producers would no longer be forced to sell their production to one national wholesale gas buyer. The EU hydrocarbons directive stated that producers were to be given the right to sell their gas to anybody interested and acceptable to them, including large consumers, intermediate traders, and retail sellers. To achieve this, a second important requirement was that the existing transport and distribution network had to be made accessible to third parties. Producers, traders, and consumers should be given the right to contract transport capacity in these systems to transport the gas to the customer at the end-use location. Instead of selling the gas to the network owners and traders, the producers and independent traders had to acquire transport and distribution services for a fee.

By 2003, the gas directive has been implemented by the EU member countries only partially and to a varying extent. In part, this is a consequence of the fact that the EU countries are free to choose how to implement the provisions of the directive. It provides a framework, leaving the actual instruments of regulation, including the way in which regulatory oversight is created, to the discretion of the member states. This has created a situation in which some of the countries, such as the United Kingdom, The Netherlands, Spain, Austria, and Denmark, are fairly advanced in opening up their markets. Also, the trading instruments referred to in the discussion of the U.S. situation are being developed in these countries.

However, other countries are delaying the implementation of the gas directive or the national adoption of an effective regulatory framework. Thus, access provisions to the several national EU markets are quite different and the market positions of the several companies involved are unequal. In addition, there are important differences between countries' perspectives on issues such as security of supply, in light of their dependence on a limited number of suppliers, and the recognition of the interests of gas-producing countries, such as The Netherlands and Norway, vis-à-vis the consumers.

Unless Europe radically changes its position on nuclear energy, increased dependence on gas imports will be unavoidable for the foreseeable future, with a possible share of gas in the EU primary energy portfolio of approximately 30%. In principle, there is enough gas in the area to meet Europe's additional demand for energy for the foreseeable future. However, doubts have been cast over this supply perspective because it is maintained that policy-makers in the consumer countries fail to accept the conditions necessary to bring this gas to the EU markets. The size of the investments required, their timing, and the quantities of gas involved in the new developments are such that producers and suppliers from outside the EU cannot undertake these ventures without a high degree of certainty that the gas will be taken by the market at the time it is produced. This implies that such supplies need to be carefully arranged with the contractual (and other) support of those buyers in the market that have the ability to evaluate and aggregate the many small parcels of demand in local markets.

During the past 40 years, the European gas industry obtained security of demand by means of long-term supply contracts with provisions on a take-or-pay basis, providing limited flexibility, and price indexation ensuring the competitiveness of gas in the specific markets. As noted previously, this industry structure is being dismantled and a new system is being introduced that focuses on short-term rather than long-term deals. This is causing significant uncertainty regarding the market and the business environment. Gas-producing countries, particularly Russia, Algeria, and Norway, and some of the large oil companies and financing groups involved are showing signs of confusion and discontent with these developments. For example, at conferences in Algiers in May and September 2002, with the exception of Norway and The Netherlands, it was made clear that the concepts that have emerged from the process of liberalization are not helpful in creating an environment for an expansion of the gas market.

4. THE FUTURE OF NATURAL GAS: NEW PATTERNS OF SUPPLY AND DEMAND?

This examination of the structuring of markets for natural gas has illustrated the importance of the way in which governments, alone or together, are involved in the organization of these markets. Although natural gas has a number of inherent advantages as a clean, versatile, and easily controllable fuel, its development also involves fundamental problems. In the past, the establishment and management of the newly developing markets for gas in the United States and Europe occurred essentially under a system of tight coordination of the economic, technical, and institutional components of these systems. The rationale for this was clear: Private investors simply could not afford to take the technical, financial, and market risks.

During the past 25 years, it has been argued that times have changed and that the state should withdraw from such ventures. However, as argued in this article, a number of experiences have posed new questions with respect to the feasibility of markets as a coordinating device for large-scale, technologically and (geo)politically complex systems such as international gas markets. It remains to be seen to what extent the development of increasingly advanced risk sharing and trading mechanisms, and regulatory instruments to mimic market behavior in parts of such systems, may be sufficient to overcome the need for public regulation. Although the pseudomarket may be a forceful means to slim down a sector, it may not necessarily be helpful in establishing and expanding the complex gas production, transmission, and distribution systems that will be required to support growth of gas use in North America, Europe, Asia, and Latin America.

SEE ALSO THE FOLLOWING ARTICLES

Market-Based Instruments, Overview • *Market Failures in Energy Markets* • *Markets for Biofuels* • *Markets for Coal* • *Markets for Petroleum* • *Natural Gas Industry, Energy Policy in* • *Natural Gas Processing and Products* • *Natural Gas Resources,*

Global Distribution of • Natural Gas Transportation and Storage • Oil and Natural Gas Drilling • Oil and Natural Gas: Economics of Exploration • Oil and Natural Gas Leasing • Oil and Natural Gas Resource Assessment: Classifications and Terminology

Further Reading

Adelman, M. A. (1962). The supply and price of natural gas. *J. Ind. Econ.* (Suppl.).

Arentsen, M. J., and Künneke, R. W. (2003). "National Reforms in European Gas." Elsevier, Amsterdam.

BP (2003). "BP Statistical Review of World Energy, June 2003: A Consistent and Objective Series of Historical Energy Data." BP, London.

Correljé, A., Vander Linde, C., and Westerwoudt, T. (2003). "Natural Gas in The Netherlands: From Cooperation to Competition?" Oranje Nassau Groep/Clingendael, Den Haag, The Netherlands.

Davis, J. D. (1984). "Blue Gold: The Political Economy of Natural Gas. World Industries Studies 3." Allen & Unwin, London.

Estrada, J., Moe, A., and Martinsen, K. D. (1995). "The Development of European Gas Markets: Environmental, Economic and Political Perspectives." Wiley, Chichester, UK.

Juris, A. (1998a). "The Emergence of Markets in the Natural Gas Industry, Policy Research Working Paper No. 1895." The World Bank, Private Sector Development Department, Private Participation in Infrastructure Group, Washington, DC.

Juris, A. (1998b). "Development of Natural Gas and Pipeline Capacity Markets in the United States, Policy Research Working Paper No. 1897." World Bank, Private Sector Development Department, Private Participation in Infrastructure Group, World Bank: Washington, DC.

Mabro, R., and Wybrew-Bond, I. (eds.). (1999). "Gas to Europe: The Strategies of the Four Major Suppliers." Oxford Univ. Press, Oxford, UK.

Newbery, D. M. (2001). "Privatization, Restructuring and Regulation of Network Utilities." MIT Press, Cambridge, MA.

Odell, P. R. (2001). "Oil and Gas: Crises and Controversies 1961–2000. Vol. 1: Global Issues." Multi Science, Brentwood, UK.

Odell, P. R. (2002). "Oil and Gas: Crises and Controversies 1961–2000. Vol. 2: Europe's Entanglement." Multi Science, Brentwood, UK.

Peebles, M. W. H. (1980). "Evolution of the Gas Industry." Macmillan, London.

Stern, J. P. (2001, November). Traditionalists versus the new economy: Competing agendas for European gas markets to 2020, Briefing Paper No. 26. Royal Institute of International Affairs, Energy and Environmental Program, London.

Wybrew-Bond, I., and Stern, J. (eds.). (2002). "Natural Gas in Asia: The Challenges of Growth in China, India, Japan and Korea." Oxford Univ. Press, Oxford, UK.

Markets for Petroleum

M. A. ADELMAN
Massachusetts Institute of Technology
Cambridge, Massachusetts, United States

MICHAEL C. LYNCH
Strategic Energy and Economic Research, Inc.
Winchester, Massachusetts, United States

1. Oil Markets
2. Natural Gas

Glossary

forward market A market in which physical crudes are traded for future delivery, usually only a few months into the future.

integration (vertical) The degree to which companies participate in upstream and downstream aspects of the industry (crude production, shipping, refining, and marketing).

liquefied natural gas (LNG) Methane that is cryogenically cooled to form a liquid for easier storage and transport.

marker crudes Large-volume crudes that are widely traded and referenced in other contracts.

middle distillates Light petroleum products that are heavier than gasoline, primarily used as diesel fuel and heating oil.

methyl tertiary butyl ether An octane enhancer used in gasoline, manufactured from methanol.

New York Mercantile Exchange (NYMEX) The largest futures market for petroleum.

resid Residual fuel oil; also called distillate fuel oil 6 or C (in Japan).

sour/sweet Degree of sulfur content, where high sulfur means more sour.

take-or-pay A contract requirement that the buyer accept a specified amount of the contracted volume or pay for it.

upgrade To improve the value of the crude or product by cleaning or lightening it through a variety of processes.

Very Large Crude Carriers Supertankers.

West Texas Intermediate (WTI) The U.S. marker crude.

Two major changes characterize energy markets of the early 21st century: globalization and growing competitiveness. Where markets used to be fragmented and opaque, in part due to heavy regulation, they have become increasingly connected to one another at the same time that transparency and competition have increased. This reflects a combination of falling transportation costs and policy changes as many governments pursue economic reform. In addition, as economic growth spreads around the world, there are many more companies large enough to operate on a global or at least regional basis.

1. OIL MARKETS

The world petroleum market remains the most important of the energy markets, in terms of both volume and global penetration. The ability of oil to provide energy in a wide variety of uses, combined with the high energy density that most petroleum products contain, makes it the most flexible and useful of fuels. Although prices are relatively high in 2003, oil has long been one of the cheaper fuels, especially after the discovery of the supergiant oil fields in the Middle East.

Until 1970, oil consisted of the U.S. domestic market, which had many small producers, refiners, and marketers, and the international market, which was dominated by a small number of multinational corporations, known as the Seven Sisters. Before World War II, and for nearly two decades afterward, those companies controlled almost all oil that was traded on the international market, producing it, shipping and refining it, and marketing it in the major consuming countries. Although some oil was sold on long-term contracts to independent companies or partly owned affiliates, most oil remained within the systems of these large companies.

The situation has completely changed in the past two decades, especially since the oil crises of the

1970s. A large portion of the world's oil is now produced by national oil companies, especially in the Organization of Petroleum Exporting Countries (OPEC) nations. Some of that oil is sold on the world market, some is sold on longer-term contracts to individual companies, and some is sold to their former concessionaires. In addition, a host of smaller companies, some privately held and others controlled by governments—mostly in the Third World—produce and sell oil on a national, regional, and occasionally global basis.

1.1 Spot Markets

One of the major changes in the world oil market in the past two decades, compared to most of the 20th century, is the rise of the spot market. Although the United States has long had a very active internal trade in petroleum, given the abundance of small producers, on a global level the amount of crude traded on the spot market has been relatively small. Oil products usually are traded only regionally, because of the higher cost of delivery. Gasoline and middle distillates especially require clean tankers so that they do not become contaminated and this has relegated their trade to local and regional markets. Rotterdam, the Mediterranean, and to a lesser extent Singapore have long seen a relatively active spot trade in oil products.

However, because the number of small producers has increased so much, more oil has entered the world's spot market. Additionally, many of the larger oil companies no longer keep their production within their own systems. A major company might sell African production into the spot market and buy spot crude in Asia rather than ship their own production between those markets, depending on the respective prices of crude and shipping. Also, advances in refinery technology have made it much easier for refiners to handle different types of crude, so that they are much more able to utilize many types of crude.

As a result, where spot crude trade was 3 to 5% of world oil markets until the 1970s, it is now estimated to be 30 to 35% on a regular basis. This could change over time with the rise of the new players in the market, such as China National Petroleum Co. and Petrobras, in Brazil, which have different goals and behavior patterns from the U.S. and European privately held oil companies. Increasing integration on the part of OPEC countries might also eventually reduce their sales on the open market.

1.2 Futures Market

One new development (actually a revival of a 19th century market) is the use of financial futures to hedge and/or speculate on the price of oil. This resumed in 1978, with the trading of heating oil in the United States on the New York Mercantile Exchange (NYMEX). Most futures trading takes place on the NYMEX and consists primarily of contracts for sweet crude [West Texas Intermediate (WTI)], gasoline, and middle distillates. Although this is a very active market, it is much less developed than those for many other commodities. The amount of open contracts on a given day is only a few times daily physical trade, whereas for many agricultural commodities, the trade is several times higher. Additionally, trading beyond 3 to 6 months is very thin.

In part, this reflects the fact that this market is still, relatively speaking, a new one and many of those active in the physical market are very large and thus are reluctant to use futures to hedge their trading, for fear their volumes would overwhelm other traders and thus drive the market. Actors such as OPEC producers also have particular foresight into future oil prices, since their decisions directly affect prices, so that any trading they undertook would influence the market. If the Saudis were to go short, for example, traders would take that as evidence that they were planning to raise production and would sell along with them. This is different from that of many other commodities, where much of the uncertainty reflects factors that are beyond the control of the players in the market, such as weather and economic uncertainty, rather than policy choices by a cartel.

However, the futures market does provide substantial transparency, especially combined with the Internet. Virtually everyone in the world can have access to oil prices on an instantaneous basis. Compared to the 1970s, when world oil prices were rarely published, this is a huge difference. It has been argued that the existence of a futures market can, by itself, influence oil prices, as speculators will tend to behave in a herd and often move in illogical ways, that is, reacting to psychological events such as headlines rather than fundamental indicators such as oil inventories. There is some truth to this, but the impact tends to be exaggerated; foresight is still rewarded and mistakes penalized, on average. Deviations from the "true" price of oil tend to be relatively small and brief and it is usually the case instead that observers are attempting to rationalize oil price movements that they cannot explain otherwise.

1.3 Price Setting and Volatility

For most of the history of the world oil market, the price of oil has been set or at least strongly influenced by one or another body. This began in the late 19th century with Rockefeller's Standard Oil Trust, which effectively monopolized the market by setting the price at which it would buy crude. After the Trust was broken up by the U.S. government, prices became more volatile, until two particular steps were taken to stabilize them. In Texas in the 1930s, the state government gave control over production to the Texas Railroad Commission, which set quotas for individual producers based on its estimate of the market situation. Other producing states in the United States followed their lead and imposed similar quotas. Since the United States represented the bulk of the world oil market through the middle of the century, this had the effect of stabilizing the world crude oil price, as the export price equaled the domestic price.

This was especially true since the world oil trade was dominated by the Seven Sisters, or major oil companies. They established a price known as "Gulf plus," which meant they would pay the price of crude oil in the Gulf of Mexico plus the cost of transportation to the point of sale. Prior to World War II, this was economically logical as Gulf crude represented the marginal barrel. In order to protect this price, the majors agreed to limit their production to avoid undermining crude prices. Since most production growth after the 1930s came from the Middle East, where consortia consisted of interlocking major oil companies, it was very easy for them to cooperate in this manner. (Earlier, they had been inclined to poach on one another.) In fact, the consortia had the power to penalize any partner who wanted additional oil that would allow them to take market shares from a competitor. It was only in the 1960s, as independent oil companies began to break into new areas such as Libya, that the system began to erode.

In the 1950s, the world oil price was often referred to as the posted price; however, this was increasingly a tax reference price. Companies paid taxes to the governments based on the posted price, but the actual price was determined by their tax payment plus the cost of production. As the cost of production was declining, actual prices increasingly diverged from the posted price. But since most crude stayed within single-company systems and was not traded openly, there was very little transparency in the market.

By the middle of the 1970s, however, most OPEC nations had completely nationalized the producing operations in their own territories and thus the posted price no longer represented the tax on production, but rather the actual sales price set by the governments. At that time, the official sales price came into wide use. OPEC nations would hold meetings to agree on a standard price, which they would then offer to all buyers. Production then fluctuated according to changes in sales, but prices remained relatively steady. Thus, Fig. 1 shows that

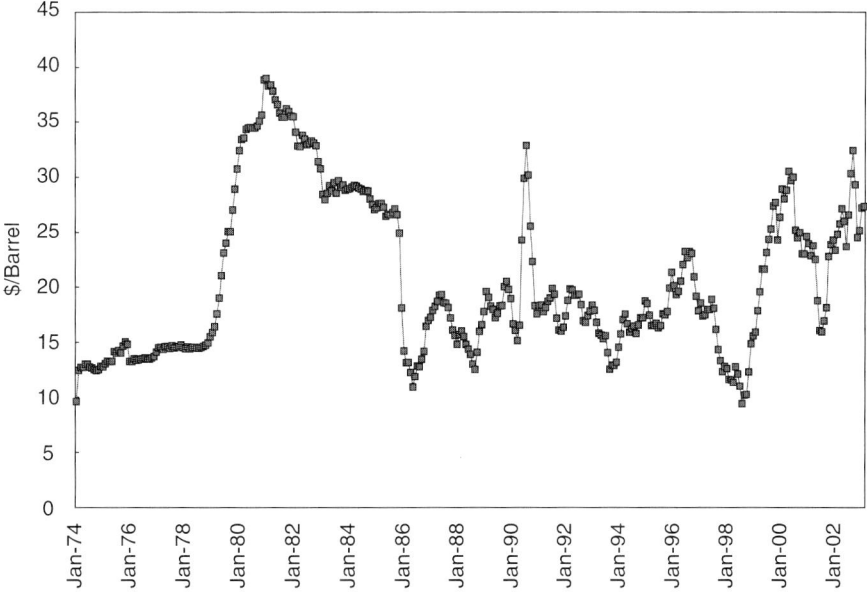

FIGURE 1   U.S. refiners' acquisition cost for imported crude (nominal dollars). From Monthly Energy Review, U.S. Department of Energy.

the world price moved very little from month to month during this period, except for the Iranian oil crisis in 1979.

The system broke down as a result of OPEC setting the price too high for too long a period after 1979, causing a substantial drop in the demand for OPEC oil, the impact of which was heaviest on Saudi Arabia, whose cash sales had practically vanished by 1985. During this period, the Saudis had been acting as the cartel's swing producer, meaning that they would absorb seasonal and short-term fluctuations in demand for OPEC oil. But they found themselves actually absorbing the long-term decline in demand for OPEC oil instead and had to initiate a price war to force the other members to both lower the price of crude oil in the long term and accept a greater share of the burden.

Since that time, prices have fluctuated much more on the world market, as Fig. 1 shows. The cartel is now allowing price to balance the market rather than their production. Since there is substantial uncertainty about global oil supply demand and inventories in the short term, the cartel is always setting a target and achieving it imperfectly.

In the past few years, prices have become even more volatile, reflecting changes in the overall market. The oil crises of the 1970s resulted in a huge amount of surplus capacity in OPEC, as well as in the global refining and shipping industries. Over time, this declined (Fig. 2), and there has been less surplus capacity to deal with short-term market tightness. Instead, prices have had to balance the market on several occasions, including 1996 and 2000. In both of those cases, most OPEC nations were producing at full capacity and the market did not balance. (Of course, the Saudis could have raised production further in both cases, but chose not to do so.)

1.4 Marker Crudes

Although crude petroleum is thought of as a fungible commodity, that is, one within which units are highly interchangeable, there are substantial quality differences in different crude types, which translates into differences in value. There are three primary aspects of quality: weight, indicating the amount of carbon relative to the amount of hydrogen; sulfur content, which is often referred to as sweetness (low sulfur) or sourness (high sulfur); and contaminants, particularly heavy metals, such as nickel and vanadium. Typically, a given field will contain one particular type of crude whose qualities are well known to buyers and even a region can produce sweet or sour crude, heavy or light crude, and so forth. Also, it is

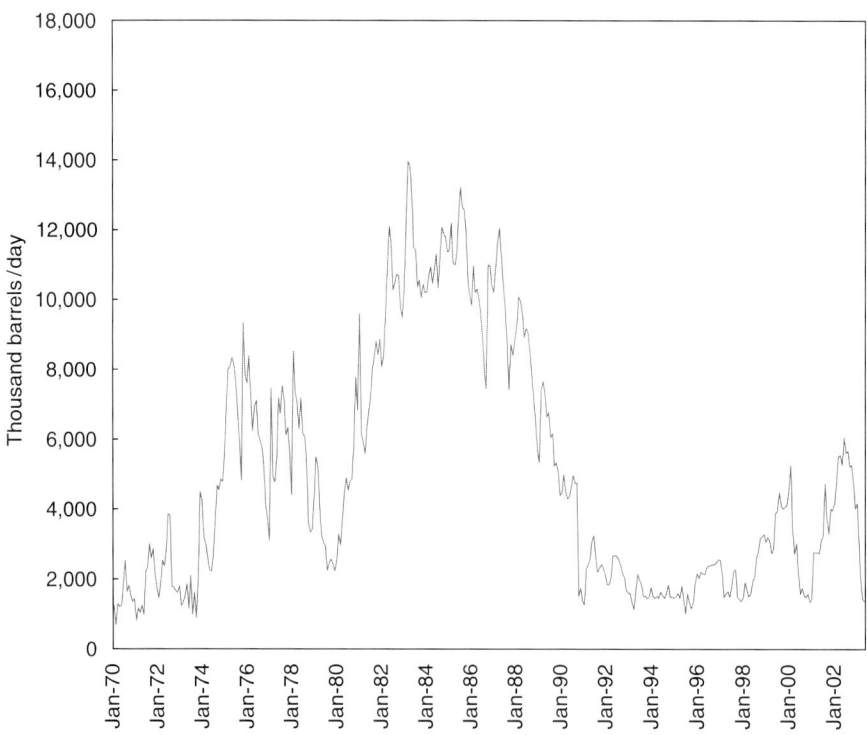

FIGURE 2 OPEC surplus capacity (thousand barrels/day). From U.S. Central Intelligence Agency and Energy Intelligence Group.

common for crudes produced in an area to be blended so that buyers do not have to differentiate between minor differences in small volumes.

Since it is difficult to track all of the many prices for crudes according to quality differences, many contracts rely on what is called a marker crude. In Europe, Brent crude is used, whereas in the United States, WTI is the primary marker crude and Asian buyers and sellers tend to rely on Dubai crude. All are traded in the open market and have well-publicized prices. Most contracts will fix prices relative to those crudes, adjusted according to differences in the quality of the crude being traded.

However, these crudes are increasingly suffering from problems of declining volumes, which leaves them vulnerable to squeezes that can cause their prices to diverge from the overall market. This has been a problem particularly for Brent crude produced in the British North Sea, which is actively traded in the forward market, and where the number of cargoes is approximately 24 per month. It is thus possible for a company to corner the market in a given month, leaving players in the forward market unable to find physical crude cargoes.

Some have suggested that the Saudis once again accept the role as a marker crude for its Arab Light, which is one of the largest volume crudes produced, but the Saudis have refused and do not allow their crudes to be traded in the spot market as a rule. They fear that Arab Light would become the target of manipulation, possibly causing their prices to rise relative to the market, thereby hurting sales, as well as making in their role in the price-setting more obvious and thus subject to political criticism.

1.5 Trade Flows

Over time, the trade in petroleum has changed significantly. In the 1970s, for example, the United States began importing increasing amounts of crude from the Arabian Gulf. However, rising production in the North Sea and the Atlantic Basin caused a sharp drop-off in that trade. Europe relies heavily on Russian and African production for its import needs and Asian imports are primarily from the Middle East. The United States still imports significant amounts of Middle Eastern crude, but most of its import needs are met by short-haul Atlantic basin producers: Canada, Mexico, Venezuela, and West Africa.

This primarily reflects the difference in transportation costs, where differences as little as pennies per barrel will shift trade between regions. Although countries such as Mexico and Venezuela may prefer to avoid being overly dependent on sales to the U.S. market, they receive much more for their production by not having to pay the higher transportation costs into more distant markets. At the same time, some oil-exporting nations continue to have an almost political concern about their market shares in certain areas, especially the U.S. market, where the Saudis sell approximately 1 million barrels of crude a day at a loss (compared to prices in Asia) to retain a foothold. It is hard to know how much this reflects a marketing strategy, such as diversification, versus a diplomatic tactic, maintaining the so-called "special relationship" with the United States.

At the same time, differences in demand and a variety of market fluctuations mean that the price in different markets can vary over time by as much as $2 per barrel for similar crudes. This then encourages balancing shifts in the trade flows to offset the differential. However, the value of crude in a given market is a reflection of a combination of factors, such as type of product demanded in the region, refinery sophistication, and openness of markets. However, isolated markets such as California are more likely to face higher prices on a regular basis.

Figure 3 shows the oil trade flows in the year 2002.

1.6 OPEC Integration

The tendency of OPEC national oil companies to move downstream is also an important recent change. Especially since the early 1950s, most crude oil was shipped to consuming nations and refined there. This was partly in response to the problems that resulted from the Iranian nationalization of British Petroleum holdings, including the Abadan refinery, in 1951, which led to a policy of "refining on the consumer's doorstep" in many industrialized nations. Also, the development of Very Large Crude Carriers made it cheap to ship very large quantities of crude.

However, as the OPEC nations gained control over their own operations, and especially as their income rose, they decided to refine their own oil where possible, partly as a corporate strategy and partly as an economic development strategy. Seeking to capture the larger share of the value added to their raw materials, OPEC countries first sought to build export refineries on their own territory and then to buy refineries overseas in consuming nations. As Fig. 4 shows, their own refinery capacity rose sharply in the late 1970s and early 1980s and then leveled off as their oil revenues weakened. Subsequently, they

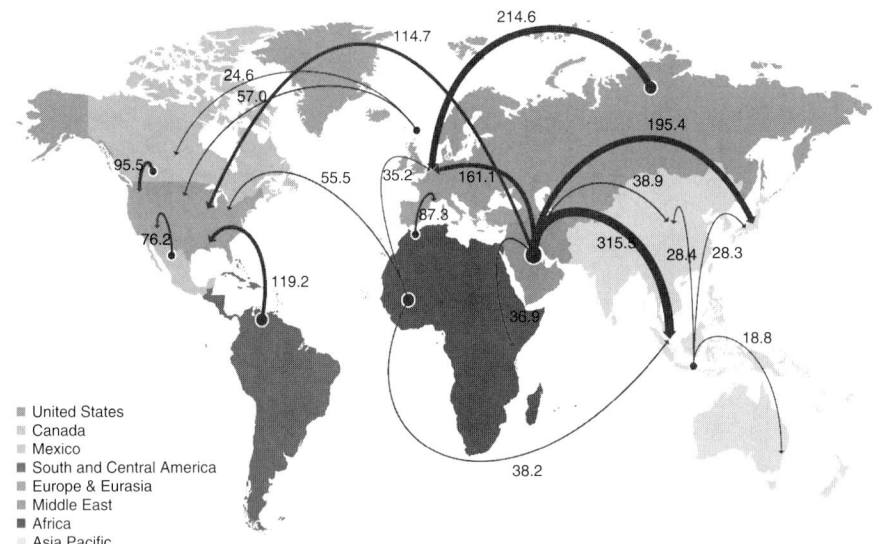

FIGURE 3 World oil trade flows (in millions of tons). From BP Statistical Review of World Energy Industry.

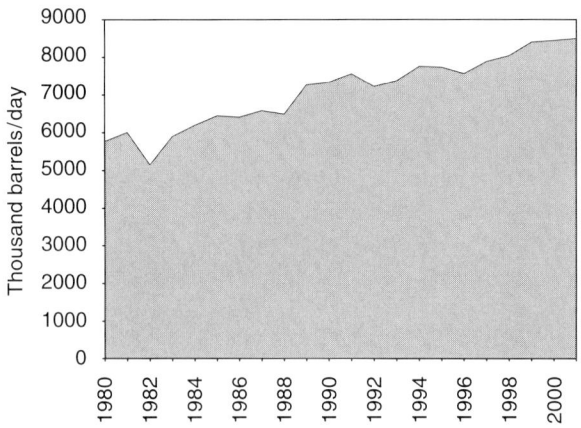

FIGURE 4 OPEC refinery capacity (thousand barrels/day). From OPEC.

realized that it was cheaper to buy into existing, underutilized assets in consuming nations, and Venezuela, Kuwait, and Saudi Arabia in particular have led in this regard, owning part or all of numerous refineries, especially in Europe and the United States.

1.7 Regional Differences

The wide variety of differences in energy usage around the world reflect a combination of economic circumstances and geography. Most important has been the abundance of commercial hydrocarbons in the United States, which meant that most stationary energy needs, such as for power generation and industrial boilers, were met by coal and later natural gas, so that petroleum was used more in the transportation sector than in many other countries. This trend continued when nuclear power became an important source of electricity beginning in the 1970s. Many other parts of the world switched from coal to oil for large users after World War II and have only just begun moving toward a more balanced mix of energy (Fig. 5).

The ready availability of cheap energy has also meant that U.S. industry and consumers have come to anticipate the ready availability of cheap energy. As Fig. 6 shows, this still remains the case. In most parts of the world, most types of energy are more expensive than they are in the United States, with a few exceptions.

Certainly, it is true that taxes play a role in raising energy prices in many parts of the world. Many countries tax gasoline at much higher levels than the United States, with gasoline taxes often higher than the final price of gasoline in the United States. But also, subsidies for coal long made it expensive in some parts of the world, such as Germany, although they are diminishing, and the high cost of natural gas transportation as well as past contracting practices have made that fuel expensive in Europe and much of Asia.

The higher rate of taxes on gasoline stems from a combination of reasons: energy security, fiscal policy, and simple protectionism. At the same time, diesel fuel is often taxed at much lower rates, which is why diesel automobiles are much more common in Europe than the United States. One side effect of this is that Europe has a surplus of gasoline produced, which it ships to the United States, whereas

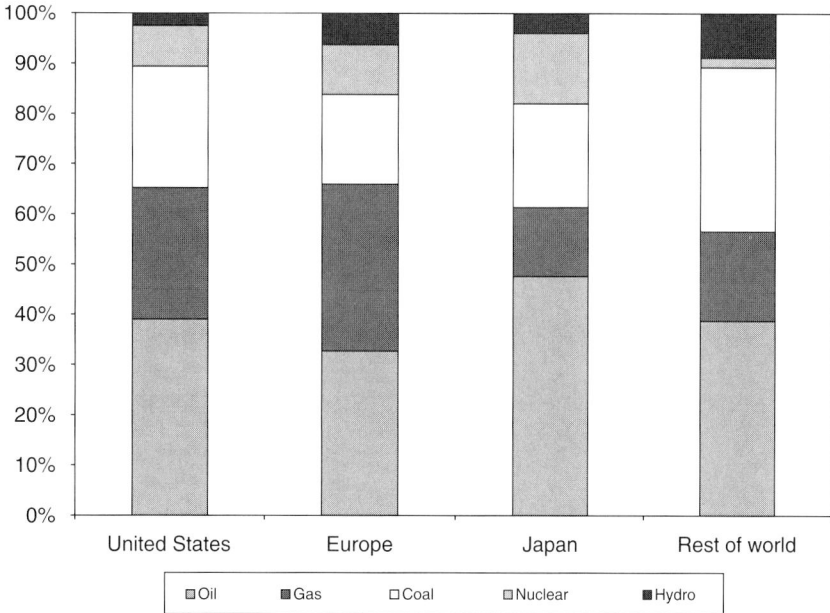

FIGURE 5 Primary energy share by region, in the year 2002. From BP Statistical Review of World Energy Industry.

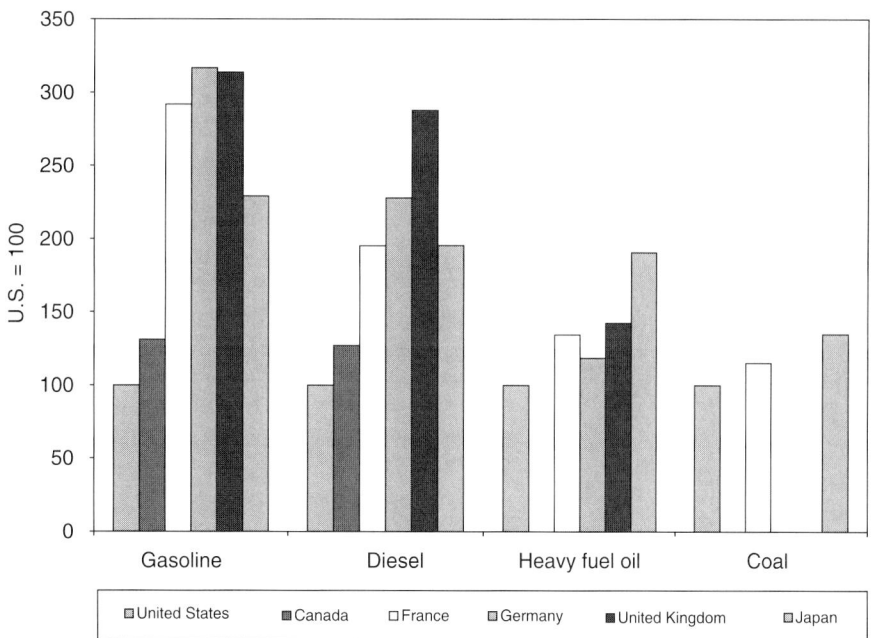

FIGURE 6 End-user prices in various countries (indexed to the United States). From International Energy Agency and Monthly Energy Review, U.S. Department of Energy.

the United States produces a surplus of middle distillates, which it exports to Europe.

1.8 Product Markets

There are a large number of products derived from crude petroleum, including many minor specialty chemicals that are not traded in quantity. But the three primary products, gasoline, middle distillates—including diesel fuel and heating oil—and residual fuel oil, make up the bulk of world oil traded, $\approx 75\%$. Each of these serves somewhat different markets, given their chemical nature and the competition they face.

Gasoline is used almost exclusively in transportation. Its volatility makes it explosive and unusable in many applications and its high price makes it less competitive with other fuels, such as coal and natural gas. Thus, it is rarely used in stationary applications, such as boiler fuel or to operate backup power generators, as diesel is. Because of its use in internal combustion engines, it is necessary for gasoline to remain very clean and uncontaminated and as a result it can be moved only in specialized vehicles, which means that bulk transport is somewhat expensive. Some gasoline moves internationally, but at volumes that are much smaller than those for crude oil.

Middle distillates play a dual role because of their nature. They are less volatile than gasoline and thus they can be used in diesel engines, which rely on high compression to ignite the fuel rather than a spark plug. But also, they can be used in small boilers, especially as a heating fuel for households. Typically, higher sulfur middle distillates are used as heating fuel, a practice that will increase as more advanced engines require lower and lower sulfur fuels.

Residual fuel oils, which are the heaviest of the major products, are named after their original origin, that is, the part of the crude that is left in the barrel after the higher value products have been refined. These fuels tend to be thicker, are harder to handle, and thus are not used by individual consumers but instead are used in factories, power plants, and large vessels, in which case they are referred to as bunker fuels. It is also the case that many of the contaminants in the original crude oil are left behind in the residual fuel oil, so that they are the least wanted of the fuels, although additional processing can improve their value by upgrading a portion to light fuels, or removing pollutants.

Prices for the different fuels are determined in large part by their competition, where gasoline faces virtually no competition in the transportation sector, automobiles being unable to switch fuels. And although diesel engines cannot readily switch fuels, some users are capable of switching, such as utilities running diesel-fired turbines. Some turbines can be switched to natural gas, should it be cheaper, but most utilities also have the option of running different power plants depending on relative cost, in which case a diesel turbine might even be competing with coal plants. Residual fuel oil, or resid, competes directly with natural gas, as many boilers are capable of using either fuel, as well as coal, typically the cheapest fuel available for power plants.

However, the use of both residual fuel oil and middle distillates is somewhat limited in the United States, as many power plants face pollution regulations that make it difficult for them to reduce usage of natural gas in favor of those fuels, which tend to give off higher levels of sulfur and other contaminants. Even so, in recent years the United States has seen as much as a half million barrels a day increase in the usage of middle distillates and residual fuel oil when natural gas prices were high.

Other major petroleum products include jet fuel, which is obviously used primarily in the airline industry and as such is very sensitive to the economic situation rather than prices of competing fuels. Liquefied petroleum gas (LPG) is another fuel used in many parts of the world for heating and cooking, especially when there is no distribution network for natural gas. Rural areas rely on this quite a bit in developed nations and less developed nations often use it even in urban areas as a convenient source of fuel for households.

Petroleum products are not completely interchangeable by any means but there is a degree of flexibility at the refinery level. By use of catalysts, cracking, blending, and other processes, refiners have the ability to alter their product streams, thereby accommodating changes in the market somewhat. For example, U.S. gasoline production increases during the summer and the production of heating oil rises in the fall and winter.

Similarly, some products can be altered when markets dictate. A variety of products can be used in petrochemical plants, including natural gas, LPG, and naphtha (which is an ingredient of gasoline), and refiners can moderate the amount of jet fuel or middle distillate they produce when conditions warrant.

1.8.1 Product Quality

The past decade has seen a major change in the impact of the environmental regulations on product formulas, especially for gasoline and diesel fuel. Historically, there have been tight regulations on the sulfur content of residual fuel oil in some parts of the industrialized world since the early 1970s, and lead in gasoline was phased out in the United States beginning at that time. These measures were in response to local pollution and had a relatively small impact on oil product trade inasmuch as low-sulfur crudes produced residual fuel oil with a low sulfur content, so that it was regularly available. As demand for low-sulfur crude began to outstrip availability, desulfurization units had to be installed at refineries. On the other hand, lead was an additive in gasoline rather than an inherent part of the product and could be replaced with other octane enhancers.

However, regulations have become both stricter and more widespread in the past decade, covering many parts of Europe and increasingly the less developed countries, requiring both lower sulfur content and also specific formulas for gasoline and diesel fuel. In the United States, there are numerous formulas for gasoline depending on the location and season, that is to say, the particular environmental problem being addressed. Factors such as benzene content, Reid vapor pressure, and sulfur content face more rigorous regulations.

This has required refiners to make two types of changes. First, they must reduce the amount of sulfur in the products been produced, but in many cases, additives are required to enable them to meet both octane needs and specific formula requirements. Also, one of the most popular additives, methyl tertiary butyl ether, is being banned in many places as a water pollutant, requiring refiners to search for other additives, such as ethanol, to replace it, although it is more expensive and difficult to handle.

Aside from requiring substantial investments in upgrading capacity, the effect has been to segregate markets in many places, as some products are not clean enough to sell legally in some markets. Most Third World refineries are not capable of producing the kinds of products needed in some parts of the industrialized world and, depending on the specific product requirements, product trade cannot take place even within a country at certain times of the year.

As a result, there have been a number of localized problems with product supply, as any difficulties at the refinery level in an area cannot be offset by nearby supplies because they do not meet the specific environmental regulations in that location. California is in the worst situation because of the severe air pollution problems and the lack of connections to other markets—products must be shipped in via the Panama Canal or across the Pacific Ocean, if they can meet the environment rules. The problem overall has been exacerbated by the rise in capacity utilization in the downstream industry (Fig. 7), which has reduced the amount of capacity available to meet unexpected supply problems.

2. NATURAL GAS

Natural gas is unusual, a premium fuel with high value to customers but low prices at the wellhead. Because of its very low density, transportation costs are extremely high. The result is that natural gas markets developed much later than those for other fuels and regional markets remain relatively isolated from one another, although that is slowly changing.

Before the development of welding technologies that allowed high-pressure steel pipelines to be constructed, global natural gas markets consisted primarily of synthetic natural gas manufactured from coal, called town gas, and the exploitation of local deposits particularly for industrial purposes. Synthetic gas had a relatively low energy content and

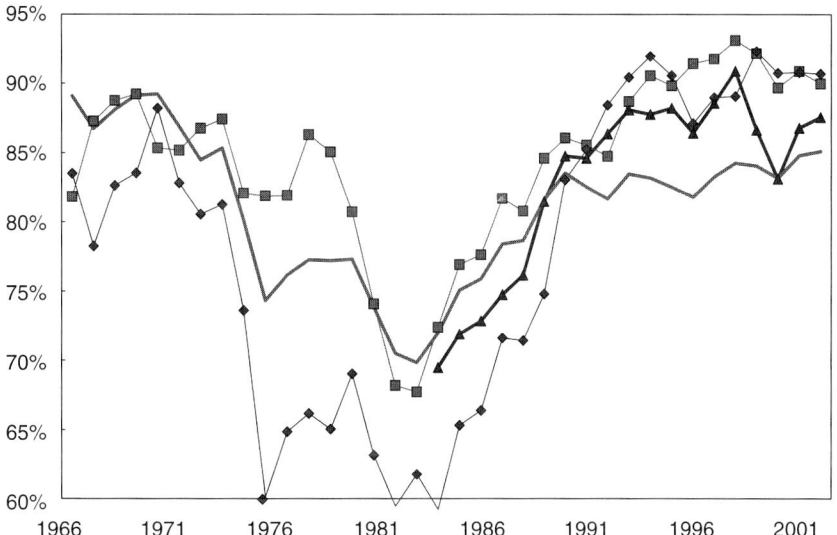

FIGURE 7 Refinery capacity utilization by region. Boxes, United States; diamonds, Europe; triangles, Asia/Australia/New Zealand; boldface line, world. From BP Statistical Review of World Energy Industry.

produced a large amount of waste products. It was used mainly in lighting in urban areas during the 19th century and began to be replaced by electricity in the early 20th century.

The early downstream petroleum industry—refining and petrochemicals—often relied on natural gas deposits as a source of cheap energy. Since long-distance transportation was impossible until the postwar era, most gas could be sold only locally and the lack of demand kept prices very low. Given this fact, it was usually not sought after intentionally, but found by those seeking oil, and often gas discoveries were abandoned unless there was a nearby facility that could use it. Some industrial centers developed specifically to take advantage of existing natural gas deposits that had no market, including the chemical industry in West Virginia and later Texas.

The U.S. natural gas market began to develop after World War II with the construction of numerous long-distance pipelines to bring the gas to consumers. European markets developed later, especially after the 1962 discovery of the supergiant Groningen field in the Netherlands, a low Btu gas that was extremely cheap to produce. Asian markets did not begin to develop until the 1970s, as high oil prices made liquefied natural gas (LNG) trade economically feasible, with the prices for gas and/or services set on a cost-plus basis. Fig. 8 shows the trade flows in 2002.

2.1 U.S. Gas Markets

The United States was the first to develop a large, well-articulated natural gas market because of the ready abundance of large amounts of gas resources and the early development of long-distance pipelines. The construction of oil pipelines from the Gulf of Mexico to the East Coast during World War II was a boon to the industry, as the two major pipelines were converted to natural gas after the war. But the ownership of large amounts of gas reserves by major oil companies and the nature of gas distribution as a natural monopoly raised concerns about possible abuse by reserve owners. In response, the U.S. government insisted that long-distance gas pipelines be owned by independent companies and that they be treated as regulated monopolies.

As a result, the U.S. gas market consisted of producers particularly in the Southwest who sold their production to pipeline companies that delivered and resold the gas to local distribution companies in the consuming areas. Since wellhead natural gas prices for the interstate market were regulated after 1954, this meant that the production, transportation, and distribution of natural gas were regulated by a combination of the federal and state governments.

However, since wellhead prices were allowed to rise only in response to inflation, the oil price boom of the 1970s made natural gas very competitive with oil in many areas and resulted in rising demand. At the same time, low prices discouraged drilling and supplies began to become scarce. The Natural Gas Policy Act of 1978 deregulated prices of some supplies and some pipeline companies signed grievously high-priced contracts for them, planning to roll the expensive gas in with their cheap, price-controlled gas. This proved disastrous, as weak demand meant that pipelines cut back on cheap gas in favor of high-

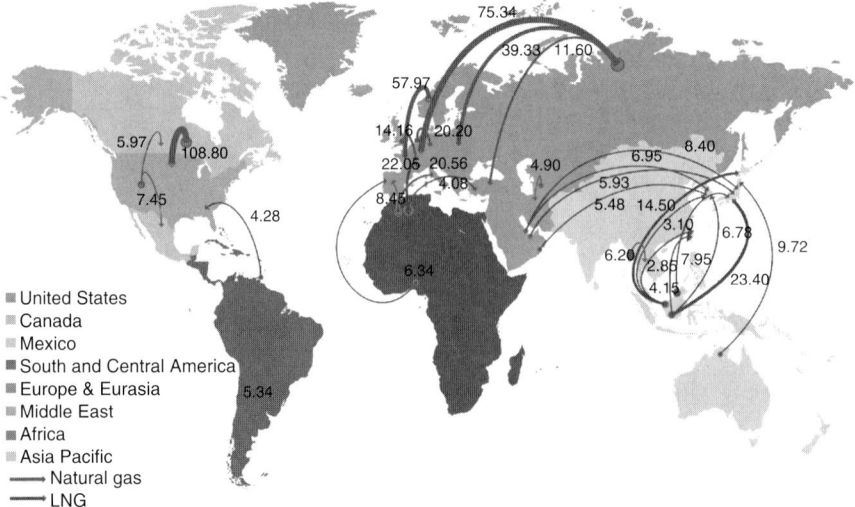

FIGURE 8 International trade flows of natural gas. From BP Statistical Review of World Energy Industry.

priced gas, for which they had guaranteed the volumes they would lift ("take-or-pay" clauses). The government deregulated all wellhead prices as a result, and the average price fell (Fig. 9).

By the 1980s, the long-distance pipeline business had also grown to the point where there were few areas that did not have multiple sources of supply, making it much more difficult to monopolize supply in an area. This led to the deregulation of the pipeline business and caused it to turn into a common carrier, rather than buyers and sellers of natural gas. It also led to the rise of natural gas traders, who located buyers and sellers of gas and then procured the necessary pipeline capacity to provide delivery.

At this point, the U.S. gas market split into the competitive, large end-user sector made up of power plants and other industrial facilities and the captive residential market, which remained regulated. The former could bid for natural gas on the open market and, having the option of switching to cheap residual fuel oil, could see its prices capped by that competing fuel. But residential users remained captive to the local distribution companies, having to pay their capital cost plus the cost of fuel.

The past few years have seen a sudden departure from the previous two decades of relatively low and stable prices, as rising demand for gas in the power sector and poor upstream performance combined with cold weather to yield extraordinarily high prices. Most analysts think that U.S. gas prices will remain well above those observed in the 1990s, but the ultimate outcome is unclear.

2.2 Canada and Mexico

The Canadian market resembles the U.S. gas market in some respects, including the presence of many gas-producing companies and the need to transport gas long distances to the major consuming areas. These include the major cities of eastern Canada, served by TransCanada Pipeline (TCPL), as well as the exports to the U.S. Midwest and Northeast, where Canada is a major source of supply.

Historically, TCPL was a regulated monopoly with partial government ownership that used its systemwide revenues to expand its pipeline further eastward, including projects that might not have been economical on their own. The province of Alberta contains the bulk of Canadian gas resources and production is gathered into a number of large pipelines and then sent south and east. There is also a major chemical industry in Alberta, which acts as a primary consumer, and the booming tar sands fields, which require heat to produce their heavy oil.

In Mexico, natural gas is produced and shipped by Pemex, the government's national oil company. Most of the gas is associated gas that is found with oil fields and only recently has Pemex been seeking to find and develop gas fields. The gas is primarily used to displace residual fuel oil in industrial and utility

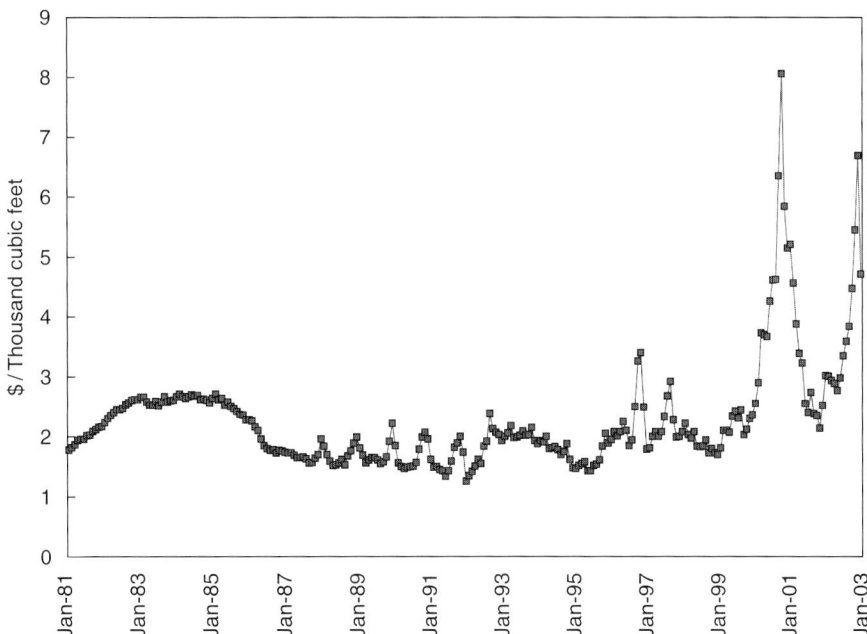

FIGURE 9 U.S. average wellhead natural price. From Monthly Energy Review, U.S. Department of Energy.

uses, which rely heavily on it. (The government monopoly on oil and gas has had the perverse effect of raising oil consumption by causing underinvestment in natural gas.)

2.3 European Gas Markets

In 1959, the discovery of the supergiant Groningen field in the Netherlands led to the beginning of the modern gas era in Western Europe. Before that time, countries such as Germany, France, and Italy had minor amounts of domestic gas production and many cities had town gas facilities. The Dutch began the large-scale international trade of natural gas in Europe and encouraged the building of national pipeline systems, by initially offering their gas at prices designed to be competitive with coal, which was the main competing fuel at the time. Their strategy was to build markets and encourage customers to make the necessary expenditures to put in the infrastructure needed for their gas to be exploited.

With the 1970s, Norwegian gas discoveries began to be marketed into Western Europe and the development of liquefied natural gas shipping also brought supplies from Algeria and Libya to the Continent. The Russians, who have the largest natural gas reserves in the world, developed long-distance pipelines from Siberia into Western Europe in order to gain hard currency. With the oil price shocks of the 1970s, these suppliers became more aggressive in seeking higher gas prices, with the Algerians going so far as to demand prices above those for crude oil, arguing that natural gas was a superior fuel. Many of these price demands did not stick, although exporters succeeded in raising prices to approximately those of crude oil equivalence, as Fig. 10 indicates.

The European gas market remains dominated by a few suppliers on one side and a few consumers on the other. In France, the buyer is Gaz de France, which delivers and resells to other customers. In Germany, Ruhrgas continues to control most of the national pipeline system, and in Italy, Snam is the main pipeline company. The heavy investment needed for the development of the gas infrastructure encouraged governments to maintain state-owned monopolies to run their gas pipeline systems and these companies have been particularly willing to accept long-term contracts with the suppliers paying prices indexed to oil prices. The European Union has been trying for some years to open up the closed markets.

The British market is somewhat unique, in that it was separated from the Continental gas market until recently, and it produced most of its own supplies in the North Sea for the past several decades. Initially, British Gas served as a monopsony buyer, purchasing gas produced by oil companies in the North Sea at fixed prices and transferring it through its pipeline system to companies and residential users. For a time, extremely low prices were paid to domestic gas

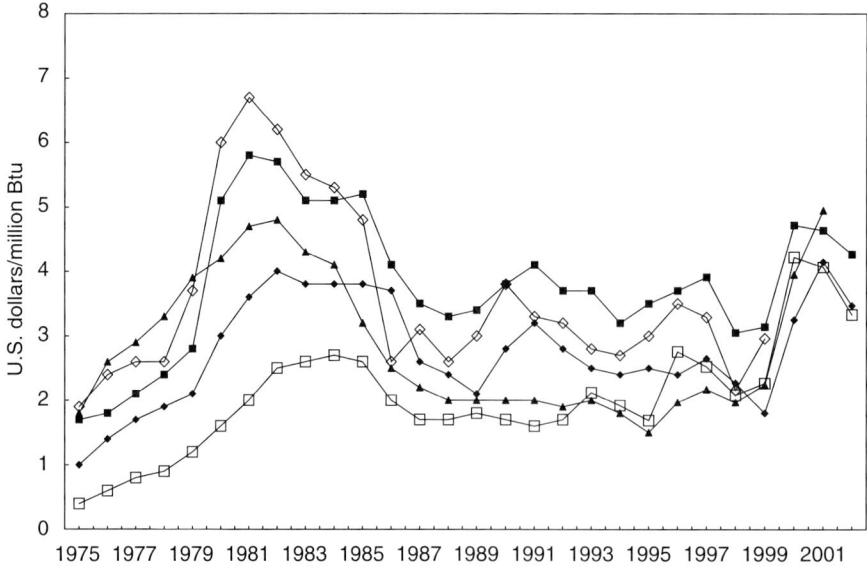

FIGURE 10 World gas and oil prices, $/million Btu (nominal). Closed squares, Japan LNG; diamonds, Western Europe gas imports; triangles, U.S. gas imports; open squares, U.S. wellhead gas; open diamonds, crude oil CIF. From BP Statistical Review of World Energy Industry and Monthly Energy Review, U.S. Department of Energy.

producers, which discouraged drilling and led to British Gas importing gas from Norway at much higher prices. The Thatcher government discontinued that practice, privatizing British Gas and allowing producers to sell their gas freely.

Norway has become a primary exporter to the European Continent, with a very mixed heritage. Although most of its natural gas has been discovered and produced by private companies, the nation relies on a national oil company, Statoil, as a primary participant and operator in many fields and used the Gas Fuel Committee to act as a monopoly marketer of its gas until recently. This reflected partly a desire to gain oligopoly rents for its gas by not having different producers bid the export price down to competitive levels and partly the fact that much of its gas exports went to areas that were dominated by monopsony buyers, particularly in France and Germany.

Norwegian sales did not begin until the 1970s, when the oil crises created a buyers' market for non-OPEC energy supplies. Thus, the Norwegians were able to achieve premium prices for their gas, selling the gas reserves in given fields and gearing their development plans according to the contracts they signed. However, after selling the gas from the supergiant Statfjord field at a significant premium in 1981, they found substantial buyer resistance in the weakening market, worsening as the price of oil collapsed in 1986. This led to a reversal of policy and an effort to price at competitive levels in order to lock in buyers.

The natural gas market is growing strongly, particularly in southern Europe, and new LNG trade is being arranged from West Africa and potentially the Middle East. At the same time, new pipelines from North Africa and Russia are being built and the Caspian producers are attempting to place their reserves into the lucrative European market.

2.4 Russian Gas

Russia contains the largest natural gas reserves in the world and is the world's largest producer and exporter. Natural gas provides the country with approximately one-fifth of its foreign currency earnings and accounts for a similar proportion of the federal government's budget. The domestic and export markets are dominated by the company Gazprom, a privatized firm that is little changed from the Soviet days (including continued partial government ownership). Gazprom produces approximately 80% of the gas in Russia and the relatively new marketing firm Itera accounts for most of the rest, with a small amount produced from oil deposits by the various oil companies. Itera also buys gas from Caspian producers and resells it to other customers, principally the Ukraine.

Gazprom continues to be the primary exporter to the West, accounting for nearly all sales outside the country, including to most of the former Soviet Bloc nations as well as the daughter republics of the Soviet Union. Sales are essentially carried out on a monopoly basis, primarily to large buyers such as Gaz de France and Ruhrgas, typically on a long-term basis with prices fixed relative to petroleum prices, usually on a moving average basis.

Domestic gas sales are a different matter, where many areas have their own local distribution companies and buy gas from either Gazprom or Itera. However, difficulties in obtaining payments from many customers have meant that many local companies have gone bankrupt and some of them have been bought out by Gazprom. The domestic price remains fixed at a fraction of the international price, although the government hopes to raise that over time. As of 2003, it is approximately 20% of the European Border price, although residential customers pay a flat monthly fee, with no metering of consumption.

Russia also holds huge reserves in the Far East, particularly in the Yakut area and near Lake Baikal, where British Petroleum holds a major interest in the supergiant Kovytka field. Hopes are that exports will be begin soon to China and other Far Eastern countries, which would require the construction of large-scale, long-distance pipelines. The primary delay relates to questions of pricing and ownership of the pipelines, although most of the major players appear to be very interested in beginning such trade, as it would have the effect of reducing coal consumption in China—a major source of pollution both locally and regionally—and reducing reliance on Middle Eastern oil imports in Korea and Japan.

2.5 Asian Gas Markets

The Asian gas market has been one of the last to develop, primarily because most gas supplies are at a great distance from the major consuming areas. Aside from minor amounts of consumption in countries such as India and Pakistan, which have exploited small domestic deposits for some time, the primary consumption of natural gas has been in the form of LNG, transported across oceans. Unlike North America and Europe, much of the gas has been consumed by the electric utility sector rather than by households.

The LNG trade began in the 1970s when Japanese utilities, looking to reduce their emissions resulting from direct consumption of crude oil, began to arrange for small quantities of LNG to be shipped from Alaska, Brunei, and other areas. Initial prices were relatively low to be competitive with crude oil prices. As crude oil prices soared in the 1970s, and many LNG exporters became more aggressive in seeking higher prices, the Japanese buyers agreed to pay prices equivalent to delivered crude oil prices in Japan. This practice has remained for some time and still dominates the trade in Asia.

Because of the greater expense of transporting natural gas in the liquefied form, the market has been much slower to develop than pipeline markets in North America and Europe. Natural gas's market share has grown slowly in Asia but remains far behind that of other regions and still does not penetrate the residential market significantly. In part, this reflects warmer climates in some areas, but it also reflects the expense of building residential distribution systems. Large consumers, such as power companies, require minimal investment in pipelines, which makes them ideal customers for a new regasification terminal, as it maximizes consumption relative to initial outlays.

However, the fact that price is not set by the market (i.e., long-run marginal costs) but rather by the price of crude oil has left prices for LNG usually well above costs and made the business fairly attractive. As a result, there are a number of projects unable to find customers. In many markets, this would mean downward pressure on prices, but such has been slow to develop in Asia for two reasons. First, many of the sellers already have large volumes committed at the higher prices and they do not wish to sacrifice the revenues from existing sales, since these customers would almost certainly demand lower prices for their contracts. Additionally, the high price of transportation means that supplies from more distant regions are not particularly competitive.

The Asian gas market is slowly changing due to a combination of falling liquefaction costs and a slowly developing gas pipeline system in the region. The construction of pipelines from Russia to Japan, China, and Korea would change the marketing of natural gas. Also, the slow growth of domestic gas usage in India, Pakistan, and Indonesia as well as development of pipeline connections in Southeast Asia mean that LNG is no longer as dominant as it once was.

2.6 Other Markets

Gas consumption is sparse in other parts of the world, but some usage occurs in areas with large resources, such as the Middle East, as the cheap gas attracts energy-intensive industry, and in areas where there is some gas and at least a local market, as in parts of Africa and Latin America. Nigeria has large amounts of associated gas and in addition to exporting LNG, it is trying to develop a local pipeline network, as are other countries with nonassociated gas, such as Bolivia and Tanzania. Power plants dominate the customer list, but this should change as the systems mature.

2.7 LNG Contracting

The LNG business is somewhat unusual, in that it involves a basic commodity that is traded almost exclusively on long-term contracts. The particularly high capital costs necessary for building liquefaction plants, combined with the small number of importers with the necessary facilities for unloading LNG, have made producers very sensitive to demand-side risk. In most cases, they have not begun construction until they have located enough gas reserves to supply their plant for its lifetime and sold nearly all of its planned production on fixed contracts.

Typically, plants would line up contracts to sell 80 to 100% of their output before beginning construction and most of those contracts would have very strict take-or-pay clauses, requiring that buyers accept most or all of their contract amounts on a regular basis. Buyers were usually not allowed to resell any of their gas either, which tended to make them hesitant to be aggressive in signing up new volumes.

Because most LNG, especially in Asia, was sold into markets without directly competing fuels, there was no real market price for the natural gas. As a result, nearly all contracts called for prices to be set equivalent to the crude oil price level, converted on a Btu basis. This has reduced the price risk, at least to some degree since crude oil prices are also volatile and uncertain, but it has generally meant that prices are adequate to provide sufficient if not quite hefty profits to the producers and the exporting governments.

However, this has also meant that market penetration has been constrained by the combination of strict contracts and high prices. The high storage costs for LNG meant that buyers did not wish to hold large inventories and thus have been reluctant to sign up for volumes that might not be used in a short period of time, reducing sales on a spot basis or for seasonal purposes. Also, where the gas competes in the industrial sector with fuel oil, usually sold at a discount to crude oil, it finds itself unable to gain a large market share.

SEE ALSO THE FOLLOWING ARTICLES

Clean Air Markets • Market-Based Instruments, Overview • Market Failures in Energy Markets • Markets for Biofuels • Markets for Coal • Markets for Natural Gas • Oil and Natural Gas: Economics of Exploration • Oil and Natural Gas Leasing • Oil and Natural Gas Resource Assessment: Classifications and Terminology • OPEC Market Behavior, 1973–2003 • Petroleum Property Valuation • Stock Markets and Energy Prices • Strategic Petroleum Reserves

Further Reading

Adelman, M. A. (1993). "The Economics of Petroleum Supply." MIT Press, Cambridge, MA.

Bradley, R. L., Jr. (1996). "Oil, Gas and Government: The U.S. Experience." The Cato Institute, Washington, DC.

British Petroleum (BP). (2003). "BP Statistical Review of the World Energy Industry." British Petroleum, London.

Hartshorn, J. E. (1993). "Oil Trade: Politics and Prospects." Cambridge University Press, Cambridge, UK.

Horsnell, P. (1997). "Oil in Asia: Markets, Trading, Refining and Deregulation." Oxford University Press, London.

Horsnell, P., and Mabro, R. (1993). "Oil Markets and Prices: The Brent Market and the Formation of World Oil Prices." Oxford University Press, London.

International Energy Agency. (2002). "World Energy Outlook." International Energy Agency, Paris.

Leffler, W. L. (2000). "Petroleum Refining in Nontechnical Language," 3rd ed. Pennwell, Tulsa, OK.

Lynch, M. C. (2002). Causes of oil price volatility. *J. Energy Dev.* 42, 373–389.

Mitchell, J. V., Beck, P., and Grubb, M. (1996). "The New Geopolitics of Energy." Royal Institute of International Affairs, London.

Razavi, H., and Fesharaki, F. (1991). "Fundamentals of Petroleum Trading." Praeger, New York.

Tussing, A., and Tippee, R. (1995). "The Natural Gas Industry: Evolution, Structure, and Economics," 2nd ed. Pennwell, Tulsa, OK.

Marx, Energy, and Social Metabolism

JOAN MARTINEZ-ALIER
Universidad Autónoma de Barcelona
Barcelona, Spain

1. Social Metabolism
2. Podolinsky's Agricultural Energetics
3. Engels's Reaction to Podolinsky
4. Otto Neurath's Naturalrechnung

Glossary

Otto Neurath (1882–1945) A famous logical empiricist philosopher of the Vienna Circle, he was also an economist or economic historian, and a Marxist. Neurath introduced the idea of incommensurable values in the economy, a concept now embraced by some ecological economists.

Sergei Andreyevich Podolinsky (1850–1891) A Ukrainian physician, economist, public and political figure, and a contemporary of Karl Marx and Fredrick Engels, who performed early detailed studies of energy flows in agriculture.

social metabolism An idea adopted by Karl Marx and others in the 19th century that nature and society were linked by material floes. Marx largely ignored energy flows, while others embraced them as a key concept.

Marx (1818–1883) and Engels (1820–1895) were contemporaries of the physicists and physiologists who established the first and second laws of thermodynamics in the 1840s and early 1850s (J. R. Mayer, 1814–1878, J. P. Joule, 1818–1889, R. Clausius, 1822–1888, W. Thomson, later Lord Kelvin, 1824–1907). Marx and Engels followed with a few years' delay many of the remarkable scientific and technical novelties of their time including the developments in thermodynamics or energetics. Engels wrote in 1888 that the nineteenth century would be remembered as the century of Darwin, Mayer, Joule, and Clausius. It had been not only the century of the theory of evolution but also of the theory of the transformation of energy. However, Marx and Engels never expressed or translated their analytical concepts (productive forces, surplus value, exploitation) into the language of energy as they could have done in their mature work after the 1850s. They never said that the productivity of labor in agriculture and in industry would depend on the energy subsidy to the economic process. After all, they were historians and economists, not yet environmental historians, ecological economists, or industrial ecologists. Increases in productivity depended on the development of the productive forces (*Produktivkr...fte*), where "Kraft" (force) was not used with the physical meaning of energy. The productive forces could be furthered or could be fettered by the social relations of production. Capitalism had meant an enormous development of the productive forces but it provoked economic crises by its very nature—that is, the exploitation of wage labor. Malthusian crises of subsistances were not relevant. These were essentials points of their theory.

1. SOCIAL METABOLISM

The first volume of *Capital* was published in 1867. There were certainly precursors of a biophysical approach to the economy from the 1880s onward, and at least one of them was close to Marxism (such as S. A. Podolinsky, 1850–1891). These proto-ecological economists did not form a school. In the 1970s, the study of the human society and economy from a physical point of view (flows of materials and energy) started to be practiced by coherent research groups with little relation to Marxism. Histories of the use of energy in the economy are not connected to Marxism (even if their authors might sometimes have Marxist pasts). The notion of energy return on energy input (EROI) was applied in the 1970s to the economy by Charles Hall and other ecologists with no relations to Marxism. Explanations of economic growth model the use of energy in the economy (or rather physical work output as distinguished from energy [exergy]). They criticize neoclassical production functions and growth theories but draw no inspiration from Marxism.

Marx and Engels had a profound interest in the interactions between the human economy and the natural environment, particularly as regards capitalist agriculture. This was expressed in Marx's use in his own drafts after 1857–1858 and in *Capital*, of the notion of metabolism between humans and nature. Marxist use of "metabolism" (by Marx himself, also by Bukharin) is well known. Alfred Schmidt was the first author to insist in the role of the concept of *Stoffwechsel* in Marx's work on the development of capitalism, noting Moleschott's and Liebig's influence, noting also its substantive use for the cycling of plant nutrients. Marx became so keen on the concept of metabolism that in a letter to his wife, he nicely wrote that what made him feel like a man was his love for her, and not his love for Moleschott's metabolism or for the proletariat. There is a rediscovery of Marx's social metabolism. Authors working on industrial metabolism or in social metabolism look at the economy in terms of flows of energy and materials. They insist that the economy should be seen as a subsystem embedded in a larger physical system. Is this perspective to be found in Marx already?

Marx and Engels were one generation younger than the agricultural chemists who published from 1840 onward their researches on the cycles of plant nutrients (phosphorous, nitrogen, potassium), influenced by the debates on the threat of decreasing agricultural yields and the wholesale imports of guano, mainly from Peru, which became an essential bulk commodity for agricultural production. About 11 million tons of guano were exported from Peru in four decades. The analyses of the composition of imported guano, and also of other manures and fertilizers (bones, for instance) already well known to farmers, laid the foundations for agricultural chemistry. Liebig's name was associated, by his own wish, to a new future leading sector of the economy, the fertilizer industry. It may also be associated to an ecological vision. He is recognized as one of the founders of ecology, before the name itself was invented. Politically he developed an argument against latifundist agriculture and agricultural exports because the plant nutrients would not return to the soil. He was in favor of small-scale agriculture and dispersed settlements. Marx quoted this opinion very favorably on several occasions.

Before proceeding further, it must be emphasized that Marx and Engels were also contemporaries of the physiologist Jacob Moleschott (1822–1893). To the present day, Marxists call him a "vulgar materialist" (meaning a reductionist materialist, in opposition to historical materialists who favor dialectics and coevolution). However "vulgar," Moleshott was a main figure in the introduction of the concept of metabolism in physiology and in what now is called ecology. Moleshott had been dismissed after 1848 from Heidelberg, he taught in Zurich and in Rome, he became a well-known professor and editor of a scientific journal, and he had been an activist. Materialism was used against the doctrine of vitalism and against religion. His book on the theory of human nutrition of 1850 had the subtitle *Ffr das Volk*. This was certainly different in intent from the studies by other authors on how to feed cheaply and efficiently soldiers, industrial workers or slaves (in America). In 1851 he published *Physiology of Metabolism in Plants and Animals*, one year later, *Der Kreislauf des Lebens* (The Cycle, Circuit or Circle of Life). Engels, in a letter to Piotr Lavrov of 1875 discussing Darwinism, summarized the idea of the "cycle of life" in "the fact that the vegetable kingdom supplied oxygen and nutriment to the animal kingdom and conversely the animal kingdom supplied plants with carbonic acid and manure." This had been interpreted as cooperation in nature by that trinity of "vulgar materialists" (Vogt, B*f*chner, and Moleschott, who, incidentally, now were talking loudly about "struggle for existence") and had been stressed by Liebig.

Indeed, Marx's use of *Stoffwechsel* (metabolism) in *Capital* (vol. I) and, before that, in some drafts of the late 1850s was influenced by Liebig, probably by other authors, and certainly by Moleschott. Marx used "metabolism" in two senses. First, as a biological analogy or metaphor to describe the circulation of commodities. Second, more relevantly for the present article, he used the expression "metabolism between man and earth, or between society and nature" to refer specifically to the cycles of plant nutrients, quoting, and praising Liebig. Such metabolism was broken by an agriculture of spoliation.

Marx found Liebig supremely relevant because he talked on the natural conditions of agricultural fertility and on the possibility of development of the productive forces by the fertilizer industry. This was useful for the polemics against Malthus and for the theory of land rent. Foster has analyzed in depth Marx's debt to Liebig, and he has wrongly dismissed Moleshott's influence on Marx's use of "metabolism." Foster does not quote Moleshott's books of 1851 and 1852 on the "circle of life" and on the physiology of metabolism in plants and animals.

In conclusion, Marx was a historian and an economist, he was also a student of agriculture and he read on agricultural chemistry and adopted the notion of "metabolism" between man and nature. The material flows between humans and nature were mobilized by human labor, except perhaps in very primitive societies that lived from gathering. The development of tools by humans was essential for the metabolism. Marx did not include the metabolic flow of energy in agriculture and in industry in his analysis, so he could not trace a fundamental distinction (as Lotka was to do 40 years later) between the use of energy in nutrition (endosomatic) and the use of energy by tools (exosomatic). This difference between biometabolism and technometabolism is crucial for the understanding of human ecology. We as a species have genetic instructions regarding our use of endosomatic energy but not our use of exosomatic energy, which must be explained by history, politics, economics, culture, and technology. This is not to be found in Marx. Not only he left energy flow aside, he did not count material flows in detail or even roughly, he was not a protoindustrial ecologist. But on the positive side, he did not see the economy as a closed system, or nature as an immutable purveyor of free gifts, as the neoclassical economists were to do.

Humans were part of nature, humans used nature's materials, we could increase its produce by the development of the so-called productive forces but we could also undermine the natural conditions of production. This was the case with capitalist agriculture. Marx wrote:

Capitalist production disturbs the metabolic interaction between man and the earth, i.e. it prevents the return to the soil of its constituent elements consumed by man in the form of food and clothing, hence it hinders the operation of the eternal natural conditions for the lasting fertility of the soil... Moreover, all progress in capitalist agriculture is a progress in the art, not only of robbing the workers, but of robbing the soil. (Capital, I)

He added that the separation of town and country, caused by latifundist agriculture and by the concentration of sources of energy in cities, provokes an "irreparable rift" in the process of social metabolism. The result of this was a squandering of the soil, aggravated by trade, undermining the conditions of agricultural production. Engels wrote, "As the individual capitalists are engaged in production and exchange for the sake of the immediate profit, only the nearest, most immediate results must first be taken into account." In modern-day parlance, Engels was saying that individual capitalists discounted the future and did not include externalities in their accounts. He continued:

As long as the individual manufacturer or merchant sells a manufactured or purchased commodity with the usual coveted profit, he is satisfied and does not concern himself with what afterward becomes of the commodity and its purchasers. The same thing applies to the natural effects of the same actions. What cared the Spanish planters in Cuba, who burned down forests on the slopes of the mountains and obtained from the ashes sufficient fertilizer for one generation of very profitable coffee tress—what cared they that the heavy tropical rainfall afterward washed away the unprotected upper stratum of the soil, leaving only the bare rock! (Dialectics of Nature)

Such quotations, which could be multiplied, should not be seen as anecdotal, they arose, as Foster has shown, from a theory of "metabolic rift." So Marx was "greener" than usually thought. However, what Greens can learn in Marx is not new to them, the knowledge of the cycles of plant nutrients and the damage to soil and forests.

"Metabolic rift" is not equivalent to "decreasing returns." Marx wrote to Engels on February 13, 1866, saying that Liebig's agricultural chemistry was more important for the discussion on decreasing returns than all the economists put together. Marx dismissed the notion of decreasing returns in agriculture, pointing out in the context of his praise for Liebig's agricultural chemistry and its promise of artificial fertilizers, that it did not make sense to assume in Britain that the produce of the land would increase in a diminishing ratio to the increase of the laborers employed, because in practice there was at the time both an increase in production and an absolute decrease (already) in the number of laborers. Marx was not worried about crisis of subsistances. Marxists attacked Malthus, which could be expected because Malthus had made the point that improving the situation of the poor was counterproductive because they would have more children. There were many debates on how many people could the earth feed around 1900. Some Marxists not only attacked Malthus, they also attacked the Neomalthusians of the late 19th century and early 20th century who were often political radicals and feminists (Paul Robin, Emma Goldman).

2. PODOLINSKY'S AGRICULTURAL ENERGETICS

The link between material metabolism (*Stoffwechsel*, literally exchanges of materials) and the flow of energy at the level of cells and organisms was made in the 1840s. It was then also understood that agriculture might be represented in terms of changes in the flow of energy and not only as an intervention in the cycling of plants nutrients (in 1845, Mayer used *Stoffwechsel* for energy flow). Metabolism was therefore used no only for materials, also for energy. Of course, materials could be cycled, energy could not. The theory of the direction of the flow of energy was developed after 1850 and the establishment of the second law.

Marx and Engels were interested in energy. For instance, Engels wrote to Marx on July 14, 1858, commenting on Joule's work on the conservation of energy (of 1840) as something of well known to them. Marx was of course interested in new sources of energy. One example will suffice: it was already discussed at the time whether hydrogen could be a net source of energy, depending on the energy requirement for electrolysis. Marx wrote to Engels on April 2, 1866, that a certain M. Rebour had found the means of separating the oxygen from hydrogen in water for very little expense. However, Marx did not refer to the flow of energy as a part of metabolism throughout his lifetime.

At a more general level, one interesting point arises from Engels's unwillingness to understand how the first and second laws could apply at the same time: the "dialectics of Nature" failed him there. As Engels became aware of Clausius's concept of entropy, he wrote to Marx on March 21, 1869:

> In Germany the conversion of the natural forces, for instance, heat into mechanical energy, etc. has given rise to a very absurd theory—that the world is becoming steadily colder... and that, in the end, a moment will come when all life will be impossible... I am simply waiting for the moment when the clerics seize upon this theory.

Indeed, not only the clerics but also W. Thomson (Lord Kelvin) brandished the second law in his religious tirades about the "heat death," although he had and could have no inkling of the source of energy in the sun in nuclear fusion. One understand Engels's dislike for the uses to which the second law was put. Josef Popper-Lynkeus (1838–1921), who with Ernst Mach became one of the sources of inspiration for the analytical, antimetaphysical philosophy of the Vienna Circle, complained since 1876 about W. Thomson's theological handling of Carnot's law. However, Engels's dislike of the second law was not only motivated only by its religious use, or abuse. He believed that ways would be found to reuse the heat irradiated into space.

Another interesting point is the reaction by Engels in 1882 (in letters to Marx) regarding Podolinsky's work on energy flows in agriculture. Podolinsky had studied, we might say, the entropy law and the economic process, and he tried unsuccessfully to convince Marx that this could be brought into the Marxist analysis. Podolinsky's work has been seen as a first effort to develop ecological economics. Politically he was not a Marxist, he was a Ukrainian federalist narodnik. He had criticized Marx's overpowering behavior at the congress of the International of 1872, praising the anarchist James Guillaume. However, Podolinsky himself saw his work on the energetics of agriculture as a contribution to Marxist theory, and he would not have objected to appear under "Marxism and Energy" in an encyclopedia. Writing to Marx on April 8,1880, he said, "With particular impatience I wait for your opinion on my attempt to bring surplus labor and the current physical theories into harmony."

In his article (published in long versions in Russian in 1880 and in German in 1883, and in short French and Italian versions in 1880 and 1881), Podolinsky started by explaining the laws of energetics, quoting from Clausius that although the energy of the universe was a constant, there was a tendency toward the dissipation of energy or, in Clausius's terminology, there was a tendency for entropy to reach a maximum. Podolinsky did not discuss explicitly the difference in thermodynamics between open, closed and isolated systems, although he stated explicitly, as the starting point of his analysis, that at the present time the earth was receiving enormous quantities of energy from the sun, and would do so for a very long time. All physical and biological phenomena were expressions of the transformations of energy. He did not enter into the controversies regarding the creation of the universe and its "heat-death," nor did he discuss the relations between thermodynamics and the theory of evolution. In March 1880 he published an article against social Darwinism. He certainly realized that the availability of energy was a crucial consideration for the increase (or decrease) of population. However, he thought that the distribution of production was explained by the relations between social classes: "in the countries where capitalism triumphs, a great part of work goes towards the production of luxury goods, that is to say, towards a gratuitous dissipation of energy instead of towards increasing the availability of energy."

He explained that plants assimilated energy, and animals fed on plants and degraded energy. This formed the *Kreislauf des Lebens*:

> *We have in front of us two parallel processes which together form the so-called circle of life. Plants have the property of accumulating solar energy, but the animals, when they feed on vegetable substances, transform a part of this saved energy and dissipate this energy into space. If the quantity of energy accumulated by plants is greater than that dispersed by animals, then stocks of energy appear, for instance in the period when mineral coal was formed, during which vegetable life was preponderant over animal life. If, on the contrary, animal life were preponderant, the provision of energy would be quickly dispersed and animal life would have to go back to the limits determined by vegetable wealth. So a certain equilibrium would have to be built between the accumulation and the dissipation of energy.*

Not only plants, also human labor had the virtue of retarding the dissipation of energy. Human labor achieved this by agriculture, although the work of a tailor, a shoemaker, or a builder would also qualify, in Podolinsky's view, as productive work, since they afford "protection against the dissipation of energy into space." The energy available for humankind came mainly from the sun. Podolinsky gave figures for the solar constant. He explained how coal and oil, wind energy, and water power were transformations of solar energy. He mentioned tides as another possible source of energy. He then started his analysis of the energetics of agriculture, remarking that only a very small proportion of sun energy was assimilated by plants. Human work, and the work of animals directed by humans, were able to increase the availability of energy by agricultural activity. This he showed by comparing the productivity of different types of land use taking statistics from France (he was living at the time in Montpellier). Table I summarizes his data.

Podolinsky then compared wheat agriculture and sown pastures and forest, concluding that production was higher when there was an input of human and animal work. Thus, comparing wheat agriculture to natural pastures, each kcal put in contributed to an increase of 22 kcal of production. If forest were taken as the terms of comparison, the energy productivity of human and domestic animals work was even higher. Notice that Podolinsky was counting human and animal work—that is, not the food intake but the work done. He did not include solar radiation in the input of energy, because he was writing as an ecological economist. Solar radiation is indeed a free

TABLE I

Annual Production and Energy Input (Only Work by Humans and Domestic Animals) per Hectare, Averages for France in 1870, According to Podolinsky

System	Production (kg)	Production (kcal) Energy input (kcal)
Forest	900 (dried wood)	2,295,000 Nil
Natural pastures	2,500 (hay)	6,375,000 Nil
Sown pastures	3,100 (hay, excluding seed)	7,905,000 37,450 (50 horse-hours and 80 human-hours)
Wheat	800 (wheat) and 2,000 (straw) (excluding seed)	8,100,000 77,500 (100 horse hours and 200 human-hours)

gift of nature (moreover, without an owner, so that there is no payment of rent).

Energy values of wood, ha,y and straw, 2550 kcal/kg, of wheat 3750 kcal/kg. Hours of work converted into kcal: 645 kcal/hour of horse-work, 65 kcal/hour of human-work.

The conclusion was that work could increase the "accumulation of energy on earth." Although Podolinsky mentioned guano, and although he must been keenly aware of the war then raging for Peruvian or Chilean saltpeter, he did not subtract from the output, or include in the input, the energy contents and cost of fertilizer. Nor did he consider the energy spent for steam engines for threshing. His methodology is nevertheless basically the same as that used later for establishing the energy balance of particular crops, or for small scale societies, or for the entire agricultural sector of particular countries.

Podolinsky then went on to explain the capacity of the human organism to do work. Otherwise "it would be difficult to explain the accumulation of energy on the surface of the earth under the influence of labour." Quoting from Hirn and Helmholtz, he concluded correctly that "man has the capacity to transform one-fifth of the energy gained from food into muscular work," giving to this ratio the name of "economic coefficient," remarking that man was a more efficient transformer of energy than a steam engine. He then used a steam engine metaphor to put across a general theoretical principle on the minimum natural conditions of human existence, from an energy point of view. He wrote that humanity was a "perfect machine" in Sadi Carnot's sense: "Humanity is a machine that not only turns heat and other physical forces into work, but succeeds also in carrying out the inverse cycle, that is, it turns work into heat and other physical forces which are necessary to satisfy our needs, and, so to speak, with its own work turned into heat is able to heat its own boiler."

Now, for humanity to ensure its elementary conditions of existence, each calorie of human work must then have a productivity of 5 calories. Taking into account that not everybody is able to work (children, old people), and that there other energy needs apart from food energy, the necessary minimum productivity would be more like ten or more. If that minimum is not achieved, then, of course, "scarcity appears and, many times, a reduction of population." Podolinsky then established the basis for a view of the economy in terms of energy social metabolism, looking at the energy return to energy input in a framework of reproduction of the social system. He thought than he had reconciled the Physiocrats with the labor theory of value, although the Physiocrats (in the 18th century) could not have seen the economy in terms of energy flow.

Podolinsky also emphasized the difference between using the flow of solar energy and the stock of energy in coal. The task of labor was to increase the accumulation of solar energy on earth, rather than the simple transformation into work of energy already accumulated on earth, more so since work done with coal was inevitably accompanied by a great dissipation of heat-energy into space. The energy productivity of a coal miner was much larger than that a primitive farmer could obtain, but this energy surplus from coal was transitory, and moreover (Podolinsky added in a footnote) there was a theory that linked climatic changes to concentrations of carbon dioxide in the atmosphere, as Sterry Hunt explained at a meeting of the British Society for the Advancement of Science in the autumn of 1878. Notice here that the emphasis here was not on capital accumulation but on increasing the availability of energy and certainly also its dissipation. Podolinsky was not, however, at all pessimistic about the prospects for the economy, and he was hopeful for the direct use of solar energy for industrial purposes, referring to the "solar engine of M. Mouchot." One could envisage that one day solar energy would be used directly to make chemical syntheses of nutritive substances, by-passing agriculture. Thus, a proper discussion of the demographic question had to take into account "the relations between the general quantity of energy on earth and the quantity of energy who live on it," and this was a more relevant consideration, in his explicit view, than the Malthusian prognosis. He was also hoping (as he had written to Marx on March 30, 1880, sending his work to him) to develop applications of his energy accounts to different forms or modes of production. Marx died in 1883 and apparently never commented on Podolinsky's work, beyond a letter of acknowledgement at the beginning of April 1880. He asked Engels to write a comment, which Engels did not do until December 1882. Podolinsky himself became ill in 1881 or 1882 and never recovered until his death ten years later, back in Kiev.

3. ENGELS'S REACTION TO PODOLINSKY

Engels had some knowledge of the energetics of human physiology, and in a note of 1875 (later included in the *Dialectics of Nature*) he referred to Fick's and Wislicenus's experiment in climbing the Faulhorn in 1865, which became popularized under the name "a day of hunger for science." Fick (1829–1901) had written in 1857 and 1958 on the amount of kilocalories (2700) that a man would need and spend per day, when not working, and how different types of work implied different energy expenditures over that rate. An idea being circulated at the time (for instance, in an article in *Das Ausland* in 1877) was that the economic values of different types of work could be established in physical terms. Engels already explicitly rejected this notion in 1875.

On December 19 and 22, 1882, Engels wrote two letters to Marx (about 1000 words) explaining that Podolinsky had "discovered" the following facts, already well known. If the food intake of one person per day were equal, say, to 10,000 kcal, then the physical work done would be fraction of this energy. This physical work could become economic work if employed in fixing solar energy, for instance through agricultural activity. With a daily food intake of, say, 10,000 kcal, a man will work in agriculture. Podolinsky's discovery was that human labor had the power of detaining solar energy on the earth's surface and permitting its activity longer than would be the case without it. All the economic conclusions he drew from this were wrong. The economic labor, which a person working in agriculture performed by the employment of these 10,000 kcal, consisted in the fixation for a greater or less time of new energy radiated to that person from the sun. Whether, however, the *new* quantity of energy fixated by the application of the original 10,000 kcal of daily nourishment "reaches 5,000, 10,000, 20,000, or 1,000,000 kcal depends solely on the degree of development attained by the means of production." In agriculture, "the energy value of the auxiliary materials, manures, etc. also enters into the calculation"—a perceptive comment that directly makes one think of the agricultural energy balances of the 1970s, which showed that the energy return to energy input of modern agriculture was on the decline although the economic productivity was increasing. "In industry," wrote Engels, all calculation comes to an end; in most cases the work added to the product (he wrote on December 19, 1892) "can no longer be expressed in kcal." Engels was right that in industrial production, there would be an economic added value although not a net gain of energy—on the contrary, there would be more available energy in the raw materials than in the final product. Actually, also in agriculture the energy in the product does not come from human work but it comes really from photosynthesis, human labor helping and guiding the process. In fact, Engels wrote on December 22, 1982:

> *Storage of energy through work really takes place only in agriculture; in cattle raising the energy accumulated in the plants is simply transferred as a whole to the animals, and one can only speak of a storage of energy in the sense that without cattle-raising, nutritious plants wither uselessly, whereas with it they are utilised. In all branches of industry, on the other hand, energy is only expended…. So that if one chooses one can translate into a physical language [ins Physikalische fbersetzen] the old economic fact that all industrial producers have to live from the products of agriculture, cattle raising, hunting, and fishing—but there is hardly much to be gained from doing so.*

Engels also wrongly wrote that "the energy value of a hammer, a screw or a needle calculated according to the costs of production is an impossible quantity," concluding, "In my opinion it is absolutely impossible to try and express economic relations in physical magnitudes." We are here as far away as possible from the "social metabolic" perspective.

Engels was most unfair when he wrote, "What Podolinsky has entirely forgotten is that man as a worker is not merely a fixer of *present* solar heat but a still greater squanderer of *past* solar heat. The stores of energy, coal, ores, forests, etc. we succeed in squandering you know better than I." Podolinsky had not forgotten this at all. Finally, Engels also wrote, "Podolinsky has strayed away from his very valuable discovery into mistaken paths because he was trying to find in natural science a new proof of the truth of socialism, and has therefore confused physics and economics."

Engels's negative reaction to Podolinsky's work, and Marx's 3-years silence from 1880 to 1883 (when he was still intellectually active, much involved in discussions on Russian peasant communes), may be seen as a missed change for an ecological Marxism. Actually Podolinsky's work is relevant not only in the Marxist context. His work and life has an entity of its own, apart from his encounters with Marx and Engels. He had a short life (he was severely ill the last 10 years of it), but he left a strong trace in Ukrainian federalist politics (as a friend of Drahomanov) and also in the Narodnik movements against the Russian autocracy (as a young colleague of Piotr Lavrov but also connected to the Narodnaya Volya group). He was trained as a medical doctor and physiologist. His work on energy and the economy received Vernadsky's approval. In a section of *La Geochimie* (1924) Vernadsky wrote about several authors (Felix Auerbach with his notion of *Ektropismus*, John Joly) who had explained that life was a process which reversed or slowed down the dissipation of energy. He then wrote a memorable phrase: Podolinsky had studied the energetics of life and tried to apply his findings to the study of the economy.

The link between the use of energy and human culture, in the form of "social energetics" (without historical statistical work), became well established and debated in European culture around 1900. Some Marxist authors adopted this outlook, and their work has been seen as an anticipation of Bertalanffy's systems theory, which grew out of the links between thermodynamics and biology. But, to repeat, there is no line of ecological Marxist history based on quantitative studies of material and energy social metabolism. Moreover, Lenin wrote a diatribe against Ostwald's social energetics in the context of his polemics against Bogdanov and his tirades against Mach's empiriocriticism. This was before the October Revolution of 1917. Afterward, Lenin's ill-considered remarks became sacred to the faithful. Thus, on the occasion of the publication of Engels's *Dialectics of Nature* in 1925 (out of drafts and notes he had left behind) and the 30th anniversary of his death, Otto Jenssen printed once again Engels's letters to Marx on Podolinsky (which were first published in 1919) and explained that Engels had anticipated a critique of Ostwald's social energetics even before Ostwald himself appeared on the scene.

4. OTTO NEURATH'S NATURALRECHNUNG

Marx wrote that wealth was undoubtedly created both by human labor and by nature, quoting William

Petty: "Labour is the Father, Nature is the Mother." Beyond facile jokes on whether fathers or mothers are more important, the interesting point is the distinction between value and wealth. Marx's objective was to explain in a capitalist wage economy the appropriation of surplus labor. Such exploitation was not so obvious as in a slave economy or a feudal (serf) economy, since wages were apparently freely established in a market. Surplus labor became surplus value.

An important contribution to the links between Marxism and a biophysical approach to the economy was made later, in the first half of the 20th century, by Otto Neurath (1882–1945). A famous logical empiricist philosopher of the Vienna Circle, he was also an economist or economic historian, and a Marxist in at least two senses. First, he defended a democratically planned economy based on physical accounting in energy and material terms (*Naturalrechnung*) influenced by Popper-Lynkeus's and Ballod-Atlanticus's quantitative, realistic "utopias," in the so-called Socialist Calculation debate of 1919 and following years. Neurath introduced the idea of incommensurable values in the economy, hence the interest in his work by ecological economists. Second, some years later, in the context of the Vienna Circle' project of the Encyclopedia of Unified Science of the 1930s and 1940s, he defended a dialectical view of history (although he did not like the word "dialectics") as the putting together the findings of the different sciences regarding concrete processes or events. The findings of one science, collected in the encyclopedia, with regard to one particular process or event, should not be contradicted by the findings of another science also present in the encyclopedia, and leave things at that. At attempt at reconciliation or removal of the contradiction should be attempted. "Consilience" (to use Edward Wilson's word) should be the rule of the encyclopedia.

Hayek's strong criticism against "social engineering" (1952) was directed, as John O'Neill has put it, "at the whole tradition which attempts to understand the ways in which economic institutions and relations are embedded within the physical world and have real physical preconditions, and which is consequently critical of economic choices founded upon purely monetary valuation." Translated into Foster's language: Hayek critized the view of the economy in terms of social metabolism. While Patrick Geddes, Wilhelm Ostwald, Frederick Soddy, and Lewis Mumford were all rudely dismissed by Hayek, it is the work of Otto Neurath that was the primary target of Hayek's criticism.

One is also reminded not only of the comments of Max Weber against Otto Neurath in *Economy and Society* but even more of his polemics against Wilhelm Ostwald in 1909. Ostwald (who was a well known chemist) was trying to interpret human history in terms of the use of energy. He proposed two simple laws, which are not untrue, and which might act or not in opposite directions. First, the growth of the economy implied the use of more energy and the substitution of human energy by other forms of energy. Second, this came together with a trend toward higher efficiency in the transformation of energy. Max Weber (1909) wrote a famous, ironic review of Ostwald's views, where he insisted on the separation between the sciences. Chemists should not write on the economy. The review was much praised by Hayek in the 1940s. Max Weber's basic point was that economic decisions by entrepreneurs on new industrial processes or new products were based on prices. It could so happen that a production process was less efficient in energy terms and nevertheless it would be adopted because it was cheaper. Energy accounting was irrelevant for the economy. Max Weber did not question energy prices as we would today when externalities such as the enhanced greenhouse effect or when the intergenerational allocation of exhaustible resources are taken into account.

Ostwald influenced many authors, among them Henry Adams (1838–1918) who proposed a "law of acceleration" of the use of energy: "the coal output of the world, speaking roughly, doubled every ten years between 1840 and 1900, in the form of utilized power, for the ton of coal yielded three or four times as much power in 1900 as in 1840." Since Marxism is in this article under critical review, it is worthwhile pointing out that from the opposite camp, Karl Popper, in *The Poverty of Historicism*, which is a polemic against Marxist laws of history, thought it worthwhile to attack also Henry Adams in a footnote, without even mentioning the word "energy" because he dared propose a historical law. Henry Adams was certainly not a Marxist; on the contrary he was a Boston aristocrat who thought the world would probably come to a bad end socially and technologically.

When the anthropologist Leslie White (1943) wrote on energy and the evolution of culture inspired by Morgan and Taylor as evolutionary anthropologists, he realized that Ostwald and Soddy were precursors of his views. He looked for stages of evolution in terms of harnessing energy in increasing magnitudes and in various forms, in a dialectical framework between the technological level (defined by the availability and the efficiency in the transformation of energy), the social system (in terms of Marxist "relations of production"), and the cultural-symbolic system.

Marx's doubts on the benefits of economic growth, clearly expressed in his ecological critique of capitalist agriculture, was not forgotten within Marxism (think of Walter Benjamin) although the technological optimists, who believed in the development of the productive forces, predominated. One of most influential Marxist technological optimists of the 20th century was the historian of science J. D. Bernal. He was in the 1950s totally in favor of the "civil" use of nuclear energy, which Lewis Mumford was strongly criticizing already at the time. Mumford, described by Ramachandra Guha as the "forgotten American environmentalist" (forgotten in comparison to G. P. Marsh, John Muir, Gifford Pinchot, Aldo Leopold, and Rachel Carson), was a heir to Patrick Geddes, William Morris, and John Ruskin; he does not belong in the Marxist tradition.

Mumford's mentor, Patrick Geddes (1854–1932), a biologist and urban planner, early on attacked neoclassical economists such as Walras because they did not count flows of energy, materials, and waste, but he did not discuss Marx. Geddes proposed the construction of a sort of input-output table inspired by the *Tableau Economique* of the Physiocrat F. Quesnay. The first column would contain the sources of energy as well as the sources of materials that are used, not for their potential energy, but for their other properties. Energy and materials were transformed into products through three stages: extraction, manufacture, transport and exchange. Estimates were needed of the losses (dissipation and disintegration) at each stage. The quantity of the final product (or "net" product, in physiocratic terms) might seem surprisingly small in proportion to the gross quantity of potential product. Now, however, the losses at each stage were not accounted for in economic terms. The final product was not added value at all. It was the value remaining from the energy and materials available at the beginning once they had been through all three stages.

Geddes's scheme is relevant to the attempt by several authors to develop a theory of ecologically unequal exchange between the metropolitan centers and the world peripheries. In neoclassical economics, provided that markets are competitive and ruled by supply and demand, there cannot be unequal exchange. This could only arise from monopoly or monopsony conditions, or because of noninternalized externalities (or excessive discounting of the future). In an ecological-economics theory of unequal exchange, one could say that the more of the original exergy (available energy or "productive potential" in the exported raw materials) has been dissipated in producing the final products or services (in the metropolis), the higher the prices of these products or services. This was indeed implied by Geddes with different words. Thus, Hornborg concluded, "market prices are the means by which world system centers extract exergy from the peripheries," sometimes helped, one must say, by military power.

A social metabolic perspective shows that capital accumulation does not take place by itself, and it is not only based on the exploitation of labor. Capitalism must advance into commodity frontiers because it uses more materials and energy, therefore it produces more waste, and it undermines not only its own conditions of production but the conditions of livelihood and existence of peripheral peoples, who complain accordingly. Such ecological distribution conflicts are more and more visible. They cannot be subsumed under the conflict between capital and labor.

SEE ALSO THE FOLLOWING ARTICLES

Aggregation of Energy • *Conversion of Energy: People and Animals* • *Ecological Footprints and Energy* • *Economic Thought, History of Energy in* • *Ecosystem Health: Energy Indicators* • *Ecosystems and Energy: History and Overview* • *Energy in the History and Philosophy of Science* • *Heterotrophic Energy Flows* • *Industrial Agriculture, Energy Flows in* • *Industrial Ecology* • *Photosynthesis and Autotrophic Energy Flows* • *Thermodynamics and Economics, Overview* • *Thermodynamic Sciences, History of*

Further Reading

Ayres, R. U. (1989). Industrial metabolism. *In* "Technology and Environment," (J. Ausubel, Ed.). National Academy Press, Washington, DC.

Ayres, R. U., Ayres, L., and Warr. B. (2002). Exergy, power and work in the US economy, 1900–1998. CMER-INSEAD, working paper, 2002.

Cipolla, C. (1974). "The Economic History of World Population." 6th ed. Penguin, London.

Cleveland, C. J. (1987). Biophysical economics: Historical perspectives and current recent trends. *Ecol. Modelling.*

Cohen, J. (1995). "How Many People Can the Earth Support?" Norton, New York.

Debeir, J. C., Del/age, J. P., H/mery, D. (1986). Les servitudes de la puissance Une histoire de l'energie, Flammarion, Paris.

Engels, F. (1972). Dialektik der Natur (1925), MEW, vol. 20, Dietz Verlag, Berlin.

Fischer-Kowalski, M. (1998). Society's metabolism: The intellectual history of materials flow analysis, Part I: 1860–1970, Part II (with W. Huettler): 1970–98. *J. Industrial Ecol.* 2(1) and 2(4).

Foster, J. B. (2000). Marx's ecology. Materialism and nature. *Monthly Rev. Press.*

Fluck, R. C., and Baird, D. C. (1980). "Agricultural Energetics." Avi, Westport, CT.

Geddes, P. (1885). An analysis of the principles of economics. Proceedings of the Royal Society of Edinburgh, 17 March, 7 April, 16 June, 7 July 1884, repr. Williand and Norgate, London.

Gootenberg, P. (1993). "Imagining Development: economic ideas in peru's "Fictitious Prosperity" og guano, 1840–80." University of California Press, Berkeley, CA.

Haberl, H. (2001). The energetic metabolism of societies, Parts I and II. *J. Industrial Ecol.*

Hall, C. h., Cleveland, C. J., and Kaufman, R. (1986). "Energy and Resources Quality: The Ecology of the Economic Process." Wiley, New York.

Hornborg, A. (1986). Toward an ecological theory of unequal exchange: Articulating world system theory and ecological economics. *Ecol. Econ.* 25(1), 127–136.

Jenssen, O. (ed.) (1925). Marxismus und Naturwissenschaft: Gedenschrifts zum 30. Todestage des Naturwissenschaftlers Friedrich Engels, mit Beitr...gen von F. Engels, Gustav Eckstein und Friedrich Adler, Verlagges. des Allgemeinen Deutschen Gewerkschaftsbundes, Berlin.

Kormondy, E. J. (1965). "Readings in Ecology." Prentice-Hall, Englewoods Cliffs, NJ.

Leach, G. (1975). "Energy and Food Production." IPC Science and Technology Press, Guildford.

Lenin, V. I. (1913). The working class and Neomalthusianism. *Pravda* 137. (Reprinted in Collected Works, vol. 19.)

Martinez-Alier, J. (2002). "The Environmentalism of the Poor: A Study of Ecological Conflicts and Valuation." Edward Elgar, Cheltenham UK, Northampton, MA.

Martinez-Alier, J., Munda, G., and O'Neill, J. (1998). Weak comparability of values as a foundation for ecological economics. *Ecological Economics* 26, 277–286.

Martinez-Alier, J., and Schlfpmann, K. (1987). "Ecological Economics: Energy, Environment and Society." Blackwell, Oxford.

Marx, K. (1969). Das Kapital, vol. I 1867, vol. III, 1894, Ullstein Verlag, Frankfurt-Vienna-Berlin.

Marx, K., and Engels, F. (1976). Lettres sur les sciences de la nature et les mathematiques, Mercure de France, Paris.

Masjuan, E. (2000). La ecolog a humana y el anarquismo ibrico: el urbanismo "orgnico" o ecol gico, el neo-malthusianismo y el naturismo social, Icaria, Barcelona.

Masjuan, E. (2003). Neomalthusianesimo e anarchismo in Italia: un capitulo della storia del pensiero ecologico dei poveri? *Meridiana* (Rome).

Mayer. J. R. (1893, 1911). Die organische Bewegung in ihrem Zusammenhang mit dem Stoffweschsel, Heilbronn, published also in Die Mechanik der W...rme: gesammelte Schriften, Stuttgart, 1893, and in W. Ostwald, Klassiker der exacten Naturwissenschaften, Akademische Verlag, Leipzig, 1911.

McNeill, J. R. (2000). "Something New under the Sun: An Environmental History of the Twentieth-Century World." Norton, New York.

Moleschott, J. (1850). "Lehre der Nahrungsmittel." F*f*r das Volk, Enke, Erlangen.

Moleschott, J. (1851). Physiologie des Stoffwechsels in Planzen und Thieren, Erlangen.

Moleshott, J. (1852). "Der Kreislauf des Lebens." Von Zabern, Mainz.

Mouchot, A. (1869, 1879). La chaleur solaire et ses applications industrielles. Gauthier-Villars, Paris, 1869; 2nd ed., 1879.

Naredo, J. M., and Valero, A. (1999). "Desarrollo econ mico y deterioro ecol gico." Argentaria-Visor, Madrid.

O'Neill, J. (1993). "Ecology, Policy and Politics." Routledge, London.

O'Neill, J. (2002). Socialist calculation and environmental valuation: Money, markets and ecology. *Science and Society*, spring.

Odum, H. T. (1971). "Environment, Power and Society." Wiley, New York.

Pfaundler, L. (1902). Die Weltwirschaft im Lichte der Physik, Deutsche Revue 22, April–June, 29–38, 171–182.

Pimentel, D., et al. (1973). Food production and the energy crisis. *Science* 182, 443–449.

Pimentel, D., M. (1979). Food, Energy and Society. Arnold, London.

Rappaport, R. (1967). "Pigs for the Ancestors: Ritual in the Ecology of a New Guinea People." Yale University Press.

Ronsin, F. (1980). La grve des ventres. Propagande no-malthusienne et baisse de la natalit/en France, 19–20 sicles, Aubier-Montagne, Paris.

Schmidt, A. (1978). "Der Begriff der Natur in der Lehre von Marx." 3rd ed. EVA, Frankfurt-Cologne.

Sieferle, R. P. (1982). Der unterirdische Wald. Energiekrise und industrielle Revolution, Beck, Munich (English trans., White Horse Press, Cambridge, United Kingdom, 2001).

Susiluoto, I. (1982). "The origins and development of systems thinking in the Soviet Union." Political and philosophical controversies from Bogdanov and Bukharin to present-day reevaluations, Suomalainen Tiedeakatemia, Helsinki.

Vernadsky, V. (1924). La G/ochimie. Alcan, Paris.

Weber, M. (1968). Energetische Kulturtheorien, *Archiv ffr Sozialwissenschaft und Sozialpolitik*, 29, 1909, repr. in Max Weber, Gessamelte Aufs...tze zur Wissenschatslehre, 3rd ed., J. C. B. Mohr (Paul Siebeck), T/bingen.

White, L. (1943). Energy and the evolution of culture. *American Anthropol.* 45(3).

White, L. (1959). The energy theory of cultural development. *In* "Readings in Anthropology" (M. H. Fried, Ed.), vol. II, pp. 139–146. Thomas Y. Cromwell, New York.

Material Efficiency and Energy Use

EBERHARD JOCHEM
Swiss Federal Institute of Technology
Zurich, Switzerland
Fraunhofer Institute for Systems and Innovation Research (FH-ISI)
Karlsruhe, Germany

1. The Conceptual Approach
2. Energy Saving Potentials by Reduced and Changed Demand for Materials
3. Joint Impact of Material Efficiency, Substitution, and Intensified Product Use and Plants

Glossary

dematerialization The declining trend of material use per capita in highly industrialized countries. This is due to the facts that a large capital stock and infrastructure are already in place and that reinvestments tend to need less material for the same service (material efficiency) or the capital stock is used more efficiently (intensified use of durables and plants).

energy intensity The relation between primary energy use and the gross domestic product (GDP) of a country in a given year.

intensified use of durables and plants The same quantity of material services is able to be carried out with fewer durables (goods, vehicles, machinery, and plants) because of more intensive use. For this purpose, additional services (e.g., car sharing, solvent service, and harvesting service) and/or additional organization (e.g., contracting and just-in-time delivery) are needed.

material efficiency The same function of a product, a material, or a technical system is performed with less material due to more intelligent design or improved properties of the materials used. Material efficiency has the largest relevance for energy use with regard to vehicles and moving parts.

material intensity The relation between primary material demand and the GDP of a country in a given year. Sometimes, the expression is defined as the relation between primary material demand and the population of a country.

material substitution The substitution of a material by another material or a different system that delivers the same or similar functions to the application considered (e.g., e-mail instead of mailing letters or books). In this context, the energy use for producing the two optional materials and for their operation during their lifetimes is compared; a net energy gain is expected in the case of material substitution.

The more efficient use of materials by improved construction design and material properties has a long success story with regard to reducing the energy demand of energy-intensive materials, vehicles, and moving parts of machines and plants. However, this more efficient use of materials, which can reduce the specific material and energy demand by approximately 1% per year, is scarcely known or perceived as a challenging development. Until recently, it was not even considered to be a promising option for energy and climate change policy in many countries. In addition to the technological option of material efficiency, the substitution of energy-intensive materials by less energy-intensive materials and the intensification of product and plant use are other options that may play a more important role in reducing the quantity of materials produced and hence energy demand. Entrepreneurial innovations will support these technical options to achieve intensified uses of machinery and vehicles by pooling (leasing, renting, and car sharing). The impact on reduced primary energy demand may range from approximately 0.3% per year (autonomous technical progress in material efficiency without specific policies) to more than 0.6% per year if supported by various policies oriented toward sustainable development.

1. THE CONCEPTUAL APPROACH

Considerations of future improvements in energy efficiency and energy demand often focus on energy-converting technologies and the distribution of grid-based energies where the energy losses amount to approximately 60% of primary energy in most economies. However, there are two additional areas for reducing future energy demand to which little attention is currently being paid (Fig. 1):

1. Energy losses at the level of useful energy (currently approximately 40% of the primary energy demand of industrialized countries) could be substantially reduced or even avoided through technologies such as low-energy buildings, membrane techniques or biotechnological processes instead of thermal processes, and lighter vehicles or reuse of waste heat.

2. The demand for energy-intensive materials could be reduced by recycling or substituting these materials with less energy-intensive ones; reusing products or product components; improving their design or material properties; and the intensified use of products, plants, and vehicles through pooling (e.g., car sharing and leasing or sharing of machines and energy-intensive plants).

This article focuses on some of the latter aspects (excluding material recycling and product reuse). Many energy demand analyzes and projections implicitly assume that the demand for energy services is a "given market demand" that has to be fueled by the various forms of useful and final energies. However, the demand for energy-intensive products and materials is not a fixed, noninfluenceable energy service; on the contrary, the demand reflects a certain status of consumer preferences, choices, and behavior; market traditions and policy boundary conditions; the division of labor in production; and technology, which can be positively influenced by many technical options and entrepreneurial innovations without reducing the services delivered by the (technically changed) products, machinery, or vehicles. Examples and their impact on energy demand are given in the following sections.

The concept of dematerialization, reducing energy demand by using less primary material, was developed more than 20 years ago. This concept integrates the different options of reducing energy services and is often measured by the indicator of material intensity, the ratio of material use in physical terms to gross domestic product (GDP) in deflated constant terms. Malenbaum was one of the first researchers to use this indicator at a global level and in 1978 made preliminary projections for the following two decades. Dematerialization is foremost a long-term concept, tied to long-term changes in the nature of technology, infrastructure, and the economy of a country. Care is required to distinguish it from short-term changes in material intensity resulting from business cycle effects. Empirical dematerialization analyses published by Bernadini and Galli in 1993 concluded two basic postulates:

1. The intensity of use of a given material follows the same bell-shaped pattern for all economies. At first, in its industrializing phase, it increases with per capita GDP, reaches a maximum, and eventually declines afterwards; the growth in material use declines to lower rates than GDP. The decline is

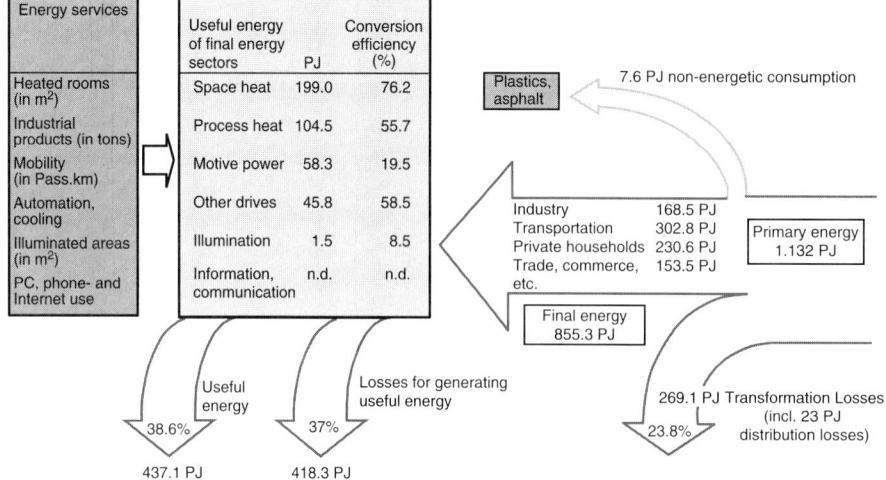

FIGURE 1 More efficient use of materials to reduce energy services demonstrated by the Swiss energy system from energy services to primary energy demand. Source: ISI, Karlsruhe Germany.

frequently accelerated by the growth of competing new materials with process, product quality, and cost advantages (e.g., aluminum or plastics instead of steel or glass).

2. The maximum intensity of material use declines the later it is attained by a given country, reflecting the technological learning and globalization of technical know-how over time.

These two postulates have been found to be valid for individual materials, and they also apply to primary energy intensity, which shows a very similar bell-shaped development. Integrating the substitution and life cycle concept of materials into a total material intensity, a further contribution to the decline of the total material intensity stems from saturation and the development of new markets based on more sophisticated intermediates and commodity products with a relatively higher value added per weight of material input.

For the total material use per capita (t/cap.) and year (including material demand for imported raw materials) of mature economies, it seems that its maximum level of approximately 80–90 t/cap. at an annual per capita GDP level of $25,000 has been reached. It is at this stage that per capita total material use may level off and even begin to decline, as Biondi and Galli found in 1992. This was confirmed by Cleveland and Ruth in 1999; see also declining U.S. figures from 100 t/cap. in the 1970s to 85 t/cap. in the 1990s published by Bringezu. This author and others, such as Rogers, stress that this trend of material use can be influenced by relevant product, recycling, and material-related policies of governments and companies in industry, trade, and other service sectors.

These changes in material intensity and their impact on energy demand were collectively treated in energy demand modeling and forecasts as intra- and interindustrial structural changes. Only recently have they been explicitly analyzed at technical levels of reduced energy and avoided greenhouse gas emissions in the national communications of the reporting process within the Framework Convention of Climate Change.

2. ENERGY SAVING POTENTIALS BY REDUCED AND CHANGED DEMAND FOR MATERIALS

The close relationship between the efficient use of energy-intensive materials and energy demand did not receive much attention from energy analysts and policy makers in the past, even though approximately one-third of total industrial energy demand is required for material production and it is obvious that moving parts and vehicles would need less energy during their lifetime if they weighed less. In most energy demand projections to date, more efficient use of materials, products, vehicles, and production plants or the substitution of materials were not seen as flexible and policy relevant but, rather, as a fixed level of activity. The following are the options to reduce energy demand by reduced material intensity discussed in this article.

1. More efficient use of materials by better design and construction; improved properties of materials, oils, or solvents, or even foamed plastics or metals; or reduced use of energy-intensive products such as nitrogen fertilizers. This strategy is particularly important in the case of moving parts or vehicles because lighter construction may contribute to significantly lower energy demand over the lifetime of the particular application (e.g., cars).

2. Substitution of highly energy-intensive materials by less energy-intensive materials or even by other technologies (e.g., substitution of steel and cement/concrete by wood; synthetic fibers by biobased fibers; and newspapers, journals, or books by electronic news and information).

3. Intensified use of products, machinery, and vehicles through short-term leasing, sharing, or renting, reducing the demand for materials by diminishing the nonused capital stock that may even include more efficiently used floor areas of residential and office buildings. This option of the strategy "renting instead of owning" seems to be a purely entrepreneurial and organizational issue at first, but it may also involve new supporting technologies (e.g., clear documentation by electronic devices on the use of the product or vehicle to identify responsibilities during its operation and to develop adequate rental tariffs).

All three elements contribute to structural changes within industrial production, mostly in the direction of lower energy intensity of total industrial production, and they offer numerous opportunities in the service sector to sell product services instead of products.

2.1 More Efficient Use of Materials

Energy-intensive materials as semifinished or final mass products make up a substantial part of industrial energy demand; of particular interest are paper, glass, steel, aluminum, plastics, rubber products, solvents, and fertilizers. When used in mobile applications, they also contribute to high energy demand, particularly in transportation. Material efficiency means that less material per product function is needed in the future. The direct energy use for primary production of energy-intensive materials is more than 5% of the primary energy demand of an industrialized country, and the indirect energy use for transportation and moving parts (e.g., lifts, moving staircases, and machine components) amounts to more than 25–30% of primary energy demand. In total, approximately 35–40% of the global primary energy use is related to the production of materials, according to estimates of Hekkert and colleagues made in 2000. Examples of past and future increased material efficiency were published by the Enquête Commission of the German Parliament in 2002 and include the following:

• Lower weight of mass papers per square meter: This has been reduced by 15% during the past 30 years.
• Lower glass weight per bottle: This has been reduced by 30% during the past 30 years (ultralight glass bottles with PET films for safety reasons are entering the market).
• Lower weight steel products [e.g., cans (reduced by 50% during the past 30 years), vehicle steel sheets, machines, and appliances]: Competition between steel, aluminum, and plastics has resulted in revolutionary changes in steel shaping and finishing technologies. Powder metallurgy, thin casting, ion beam implantation, directional solidification, cold forging, cold automatic heating, and extruding have allowed savings up to 50% of steel input during the past 30 years;
• Lower specific weight of aluminum per application [e.g., packaging (reduced by 40% during the past 20 years), facades, window frames, vehicle sheets, and construction (airplanes, packing, and cans)].
• Lower specific weight of plastics through better quality plastics produced by improved catalysts and copolymers (e.g., the development of radial tires in the 1970s led to a 25% reduction in tire weight and doubled its lifetime; linear low-density polyethylene allows the use of thinner films in plastic products and savings up to 50%; in the future, the use of foamed plastics will reduce the weight of parts in vehicles, window frames, floor materials, and home and office textiles): Supercellular alloys of magnesium and aluminum and foamed nonferrous metals may contribute to less material input required for many applications.

Worrell and colleagues made the interesting observation in 1995 that the amount of nitrogen fertilizer used can be reduced by several measures applicable to many crops; the reduction potential is between 30 and 50% (e.g., for The Netherlands, from 230 to 130 kg/ha) within one decade.

2.1.1 Potentials and Drivers of Material Efficiency

Besides cost reduction potentials, material efficiency will also be enhanced by increasing the useful life, for example, by improving surface layers protecting buildings and infrastructure against ultraviolet radiation, acid rain, algae growth, and pollution. Surface protection is of particular interest in coastal regions with salty air. Other technological developments rely on the progress in microelectronics, information, and communication technology—miniaturization, design and quality control, and improved logistics. Computer-aided design and manufacturing has allowed the optimization of the number of components that can be produced by a given quantity of material, leading to a significant reduction of waste or to fewer unfinished or rejected goods.

As indicated by these examples, the strategy of reducing a product's material use while maintaining identical functionality is not a new development. In particular, much attention has been paid to postconsumer wastes by national policies or by the European Union regarding the reduction of packaging wastes. The first analytical work by Gielen in 1999 suggests that material efficiency has improved in the past decades by approximately 1% per year on average, which indicates a comparable rate of energy efficiency improvement. The question is whether these trends can be maintained or even accelerated. Competition among the materials and legal frameworks that strive for the internalization of the external costs of material production, waste disposal, and energy conversion will be a major driver for maintaining progress in material efficiency.

The future energy saving potential stemming from improved material efficiency has not been systematically analyzed. However, the methodological approach used to analyze the impact of material efficiency on energy demand (and energy-related emissions) has been developed and described by several authors, such as Worrell and colleagues in 1995 and Hekkert and colleagues in 2000. These authors have also performed a number of case studies for several basic materials, products, and packaging materials. Tentative analyses suggest two major conclusions:

1. The energy savings during the lifetime of vehicles and moving components is approximately three times higher than the savings for reduced material demand.

2. The direct and indirect primary energy savings from the more efficient use of materials range between 0.1% per year (autonomous technical progress) and, according to the findings of the Enquête Commission, approximately 0.25% per year if technical progress is supported by relevant policies.

2.1.2 Obstacles and Issues of Research and Development

Many of the new materials and their specific properties and production processes are known today, but, as for energy efficiency potentials, various obstacles hinder the rapid diffusion of many new materials and the related benefit of reduced energy demand. Because many of the new material technologies are based on expensive, high-quality materials, price is often the most common barrier to their widespread implementation. For this reason, apart from research and development (R&D), Ayres stresses the fact that the promotion must also integrate favorable economic and policy conditions, including innovation and market introduction policies.

In some cases, there are conflicts of interest that have to be taken into account for future market introduction and acceptance. There may also be problems in the transition phase from today's products to light material products (e.g., safety aspects of heavy traditional cars and new light cars and road vehicles in cities).

In order to realize the high-end energy saving potential of material efficiency in the coming decades, some of the research involved is very basic applied research (e.g., foamed plastics or even metals and alloys of magnesium or aluminum). The R&D for the different applications seems to be very dispersed but is often very advanced applied research. In many cases, the higher quality of the improved materials and their applications require high-performance production machines and high production expertise; for example,

- R&D on new polymer catalysts, particularly for polyethylene, polypropylene, polystyrene, polyester, and interesting copolymers, to improve their properties and to substitute specialized, more energy-intensive plastics (polycondensates and polyadducts).

- R&D on improved surface protection to avoid corrosion and/or to increase the life span of the products, facades, or vehicles, including economic optimization over the product's and component's lifetime.

- R&D on foamed plastics and nonferrous metals for lightweight construction, particularly for moving parts and vehicles.

2.2 Substitution of Energy-Intensive Materials

The specific energy demand for producing materials can vary significantly. Replacing mass-produced, energy-intensive materials with less energy-intensive materials can generate a large energy saving potential. This depends on the specific substitution involved, where different product life spans may also have an influence. Substantial energy savings can be achieved by substitution of

- Heavy materials with lightweight materials in the transport sector. Conventional steel constructions, for example, may be replaced with lighter aluminum or magnesium constructions. Although the production of aluminum or magnesium components is more energy-intensive, this additional energy consumption can be compensated for during the components' lifetime. The average weight of U.S. cars declined approximately 20% between 1976 and 1990; most of this decline was accounted for by the use of aluminum, plastics, and composites as substitutes. Schön and colleagues concluded that this past achievement is likely to be repeated in the next 20 years by further substitution using nonferrous metals and plastics.

- Energy-intensive metals, paper, and glass with plastics; their advantages are often lighter weight, greater resistance to tear, relative inertness to moisture and organic environments, and less storage space requirements. Not only do plastics continue to substitute for metals, glass, and paper in traditional applications but also new polymers with improved properties (e.g., high strength-to-weight ratio, transparency, greater durability, and specific surface properties) are increasingly replacing older generation plastics widely in use today. These new polymers often have a lower manufacturing energy demand than the materials they displace when the comparison is made in terms of end-use qualities rather than on a unit weight basis.

- Petrochemically based plastics with biobased polymers or plastics or natural materials. Well-established substitution routes are oleochemical surfactants and lubricants from vegetable oils, as Patel and colleagues reported in 2000. Future examples are the increased use of wood as a construction material, the renaissance of flax and hemp as natural fibers instead of the use of chemical fibers, manufacturing and cooling oils, and bioplastics produced from starch. In the long term, Hüsing and colleagues expect major innovations such as polyhydroxyalkanoates from gene-manipulated plants, which have similar properties as those of the petrochemically based polypropylene.

There are many other small substitutions of materials, such as the substitution of copper in the telecommunication market with optical fibers with 30–40 times the transmission capacity of conventional wiring and by mobile telecommunications, substitution of metals with ceramics (e.g., spark plugs, insulation layers, titan oxide-based solar cells, superconductors, and high-temperature protection layers), and substitution of printed media with electronic information and communication.

A fundamental connotation of substituting old materials with new ones is the trend from commodities to specialty materials adapted for specific products, such as car frames or other car components and biodegradable packaging materials, specific functions of surfaces (e.g., zero corrosion and improved heat transfer), or specific mechanical properties per weight of material. These advanced materials (e.g., produced from nano powder) are becoming more important as they substitute for basic traditional metallic materials or plastics.

2.2.1 Potentials and Opportunities

The future energy saving potentials stemming from material substitution have not been systematically analyzed; the evidence for the energy saving effect of material substitution is restricted to a limited number of metals and plastics applied to a limited number of products. The impact of bioengineered basic natural materials, for example, is an unresolved issue but theoretically implies high energy saving potentials. In addition, the competition between different materials within the substitution process and the simultaneous process of material efficiency makes the analysis of energy saving potentials extremely complex. Initial tentative and partial analyses suggest results similar to those of the impact studies on material efficiency:

1. The energy savings due to substituted materials during the lifetime of vehicles and moving components are approximately three times higher than the energy savings due to material substitution in other fields of application.
2. The direct and indirect primary energy savings from material substitution are estimated to be less than 0.1% per year (autonomous substitution) and 0.2% per year or more if the technical progress by material substitution is supported by relevant policies. This potential may increase during the next decades due to new opportunities resulting from technologies such as nano- and biotechnology and genetic engineering.

2.2.2 Obstacles and Policy Issues

Traditional consumer wishes, the traditional material choice of producers, a certain "image" of a given material in its societal context, and sometimes the higher cost of the substituting and less energy-intensive material can adversely affect the realization of the energy saving potentials in various applications. Early R&D on biobased polymers and products, including applied genetic engineering to improve properties and yields (e.g., natural fibers, starch, and wood), may be important given the perspective of changing boundary conditions in the 2000 "World Energy Assessment," edited by Goldemberg and Johannson: increasing prices of oil and natural gas within the next two decades when global crude oil production is likely to pass its middepletion point. Reducing the production costs of the less energy-intensive materials by learning and economy-of-scale effects may also be an issue of public policy through limited financial incentives, procurement schemes, information and training, as well as by setting technical standards to be met by the new materials.

2.3 Intensified Use of Products, Machinery, and Vehicles

Durables and investment goods are generally purchased but could often be leased, rented, or hired, particularly in cases of small annual usage times (e.g., cars, 200–500 h per year; harvesting machines, <50 h; construction machines, <500 h; and many high-temperature industrial processes in small and medium-sized companies operating on one-shift schedules or <2500 h per year). Regarding material use (and the imbedded energy to produce durables and investment goods), Friend stated in the mid-1990s that the low annual usage time implies a huge idle capital and material stock of an economy due to the concept of selling and owning products instead of selling and buying product-based services. In the case of high-temperature industrial processes, the energy losses for daily startup and shutdown may be substantial.

The strategy of product-based services of manufacturers or service companies is based on the concept of selling the total service (e.g., harvesting, partial outsourcing of energy-intensive production steps, using centralized voice mail instead of answering machines, and renting solvents from chemical suppliers) or pooling durables and investment goods. Mont describes many cases in great detail in which the intensified use of the durables and investment goods reduces their capital cost and compensates for the additional service cost for either the total service or contracting, operating, reservation, billing, controlling, and additional insurance. In many cases, ecological or environmental arguments and considerations contribute to the acceptance of these new services, which are labeled ecoefficient services by Popov and DeSimone. However, very seldom are these product-based services examined from the perspective of material efficiency and indirectly reduced energy use.

At first glance, using products, machinery, and vehicles more intensively through short-term leasing, sharing, or renting and its technical dimension of material efficiency (and, indirectly, of energy efficiency) seems to be a research area more related to the social sciences. In order to realize the strategy "renting instead of owning" at the commercial level, however, experience strongly suggests the need for supporting technologies in many cases. These technologies would offer the user clear documentation on the use of the product or vehicle, identify responsibilities during its operation, and develop adequate rental tariffs. These predominantly information and communication technologies require research focused on small, inexpensive, and reliable electronic systems.

2.3.1 Potentials and Opportunities

The realizable direct energy efficiency potential of these numerous options of intensified use of products, vehicles, and production plants has scarcely been analyzed and depends very much on the future acceptance and market diffusion of services such as car sharing (studied by Meijkamp in The Netherlands), pooling public (municipal) or company vehicle fleets or machinery (studied by Fleig), and moving elderly people into adequate but appropriately sized apartments (first analyzed by Zanger and colleagues in 1999). Assuming moderate market shares of two of these new services in 2050 (e.g., 10% of car-using households opt for car sharing, and 2% of elderly people move to smaller apartments with an average gain of $20\,m^2$ per move and apartment), the energy savings in industrialized countries may be on the order of 3% of current levels of primary energy demand. Increased renting of products and vehicles may indirectly affect energy use through, for example, the higher energy efficiency of "younger" capital stock or vehicle fleets, walking or cycling displacing the car for short trips, and a more sensible use of machines or vehicles due to the accompanying documentation system. The indirect effects are estimated to be approximately 10% of the direct effects.

The impact of these numerous possibilities of intensifying the use of products and production machinery and plants on energy demand is almost unknown because very few existing services have been assessed under energy aspects and many conceivable services to intensify the use of products, vehicles, or production facilities have not been offered on suitable markets. The estimates of known and offered services thus assume yearly primary energy savings of less than 0.02% per year (autonomous substitution), but in the longer term the Enquête Commission estimated that they may contribute to the annual energy savings of 0.1–0.2% of total primary energy demand if relevant incentives and policies are in place.

2.3.2 Obstacles and Policy Issues

To realize the full impact of intensifying the use of products, vehicles, and apartments, new entrepreneurial forms are also required that reduce the capital stock for those applications in which intensified use is being targeted.

Obviously, much research must be undertaken to develop a clearer understanding of the possible potentials of energy savings:

- Research on customer acceptance: In many cases, the various services will be innovative—for example, incentives for machinery and car sharing or improved awareness about the fixed and capital costs of owning machinery and vehicles with low annual operating hours.
- Research on the willingness of innovative companies to invest in pooling services: R&D on inexpensive accompanying information and technology-based monitoring and control techniques for easy and fair pricing, easy scheduling and access, and protection against theft.

New services may first require pilot projects with socioeconomic evaluations in order to design the necessary policy and financial boundary conditions, clarify legal questions, and design professional training and educational programs.

3. JOINT IMPACT OF MATERIAL EFFICIENCY, SUBSTITUTION, AND INTENSIFIED PRODUCT USE AND PLANTS

Although the relation between these three elements of sustainable material use on energy demand has only been partially analyzed, the results of analytical technology-specific studies suggest substantial energy saving potentials. These potentials may materialize less quickly if policy action is not taken by governments, the scientific community, and the industrial players (Table I). It seems that the energy saving potential of the strategy of material efficiency may be most important during the next few decades but will have decreasing importance thereafter, whereas the potentials of material substitution and intensification of the use of products, vehicles, and production plants may require much additional research (e.g., on biomass-based plastics) and changes in social values and decision making (renting instead of owning) before they can substantially contribute to a reduced use of energy. However, the energy saving potentials of these two options may be quite high in the long term (i.e., over many decades during this century).

Preliminary estimates of the energy saving potentials for the next three decades have been made for an industrialized country such as Germany. The results showed approximately 450 PJ saved in the reference scenario, which is expected to occur without additional policy activities and is reflected in the assumptions on structural change in the economy (Table II). This energy saving potential corresponds to an annual decrease in the total primary energy demand of Germany of 0.12% per year. In the policy scenario, called the sustainability scenario, the energy saving potential was assumed to double. Approximately half the savings are estimated to come from improvements in road vehicles [material efficiency (light vehicles) and car sharing/truck leasing]. Other major savings are expected to result from the lighter construction of machinery and plants and the substitution of primary aluminum, which is a light but energy-intensive material.

There are indications that recycling seems to play a smaller role compared to material efficiency and material substitution mainly due to the savings in vehicles and moving parts of machinery and to the substitution of primary aluminum. The contribution of the intensified use of products, vehicles, and production plants is certainly underestimated, but analytical and empirical work still has to be done in many product areas to obtain reliable estimates of energy savings in this area of activity and entrepreneurial innovations. The results do not reflect changes in exports and imports of energy-intensive products, which may affect the saving potentials at the national level of an economy. Finally, the integration of the four options cited in Table II is highly dependent on socioeconomic conditions; hence, the joint effect on energy savings will depend on the constellation of the four options.

There is no doubt that the energy saving potentials will very much depend on future progress in the R&D of improved properties of existing materials, such as steel, nonferrous metals, cement, glass, plastics, and fibers, as well as the invention, market introduction, and diffusion of new materials (e.g., plastics or fibers based on biomass that may be genetically designed or modified). In addition, the intensified use of products, vehicles, and production plants will be dependent on changes in social values and entrepreneurial innovations that must

TABLE I

Estimates of Annual Primary Energy Saving Potential from Material Efficiency, Substitution, and Intensified Use of Products, Vehicles, and Production Facilities

Energy-related material aspect	Annual primary energy saving potential (%)		Comment
	Autonomous development	Policy-induced development	
More efficient use of materials	0.1	Up to 0.25	Relatively well analyzed, but still unknown potentials
Substitution of materials	<0.1	Up to 0.20	Open to major technical breakthroughs (biomass, nanotechnology)
Intensified use of durables, vehicles, etc.	<0.02	Up to 0.20	Unknown potential due to changes in societal values
Total effect	<0.2	Up to 0.5 or 0.6	

TABLE II

Impact of Material Recycling, Efficiency, Substitution, and Intensified Use of Products and Production Facilities on Primary Energy Demand: Reference and Sustainability Scenario for Germany, 2000–2030[a]

Material/group of product	Additional saving potential of primary energy until 2030, reference and sustainability scenario (PJ)									
	Intensified recycling of waste materials		Improved material efficiency		Material substitution		Intensified use of products, vehicles, etc.		Total	
	Ref.	Sustainability	Ref.	Sustainability	Ref.	Sustainability	Ref.	Sustainability	Ref.	Sustainability
Steel	30	60	5	30			n.a.	n.a.	35	60
Aluminum	23	43			100[b]	240[b]	n.a.	n.a.	83	163
Plastics	30	90			10	20	n.a.	n.a.	40	110
Cement/concrete, bricks			n.a.	n.a.	2	10	0	0	2	10
Bitumen, asphalt	5	11			n.a.	n.a.	0	0	5	11
Glass	5	10	5	25	5	10			5	45
Paper	11	40	11	55	10	50			32	145
Road vehicles										
Production	n.a.	n.a.	30	50	n.a.	n.a.	n.a.	n.a.	30	50
Use	n.a.	n.a.	250[b]	500[b]			45	110	170	360
Construction	n.a.	n.a.	80	120	n.a.	n.a.	n.a.	n.a.	80	120
Machinery/plants, office and home appliances	n.a.	n.a.			n.a.	n.a.				
Dwellings	0	0	n.a.	n.a.	n.a.	n.a.	4	10	4	10
Total	104	254	256	530	77	210	49	120		
Weighted by 0.9	94	230	230	480	70	190	45	110	430	1020

[a] *Sources.* Enquête Commission (2002); and Schön *et al.* (2003). Ref, reference; n.a., not applicable.
[b] In total, only 50% of the value considered.

occur to realize the potentials expected in the coming decades.

The following conclusions can be made with regard to the preliminary findings of the relationship between the various options of sustainable material use and energy demand:

- The impact of material efficiency, material substitution, and the intensified use of durables has not been explicitly treated as an energy policy option in the past, nor has it been explicitly analyzed in energy demand projections. It has only been treated as an unexplained aggregate of structural change.
- If consciously adopted as a policy of sustainable development by interested companies in industry and the new service sectors, the impact of the three options for energy demand may be significant—an approximately 0.5% reduction in annual primary energy demand in the coming decades.
- The obstacles to these potentials are manifold, but the material and energy savings provide opportunities to recover the R&D expenses and additional investments; the value added in the new materials and systems may also be an incentive for companies as well as for governments.

The close relationship between material use and energy demand opens up an entirely new area of energy systems analysis and, more important, offers major entrepreneurial opportunities for a lighter, more efficient, and less polluting industrial society. It opens up new jobs in many service areas by substituting the use of natural resources by know-how-intensive products and services. These effects have been studied and calculated by Walz and colleagues for Germany, with the conclusion that more efficient use of material and capital stock is likely to contribute to a sustainable development in the long term.

SEE ALSO THE FOLLOWING ARTICLES

Cement and Energy • Consumption, Energy, and the Environment • Economics of Energy Efficiency • Energy Efficiency and Climate Change • Energy Efficiency, Taxonomic Overview • Industrial Energy Efficiency • Material Use in Automobiles • Obstacles to Energy Efficiency • Plastics Production and Energy • Recycling of Metals • Recycling of Paper • Steel Production and Energy

Further Reading

Ayres, R. U. (2001, November). The energy we overlook. *World Watch* **14**(6), 30–39.

Bernadini, O., and Galli, R. (1993). Dematerialization: Long-term trends in the intensity of use of materials and energy. *Futures* **25**(5), 431–448.

Biondi, L., and Galli, R. (1992). Technological trajectories. *Futures* **24**(7), 580ff.

Bringezu, St. (2000). "Use of Resources in Economic Regions [in German]." Springer, Berlin.

Enquête Commission (2002). Sustainable energy supply under the conditions of globalization and liberalization, Final report, No. BtgsDs. 14/0400, pp. 311–327. German Bundestag, Berlin.

Flaig, F. (ed.). (2000). "Sustainable Cycling Economy [in German]." Schäffer-Poeschel, Stuttgart.

Friend, G. (1994). The end of ownership? Leasing, licensing, and environmental quality. *New Bottom Line* **3**(11).

Gielen, D. J. Materialising dematerialisation: Integrated energy and materials system engineering for greenhouse gas emission mitigation. Thesis. Delft Technical University. The Netherlands

Hekkert, M. P., Joosten, L. A. J., Worrell, E., and Turkenburg, W. C. (2000). Reduction of CO_2 emissions by improved management of material and product use: the case of primary packaging. *Resources, Conservation and Recycling* **29** Issues 1–2, 33–64.

Hekkert, M. P., van den Reek, J., Worrell, E., and Turkenburg, W. C. (2002). The impact of material efficient end-use technologies on paper use and carbon emissions. *Resour. Conserv. Recycling* **36**(3), 241–266.

Hüsing, B., Angerer, G., Gaisser, S., and Marscheider-Weidemann, F. (2002). "Bio-technological Production of Basic Materials under Specific Aspects of Energy and Bio-polymers from Residues [in German]." Federal Environmental Protection Agency, Berlin.

Jochem, E., *et al.* (2000). End-use energy efficiency. *In* "World Energy Assessment" (J. Goldemberg and Th. Johannson, Eds.), pp. 173–217. UNDP/World Energy Council/UNDESA, New York.

Jochem, E., Favrat, D., Hungerbühler, K., Rudolph, V., Rohr, P., Spreng, D., Wokaun, A., and Zimmermann, M. (2002). "Steps Towards a 2000 Watt Society. Developing a White Paper on Research and Development of Energy-Efficient Technologies." CEPE/ETH, Zurich.

Maijkamp, R. (1998). Changing consumer behaviour through eco-efficient services: An empirical study on car sharing in The Netherlands. *Business Strategy Environ.* **7**, 234–244.

Malenbaum, W. (1978). "World Demand for Raw Materials in 1985 and 2000." McGraw-Hill, New York.

Mont, O. (2000). "Product-service systems. Shifting corporate focus from selling products to selling product-services: A new approach to sustainable development, AFR Report No. 288." Natur Vards Verket, Stockholm, Sweden.

Nakicenovic, N., Grübler, A., and MacDonald, A. (1998). "Global Energy Perspectives." Cambridge Univ. Press, Cambridge, UK.

Popov, F., and DeSimone, D. (1997). "Eco Efficiency—The Business Link to Sustainable Development." MIT Press, Cambridge, MA.

Rogers, W. M. (2000). "Third Millennium Capitalism: Convergence of Economic, Energy, and Environmental Forces." Quorum, Westport, CT.

Schön, M., *et al.* (2003). "System Analysis of Recycling Potentials of Energy-Intensive Materials and Their Impact on Efficient Energy Use [in German]." Fh-ISI, Karlsruhe, Germany.

Walz, R., *et al.* (2001). "The World of Labour in a Sustainable Economy—Analysis of the Impact of Environmental Strategies on the Economy and Labour Structures [in German]." Federal Environmental Protection Agency, Berlin.

Worrell, E., Meuleman, B., and Blok, K. (1995a). Energy savings by efficient application of fertilizer. *Resour. Conserv. Recycling* **13**(3/4), 233–250.

Worrell, E., Faaij, A. P. C., Phylipsen, G. J. M., and Blok, K. (1995b). An approach for analysing the potential for material efficiency improvement. *Resour. Conserv. Recycling* **13**(3/4), 215–232.

Zanger, C., Drengner, J., and Gaus, H. (1999). Consumer acceptance of increased life times of products and intensified use of products [in German]. *UWF* **1**, 92–96.

Materials for Solar Energy

CLAES G. GRANQVIST
Uppsala University
Uppsala, Sweden

1. Solar Energy Materials
2. Fundamental Optical Properties of Materials
3. Transmitting and Reflecting Materials
4. Thin Films
5. Transparent Thermal Insulators
6. Solar Thermal Converters
7. Radiation Cooling
8. Solar Cells
9. Coatings for Glazings: Static Properties
10. Coatings for Glazings: Dynamic Properties
11. Solar Photocatalysis

Glossary

cermet A mixture of ceramic and metal.
chromogenic Pertaining to the ability to change optical properties persistently and reversibly by the action of an external stimulus.
electrochromic Pertaining to the ability to change optical properties reversibly and persistently by the action of a voltage.
float glass Solidified glass floated on a bed of molten metal so that the glass surface is nearly atomically smooth.
infrared Referring to wavelengths of light longer than 0.7 μm.
iridescence A rainbowlike exhibition of color.
photocatalysis Chemical reactions on an "active" surface, stimulated by light.
photochromic Pertaining to the ability to change optical properties reversibly and persistently by the action of light irradiation.
Planck's law A relationship describing the spectral distribution of thermally emitted radiation.
smart window A window incorporating a chromogenic material.
spectrally selective Having distinctly different optical properties in different wavelength ranges.
thermochromic Pertaining to the ability to change optical properties reversibly and persistently by the action of temperature.
ultraviolet Referring to wavelengths of light shorter than 0.4 μm.

This article summarizes work on materials with properties tailored to meet the specific parameters of the spectral distribution, angle of incidence, and intensity of the electromagnetic radiation prevailing in our surroundings. Spectral and angular selectivity are introduced to specify many of the desired properties. Materials with optimized transmittance, reflectance, absorptance, and emittance are discussed. A general introduction of the properties of glasses, polymers, and metals is followed by an account of thin film technology, which is of importance for solar energy materials of many different kinds. The article treats materials for transparent thermal insulation, solar thermal converters, radiative cooling devices, solar cells, architectural windows with static as well as dynamic properties, and solar photo-catalysis. The aim is to introduce basic concepts and commonly used materials, and also to bring the exposition up to the work in today's scientific laboratories.

1. SOLAR ENERGY MATERIALS

Solar energy materials have properties tailored to meet the specific parameters of the spectral distribution, angle of incidence, and intensity of the electromagnetic radiation prevailing in our natural surroundings. Materials for thermal and electrical conversion in man-made collectors, as well as for energy-efficient passive design in architecture, are typical examples.

Figure 1 introduces the ambient electromagnetic radiation of Earth's environment in a unified manner. The most fundamental property of this radiation, which ensues from the fact that all matter emits electromagnetic radiation, is conveniently introduced

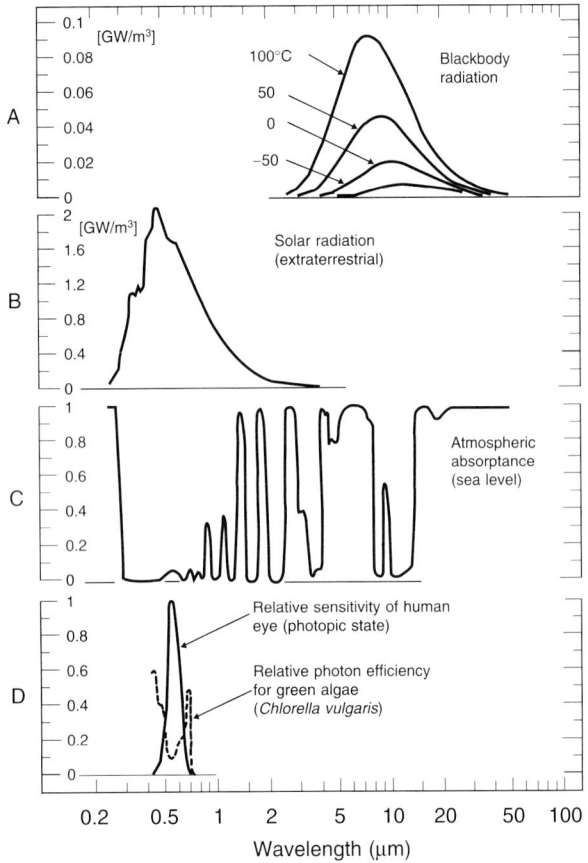

FIGURE 1 Spectra for (A) black-body radiation pertaining to four temperatures, (B) solar radiation outside Earth's atmosphere, (C) typical absorptance across the full atmospheric envelope, and (D) relative sensitivity of the human eye and relative photosynthetic efficiency for green algae.

by starting with the ideal blackbody. The emitted spectrum–known as the Planck spectrum–of blackbody radiation is uniquely defined if the temperature is known. Planck's law is a consequence of the quantum nature of the radiation. Such spectra are depicted in Fig. 1A for four temperatures. The vertical scale denotes power per unit area and wavelength increment (hence the unit GW/m^3). The spectra are bell-shaped and confined to the $2<\lambda<100$-µm wavelength range. The peak in the spectrum is displaced toward shorter wavelengths as the temperature increases. At room temperature, the peak lies at about 10 µm. Thermal radiation from a material is obtained by multiplying the Planck spectrum by a numerical factor (the emittance), which is less than unity. In general, the emittance is wavelength dependent.

A solar spectrum is depicted in Fig. 1B for radiation just outside Earth's atmosphere. The curve has a bell shape, defining the sun's surface temperature ($\sim 6000°C$). It is important to observe that the solar spectrum is limited to $0.25<\lambda<3$ µm, so that there is almost no overlap with the spectra for thermal radiation. Hence, it is possible to have surfaces with properties that are entirely different with regard to thermal and solar radiation. The integrated area under the curve gives the solar constant (1353 ± 21 W/m^2); this is the largest possible power density on a surface oriented perpendicular to the sun in the absence of atmospheric extinction.

Most solar energy conversion systems are located at ground level, and it is of obvious interest to consider to what extent atmospheric absorption influences solar irradiation and net thermal emission. Figure 1C illustrates a typical absorption spectrum vertically across the full atmospheric envelope under clear weather conditions. The spectrum is complicated with bands of high absorption–caused mainly by water vapor, carbon dioxide, and ozone–and bands of high transparency. It is evident that most of the solar energy can be transmitted down to ground level and only parts of the ultraviolet ($\lambda<0.4$ µm) and infrared ($\lambda>0.7$ µm) radiation are strongly damped. The maximum power density perpendicular to the sun is about 1000 W/m^2. Thermal radiation from a surface exposed to the clear sky is strongly absorbed except in the $8<\lambda<13$-µm range, where the transmittance can be large, provided that the humidity is moderately low.

Two biological conditions of relevance for solar-energy-related applications are illustrated in Fig. 1D. The solid curve shows the relative sensitivity of the human eye in its light-adapted state. A bell-shaped curve extends across the $0.4<\lambda<0.7$-µm interval, with its peak at 0.555 µm. Clearly, a large part of the solar energy comes as invisible infrared radiation. The dashed curve indicates that photosynthesis in plants operates at wavelengths in approximately the same range as those for the human eye, which is relevant for greenhouse applications.

Figure 1 depicts an important fact about ambient radiation: that it is spectrally selective, i.e., confined to specific and usually well-defined wavelength ranges. This property is of great importance for most types of solar energy materials. Another type of selectivity–known as angular selectivity–results from the fact that different angles may apply for different types of radiation; for example, solar radiation comes from a point far from the horizon during most of the day, whereas the visual contact between a person and his or her surroundings is often at near-horizontal lines-of-sight.

2. FUNDAMENTAL OPTICAL PROPERTIES OF MATERIALS

When electromagnetic radiation impinges on a material, one fraction is transmitted, a second fraction is reflected, and a third fraction is absorbed. Energy conservation yields, at each wavelength, that

$$T(\lambda) + R(\lambda) + A(\lambda) = 1, \qquad (1)$$

where T, R, and A denote transmittance, reflectance, and absorptance, respectively.

Another fundamental relationship, also following from energy conservation and referred to as Kirchhoff's law, is

$$A(\lambda) = E(\lambda) \qquad (2)$$

with E being the emittance, i.e., the fraction of blackbody radiation (cf. Fig. 1A) that is emitted at wavelength λ. It is evident that Eq. (2) is of relevance mainly for $\lambda > 3\,\mu m$.

It is frequently convenient to average the spectral data over the sensitivity of the eye, a solar spectrum (with or without accounting for atmospheric effects), or a blackbody emittance spectrum (for a certain temperature). Luminous, solar, or thermal values pertinent to the respective optical properties are then obtained; these are denoted X_{lum}, X_{sol}, and X_{therm}, with X being T, R, A, or E.

3. TRANSMITTING AND REFLECTING MATERIALS

3.1 Glasses

Glass can be used for protection against an unwanted environmental impact, for convection suppression, and as a substrate for surface coatings (normally referred to as thin films). Float glass, a highly standardized product used in a great majority of windows in buildings, is characterized by uniformity and flatness almost on the atomic scale. The latter feature is an outcome of the production process in which the molten glass is solidified on a surface of molten tin. The middle curve in Fig. 2 shows the spectral transmittance of 6-mm-thick standard float glass within the wavelength range relevant to solar radiation. A characteristic absorption feature at $\lambda \approx 1\,\mu m$, due to the presence of Fe oxide, limits T_{sol} to some extent. Special float glass is available with varying amounts of Fe oxide. Fig. 2 shows that low-Fe content leads to glass with very high T_{lum} and T_{sol} values, and also with substantial transmittance in the ultraviolet region.

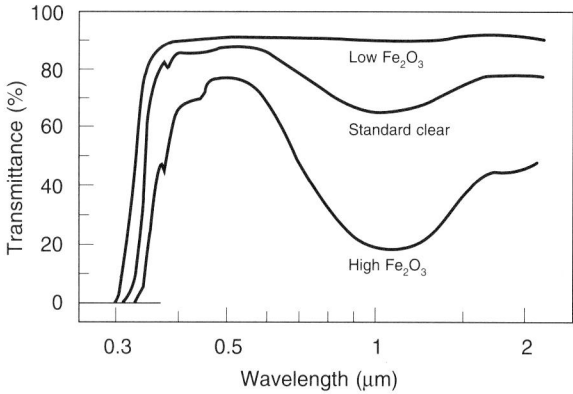

FIGURE 2 Spectral transmittance for float glass with three different amounts of Fe_2O_3.

Glass with a large Fe content limits T_{sol} whereas T_{lum} remains rather large; such glass has a distinctive greenish tint. The reflectance of each interface between glass and air is about 4% in the $0.4 < \lambda < 2\,\mu m$ range, implying that the maximum transmittance for a glass pane is 92%. Glass is strongly absorbing for $\lambda > 3\,\mu m$, and E_{therm} is as large as about 85%.

Laminated glass may be used for safety and other reasons. This glass comprises a layer of polyvinyl butyral sandwiched between two glass panes and bonded under heat and pressure. The laminate is an efficient absorber of ultraviolet light, and the transmittance at $\lambda < 0.38\,\mu m$ is almost zero for a layer thickness of the order of 1 mm. Photochromic glass is able to darken under irradiation of ultraviolet light from the sun and clears in the absence of such irradiation. Fatigue-free photosensitivity is normally accomplished by adding metal halides–particularly silver halide–to the vitreous matrix of the glass. A number of other additives are important as well. Figure 3 shows T_{lum} as a function of time for darkening and clearing of two photochromic glass plates. Darkening progresses rapidly and reaches about 80% after 1 minute. Clearing is much slower, though, and is still incomplete after 1 hour. The photosensitivity occurs almost entirely in the $0.4 < \lambda < 1.0\,\mu m$ interval, and hence the modulation of T_{lum} is much more pronounced compared to the modulation of T_{sol}. Photosensitivity exerts no effect on the reflectance or on E_{therm}.

3.2 Polymers

Many polymers can be transparent and can, in principle, replace glass. However, polymers degrade more easily than glass and hence have fewer

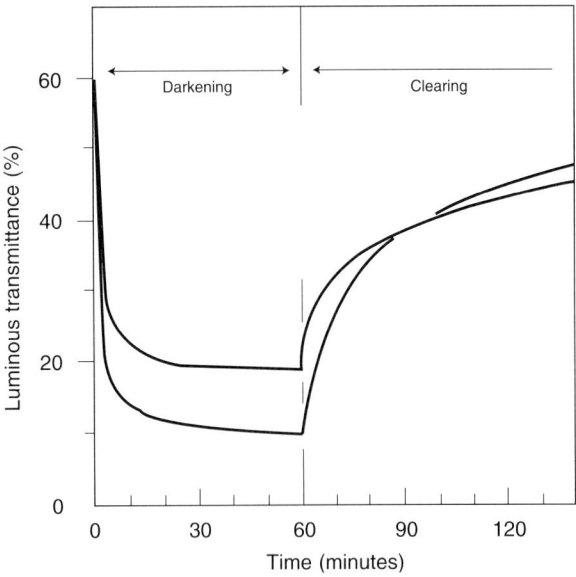

FIGURE 3 Luminous transmittance vs time for two photochromic glasses.

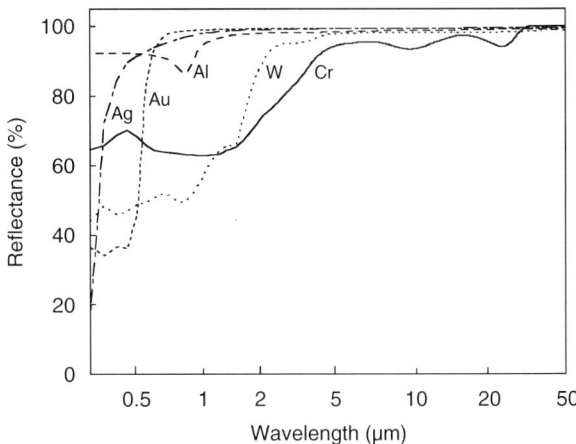

FIGURE 4 Spectral reflectance for some metal surfaces.

applications. Polyester foil deserves special mention because it can serve as a flexible and inexpensive substrate for thin films of many different kinds, transparent as well as reflecting. Another alternative is ethylene tetrafluoroethylene. Photochromism is known in a number of polymers, and detailed information is available for spirooxazine in a host of cellulose acetate butyrate. The coloration dynamics are similar to those for photochromic glass, but the bleaching is faster for the polymer and takes place in about 20 minutes. The photo-effects are limited to T_{lum}; they vanish at elevated temperatures.

3.3 Metals

Metal surfaces can be used to accomplish high R_{lum} and R_{sol} (for example, in mirrors of various types), or for providing low E_{therm}. Figure 4 shows spectral reflectance for five different metals. The highest values of R_{lum} and R_{sol} are found in silver (Ag) and aluminum (Al), and these metals are used in bulk form or as thin films supported by substrates in several solar-energy-related applications. Surface protection of such surfaces is normally needed to obtain long-term durability. More corrosion-resistant metals, such as platinum or rhodium, are less reflective. All of the metals can yield a low magnitude of E_{therm}.

4. THIN FILMS

4.1 Thin Film Deposition

Surface treatments can be used to modify the optical properties of an underlying material and thus enhance its ability to adapt to the conditions of the ambient radiation (cf. Fig. 1). Many specific types of solar energy materials are used in this way. Thin film deposition, a particularly important application, represents a vast technological field, and only the most common methods are mentioned here.

Sputter deposition is widely used to make uniform coatings on glass, polyester, metal, etc. Typical film thicknesses are 0.1 to 1 μm. Essentially, a plasma of inert and/or reactive gases is set up under low pressure, and energetic ions in the plasma dislodge atoms from a solid plate of the raw material of the film (known as the "target"); these atoms are deposited as a uniform film on an adjacent surface (the substrate).

Evaporation can be an alternative to sputter deposition. Here, the raw material of the film is heated in vacuum so that a condensing vapor can transfer material to the substrate at a sufficient rate. Other technologies can be applied without recourse to vacuum: For example, sol-gel deposition involves dipping the substrate in a chemical solution, withdrawing it at a controlled rate, and subsequent annealing. Alternatively, the chemical solution can be applied by spray coating.

Chemical vapor deposition uses heat to decompose a vapor of a "precursor" chemical to produce a thin film of a desired composition. Electrochemical techniques include cathodic electrodeposition from a chemical solution and anodic conversion of a metallic surface to form a porous oxide. Numerous alternative techniques exist.

4.2 Antireflection Treatment

Antireflection treatment (for example, to obtain high T_{lum} through glass) can be accomplished by the application of a layer with a refractive index that is close to the square root of the refractive index of the underlying glass (i.e., 1.23) and with a thickness corresponding to one-quarter of the wavelength of visible light (i.e., ~100 nm). The condition imposed on the refractive index allows for good durability for only a few materials. Figure 5 shows an example of a glass slab of which two surfaces have been coated with a layer of porous silica obtained via sol-gel technology. It is seen that T_{lum} exceeds 99% in the middle of the luminous spectrum. Other antireflection treatments of practical interest include thin films of aluminum oxyfluoride or magnesium fluoride, and porous or microstructured surface layers produced by etching the glass surface in fluorosilicic acid or by application of a submicrometer-structured layer.

Simple antireflection treatments are able to lower the reflectance within a rather narrow wavelength range, but if a low magnitude of R_{sol} is required, a multilayer film with carefully adjusted refractive indices and thicknesses could be used. It should be noted that antireflection is efficient only for a specific range of incidence angles, especially in the case of multilayer films.

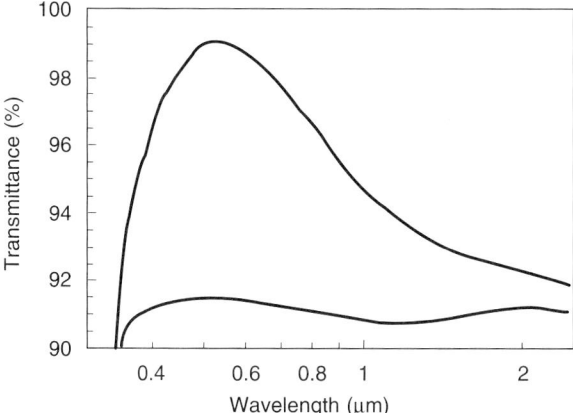

FIGURE 5 Spectral transmittance of a glass plate coated on both sides by antireflecting thin films of porous silica.

5. TRANSPARENT THERMAL INSULATORS

5.1 Principles

Heat transfer occurs via the additive effects of radiation, convection, and conduction. Cutting back this transfer is of concern for many different applications of solar energy. Radiation transfer can be diminished by having surfaces with low values of E_{therm}. Convection, which is of interest for gases, can be lowered by breaking up the gas into cells with dimensions of the order of 1 cm or less (or, possibly, by decreasing the gas pressure), and conduction heat transfer can be minimized by suitable constructions using low-conductance materials.

5.2 Solid Materials

Figure 6 shows four types of solid transparent insulation materials, including flexible polymer foils; polymer honeycomb materials; bubbles, foam, and fibers; and inorganic microporous materials such as silica aerogels. If the honeycomb cross-section is small compared to the cell length, a capillary structure exists. Foils and aerogels can be almost invisible to the eye, whereas honeycombs, bubbles, foam, and fibers cause reflection and scattering and hence a limited direct transmittance; thus, only the former group of materials is of interest when unperturbed vision is required–such as in many windows–and the other materials find applications in translucent wall claddings and solar collectors. The materials can be used in vertical, horizontal, or inclined positions, depending on the type of application.

Flexible polyester foils can be suspended to break up a thick gas slab into ~1-cm-thick layers with

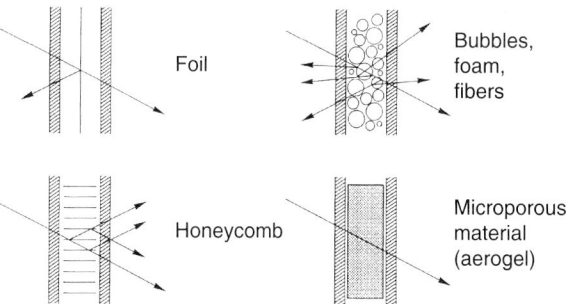

FIGURE 6 Principles underlying performance of four different types of solid transparent insulation materials placed between glass panes. Arrows signify light rays. Reflections at the glass surfaces are not shown.

diminished convection, and thin films can be applied to adjust the magnitudes of T_{lum} and T_{sol}, and to minimize E_{therm}. Similarly, the macroporous materials–typically made of polystyrol, polyamide, polyvinylchloride, or polycarbonate–produce gas-filled cells of a sufficiently small size to practically eliminate convection. A characteristic value of the heat transfer of the honeycombs, capillaries, bubbles, foam, and fibers is $1 \text{ W/m}^2 \text{K}^{-1}$ for a material thickness of 10 cm.

Silica aerogel can be made by supercritical drying of colloidal silica gel under high temperature and high pressure. The ensuing material consists of ~ 1-nm-diameter silica particles interconnected so that a loosely packed structure with pore sizes of 1 to 100 nm is formed. The relative density can be as low as a few percent of the bulk density. The material can be produced as translucent granules or as transparent tiles with a certain degree of haze. Figure 7 shows spectral transmittance through a 4-mm-thick tile. The low transmittance at short wavelengths is due to optical scattering. Heat transfer lower than $1 \text{ W/m}^2 \text{K}^{-1}$ can be achieved in 1-cm-thick slabs of aerogel under favorable conditions (positioning between surfaces with low E_{therm} and at low gas pressure).

5.3 Gases

The heat transfer between two surfaces can be lowered by use of a gas other than air, as is common in modern fenestration technology involving hermetically sealed glazing units. The heat transfer across a typical 1.2-cm-wide gas slab is diminished on the $\sim 10\%$ level if air is replaced by argon or CO_2, whereas larger effects are found with krypton or xenon. The relative role of the gas filling is enlarged (in practice, to the $\sim 20\%$ level) if the radiation component to the heat transfer is decreased by the use of surfaces with a low E_{therm}. Infrared-absorbing gases can diminish the heat transfer across a given distance, but the limited availability of suitable gases makes this possibility less interesting. Among other options related to gases, some improvement of the thermal properties can be combined with sound insulation by the use of SF_6 gas. Finally, minimized heat transfer can be achieved by eliminating the gas, so that both the conduction and the convection components vanish. Obviously, the mechanical strength of the device must be sufficient to allow a vacuum. This necessitates inconspicuous spacers for "vacuum windows."

6. SOLAR THERMAL CONVERTERS

6.1 Principles

Photothermal conversion of solar energy, to produce hot fluid or hot air, takes place in solar collectors. Efficient conversion requires that a solar-absorbing surface is in thermal contact with the fluid or gas, and that thermal losses to the environment are minimized. The absorber should be positioned under a transparent cover, and a transparent thermal insulation material may be used as well. The back and sides of the solar collector should be well insulated using conventional techniques. The most critical part for the photothermal conversion is the actual solar absorber surface, which obviously should have maximum A_{sol}. Its radiation heat losses should be minimized, and this requires low E_{therm}. The absorber is nontransparent, and the conditions can be written in terms of an ideal spectral reflectance as

$$R(\lambda) = 0 \text{ for } 0.3 < \lambda < \lambda_c \text{ μm} \quad (3)$$

$$R(\lambda) = 1 \text{ for } \lambda_c < \lambda < 50 \text{ μm}, \quad (4)$$

where λ_c denotes a "critical" wavelength lying between 2 and 3 μm, depending on the temperature (cf. Fig. 1). Clearly, this is one example of how spectral selectivity is used to obtain optimized properties of a solar energy material.

A number of different design principles and physical mechanisms can be used in order to create a spectrally selective solar-absorbing surface. Six of these are shown schematically in Fig. 8. The most straightforward approach would be to use a material with suitable intrinsic optical properties. There are

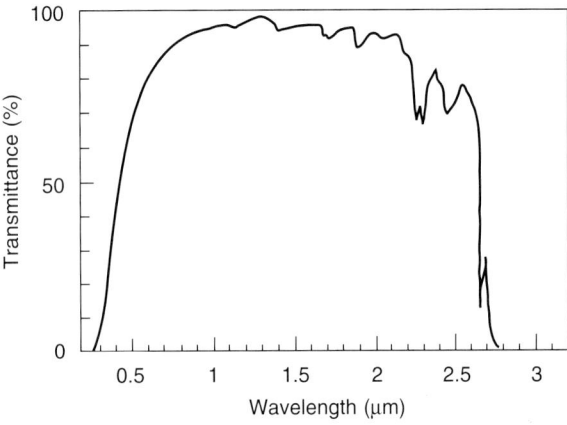

FIGURE 7 Spectral transmittance of a silica aerogel tile.

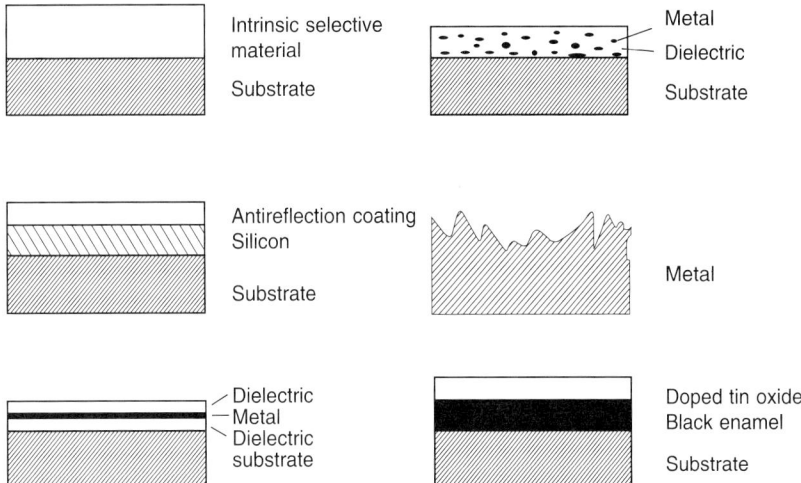

FIGURE 8 Cross-sectional schematic designs of six types of thin films and surface treatments for selective absorption of solar energy.

few such materials, though, and this approach has not been particularly fruitful. Semiconductor-metal tandems can give the desired spectral selectivity by absorbing short-wavelength radiation in a semiconductor with a bandgap corresponding to a wavelength of ~2 µm and having low E_{therm} as a result of the underlying metal. The pertinent semiconductors all have large refractive indices, which tend to give strong reflection losses and lower A_{sol}, implying that complex antireflection treatments must be invoked. Multilayer-metal tandems can be tailored so that the thin film stack becomes an effective selective absorber of solar energy. Metal/dielectric composite-metal tandems contain metal nanoparticles in a dielectric (often oxide) host. The composite is often referred to as a cermet. This design principle offers a large degree of flexibility, and the optimization of the spectral selectivity can be made with regard to the choice of constituent materials, film thickness, particle concentration, and grading, as well as the shape and orientation of the particles. The particles are much smaller than any relevant wavelengths of solar (or thermal) energy, and the composite behaves as an "effective medium" with properties intermediate between those of the metal and the dielectric. Textured metal surfaces can produce a high A_{sol} by multiple reflections against metal dendrites that are ~2 µm apart, but the low E_{therm} is rather unaffected by this treatment because the relevant wavelengths are much larger that the dendrite separation. Finally, selectively solar-transmitting films on blackbody-like absorbers can be used to impart spectral selectivity. The relevant thin films are discussed in the following section.

6.2 Spectrally Selective Thin Films

A number of thin films, coated onto metallic substrates by one of the thin film technologies discussed in the preceding section, have been developed in the past, and some of these have reached commercialization. These films normally exploit several of the design principles and physical mechanisms mentioned previously. Electrochemical techniques have been used for several decades to make spectrally selective surfaces (suitable for flat-plate solar collectors) based on electrodeposited films containing chrome or nickel (referred to as black chrome and black nickel, respectively) of complex compositions. Another technique employs anodic conversion of aluminum to make a porous surface layer, followed by electrolysis to precipitate out nickel particles inside the pores. The solid curve in Fig. 9 illustrates a spectral reflectance curve. $A_{sol} \approx 96\%$ and $E_{therm} \approx 15\%$ at 100°C are typical performance data. The durability of these films has been modeled in detail.

The electrochemical techniques involve the handling of large quantities of environmentally hazardous chemicals, and this has led to a current trend toward thin film techniques based on deposition in vacuum. The dotted curve in Fig. 9 shows a spectrum for an optimized sputter-deposited film based on nickel (Ni); the design includes an antireflection layer that boosts A_{sol}. The data correspond to $A_{sol} \approx 95\%$ and $E_{therm} \approx 10\%$ at 100°C. Other sputter-deposited and evaporated films of practical use include a number of cermets with stainless-steel and metal nitride, and compositionally graded films containing chrome oxide and chrome nitride. These films can be used

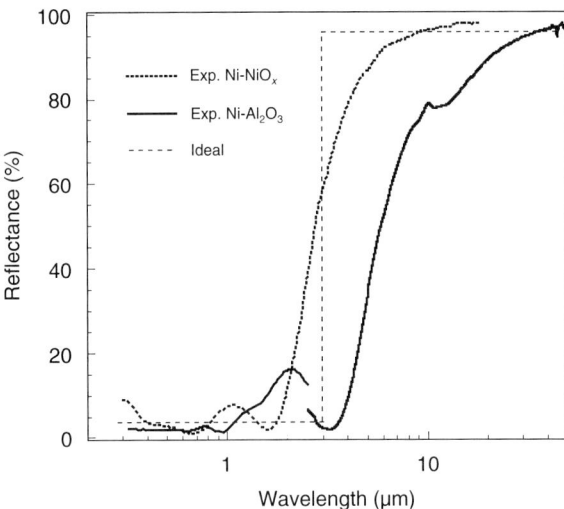

FIGURE 9 Experimental spectral reflectance results for two types of selectively solar-absorbing surfaces, compared to the ideal performance.

for different types of solar collectors, including concentrating ones. Advanced selectively solar-absorbing surfaces have been developed for solar thermal power stations with parabolic trough reflectors and tubular solar collectors. The absorber surfaces used in this case employ sputter deposition to manufacture a complex multilayer film stack incorporating graded molybdenum (Mo)/oxide films, antireflection treatment, and an emittance-suppressing Mo film on the stainless-steel tube serving as a substrate. These absorbers have $A_{sol} \approx 97\%$ and $E_{therm} \approx 17\%$ at 350°C.

6.3 Spectrally Selective Paints

Paints have an obvious advantage in that they can be applied by commonly available techniques. Typically, a strongly absorbing pigment (for example, an oxide of FeMnCu) is mixed in an infrared-transparent polymeric binder such as a silicone or a siloxane. The paint is applied in a layer 2 to 3 μm thick onto a metallic substrate with low E_{therm}. Characteristic data are $A_{sol} \approx 90\%$ and $E_{therm} \approx 30\%$; these data are not as good as those for the thin films. Possibilities to prepare thickness-insensitive spectrally selective paints by including metallic flakes in the binder may lead to interesting products in the future. Visibly colored solar-absorbing surfaces, which may be desirable for aesthetic reasons, can be obtained by relaxing the requirement on A_{sol} somewhat.

7. RADIATION COOLING

As was shown in Fig. 1, the atmosphere can be transparent to thermal radiation in the $8 < \lambda < 13$-μm "window" range. This effect can lead to radiation cooling of surfaces oriented toward the sky. The resource for this cooling has been evaluated through detailed calculations showing that the cooling power for a blackbody surface is 80 to 120 W/m², depending on the atmospheric conditions, when the surface is at ambient temperature. If nothing but radiation exchange is considered, the surface can drop to 14–26°C below the air temperature. In the presence of a nonradiative heat transfer equal to 1 W/m² K⁻¹, the maximum temperature drop is expected to be 10–20°C. Practical tests, using a multistage radiation cooler placed under an exceptionally dry atmosphere, have demonstrated that a temperature decrease as large as ~33°C can be achieved in practice by purely passive means.

The temperature drop can be enhanced if radiation exchange takes place only within the atmospheric "window." This calls for surfaces with infrared selective emittance, with an ideal reflectance of

$$R(\lambda) = 1 \text{ for } 3 < \lambda < 8 \text{ μm and } 13 < \lambda < 50 \text{ μm} \quad (5)$$

$$R(\lambda) = 0 \text{ for } 8 < \lambda < 13 \text{ μm}. \quad (6)$$

If cooling should take place also in the day, it is desired to have $R(\lambda) = 1$ for $0.3 < \lambda < 3$ μm. Calculations have shown that infrared-selective surfaces can have a cooling power between 60 and 95 W/m², depending on the atmosphere. Maximum temperature drops of 26–60°C with nothing but radiation exchange, and of 18–34°C when the nonradiative heat influx is 1 W/m² K⁻¹, have been predicted. The significance of spectral selectivity for reaching low temperatures is hence evident. Spectral selectivity approximately according to Eqs. (5) and (6) has been achieved with ~1-μm-thick silicon oxynitride films backed by aluminum and by some aluminized polymer foils, including Tedlar. Another possibility is to use confined gas slabs of NH_3, C_2H_4, C_2H_4O, or mixtures of these, with thicknesses of a few centimeters. In both cases, the low reflectance in the atmospheric "window" range is associated with molecular vibrations, and the high reflectance outside this range is caused by the aluminum.

Practical tests of the cooling ability of these materials have been carried out with devices in which the cooling surface, or the gas, is located under an infrared-transparent convection shield. High-density polyethylene foil has suitable optical properties but requires stiffening in order not to cause convection heating due to mechanical movements caused by wind. Using a radiation-cooled gas, the coolant can be circulated and heat exchanged so that it provides a continuous source for cooling under the clear sky. Radiation cooling can be used for condensation of water from the atmosphere. Preliminary measurements have demonstrated that significant amounts can be collected even under unfavorable conditions (during drought months in a semidesert part of Tanzania). The material used in this case was a 0.4-mm-thick polymer foil pigmented by white TiO_2 and $BaSO_4$.

8. SOLAR CELLS

Solar cells for generating electricity depend on certain materials. In principle, all that is needed to generate electricity is a carrier in an excited state or states; a carrier can be excited from a ground state to an excited state by photon absorption, and to generate electricity requires some means of extracting or supplying carriers to or from these states. There are many options but, despite much research and development over decades, the attention has been focused almost entirely on a few materials.

Silicon (Si) is an excellent material for solar cells and is by far the most widely used one. It is nontoxic, very abundant, and has a well-established technological base due to its ubiquitous uses in microelectronics. Essentially, a slab or film of Si is n-doped and p-doped in the two film surface regions, and this structure is positioned between a transparent front electrode arrangement and an opaque metallic back electrode. (Elements that serve as dopants are either p-type acceptors or n-type donors; adjoining n and p regions form a junction that passes current from p to n.) The Si can be of three types: crystalline (cut from a single crystal ingot), polycrystalline (made from a multicrystalline ingot or prepared by "ribbon growth"), or amorphous and hydrogenated (typically a ~0.3-μm-thick film made by glow discharge deposition of a silane gas). High photoelectric conversion efficiency demands low reflection losses, which can be achieved by an antireflecting film or–in the case of a single crystalline material–by using anisotropic etching to make a pyramid-type texturing.

Compound semiconductors of the III–V type (gallium, arsenic, aluminum, indium), based on GaAs, (Al,Ga)As, InP, or (In,Ga)P, can show high efficiency as well as good resistance to ionizing radiation; they are costly, though, and mainly used for space applications. Among the II–IV compounds (cadmium, tellurium), CdTe and Cd(S,Te) are known for their robust manufacturability and are of interest for ground-based solar cells, although environmental concerns regarding large-scale use of a technology including cadmium should not be overlooked. Among the I–III–IV$_2$ compounds (copper, indium, selenium), CuInSe$_2$, Cu(In,Ga)Se$_2$, and Cu(In,Ga) (S,Se)$_2$ are notable for their possibility to reach high photothermal conversion efficiencies in potentially low-cost thin film solar cells. Numerous other materials–including polymers–and approaches to solar cells are possible, and many other compound semiconductors may well turn out to have favorable properties. Some of those may find applications in thermophotoelectricity, i.e., cells operating with thermal irradiation. Others may be used in tandem cells, encompassing superimposed cells with optical bandgaps that are large on the exposed surface and gradually shrink toward the underlying substrate.

Alternatively, photoelectrochemical techniques can be employed to generate electricity, and studies of nanocrystalline dye cells are presently of interest. These cells absorb light in the dye molecules–normally containing a ruthenium-based compound–coated onto nanocrystals of TiO_2 made by colloidal technology or other means. Electrons are excited in the dye and can flow into the conduction band of the adjacent n-type TiO_2. The electron is transferred to a transparent front surface electrode, through the load, and to a counterelectrode having a thin layer of platinum. Here it reduces triiodine to iodide, which then diffuses to the dye molecules and reduces them back to their original state. Sealing and durability are issues that need to be resolved before widespread applicability is possible. Also, the fact that the fundamental principles of functionality remain obscure may hamper rapid progress.

Generally speaking, solar cells lose efficiency when they become warm. It is therefore of interest to devise cells and cell systems that minimize heating. One possibility is to use external surfaces that reflect the subbandgap wavelengths useful for photothermal conversion onto the solar cell while absorption prevails at longer wavelengths. Mirrors of SnO_2-coated aluminum have been developed for this purpose. One aspect of today's solar cell technology that requires particular attention, because it proliferates, is the global availability of the raw materials. Gallium, selenium, and ruthenium are all rare components of Earth's crust. Indium is also widely believed to be rare, but recent assessments indicate that availability is not a serious issue.

9. COATINGS FOR GLAZINGS: STATIC PROPERTIES

9.1 Principles

Architectural windows and glass facades are problematic from an energy perspective. On the one hand, their primary function, to provide visual contact between indoors and outdoors, as well as day lighting, should not be compromised. On the other hand, there are often unwanted energy flows, with too much thermal energy leaving or entering the building via the window and a concomitant demand for space heating and cooling; furthermore excessive solar energy may be admitted, which puts additional demand on air conditioning. Present architectural trends tend to increase the window areas, so that the energy issue becomes even more pressing.

The radiation part of the heat transfer can be controlled by thin films with a low E_{therm}, which, in a multiply glazed window, should face one of the confined air (or gas) spaces. In this way, the heat transfer across a vertically mounted window can drop from ~ 3 to $\sim 1.5\,W/m^2\,K^{-1}$ for double glazing and from ~ 1.8 to $\sim 1.0\,W/m^2\,K^{-1}$ for triple glazing. Obviously, T_{lum} must be large for these films. The infrared part of the solar radiation, which transmits energy through the glazing but is not needed for vision, can be stopped by a thin film having low transmittance at $0.7 < \lambda < 3$ µm. These demands have led to the development of two types of thin films, i.e., "low-emittance coatings" (low-E coatings), characterized by

$$T(\lambda) = 1 \text{ for } 0.4 < \lambda < 0.7 \text{ µm} \quad (7)$$

$$R(\lambda) = 1 \text{ for } 3 < \lambda < 50 \text{ µm} \quad (8)$$

and "solar control coatings," characterized by

$$T(\lambda) = 1 \text{ for } 0.4 < \lambda < 0.7 \text{ µm} \quad (9)$$

$$R(\lambda) = 1 \text{ for } 0.7 < \lambda < 50 \text{ µm}. \quad (10)$$

The spectral selectivity inherent in these relationships is of course idealized.

Another way to achieve energy efficiency is to use angular selective thin films on the glass. In this way, it is possible to take advantage of the fact that the indoor–outdoor contact across a window normally occurs along near-horizontal lines-of-sight, whereas solar irradiation comes from a small element of solid angle high up in the sky during most of the day. It is also possible to combine spectral and angular selectivity.

9.2 Spectrally Selective Thin Films

Very thin metal films can have properties resembling those defined in Eqs. (7)–(10). Calculations for extremely thin bulklike films of silver, gold, and copper predict that large values of T_{lum} and T_{sol} can be combined with low values of E_{therm}. However, such films cannot be made on glass or polyester. Rather, the thin film goes through a number of distinct growth phases, and large-scale coalescence into a metallic state with low E_{therm} requires thicknesses exceeding 10 nm, and then T_{lum} and T_{sol} are limited to about 50%. Most of the transmittance loss is due to reflectance at the film interfaces, implying that the transmittance can be boosted by embedding the metal film between dielectric layers, serving, essentially, for antireflection purposes.

Silver (Ag) has the best optical properties and is used in most applications. Gold (Au) and copper (Cu) tend to produce colored films that are undesirable for many architectural applications. Titanium nitride (TiN) can serve as a replacement for Au. The dielectric layer can be of several different kinds, such as Bi_2O_3, SnO_2, TiO_2, ZnO, and ZnS. The multilayer structure may also include corrosion-impeding layers, such as Al_2O_3. Figure 10 illustrates $T(\lambda)$ for three types of thin films based on $TiO_2/Ag/TiO_2$ and

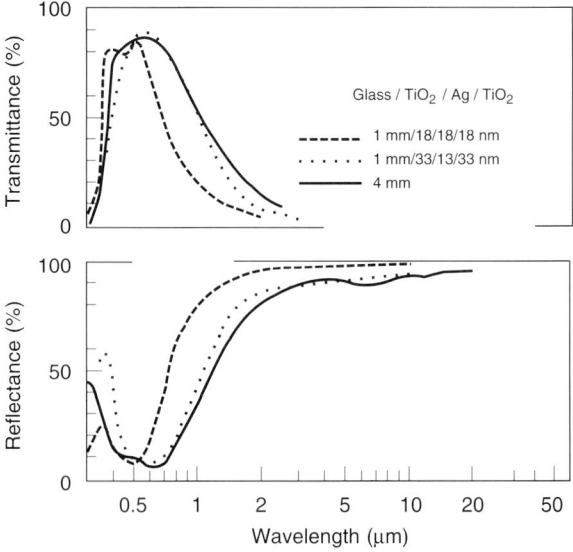

FIGURE 10 Spectral transmittance and reflectance results for glass having thin films based on $TiO_2/Ag/TiO_2$. Thicknesses are given for two of the film stacks.

deposited onto glass. Depending on the film thicknesses, it is possible to optimize T_{lum} or T_{sol}, whereas E_{therm} is invariably low. Thus, low-E coatings as well as solar control coatings can be made from metal-based thin films. Much of the glass that is used in modern windows and glass facades employs sputter-deposited films based on Ag. The films can be used only under protected conditions, such as inside sealed, double-glazed windows. The film thicknesses are small enough that iridescence, i.e., a rainbow-like exhibition of colors, is not a problem.

Doped oxide semiconductor coatings offer an alternative to the spectrally selective metal-based films just discussed. The practically useful materials include In_2O_3:Sn, SnO_2:F, SnO_2:Sb, and ZnO:Al. Many alternatives can be found among the ternary and quaternary oxides as well as among mixtures of the mentioned oxides. Nevertheless, the binary oxides have remained popular. The indium-based oxide has the lowest E_{therm}, the selenium-based oxides are most durable, and the zinc-based oxide is most popular for some applications (such as in many solar cells). The films are electrically conducting with a minimum resistivity of $\sim 10^{-4}\,\Omega\,cm$ for optimally prepared In_2O_3:Sn. Achieving such properties is notoriously difficult, though, and typically requires thermal treatment at $\sim 300\,°C$ or above.

Figure 11 shows $T(\lambda)$ and $R(\lambda)$ in the full $0.3 < \lambda < 50$-µm range for a 0.2-µm-thick film of In_2O_3:Sn. The parameter denoted n_e is the density of free electrons, which is directly associated with the amount of tin (Sn) doping; in practice, this amount is of the order of a few percent. The data in Fig. 11 were computed from a fully quantitative theoretical model of heavily doped semiconductors and give a clear representation of the salient features developing on increased doping: an onset of high reflectance in the near-infrared (i.e., a lowering of E_{therm}), bandgap widening tending to enhance T_{sol}, and disappearance of absorption features in the thermal infrared (approximately at $20 < \lambda < 30\,\mu m$). The doped semiconductor oxides can be very hard and resistant to corrosion, and they can be applied on glass surfaces exposed to ambient conditions. The practically useful thicknesses are larger than $\sim 0.2\,\mu m$, implying that thickness variation, which may be difficult to eliminate in practical manufacturing, may lead to aesthetically unpleasing iridescence.

The refractive index of the doped oxide semiconductors is about 2 for luminous radiation, which tends to limit T_{lum} to $\sim 80\%$. Higher transmittance values can be obtained by antireflection treatment,

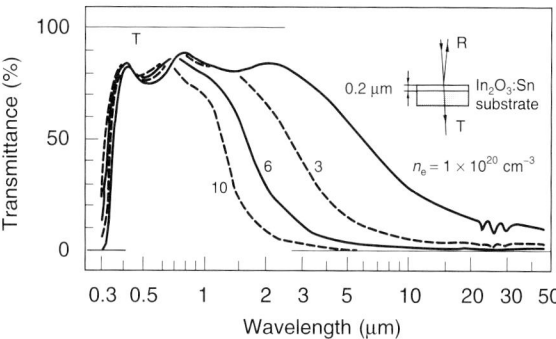

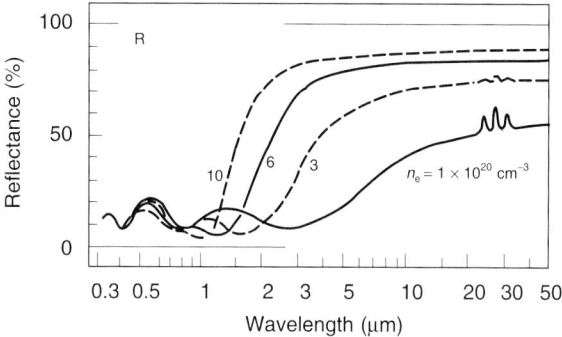

FIGURE 11 Spectral transmittance (T) and reflectance (R) computed from a fully quantitative model for the optical properties of In_2O_3:Sn, using the indicated values of film thickness and free-electron density (n_e).

and T_{lum} can then even exceed the value for uncoated glass. The antireflection layer lowers E_{therm}, but on an insignificant level of a few percent only.

9.3 Angular Selective Thin Films

Pronounced angular properties can be used to invoke energy efficiency for glazings, as mentioned previously. Figure 12 illustrates this feature as computed for a five-layer stack of thin films, specifically with two 12-nm-thick silver films embedded between three SiO_2 films with thicknesses of 120 or 170 nm. Strong angular effects prevail at the larger of these thicknesses, and T_{sol} is 23% for normally incident light and as large as 58% at an incidence angle of 60°. The data in Fig. 12 are symmetrical around the surface normal of the film. However, angular properties that are distinctly different at equal angles on the two sides of the normal can be obtained by the use of films with inclined columnar microstructures. Figure 13 shows a polar plot of T_{lum} recorded for a chromium-based film prepared by evaporation under conditions so that the incident flux arrives at the substrate at an almost glancing angle. The columns needed for the angular selectivity then

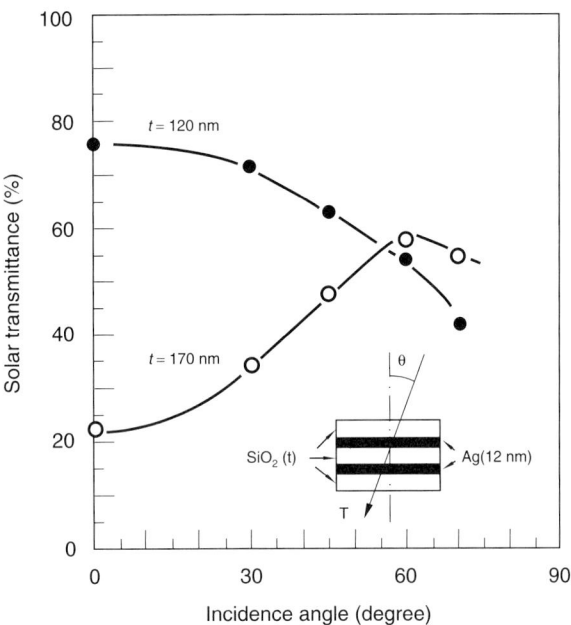

FIGURE 12 Solar transmittance for thin films of $SiO_2/Ag/SiO_2/Ag/SiO_2$ as a function of SiO_2 thickness (t) and incidence angle (θ).

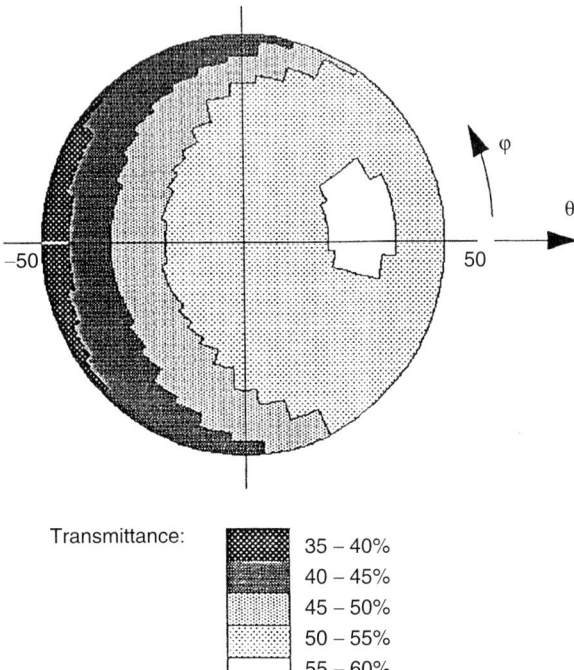

FIGURE 13 Polar plot of luminous transmittance through an obliquely evaporated Cr-based thin film. Data are given as a function of incidence angle (θ) and azimuthal angle (φ).

tend to form an array inclined in the direction toward the evaporated flux. Sputter deposition can also be used for making angular selective films.

10. COATINGS FOR GLAZINGS: DYNAMIC PROPERTIES

Windows and glass facades with variable properties have been the dream of architects for years. Such products are now becoming possible by exploiting "chromogenic" materials. Electrochromic devices are now being used in windows, albeit as yet on a small scale. A range of other options for dynamic fenestration also exists.

10.1 Electrochromic Devices

Electrochromic devices are able to vary their throughput of visible light and solar radiation on electrical charging and discharging, using a low voltage. Thus, they can regulate the amount of energy entering through a "smart window" so that the need for air conditioning in a cooled building becomes lessened. The energy efficiency inherent in this technology can be large, provided that the control strategy is adequate. Additionally, the transmittance regulation can impart glare control as well as, more generally, user control of the indoor environment. The absorptance, rather than the reflectance, is modulated so that the electrochromic devices tend to heat up in their low-transparent state, which has to be considered in a practical window construction. Modulation of E_{therm} is possible in principle, and development of variable-emittance surfaces for temperature control of space vehicles, as well as for other applications, is in progress.

Figure 14 shows an electrochromic five-layer prototype device that introduces basic design concepts and types of materials. A kinship, in principle, to an electrical battery is noteworthy. The central part of the construction is a purely ionic conductor (i.e., an electrolyte), either a thin film or a polymer laminate material; it must be a good conductor for small ions such as H^+ or Li^+. The electrolyte is in contact with an electrochromic layer and a counterelectrode. For a transparent device, the latter must remain nonabsorbing, irrespective of its ionic content, or it should exhibit electrochromism in a sense opposite to that of the base electrochromic film. This three-layer configuration is positioned between transparent electrical conductors, normally being doped oxide semiconductors of the kinds previously discussed; they are backed by glass or polyester foil. By applying a voltage of a few volts–conveniently obtained from solar cells–between the outer layers, ions can be shuttled into or out of the electrochromic

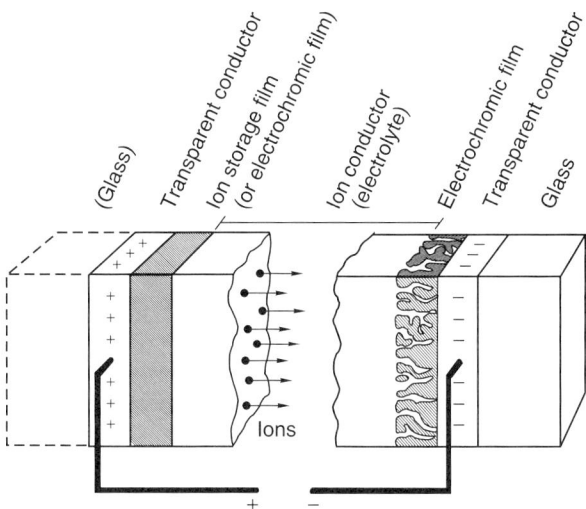

FIGURE 14 Basic design of an electrochromic device, indicating transport of positive ions under the action of an electric field.

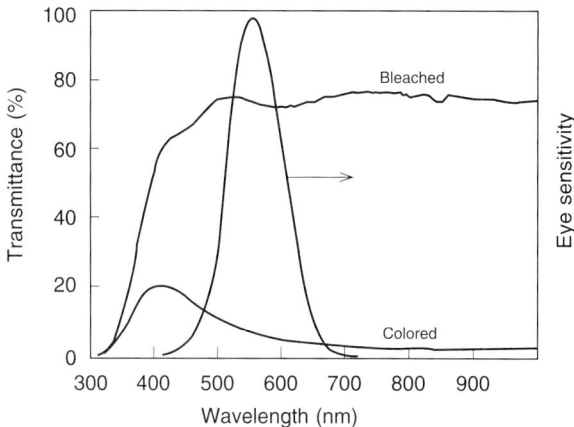

FIGURE 15 Spectral transmittance through an electrochromic foil including thin films of tungsten-based and nickel-based oxides. Data are shown for fully bleached and colored states.

film(s), the optical properties of which are thereby changed so that the overall optical performance is changed. The voltage needs to be applied only when the transmittance is altered, i.e., the electrochromic devices exhibit open-circuit memory. The time for going between bleached and colored states is largely governed by the electrical resistance of the transparent conducting films. Smart windows of the scale of a square meter require about ten minutes to change from a bleached state to a moderately colored state. Smaller devices can display much faster dynamics.

There are many materials and design issues for practical electrochromic smart windows. Electrochromism is found in numerous organic and inorganic materials, the latter mainly being transition metal oxides. Films of WO_3 are used as the base electrochromic layer in almost all devices studied until now. However, many options exist for the counterelectrode, with good results having been documented for oxides based on Ce, Co, Ir, Ni, and V, whereas other designs have used $KFe^{3+}Fe^{2+}(CN)_6$ (known as "prussian blue"). Additions of Al and Mg to oxides of Ir or Ni can enhance the bleached-state transmittance of the devices, which is important for architectural applications. In the case of Ir oxide, the additions lead to significant cost reduction. Many alternatives exist for the central electrolyte, which can be an adhesive polymer with high ionic conductivity or–in an all-solid-state approach with superimposed thin films–a hydrous layer of Ta_2O_5, ZrO_2, or SiO_2 serving as a conductor for H^+.

Figure 15 illustrates the optical performance of a laminated polyester-based electrochromic foil device comprising films based on WO_3 and NiO in the bleached state and after maximum coloration. It is seen that T_{lum} can be modulated between 74 and 7%. There is a trade-off between durability and maximum coloration, though, and a minimum T_{lum} of ~30% may be more adequate for practical applications.

10.2 Some Alternative Chromogenic Technologies

"Gasochromic" thin films represent an alternative to the electrochromic devices, with specific pros and cons. Glass is coated with a WO_3-based thin film having an extremely thin surface layer of catalytically active Pt. The coated side of the glass is in contact with the gas confined in a carefully sealed double-glazed window. By changing the amount of H_2 in contact with the thin film, it is possible to insert a variable amount of H^+ and the transmittance is then modulated in analogy with the case of the electrochromic device. Among the many other possibilities to control T_{lum} and T_{sol} electrically, of particular note are suspended-particle devices and phase-dispersed liquid crystal devices. These have much swifter dynamics compared to electrochromic devices, but require higher voltages. Rapid technological progress, as well as the proprietary nature of much basic information, make a detailed comparison of the various techniques difficult.

Thermochromism represents another possibility to control T_{sol} in order to create energy efficiency in buildings. The most interesting material, a thin film based on VO_2, undergoes a structural transition at a certain "critical" temperature τ_c, below which the material is semiconducting and relatively nonabsorbing in the infrared. Above τ_c, it is metallic and

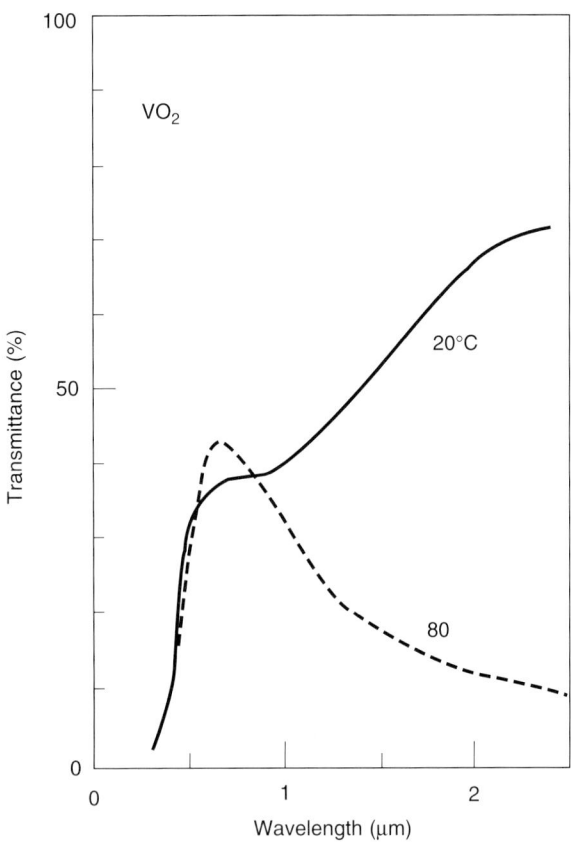

FIGURE 16 Spectral transmittance through a thermochromic VO$_2$ film at two temperatures.

infrared reflecting. Figure 16 shows typical data on $T(\lambda)$ above and below τ_c. Bulklike VO$_2$ has $\tau_c \approx 68°C$, which obviously is too high for architectural applications in buildings, but τ_c can be decreased to room temperature by additions of Mo, Nb, or W, for example.

Still another possibility to vary T_{lum} as a result of temperature changes is found among the "cloud gels," which can be interposed between the panes in a double-glazed window. These are polymers for which "clouding"–i.e., transition to a diffusely scattering state–can set in above a certain τ_c due to a reversible thermochemical dissolution and a thermally induced modification in the lengths of the polymer molecules.

11. SOLAR PHOTOCATALYSIS

Under the action of solar irradiation, photocatalytically active surfaces can cause the breakdown of organic molecules as well as microorganisms adsorbed on these surfaces. Thus hazardous or otherwise unpleasant pollutants in air or water can be decomposed into tasteless, odorless, or at least less toxic compounds. Well-known photocatalysts are oxides of Cd, Ce, Fe, Sb, Sn, Ti, W, Zn, and Zr, with most developmental work having been devoted to TiO$_2$. All of these oxides are able to use ultraviolet light only. This is an advantage for some applications, and, for example, a double-glazed window may be transformed into a photocatalytic reactor with an outer ultraviolet-transparent pane (cf. Fig. 2) and an inner pane with a TiO$_2$ film. In other cases, it is also an advantage to employ visible light for inducing photocatalysis; transition metal additions to the oxides previously mentioned, as well as nitrogen incorporation, are able to widen the spectral range for photocatalysis.

Photocatalytic decomposition of water into hydrogen and oxygen has been investigated for many years, but it has turned out to be difficult to accomplish effectively using solar irradiation. Present research includes studies of noble-metal-loaded oxide catalysts of many kinds, sometimes covered with NaOH or Na$_2$CO$_3$. Ultrasonic irradiation of the catalyst can improve the efficiency of water splitting.

SEE ALSO THE FOLLOWING ARTICLES

Lunar–Solar Power System • Solar Cells • Solar Cookers • Solar Energy, History of • Solar Fuels and Materials • Solar Heat Pumps • Solar Ponds • Solar Thermal Energy, Industrial Heat Applications • Sun, Energy from

Further Reading

Gordon, J. (ed.). (2001). "Solar Energy: The State of the Art." James & James Science Publ., London.

Granqvist, C. G. (ed.). (1993). "Materials Science for Solar Energy Conversion Systems." Pergamon Press, Oxford, U.K.

Granqvist, C. G. (1995). "Handbook of Inorganic Electrochromic Materials." Elsevier Science, Amsterdam.

Kaneko, M., and Okura, I. (eds.). (2002). "Photocatalysis: Science and Technology." Springer-Verlag, Berlin and New York.

Lampert, C. M., Granqvist, C. G. (eds.). (1990). "Large-Area Chromogenics: Materials and Devices for Transmittance Control." SPIE—The International Society for Optical Engineering, Bellingham, Washington.

Mbise, G. W., Le Bellac, D., Niklasson, G. A., and Granqvist, C. G. (1997). Angular selective window coatings: Theory and experiment. *J. Phys. D: Appl. Phys.* **30**, 2103–2122.

Material Use in Automobiles

SUJIT DAS
Oak Ridge National Laboratory
Oak Ridge, Tennessee, United States

1. Material Usage Trends in Automobiles
2. Factors Impeding Lightweight Materials Growth
3. Role of Lightweight Materials in Energy Consumption
4. Automotive Lightweight Materials Outlook
5. SAFETY Considerations in Automobile Lightweighting

Glossary

automobile An average light-duty vehicle representing passenger cars, sport-utility vehicles, minivans, and light trucks.
composites A material composed of a reinforcing material (e.g., glass or carbon fibers) in a matrix material (e.g., polymers).
lightweighting Using lighter materials or efficient system design to reduce vehicle weight.
monocoque A type of motor vehicle design in which the body and frame are integrated into eggshell shape, resulting in a small number of relatively large parts.
recovery rate The percentage of usable recycled materials recovered from discarded vehicles.
recycling rate The percentage of vehicles from which materials are collected, separated, or processed and returned to the economic mainstream to be reused in the form of raw materials or finished goods.
thermoplastic A polymer in which the molecules are held together by weak secondary bonding forces that can be softened and melted by heat, then shaped or formed before being allowed to freeze again.
thermoset A polymer that solidifies irreversibly when heated due to a chemical reaction involving cross-linking between chains.
unibody A conventional body on frame motor vehicle body structure design composed of a relatively large number of hollow sheet metal components onto which body panels are welded by welding robots or in multispot welding units.

Material composition continues to play an important role in automobile engineering, particularly with the renewed importance of fuel economy standards and advanced vehicle designs such as hybrid and fuel cell vehicles. Of the numerous types of materials currently in use in automobiles, lightweight materials, particularly aluminum, polymer composites, and magnesium, have received increasing attention because of their potential to reduce the weight of vehicles. Most of these lightweight materials are more energy intensive to produce than conventional steel, the predominant material in current vehicles. A life cycle energy comparison that addresses the vehicle-use phase, which contributes the major share of a vehicle's life cycle energy use, is appropriate to examine the materials' energy implications. The effect of vehicle weight on its safety is an overriding sensitive issue to consumers, and it remains a hotly debated issue in vehicle lightweighting.

1. MATERIAL USAGE TRENDS IN AUTOMOBILES

Automobiles play an important role in the world economy with a share of about 50% of total transportation energy demand. The transportation sector accounted for about 27% of total U.S. energy consumption in 2000. Growth in U.S. energy consumption of the past decade averaged about 2% annually, with automobiles accounting for about 57% of total transportation energy in 2000. Between 1978 and 2001, the average U.S. automobile weight has decreased only by 7.3% (3570 lbs versus 3309 lbs), whereas sales-weighted fuel economy improvements in automobiles have been more than 40% (19.9 miles per gallon versus 28.6 miles per gallon). As the share of heavier SUVs, minivans, and light trucks continued to grow during the past decade, the percentage increase in average vehicle weight in the fleet is significantly higher. A combination of technology improvements and weight reduction brought about this fuel-economy gain. The technology improvements range from reduced internal engine friction to reduced aerodynamic drag. Major improvements were provided by the transition from rear- to front-wheel drive, which allowed interior volume to be maintained while weight and exterior size were reduced, and by conversion from carburetors to electronically controlled fuel injection. There were many other improvements as well: more efficient automatic transmissions, improved lubricants, lower rolling resistance tires, reduced accessory loads, material substitution, and so forth.

The technology choices in U.S. automobiles have been a concern since the environmental movement of the 1960s, the oil crisis of the 1970s, the Clean Air Act of 1970 mandating standards for emissions, and the Energy Policy and Conservation Act of 1975 mandating national standards for automotive fuel economy (popularly known as corporate average fuel economy). These events and policies prompted automakers to address fuel economy, formerly a minor factor in automotive design, as a vital engineering requirement. The 1994 Partnership for a New Generation of Vehicles (PNGV) program (lately superseded by Freedom Cooperative Automotive Research (FreedomCAR), which is unlike independent of any specific vehicle platform with a focus on fuel cell vehicles without any specific fuel economy goal), has a formidable goal of developing production-ready mid-sized vehicles that are 40% lighter than baseline vehicles and achieve 80 miles per gallon fuel economy. This program considered many new research areas, including advanced powertrain concepts with as much as 40 to 45% powertrain thermal efficiency, body and chassis weight reduction technologies, alternate design concepts, and more efficient accessories development. To achieve this goal, the focus was on hybrids and fuel cells for propulsion systems; aluminum, magnesium, and glass-fiber composites for body and chassis; and a range of improvements in accessories.

With improved vehicle performance and emission requirements, vehicle weight is becoming increasingly important to achieving better fuel economy. It is estimated that each 1% reduction in vehicle weight provides a 0.66% improvement in fuel economy, after a vehicle has been fully redesigned to account for the reduced power demands of a lower mass. The primary weight savings leads to secondary savings with lighter engine and powertrain configuration. Although more than 85% of fuel economy gain observed during the past three decades can be credited to technology improvements, the use of lightweight materials has played a key role in maintaining the vehicle weight, countering the weight added by the technology gains and passenger safety and comfort features. Furthermore, since 75% of a vehicle's energy consumption is directly related to factors associated with vehicle weight, the need to produce lightweight vehicles that also are safe and cost-effective is critical.

An average U.S. automobile weighs about 3240 lbs, of which body, chassis, and powertrain vehicle subsystems contribute 35%, 34%, and 27%, respectively, of its total weight. Compared to a U.S. automobile, European and Japanese automobiles are lighter in total vehicle weight due to a significant growth in the use of nonferrous metals and plastics/composites in the past. A European automobile has the least share of ferrous materials content (i.e., about 60%), and its current share of plastics/composites is estimated to be more than 10% of its vehicle weight. Despite the competition from other materials, ferrous metals continue to be the dominant material in an automobile, as shown for the United States in Fig. 1. Ferrous materials are flexible in use and relatively cheap, and their properties including its crash behavior and recyclability, are well understood and established. Although there has been a steady decline in conventional steel usage, the increased use of lighter, high-strength steel, with its improved formability in combination with efficient system design with lower weight, has helped to maintain steel's >50% share of total materials. Tailored blanks, hydroforming, and various metallurgical improvements have also kept the steel content of vehicles relatively constant. The decrease in iron content is due mainly to the switch to aluminum engine blocks and heads. The use of stainless steel is very limited to exhaust systems and engine gaskets applications today. The use of lightweight materials is largely responsible for the decline of 7.3% in total vehicle weight during the 1978–2001 period, as shown in Fig. 1.

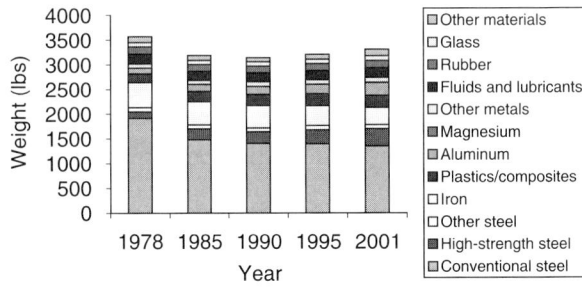

FIGURE 1 Material composition trend of a U.S. automobile.

Of the lightweight materials, aluminum, plastics/composites, and (to a far lesser extent) magnesium continue to have significant roles in automotive materials usage. The new Lincoln LS represents a current example of lightweight materials on a high-volume production vehicle. Aluminum, plastics, and magnesium total more than 20% of the vehicle's weight. Between 20 and 25% of the ferrous materials are high-strength steels selected to achieve weight reduction. Aluminum, composites, and magnesium have weight reduction potentials ranging from 25 to 75%, depending on the material replaced and automobile part application, as shown in Table I. Due to some inferior mechanical properties and manufacturing technology limitations, the mass reduction achieved using some of these lightweight materials in commercial applications is generally lower than the theoretical densities shown in this table. Consumer preferences—such as performance, style, comfort, economy, and the brand image—are relatively less important than many other hard-headed criteria—such as cost—which dictate the material choice selection by the vehicle manufacturer. Life cycle (e.g., recyclability, manufacturability, environmental issues, etc.) and engineering needs are also gaining momentum in material selection criteria.

Of the nonferrous lightweight metals, aluminum, having a density one-third that of steel, has seen the most use, with applications ranging from body closure panels to wheels and engines. Although aluminum parts are generally 1.5 times as thick as steel to achieve comparable stiffness, they still offer a significant weight savings. Aluminum use in automobiles has grown steadily from an average of 180 lbs in 1978 to 257 lbs in 2001, with roughly 80% of that in cast components, mostly engines and transmissions. This substitution of steel has been on a part-by-part basis, not the result of any radical design change. Aluminum is rapidly becoming the material of choice for vehicle's cylinder blocks.

Use of wrought aluminum sheet, however, is limited to air conditioning units and a few closure panels for the car body. Both panels and structures are still largely an unfilled dream for aluminum. The most popular body panel application is the hood (e.g., F-series pickup truck) due to weight balancing advantages. Liftgates are a rapidly growing application. Currently, only 5 to 10% of body panels are made of aluminum. A significant displacement of steel by aluminum would occur if and when steel as the primary material is being replaced in the chassis or the body of the car, since the latter contributes more than 30% of the average total vehicle weight. Aluminum panels on an extruded aluminum spaceframe were first used in the 1994 Audi A8—a low-volume production car. Price is a major problem presently, but the price (and its variability) is likely to be reduced as more aluminum scrap becomes available.

Composites have been the important peripheral contender in the material war between steel and aluminum with magnesium. Automotive composites consist of a reinforcing fiber in a polymer matrix, with polyester, vinyl ester, and epoxy resins the most often used matrixes. These polymer-based composites—having densities one-sixth to one-third that of steel, depending on the fiber type, fiber loading, and resin type—are the most common type of composites used in automotive applications. The fibers have densities ranging from 1.8–2.5 gm/cm^3, as shown in Table I; the commonly used polymer matrix materials have densities ranging from 1.1 to 1.4 gm/cm^3. About two-thirds of all composites currently use glass fiber and polyester or vinyl ester resins and provide a wide range of potential automotive applications, including body panels, suspension, steering, brakes, and other accessories. Demand varies widely by application. The composite materials have not changed significantly over the past 30 years, with glass fibers being the mainstay among reinforcement materials, while in some select applications carbon and aramid fiber materials are used. Today's average automobile is about 8% plastics and composites, combined, and their use has grown slowly, increasing by only 73 lbs during the 1978–2001 period, as shown in Fig. 1.

Polymer composites have been used for applications with low production volumes because they

TABLE I

Weight Savings and Relative Cost of Lightweight Materials

Lightweight material	Density (gm/cm^3)	Material replaced	Mass reduction (%)	Relative cost (per part)
High strength steel	7.8	Mild steel	10–24	1
Aluminum	2.7	Steel or cast iron	40–60	1.3–2
Magnesium	1.8	Steel or cast iron	60–75	1.5–2.5
Magnesium	1.8	Aluminum	25–35	1–1.5
Glass FRP	2.5 (fiber)	Aluminum	25–35	1–1.5
Carbon FRP	1.8 (fiber)	Mild steel	50–65	2–10+
Titanium	4.5	Alloy steel	40–55	1.5–10+
Stainless steel	7.9	Mild steel	25–40	1.2–1.7

require less lead times and lower investment costs relative to conventional steel fabrication. Important drivers of the growth of polymer composites have been the reduced weight and parts consolidation opportunities the material offers, as well as design flexibility, corrosion resistance, material anisotropy, and mechanical properties. Composites use is still predominantly in nonstructural elements of the vehicle because they are highly competitive for bolt-on exterior panels such as hoods, decklids, and fenders, especially in low- and mid-volume light-duty vehicles such as Corvette, Dodge Viper, and AP minivans. In addition, composite cross vehicle beams, radiator housings, fascia supports, bumper systems, and floor trays have contributed to weight savings. Corvette's body has been made largely of composites since the 1950s, and General Motor's Saturn and other vehicles use composite side panels. The latest growth in automotive composites has been mostly in exterior parts such as fenders, tailgates, and pickup beds.

Automotive manufacturers are becoming increasingly interested in carbon-fiber-based polymer composites because they offer high stiffness and low density for structural applications. They can also be made lighter than their glass-reinforced counterparts, providing a significantly higher weight-savings potential. Its most significant use to date has been in concept cars (e.g., about two decades ago in the Ford Granada and in General Motors' 1991 Ultralite). Carbon fiber composites became part of the government's research plans with the PNGV goal of achieving by 2011 a production-ready, 100-mpg, mid-sized vehicle, which is expected to contain at least 10 lbs of carbon fiber. In addition, there has been interest in fiber-reinforced thermoplastic composites that have the typical advantages of polymer matrix composites—such as weight savings, high strength, high stiffness, corrosion resistance, parts integration, and energy absorption—and have an indefinite shelf life, are recyclable, and are feasible for automated, high volume processing with a potential for rapid and low-cost fabrication. The most-used thermoplastics today are glass-filled thermoplastics developed for a variety of applications from intake manifolds to engine covers and, to a lesser extent, for body panels.

Magnesium is becoming a significant automotive material in varied applications such as engine blocks, seat frames, steering column components, transmission housings, wheels, and valve covers. Despite density that is 25% and 67% lower than aluminum and steel, respectively, magnesium use in automobiles has been limited to die castings because of its high cost and limited supply. The lower cost of producing die cast parts helps to offset its high cost. The market for automotive magnesium parts has grown rapidly, nearly 15% annually during the 1990s, achieving a level of about 8.5 lbs per vehicle in 2001. Some high-end models, such as the 2003 Dodge Viper, are slated to contain over 44 lbs of magnesium, mostly in the one-piece instrument panel that has been incorporated into the design. Magnesium use in wrought automotive products, such as seat frames, wheels, and body panels, has been very limited because of the lower process throughput and the metal's cost. Magnesium's creep resistance is inferior to aluminum's and limits its application in automotive powertrains today.

The "other metals" category in Fig. 1 mainly includes copper, brass, powder metal parts, and zinc die castings, where the latter represents only about 3% of this category. Zinc is also used for the coating of steel and being considered to protect automotive underbody parts. Nonferrous metals are important in powder metallurgy and metal matrix composites, both of which are being used increasingly in components such as camshafts, connecting rods, pistons, gears, and pulleys. The use of ferrous powder metallurgy parts in applications such as connecting rods has been trending upward primarily because it is less expensive than other methods. Metal matrix composites are gaining wide acceptance in niche applications such as brake discs, drums and calipers, pistons, prop shafts, and tire studs. Copper use in automobile radiators has been declining as manufacturers shift to aluminum, which is lighter and perceived to have a stable market price.

Another promising lightweight material that has only recently seen even limited use is titanium. It has potential application in suspension springs, exhaust systems, and a number of engine and chassis parts. Passenger car components that could benefit from titanium include engine valves, connecting rods, valve-spring retainers, and valve springs. Titanium has the benefits of being half the weight and twice the strength of steel (as shown in Table I), with great potential for a reduction in engine noise and vibration and improved durability due to reduced component loads, vehicle weight reduction, and improved powertrain efficiency. However, the cost (both raw material as well as component fabrication) is prohibitive, four times higher than even stainless steel. Titanium's high strength, low density, excellent corrosion resistance, and the relative worldwide abundance of its ore make it a long-term viable lightweight alternative to steel in engine and body/chassis applications. Because of its high corrosion resistance at elevated temperatures, titanium is the only lightweight substitute for stainless steel in exhaust systems. The first use of titanium in a production automobile was for connecting rods in the Acura NSX V-6, and the use of titanium springs in the 2001 model year Volkswagen Lupo FSI has contributed to about 180 lbs reduction in weight.

2. FACTORS IMPEDING LIGHTWEIGHT MATERIALS GROWTH

Although lightweight materials have significantly penetrated the aerospace and defense industry, their cost has been the major barrier to their large-scale penetration in the automotive industry. A part-cost comparison (which includes both materials and manufacturing) of these materials indicates that relative part costs are anywhere from 1.3 to 10+ times higher than the materials they replace (see Table I). At the higher end of these cost differences are carbon fiber reinforced polymer composites, titanium, and magnesium, which are very little used in automobiles today. In most cases, the cost of the primary material is significantly higher than conventional steel's cost, and the manufacturing-process cost is only a small share of the overall part cost. Thus, any automotive system application with a greater potential for parts integration improves the economic viability of parts made of lightweight materials. These expensive lightweight materials have comparatively penetrated in Europe and Japan more than the United States due to lower fuel cost in the United States. In addition, there is a lack of economies of scale due to these materials' relatively smaller industry base. The current consumption of these materials is relatively small compared to steel, and in some cases (e.g., magnesium and rubber) the automotive industry accounts for a major share of the materials' total consumption, as shown in Table II. Most of the lightweight material industries are small, causing instability in material price that promotes substitution among the competing materials. Plastic in Table II does not include polymer composites, whose total consumption by the automotive industry is estimated to be 156 K tonnes in 2001.

Factors in addition to cost that have dampened the growth in lightweight materials use include relatively less pressure for improved fuel economy and from environmental pollution concerns, along with the availability of inexpensive gasoline and to some extent recent subsidies available to traditional materials such as steel. The existing well established manufacturing base is always a big positive factor for steel—there is always reluctance in the industry to avoid the enormous potential capitalization costs of large-scale conversion from steel to lightweight materials. All of the industries lack experience and play a small role in the automotive industry, situations that will only change with the materials' improved economic viability.

A widespread use of aluminum in automobiles would require a dramatic shift in wrought applications. Aluminum stamped parts are more expensive than extruded and cast parts because production-line rates are slower, there are difficulties in welding and assembly, aluminum sheet cost is higher as are die costs because they require special coatings to compensate for springback. The tendency of aluminum sheets to tear requires low stamping rates and extra hits for the stampings. Most aluminum used in outer-body panels is heat-treatable, 6000-series aluminum alloys, which are still relatively expensive to produce. The potential exists to produce 5000-series aluminum alloy for inner body panels at $1/lb using continuous casting and exploiting large economies of scale. Under existing fabrication conditions, only a drastically consolidated design such as a spaceframe is even slightly competitive with the conventional unibody design. A cost comparison indicates that aluminum body structures made of continuous-cast, coated aluminum at $1/lb (unlike ingot and stamped sheet rolling cost at $1.30/lb) cost 1.6 times more than body structures made of conventional steel. It is not only raw material cost difference, but also the rivet bonded assembly for aluminum structures that causes this large cost difference.

TABLE II
2001 Material Usage by the Automotive Industry

Material	U.S. total consumption (tonnes)	Automotive share (%)
Steel	90 M	14
Iron	9 M	31
Aluminum	11 K	32
Plastic	46 K	4
Magnesium	96 K	47
Zinc	1 M	23
Rubber	3 M	68

As with other lightweight materials, the price of resins and glass fiber reinforcements poses a major barrier to the growth of automotive composites. The long cure-cycle times required for automotive part applications make it impossible to maintain high automotive production rates. For composites to be cost competitive on a part-by-part substitution, improvements are necessary in cycle times and material utilization, which in some cases have been estimated to currently contribute 21% and 60%, respectively, of the total cost of carbon-fiber-reinforced thermoplastics. On a $/lb basis, the cost of polymer composites is about two to three times higher than steel, but a recent study comparing the monocoque designs indicates that the cost of glass-fiber reinforced thermosets and carbon-fiber-reinforced thermoplastics are about 62% and 76% higher than the conventional steel unibody. The material cost plays a key role in the economic viability of polymer composites, particularly at higher production volumes for glass-fiber-reinforced thermosets and for carbon fiber-reinforced thermoplastics composites. A 50% reduction in the carbon fiber cost (to prices in the $3 to $5/lb range) and smaller cost reductions in other thermoplastic materials are needed for the composites to be economically viable. Also impeding large-scale use of composites is a curious mixture of concerns about material issues such as crash energy absorption, recycling challenges, competitive cost pressures, the industry's general lack of experience and comfort with the material, and industry concerns about its own capabilities.

Technical and institutional drawbacks, in addition to cost, have limited magnesium use in vehicles, especially for wrought magnesium products. Because magnesium is very reactive, corrosive, hard to form at room temperature, and has a relatively high percentage of scraps and refuses, its part production cost is significantly higher than that of competing materials. In addition, magnesium has a small supply base, with annual production about 1400 times and 45 times less than steel and aluminum, respectively. Because it is a small industry and not traded on the London Metal Exchange, its price is more prone to swings as demand grows and absorbs available production. In addition, past supply expansions in magnesium production capacity have not been systematic enough to harness potential demand for low-cost material in the auto industry. Automakers' demand surpassed industry capacity, causing wild price swings, which caused parts manufacturers to switch from magnesium to other competing materials. Recent cheap imports, particularly from Russia and China, have caused a severe blow to the industry, forcing several magnesium production plants worldwide to close down, including one of the two remaining plants in the United States.

Several current governmental R&D programs (e.g., the Automotive Lightweighting Materials Program by the United States Department of Energy in collaboration with the automotive industry) focus on the viability of primary material production and manufacturing technologies for significantly reducing automotive vehicle body and chassis weight without compromising other attributes such as safety, performance, recyclability, and cost. Aluminum and glass-reinforced polymer composites have been considered as the near-term automotive materials, whereas, magnesium, titanium, and carbon-reinforced polymer composites are considered for the long term. Similarly, the Canadian Lightweight Materials Research Initiative (CLiMRI) coordinates research and development of materials and processes for weight reduction in transportation applications, besides the multidisciplinary auto-related research and development initiative "Auto21" to influence the vehicle in the 21st century. The Industrial and Materials Technologies program of the European Union indicates similar activities in this area of research. The resulting ongoing research is expected to address some of the aforementioned impediments to lightweight material use.

3. ROLE OF LIGHTWEIGHT MATERIALS IN ENERGY CONSUMPTION

Due to the increasing use of automotive lightweight materials, concerns have surfaced regarding the fuel savings, lower emissions, and weight savings, particularly with respect to aluminum and magnesium. Life cycle analysis has shown that high energy consumption during production of virgin aluminum and magnesium is offset by the reduced energy consumption in the use phase, bringing the breakeven point to within the lifetime of the car. The point at which a breakeven is reached depends on the amount of lightweight materials included, the amount of recycled material used for production of the part, and the system and conditions assumed in the study. Life cycle assessments with the consideration of "cradle to grave" aspects of a material used in automobiles have been instrumental in providing a more holistic view of energy consumption, considering energy consumption beyond the materials production stage and demonstrating a significant potential for energy savings.

The major stages of "life" included in the life cycle analysis of automotive lightweight materials are material extraction, part production, use phase, and end-of-vehicle stage. The second and last life cycle stages consume considerably less energy than the first and third stages and have only sparse data about them available in the literature today. The use phase energy consumption has the biggest impact on the life cycle analysis, contributing 60 to 80% of the total energy consumption. The next two largest shares, in order, are material production and part manufacturing. The transportation necessary within each of the phases has often been included in the past studies. The life cycle energy benefit in switching to a lighter weight material is not just a function of weight saved; it is also dependent on material production energies and substitution factors and vehicle operational efficiency.

Table III shows the primary- and secondary-production energy of various materials used in automobiles today. There is considerable variation—as much as 40%—in reported values of material production energies, depending on the estimation methodology used and assumptions about production technology and its efficiency. The primary production energy is the sum of all energies required to extract raw materials from the earth (e.g., mine ore or pump oil) and to process (wash, concentrate, or refine) it into a usable form (ingot or rolled sheet). Much of this production energy is avoided by recycling, although getting a recycled material of a comparable purity to virgin stock is often not readily achieved due to the difficulty in removal of impurities. The recycled content is an important factor in the overall life cycle analysis, as less than 15% of the energy used for production of virgin material is needed for remelting in case of aluminum and magnesium, compared to 30% in the case of steel as shown in Table III. Recycling is gaining momentum with the End-of-Life directive in the European Union affecting the design phase and material decision making processes of automobile manufacturers as well as influencing other industrial activities such as automotive dismantlers and shredders. The directive's recycling targets are 80% and 85% by 2006 and 2015, respectively, with a considerable higher recovery rate of 85% and 95%, respectively, for the corresponding years. This is in contrast with 75% recycling rate today, mostly limited to ferrous materials.

The life cycle energy consumption of a generic family U.S. sedan based on 120,000 lifetime mileage is estimated to be about 974 GJ, which corresponds to the consumption of 46,000 pounds of hydrocarbons (coal, oil, and natural gas). Weight reduction through the use of lightweight materials is shown to significantly reduce the life cycle energy of a vehicle. Several life cycle analyses on a single product basis indicate that although aluminum production is about ten times more energy intensive than steel production, aluminum in automotive applications is less energy intensive during its life cycle than conventional steel. Since vehicle operation dominates overall life-cycle energy use, "lightweighting" with aluminum increases fuel efficiency enough to reduce overall energy use.

For example, a radical use of carbon fiber composites and aluminum (i.e., replacement of all steel and iron with carbon fiber composites and aluminum, respectively, resulting in overall 40% vehicle weight savings) is predicted to yield a life cycle energy savings of 16% for a conventional mid-sized vehicle. The use of recycled materials improves life cycle energy consumption considerably—for example, a 7.5% life cycle energy reduction for a particular aluminum substitution example becomes 12% when the part is 50% recycled aluminum. If aluminum-intensive light-duty vehicles were commercialized on a mass scale by 2005, the reported national petroleum energy savings would be quite significant, about 4% and 5% by 2020 and 2030, respectively. The impacts of lightweight materials on non-conventional vehicles such as electric and hybrid are comparatively less due to the use of highly energy

TABLE III

Primary and Secondary Production Energy for Automotive Materials

Material	Energy (MJ/kg)	
	Primary	Secondary
Steel	30	9
Cast iron	34	24
Wrought aluminum	201	29
Cast aluminum	189	26
Cast magnesium	284	27
Reinforced plastic	56	37
Unreinforced plastic	79	14
Copper	140	35
Zinc	53	16

efficient power-plants that reduce energy consumption in the vehicle use stage.

Life cycle energy comparisons on a single product basis are quite sensitive to the underlying assumptions made for major input parameters. For example, a comparative assessment of material substitution examining automatic transmission cases made from die-cast magnesium and aluminum finds aluminum use more favorable, despite magnesium's significantly higher weight reduction potential. This result is assumed to be due to the different recycling rates for the two materials. On the other hand, magnesium substitution ranging from 10 to 30% in automotive parts like gearbox housing and dashboards, indicated a breakeven point around 40,000 miles.

Recent studies argue that since new product designs are introduced progressively in a fleet, only a fleet level analysis can account for the temporal effects of use and the distribution of energy over a vehicle lifetime. Since the product manufacture and use are distributed over long periods of time that are not simple linear combinations of single product life cycles, it is all the products in use over a period of time, rather than a single product that are more appropriate for the life cycle analysis. This approach allows the prediction of time when benefits are realized. A recent study looking at the rolled products for body-in-white applications (in which aluminum provides the greatest weight saving potential compared to steel and ultra light steel autobody) found that it would take about 4 and 10 years, respectively, for aluminum vehicles to achieve a life cycle energy equivalence with steel and ultra light steel autobody on a single vehicle basis. It would take twice as long to realize the benefits of aluminum in the fleet.

4. AUTOMOTIVE LIGHTWEIGHT MATERIALS OUTLOOK

During the next two decades, a combination of customer expectations, environmental stewardship, and tightened regulations will lead to significant changes in the automobile technologies, where materials will be at the forefront. As technological improvements continue to address the cost and manufacturability issues of automotive lightweight materials, a shift from current high-end and niche applications to widespread applications on a part-by-part basis will be seen. The pressure to boost the light-truck corporate fuel economy with an anticipation of higher future sales in this vehicle category will provide impetus to the growth in the automotive lightweight materials during this decade. As long as the pressure of the 'hypercar" strategy based on a conceptual vehicle combining very low weight with a highly aerodynamic design and a electric drive system capable of achieving a fuel economy 80 to 100 mpg continues, a much greater role for composites, aluminum, magnesium, and possibly titanium in automotive use will be seen. These materials would enable less powerful engines to be used for equivalent performance, compounding the weight reduction throughout the vehicle. This higher usage of lightweight materials trend can be seen in the concept cars developed by the three automobile manufacturers (i.e., Precept by General Motors, ESX3 by DaimlerChrysler and Prodigy by Ford) during the end of the past decade to meet the three-times-greater fuel efficiency goal of the PNGV program, as shown in Table IV. There is a significant increase in aluminum use for all of these concept vehicles, whereas composites use increases significantly in the case of ESX3 only. Carbon fiber and titanium, now rarely used in vehicles, had a significant share in the material composition of these concept vehicles. Development of more efficient design and manufacturing initiatives, as well as parts consolidation, will be necessary for the widespread use of lightweight material

TABLE IV

Material Composition of PNGV Concept Vehicles by Three Automobile Manufacturers

Material	Baseline	Prodigy	ESX3	Precept
Wrought aluminum	47	462	330	304
Cast aluminum	159	271	120	820
Magnesium	6	86	122	7
Titanium	0	11	40	33
Platinum	0.0033	0.01	0.01	0.01
Ferrous	2168	490	528	487
Plastics	193	209	52	187
Resins (for composites)	28	40	428	86
Carbon fiber	0	8	24	22
Glass fiber	19	19	60	35
Lexan	0	30	20	0
Glass	97	36	70	57
Rubber	139	123	148	77
NiMH or LiIon batteries	0	138	88	138
Other	391	83	212	338
Total	3248	2010	2250	2591

applications. Lightweighting will become more important as the trend for comfort and safety related system gadgets continue to add weight to the vehicle. With the European directive establishing considerably higher recycling goals, recyclability will pose major barriers to these lightweight materials unless viable recycling technologies and infrastructure are solved and in place.

The development of high-strength steels and relevant design opportunities mean major weight savings can be made without resorting to other materials. Hence, steel is likely to be a substantial automotive material for at least the first third of the century. The $22 million ultra light steel auto body (ULSAB) project recently completed by American Iron and Steel Institute (AISI) demonstrated a 36% mass reduction using high-strength steels based on drastic design changes on benchmark vehicles using tailor-welded blanks, lighter material, and advanced forming processes. A similar ultra light steel auto closures (ULSAC) project demonstrated a number of weight-saving vehicle closure elements such as doors, hoods, decklids, and hatchbacks. Faster growth is predicted for high-strength steel than for aluminum in body panel applications. High-strength steels in optimized designs enable substantial weight reductions when made by innovative manufacturing and assembly technologies. In addition, an extensive use (i.e., about 80%) of advanced high-strength steels (AHSS)—a new category of steels that combine very high-strength with excellent formability—defines the latest ULSAB-Advanced Vehicle Concept (AVC) project by AISI.

Growth in the use of aluminum castings is anticipated to continue, particularly for engines and wheels. Another aluminum-casting growth area is anticipated to be suspension (e.g., suspension knuckles and control arms, as well as structural components, drivelines, and interiors). The possibility exists for other lightweight materials to capture some of these substitution opportunities (e.g., for intake manifolds to be made with reinforced nylon and other polymers and for transmission transfer cases made by magnesium casting). Panel and structures applications are still largely an unfilled dream for aluminum. Aluminum still has to overcome significant technological and economic hurdles before it can replace steel in the car body, and competing with recent advancements in high-strength steel will be difficult. The shift of aluminum producers from being merely material suppliers to being partners with the automakers will facilitate larger-scale use of aluminum. Developments in aluminum stamping, legislative pressures to improve of fuel economy, and recycling targets might make aluminum the primary material in the auto body.

Glass-fiber-based polymer composites using thermosets as the matrix material will likely hold their own, but usage will grow at a much lower rate than expected for reinforced thermoplastics, which are likely to capture nearly all the growth anticipated later this decade. These materials will play a significant role where part consolidation is great, at a system level rather than on part-by-part substitution. Low tooling cost and lightweighting potential will continue to be the driving force for increased penetration into the light truck area. Due to the cost and toughness advantages of thermoplastic composites, automotive underbody shield application areas look promising. Carbon-fiber-based polymer composites use will be limited to high-end and niche applications demanding substantially higher weight reduction; further growth will be impeded by the significant cost of carbon fiber.

There appears to exist a general commitment towards the use of magnesium by the automotive manufacturers as evidenced by the partnerships formed between them and major magnesium suppliers. These partnerships include Volkswagen and the Dead Sea Works, Ford and Australian Magnesium Company, and General Motors and Norsk-Hydro. Although magnesium remains a major competitor of aluminum, its growth will be limited mainly to niche automotive applications from the die casting industry until major breakthroughs are achieved in wrought automotive applications. The historical, worldwide growth rate for magnesium die casting has been 13 to 14%, a rate that appears to be sustainable with continued pressure to increase fuel economy and as more vehicles use magnesium dashboards, seats, transfer cases, knee bolsters, and so on. Since the largest current use of magnesium is for aluminum alloying, growth in use of these materials are complementary in the context of lightweighting the automotive fleet. If magnesium's wild price swings observed during the past decade—resulting from the mismatch between supply and demand and the more recent market oversupply with cheap imports from non-Western countries—continue, they will be detrimental to the industry's viability. It is anticipated that a slow expansion in the supply base and prices maintained at above $1.40/lb can lead to a more stable demand and supply growth—in the range of 20 lbs to 80 lbs of magnesium per vehicle by 2015 compared to less than 10 lbs today.

5. SAFETY CONSIDERATIONS IN AUTOMOBILE LIGHTWEIGHTING

It has been strongly argued that improving fuel economy by reducing vehicle weight reduces vehicle safety. Technological improvements that have played a key role in fuel economy improvements appear to be neutral or beneficial to safety. The relationship between vehicle weight and safety is important by itself. Vehicle safety concerns have been gaining momentum lately with the growth of light-duty trucks and sport utility vehicles (SUVs) and the pending legislation to boost the CAFÉ for light duty trucks by 1.5 mpg in model year 2007 vehicles. In 1988, light trucks constituted roughly 30% of the vehicle fleet; by 1994 they were more than 40%; and by 2000 they reached an estimated 45% of the fleet. As the number of car-light truck collisions increased, it raised the question whether the growth in the number and weight of light trucks is having an adverse impact on the safety of passenger car occupants and other road users, possibly exceeding any safety benefits of the vehicle-weight increases for the occupants of the trucks.

Safety issues surrounding a general downsizing are concerned with the details of how vehicle designs may change, differences in the performance of lighter weight materials, the precise distribution of changes in mass and size across the fleet, and interactions with other highway users. Of the driver, environment, and vehicle, the vehicle is the least important factor in highway fatalities. To isolate the effects of a less important factor from the effects of more important yet related factors is often not possible. Measure of vehicle exposure with which to control for confounding influences of drivers, environment, and other vehicle characteristics are almost always inadequate—resulting in biased inferences in the literature today.

In a collision between two vehicles of unequal weights, the occupants of the lighter vehicle are at a greater risk whereas occupants of the heavier vehicles benefit, and so there may not be a win-win situation from a societal perspective. In addition, because of the powerful heavier light-truck designs, lighter vehicles are at a serious safety disadvantage. A clear pattern has been observed: reducing a vehicle's weight increases net risk in collisions with substantially larger and heavier entities, reduces net risk in collisions with much smaller and more vulnerable entities (e.g., motorcyclists, bicyclists, and pedestrians) and has little effect on net risk in collisions with vehicles of about same size. Overall, it has been found that the variation of weight among cars results in a net increase of fatalities in collisions, as the reduction in the fatality risk for the driver of the heavier car is less than the increase of the fatality risk for the driver of the lighter car. This increase of fatalities in collisions is offset some by benefits to smaller, lighter highway users (pedestrians and cyclists) in collisions with lightweighted vehicles. The relative weight of the vehicles rather than their absolute values, as well as the distribution of vehicle weights, is what leads to the adverse consequences for the occupants of the lighter vehicle. In fact, some evidence exists, such as the 1997 National Highway Traffic safety Administration study, that proportionately reducing the mass of all vehicles would have a beneficial safety effect in vehicle collision. However, if cars are downweighted and downsized more than light trucks, the increased disparity in weights would increase fatalities.

There is also evidence that smaller, lighter cars have historically had a greater propensity to roll over, and single-vehicle rollover crashes are a major component of all traffic fatalities. The stability of vehicles depends on their dimensions, especially track width, and the height of their center of gravity. The correlation between vehicle weight and the stability has not been strong particularly for light trucks, given the historical tendency that reduced mass means narrower, shorter, less stable cars. It has been argued with some evidence that increase in fatalities due to downweighting and downsizing in single vehicle accidents is a matter of vehicle design rather than mass and that smaller and lighter vehicles have better crash avoidance capabilities. The effect of weight reductions on fatalities in passenger car rollovers might be smaller if weight could be reduced without changing track width.

A recent ACEEE study provides additional evidence to contradict conventional wisdom. This study was based on a measure of risk comparing driver fatalities to the number of vehicles on the road, associated with recent vehicle models sold between 1995 and 1999 with up-to-date safety designs and restraint technologies. The study found that many small cars have lower fatality rates among their drivers than do SUVs or pickups. SUVs have been the major concern, as they are the fastest growing segment of new vehicles, today comprising 21% of the new vehicle market, up from 6% just 13 years ago. This study further illustrated the importance of design in determining a vehicle's safety. Among all major vehicle groups, minivans have the lowest fatality rates among their drivers, while pickups have the highest. Minivans generally are built on car rather than pickup truck platforms, which may reduce the risk to their drivers.

Overall, there does not appear to be any direct correlation between improved fuel economy and vehicle safety, perhaps largely because fuel economy improvement to date have been the result of technology advances rather than decreased weight. The relationships between vehicle weight and safety are complex and not measurable with any reasonable degree of certainty at present. The distribution of vehicle weights is an important safety issue. It is important to know how vehicle weight and size changes will affect the size/weight distribution of the fleet and, finally, how this distribution will affect different fatality rates (i.e., single-vehicle versus vehicle-to-vehicle rates). Vehicles of uniform weight would correct for the negative social externality that encourages individuals to transfer safety risks to others by buying ever larger and heavier vehicles. It appears that in certain kinds of accidents, reducing weight will increase safety risk, while in others it may reduce it. Reducing the weights of light-duty vehicles will neither benefit nor harm all highway users; there will be winners and losers. Lastly, there appears to be no definitive answer to the fuel-economy and safety question, but to the extent that fuel economy can be improved with advanced technology rather than size and weight reductions, adverse safety impacts, if any, can be minimized.

SEE ALSO THE FOLLOWING ARTICLES

Aluminum Production and Energy • Fuel Economy Initiatives: International Comparisons • Glass and Energy • Industrial Energy Use, Status and Trends • Internal Combustion Engine Vehicles • Material Efficiency and Energy Use • Motor Vehicle Use, Social Costs of • Plastics Production and Energy • Recycling of Metals • Steel Production and Energy • Transportation Fuel Alternatives for Highway Vehicles

Further Reading

Das, S. (2000). The life cycle impacts of aluminum body-in-white automotive material. *J. Metals* 52(no. 8), 41–44.

Faller, K., and Froes, F. H. (2001). The use of titanium in family automobiles: Current trends. *J. Metals* 53(7), 27–28.

Kahane, C. J. (1997). "Relationships between Vehicle Size and Fatality Risk in Model Year 1985–93 Passenger Cars and Light Trucks." NHTS Technical report, DOT HS 808 570. National Technical Information Services, Springfield, VA.

Kramer, D. A. (2002). "U.S. Geological Survey Minerals Yearbook – 2001." U.S. Geological Survey. Also available at http://minerals.er.usgs.gov/.

National Academy Press (2002). "Effectiveness and Impact of Corporate Average Fuel Economy (CAFÉ) Standards." U.S. Government Printing Office, Washington, DC.

Paxton, H. W., and DeArdo, A. J. (1997). What's driving the change in materials. *New Steel* 1(2), 64–74.

Pinkham, M. (2001). Will the 21st century be the aluminum age? *Metal Center News* 41(8), 32–35.

Powers, W. F. (2000). Automotive materials in the 21st century. *Adv. Mat. Proc.* 157(5), 38–41.

Ross, M., Wenzel, T. (2001). "An Analysis of Traffic Deaths by Vehicle Type and Model," Report No. T021. American Council for an Energy Efficient Economy, Washington, DC.

Schultz, R. A. (1999). "Aluminum for Lightweight Vehicles: An Objective Look At the Next 10 to 20 Years." Presented at the Metal Bulletin 14th International Aluminum Conference, Montreal, Canada, September 15.

Sullivan, J. L., and Hu, J. (1995). "Life Cycle Energy Analysis for Automobiles." SAE Paper no. 951829, Warrendale, PA.

Ward's Communications (2002). "Ward's Motor Vehicle Facts & Figures." Southfield, MI.

ISBN 0-12-176483-4